ENVIRONMENTAL ECONOMICS & MANAGEMENT

Theory, Policy, and Applications

Scott J. Callan

Bentley University

Janet M. Thomas

Bentley University

Australia • Brazil • Japan • Korea • Mexico • Singapore • Spain • United Kingdom • United States

***Environmental Economics & Management: Theory, Policy, and Applications*, Sixth Edition**
Scott J. Callan and Janet M. Thomas

Senior Vice President, LRS/Acquisitions & Solutions Planning: Jack W. Calhoun

Editorial Director, Business & Economics: Erin Joyner

Editor-in-Chief: Joe Sabatino

Senior Acquisition Editor: Steven Scoble

Developmental Editor: Clara Goosman

Editorial Assistant: Elizabeth Beiting-Lipps

Senior Market Development Manager: John Carey

Senior Brand Manager: Robin LeFevre

Marketing Coordinator: Ilyssa Harbatkin

Media Editor: Anita Verma

Senior Manufacturing Planner: Kevin Kluck

Rights Acquisition Director: Audrey Pettengill

Senior Rights Acquisition Specialist, Text/Image: Deanna Ettinger

Art and Cover Direction, Production Management, and Composition: PreMediaGlobal

Cover Images:
Black Splash Background: © Comstock/thinkstock
Globe with Photos: © iStockphoto/thinkstock

Library of Congress Control Number: 2012935809

ISBN 13: 978-1-111-82667-3

ISBN 10: 1-111-82667-6

South-Western
5191 Natorp Boulevard
Mason, OH 45040
USA

Cengage Learning products are represented in Canada by Nelson Education, Ltd.

For your course and learning solutions, visit **www.cengage.com**

Purchase any of our products at your local college store or at our preferred online store **www.cengagebrain.com**

Printed in the United States of America
1 2 3 4 5 6 7 16 15 14 13 12

DEDICATION

To our students.

Preface

How fascinating it was to learn in 2011 that an earth-like planet was identified some 600 light-years away as part of NASA's Kepler project. Called *Kepler-22b,* this newly discovered planet orbits a sun similar to our own and has an estimated surface temperature of 72° Fahrenheit.[1] The excitement associated with this finding, however remote, implicitly conveys the value society places on our own planet and its unique capacity to support life as we know it. Such awareness reminds us of the responsibility to maintain or improve the earth for future generations.

Another vivid reminder came in 2012 with the fiftieth anniversary of the publication of Rachel Carson's *Silent Spring,* the influential best seller that called attention to the effects of pesticides and other chemicals on the ecology. Although environmental quality is an ongoing and sometimes elusive objective, most would agree that at least in some regions of the world, the environment is far cleaner than the one about which Rachel Carson wrote in the early 1960s.

Still, society must continue to be mindful of its responsibility to preserve and protect the earth, even in the face of twenty-first-century challenges. In the developed world, the recent decline in economic growth and the European crisis have placed difficult burdens on countries around the world, which in turn constrain public and private initiatives, including those aimed at the environment. At the same time, high economic growth rates in developing nations, although moderating of late, stress already depleting natural resources, threaten water supplies, and add to carbon emissions and other pollution releases. Then too, world population continues to climb, estimated to have reached 7 billion in 2011, adding to production and consumption levels that further stress the planet's resources. Beyond these issues are unpredictable events that damage the environment, such as the *Deepwater Horizon* blowout of 2010, the Australian floods in 2010 and 2011, and the earthquake and tsunami that devastated Japan in 2011. Without question, achieving a cleaner environment is a tough challenge and remains a work in progress.

As a world community, we have learned some difficult lessons about the costs and implications of pollution and the irresponsible use of scarce natural resources, and some believe that we still have much to learn. In any case, we do seem to be moving in the right direction. Consider the following. In the United States and other developed nations, air quality has improved dramatically, thanks to marked reductions in damaging emissions, such as carbon monoxide and sulfur dioxide. Technological advances and government initiatives have promoted greater use of hybrid and zero-emission vehicles. Nearly 200 nations meet every year at the United Nations Climate Change Convention to seek solutions to rising carbon emissions, which recently involved efforts to extend the Kyoto Protocol at the 2011 conference in Durbin, South Africa. Countries around the world are promoting resource conservation and cleaner energy use, and most have banned lead in gasoline. Along the way, environmentally responsible trends

---

[1] Potter (2011).

continue to evolve aimed at sustainable development, pollution prevention, and market-based solutions. In sum, we are making important strides.

At a micro level, concern for the environment and natural resource depletion has become part of consumer decision making and has been integrated into virtually every firm's business model. Consumers look for products offered with minimal packaging and those made from recycled materials. They read labels more carefully to avoid goods with large carbon footprints or those made with damaging contaminants. On the supply side, producers respond by conserving natural resources, reducing emissions, promoting product recovery and recycling, and becoming more environmentally responsible in how they design, manufacture, and market their goods and services. Such socially responsible activities are part of what is known as corporate social responsibility, or CSR, which has become an integral element of contemporary business strategy and tactical planning. Taken together, society has come to recognize that economic activity and the natural environment are inexorably linked, and this profound relationship is at the core of environmental economics and management.

Few would debate that environmental economics is a dynamic and relevant field. Hence, as authors, we continue to work diligently to keep the content of this text current and, at the same time, accessible to students. In this sixth edition of our book, we again have had the opportunity to integrate suggestions offered by survey respondents, insightful comments made by our adopters, plus some new ideas of our own. We continue to believe that teaching environmental economics is an exciting opportunity to show students the broad applicability of economic thinking. Students are more environmentally literate and aware than ever before, and most are eager to understand how the market process can help explain and even solve environmental problems. It is, to say the least, an energizing challenge to present this evolving field to what typically is a diverse audience of students.

What hasn't changed over the years is our underlying purpose in writing this textbook—to offer undergraduate students and selected graduate students a clear perspective of the relationship between market activity and the environment. Although we generally assume that students have been exposed to principles of microeconomics, we offer a good review of basic microeconomic fundamentals in Chapter 2 and the major concepts of public goods and externality theory in Chapter 3. Together, these two chapters provide students with the necessary foundation for the course.

Common to the previous editions, our overall motivation is to illustrate in a practical manner how economic tools such as market models and benefit-cost analysis can be used to assess environmental problems and to evaluate policy solutions. Along with traditional discussions, we incorporate contemporary examples of business and consumer practices that are part of environmental decision making. The presentation does not compromise economic theoretical concepts. Instead, it complements the theory with timely, real-world applications. In so doing, even seemingly abstract concepts are given relevance through actual cases about consumers, industry, and public policy.

Content: A Modular Approach

Without question, it is challenging to organize the vast amount of material that an environmental economics course generally covers. Mindful of the usual time constraint in a one-semester course and the fact that the student audience can be highly varied, we developed a **modular structure** for the text. This approach not only organizes the presentation by major topic but also provides a format that facilitates customizing the material to suit a variety of course objectives. At the instructor's discretion, certain chapters within a given module can be omitted or covered less thoroughly without loss of continuity in the overall presentation. Likewise, the order in which the modules are covered can be varied to suit instructor preferences or student interests.

The first three modules form the foundation for the course. These are:

- **Module 1. Modeling Environmental Problems:** A three-chapter module illustrating how environmental problems are modeled from an economic perspective. Primary topics are the materials balance model, a review of market theory and price determination in an environmental context, and the market failure of pollution using both a public goods model and externality theory.
- **Module 2. Modeling Solutions to Environmental Problems:** A two-chapter module on environmental regulatory approaches—one on the command-and-control approach and one on the market approach. Allocative efficiency and cost-effectiveness are used to analyze these, and models are developed to study various control instruments such as technology-based standards, pollution charges, deposit/refund systems, and tradeable pollution permits.
- **Module 3. Analytical Tools for Environmental Planning:** A four-chapter module introduced by an in-depth investigation of risk assessment, risk management, and benefit-cost analysis. Included is a thorough presentation of benefit estimation procedures such as the contingent valuation method and the averting expenditure approach.

Following these are three media-specific modules, which are actually comprehensive case studies of major environmental problems and policy solutions. Using economic modeling and analytical tools, each module assesses the associated environmental risk, evaluates the policy response, and presents a benefit-cost analysis of major legislation and international agreements. These three modules can be covered in any sequence following the foundational material presented in the first half of the text. Some instructors might elect to focus on only one or two of these modules.

- **Module 4. The Case of Air:** A four-chapter module assessing major air pollution problems and the policy initiatives aimed at controlling them. The first chapter lays the groundwork by discussing air quality policy in general and the standard-setting process. The next two chapters cover policy implementation—one aimed at mobile sources and the other at stationary sources, with accompanying discussions of urban smog and acid rain. The last chapter in this module deals with global air quality, specifically ozone depletion and climate change, as well as domestic and international policies that address these problems.
- **Module 5. The Case of Water:** A three-chapter module covering the problems of groundwater and surface water contamination and specific policies aimed at point and nonpoint polluting sources. Two chapters are devoted to an economic analysis of controls on point and nonpoint sources under the Clean Water Act, and a third conducts an analogous investigation of safe drinking water policy.
- **Module 6. The Case of Solid Wastes and Toxic Substances:** A three-chapter module analyzing the solid waste cycle and the use of pesticides and other toxic

substances. Among the primary topics discussed are risk management of the hazardous waste stream, the Superfund controversy, market solutions to controlling municipal solid waste, and risk-benefit analysis in pesticide control.

Each of these media modules utilizes the analytical tools presented in Modules 1 through 3, such as economic models, risk management, and benefit-cost analysis.

The concluding module covers topics in global environmental management. Although international issues are integrated throughout the text, this module concentrates on environmental objectives, policies, and strategies that involve the global community. We focus here on sustainable development, international trade and environmental protection, industrial ecology, and pollution prevention.

- **Module 7. Global Environmental Management:** A two-chapter module examining sustainable development as a worldwide objective and various efforts underway to achieve it. The first chapter in the module addresses the effect of economic growth on environmental quality, international agreements aimed at transboundary pollution, and the effect of environmental protection on international trade. In the second chapter, the focus is on approaches, specifically industrial ecology and pollution prevention. Beyond explaining these concepts at a fundamental level, we also illustrate how these ideas are put into practice through various programs and partnerships in nations around the world.

In essence, this module "closes the loop" of the text by revisiting the materials balance model introduced originally in Chapter 1. Here, we use this model to illustrate the importance of long-run environmental planning and global policy initiatives that go beyond traditional command-and-control abatement efforts.

Featured in the Sixth Edition

In response to both formal and informal input from our adopters and reviewers, we have maintained the fundamental organization and modular structure that have become the hallmark of our text. From a practical perspective, this consistency of style means that our adopters face minimal adjustment in transitioning from the fifth to sixth edition. For new adopters, it is our sincere hope that the flexibility of this modular structure makes it easy to adapt this text to virtually any course in environmental economics. Within this time-tested structure is content that supports theory with real-world examples, current data from the private and public sectors, and boxed applications that link theory to reality both domestically and internationally.

Although the basic modular structure remains consistent with the prior edition, we did revise the text in substantive ways that reflect domestic and international policy changes, evolving corporate environmentalism, and findings of recent empirical studies. We also integrated suggestions offered by survey respondents, reviewers, adopters, and students.

New Topics, New Applications, and Other Updates

To assure that instructors and students have ready access to current science, government initiatives, and private sector environmentalism, we searched diligently for **updates** on public policy, data, applications, and examples. Simply put, we want our text to be not only a complementary textbook to a course, but also a reliable resource for any educator or student interested in environmental economics.

Wherever data are presented in tables or within the text content, we looked for more recent values to update the presentation. Adopters and students will discover the most recent emissions data available have been integrated throughout Module 4, *The Case of Air*. For example, in Chapter 10, *Defining Air Quality: The Standard-Setting Process,* the National Ambient Air Quality Standards are current through 2011, as are data on nonattainment areas. Chapter 12, *Improving Air Quality: Controlling Stationary Sources*, includes new information from the most recent annual spot auction for SO_2 allowances, including prices through 2011. In Module 5, *The Case of Water*, use-support status of water bodies in the United States reflects findings from the latest available *National Water Quality Inventory Report to Congress*, and construction funding for publicly owned treatment works is shown through 2009. Similarly, in Module 6, *The Case of Solid Wastes and Toxic Substances*, we report data from the most recently available sources. In Chapter 18, *Managing Municipal Solid Waste*, for example, we report U.S. waste data through 2010. We also present information on state-level deposit/refund programs as of 2011.

As part of this updating process, we researched all major environmental policies and programs to determine if any revisions had been passed or new initiatives launched. Among the new policies discussed in this edition are: recent NESHAP revisions for power plants; the Cross–State Air Pollution Rule (CSAPR); President Obama's Clean Water Framework; the Integrated Cleanup Initiative (ICI); the Pesticide Registration Improvement Renewal Act (PRIA 2); and the Border 2020 U.S.-Mexico Environmental Program draft document.

We also added **new topics** and offered **expanded coverage** of others. Among these are the following:

- EJ2014, a new U.S. environmental justice strategy
- Obama's Executive Order 13563, supporting benefit-cost analysis for policy
- Cross–State Air Pollution Rule (CSAPR)
- Proposals to tighten certain National Ambient Air Quality Standards (NAAQS)
- Revisions to the National Emission Standards for Hazardous Air Pollutants (NESHAP)
- Market assessment of the new greenhouse gas (GHG) controls on motor vehicles
- Analysis of tax incentives for plug-in hybrids
- Carbon tax initiatives, including the EU's carbon tax proposal
- Conference of the Parties (COP17) in Durbin, South Africa
- National Commission on the BP *Deepwater Horizon* Oil Spill and Offshore Drilling
- EPA's new Drinking Water Strategy (DWS)
- Integrated Cleanup Initiative (ICI) for contaminated sites
- Rio+20 summit for sustainable development
- EU's Integrated Pollution Prevention and Control (IPPC) Directive
- China's Cleaner Production Promotion Law
- U.S. Pollution Prevention Program's five-year (2010–2014) strategic plan

Throughout our text are boxed applications of real-world issues related to environmental economic theory and concepts. These have been well received by adopters and their students. For this edition, all applications were reviewed, updated, and in some cases replaced with more current cases. Overall, we added 8 entirely new applications, completely revised and renamed more than 12, and updated the remainder, for a total of 60 applications placed throughout the text. The **new applications** deal with recent issues that have affected or are continuing to influence policy decisions or business practices. The new titles are:

- *The Gulf Oil Spill of 2010*
- *Command-and-Control Under Fire: Will New Environmental Regulations Hurt the Economic Recovery?*
- *Benefit-Cost Analysis of the Clean Air Act Amendments (CAAA): The Second Prospective Study*
- *The Cross–State Air Pollution Rule (CSAPR): Successor to the CAIR*
- *Renewable Energy Certificates (RECs): Currency for Renewable Energy Markets*
- *Abating the Deepwater Horizon Blowout: Benefits and Costs*
- *Creating Environmental Markets in the Chesapeake Bay Watershed*
- *Consumer Electronics and the Growth of e-waste*

Among the **revised and renamed applications** are:

- *Environmental-Economic Accounting and GDP: A Worldwide Initiative*
- *Environmentally Driven Demand: A 20-Year Look Back*
- *An Update on Boston Harbor: Restoring an Environmental Public Good*
- *Responding to a Negative Externality: From the CD Long Box to the Carbon-Neutral Digipak™*
- *Undefined Property Rights to China's Pearl River Delta (PRD) Region Leads to Environmental Damage*
- *Costs of Remediating Hazardous Waste and Oil Spills*
- *A Benefit-Cost Analysis of the UK's Climate Change Act 2008*
- *Defining Air Quality Standards in the United Kingdom*
- *The Car Allowance Rebate System (CARS): Cash for Clunkers Revisited*
- *Saving the Chesapeake Bay: A Renewed Effort*
- *The Aircraft Drinking Water Rule (ADWR): Ensuring Safe Drinking Water on Aircraft*
- *Kalundborg Symbiosis in the Twenty-First Century*
- *An Update on Remanufacturing: A Lucrative Approach to Pollution Prevention*

A Streamlined, Accessible Presentation

Over the years, many reviewers recommended at least one or two topics to be added to the text, yet others suggested that we make the content more concise and more accessible to students. Based on our own work in environmental economics, we fully understand the spirit of these comments. Environmental economics is a multidisciplinary, extensive, and growing field. We, too, want to offer as much as possible to our students, but we also know that a one-semester course is limiting. Striving to achieve a balance of these seemingly mutually exclusive ideas, we arrived at what we believed was a workable solution: a **topic-rich text with a streamlined presentation**.

With each new edition, we find new policies, initiatives, or events to be added, which in turn means that we must make careful edits to maintain a clean and concise presentation. While we continue to believe that these edits have been worthwhile, they did not take precedence over the integrity of content and the text's attention to economic theory and public policy. The result, according to our adopters and students, has been a success.

We hope that our adopters and their students find this sixth edition to be contemporary, relevant, and accessible, capturing the rich, policy-based content of environmental economics in a way that motivates and informs. We remain dedicated to our goal of preserving the careful exploration of environmental issues and economic analysis that has become the hallmark of our text, while integrating revisions that keep the material accessible to students and accommodative of instructor time constraints. The table of contents attests to the topic-rich character of the text, but the presentation continues to be concise and approachable. Where appropriate, we convert values to more current dollars so that the data can be more easily understood by today's students. For example, most data values, such as in the benefit-cost chapters and selected media chapters, have been restated in 2010 dollars. We also use well-known companies and familiar corporate practices to illustrate such issues as green markets and corporate social responsibility. Among the representative business contexts used in this edition are BMW Group, Caterpillar, Inc., McDonald's Corporation, and The Walt Disney Company.

More Graphs and Quantitative Problems

In response to input from adopters and reviewers, we continue to integrate **quantitative problems and models** throughout every chapter. These learning tools reinforce the economics and the analytical approaches introduced in the first three modules. Where relevant and appropriate, algebraic models are integrated in the text, and most are accompanied by graphical figures. This allows students to link graphs to simple quantitative models more readily. Not only does this help students develop good quantitative skills, but it also gives them two avenues to pursue when studying key concepts in the discipline. In some cases, we have edited existing models to make the set-up and the solution more explicit.

At the end of virtually every chapter, the Review Questions include at least one problem using algebra and/or a graphical analysis. And for this sixth edition, we added at least one new question to each chapter, which is in direct response to reviewers and adopters. As in past editions, solutions to all Review Questions are found in the available Instructor's Manual.

Integrated International Coverage

As in previous editions, we continue our practice of offering instructors and students an integration of international issues throughout the text. Reinforcing this approach is the inclusion of **at least one boxed application with an international context** in every chapter. A quick scan of the table of contents, including the new "List of Applications," will convey the depth of coverage and the interesting content in these supporting applications. These provide real-world examples that illustrate how environmental economics is broad in scope and applicability.

With the same motivation, we made certain to retain the focus and relevance of Module 7, *Global Environmental Management,* to which reviewers and adopters have responded positively. This module gives instructors the option of devoting a class or two strictly to **international issues** in environmental economics across air, water, and land contexts.

To keep this module relevant, we researched the status of key international agreements and any revisions in progress so that the discussion would be as current as possible. In this new edition, instructors and students will find information on the Integrated Pollution Prevention and Control (IPPC) Directive 2008/1/EC; the 2011 Conference of Parties (COP) in Durban, South Africa; and the draft framework for the U.S.–Mexico Border 2020 Program. In addition, updated information is provided on such international agreements as the United States–Canada Air Quality Agreement, the London Convention, and the North American Free Trade Agreement (NAFTA). As in the past, we made best efforts to update such matters as ratification status and international funding commitments and to report on pending negotiations. This module also highlights corporate programs in pollution prevention and sustainable development and reports on the accomplishments of such firms as Apple, Inc., Ford Motor Company, and Johnson & Johnson.

Updated Internet Links and Icons

Throughout the book, we provide **Internet links** to Web sites that support and enhance the text presentation. As in previous editions, every effort has been made to update all URLs and to add new Web sites that are relevant and useful. Of course, all such changes were carried through to the textbook's Companion Sites, where these URLs are also provided. These links are integrated within the text and in footnotes. To facilitate using the Internet as a complement to instruction and study, an icon has been placed next to each link within the main text, which points out the link clearly to instructors and students.

Pedagogical Features

This edition continues to offer various features designed to help instructors prepare lectures and class materials and to make the material interesting and accessible to students. There are both end-of-chapter and end-of-text pedagogical tools, including chapter summaries, review questions, a comprehensive glossary, and a complete list of references. In every chapter, important definitions are given in the margins, key concepts are shown in boldface, and an extensive offering of real-world applications are provided in shaded boxes.

Applications

60 boxed applications complement the text presentation by illustrating the relevance of economic theory, environmental risk, and public policy. The content has been drawn from many sources—the business press, domestic and international government reports, economic research, and the environmental science literature. Topics range from corporate strategies to international policy initiatives. We believe that these real-world cases motivate learning because they illustrate fundamental concepts in relevant, contemporary settings. They also might stimulate more in-depth study in a term paper or course project. In addition to the new applications listed previously, other titles include:

- *Taxing Gasoline Consumption: An International Comparison*
- *Fighting Acid Rain with Pollution Rights: The First Annual Auction*
- *Incremental Benefits of the European Union's Climate Change Initiatives*

- *Hybrids and Plug-In Hybrid Electrics*
- *China's Power Plants and Acid Rain*
- *What Is a Carbon Footprint? Some Firms Are Finding Out*
- *Worldwide Demand for Bottled Water: Is It Safer Than Tap Water?*
- *Using the Market to Control Hazardous Wastes: An International Approach*
- *The Environmental Kuznets Curve*
- *ISO 14000 International Standards on Environmental Management*

Margin Definitions

Throughout each chapter, **Margin Definitions** of key concepts, terms, and relationships are placed adjacent to the associated text presentation. Doing so calls attention to important points in the text, helps familiarize students with new terminology, and assists them in reviewing and self-testing their comprehension. Each of these terms is repeated in the end-of-text glossary for easy reference, and each also appears in the text's Companion Sites.

End-of-Chapter Learning Tools

We conclude each chapter with a **Summary** to help students review and assimilate what they have read. Instructors may find these summaries useful in organizing their courses and in preparing classroom presentations. We also provide both conceptual and analytical **Review Questions** that can be used for regular assignments, in-class discussions, or sample test questions, and as noted, for this new edition, new questions have been added to each chapter. Solutions are provided in an Instructor's Manual, which is available online at the Instructor's Companion Site. We also offer a selection of **Additional Readings** (beyond those cited in the chapter), which are useful for supplementing reading assignments or supporting student projects. In each chapter, these readings have been updated to include recently published articles and books. We also provide a reference list of commonly used **Acronyms** at the end of the chapters in the media-specific modules, i.e., Modules 4, 5, and 6. These also are provided online at the Companion Sites.

End-of-Text Learning Tools

At the end of the text is a list of **References**, which gives complete information on all sources cited in short form throughout the book. Both instructors and students should find this collection of resources helpful in conducting independent investigations of selected topics. Many include URLs to facilitate online access. There is also a convenient **Glossary** of all defined terms given in the chapters.

Text Ancillaries

The following ancillary materials are available:

- An **Instructor's Companion Site**, accessible at **http://login.cengage.com**. At this site, password-protected instructor resources can be downloaded, and student resources can be reviewed. The instructor resources include the following:
  - An **Instructor's Manual** that provides quantitative solutions and suggested responses to end-of-chapter Review Questions. New to this edition of the Instructor's Manual

are Paper Topics for each chapter, and two appendices—one entitled *Graphing Tools and Quantitative Techniques* and another called *Guidelines for Writing a Research Paper*.
  - An updated collection of **PowerPoint® slides** with animated graphs and active links to relevant Internet sites.
  - A new **Test Bank** that offers True-False and Multiple Choice questions, Short Problems, and Case Studies.
- A **Student's Companion Site**, accessible at **www.cengagebrain.com**. At the home page, students can search for the ISBN of this title (shown on the back cover). This will take them to the product page where the following resources can be found: a list of commonly used acronyms; a set of flashcards with terms and definitions; links to all Internet addresses cited in the book; and more.
- **Cengage Learning's Global Economic Watch**. The Watch, a ground-breaking resource accessible at **www.cengage.com/thewatch**, stimulates discussion and understanding of the global downturn with easy-to-integrate teaching solutions. Included is a content-rich blog of breaking news, expert analysis and commentary, a real-time database of hundreds of relevant and vetted journal, news, and periodical articles, videos, podcasts, and much more.

Acknowledgments

As much of own energy and effort is on every page of this book, we never could have done this work without the support and help of others. Countless people supported and contributed to this textbook from its initial prospectus in the early 1990s to the production and publication of this sixth edition. Over the years, we have been fortunate to hear from adopters, giving us the opportunity to integrate their ideas into our revision plans. Similarly, we have received useful input from our students, as they used the text and worked with the review problems, boxed applications, and other pedagogical features. We are most grateful for this important feedback from those who have read and used the textbook.

We also must acknowledge the many economists who reviewed our book and contributed to its development and evolution over time. They provided numerous suggestions, comments, and constructive criticism for which we are grateful.

Mark Aldrich, Smith College
Bill Ballard, College of Charleston
Laurie Bates, Bryant College
John Braden, University of Illinois-Urbana
Michael Carew, Baruch College
Joni Charles, Texas State University—San Marcos
Xudong Chen, Baldwin-Wallace College
Ariaster B. Chimeli, Ohio University
Susan Christoffersen, Philadelphia University
Alan R. Collins, West Virginia University
Michelle Correia, Florida Atlantic University
Jeffrey Czajkowski, Austin College
Jean W. Dupon, Menlo College
Finley Edwards, Colby College
Warren Fisher, Susquehanna University
Juergen Fleck, Hollins University
Per G. Fredriksson, University of Louisville
Shelby Frost, Georgia State University
Craig A. Gallet, California State University—Sacramento
Jacqueline Geoghegan, Clark University
Joyce Gleason, Nebraska Wesleyan College
Douglas F. Greer, San Jose State University
Gauri-Shankar Guha, Arkansas State University
Darwin C. Hall, California State University
Sue Eileen Hayes, Sonoma State University
Pablo Hernandez, Hollins University
Aref A. Hervani, Chicago State University
Naphtali Hoffman, Elmira College and Binghamton University
John Homer, Olivet College
Richard D. Horan, Michigan State University
Robert G. Houston Jr., Eastern Kentucky University
Charles Howe, University of Colorado
Debra Israel, Indiana State University
Tulin Johansson, Lasell College
Donn Johnson, University of Northern Iowa
Mahbubul Kabir, Lyon College
Stanley R. Keil, Ball State University
Derek K. Kellenberg, University of Colorado—Boulder
Katherine Kiel, College of the Holy Cross
David C. Kingsley, Westfield State College
Douglas Kinnear, Hastings College
Janet E. Kohlhase, University of Houston
Supriya Lahiri, University of Massachusetts
Franklin A. Lopez, University of New Orleans
Donald Marron, University of Chicago
Gary McDonnell, Northern Michigan University

Warren Matthews,
Houston Baptist University
Fredric Menz, Clarkson University
J. Todd Ondell, Viterbo University
Brian J. Peterson, Manchester College
Arden Pope, Brigham Young University
Zeyuan Qiu,
New Jersey Institute of Technology
Thomas Rhoads, Towson University
Robert B. Richardson,
Michigan State University
H. David Robison, La Salle University
Frederick B. Rodgers, Medaille College
Duane J. Rosa,
West Texas A&M University
Richard Rosenberg,
Pennsylvania State University
Matthias Ruth, University of Maryland
Farhad Saboori, Albright College
George D. Santopietro,
Radford University
Peter Schuhmann,
University of North Carolina
Wilmington
Peter Schwarz,
University of North Carolina—Charlotte
Joseph J. Seneca, Rutgers University
Hilary A. Sigman, Rutgers,
The State University of New Jersey
Robert Sonora, Fort Lewis College
Marcia S. Snyder, College of Charleston
Douglas D. Southgate Jr.,
Ohio State University
Philip Sprunger, Lycoming College
Tesa Stegner, Idaho State University
Albert Sumell, Youngstown State University
Jeffrey Sundberg, Lake Forest College
David Terkla, University of Massachusetts
Nora Underwood,
University of Central Florida
Charles S. Wassell, Jr., Central Washington
Thomas White, Assumption College
Mark Witte, Northwestern University
John Whitehead, East Carolina University
Keith Willett, Oklahoma State University
David Yoskowitz,
Texas A&M International University

To all these teachers and scholars, we offer our sincere appreciation for their supportive commentary, constructive criticism, and many useful ideas. By incorporating many of their suggestions along the way, we believe our text has been strengthened in every dimension.

We also thank our entire project team at South-Western, Cengage Learning, for their support and expertise in developing, designing, and producing this edition. This team includes:

Joe Sabatino, Editor-in Chief
Steven Scoble,
Senior Acquisitions Editor
Clara Goosman, Developmental Editor
Elizabeth Beiting-Lipps,
Editorial Assistant
John Carey,
Senior Market Development Manager
Robin LeFevre, Senior Brand Manager
Ilyssa Harbatkin, Marketing Coordinator
Kim Kusnerak, Production Manager
Jerusha Govindakrishnan,
Content Project Manager
Karunakaran Gunasekaran,
Manager, Project Management
Anita Verma, Media Editor
Deanna Ettinger,
Senior Rights Acquisitions Specialist
Audrey Pettengill,
Rights Acquisitions Director
Kristina Mose-Libon, Art Director
Sandy Reutelshofer, Inventory Manager
Terina Bradley, Inventory Analyst
Kevin Kluck,
Senior Manufacturing Planner

Lastly, we extend our gratitude to our respective families for their continued support of our work on this project and of all our professional endeavors.

Scott J. Callan
Janet M. Thomas

About the Authors

Dr. Scott J. Callan

Scott J. Callan

Dr. Scott J. Callan is a professor of economics at Bentley University. He received his M.S. and Ph.D. degrees from Texas A&M University. His teaching areas of interest focus on quantitative methods and applied microeconomic topics, such as environmental economics, managerial economics, and sports economics. He has taught courses in environmental economics and the economics of natural resources at both the undergraduate and graduate levels.

In addition to this textbook, Dr. Callan has authored numerous applied microeconomic articles that have been published in a variety of economic journals, including the *Southern Economic Journal*, *Review of Industrial Organization*, *Journal of Business and Economic Statistics*, and *Journal of Sports Economics*. His environmental economics research investigates the demand and supply characteristics associated with the market for municipal solid waste (MSW). Demand-side topics include the effect of pay-as-you-throw programs on waste generation, disposal, and recycling activities. Supply-side issues focus on the extent of economies of scale and scope in the provision of MSW services. These research findings have appeared in *Land Economics* and *Environmental and Resource Economics*, among other academic journals. Recent research efforts have been spent exploring the many facets of corporate social performance and its effect on firm behavior. Articles dealing within this area of research have been published in *Corporate Social Responsibility and Environmental Management*. In addition to his many publications, Dr. Callan has reviewed scholarly articles for a wide variety of academic publications as well as research grant proposals for the U.S. Environmental Protection Agency.

About the Authors

Dr. Janet M. Thomas

Courtesy of Framingham State University

Dr. Janet M. Thomas earned her M.A. and Ph.D. degrees in economics at Boston College. After completing her doctorate, she was appointed to the faculty of Bentley University. She is currently a full professor at Bentley, teaching at both the undergraduate and graduate levels and in Bentley's Honors Program. In addition to environmental economics, Dr. Thomas has taught intermediate microeconomics, industrial organization, public finance, and principles of economics. She has been actively involved in course and curriculum development in environmental economics and served as coordinator of the MBA, Environmental Management Concentration Program at Bentley. In 2009, she was appointed to the Board of Directors for the Framingham State University Foundation, Inc.

Dr. Thomas also is an active researcher in environmental economics, corporate social responsibility, sports economics, and other fields in applied microeconomics. Her present research agenda centers on the relationships among corporate social performance, corporate financial performance, and executive compensation. In addition to this textbook, she has published her research results in such academic journals as *Land Economics*, *Corporate Social Responsibility* and *Environmental Management, Southern Economic Journal, Environmental and Resource Economics, Journal of Sports Economics, Review of Industrial Organization,* and *Eastern Economic Journal.* Dr. Thomas was named Faculty Member of the Year by Bentley's Student Government Organization in 1991. In 1993, she received the Gregory H. Adamian Award for Teaching Excellence. In 1996, she received the Bentley University Scholar of the Year Award, and she received the Outstanding Scholarly Contribution Award in 2005.

Brief Contents

Contents

List of Applications

MODULE 1 Modeling Environmental Problems

Over the past several decades, environmental pollution and other problems have taken on a more significant and visible role in business decisions and corporate planning. It's safe to say that the world has become more aware of the natural environment and more sensitive to the implications of ecological damage. People are changing their consumption patterns to incorporate environmental responsibility into their market decisions. Many are reordering their preferences in favor of biodegradable detergents, non–ozone-depleting products, and recyclable packaging. Similarly, governments are responding by enacting environmental legislation and establishing pollution-monitoring networks to protect the ecology. Firms are adding environmental concerns to their list of business priorities. This corporate response is necessary, not only to comply with regulations on production and product design, but also to remain competitive in a marketplace in which many consumers specifically seek out environmentally responsible producers.

To comprehend this changing marketplace, it is necessary to understand the fundamentals of how markets work and the relationship between market activity and nature. Economic analysis uses models to explain the strategic decision making and economic conditions that define the marketplace. By eliminating unnecessary detail, models allow economists to test theories about economic relationships and to make predictions about behavioral and institutional reactions to changes in market conditions. For these reasons, modeling is a fundamental tool in environmental economics.

In Module 1 of the text, we devote three chapters to basic models that are useful for understanding environmental issues. In Chapter 1, we develop the materials balance model, which illustrates the linkages between the circular flow of economic activity and nature. In so doing, we show how environmental damage and resource depletion occur as a result of market decision making. In Chapter 2, we focus on the fundamentals of the market process, reviewing such key concepts as supply and demand, economic efficiency, and measures of social welfare. All of this lays the foundation for Chapter 3, in which we explain how environmental problems occur when the market fails. Here, too, we use models to illustrate the sources of market failure and the conditions under which such an outcome arises.

The Role of Economics in Environmental Management

"The supreme reality of our time is ... the vulnerability of our planet."
—John F. Kennedy (1917–1963)

In the twenty-first century, society must understand and accept the challenge of protecting and preserving the earth's resources while continuing to develop economically. The rapid growth and advancing technology that began in earnest with the Industrial Revolution have taken a toll on the natural environment. Manufacturing processes, mass transit systems, telecommunications, and synthetic chemicals are responsible both for the highly advanced lifestyle that society enjoys *and* for much of the environmental damage that has occurred. With 20/20 hindsight, we now recognize that the trade-off between economic growth and environmental quality has been significant.

An important objective, therefore, is to understand the critical relationship between economic activity and nature and to use that knowledge to make better and wiser decisions. Of course, there will always be some amount of trade-off—precisely what economic theory conveys. We cannot expect to have perfectly clean air or completely pure water, nor can we continue to grow economically with no regard for the future. But there *is* a solution, although it is a compromise of sorts. We first have to decide what level of environmental quality is acceptable and then make appropriate adjustments in our market behavior to sustain that quality as we continue to develop as a society.

The adjustment process is not an easy one, and it takes time. As a society, we are still learning—about nature, about market behavior, and about the important relationships that link the two together. What economics contributes to this learning process are analytical tools that help to explain the interaction of markets and the environment, the implications of that relationship, and the opportunities for effective solutions.

In this chapter, we support these assertions with a simple but powerful model that illustrates the link between economic activity and nature. As we will discover, the underlying relationships motivate economic analysis of environmental issues, which is formally defined through two disciplines: natural resource economics and environmental economics, the latter of which is our focus in this book. With this model as a foundation, we lay the groundwork for our course of study, starting with an introduction of basic concepts. From there, we identify and discuss major objectives in environmental economics and then present an overview of public policy development and the role of economics in that process.

ECONOMICS AND THE ENVIRONMENT

One of the most pervasive applications of economic theory is that it logically explains what we observe in reality. For example, through microeconomic analysis, we can understand the behavior of consumers and firms and the decision making that defines the marketplace. This same application of economic theory can be used to analyze environmental problems—why they occur and what can be done about them. Stop to consider how pollution or resource depletion comes about—not from a sophisticated scientific level, but from a fundamental perspective. The answer? Both arise from decisions made by households and firms. Consumption and production draw on the earth's supply of natural resources, and both activities generate by-products that can contaminate the environment. This means that the fundamental decisions that comprise economic activity are directly connected to environmental problems. To illustrate this relationship, we begin by presenting a basic model of economic activity. Then, we expand the model to show exactly how this connection arises.

Circular Flow Model

circular flow model Illustrates the real and monetary flows of economic activity through the factor market and the output market.

The basis for modeling the relationship between economic activity and the environment is the same one that underlies all of economic theory—the **circular flow model**, shown in Figure 1.1. Typically, this is the first model students learn about in introductory economics.

First, consider how the flows operate, holding all else constant. Notice how the **real flow** (i.e., the nonmonetary flow) runs counterclockwise between the two market sectors, households (or consumers) and firms (or producers). Households supply resources or factors of production to the **factor market**, where they are demanded by firms to produce goods and services. These commodities are then supplied to the **output market**, where they are demanded by households. Running clockwise is the **money flow**. The exchange of inputs in the factor market generates an income flow to

FIGURE 1.1 Circular Flow Model

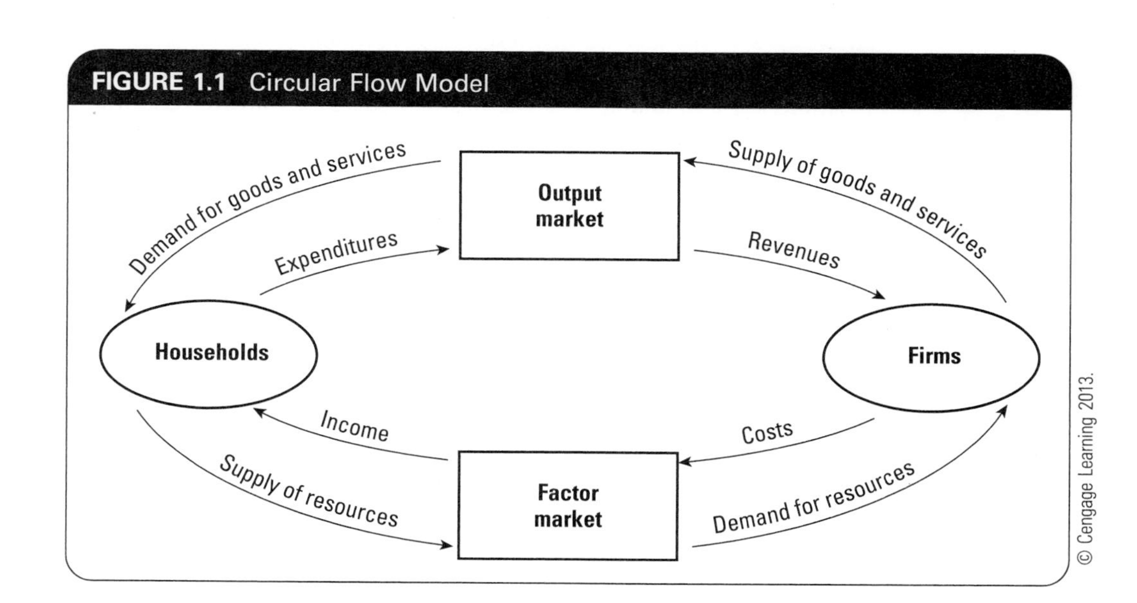

households, and that flow represents costs incurred by firms. Analogously, the money flow through the output market shows how households' expenditures on goods and services are revenues to firms.

Now, think about how the volume of economic activity and, hence, the size of the flow are affected by such things as population growth, technological change, labor productivity, capital accumulation, and natural phenomena such as drought or floods. For example, holding all else constant, technological advance would expand the productive capacity of the economy, which in turn would increase the size of the flow. Similarly, a population increase would lead to a greater demand for goods and services, which would call forth more production and lead to a larger circular flow.

Notice that by analyzing how the flows operate and how the size of an economy can change, we can understand the basic functioning of an economic system and the market relationships between households and firms. However, the model does not explicitly show the linkage between economic activity and the environment. To illustrate this interdependence, the circular flow model must be expanded to depict market activity as part of a broader paradigm, called the materials balance model.

Materials Balance Model

The explicit relationship between economic activity and the natural environment is illustrated by the **materials balance model** shown in Figure 1.2.[1] Notice how the real flow of the circular flow model is positioned within a larger schematic to show the connections between economic decision making and the natural environment.

materials balance model
Positions the circular flow within a larger schematic to show the connections between economic decision making and the natural environment.

Flow of Resources: Natural Resource Economics

Look at the linkages between the upper block representing nature and the two market sectors (households and firms), paying particular attention to the direction of the arrows. Notice that one way an economic system is linked to nature is through a flow of materials or natural resources that runs *from* the environment *to* the economy, specifically through the household sector. (Recall that, by assumption, households are the owners of all factors of production, including natural resources.) This flow describes how economic activity draws on the earth's stock of natural resources, such as soil, minerals, and water. It is the primary focus of **natural resource economics**, a field of study concerned with the flow of resources from nature to economic activity.

natural resource economics
A field of study concerned with the flow of resources from nature to economic activity.

Flow of Residuals: Environmental Economics

A second set of linkages runs in the opposite direction, *from* the economy *to* the environment. This flow illustrates how raw materials entering the system eventually are released back to nature as by-products or **residuals**. Most residuals are in the form of gases released into the atmosphere, and in the short run, most are not harmful. In fact, some are absorbed naturally through what is called the **assimilative capacity** of the environment. For example, carbon dioxide emissions from the combustion of fossil fuels (i.e., oil, coal, and natural gas) can be partially absorbed by the earth's oceans and forests. Other released gases are not easily assimilated and may cause harm, even in the

residual
The amount of a pollutant remaining in the environment after a natural or technological process has occurred.

[1]Kneese, Ayres, and D'Arge (1970). See Chart 1, p. 9, in this source for a more detailed depiction of this model.

FIGURE 1.2 Materials Balance Model

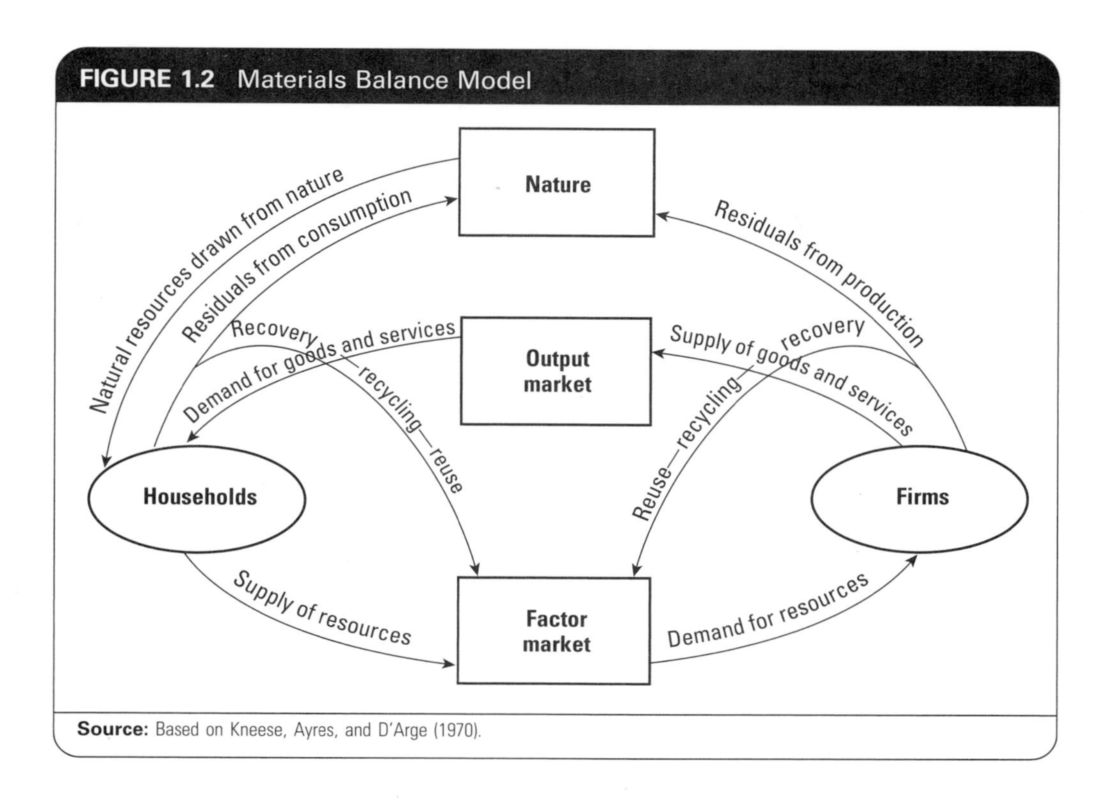

Source: Based on Kneese, Ayres, and D'Arge (1970).

short term. There are also liquid residuals, such as industrial wastewaters, and solid residuals, such as municipal trash and certain hazardous wastes—all of which are potential threats to health and the ecology. Notice in Figure 1.2 that there are two residual outflows, one leading from each of the two market sectors, meaning that residuals arise from both consumption and production activity. This set of flows is the chief concern of **environmental economics**.

environmental economics A field of study concerned with the flow of residuals from economic activity back to nature.

It is possible to delay, though not prevent, the flow of residuals back to nature through **recovery**, **recycling**, and **reuse**. Notice in the model that there are inner flows running from the two residual outflows back to the factor market. These inner flows show that some residuals can be recovered from the stream and either recycled into another usable form or reused in their existing form. For example, Application 1.1 discusses how Germany's BMW Group has made advances in automobile design to facilitate recycling once a vehicle has reached the end of its economic life.

Although recycling efforts are important, keep in mind that they are only short-term measures, because even recycled and reused products eventually become residuals that are returned to nature. Indeed, what the materials balance model shows is that all resources drawn from the environment ultimately are returned there in the form of residuals. The two flows are balanced, a profound fact that is supported by science.

Using Science to Understand the Materials Balance

first law of thermodynamics Matter and energy can neither be created nor destroyed.

According to the **first law of thermodynamics**, matter and energy can neither be created nor destroyed. Applying this fundamental law to the materials balance model means that in the long run, the flow of materials and energy drawn from nature into consumption and production must equal the flow of residuals that run from these

APPLICATION 1.1 BMW Group's Sustainable Decisions and Design for Recycling

In addition to BMW Group's venerable reputation for well-engineered automobiles, it is also recognized for another distinction. The German automaker is firmly committed to sustainable corporate decisions and therefore to developing and building BMWs in an environmentally responsible manner.

Thanks to disassembly analyses conducted at its Munich Recycling and Dismantling Center in Lohhof, Germany, BMW Group is developing what it calls "solutions for environmentally and economically sensible recycling." These analyses determine the time and resources needed to dismantle a vehicle at the end of its useful life. This information is then integrated into vehicle construction plans. Such an approach is called "Design for Disassembly (DFD)"—a manufacturing method aimed at building a product to facilitate end-of-life recycling. Along with other major corporations such as Volkswagen, 3M, and General Electric, BMW Group is investigating ways to manufacture a DFD product that is economically competitive and that stands up to the company's high standards for quality engineering. As a consequence of these and other efforts, today 85 percent of the materials in a BMW Group vehicle are recyclable, with another 10 percent available for use in energy generation.

BMW's research in DFD is part of a long-term commitment to environmentally responsible production decisions. The company has been recycling its catalytic converters since 1987. A year later, it introduced its limited-production Z1 roadster. The two-seater is totally recyclable and is considered the first DFD product ever made. Subsequently, BMW Group built a pilot plant in Bavaria dedicated solely to researching the DFD approach to manufacturing. Teams of workers at the facility systematically dismantle cars, beginning with the fluids and oils and ending with the removal and sorting of interior materials.

A major objective is to build an automobile that can be dismantled at a relatively low cost. Long hours for disassembly elevates costs, which ultimately forces up car prices and reduces competitiveness—an outcome no carmaker can afford. Another key goal is to ensure that parts can be readily sorted. This is particularly critical for plastics, which are more complex to recycle but are increasingly used to lower vehicle weight and to improve fuel efficiency. Using materials flow management, BMW Group and various partners have developed separation techniques for polyurethane foam and pure-grade plastics. Taking this a step further, the firm also uses plastic recyclables to build its new cars. In fact, plastic recyclables in BMW Group vehicles account for 15 percent of the total weight of plastic components, a proportion that the company intends to increase to 20 percent.

Closing the loop, BMW Group and selected recycling partners have established a network of automobile take-back and recycling centers throughout the European Union (EU). This infrastructure is aimed at making it more convenient for BMW, MINI, and Rolls-Royce owners to dispose of their cars at the end of their economic life. In fact, beginning in January 2007, the return of end-of-life vehicles has been available to consumers free of charge. Although an EU Directive called for such networks in 2000, BMW Group had developed its own in the early 1990s, long before the directive was in place. Today, approximately 2,000 recovered automobiles are dismantled and recycled at the Munich Recycling and Dismantling Center, which is the leading center of its kind in the world. Further development of this vehicle return process is among BMW Group's sustainable objectives in the area of product responsibility. More detail is available at **www.bmwgroup.com/recycling**.

Sources: BMW Group (2009; 2007; 2003; 2001); Protzman (July 4, 1993); Knepper (January 1993); Nussbaum and Templeman (September 17, 1990).

activities back into the environment. Put another way, when raw materials are used in economic activity, they are converted into other forms of matter and energy, but nothing is lost in the process. Over time, all these materials become residuals that are returned to nature. Some arise in the short run, such as waste materials created during production. Other resources are first transformed into commodities and do not enter the residual flow until the goods are used up. At this point, the residuals can take various forms, such as carbon monoxide emissions from gasoline combustion or trash disposed in a municipal landfill. Even if recovery does take place, the conversion of residuals into recycled or reused goods is only temporary. In the long run, these too end up as wastes.

second law of thermodynamics
Nature's capacity to convert matter and energy is not without bound.

There is one further point. Because matter and energy cannot be destroyed, it may seem as though the materials flow can go on forever. But the **second law of thermodynamics** states that nature's capacity to convert matter and energy is not unlimited. During energy conversion, some of the energy becomes unusable. It still exists, but it is no longer available to use in another process. Consequently, the fundamental process on which economic activity depends is finite.

These scientific laws that support the materials balance model communicate important, practical information to society. First, we must recognize that every resource drawn into economic activity ends up as a residual, which has the potential to damage the environment. The process can be delayed through recovery but not stopped. Second, nature's ability to convert resources to other forms of matter and energy is limited. Taken together, these assertions provide a comprehensive perspective of environmental problems and the important connections between economic activity and nature. It is the existence of these connections that motivates the discipline of environmental economics.

UNDERSTANDING ENVIRONMENTAL DAMAGE

Although it is not necessary to master the rigors of environmental science, it is important to become familiar with the basic concepts used to identify environmental damage and to describe policy solutions. To that end, what follows is a brief overview of selected terms and concepts in environmental economics. A useful online glossary of environmental terms is available at the Environmental Protection Agency's (EPA) Web site at **www.epa.gov/OCEPAterms**.

pollution
The presence of matter or energy whose nature, location, or quantity has undesired effects on the environment.

Environmental economics is concerned with identifying and solving the problem of environmental damage, or **pollution**, associated with the flow of residuals. Although pollution is defined differently in different contexts, it can be thought of generally as the presence of matter or energy whose nature, location, or quantity has undesired effects on the environment. Virtually any substance can cause pollution solely on the basis of a single characteristic, such as its fundamental constituents, its location, or its quantity. What this implies is that finding solutions to environmental damage depends critically on identifying the causes, sources, and scope of the damage.

Causes of Environmental Damage

natural pollutants
Contaminants that come about through nonartificial processes in nature.

anthropogenic pollutants
Contaminants associated with human activity.

How do we identify which substances are causing environmental damage? One way is to distinguish them by their origin—that is, whether they are **natural pollutants** arising from nature or **anthropogenic pollutants** resulting from human activity.

- **Natural pollutants** arise from nonartificial processes in nature, such as particles from volcanic eruptions, salt spray from the oceans, and pollen.
- **Anthropogenic pollutants** are human induced and include all residuals associated with consumption and production. Examples include gases from combustion and chemical wastes from certain manufacturing processes.

Of the two, anthropogenic pollutants are of greater concern to environmental economists, particularly those for which nature has little or no assimilative capacity.

Sources of Environmental Damage

stationary source A fixed-site producer of pollution.

mobile source Any nonstationary polluting source.

point source Any single identifiable source from which pollutants are released.

nonpoint source A source that cannot be identified accurately and degrades the environment in a diffuse, indirect way over a broad area.

Once pollutants have been identified, the next step is to determine the sources responsible for their release. Polluting sources are many and varied, ranging from automobiles to waste disposal sites. Even a seemingly pristine setting such as farmland can be a polluting source if rainwater transports chemical pesticides and fertilizers to nearby lakes and streams. Because polluting sources are so diverse, they are usually classified into broad categories that are meaningful to policy development. Depending on the environmental media (air, water, or land), sources of pollution are generally grouped by (1) their mobility (**stationary source** or **mobile source**) or (2) their identifiability (**point source** or **nonpoint source**). See Table 1.1 for detail.

Scope of Environmental Damage

Although environmental damage is a universal concern, some types of pollution have detrimental effects that are limited to a single community, whereas others pose a risk over a large geographic region. The point is, the extent of the damage associated with pollution can vary considerably—an observation vitally important to policy formulation. Consequently, environmental pollution is often classified according to the relative size of its geographic impact as local, regional, or global.

Local Pollution

local pollution Environmental damage that does not extend far from the polluting source.

Local pollution refers to environmental damage that does not extend far from the polluting source and typically is confined to a single community. Although the negative effects are limited in scope, they nonetheless pose a risk to society and can be difficult to control. A common local pollution problem is **urban smog**. Visible as a thick yellowish haze, smog is caused by pollutants that chemically react in sunlight. It is particularly severe in major cities like Beijing, Los Angeles, and Mexico City, as Figure 1.3 illustrates.

Another local pollution problem that is receiving increasing attention is **solid waste pollution**. Poor waste management practices can allow contaminants such as

TABLE 1.1 Sources of Pollution

SOURCES GROUPED BY MOBILITY		
Type	**Description**	**Examples**
Stationary	A fixed-site producer of pollution.	Coal-burning power plants, sewage treatment facilities, manufacturing plants
Mobile	Any nonstationary polluting source.	Automobiles, trucks, airplanes
SOURCES GROUPED BY IDENTIFIABILITY		
Type	**Description**	**Examples**
Point	Any single identifiable source from which pollutants are released.	Factory smokestack, effluent pipe, a ship
Nonpoint	A source that cannot be identified accurately and degrades the environment in a diffuse, indirect way over a relatively broad area.	Agricultural runoff, urban runoff

FIGURE 1.3 Pollutants Contributing to Urban Air Pollution in Major Cities

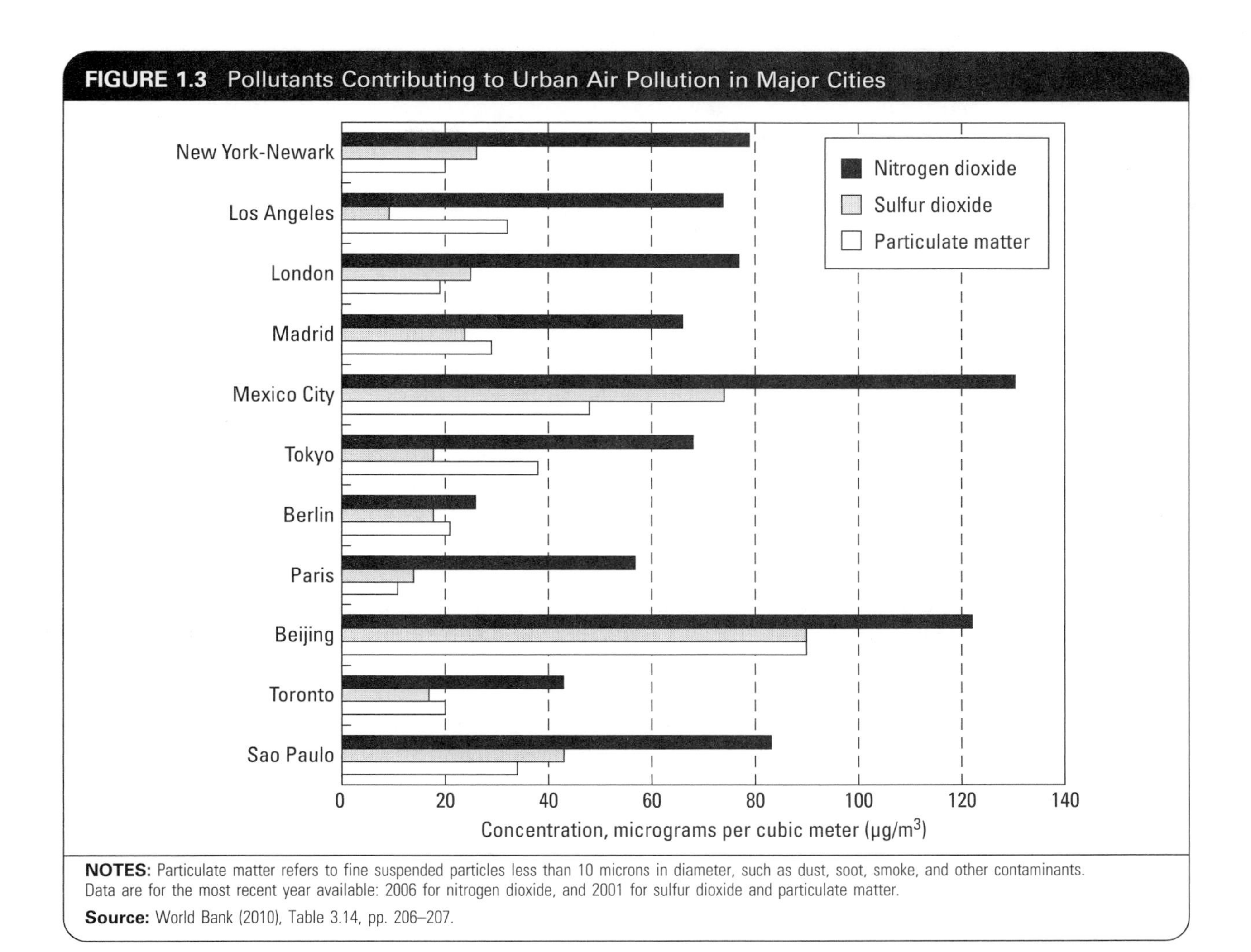

NOTES: Particulate matter refers to fine suspended particles less than 10 microns in diameter, such as dust, soot, smoke, and other contaminants. Data are for the most recent year available: 2006 for nitrogen dioxide, and 2001 for sulfur dioxide and particulate matter.

Source: World Bank (2010), Table 3.14, pp. 206–207.

lead and mercury to leach into soil and water supplies. Beyond measures aimed at improving waste management are efforts to reduce the amount of waste being generated in the first place. Look at the per capita estimates of municipal waste generation and gross domestic product (GDP) for selected countries given in Table 1.2. Note that these data generally suggest a positive relationship between waste generation and industrialization.

Regional Pollution

regional pollution Degradation that extends well beyond the polluting source.

Environmental pollution that poses a risk well beyond the polluting source is called **regional pollution**. An important example is **acidic deposition**, which arises from acidic compounds that mix with other particles and fall to the earth either as dry deposits or in fog, snow, or rain. Acidic deposition is commonly known as acid rain. Acid rain is characterized as regional pollution because the harmful emissions can travel hundreds of miles from their source. Another example of

TABLE 1.2 Per Capita Municipal Solid Waste Generation for Selected Countries

Country	Waste (kilograms per capita)	2008 GDP ($ per capita)
United States	745	43,250
Netherlands	623	38,035
Germany	581	33,663
Spain	575	27,747
United Kingdom	546	34,356
Italy	545	28,245
France	543	30,624
Sweden	515	33,744
Norway	490	49,416

NOTES: Waste and GDP figures shown are for 2008.
GDP values are in 2005 U.S. dollars.
Sources: European Commission, Eurostat (January 17, 2011), Table 4; U.S. Census (2011), Table 373, p. 229, Table 1348, p. 847.

regional pollution is the Gulf Oil Spill that occurred in 2010, which is discussed in Application 1.2.

Global Pollution

global pollution Environmental effects that are widespread with global implications.

Some environmental problems have effects so extensive that they are called **global pollution**. Global pollution is difficult to control, both because the associated risks are widespread and because international cooperation is needed to achieve effective solutions. Consider, for example, the problem of **global warming**. Also known as the greenhouse effect, global warming occurs as sunlight passes through the atmosphere to the earth's surface and is radiated back into the air where it is absorbed by so-called greenhouse gases (e.g., carbon dioxide). Although this warming process is natural, activities such as fossil fuel combustion add to the normal level of greenhouse gases, which in turn can raise the earth's natural temperature. These climate disruptions may affect agricultural productivity, weather conditions, and the level of the earth's oceans—all effects that are worldwide in scope.

Global warming falls under the broader heading of **climate change**, although the two are often used interchangeably. Climate change refers to a major alteration in any climate measure, including temperature, wind, and precipitation, that is prolonged (decades or longer).[2] Such an alteration in climate can arise from natural phenomenon, such as changes in oceanic circulation or variances in the sun's intensity; it can also be the result of human activities like fossil fuel combustion or deforestation. For further detail, visit the EPA's site on climate change, **www.epa.gov/climatechange/index.html**.

Similarly widespread are the risks of **ozone depletion**, a thinning of the earth's ozone layer. The ozone layer protects the earth from harmful ultraviolet radiation, which can weaken human immune systems, increase the risk of skin cancer, and harm ecosystems.

[2] U.S. EPA, Office of Air and Radiation, Office of Atmospheric Programs, Climate Change Division (April 14, 2011b).

APPLICATION 1.2 The Gulf Oil Spill of 2010

In a speech to the American people in June 2010, President Obama called the *Deepwater Horizon* blowout "...the worst environmental disaster America ever faced." *Deepwater Horizon* is the name of the offshore drilling rig that was positioned in the Gulf of Mexico in the Macondo prospect, which is owned and operated by British Petroleum (BP). On April 20, 2010, an explosion and fire took place on the rig, which damaged the wellhead, causing millions of gallons of crude oil to leak into the Gulf. There were significant human losses. Eleven workers on the rig lost their lives, and another 17 were injured. Damage to the environment extended well beyond the rig site, resulting in nontrivial, regional pollution that has not yet been fully identified.

Crude oil spilled out of the well into the Gulf of Mexico for three months before the wellhead was capped. Over that period, officials estimate that more than 170 million gallons of crude oil despoiled the gulf, making it the largest marine oil spill accident in the history of the United States. The massive spill is responsible for both environmental and economic damage, but scientists have not yet identified the magnitude and extent of the losses. Part of the difficulty is the absence of comprehensive data on ecological conditions in the area *prior to* the accident to serve as a benchmark to which post-spill conditions could be compared. Another challenge is the limited scientific knowledge about deepwater ecosystems, which were subjected to the oil pollution.

At risk are all types of marine life, including fish, sea turtles, dolphins, seaweed, and plankton. Of particular concern are those that are endangered based on the Endangered Species Act. The National Oceanic and Atmospheric Administration (NOAA) reports that all 28 species of marine mammals living in the Gulf of Mexico are protected, and 6 of these are endangered. Also at risk are seabirds and shorebirds that rely on the ocean and coastal estuaries for food, such as bald eagles and falcons. Oil that reaches shorelines threatens ecosystems living in such coastal habitats as salt marshes and beaches. Reportedly, more than 650 miles of coastal habitats along the Gulf were oiled, and of these, over 130 miles have been identified as heavily oiled. These shorelines affected four states—Louisiana, Mississippi, Alabama, and Florida. Less apparent but nonetheless cause for concern is the possibility of a deepwater plume, defined as "clouds" of tiny oil droplets and dissolved gases moving far below the ocean's surface. Scientists are concerned that these substances may detrimentally affect oxygen levels that support marine species.

Beyond the damage to the natural environment, the Gulf spill also caused economic losses, particularly to the fishing and tourist industries. In addition to negative effects from actual damages to the region were those arising from public perception and uncertainty about the ocean and area beaches as well as the safety of fish and shellfish caught in the Gulf. Diminished hotel bookings and restaurant reservations served as anecdotal evidence of reduced profits to local economies because of the well blowout.

Taken together, the human, environmental, and economic losses linked to the Gulf spill are significant. But testing and monitoring of ecosystems over the long term are necessary to more accurately identify the magnitude and extent of the associated damage, which will take time. Nonetheless, officials are reporting that the damages might not be as extensive as originally feared. For example, initial reports indicate that spill-related bird kills were less than 1 percent of those linked to the Exxon *Valdez* spill. Why? For one thing, ocean currents and winds prevented most of the oil from reaching shorelines. For another, the warmer temperatures in that region are said to have facilitated evaporation and degradation of the spilled oil, in part because the oil-eating bacteria used to combat the spill function better in warmer waters. Also, the oil in this spill was much lighter and degradable than that released in the Exxon *Valdez* spill.

Scientists and public officials continue to monitor the region, collecting samples and gathering data, to better understand and assess the magnitude of loss linked to this unprecedented environmental disaster. Hence, a full assessment of the associated effects will not be known for some time. In the interim, ongoing data and reports are available at various Web sites, including the U.S. government's official site on the spill, **www.restorethegulf.gov**, and the site of the National Commission on the BP Deepwater Horizon Oil Spill and Offshore Drilling at **www.oilspillcommission.gov**.

Sources: National Commission on the BP Deepwater Horizon Oil Spill and Offshore Drilling (January 2011); Grunwald (August 9, 2010).

In 1985, scientists discovered that a previously observed thinning of the ozone layer over the Antarctic region had become an "ozone hole" the size of North America. Ozone depletion is caused mainly by a group of chemicals known as chlorofluorocarbons (CFCs), which had been commonly used in refrigeration, air conditioning, packaging, insulation, and aerosol propellants. Although national governments have controlled CFC usage, the main policy thrust has arisen through international agreements because of the global nature of the problem.

IDENTIFYING ENVIRONMENTAL OBJECTIVES

Just as fundamental environmental problems are universal, so too are the overall objectives. However, articulating the specifics of these objectives and accepting the trade-offs that such goals imply is a process that is not without debate. Indeed, such is the substance of environmental summits, where national leaders, industry officials, and environmentalists gather to exchange ideas about appropriate objectives and to garner cooperation from one another. A case in point was the 2011 United Nations climate conference in South Africa at which negotiations continued on the Kyoto Protocol. A more comprehensive environmental agenda was addressed at the United Nations Conference on Environment and Development (UNCED) held in Rio de Janeiro, Brazil, in 1992. Known as the Earth Summit, the event marked the twentieth anniversary of the first worldwide environmental conference in Stockholm and was attended by 6,000 delegates from more than 170 countries.[3]

Although the objective-setting process has been difficult at times and is often immersed in political debate, it is nonetheless moving forward. Today, virtually every environmental decision is guided by what have become worldwide objectives: **environmental quality**, **sustainable development**, and **biodiversity**.

Environmental Quality

environmental quality
A reduction in anthropogenic contamination to a level that is "acceptable" to society.

Given the pervasive problems of local, regional, and global pollution, few would debate including **environmental quality** among the world's objectives. However, there is a lack of consensus about how to define this concept in practice. Most of us consider environmental quality to mean clean air, water, and land. However, when environmental quality is being defined to guide policy, we have to decide just how clean is clean.

The debate usually begins with asking why environmental quality should not mean the absence of all pollution. The answer is that such an objective is impossible, at least in a pure sense. Recall that some pollution is natural and therefore not controllable. Furthermore, the absence of all anthropogenic pollutants could be achieved only if there were a prohibition on virtually all the goods and services that characterize modern living. This means that a more rational perception of environmental quality is that it represents a reduction in anthropogenic contamination to a level that is "acceptable" to society. This acceptable level of pollution will, of course, be different for different contaminants, but in each case, certain factors are considered in making the determination.

[3]Council on Environmental Quality (January 1993), p. 140.

Among these factors are the gains to human health and ecosystems, expenditures needed to achieve the reduction, availability of technology, and the relative risk of a given environmental hazard.

Although the world has made progress toward achieving environmental quality, there is still work to be done. In some parts of the world, such as Eastern Europe and developing nations, environmental pollution is extreme, and progress toward reducing the effects is slow. Application 1.3 discusses some of the environmental problems China is experiencing as it strives to advance economically.

Recognizing environmental quality as a worldwide objective has triggered an awareness of its importance over a longer time horizon. Society has begun to realize that pursuing economic growth could so adversely affect the natural resource stock that the productive capacity and welfare of future generations could be threatened. The potential of such an *intertemporal* trade-off has prompted a sense of obligation to the future that has materialized into two related objectives: **sustainable development** and **biodiversity**.

Sustainable Development

sustainable development Management of the earth's resources such that their long-term quality and abundance are ensured for future generations.

Economic growth is defined as an increase in real gross domestic product (GDP). Although growth is a favorable outcome, there are long-term environmental implications, as the materials balance model suggests. Achieving an appropriate balance between economic growth and the preservation of natural resources is the essence of the objective known as **sustainable development**, which calls for managing the earth's resources to ensure their long-term quality and abundance.[4] This reminds us that the circular flow of economic activity cannot be properly understood without recognizing how it fits into the larger scheme of the natural environment. Yet only in the recent past have economists and society at large begun to accept this broader and more realistic view. For example, new initiatives have been proposed to capture the ecological effects of growth in macroeconomic performance measures—an issue discussed in Application 1.4. More on sustainability is available at **www.epa.gov/sustainability**.

On a much broader scale, the *Rio Declaration*, drafted at the Rio Summit, outlines 27 principles to serve as guidelines for global environmental protection and economic development. Similar commitments are given in the summit's 40-chapter document, *Agenda 21*, an international agenda of comprehensive environmental goals. *Agenda 21* is dedicated in large part to sustainable development, with an emphasis on regions where achieving this objective is particularly critical, such as in developing nations. In celebration of the Rio Summit's tenth anniversary, another worldwide event was held in South Africa—the Johannesburg Summit 2002. As in Rio, thousands of people attended, including heads of state, national delegates, and business leaders, all coming together to discuss issues relating to the goal of sustainable development.

Biodiversity

biodiversity The variety of distinct species, their genetic variability, and the variety of ecosystems they inhabit.

Another environmental objective that addresses the legacy left to future generations is **biodiversity**. This refers to the variety of distinct species, their genetic variability, and

[4]Council on Environmental Quality (January 1993), p. 135.

APPLICATION 1.3 The High Price of China's Economic Advance

China's rapidly growing economy has become a double-edged sword. Although China's 1.3 billion residents are enjoying greater prosperity, the quality of the nation's air, water, and land resources has severely deteriorated. Despite significant sums dedicated to environmental clean-up, some ecological damage is going virtually unchecked. In fact, much of the abatement effort to date has been aimed at highly visible, large urban centers, such as Beijing and Shanghai, leaving smaller cities and rural communities to bear a disproportionate burden of the nation's contaminated water and severe air pollution. Beijing had become a particular focus as the designated host city for the 2008 Olympic Games. Unfortunately, much of the environmental success associated with Olympics' preparation has been short-lived, as air quality in the city has begun to deteriorate back to pre-Olympic levels.

As is sometimes the case in developing economies, environmentalism is perceived as an obstacle to economic advance that can frustrate industrial development. But in China, the environmental damage has become so severe that its economic advance is being diminished by a lack of clean water, reduced productivity associated with pollution-induced health problems, and other damages that limit production. Official estimates suggest that this drag on China's growth may be as much as 5.8 percent of GDP annually, a considerable increase from the 3 percent estimate in the mid-1990s. Such a backlash has prompted a more aggressive response to the nation's environmental decline.

Over the 2006–2010 period, Chinese officials spent 1.4 trillion ¥ (yuan), or about $207 billion, on environmental protection and will likely spend an estimated 3.1 trillion ¥ over the next five-year period through 2015 (almost $460 billion). Nonetheless, it is not clear that the Chinese government is doing enough to protect the environment. The agency responsible for putting these sums of money to work is the Ministry of Environmental Protection (formerly the State Environmental Protection Agency [SEPA]). Reportedly, the SEPA lacked the power and influence to effect significant improvements, and it was thought that most of the designated funds might have gone to other national agencies, such as the State Forestry Administration, or local environmental bureaus, which generally suffer from conflicts of interest. Unless the Ministry of Environmental Protection corrects these problems, grassroots efforts may play a greater role in effecting environmental change than the Chinese government. According to ministry officials, 51,000 protests relating to the environment took place in 2005, and more than 600,000 environmental complaints were made in 2006.

Poor air quality in China is pervasive, accounting for serious respiratory problems and poor visibility. Although both industrial and automobile emissions are contributors, China's air pollution can be traced mainly to the country's heavy reliance on coal. During the 1980s, its coal consumption increased from 620 million tons to more than 1 billion tons. Industrial centers, such as Chongqing, suffer the effects of severe acid rain linked to the use of high-sulfur coal. Documented damages range from erosion of buildings to the destruction of crops and other plant life. Reportedly, trees along city streets have had to be replaced three times in a 30-year period.

China's water resources also are at risk. Sewage treatment is often inadequate, even nonexistent in some locations, and industrial wastes are contaminating many of the country's rivers and streams. These are nontrivial problems, as evidenced by a recent estimate suggesting that about 700 million people in China drink some form of contaminated water.

For a time, Chinese officials had been reluctant to allocate resources away from economic development and toward environmental cleanup and protection. Instead, China relied on financial support from other countries and organizations. For example, Denmark has helped Chongqing build a modern sewage treatment plant, and the World Bank has provided loans to fund major environmental efforts, including a multimillion-dollar project to clean up Beijing. Despite this international support, recent estimates indicate that China must spend between 2 and 4 percent of its GDP to clean up its environment.

What does China stand to gain economically from intensified environmental policy development? If China can halt its environmental deterioration, production and productivity should improve, allowing it to recapture the nearly 6 percent of GDP it now loses to pollution. Moreover, its stepped-up environmental investments are estimated to increase 14.5 percent per year, reaching 2.3 trillion ¥ (about $340 billion) by 2020. Such investments in emissions testing, water purification, and the development of cleaner energy will also add to China's national production and overall job growth.

Sources: Sim and Rong (September 17, 2010); "Asia: Don't Drink the Water and Don't Breathe the Air; Environmental Protection in China" (January 26, 2008); Roberts (October 27, 2003); Biers (June 16, 1994).

APPLICATION 1.4 Environmental-Economic Accounting and GDP: A Worldwide Initiative

Without question, measuring a nation's macroeconomic performance is a complex, yet important, undertaking. The objective is to monetize total production and use the result to assess and monitor a nation's growth. Currently, the accepted method for measuring economic activity follows the System of National Accounts (SNA) endorsed by the United Nations (UN), a universal accounting framework for calculating such performance measures as GDP.

Conventionally, GDP is defined as the monetized value of all final goods and services produced in a country each year. Although designed to be a comprehensive measure of productive activity, GDP is admittedly flawed. For example, difficulties in attempting to capture the so-called underground economy or the value of nonmarketed goods, such as "do-it-yourself" projects, necessarily bias the measure. Beyond these well-known flaws, other inherent biases have caused some to question the validity of GDP as a measure of economic welfare. Among these is the absence of consideration for ecological damage and natural resource depletion associated with economic activity.

Recognizing this shortcoming, many have argued for making environmental adjustments to the GDP measure. The justifications for such revisions are based on the following considerations. Just as the SNA allows for the depreciation of physical capital, so too should it recognize the devaluation of natural resources associated with economic activity. If pesticides leach into underground springs, the damage to drinking water should be monetized and reflected in the GDP measure. Likewise, if a forest is destroyed to make way for urban development, then that loss should be recorded. In both cases, natural assets would be depreciating, and that depreciation should be captured in the SNA to avoid a serious bias.

Proponents of environmentally adjusted performance measures also claim that ignoring the economic value of natural resources perpetuates society's failure to acknowledge the effect of economic growth on future generations. In fact, in some cases, current accounting practices falsely record environmental deterioration as a contribution to economic welfare. An example would be increased medical spending associated with the effects of a toxic chemical leak, which ironically *elevates* a nation's GDP. An explicit accounting of resource depletion and environmental damage would correct this outcome and restore accuracy to national product and income accounts.

Officials in many nations are aware of this environmental bias, and many have undertaken initiatives to address the problem. An important case in point is the 2003 revision to the *Handbook of National Accounting: Integrated Environmental and Economic Account*, known as the SEEA (for the System of Integrated Environmental and Economic Accounting) published by the UN. This revision was a collaborative effort of the UN, the European Commission, the International Monetary Fund, the Organization for Economic Co-operation and Development, and the World Bank, and provided a detailed set of guidelines for countries to integrate environmental assets into their national economic accounting methods. These guidelines have been recognized as an important step toward reducing the environmental bias inherent in GDP measures. In fact, the United Nations Statistical Commission decided in 2005 that these guidelines should become an international statistical standard, which will become known as the revised SEEA when completed.

To learn more about the status of this effort and specific country applications, visit the UN's environmental-economic accounting site at **http://unstats.un.org/unsd/envaccounting/default.asp**.

Sources: United Nations Statistics Division (2011); United Nations, European Commission, International Monetary Fund, Organization for Economic Co-operation and Development (OECD), and World Bank (2003); Banzhaf (2003); Nordhaus (November 1999); Repetto (September 1992); Repetto (June 1992).

the variety of ecosystems they inhabit.[5] There is still much that scientists don't know about the diversity of life on earth. Although approximately 1.8 million species have been identified, most estimates suggest that the actual number may be at least 5 to 10 million, and some biologists believe there may be as many as 100 million.[6] Although

[5]Council on Environmental Quality (January 1993), p. 135.

[6]Raven, Berg, and Hassenzahl (2008), p. 377.

there are many unknowns, consensus within the scientific community is that the variety of species on earth is important to the ecology. Beyond the preservation of a species for its own sake, it is also the case that all life on earth is inexorably connected. Hence, the loss of one species may have serious implications for others, including human life.

The longevity of any biological species can be directly threatened by exposure to pollutants or by other human actions, such as commercial or sport hunting. The major threat to biodiversity, however, is natural habitat destruction linked to changes in land use, which affects entire ecosystems. Population growth and economic development are primarily responsible for this destruction, which includes the harvesting of tropical forests and the engineered conversion of natural landmasses into alternative uses.[7] For example, of the more than 220 million acres of wetlands believed to have existed in the contiguous United States, less than half remain today, estimated to comprise about 107.7 million acres. Indeed, over the 1998–2004 period, an estimated 361,000 acres of coastal wetlands in the eastern United States were lost, despite an overall net gain in U.S. wetlands acreage of about 192,000 for the period.[8] Biodiversity can also be jeopardized by habitat alteration, which is often attributed to environmental pollution. For example, acid rain has been linked to changes in the chemical composition of rivers and lakes as well as to forest declines in Europe and North America. Such disturbances in the natural conditions to which biological life has become adapted can pose a threat to the longevity of these species.

The extent of biodiversity loss is not known. There are, however, indications that concern about diversity loss is warranted. As of 2012, over 1,380 plant and animal groups have been classified as endangered or threatened in the United States and more than 600 in other nations, as shown in Table 1.3. That this issue is one of global

TABLE 1.3 Threatened and Endangered Species as of 2012

	ENDANGERED		THREATENED		TOTAL	
	U.S.	International	U.S.	International	U.S.	International
Mammals	70	256	13	20	83	276
Corals	0	0	2	0	2	0
Birds	76	204	16	14	92	218
Herptiles (reptiles, amphibians)	28	74	33	17	61	91
Fish	77	11	68	1	145	12
Invertebrates (snails, clams, crustaceans, insects, arachnids)	171	7	33	0	204	7
Plants	644	1	150	2	794	3
TOTAL	1,066	553	315	54	1,381	607

Source: U.S. Department of the Interior (DOI), Fish and Wildlife Service (FWS) (January 6, 2012).

[7]Raven, Berg, and Hassenzahl (2008), pp. 385–86.

[8]Stedman and Dahl (2008); Dahl (2006); U.S. EPA, Office of Water, Office of Wetlands, Oceans, and Watersheds (February 22, 2006).

concern is evidenced by the *Convention on Biodiversity*, executed by 153 nations at the Rio Summit. Among the convention's mandates are devising measures that identify which species are in decline and discovering why the declines have occurred.

Collectively, the goals of environmental quality, sustainable development, and biodiversity set an ambitious agenda. This means that all of society must work toward developing effective environmental policy initiatives. Central to this effort is a planning process in which public officials, industry, and private citizens participate. In the context of environmental problems, this process involves a series of decisions about assessing environmental risk and responding to it, as the following overview explains.

ENVIRONMENTAL POLICY PLANNING: AN OVERVIEW

An important observation to make about environmental policy planning is that it involves the interdependence of many segments of society, including government agencies, private industry, the scientific community, and environmentalists. Each group of participants, albeit from a different vantage point, plays a significant role in formulating policy, and each offers expertise to the outcome.

Policy Planning in the United States

We can categorize the individuals who are instrumental to environmental policy planning in various ways. Using the broadest classification, there are two major groups: the public sector and the private sector. In the United States, the EPA acts as a sort of liaison between the various constituents of each sector, as shown in Figure 1.4. Established in 1970, the EPA was created by President Nixon from various components of existing federal agencies and executive departments. Today, the EPA operates as an independent agency headed by a president-appointed administrator who oversees its

FIGURE 1.4 Parties Involved in Environmental Policy Planning

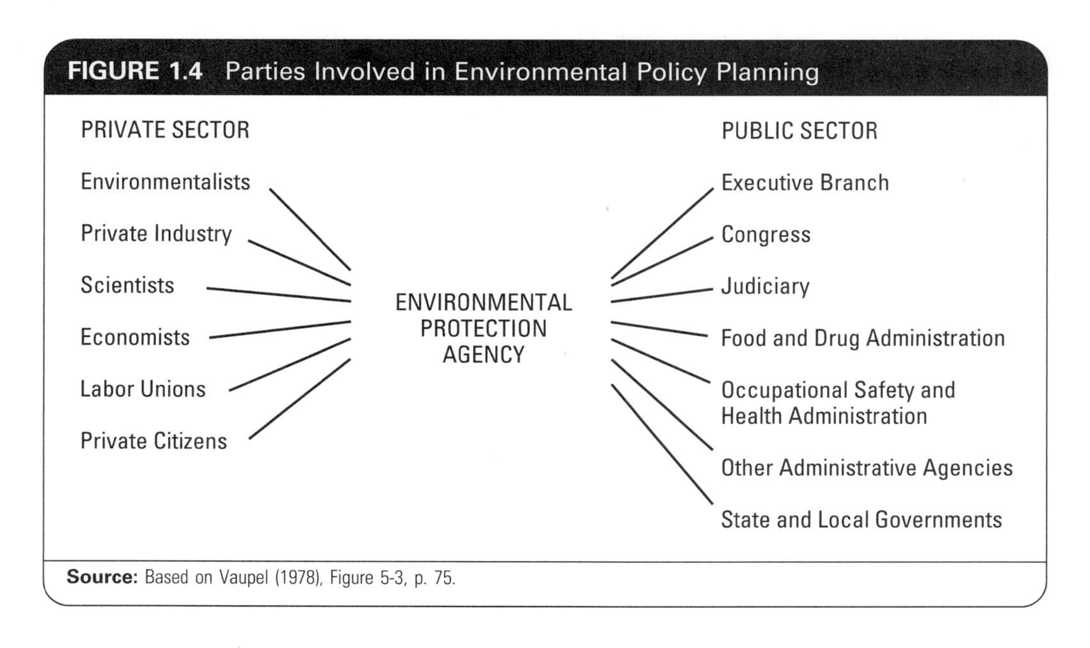

Source: Based on Vaupel (1978), Figure 5-3, p. 75.

vast infrastructure. To better understand this vast federal agency, read about the EPA's mission statement and its strategic plan online at **www.epa.gov/aboutepa/index.html**.

The National Environmental Policy Act (NEPA)

As Figure 1.4 implies, some environmental issues fall under the jurisdiction of federal agencies other than the EPA. For example, issues relating to the contamination of foods or the use of food additives are the responsibility of the Food and Drug Administration (FDA). To coordinate these efforts, the National Environmental Policy Act (NEPA) of 1969 directs the integration of tasks across agencies, executive departments, and branches of government. This act guides the formulation of all U.S. federal environmental policy and requires that the environmental impact of public policy proposals be formally addressed. Among NEPA's provisions is a requirement for an Environmental Impact Statement (EIS) on proposals for legislation or major federal actions. Since the NEPA was passed in 1969, literally thousands of EISs have been completed by a variety of federal agencies.[9] For more information, visit the EIS database at **www.epa.gov/compliance/nepa/eisdata.html**.

In the United States and around the world, environmental policy planning relies on careful research and analysis, which in turn depends on individuals with expertise in many disciplines, among them biology, chemistry, economics, law, and medicine. Input from these and other fields is used to evaluate data and make decisions that ultimately lead to specific policy prescriptions. The underlying tool that guides this policy-planning process is **risk analysis**, which comprises two decision-making procedures: **risk assessment** and **risk management**.

Risk Assessment

At any point in time, a number of environmental objectives must be met with a limited amount of economic resources. This means that as problems are identified, they have to be prioritized. In general, this is done through scientific assessment of the relative risk to human health and the ecology of a given environmental hazard—a procedure known as **risk assessment**. The assessment must determine whether or not a causal relationship exists between the identified hazard and any observed health or ecological effects. If causality is determined, scientists then attempt to quantify *how* the effects change with increased exposure to the hazard. These findings are critical, because they determine whether or not a policy response is necessary and, if so, how immediate and how stringent that policy should be.

risk assessment Qualitative and quantitative evaluation of the risk posed to health or the ecology by an environmental hazard.

risk management The decision-making process of evaluating and choosing from alternative responses to environmental risk.

Risk Management

Assuming that the risk assessment findings warrant it, the planning process enters its next phase: **risk management**. This refers to the decision-making process of evaluating and choosing from alternative responses to environmental risk. In a public policy

[9]Council on Environmental Quality (1997), pp. 47–53.

context, risk responses refer to various types of control instruments, such as a legal limit on pollution releases or a tax on pollution-generating products. The objective of risk management is clear—to choose a policy instrument that reduces the risk of harm to society. What is less obvious is how public officials determine the level of risk society can tolerate and on what basis they evaluate various policy options.

Policy Evaluation Criteria

A number of risk management strategies have been devised to guide these important policy decisions. These strategies use criteria to evaluate policy options. In general, these criteria are based on measures of risk, costs, or benefits—either singularly or in comparison to one another. Two criteria that are economic in motivation are **allocative efficiency** and **cost-effectiveness**.

allocative efficiency
Requires that resources be appropriated such that the additional benefits to society are equal to the additional costs.

- **Allocative efficiency** requires that resources be appropriated such that the additional benefits to society are equal to the additional costs incurred.
- **Cost-effectiveness** requires that the least amount of resources be used to achieve an objective.

cost-effectiveness
Requires that the least amount of resources be used to achieve an objective.

In practice, the choice of criteria is mandated, or at least implied, by law. For example, some provisions in U.S. legislation prohibit cost considerations in policy formulation, implicitly blocking either an efficient or a cost-effective outcome—a result that is frequently debated in the literature and one that we will analyze in later chapters.

Although both efficiency and cost-effectiveness are rooted in resource allocation, **environmental justice** has a different perspective. It is an environmental equity criterion that considers the fairness of the risk burden across geographic regions or segments of the population. More information about this criterion is available at **www.epa.gov/compliance/environmentaljustice**. U.S. policy officials have recognized the importance of environmental justice. In 1994, President William Clinton signed Executive Order 12898, which directed all federal agencies to incorporate environmental justice into their decision-making process. Since that initial signing, however, implementation of that directive has been slow. In fact, according to the Inspector General's office, the EPA under George W. Bush's administration failed to fully incorporate and implement the requirements under Clinton's Executive Order.[10]

environmental justice
Fairness of the environmental risk burden across segments of society or geographical regions.

Recognizing the need for further progress, the EPA under the Obama administration drafted a new strategy in 2010 aimed at improving the integration of environmental justice into EPA programs. The four-year plan, called EJ2014 to mark the twentieth anniversary of Executive Order 12898, outlines specific goals to protect and empower overburdened communities and to establish partnerships at all levels of government to improve these populations. For more detail on the plan and its implementation, visit **www.epa.gov/compliance/environmentaljustice/plan-ej/index.html**.

[10]U.S. EPA, Office of Inspector General (2004).

Government's Overall Policy Approach

An important element of risk management is the regulatory approach used by government. Policies calling for direct regulation of pollution are indicative of a **command-and-control approach**. This form of regulation uses rules or standards to control the release of pollution. In practice, standards either set a maximum on the amount of residuals polluters may release or designate an abatement technology that all sources must use. In either case, polluters have little or no flexibility in deciding how they comply with the law. The command-and-control approach has been the predominant one used in the United States over the past several decades, but in recent years, more economic incentives have been integrated into strategic policy plans. This suggests a shift toward a **market approach** to policy.

command-and-control approach
A policy that directly regulates polluters through the use of rules or standards.

The **market approach** is incentive-based, meaning it attempts to *encourage* conservation practices or pollution-reduction strategies rather than force polluters to follow a specific rule. Many policy instruments can achieve this result, such as a fee on pollutant releases or a tax levied on pollution-generating commodities. What all such instruments have in common is that they tap into natural market forces so that polluters' optimizing decisions will benefit the environment.

market approach
An incentive-based policy that encourages conservation practices or pollution-reduction strategies.

Think about how private firms are motivated by profit. What a market approach does is strategically use this motivation to design environmental policy. For example, if a profit-maximizing firm were discharging a chemical into a river, a market approach might be to charge that polluter a fee for every unit of chemical released. In so doing, the firm would have to pay for the damage it caused, which would erode its profits. This tactic is sometimes called the "**polluter-pays principle**." The expected outcome is that the profit-maximizing firm will reduce the chemicals it releases, using the least-cost method available. The favorable outcomes are that society enjoys the benefit of a cleaner environment and that the associated costs to achieve that gain are minimized.

That such market-based strategies can be effective is validated by experience all over the world. In the United States and other industrialized nations, more market incentives are being integrated within the conventional command-and-control policy approach. In cooperation with the European Environment Agency (EEA), the Organisation for Economic Co-operation and Development (OECD), which supports the polluter-pays principle, has developed a searchable database that describes economic instruments currently in use by its member countries. Covering over 40 nations, the database lists hundreds of examples that illustrate how market instruments can facilitate decision making. To learn more about specific country applications, visit **www2.oecd.org/ecoinst/queries/index.htm**.

Setting the Time Horizon

Another element of risk management decision making is determining the most effective time plan for policy initiatives. One approach is to target policy at more immediate, or *short-term*, problems. These types of initiatives are called **management strategies**, since their purpose is to manage an existing problem. Here, the intent is ameliorative. In terms of the materials balance model, such strategies attempt to reduce the damage from the residual flow.

management strategies
Methods that address existing environmental problems and attempt to reduce the damage from the residual flow.

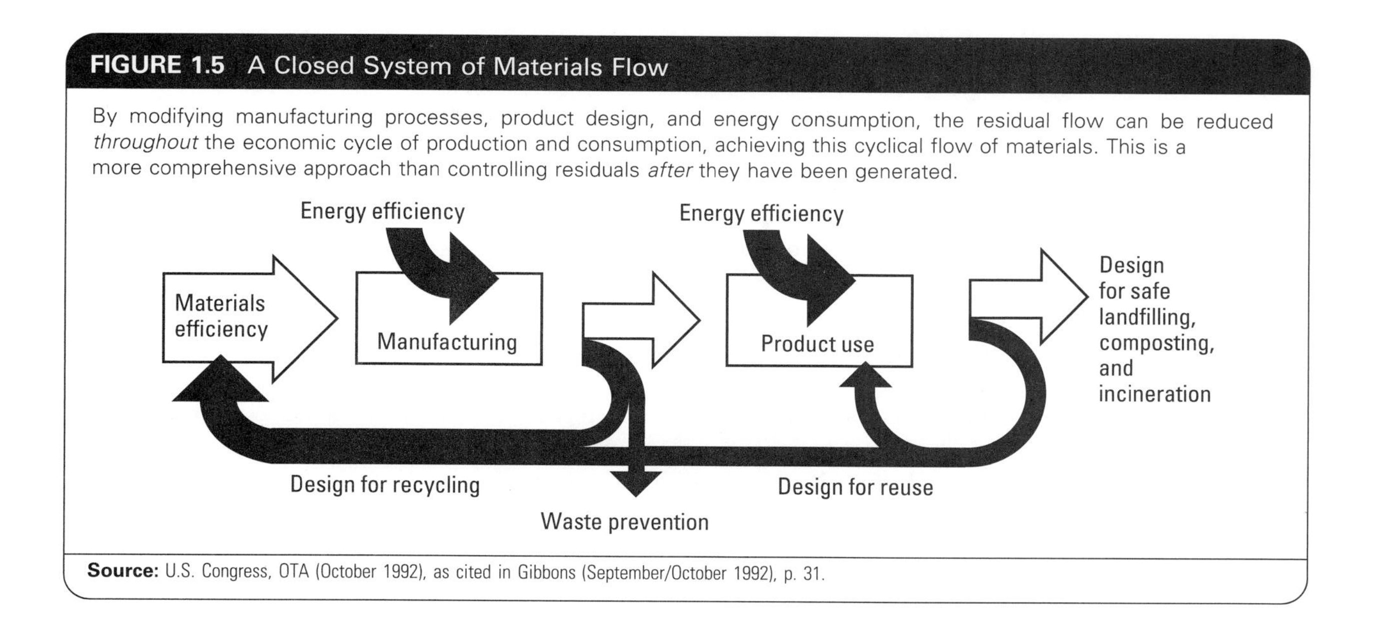

FIGURE 1.5 A Closed System of Materials Flow

By modifying manufacturing processes, product design, and energy consumption, the residual flow can be reduced *throughout* the economic cycle of production and consumption, achieving this cyclical flow of materials. This is a more comprehensive approach than controlling residuals *after* they have been generated.

Source: U.S. Congress, OTA (October 1992), as cited in Gibbons (September/October 1992), p. 31.

pollution prevention (P2) A long-term strategy aimed at reducing the amount or toxicity of residuals released to nature.

An alternative approach addresses the potential of future deterioration and is therefore preventive in purpose. This *long-term* strategy is referred to as **pollution prevention (P2)**. It is implemented by reducing the residual flow and/or minimizing harmful components of residuals, such as toxic chemicals. By modifying energy consumption, manufacturing processes, and product design, as BMW Group does in its DFD method (discussed in Application 1.1), the residual flow can be reduced *throughout* the economic cycle of production and consumption. We model this as a cyclical flow of materials, as shown in Figure 1.5. Notice that this is a more comprehensive approach than controlling residuals after they have been generated. Preventive strategies are becoming more prevalent in U.S. policy. In fact, the nation made a formal commitment to pollution prevention when Congress enacted the Pollution Prevention Act of 1990.

CONCLUSIONS

Concerns about the risks of pollution and the threat of natural resource depletion have been expressed by private citizens, the business community, and governments all over the world. In large part, this perspective has come from a growing awareness of the delicate balance between nature and economic activity. The materials balance model illustrates the strength of this relationship and the consequences of naive decision making that ignores it.

Recognizing the implications, many nations have made measurable progress in identifying problems and setting an agenda to address them. As that work continues, comparable efforts are under way to develop and implement solutions. To that end, policy reform, collaborative arrangements between government and industry, and international summits are being aimed at resolving environmental problems. Scientists are working to learn more about the ecology, the diversity of species, and environmental risks. Laws are being changed to incorporate more preventive measures as well as incentive-based instruments that encourage pollution reduction and resource conservation. As this process evolves, society is changing the way it thinks about the earth's resources, the long-term consequences of its decisions, and its obligation to the future.

Economics has much to contribute to this evolution, in large part because of the interdependence between market decisions and nature. The fundamental concepts of price and optimizing behavior can be used to analyze the effectiveness of environmental policy and to develop alternative solutions. As we explore the discipline of environmental economics, we will use these same concepts to study the effects of pollution as well as the public policy and private responses to the associated risks.

As this chapter suggests, there is a lot of ground to cover, but the importance and relevance of the issues justify the effort. Environmental economics, much like the problem it examines, presents both a challenge and an opportunity—a characterization drawn from the opening address to the Rio Summit:

> The Earth Summit is not an end in itself, but a new beginning.... The road beyond Rio will be a long and difficult one; but it will also be a journey of renewed hope, of excitement, challenge and opportunity, leading as we move into the 21st century to the dawning of a new world in which the hopes and aspirations of all the world's children for a more secure and hospitable future can be fulfilled.[11]

[11]Maurice F. Strong, UNCED secretary-general, in his opening address to the UNCED Conference on Environment and Development, Rio de Janeiro, Brazil, June 3, 1992, as cited in Haas, Levy, and Parson (October 1992), p. 7.

Summary

- The circular flow model is the basis for modeling the link between economic activity and nature.
- The relationship between economic activity and the natural environment is illustrated by the materials balance model.
- The first law of thermodynamics asserts that matter and energy can be neither created nor destroyed. The second law of thermodynamics states that the conversion capacity of nature is limited.
- Pollution refers to the presence of matter or energy, whose nature, location, or quantity produces undesired environmental effects. Some pollutants are natural; others are anthropogenic.
- Sources of pollution are sometimes grouped into mobile and stationary sources. Another common classification is to distinguish point sources from nonpoint sources.
- Local pollution problems are those whose effects do not extend far from the polluting source.
- Regional pollution has effects that extend well beyond the source of the pollution.
- Global pollution problems are those whose effects are so extensive that the entire earth is affected.
- Among the most critical environmental objectives are environmental quality, sustainable development, and biodiversity.
- The National Environmental Policy Act (NEPA) guides the formulation of U.S. federal environmental policy and requires that the environmental impact of all public policy decisions be formally addressed.
- The underlying tool that guides policy planning is risk analysis, which comprises two decision-making procedures: risk assessment and risk management.
- Risk assessment is a scientific evaluation of the relative risk to human health or the ecology of a given environmental hazard. Risk management refers to the process of evaluating and selecting an appropriate response to environmental risk.
- Two economic criteria used in risk management are allocative efficiency and cost-effectiveness.
- Environmental justice, an environmental equity criterion, considers the fairness of the risk burden across geographic regions or across segments of the population.
- A command-and-control policy approach uses limits or standards to regulate environmental pollution.
- A market approach uses economic incentives to encourage pollution reduction or resource conservation.
- Management strategies have a short-term orientation and are ameliorative in intent.
- Pollution prevention strategies have a long-term perspective and are aimed at precluding the potential for further environmental damage.

Review Questions

1. **a.** State how each of the following factors affects the materials balance model: (i) population growth; (ii) income growth; (iii) increased consumer recycling; (iv) increased industrial recycling; (v) increased use of pollution prevention technologies.

   b. Assume that stringent pollution controls are placed on the flow of residuals released into the atmosphere. According to the materials balance model, what does this imply about the residual flows to the other environmental media and/or the flow of inputs into the economy?

2. **a.** Why is design for recycling, as practiced by BMW Group and discussed in Application 1.1, important to the environment?

   b. Explore the design for recycling approach employed by another automobile manufacturer, such as Toyota or another manufacturer, and compare it to that of BMW Group.

3. Faced with the oil crisis of the mid-1970s, the U.S. Congress instituted Corporate Average Fuel Economy (CAFE) standards. These have been increased over time, most recently in 2010 for the 2012–2016 model years. (For detail on the CAFE

standards, visit the Web site of the National Highway Traffic Safety Administration, **www.nhtsa.dot.gov/fuel-economy**.) These standards are aimed at increasing the fuel economy, or miles per gallon (MPG), of automobiles.

a. Briefly describe the expected environmental effect of increasing the MPG of automobiles, holding all else constant.

b. Serious criticism has been lodged against the CAFE standards because U.S. automakers responded by using more plastics in automobiles (to make the cars lighter in weight) to meet the more restrictive CAFE standards. Explain how the use of this technology affects your answer to part (a). Are there any other relevant issues associated with this manufacturing decision?

4. Using the data in Table 1.2, graphically illustrate the relationship between a country's per capita GDP and its per capita municipal waste generation. What conclusion can you draw from your graphical analysis?

5. Use your knowledge of economic principles to discuss how the market premise operates under the "polluter-pays principle."

6. Reconsider the problem of U.S. wetlands loss and the implications for biological diversity. Briefly contrast how a command-and-control policy approach to this problem would differ in intent and implementation from a market approach.

Additional Readings

Anonymous. "Money Can Grow on Trees." *The Economist* (September 25, 2010), p. 6.

Aronsson, Thomas, and Karl-Gustaf Löfgren. *Handbook of Environmental Accounting*. Northampton, MA: Elgar, 2011.

Bartelmus, Peter, and Eberhard K. Seifert. *Green Accounting*. Burlington, VT: Ashgate, 2003.

Beierle, Thomas C., and Jerry Cayford. *Democracy in Practice: Public Participation in Environmental Decisions*. Washington, DC: Resources for the Future, 2002.

Bennett, Jeff, ed. *The International Handbook on Non-Market Environmental Valuation*. Northampton, MA: Elgar, 2011.

Cohen, Mark A. "Deepwater Drilling: Recommendations for a Safer Future." *Resources, No. 177* (Winter/Spring 2011), pp. 23–27.

Commoner, Barry. "Economic Growth and Environmental Quality: How to Have Both." *Social Policy* (Summer 1985), pp. 18–26.

Gertner, John. "The Rise and Fall of the G.D.P." *New York Times Magazine* (May 13, 2010), pp. 60–71.

Goldman, Rebecca L. "Ecosystem Services: How People Benefit From Nature." *Environment 52(5)* (September/October 2010), pp. 15–23.

Hahn, Robert W. "The Impact of Economics on Environmental Policy." *Journal of Environmental Economics and Management 39(3)* (May 2000), pp. 375–99.

Hecht, Joy E. *National Environmental Accounting: Bridging the Gap Between Ecology and Economy*. Washington, DC: Resources for the Future, 2005.

Klyza, Christopher McGrory, and David Sousa. *American Environmental Policy, 1990–2006: Beyond Gridlock*. Cambridge, MA: MIT Press, 2008.

Landy, Marc K., Marc J. Roberts, and Stephen R. Thomas. *The Environmental Protection Agency: Asking the Wrong Questions: Nixon to Clinton*. New York: Oxford University Press, 1994.

Metrick, Andrew, and Martin L. Weitzman. "Conflicts and Choices in Biodiversity Preservation." *Journal of Economic Perspectives 12(3)* (Summer 1998), pp. 21–34.

Najam, Adil, Janice M. Poling, Naoyuki Yamagishi, Daniel G. Straub, Jillian Sarno, Sara M. DeRitter, and Eonjeong Michelle Kim. "From Rio to Johannesburg: Progress and Prospects." *Environment 44(7)* (September 2002), pp. 26–37.

Nordhaus, William D. "New Directions in National Economic Accounting." *American Economic Review, Papers and Proceedings 90(2)* (May 2000), pp. 259–63.

Parry, Ian W. H., and Felicia Day. *Issues of the Day: 100 Commentaries on Climate, Energy, the Environment, Transportation, and Public Health Policy*. Washington, DC: Resources for the Future, 2010.

Polasky, Stephen. *The Economics of Biodiversity Conservation*. Burlington, VT: Ashgate, 2002.

Portney, Paul R. "EPA and the Evolution of Federal Regulation." In Paul R. Portney and Robert N. Stavins, eds. *Public Policies for Environmental Protection*. Washington, DC: Resources for the Future, 2000, pp. 11–30.

Solow, Robert M. "Sustainability: An Economist's Perspective." In Robert N. Stavins, ed. *Economics of the Environment: Selected Readings*. New York: Norton, 2005, pp. 505–13.

Stavins, Robert N. *Environmental Economics and Public Policy*. Northampton, MA: Elgar, 2001.

Stroup, Richard L. *Economics: What Everyone Should Know About Economics and the Environment*. Washington, DC: Cato Institute, 2003.

Venkataraman, Bhawani. "Why Environmental Education?" *Environment 50*(2) (September/October 2008), p. 8.

Wonacott, Peter. "Polluters in China Feel No Pain." *Wall Street Journal Online*, March 24, 2004.

______. "Green Groups Move To Clean Up China." *Wall Street Journal Online*, June 14, 2004.

To access additional course materials, visit www.cengagebrain.com. At the home page, search for the ISBN of this title (shown on the back cover). This will take you to the product page where these resources can be found.

CHAPTER

2

Modeling the Market Process: A Review of the Basics

"There exists no more democratic institution than the market."

—J. A. Schumpeter (1883–1950)

As conveyed through the materials balance model, environmental problems are linked to market activity. Decisions made by consumers and firms affect the abundance and quality of the earth's natural resource stock. Because environmental economics is concerned with resource damage, we need to develop a thorough understanding of how market activity gives rise to polluting residuals and why market forces alone cannot solve the problem. From an economic perspective, environmental pollution is characterized as a market failure. Hence, environmental economics uses market failure models to analyze the problem and to identify solutions. But these models rely on a solid understanding of the market process itself.

To that end, in this chapter we present the essential components of a market and the basic concepts used in microeconomics. We focus on the circular flow model, which is central to the materials balance paradigm. Supply and demand fundamentals are reviewed to reestablish a good grasp of market behavior, the motivations for consumer and firm decision making, and price determination.

The context for our review is a hypothetical market for bottled water in which competitive conditions are assumed. Through an analysis of market equilibrium, we discuss allocative efficiency, an economic criterion used throughout the study of environmental economics to evaluate public and private responses to pollution problems. From there, we develop welfare measures, which are useful in evaluating the effect of environmental policy on society. The analytical and modeling tools presented here will serve as the foundation for our study of market failure in Chapter 3.

MARKET MODELS: THE FUNDAMENTALS

market
The interaction between consumers and producers to exchange a well-defined commodity.

Defining the Relevant Market

In economics, the concept of a market has a broader definition than in everyday usage. Specifically, a **market** refers to the interaction between consumers (or buyers) and producers (or sellers) for the purpose of exchanging a well-defined commodity. This definition is purposefully general and abstract, since an economic market refers to the process of exchange and the conditions underlying that exchange for a broad range of economic activities. For example, this definition is as relevant to hiring workers in the labor market as it is to buying groceries at a supermarket. As we will discover in upcoming chapters, it can even be applied to an analysis of pollution control in the "market for environmental quality." Therefore, one of the more critical steps in economic analysis is defining the market context for the good or service under investigation.

Specifying the Market Model

Once the relevant market has been defined, a model of that market and its characteristics must be specified. The form of the model varies with the objective of the study and its level of complexity. Simple qualitative relationships among economic variables can be modeled using a two-dimensional graph. To quantify these relationships, models are refined through the use of equations or functions. Formal testing is accomplished through empirical analysis of these theorized relationships using real-world data.

MODELING SUPPLY AND DEMAND: AN OVERVIEW

By definition, a market exchange for any commodity comprises two sets of independent decision makers: buyers and sellers. Each is motivated by different objectives, and each is influenced and even constrained by different factors. The decisions of producers are modeled through a **supply** function, while consumers' decisions are modeled through a **demand** function. When considered simultaneously, the resulting market model of supply and demand determines equilibrium output and price.

Purpose of the Model

The main objective of the supply and demand model is to facilitate an analysis of market conditions and any observed price changes. An investigation of price movements can identify shortages and surpluses, the existence of resource misallocations, and the economic implications of government policy. For example, environmental economists can use supply and demand models to investigate the effectiveness of a gasoline tax on reducing gas consumption to help improve urban air quality. By studying the associated changes in market conditions and movements in gasoline prices, economists can determine how consumption patterns are affected, how the tax burden is shared between the consumer and the producer, and how income distribution is affected.

More complex analysis is necessary when the market system fails to operate properly. In these instances, the conventional model of supply and demand must be modified to account for those conditions that weaken the operation of market forces. Economic theory suggests that the persistence of environmental problems such as urban smog and water pollution is the result of failures or breakdowns in the market system. Of course, to understand why this is so and to begin to find solutions, we must have a good command of the market process, the underlying supply and demand conditions, and the mechanisms of the price system.

Building a Basic Model: Competitive Markets for Private Goods

To develop a basic model of supply and demand, we make a number of assumptions. First, a competitive *goods* market is assumed, which is characterized primarily by (1) a large number of independent buyers and sellers with no control over price, (2) a homogeneous product, (3) the absence of entry barriers, and (4) perfect information. Second, we assume that the market for *resources* also is competitive. This implies that the individual firm has no control over input prices—a result that will simplify the model. Finally, the output being exchanged in the market is assumed to be a private good. A **private good** is one that has two characteristics—rivalry in consumption and excludability. This means that consumption of the good by one person precludes that of another, and the benefits of consumption are exclusive to that single consumer. This assumption of a private good (as opposed to a public good[1]) is critical to a conventional analysis of quantity and price determination.

private good
A commodity that has two characteristics—rivalry in consumption and excludability.

MARKET DEMAND

Demand refers to the market response of consumers, who adjust their purchasing decisions to maximize their satisfaction, or what economists term **utility**. Many factors influence consumers' decisions. However, because a key objective of market analysis is price determination, the demand function is specified as the relationship between quantity demanded and price, holding constant all other variables that influence this decision. Economists use the Latin phrase *ceteris paribus*, abbreviated *c.p.*, to mean "holding all else constant." Hence, **demand** is formally defined as the quantities of a good the consumer is willing and able to purchase at some set of prices during some time period, *c.p.* The consumer's "ability to pay" refers to the income constraint that limits consumer choice. The "willingness to pay" is the value or benefit the consumer expects to receive from consumption of the commodity. In fact, this willingness to pay, or **demand price**, is considered a measure of the **marginal benefit (*MB*)** associated with consuming another unit of the good.

demand
The quantities of a good the consumer is willing and able to purchase at a set of prices during some time period, *c.p.*

The key economic variables that are held constant when specifying demand are the wealth and income of the consumer, the prices of related goods (i.e., substitutes and complements), preferences, and price expectations. A change in any of these variables alters the entire price-quantity relationship, which represents a *change in demand*. This is distinct from the effect on consumption of a change in price, which causes only a *change in quantity demanded*. For example, consumers typically buy more of a good when it goes on sale. This is simply a change in the quantity demanded of the product

[1]Public goods are excluded from our analysis because they represent a type of market failure. These are commodities such as national defense and clean air, whose consumption is both nonrival and nonexclusive. That is, once a public good is provided to one consumer, it is difficult or very costly to prevent others from sharing in its consumption. A more formal development of the theory of public goods as it pertains to environmental issues is presented in Chapter 3.

in response to a price change. On the other hand, if, for example, consumers' tastes change such that they want more of a product at *all* possible prices, the result would be a change in demand. A case in point is the observed shift in consumer preferences toward "environmentally friendly" products. Application 2.1 discusses survey data that speak to this phenomenon and its effect on market demand.

APPLICATION 2.1 Environmentally Driven Demand: A 20-Year Look Back

Over two decades ago, a 1991 survey of New York adults revealed the development of a trend in consumers' spending decisions that reflected brand-switching in favor of environmentally safer products. Such a change in preferences implied a shift in market demand toward these products and away from less safe substitutes. According to the survey, some of the nondurable consumer products affected by this phenomenon were identified as detergents, diapers, aerosol sprays, and cleaners. Overall, 44.6 percent of those survey participants said they had switched brands for environmental reasons, with a higher proportion reported for the 31- to 45-year-old age group. Gender differences also were identified with respect to this switching phenomenon. Some 49.4 percent of females in the survey expressed a willingness to substitute toward more environmentally safe brands, whereas the comparable proportion for males was 37.9 percent.

On a broader level, the 1991 survey also suggested that environmental safety had become an important influence on consumer decision making. About 93 percent of survey respondents asserted that protecting the environment was a "very important factor" in their purchasing decisions. Yet, only 1 in 10 respondents placed environmental safety as a top priority. What factors were identified as more important? It turns out that product price and quality were shown to be the most significant determinants of consumers' buying decisions.

A decade later, a 2001 Gallup poll showed that almost 75 percent of Americans would in fact be willing to pay higher prices to cover the added costs associated with tougher emission standards. Several years later, a 2007 Gallup poll revealed similar sentiments—this time in the context of global warming. This 2007 poll found that 78 percent of Americans believed that individuals should spend more to reduce home energy usage, and over 60 percent of respondents supported the purchase of pricier hybrid cars.

Interestingly, more recent survey results suggest a slight shift in consumer attitudes about the overall environment. Specifically, a 2010 Gallup poll shows relatively low levels of public concern about the environment, particularly global warming, and limited apprehension about environmental protection. These results are believed to be linked to difficult economic conditions as well as political divisiveness about the environment. However, the poll also indicates that environmental activism has remained fairly stable over the 10-year period from 2000 to 2010. Five of eight pro-environmental behaviors surveyed, including recycling, product selection, and energy conservation, show either no change or slight increases in participation over the period. For example, 90 percent of respondents reported participating in recycling activity on a voluntary basis, which is the same proportion reported 10 years prior. And 76 percent of Americans reported having purchased a product specifically because it was perceived to be less harmful to the environment, which was a 3-percentage-point increase from 2000.

Taken together, it is no surprise that manufacturers continue to respond to consumers' environmentally driven product preferences by promoting their brands as "environmentally safe." The 1991 New York survey provided early information about the relative success of this effort across competitive commodities. Brand names such as Clorox, Arm & Hammer, Tide, L'Oréal, and Revlon were among those perceived by consumers as associated with environmentally superior products. Similar observations can be made about many of today's consumer goods and services. Individuals routinely examine product labels for recycled content or energy efficiency ratings as a way to differentiate among the multitude of products for sale in the marketplace.

So, just as demand theory predicts, price is a major determinant of demand. But, many other factors influence consumer decisions. Survey findings over the past 20 years along with anecdotal evidence suggest that environmental concerns have an observable effect on consumers' buying habits. And logically, profit-maximizing firms respond to these consumer preferences by offering and marketing environmentally safer goods and services.

Sources: Dunlap (2010); Carroll (2007); Dunlap and Saad (April 16, 2001); Manly (March 23, 1992).

Law of Demand

Law of Demand
There is an inverse relationship between price and quantity demanded of a good, *c.p.*

Under conventional circumstances, the relationship between quantity demanded and price is an inverse one, which is referred to as the **Law of Demand**. This means that a *rise* in price is associated with a *fall* in quantity demanded, *c.p.* This is a highly intuitive theory, because we expect consumers to view price as an obstacle that limits consumption, given their income constraints.

Modeling Individual Demand

To illustrate the Law of Demand, Table 2.1 presents hypothetical data for an individual's demand for one-liter bottles of water. The values show the utility-maximizing quantity decisions in a one-month period for prices ranging from \$0.50 to \$11.50 per bottle. Notice how the inverse relationship between quantity demanded and price holds throughout. Although these data give only a sampling of price-quantity pairs, an equation of the same relationship models *all* possible price-quantity responses for the consumer. In this case, the demand function is the linear relationship $q_d = -2P + 23$, where q_d signifies the quantity demanded of a single individual, and P is the price per one-liter bottle. Notice that if any of the P values from Table 2.1 are substituted into the right-hand side of the equation, the corresponding q_d values are obtained. For example, if $P = \$3.50$, then $q_d = -2(3.50) + 23 = 16$.

By convention, the graph of demand uses the *inverse* form of the function, that is, $P = f(q_d)$. In this case, solving the demand equation for P in terms of q_d gives $P = -0.5q_d + 11.50$. The graph of this single consumer's demand curve (d) is shown in Figure 2.1.

TABLE 2.1 Single Consumer's Demand Data for Bottled Water

Price (\$) P	Quantity Demanded (bottles/month) $q_d = -2P + 23$
0.50	22
1.50	20
2.50	18
3.50	16
4.50	14
5.50	12
6.50	10
7.50	8
8.50	6
9.50	4
10.50	2
11.50	0

FIGURE 2.1 One Consumer's Demand (*d*) for Bottled Water

This model of one consumer's demand (*d*) uses the ***inverse*** form of the function $P = -0.5q_d + 11.50$, where -0.5 is the slope and $+11.50$ is the vertical intercept. The negative slope illustrates the Law of Demand.

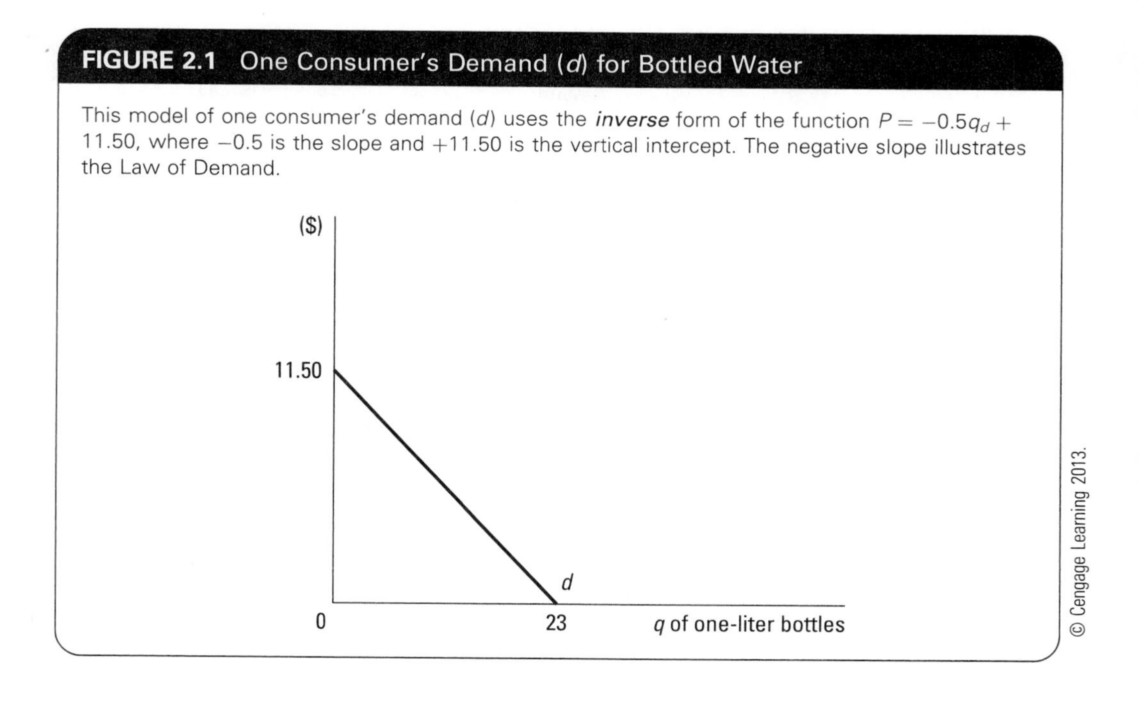

Deriving Market Demand From Individual Demand Data

market demand for a private good
The decisions of all consumers willing and able to purchase a good, derived by *horizontally* summing individual demands.

For most applications in economics, the collective decision making of all consumers in a market is more relevant than that of a single consumer. Thus, the more appropriate concept is **market demand**, representing all consumers who are willing and able to purchase the commodity. Market demand is found by summing over the individual demand values. For private goods, we sum the quantities at each demand price, a procedure called "horizontal summing," because quantity is conventionally plotted on the horizontal axis.[2] Even though price is the same for all consumers, quantity decisions vary because of differences in other factors such as income, wealth, tastes, and expectations.

For simplicity, we first illustrate this summing procedure by adding the demand of only one other consumer to the model specified previously. Individual demand data for two hypothetical consumers, Consumer 1 and Consumer 2, are given in Table 2.2. Notice that each consumer makes unique decisions about quantity because each has unique preferences, income, wealth, and so forth. The aggregate demand for these two individuals is found by summing the two quantity columns at each price level.

[2]Horizontal summing is characteristic of **private goods**. Because consumption of such goods is excludable, each individual is able to choose his or her own quantity. As we discuss in Chapter 3, this is *not* true for public goods, an outcome with important environmental economic implications.

TABLE 2.2 Combined Demand Data for Bottled Water for Two Consumers

Price ($) P	Quantity Demanded by Consumer 1 (bottles/month) $q_{d1} = -2P + 23$	Quantity Demanded by Consumer 2 (bottles/month) $q_{d2} = -4P + 46$	Combined Quantity Demanded by Both Consumers (bottles/month) $q_{d(1+2)} = (q_{d1} + q_{d2}) = -6P + 69$
0.50	22	44	66
1.50	20	40	60
2.50	18	36	54
3.50	16	32	48
4.50	14	28	42
5.50	12	24	36
6.50	10	20	30
7.50	8	16	24
8.50	6	12	18
9.50	4	8	12
10.50	2	4	6
11.50	0	0	0

The result is shown in the far right-hand column of the table. The same method can be applied to an algebraic model by adding each pair of corresponding terms in the two demand equations, as shown below.

$$\begin{array}{lrl}
\text{Demand for Consumer 1:} & q_{d1} & = -2P + 23 \\
+ \quad \text{Demand for Consumer 2:} & q_{d2} & = -4P + 46 \\
\hline
\text{Demand for Consumers 1 and 2:} & q_{d(1+2)} = (q_{d1} + q_{d2}) & = -6P + 69
\end{array}$$

To derive the market demand, we use the same approach except we aggregate across *all* consumers. Maintaining the characteristics of a competitive market, we assume there are 100 consumers in the market for bottled water. The hypothetical data are shown in Table 2.3. For each price, we show the combined quantity values for Consumers 1 and 2 in the second column, the aggregated data for the remaining 98 consumers in the third column, and the market quantity demanded, $Q_D = \Sigma q_d$, in the fourth column. The corresponding equations are given below.

$$\begin{array}{lrl}
\text{Demand for Consumers 1 and 2:} & q_{d(1+2)} & = -6P + 69 \\
+ \quad \text{Demand for Consumers 3 through 100:} & q_{d(3,\ \ldots\ ,100)} & = -94P + 1{,}081 \\
\hline
\text{Market Demand:} & Q_D & = -100P + 1{,}150
\end{array}$$

The graph of the *inverse* market demand curve (D), or $P = -0.01Q_D + 11.5$, is shown in Figure 2.2.

TABLE 2.3 Market Demand Data for Bottled Water

Price ($) P	Quantity Demanded by Consumers 1 and 2 (bottles/month) $q_{d(1+2)} = -6P + 69$	Quantity Demanded by Consumers 3 through 100 (bottles/month) $q_{d(3,\ldots,100)} = -94P + 1{,}081$	Market Demand (bottles/month) $Q_D = -100P + 1{,}150$
0.50	66	1,034	1,100
1.50	60	940	1,000
2.50	54	846	900
3.50	48	752	800
4.50	42	658	700
5.50	36	564	600
6.50	30	470	500
7.50	24	376	400
8.50	18	282	300
9.50	12	188	200
10.50	6	94	100
11.50	0	0	0

FIGURE 2.2 Market Demand (*D*) for Bottled Water

This model of the market demand (D) for bottled water uses the *inverse* market demand function $P = -0.01Q_D + 11.5$, which represents the decisions of all consumers in the market. The market demand curve is derived using a horizontal summing procedure across all individual demand curves.

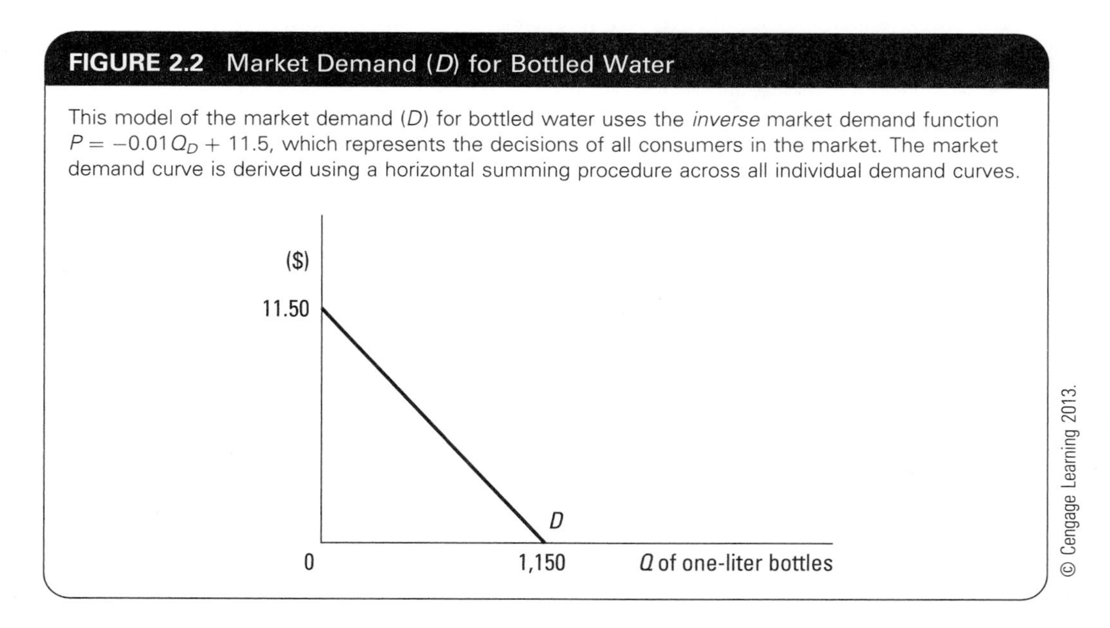

MARKET SUPPLY

supply
The quantities of a good the producer is willing and able to bring to market at a given set of prices during some time period, *c.p.*

On the opposite side of the market, we derive a supply relationship based on the decisions of producers who are motivated by profit. Each firm's supply decision is modeled as a function of price, even though this decision is influenced by many other variables. Therefore, we say that **supply** refers to the quantities of a good the producer is willing and able to bring to market at a given set of prices during some time period, *c.p.* Among the variables that potentially affect the firm's supply decision are production technology, input prices, taxes and subsidies, and price expectations. Analogous to the demand side of the market, changes in these determinants affect the entire price-quantity relationship, causing a *change in supply*, whereas a movement in price is associated with a *change in quantity supplied.*

Law of Supply

Law of Supply
There is a direct relationship between price and quantity supplied of a good, *c.p.*

The relationship between quantity supplied and price is generally a positive one, expressed as the **Law of Supply**. This means that a *rise* in price is associated with a *rise* in quantity supplied, *c.p.* The conventional assumption that firms are profit maximizers suggests that a higher price is an incentive for firms to produce more output. A contemporary example of how powerful this profit motive can be is in the context of so-called green markets, discussed in Application 2.2.

APPLICATION 2.2 International and Domestic Firms Discover Profit in Environmentalism

For a time, the common perception of how environmental goals would affect the business sector was that such efforts only added to costs, which in turn reduced profitability. Although it is true that meeting new environmental regulations can be costly, it is also the case that today's companies are finding ways to reduce pollution *and* enhance profits. How? The approach varies widely across firms, industries, and nations. This is illustrated with a few examples.

One obvious way that firms are enjoying profits tied to environmentalism is through recycling. A commonly cited example is the use of sludge from electric utilities' smokestacks to produce gypsum, an input in the manufacture of such products as wallboard. Although corporate recycling may seem like old news, what *is* new is that such efforts are becoming more pervasive, showing up in entirely new contexts. An innovative example is McDonald's experimental use of a Big Mac package made from discarded juice boxes. Another is Airbus's dismantling project, "Process for Advanced Management of End-of-Life Aircraft (PAMELA)," launched in 2005 to develop an efficient recycling procedure for jetliners. Competing with Boeing in this effort just as it does in sales, Airbus's goal is to recycle 95 percent of every aircraft by 2015. In every case, the environment is saved from excess waste and pollution, and the company enjoys lower costs and a better bottom line. For more on Airbus's aircraft end-of-life efforts, visit **www.airbus.com/innovation/eco-efficiency/aircraft-end-of-life/**.

A fairly new corporate approach to "profitable environmentalism" is something called remanufacturing. Unlike recycling, where used products or wastes are recovered and formed into alternative raw materials, remanufacturing involves the collection, reconditioning, and reselling of the same product over and over again. A case in point is the remanufacturing of used printer cartridges, which has grown to a $3 billion industry. This remanufacturing trend is a win-win situation. The gains to the environment are energy conservation and reduced waste disposal. For the firm, the gain is cost savings, assuming disassembly and reconditioning are cheaper than starting from scratch, where the cost difference is generally 40–65 percent.

Other environmentally based profits arise in an entirely different way. Sometimes, companies launch a project solely for environmental reasons, but it ends up being a lucrative source of new profit that is entirely unexpected. Starbucks Corporation experienced such an outcome in late 1998. The company had received negative commentary about coffee production causing the destruction of rain forests. In response, Starbucks subsidized Mexican farmers to find ways to improve the production of coffee beans grown in the shade. Quite unexpectedly, the new beans were a market success, and the coffee-making giant expanded the project to four other countries. More on Starbucks's environmental policies is available at **www.starbucks.com/responsibility/environment**.

No matter the source or the motivation, in every case and in numerous others like these, corporations are proving that profits and environmental responsibility need not be mutually exclusive.

Sources: Kingsley-Jones (May 27–June 2, 2008); Michaels (June 1, 2007); Deutsch (September 9, 2001); Arndt (April 16, 2001); Ginsburg (April 16, 2001).

A more formal justification for the Law of Supply is based upon the nature of firms' costs as production is carried out. As the firm produces more output (Q), its **total costs (TC)** rise proportionately faster, meaning that the ratio of the change in TC (ΔTC) to the change in Q (ΔQ) is increasing. This ratio ($\Delta TC/\Delta Q$) defines the firm's **marginal cost (MC)** of production, the additional cost of producing another unit of output. Because MC rises as Q rises, firms need to charge a higher price for each extra unit of output they produce. Hence, the existence of rising MC supports the positive price-quantity relationship given by the Law of Supply.

Modeling Individual Supply

The Law of Supply is illustrated by using the hypothetical data for a single producer of bottled water given in Table 2.4. The quantity column shows a single firm's profit-maximizing output decisions associated with prices ranging from \$0.50 to \$11.50. Notice how the positive relationship between the **supply price** and the quantity supplied holds throughout. The linear equation associated with these data is $q_s = +16P - 4$, where q_s is used to signify the quantity supplied by a single firm. Each of the price-quantity pairs given in Table 2.4 satisfies this equation.

The graphical model of the single firm's supply curve (s) is illustrated in Figure 2.3. Again, the convention is to plot price on the vertical axis and quantity on the horizontal axis, using the inverse form of the equation. In this case, the inverse supply function is $P = +0.0625q_s + 0.25$. Given the absence of extremes, the supply curve has the expected positive slope in accordance with the Law of Supply and the underlying theory of rising MC. In fact, under the assumption of competitive output markets, the firm's short-run supply curve *is* its MC curve.

market supply of a private good The combined decisions of all producers in a given industry, derived by *horizontally* summing individual supplies.

Deriving Market Supply From Individual Supply Data

To analyze production at an aggregated level, **market supply** of a private good is derived using a *horizontal* summing procedure over the individual quantity decisions of

TABLE 2.4 Single Producer's Supply Data for Bottled Water

Price (\$) P	Quantity Supplied (bottles/month) $q_s = +16P - 4$
0.50	4
1.50	20
2.50	36
3.50	52
4.50	68
5.50	84
6.50	100
7.50	116
8.50	132
9.50	148
10.50	164
11.50	180

FIGURE 2.3 One Producer's Supply (s) of Bottled Water

This graph shows the supply (s) of one producer in the bottled water market. It is based on the inverse supply function $P = +0.0625q_s + 0.25$. Because this market is assumed to be competitive, this supply curve is the firm's *MC* curve. Notice that this function has a slope of +0.0625 and a vertical intercept of +0.25. The positive slope reflects the Law of Supply and the theory of rising *MC*.

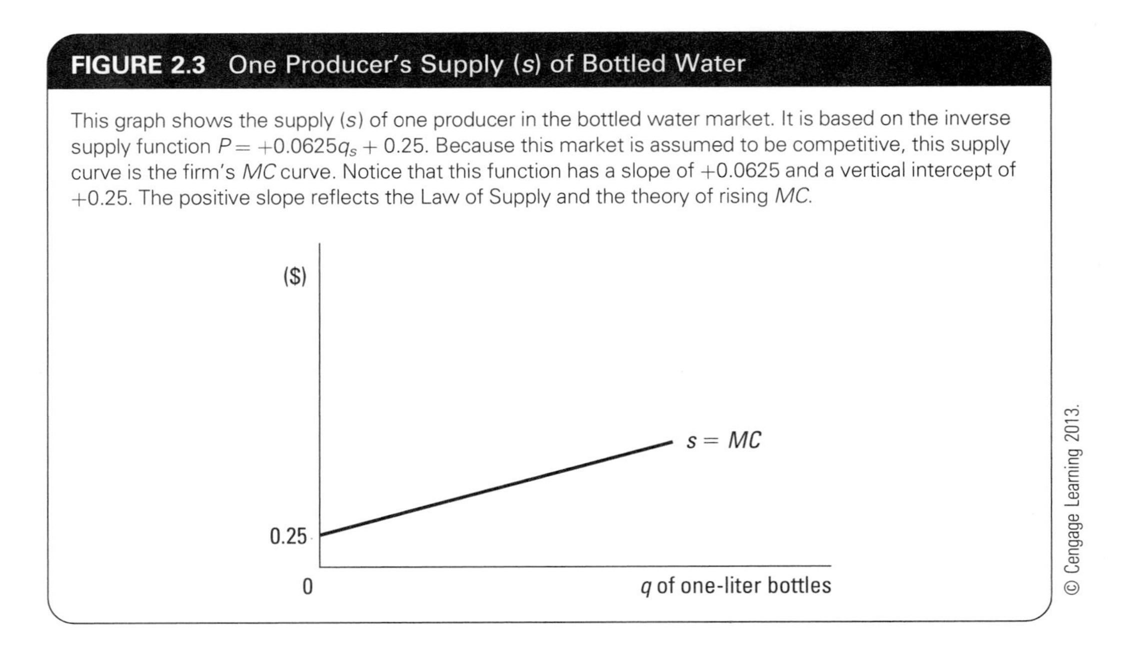

firms, analogous to what is done on the demand side. However, because the output market is assumed to be competitive, the adding-up process on the supply side is simpler, since all firms are identical under such market conditions.

Just as on the demand side, we begin by illustrating the aggregation procedure using only two firms. Hypothetical data for two representative producers of bottled water and their combined supply decisions are given in Table 2.5. The algebraic counterpart is shown next.

TABLE 2.5 Combined Supply Data for Bottled Water for Two Producers

Price ($) P	Quantity Supplied by Producer 1 (bottles/month) $q_{s1} = +16P - 4$	Quantity Supplied by Producer 2 (bottles/month) $q_{s2} = +16P - 4$	Combined Quantity Supplied by Both Producers (bottles/month) $q_{s(1+2)} = q_{s1} + q_{s2} = +32P - 8$
0.50	4	4	8
1.50	20	20	40
2.50	36	36	72
3.50	52	52	104
4.50	68	68	136
5.50	84	84	168
6.50	100	100	200
7.50	116	116	232
8.50	132	132	264
9.50	148	148	296
10.50	164	164	328
11.50	180	180	360

	Supply by Producer 1:	$q_{s1} = +16P - 4$
+	Supply by Producer 2:	$q_{s2} = +16P - 4$
	Combined Supply by Producers 1 and 2:	$q_{s(1+2)} = (q_{s1} + q_{s2}) = +32P - 8$

To bring the supply relationship up to the market level, we sum output levels over *all* firms in the industry, assumed to total 25. The hypothetical **market supply** data are shown in Table 2.6. For each price, the combined supply decisions for Producers 1 and 2 are shown in the second column, the aggregated data for the remaining 23 producers are given in the third column, and the market quantity supplied ($Q_S = \Sigma q_s$) is given in the last. The corresponding supply equations are given below.

	Supply by Producers 1 and 2:	$q_{s(1+2)} = +32P - 8$
+	Supply by Producers 3 through 25:	$q_{s(3, \ldots ,25)} = +368P - 92$
	Market Supply:	$Q_S = +400P - 100$

Figure 2.4 shows the graph of the market supply curve (S) in inverse form, which is $P = +0.0025Q_S + 0.25$. At each price, the horizontal distance between the vertical axis and S is exactly equal to the sum of the analogous horizontal intervals for each of the 25 individual supply (s) curves. Furthermore, because the competitive firm's supply (s) is its MC curve, the market supply (S) curve is in turn the horizontal sum of the individual firms' MC curves.

TABLE 2.6 Market Supply Data for Bottled Water

Price ($) P	Quantity Supplied by Producers 1 and 2 (bottles/month) $q_{s(1+2)} = +32P - 8$	Quantity Supplied by Producers 3 through 25 (bottles/month) $q_{s(3, \ldots ,25)} = +368P - 92$	Market Supply (bottles/month) $Q_S = +400P - 100$
0.50	8	92	100
1.50	40	460	500
2.50	72	828	900
3.50	104	1,196	1,300
4.50	136	1,564	1,700
5.50	168	1,932	2,100
6.50	200	2,300	2,500
7.50	232	2,668	2,900
8.50	264	3,036	3,300
9.50	296	3,404	3,700
10.50	328	3,772	4,100
11.50	360	4,140	4,500

FIGURE 2.4 Market Supply (S) of Bottled Water

This model depicts the market supply curve (S) for bottled water based on the inverse market supply function $P = +0.0025Q_S + 0.25$. The market supply is found by summing all individual supply curves (s) horizontally. Because each individual supply (s) curve is the firm's MC curve, the market supply (S) represents the horizontal sum of these MC curves.

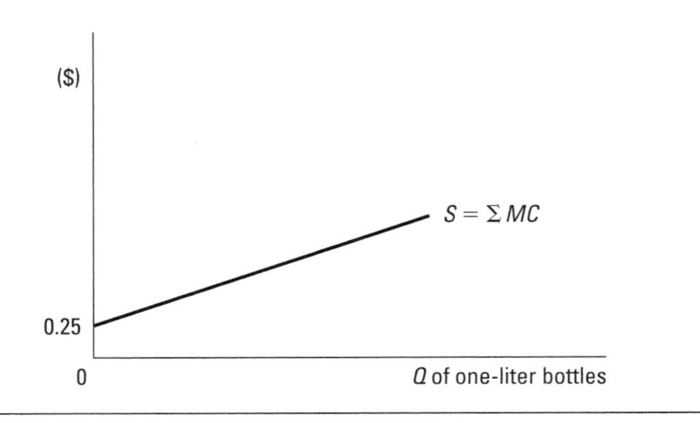

MARKET EQUILIBRIUM

Thus far, we have considered each side of the market separately to develop distinct models of economic decision making. To develop a model of price determination, we must consider supply and demand *simultaneously* to allow for the interaction of consumers and producers in the marketplace. The formal theory that price is simultaneously determined by supply and demand is one of the most significant in all of economic analysis.[3]

Equilibrium Price and Quantity

equilibrium price and quantity The market-clearing price (P_E) associated with the equilibrium quantity (Q_E), where $Q_D = Q_S$.

Together, the forces of supply and demand determine a unique **equilibrium price (P_E)** at which point the market system has no tendency for change. Equilibrium, or market-clearing, price (P_E) is the price at which quantity demanded by consumers, Q_D, is exactly equal to the quantity supplied by producers, Q_S (i.e., where $Q_D = Q_S$). Only at P_E will the associated **equilibrium quantity (Q_E)** be both the profit-maximizing production level for firms and the utility-maximizing consumption level for consumers.

Market equilibrium for our hypothetical bottled water market is shown in Table 2.7, where market supply and demand data are presented together. Notice that equilibrium price is \$2.50, the only price at which Q_S equals Q_D of 900 units. This result also can be determined algebraically by solving the market demand and market supply equations simultaneously, as follows:

At equilibrium: $Q_S = Q_D$

Substituting: $+400P - 100 = -100P + 1{,}150$

Solving: $500P = 1{,}250$ or $P_E = \$2.50$

Substituting into either equation: $Q_E = 400(2.50) - 100$, or $-100(2.50) + 1{,}150 = 900$

[3]This theory is attributable to Alfred Marshall, who first published this theory in his *Principles of Economics* in 1890.

TABLE 2.7 Market Supply and Demand Data for Bottled Water

Price ($) P	Market Quantity Supplied $Q_S = +400P - 100$	Market Quantity Demanded $Q_D = -100P + 1,150$	Market Surplus or Shortage
0.50	100	1,100	Shortage = 1,000
1.50	500	1,000	Shortage = 500
2.50	**900**	**900**	**Equilibrium**
3.50	1,300	800	Surplus = 500
4.50	1,700	700	Surplus = 1,000
5.50	2,100	600	Surplus = 1,500
6.50	2,500	500	Surplus = 2,000
7.50	2,900	400	Surplus = 2,500
8.50	3,300	300	Surplus = 3,000
9.50	3,700	200	Surplus = 3,500
10.50	4,100	100	Surplus = 4,000
11.50	4,500	0	Surplus = 4,500

FIGURE 2.5 Equilibrium in the Market for Bottled Water

Market demand (D) and market supply (S) determine equilibrium price (P_E) and quantity (Q_E) at their point of intersection.

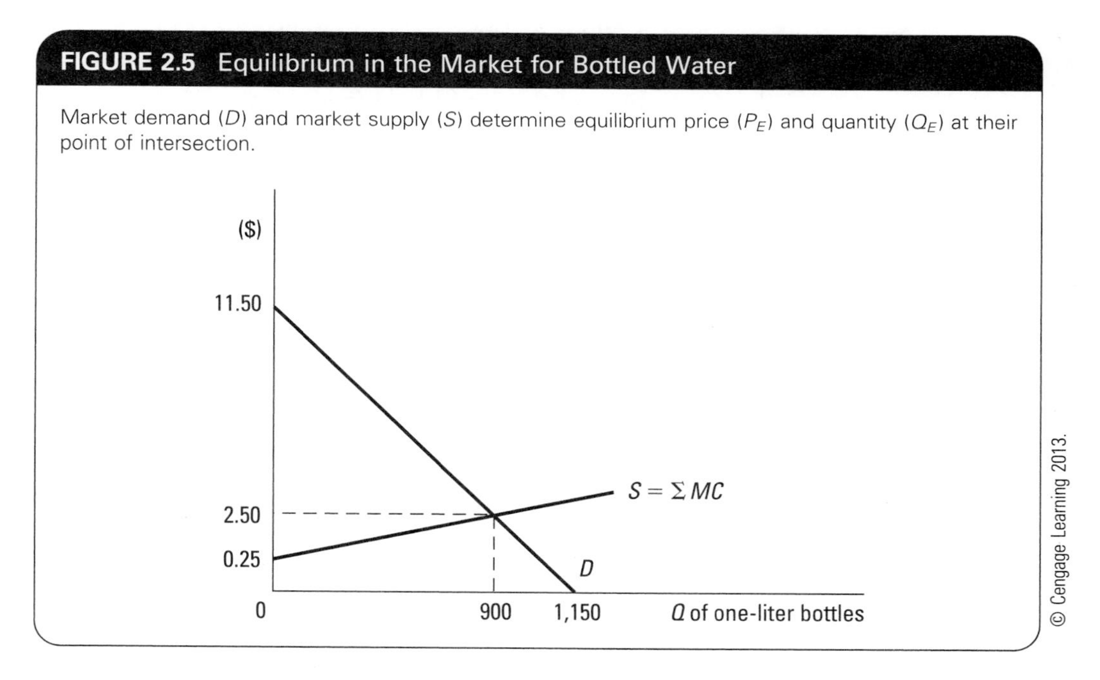

Graphically, we model market equilibrium by diagramming the market demand (D) and market supply (S) curves together. This is shown in Figure 2.5. Notice that equilibrium is shown as the point where D and S intersect, or where $Q_D = Q_S$.

Market Adjustment to Disequilibrium

If the prevailing market price is not at the equilibrium level, it must be the case that $Q_D \neq Q_S$, and the market is said to be in **disequilibrium**. As a consequence, consumers and producers have an incentive to make some adjustment that will restore market equilibrium. The motivation for this adjustment and the process by which it is implemented depend on whether there is excess demand or excess supply.

Shortage

shortage
Excess demand of a commodity, equal to $(Q_D - Q_S)$, that arises if price is *below* its equilibrium level.

If the actual price P is *below* the equilibrium price P_E, then there is **excess demand**, meaning that Q_D exceeds Q_S at that price. The result is a **shortage** of the good equal to $(Q_D - Q_S)$. Refer back to Table 2.7 to see the shortages in the bottled water market at every price below \$2.50. In such a case, consumers want more of the good at the existing price than firms are willing and able to sell. In response, firms would be willing to increase quantity supplied, moving up the market supply curve and elevating price as they do so. This process would continue until price is increased to its equilibrium level.

Surplus

surplus
Excess supply of a commodity, equal to $(Q_S - Q_D)$, that arises if price is *above* its equilibrium level.

When the actual price P is *above* the equilibrium price P_E, then firms are unable to sell all their output, and they begin to observe their inventories accumulating with unsold stock. In this case, there is **excess supply**, meaning that Q_S exceeds Q_D. The result is a **surplus** equal to $(Q_S - Q_D)$. Look back at Table 2.7 to see the surpluses in the bottled water market at all prices above \$2.50. Observing unsold supplies, firms would have an incentive to lower price to eliminate the surplus. This price reduction would continue until all unsold supplies are purchased, which occurs at the equilibrium price.

Notice how price movements serve as a signal that a shortage or a surplus exists, whereas the stability of price suggests equilibrium. The adjustments that occur when the market is in disequilibrium happen because of the market's internal forces. As long as the assumptions of the model hold (and this is an important point), there is no need for government intervention or other third-party mediation to achieve equilibrium.

Thus far, we have examined the fundamental elements of a competitive market and illustrated how the simultaneous effect of demand and supply decisions determines equilibrium price and quantity. However, we need to go beyond this point to understand the implications of a competitive equilibrium—first in terms of resource allocation and then in terms of society's well-being.

EFFICIENCY CRITERIA

Economic analysis is guided by specific criteria. One criterion deals with the proper allocation of resources among alternative uses, referred to as **allocative efficiency**. Another is concerned with economizing on resources used in production, called **technical efficiency**. Each of these criteria is relevant to all applied economic disciplines, including environmental economics. We begin with a discussion of allocative efficiency.

Allocative Efficiency

As illustrated by the materials balance model, the way a market system uses resources is critical, not only for production and consumption, but also for the environment. But how is the resource allocation to be evaluated? The answer is through a procedure that involves two elements:

- an assessment of benefits and costs
- the use of marginal analysis

We have already introduced these elements in our previous discussion of demand and supply. Now, we use them to draw important conclusions about decision making and equilibrium under competitive conditions—first at the *market* level and then at the *firm* level.

Resource Allocation at the Market Level

Competitive markets are considered an ideal, a standard by which other market structures are evaluated. If we look carefully at what is conveyed by a competitive equilibrium, we can better appreciate this characterization. At equilibrium, we know that market demand intersects market supply. What does this mean in terms of resource allocation?

Recall that prices along a demand curve are measures of marginal benefit (MB). Each demand price communicates the value consumers place on the next, or *marginal*, unit of the good based on the added benefit they expect to receive from consumption. On the supply side, there is also a marginal interpretation, but here the prices are measures of economic cost. Because the market supply in a competitive market is the horizontal sum of firms' marginal cost (MC) curves, each supply price represents the additional cost of resources needed to produce another unit of the good. Economic costs include both the **explicit**, or out-of-pocket, costs associated with production and all **implicit** costs based on the highest valued alternative use of any economic resource.

allocative efficiency
Requires that resources be appropriated such that the additional benefits to society are equal to the additional costs.

Taking these two interpretations together yields an important result. At the competitive equilibrium, the value society places on the good is equivalent to the value of the resources given up to produce it, or, more simply, $MB = MC$. By definition, this result ensures that **allocative efficiency** is achieved. **Allocative efficiency** requires that the additional value society places on another unit of the good is equal to what society must give up in resources to produce it. Recognizing this outcome as part of the competitive model is important, but it is just as critical to understand why it arises. For that, we need to focus on the decision making of individual firms under these market conditions.

Resource Allocation at the Firm Level

Starting from a general perspective, the assumed motivation of a firm's decision making is profit maximization. We further assume that the choice variable for producers is output. Therefore, all firms, regardless of competitive conditions, choose the output level that maximizes profit. **Total profit (π)** is defined as **Total revenue (TR) – Total costs (TC)**. TR is simply the dollar value of the firm's sales, which equals market price (P) multiplied by the quantity of output sold (q) (i.e., $TR = Pq$). Total costs (TC) include all economic costs associated with production.

total profit
Total profit (π) = Total revenue (TR) – Total costs (TC).

To find the q that achieves the highest possible π, the firm makes its decisions *at the margin*, taking into account the relative benefits and costs of producing each additional unit of output. From the firm's perspective, the benefit is measured by TR and the cost by TC. Thus, it considers the profit implications of each successive unit of output by looking at the associated changes in TR and TC. If producing the next unit of output adds more to TR than it does to TC, then the firm increases production. If production adds more to TC than to TR, production is decreased. This process goes on until there is no incentive to continue, or when the *change in TR* (ΔTR) from increasing output is equal to the *change in TC* (ΔTC). At this point, the *change in π* ($\Delta\pi$) from producing the last unit of output is zero, and any further increase in output would cause π to decline. At this precise point, π is at its maximum.[4]

[4]This, of course, assumes that in the short run, the firm's revenues cover its variable costs. If not, it will shut down. In the long run, the firm produces as long as its total revenues cover all its total costs. If not, it will exit from the market.

Notice that this decision-making process relies on *changes*, which is the definition of marginal analysis. In this context, the relevant marginal variables are as follows:

- **Marginal revenue (*MR*)** is the change in total revenue (TR) associated with a change in output (q), or $MR = \Delta TR/\Delta q$.
- **Marginal cost (*MC*)** is the change in total costs (TC) associated with a change in output (q), or $MC = \Delta TC/\Delta q$.
- **Marginal profit (*M*π)** is the change in total profit (π) associated with a change in output (q), or $M\pi = \Delta\pi/\Delta q$.

Thus, the firm implicitly makes its profit-maximizing decisions according to the following rules:

- The firm increases production as long as $MR > MC$ or as long as $M\pi > 0$.
- The firm contracts production as long as $MR < MC$ or as long as $M\pi < 0$.
- The firm achieves **profit maximization** where $MR = MC$ or where $M\pi = 0$.

profit maximization
Achieved at the output level where $MR = MC$ or where $M\pi = 0$.

Observe that the firm's optimal output level occurs at precisely the point where the marginal benefit *to the firm* of doing so (MR) is exactly offset by the marginal cost of the resources it uses (MC). Although this outcome validates the use of benefits and costs at the margin, it does not necessarily result in an allocatively efficient outcome. Why not? Because MR is the marginal benefit *to the firm*, which is not necessarily equal to the marginal benefit *to society*, measured by price P. In fact, the firm's optimizing behavior achieves allocative efficiency only if MR equals P, and this equality occurs only under competitive conditions. To understand this, we need to reconsider how the firm's profit-maximizing decision is affected by a competitive market.

As a price taker, each firm in a competitive market accepts the market-determined price as a given. It is unable to charge a higher price, because by assumption its product is identical to that of all its competitors. Thus, if a firm were to raise price, consumers would demand none of its output and instead buy from other suppliers. Moreover, a firm has no incentive to lower its price, because it can sell all it wants at the market-determined price.

Because the competitive firm has no control over prevailing market conditions, it faces a constant price. Therefore, each additional unit sold raises total revenue by an amount exactly equal to the price of the good. This outcome translates to an important equality unique to competitive markets, namely, that $P = MR$. Thus, although competitive firms follow the profit-maximizing rules derived earlier, the equilibrium is markedly different in that allocative efficiency is automatically ensured. This is summarized below.

Profit maximization requires:	$MR = MC$
Competitive markets imply:	$P = MR$
Thus, profit maximization for competitive firms requires:	$P = MC$

Now it should be clear why the competitive *market* equilibrium achieves allocative efficiency—because every firm in that market independently produces where $P = MC$.

To illustrate this outcome, Table 2.8 presents revenue and cost data for a representative firm in our hypothetical bottled water market. First, let's confirm that $P = MR$ at all output levels. In keeping with the competitive model, notice that the prices faced by the firm are constant at the market-determined equilibrium price (P_E) of \$2.50. The MR values are found by dividing the change in TR by the change in q. So, for example, when output rises from 28 units to 36 units, generating an increase in TR from \$70 to \$90,

TABLE 2.8 Revenue and Cost Data for a Representative Firm in the Bottled Water Market

Price ($) P_E	Quantity (bottles/month) q	Total Revenue ($) TR	Marginal Revenue ($) MR	Marginal Cost ($) MC
2.50	4	10.00	2.50	0.50
2.50	12	30.00	2.50	1.00
2.50	20	50.00	2.50	1.50
2.50	28	70.00	2.50	2.00
2.50	36	90.00	2.50	2.50
2.50	44	110.00	2.50	3.00
2.50	52	130.00	2.50	3.50
2.50	60	150.00	2.50	4.00
2.50	68	170.00	2.50	4.50
2.50	76	190.00	2.50	5.00

$MR = \Delta TR/\Delta q = (\$90 - \$70)/(36 - 28) = \$20/8 = \$2.50$. Repeating this calculation for all changes in q shown in the table results in $MR = \$2.50$ in every case—exactly equal to P_E.

Now let's examine how the firm chooses the output level that maximizes profit. As predicted by the competitive model, the MC values in the last column of the table are the same as the prices associated with the firm's supply schedule. To guide its production decisions, the firm considers MC relative to MR at each output level. In this case, the profit-maximizing equilibrium output occurs at $q_E = 36$, where $MR = MC = \$2.50$. And because $P = MR$ at all q levels, this decision also ensures that $P = MC$, indicating that resources are being allocated efficiently. Figure 2.6 illustrates this equilibrium.

FIGURE 2.6 Competitive Firm's Profit-Maximizing Equilibrium

The profit-maximizing equilibrium for a competitive firm occurs at $q_E = 36$, where $MR = MC = \$2.50$. Because $P = MR$ at all q levels, it must be the case that $P = MC$ at equilibrium, indicating that resources are allocated efficiently. At any q *below* q_E, $MR > MC$, or $M\pi > 0$, which means that the firm can increase π by expanding production. The converse is also true. Only at $q_E = 36$ is $M\pi = 0$, meaning that further additions to π are not possible, and π is at its maximum.

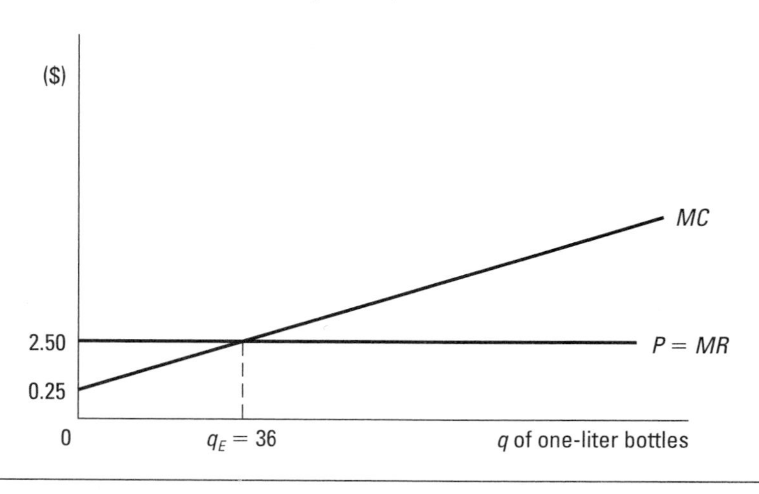

Technical Efficiency

technical efficiency
Production decisions that generate maximum output, given some stock of resources.

Another important economic criterion is **technical efficiency**, which refers to production decisions that generate maximum output, given some stock of resources, or (saying the same thing from a slightly different perspective) decisions to produce a given output level using minimum resources. In the context of the materials balance model, achieving technical efficiency preserves natural resources and minimizes the generation of residuals arising from resource use. Furthermore, given the relationship between production and costs, technical efficiency implies that economic costs are minimized when producing a given level of output. Viewed from this perspective, we observe that technical efficiency is an application of the more general cost-effectiveness criterion introduced in Chapter 1.

A key point is to understand that market forces can achieve technical efficiency as long as competitive conditions prevail. To remain viable, the competitive firm must minimize costs, because it cannot raise its price to cover the added expense of inefficient production. If it attempted such a strategy, demand for its product would fall to zero, because many other firms can sell the same product at a lower price. Recognizing how technical efficiency is achieved under such ideal market conditions helps economists determine *why* it is not being met in other market contexts. More importantly, the magnitude of a technically inefficient decision can be assessed by comparing the resulting costs to what they *would* have been if the market had been allowed to operate freely.

WELFARE MEASURES

In economics, an important objective is to assess the gains and losses to society associated with any event that alters market price. The supply and demand model provides the information necessary to conduct these analyses, using concepts known as **consumer surplus** and **producer surplus**. By comparing these measures before and after a market disturbance, it is possible to quantify how society has been affected.

Consumer Surplus

consumer surplus
Net benefit to buyers estimated by the excess of marginal benefit (*MB*) of consumption over market price (*P*), aggregated over all units purchased.

To get a sense of the logic of consumer surplus, we start with a working definition. **Consumer surplus** is a measure of net benefit accruing to buyers of a good estimated by the excess of what they are willing to pay over what they must actually pay, aggregated over all units of the good purchased. Notice that consumer surplus depends on two distinct notions of price—one that measures *willingness to pay* and one that measures what is *actually paid*. The series of prices consumers are willing to pay for various quantities of a good are those that define the demand curve. As discussed, each demand price measures the marginal benefit (*MB*) of consumption. Conversely, the price that consumers must actually pay is the prevailing market price (*P*) determined by demand and supply.

There are two major differences between these prices. First, they have different determinants. The demand price (*MB*) is determined *solely* by demand, a sort of psychic price based on how consumers value a good. On the other hand, market price (*P*) arises from the forces of supply and demand and is driven by *both* producer and consumer incentives. Second, although there is a whole series of demand prices, there is only one market price charged for *all* units sold. The result? Once the market price is determined, *all* units are sold for that single price, even those for which the demand price is much higher. Consequently, consumers receive a surplus benefit for every unit purchased at a demand price that exceeds the market price.

To graphically illustrate consumer surplus, we reproduce in Figure 2.7 the market demand for bottled water derived previously. Added to the diagram is a reference price line drawn horizontally at the equilibrium price level of $2.50. Notice that for every output level up *to* the equilibrium quantity of 900, the demand price is higher than the market price. So each unit purchased gives consumers a surplus benefit over and above what they had to pay for it. For example, for the first bottle of water, consumers are willing to pay $11.49 based on the market inverse demand function,[5] but they actually have to pay only $2.50. That means they receive a net benefit from consuming this first unit of water equal to the excess of $11.49 over $2.50, or $8.99. Geometrically, this is measured as the vertical distance from the demand curve to the price line at $Q = 1$, shown in Figure 2.7 as distance *ab*.

Since consumers receive a net benefit for every unit purchased up to the equilibrium point, this value must be aggregated over all units consumed to derive the measure of **consumer surplus**. Graphically, this is the triangular area above the price line and below the demand curve. In Figure 2.7, the consumer surplus for bottled water is the area of the triangle WXY. This makes sense because the entire area under the demand curve is a measure of total benefit from consumption (i.e., the aggregation of the marginal benefit from each unit of the good), and the rectangular area under the price line is the total expenditure on the good. Thus, the difference between the two is the net benefit to the consumer.

FIGURE 2.7 Consumer Surplus

For the first bottle of water, consumers' demand price is $11.49, but the market-determined price is only $2.50. Therefore, consumers receive a net benefit of $11.49 − $2.50 = $8.99, shown as vertical distance *ab*. Aggregating this net benefit over all units of water consumed gives the measure of consumer surplus shown as triangular area WXY. The dollar value of this area is ½(900)($9.00) = $4,050.

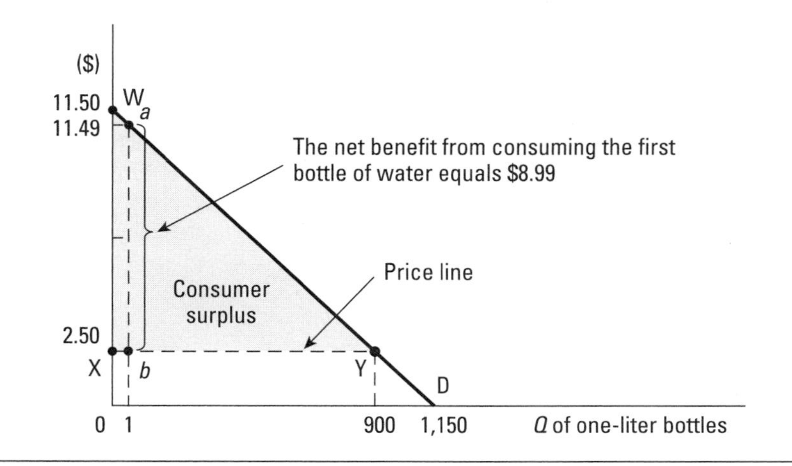

[5]The price of $11.49 is found by substituting $Q_D = 1$ into the inverse demand function $P = -0.01Q_D + 11.5$.

The dollar value of consumer surplus can be found by calculating the area of the triangle that represents it in the graphical model.[6] In Figure 2.7, the base of the triangle WXY is the horizontal distance from the vertical axis to the equilibrium quantity, or 900. The height is \$9.00, which is the difference between the vertical intercept of the demand curve (\$11.50) and the price line (\$2.50). Thus, the dollar value of consumer surplus in this market is ½(900)(\$9.00) = \$4,050.

The applicability of consumer surplus stems from the fact that its magnitude is related to equilibrium price and quantity. Any disturbance to market equilibrium will change the size of consumer surplus. And because this surplus measures consumer benefit, any change in its value can be used to assess the associated gain or loss to consumer welfare.

Producer Surplus

producer surplus
Net gain to sellers of a good estimated by the excess of market price (*P*) over marginal cost (*MC*), aggregated over all units sold.

On the supply side, the comparable welfare measure is **producer surplus**. It measures net gain to sellers, estimated by the excess of market price (*P*) over marginal cost (*MC*), aggregated over all units sold.

Based on our discussion of supply decisions, we know that $P = MC$ at a competitive equilibrium. However, at every output level below equilibrium, *MC* is *lower* than *P*. So firms are actually willing to supply these smaller quantities at prices below what the market dictates. The price that firms are "willing to accept" for each output level is their supply price, and it is this price that is reflected in the *MC* curve. Thus, at each quantity below equilibrium, firms accrue a net gain measured by the excess of *P* over *MC*. This net gain is illustrated in Figure 2.8.

FIGURE 2.8 Producer Surplus

For the first bottle of water produced, the *MC* is \$0.2525, but firms can sell this bottle for \$2.50. Therefore, they receive a net gain of \$2.50 – \$0.2525, or \$2.2475, shown as the vertical distance *cd*. Aggregating this gain over all units sold yields the measure of producer surplus represented by triangular area XYZ. The dollar value of this area is ½(900)(\$2.25) = \$1,012.50.

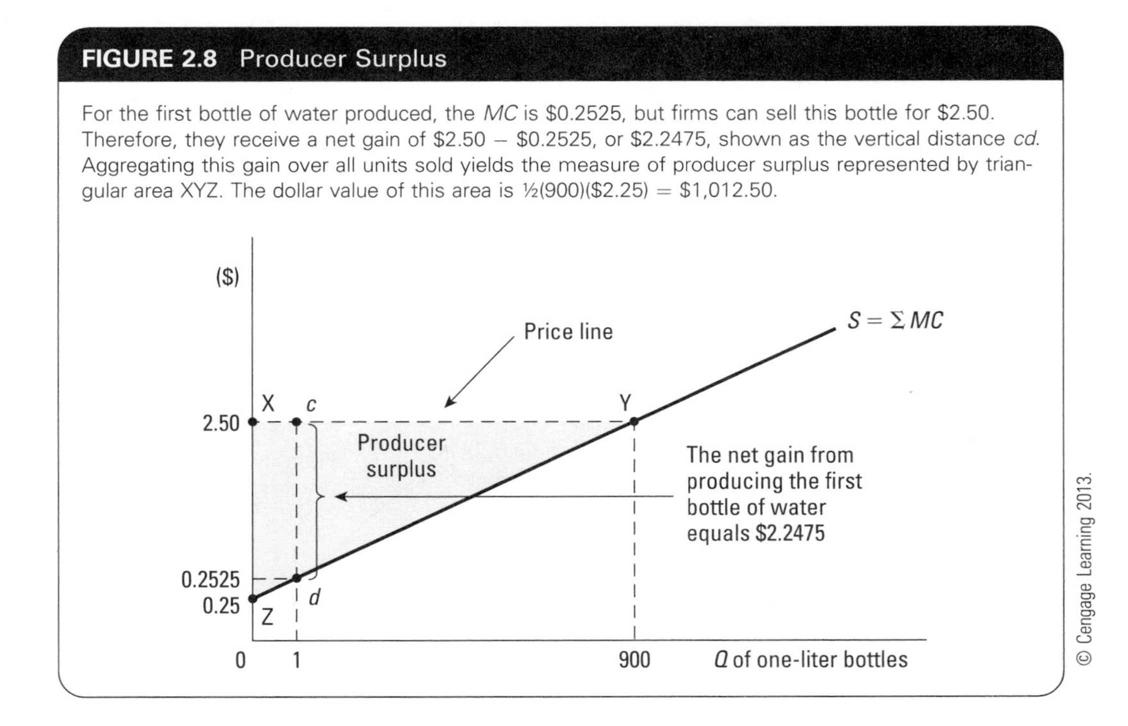

[6]Recall that the formula for finding the area of a triangle is ½(base)(height).

The diagram shows the competitive market supply for bottled water, which is also the *MC* curve, and a reference line drawn horizontally at the equilibrium price of $2.50. For every unit of output supplied up *to* the equilibrium point at 900 units, producers enjoy a surplus equal to the excess of *P* over *MC*. For example, for the first unit of bottled water produced, the *MC* is $0.2525 based on the market supply function.[7] However, firms can sell this first bottle for $2.50. Thus, the net gain associated with this unit of output is the excess of $2.50 over $0.2525, or $2.2475. Geometrically, this gain is the vertical distance from the supply curve to the price line at $Q = 1$, labeled as distance *cd* in Figure 2.8.

Just as on the demand side, this net gain must be aggregated over all units sold up to the equilibrium quantity to find producer surplus. Graphically, this is the sum of all the vertical distances between the *MC* curve and the price line, or the triangular area bounded by the *MC* curve and the price line up to the equilibrium point. In Figure 2.8, this is the area of the triangle XYZ.

By using the same method as described for consumer surplus, we can quantify producer surplus by calculating its representative triangular area depicted in the graph. In Figure 2.8, the base of the triangle XYZ is 900, and its height is $2.25, found as the difference between the $2.50 price line and the vertical intercept of the supply curve, $0.25. Thus, producer surplus at equilibrium in the bottled water market is ½(900)($2.25) = $1,012.50. Notice that the magnitude of producer surplus, just like consumer surplus, is based on equilibrium price and quantity. Hence, any market disturbance will change its value and thus provide a way to assess any associated welfare gain or loss to firms.

Society's Welfare: Sum of Consumer and Producer Surplus

society's welfare
The sum of consumer surplus and producer surplus.

Economists use the sum of consumer and producer surplus to capture the gains accruing to both sides of the market, or **society's welfare**. By applying this concept to the bottled water market, we see that society enjoys a surplus valued at $4,050 + $1,012.50, or $5,062.50. Because this surplus is based on a competitive equilibrium, its value is maximized. This is so because of the allocative efficiency that characterizes a competitive market. Put another way, it is not possible to reallocate resources to improve society's welfare. By default, this implies that any market outcome not meeting the criterion of allocative efficiency has a negative effect on society's well-being. In such an instance, the loss can be monetized by comparing the resulting sum of consumer and producer surplus to what it *would* have been if allocative efficiency had been achieved.

Measuring Welfare Changes

By way of illustration, let's consider a hypothetical policy initiative in the bottled water market that forces price above *MC* up to $6.50 per unit. To measure the effect on society's welfare, we need to compare the post-policy level of consumer and producer surplus to the benchmark competitive level of $5,062.50. We begin by reproducing the model of the bottled water market in Figure 2.9, adding to the diagram the policy price and quantity of $6.50 and 500, respectively.[8] Capital letters A through F have been

[7]The supply price, or *MC*, of producing the first unit of output is found by evaluating the market supply function at Q_S.

[8]At a price of $6.50, $Q_S > Q_D$, so the quantity exchanged in the market is Q_D. Thus, we find quantity by substituting $6.50 into the demand equation: $Q_D = -100(6.50) + 1{,}150 = 500$.

FIGURE 2.9 Deadweight Loss to Society Under a Pricing Regulation

At the competitive equilibrium, area (A + B + C) is consumer surplus, valued at $4,050, and area (D + E + F) is producer surplus valued at $1,012.50, for a total welfare measure of $5,062.50. Under a policy that sets price at $6.50, consumer surplus falls to $1,250 (shown as area A), and producer surplus increases to $2,812.50 (shown as area B + D + F), for a new total welfare measure of $4,062.50. Consumers incur a net loss of −(B + C), valued at $2,800, whereas producers gain +(B − E) valued at $1,800. So as a result of this policy, there is a **deadweight loss to society** of −(C + E), valued at $1,000.

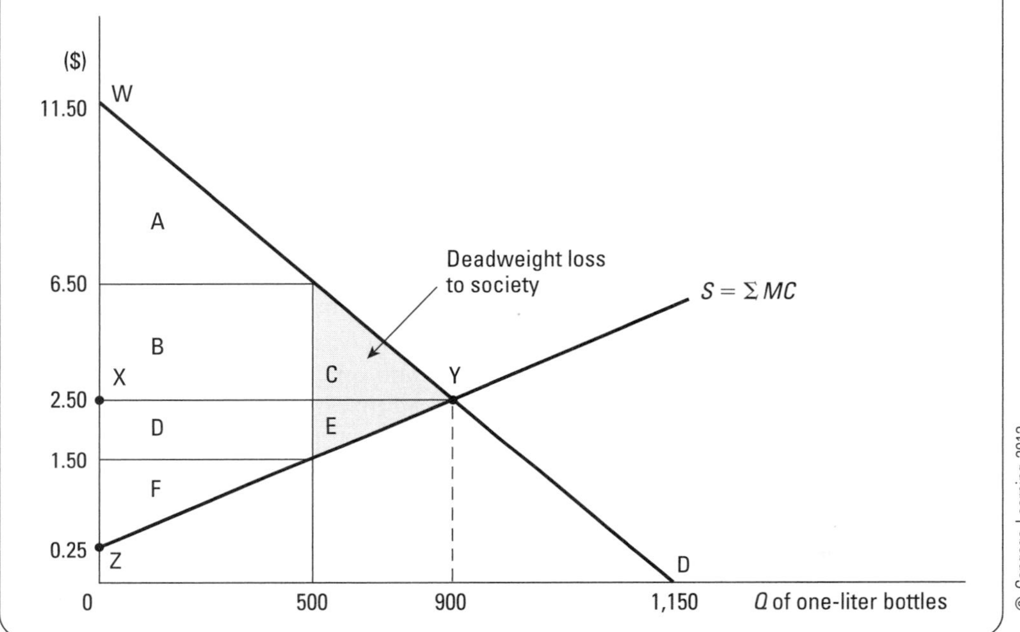

added to facilitate our discussion. The benchmark level of consumer surplus is shown as area (A + B + C), and producer surplus is area (D + E + F).

Now, we determine the comparable surplus values under the policy-determined price and quantity. At the $6.50 price, consumer surplus is reduced to triangular area A, valued at $1,250. This new value is $2,800 *lower* than it was before the policy, so we know that the policy generates a net *loss* to consumers. The producer surplus at the $6.50 price is area (B + D + F) valued at $2,812.50, found by summing the area of the rectangle (B + D) and the area of the triangle (F). Because this magnitude is $1,800 *higher* than the original surplus, we know that producers enjoy a net *gain* as a result of the policy.

Finally, consider the overall effect. The total surplus under the new policy is $1,250 + $2,812.50, or $4,062.50, which is $1,000 less than the original value. Although producers enjoy a net gain of $1,800, this gain is outweighed by the loss to consumers of $2,800. So the new pricing policy causes a decline in society's welfare of $1,000. This change can be confirmed geometrically by looking at the areas representing surplus before and after the policy:

	Change in consumer surplus:	(A) − (A + B + C)	= −(B + C)
+	Change in producer surplus:	(B + D + F) − (D + E + F)	= +(B − E)
	Net loss to society:		= −(C + E)

deadweight loss to society
The net loss of consumer and producer surplus due to an allocatively inefficient market event.

Area (C + E), valued at $1,000, is called the **deadweight loss to society**. Why? Because it was once a part of the surplus accruing to producers and consumers under allocatively efficient conditions, but as a result of the policy, it has been lost or unaccounted for. Notice that area B, although a loss to consumers, is a gain to firms. Thus, the value is redistributed from one market sector to another. Some may view this outcome as unfair, but the relevant point is that such transfers are not inefficient, because the amount is captured somewhere within the market system. What *is* problematic is that the policy generates a loss to society as a whole because it forces *P* above *MC*, violating the allocative efficiency criterion.

CONCLUSIONS

Understanding the fundamentals of how markets operate is an important basis for the study of environmental economics. To that end, we have limited our discussion in this chapter to the circular flow model—reviewing the mechanics of supply and demand, the signaling mechanism of price, and marginal analysis—all within the context of a classical competitive market. Competitive markets establish a benchmark that helps us to evaluate the effects of market failures and market disturbances, both of which are important to environmental economics. From a practical perspective, we can assess the effects of environmental pollution or any policy initiative using allocative efficiency as a criterion. Furthermore, these effects can be measured by quantifying the associated changes in consumer and producer surplus.

The competitive model also illustrates how an economic system operates in the absence of any condition that impedes natural market forces. Recognizing how a fully functioning market performs is necessary to understanding the economic perception of environmental pollution as a market failure—the subject of our next chapter. To make this transition, we will expand our analysis to the full materials balance model, allowing for the interdependence of the circular flow with the environment. By using the modeling tools shown in this chapter and expanding on marginal benefits and costs, we will develop more elaborate market models that explicitly account for this interdependence. These models will show *how* and *why* the market fails to correct environmental damage, which in turn will suggest approaches to finding effective policy solutions.

Summary

- A market refers to the interaction between consumers and producers for the purpose of exchanging a well-defined commodity.
- A competitive market model is characterized by a large number of independent buyers and sellers with no control over price, a homogeneous product, the absence of entry barriers, and perfect information.
- Demand is a relationship between quantity demanded (Q_D) and price (P), holding constant all other factors that may influence this decision, such as wealth, income, prices of related goods, preferences, and price expectations.
- According to the Law of Demand, there is an inverse relationship between quantity demanded and price, *c.p.*
- Market demand for a private good is found by horizontally summing individual demands.
- Supply is a relationship between quantity supplied (Q_S) and price (P), holding constant all other supply determinants, such as technology, input prices, taxes and subsidies, and price expectations.
- The Law of Supply states that there is a direct relationship between quantity supplied and price, *c.p.*
- Market supply for a private good is found by horizontally summing individual supplies.
- The equilibrium, or market-clearing, price (P_E) is the price at which $Q_D = Q_S$. If price is above (below) its equilibrium level, there is a surplus (shortage) of the commodity, which puts pressure on the prevailing price to fall (rise) toward equilibrium.
- Allocative efficiency requires that the additional value society places on another unit of a good is precisely equivalent to what society must give up in resources to produce it.
- All profit-maximizing firms expand (contract) output as long as the additional revenue (MR) is greater (lower) than the increase in costs (MC). Profit-maximizing output occurs where $MR = MC$, or where marginal profit ($M\pi$) = 0.
- Competitive firms are price takers. Because $P = MR$ for competitive firms, the profit-maximizing output level at $MR = MC$ is also the point where $P = MC$, which signifies allocative efficiency.
- Technical efficiency arises when the maximum output is produced from some fixed stock of resources, or, equivalently, when minimal resources are used to produce a given output level.
- Consumer surplus measures the net benefit to buyers, measured as the excess of what consumers are willing to pay (MB) over what they must actually pay (P), aggregated over all units purchased.
- Producer surplus measures the net gain to sellers, estimated as the excess P over MC, aggregated over all units sold.
- Society's welfare is measured as the sum of consumer and producer surplus, which is maximized when allocative efficiency is achieved.
- The deadweight loss to society measures the net change in consumer and producer surplus caused by an allocatively inefficient market event.

Review Questions

1. Suppose $Q_D = 200 - 4P$ and $Q_S = 100$ describe market demand and market supply in a given market.
   a. Algebraically find equilibrium price and quantity, and support your answer graphically.
   b. What is unusual about this market? Give an example of a good or service that might be characterized in this way.
2. In 1995, the Food and Drug Administration (FDA) published new labeling standards for bottled water. (The full text of the final rule can be found at **http://cfr.vlex.com/vid/165-110-bottled-water-19705533**.) Prior to that time, bottlers could sell regular tap water under a bottled water label. In fact, the FDA estimated that approximately 25 percent of the supply of bottled water was nothing more than ordinary tap water. Consider how these tougher standards eliminated 25 percent of the supply of bottled water. If market demand is unaffected, what qualitative impact

would this labeling change have on equilibrium price and quantity for bottled water? Support your answer with a graphical model.

3. Suppose the market for organically grown wheat is modeled through the following market supply and demand functions:

$$P = 10 + 0.5Q_S \text{ and } P = 22 - 2.5Q_D,$$

where Q_S and Q_D are in millions of bushels, and P is price per bushel.

   a. Find the market equilibrium price, P_E, and market equilibrium quantity, Q_E.

   b. Now determine the value of producer surplus and consumer surplus at equilibrium.

4. Reconsider the implications of the revised labeling standards discussed in Question 2 in the context of the hypothetical market for bottled water modeled in the text. Recall that the market demand and market supply equations are

$$Q_D = -100P + 1,150 \text{ and } Q_S = 400P - 100,$$

where P_E = \$2.50 and Q_E = 900.

Now, suppose the change in standards results in a new market supply of $Q_S{}' = 400P - 350$, with no change in market demand.

   a. Determine the new $P_E{}'$ and $Q_E{}'$ for bottled water. Do your results agree with your intuitive answer to Question 2?

   b. Graphically illustrate the market for bottled water before and after the change in labeling standards. Be sure to label all relevant points.

   c. Compare the values of consumer surplus and producer surplus before and after the change in labeling standards. Is this result expected? Why or why not?

5. **a.** Describe a real-world government policy that creates a market surplus. Be sure to carefully define the relevant market.

   b. Explain the efficiency implications of such a policy. Be specific.

   c. In the instance you have described, what is the government's motivation for intervening in the market in this way?

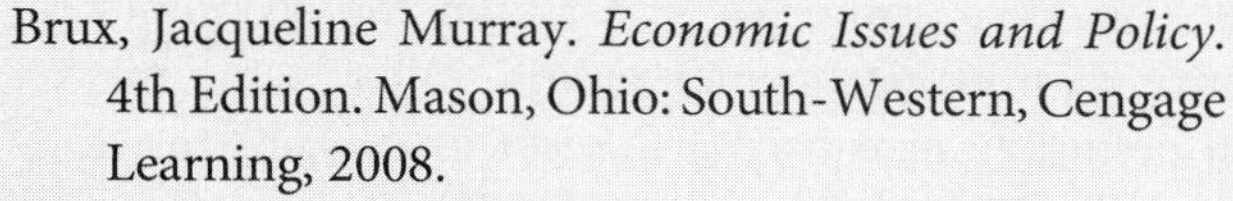

Additional Readings

Brux, Jacqueline Murray. *Economic Issues and Policy.* 4th Edition. Mason, Ohio: South-Western, Cengage Learning, 2008.

Cortese, Amy. "Can Entrepreneurs and Environmentalists Mix?" *New York Times*, May 6, 2001, p. 3.

Heilbroner, Robert L. *The Worldly Philosophers.* New York: Simon & Schuster, 1980.

Mankiw, N. Gregory. *Principles of Microeconomics.* 6th Edition. Mason, Ohio: South-Western, Cengage Learning, 2012.

Nicholson, Walter, and Christopher Snyder. *Intermediate Microeconomics and Its Application.* 11th Edition. Mason, Ohio: South-Western, Cengage Learning, 2010.

Pindyck, Robert S., and Daniel L. Rubinfeld. *Microeconomics.* 8th Edition. Upper Saddle River, NH: Pearson Prentice Hall, 2013.

To access additional course materials, visit www.cengagebrain.com. At the home page, search for the ISBN of this title (shown on the back cover). This will take you to the product page where these resources can be found.

CHAPTER 3

Modeling Market Failure

"The law of property determines who owns something, but the market determines how it will be used."
—Ronald Coase (1910–)

Based on the circular flow model, we know that free markets provide desired goods and services to the market, resolve shortages and surpluses, and eliminate inefficiency through the pricing mechanism—all without government intervention. This is a remarkable result, given that consumers and producers are not motivated by philanthropic goals but, rather, are driven by their own self-interest. As first described through Adam Smith's metaphor of the "invisible hand," this market outcome emerges as though consumers and firms are guided to make decisions that enhance society's well-being.[1] Recognizing the efficiency and welfare implications of a competitive equilibrium underscores what is at stake when something impedes the market process that underlies it. A case in point is the persistence of pollution.

When we consider the circular flow in the fuller context of the materials balance model, we become aware of how economic activity generates residuals that can damage natural resources. However, we need to look further to understand the economics of why pollution persists in the absence of third-party intervention. Why is the market unable to respond to environmental pollution, or can it? The most immediate answer is that pollution is a **market failure** that distorts the classical market outcome.

From an economic perspective, environmental problems persist because they implicitly violate the assumptions of a fully functioning market. The incentive mechanisms that normally achieve an efficient solution are unable to operate, and government has to intervene. However, if the market failure is understood, incentives can be restored through environmental policy. Conceptually, the idea is to ferret out the conditions that cause the pricing system to break down, make the needed adjustments to the underlying conditions, and then let the power of the market work toward a solution.

In this chapter, we provide the analytical tools necessary to understand market failure in the context of environmental problems. Our discussion centers on the development of two economic models. The first is based on the public goods characteristics of environmental quality. The second uses what is known as externality theory to show how market incentives fail to capture the effects of pollution associated with production or consumption. Finally, we link the two models through a discussion of property rights and their role in environmental market failures.

[1] Smith (1937), p. 423 (originally published in 1776).

ENVIRONMENTAL PROBLEMS: A MARKET FAILURE

market failure
The result of an inefficient market condition.

Classical microeconomic theory predicts an efficient outcome under certain assumptions about pricing, product definition, cost conditions, and entry barriers. If any of these assumptions fails to hold, market forces cannot operate freely. Depending on which assumption is violated, the result will be any of a number of inefficient market conditions, collectively termed **market failures**. These include imperfect competition, imperfect information, public goods, and externalities. For example, if we relax the assumption of freedom of entry in the competitive model, some degree of market power will develop. As this occurs, society's welfare declines, and resources are allocated inefficiently.

Economists model environmental problems as market failures using either the theory of public goods or the theory of externalities. Each is distinguished by how the market is defined.

- If the market is defined as "environmental quality," then the source of the market failure is that environmental quality is a **public good**.
- If the market is defined as the good whose production or consumption generates environmental damage, then the market failure is due to an **externality**.

Although each of these models suggests a different set of solutions, the theories are not totally unrelated. In the context of environmental problems, both are exacerbated by a third type of market failure: **imperfect information**. We begin by analyzing the market failure aspect of public goods.

ENVIRONMENTAL QUALITY: A PUBLIC GOOD

public good
A commodity that is nonrival in consumption and yields benefits that are nonexcludable.

Economists distinguish public goods from private goods by examining their inherent characteristics—*not* by whether they are publicly or privately provided.[2] A **public good**, or more technically a *pure* public good, is one that possesses the following characteristics: it is **nonrival** in consumption, and its benefits are **nonexcludable.**[3] At the other extreme is a pure private good, which is characterized by rivalness and excludability.

Characteristics of Public Goods

nonrivalness
The characteristic of indivisible benefits of consumption such that one person's consumption does not preclude that of another.

Nonrivalness refers to the notion that the benefits associated with consumption are *indivisible*, meaning that when the good is consumed by one individual, another person is not preempted from consuming it at the same time. Put another way, the marginal cost (MC) of another individual sharing in the consumption of the good is zero. Consider, for example, the network television broadcast of the NBA championship finals. The benefits to the existing television audience are completely unaffected when another person tunes in to view the broadcast. Contrast this with what happens when a private good, such as a personal computer, is consumed. Once someone is using the computer, that consumption prohibits another person from using it at the same time.

---

[2]For a good discussion of this distinction, see Rosen and Gayer (2008), Chapter 4.

[3]Much of the important work in the theory of public goods is credited to Samuelson (1954, 1955, 1958).

nonexcludability
The characteristic that makes it impossible to prevent others from sharing in the benefits of consumption.

Nonexcludability means that preventing others from sharing in the benefits of a good's consumption is not possible (or prohibitively costly in a less strict sense). An example of a good with this characteristic is a jogging path. It would be virtually impossible to ration the use of the path to a select group of runners. In contrast, consider the inherent excludability of a conventional private good, such as hotel lodging. Exclusive rationing of hotel services to the consumer paying for them is easily accomplished, and the associated benefits accrue solely to that single consumer.

Although nonrivalness and nonexcludability may seem similar, they are not identical. A good way to distinguish them is as follows. Nonrivalness means that rationing of the good is not *desirable*, whereas nonexcludability means rationing of the good is not *feasible*.[4] In fact, it is possible for a good to possess one of these attributes but not the other. Reconsider the examples for each characteristic. Although a televised NBA game is a nonrival good, it can be made excludable by broadcasting it over a cable network only to subscribers who have paid for it and by scrambling the signal to everyone else. Likewise, provision of the jogging path, which is nonexcludable absent prohibitively high costs, does not possess the nonrivalry attribute. Use of the path by more and more runners would lead to congestion, which would affect every consumer of the jogging path.

Two classic examples of public goods are a lighthouse and national defense. Contemplate the services provided by these goods to see that the benefits of each are both nonrival and nonexcludable. A more contemporary and, from our perspective, more relevant example of a public good is environmental quality.[5] Just like the lighthouse, cleaner air, for example, is both nonexcludable and indivisible. Consider the futility of trying to restrict the benefits of air quality to a single person. It's unreasonable to think that others could be excluded from the consumption of a cleaner air supply just because another person has paid for it. Moreover, once the air is made cleaner for one person, others could simultaneously enjoy the benefits of breathing healthier air.

So what, then, is the problem? Having accepted the idea that environmental quality is a public good, we still need to explain why this, or any public good for that matter, is a market failure. To do this, we must develop a model of a public good.

MODELING A PUBLIC GOODS MARKET FOR ENVIRONMENTAL QUALITY

Public goods generate a market failure because the nonrivalness and nonexcludability characteristics prevent natural market incentives from achieving an allocatively efficient outcome. To illustrate this assertion, we reintroduce the supply and demand model but redefine the market as the public good, air quality.[6] It turns out that by changing the

[4]Stiglitz (1988), pp. 119–23.

[5]Alternatively, we could define environmental pollution as a public "bad."

[6]To direct attention solely to the public goods aspect of the model, we continue to assume competitive markets. This prevents confounding the market failure of public goods with that of imperfect competition. Further, the assumption of competitive markets allows the market supply curve to be modeled as the horizontal sum of all producers' marginal cost curves, just as was done for the private goods case in Chapter 2.

market definition from a private good to a public good, the conventional derivation of market demand is no longer viable. It is this modeling dilemma that is at the root of the public goods problem.

Allocative Efficiency in a Public Goods Market

Just as in the private goods case, achieving an allocatively efficient equilibrium in a public goods market depends on the existence of well-defined supply and demand functions. To develop these functions for air quality, we adjust the market definition so that output can be quantified. Air quality can be defined as "an acceptable level of pollution abatement," which for this discussion is assumed to be some percentage reduction in sulfur dioxide (SO_2) emissions.

Market Supply for Air Quality

Although the market supply of a public good often represents government's decisions about production, as opposed to those of private firms, the general derivation of the supply function is analogous to what we developed in Chapter 2 for a private good. We begin by assuming that there is some number of hypothetical producers, each of which is willing and able to supply various reductions in SO_2 at different price levels, *c.p.* The aggregation of these production decisions gives rise to market supply, which we assume is represented by the data in Table 3.1. Price (P) is measured in millions of dollars, and quantity supplied (Q_S) is measured as a percentage of SO_2 abatement. The algebraic counterpart for these data is

$$\text{Market Supply:} \quad P = 4 + 0.75Q_S$$

Market Demand for Air Quality

On the demand side, the model for a public good is quite different from that for a private commodity. Recall from Chapter 2 that the market demand for a private good is found by *horizontally* summing the demands of individual consumers. It is as though each consumer were asked, "What *quantity* of this good would you consume at each of the following prices?" But this question is not relevant to the demand for a public good, because once such a commodity is provided, it is available at the same quantity to all

TABLE 3.1 Hypothetical Supply Data in the Market for Air Quality

Quantity Supplied Q_S (% of SO_2 abatement)	Market Supply Price $P = 4 + 0.75Q_S$ ($ millions)
0	4.00
5	7.75
10	11.50
15	15.25
20	19.00
25	22.75
30	26.50

market demand for a public good The aggregate demand of all consumers in the market, derived by *vertically* summing their individual demands.

consumers—a direct consequence of the **nonrivalness** characteristic. How, then, *is* market demand determined for a public good if quantity is not a decision variable?

The key is to recognize that the demand price for a public good is variable, even though the quantity is not. So the relevant question in deriving this demand must be, "What *price* would you be willing to pay for each quantity?" In theory, each consumer should express a unique "**willingness to pay (WTP)**" for the public good based on the benefits each expects to derive from consumption. The **market demand for a public good** is the aggregate demand for all consumers in the market. It is derived by summing each individual demand *vertically* to determine the market price ($P = \Sigma p$) at each and every possible market quantity (Q), *c.p.*

To illustrate this procedure in the market for SO_2 abatement, we initially focus on only two consumers. The scenario is that we conduct a survey, asking the two consumers how much each would be willing to pay each year for various amounts of SO_2 abatement, *c.p.* The results are given in Table 3.2. The quantity demanded (Q_D) column shows a selection of possible abatement levels to be provided to all consumers. The demand prices, or WTP responses, are labeled p_1 and p_2 for Consumer 1 and Consumer 2, respectively. These responses are based on the following demand equations, which are expressed in inverse form to signify that price is the decision variable:

$$\text{Demand for Consumer 1:} \quad p_1 = 10 - 0.12Q_D$$
$$\text{Demand for Consumer 2:} \quad p_2 = 15 - 0.18Q_D$$

Notice that each consumer's WTP response for a given Q_D is unique. For example, for a 5 percent reduction in SO_2, Consumer 1 is willing to pay \$9.40 per year, whereas Consumer 2 is willing to pay \$14.10 per year. These responses are different because each consumer has a unique level of income, wealth, preferences, and so forth. In this context, we might expect the two consumers to have distinct preferences for air quality and even disparate views about the associated benefits. For example, perhaps Consumer 2's WTP is higher because he or she is a member of an environmental group and is more aware of the benefits of cleaner air.

To find the combined demand for air quality, the individual price responses at each quantity level are added together. The far right-hand column of Table 3.2 shows

TABLE 3.2 Hypothetical Demand Data in the Market for Air Quality

Quantity Demanded Q_D (% of SO_2 abatement)	Consumer 1's WTP $p_1 = 10 - 0.12Q_D$ (\$)	Consumer 2's WTP $p_2 = 15 - 0.18Q_D$ (\$)	Combined Demand Price for Consumers 1 and 2 $p_1 + p_2 = 25 - 0.3Q_D$ (\$)
0	10.00	15.00	25.00
5	9.40	14.10	23.50
10	8.80	13.20	22.00
15	8.20	12.30	20.50
20	7.60	11.40	19.00
25	7.00	10.50	17.50
30	6.40	9.60	16.00

NOTE: WTP = willingness to pay.

the results. The algebraic counterpart is found by summing each pair of corresponding terms on the right-hand side of each equation:

$$\begin{array}{lllr} & \text{Demand for Consumer 1:} & & p_1 = 10 - 0.12Q_D \\ + & \text{Demand for Consumer 2:} & & p_2 = 15 - 0.18Q_D \\ \hline & \text{Demand for Consumers 1 and 2:} & & p_1 + p_2 = 25 - 0.3Q_D \end{array}$$

The graphical model is shown in Figure 3.1. For any given quantity on the combined demand curve ($d_1 + d_2$), the corresponding measure of price is equal to the vertical sum of the individual price responses for that same quantity based on the demand curves d_1 and d_2.

Having developed the general procedure, we now bring the demand relationship up to the market level by assuming that the combined demand of the two consumers represents the decisions of one million consumers. Notice that this conversion means that price (P) is now denominated in millions of dollars, which corresponds to the scale of the market supply function. More formally, the market demand function is specified as

$$\text{Market Demand:} \quad P = 25 - 0.3Q_D$$

Equilibrium in the Air Quality Market

To find equilibrium in the market for SO_2 abatement, we begin by combining the market demand and supply data in Table 3.3. We observe that Q_E is equal to 20 percent, the point at which both the supply price and the demand price are equal to $19 million,

FIGURE 3.1 Combined Demand of Two Consumers for Air Quality

The demand curve labeled ($d_1 + d_2$) represents the combined consumption decisions of two consumers for air quality. For any quantity on this curve, the corresponding price is equal to the vertical sum of the individual price responses for that same quantity as given on curves d_1 and d_2.

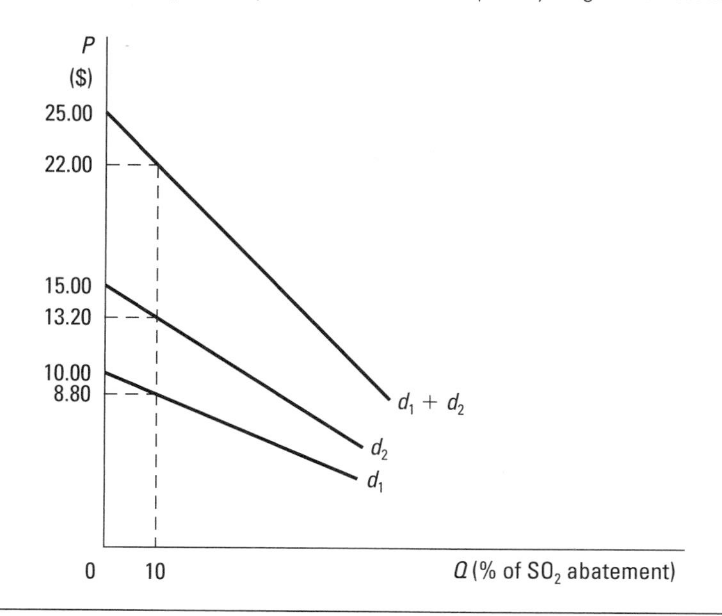

TABLE 3.3 Hypothetical Market Demand and Market Supply Data for Air Quality

Quantity Q (% of SO_2 abatement)	Market Demand Price $P = 25 - 0.3Q_D$ ($ millions)	Market Supply Price $P = 4 + 0.75Q_S$ ($ millions)
0	25.00	4.00
5	23.50	7.75
10	22.00	11.50
15	20.50	15.25
20	**19.00**	**19.00**
25	17.50	22.75
30	16.00	26.50

which is P_E. This result can also be found algebraically by solving the equations for market supply and demand simultaneously, as shown below.

At equilibrium:	Market Supply	= Market Demand
Substituting:	$4 + 0.75Q_S$	$= 25 - 0.3Q_D$, where $Q_S = Q_D$ at equilibrium
Solving:	$1.05Q$	$= 21$, or $Q_E = 20$ percent
Substituting into either equation:	P_E	$= 4 + 0.75(20)$ or $25 - 0.3(20) = \$19$ million

The corresponding graphical model is shown in Figure 3.2, where the intersection of the market supply (S) and market demand (D) curves identifies P_E and Q_E.

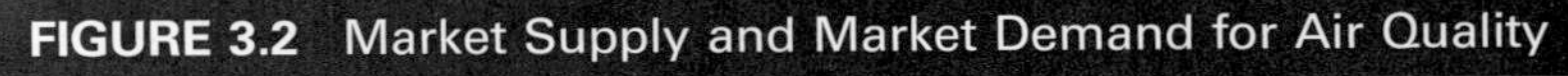

FIGURE 3.2 Market Supply and Market Demand for Air Quality

Assuming the market for SO_2 abatement is competitive, equilibrium price (P_E) of $19 million and equilibrium quantity (Q_E) equal to 20 percent abatement represent an allocatively efficient solution. Q_E represents the optimal level of *abatement* measured from *left to right* and implicitly the optimal level of *air pollution* measured from *right to left*. Notice that the optimal level of pollution is not zero.

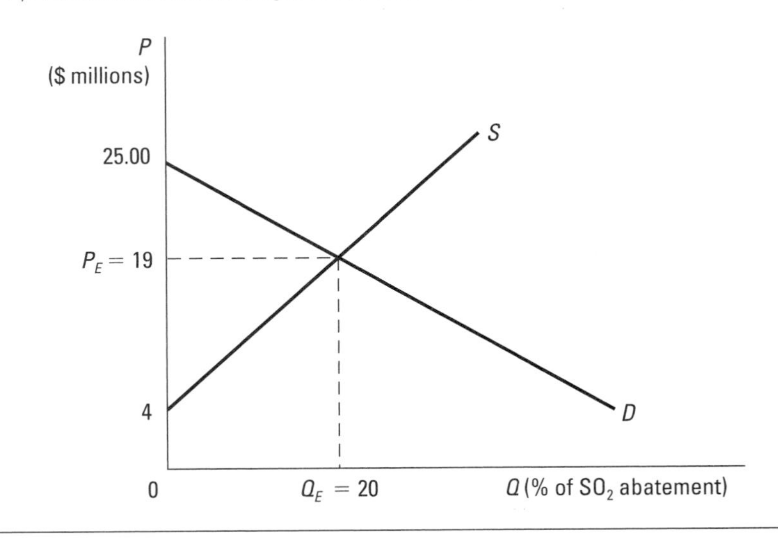

What Are the Implications?

Determining P_E and Q_E for the abatement of any kind of pollution is a significant result. As will become apparent in upcoming chapters, it is this result to which economists refer when addressing market-based solutions to environmental pollution. Notice in Figure 3.2 that Q_E represents the efficient or optimal level of *abatement* measured from *left to right* and implicitly the optimal level of *pollution* measured from *right to left*. As the model suggests, this optimal level is not necessarily zero.[7]

From a general perspective, abating at the 100 percent level to reduce pollution to zero involves prohibitive opportunity costs. These include the forgone production and consumption of any good generating even the smallest amount of pollution. Given our present technology, a zero-pollution world would be one without electricity, advanced transportation systems, and virtually all manufactured products. Thus, it makes little sense to argue for the elimination of all pollution in our environment.[8]

Understanding the Market Failure of Public Goods Markets

As we stated at the outset, the achievement of an allocatively efficient outcome in a public goods market depends on the identification of well-defined demand and supply functions. Although we developed both functions for our hypothetical market for air quality, market demand was identified only because we implicitly made one critical assumption—that consumers would reveal their willingness to pay (WTP) for SO_2 abatement. However, without third-party intervention, the nonexcludability of this or any public good makes it difficult, if not impossible, to ascertain such information. If consumers' WTP responses are unknown, market demand cannot be identified, and an efficient outcome cannot be obtained. It is precisely the inability of free markets to capture the WTP for a public good that causes the market failure.

nonrevelation of preferences
An outcome that arises when a rational consumer does not volunteer a willingness to pay because of the lack of a market incentive to do so.

free-ridership
Recognition by a rational consumer that the benefits of consumption are accessible without paying for them.

Consider the meaning of demand in the context of private versus public goods. In general, market demand captures the expected benefits associated with consumption. If the good is private (i.e., excludable), its benefits can be obtained only through purchase. Therefore, the consumer's WTP for a private good is a suitable proxy for the anticipated marginal benefits of consuming it. However, in the case of a public good that is nonexcludable, the consumer can share in its consumption even when it is purchased by someone else. Thus, there is no incentive for a rational consumer to volunteer a WTP for something he or she can consume without having to pay for it.

Formally, this problem is known as **nonrevelation of preferences**, which in turn is due to the more basic dilemma of **free-ridership**. The rational consumer recognizes that the benefits of a public good are accessible simply by allowing someone else to purchase it. The consumer becomes a free rider. Individual preferences about the public good remain undisclosed, and thus market demand is undefined. When the public good is environmental quality, the consequence can be serious ecological harm. A case in point is the pollution that damaged Boston Harbor, discussed in Application 3.1.

---

[7] Of course, a zero optimal level of pollution is *possible* under certain conditions, although this is the exception rather than the rule.

[8] One of the classic research papers that discusses this economic view of pollution is Ruff (Spring 1970).

APPLICATION 3.1 An Update on Boston Harbor: Restoring an Environmental Public Good

During the 1988 presidential campaign, the pollution of Boston Harbor became a highly charged political issue. Wanting to promote an image as the prospective "environmental president," then–Vice President George H. W. Bush focused the nation's attention on what he called the "dirtiest harbor in America." Political motivations aside, Bush was, at least in spirit, accurate in his assessment of the famous harbor that for a century had suffered the ill effects of public and industrial waste dumping. Boston Harbor and its nine beaches had been fouled by pollutants ranging from raw sewage to heavy metals, such as lead and mercury.

Why had the condition of this vital natural resource declined to such a deplorable state? The specifics are complex and entangled with bureaucracy and local politics, but the bottom line is that the harbor became seriously polluted because clean waterways are a public good. A cleaner Boston Harbor possesses the requisite characteristics of a public good, namely, nonexclusivity and nonrivalness in consumption. The result is the typical free-ridership problem that characterizes the market failure aspect of all public goods. No one is willing to reveal their preferences for a cleaner harbor and accept a share of the financial responsibility as long as the associated benefits can be enjoyed without having to pay for them. In addition, most harbor users are not fully aware of the benefits of cleaning up the harbor. Therefore, no one accepts responsibility for the preservation and protection of a waterway that is, in a real sense, commonly owned. The unfortunate outcome was that industries used the harbor as a dumping ground for chemical discharges, and residents left their trash on the harbor's beaches.

As predicted by theory, third-party intervention—implemented in this case through a federal court order—was necessary to initiate corrective action to save the harbor. The federally mandated plan, spearheaded by the Massachusetts Water Resources Authority (MWRA) (see **www.mwra.state.ma.us**), was completed in 2001 with a price tag of $4.1 billion. The monies were spent on an extensive reconstruction of the area's sewage treatment facilities. Wastes from Boston and nearby communities that formerly had been dumped into the harbor after receiving only limited treatment now are treated with state-of-the-art technologies and disposed of some 9.5 miles out into deeper waters. Other victories include a dramatic reduction in the discharge of toxic metals and sludge. In fact, metals in the discharges from the MWRA treatment facility declined by over 80 percent from 1990 to 2006.

Over $30 million has been invested by the MWRA in ongoing monitoring efforts and studies of water quality and marine life health, work that has been reviewed by regulatory officials and independent scientists. As this work continues, the public authority has witnessed a rebound of harbor beaches and associated ecological systems. Although the massive plan took years to implement, the results have been impressive. In fact, the Boston Harbor Project is considered to be "*... one of the nation's greatest environmental achievements*."

Today, the original effort is being supported by another public sector undertaking: the construction of the South Boston Combined Sewer Overflow (CSO) Project. This $148 million CSO venture is intended to hold stormwaters and combined sewer overflows until they can be processed by the treatment facility. In so doing, overflows from unusual weather conditions, up to a five-year storm event, are diverted from local beaches. This underground storage facility, which includes a 2.1-mile-long tunnel under South Boston that holds 19 million gallons of overflows, should make beach closings relatively rare.

To review the environmental status of the Boston Harbor and related data and information, visit **www.mwra.state.ma.us/harbor/html/bh_wq.htm**.

Sources: MWRA (June 3, 2009; January 30, 2009); Daley (September 10, 2001); Allen (September 6, 1992); Dolin (July/August 1992); Doneski (Spring 1985).

If we relax the conventional assumption of perfect information, adding more realism to the model, the identification of market demand becomes even more obscure. In many public goods markets, consumers are not fully aware of the benefits associated with consumption. This is certainly the case for environmental quality. Most people are not aware of all the health, recreational, and aesthetic benefits associated with pollution abatement. So even if consumers could be induced to express their WTP for a cleaner environment, it is highly likely that the resulting demand price would underestimate the true benefits. This added complication is due to **imperfect information**, which is another source of market failure.

What can be concluded from all this? As we initially claimed, market forces alone cannot provide an allocatively efficient level of a public good. This realization helps to explain what we observe in public goods markets—intervention by some third party, typically government. But to what degree should government become involved in the market process, and what tasks should government perform to achieve an efficient solution? These are difficult questions, but the underlying theories that identify the public goods problem suggest approaches to solving it.

The Solution: Government Intervention

In practice, a common means by which government responds to the dilemma of free-ridership and nonrevelation of preferences is through **direct provision of public goods.** Simple observation reveals that many public goods, such as fire protection, parks, and roadways, are provided by government. Similarly, government is involved in the preservation of natural resources and the provision of environmental quality. Among the government agencies whose responsibilities include the environment are the Army Corps of Engineers, the Federal Energy Regulatory Commission, and, of course, the Environmental Protection Agency (EPA).

An alternative government response is the use of **political procedures** and **voting rules** designed to identify voters' preferences about public goods. For example, members of Congress are responsible for determining their constituents' views on environmental issues and representing those interests in legislation.

ENVIRONMENTAL PROBLEMS: EXTERNALITIES

externality
A spillover effect associated with production or consumption that extends to a third party outside the market.

Another way to model environmental problems is through externality theory. Instead of defining the market as environmental quality or pollution abatement, this approach specifies the relevant market as the good whose production or consumption generates environmental damage *outside* the market transaction. (Recall from the materials balance model that residuals are generated by both producers and households.) Any such effect that is *external to* the market is aptly termed an **externality**.

Basics of Externality Theory

negative externality
An external effect that generates costs to a third party.

positive externality
An external effect that generates benefits to a third party.

Microeconomic theory argues that price is the most important signaling mechanism in the market process. Equilibrium price communicates the marginal value that consumers assign to a good and the marginal costs incurred by firms in producing it. Under ordinary conditions, this theory predicts the realities of the market remarkably well. Sometimes, however, price fails to capture *all* the benefits and costs of a market transaction. Market failures such as these occur when a third party is affected by the production or consumption of a commodity. Such a third-party effect is called an **externality**. If the external effect generates *costs* to a third party, it is a **negative externality**. If the external effect generates *benefits* to a third party, it is a **positive externality**.

Although the notion of an externality may seem obscure, it is nonetheless familiar conceptually. If a family purchases an unsightly satellite dish and installs it in their front yard, that action imposes costs to neighbors in the form of declining property values, a negative externality not reflected in the price of the satellite dish. Conversely, if one firm conducts research that advances a production process, there is a benefit to the entire industry, a positive externality not accounted for in the research investment decision.

Common to both examples is a spillover effect that occurs outside the market transaction. This effect is not captured by the price of the commodity being exchanged. If price does not reflect *all* the benefits and costs associated with production and consumption, it is unreliable as a signaling mechanism, and the market fails. An important consequence is that scarce resources are misallocated. If consumption generates external benefits, the market price undervalues the good, and *too little* of it is produced. If there is a negative externality, the market price does not reflect the external costs, and *too much* of the commodity is produced.

Environmental Externalities

Of interest to environmental economists are externalities that damage the atmosphere, water supply, natural resources, and the overall quality of life. The classic case is the negative externality associated with *production*. For example, the provision of air transportation causes noise pollution, damages air quality, and reduces the value of nearby residential properties. These are real costs that are not absorbed by airlines or by air travelers. Because these costs are incurred by parties outside the market transaction, they are not captured in the price of airline tickets. Environmental externalities also can be associated with *consumption*. A good example is the cost of waste disposal associated with consuming products having excess or nonbiodegradable packaging. Application 3.2 discusses how this type of externality affected the production of CDs.

Positive externalities also can help to explain the persistence of environmental problems. Consider the market for pollution abatement equipment, such as scrubbers. Scrubbers are elaborate systems used to clean emissions from producers' smokestacks. When an electricity producer, for example, purchases and installs a scrubber system, the benefits of cleaner air accrue to all people living in the nearby area. Because these individuals are not a part of the market transaction, the external benefits are not captured in the scrubber's price. Resources are misallocated, and too few scrubbers are exchanged in the marketplace.

Notice that there is a qualitative relationship between the external benefits associated with pollution abatement and the external costs of pollution-generating commodities. They are just the inverse of one another. If the market is defined as the abatement equipment industry, there is a *positive* externality, and the external benefits are improved health, natural resources, aesthetics, etc. On the other hand, if the market is defined as electricity production, for example, there is a *negative* externality, and the external costs are the damages to health, natural resources, and aesthetics. Which of the two models is relevant depends solely on which market is specified.

APPLICATION 3.2 Responding to a Negative Externality: From the CD Long Box to the Carbon-Neutral Digipak™

Despite consumer reliance on digital downloads, compact disks (CDs) continue to play an important role within the music industry. In fact, according to Oasis Disc Manufacturing, the leading CD and DVD manufacturer to independent musicians and filmmakers, the great majority of retail music revenues are generated from physical CD media. Moreover, some argue that CDs have a lasting appeal because their packaging is an art form that makes them true collectibles. Yet, it has been the packaging of CDs that has generated concern among environmentally-conscious consumers and other groups for over two decades.

Compact disks (CDs), introduced to the U.S. market in 1982, were originally packaged in two boxes. The outer box was a 6-by-12-inch cardboard package, known in the industry as the long box. Inside the long box was a 5-by-5½-inch jewel box constructed of clear plastic, which housed the CD. The jewel box, which continues in use today, generally is saved by the consumer as a protective container for the CD when not in use. However, the long box had no practical value beyond identifying the contents at the point of sale. In fact, most consumers discarded the long box immediately after purchase, so this part of the packaging ended up in landfills or burned in municipal incinerators.

Of the 250 million CDs sold in 1990, an estimated 23 million pounds of CD packaging was discarded. Think about the costs associated with the generation and disposal of this much waste, costs that are external to the purchase and sale of CDs. The producer of the CD does not consider these external costs as part of its production decision, so they are not reflected in the CD's price. Who, then, bears the costs? Society as a whole has to pick up the tab, the classic symptom of a negative externality, in this case, one associated with consumption.

The obvious question to ask is why the long box was used at all in packaging CDs. Music company executives defended this packaging with two arguments. First, its large size helped to discourage shoplifting. Second, the long box helped CD sales by allowing retailers to use existing display racks originally designed for 12-by-12-inch record albums. According to one industry official, the long box saved retailers an estimated $100 million in redesign costs for display units.

Despite the industry's arguments, followers of the green movement pressured the industry to find an alternative to the wasteful packaging. In 1991, a "Ban the Box" campaign materialized to eliminate the long box from CD packaging. Environmentally minded recording artists joined the crusade, demanding that a more environmentally friendly package design be developed.

Industry officials searched for a new design that would facilitate retailers' display needs *and* ameliorate environmentalists. After weighing all the options, the recording industry agreed in 1992 to a voluntary ban on the long box, and since April 1993, CDs have been shipped without the objectionable packaging. As a result, the standard CD package used by most American companies is the shrink-wrapped jewel box or the paperboard Digipak, which was subsequently introduced.

Today, the evolution of CD packaging has reached a new level. Oasis Disc Manufacturing is the first to offer a totally carbon-neutral CD. Compared to the jewel case, production of the Zero Carbon Footprint Digipak™ generates 88 percent less carbon emissions, which, when coupled with carbon offsets, renders the CD and its packaging carbon-neutral. The packaging also features a CD flexible tray manufactured from 100 percent recycled water bottles. Visit **www.oasiscd.com** for more information.

The CD packaging controversy is an important example of how externalities persist in the marketplace. Wasteful CD packaging was used for a full decade before any action was taken. In this case, the third-party intervention came not from government but from environmentalists who made market participants aware of how their decisions adversely affected the environment, which in turn prompted a positive response.

Sources: Oasis Disc Manufacturing (February 2, 2011); "New Carbon-Neutral Packaging Is Music to Your Ears." (February 1, 2011); Block (June 1993); Cox (July 25, 1991); Newcomb (May 13, 1991).

Relationship Between Public Goods and Externalities

Environmental externalities are those affecting air, water, or land, all of which have public goods characteristics. What this implies is that, although public goods and externalities are not the same concept, they are closely related. In fact, if the externality affects a broad segment of society and if its effects are nonrival and nonexcludable, the externality is itself a public good.[9] If, however, the external effects are felt by a more narrowly defined group of individuals or firms, then those effects are more properly modeled as an externality.

MODELING ENVIRONMENTAL DAMAGE AS A NEGATIVE EXTERNALITY

Having established the basics of what externalities are conceptually, we now develop a formal model of a negative environmental externality. We elect to model a production externality, because this approach addresses the most prevalent source of environmental pollution.

Defining the Relevant Market

As always, one of the first steps in model building is to define the relevant market. In this case, we define the market as refined petroleum products. This is a fitting choice, because refined petroleum plants are major water polluters. Among the associated external costs are serious health risks for people using the rivers and streams.

Modeling the Private Market for Refined Petroleum

To avoid complicating our analysis with the market failure of imperfect competition, we assume that the private market for refined petroleum is competitive. At the industry level, the hypothetical supply and demand relationships for refined petroleum products are

$$\text{Supply:} \quad P = 10.0 + 0.075Q$$
$$\text{Demand:} \quad P = 42.0 - 0.125Q$$

where Q is measured in thousands of barrels per day and P is the price per barrel.

Recall from Chapter 2 that supply represents the marginal costs of production, and demand represents the marginal benefits of consumption, both based on *private* or *internal* decision making. In markets such as this one where there are also external costs, it is necessary to distinguish internal, or private, costs from external costs. To do this, we formally refer to the supply function as the **marginal private cost (*MPC*)** of production. For consistency, we also will refer to the demand relationship as the **marginal private benefit (*MPB*)** function, even though it is assumed that there are no external benefits in this market. Thus, the two functions are restated as

$$MPC = 10.0 + 0.075Q$$
$$MPB = 42.0 - 0.125Q$$

[9]Technically, if the externality provides *benefits* to a large component of society, it is a public *good*; if the opposite is true, the externality is a public "*bad*."

Inefficiency of the Competitive Equilibrium

Given the usual assumptions about the underlying motivations of supply and demand, the competitive market clears where $MPB = MPC$, or, equivalently, where marginal profit ($M\pi$) or ($MPB - MPC$) = 0. Solving the MPB and MPC equations simultaneously yields a competitive market price (P_C) of \$22 per barrel and a market quantity (Q_C) of 160,000 barrels. The graph is shown in Figure 3.3, where equilibrium occurs at the intersection of the MPB and MPC curves.

The problem with this equilibrium is that it ignores the external costs to society of contaminated water supplies caused by refined petroleum production. Remember, natural market forces motivate firms to satisfy their own interests, not those of society. The costs of the water pollution are *external* to the market exchange and consequently are not factored into private market decisions. The implications are serious, because allocative efficiency requires that marginal benefits be equal to *all* marginal costs of production. Because the external costs are not included in private decision making, the MPC undervalues the opportunity costs of production, and the resulting output level is too high.

From a practical perspective, economists want to identify and monetize these external costs, but assigning a dollar value to negative externalities is difficult. Think about trying to monetize the damage to aquatic life from water pollution or the increased health risks from swimming in a polluted lake. Although there are economic methods that estimate these costs, they are not straightforward. Therefore, we will defer discussion of these methods to later chapters. For now, we will simply argue that these costs do exist and assume for the present that they can be quantified.

FIGURE 3.3 Competitive Equilibrium in the Refined Petroleum Market

The intersection of the *MPB* and *MPC* curves identifies the competitive equilibrium in the refined petroleum market, where $P_C = \$22$ per barrel and $Q_C = 160{,}000$ barrels. This equilibrium ignores the external costs of contaminated water supplies caused by refined petroleum production. Therefore, the *MPC* undervalues the true opportunity costs of production, and the competitive output level is too high.

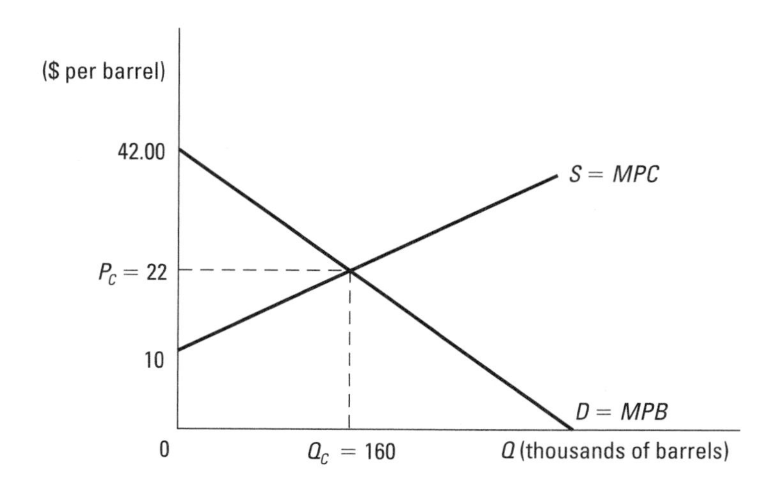

Modeling the External Costs

To complete our analysis of the refined petroleum market, we model the hypothetical **marginal external cost (*MEC*)** function as

$$MEC = 0.05Q$$

Think about the economic interpretation of this equation. Based on the constant slope of 0.05, the *MEC* resulting from water pollution is increasing at a constant rate of 0.05 with respect to oil production. Because Q is measured in thousands of barrels, this value implies that for every additional 1,000 barrels of refined oil produced, the marginal external costs of pollution rise by $0.05 per barrel.

Modeling the Marginal Social Costs and Marginal Social Benefits

To achieve allocative efficiency, we must consider the external costs when finding equilibrium price and quantity. To accomplish this, we add the *MEC* to the firm's *MPC* to derive the **marginal social cost (*MSC*)** equation:

marginal social cost (*MSC*) The sum of marginal private cost (*MPC*) and marginal external cost (*MEC*).

$$\begin{aligned} MSC &= MPC + MEC \\ &= 10.0 + 0.075Q + 0.05Q \\ MSC &= 10.0 + 0.125Q \end{aligned}$$

The *MSC* is relevant to production decisions because it captures *all* costs of producing refined petroleum, that is, the private costs of production *and* the external costs of environmental damage to society.

marginal social benefit (*MSB*) The sum of marginal private benefit (*MPB*) and marginal external benefit (*MEB*).

On the demand side, there is an analogous benefit relationship, called the **marginal social benefit (*MSB*)**, which is the sum of the *MPB* and any **marginal external benefit (*MEB*)**. Because we have assumed that there are no positive consumption externalities, the *MEB* is zero; therefore, the *MPB* equals the *MSB* in this case.

Efficient Equilibrium

Once determined, *MSC* must be set equal to *MSB* to solve for the efficient equilibrium price (P_E) and quantity (Q_E). In this case, the efficient level of refined petroleum products is 128,000 barrels per day sold at a market price of $26 per barrel. Compare this to the competitive equilibrium of $Q_C = 160{,}000$ and $P_C = \$22$. As we asserted, the competitive equilibrium in the presence of a negative externality is characterized by an overallocation of resources to production. Furthermore, the competitive price is too low, because the *MEC* is not captured by the market transaction.

The results are shown in Figure 3.4. Geometrically, the *MSC* curve is the vertical sum of the *MEC* and *MPC* curves. The intersection of *MSC* and *MSB* identifies the efficient equilibrium at $P_E = \$26$ and $Q_E = 128{,}000$. The graph also shows the competitive equilibrium at $P_C = \$22$ and $Q_C = 160{,}000$, which corresponds to the intersection of *MPC* and *MPB*. Notice that at Q_C, *MSB* is *below* *MSC*. This signifies that society is giving up more in scarce resources to produce petroleum than it gains in benefits from consuming it. To restore the equality of *MSB* = *MSC*, signifying allocative efficiency, output must be decreased—precisely what is predicted by theory.

FIGURE 3.4 Competitive and Efficient Equilibria in the Presence of a Negative Externality

The intersection of *MSC* and *MSB* identifies the efficient equilibrium at $P_E = \$26$ and $Q_E = 128{,}000$. Notice how this compares to the competitive equilibrium where $P_C = \$22$ and $Q_C = 160{,}000$, which occurs at the intersection of *MPC* and *MPB*. At Q_C, *MSB* is *below MSC*, which means that society is giving up more in scarce resources to produce petroleum than it gains in benefits from consuming it.

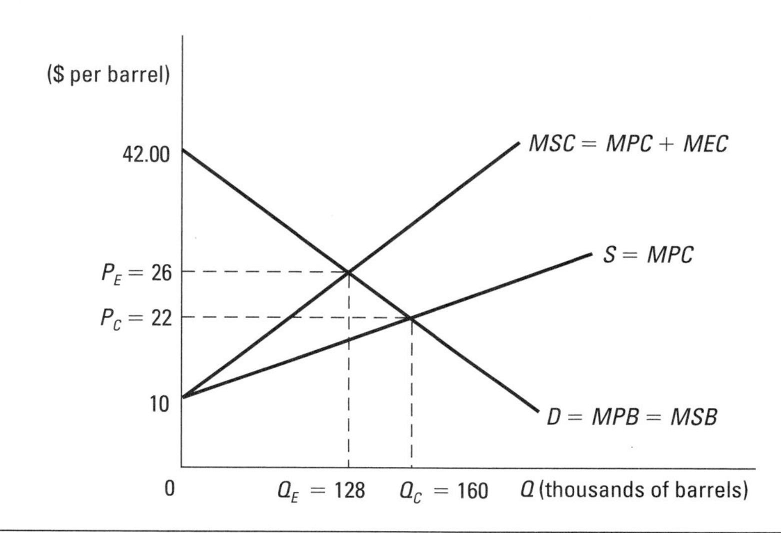

An alternative way to analyze the two equilibria is to examine the corresponding levels of $M\pi$. At the competitive equilibrium, we know that $M\pi = 0$, because that point is defined where $MPB = MPC$. At the efficient equilibrium, we know that $MSB = MSC$, which can be restated as $MSB - MSC = 0$, or equivalently, where $MPB - MPC = MEC$. We conclude that at the efficient equilibrium point, $M\pi = MEC$. These derivations are summarized below.

competitive equilibrium The point where marginal private benefit (*MPB*) equals marginal private cost (*MPC*), or where marginal profit ($M\pi$) = 0.

efficient equilibrium The point where marginal social benefit (*MSB*) equals marginal social cost (*MSC*), or where marginal profit ($M\pi$) = marginal external cost (*MEC*).

Competitive equilibrium:

$$MPB = MPC$$
$$MPB - MPC = 0$$
$$M\pi = 0$$

Efficient equilibrium:

$$MSB = MSC$$
$$MPB + MEB = MPC + MEC$$
$$MPB - MPC = MEC \quad (\text{since } MEB = 0)$$
$$M\pi = MEC$$

Measuring the Welfare Gain to Society

One important conclusion of the preceding analysis is that efficiency in the market for refined petroleum would improve if output were reduced by 32,000 barrels per day (i.e., 160,000 − 128,000). This output adjustment would increase society's welfare. To illustrate this, the *MPC*, *MSC*, and *MPB* curves are reproduced in Figure 3.5 with some added notation. Use this model to consider the separate effects on the refineries and on society associated with this output restriction.

FIGURE 3.5 Net Gain to Society of Restoring Efficiency

To restore efficiency, Q would have to be reduced from 160,000 to 128,000 barrels. As this output restriction occurs, refineries lose profit, represented as area WYZ. At the same time, society gains the associated decline in damages to health and the ecology, shown as area WXYZ. On net, society gains by area WXY as a result of efficiency being restored.

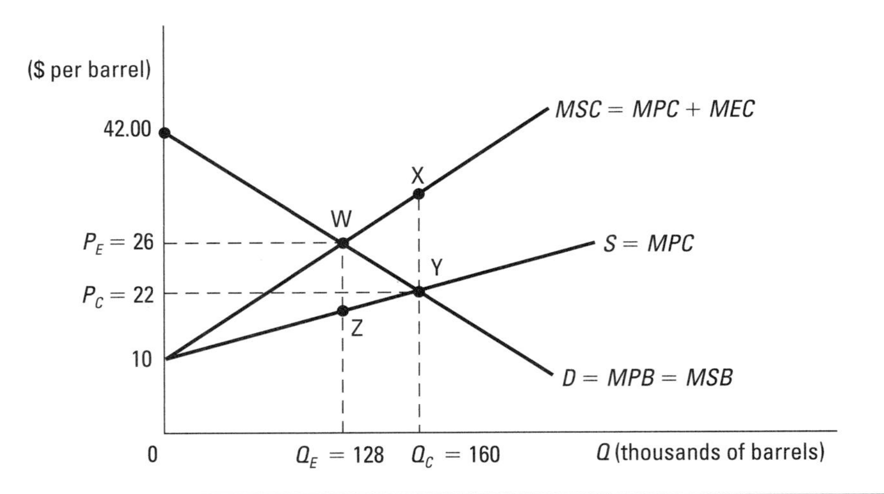

From the firm's perspective, there is a loss in profit. Notice that as Q falls from 160,000 to 128,000, refineries lose profit, measured as the excess of *MPB* over *MPC* for each unit of output. Adding all the $M\pi$ values between $Q_E = 128{,}000$ and $Q_C = 160{,}000$ defines the triangular area WYZ, which represents the total loss in profits. However, from society's vantage point, there is a gain equal to the accumulated *reduction* in *MEC* associated with the output decline. This reduction in external costs represents the decrease in health and ecological damage. Geometrically, this gain is area WXYZ. On net, society gains the triangular area WXY as efficiency is restored. Notice that, although the movement to the efficient output level achieves a welfare gain, it does not entirely eliminate the externality. Firms are still generating pollution at Q_E, but the amount is lower than it would be at Q_C.

In sum, if production of a good generates a negative externality, the market will yield an inefficient solution with too many resources allocated to production. If that externality were somehow accounted for within the market, society as a whole would gain. Of course, the operative issue is *how* to account for externalities such that efficiency can be restored.

Market Failure Analysis

It is important to understand the lack of incentives in the natural market process to explicitly account for external costs. Petroleum refineries are motivated by *private* gain, not by *social* gain. Although these firms may be aware of the environmental damage associated with their production, there is no incentive—in fact, there is a disincentive—for them to absorb these costs. Doing so would negatively affect profits. It would be as though firms offered to pay for the external costs on society's behalf. However, there is no market incentive for a rational firm to incur higher costs than it has to, even if it is for the good of society.

These assertions should not deter efforts to solve the problems of environmental damage; quite the contrary. Market failure models give us a better understanding of *why* we observe increasing damage to the physical environment as industrial production has intensified throughout the world. The theory also explains the persistence of environmental problems from a market perspective and the need for government to intervene where such problems arise. Finding appropriate policy solutions is not easy, but the process is facilitated by an understanding of how and why markets fail. On that point, if we consider both the public goods problem and the externality model we have examined, an important common element leads us to the source of virtually all environmental problems—the absence of property rights.

THE ABSENCE OF PROPERTY RIGHTS

In every model presented thus far in this chapter, the market failed to provide an efficient solution. Using basic economic theory, we demonstrated *how* the market fails and even quantified the failure in terms of the overproduction of output. Although in each case the failure was examined conceptually, we have not yet focused on the underlying root of the problem.

Reconsider the negative externality model, and recall the assertion that an externality is a public good if it affects a broad segment of society. In our model, the external cost is damage to water supplies, which does fit that characterization. Furthermore, clean water possesses the two distinguishing characteristics of a public good. Now recall our public goods model in which the relevant market was defined as air quality. What do these two public goods have in common? The answer is that in each case, the property rights of the good are undefined, and as a result, markets for these goods are virtually nonexistent. **Property rights** are the set of valid claims to a good or resource that permits its use and the transfer of its ownership through sale. These rights are generally limited by law and/or social custom.

property rights
The set of valid claims to a good or resource that permits its use and the transfer of its ownership through sale.

In the context of environmental public goods, it is unclear who "owns" the rights to water supplies, or who "owns" the rights to the air. For example, do swimmers own the right to clean water, or do refineries hold the rights to pollute it? Do individuals own the right to breathe clean air, or do polluting firms own the right to contaminate it? Because the answers to these questions are not clear-cut, there is no built-in market mechanism to solve environmental problems. Indeed, the absence of property rights to rivers in the Pearl River Delta (PRD) region of southern China is the underlying reason for serious environmental damage to that region, an issue discussed in Application 3.3. It turns out that property rights are critically important to the sound functioning of the market system. In fact, as pointed out by Nobel laureate Ronald Coase, the assignment of property rights alone can provide for an efficient solution even in the presence of an externality.[10] To learn more about Coase, visit the Ronald Coase Institute at **www.coase.org**.

[10]In 1991, Ronald Coase received the Nobel Prize in economics for his pioneering work in the theory of transaction costs and other concepts that link economic theory and the law (Coase, October 1960). Much of his research is relevant to the economics of environmental problems.

APPLICATION 3.3 Undefined Property Rights to China's Pearl River Delta (PRD) Region Leads to Environmental Damage

The Pearl River Delta (PRD) in southern China is an enormously important economic region. Located in the southern part of Guangdong Province, the region is responsible for a third of China's exports, explaining why it is sometimes called the "world's factory floor." Over the last two decades, much of Hong Kong's manufacturing sector moved there to take advantage of lower-cost land and labor, a reality not overlooked by importers of Chinese goods. Cases in point are Chinese textile mills, which face strong demand by apparel retailers needing inexpensive fabrics to produce clothing. Among these retailers are well-known U.S. firms, such as Wal-Mart, Lands' End, and Nike. To satisfy demanders, these export-dependent textile factories attempt to keep prices low, which in turn means that low-cost operations must be maintained.

None of this market pressure would have been problematic had it not been for the fact that textile manufacturing is among the more environmentally damaging industries in China. Wastes from textile production include fabric dyes, which comprise such pollutants as heavy metals, carcinogens, and organic materials. Such wastes can be treated prior to release, but doing so adds to production costs, which in turn puts upward pressure on prices. Hence, Chinese textile manufacturers had no market incentive to internalize these external costs. In fact, the Ministry of Commerce in China asserted that export prices are artificially low precisely because textile plants do not absorb the costs of damages to waterways. Instead, wastes are released into nearby rivers of the PRD—rivers for which property rights are undefined. The resulting pollution is a negative externality affecting a large segment of Chinese society, which identifies the externality as a public "bad." In the absence of well-defined property rights, markets do not function properly, and too many resources are allocated to production.

A much publicized example was the dumping of textile wastes by Fuan Textile Mill, a supplier to such American firms as Eddie Bauer, Gap, and Target. A subsidiary of Hong Kong's Fountain Set Holdings, Ltd., the world's largest maker of cotton knit fabric, Fuan Textile was releasing some 22,000 tons of dye-contaminated water each day into an adjacent river allegedly through a pipe under the plant's floor. The violation was discovered by Chinese authorities in 2006.

In the past, fines for polluting activity, which ranged from $700 to $1,300, had little effect on violators, because the fines were sometimes lower than firms' abatement costs. Reportedly, wastewater treatment can cost hundreds of thousands of dollars annually for large plants. As long as such a cost imbalance held, profit-maximizing firms would continue to damage waterways.

In response, China's State Environmental Protection Agency (SEPA), in collaboration with the Ministry of Commerce, introduced tougher regulations aimed at the country's export industries, which account for about 20–30 percent of China's water pollution. New initiatives were launched, including market-based instruments aimed at encouraging polluters to internalize the negative externality.

As local factories faced these tougher and costly regulations, they elevated prices, making them less competitive globally. In some cases, factory closures resulted. According to the Federation of Hong Kong Industries, 10 percent of Hong Kong-owned factories in the PRD region were expected to close in 2008. The associated decline in production, according to economic theory, should bring output closer to its efficient level. Moreover, because the production of exports accounts for a large proportion of China's pollution, the regulatory changes, if successful, should eventually improve water quality in the PRD.

That the situation is dire and continuing is validated in part by a 2009 investigation by Greenpeace China, which examined the environmental status of the PRD region, collecting wastewater and sediment samples from five manufacturing plants. Among their findings was that all five facilities were releasing wastewater polluted with proven or suspected hazardous chemicals, and that effluents from three of the five plants had chemical concentrations above the standards set by the Guangdong province. Perhaps more troubling is the finding that some effluents contained hazardous organic chemicals not even regulated by existing standards.

Taken together, these results indicate that more must be done to improve water quality in the PRD region, as outlined in the Greenpeace China report on this investigation. To access the report and the associated recommendations for industry and the Chinese government, visit **www.greenpeace.org/eastasia/publications/reports/toxics/2010/pearl-river-report-2009-2nd/**.

Sources: Greenpeace China (2010); U.S. Department of Commerce, International Trade Administration, U.S. Commercial Service (October 3, 2008); Fong and Canaves (February 22, 2008); Spencer (November 1, 2007; August 22, 2007).

The Coase Theorem

Coase Theorem Assignment of property rights, even in the presence of externalities, will allow bargaining such that an efficient solution can be obtained.

Coase's exposition about property rights and their relevance to externality problems is so significant that it has come to be known as the **Coase Theorem**. This theorem states that proper assignment of property rights to any good, even if externalities are present, will allow bargaining between the affected parties such that an efficient solution can be obtained, no matter which party holds the rights. Two important underlying assumptions of this theory are noteworthy:

- Transactions are costless.
- Damages are accessible and measurable.

To illustrate the Coase Theorem, let's revisit our market model of refined petroleum to see how the assignment of property rights and bargaining can restore efficiency. We impose the assumptions given above and test the theory's prediction by examining the outcome under two different assignments of rights to some hypothetical river. One is that private individuals hold the rights to the river for recreational use, and the other is that refineries hold the rights to release toxic chemicals into the river.[11]

Bargaining When Property Rights Belong to the Refineries

We begin by assigning the property rights to the refineries. This means that the refineries have the right to pollute the river as part of their production processes. Although this assignment of rights may appear to be perverse, recall that according to the Coase Theorem, this should have no effect on the outcome at all.

Remember that the refineries are interested in producing petroleum products up to the level that maximizes profit—by itself, not an objectionable motive in a free enterprise system. The pollution they cause is an unintended by-product of production. Nonetheless, the recreational users of the river are harmed by the associated pollution, or more formally, their utility is negatively affected. Knowing that the refineries have the right to pollute, and given their own motivation to maximize utility, these recreational users have an incentive to negotiate. For each unit change in output, they would be willing to pay the refineries *not* to pollute up to an amount equal to the associated negative effect on their utility. The refineries, on the other hand, would be willing to accept payment *not* to pollute as long as that payment is greater than the loss in profits from cutting back production.

The terms of the negotiation can best be understood by examining Figure 3.6, which replicates the negative externality model. As before, Q_C represents the profit-maximizing competitive equilibrium, and Q_E is the efficient equilibrium. We assume that the market is at Q_C when the bargaining begins, because this is the point that the owners of the rights, the refineries, would choose. Each party's bargaining position is given in the context of the cost and benefit curves in the diagram:

Recreational users:	Willing to offer a payment (ρ) such that $\rho < (MSC - MPC)$
Refineries:	Willing to accept a payment (ρ) such that $\rho > (MPB - MPC)$

[11]Although it would be more realistic to assume that only a single refinery is involved in a bargaining scenario about the use of a river, it is simpler to approach the problem from an industry perspective so that we can use the same data and graphical models developed in the negative externality discussion.

FIGURE 3.6 Bargaining in the Refined Petroleum Market

If the refineries own the rights to the river, bargaining begins at Q_C and continues to Q_E, because at all output levels between these points, the payment ρ satisfies the condition $MEC > \rho > M\pi$, which is acceptable to both parties. If the recreational users own the rights, bargaining begins at $Q = 0$ and continues up to Q_E. This occurs because at all output levels between these points, the payment ρ satisfies the condition $M\pi > \rho > MEC$, which is acceptable to both parties.

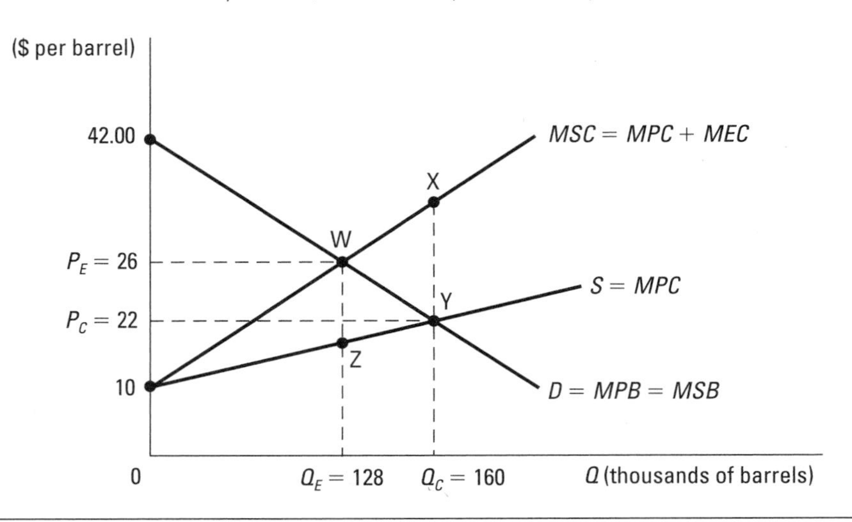

Note that the value of (*MSC* – *MPC*) measured vertically is simply the *MEC*, the value of the marginal damage incurred by the recreational users for each added unit of production. The vertical distance measured by (*MPB* – *MPC*) is the $M\pi$ earned by the refineries for each extra unit of output produced. In theory, the bargaining between the two groups should continue as long as the payment is greater than the refineries' loss in profit but less than the recreational users' damage. In the context of the model, payment ρ will be acceptable to both parties as long as the following condition holds:

$$(MSC - MPC) > \rho > (MPB - MPC), \text{ or equivalently,}$$
$$MEC \quad > \rho > \quad M\pi$$

Find these vertical distances in Figure 3.6. Note that at the competitive equilibrium (Q_C), the refineries' $M\pi$ equals zero, or $(MPB - MPC) = 0$. On the other hand, the *MEC* at Q_C is some positive value, or $(MSC - MPC) > 0$, represented as distance XY. Thus, bargaining between the parties is feasible at the competitive equilibrium, because $MEC > M\pi$ at Q_C. Moreover, because this condition holds for all output levels between Q_C and Q_E, the parties will continue to bargain all the way to Q_E, where negotiation ceases. The reason? Because at Q_E, (*MSC* – *MPC*) is exactly equal to (*MPB* – *MPC*), or $MEC = M\pi$, each measured as distance WZ. Output reductions beyond that point would generate a loss in profit to the refineries greater than what the recreational users would be willing to pay, or $M\pi > MEC$, so bargaining would break down.

This is a profound result. Under the assumptions of Coase's model, assigning the property rights to the refineries will lead to bargaining between the parties, which ultimately generates an efficient outcome without any third-party intervention.

Bargaining When Property Rights Belong to the Recreational Users

Now reconsider the model with the property rights assigned to the recreational users. Remember that according to the Coase Theorem, an efficient outcome can be achieved regardless of which of the affected parties controls the property rights. We should expect to obtain the same result as before. The starting point at which bargaining begins, however, is different. If the recreational users hold the rights to the river, then technically, the refineries cannot produce at all unless they pay for the rights to pollute. Thus, we assume that the point at which bargaining begins is where $Q = 0$.

The recreational users pursue their rights to a clean river, not to obstruct production of the refineries but to maximize their utility. Nonetheless, the refineries' profit objective is negatively affected. Hence, it is they who have a market-based motivation to bargain with the recreational users in an attempt to get them to accept payment for their rights to the river. The terms of the bargaining are analogous to the previous scenario, except that now the refineries are in the offering position, and the recreational users are in the accepting position.

Refer back to Figure 3.6, and consider the two sides' relative positions at $Q = 0$. For each unit change in output, the refineries are willing to pay for the right to pollute (and thus the right to produce output), up to an amount equal to the $M\pi$ received from production, measured as the vertical distance between *MPB* and *MPC*. The recreational users are willing to trade off their rights to clean water only if they receive payment greater than the damage they incur as the river becomes polluted. This damage is the *MEC*, measured as the distance between *MSC* and *MPC*. So each party's bargaining stance is as follows:

Refineries:	Willing to offer a payment ρ such that $\rho < (MPB - MPC)$
Recreational users:	Willing to accept a payment ρ such that $\rho > (MSC - MPC)$

Therefore, there is opportunity for bargaining to proceed as long as the following condition holds:

$$(MPB - MPC) > \rho > (MSC - MPC), \text{ or equivalently,}$$
$$M\pi > \rho > MEC$$

The model presented in Figure 3.6 confirms that this condition holds at $Q = 0$ and continues to hold for all output levels up to Q_E. At Q_E, bargaining ceases, because at that point $(MPB - MPC)$ is exactly equal to $(MSC - MPC)$, or $M\pi = MEC$, each represented by distance WZ. Note that beyond Q_E, the marginal damage associated with increased production exceeds the addition to profit that the refineries would receive, or $MEC > M\pi$, so bargaining would break down. Once again, the assignment of property rights, this time to the recreational users, leads to an efficient outcome without any government intervention.

Limitations of the Coase Theorem

Coase's model yields a powerful result and one that underscores the importance of property rights to the market process, regardless of who is assigned those rights. However, as was stated at the outset, the model's prediction of an efficient outcome depends on two very limiting assumptions: that transactions are costless and that damages are accessible and measurable. Thus, for the theory to hold in practice, at minimum, it must be the case that few individuals are involved on each side of the market.

The reality of the refinery market and, in fact, of most markets is that there are many affected parties on both sides of the market. Nontrivial costs would be associated with trying to reach a consensus about the bargaining terms within each group even before negotiations could begin. Undoubtedly, legal counsel would be necessary, adding still more to the transaction costs. Then there is the difficulty of identifying the sources of the damage and assigning a value to that damage. As more parties get involved, the task becomes increasingly difficult.

Common Property Resources

It turns out that property rights need not be completely missing for a market problem to exist. If property rights exist in some form but are ill defined, the outcome will also be an inefficient one. Such is the case for common property resources, which represent another source of externalities. **Common property resources** are those for which property rights are shared by some group of individuals.

common property resources
Those resources for which property rights are shared.

Notice that common property resources fall somewhere on a continuum between the extremes of pure public goods and pure private goods. Unlike pure public goods, common property resources are not accessible to everyone, meaning there is some measure of excludability.[12] However, because the property rights extend to more than one individual, they are not as clearly defined as they are for pure private goods. Some classic examples of common property resources are fisheries and oil pools.

With common property resources, the problem is that public access without any control leads to resource exploitation, which in turn generates a negative externality. The problem arises because each co-owner makes decisions about using the resource based only on private costs and benefits, ignoring how that decision would affect other owners. For example, consider a lake stocked with fish for exclusive use by members of a local community. Unless the catch is limited in some way, each person would fish from the lake based on private motivations, ignoring how the associated decline in the fish population would negatively affect others' ability to share in the benefits. Furthermore, each individual would have no incentive to consider the ultimate cost of restocking the lake, because their personal share of that cost would be very small. The result is overuse or depletion of the scarce resource.

[12]Put another way, one can think of public goods as an extreme case of common property resources, when the group that shares the property rights consists of all individuals.

The Solution: Government Intervention

From an economic perspective, the general solution to externalities, including those affecting the environment, is to **internalize the externality**, that is, to force the market participants to absorb the external costs or benefits. One way this can be done is through the assignment of property rights. In our model, when the refineries owned the rights to pollute, the recreational users internalized the externality through their payment offer. Conversely, when the recreational users owned the rights to clean water, the refineries internalized the external cost by paying for the right to pollute. But how do these rights get assigned in the first place? In practice, the government would have to make this determination as well as enforce limitations on these rights for the good of society.

Other approaches to internalizing environmental externalities are policies that change the effective product price by the amount of the associated external cost or benefit. In the petroleum refinery market, for example, the price per barrel of oil could be forced up by the *MEC*, perhaps through a unit tax. More recent policy prescriptions involve the most direct form of internalizing environmental externalities, which is to establish a market and a price for pollution. These approaches will be investigated in later chapters, all of which are rooted in market failure theory.

CONCLUSIONS

Economic theory supports the argument that environmental problems persist because they are market failures. Whether we model pollution as a negative externality or as damage to environmental public goods, we observe conditions that impede natural market forces. At the root of the dilemma is the absence of property rights. Because no one owns the atmosphere or the earth's water bodies, there are no market incentives to pay for the right to protect these resources or for the right to pollute them. The result is a misallocation of economic resources and a decline in society's welfare. Some third-party mediation, typically government, is needed to correct the market failure and reach an efficient equilibrium.

How strong a presence should government have in affected markets, and how should government go about the difficult task of developing effective policy solutions? Although market failure models identify what the overall approach should be, many practical issues have to be addressed. In theory, we know that the value of an environmental externality must be internalized so that the efficient solution can be identified. But *how* is this accomplished in practice, and is this a practical solution? Are there methods that determine some proxy measure of the costs and benefits of pollution abatement? Or are there better, alternative solutions or more reasonable objectives that should be pursued?

In Chapters 4 and 5, we will begin to answer these questions by investigating various policy solutions, ranging from legislated regulations to market-based initiatives. By using the analytical modeling tools developed thus far, we will critically evaluate different policy approaches to reducing environmental pollution.

Summary

- There are two basic explanations for the economic assessment of environmental problems as market failures: environmental quality is a public good, and pollution-generating products are associated with externalities.
- A pure public good is one that is both nonrival and nonexcludable in consumption.
- Market demand for a public good is found by vertically summing individual demand curves.
- The market failure of public goods exists because demand is not readily identified. The market failure arises because of nonrevelation of preferences, which in turn is due to free-ridership.
- Even if consumers revealed their willingness to pay, the resulting price would likely underestimate the good's true value because of imperfect information.
- Governments respond to the public goods problem through direct provision of public goods or through political procedures and voting rules.
- An externality is a third-party effect associated with production or consumption. If this effect generates costs, it is a negative externality; if it yields benefits, it is a positive externality.
- In the presence of a negative (positive) externality, the competitive equilibrium is characterized by an overallocation (underallocation) of resources such that too much (too little) of the good is produced.
- In a negative externality model, the competitive price is too low because the marginal external cost (MEC) is not captured by the market transaction.
- To identify the efficient equilibrium, the MEC is added to the marginal private cost (MPC) to derive the marginal social cost (MSC), which must be set equal to the marginal social benefit (MSB).
- The source of the public goods problem and of externalities in private markets is that property rights are not defined.
- The Coase Theorem argues that under certain conditions the assignment of property rights will lead to bargaining between the affected parties such that an efficient solution can be obtained.
- If property rights exist but are ill defined, such as in the case of common property resources, the market solution is inefficient because of externalities.
- Solutions to market failures typically involve government intervention, which may include regulation, tax policy, or market-based solutions.

Review Questions

1. Use economic analysis to evaluate the following statement: *The only amount of acceptable pollution is no pollution at all.*
2. Using the theory of public goods, explain the logic of why in some resort communities the ownership of waterfront homes also includes some defined area along the beach.
3. A chemical pesticide producer releases wastes into the Ohio River. Because the negative effects are not captured by the private market, there is a market failure, which is illustrated by the following functions, where Q is the amount of liquid chemical pesticides in thousands of barrels, and P is the price per barrel:

   $$MSB = 70 - 0.1Q \qquad MPC = 10 + 0.4Q$$
   $$MEC = 0.1Q$$

   a. Find the competitive equilibrium and the level of $M\pi$ at that point.
   b. Find the efficient equilibrium and the $M\pi$ function at that point.
4. Assume that a small town uses a referendum to overcome the free-ridership problem and determine how its residents might value a new water

filtration system for its public water supply. The voting results are aggregated by the town's two districts, yielding the following demand estimates:

District 1: $Q = 160 - 20P_1$

District 2: $Q = 60 - 5P_2$,

where Q is the expected percent of copper to be filtered by the system, and P is the price in millions of dollars.

a. Based on these estimates, determine the town's market demand for this public good, the new filtration system.

b. If the market supply for the system were $P = 6 + 0.15Q$, what would be the equilibrium price and quantity for the town?

5. Suppose you serve on an environmental policy planning board for the federal government. Your task is to propose a policy initiative aimed at reducing urban air pollution, using the assignment of property rights according to the Coase Theorem. Assume that the major polluters to be targeted are commuter bus companies and that the major parties affected are the collection of city dwellers—residents and workers. Describe your proposal in detail, and include the following:

a. Whether the rights are to be assigned to city dwellers or to bus companies.

b. How these rights are to be defined.

c. How the rights are to be distributed to the parties.

d. Whether the rights are to be auctioned or sold and, if so, at what price.

6. A New Hampshire textile mill releases pollution into nearby wetlands, and the associated health and ecological damages are not considered in the private market. Suppose you are an environmental economist working with the following marginal benefits and costs for this market, where Q is thousands of pounds and P is price per pound.

$MPB = 800 - 0.5Q$ $\quad MPC = 20 + 0.3Q$

$MEB = 0$ $\quad MEC = 0.4Q$

a. Find the competitive equilibrium, Q_C and P_C, and the efficient equilibrium, Q_E and P_E.

b. Suppose the textile mill owned the rights to the wetlands, and it is negotiating with a private environmental group that is willing to pay the mill to produce less output. For the 800th unit of output, determine the range within which a payment would be acceptable to both parties.

Additional Readings

Braden, John B., and Charles D. Kolstad, eds. *Measuring the Demand for Environmental Quality.* Amsterdam: North Holland, 1991.

Cornes, Richard, and Todd Sandler. *The Theory of Externalities, Public Goods, and Club Goods.* 2nd Edition. Cambridge: Cambridge University Press, 1996.

Fullerton, Don, and Robert N. Stavins. "How Economists See the Environment." *Nature 395: 6701* (1998), pp. 433–34.

Hardin, Garrett. "The Tragedy of the Commons," *Science 162* (December 13, 1968), pp. 1243–48.

Hyman, David N. *Public Finance: A Contemporary Application of Theory to Policy.* 10th Edition. Mason, OH: South-Western, Cengage Learning, 2011.

Kim, A. M. "A Market without the 'Right' Property Rights." *Economics of Transition 12(2)* (June 2004), pp. 275–305.

King, Andrew. "Cooperation between Corporations and Environmental Groups: A Transaction Cost Perspective." *The Academy of Management Review 32(3)* (July 2007), pp. 889–900.

Ostrom, Elinor. "The Challenge of Common-Pool Resources." *Environment 50*(*4*) (July/August 2008), pp. 8–20.

Popp, David. "Altruism and the Demand for Environmental Quality." *Land Economics 77*(*3*) (August 2001), pp. 339–49.

Raymond, Leigh. *Private Rights in Public Resources: Equity and Property Allocation in Market-Based Environmental Policy.* Washington, DC: Resources for the Future, 2003.

Sagoff, Mark. *The Economy of the Earth: Philosophy, Law and the Environment.* 2nd Edition. Cambridge: Cambridge University Press, 2007.

To access additional course materials, visit www.cengagebrain.com. At the home page, search for the ISBN of this title (shown on the back cover). This will take you to the product page where these resources can be found.

MODULE 2 Modeling Solutions to Environmental Problems

Once we recognize the risks of environmental pollution, we can better appreciate what is at stake if the problem goes unchecked. Knowing how the market functions helps us to understand its inability to correct the problem on its own. Solutions to reduce the effects of polluting residuals must come from outside the market, generally in the form of government policies and programs. But how does government go about designing environmental policy? What objectives does it establish to protect society's health and welfare, and what kinds of policy instruments does it use to achieve those objectives? These are tough questions, and in this two-chapter module, we'll begin to answer them.

As a starting point, we know that government solutions will not eliminate pollution entirely. The laws of science and the materials balance model tell us that much. However, determining exactly where environmental targets should be set is one of the more difficult and even controversial elements of policy development, as is the approach undertaken by government to intervene in pollution-generating markets. Governments all over the world use a variety of policy instruments aimed at reducing environmental risks, some more successfully than others. These range from regulations that directly control polluters' activities to incentive-based plans that use market forces and the price mechanism to achieve a cleaner environment.

In Module 2, we introduce this array of policy alternatives by exploring both conventional and economic solutions to environmental degradation. Conventional policy is the subject of Chapter 4, where we study the traditional use of standards to define environmental objectives, along with the command-and-control approach to implementing these standards. In Chapter 5, we examine the market approach, which has recently been adopted by the United States and other nations as a secondary form of environmental control policy. By using economic modeling and the criteria of allocative efficiency and cost-effectiveness, we evaluate the implications of using these different regulatory strategies to solve the complex problem of environmental pollution.

CHAPTER 4

Conventional Solutions to Environmental Problems: The Command-and-Control Approach

"That government is best which governs the least, because its people discipline themselves."

—Thomas Jefferson (1743–1826)

In Chapter 3, the public goods model and the theory of externalities were used to illustrate how pollution is the result of market failures—failures that arise because of the absence of property rights. Because no one owns the atmosphere or the earth's rivers and streams, there is no market incentive to prevent or correct contamination of these resources. According to the Coase Theorem, assigning property rights would solve the dilemma, but only under certain limiting conditions. The bottom line is that government must act as a third-party mediator in those markets where pollution problems arise.

Recognizing the need for government to correct environmental market failures is an important observation gained through economic modeling, but it is only a first step. We can also use these models to determine *how* government should respond to achieve effective policy solutions. For example, economic theory maintains that government should set objectives to achieve **allocative efficiency**, balancing social benefits and costs at the margin. However, government is generally not motivated by efficiency, and even when it is, it is unlikely that these benefits and costs can be clearly identified. Nonetheless, the efficiency criterion is useful in assessing whatever policy objectives are set relative to their optimal level. Also, the criterion of **cost-effectiveness** can be used to evaluate *how* these objectives are being implemented, even an objective set at something other than its efficient level.

In this chapter, our goal is to analyze government's use of conventional policy solutions to respond to environmental market failure. We begin by providing an overview of environmental standards and their role in policy development. We use allocative efficiency to assess the level at which standards are set to define environmental objectives. Next, we provide an overview of the two broadly defined approaches to implementing these standards-based objectives: the **command-and-control approach** and the **market approach**. Finally, we investigate the cost-effectiveness of the command-and-control approach, the more conventional of the two, deferring an analysis of the market approach to Chapter 5.

USING STANDARDS IN ENVIRONMENTAL POLICY

Standards form the fundamental basis of most environmental policies. In the United States, setting standards follows a lengthy set of procedures involving scientific research and a series of formal reviews. The Environmental Protection Agency (EPA) is charged with oversight of these tasks and for making a formal recommendation about how these standards are to be defined. Ultimately, the standards are legislated by Congress and subsequently monitored for compliance and enforced by the EPA. Information on environmental regulations and rulings is available online at **www.epa.gov/lawsregs/index.html**.

Types of Environmental Standards

ambient standard A standard that designates the quality of the environment to be achieved, typically expressed as a maximum allowable pollutant concentration.

When environmental standards are defined in the law, they can be specified as **ambient standards**, **technology-based standards**, or **performance-based standards**. **Ambient standards** designate the desired quality level of some element of the environment, such as the outdoor air or a body of water. These typically are expressed as a maximum allowable concentration of some pollutant in the ambient environment. For example, the United States and most major industrialized countries use ambient standards to define air quality and water quality. In each case, the ambient standard is not directly enforceable but serves as a target to be achieved through a pollution limit, which in turn is implemented through one of the other types of standards.

technology-based standard A standard that designates the equipment or method to be used to achieve some abatement level.

As its name implies, a **technology-based standard** stipulates the type of abatement control that must be used by all regulated polluting sources. In the United States, the EPA is responsible for researching available technologies and evaluating their relative effectiveness in accordance with certain criteria outlined in the law. It then selects the "best" technology, which subsequently must be adopted by all regulated polluters.[1] The motivation is straightforward—to ensure a specific limit on pollution releases by controlling *how* that limit is to be achieved. For example, to reduce sulfur dioxide emissions, the EPA might require all coal-burning power plants to use a scrubber system, forcing each one to achieve the same level of abatement in precisely the same way.

performance-based standard A standard that specifies a pollution limit to be achieved but does not stipulate the technology.

An alternative type of environmental standard is performance-based. A **performance-based standard** specifies an emissions limit to be achieved by every regulated polluter but does not stipulate the technology to be used to achieve that limit. By definition, performance-based standards are more flexible than their technology-based counterparts. They implicitly allow polluting sources to choose *how* they will reduce pollution releases, as long as they meet the statutory emissions limit.

[1]The meaning of "best" in this context is one that is often the subject of debate, an issue we will investigate in upcoming chapters.

Economic Implications of Using Standards

Although the use of standards sounds straightforward enough, there are important economic implications to be considered. We can investigate these through a two-part economic evaluation that centers on the following questions:

- Are the standards being used to define environmental objectives set at a level that is **allocatively efficient**? That is, does the marginal social cost of pollution abatement equal the marginal social benefit?
- Given some predetermined environmental objective, is the implementation of that objective conducted in a **cost-effective** manner?

ARE ENVIRONMENTAL STANDARDS ALLOCATIVELY EFFICIENT?

allocatively efficient standards Standards set such that the associated marginal social cost (*MSC*) of abatement equals the marginal social benefit (*MSB*) of abatement.

Because environmental objectives are defined by standards, it is important to determine whether they are **allocatively efficient standards**. This criterion is met if economic resources are allocated such that the associated benefits and costs to society are equal at the margin. Therefore, we need to develop these benefit and cost concepts specifically for the pollution abatement market, expanding on what we presented in Chapter 3. The goal is to learn precisely what is required to identify an allocatively efficient abatement level so that we can assess the likelihood of government achieving such an outcome using a standards-based environmental objective.

Marginal Social Benefit of Abatement

marginal social benefit (*MSB*) of abatement A measure of the additional gains accruing to society as pollution abatement increases.

As pollution is abated, the social gains are all the benefits associated with a cleaner environment, such as improvements in health, ecosystems, aesthetics, and property. If we measure how these benefits increase relative to increases in abatement, we arrive at the **marginal social benefit (*MSB*) of abatement**. It is equally correct to think of this *MSB* as a measure of the *reduction in damages or costs* caused by pollution.[2] In theory, if we were to add up all the marginal reductions in environmental external costs across every market where pollution is reduced, we would arrive at the *MSB*. We have actually modeled damage reduction, although in a limited context, in our discussion of bargaining in Chapter 3. By paying refineries to pollute less, recreational users of the river gained the *reduction* in the marginal external cost (*MEC*) associated with refined petroleum production. A real-world case in which damages were shown to be directly attributable to industrial pollution is in the town of Catano, Puerto Rico, the subject of Application 4.1.

From a market perspective, the *MSB* of abatement is society's *demand* for pollution abatement, or, equivalently, its demand for environmental quality.[3] Just as the

[2]Technically, the *MSB* also includes the reduction in social costs from attempting to avoid the effects of pollution, such as the costs of air purification systems, bottled water, or water filtration systems.

[3]Although we generally model demand as the *MPB*, which differs from the *MSB* by the amount of any marginal external benefit (*MEB*), in this context, there is no *MEB*, because the demanders are *all* of society. Thus, there are no third parties to which any external benefits could accrue.

APPLICATION 4.1 Catano, Puerto Rico: Reducing the Damages of Industrial Pollution

For years, the people of Catano, Puerto Rico, blamed their persistent health problems on the town's severe air pollution. The residents had been living with a barrage of emissions released by a nearby oil refinery, a sewage sludge incinerator, ships in San Juan Bay, and a parade of 18-wheelers transporting goods from nearby docks. According to the island dwellers, however, most of the problem has been caused by two giant power plants operated by the Puerto Rico Electric Power Authority (PREPA). Together, the two facilities release an average of 100 million pounds of sulfur dioxide emissions into Catano's atmosphere each year. One of the town's residents leading the charge against the utility says that Catano's air can be likened to a "toxic soup."

Acting on the residents' complaints, several agencies conducted health studies. Findings by a Puerto Rico Medical Association study showed that cancer rates in Catano were nearly twice the national average. Further disturbing evidence came from a U.S. Public Health Service investigation, which found an alarming rate of respiratory disease among the people of Catano. Because it is well known that long-term exposure to sulfur dioxide causes respiratory ailments, this report validated the suspicion that the utility's emissions were the primary cause of health damages in the town.

Responding to these disturbing medical reports, the EPA began its own study of the area, focusing on the power plants. What they found confirmed the accusations of Catano's residents. Pollutant releases from the plants were in violation of air quality regulations. In an effort to bring the power authority into compliance, the United States filed suit against PREPA in 1993, citing numerous violations. In 1999, the legal action was settled, and a consent decree was entered, which was disputed in part by PREPA. Following a protracted period of negotiations, a final agreement was reached in 2004. As part of the judicial settlement, PREPA is required to use low-sulfur fuel to reduce sulfur dioxide emissions. This mandated change in fuel, together with fines and other corrective measures, was aimed at bringing PREPA's plants into EPA compliance. More on the agreement is available online at **www.justice.gov/opa/pr/2004/June/04_enrd_433.htm**.

As an interesting side note, PREPA also was fined $7 million in punitive damages for its violations. Originally, these funds were to be paid directly to the federal government. However, at the urging of a local environmental group and its leading activist, Rosa Hilda Ramos, the EPA agreed to use $3.4 million of the $7 million fine to purchase and protect a significant area of marshland that borders the San Juan Bay Estuary. Taken together, these decisions should provide the residents of Catano with cleaner air to breathe and protection of an important water body for years to come.

Sources: The Goldman Environmental Prize (2008); U.S. Department of Justice (June 23, 2004); Ross (January 7, 1993).

recreational users were *willing to pay* the refineries for a cleaner river, society is *willing to pay* for a cleaner environment. We expect this willingness to pay to decline with increasing levels of abatement, just as the Law of Demand predicts. Therefore, the *MSB* is negatively sloped.

marginal social cost (*MSC*) of abatement
The sum of all polluters' marginal abatement costs plus government's marginal cost of monitoring and enforcing these activities.

Marginal Social Cost of Abatement

On the supply side, we need to model the costs to society as polluters reduce their releases of contaminating residuals. This relationship is called the **marginal social cost (*MSC*) of abatement**. To gain a better sense of what the *MSC* represents, it's easier to think of it as comprising two parts: an aggregation of the marginal costs of every polluter's abatement activities, and the marginal costs government incurs to monitor and enforce those activities.

Let's begin with the first part by considering a representative polluter and how it goes about the task of abating pollution.

Firm-Level Marginal Abatement Cost

Referring again to our discussion of bargaining under the Coase Theorem, recall that refineries reduced their toxic releases by decreasing output. In that case, the marginal cost to refineries of abating pollution was forgone profit, modeled as a movement from right to left along the marginal profit ($M\pi$) curve. Therefore, if the decision variable is output, each polluting firm faces a marginal abatement cost equal to its forgone $M\pi$. However, such a model implicitly assumes that polluters can meet an environmental standard *only* by reducing output, a very limiting assumption. Because other methods are available, we need to expand this output-based specification to something more general.

Remember that as a profit maximizer, the polluting firm is implicitly a cost minimizer. To meet an environmental standard, the firm will consider all available abatement options and select the least-cost method. While it certainly could reduce production to meet the standard, it has other options, such as installing new abatement technology or changing a production process. To further understand this issue, read Application 4.2, which discusses a relatively low-cost abatement technology used to clean up oil spills and hazardous wastes. To allow for the fact that polluters choose from a menu of available abatement methods, we model what is conventionally called a **marginal abatement cost (*MAC*)** function. The *MAC* measures the change in economic costs associated with increasing pollution abatement (A), using the least-cost method.

marginal abatement cost (*MAC*)
The change in costs associated with increasing abatement, using the least-cost method.

Each polluting source likely faces a unique *MAC* curve. Among the factors that affect the shape and position of the *MAC* curve are the firm's location, the type of contaminants it releases, the nature of its production, and the availability of technology. However, a typical *MAC* curve is positively sloped and increases at an increasing rate, as shown in Figure 4.1 (page 86). Think about what this means. When pollution levels are high, the addition of virtually any type of abatement technology will likely be quite effective. So, although costs are rising, they do so at a fairly slow rate relative to the abatement being accomplished. As this process continues and the environment becomes cleaner, however, it becomes difficult to remove more pollution. Thus, the added costs relative to the abatement achieved increase at a much faster rate.

Should a firm's abatement options change, the position of its *MAC* curve would be affected. For example, the introduction of a cost-saving abatement technology would pivot the *MAC* curve downward, as shown in Figure 4.2 (page 86).[4] Likewise, a polluter's *MAC* for one pollutant might be lower than that of another. Consequently, an *MAC* function generally is defined for a particular contaminant at a given level of technology.

market-level marginal abatement cost (MAC_{mkt})
The horizontal sum of all polluters' *MAC* functions.

Market-Level Marginal Abatement Cost

The aggregation of all polluters' *MAC*s represents the **market-level marginal abatement cost (MAC_{mkt})** defined as the horizontal sum of each polluter's *MAC*, or $MAC_{mkt} = \Sigma MAC_i$ for all i firms. This is exactly the same procedure used to derive market supply

[4] A firm's *total* abatement costs are represented as the area under the *MAC* curve up to the abatement level required by the standard, assuming no fixed costs. Thus, the effect of the cost-saving technology on total abatement costs is implicitly shown as the reduction in the area under the firm's *MAC* curve.

APPLICATION 4.2 Costs of Remediating Hazardous Waste and Oil Spills

Environmental damage linked to hazardous waste disposal or accidents is a problem that has captured world attention in large part because of the contamination caused by major oil spills. The 1989 *Valdez* incident, for example, despoiled Alaskan shorelines with 11 million gallons of oil. More recently, the 2010 British Petroleum (BP) oil well blowout, which released over 200 million gallons of crude oil into the Gulf of Mexico, renewed public awareness of and concern about such accidents. These disasters are international news both because the environmental implications are severe *and* because the abatement costs are so staggering. Adding to these concerns is the well-publicized problem of deliberate hazardous waste dumping. Many landfill sites have become contaminated with cancer-causing substances, ranging from polychlorinated biphenyls (PCBs) to heavy metals.

Based on a recent report released by the EPA, there are currently more than 22,000 contaminated sites needing attention and over 62,000 known and suspected sites throughout the United States. Associated abatement costs are about $6 billion to $8 billion annually. Oil spill abatement expenses are no less disturbing. Exxon incurred over $3 billion to clean up the damage linked to the *Valdez* oil spill, not counting fines and legal settlements. And at the end of 2010, BP publicly estimated that the costs to clean up the Gulf oil spill, along with such cost as fines and legal fees, would run to $40 billion, an amount disputed by some to be too conservative. Some analysts argue the costs could be closer to $200 billion. These and other cleanup expenditures have prompted new and ongoing research to find lower-cost abatement technologies.

Recently, scientists have made important strides in abating contaminated soil and oil spills using a technology called **bioremediation**. Bioremediation is a relatively low-cost technique that relies on microbes to consume waste materials. In some contexts, fertilizer sprays are used to stimulate the feeding of these microorganisms, which accelerates the usually lengthy process of returning soil to its original state. Used for some 50 years to clean wastewater treatment plants, the process is now being used to abate toxic contaminants. In one of its most successful applications to date, bioremediation helped to clean up the oily mess left by the *Valdez* on Alaskan beaches. Within three weeks, the soil was restored down to a foot below the surface.

In 2010, scientists reported similar success in the cleanup of the BP oil spill. Naturally occurring bacteria and fungi that live in the ocean consume the hydrocarbons present in the released oil. Dispersants used to break down the oil into smaller droplets should aid in the process, although scientists are studying whether the toxicity of some dispersants deployed in that accident may actually have run counter to their intended effect. Remediation of oil from the deep ocean waters is totally dependent on these microorganisms despite the fact that the colder temperature of those waters inhibits their growth. At the shoreline, fertilizers can be added to stimulate microbial activity, as was done to abate the oil from the *Valdez*.

Although there is much more to study and learn, many believe that bioremediation holds great promise for the environment. Such treatment processes provide decision makers with a wider range of alternatives to abate pollution and greater opportunities to find cost-effective solutions. For more information on bioremediation and other remediation technologies, visit **www.epa.gov/superfund/remedytech/remed.htm**.

Sources: Biello (2010); U.S. EPA, Office of Solid Waste and Emergency Response (September 2004); Environmental Law Institute (2002); Hof (June 4, 1990).

in Chapter 2. In this context, the horizontal summing ensures that the MAC_{mkt} represents least-cost decisions, because it effectively sets each MAC equal at every abatement level.

Marginal Cost of Enforcement

marginal cost of enforcement (*MCE*) Added costs incurred by government associated with monitoring and enforcing abatement activities.

Now, consider the second element of the marginal social cost (MSC) of abatement. To the MAC_{mkt} function we need to add the marginal costs incurred by government for enforcing and monitoring abatement activities. This component is commonly referred to as the **marginal cost of enforcement (*MCE*)**. Figure 4.3 (page 87) illustrates how the MCE is added vertically to the MAC_{mkt} to derive the MSC function. (For simplicity, all functions are assumed to be linear.) At any abatement level (A), the MCE is the vertical distance between MAC_{mkt} and MSC. Notice that this distance is increasing with

FIGURE 4.1 Single Polluter's Marginal Abatement Cost (*MAC*)

A typical *MAC* curve is positively sloped and increases at an increasing rate. Notice that as the firm continues to abate from A_1 to A_2, the *MAC* increases by a proportionately greater amount from MAC_1 to MAC_2. This reflects the fact that as the abatement process continues and the environment becomes cleaner, it becomes increasingly difficult and therefore more costly to remove each additional unit of pollution.

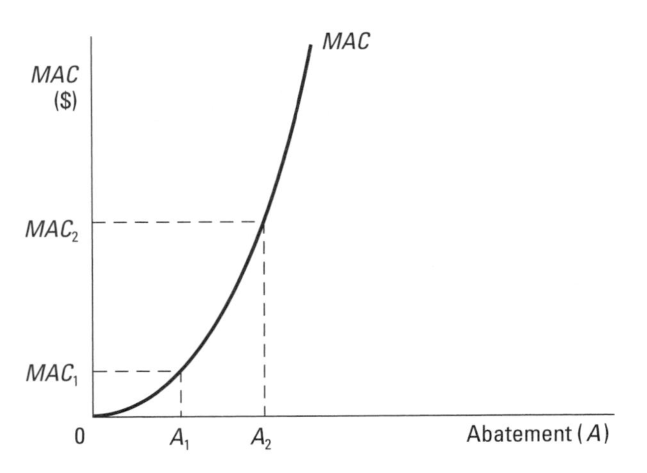

FIGURE 4.2 Effect of Cost-Saving Technology on the *MAC* Curve

Changes in a firm's abatement options change the position of the *MAC* curve. For example, this model shows how the introduction of cost-saving abatement technology would pivot the *MAC* downward to *MAC*′.

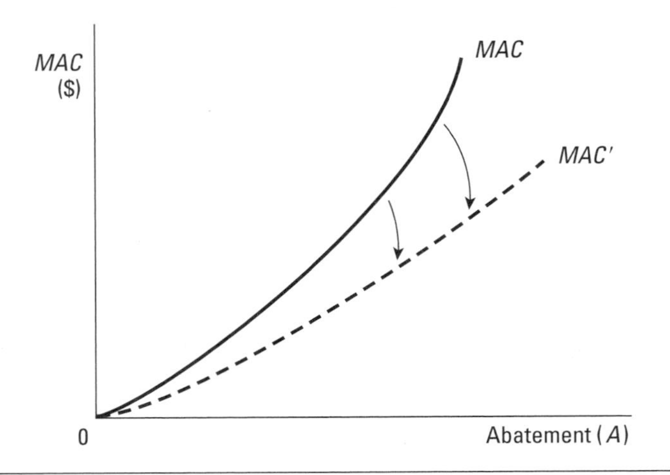

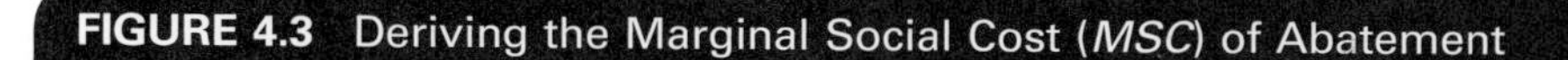

FIGURE 4.3 Deriving the Marginal Social Cost (*MSC*) of Abatement

To derive the *MSC*, the *MCE* is vertically added to the MAC_{mkt}. At any abatement level *A*, the *MCE* is the vertical distance between MAC_{mkt} and *MSC*. Notice that this distance increases with higher abatement levels. As pollution standards become more stringent, polluters have a greater tendency to evade the law, which calls for more sophisticated, and thus more costly, monitoring and enforcement programs.

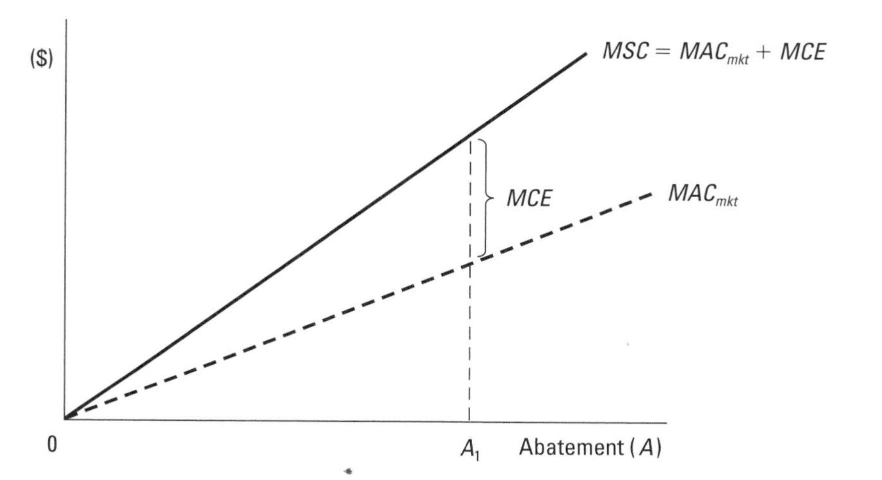

higher abatement levels. As pollution standards become more stringent, polluters have a greater tendency to evade the law, which in turn calls for more sophisticated, and thus more costly, monitoring and enforcement programs. To learn more about enforcement in the United States, visit the EPA's Office of Enforcement and Compliance Assurance at **www.epa.gov/compliance/index-e.html**.

Are Abatement Standards Set Efficiently?

From our discussion in Chapter 3, we know that *MSB* and *MSC* simultaneously determine the efficient level of abatement (A_E), which occurs at the intersection of the two functions. Whether or not the government sets environmental standards to achieve this level depends on various considerations. Four factors in particular suggest that this outcome is highly unlikely: (1) legislative constraints, (2) imperfect information, (3) regional differences, and (4) nonuniformity of pollutants.

Legislative Constraints

benefit-based standard
A standard set to improve society's well-being with no consideration for the associated costs.

The reality of a standards-based approach is that it does not necessarily set pollution limits to account for the associated benefits and costs. In fact, under U.S. law, many standards are said to be **benefit-based**, meaning that they are set to improve society's well-being with no allowance for balancing the associated costs. For example, under the U.S. Clean Air Act, air quality standards are motivated solely by the expected benefits of improved health and welfare. If costs are not accounted for in the standard-setting process, resources likely will be overallocated to abatement.

Imperfect Information

Even when a cost-benefit balancing *is* called for by law, the absence of full information would likely prevent the government from identifying the *MSB* and *MSC* of abatement. Let's consider the *MSB* relationship first. Recall from our discussion of market failure in Chapter 3 that pollution abatement is a public good. As such, its demand, which is the *MSB* curve, is not readily identified because of **nonrevelation of preferences**. In practice, there are methods to estimate the value society places on damage reductions associated with abatement. However, given the difficulty of trying to monetize such intangibles as health improvements and longevity of life, the probability of accurately modeling *MSB* is low.

There are similar problems in identifying the *MSC*. In addition to estimating the *MCE*, the government also would have to know the *MAC* for every polluter. Obtaining this firm-level information would be virtually impossible, given the diversity of production and abatement techniques across polluting sources. Furthermore, the *MAC* must also account for the **implicit costs** of abatement, which are difficult to quantify. In this context, implicit costs would include any unemployment associated with production declines, the potential loss of consumer choice if products were eliminated or altered, and any price and income effects arising from abatement requirements.[5]

In the absence of perfect information, it is highly probable that the government will unknowingly establish the abatement standard at some level other than the allocatively efficient one, even if that was the legislated intent. Figure 4.4 illustrates the market for

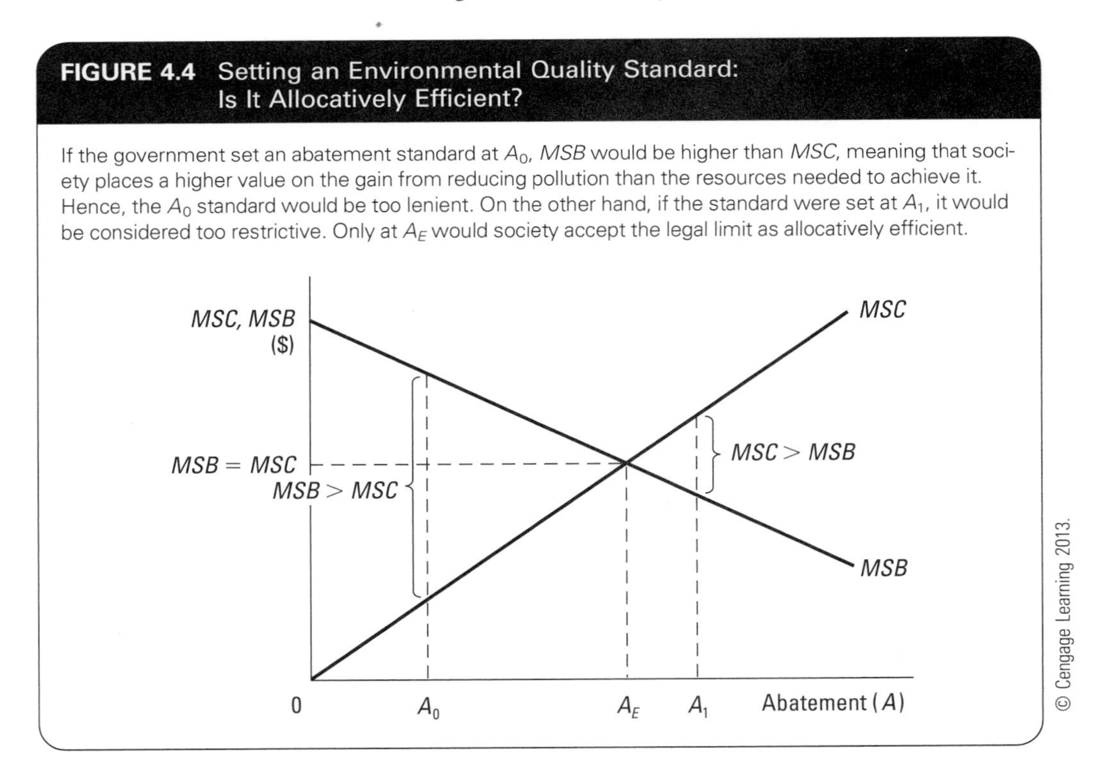

FIGURE 4.4 Setting an Environmental Quality Standard: Is It Allocatively Efficient?

If the government set an abatement standard at A_0, *MSB* would be higher than *MSC*, meaning that society places a higher value on the gain from reducing pollution than the resources needed to achieve it. Hence, the A_0 standard would be too lenient. On the other hand, if the standard were set at A_1, it would be considered too restrictive. Only at A_E would society accept the legal limit as allocatively efficient.

[5]In Chapters 7 and 8, we will investigate the methods used in practice to estimate the benefits and costs of improving environmental quality.

abatement at equilibrium, comparing the allocatively efficient outcome (A_E), with two other possible levels (A_0 and A_1). If the standard is set at A_0, MSB would be greater than MSC, meaning that society places a higher value on reducing pollution than it must give up in resources to achieve it. Hence, the A_0 standard would be considered too lenient. Conversely, a standard set at A_1 would be too restrictive. Only at A_E would society consider the legal limit to be allocatively efficient. However, the information needed to find this optimal level of pollution abatement is immense.

Regional Differences

Even if the law permits a balancing of costs and benefits *and* even if full information were available, there is a qualifier on the use of A_E as a national standard across all polluting sources. Why? Because this optimal level is determined from MSB and MSC, both of which assume the absence of region-specific abatement benefits and costs. The only way that A_E would be allocatively efficient in all regions is if the respective MSB and MSC functions defined for those locations were identical.

By way of example, consider two hypothetical regions, X and Y, that have identical MSC functions (i.e., $MSC_X = MSC_Y$) but different MSB functions, such that MSB_X is *lower* than MSB_Y at all abatement levels. Such a disparity might be due to differences in income, education, or population across the two locations. In any case, at most, only one of the two locations would consider a nationally determined A_E as efficient. Look at Figure 4.5, which superimposes the MSB and MSC for each region on the same diagram. The allocatively efficient level of abatement in region X (A_X) is much lower than that for region Y (A_Y). So there is no way that a single national standard of abatement—even one that *is* efficient on a national level—would be optimal for both regions.

FIGURE 4.5 Effect of Regional Differences on Achieving Allocative Efficiency

This model shows how region-specific conditions can give rise to different optimal levels of abatement. Although the *MSC* curves are identical (i.e., $MSC_X = MSC_Y$), the *MSB* in region X is *lower* than that in region Y, i.e., $MSB_X < MSB_Y$. Therefore, the allocatively efficient abatement level for region X (A_X) is lower than that for region Y (A_Y). This means that a national abatement standard cannot be optimal for both regions.

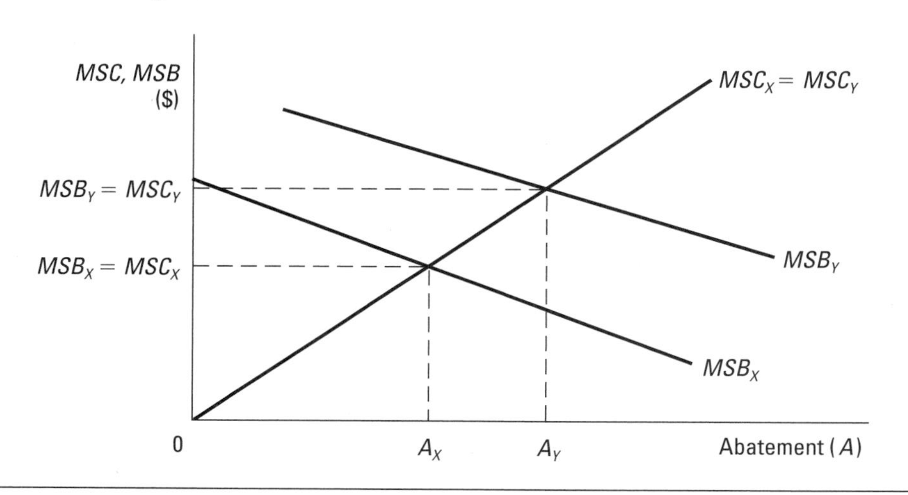

Nonuniformity of Pollutants

An inefficient outcome can also arise within the same region, if changes in releases from polluting sources do not have a uniform impact on the environment. This can occur if the relationship between the change in pollutant releases and the change in exposure is nonlinear, or not directly proportional. The nonuniformity can also arise when polluters are located at varying distances from an exposed population or ecosystem, even if their pollution releases are identical.[6] In general, the farther away from a source an affected population is, the lower the associated damage, because there is greater opportunity for dilution of contaminants.[7] This in turn would mean that the *MSB* of abatement varies inversely with the distance between a source and the affected population or resource. Thus, even with equal *MSC* of abatement, the efficient level of abatement would not be the same for all polluting sources. Consequently, just as for regional differences, a nationally determined abatement standard would not be optimal for all sources.

What we can surmise from this assessment is that, in most real-world settings, at least one of the factors—legislative constraints, imperfect information, regional differences, or nonuniformity of pollutants—will be present. Therefore, there is a low probability that abatement standards will be set at an allocatively efficient level. In accepting this, we must rely on a different criterion to evaluate not where the standards are set but how they are implemented.

GENERAL APPROACHES TO IMPLEMENTING ENVIRONMENTAL POLICY

cost-effectiveness Requires that the least amount of resources be used to achieve an objective.

command-and-control approach A policy that directly regulates polluters through the use of rules or standards.

market approach An incentive-based policy that encourages conservation practices or pollution-reduction strategies.

Our preceding analysis makes a strong case that a government-established environmental standard will be set at something other than its allocatively efficient level. Yet, it is possible that even a nonoptimal environmental standard can be implemented using the least amount of resources. If so, the policy is said to meet the **cost-effectiveness** criterion. Whether or not this "second-best" criterion is met depends on the method used to bring about the desired reduction in pollution.

Most governments, including that of the United States, use a number of different policy tools to achieve environmental quality. The majority of these fall into two broad-based categories or approaches. One is the **command-and-control approach**, which uses pollution limits or technology-based restrictions to directly regulate polluting sources. The second is the **market approach**, which uses incentive-based policy tools to motivate abatement through market forces.

Of the two, command-and-control is the more conventional approach, and it dominates environmental policy in most countries. This nearly universal reliance on direct regulation seems to have evolved from an attempt to gain immediate control of what was initially an unfamiliar and urgent dilemma. Although well intentioned, the use of inflexible regulations and pollution limits, often imposed uniformly across

[6]National Acid Precipitation Assessment Program (NAPAP) (November 1991), Chapter 3, Sec. 3.1.

[7]This relationship can be captured by what is called a transfer coefficient. For details, see Tietenberg and Lewis (2009), pp. 371–79.

all polluters, has not met with consistent success. By way of example, Application 4.3 discusses how EPA proposals for tougher environmental regulations have sparked concerns about the implications for job creation and economic recovery in the United States. This example and numerous others help to explain why, over time, policymakers began to look for alternatives.

The United States and other industrialized countries have gradually integrated market-based solutions into their environmental policy programs. Combining incentive-oriented policy instruments with more conventional methods seems to be an evolving trend. However, the *relative* gains of market-based solutions cannot be fully appreciated without assessing the cost-effectiveness of the command-and-control approach.

APPLICATION 4.3 Command-and-Control Under Fire: Will New Environmental Regulations Hurt the Economic Recovery?

Under the Obama administration, the EPA has been actively promoting tighter environmental regulations and more stringent emissions limits. Among the EPA's regulatory proposals are new standards to control carbon dioxide and other greenhouse gas emissions; tighter ambient air quality standards on ground-level ozone, nitrogen dioxide, and sulfur dioxide; and new controls on industrial boilers. Because these regulations are based on setting limits on pollution or establishing rules affecting production, they are considered part of a **command-and-control approach** to environmental policy.

Notwithstanding widespread support for abating pollution and assuring a clean environment, these rules-based EPA proposals have prompted resistance from a number of stakeholders, particularly the private business sector. Why? Because tougher controls on emissions, production, and industrial equipment, holding all else constant, add to operating costs and capital investment, which in turn reduces profits. Attention to this reality has been magnified by the slow economic recovery that followed the collapse of the housing and financial sectors. The opposition voiced by constituents from all segments of the economy is linked in large part to concerns that tougher environmental rules and regulations may deter job creation and slow the decline in the national unemployment rate.

By way of example, critics predict that the more stringent ozone standards proposed by the EPA will bring hundreds of counties into nonattainment status, which in turn could bring about plant closures and elevated fuel and energy costs. This negative view has been adopted by Unions for Jobs and the Environment (UJAE), a collaborative of a dozen unions from around the world, which include such major labor organizations as the Teamsters and the United Mine Workers. Others point to the economic implications of new controls on greenhouse gases, which, according to the U.S. Chamber of Commerce, are expected to give rise to stiffer regulations on more than 250,000 office buildings, 150,000 warehouses, and tens of thousands of other facilities around the country.

Following the mid-term elections in 2010, House Oversight and Government Reform Committee Chairman Darrell Issa asked the business community to identify those government regulations viewed as most burdensome. The responses, which were made public in February 2011, underscored the dilemma: EPA rules were identified more than those administered by any other government agency. The president called for a review of government regulations early in 2011, with the goal of eliminating any that were outdated or particularly burdensome to firms. However, the initiative seemed to be aimed more at older rulings than at the proposed controls on greenhouse gases, which have been the source of broad-based criticism. In fact, in early 2011, the House Energy and Commerce Committee proposed the Energy Tax Prevention Act of 2011, which would repeal the EPA's authority to control greenhouse gases.

It is clear that the debate both in Washington and in the private sector about the command-and-control approach to new environmental policy under the Obama administration is expected to continue for some time. In the interim, to learn more about regulations and rules proposed or administered by the EPA, visit **www.epa.gov/lawsregs/index.html**.

Sources: Ackley (2011); Banerjee (2011); Radnofsky (2011); Inhofe (2010).

IS THE COMMAND-AND-CONTROL APPROACH COST-EFFECTIVE?

The practical way to assess whether the command-and-control approach is cost-effective is to determine whether society is incurring higher costs than necessary to achieve a given level of environmental quality. For discussion purposes, let's consider some abatement standard as the socially desirable (as opposed to efficient) outcome, perhaps motivated to protect human health. In theory, to achieve cost-effectiveness, policymakers must identify the relative costs of all control instruments that can achieve this objective and then select the one that minimizes costs. Given this general premise, we can identify two command-and-control decisions that may violate the cost-effectiveness criterion. The first is the use of a technology-based standard, and the second is the use of uniform standards. We discuss each of these in turn.

Cost-Ineffectiveness of Technology-Based Standards

Recall from the beginning of this chapter that there are three types of standards: **ambient standards**, **performance-based standards**, and **technology-based standards**. Take a minute to reread the definitions of these control instruments, and think about the cost implications of each. What should be apparent is that the technology-based standard potentially prevents the polluter from minimizing the costs of achieving a given abatement level. Remember that the *MAC* curve is defined under the assumption that the polluter selects the *least-cost* available method. If the government forces polluters to use a specific technology to meet an emissions limit, it is impeding the firm's incentive to abate in a cost-effective manner. Unless the mandated technology happens to be the least-cost approach for *all* polluters, at least some will be forced to operate above their respective *MAC* curves. This in turn means that society is incurring costs higher than the *MSC* of abatement. The result is a waste of economic resources with no additional benefits to society.

If instead a performance-based standard were used, each polluter could select the means by which it achieves that standard. Without further guidance, it would follow its self-interest and choose the least costly abatement method. Society would still gain the benefits of a cleaner environment, but fewer resources would be used to achieve that gain. A word of caution is in order, however. Although using performance-based standards has potential cost advantages over using technology-based standards, this selection does not by itself ensure a cost-effective solution. In fact, regardless of which type of standard is used, resources will be wasted if they are imposed *uniformly* across polluters. Let's consider this issue next.

Cost-Ineffectiveness of Uniform Standards

Under a strict command-and-control framework, standards are often imposed *uniformly* across groups of polluting sources. The operative question is whether such a policy approach is cost-effective. The answer? The use of uniform standards across polluting sources will waste economic resources as long as abatement cost conditions differ among those sources. Of course, the reality is that there are many factors that might give rise to such differences. One is the age of the polluter's physical plant. Newer facilities typically are designed and built with advanced pollution control equipment, making them capable of meeting an abatement standard at a much lower marginal cost than their less modern counterparts. Another relevant factor is regional differences in input prices. There is no reason to expect firms in different locations to face the same costs of labor, land, and capital. As long as input prices vary, so too will polluters' costs to achieve a given standard.

By accepting that abatement costs will likely differ among polluters, we need to explore why this makes the use of uniform standards cost-ineffective. The problem is that uniform standards force high-cost abaters to reduce pollution as much as low-cost abaters, so more resources than necessary are used to achieve a cleaner environment. Cost savings could be realized by having more of the abatement accomplished by polluters who can do so at a lower cost. A simple model illustrates this assertion.

To begin, assume there are only two polluting sources in a given region, each of which generates 10 units of pollution for a total of 20 units released into the environment. The government determines that emissions must be reduced by 10 units across the region to achieve the "socially desirable level of pollution." Each firm faces different abatement cost conditions, modeled as follows:[8]

Polluter 1's Marginal Abatement Cost (MAC_1): $MAC_1 = 2.5A_1$
Polluter 1's Total Abatement Costs (TAC_1): $TAC_1 = 1.25(A_1)^2$

Polluter 2's Marginal Abatement Cost (MAC_2): $MAC_2 = 0.625A_2$
Polluter 2's Total Abatement Costs (TAC_2): $TAC_2 = 0.3125(A_2)^2$

where A_1 is the amount of pollution abated by Polluter 1, and A_2 is the amount of pollution abated by Polluter 2.

Now, assume the government implements the 10-unit standard *uniformly*, requiring each polluter to abate by 5 units (i.e., $A_1 = A_2 = 5$). At this level, the *MAC* for Polluter 1 is \$12.50 ($MAC_1 = 2.5[5] = \$12.50$), and its *TAC* is \$31.25 ($TAC_1 = 1.25[5]^2 = \$31.25$). For Polluter 2, MAC_2 is \$3.13, and TAC_2 is \$7.81. Therefore, the total abatement costs for the region (absent the costs of monitoring and enforcement) equal \$39.06, which represents the value of resources used to meet the standard. The question is, could the same standard be achieved at a lower cost?

Notice that Polluter 2 has an abatement cost advantage over Polluter 1. For example, the fifth unit of abatement costs Polluter 2 \$9.37 less than Polluter 1. Therefore, it would be cheaper if Polluter 2 were to do more of the abating. Of course, it would need to have an incentive to do this, and the two firms would have to negotiate to arrive at some mutually beneficial agreement.[9] However, no such opportunity is allowed when the government forces every polluter to abate by the same amount. Thus, we conclude that the use of uniform standards under a command-and-control approach does not achieve the cost-effectiveness criterion as long as *MAC* conditions differ across polluters.

cost-effective abatement criterion
Allocation of abatement across polluting sources such that the *MAC*s for each source are equal.

Could the government reallocate abatement levels across the two polluters to achieve a cost-effective solution? The answer is yes, and economic theory conveys exactly how this result could be achieved. If each polluter were to abate to the point where the corresponding level of *MAC* is equal across firms, the **cost-effective abatement criterion** would be achieved. This means that the environmental standard would be met at minimum cost. This result is one application of what microeconomic theory calls the **equimarginal principle of optimality**.

To illustrate the cost savings of such an approach, let's return to our two-polluter model. We need to find the abatement levels for each polluter at which their respective

[8]Assuming no fixed costs, the total abatement costs (*TAC*) for a given abatement level *A* are found as the aggregation of the *MAC* at each abatement level up to *A*. Graphically, this means that the *TAC* for abatement level *A* is the area under the *MAC* curve up to point *A*.

[9]In a real-world setting, there would also be transaction costs associated with the negotiations between the two firms.

*MAC*s are equal, holding the combined abatement level at 10 units to meet the standard. Algebraically, the steps in the solution are:

Step 1: Set $MAC_1 = MAC_2$: $2.5A_1 = 0.625A_2$

Step 2: Set $A_1 + A_2 =$ abatement standard: $A_1 + A_2 = 10$

Step 3: Solve the equations simultaneously: $A_1 = 2; A_2 = 8$

This same result is shown in Figure 4.6. Both firms' *MAC* curves are plotted on the same diagram. For Polluter 1, A_1 is measured horizontally *left to right*, whereas for Polluter 2, A_2 is measured *right to left*. The horizontal axis measured in either direction ranges from 0 units of abatement up to the 10-unit requirement imposed by the regulatory authority. Thus, every point on this axis represents a *combined* abatement level that satisfies the standard. Notice that the intersection of the two *MAC* curves yields the cost-effective solution ($A_1 = 2$ and $A_2 = 8$). At this point, $MAC_1 = MAC_2 = \$5.00$.

If each polluter were given these firm-specific abatement targets, the total costs of achieving the environmental objective would be minimized. These costs can be calculated from each firm's *TAC* equation, substituting in the cost-effective abatement levels:

$$TAC_1 = 1.25(2)^2 = \$5.00$$
$$TAC_2 = 0.3125(8)^2 = \$20.00$$

By following the equimarginal principle of optimality, the total cost to society of achieving the 10-unit abatement standard is $25.00, which is a $14.06 savings over the uniform standards approach. Equivalently, this $14.06 represents the unnecessary costs incurred by society when a uniform standard is imposed across nonidentical polluters.

FIGURE 4.6 Cost-Effective Solution in a Two-Polluter Model

The model shows the *MAC* curves for Polluter 1 and Polluter 2. Polluter 1's abatement level (A_1) is measured left to right on the horizontal axis, and Polluter 2's abatement (A_2), right to left. The horizontal axis measures from 0 up to the 10-unit abatement requirement imposed by government, so every point represents a *combined* abatement level that satisfies the standard. In accordance with the equimarginal principle of optimality, the intersection of the two *MAC* curves yields the cost-effective solution, where $A_1 = 2$ and $A_2 = 8$. At this point, notice that $MAC_1 = MAC_2 = \$5.00$.

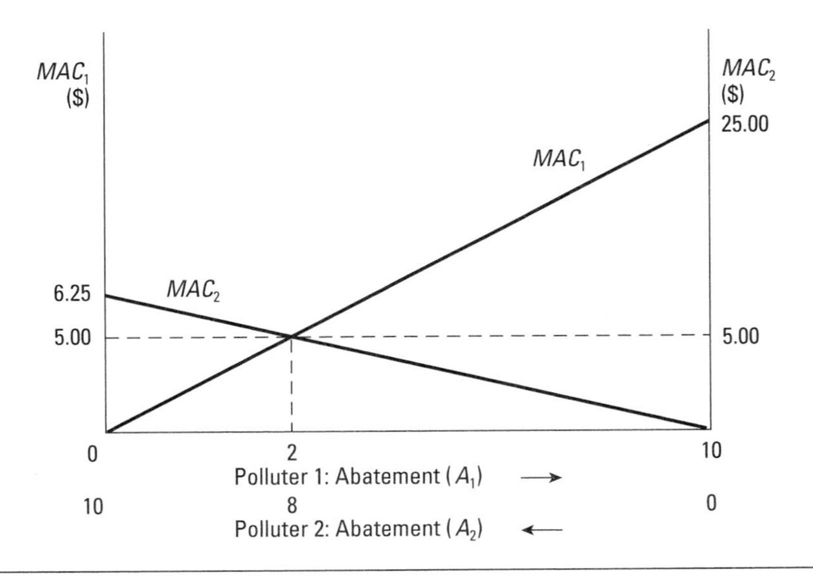

As a final point, it's reasonable to ask how in practice the government could arrive at these firm-specific abatement standards within a command-and-control framework. The answer is that it would have to know the abatement cost conditions for every firm it was regulating. Of course, this kind of information would be virtually impossible to determine, particularly when thousands of individual sources are being controlled. But there *is* a way around the problem, although not within the command-and-control approach. As we will discover in Chapter 5, the market approach can arrive at this same cost-effective solution without specific knowledge of polluters' costs. How? By using market incentives and the price mechanism in place of inflexible rules.

CONCLUSIONS

In this chapter, we have begun to evaluate solutions to environmental problems by focusing on the more conventional policy tools used in practice—the use of standards to define environmental objectives and the use of a command-and-control approach to implement those objectives. Even at this introductory level, we were able to reach important conclusions. It is apparent, for example, that government-mandated environmental standards are not likely to be set at an efficient level. Beyond those instances when the law does not allow for the requisite balancing of benefits and costs, there is still an information problem. Policymakers would need extensive data to measure the marginal costs and benefits of abatement. Hence, in all likelihood, the level at which environmental objectives are set will not be allocatively efficient.

Accepting this realization, we find that cost-effectiveness becomes the relevant criterion by which to assess the command-and-control approach and compare it to the market approach. Thus far, we have shown that using uniform standards under a command-and-control framework likely wastes resources. Cost savings can be realized if polluters reduce emissions up to the point where their marginal costs of doing so are equal. But how can such a result be achieved in practice? The answer requires an investigation of alternative control instruments, in particular those that are part of the market approach to environmental policy.

In sum, although our findings in this chapter are important, they are incomplete. We still must learn how market-based initiatives are designed and how they compare to their command-and-control counterparts—precisely the agenda of Chapter 5. This part of our analysis is particularly relevant, given the recent trend in the United States and other nations to integrate market-based instruments into what had been an exclusively command-and-control approach to environmental policy. To appreciate this trend and what it means for society, we must understand how market instruments operate and evaluate their effectiveness in achieving environmental goals.

Summary

- There are three basic types of standards used in environmental control policy: ambient standards, which designate the level of environmental quality as a maximum allowable pollutant concentration; technology-based standards, which indicate the abatement method to be used; and performance-based standards, which specify an emissions limit to be achieved.
- An environmental standard achieves allocative efficiency if resources are allocated such that the marginal social benefit (*MSB*) of abatement equals the marginal social cost (*MSC*) of abatement.
- The *MSB* measures the additional gains to society associated with the reduction in damages caused by pollution.
- The *MSC* is the horizontal sum of the market-level *MAC* (MAC_{mkt}) and the government's marginal cost of enforcement (*MCE*).
- Four factors suggest that a government-mandated abatement standard is not likely to meet the allocative efficiency criterion: (1) the existence of legislative constraints, (2) imperfect information, (3) regional differences, and (4) nonuniformity of pollutants.
- Governments generally use one of two approaches to implement environmental policy: the command-and-control approach or the market approach.
- Two aspects of the command-and-control approach may violate the cost-effectiveness criterion: the use of technology-based standards and the use of uniform standards.
- Because technology-based standards dictate a specific abatement method to polluting sources, they prevent the polluter from minimizing costs.
- Uniform standards force high-cost abaters to reduce pollution as much as low-cost abaters, so more resources than necessary are used to achieve the benefits of a cleaner environment.
- To achieve a cost-effective outcome, abatement responsibilities across polluting sources must be allocated such that the level of *MAC* is equal across polluters.

Review Questions

1. One of the major problems in applying the Coase Theorem in practice is the existence of high transaction costs. Propose an approach that a third party could use that would reduce these costs sufficiently so that bargaining could proceed. How likely is the solution to be efficient, and why?
2. Using a graph of the pollution abatement market, model a situation in which the allocatively efficient level of abatement occurs at 100 percent, or equivalently, where pollution is zero. Referring to the relative position of the *MSC* and *MSB* curves, explain this outcome intuitively.
3. Suppose that the state of Connecticut is attempting to set a water quality standard, where water quality is measured as the percent of mercury abated (*A*), and the marginal social benefit (*MSB*) and marginal social cost (*MSC*) of abatement have been estimated as follows:

   $$MSB = 40 - 0.1A \quad MSC = 36 + 0.25A$$

   The state's department of environmental protection sets the standard at 20 percent. Is this standard set efficiently, too stringently, or too leniently? Explain briefly.
4. **a.** Under a strict command-and-control framework, suppose abatement standards are set equally across polluters. Assume the total abatement target is 30 units. Show the cost implications using three graphs, each of a different polluter with a unique *MAC* curve drawn to depict a low-cost abater, a moderate-cost abater, and a high-cost abater. On each graph, identify the abatement level corresponding to a uniform standards approach, and show the level of *MAC* at that point and the area corresponding to *TAC*.

   b. Now, refer directly to your model, and summarize what would happen *qualitatively* to the abatement levels of each firm if the equimarginal principle of optimality were used. Explain intuitively why this method would be cost-effective.
5. It is well documented that the carbon monoxide (CO) emissions from combustible engines increase in colder climates. This implies that the associated damages are expected to be less severe in summer months than in winter. Nonetheless, air quality control authorities use a standard for CO that is uniform throughout the year with no allowance for seasonal effects. Use this information and the following model to answer the questions:

MSB of CO abatement in winter $= 350 - 0.5A$
MSB of CO abatement in summer $= 140 - 0.2A$
MSC of CO abatement $= 0.2A$

where A is the level of CO abatement.

a. Graph the *MSB* and *MSC* functions on the same diagram.

b. Assume the government sets a uniform standard for winter and summer at $A = 500$. Support or refute this policy based on the criterion of allocative efficiency, using your model to explain your response.

c. If you were in charge of setting policy for CO emissions, what action would you recommend to ensure an allocatively efficient outcome across the two seasons?

6. Assume that two power plants, Firm 1 and Firm 2, release sulfur dioxide (SO_2) in a small urban community that exceeds the emissions standard. To meet the standard, 30 units of SO_2 must be abated in total. The two firms face the following abatement costs:

$$MAC_1 = 16 + 0.5A_1 \quad MAC_2 = 10 + 2.5A_2$$

where costs are measured in thousands of dollars.

a. Prove that a uniform standard will not meet the cost-effectiveness criterion.

b. Determine how the abatement levels should be reallocated across the two firms to minimize costs.

Additional Readings

Coglianese, Cary, and Laurie K. Allen. "Does Consensus Make Common Sense? An Analysis of EPA's Common Sense Initiative." *Environment 46*(1) (January/February 2004), pp. 10–25.

Farzin, Y. H. "The Effects of Emission Standards on Industry." *Journal of Regulatory Economics 24*(3) (November 2003), pp. 315–27.

Gray, Wayne B., and Ronald J. Shadbegian. "'Optimal' Pollution Abatement—Whose Benefits Matter, and How Much?" *Journal of Environmental Economics and Management 47*(3) (May 2004), pp. 510–34.

Harrington, Winston, and Richard D. Morgenstern. *Choosing Environmental Policy: Comparing Instruments and Outcomes in the United States and Europe.* Baltimore, MD: Johns Hopkins University Press, 2004.

———. "Economic Incentives versus Command and Control: What's the Best Approach for Solving Environmental Problems?" *Resources 152* (Fall/Winter 2004), pp. 13–17.

Helfand, Gloria E., and Peter Berck. *The Theory and Practice of Command and Control in Environmental Policy.* Burlington, VT: Ashgate, November 2003.

Hutchinson, Emma, and Peter W. Kennedy. "State Enforcement of Federal Standards: Implications for Interstate Pollution." *Resource and Energy Economics 30*(3) (August 2008), pp. 316–44.

Karp, Larry, and Jiangfeng Zhang. "Regulation with Anticipated Learning about Environmental Damages." *Journal of Environmental Economics and Management 51*(3) (May 2006), pp. 259–79.

Keohane, Nathaniel O., Richard L. Revesz, and Robert N. Stavins. "The Choice of Regulatory Instruments in Environmental Policy." *Harvard Environmental Law Review 22* (1998), pp. 313–67.

Koontz, Tomas M., Toddi A. Steelman, JoAnn Carmin, Katrina Smith Korfmacher, Cassandra Moseley, and Craig W. Thomas. *Collaborative Environmental Management: What Roles for Government?* Washington, DC: Resources for the Future, 2004.

Portney, Paul R. "EPA and the Evolution of Federal Regulation." In Paul R. Portney, ed. *Public Policies for Environmental Protection.* Washington, DC: Resources for the Future, 2000, pp. 11–30.

Van Rooji, Benjamin. "Greening Industry Without Enforcement? An Assessment of the World Bank's Pollution Regulation Model for Developing Countries." *Law & Policy 23*(1) (January 2010), pp. 127–52.

Westmoreland, Joshua K. "Global Warming and Originalism: The Role of the EPA in the Obama Administration." *Boston College Environmental Affairs Law Review 37*(1) (2010), pp. 225–56.

To access additional course materials, visit www.cengagebrain.com. At the home page, search for the ISBN of this title (shown on the back cover). This will take you to the product page where these resources can be found.

CHAPTER 5

Economic Solutions to Environmental Problems: The Market Approach

"The government's view of the economy could be summed up in a few short phrases: If it moves, tax it. If it keeps moving, regulate it. And if it stops moving, subsidize it."

—Ronald Reagan (1911–2004)

It is true that the market fails to correct environmental problems on its own, but the incentives that define the market process can nonetheless be put to work by policy-makers. The **market approach** to environmental policy, recommended for some time by economists, has begun to be adopted by governments as part of their overall response to the risks of pollution. Distinct from the use of more traditional command-and-control instruments, the market approach uses price or other economic variables to provide incentives for polluters to reduce harmful emissions.

Economists are strong proponents of the market approach because it can achieve a cost-effective solution to environmental problems. How? By designing policy initiatives that allow polluters to respond according to their own self-interests. Market instruments are aimed at bringing the external costs of environmental damage back into the decision making of firms and consumers. Taking its cue directly from market failure theory, the market approach attempts to restore economic incentives by assigning a value to environmental quality, or, equivalently, by pricing pollution. Once done, firms and consumers adjust their behavior to the resulting change in market conditions.

Recognition of the gains associated with market-based instruments is growing in both the private and public sectors. Consequently, it is important to understand this alternative policy approach and its advantages over more traditional forms of regulation. To that end, in this chapter, we examine the theory and the practical implications of environmental policy based on market incentives. Our primary tool of analysis will be economic modeling, and we will use the criteria of allocative efficiency and cost-effectiveness to make policy assessments.

To provide a framework for our investigation, we begin with a brief overview of the major categories of market-based instruments: pollution charges, subsidies, deposit/refund systems, and pollution permit trading systems. We then use these categories to structure our economic analysis. Taking each category in turn, we develop models of specific instruments, assess the results, and provide examples of how each is used in practice.

DESCRIPTIVE OVERVIEW

market approach
An incentive-based policy that encourages conservation practices or pollution-reduction strategies.

The primary distinction between the **market approach** and the command-and-control approach is the way in which environmental objectives are implemented, as opposed to the level at which those objectives are set. From a practical perspective, standards-based objectives are set at a socially desirable level rather than at an efficient level. Where the market approach parts company with the command-and-control approach is in *how* it attempts to achieve those objectives, that is, in its design of policy instruments.

Types of Market Instruments

Because many policy instruments use market incentives, it is helpful to classify these instruments into major categories: **pollution charges**, **subsidies**, **deposit/refund systems**, and **pollution permit trading systems**. Table 5.1 offers a brief description of each.

Nations all over the world use market-based instruments to help control pollution. In fact, all Organisation for Economic Co-operation and Development (OECD) member nations[1] use approximately 375 different environmental taxes as well as some 250 environmentally based fees and charges.[2] Although the market approach continues to be a secondary form of control, its use in national policy prescriptions speaks to its importance as part of the range of available solutions to environmental problems. To further investigate international market-based policy, access the OECD/EEA database on instruments used for environmental policy and natural resources management, which is available online at **www2.oecd.org/ecoinst/queries/index.htm**.

TABLE 5.1 Categories of Market-Based Instruments

Market Instrument	Description
Pollution charge	A fee charged to the polluter that varies with the quantity of pollutants released.
Subsidy	A payment or tax concession that provides financial assistance for pollution reductions or plans to abate in the future.
Deposit/refund	A system that imposes an up-front charge to pay for potential pollution damages that is returned for positive action, such as returning a product for proper disposal or recycling.
Pollution permit trading system	The establishment of a market for rights to pollute, using either credits or allowances.

NOTE: A good overview of market-based instruments used in the United States and internationally is found at **http://yosemite.epa.gov/ee/epa/eed.nsf/webpages/EconomicIncentives.html**.

Source: U.S. EPA, Office of Policy, Economics, and Innovation (January 2001).

[1]For a complete list of the 34 OECD member nations, visit **www.oecd.org**, and click on "About OECD" and then on "Members and Partners."

[2]OECD (2006), p. 1.

POLLUTION CHARGES

pollution charge
A fee that varies with the amount of pollutants released.

The theoretical premise of a **pollution charge** is to internalize the cost of environmental damages by pricing the pollution-generating activity. By definition, a pollution charge is a fee that varies with the quantity of pollutants released. It can be implemented as a product charge or as an effluent or emission charge. The motivation follows what's known as the "**polluter-pays principle**," a position rooted in the belief that the polluter should bear the costs of control measures to maintain an acceptable level of environmental quality.[3] Let's begin by modeling a **product charge** implemented as a tax—the classical solution to negative externalities.

Modeling a Product Charge as a Per-Unit Tax

Consider a good in a competitive market whose production generates a negative environmental externality. Because producers base their decisions solely on the marginal private cost (*MPC*) of production, ignoring the marginal external cost (*MEC*) of the environmental damage, too many resources are allocated to production. As depicted in Figure 5.1, firms produce output level Q_C, where the marginal social benefit (*MSB*) of consuming the good is equal to the marginal private cost (*MPC*) of producing it.[4] Notice that the competitive equilibrium output (Q_C) is higher than the efficient level

FIGURE 5.1 A Pigouvian Tax

This model illustrates how a Pigouvian tax achieves efficiency in the market for some pollution-generating product (*Q*). By setting the unit tax equal to the *MEC* at Q_E, shown as distance *ab*, the *MPC* curve shifts up to MPC_t. Equilibrium output is then determined by the intersection of MPC_t and *MSB*, which is the efficient production level, Q_E.

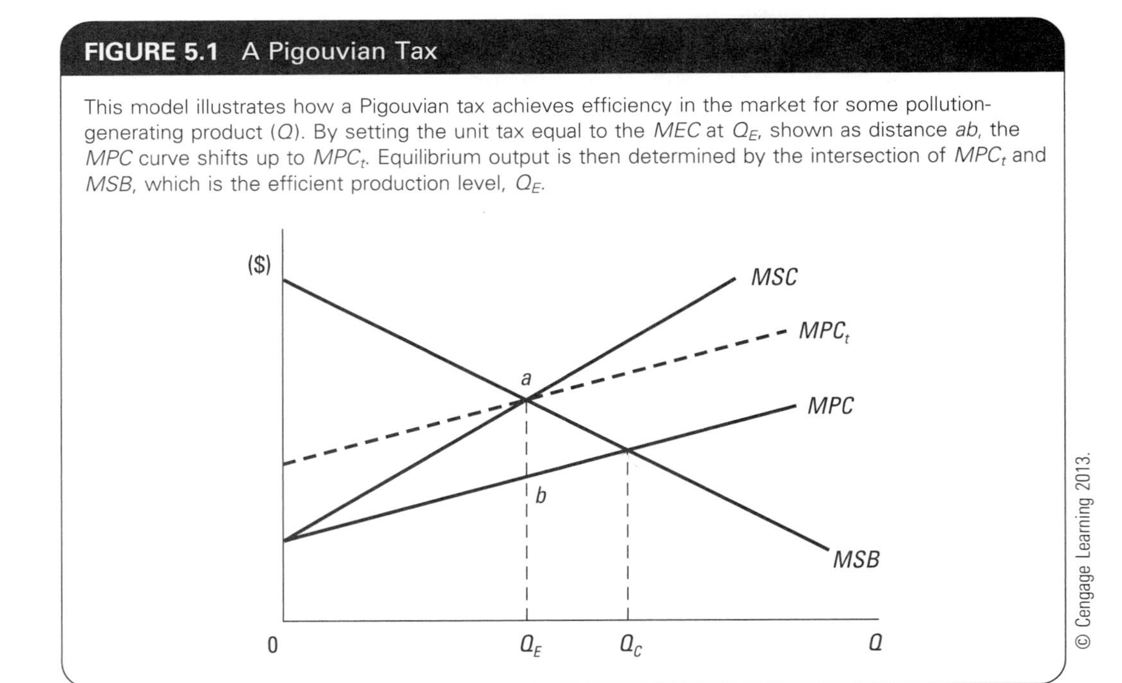

[3] OECD (1989), p. 27.

[4] As we have done in previous discussions, we are implicitly assuming that there is no marginal external benefit (*MEB*) in this market, so *MPB* = *MSB*.

(Q_E), which corresponds to the point where the *MSB* equals the marginal social cost (*MSC*).

product charge
A fee added to the price of a pollution-generating product based on its quantity or some attribute responsible for pollution.

The motivation of a **product charge** is to induce firms to internalize the externality by taking account of the *MEC* in their production decisions. One way this can be done is by imposing a unit tax on the pollution-generating product equal to the *MEC* at the efficient output level (Q_E). This type of tax is called a **Pigouvian tax**, named after English economist A. C. Pigou, who initially formulated the theory. As illustrated in Figure 5.1, this policy instrument effectively shifts up the *MPC* curve by distance *ab* to MPC_t, which generates an equilibrium at the efficient output level.

Pigouvian tax
A unit charge on a good whose production generates a negative externality such that the charge equals the *MEC* at Q_E.

Assessing the Model

In theory, the Pigouvian tax forces firms to lower production to the efficient level. Although theoretically pleasing, this instrument is difficult to impose in practice and is not commonly used. Why? One problem is the difficulty in identifying the dollar value of *MEC* at Q_E and, hence, the level of the tax. A second problem is that the model implicitly allows only for an output reduction to abate pollution—an unrealistic restriction. To address both of these reservations, let's consider a more practical alternative. The pollution charge can also be implemented as an **emission charge**, which is a tax levied on the pollution, instead of as a product charge. By moving out of the product market, the polluter's response is not restricted to an output reduction.

Modeling an Emission Charge: Single-Polluter Case

emission or effluent charge
A fee imposed directly on the actual discharge of pollution.

An **emission** or **effluent charge** assigns a price to pollution, typically through a tax. Once this price mechanism is in place, the polluter can no longer ignore the effect of its environmental damages on society. Instead, the pollution charge forces the polluter to confront those damages, pay for them, and in so doing, consider them as part of its production costs. Faced with this added cost, the polluting firm can either continue polluting at the same level and pay the charge or invest in abatement technology to reduce its pollutant releases and lower its tax burden. Based on normal market incentives, the firm will choose whichever action minimizes its costs.

It is fairly simple to model an emission charge that allows polluters to make cost-minimizing decisions. To begin, we assume that the government sets an abatement standard at some "acceptable level," A_{ST}. Now, we consider a policy that presents the polluter with the following options to be undertaken singularly or in combination:

- The polluter must pay a constant per-unit tax (t) on the difference between its existing abatement level (A_0) and the standard (A_{ST}) such that Total Tax $= t(A_{ST} - A_0)$; and/or
- The polluter incurs the cost of abating.

In Figure 5.2, we graph these options from a single firm's perspective using marginal curves. Because the per-unit tax is constant at t, the marginal tax (*MT*) curve is a horizontal line at that tax level. The cost of abating at the margin is shown as the marginal abatement cost (*MAC*) curve. At each unit of *A*, the cost-minimizing firm will compare *MAC* to *MT* and choose whichever is lower.

In our model, the firm will abate up to A_0, because up to that point *MAC* is below *MT*. Assuming no fixed costs, total abatement costs (*TAC*) are represented by the area

FIGURE 5.2 Modeling an Emission Charge for a Single Polluter

This model illustrates the decisions of a single polluter, faced with the options of paying a tax $[t\ (A_{ST} - A_0)]$ or incurring the cost of abating. In this case, the firm will abate up to A_0, because up to that point $MAC < MT$. Its total abatement costs (TAC) are shown as area $0aA_0$. Beyond A_0 and up to the standard, A_{ST}, the firm will pay the taxes, because $MT < MAC$. Its tax burden is represented by area A_0abA_{ST}. Hence, the total costs to the polluter of complying with this policy are area $0abA_{ST}$.

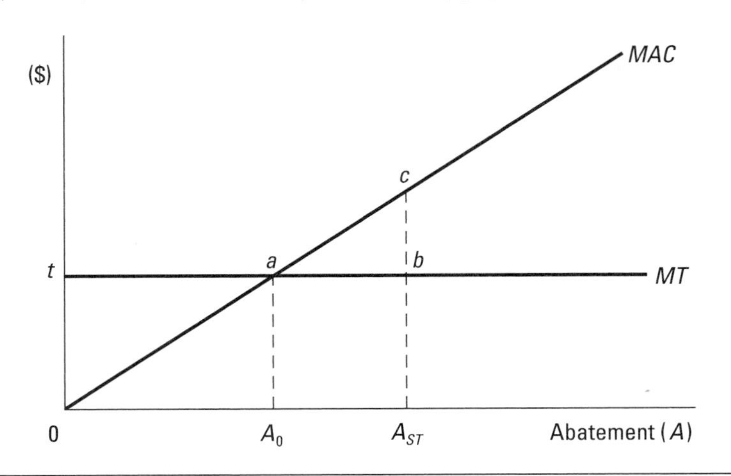

under the MAC curve up to A_0, or area $0aA_0$. Notice that these costs are lower than what the taxes would be up to A_0, shown as area $0taA_0$. Beyond A_0 and up to A_{ST}, the firm will opt to pay the tax, because MT is lower than MAC in that range. The firm's total tax payment for not abating between A_0 and A_{ST} is represented by area A_0abA_{ST}, which is smaller than the cost of abating that amount, which is area A_0acA_{ST}.

In sum, the total costs to the polluter of complying with this policy are area $0abA_{ST}$, which comprises the following two elements:

- Area $0aA_0$, the total cost of abating A_0 units of pollution
- Area A_0abA_{ST}, the tax on pollution not abated up to A_{ST}

Assessing the Model

Be sure to recognize that the emission charge stimulates the natural economic incentives of the polluter. At any point in time, *static* incentives motivate the firm to choose among the available options, given its existing technology. Seeking to satisfy its own self-interest to maximize profit, the polluter makes a least-cost decision between paying the tax and abating. The result is that the externality is internalized, using the least amount of resources.

There are also *dynamic* incentives that encourage the firm to advance its abatement technology. More efficient abatement techniques would allow the firm to reduce pollution more cheaply and enjoy the associated cost savings. Furthermore, the lower abatement costs might even allow the firm to avoid paying any emission charges. Consider the effect of a hypothetical technological advance, shown in Figure 5.3 as a downward pivot of MAC to MAC'. In this case, if the firm were faced with the same set of policy options based on A_{ST}, it would be better off abating all the way up to that standard. The firm would incur total abatement costs equal to $0bA_{ST}$ and would pay no emission

FIGURE 5.3 Effect of Technology Improvement on a Firm's Decision Making

A technological advance causes the *MAC* curve to pivot downward to *MAC′*. As a result, if the firm is faced with the option of abating or paying the unit tax at each abatement level up to A_{ST}, it would be better off abating up to A_{ST}. In this case, the technological change helped the firm avoid paying any emission charges.

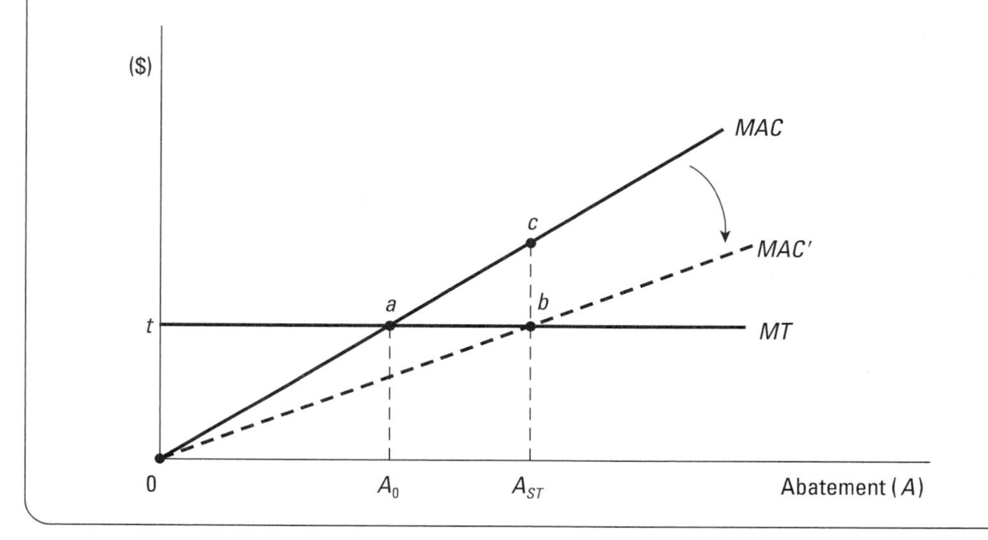

charges. Relative to its expenditures before the technological advance, which were $0abA_{ST}$, it saves an amount equal to $0ab$.

Modeling an Emission Charge: Multiple-Polluters Case

To evaluate the cost-effectiveness of an emission charge across multiple polluters, we return to the two-polluter model used in Chapter 4. As before, we assume that the government imposes a 10-unit abatement standard for the region. We repeat the cost functions from Chapter 4 to facilitate our discussion:

Polluter 1's Marginal Abatement Cost (MAC_1): $MAC_1 = 2.5A_1$
Polluter 1's Total Abatement Costs (TAC_1): $TAC_1 = 1.25(A_1)^2$

Polluter 2's Marginal Abatement Cost (MAC_2): $MAC_2 = 0.625A_2$
Polluter 2's Total Abatement Costs (TAC_2): $TAC_2 = 0.3125(A_2)^2$

where A_1 is the amount of pollution abated by Polluter 1, and A_2 is the amount of pollution abated by Polluter 2.

Assume that the government imposes the same emission charge as in the single-polluter case, that is, Total Tax $= t(A_{ST} - A_0)$. In this case, because $A_{ST} = 10$, the emission charge becomes $t(10 - A_0)$. We also assume that the tax rate (t) is set at \$5, so Total Tax $=$ \$5$(10 - A_0)$ for each polluter.

Now consider each firm's response to the tax. As we proceed, refer to Figure 5.4, which reproduces the model used in Chapter 4, adding a horizontal line (MT) at \$5 to represent the emission charge. When faced with the \$5-per-unit charge, Polluter 1

FIGURE 5.4 Modeling an Emission Charge in a Two-Polluter Model

In this model, each firm abates as long as its $MAC < MT$ and pays the emission charge on all pollution not abated. Polluter 1, the high-cost abater, abates 2 units and pays $40 in taxes on the remaining 8 units. Polluter 2, the low-cost abater, abates 8 units and pays taxes of $10 on the remaining 2 units. At this point, $MAC_1 = MAC_2 = \$5$, which indicates the least-cost abatement allocation across the two firms.

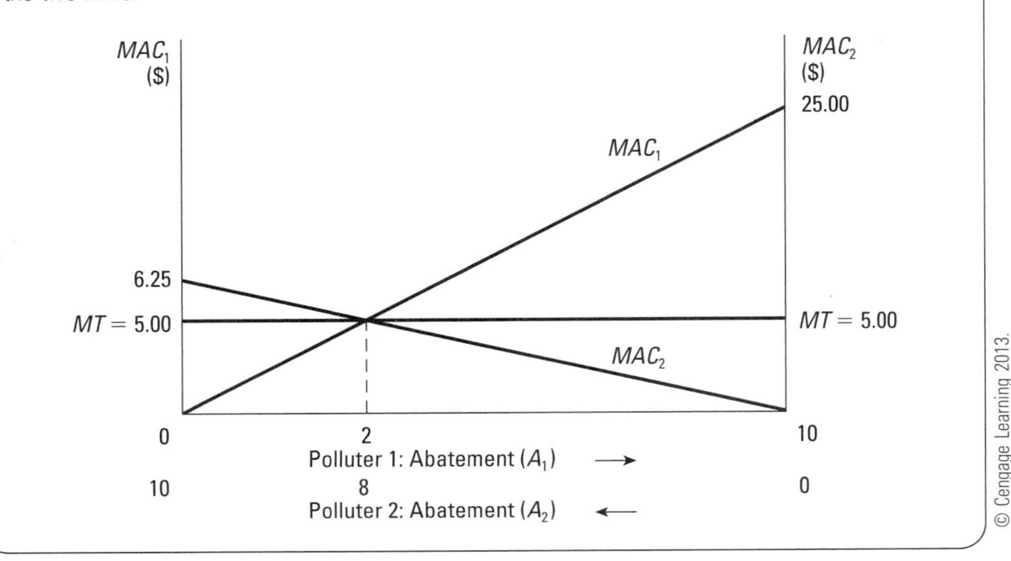

would compare the relative costs of MT and MAC_1 for each incremental unit of abatement, just as in the single-firm case. It would abate as long as $MAC_1 < MT$ and pay the tax when the opposite is true. Thus, Polluter 1 would abate up to the point where $MAC_1 = MT$, which occurs at $A_1 = 2$, and pay the tax on the remaining 8 units. Polluter 2 would proceed in the same way, abating to the point where $MAC_2 = MT$ at $A_2 = 8$ and paying the tax on the remaining 2 units. The corresponding calculations are as follows:

Polluter 1

Abates up to the point where $MAC_1 = MT$:	$2.5(A_1) = \$5$, or $A_1 = 2$
Incurs Total Abatement Costs of:	$TAC_1 = 1.25(2)^2 = \$5$
Incurs Total Tax Payment of:	Total Tax $= 5(10 - 2) = \$40$

Polluter 2

Abates up to the point where $MAC_2 = MT$:	$0.625(A_2) = \$5$, or $A_2 = 8$
Incurs Total Abatement Costs of:	$TAC_2 = 0.3125(8)^2 = \$20$
Incurs Total Tax Payment of:	Total Tax $= 5(10 - 8) = \$10$

Combined Values for the Region

Total Abatement Level $= 10 = A_{ST}$

Total Abatement Costs $= \$25$

Total Tax Payments $= \$50$

There are two important observations to make about these results. First, the $5 unit tax achieves the 10-unit abatement standard. Second, this objective has been achieved using the cost-effective allocation of abatement resources across polluters.[5] Because each polluter abates to the point where its *MAC* equals *MT*, all *MAC*s are equal, precisely in accordance with the cost-effective abatement criterion. Notice that most of the abating is done by the low-cost abater, Polluter 2. The high-cost polluter, Polluter 1, abates less but pays much higher emission charges in the form of taxes.

Assessing the Model

The emission charge exploits each polluter's natural incentive to pursue a least-cost strategy. As a result, the low-cost abaters do most of the cleaning up, and the high-cost abaters pay more in taxes to cover the greater damages they cause. The benefits to society are not affected by *who* does the abating, but the costs are. In this case, costs are minimized because Polluter 2 does most of the abating. It does so, not because it is motivated by society's objectives, but because doing so is in its own best interest. An added advantage is that the tax generates revenues, which can be used by the government to help finance the costs of enforcement and monitoring.

Notwithstanding the advantages of the emission charge, there are several caveats. The first concerns the setting of the emission charge. Realistically, the government will not know the tax rate at which polluters' abatement levels collectively meet the standard and therefore will have to adjust the tax until the environmental objective is achieved.[6] This adjustment process can be time intensive. Another consideration is that monitoring is likely to be more complex, and therefore more costly, when each polluter responds to a policy based on its own internal operations. Distributional implications also have to be considered. Because polluting firms pay higher taxes, part of the tax burden is shared with consumers in the form of higher prices. Job losses also may occur as firms adjust to the tax or change technologies to increase abatement. Finally, firms may try to evade the tax by illegally disposing of pollutants. To minimize that potential, the government may have to strengthen its monitoring programs, which adds to costs.

Pollution Charges in Practice[7]

Internationally, the pollution charge is the most commonly used market-based instrument. Several countries, including Australia, France, Germany, Italy, and Japan, use **fees or taxes** to control the noise pollution generated by aircraft. Australia, Canada, France, and Poland are among the nations using **effluent charges** to protect water resources.

A real-world application of a **product charge** is one levied on batteries, which is done in Austria, Denmark, Hungary, and Italy. Another example is a fee charged on

---

[5]Recall from Chapter 4 that using a uniform standard under a command-and-control approach generates total abatement costs of $39.06.

[6]For example, if the government had set the tax at too low a level, say $4, each polluter would have abated to where its *MAC* equals $4. The result would have been $A_1 = 1.6$ and $A_2 = 6.4$ for a combined abatement level of 8.0 units—too low to satisfy the objective. So, the government would have had to raise the tax until the combined total reached the 10-unit abatement standard.

[7]Drawn from OECD/EEA (2011).

automobile tires. Finland uses this particular product charge to help cover the costs of collecting and recycling used tires. Other targeted products used by various countries include electronics, lubricant oil, packaging, paint, paint containers, and gasoline. To access current data on gasoline tax rates in the United States, both at the federal and state levels, visit the American Petroleum Institute at **www.api.org/Oil-and-Natural-Gas-Overview/Industry-Economics/Fuel-Taxes.aspx**. For an overview of the international experience with gasoline taxes, read Application 5.1.

APPLICATION 5.1 Taxing Gasoline Consumption: An International Comparison

According to Kazuo Aichi, former chief of Japan's environmental agency, "Gasoline is too cheap in the United States and should be taxed more to cut energy use." Although Aichi's comment was made over a decade ago, it remains relevant today. The argument is motivated in part by ongoing disagreements between the United States and some of its industrialized counterparts about the appropriate response to such problems as natural resource depletion and global air pollution. This criticism of U.S. policy on fuel taxes underscores the importance of price in encouraging conservation.

Consider how the U.S. tax on gasoline compares to what is imposed by other industrialized countries by examining the data in the accompanying table.

These data validate the claim that the price of gasoline in the United States *is* cheap in a relative sense and that much of the difference is due to a lower tax rate. Notice that the tax rate is between 50 percent and 62 percent of the unit price in European nations, about 32 percent in Canada, and just under 46 percent in Japan. The comparatively low tax rate of 13.2 percent in the United States reflects its fuel tax policy.

Beyond this brief comparison of tax rates, there is the more important issue of the economic intent of a gasoline tax. In addition to boosting government revenues, taxing a commodity such as gasoline is designed to internalize the negative externalities of consumption, including air pollution caused by operating gasoline-powered vehicles. Other negative externalities are highway congestion and the increased risk of traffic accidents. These adverse effects have one thing in common—they extend to parties beyond those engaged in the market transaction. Consequently, the costs of these damages are not reflected in the price, and too much gasoline is brought to market. As other nations have apparently learned, raising the tax on gasoline can reduce consumption and bring the associated social costs and benefits closer together.

Gasoline Prices and Taxes by Country (January 2011)

Country	Total Gasoline Price per Liter (U.S. dollars)	Tax Rate % of Price
United States	0.818	13.2
Canada	1.150	31.8
Japan	1.645	45.8
Spain	1.677	50.1
France	1.923	58.1
Italy	1.930	55.4
Germany	1.949	60.5
United Kingdom	1.958	61.6

Source: International Energy Agency (2011).

Sources: International Energy Agency (2011); Reifenberg and Sullivan (May 1, 1996); Tanner (June 9, 1992); "Japanese Environmentalist Says Gas Too Cheap in U.S." (February 2, 1992).

ENVIRONMENTAL SUBSIDIES

An alternative market approach to reducing environmental damage is to pay polluters *not* to pollute through an **environmental subsidy**. There are two major types of subsidies—**abatement equipment subsidies** and **pollution reduction subsidies**. We discuss each of these in turn.

Modeling an Abatement Equipment Subsidy

abatement equipment subsidy A payment aimed at lowering the cost of abatement technology.

Abatement equipment subsidies are aimed at reducing the costs of abatement technology. Because subsidies are "negative taxes," they have a similar incentive mechanism to pollution charges except that they reward for *not* polluting as opposed to penalizing for engaging in polluting activities. In practice, abatement equipment subsidies are implemented through grants, low-interest loans, or investment tax credits, all of which give polluters an economic incentive to invest in abatement technology.

From a theoretical perspective, subsidies are used to internalize the *positive* externality associated with the consumption of abatement activities. If a subsidy were offered for installing specific abatement equipment, such as scrubbers, quantity demanded would increase because the effective price would be lower. To achieve an efficient equilibrium, the subsidy would have to equal the marginal external benefit (*MEB*) of scrubber consumption measured at the efficient output level. Notice that this is analogous to a Pigouvian tax, and in fact, this type of subsidy is known as a **Pigouvian subsidy**.

Pigouvian subsidy A per-unit payment on a good whose consumption generates a positive externality such that the payment equals the *MEB* at Q_E.

A model of a hypothetical competitive market for scrubbers is as follows:

$$MSC = 70.0 + 0.5Q, \qquad \begin{aligned} MPB &= 350.0 - 0.9Q \\ +MEB &= 56.0 - 0.2Q \\ \hline MSB &= 406.0 - 1.1Q \end{aligned}$$

where Q is the number of scrubber systems produced in a year, and *MSC*, *MPB*, and *MEB* are denominated in millions of dollars.[8] The corresponding graph is shown in Figure 5.5. The competitive equilibrium arises where $MPB = MSC$, or at $Q_C = 200$ and $P_C = \$170$ million. However, the efficient equilibrium occurs where $MSB = MSC$, or at $Q_E = 210$ and $P_E = \$175$ million. In an unregulated competitive market, too few scrubbers are exchanged at too low a price because the external benefits of a cleaner environment are not recognized by the market participants. If a subsidy (s) equal to the *MEB* at the efficient output level were provided to demanders (i.e., polluters), it is as if *MPB* shifts up to MPB_S, and more scrubbers would be traded. In this case, the Pigouvian subsidy would equal $MEB = 56.0 - 0.2(210) = \$14$ million, shown as distance KL in Figure 5.5. The effective price to polluters would be the efficient market price *less* the subsidy, or ($P_E - s$), which in this case would be (\$175 million − \$14 million), or \$161 million.

Assessing the Model

Just as in the case of a Pigouvian tax, one problem with implementing a Pigouvian subsidy is measuring the *MEB*. Monetizing the marginal external benefits of such

[8]For simplicity, we assume there are no marginal external costs (*MEC*), so that $MPC = MSC$.

FIGURE 5.5 A Pigouvian Subsidy

Because of the positive externality in the scrubber market, the true measure of benefits is given by *MSB*, the vertical sum of *MPB* and *MEB*. If a Pigouvian subsidy equal to the *MEB* at Q_E were provided to purchasers, it is as if the *MPB* shifts up to MPB_S, and 210 scrubbers would be traded instead of the competitive output level of 200. In this case, the subsidy (s) is $14 million, labeled as distance KL. The effective price to polluters would be ($P_E - s$), which in this model is ($175 million – $14 million), or $161 million.

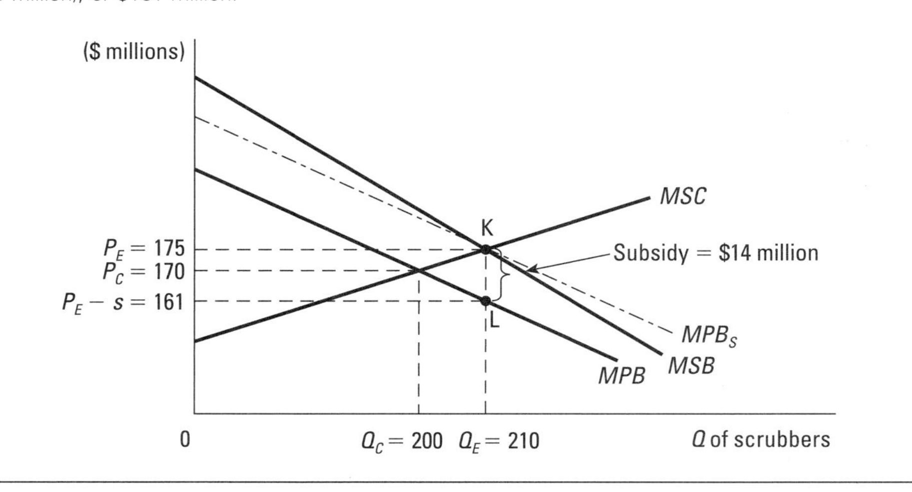

intangibles as better health and more stable ecosystems is difficult at best. So it is not likely that a subsidy of abatement equipment will achieve allocative efficiency. However, its associated effect of encouraging greater consumption because of an effectively lower price should still occur.

Even setting aside the difficulty in achieving efficiency, equipment subsidies may have other drawbacks. One commonly cited criticism is that this type of control instrument biases polluters' decisions about how best to abate. Subsidies affect relative prices, making other alternatives less attractive from a financial perspective. However, some of these abatement alternatives might be more effective in reducing pollution. For the same reason, innovation of a potentially superior abatement system could be discouraged as long as the government is subsidizing existing equipment. Finally, subsidies must be financed through taxes or government borrowing. Thus, they effectively redistribute income from society to polluters, an outcome some view as unacceptable despite the associated gain of a cleaner environment.

Modeling a Per-Unit Subsidy on Pollution Reduction

An alternative type of subsidy is one based on emission or effluent reductions, called a **per-unit subsidy on pollution reduction**. In this case, the government agrees to pay the polluter a subsidy (s) for every unit of pollution removed below some standard (Z_{ST}). This is modeled as:

per-unit subsidy on pollution reduction A payment for every unit of pollution removed below some predetermined level.

$$\text{Total Subsidy} = s(Z_{ST} - Z_0)$$

where Z_0 is the actual level of pollution. Suppose, for example, that Z_{ST} is set at 200 tons of emissions per month, and the subsidy (s) is set at $100 per ton per month. Then, if a polluter reduces its emissions to 180 tons per month, it would receive a subsidy of \$100(200 − 180), or \$2,000.

Assessing the Model

On the plus side, a per-unit pollution reduction subsidy *might* be less disruptive than an equipment subsidy, because it is established independent of the abatement method used and thus avoids any technological bias. On the other hand, these subsidies can have the perverse effect of elevating pollution levels in the aggregate. How does such a paradox arise? Because a per-unit subsidy effectively lowers a polluter's unit costs, which in turn raises its profits. If the industry has limited entry barriers, these profits would signal entrepreneurs to enter the industry. In the long run, although each individual polluter reduces its emissions, the subsidy may cause the market to expand such that *aggregate* emissions end up higher than they were originally.[9] The dilemma could be solved if entry were prohibited or at least limited in some way. Whether or not this is feasible or even desirable depends on the industry structure, the extent of environmental damage, and the associated costs.

Environmental Subsidies in Practice

Internationally, a common application of environmental subsidies is in the form of grants or low-interest loans for energy efficient automobiles and other products, which are being used in many countries, including Belgium, Denmark, Finland, Japan, and Turkey.[10] In the United States, the most common use of subsidies is federal funding for such projects as publicly owned treatment works. Federal subsidies are also used to promote the use of pollution control equipment and to encourage the use and development of cleaner fuels and low-emitting vehicles. Federal subsidies are implemented in a variety of ways, such as through grants, rebates, tax exemptions, and tax credits. At the state level, a common application of environmental subsidies is in the form of tax incentives to encourage recycling activities.[11] See Table 5.2 for an overview of some of these state programs.

DEPOSIT/REFUND SYSTEMS

deposit/refund system
A market instrument that imposes an up-front charge to pay for potential damages and refunds it for returning a product for proper disposal or recycling.

The potentially perverse consequences of abatement subsidies suggest that pollution charges might be a better alternative. In some contexts, however, pollution charges can be costly to administer because of the associated expense of monitoring and enforcement. Recall that one of the drawbacks of pollution charges is that they may encourage illegal disposal of contaminants. This potential problem is an important motivation for using a **deposit/refund system**.

Operationally, deposit/refund systems attach a front-end charge (the deposit) for the *potential* occurrence of a damaging activity and guarantee a return of that charge (the refund) upon assurance that the activity has not been undertaken. This market

[9]For further details, see Baumol and Oates (1975), Chapter 12.
[10]OECD/EEA (2011).
[11]U.S. EPA, Office of Policy, Economics, and Innovation (January 2001), Chapter 7.

TABLE 5.2 Selected State Subsidies for Recycling

State	General Description	Eligible Applicants
Arizona	Income tax credit on recycling equipment equal to 10 percent of the installed cost.	Individuals and corporations.
Delaware	Recycling investment tax credit of $500 for every $100,000 invested. Recycling employment income tax credit of $500 for every new employee associated with adding recycled products into the process.	Recycling companies using at least 25 percent of recycled or recovered materials.
Georgia	Personal income tax credit for investing in recycling facilities, equipment, or machinery equal to 3, 5, or 8 percent of the investment.	Manufacturing industries.
Hawaii	Sales tax reduction on recycling equipment of between 0.5 and 4.0 percent.	Solid waste processing facilities.
Idaho	Income tax credit on recycling equipment of up to 20 percent of equipment costs up to $30,000 per year.	Recycling firms handling post-consumer paper, glass, and plastic.
Maryland	Personal property tax exemption on tools, machinery, and manufacturing apparatus or engines for certain counties.	Recycling firms.
New Jersey	Exemption from sales tax on recycling equipment.	Recycling firms.
South Carolina	Sales and use tax exemption for recycling equipment.	Recycling and manufacturing firms.
Virginia	Income tax credit equal to 10 percent of recycling equipment purchase price.	Recycling firms.
Wisconsin	Property tax exemption for machinery and equipment used exclusively and directly in recycling or reducing waste.	Recycling firms.

NOTE: For further detail and other examples, visit **www.epa.gov/epawaste/conserve/rrr/rmd/bizasst/rec-tax.htm**.

Source: U.S. EPA, Office of Solid Waste and Emergency Response (March 12, 2012).

instrument combines the incentive element of a pollution charge with a built-in mechanism for controlling monitoring costs. Its intent is to capture the difference between the private and social costs of improper waste disposal, with its most common targets being beverage containers and lead-acid batteries.

Economics of Deposit/Refund Systems

Improper or illegal waste disposal gives rise to a negative externality. The external costs include health damages, such as lead contamination from discarded lead-acid batteries, and aesthetic impairment from litter and trash accumulation. Deposit/refund systems are designed to force the potential polluter to account for both the marginal private cost (*MPC*) and the marginal external cost (*MEC*) of improper waste disposal, should that activity be undertaken.

As with the pollution charge, the deposit is intended to capture the *MEC* of improper waste disposal. The deposit forces the polluter to internalize the cost of any damage it may cause by making it absorb this cost *in advance*. Unique to the deposit/refund system is the refund component, which introduces an incentive to properly dispose of wastes and hence prevent environmental damage from taking place at all.

Taken together, the deposit/refund system targets the *potential* polluter instead of penalizing the actual polluter, using the refund to reward appropriate behavior.

Modeling a Deposit/Refund System

A model of a deposit/refund system is presented in Figure 5.6. From left to right, the horizontal axis measures *improper* waste disposal (*IW*) as a percentage of all waste disposal activity. Implicitly, then, the percentage of *proper* waste disposal (*PW*) is measured right to left. Thus, if 25 percent of all wastes is improperly disposed of, then by default, 75 percent is disposed of appropriately and safely.

The MPC_{IW} includes expenses for collecting and illegally dumping wastes plus the costs of improperly disposing recyclable wastes, such as the expense of trash receptacles, collection fees paid to refuse companies, and the opportunity costs of forgone revenue associated with recycling. The MSC_{IW} includes the MPC_{IW} plus the MEC_{IW}, represented implicitly as the vertical distance between MSC_{IW} and MPC_{IW}. The MPB_{IW} is the demand for improper waste disposal. It is motivated by the avoidance of time and resources to collect wastes, bring nonrecyclables to a landfill, and haul recyclables to a collection center.[12] Because we assume no external benefits in this case, $MPB_{IW} = MSB_{IW}$.

In the absence of environmental controls, equilibrium is determined by the intersection of MSB_{IW} and MPC_{IW}, or Q_{IW}. The efficient equilibrium occurs where MSC_{IW}

FIGURE 5.6 Modeling a Deposit/Refund System

To correct the negative externality associated with improper waste disposal, a deposit/refund system can be instituted with the deposit set equal to MEC_{IW} measured at Q_E. The deposit elevates MPC_{IW} by distance *ab*, forcing market participants to establish a new equilibrium at the socially optimal level. The result is an increase in the proportion of proper waste disposal from Q_{IW} to Q_E.

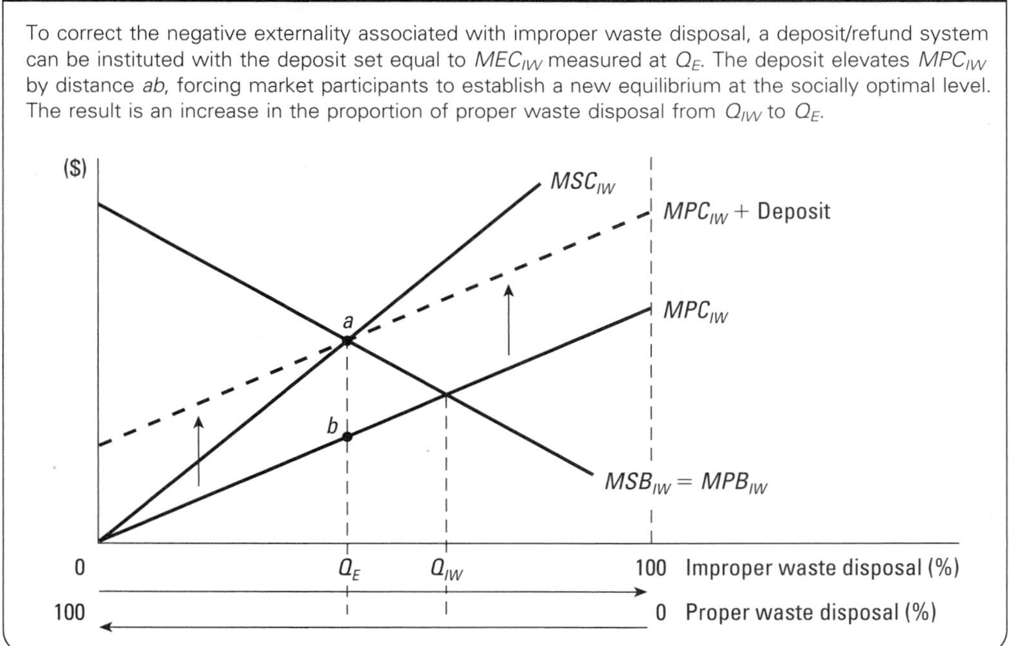

[12]The MPB_{IW} measured left to right is equivalent to the MPC_{PW} measured right to left.

equals MSB_{IW}, or at Q_E, which is smaller than Q_{IW}. Predictably, we observe that in the presence of a negative externality, too much improper waste disposal is produced because market participants do not consider the full impact of their actions.

To correct the negative externality, assume that a deposit/refund system is instituted whereby the deposit equals the MEC_{IW} at Q_E. This is labeled as distance *ab* in Figure 5.6. Once imposed, the deposit effectively elevates MPC_{IW} by distance *ab*, forcing the market participants to a new equilibrium at Q_E. In so doing, a percentage of waste disposal is converted from improper methods to appropriate ones, measured by distance $(Q_{IW} - Q_E)$. Notice that the deposit serves the same function as a pollution charge. The critical difference is that the refund helps to deter improper waste disposal. The potential polluter has an explicit incentive to properly dispose of wastes, because doing so allows it to reclaim the deposit. Should disposers choose instead to illegally discard waste, at least they will have paid for the external costs in advance. Authorities also have the flexibility to adjust the deposit or refund amounts to enhance the built-in incentives.

Assessing the Model

The value added of the deposit/refund system is that the refund encourages environmentally responsible behavior without adding significantly to government's monitoring and compliance costs. What makes this instrument unique is that once established, the incentives operate with limited supervision.

Another advantage of the deposit/refund instrument is that it can be used to encourage more efficient use of raw materials. An inordinate amount of used products and materials ends up in landfills or burned in incinerators, when they could be recycled. The availability of recycled products and wastes can help slow the depletion of such virgin raw materials as aluminum and timber and may result in associated price declines as well. Charging firms a deposit on raw materials acts as a tax, encouraging more efficient use of resources *during* the production process. The refund encourages proper disposal or recycling of raw material waste at the *end* of the production phase. Firms that elect to ignore this incentive face not only conventional disposal costs but also the opportunity cost of the forgone refund.

Deposit/Refund Systems in Practice

Perhaps the best known applications of deposit/refund systems are those used to encourage proper disposal of beverage containers. In the United States, such programs typically are initiated through state bottle bills. As of 2011, 10 states have passed legislation requiring deposits on beer and soft drink containers: California, Connecticut, Hawaii, Iowa, Maine, Massachusetts, Michigan, New York, Oregon, and Vermont. As noted in Table 5.3, deposits range from 2 cents to 15 cents per container.

Attesting to the effectiveness of these programs, Oregon, the first state to enact a bottle bill, reported that roadside litter was reduced by 75–85 percent just two years after its bill became law.[13] Similar results also have been observed in other countries. Examples include Canada, Hungary, and the Netherlands. As might be expected, success

[13]U.S. EPA, Office of Policy, Economics, and Innovation (January 2001), pp. 59 and 63.

TABLE 5.3 State Deposit/Refund Programs Under Bottle Bills

State	Beverage Containers Covered	Amount of Deposit
California	Beer, malt, wine and distilled spirits coolers, all nonalcoholic beverages except milk	5¢ for < 24 oz. 10¢ for > 24 oz.
Connecticut	Beer, malt, carbonated soft drinks, bottled water	5¢
Hawaii	All nonalcoholic drinks, except milk or dairy products, beer, malt beverages, mixed spirits, mixed wine	5¢
Iowa	Beer, carbonated soft drinks, mineral water, wine coolers, wine, liquor	5¢
Maine	All beverages except dairy products and unprocessed cider	Wine, liquor: 15¢ All others: 5¢
Massachusetts	Beer, malt, carbonated soft drinks, mineral water	5¢
Michigan	Beer, soft drinks, carbonated and mineral water, wine coolers, canned cocktails	10¢
New York	Beer, malt, carbonated soft drinks, water, wine coolers	5¢
Oregon	Beer, malt, carbonated soft drinks, bottled water	Standard refillables: 2¢ All others: 5¢
Vermont	Beer, malt, carbonated soft drinks, mixed wine drinks, liquor	Liquor: 15¢ All others: 5¢

Sources: Container Recycling Institute (2011; 2009); U.S. EPA, Office of Policy, Economics, and Innovation (January 2001), Table 5.1, p. 59; pp. 57–66.

of these programs varies greatly across countries. For example, across the Canadian Provinces, return rates range between 47 and 96 percent of targeted bottles, and for Hungary, the return rate is between 70 and 80 percent. While impressive, these rates pale in comparison to the Netherlands, which has been able to achieve a 95–99 percent return rate for bottles and cans targeted under its deposit/refund system.[14] Figure 5.7 presents the mechanics of these types of deposit/refund systems.

Other deposit/refund initiatives are aimed at encouraging responsible disposal of such products as used tires, car hulks, and lead-acid batteries. Proper discard of lead-acid batteries is of particular concern because of the health risks linked to lead exposure. Consequently, in the United States, some states have imposed mandatory deposit/refund systems for lead-acid batteries. Typically, the deposit is $5 or $10 per battery, and the consumer can obtain a refund by returning a used battery within a specified period along with proof that the deposit was paid.[15] Denmark, Mexico, and Poland have set up similar systems for lead-acid batteries with return rates ranging between 70 percent and over 99 percent. Other applications of deposit/refund systems include Poland and Spain for product packaging; Denmark for used tires; and Denmark, Norway, and Sweden for scrapped vehicles. In this latter instance, Denmark and Norway charge a

[14]OECD/EEA (2011).

[15]U.S. EPA, Office of Policy, Economics, and Innovation (January 2001), p. 64.

FIGURE 5.7 Typical Phases of a Deposit/Refund System for Beverage Containers

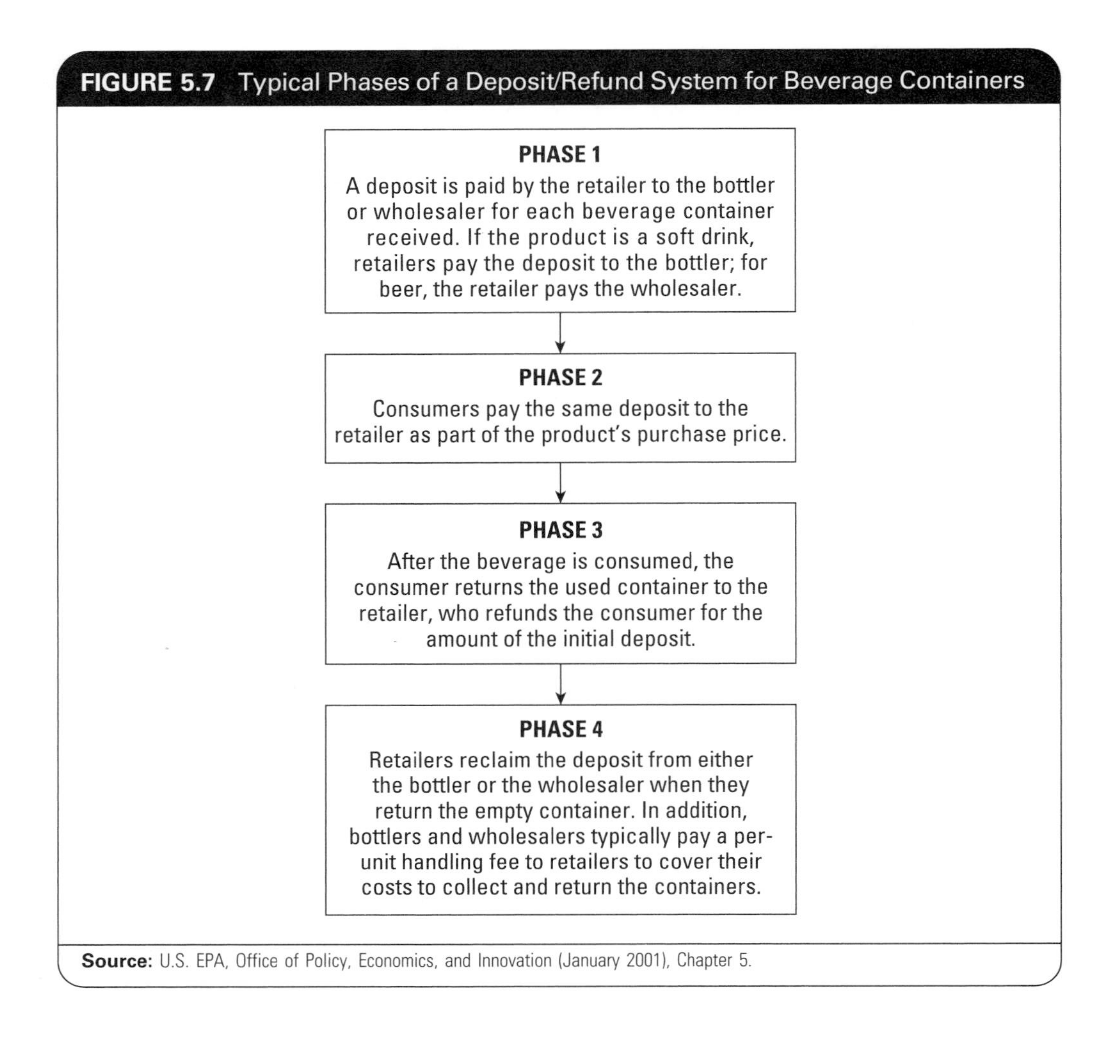

PHASE 1
A deposit is paid by the retailer to the bottler or wholesaler for each beverage container received. If the product is a soft drink, retailers pay the deposit to the bottler; for beer, the retailer pays the wholesaler.
PHASE 2
Consumers pay the same deposit to the retailer as part of the product's purchase price.
PHASE 3
After the beverage is consumed, the consumer returns the used container to the retailer, who refunds the consumer for the amount of the initial deposit.
PHASE 4
Retailers reclaim the deposit from either the bottler or the wholesaler when they return the empty container. In addition, bottlers and wholesalers typically pay a per-unit handling fee to retailers to cover their costs to collect and return the containers.

Source: U.S. EPA, Office of Policy, Economics, and Innovation (January 2001), Chapter 5.

deposit that is actually less than the refund, presumably to further enhance the incentive to return unwanted vehicles for proper disposal or recycling.[16]

POLLUTION PERMIT TRADING SYSTEMS

Thus far, we have illustrated that market instruments can be used to set prices for polluting and abatement activities. It is also possible for government to use the price-quantity relationship in the opposite direction—by establishing the quantity of pollution or abatement to be achieved and letting the market determine the price. With perfect information, either approach is viable, and both will lead to the same outcome. However, pricing instruments can be problematic in that government does not know in advance what price will achieve a quantity-based environmental objective. That means

[16]OECD/EEA (2011).

it has to monitor the quantity response to some initially established price and continually make adjustments until the proper pollution level is achieved, essentially a trial-and-error process.

It may be more efficient to use a policy instrument that operates from the *known* variable, that is, the socially desirable quantity of pollution or abatement, and let the market establish the price. This is the underlying premise of a **pollution permit trading system**, which can be implemented through the use of credits or allowances. Under a system of **pollution credits**, a polluter earns marketable credits only if it emits below an established standard. If instead the trading system uses **pollution allowances**, each permit gives the bearer the right to release some amount of pollution. These too are marketable, so that polluters can buy and sell allowances as needed, based on their access to abatement technologies and their costs.

pollution permit trading system
A market instrument that establishes a market for rights to pollute by issuing tradeable pollution credits or allowances.

pollution credits
Tradeable permits issued for emitting below an established standard.

pollution allowances
Tradeable permits that indicate the maximum level of pollution that may be released.

Structure of a Pollution Permit Trading System

A system of marketable pollution permits has two key components:

- the issuance of some **fixed number of permits** in a region
- a provision for **trading** these permits among polluting sources within that region

The fixed number of permits issued is bound by whatever pollution level is mandated by law, capping emissions to meet that regulated level. For example, if the level were set at 200 units of emissions, a maximum of 200 one-unit permits could be issued. Any polluter releasing emissions not authorized by permits would be in violation of the law. Once the limited permits are distributed, polluters may trade them with one another, hence the common description of such a program as a **cap-and-trade system**.

A bargaining process should develop, which gives rise to a **market for pollution rights**. Following their own self-interest, polluters either purchase these rights to pollute or they abate, whichever is the cheaper alternative. High-cost abaters have an incentive to bid for available permits, whereas low-cost abaters have an incentive to abate and sell their permits on the open market. The result is a cost-effective abatement allocation. But notice that trading is critical to the cost-effective outcome. For example, if the permits were allocated equally across all polluters and no trading were allowed, the result would be no different than a command-and-control system of uniform standards.

The tradeable permit system accommodates environmental objectives, defined at an aggregate level. For example, in the United States, air pollution policies are designed to achieve national ambient air quality standards within well-defined regions. Within any region, however, some polluters might perform above the standard and others below it, which is acceptable as long as in the aggregate the region is in compliance. This is exactly how the permit system operates—controlling the total amount of emissions in a region but not the releases for each source within that region.[17]

The trading component of the permit system capitalizes on differences in polluters' abatement technologies and opportunities. Sources that can abate efficiently are given the incentive to do so because they can sell their unused permits to their less efficient

[17]Contrast this scenario with a command-and-control instrument that forces every polluting source to meet identical standards of emission or effluent levels. Such instruments equalize the *level of control* across polluters rather than the *marginal costs of control.*

counterparts. As long as the environmental goal is achieved in the aggregate, the benefit to society is the same whether the task is undertaken by a select few or by all firms doing an equal amount of abating. The costs, however, will be markedly lower if abatement is done by more efficient polluters.

Modeling a Pollution Permit System for Multiple Polluters

To illustrate the operation of a permit system, we return to our two-polluter model, where each firm faces distinct abatement costs:

$$\text{Polluter 1:} \quad TAC_1 = 1.25(A_1)^2$$
$$MAC_1 = 2.5A_1$$
$$\text{Polluter 2:} \quad TAC_2 = 0.3125(A_2)^2$$
$$MAC_2 = 0.625A_2$$

Before any government intervention, we assume that each firm releases 10 units of pollution for a total of 20 units in their region. The government has determined that the "acceptable" level of pollution for this region is 10 units, and it decides to meet this objective by using a tradeable permit system. It therefore issues 10 permits, each of which allows the bearer to emit 1 unit of pollution. For simplicity, assume that the government allocates 5 permits to each polluter.[18] Under the permit system, each firm is required to hold a permit for each unit of pollution released and to undertake abatement on all remaining units.

Based on the *initial* allocation of permits, each polluter must abate 5 units of pollution. This initial condition, termed Round 1 of the permit system, is summarized as follows:

ROUND 1: Government Issues Five Permits to Each Polluter

Polluter 1:

Current pollution level:	10 units
Number of permits held:	5
Abatement required:	5 units

$$MAC_1 = 2.5A_1 = 2.5(5) = \$12.50$$
$$TAC_1 = 1.25(A_1)^2 = 1.25(5)^2 = \$31.25$$

Polluter 2:

Current pollution level:	10 units
Number of permits held:	5
Abatement required:	5 units

$$MAC_2 = 0.625A_2 = 0.625(5) = \$3.125$$
$$TAC_2 = 0.3125(A_2)^2 = 0.3125(5)^2 = \$7.81$$

If the permit system did not allow for trading, each firm would have no choice but to abate 5 units each. Although the environmental objective would be met, it would not

---

[18]The government could have introduced the permits through a direct sale, assigning a price to each permit, or through an auction. Either method has the advantage of generating revenue to the government, and the revenue could help absorb some of the administrative costs.

be achieved in a cost-effective manner. The combined abatement cost for both sources without trading is \$39.06.[19]

Now, consider how the result changes when permit trading is allowed. Because the two firms face different *MAC* levels at the end of Round 1, there is an incentive for trade. Polluter 1 has an incentive to buy permits from Polluter 2 as long as the price of each permit is less than its MAC_1. Likewise, Polluter 2 has an incentive to sell permits to Polluter 1 as long as it can obtain a price greater than its MAC_2.

Suppose that in Round 2 of the trading process, the two firms agree on the purchase and sale of one permit at a price of \$8.00.[20] Polluter 1 purchases one permit from Polluter 2, giving Polluter 1 the right to pollute 6 units and the obligation to abate 4 units. Polluter 2 now possesses the right to release 4 units of pollution, which means it must abate 6 units. Round 2 is summarized as follows:

ROUND 2:	***Polluter 1 Purchases One Permit From Polluter 2***	
Polluter 1:	Current pollution level:	10 units
	Number of permits held:	6
	Abatement required:	4 units
	$MAC_1 = 2.5A_1 = 2.5(4)$	$= \$10.00$
	$TAC_1 = 1.25(A_1)^2 = 1.25(4)^2$	$= \$20.00$
	Cost of one permit purchased	= \$8.00
Polluter 2:	Current pollution level:	10 units
	Number of permits held:	4
	Abatement required:	6 units
	$MAC_2 = 0.625A_2 = 0.625(6)$	$= \$3.75$
	$TAC_2 = 0.3125(A_2)^2 = 0.3125(6)^2$	$= \$11.25$
	Revenue from one permit sold	= \$8.00

This outcome can be analyzed from two perspectives—that of society and that of the firm. From the vantage point of society, the total costs of abating 10 units of pollution are now \$31.25, which is \$7.81 *less* than the costs without permit trading. Qualitatively, this is exactly what should happen, because trading has brought the two firms' *MAC* values closer together. Polluter 1 now faces a lower MAC_1 of \$10 (compared to \$12.50 in Round 1), and Polluter 2 now has a higher MAC_2 of \$3.75 (compared to \$3.125 in Round 1).

Next, consider the gains that accrue to each firm as a result of the trade. Polluter 1 is better off, because its total expenditures have decreased. Its outlay for abating plus the cost of the added permit is \$28.00 (i.e., TAC_1 of \$20.00 plus the cost of the additional permit, \$8.00), which is \$3.25 less than its TAC_1 in Round 1. Likewise, Polluter 2 is better off because its net expenditures on abating and trading are \$3.25 (i.e., TAC_2 of

[19]Recall from Chapter 4 that this is precisely the same expenditure incurred by the two polluters if a uniform standard is used under a command-and-control approach.

[20]Any negotiated price between MAC_1 and MAC_2 would be acceptable. The ultimate selling price within that range would be determined by the two firms' relative bargaining strengths.

\$11.25 minus the revenue received from selling one permit, \$8.00), which is \$4.56 less than its TAC_2 at the end of Round 1.

Because there is an incentive for trade as long as the two firms face different *MAC* levels, it should be apparent that Round 2 does not represent a cost-effective solution. After the exchange of one permit, Polluter 1 still faces a higher *MAC* level than Polluter 2 (i.e., \$10.00 for Polluter 1 versus \$3.75 for Polluter 2). Thus, it is in each polluter's best interest to continue to trade. The rule of thumb is that in the presence of differing *MAC* levels among polluting sources, high-cost abaters have an incentive to purchase permits from low-cost abaters, and low-cost abaters have an incentive to sell them. The result? Low-cost abaters will do what they do best—clean up the environment—and high-cost abaters will pay for the right to pollute by buying more permits. Trading will continue until the incentive to do so no longer exists, that is, when the *MAC* levels across both firms are equal. At precisely this point, the cost-effective solution is obtained.

By applying this equimarginal principle to our model, we come to the final round, or equilibrium.

FINAL ROUND:	***Polluter 1 Purchases Three Permits From Polluter 2***		
	Equalization of MAC Levels Across Polluters Is Achieved[21]		
Polluter 1:	Current pollution level:		10 units
	Number of permits held:		8
	Abatement required:		2 units
	$MAC_1 = 2.5A_1 = 2.5(2)$	=	\$5.00
	$TAC_1 = 1.25(A_1)^2 = 1.25(2)^2$	=	\$5.00
	Cost of three permits purchased	=	\$20.00
Polluter 2:	Current pollution level:		10 units
	Number of permits held:		2
	Abatement required:		8 units
	$MAC_2 = 0.625A_2 = 0.625(8)$	=	\$5.00
	$TAC_2 = 0.3125(A_2)^2 = 0.3125(8)^2$	=	\$20.00
	Revenue from three permits sold	=	\$20.00

At this point, each polluter faces an *MAC* of \$5, and society's total cost to achieve the environmental objective is \$25. (Notice that the \$20 payment for permits is not included in society's abatement costs, because this amount is just a transfer from one firm to another.) As predicted, the low-cost abater, Polluter 2, is doing most of the abating at 8 units, whereas the high-cost abater, Polluter 1, abates only 2 units.

Assessing the Model

It is no coincidence that the final abatement allocation for these firms is identical to what happens if a \$5 pollution charge is used. Logically, the outcome is the same

[21]We assume that the second permit is sold for \$7 and the third for \$5. These values plus the \$8 price for the first permit results in a total payment of \$20 for permits.

because both instruments use incentives linked to the firm's *MAC*. There are, however, three important differences.

First, with a pollution charge, the government has to search for the price that will bring about the requisite amount of abatement. In the permit system, trading establishes the price of a right to pollute without outside intervention. Second, the trading system is more flexible; the number of permits can be adjusted to change the environmental objective. If the objective is too stringent, more permits can be introduced. If it is too lenient, the government, environmental groups, or concerned citizens can buy up permits, effectively reducing the amount of pollution allowed in the affected region. Third, the pollution charge generates tax revenues on all units of pollution not abated, whereas no revenues are generated from the permit system. This distinction may be critical in jurisdictions with tight fiscal budgets. However, a trading system can be designed to generate revenues if the government sells or auctions off the initial allocation of permits.

Some measure of controversy surrounds the use of trading systems. Economists typically tout the advantages of a system that so explicitly uses the market. On the other side of the coin, opponents argue that trading systems can create pollution hot spots, localized areas facing high concentrations of pollutants where most of the permit buying takes place. Another objection is the potential for elevated administrative costs to keep records of trades and the emissions of buyers and sellers.[22] Hypotheticals aside, the true test will be in observing how these permit systems perform in practice.

Pollution Permit Trading Systems in Practice

Of all the available control instruments, pollution permit trading systems are by far the most market-oriented. Much of the initial development of these programs has occurred in the United States at the federal level. In attempting to combat the adverse effects of acid rain, the Clean Air Act Amendments of 1990 established an allowance-based trading program to control sulfur dioxide (SO_2) emissions. Application 5.2 describes the major events that characterized the first day of official trading on March 31, 1993.

More recent developments in U.S. trading programs have occurred at the regional and state levels. For example, in an effort to reduce urban smog, a group of northeastern states organized the Ozone Transport Commission, which ultimately designed and implemented a tradeable permit system to reduce nitrogen oxide (NO_x) emissions. An analogous state-level system was devised by California's Regional Clean Air Incentives Market (RECLAIM) to combat urban smog in the Los Angeles area.[23] Visit RECLAIM's Web site at **www.aqmd.gov/reclaim/reclaim.html** for more information.

On a global scale, a trading program that has received considerable worldwide attention is the tradeable permit system for greenhouse gases (GHGs) established by the Kyoto Protocol, an international accord that addresses global warming.[24] To meet their obligations under the Protocol, member states of the European Union executed a formal directive to establish the European Union Greenhouse Gas Emission Trading

[22]U.S. EPA, Office of Policy, Economics, and Innovation (January 2001), p. 68.

[23]OECD/EEA (2011).

[24]More detail on the Kyoto Protocol, including the trading program, is provided in Chapter 13.

APPLICATION 5.2 Fighting Acid Rain With Pollution Rights: The First Annual Auction

In March 1993, the first annual auction of sulfur dioxide (SO_2) emission permits was held. The event was administered by the Chicago Board of Trade (CBOT), the largest commodity exchange in the world. Although most of the available permits are allocated by the EPA directly to the largest polluting sources, a relatively small number (150,000 one-ton permits per year) are set aside in an auction subaccount for direct sale.

Most of the bids in the 1993 auction came from the nation's major utilities, who are the largest SO_2 polluters. A case in point is Illinois Power, a utility that releases about 240,000 tons of SO_2 emissions each year. This facility was unable to operate on the 171,000 permits issued by the EPA, so it bought more from other utilities at about $225 each. It subsequently submitted bids for another 5,000 permits at the 1993 auction. For Illinois Power and others like it, the costs to abate were apparently greater than the expected outlay to purchase permits.

Despite the predominance of utilities in the bidding, there was at least one important exception—a not-for-profit environmental group called National Healthy Air License Exchange. According to the group's president, any permits bought in the auction would be retired and kept off the market. That private citizens can exercise such a tangible influence over environmental policy is one of the advantages of an emissions trading program. National Healthy Air License Exchange submitted bids for 1,100 permits but came away with only one, for which it paid $350. Nonetheless, the organization was able to participate in the auction and eliminate some emissions from the atmosphere.

In an apparently philanthropic move, Northeast Utilities of Connecticut donated 10,000 of its permits to the American Lung Association just before the auction. The association retired the permits to keep them out of the bidding process. The utility, which did not need the permits to operate, could have sold them for an estimated value of $3 million on the open market. However, the gesture was not totally without financial incentive. Northeast Utilities was expected to enjoy tax deductions for the contribution that would offset any sacrifice of pollution rights' revenues. Hoping to encourage other such donations, Northeast Utilities and the American Lung Association established a repository for permits donated by other utilities.

How did the bidders fare in the first pollution rights auction? Rights to emit the 150,000 tons of SO_2 were purchased by utilities, brokers, and environmentalists for a total of $21 million. Permit prices ranged from $122 to $450. The largest single purchaser was Carolina Power and Light Company, a utility that bid for and won over 85,000 permits. In accordance with the law, all auction proceeds went to the EPA, which then allocated the funds to those utilities from which the permits were originally obtained.

For direct access to all the annual auction results through the present, visit the EPA's Acid Rain Program Allowance Auctions Web page at **www.epa.gov/airmarkets/trading/auction.html**.

Sources: Allen (March 20, 1993); Taylor and Gutfeld (September 25, 1992); Taylor (March 31, 1992); Taylor and Kansas (March 26, 1992).

System (EU ETS), which was launched in 2005. This multination trading program effectively limits aggregate carbon dioxide (CO_2) emissions of participating countries and grants each nation marketable emissions allowances. More on this innovative program is available at **http://ec.europa.eu/clima/policies/ets/index_en.htm.**

CONCLUSIONS

No environmental policy instrument is without flaws. In truth, this less than perfect outcome should be expected. The market process works as well as it does because it operates autonomously, without external guidance. Therefore, it should not be surprising that any attempt to correct a market problem by imposing third-party controls is likely to have its share of pitfalls.

This line of defense is precisely the motivation of market-based policy approaches. The aim is not to add more restraint but to restore the market forces that broke down in the first place. In one form or another, market-based instruments effectively assign a price to environmental goods, such as clean air and clean water. Once this signaling mechanism is in place, polluters are forced to internalize the costs of pollution damage and adjust their decisions accordingly.

Of course, not all market-based instruments are well suited to all environmental problems. Both the nature of the problem and the market context must be understood before any policy can be implemented with success. Environmental problems are complex, both in origin and in implication. Likewise, today's markets are sophisticated and dynamic. However, evidence is beginning to accumulate that the link between the two is a fundamental step toward finding solutions.

Summary

- The market approach to pollution control uses economic incentives and the price mechanism to achieve an environmental standard.
- The major categories of market-based instruments are pollution charges, subsidies, deposit/refund systems, and pollution permit trading systems.
- A pollution charge is a fee that varies with the quantity of pollutants released. It can be implemented as a product charge or as an effluent or emission charge.
- A Pigouvian tax is a unit charge on a pollution-generating product equal to the *MEC* at the efficient output level.
- An emission charge is levied directly on the actual release of pollutants. Given a choice of abating or paying the fee, the profit-maximizing polluter will choose the least-cost option.
- When multiple polluters face an abatement standard, the emission charge yields a cost-effective allocation of abatement responsibilities where the *MAC* levels for all sources are equal.
- Abatement equipment subsidies are designed to reduce the costs of abatement.
- If the subsidy equals the *MEB* at the efficient output level, it is called a Pigouvian subsidy.
- Per-unit pollution reduction subsidies pay the polluter for abating beyond some predetermined level.
- A deposit/refund system imposes an up-front charge to pay for potential pollution damage and later refunds the charge when the product is returned for proper disposal or recycling. The deposit is intended to capture the *MEC* of improper disposal, and the refund provides an incentive to properly dispose of or recycle wastes.
- A pollution permit trading system involves the issuance of tradeable pollution rights based on a given environmental objective. Following natural incentives, polluters will either purchase these rights or abate, whichever is the cheaper alternative.

Review Questions

1. Suppose that a chemical manufacturing plant is releasing nitrogen oxides into the air, and these emissions are associated with health and ecological damages. Economists have estimated the following marginal costs and benefits for the chemical market, where Q is monthly output in thousands of pounds and P is price per pound.

   $MSB = 50 - 0.4Q$ $\quad$ $MSC = 2 + 0.4Q$

   $MEB = 0$ $\quad$ $MEC = 0.2Q$

   a. Find the competitive equilibrium, Q_C and P_C, and the efficient equilibrium, Q_E and P_E.

   b. Find the dollar value of a product charge that would achieve an efficient solution.

2. Assume that the U.S. market for some inorganic chemical is modeled as follows, where Q is in thousands of kilograms, P is price per kilogram, and there are no consumption externalities:

   $MSB = 90 - 0.5Q$ $\quad$ $MSC = 30 + 0.3Q$

   $MEC = 0.2Q$

   a. Find the competitive equilibrium.

   b. Find the efficient equilibrium.

   c. Determine the dollar value of the Pigouvian tax that would ensure an efficient equilibrium.

3. Despite economists' support of a market approach to environmental policy, the command-and-control approach continues to dominate the policy of most nations. Explain why this is the case. In your response, cite and then comment on some of the common criticisms of market-based initiatives.

4. In 1996, Michigan launched a voluntary emissions trading program, which allows polluters to achieve cost-effective solutions when meeting requirements of the U.S. Clean Air Act. (For more on this program, visit **www.michigan.gov/deq**, and click on Air, Assessment and Planning, and Emissions Trading.) Suppose that Michigan's objective for two major firms in an urban area is a 16 percent reduction in carbon

monoxide (CO) emissions and that each firm faces the following costs:

Firm 1: $TAC_1 = 1{,}000 + 2.5(A_1)^2$

$MAC_1 = 5A_1$

Firm 2: $TAC_2 = 500 + 1.5(A_2)^2$

$MAC_2 = 3A_2$

where A_1 and A_2 represent the percentages of CO emission abatement achieved by firm 1 and firm 2, respectively, and *TAC* and *MAC* are measured in thousands of dollars.

a. Calculate the *TAC* and *MAC* for each firm if a uniform abatement standard were used.

b. Based on your answer to part (a), is there an economic incentive for the sources to participate in the trading program? Explain.

c. Quantify the cost savings associated with a cost-effective abatement allocation that could be achieved through trading.

d. At what price must each tradable permit be set to achieve the cost-effective solution?

Additional Readings

Bohringer, C. "Industry-Level Emission Trading between Power Producers in the EU." *Applied Economics 34*(*4*) (March 2002), pp. 523–33.

Coria, Jessica, and Thomas Sterner. "Tradeable Permits in Developing Countries: Evidence from Air Pollution in Santiago, Chile." *Journal of Environment Development 19*(*2*) (June 2010), pp. 145–70.

David, Maia, and Bernard Sinclair-Desgagné. "Pollution Abatement Subsidies and the Eco-Industry." *Environmental and Resource Economics 45*(*2*) (February 2010), pp. 271–82.

Ellerman, A. Denny, Paul L. Joskow, and David Harrison, Jr. "Emissions Trading in the U.S.: Experience, Lessons, and Considerations for Greenhouse Gases." Arlington, VA: Pew Center on Global Climate Change, May 2003.

Fullerton, Don. "A Framework to Compare Environmental Policy." *Southern Economic Journal 68*(*2*) (October 2001), pp. 224–48.

Harrington, Winston, and Richard D. Morgenstern. *Choosing Environmental Policy: Comparing Instruments and Outcomes in the United States and Europe*. Washington DC: Resources for the Future, 2004.

Hutchinson, Emma, Peter W. Kennedy, and Cristina Martinez. "Subsidies for the Production of Cleaner Energy: When Do They Cause Emissions to Rise?" *Contributions to Economic Analysis and Policy 10*(*1*) (2010), pp. 41–44.

Krysiak, Frank C., and Iris Maria Oberauner. "Environmental Policy à la Carte: Letting Firms Choose Their Regulation." *Journal of Environmental Economics and Management 60*(*3*) (November 2010), pp. 221–32.

Mandell, Svante. "Optimal Mix of Emission Taxes and Cap-and-Trade." *Journal of Environmental Economics and Management 56*(*2*) (September 2008), pp. 131–40.

Muller, Nicholas Z., and Robert Mendelsohn. "Weighing the Value of a Ton of Pollution." *Regulation 33*(*2*) (Summer 2010), pp. 20–24.

Oates, Wallace. "Green Taxes: Can We Protect the Environment and Improve the Tax System at the Same Time?" *Southern Economic Journal 61*(*4*) (April 1995), pp. 915–22.

Organisation for Economic Co-operation and Development (OECD). *Environmentally Related Taxes in OECD Countries: Issues and Strategies*. Paris: OECD, 2001.

______. *Tradeable Permits: Policy Evaluation, Design and Reform*. Paris: OECD, 2004.

Repetto, Robert, Roger C. Dower, Robin Jenkins, and Jacqueline Geoghegan. *Green Fees: How a Tax Shift Can Work for the Environment and the Economy*. Washington, DC: World Resources, November 1992.

Requate, Till. "Dynamic Incentive by Environmental Policy Instruments–A Survey." *Ecological Economics 54*(*2–3*) (August 2005), pp. 175–95.

Ruff, Larry. "The Economic Common Sense of Pollution." *The Public Interest 19* (Spring 1970), pp. 69–85.

Sartzetakis, E. S. "On the Efficiency of Competitive Markets for Emission Permits." *Environmental and Resource Economics 27(1)* (January 2004), pp. 1–19.

Segerson, Kathleen., ed. *The Economics of Pollution Control.* Northhampton, MA: Elgar, 2011.

Stavins, Robert N. "Market-Based Environmental Policies." In Paul R. Portney and Robert N. Stavins, eds. *Public Policies for Environmental Protection.* Washington, DC: Resources for the Future, 2000, pp. 31–76.

Sterner, Thomas. *Policy Instruments for Environmental and Natural Resource Management.* Washington, DC: Resources for the Future, 2003.

Tietenberg, T. H., ed. *Emissions Trading Programs, Vols. 1 and 2.* Burlington, VT: Ashgate, 2001.

______ "Cap-and-Trade: The Evolution of an Economic Idea." *Agricultural and Resource Economics Review 39(3)* (October 2010), pp. 359–67.

Wirth, Timothy E., and John Heinz. *Project 88: Harnessing Market Forces to Protect Our Environment: Initiatives for the New President.* Washington, DC, December 1988.

To access additional course materials, visit www.cengagebrain.com. At the home page, search for the ISBN of this title (shown on the back cover). This will take you to the product page where these resources can be found.

MODULE 3

Analytical Tools for Environmental Planning

Economics uses powerful models to explain environmental market failures and the policy solutions used to address them. With these models as a foundation, we can now move to the practical implications of environmental planning—the process through which government identifies environmental risks, prioritizes them, and responds with a policy plan. The planning process involves difficult decisions: determining which hazards pose the greatest threat to society, where to set policy objectives, and which control instruments to use. These decisions are guided by analytical tools designed to evaluate environmental risks and assess the costs and benefits of minimizing them. In this module, we conduct an in-depth investigation of two analytical tools used in environmental planning and decision making: risk analysis and benefit-cost analysis.

In Chapter 6, we study the two components of risk analysis: risk assessment, which is the identification of risk; and risk management, which is the formulation of a risk response. In our study of risk assessment, we explore how health and ecological risks are evaluated and characterized. Once done, several risk management strategies are examined, including benefit-cost analysis, an economic tool used to guide environmental policy decisions. Because of its potential significance, we devote the final three chapters of this module to an investigation of this important analytical tool. Chapter 7 explains how economists measure environmental benefits. Chapter 8 presents the analogous discussion for costs, and Chapter 9 shows how both sets of results are used in a comparative evaluation to guide policy decisions.

CHAPTER 6

Environmental Risk Analysis

"Risk comes from not knowing what you're doing."

—Warren Buffet (1930–)

In Module 2, we learned that there are many types of policy instruments that government can use to control pollution, ranging from uniform standards to tradeable pollution permits. Now we are prepared to investigate *how* such policies are designed and implemented in practice. It turns out that the way government becomes involved in any market failure and how it ultimately responds through policy are elements of a complex process. First, government must use scientists' evaluation of an environmental hazard to formally assess the environmental problem and the risk it poses to society—a procedure referred to as **risk assessment**. Second, public officials must decide an appropriate course of action or policy response to reduce that risk—part of the process known as **risk management**. In this chapter, we study these risk-based procedures, which are the key components of risk analysis.

Few would debate that risk assessment and risk management involve difficult and sometimes controversial decisions. The decision making is difficult because there is uncertainty about environmental hazards and the implications for human health and the ecology, particularly over the long term. The controversy arises because there is no clear consensus about how government should respond to what is known about a given hazard. Accepting that all environmental risk cannot be eliminated, policymakers have to determine how much risk society can tolerate—a decision about which there is usually much debate. They also must decide what policy to use to achieve whatever risk level is deemed acceptable. How do public officials justify choosing one policy instrument over another? What criteria guide this decision-making process, and are they appropriate?

From an economic perspective, managing environmental risks should be guided by the costs and benefits associated with abatement. Unfortunately, the data to fully assess these benefits and costs are often insufficient. So, policymakers have to rely on the best available estimates or use an alternative risk management strategy. What are these alternative strategies, and what criteria are used to motivate them? These and other issues in risk analysis frame our agenda in this chapter.

CONCEPT OF RISK

risk
The chance of something bad happening.

Risk is a somewhat obscure notion, yet all of us have some intuitive sense of what it is. After all, while most of us choose not to dwell on it, risk—or the chance of something bad happening—is part of life. Accepting that risk is a pervasive phenomenon, we also know that some risks can be minimized or even avoided, provided they have been recognized. This realization suggests that dealing with risk involves two tasks: (1) *identifying* the degree of risk and (2) *responding* to it. Individuals participate in both activities, although usually not in any systematic fashion. As a matter of course, they usually formulate a perception of risk, based not on scientific data but on a subjective or even instinctive level. These perceptions, however unscientific, are what determine how an individual responds to risk. A person might choose to accept the risk as is, find ways to reduce it, or try to avoid it entirely.

All this seems to imply that dealing with risk is strictly a private exercise, but it turns out that risk analysis is also an important part of public policy development. Why? Because at its core, policy is a formal response to risks faced by society. Because the government is devising a response for society as a whole, it cannot rely on individual perceptions of risk that are highly subjective, often uninformed, and possibly off the mark. Instead, the policymaker must use a systematic assessment of risk before devising a policy response.

Classifying Risk: Voluntary and Involuntary Risk

Attempting to impose structure on obscure concepts is always difficult. Yet it is precisely in these instances that order is needed. That researchers and other analysts classify risk in a number of ways seems to bear out this observation. One of the more common approaches is to consider two broad risk categories: **voluntary risk** and **involuntary risk**.

Voluntary Risks

voluntary risk
Risk that is deliberately assumed at an individual level.

As the label suggests, **voluntary risks** are those that are deliberately assumed at an individual level, that is, the result of a conscious decision. Every day, we make private decisions to engage in activities that implicitly change the amount of risk that we elect to accept.

Most voluntary risks arise from personal activities, such as driving a car, flying in an airplane, or drinking coffee. Because these risks are self-imposed, individuals can and do make decisions to respond to them. In particular, they adjust their personal exposure level to the underlying hazards. For example, an individual might choose to respond to the risks of skydiving by avoiding the activity entirely. In so doing, exposure to the hazard and the risks to that individual are reduced to zero.

Since voluntary risks are self-imposed and the potential outcome is confined to a single person, the public sector typically is not involved or at most plays a limited role. When government does intervene, it is usually confined to identifying potentially dangerous conditions or products and communicating that information to society. In so doing, the government helps people identify and assess risk, but it does not impose a response decision. This communication might be accomplished through public service announcements or through product labeling. The Environmental Protection Agency (EPA) has instituted a Consumer Labeling Initiative to improve understanding of household

product labels, such as on pesticides and household surface cleaners. To learn more about this project, visit **www.epa.gov/pesticides/regulating/labels/consumer-labeling.htm**.

Involuntary Risks

involuntary risk Risk beyond one's control and not the result of a willful decision.

People are also exposed to hazards that are beyond their control. Here, the risks are **involuntary**, since they do not arise from a willful decision. A classic example of involuntary risk is the likelihood of property damage and personal injury caused by a natural disaster. The risk of being harmed by a hurricane or an earthquake is not self-imposed, yet the chance of such an event, while relatively small, exists nonetheless. Environmental hazards, such as air pollution or toxic waste sites, are another source of involuntary risk. Here, the hazard is often chemical exposure that arises as an externality of production. In such cases, the risk is considerable because the effects are pervasive and because the potential harm extends to human health and the ecology.

Characteristic of involuntary risk is that individual responses are limited. Personal exposure to the hazard can be adjusted, but absent extreme behavior, the risk cannot be reduced to zero. For example, to reduce personal risk of the effects of dirty air, a person could avoid living in such cities as Los Angeles or Mexico City. Such a response would decrease that individual's risk of harm, but the risk still exists. Because the sources of involuntary risk are beyond the control of private individuals, the associated threat is a *public* problem. Consequently, government tries to control society's exposure to some involuntary risks.

Distinguishing Between Voluntary and Involuntary Risks

To illustrate the difference between voluntary and involuntary risks, think about the voluntary risk of smoking cigarettes. As researchers learned more about the associated health risks, the government began to disseminate this information through public service announcements and by placing warnings on cigarette packages. However, the individual decision to smoke, or, in this context, the response to the voluntary risk, remained a private decision because the risk was believed to extend *only* to the smoker. As research progressed, it became apparent that passive, or secondhand, smoke presents a risk to persons *other than* the smoker. This discovery meant that what was originally perceived as a voluntary risk turned out to pose a threat to others that was beyond their control—an involuntary risk. As such, passive smoke was considered a public problem, and government assumed a more aggressive position on smoking. For more on the risks of passive smoke, see Application 6.1.

environmental risk Involuntary risk of exposure to an environmental hazard.

Defining Environmental Risk

hazard Source of the environmental damage.

exposure Pathways between the source of the damage and the affected population or resource.

One of the more important concerns of environmental decision makers is determining **environmental risk**, the involuntary risk of exposure to hazards such as pollutant emissions and toxic substances. Notice that two elements determine the extent of environmental risk: the hazard itself and exposure to that hazard. The **hazard** is the source of the damage or the negative externality, such as poisonous factory emissions or toxic chemicals dumped into a river. **Exposure** refers to the pathways between the source of the damage and the affected population or resource. Although both hazard and exposure define environmental risk, each can independently affect the outcome. That is, some hazards are relatively minor but affect a large part of the population; others, like certain chemicals, are dangerous, but exposure to them is limited.

APPLICATION 6.1 EPA Declares Secondhand Smoke a Carcinogen

In January 1993, the EPA released a long-awaited report, presenting the agency's conclusion that environmental tobacco smoke, also known as passive, or secondhand, smoke, is a human carcinogen responsible for the lung cancer-related deaths of 3,000 nonsmokers each year.

The EPA's announcement alarmed the general public, but it also stimulated debate about the agency's treatment of the matter. Harsh criticism was mounted about the two-year delay between the EPA's initial draft report on environmental smoke and its formal announcement. The EPA had completed its risk assessment on passive smoke in 1990 and prepared a first draft report. However, the final report was delayed while the tobacco industry disputed the validity of the EPA's findings. At the same time, the EPA's indoor air research program terminated its work on tobacco smoke.

According to the agency's critics, this halt in tobacco research *and* the delay in announcing its conclusions were the result of the EPA's succumbing to pressure from the tobacco industry. The EPA denied the allegation, asserting that its research on tobacco ended because the work had been completed. It further argued that the 1993 announcement was connected to their risk assessment division, whose research was not affected by any changes within the agency's indoor air program.

Other questions were raised about the methods the EPA used to arrive at the risk estimates. In fact, soon after the report was issued, tobacco growers and cigarette manufacturers jointly filed suit against the EPA. The objective was to obtain a declaration that the agency's report was null and void because it was based on faulty scientific evidence, invalid procedures, and manipulated data. Notwithstanding the motivation of the plaintiffs, the agency *did* adjust the statistical confidence level of the studies downward, which allowed one study to show a statistically significant increased risk of lung cancer in nonsmokers from exposure to passive smoke. Without the adjustment, no single study arrives at such a conclusion. In any case, in 1998, a North Carolina district court ruled to vacate several chapters in the EPA's report, including its classification of secondhand smoke as a human carcinogen. The judge argued that the methods used in the study were questionable and that its conclusion was motivated by the EPA's desired outcome rather than by science and standard statistical tests.

The following year, the Clinton administration countered with an appeal. In December 2002, the Fourth U.S. Circuit Court of Appeals overturned the lower court's ruling, asserting that the EPA's report was advisory and therefore not subject to court review. As a consequence, the EPA's report and its findings linking secondhand smoke and lung cancer incidence have been vindicated, absent further legal action by the tobacco industry.

A brief summary of the EPA's results and a link to the full report are available at **http://cfpub2.epa.gov/ncea/cfm/recordisplay.cfm?deid=2835**.

Sources: "Tobacco Firms Fail in Appeal of Ruling on Second Hand Smoke" (December 12, 2002); Schlesinger (July 20, 1998); Shalal-Esa (June 23, 1993); "Quietly, EPA Drops Some Tobacco Research" (January 7, 1993); Noah (January 6, 1993); U.S. EPA, Office of Air and Radiation (December 1, 1992).

Because risk analysis is central to environmental decision making, policymakers have devised methods to assess, characterize, and respond to environmental risk. These interdependent methods are referred to as **risk assessment** and **risk management**. We investigate each in some detail.

RISK ASSESSMENT[1]

risk assessment Qualitative and quantitative evaluation of the risk posed to health or the ecology by an environmental hazard.

Risk assessment refers to the qualitative and quantitative evaluation of the risk posed to health or the ecology by an environmental hazard. In practice, environmental risk assessment is conducted by scientists who gather, analyze, and interpret data about a

[1]Much of the following discussion is drawn from U.S. EPA, Risk Assessment Portal (August 19, 2010a); National Research Council of the National Academies (2008); U.S. EPA, Office of the Science Advisor, Risk Assessment Forum (1998); National Research Council (1983).

given contaminant. In the United States, the EPA uses a paradigm of this process based on a model first presented by the National Academy of Sciences in 1983 and later amended in 2008 by an independent committee selected by the National Research Council. The process differs somewhat for human health risks and ecological risks in response to a reexamination and definition of the latter in 1998. We briefly examine each of these assessment processes in turn. Visit **www.epa.gov/riskassessment/** for more information.

Human Health Risk Assessment[2]

The accepted model of risk assessment in the United Stated is used to conduct evaluations of human health risks. To provide context for this process, it helps to know that this risk assessment model is part of a three-phase framework for risk-based decision making: Phase I, Problem Formulation and Scoping; Phase II, Risk Assessment; and Phase III, Risk Management. The risk assessment phase, which is our present focus, is comprised of four steps: **hazard identification**, **dose-response assessment**, **exposure assessment**, and **risk characterization**. Figure 6.1 shows the information flow through the three phases of the overall framework as well as through the four assessment stages. Let's take a look at these four stages in turn.

Hazard Identification

hazard identification Scientific analysis to determine whether a causal relationship exists between a pollutant and any adverse effects.

The first step of risk assessment is known as **hazard identification**. In this step, a determination is made whether a causal relationship exists between a pollutant and an increased incidence of adverse effects on human health and whether these effects are likely to arise. This determination is based on available data, using several scientific methods to gather evidence about any association between human health and exposure to a pollutant. Three common methods are **case clusters**, **animal bioassays**, and **epidemiology**, which are described in Table 6.1. Specific information about hazard identification approaches is detailed in a set of human health guidelines recommended by the National Academy of Sciences and available online at **http://cfpub.epa.gov/ncea/cfm/nceaguid_human.cfm**.

Dose-Response Assessment

dose-response relationship A quantitative relationship between doses of a contaminant and the corresponding reactions.

threshold The level of exposure to a hazard up to which no response exists.

Once a substance has been identified as a hazard, scientists must investigate its potency by quantifying the human response to various doses. This element of risk assessment determines the **dose-response relationship**. Using data collected in the hazard identification stage, dose-response assessment attempts to develop a complete profile of the effects of an environmental pollutant. An important aspect of this assessment is determining whether some range of exposures to the hazard from zero to some specific level is tolerated. Scientists call this a **threshold** level of exposure, which is the point up to which no response exists based on scientific evidence.

As in any statistical study, estimating a dose-response relationship requires the researcher to make initial assumptions. These include which factors are being controlled when defining the relationship and what the underlying relationship

[2]Drawn from U.S. EPA, Risk Assessment Portal (August 19, 2010b).

FIGURE 6.1 Human Health Risk Assessment Process

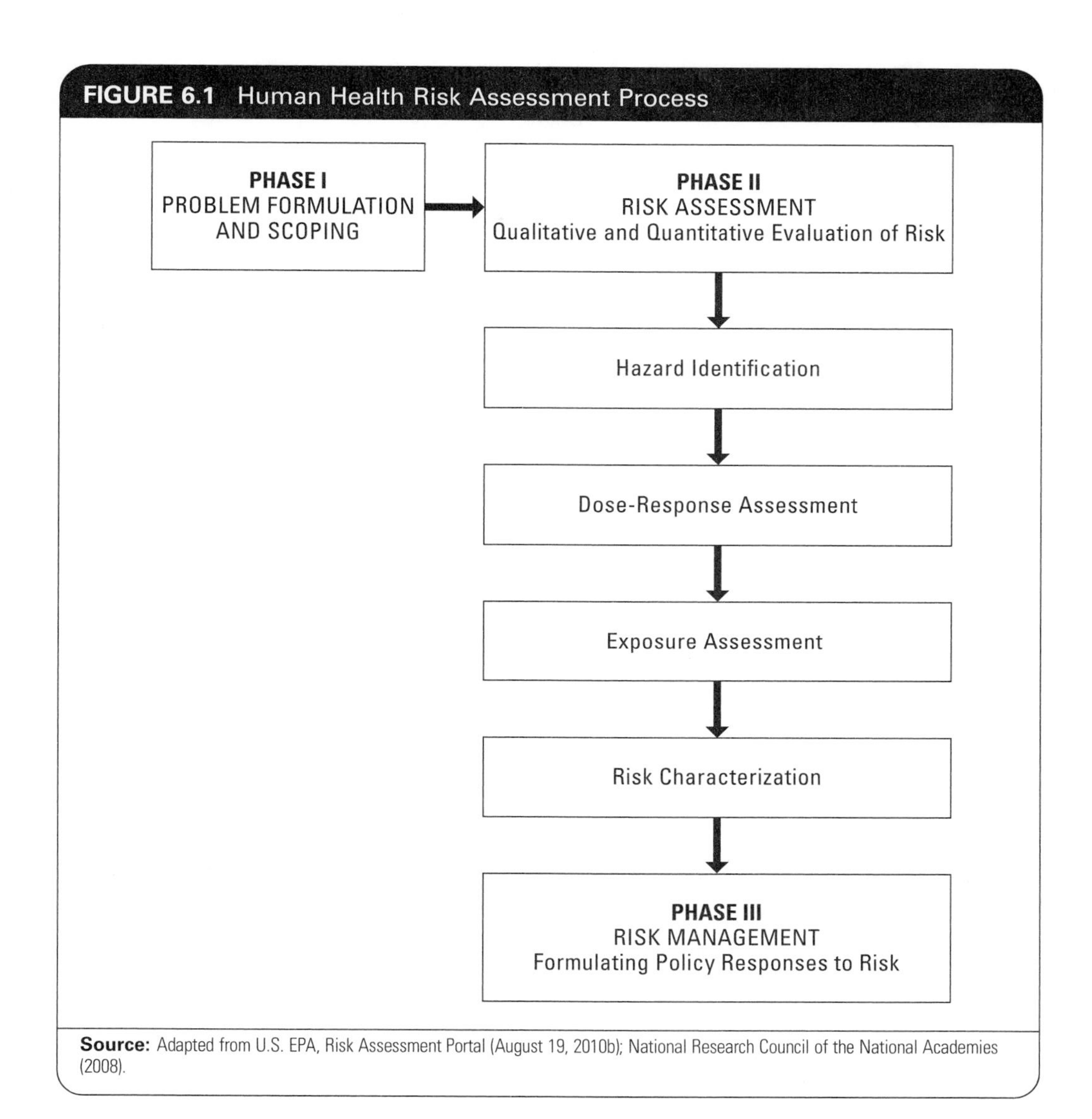

Source: Adapted from U.S. EPA, Risk Assessment Portal (August 19, 2010b); National Research Council of the National Academies (2008).

TABLE 6.1 Scientific Methods to Identify an Environmental Health Hazard

Scientific Method	Definition
Case Cluster	A study based on the observation of an abnormal pattern of health effects within some population group.
Animal Bioassay	A study based on the comparative results of laboratory experiments on living organisms both before and after exposure to a given hazard.
Epidemiology	The study of the causes and distribution of disease in human populations based on characteristics such as age, gender, occupation, and economic status.

Source: Lave (1982).

looks like.[3] Figure 6.2(a), (b), and (c) illustrates three hypothetical dose-response functional forms. Common to each graph is a positive relationship between dose level and response. What differs among them is the *rate* at which the response increases with the dosage.

In Figure 6.2(a), the relationship is linear, meaning that the rate of increase between dose and response is constant. Notice also that this curve begins at the origin. This means that a response is expected no matter how small the dose, or, equivalently, that no threshold level has been observed. Contrast this with the relationship shown in Figure 6.2(b). In this case, there is a constant rate of change between dose levels and response, but only beyond dose level D_t. Up to and including that point, there is no response at all, meaning that the hazard has an identified threshold at D_t. Finally, Figure 6.2(c) shows a cubic (or S-shaped) relationship drawn from the origin. Here, the response level increases at an increasing rate up to dose level D_0 and then increases at a decreasing rate beyond that point.

The scientific results derived from hazard identification and dose-response assessment provide general information about an environmental risk based on some known population defined by the laboratory or test conditions. This information can then be used as a basis for assessing the risk to a potentially exposed population in a specific context. The EPA has established a database of identified environmental hazards and their estimated dose-response relationships for use by both researchers and the general public. Referred to as the **Integrated Risk Information System (IRIS)**, this database is

FIGURE 6.2 Dose-Response Relationships

Panel (a) shows a linear dose-response function, meaning that the rate of increase between dose and response is constant. Because the function starts at the origin, no threshold level is observed. The function in panel (b) shows a constant rate of change between dose levels and response beyond dose level D_t. Up to and including that point, there is no response, meaning that there is a threshold at D_t. Panel (c) depicts a cubic relationship drawn from the origin, showing that the response level increases at an increasing rate up to D_0 and then increases at a decreasing rate thereafter.

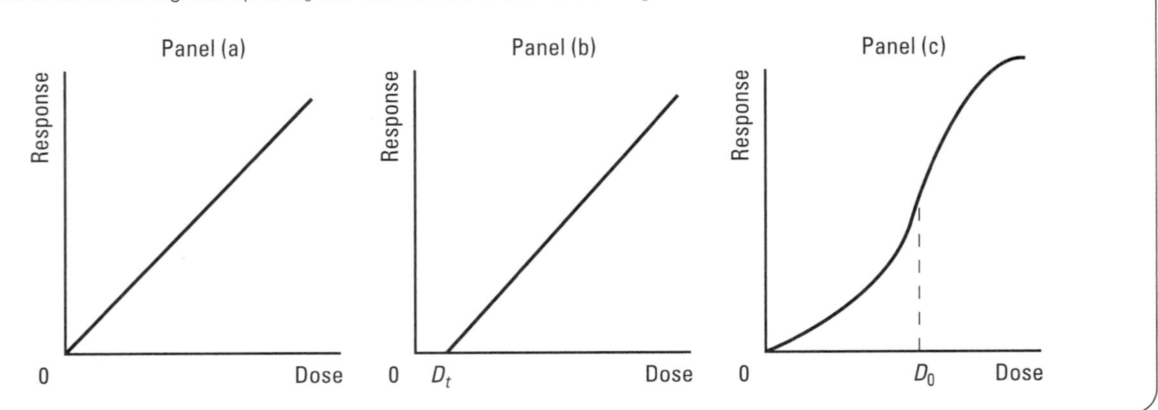

[3] See Lave (1982), pp. 43–47, for further information.

designed to improve risk assessment by lending consistency to what is a difficult and time-intensive process. IRIS can be accessed via the Internet at **www.epa.gov/iris/index.html**.

Exposure Assessment

exposure assessment Measures the magnitude, frequency, and duration of exposure, pathways and routes, and any sensitivities.

The process through which a generalized dose-response relationship is applied to specific conditions for an affected population is called **exposure assessment**. This evaluation measures or estimates the following:

- the magnitude, frequency, and duration of exposure
- the pathways from the source to the affected population and the routes into the body
- any sensitivities within the population group

To illustrate, consider conducting an exposure assessment for lead. The potentially affected target group is the general population. The sources of lead are many, including painted surfaces, factory emissions, lead-acid batteries, and lead water pipes. Because lead is ubiquitous, it can reach the population through many pathways. Moreover, it can enter the body through inhalation of contaminated air, ingestion from drinking contaminated water or eating tainted food, or direct ingestion of lead particles. Existing research also shows that sensitivity to lead exposure is greater for developing fetuses, infants, and young children.

Risk Characterization

risk characterization Description of expected risk, how the risk was assessed, and areas in need of policy decisions.

The final step of risk assessment is called risk characterization, which is the objective of the entire process. **Risk characterization** communicates the nature of the expected risk, information on how risk was assessed along with underlying assumptions and uncertainties, and areas where policy decisions are needed. Included are individual characterizations of the three prior steps as well as an integrated analysis of the overall process. More than just an assimilation of the previous steps, the description includes both a **quantitative** and a **qualitative** risk evaluation.

The **quantitative component** identifies the magnitude of the risk and provides a way to compare one risk with another. Drawn from the dose-response assessment, risk can be measured as a probability that an event will occur or as a reference dose.

Some probabilities are based on what are called actuarial risks, or those determined from factual data. Actuarial risk measures are found by calculating the number of victims of a given hazard relative to the total number exposed. For example, the actuarial risk of death per year from driving an automobile has been estimated at 24 in 100,000, or 0.024 percent. The likelihood of premature death from being struck by lightning is 0.00005 percent, or 5 people for every 10 million. Other probability measures, such as the carcinogenic risks associated with chemical exposure, are based not on actual data but on inferences derived from animal bioassays or epidemiological studies. For example, the risk of getting cancer in a year from drinking chlorinated water has been estimated to be 0.0008 percent, or 8 in 1 million persons exposed.[4]

---

[4]The estimates of risk presented in this section are given in Scheuplein (January–March 1993).

Environmental risks, such as noncarcinogenic health risks, are quantified as the exposure level to a hazard that can be tolerated over a lifetime without harm. This is communicated as a **reference dose (RfD)** expressed as:

RfD = milligrams of a pollutant per body weight (in kilograms) per day

For example, an RfD for pollutant X of 0.005 milligrams/kilogram/day means that exposure to 0.005 milligrams of pollutant X per kilogram of body weight each day over a lifetime should cause no harm.

The **qualitative component** of risk characterization gives context to the numerical risk value. It gives a description of the hazard, an assessment of exposure that notes any susceptible population groups, an identification of the data used, the scientific and statistical methods employed, and all underlying assumptions. Any scientific uncertainties, data gaps, or measurement errors that distinguish the findings are pointed out as well. All this information characterizes the reliability of the results and facilitates further research.

Ecological Risk Assessment[5]

An ecological risk assessment evaluates the probability of changes to the natural environment that are linked to such stressors as pollution exposure or climate change. Examples include crop damage and soil contamination. Application 6.2 offers some insight into the ecological risks of climate change. Under most public policies, these ecological changes are viewed as secondary to human health effects. However, ecological effects and human health effects are not independent. Over time, human health is adversely affected if ecological health deteriorates. For example, damage to soil and crops may negatively affect economic productivity, human fitness, and the quality of life. In fact, a report on environmental risk conducted by the EPA's Science Advisory Board (SAB) recommended that the EPA devote more attention to reducing ecological risks and to recognizing the link between ecological health and human health.[6]

In response, the EPA developed guidelines aimed specifically at ecological risk assessment.[7] These can be accessed online at **www.epa.gov/raf/publications/guidelines-ecological-risk-assessment.htm**. Today, these guidelines help with the implementation of three phases comprising ecological risk assessment. Following an initial planning process, these phases are identified as: **Problem Formulation**, **Analysis**, and **Risk Characterization**, as shown in Figure 6.3 (page 136). Let's examine each of these briefly.

Problem Formulation

problem formulation Identifies the ecological entity that is potentially at risk.

In **problem formulation**, the ecological entity, or assessment endpoint, that is potentially at risk is identified. This entity might be an individual plant or animal species, an ecological community, an entire ecosystem, or a habitat. Once done, the assessment must determine which characteristic of the entity may be at risk and the entity's overall importance or relevance, such as its role in the ecosystem. Ultimately, a model is developed to show possible links between the ecological entity and the environmental stressor with an accompanying description that includes the assessment endpoint, i.e., the predicted effects on the ecological entity.

---

[5]Drawn from U.S. EPA, Risk Assessment Portal (January 5, 2011).

[6]U.S. EPA, Science Advisory Board (1990).

[7]U.S. EPA, Office of the Science Advisor, Risk Assessment Forum (1998).

APPLICATION 6.2 Climate Change: Assessing Ecological Risk by International Region

In recent years, increased attention has been directed toward climate change and the associated implications for the ecology. In part, the intense and often contentious discussion is linked to the uncertainty associated with climate change. Scientists do agree on some issues. Specifically, they know that greenhouse gases (GHGs), such as carbon dioxide (CO_2), occur naturally in the atmosphere and are responsible for warming the earth to support life on the planet; however, human activity, specifically the burning of fossil fuels, elevates GHGs beyond their natural levels, causing a rise in the earth's temperature. In fact, scientists have documented a warming trend of between 1 and 1.7 degrees Fahrenheit over the 100-year period up to 2005.

What *is* uncertain about GHG accumulation is the extent of warming likely to occur, how quickly it will arise, and how the predicted elevated temperature will affect other climate-sensitive systems. Adding further to the difficulty of prediction is that the associated risks are expected to vary by region throughout the world. To illustrate, we offer a brief overview of the ecological risks of climate change for selected international regions.

According to the Intergovernmental Panel on Climate Change (IPCC), Africa is considered to be among those continents most vulnerable to climate change because of its low capacity to adapt and the numerous stresses it faces. Among the effects Africa may confront are increased water stress, decreased agricultural yields, and rising sea levels. Reduced fish stocks are also likely due to elevated water temperatures, which could be worsened by overfishing.

IPCC scientists further predict that glacier retreat in the Himalayas of Asia will increase the risk of flooding and rock avalanches. Coastal areas, too, are expected to experience flooding from both the ocean and nearby rivers. Predicted changes in crop yields vary by location, with rising yields anticipated in the east and southeast, and declines expected in the south and central areas.

Latin America is expected to face reduced crop yields and diminished livestock productivity as its agricultural areas degrade to desert-like regions. Predicted rising sea levels should increase the risk of flooding in low-lying areas, and the higher water temperatures will likely threaten coral reefs and cause dislocation of fish stocks.

Expected ecological risks to North America will vary within the region. For example, agricultural yields may improve in some areas, but crops grown at the higher range of tolerated temperature levels may be challenged. Urban locations already stressed by heat waves likely will witness greater intensity, frequency, and numbers of these events. Warmer mountainous areas in the west are expected to face greater frequency of winter flooding and decreased snowpack, and forested regions may be at greater risk from fires, pests, and disease.

Significant biodiversity loss by 2020 is among the risks predicted for Australia and New Zealand, including the Great Barrier Reef in northeast Australia. The region also may experience declines in agricultural and forestry production because of increased risks of drought and fire. Interestingly, however, western and southern parts of New Zealand could incur initial benefits because of a longer growing season and higher rainfall levels.

In Europe, expected climate-linked changes include a significant heat wave, receding glaciers, and longer growing seasons. Ecological risks are expected to vary considerably among the southern, central and eastern, and northern regions. Most, however, are expected to witness negative effects, including more frequent coastal floods, greater erosion linked to storms and rising sea levels, and in mountainous areas, retreating glaciers, diminished snow cover, and species loss.

For full information, consult the research of the Intergovernmental Panel on Climate Change (IPCC), which is accessible online at **www.ipcc-wg2.gov/publications/AR4/index.html**.

Sources: U.S. EPA, Office of Air and Radiation, Office of Atmospheric Programs, Climate Change Division (December 17, 2008; December 20, 2007); Intergovernmental Panel on Climate Change (IPCC) (2007a).

Analysis

analysis phase Identifies information to predict ecological responses to environmental hazards under various exposure conditions.

The **analysis phase** identifies all the information necessary to predict ecological responses to environmental hazards under various exposure conditions. Included is a determination of which ecological entities are exposed, the degree of the exposure, and the likelihood that harmful effects will arise because of that exposure. To support the analysis, calculations are made to quantify the risk, such as a hazard quotient, which forms the ratio of a contaminant concentration to some screening benchmark, and a bioaccumulation rate, which measures how pollutants are taken up by an ecological species.

FIGURE 6.3 Ecological Risk Assessment Process

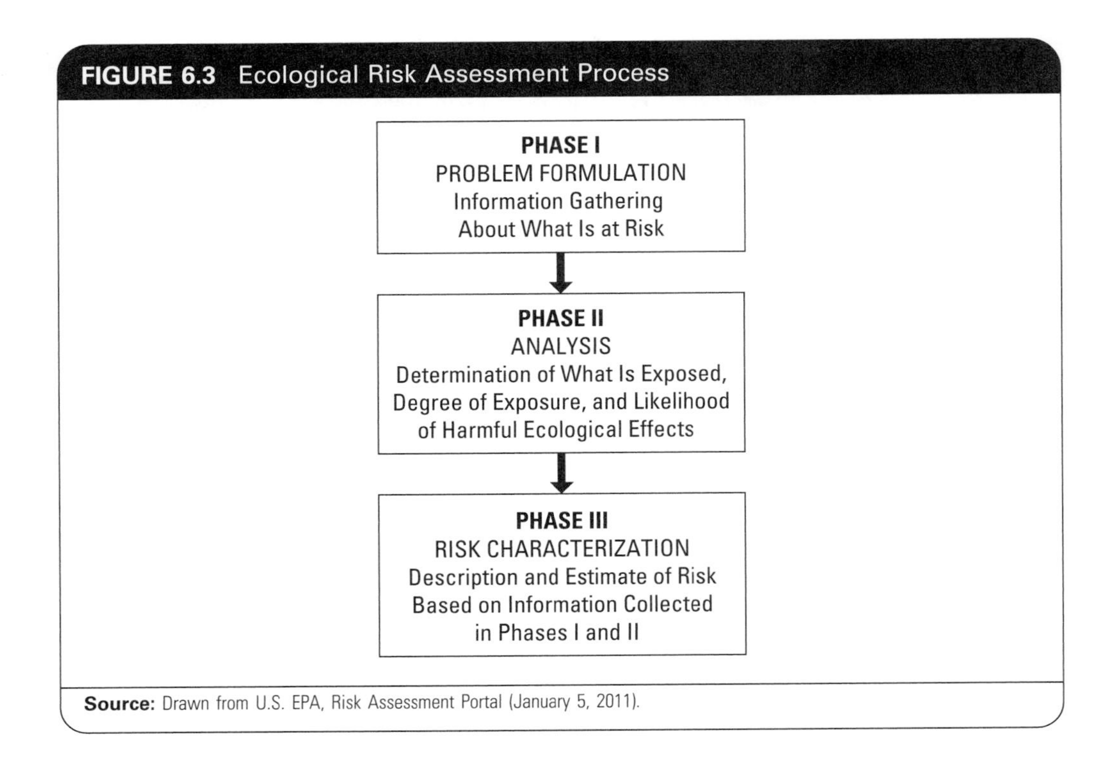

Source: Drawn from U.S. EPA, Risk Assessment Portal (January 5, 2011).

Risk Characterization

Just as in health risk assessments, this final phase of ecological risk assessments is a description of risk based on the information gathered in the previous phases. This description provides the degree of confidence in the risk estimates, any evidence that supports the findings, and an interpretation of the predicted ecological effects. Included are such risk descriptors as the severity of the damage, the time over which the damage occurs, and the extent of the damage in terms of numbers and types of species.

RISK MANAGEMENT

risk management The decision-making process of evaluating and choosing from alternative responses to environmental risk.

While the objective of risk assessment is to *identify* risk, it is the goal of **risk management** to *respond* to it. More to the point, risk management is concerned with evaluating and selecting from alternative policy instruments to reduce society's risk of a given hazard. To evaluate various policy options, the decision maker must consider not only the information given by the risk characterization but also such factors as technological feasibility, implementation costs, and other economic implications.

Implementation of the risk management process involves a series of decisions aimed at two major tasks:

1. To determine what level of risk is "acceptable" to society
2. To evaluate and select the "best" policy instrument to achieve that risk level

None of the underlying decisions are unidimensional, and realistically, none can be made with complete objectivity. However, there are strategic approaches used to guide the decision making, some of which are mandated by law. We begin with a brief discussion of these two major tasks of risk management, which lays the groundwork for an

analysis of risk management strategies. Further detail and links to various resources are available at **www.epa.gov/riskmanagement/**.

Tasks of Risk Management

Determining "Acceptable" Risk

The universal objective of all risk management strategies is to reduce risk. However, for each policy proposal, the public official must decide how much of a reduction is appropriate. Although risk is a function of both hazard and exposure, only one of these–exposure–can be controlled. So when the risk manager decides on the risk reduction to be achieved, the exposure level is implicitly determined. This in turn dictates how stringent public policy must be. In setting the level of **"acceptable" risk**, the policymaker confronts a difficult but necessary question: Should the policy eliminate the risk by reducing exposure to zero, or should some compromise be struck at some positive risk level, and if so, where?

"acceptable" risk
Amount of risk determined to be tolerable for society.

If the acceptable risk level is set at zero, the policy must reduce society's exposure to zero (assuming no threshold level). Although this action eliminates the associated health and ecological damages, such a stringent policy is likely to cause economic problems. For example, if the hazard is a chemical used in manufacturing, reducing exposure to zero means that its use must be prohibited. Such a ban may force plant closings and create job losses.

Conversely, if some positive level of risk is deemed "acceptable," then the decision maker is setting policy that allows exposure and therefore some amount of damage. To help guide this difficult decision, policymakers sometimes use the concept of *de minimis* risk. ***De minimis* risk** refers to a negligible level of risk such that reducing it further would not justify the costs of doing so. This concept is sometimes equated to the risk of a natural hazard, such as a lightning strike or an earthquake. Once the baseline is established, the decision maker might use **comparative risk analysis** to evaluate how the positive risk level compares to other risks currently faced and accepted by society. This sort of analysis has been used to communicate the relatively unfamiliar risks associated with exposure to radon, a naturally occurring gas that can be harmful when trapped indoors. See Application 6.3 for a discussion of how the risks of radon exposure are expressed in terms of the risks from more familiar activities such as cigarette smoking.

***de minimis* risk**
A negligible level of risk such that reducing it further would not justify the associated costs.

Evaluating and Selecting a Policy Instrument

Once the degree of risk and hence the stringency of policy have been determined, the second task of risk management is to decide what type of policy to use. Here, the decision maker has to first evaluate alternative policies that can achieve the "acceptable" risk level and then select the "best" option from among them. Recall from Chapters 4 and 5 that the options likely will include some that are command-and-control and others that are market-based. In making the decision, the risk manager considers the magnitude of risks, benefits, and/or costs associated with each available control instrument. There are risk management strategies that define exactly how this evaluation is to be done.

Risk Management Strategies

Executing the two tasks of risk management—determining the "acceptable" risk level and choosing the appropriate policy instrument—requires a systematic evaluation of available options. From an economic perspective, the most important considerations are:

- the level of risk established
- the benefits that accrue to society from adopting the policy
- the associated costs of implementing the policy

APPLICATION 6.3 Using Comparative Risks to Communicate the Dangers of Radon

Over time, public awareness of radon has improved, but many people are still unaware of the potential risks. The result? Government is faced with a public policy problem. Although the risk of exposure to radon is involuntary, government cannot directly intervene, because radon pollutes a part of the environment that is beyond its jurisdiction—the *indoor* air.

Exposure to radon is an involuntary risk because it is a natural hazard. It is a radioactive gas caused by the decaying of uranium found in soil deposits and rocks. Outdoors, as radon is released, it dissipates quickly and poses no adverse health effects. It becomes a potential problem only if it enters homes and buildings through cracks in foundations, gaps around pipes, and the water supply. If radon is trapped indoors, it can accumulate to dangerous levels. It is odorless and colorless, so people are unaware of its presence and are exposed to it quite involuntarily. Research studies suggest that long-term exposure to radon causes lung cancer. According to the EPA's 2003 risk assessment of radon in homes, chronic exposure may be responsible for about 21,000 lung cancer deaths per year, ranking it second only to smoking as a cause of lung cancer.

The EPA has joined forces with the surgeon general and various national health associations to educate society about the little-known but potentially dangerous problem of radon. People have been encouraged to measure the radon level in their homes using readily available test kits. If the indoor radon level is 4 picocuries per liter (pCi/L) or higher, the EPA recommends that some action be taken to correct the problem. Brochures have been prepared by government and other associations to disseminate facts about radon to homeowners. Yet all of this well-intentioned information is useful only if it is communicated in a way that is meaningful to the general public. To accomplish this, the government uses **comparative risks**.

Risk estimates for lifetime exposure to various levels of radon are given as the number of people who get lung cancer in a given population. To communicate these estimates more effectively, comparable lifetime risks of more familiar activities are provided, as shown in the accompanying table.

Radon Level (pCi/L)	Risk of Getting Lung Cancer (If You've Never Smoked) (Number of persons out of 1,000)	Comparative Risk*
20	36	35 times the risk of drowning
10	18	20 times the risk of dying in a home fire
8	15	4 times the risk of dying in a fall
4	7	Risk of dying in a car crash
2	4	Risk of dying from poison
1.3	2	(Average indoor radon level)

*These risks are calculated using the Centers for Disease Control and Prevention's 1999–2001 National Center for Injury Prevention and Control Reports.

Source: U.S. EPA, Indoor Environments Division (January 2009). (An online version of this document can be accessed at **www.epa.gov/radon/pubs/citguide.html**.)

Sources: U.S. EPA, Indoor Environments Division (January 2009); U.S. EPA, Office of Air and Radiation (June 2003).

Several risk management strategies have been developed over time, each of which outlines *how* these factors are to be evaluated. The most prevalent of these are **comparative risk analysis**, **risk-benefit analysis**, and **benefit-cost analysis**.

comparative risk analysis
An evaluation of relative risk.

Comparative Risk Analysis

Just as **comparative risk analysis** helps the risk manager select an "acceptable" risk level, it also can be used to help officials identify which risks are most in need of an official response. The EPA's Science Advisory Board (SAB) has prepared a ranking of environmental problems by degree of risk, shown in Table 6.2. The ranking was based on the best available data and scientific information and considered such factors as severity of effects and number of people exposed. The SAB specifically advises that EPA programs should be guided by the principle of **relative risk reduction**, meaning the agency should order its policy decisions to reduce the most severe environmental risks first.

One difficulty associated with setting risk-based priorities is that the rankings set by government are often different from how society *perceives* environmental risks. Look at Table 6.3 to see a ranking of environmental problems based on general public perception in the United States. Figure 6.4 (page 141) offers similar and more current information from the European Union (EU). Notice that public perception of environmental risk is generally not aligned with the SAB's ranking given in Table 6.2. Such a dichotomy of views between public perception and scientific evidence presents a dilemma to officials attempting to gain support for policy proposals. Therefore, it is important that government communicate scientific findings to the public to improve its understanding of environmental risk.

Comparative risk analysis also can be used to select from among alternative control instruments. Used in this context, the approach is often called **risk-risk analysis**. This strategy involves a comparison of the estimated risk probabilities or risk ranking scores from two or more policy options. For example, the decision maker might compare the relative risks of two different mandates for hazardous waste treatment, such as land disposal versus incineration, and propose whichever approach is more effective in reducing risk. Implicitly, the objective of a risk-risk strategy is risk minimization, with no explicit consideration allowed for associated costs.[8]

TABLE 6.2 Scientific Rankings of Environmental Problems

Relative Risk Ranking	Environmental Problem
High risk to human health	Ambient air pollutants Worker exposure to chemicals in industry and agriculture Indoor pollution Contamination of drinking water
High risk to natural ecology and human welfare	Habitat alteration and destruction Species extinction and loss of biological diversity Stratospheric ozone depletion Global climate change
Medium risk to natural ecology and human welfare	Herbicides/pesticides Contamination of surface waters Acid deposition Airborne toxics
Low risk to natural ecology and human welfare	Oil spills Groundwater contamination Radionuclides Thermal pollution Acid runoff to surface waters

Source: U.S. EPA, Office of Communications, Education, and Public Affairs (April 1992), p. 9; U.S. EPA, Science Advisory Board (September 1990).

[8]U.S. Congress, Office of Technology Assessment (1983), pp. 226–27.

TABLE 6.3 Public Perceptions of Environmental Problems in the United States

Environmental Problem	Percentage Responding That Problem Is Very or Extremely Serious
Hazardous waste	89
Oil spills	84
Air pollution	80
Solid waste disposal	79
Atmospheric damage	79
Nuclear waste	78
Contaminated water	77
Forest destruction	76
Ocean pollution	75
Endangered species	67
Threats to wildlife	65
Pesticide use	60
World population	57
Poor energy use	56
Global warming	56
Reliance on coal/oil	53
Wetland development	50
Radon gas	35
Indoor air pollution	27
Electromagnetic fields	19

Source: U.S. EPA, Office of Communications, Education, and Public Affairs (April 1992), p. 9.

Risk-Benefit Analysis

risk-benefit analysis
An assessment of risks of a hazard along with the benefits to society of not regulating that hazard.

An alternative risk management strategy, called **risk-benefit analysis**, simultaneously considers the benefits to society of *not* regulating an environmental hazard along with the level of associated risk. Here, the objective is to simultaneously maximize the expected benefits *and* minimize the risk. Although it may seem perverse, it is true that the source of some environmental hazards offers benefits to society. Think about gasoline, which during combustion gives off emissions that present a health risk. Nonetheless, gasoline *does* benefit society by fueling motor vehicles. Hence, if the risk manager were to assess only the risks in this case, the solution might be to ban gasoline. Yet, if the risk-reduction strategy were balanced with a consideration of benefits, the risk manager would have to consider how a reduction in gasoline usage would diminish society's well-being.

The use of a risk-benefit strategy is commonly mandated in environmental law. An example is found in the Toxic Substances Control Act (TSCA). This law requires the EPA to simultaneously consider the health and environmental effects, the degree of exposure to the substance, *and* the benefits the substance provides to society in use. A similar mandate is called for in the Federal Insecticide, Fungicide, and Rodenticide Act (FIFRA).[9]

[9] *"Some Statutory Mandates on Risk"* (January–March 1993). We will have more to say about TSCA and FIFRA in Chapter 19.

FIGURE 6.4 Public Perceptions of Environmental Problems From the European Union (EU)

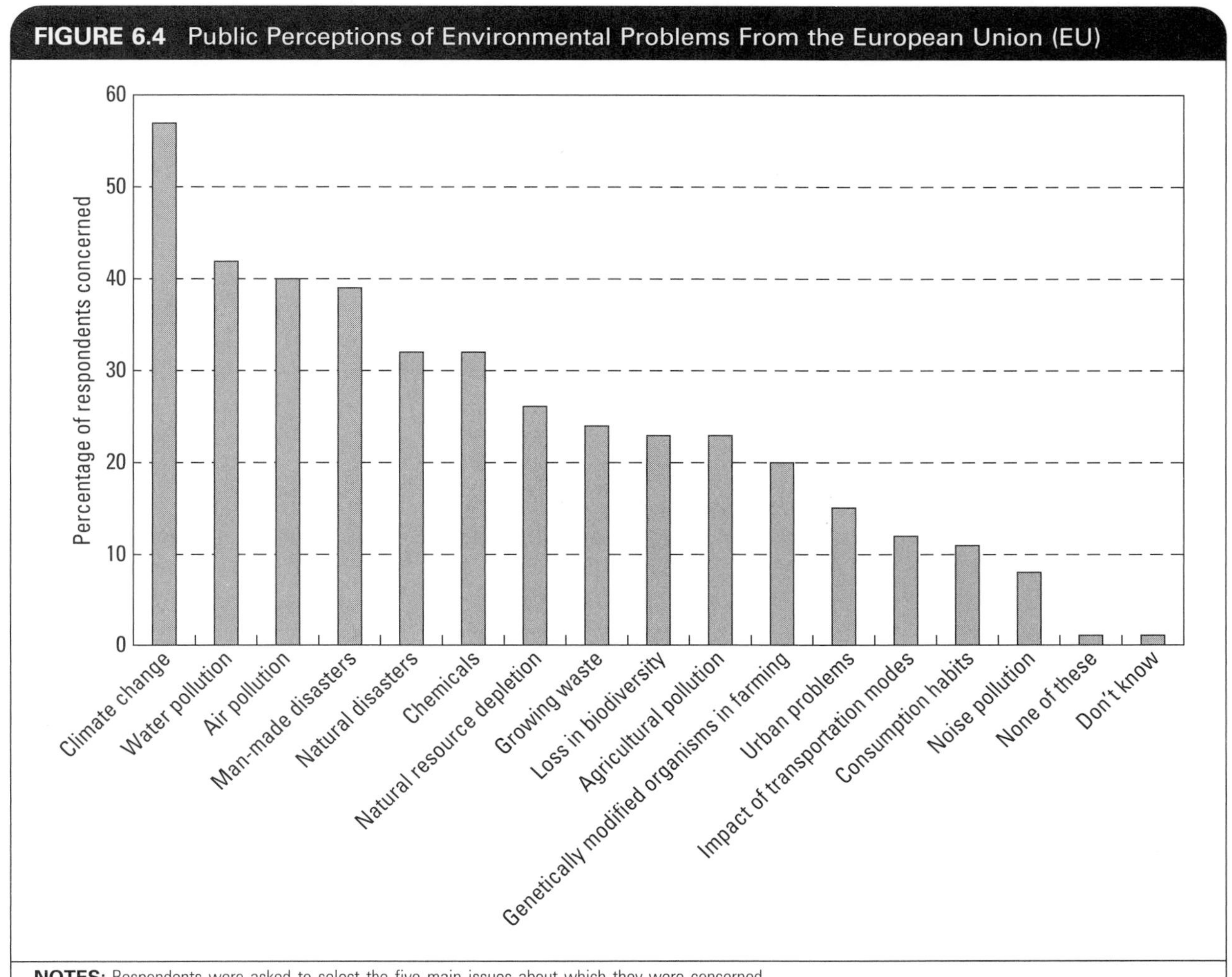

NOTES: Respondents were asked to select the five main issues about which they were concerned.
Man-made disasters include such events as major oil spills and industrial accidents.
Natural disasters include events such as earthquakes and floods.
Chemicals refers to the effect of chemical use in everyday products.
Urban problems refer to traffic congestion, lack of green space, etc.

Source: European Commission (2008). *Attitudes of European Citizens Towards the Environment*, Special Eurobarometer 295, **http://ec.europa.eu**

Benefit-Cost Analysis

President Reagan's Executive Order 12291 is largely responsible for the more intensified use of **benefit-cost analysis** in formulating environmental policy. President Clinton's Executive Order 12866 continued to support this risk management strategy, and President George W. Bush's Executive Orders 13258 and 13422 amended but adhered in substance to Executive Order 12866. In 2011, President Obama revoked the two Bush mandates and issued Executive Order 13563, which reaffirmed and extended President Clinton's Executive Order 12866. To access Executive Orders 12866 and 13563 online, visit **www.archives.gov/federal-register/executive-orders/pdf/12866.pdf** and **www.gpo.gov/fdsys/pkg/FR-2011-01-21/pdf/2011-1385.pdf**, respectively.

benefit-cost analysis
A strategy that compares the *MSB* of a risk reduction policy to the associated *MSC*.

Benefit-cost analysis can identify an "acceptable" risk level based on the criterion of **allocative efficiency**. For incremental risk reductions, the decision maker would compare the monetized value of social benefits with the associated costs to find the efficient risk level, where the marginal social benefit (*MSB*) and the marginal social cost (*MSC*) of risk reduction are equal. Equivalently, this corresponds to the risk level that maximizes the difference between total social benefits (*TSB*) and total social costs (*TSC*).

Often, environmental law establishes the risk reduction to be achieved, which means that the level of associated benefits has been predetermined. In such cases, the risk manager still can use benefit-cost analysis but with a different objective in mind. Here, the goal would be to select a policy instrument that meets the legislated risk objective at least cost. If the selection is made properly, the initiative will achieve the economic goal of **cost-effectiveness** rather than allocative efficiency.

Finally, benefit-cost analysis can be used in the policy appraisal stage much in the same way it is used to find an "acceptable" risk level. At this stage, allocative efficiency is used to evaluate the effectiveness of an ongoing initiative. In practice, estimates of the *MSB* and *MSC* at the risk level achieved by policy are compared to see if they are equivalent. If not, the risk manager knows that the policy needs to be amended to correct the resource misallocation.

CONCLUSIONS

While most policy decisions are difficult and even controversial, those made in the context of environmental issues are particularly so. Indeed, much of what environmental policymakers struggle with is how to deal objectively and fairly with the risks posed by environmental hazards. Such is the purpose of risk analysis and its two components, risk assessment and risk management.

Scientists provide the data and analysis needed for risk assessment. As a result of their research, public officials gain valuable information about the nature of environmental hazards and the risks of exposure. Armed with a characterization of the risks involved, government can make better and more informed policy decisions. Through risk management strategies, "acceptable" risk levels can be determined and alternative policy instruments can be evaluated objectively using well-defined criteria. Independent of how the "acceptable" risk level is determined, estimates of the social benefits and costs of a policy are useful in evaluating the effectiveness of policy initiatives after they have been adopted into law. These data can guide proposals for legislation that characterizes the dynamic process of environmental policy development.

The use of benefit-cost analysis as a decision rule is becoming more prevalent in public policy decision making. Monetizing environmental costs and benefits is an attempt to provide an impartial guideline to the risk manager, but the task can be difficult to execute in practice. It also has been the source of some controversy. Think about the dilemma of assigning a dollar value to saving a life or restoring the quality of coastal waters. Consequently, economists are continuing to research better methods and more comprehensive data to improve their estimates of benefits and costs. In Chapters 7, 8, and 9, we will explore the fundamental steps of benefit-cost analysis and assess the contribution of this risk management strategy to environmental decision making.

Summary

- Voluntary risks are deliberately assumed at an individual level. Involuntary risks arise from exposure to hazards beyond the control of individuals.
- Environmental risk measures the likelihood that damage will occur due to exposure to an environmental hazard. The hazard is the source of the damage, and exposure refers to the pathways between this source and the affected population or resource.
- Risk assessment is the qualitative and quantitative evaluation of the health or ecological risk posed by an environmental hazard. The process differs somewhat for human health risks and ecological risks, in response to a reexamination and definition of the latter in 1998.
- The accepted model of risk assessment in the United States is used to conduct evaluations of human health risks. This model is part of a three-phase framework for risk-based decision making: Phase I, Problem Formulation and Scoping; Phase II, Risk Assessment; and Phase III, Risk Management.
- Within the risk assessment phase are four steps: hazard identification, dose-response assessment, exposure assessment, and risk characterization.
- In hazard identification, a determination is made as to whether a causal relationship exists between a pollutant and an increased incidence of adverse effects on human health and whether these effects are likely to arise. Several methods are used, including case clusters, animal bioassays, and epidemiology.
- Using data collected in the hazard identification stage, a dose-response assessment attempts to develop a complete profile of the effects of an environmental pollutant. An important objective is to identify whether there is a threshold level of exposure, the point up to which no response is observed.
- Exposure assessment is the process through which a generalized dose-response relationship is applied to specific conditions for an affected population.
- Risk characterization is a quantitative and qualitative description of expected risk.
- The quantitative component of risk characterization provides a means to gauge the relative magnitude of the risk. Risk might be measured as a probability or as a reference dose (RfD).
- The qualitative component of risk characterization gives context to the numerical measure of risk and includes a description of the hazard, an assessment of exposure, an identification of the data, the scientific and statistical methods used, and any uncertainties in the findings.
- An ecological risk assessment evaluates the probability of changes to the natural environment that are linked to such stressors as pollution exposure or climate change. The three phases comprising ecological risk assessment are: Problem Formulation, Analysis, and Risk Characterization.
- In problem formulation, the ecological entity, or assessment endpoint, that is potentially at risk is identified.
- The analysis phase identifies all the information necessary to predict ecological responses to environmental hazards under various exposure conditions.
- Just as in health risk assessments, risk characterization is the final phase of ecological risk assessments, aimed at providing a description of risk based on the information gathered in the previous phases.
- Risk management is concerned with evaluating and selecting from alternative policy instruments to reduce society's risk of a given hazard. Several risk management strategies are used in practice, including comparative risk analysis, risk-benefit analysis, and benefit-cost analysis.
- Comparative risk analysis, known in some contexts as risk-risk analysis, involves an evaluation of relative risk. This can be used to help officials identify which risks are most in need of an official response. It also can be used to select among alternative control instruments.
- Risk-benefit analysis is aimed at simultaneously maximizing expected benefits *and* minimizing risk.
- Benefit-cost analysis evaluates alternative risk levels by comparing the value of the expected gains with the associated costs. If the "acceptable" risk level maximizes the difference between total social benefits (*TSB*) and total social costs (*TSC*), the outcome will be allocatively efficient. If the law establishes the risk level to be achieved, a cost-effective solution can be realized by selecting the least-cost policy instrument that achieves the risk objective.

Review Questions

1. Refer to Tables 6.2 and 6.3, and investigate how the public's perception of the risk of indoor pollution compares with the ranking given by the EPA's Science Advisory Board. How is this comparison relevant to the use of comparative risk analysis in communicating the hazards of radon, as discussed in Application 6.3? (Or access the radon report directly at **www.epa.gov/radon/pubs/citguide.html**.)
2. Comment on the following statement: "Without exposure, there is no risk."
3. Other than those mentioned in this chapter, give several real-world examples of how government has provided public information to enhance the identification of a voluntary risk.
4. **a.** Interpret the shape of the dose-response function in Figure 6.2(c).
   b. Does this dose-response relationship suggest the presence of a threshold level? If so, where is it? If not, why not?
5. **a.** Verbally describe what an RfD of 0.002 for some pollutant Z means.
   b. Graphically sketch a dose-response function for pollutant Z, assuming that the dose-response relationship increases at a decreasing rate throughout. Label the RfD on your diagram.
6. Suppose a dose-response function has been estimated to be $R = -0.2 + 1.6D$, where D is emissions of sulfur dioxide in parts per million (ppm), and R is the response measured as percent of agricultural crop decline. Determine if there is a threshold, and if so, what is the threshold? If there is none, explain how you know.

7. What is a hazard quotient, and how is it used in risk assessment?
8. Suppose you are using risk-benefit analysis to evaluate a policy aimed at limiting the use of a pesticide applied to grain crops. Describe the risks and benefits that would have to be estimated to conduct this analysis properly.

Additional Readings

Carnegie Commission on Science, Technology, and Government. *Risk and the Environment: Improving Regulatory Decision Making*. New York: Carnegie Commission, June 1993.

Davies, Terry. "Congress Discovers Risk Analysis." *Resources* (Winter 1995), pp. 5–8.

Dionne, Georges, and Sandrine Spaeter. "Environmental Risk and Extended Liability: The Case of Green Technologies." *Journal of Public Economics 87*(5–6) (May 2003), pp. 1025–60.

Eyles, J., and N. Consitt. "What's at Risk? Environmental Influences on Human Health." *Environment 46*(8) (October 2004), pp. 24–39.

Farrell, Alexander E., and Jill Jäger. *Assessments of Regional and Global Environmental Risk*. Washington, DC: Resources for the Future, 2005.

Gerrard, Simon, R. Kerry Turner, and Ian J. Bateman, eds. *Environmental Risk Planning and Management*. Northampton, MA: Elgar, 2001.

Hill, Marquita K. *Understanding Environmental Pollution*. NY: Cambridge University Press, 2010.

Koop, Gary, and Lise Tole. "Measuring the Health Effects of Air Pollution: To What Extent Can We Really Say That People Are Dying from Bad Air?" *Journal of Environmental Economics and Management 47*(1) (January 2004), pp. 30–54.

Linkov, Igor, and Abou Bakr Ramadan, eds. *Comparative Risk Assessment and Environmental Decision Making*. Dordrecht, The Netherlands: Kluwer, 2004.

Lorenzoni, Irene, Nick F. Pidgeon, and Robert E. O'Connor. "Dangerous Climate Change: The Role for Risk Research." *Risk Analysis 25*(6) (December 2005), pp. 1387–98.

Robson, Mark G., and William A. Toscano. Risk *Assessment for Environmental Health*. San Francisco: Jossey-Bass, a Wiley Imprint, 2007.

Scarlett, Lynn, Arthur Fraas, Richard Morgenstern, and Timothy Murphy. "Managing Environmental,

Health, and Safety Risks: A Closer Look at How Three Federal Agencies Respond." *Resources, No. 177* (Winter/Spring 2011), pp. 38–43.

Sinclair-Desgagné, Bernard. *Corporate Strategies for Managing Environmental Risk*. Burlington, VT: Ashgate, 2004.

Smith, V. Kerry. *The Economics of Environmental Risk: Information, Perception and Valuation (New Horizons in Environmental Economics)*. Northampton, MA: Elgar, 2005.

Sorensen, M. T., W. R. Gala, and J. A. Margolin. "Approaches to Ecological Risk Characterization and Management: Selecting the Right Tools for the Job." *Human and Ecological Risk Assessment 10(2)* (April 2004), pp. 245–69.

U.S. Environmental Protection Agency (EPA). *Setting the Record Straight: Secondhand Smoke Is a Preventable Health Risk*. Washington, DC: U.S. EPA, June 1994.

U.S. Environmental Protection Agency (EPA), Office of Health and Environmental Assessment, Office of Research and Development. *Respiratory Health Effects of Passive Smoking: Lung Cancer and Other Disorders*. Washington, DC: U.S. EPA, December 1992.

Viscusi, W. Kip. "The Value of Risks to Life and Health." *Journal of Economic Literature 31(4)* (December 1993), pp. 1912–46.

Viscusi, W. Kip, and Ted Gayer. *Classics in Risk Management*. Northampton, MA: Elgar, 2004.

Vose, David. Risk Analysis: *A Quantitative Guide*, 3rd Edition, West Sussex, England: John Wiley and Sons, 2008.

Wiener, Jonathan B., Michael D. Rogers, James K. Hammitt, and Peter H. Sand, eds. *The Reality of Precaution: Comparing Risk Regulation in the United States and Europe*. Washington, DC: RFF Press, June 2010.

To access additional course materials, visit www.cengagebrain.com. At the home page, search for the ISBN of this title (shown on the back cover). This will take you to the product page where these resources can be found.

CHAPTER 7

Assessing Benefits for Environmental Decision Making

"Write injuries in dust, benefits in marble."

—Benjamin Franklin (1706–1790)

Risk assessment and risk management are central to the decision-making process that guides environmental policy. Once the degree of environmental risk has been identified, public officials begin the critical task of formulating policy. Ultimately, the objective is to minimize risk, which is a benefit to society. Meeting this objective is not an unconstrained decision, however. There are opportunity costs. Public officials must consider that resources used to reduce smog, for example, are no longer available to clean the Great Lakes, to save the California condor, or to improve public education. How do policymakers come to grips with such tough decisions? There is no simple answer, and, in fact, most would argue that the public sector wrestles with this problem on an ongoing basis. Yet there are strategies that can be effective in environmental policy development, among them, **benefit-cost analysis**.

Benefit-cost analysis underlies much of economic theory. For example, the balancing of revenues and costs at the margin to maximize profit is an application of benefit-cost analysis. In the broader context of policy decisions, benefit-cost analysis is used to evaluate the associated gains and losses to society. In every case, the principle is the same: An efficient solution results if benefits and costs are balanced at the margin. However, applying this theory in practice is not so clear-cut.

In order to use benefit-cost analysis to guide environmental decisions, policymakers must quantify the associated social benefits and costs. Yet many intangibles that are difficult to quantify are involved, for example, longevity of human life, improved aesthetics, and the preservation of ecosystems. The process is difficult but critically important.

In this chapter, we explore the motivation for valuing **environmental benefits** and the methods used to measure them. We start by presenting the conceptual issues of how and why society values natural resources and environmental quality. This theoretical framework supports our subsequent investigation of various benefit estimation methods.

ENVIRONMENTAL BENEFITS: CONCEPTUAL ISSUES

As a starting point, we need to establish the appropriate level of analysis for assessing policy-induced environmental benefits. From previous chapters, we know that these health and ecological gains can be assessed as damage reductions. The key is to recognize that the relevant measure is the change in damage reductions brought about by policy. In practice, these changes are called incremental benefits.

Identifying Incremental Benefits

incremental benefits
The reduction in health, ecological, and property damages associated with an environmental policy initiative.

To assess the social benefits attributable to environmental policy, decision makers must identify **incremental benefits**, i.e., how health, ecological, and property damages change as a consequence of that policy. This focus on the *change* in damages instead of on their absolute level is not new. Economic theory is concerned with effects that occur *at the margin*. These, too, are changes, although they are infinitesimal, measured at a point. When the relevant change is over a discrete range, it is referred to as *incremental* rather than *marginal*. Because policy evaluation is concerned with identifying damage reductions over some discrete time period, the appropriate measure of benefits is incremental.

To identify incremental benefits, analysts must compare the actual or expected benefits to society after some policy is implemented to a baseline measure of current conditions. Environmental benefits are commonly separated into categories, such as improvements in human health, aesthetics, the economy, recreation, property, and the ecology. A timely example of incremental environmental benefits are those gains anticipated by the European Union (EU) as it works toward reducing greenhouse gases (GHGs), an international policy initiative discussed in Application 7.1.

Within the broad category of incremental benefits are two types of damage-reducing effects: primary environmental benefits and secondary environmental benefits. A **primary environmental benefit** is a damage-reducing effect that is the direct consequence of implementing policy. Examples include a lower incidence of respiratory ailments, more stable ecosystems, and economic benefits, such as a more prosperous fishing industry resulting from clean water regulations.

In contrast, a **secondary environmental benefit** is an indirect gain to society associated with policy implementation. One source might be the stimulative effect of a primary benefit, such as higher worker productivity resulting from the primary benefit of improved health. Another would be a demand-induced change, such as the increased demand for labor to implement new policy. By way of example, Table 7.1 (page 149) gives annual benefit estimates for several key titles of the Clean Air Act. As noted in the table, the welfare category includes secondary benefits, such as improved worker productivity and enhanced agricultural productivity.

Conceptually Valuing Environmental Benefits

What is the value to society of cleaner air or cleaner water? What value does society place on cleaning up a hazardous waste site? As discussed in previous chapters, both questions could be answered directly if the commodity in each case were a private good traded in the open market. In that case, the demand price would convey the marginal benefit of each additional unit of the good. The problem is that environmental quality

APPLICATION 7.1 Incremental Benefits of the European Union's Climate Change Initiatives

Like other regions in the world, Europe stands to enjoy environmental benefits from implementing successful policies aimed at reducing air emissions. In the context of climate change, the targeted emissions are carbon dioxide (CO_2) and other greenhouse gases (GHGs), and the policy directives are based mainly on the commitment of the European Union (EU) to the Kyoto Protocol. This protocol is the international agreement on climate change, which entered into force in 2005. To help achieve its commitment under the protocol, the EU established various policy initiatives, including its own Emissions Trading Scheme (ETS) for GHGs. In so doing, the EU became the only major power to price carbon.

Not surprisingly, much has been made of the EU's ETS and Europe's progress toward achieving cleaner air, particularly in light of the controversy surrounding the Kyoto Protocol and the uncertainties associated with climate change predictions. Among the issues of interest are the incremental environmental benefits expected to accrue to the EU as a result of its Kyoto-linked policies. From a qualitative perspective, these expected benefits should include ecological, geophysical, meteorological, and human health gains.

In the Arctic, for example, a significant incremental benefit associated with controlling global warming would be a diminished rate of ice loss in Greenland and in the Arctic Sea. In the Mediterranean region, crop yield declines linked to rising temperatures should lessen, fewer forest fires should occur, and the risk of biodiversity loss and desertification should be diminished. In the mountainous regions, the risks of species extinction, soil erosion, and rock falls should decrease, and in Northwestern areas, the risk of coastal flooding and increased winter precipitation should decline. Although monetizing these benefits would be difficult, most would agree that these gains would be significant and highly valued by society; yet they are only part of the story.

Interestingly, because the reduction of GHGs necessitates reduced use of fossil fuels, emissions of other air pollutants like nitrogen oxides and particulate matter would also decline. This outcome would generate ancillary environmental benefits. For example, estimates suggest that abatement costs for sulfur dioxide and nitrogen oxides would be lower by 12 percent and 20 percent, respectively. In the aggregate, the cost savings to abate these and other pollutants to the level achieved as part of the climate change initiative are estimated to be about €12 billion (or about \$14.2 billion). Moreover, premature deaths from exposure to ozone and particulate matter are expected to fall by more than 20,000 by 2030, which adds markedly to expected benefits. In addition, avoided annual health expenditures should be in the range of €16 billion to €48 billion (or about \$18.9 billion to \$56.8 billion).

Although monetary estimates have been assigned to some of these incremental benefits, doing so for all identified gains would be difficult at best. Not only is the extent of the identified gains uncertain, but the time period over which they might be realized is also tentative. Because of these realities, policymakers face a daunting task in designing and implementing climate change policy that can be justified by benefit-cost analysis.

For further information about environmental benefits of climate change policy in the EU, visit **www.eea.europa.eu/data-and-maps/figures/benefits-of-climate-policy**.

Sources: European Environment Agency (March 2008; 2006); Intergovernmental Panel on Climate Change (IPCC) (2007a); "International: How to Cool the World; Climate Change" (2007).

is a public, nonmarketed good. The absence of prices and the dilemma of nonrevelation of preferences cloud a determination of how society values a cleaner environment. In theory, if we could infer society's demand for environmental quality, we could then measure the incremental benefits of environmental policy.

To illustrate this assertion, we return to our model of air quality based on SO_2 pollution abatement, which we introduced in Chapter 3. Recall that since demand for this public good represents society's decisions, it is both the marginal private benefit (MPB) and the marginal social benefit (MSB) of air quality. Thus, we can refer to the demand for SO_2 abatement as $MSB = 25 - 0.3A$, where MSB is measured in millions of dollars, and A is the percentage of SO_2 abated.

TABLE 7.1 Annual Benefits Associated With Titles I Through V of the 1990 Clean Air Act

Source Category	2000 Benefits (in millions)	2002 Benefits (in millions)
Mortality	$113,556	$126,894
Chronic Illness	6,742	7,476
Hospitalization	502	589
Minor Illness	1,864	2,084
Welfare	5,245	5,696
TOTAL	$127,909	$142,739

NOTES: All values are in 2010 dollars. Mortality benefits include only deaths of those who are at least 30 years of age. Welfare benefits include enhanced worker productivity, enhanced agricultural productivity, and increased recreational activity. Ecological benefits are not quantified and therefore are excluded.

Source: U.S. EPA, Office of the Chief Financial Officer (September 30, 2003), Table 3, p. 180.

The graph of this relationship is shown in Figure 7.1. At each level of abatement, *MSB* is measured as the vertical distance from the horizontal axis up to the demand curve. The total social benefits (*TSB*) for any abatement level are measured as the aggregation of these vertical distances, or the area under the *MSB* curve up to that point. In Figure 7.1, the *TSB* for some hypothetical abatement level A_1 are shown as the shaded

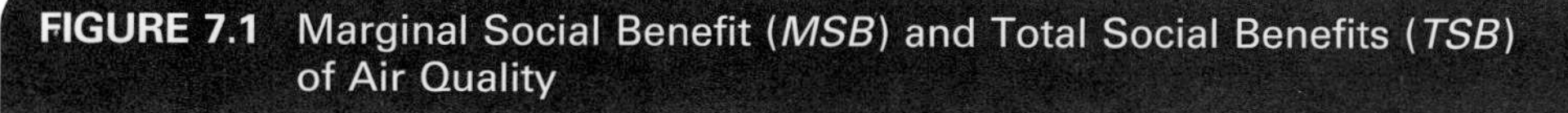

FIGURE 7.1 Marginal Social Benefit (*MSB*) and Total Social Benefits (*TSB*) of Air Quality

The market demand curve for SO_2 abatement represents the marginal social benefit (*MSB*) of air quality. At the hypothetical level of abatement, A_1, the corresponding *MSB* is the vertical distance from the horizontal axis at that point up to the curve. The total social benefits (*TSB*) associated with A_1 are shown as the shaded area under the *MSB* up to that point.

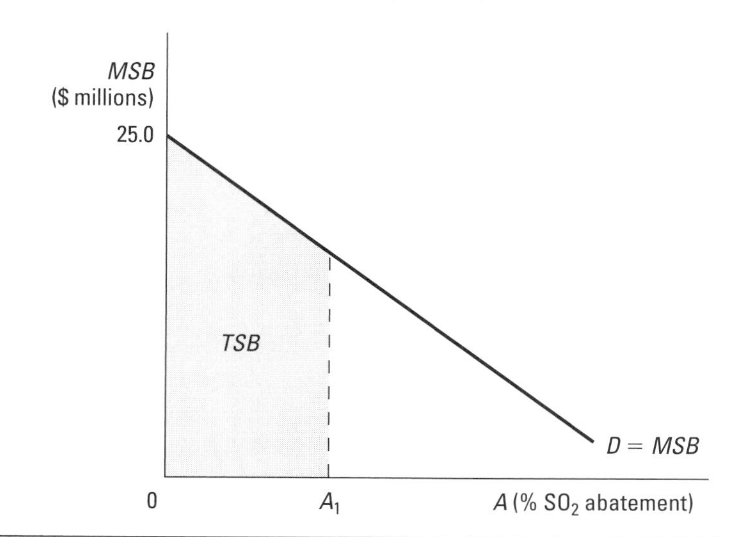

area under the demand curve up to A_1. Using this model, we can measure the incremental benefits from a policy-induced increase in SO_2 abatement in three steps:

1. Find the baseline level of *TSB* *before* the policy is undertaken.
2. Find the new level of *TSB* *after* the policy is implemented.
3. Subtract the baseline *TSB* from the post-policy *TSB*.

Suppose that the current level of SO_2 abatement is 20 percent, and the objective is to find the incremental benefits of a policy that increases abatement to 25 percent. How should this be measured? First, find the baseline *TSB*. Referring to Figure 7.2, notice that the *MSB* at the 20 percent abatement level is \$19 million. The *TSB* at this level are shown as the area under the *MSB* curve up to that point, or \$440 million. This value represents society's **willingness to pay (WTP)** for the benefits achieved when 20 percent of SO_2 emissions are abated. Next, complete the analogous calculations for the proposed abatement increase to 25 percent. At this post-policy abatement level, *MSB* is \$17.5 million, and *TSB* equal \$531.25 million. Finally, the incremental benefits are found as the difference between the two *TSB* values, or \$91.25 million. This is shown as the shaded area in Figure 7.2.

An alternative model of the same result is shown in Figure 7.3, where the *TSB* are graphed directly with SO_2 abatement. In this model, the *TSB* associated with a given abatement level are measured simply as the vertical distance up to the curve. Identify on the graph the pre- and post-policy levels of the *TSB* corresponding to the 20 percent and 25 percent abatement levels. Notice that the incremental benefits of \$91.25 million are measured as the vertical distance between these two *TSB* levels.

FIGURE 7.2 Modeling Incremental Social Benefits for Air Quality Using the *MSB* Function

The *MSB* at the baseline abatement level of 20 percent is \$19 million, and the *TSB* are shown as the area under the *MSB* curve up to that point, or \$440 million. If a policy increased SO_2 abatement to 25 percent, *MSB* would be \$17.5 million, and *TSB* would rise to \$531.25 million. Thus, incremental benefits are the difference between the two *TSB* values, or \$91.25 million, shown as the shaded area under the *MSB* curve between the two abatement levels.

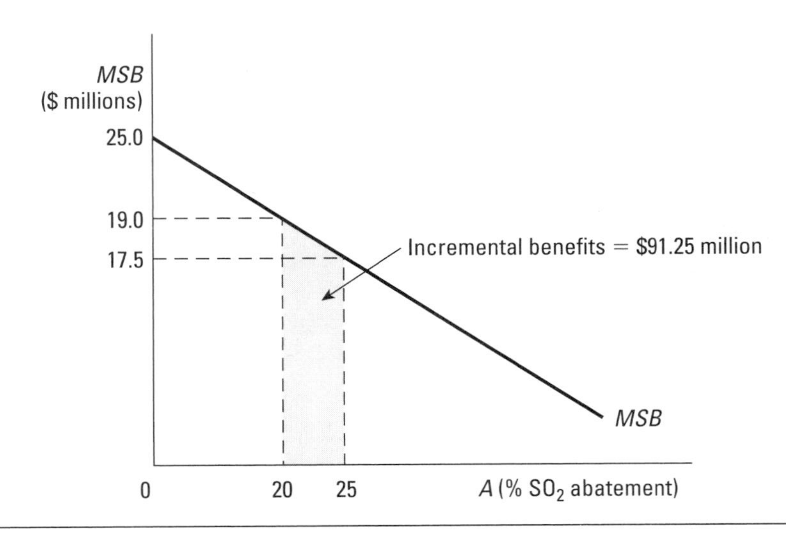

FIGURE 7.3 Modeling Incremental Social Benefits for Air Quality Using the *TSB* Function

At each abatement level, the *TSB* are shown as the vertical distance from the horizontal axis up to the curve. The *TSB* for the 20 percent baseline abatement level and for the 25 percent post-policy abatement level are $440 million and $531.25 million, respectively. Incremental benefits are shown as the vertical distance between the two points on the *TSB* curve, or $91.25 million.

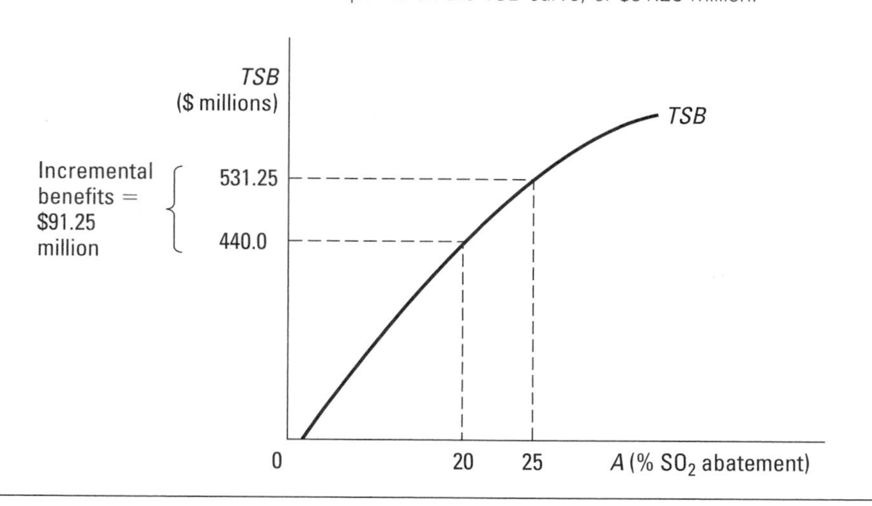

Both models assume that society's valuation of environmental quality, or equivalently, its demand for pollution abatement, can be identified. However, because there is no explicit market for this commodity, the valuation cannot be obtained from observing market-determined prices. Instead, inferences must be made about *how* society derives value or utility from various levels of environmental quality.

User Versus Existence Value[1]

Discovering how society values a good is difficult in the absence of market prices. Even if the dilemma of nonrevelation of preferences could be overcome, the value of a good like clean air or water is difficult to quantify because of the many intangibles involved. Although economists recognize that some intangibles are immeasurable, they still need some sense of how the benefits of environmental quality are perceived by society. Fortunately, theories that can help have been advanced.

user value
Benefit derived from physical use of or access to an environmental good.

existence value
Benefit received from the continuance of an environmental good.

From a purely conceptual vantage point, it is generally recognized that society derives utility from environmental quality through two sources of value: **user value** and **existence value**. **User value** refers to the utility or benefit received from usage of or access to an environmental good. In contrast, **existence value** is the utility or benefit received from an environmental good simply through its continuance as a good or service. Collectively, these components measure society's total valuation of an environmental good, expressed as follows:

Total value of environmental quality = User value + Existence value

[1]Much of the following discussion is drawn from Mitchell and Carson (1989), Chapter 3.

User Value

To better understand user value, consider the benefits of using a lake. If an individual swims in a lake, that person derives utility from physically using the natural resource. Likewise, a commercial fishing fleet derives user benefit from catching fish from the lake. In both cases, benefits are derived from directly consuming services provided by the resource. These activities, both recreational and commercial, generate benefits that yield **direct user value**. This valuation helps to determine what the individual or the fishing fleet would be willing to pay to maintain or improve the lake's quality.

direct user value Benefit derived from directly consuming services provided by an environmental good.

Continuing with the same example, another individual might receive utility from simply looking at a view of the lake. Here, the utility is derived from the lake's aesthetic qualities. This type of activity involves using the lake in a less immediate way than swimming in it or fishing from it. Hence, the lake is said to yield **indirect user value**. Both direct and indirect user value are elements of society's total valuation of environmental quality.[2]

indirect user value Benefit derived from indirect consumption of an environmental good.

Existence Value

Society also receives benefits from environmental goods beyond the utility associated with direct or indirect use. Think about how people value such natural resources as the great rain forests, the Grand Canyon, or the bald eagle. Consumption does not explain how or why society values these resources. Yet we know that society is willing to pay to preserve them. In such circumstances, benefits accrue to society from simply knowing that these resources exist and are being preserved. This component of total valuation is referred to as **existence value**.[3]

While seemingly abstract, existence value is an important motivation for privately funded conservation efforts and for many environmental policy initiatives. A case in point is the Endangered Species Act (accessible online at **www.nmfs.noaa.gov/pr/laws/esa/text.htm**), which provides for the protection and preservation of animals, birds, fish, and plants threatened with extinction. As discussed in Application 7.2, this act is an example of the U.S. government's recognition of existence value. Other tangible evidence is the willingness of society to support the work of environmental groups, such as the National Wildlife Federation and the Sierra Club—groups whose agendas focus on preserving resources that many of their benefactors never expect to use or even see firsthand.

vicarious consumption Utility associated with knowing that others derive benefits from an environmental good.

In a text about valuing public goods, Mitchell and Carson (1989) classify the motives for existence value as **vicarious consumption** and **stewardship**, among others. **Vicarious consumption** refers to the notion that individuals value a public good for the benefit it provides to others, whether or not these others are known personally. This suggests that the utility derived is *interdependent*, which means that an individual can receive benefit from the knowledge that others are enjoying the public good. **Stewardship** arises both from a sense of obligation to preserve the environment for

stewardship Sense of obligation to preserve the environment for future generations.

[2]Although these examples involve present period consumption, economists have also begun to examine how society expects to benefit from consumption in some future period. This valuation concept, which adds uncertainty to benefit assessment, is referred to as **option value**.

[3]One of the earlier discussions of existence value is presented by Krutilla (1967).

APPLICATION 7.2 The Endangered Species Act

In 1973, the Endangered Species Act was passed by Congress to "… provide a means whereby the ecosystems upon which endangered species and threatened species depend may be conserved (Sec. 2b)." Officially, the act protects the biodiversity of the earth: the diversity of genes, species, ecosystems, and the interaction among them. An important outcome of this act was the creation of a formal list of biological organisms in danger of extinction, *regardless of their direct or indirect use to humans*. Originally numbering 109 in 1973, the list of endangered and threatened species in the United States has grown to include over 1,380 species as of 2012.

When the Endangered Species Act was originally proposed and when it went through subsequent reauthorizations, arguments were made to justify its passage on economic grounds. One argument was that species were to be protected because they may serve a direct benefit to mankind that scientific study has not yet discovered. Some justification for this argument can be found in the discovery and use of TAXOL® (paclitaxel). This compound, found in the bark of Pacific yew trees, is used for the treatment of certain cancers. Prior to production of a semisynthetic form of TAXOL®, approximately 60 pounds of yew bark were needed to produce enough TAXOL® to treat one cancer patient. Consequently, the Pacific yew, which is a slow-growing species, was being harvested by timber companies at a rapid rate. Notice that this discourse focuses on the **user value** of environmental resources.

Just as strong in their position were those who asserted that environmental resources offer benefits that span a much broader range than their direct or indirect value in human consumption. Here, the premise is that species should be protected based on a presumed right of survival—evidence of the role of **existence value** in benefit assessment.

To learn more about the Endangered Species Act and the U.S. Endangered Species Protection Program, visit **www.epa.gov/espp/**.

Sources: U.S. Department of the Interior (DOI), Fish and Wildlife Service (FWS) (January 6, 2012); Bristol-Myers Squibb Company (October 19, 2004); Council on Environmental Quality (January 1993), pp. 17–28; Adler and Hager (April/May 1992); Endangered Species Act of 1973, 16 U.S.C. §§ 1531–44.

future generations *and* from the recognition of the intrinsic value of natural resources. In sum, we express the total valuation of environmental quality as:

Total value	=	**User value** (direct and indirect user value)	+	**Existence value** (vicarious consumption and stewardship value)

Recognizing how society values an environmental resource is important for identifying the social benefits of a policy proposal. It also helps economists decide which estimation method might be most effective in quantifying those benefits. Yet the question remains: How do economists assign dollar values to nonmarketed environmental goods, such as clean water and the spotted owl? Remember that the objective is to *monetize*, or find the **WTP** for, changes in user and existence value arising from a policy-driven increase in environmental quality.

APPROACHES TO MEASURING ENVIRONMENTAL BENEFITS: AN OVERVIEW[4]

Economists have made great strides in developing methods to estimate the benefits of environmental quality improvements. For the most part, these methods are aimed at estimating primary benefits, implicitly assuming that secondary benefits are insignificant

[4]This section is drawn mainly from Mitchell and Carson (1989), particularly pp. 74–78, and from Cropper and Oates (June 1992).

and likely offset by secondary costs. A review of the extensive literature on the subject reveals that several methods are used in practice. Some are better than others at quantifying the more intangible benefits of improved environmental quality, including the somewhat elusive concept of existence value.

To organize our discussion of benefit measurement methods, we rely upon a general classification introduced by Smith and Krutilla (1982), which places the various measurement techniques into two broad categories: the **physical linkage approach** and the **behavioral linkage approach**. A summary of selected benefit valuation methods within each approach is presented in Table 7.2.

Physical Linkage Approach

physical linkage approach
Estimates benefits based on a technical relationship between an environmental resource and the user of that resource.

The **physical linkage approach** measures benefits based on a technical relationship between an environmental resource and the user of that resource. A common estimation procedure that uses this approach is the **damage function method**. This method uses a functional relationship to capture the link between a contaminant and any associated damages. Using this function, incremental benefits are measured as the reduction in damages arising from a policy-induced decrease in the contaminant. This damage reduction is then monetized to obtain a dollar value of the benefits brought about by the policy.

TABLE 7.2 Benefit Estimation Methods

THE PHYSICAL LINKAGE APPROACH

Damage Function Method

Uses a model of the relationship between levels of a contaminant and observed (or statistically inferred) environmental damage to estimate the damage reduction arising from a policy-induced decline in the contaminant.

THE BEHAVIORAL LINKAGE APPROACH

DIRECT METHODS

Political Referendum Method

Uses the *actual* market of a public good by monitoring voting results from political referenda on proposed changes in environmental quality.

Contingent Valuation Method (CVM)

Employs surveys to inquire about individuals' willingness to pay (WTP) for environmental improvements based on *hypothetical* market conditions.

INDIRECT METHODS

Averting Expenditure Method (AEM)

Assesses changes in an individual's spending on goods and services that are *substitutes* for personal environmental quality to assign value to changes in the overall environment.

Travel Cost Method (TCM)

Values a change in the quality of an environmental resource by assessing the effect of that change on the demand for a *complementary* good.

Hedonic Price Method (HPM)

Uses the theory that a good is valued for the attributes it possesses to estimate the implicit or hedonic price of an environmental attribute and identify its demand as a means to assign value to policy-driven improvements in quality.

Behavioral Linkage Approach

behavioral linkage approach
Estimates benefits using observations of behavior in actual markets or survey responses about hypothetical markets.

In general, the **behavioral linkage approach** to quantifying benefits is based upon observations of behavior in *actual* markets or survey responses about *hypothetical* markets for environmental goods. Techniques that assess responses immediately related to environmental changes are broadly termed **direct methods**. As shown in Table 7.2, two types of direct methods are the **political referendum method**, which relies on *actual* market information, and the **contingent valuation method (CVM)**, which uses *hypothetical* market data. **Indirect methods** are those that examine responses not about the environmental good itself but about some set of market conditions related to it. Three examples are the **averting expenditure method (AEM)**, the **travel cost method (TCM)**, and the **hedonic price method (HPM)**, which also are listed in Table 7.2.

The published literature on benefit valuation methods is extensive. Many research papers present specific empirical findings, and numerous texts focus on the methodology itself. We offer only an overview of the operational issues associated with the more common methods used by researchers.

ESTIMATION UNDER THE PHYSICAL LINKAGE APPROACH

Damage Function Method

damage function method
Models the relationship between a contaminant and its observed effects to estimate damage reductions arising from policy.

When using the **damage function method**, the researcher models the relationship between an environmental contaminant and observed damage.[5] An example of a damage function is shown in Figure 7.4. The contaminant level (C) is measured horizontally, and the total damages (TD) from exposure to that contaminant are measured vertically. Once the function is specified, the analyst uses the model to estimate the damage reduction from any policy-induced decline in the contaminant. At this point, the reduction is measured in nonmonetary units, such as the number of acidified lakes, acres of damaged forests, or number of premature deaths. Ultimately, this reduction must be assigned a monetary value, either by using market prices (if available) or by using some estimation technique.

To illustrate the procedure, look at Figure 7.4, and assume that a policy initiative is expected to reduce the contaminant from C_0 to C_1. Based on this model, the proposal would reduce damage (or, equivalently, increase benefits) by the vertical distance between TD_0 and TD_1. If, for example, the damage reduction was diminished injury to wheat crops, this vertical distance might be measured as thousands of bushels of wheat. This could be monetized by multiplying the number of bushels by the market price.

Assessing the Damage Function Method

Although the damage function method is useful, it has limitations. First, by construction, it estimates only one aspect of incremental benefits. In our example, the measured

[5] If the benefit assessment is the reduction in adverse effects to a *biological organism*, the analyst would use a **dose-response function**, which is a particular type of damage function. Recall from Chapter 6 that scientists use dose-response functions in the risk assessment process.

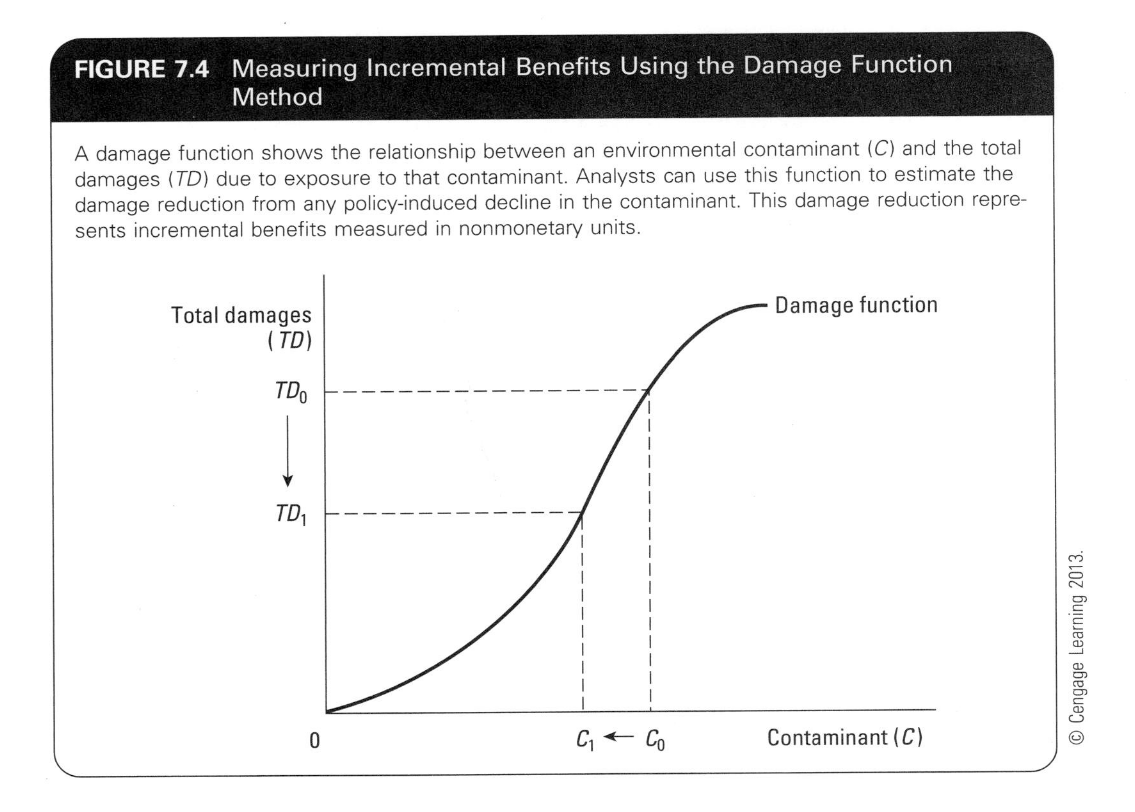

FIGURE 7.4 Measuring Incremental Benefits Using the Damage Function Method

A damage function shows the relationship between an environmental contaminant (*C*) and the total damages (*TD*) due to exposure to that contaminant. Analysts can use this function to estimate the damage reduction from any policy-induced decline in the contaminant. This damage reduction represents incremental benefits measured in nonmonetary units.

benefits are increased wheat crops. In most cases, a contaminant reduction would generate other gains, perhaps increases in other crops or improvements in human health. Hence, a full assessment of benefits using the damage function approach would require that the same estimating procedure be performed for every type of damage reduction. Second, the procedure is only a first-step approach in that it is not capable of simultaneously monetizing the benefits it quantifies.

Applications of the Damage Function Method

Recognizing the limitations of the damage function method, analysts typically use it for measuring a specific type of incremental benefit, as opposed to performing a comprehensive benefit assessment. Furthermore, the context is often one in which market-determined prices are available to monetize the gain.

To illustrate, consider a benefit assessment of the Clean Air Act provisions aimed at reducing ozone in the lower atmosphere or troposphere. According to scientific evidence, one type of associated benefit is an increase in crop yields.[6] Conceptually, such agricultural benefits could be modeled by measuring the change in consumer and producer surpluses associated with an increase in crop yields. Figure 7.5 models this effect

[6]The interested reader may wish to consult Kopp and Krupnick (December 1987) for an empirical investigation that estimates the agricultural benefits of ozone reduction.

FIGURE 7.5 Modeling Incremental Benefits of an Ozone-Reducing Policy

The increase in crop yields associated with a hypothetical ozone-reducing policy can be modeled as a shift in crop supply from S_0 to S_1. Before the policy is implemented, consumer surplus is area P_0ab, and producer surplus is area P_0be, for a total of area *eab*. After the policy is implemented and supply shifts to S_1, consumer surplus becomes area P_1ac, and producer surplus becomes area P_1ce, for a total of area *eac*. Thus, the incremental benefit is area *ebc* (i.e., *eac* – *eab*).

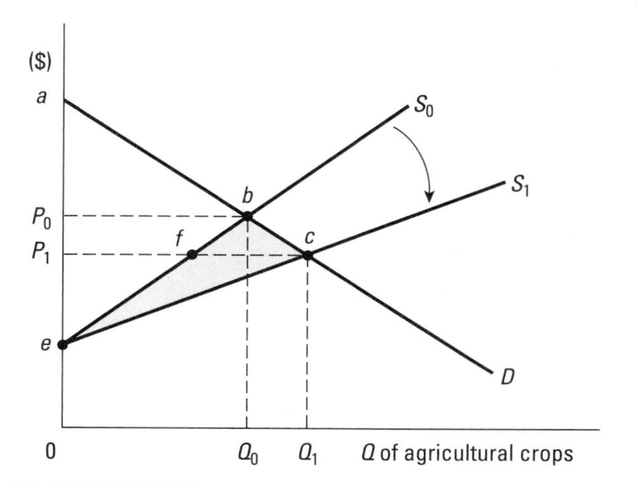

as an increase in crop supply from S_0 to S_1, which in turn causes a price decline from P_0 to P_1.

Consider first the size of the surpluses before any policy change, using supply curve S_0. Consumer surplus is the area below the demand curve and above the market price, or area P_0ab. Producer surplus is the area above the supply curve and below the market price, or area P_0be. Thus, the total surplus *before* a policy change is area *eab*. After the ozone-reducing policy is implemented and supply shifts to S_1, consumer surplus becomes area P_1ac, and producer surplus becomes area P_1ce. Thus, the total surplus *after* the policy change is area *eac*. The operative question is whether society is better off as a result. Since the total surplus increased from *eab* to *eac*, the answer is yes, and the incremental benefit can be quantified as area *ebc* (i.e., *eac* − *eab*).

Notice, however, that the *distribution* of benefits is not as easily determined. Consumers definitely gain, because consumer surplus rises from P_0ab to P_1ac, but the same assertion cannot be made about producers. Some of the original producer surplus, area P_0bfP_1, has been transferred to consumers, but there is also a gain to producers of area *efc*. Whether or not the gain of *efc* exceeds the transfer of P_0bfP_1 depends on the shapes of the supply and demand curves and the magnitude of the supply shift. Consequently, to measure the incremental benefits of policy *and* to determine the distribution of these benefits, sophisticated models must be used. Application 7.3 presents the results of such a benefit estimation conducted as part of a U.S. air quality assessment.

APPLICATION 7.3 Valuing Agricultural Benefits: The Case of Tropospheric Ozone Reductions

The National Acid Precipitation Assessment Program (NAPAP) was launched as an investigative plan to gather information for formulating policy on acidic deposition. According to NAPAP's 1990 report, scientific research has been unable to find a consistent adverse effect on crop yields caused by acidic deposition. However, there *is* well-documented evidence that elevated levels of tropospheric ozone can diminish crop yields. In fact, estimates of this damage range from 2 percent to 56 percent, depending on such factors as crop species, location, and exposure levels. Thus, a key part of NAPAP's research dealt with quantifying the **incremental benefits** that would result from an ozone-reducing policy initiative.

Modeling techniques to measure environmental benefits were used in the NAPAP study of ozone's effects on agricultural yields. The final report presents estimates of changes in consumer and producer surplus for alternative ozone policies (see accompanying table).

Under the first scenario, a 10 percent reduction in ozone concentrations is estimated to increase total surplus by approximately $1.30 billion. Of this amount, consumer surplus would increase by $1.38 billion, and producer surplus would decline by $80 million. A more restrictive ozone policy, such as a 25 percent reduction, would increase the benefits to both producers and consumers and elevate the change in total surplus to $3.05 billion.

These findings have important implications about the distribution of these incremental benefits across buyers and sellers. Notice that consumers and producers do not respond symmetrically to increases in ozone levels. In fact, if ozone levels were to increase by either 10 percent or 25 percent, society as a whole would lose but producers would actually benefit. According to the NAPAP report, this gain to producers arises because of farmers' ability to pass on the higher costs of depleted supplies to consumers in the form of higher prices.

Change in Surplus (billions of 2010 dollars)

Changes in Ozone (%)	Consumer Surplus	Producer Surplus	Total Surplus
−10	1.38	−0.08	1.30
−25	2.88	0.17	3.05
+10	−1.84	0.38	−1.46
+25	−4.68	0.80	−3.88

Source: National Acid Precipitation Assessment Program (NAPAP) (November 1991), Table 4.9-6, p. 401.

Source: National Acid Precipitation Assessment Program (NAPAP) (November 1991), pp. 55, 154–56, 398–401.

DIRECT ESTIMATION METHODS UNDER THE BEHAVIORAL LINKAGE APPROACH

The **direct methods** under the behavioral linkage approach estimate environmental benefits according to responses or observed behaviors directly tied to environmental quality. Although a number of direct methods are available to researchers, we consider only one representative example: the **contingent valuation method (CVM)**.

Contingent Valuation Method (CVM)

When market data are unavailable or unreliable, economists can use alternative estimation methods that rely on *hypothetical* market conditions. Such methods typically use surveys to inquire about individuals' willingness to pay (WTP) for some environmental initiative.

contingent valuation method (CVM)
Uses surveys to elicit responses about WTP for environmental quality based on hypothetical market conditions.

This survey approach to benefit estimation is known as the **contingent valuation method (CVM)** because the results are dependent, or *contingent*, on the devised hypothetical market. This market serves as the context for a series of survey questions. The critical assumption is that properly designed surveys can elicit responses comparable to those arising in actual situations. In some sense, the survey instrument helps to finesse the problem of nonrevelation of preferences that characterizes public goods.

Implementing this survey approach involves the following three tasks:

- Constructing a detailed model of the hypothetical market, including the characteristics of the good and any conditions that affect the market
- Designing a survey instrument to obtain an unbiased estimate of individuals' WTP
- Evaluating the truthfulness of survey respondents' answers

Assessing the CVM

The CVM is favored by researchers because it can be applied to a variety of environmental goods and because it can assess existence value as well as user value. However, because this approach makes inferences about actual markets from a hypothetical model, it is subject to the biases that typically plague a survey-dependent study, such as an individual's unwillingness to reveal a WTP because of the free-ridership problem.

Responding to the potential biases, economists continue to make improvements to the CVM. For instance, some researchers add more detail to their hypothetical models. Others improve the design of the survey instrument. Some surveys include maps to illustrate the location of the good or photographs of the commodity and the area affected by its provision.[7] Whatever the form, the objective is to make the hypothetical market situation as factual and as close to actual conditions as possible.

Applications of the CVM

Researchers have used the CVM in a variety of contexts to estimate environmental benefits. An important application is estimating the value of a statistical human life.[8] In a review of these findings, estimates were reported as falling within a range of $3.2 million to $8.0 million ($2010).[9]

Another common focus of CVM studies is to measure society's WTP for water quality improvements. Two examples are a study by Smith and Desvousges (1986), which examines a specific water body (the Monongahela River in Pennsylvania), and an analysis by Carson and Mitchell (1988), which estimates a generalized measure across all U.S. water sites. The study by Smith and Desvousges (1986) finds that the average household in five western Pennsylvania counties is willing to pay $60 ($2010) per year to improve the

---

[7] Brookshire and Crocker (1981).

[8] A statistical life saved is related to the concept of environmental risk introduced in Chapter 6. For example, if an environmental policy lowers the risk of death from 2 in 100,000 persons exposed to 1 in 100,000 exposed, the incremental benefit of that policy is one human life saved. To give this context, according to an EPA publication, a statistical life saved is valued at $8.0 million in 2010 dollars (U.S. EPA, Office of the Administrator, September 2000).

[9] Cropper and Oates (June 1992), p. 713. To facilitate comparison, we have converted all reported values to 2010 dollars, indicated by the notation ($2010).

Monongahela River from boatable to fishable quality. Carson and Mitchell's (1988) nationwide survey shows that the average respondent is willing to pay $175 ($2010) per year for water quality improvements. How can the difference be explained?

Since the valuation in the national survey is higher than the more localized finding of Smith and Desvousges (1986), the difference may be attributable to existence value. Why? Because the respondents in the more general survey are willing to pay for water quality improvements throughout the United States, even though they do not expect to use these water bodies themselves.[10]

Incremental benefits from air quality improvements also have been estimated using the CVM. In fact, some argue that the CVM is particularly useful for valuing visibility improvements at national parks, where existence value is likely to be significant. One study by Schulze and Brookshire (1983) seems to support this hypothesis. These researchers find that the user value of improving visibility at the Grand Canyon from 70 miles to 100 miles is $3.70 ($2010) per visitor per day. In contrast, they estimate the comparable existence value at $175 ($2010) per household per year to prevent diminished visibility at the Grand Canyon.

In a more recent study, Groothuis, Groothuis, and Whitehead (2007) use the CVM to estimate an individual's WTP for another visibility-related environmental improvement. The objective of this research is to monetize the improved scenic benefits associated with removal of highway billboards. Using a sample of Watauga County, North Carolina, residents and correcting for any hypothetical bias, these researchers estimate that each household would be willing to pay a one-time charge of between $28 ($2010) and $54 ($2010) to remove billboards to improve the scenic views of the nearby Appalachian Mountains.

Because the CVM is capable of capturing existence value, it has been used to value ecological benefits, such as preserving an endangered species. For example, one study estimates that individuals would be willing to pay $48 ($2010) per year to save the whooping crane.[11]

INDIRECT ESTIMATION METHODS UNDER THE BEHAVIORAL LINKAGE APPROACH

For some environmental proposals, direct estimation procedures such as the CVM might not be viable. In these cases, economists use **indirect methods**, which make inferences about markets or conditions that are linked to the environmental good under investigation. Three such methods dominate the literature: the **averting expenditure method (AEM)**, the **travel cost method (TCM)**, and the **hedonic price method (HPM)**.

averting expenditure method (AEM) Estimates benefits as the change in spending on goods that are *substitutes* for a cleaner environment.

Averting Expenditure Method (AEM): An Approach Using Substitutes

To indirectly estimate the willingness to pay (WTP) for such nonmarketed commodities as cleaner air or water, the **averting expenditure method (AEM)** uses changes in spending on goods that are *substitutes* for environmental quality. The motivation for

[10]Cropper and Oates (June 1992), pp. 716–17.

[11]Bowker and Stoll (May 1988).

this approach is quite intuitive. Exposure to pollution causes damages that negatively affect a person's utility. Consequently, people undertake averting action by purchasing goods and services that improve their *personal* environmental quality, such as the indoor air or a private drinking water supply.[12]

Table 7.3 gives some common examples of what people do to reduce the effects of pollution on their personal environment. Notice that in each case the averting action involves an expenditure on a substitute good or service. Therefore, if the general environment is improved by some policy initiative, the individual can spend *less* on these substitute goods. It is precisely this decline in averting expenditures that gives an *indirect* estimate of the individual's WTP for the associated incremental benefits. For instance, faced with a contaminated drinking water supply, an individual might purchase bottled water or install a water-filtering system. If a government policy improves the public drinking water supply, the individual can spend less on these substitute commodities. This reduction in spending identifies the incremental benefits provided by the drinking water policy.

We model the AEM in Figure 7.6, where the relevant market is defined as *personal* environmental quality (X). The demand curve (d) is also the marginal benefit (MB) function, and each supply (s) relationship is modeled as a marginal cost (MC) curve.[13] The critical assumption is that each MC curve represents the averting expenditures on environmental quality substitutes to achieve various levels of *personal* environmental quality (X), given some level of *overall* environmental quality (E). In our diagram, MC_0

TABLE 7.3 Averting Actions

Pollution	Effect	Averting Action
Air pollution	Material soiling	Clean or repaint material surfaces; use protective covers; move to new location.
	Health problems	Install air purifier or air conditioner; schedule more frequent visits for medical examinations; purchase medications to alleviate respiratory symptoms; move to new location.
Water pollution	Material soiling	Install water filtration system; purchase cleaning products and rust removers; move to new location.
	Health problems	Install water filtration system; purchase bottled water; move to new location.
Hazardous waste site	Aesthetic degradation	Install fencing or shrubbery; move to new location.
	Health problems	Test water supply for contamination; install air filtration or air conditioner; move to new location.
Noise pollution	Health problems	Install sound-deadening insulation; purchase medication to aid sleeping; move to new location.

Source: Bartik (1988), pp. 111–26, Table 1. Copyright © 1988 by Academic Press, Inc.

[12]The reference to an individual's "personal" environment was first used by Bartik (1988) to describe the use of the AEM to value nonmarginal improvements in the environment.

[13]Because we are modeling *personal* environmental quality, the marginal benefits and costs accrue to a single individual. To avoid confusion, we use the simple labels of MB and MC so as not to infer a distinction between private and social decisions, which is irrelevant in this unique situation.

FIGURE 7.6 Measuring Incremental Benefits Using the Averting Expenditure Method (AEM)

When overall environmental quality is E_0, personal environmental quality is X_0, and averting expenditures are area $0abX_0$. After a policy increases overall environmental quality to E_1, personal environmental quality increases to X_1, and averting expenditures change to area $0acX_1$. To achieve X_1 in the absence of the policy change, the individual would be willing to spend an amount equal to area $0abcX_1$. Thus, the individual's WTP for the incremental benefits is the difference between $0abcX_1$ and $0acX_1$, or triangular area *abc*.

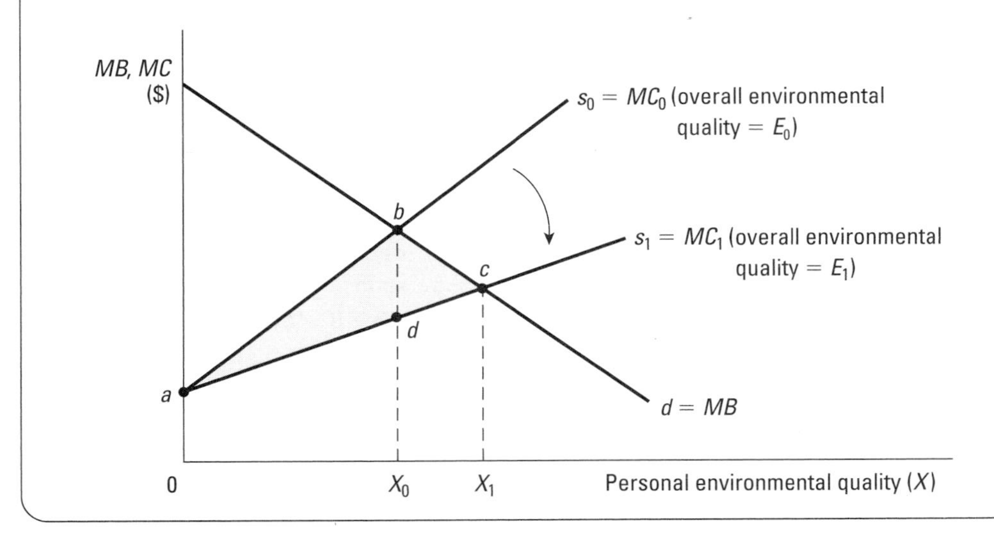

represents the marginal cost of averting expenditures at the existing level of overall environmental quality, E_0. As overall environmental quality improves to E_1, the individual spends less (or incurs lower costs) to achieve each level of personal environmental quality, and the *MC* curve moves downward to the right, becoming MC_1.

At the initial equilibrium when overall environmental quality is E_0, the individual's personal environmental quality is X_0, where *MB* and MC_0 intersect. At this point, total averting expenditures are the area under the MC_0 curve up to X_0, or area $0abX_0$. After the policy improves overall environmental quality to E_1, the individual's marginal cost curve shifts to MC_1. At the new equilibrium where *MB* and MC_1 intersect, personal environmental quality increases to X_1, and averting expenditures change to area $0acX_1$. Now, we can use this information to monetize the incremental benefits of the improvement in overall environmental quality from E_0 to E_1.

The key is to compare averting expenditure levels before and after the policy change *for the same level of personal environmental quality*. As we determined, post-policy averting expenditures for X_1 are represented by area $0acX_1$. Let's compare this to what the individual would be willing to spend to achieve X_1 without the influence of the policy. This is found by calculating the area under the original MC_0 curve up to X_1 and bounded by *MB*, or area $0abcX_1$. Thus, the individual's WTP for the incremental benefit is the difference between the two areas ($0abcX_1$ and $0acX_1$) or the triangular area *abc*.

An alternative valuation could be based on a constraint that holds personal environmental quality level at its original level X_0. This approach might be preferred because the calculation is simpler, requiring information on only the two *MC* curves rather than on both the *MC* curves and the *MB* curve. Under this scenario, the incremental benefits would be the difference between areas $0abX_0$ and $0adX_0$, or the area *abd*. This smaller area can be interpreted as a lower bound for the WTP valuation.[14]

Assessing the AEM

A drawback of the AEM arises from the phenomenon known as *jointness of production*. This refers to the fact that some averting expenditures yield benefits beyond those associated with a cleaner environment. Consider, for example, the averting expenditures on an air-conditioning system. While the system does reduce certain health risks of air pollution, it also provides comfort. Therefore, the savings in expenditures arising from a clean air policy initiative cannot be attributed solely to the incremental benefits of that policy.

Applications of the AEM

A number of studies have used the AEM to identify and estimate environmental benefits. For example, Bresnahan, Dickie, and Gerking (1997) identify three averting behaviors undertaken by Los Angeles residents to mitigate the effects of urban smog. Specifically, individuals might spend less time outdoors, alter outdoor recreational activities when air quality is poor, and/or increase their use of home air conditioners. Changes in spending on any of these activities associated with a policy-induced improvement in air quality would represent incremental benefits. Abrahams, Hubbell, and Jordan (2000) use the AEM to estimate the annual benefits to Georgia residents associated with safer drinking water. The researchers accomplish this by measuring changes in the residents' spending on bottled water and water filters. Based on the findings, an individual's WTP for safe drinking water is approximately $67 ($2010) per year, and the total benefits to all Georgia residents are estimated to be $495 million ($2010).

In an international context, Um, Kwak, and Kim (2002) survey residents in Pusan, Korea, to identify averting activities used to avoid drinking polluted tap water, including boiling tap water, buying bottled water, and installing water filtration systems. As Korea's water quality improves, spending on these activities should decline, and the associated savings are estimates of an individual's WTP for cleaner drinking water. According to this research, households of Pusan, Korea, are willing to pay between $0.09 ($2010) and $8.16 ($2010) per month for improvements in their tap water quality.

Travel Cost Method (TCM): An Approach Using Complements

travel cost method (TCM)
Values benefits by using the *complementary* relationship between the quality of a natural resource and its recreational use value.

An alternative approach to valuing environmental benefits is the **travel cost method (TCM)**, which uses the *complementary* relationship between the quality of a natural resource and its recreational use value. Simple observation suggests that the demand for the recreational use of an environmental resource, such as a lake or a national forest, increases as its quality improves. Therefore, as this demand function shifts with a change in environmental quality, the resulting change in consumer surplus approximates the associated incremental benefits.

We model the TCM in Figure 7.7, assuming that recreational demand has been properly identified.[15] Two demand curves for the recreational use of a lake are shown in the diagram,

[14]Bartik (1988).

[15]One technique used to identify a recreational demand curve is called the Clawson-Knetch method (Clawson and Knetch, 1966). This method uses travel costs to a recreational site, visitation rates, and other socioeconomic data to estimate recreational demand.

FIGURE 7.7 Measuring Incremental Benefits Using the Travel Cost Method (TCM)

D_0 is the recreational demand for a lake at some preexisting environmental quality level, E_0. D_1 is the new demand curve after a policy improves the lake's quality to E_1. The price line at P_0 is the admission fee. Before the policy is implemented, number of visits to the site is V_0, and consumer surplus is area abP_0. After the policy is put into effect, number of visits increases to V_1, and consumer surplus increases to area cdP_0. The change in consumer surplus, area *acdb*, represents the incremental benefits of improving the lake's quality from E_0 to E_1.

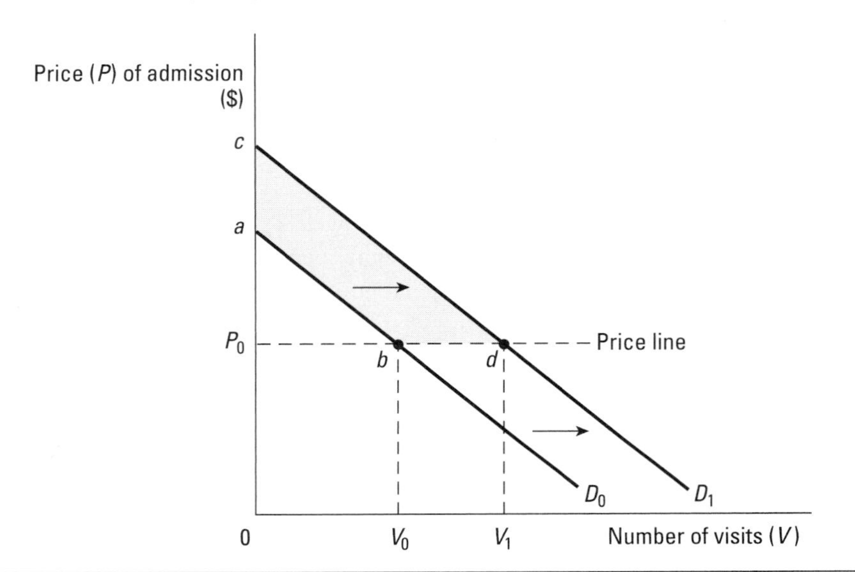

D_0 and D_1. D_0 is the relevant demand at some preexisting environmental quality level E_0. D_1 is the new demand curve after a policy has been implemented to improve the lake's quality to E_1. A price line is drawn at P_0 to represent the admission fee to use the lake. Before the policy is implemented, the number of visits to the site is V_0, where visitors enjoy a consumer surplus equal to area abP_0. After the policy is put into effect, the number of visits rises to V_1, and consumer surplus increases to area cdP_0. The resulting *change* in consumer surplus, shown as area *acdb* (i.e., area cdP_0 minus area abP_0), estimates the incremental benefits to visitors associated with improving the lake's quality.[16]

Assessing the TCM

One disadvantage of the TCM is that it estimates only user value and not existence value—an omission likely to create bias. Another is that it focuses on recreational use, making it ineffective for estimating benefits that accrue to commercial users of a resource. Finally, the TCM has been found to generate estimates that are biased downward if access to a site is deterred by congestion.[17]

[16]Notice that even if there were no entry fee to use the lake, the model would be the same, except that consumer surplus would be the entire area under each demand curve. In such an instance, the resulting change in consumer surplus would extend to the horizontal axis.

[17]Smith and Desvousges (1986), p. 220.

Applications of the TCM

Because of its limitations, the TCM is commonly used to value improvements to water bodies used mainly for recreation. For example, a study by Mullen and Menz (1985) uses the TCM to value the effect of acid rain damages to the Adirondack, New York, lake region. Other studies have used the procedure to value the benefits of improving water quality from boatable to fishable conditions. These findings tend to vary considerably. Consider the following results from three independent analyses, all measured as WTP per person per day in 2010 dollars:[18]

- Vaughan and Russell (1982): Between $10.58 and $21.17
- Smith and Desvousges (1985): Between $0.14 and $67.61
- Smith, Desvousges, and McGivney (1983): Between $2.35 and $4.86

Part of the reason for the inconsistent valuations is that TCM estimates tend to be sensitive to the site under study. Although demographic variables across regions can be controlled for, other site differences, such as aesthetics, access to major highways, and substitute recreational opportunities, are difficult to quantify and control. Consequently, it is unlikely that the TCM can determine a *generalized* value of improved water quality.

Hedonic Price Method (HPM): An Approach Using Product Attributes

hedonic price method (HPM)
Uses the estimated hedonic price of an environmental attribute to value a policy-driven improvement.

The **hedonic price method (HPM)** is based on the theory that a good or service is valued for the attributes or characteristics it possesses.[19] This perception of value suggests that *implicit* or *hedonic prices* exist for product attributes, and these prices can be determined from the explicit price of the product. In environmental economics, researchers use this technique to value the environmental attributes of certain commodities.

Housing markets have been a classic context for hedonic pricing studies. Such analyses assume that the market price of a home is determined by the implicit value of its characteristics, such as location, number of baths, lot size, and the environmental quality of the community. Therefore, changes in any of these characteristics are capitalized into the property's price.

The conventional model specifies the market price of a house, P, as a function of its attributes. A simplified version of such a model is:

$$P = f(X_1, X_2, \ldots, X_n, E)$$

where each X variable represents some housing attribute, such as lot size or number of baths, and E signifies the associated environmental quality.

As any one of these characteristics increases in magnitude (assuming each is specified as favorable), the price of the property P increases. It is this *marginal price* that is the implicit value of that attribute. Thus, as environmental quality improves, the resulting increase in property value can be used to estimate the associated incremental benefits. Once the implicit price of E is determined, the demand for environmental quality can be estimated. This in turn can be used to measure changes in consumer surplus arising from policy-driven improvements in environmental quality.

---

[18]These comparisons are discussed by Smith and Desvousges (October 1985).

[19]Lancaster (1966).

Assessing the HPM

The appeal of the HPM is that it is highly intuitive. It approaches the problem of monetizing incremental benefits in a logical way, directly using market prices. Its major disadvantage is that it requires a fairly complicated empirical model. Furthermore, the method calls for extensive data on product characteristics, which often are unavailable or incomplete. If an important product attribute is missing, it will not be possible to determine the connection between a change in an explicit price and a change in environmental quality.

Applications of the HPM

One application of the HPM is in measuring how the siting of hazardous waste facilities affects prices of nearby properties. A study by Kohlhase (1991) finds that housing prices in the Houston area are positively affected by the distance from a Superfund site up to 6.2 miles, specifically that each additional mile in distance adds $4,791 ($2010) to a property's value. In a similar study of single-family homes in Woburn, Massachusetts, Kiel (1995) estimates the analogous marginal benefit to be $3,015 ($2010) for the period when waste facilities were declared Superfund sites. A related investigation of housing values in the same city by Kiel and Zabel (2001) finds that the estimated benefit of cleaning Superfund sites is between $111.9 million and $189.6 million ($2010).

Using a more comprehensive dataset, Kiel and Williams (2007) determine that the proximity to a Superfund site does not have the same effect on housing prices across a sample of 57 sites located throughout the United States. Specifically, the WTP for clean-up of a Superfund is dependent on such factors as the size of the site and the percentage of blue-collar workers in the community. Taken together, these findings suggest that regulators should not assume that the dollar value of environmental benefits of Superfund site abatement is constant across sites.

CONCLUSIONS

As a risk management strategy, benefit-cost analysis can be used to set policy objectives and to select the best available control instrument to achieve those objectives. However, its success in guiding these important decisions hinges on the accuracy of benefit and cost measurements. In this chapter, we have focused on the conceptual issues of benefit valuation and some of the estimation methods used in practice. Of all the available benefit valuation techniques, there is no clear consensus about which is consistently superior. The diversity of approaches reflects both the complexity of the task *and* recognition of its importance in public policy decision making.

From a general perspective, our investigation in this chapter uncovered some of the difficulties inherent in any social benefit estimation—difficulties that often are magnified in an environmental context. The primary challenge is in monetizing gains that involve intangibles not traded in the marketplace. Hence, economists have had to devise methods to quantify these intangibles using something other than explicit prices.

On balance, research efforts in measuring social benefits have been fruitful. Progress has been made in fine-tuning estimation procedures, in recognizing which methods are most useful for which contexts, and in interpreting the results. These efforts are important to support the use of benefit-cost analysis as a risk management strategy and ultimately to devise better policy solutions. Of course, all of this necessitates that comparable progress be made on the cost side of the analysis, which is the focus of Chapter 8.

Summary

- To assess incremental benefits attributable to environmental policy, policymakers must determine how health, ecological, and property damages change as a result of that policy initiative.
- A primary environmental benefit arises as a *direct* consequence of implementing policy, whereas a secondary environmental benefit is an *indirect* gain arising either from the primary benefit or from some demand-induced effect.
- If we could infer society's demand for environmental quality, we could measure incremental benefits, because this demand represents the marginal social benefit of abatement.
- It is generally recognized that society derives utility from environmental quality based on its user value and its existence value.
- User value refers to the benefit received from physical utilization or access to an environmental resource. Existence value is the benefit received from the continuance of the resource based on motives of vicarious consumption and stewardship.
- There are two major types of benefit measurement techniques: the physical linkage approach and the behavioral linkage approach.
- A common procedure that uses the physical linkage approach is the damage function method. This method models the relationship between levels of a contaminant and the associated damages. Incremental benefits are estimated as the damage reduction achieved from any policy-induced decline in the contaminant.
- Two direct methods that use the behavioral linkage approach are the political referendum method and the contingent valuation method (CVM). Examples of indirect methods using this approach are the averting expenditure method (AEM), the travel cost method (TCM), and the hedonic price method (HPM).
- The contingent valuation method (CVM) is a survey approach that determines individuals' willingness to pay (WTP) for some environmental improvement based on hypothetical market conditions.
- The averting expenditure method (AEM) uses expenditures on goods that are *substitutes* for environmental quality to indirectly determine the WTP for a cleaner environment.
- The travel cost method (TCM) relies on identifying the recreational demand for an environmental resource, which is a *complementary* good to environmental quality. As environmental quality improves, recreational demand increases, and the associated benefits can be estimated as the change in consumer surplus.
- The hedonic price method (HPM) is based on the theory that implicit or hedonic prices exist for individual product attributes, including those related to environmental quality.

Review Questions

1. Is it possible for an individual's valuation of an environmental commodity to include both user value and existence value? Explain briefly.
2. Suppose the federal government is considering an air quality policy initiative that would effectively increase abatement (A) of ozone from 10 percent to 20 percent and that the marginal social benefit (MSB) of ozone abatement in millions of dollars has been estimated as $MSB = 120 - 2.5A$. Determine the dollar value of incremental benefits associated with this initiative.
3. In response to acid rain damage to Chesapeake Bay, a collaborative federal and state program has been proposed. You have been hired to evaluate the benefits of the plan as part of a formal benefit-cost analysis. Use the travel cost method (TCM) to accomplish this goal, based upon a $20 admission fee and the following pre- and post-policy recreational demand functions:

   Pre-policy: $P = 72 - 0.04V_0$

   Post-policy: $P = 90 - 0.04V_1$

where V is number of visitors in thousands and P is the admission fee.

4. Contrast the averting expenditure method (AEM) with the travel cost method (TCM), and discuss the relative strengths and weaknesses of each.
5. Refer to Application 7.3 and the estimated changes in consumer and producer surplus reported by the NAPAP for various ozone policies. Graphically model the result of a 25 percent *increase* in tropospheric ozone to *qualitatively* show the distribution of benefits reported by the NAPAP. (Do not attempt to arrive at the numerical values for the changes in surplus values. Show only how the consumer, producer, and total surpluses *change* in accordance with the reported findings.)
6. One of the strengths of the contingent valuation method (CVM) is its ability to capture existence value. How can the researcher take advantage of this, yet avoid some of the biases of such a survey-based approach?
7. **a.** Suppose you are part of a research team evaluating a proposal to clean up a hazardous waste site. You are in charge of assessing the incremental benefits. Which method would you choose to derive the estimation? Explain briefly.
   b. Based on your selection, outline your research plan for this specific estimation problem. Be sure to identify the following in your outline: a general description of your model, the relevant market for your model, the primary variables of interest, the data requirements, and any potential bias in your results.

Additional Readings

Abdalla, Charles W., Brian Roach, and Donald J. Epp. "Valuing Environmental Quality Changes Using Averting Expenditures: An Application to Groundwater Contamination." *Land Economics 68(2)* (May 1992), pp. 163–69.

Boyle, Melissa A., and Katherine A. Kiel. "A Survey of House Price Hedonic Studies of the Impact of Environmental Externalities." *Journal of Real Estate Literature* 9(2) (2001), pp. 117–44.

Bresnahan, Brian W., Mark Dickie, and Shelby Gerking. "Averting Behavior and Urban Air Pollution." *Land Economics 73(3)* (August 1997), pp. 340–57.

Brown, Gardner M., Jr., and Jason F. Shogren. "Economics of the Endangered Species Act." *Journal of Economic Perspectives 12(3)* (Summer 1998), pp. 3–20.

Carson, Richard. *Contingent Valuation: A Comprehensive Bibliography and History*. Northampton, MA: Elgar, 2004.

Carson, Richard T., Robert C. Mitchell, Michael Hanemann, Raymond J. Kopp, Stanley Presser, and Paul A. Ruud. "Contingent Valuation and Lost Passive Use: Damages from the Exxon Valdez Oil Spill." *Environmental and Resource Economics 25(3)* (July 2003), pp. 257–86.

Champ, Patricia A., Kevin J. Boyle, and Thomas C. Brown, eds. *A Primer on Nonmarket Valuation*. Norwell, MA: Kluwer Academic Publishers, 2003.

Champ, Patricia A., Nicholas E. Flores, Thomas C. Brown, and James Chivers. "Contingent Valuation and Incentives." *Land Economics 78(4)* (November 2002), pp. 591–604.

Common, M., I. Reid, and R. Blamey. "Do Existence Values for Cost Benefit Analysis Exist?" *Environmental and Resource Economics* 9(2) (March 1997), pp. 225–38.

Freeman, A. Myrick III. *The Measurement of Environmental and Resource Values: Theory and Methods*. 2nd Edition. Washington, DC: Resources for the Future, 2003.

Giraud, K. L., J. B. Loomis, and J. C. Cooper. "A Comparison of Willingness to Pay Estimation Techniques from Referendum Questions." *Environmental and Resource Economics 20(4)* (2001), pp. 331–46.

Haab, Timothy C., and Kenneth E. McConnell. *Valuing Environmental and Natural Resources: The Econometrics of Non-Market Valuation*. Northampton, MA: Elgar, 2002.

Hokby, Stina, and Tore Soderquist. "Elasticities of Demand and Willingness to Pay for Environmental

Services in Sweden." *Environmental and Resource Economics 26*(3) (November 2003), pp. 361–83.

Hoshino, Tadao, and Koichi Kuriyama. "Measuring the Benefits of Neighbourhood Park Amenities: Application and Comparison of Spatial Hedonic Approaches." *Environmental and Resource Economics 45*(3) (March 2010), pp. 429–44.

Johnstone, C., and A. Markandya. "Valuing River Characteristics Using Combined Site Choice and Participation Travel Cost Models." *Journal of Environmental Management 80*(3) (August 2006), pp. 237–47.

Muller, Nicholas Z., and Robert Mendelsohn. "Measuring the Damages of Air Pollution in the United States." *Journal of Environmental Economics and Management 54*(1) (July 2007), pp. 1–14.

Portney, Paul R. "The Contingent Valuation Debate: Why Economists Should Care." *Journal of Economic Perspectives 8*(4) (Fall 1994), pp. 3–18.

Ransom, Kevin P., and Stephen C. Mangi. "Valuing Recreational Benefits of Coral Reefs: The Case of Mombasa Marine National Park and Reserve, Kenya." *Environmental Management 45*(1) (January 2010) pp. 145–54.

Schläpfer, Felix. "Contingent Valuation: A New Perspective." *Ecological Economics 64*(4) (February 2008), pp. 729–40.

Smith, V. Kerry. "Can We Measure the Economic Value of Environmental Amenities?" *Southern Economic Journal 56*(4) (April 1990), pp. 865–78.

U.S. Environmental Protection Agency, Office of the Administrator. *Guidelines for Preparing Economic Analyses*. Washington, DC: September 2000.

Whittington, Dale. "Improving the Performance of Continent Valuation Studies in Developing Countries." *Environmental and Resource Economics 22*(1–2) (June 2002), pp. 323–67.

To access additional course materials, visit www.cengagebrain.com. At the home page, search for the ISBN of this title (shown on the back cover). This will take you to the product page where these resources can be found.

CHAPTER 8

Assessing Costs for Environmental Decision Making

"We must recognize that the goal of a cleaner environment ... will not be cheap or easy and the costs will have to be borne by each citizen, consumer and taxpayer."

—Richard Nixon (1913–1994)

In Chapter 7, we launched our formal study of benefit-cost analysis with the theory and measurement of environmental benefits. Equally critical is the analysis of **environmental costs**, that is, the costs of environmental improvement. In this phase of risk management, the policymaker must consider the value of all economic resources allocated to reducing environmental risk. Unlike the benefit side, the challenge here is not assigning a monetary value to costs, since most are already expressed in money terms. Rather, the more critical issue is identifying all the resources used to design, implement, and execute the policy prescription.

Few would debate that identifying environmental costs is a major undertaking. Think about the amount of government spending necessary to support the scientific research, the network of administrative agencies, and the labor force to implement a major environmental initiative. Add to that the billions of dollars spent by private businesses on abatement equipment and labor to comply with environmental regulations. Identifying these expenses on such a massive scale is by itself a tremendous task. But there is another element of cost analysis that adds to the difficulty—the premise that *economic* costs, and not simply *accounting* costs, are to be determined.

In this chapter, we discuss all these interesting issues, beginning with an overview of the fundamentals, much as we did on the benefit side. In particular, we start by defining incremental costs, the distinction between explicit and implicit costs, and the concept of valuing environmental costs. Once done, we address actual cost estimation methods used in practice today. Finally, we discuss the more prevalent ways in which environmental costs are classified and reported.

ENVIRONMENTAL COSTS: CONCEPTUAL ISSUES

Just as is the case for benefits, the appropriate level of analysis for evaluating the cost of an environmental initiative is **incremental costs**. The rationale is to allow a comparison between post-policy expenditures and their pre-policy level, which we call the baseline.

Identifying Incremental Costs

incremental costs
The change in costs arising from an environmental policy initiative.

Starting with the basics, we know that environmental costs must be defined in incremental terms. The motivation for using incremental variables is to capture *changes* brought about by policy. In this case, the relevant change is the increase in costs associated with policy-induced improvements in environmental quality. **Incremental costs** are calculated by first identifying the existing level of environmental expenditures, then estimating the costs after the policy is implemented, and finally finding the difference between the two. By way of example, the United States incurred millions of dollars in incremental costs to address the environmental damage associated with the terrorist attacks of September 11, 2001. An overview of these costs is described in Application 8.1.

Ideally, incremental costs should reflect changes in *economic* costs. As discussed in Chapter 2, economic costs are a more accurate measure of resource utilization than are accounting costs, since they include both **explicit** (i.e., out-of-pocket) **costs** and **implicit costs**. However, since the latter are not readily identifiable, analysts often derive incremental cost values based solely on explicit expenditures.

Explicit Environmental Costs

explicit costs
Administrative, monitoring, and enforcement expenses paid by the public sector plus compliance costs incurred by all sectors.

The **explicit costs** of implementing an environmental policy include the administrative, monitoring, and enforcement expenses paid by the public sector as well as the compliance costs incurred by virtually all sectors of the economy. See, for example, Table 8.1, which reports annual compliance cost estimates for several key titles of the Clean Air Act.

Explicit costs and their components are easier to identify than their benefit counterparts, since most of the resources used to implement pollution control policies are

TABLE 8.1 Annual Compliance Costs Associated With Titles I Through V of the 1990 Clean Air Act

Source Category	2000 Costs (in millions)	2002 Costs (in millions)
POINT SOURCES		
Nonutility Point Sources	$5,228	$ 5,407
Utility Point Sources	5,588	6,128
Permits	541	541
MOBILE AND AREA SOURCES		
Motor Vehicles and Non-Road Engines	n/a	$23,272
TOTAL	—	$35,348

NOTE: All data are in 2010 dollars.
Source: U.S. EPA, Office of the Chief Financial Officer (September 30, 2003), pp. 179–80, including Table 2, p. 179.

APPLICATION 8.1 Incremental Environmental Costs of the September 11, 2001, Terrorist Attacks

Identified as the worst terrorist attack on U.S. soil in history, the events of September 11, 2001, imposed grave harm on virtually every aspect of American living. Beyond the horrific human loss in New York, Washington, DC, and Pennsylvania, there was enormous damage to the Pentagon and to the entire infrastructure of Lower Manhattan. The explosions and structural ruin placed environmental quality in the two metropolitan areas at risk. Because of the collapse of the World Trade Center and the ensuing fires at ground zero, the environmental risks were of greater concern in the New York City area.

When the massive twin towers fell, the debris field was enormous. As a consequence, the air was filled with smoke, dust, and such substances as asbestos, chemical gases, and fiberglass. Then Mayor Giuliani and other officials donned dust masks as they made their way through the third largest business district in America. The image was nothing less than extraordinary. The EPA and other government agencies were summoned to the scene to assist with cleanup efforts and to test the air and water supplies. Of major concern was the potential exposure of rescue workers and others on the scene to elevated levels of hazardous contaminants.

The EPA's work in the wake of the disaster was immense. Working with the Occupational Safety and Health Administration (OSHA), the Centers for Disease Control and Prevention (CDC), and other government agencies, the EPA provided expertise on the abatement of hazardous materials, sampled drinking water supplies, tested runoff, and established a monitoring network to test air and water quality in New York City and in the area adjacent to the Pentagon. About 20 fixed air monitors were placed in and around the World Trade Center site, with others located in the Bronx, Brooklyn, Queens, and Staten Island. Portable equipment also was used to gather data from other locations. The results were assessed relative to standards and benchmarks established by various federal regulations. Daily updates of the environmental monitoring results were made available by the EPA through July 2002. A summary of the overall response by the EPA is available at **www.epa.gov/wtc/**.

These environmental initiatives necessitated additional spending by the EPA, which by definition represents the incremental environmental costs associated with the agency's response to the disaster. Initially, these expenditures were financed by emergency funding of $23.7 million. However, on September 18, 2001, then EPA administrator, Christine Todd Whitman, announced that the Federal Emergency Management Agency (FEMA) had given the EPA up to $83 million to cover the incremental costs of its abatement, advisory, and monitoring efforts in New York City and Washington, DC.

Because of the difficulty in identifying implicit costs, the dollar values referenced by these agencies were necessarily confined to explicit cost estimates. Clearly, if the full economic costs were considered, including all implicit costs, the dollar values cited would have been considerably higher.

In 2004, the EPA convened a World Trade Center Expert Technical Review Panel to gather input, characterize any remaining health risks at the site of the attacks, and recommend ways to minimize risks. When this review panel completed its work in 2005, a voluntary program was implemented during 2007 and 2008 called the Lower Manhattan Test and Clean Program. It was designed to test for residual contamination in buildings located close to ground zero and is considered the final phase of the EPA's response to the events of September 11. More information and access to that program's final report are available at **www.epa.gov/wtc/reports/lower_manhattan_test_&_clean_program_final_report.pdf**.

Sources: U.S. EPA (November 19, 2008; September 24, 2002a; September 24, 2002b; September 18, 2001).

traded in private markets. Those associated with the use of economic resources—land, labor, and capital—are rent, wages, and interest, respectively. Since these resources are traded on the open market, expenditures are based on market-determined input prices. However, these expenditures are not made simultaneously, and some are less controllable than others in the short run.

Recognizing these distinctions, economists classify costs into two components: (1) **fixed costs**, which are not controllable in the short run and do not depend on production levels, and (2) **variable costs**, which have the opposite characteristics. In the context of environmental policy implementation and compliance, the accounting equivalents of these categories are **capital costs** and **operating costs**, respectively.

capital costs
Fixed expenditures for plant, equipment, construction in progress, and production process changes associated with abatement.

operating costs
Variable expenditures incurred in the operation and maintenance of abatement processes.

Capital Costs and Operating Costs

According to U.S. government guidelines provided by the Department of Commerce and used by the Environmental Protection Agency (EPA), capital costs and operating costs have specific meanings. **Capital costs** are expenditures for plant, equipment, construction in progress, and the costs of changes in production processes that reduce or eliminate pollution generation. **Operating costs** are those incurred in the operation and maintenance of pollution abatement processes, including spending on materials, parts and supplies, direct labor, fuel, and research and development.

A critical distinction between these two categories is in how costs are related to the level of abatement. A capital cost is incurred regardless of the amount of pollution abated, analogous to the economic definition of fixed cost. Examples include the installed price of a scrubber system to control air pollution, construction costs of a wastewater facility, and the purchase price of acreage to use as a landfill. Conversely, operating costs are directly related to the quantity of abatement, comparable to the economic definition of variable costs. Examples include the costs of monitoring emissions and the costs of labor to run an abatement facility.

Implicit Environmental Costs

implicit costs
The value of any nonmonetary effects that negatively influence society's well-being.

Implicit costs are those concerned with any nonmonetary effects that negatively affect society's well-being. Examples include the value of diminished product variety arising from a ban on certain inputs, the time costs of searching for substitutes, and the reduced convenience that environmental control policies might impose. Although implicit costs are arguably important, identifying and measuring them in practice is another issue entirely. In fact, most analyses fail to fully capture these values. The result? Many environmental cost assessments are seriously understated, a problem that has yet to be fully resolved.

Conceptually Valuing Environmental Costs

What economic resources are used to achieve cleaner air or water, and what are the costs of these resources? Ideally, one should determine the social costs of environmental policy to answer these questions.

social costs
Expenditures needed to compensate society for resources used so that its utility level is maintained.

In theory, the **social costs** of any policy initiative are the expenditures needed to compensate society for the resources used so that its utility level is maintained. This compensation would have to account for all price, output, and income effects that arise from a given regulation. An example of a price effect is the increased price of automobiles caused by the cost of abatement equipment like catalytic converters. Others include the higher price of household products arising from the use of recyclable packaging and the costs of more highly regulated labeling. Income effects refer to real income changes, such as those caused by higher taxes imposed to support regulatory expenses. Beyond these effects are implicit or nonmonetary costs. Examples include the inconvenience of using public transportation or car pools in response to urban air quality policies or the search costs to find nontoxic substitute products.

To clarify a bit further, consider the following. The EPA identifies five basic components of total social costs linked to environmental policy. These are explained in Table 8.2 and ordered by ease of estimation. Notice the specific references to price, output, and income effects in certain of these components.

TABLE 8.2 Five Basic Components of Social Costs

Cost Component	Description
Real-resource compliance costs	Direct costs comprising the primary component of social costs associated with purchasing, installing, and operating pollution equipment, changing production processes, or capturing and selling or reusing waste products.
Government regulatory costs	Monitoring, administrative, and enforcement costs linked to new regulations.
Social welfare losses	Reduction in consumer and producer surpluses due to the rise in price (or decline in output) of goods associated with environmental policy.
Transitional costs	Value of resources displaced due to regulation-induced production declines and private costs of reallocating those resources.
Indirect costs	Adverse effects on product quality, innovation, productivity, and market effects indirectly influenced by policy, all of which may affect net levels of producer and consumer surpluses.

NOTE: These components are ordered by relative ease of estimation.

Source: U.S. EPA, Office of the Administrator (September 2000), pp. 113–14.

From a modeling perspective, we capture marginal social costs (*MSC*) by the supply of the public good, environmental quality. Although the demand for a public good can only be inferred, the supply function can be identified in the same manner as that used for a private good. Let's reconsider the market supply model presented in Chapter 3 that represents the *MSC* of sulfur dioxide (SO_2) abatement. This function is specified as $MSC = 4 + 0.75A$, where *MSC* is measured in millions of dollars, and *A* is a percentage of SO_2 abatement. In Chapter 4, we learned that the *MSC* is the vertical sum of the market-level marginal abatement costs (MAC_{mkt}) plus the government's marginal cost of enforcement (*MCE*). The graph is shown in Figure 8.1.

Analogous to the benefit side, *MSC* for each abatement level is shown as the vertical distance from the horizontal axis up to the supply curve. Total social costs (*TSC*) for any abatement level are measured as the area under the *MSC* curve up to that point.[1] The shaded area in Figure 8.1 represents the *TSC* to achieve an abatement level of A_1.

Just as on the benefit side, the incremental cost assessment follows a logical three-step process:

1. Find the baseline level of *TSC before* the policy is undertaken.
2. Find the new level of *TSC after* the policy is implemented.
3. Subtract the baseline *TSC* from the post-policy *TSC*.

To illustrate this procedure, consider the incremental costs of a policy-induced increase in SO_2 abatement from 20 to 25 percent. First, identify the baseline level of *TSC* when *A* equals 20 percent. Referring to Figure 8.2, notice that the *MSC* at this

[1]Technically, the area under any marginal cost curve represents only total *variable* costs. Hence, to argue that the area under the *MSC* curve represents the *TSC* implicitly assumes there are no fixed costs.

FIGURE 8.1 Marginal Social Cost (*MSC*) and Total Social Costs (*TSC*) of Air Quality

The market supply for sulfur dioxide (SO_2) abatement represents the marginal social cost (*MSC*) of air quality. For each abatement level (*A*), the associated *MSC* is shown as the vertical distance from the horizontal axis up to the supply curve. Total social costs (*TSC*) for any abatement level are measured as the area under the *MSC* curve up to that point. Thus, the shaded area represents the *TSC* to achieve abatement level A_1.

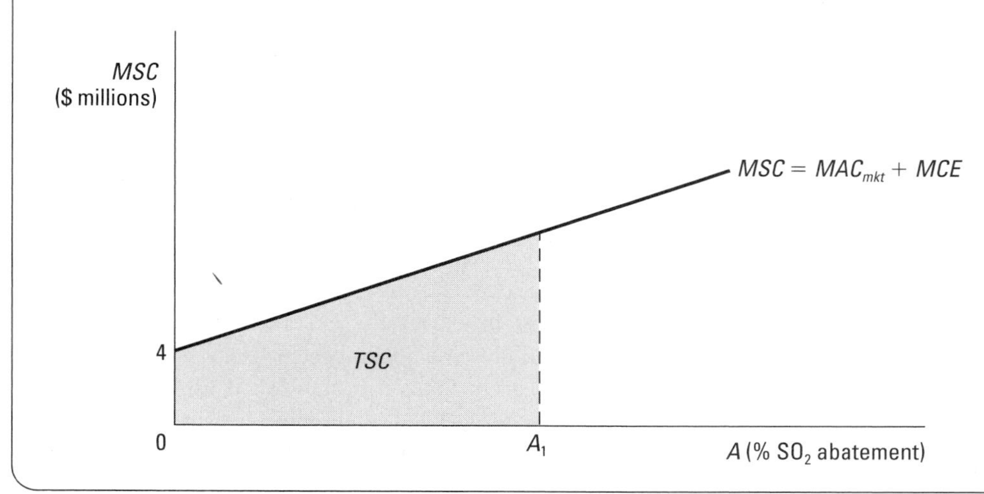

FIGURE 8.2 Modeling Incremental Costs for Air Quality Using the *MSC* Function

At the baseline abatement level of 20 percent, the *MSC* equals $19 million, and the *TSC* are the area under the *MSC* curve up to that point, or $230 million. At the post-policy abatement level of 25 percent, the *MSC* is $22.75 million, and the corresponding *TSC* are $334.375 million. Therefore, the difference between the two *TSC* values, or $104.375 million, represents the incremental costs to achieve the additional 5 percent abatement of SO_2 emissions. These incremental costs are shown as the shaded area.

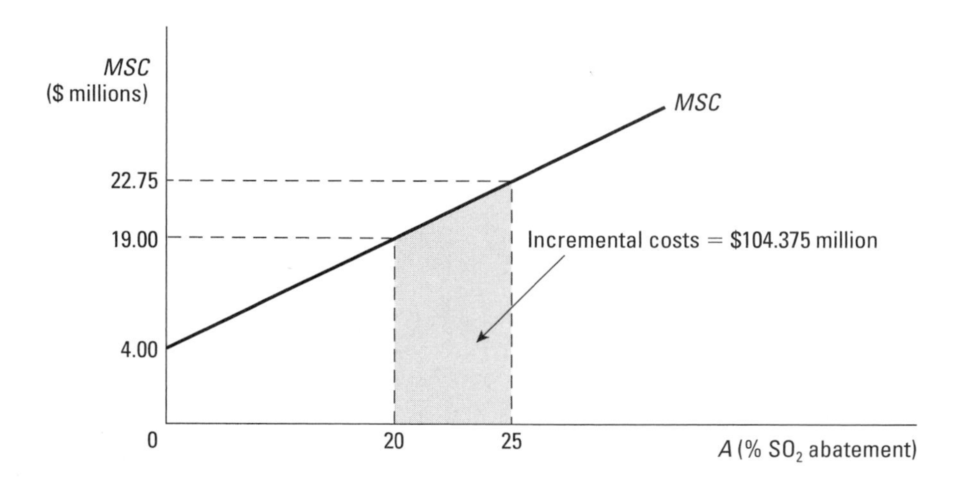

abatement level is \$19 million and the corresponding *TSC* are shown as the area under the *MSC* curve up to that point, or \$230 million. Second, determine the post-policy *TSC* to increase abatement to 25 percent. Following the same steps, the *MSC* when *A* equals 25 percent is \$22.75 million, and the *TSC* are \$334.375 million. Hence, the difference between the two *TSC* values, or \$104.375 million, represents the incremental costs to achieve the additional 5 percent abatement of SO_2 emissions.

Figure 8.3 presents an alternative model that graphs the *TSC* directly with the percentage of SO_2 abatement. In this case, the *TSC* associated with any abatement level can be found simply as the vertical distance from the horizontal axis up to the *TSC* curve. To determine incremental costs from this model, identify the pre- and post-policy abatement levels (20 percent and 25 percent, respectively) and the corresponding *TSC* levels (\$230 million and \$334.375 million, respectively). The difference, or \$104.375 million, is shown as the vertical distance between the two values.

In practice, to properly assess the social costs of an environmental initiative, the analyst would have to measure changes in social welfare, perhaps by estimating changes in consumer and producer surplus. This is a complex undertaking, but there is some evidence that doing so improves the accuracy of cost assessments. A study by Hazilla and Kopp (1990) estimates the social costs of U.S. air and water quality control policies. According to this research, there is a significant difference between social costs and the explicit cost estimates typically used in practice. In fact, these researchers recommend that policymakers adjust procedures to consider more of a general equilibrium approach to cost analysis than is currently in place. However, available methods to execute this more comprehensive cost assessment are too cumbersome and unreliable to be used as a regular practice. So at least for the present, cost analysis continues to be based solely upon explicit expenditures, although it is generally acknowledged that such an approach understates the true social costs.

FIGURE 8.3 Modeling Incremental Costs for Air Quality Using the *TSC* Function

The *TSC* associated with any abatement level are found as the vertical distance from the horizontal axis up to the *TSC* curve. Notice that at the pre- and post-policy abatement levels, 20 percent and 25 percent, respectively, the corresponding *TSC* levels are \$230 million and \$334.375 million, respectively. Thus, the difference, or \$104.375 million, is the vertical distance between the two cost values.

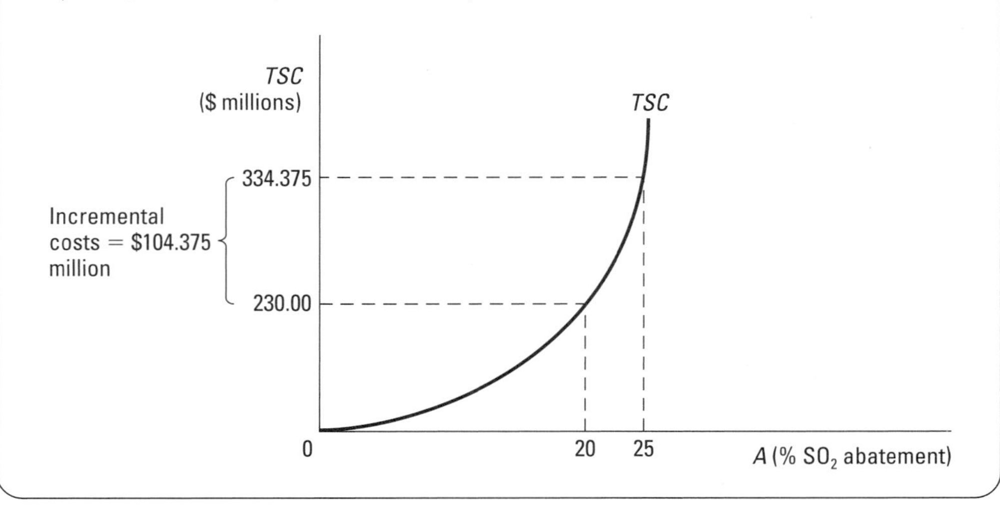

ESTIMATION METHODS FOR MEASURING EXPLICIT COSTS

Economists use two major approaches to estimate incremental environmental costs: the **engineering approach** and the **survey approach**. The **engineering approach** estimates abatement expenditures based on the least-cost available technology needed to achieve some level of pollution abatement. The **survey approach** derives estimated abatement expenditures directly from polluting sources. Although the two approaches are distinct, a common practice is to use a combination approach that draws from both. First, the survey approach is used to solicit information about available technologies and existing market conditions. Then, the engineering approach is employed to estimate dollar values based on the collected data. What follows is a brief discussion of each approach, with some mention of their respective shortcomings.

Engineering Approach

engineering approach
Estimates abatement expenditures based on least-cost available technology.

The **engineering approach** to expenditure estimation relies upon the knowledge of experts in abatement technology. Based on the state of the art in abatement, engineers and scientists are called upon to identify combinations of equipment, labor, and materials needed by polluters to comply with a policy mandate. Then, capital and operating costs for all feasible abatement designs are estimated. Finally, the analyst selects the least-cost model from among these technology-based designs and uses the result to estimate the aggregate incremental cost for all affected polluting sources.

While theoretically reasonable, the engineering approach is not without problems. For one thing, the procedure is difficult to implement for *proposed* environmental controls, since there is uncertainty about price movements, availability of raw materials, and future energy costs. In addition, most comprehensive policy initiatives affect a wide range of industries and public facilities. To properly account for these, the engineering approach would have to be customized to suit each type of production setting. If ignored, the results likely would suffer from the averaging process. Even if industry-specific values were derived, the heterogeneity of firms and the unique market conditions that each faces could not be captured in such a generalized estimate. Finally, because the approach is based on the least-cost abatement design, it explicitly assumes that all firms are cost-effective or technically efficient entities—an assumption that likely understates the true costs incurred.

Survey Approach

survey approach
Polls a sample of firms and public facilities to obtain estimated abatement expenditures.

In contrast to the engineering approach that relies on external experts, the **survey approach** relies directly upon polluting sources to provide data for the estimation. Similar to the contingent valuation method for estimating benefits, the survey approach polls a sample of firms and public facilities to inquire about existing or projected environmental expenditures.

On the plus side, the survey method is a more direct means to obtain abatement cost data than the engineering approach. There are, however, disadvantages. First, it implicitly assumes that polluting sources are sufficiently well informed to provide reasonable estimates. Second, there is an inherent bias. Polluting sources have an incentive to offer inflated values to government officials, because they recognize that higher costs will increase the probability that the proposed regulation will be rejected.

Just as in benefit assessment, no cost estimation method is perfect. In fact, this realization has motivated researchers to use a combination of the engineering and survey methods. The objective is to have the best of both worlds and minimize potential bias.

COST CLASSIFICATIONS IN PRACTICE

As part of cost assessment, analysts are interested in determining the composition of environmental control costs. Information about the composition of control costs gives policymakers a clearer sense of how resources are being allocated to achieve environmental objectives—critical information for evaluating policy on the basis of allocative efficiency. Similarly, adjustments to correct for environmental inequities require knowledge about how the cost burden is distributed across economic sectors.

Because of the diversity of objectives in cost assessment, different cost classifications are used to obtain different kinds of information. An overview of two of the more commonly used classifications follows.

Cost Classifications by Economic Sector

Policy analysts often are concerned with how environmental costs are distributed across the public and private sectors of the economy. For instance, the costs of policy revisions that strengthen monitoring and enforcement programs shift a higher cost burden to the public sector. This in turn would shift up the *MCE* curve and hence the *MSC* of abatement. Conversely, tighter abatement requirements will change the cost distribution more toward private industry. To illustrate, consider the U.S. trend data in Table 8.3. The expenditures are broken down by major function: abatement, regulation and monitoring, and research and development.

As the data in Table 8.3 indicate, the aggregate level of real spending on environmental issues has grown fairly steadily in the United States, up to $179.23 billion ($2010) over the 20-year period shown.[2] Furthermore, spending on abatement consistently represents the vast majority of total environmental expenditures over time.

TABLE 8.3 U.S. Pollution Abatement and Control Expenditures by Function

	POLLUTION ABATEMENT					
YEAR	Personal	Business	Government	Regulation and Monitoring	Research and Development	TOTAL
1975	13.17	66.51	27.93	2.63	4.46	114.70
1980	17.60	79.36	29.40	3.33	4.63	134.33
1985	24.64	85.20	29.47	2.53	2.80	144.61
1990	15.57	97.27	40.01	2.99	2.37	158.19
1994	14.36	112.75	45.95	3.24	2.93	179.23

NOTES: Data are in billions of 2010 dollars. Some totals may not agree because of independent rounding. All of personal spending is used to purchase and operate motor vehicle emission abatement devices.

Source: Vogan (September 1996).

[2]Further updates to these data (beyond 1994) were discontinued for budgetary reasons.

For the most current year shown, aggregate spending on abatement was 96.6 percent of the total, with spending on regulation and monitoring and on research and development representing less than 2 percent each.

Within the abatement category, spending by each economic sector has shown a similar growth pattern, so much so that the relative proportions borne by each have remained fairly constant. However, the cost burden is not shared equally across the economy. Over time, private business spending on pollution abatement represents about 65 percent of the total on average, public sector spending accounts for about 27 percent, with the personal private sector responsible for the remainder. Internet access to the report from which these data are drawn is available at **www.bea.gov/scb/pdf/national/niparel/1996/0996eed.pdf**.

To gain a better sense of abatement spending *within* the business sector, we can examine available abatement cost data for a more current year, 2005 to be specific. In Table 8.4, we present operating and capital abatement costs for five major manufacturing industries—chemicals, petroleum and coal products, primary metals, paper, and food. Notice that these costs are further broken down by type of abatement activity, specifically treatment, prevention, recycling, and disposal. This presentation gives us a better sense of how abatement costs are allocated within markets.

For every industry represented in Table 8.4, we observe that the highest proportion of abatement costs is associated with waste treatment. In the chemical industry, for example, this activity accounts for nearly 53 percent of operating costs and about 46 percent of capital costs. Similarly, for petroleum and coal, the proportions are 51 percent and 53 percent, respectively. Interestingly, firms in this latter industry also spend proportionately more on prevention relative to the other four industries, specifically 35 percent of operating costs and 45 percent of capital costs. Relatively high expenditures on treatment and prevention explain why petroleum and coal companies incur a proportionately lower amount of disposal costs relative to the other markets.

Aggregating across all abatement activities, total operating and capital abatement costs for each of these five industries are depicted in Figure 8.4. This illustrates at a glance how these costs vary across major manufacturing markets. Not surprisingly, total

TABLE 8.4 U.S. Pollution Operating and Capital Abatement Costs in 2005 by Activity

	OPERATING COSTS					CAPITAL COSTS				
		Activity					Activity			
INDUSTRY	TOTAL	Treatment	Prevention	Recycling	Disposal	TOTAL	Treatment	Prevention	Recycling	Disposal
Chemical	5825.1	3079.3	903.9	465.8	1376.1	1419.8	647.5	613.6	59.6	99.0
Petroleum and Coal Products	4182.6	2117.1	1444.9	305.5	315.1	1946.1	1036.1	870.9	34.5	4.6
Primary Metal	2558.1	1382.6	305.0	244.9	625.7	571.6	391.0	72.1	81.8	26.6
Paper	2005.5	1196.9	211.5	132.4	464.7	640.1	328.9	266.1	23.0	22.2
Food	1756.1	959.2	192.8	120.6	483.5	500.2	286.2	125.8	29.7	58.4

NOTE: Data are in millions of 2010 dollars.
Source: U.S. Census Bureau (2008).

FIGURE 8.4 Pollution Abatement Costs in 2005 by Industry

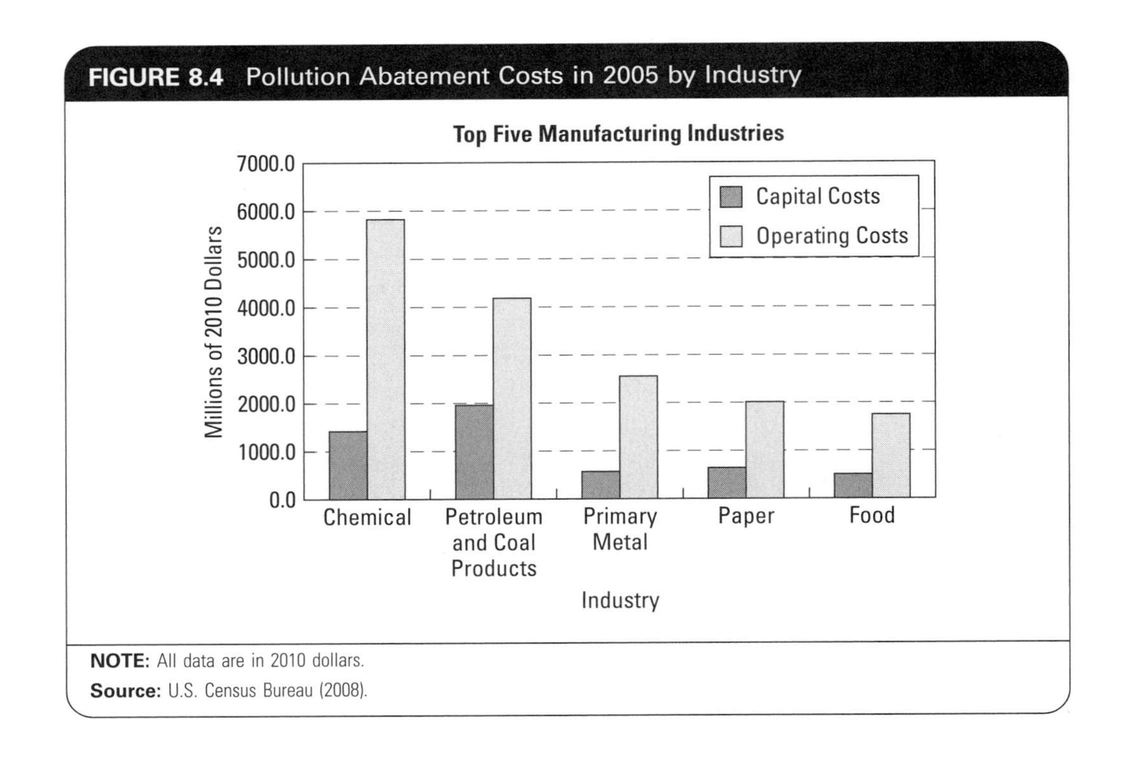

NOTE: All data are in 2010 dollars.
Source: U.S. Census Bureau (2008).

abatement spending by the chemical industry is higher than for any other industry at \$7,244.9 million (\$2010), comprising \$1,419.8 million (\$2010) in capital costs and \$5,825.1 million (\$2010) in operating costs. Moreover, total abatement spending for this market is more than \$1 billion higher than the petroleum industry, which is ranked second by total abatement expenditures. These and other related data are available online at **www.census.gov/prod/2008pubs/ma200-05.pdf**.

Cost Classifications by Environmental Media

Another important policy consideration is the distribution of expenditures by environmental media (i.e., air, water, and solid waste). Table 8.5 shows U.S. abatement and control expenditures for 1975 through 1994 disaggregated by environmental media.

TABLE 8.5 U.S. Pollution Abatement and Control Expenditures Disaggregated by Environmental Media

Year	GDP	Air	Water	Solid Waste
1975	6,637.74	48.31	48.92	18.32
1980	7,378.18	59.15	54.62	22.55
1985	8,546.94	62.17	53.03	30.76
1990	9,677.38	47.26	61.95	51.12
1994	10,424.90	55.32	62.36	61.41

NOTE: Data are in billions of 2010 dollars.
Sources: U.S. Department of Commerce, Bureau of Economic Analysis (February 25, 2011); Vogan (September 1996).

Such information helps policymakers examine how resources are allocated across environmental problems and how this allocation changes from period to period. To give context to these values, real gross domestic product (GDP) data also are provided in the table. This helps to illustrate the *relative* magnitude of environmental spending in the United States. Note that in 1994, for example, environmental spending on water resources was $62.36 billion ($2010). However, this is a relatively small percentage of GDP, at approximately 0.6%.

As in the prior discussion, we can look at similar data within the manufacturing sector for a more current period. In this case, Figure 8.5 presents 2005 abatement cost data delineated by environmental media for the same five manufacturing industries we studied previously, i.e., chemicals, petroleum and coal, primary metals, paper, and food. To facilitate comparison, these values are expressed in 2010 dollars. Notice that abatement costs for air quality policies are significantly higher than for water and solid waste initiatives. In fact, total abatement expenditures for air quality are approximately 1.5 times as great as for water policy and some 2.4 times as high as for solid waste initiatives. We will investigate media-specific policy differences that might explain these relative costs in upcoming chapters. For now, Application 8.2 offers some insight on air quality abatement spending by automobile manufacturers in response to recent global warming initiatives.

FIGURE 8.5 Pollution Abatement Costs in 2005 by Environmental Media

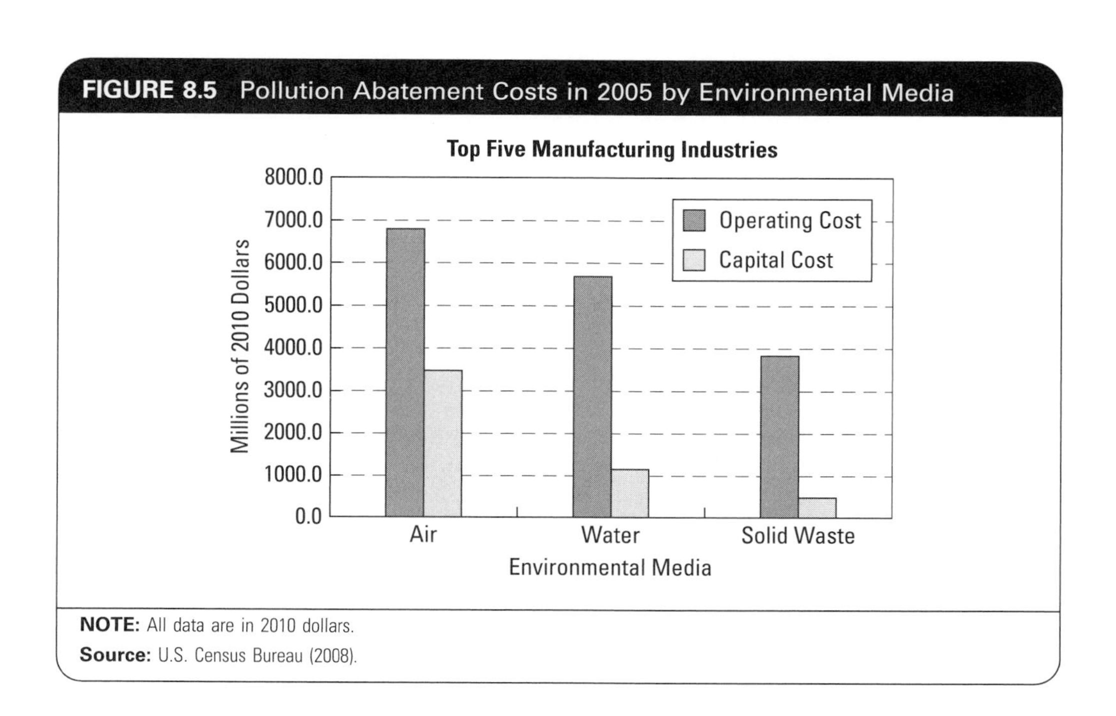

NOTE: All data are in 2010 dollars.
Source: U.S. Census Bureau (2008).

APPLICATION 8.2 Abatement Costs for the Global Auto Industry: Reducing Carbon Emissions

Heightened awareness of air pollution and climate change has brought about a renewed interest in cleaner running cars. Much of the attention to global warming began as a reaction to the 2005 ratification of the Kyoto Protocol, the international agreement aimed at reducing greenhouse gas (GHG) emissions. As a consequence, auto manufacturers around the world must respond with new technologies to reduce carbon and other GHG emissions.

Analysts, public officials, and private industry have begun to examine the economic implications of this worldwide effort to control global climate change. Of particular interest is how the requisite changes in GHG emissions translate into incremental costs for manufacturers. In a 2003 report prepared jointly by Sustainable Asset Management and World Resources Institute, such cost estimates are projected for automobile manufacturers as they adjust production technologies to meet restrictions on carbon releases.

Interestingly, the report's cost projections vary by a factor of 25 across firms in the automobile industry. At the high end is the sales-weighted average cost for BMW, projected to be $649 per vehicle. This is in stark contrast to the analogous value for Honda, estimated to be $24. DaimlerChrysler, Ford, and GM fall in line after BMW with unit costs per vehicle of $459, $403, and $377, respectively. Logically, consumer prices are expected to rise as well, as manufacturers attempt to pass on some proportion of these higher costs to buyers, with the proportion varying with the demand conditions for each individual vehicle. However, associated fuel savings are expected to offset the price increases.

Although the United States did not ratify the Kyoto Protocol, federal- and state-level initiatives to combat climate change are also expected to affect automakers' costs of production. In 2004, the California Air Resources Board (ARB) adopted the country's first GHG regulations for automobiles, and over a dozen other states along with the District of Columbia followed suit. The California law that called for these regulations was signed in 2002. Enforcement of such state-level regulations required an EPA waiver from the Clean Air Act, but the waiver was declined in 2008. California subsequently asked for a reconsideration, and, with the urging of President Obama, the EPA granted the waiver in June 2009. In 2010, the EPA along with the Department of Transportation established national GHG emission standards for cars and light-duty trucks, requiring the achievement of 250 grams of carbon dioxide per mile, on average, by model year 2016, with analogous requirements to follow for medium- and heavy-duty vehicles.

How will these new policy initiatives affect environmental abatement costs? The estimates and the projected implications vary widely. According to the EPA and the National Highway Traffic Safety Administration (NHTSA), the standards on light-duty vehicles are estimated to increase 2012 vehicle prices by $476, and 2016 prices by $1,091. However, not unlike the estimates prepared by Sustainable Asset Management and World Resources Institute, these costs are expected to be more than countered by fuel savings, which are estimated to be higher than $3,000 over the life of a 2016 model year vehicle.

At this stage, there is a lack of consensus on both the expected magnitude of the direct incremental costs and the extent to which these direct costs might be offset by associated fuel cost savings. This is due to the absence of full information both about manufacturing technologies and the actual fuel efficiencies that might be achieved. That said, it is a certainty that proper evaluation of these GHG regulations must consider both the gains to society as well as the associated environmental costs.

Sources: Pew Center on Global Climate Change (2011); U.S. EPA, Office of Transportation and Air Quality (April 2010); Power (January 27, 2009); Hakim (November 2, 2004; July 25, 2004); California Environmental Protection Agency, Air Resources Board (September 24, 2004); Alliance of Automobile Manufacturers (September 23, 2004); Austin, Rosinski, Sauer, and Le Duc (2003).

CONCLUSIONS

Environmental cost assessment and estimation is a critical element of risk management. Although more tangible than benefit assessment, the valuation of environmental costs has its share of complexities. Chief among these is the divergence between the social costs of an environmental initiative and the explicit costs estimated in practice. In the absence of better techniques that can capture implicit costs, incremental cost estimates are generally assumed to be biased downward.

As economists search for a practical solution to this problem, they also look for ways to improve their estimates of explicit costs. The combined use of the survey and engineering approaches appears to be one way that greater accuracy can be achieved. Related to these efforts are attempts to classify cost data in ways that are meaningful to policy evaluation procedures and to the formulation of new programs and initiatives. The motivation is simple. Policymakers have to make important decisions based in part on the results of environmental cost estimation. In Chapter 9, we will explain precisely how environmental policy is influenced by both benefit and cost estimates when they are systematically considered in a formal benefit-cost analysis.

Summary

- Environmental cost analysis is concerned with incremental costs, or the change in explicit and implicit costs to society, incurred as a result of government policy. Because implicit costs are not readily identifiable, analysts generally derive incremental costs based solely on explicit expenditures.
- Explicit costs of implementing environmental policy include the administrative, monitoring, and enforcement expenses paid by the public sector as well as the compliance costs incurred by virtually all sectors of the economy.
- Capital costs are expenditures for plant and equipment used to reduce or eliminate pollution.
- Operating costs are those incurred in the operation and maintenance of abatement processes.
- Implicit costs are those concerned with any nonmonetary effects that negatively affect society's well-being.
- The social costs of any policy initiative are the expenditures needed to compensate society for the resources used so that its utility level is maintained.
- Marginal social costs (*MSC*) can be modeled as the supply of the public good, environmental quality, where *MSC* is the vertical sum of the market-level marginal abatement cost (MAC_{mkt}) and the marginal cost of enforcement (*MCE*).
- The engineering approach to cost estimation is based on the least-cost available technology needed to achieve a given level of abatement.
- The survey approach to cost estimation relies on estimated abatement expenditures obtained directly from polluting sources.
- Because of the diversity of objectives in environmental cost assessment, costs are commonly communicated through various classifications, such as by environmental media or by economic sector.

Review Questions

1. Consider a policy proposal to impose more stringent controls on automobile tailpipe exhausts. Distinguish between the explicit and implicit costs of this proposal, and support your discussion with several specific examples of each.
2. Of the two approaches to cost estimation, which in your view likely produces the most reliable estimates? Explain.
3. Suppose the *MSC* of cleaning Puget Sound is modeled as $MSC = 10 + 1.4A$, where *A* is the percentage of phosphorus abated and *MSC* is measured in millions of dollars.
   a. Find the incremental costs of a policy initiative that increases the phosphorus abatement level from its baseline of 30 percent to 45 percent.
   b. Graphically illustrate using the *MSC* function, labeling clearly where incremental costs are shown.
   c. Repeat part (b) using the *TSC* function directly.
4. Suppose the *TSC* of mercury abatement has been estimated as $TSC = 20A + 2.4A^2$, where *A* is a percentage of mercury abatement, and *TSC* is in millions of dollars. Find the incremental cost of a statewide policy that would increase abatement from 5 to 10 percent, and illustrate graphically.
5. Assume you are responsible for assessing whether the air quality standard for carbon monoxide (CO) is set at the efficient level for some region. To accomplish this task, you have estimated the following marginal benefit and marginal cost functions for CO abatement:

$$MSB = 20 - 0.25A$$
$$MAC_{mkt} = 6 + 0.25A$$
$$MCE = 2 + 0.1A$$

where *A* is the percentage of CO abatement, and *MSB*, MAC_{mkt}, and *MCE* are measured in millions of dollars.

a. Find the *MSC* of abatement function.
b. Solve for the efficient level of abatement.
c. Show the importance of the *MCE* by determining the effect on the solution if these government costs were ignored. Support your answer algebraically and graphically.

Additional Readings

Becker, Randy A. "Air Pollution Abatement Costs Under the Clean Air Act: Evidence From the PACE Survey." *Journal of Environmental Economics and Management 50(1)* (July 2005), pp. 144–69.

Cohen, Mark A. "A Taxonomy of Oil Spill Costs: What Are the Likely Costs of the Deepwater Horizon Spill"? *Backgrounder* (June 2010), pp. 1–6.

Dasgupta, Susmita, Minul Huq, David Wheeler, and Chonghua Zhang. "Water Pollution Abatement by Chinese Industry: Cost Estimates and Policy Implications." *Applied Economics 33(4)* (March 2001), pp. 547–57.

Gray, Wayne B. *Economic Costs and Consequences of Environmental Regulation.* Burlington, VT: Ashgate, 2002.

Harrington, Winston, Richard D. Morgenstern, and Peter Nelson. "On the Accuracy of Regulatory Cost Estimates." *Journal of Policy Analysis and Management 19(2)* (Spring 2000), pp. 297–322.

______. "Predicting the Costs of Environmental Protection." *Environment 41(7)* (September 1999), pp. 10–19.

Hartman, Raymond S., David Wheeler, and Manjula Singh. "The Cost of Air Pollution Abatement." *Applied Economics 29* (June 1997), pp. 759–74.

McKitrick, Ross. "A Derivation of the Marginal Abatement Cost Curve." *Journal of Environmental Economics and Management 37(3)* (May 1999), pp. 306–14.

Palmer, Karen, Hilary Sigman, and Margaret Walls. "The Cost of Reducing Municipal Solid Waste." *Journal of Environmental Economics and Management 33(2)* (June 1997), pp. 128–50.

Pereira, Alfredo Marvao, and Rui Manuel Marvao Pereira. "On the Potential Economic Costs of Cutting Carbon Dioxide Emissions in Portugal." *Portuguese Economic Journal 9(3)* (2010), pp. 211–22.

Russell, Clifford. *The Economics of Environmental Monitoring and Enforcement.* Burlington, VT: Ashgate, 2003.

Vermont, Bruno, and Stephane De Cara. "How Costly Is Mitigation of Non-CO_2 Greenhouse Gas Emissions From Agriculture?" *Ecological Economics 69(7)* (May 2010), pp. 1373–86.

To access additional course materials, visit www.cengagebrain.com. At the home page, search for the ISBN of this title (shown on the back cover). This will take you to the product page where these resources can be found.

CHAPTER 9

Benefit-Cost Analysis in Environmental Decision Making

"Human capital analysis starts with the assumption that individuals decide on their education, training, medical care, and other additions to knowledge and health by weighing the benefits and costs."

—Gary Becker (1930–)

Should the national limit on sulfur dioxide emissions be tightened? Would society be better off if a tradeable permit system were used instead of technology-based standards to control water pollution? These are the kinds of questions that environmental decision makers must address. Analytical tools such as **benefit-cost analysis** can help them find answers. Benefit-cost analysis begins with identifying and monetizing environmental benefits and costs, the subject of Chapters 7 and 8. The estimates must then be adjusted and systematically compared to arrive at a decision. These critical steps that link benefit and cost estimates to a decision rule complete the strategic process of benefit-cost analysis.

Adjustments to estimated values are necessary because incremental costs and benefits are not realized immediately. Instead, they accrue over a period of years. How do these time differences affect the result? There are two considerations. The first is that future costs and benefits have to be adjusted downward to be comparable to those incurred in the present. The second is that these values must be adjusted for expected changes in the price level over time. Without these modifications, the benefit-cost analysis would yield biased results, and any policy decision based upon them would be misguided.

Once benefit and cost estimates are corrected for time differences, they must then be compared to one another. But on what basis is the comparison made? Is it sufficient for the benefits to outweigh the costs, or must the numerical difference between the two be at a particular level? What if the law precludes the analysis by predetermining the benefit level to be achieved? The answers are found in the decision rules that guide benefit-cost analysis.

The growing prominence of benefit-cost analysis as a risk management strategy is an important reason for understanding how it is undertaken, what it implies from an economic perspective, and how it has come to be required by the U.S. government in major policy decisions. To that end, this chapter presents and analyzes these important aspects of benefit-cost analysis.

TIME ADJUSTMENTS TO ENVIRONMENTAL BENEFITS AND COSTS

Critical to any environmental benefit-cost analysis is reconciling the timing of benefits and costs. First, benefits and costs do not necessarily accrue at the same time. Furthermore, even if they did, they are not realized immediately. The effects of environmental policy typically extend well into the future. Consequently, decision makers must be forward thinking in their evaluation of a policy proposal and must make projections about its future implications.

To support these forecasts, benefit and cost estimates must be adjusted to account for the fact that the value of a dollar is not constant over time. Two types of time-oriented modifications are necessary. One is **present value determination**, which accounts for the opportunity cost of money. The second is **inflation correction**, which adjusts for changes in the general price level. Let's examine each of these procedures.

Present Value Determination

Opportunity cost is one of the most pervasive concepts in economic thinking. It means that the highest valued alternative of any decision represents the full cost of that decision, whether it arises in production, consumption, or even purely financial transactions. To understand the concept in this latter context, consider the following simple scenario.

Assume a friend asks to borrow $200 today, promising to pay back the loan one year from today. What amount of money would the payback have to be to maintain your well-being or utility level? (To keep things simple, assume there is no inflation.) Economically, the answer depends on the opportunity cost of money, that is, its highest valued alternative use. If the $200 could be invested to yield a 5 percent return, then the opportunity cost of the loan is $10. Therefore, the appropriate payback should be $210, or ($200 + [$200][0.05]). Technically, this calculation represents the conversion of the loan value in the *present period* into its value in the *future period*.

Mathematically, the conversion for a one-year period is achieved using the simple formula:

$$FV = PV + r(PV) = PV(1 + r)$$

where

$$FV = \text{future value}$$
$$PV = \text{present value}$$
$$r = \text{rate of return}$$

This equation shows that the future value (FV) of a dollar is equal to its present value (PV) plus the opportunity cost of not using the dollar in the present period ($r[PV]$). If the future valuation has to account for more than one time period (t), the formula becomes:

$$FV = PV(1 + r)^t$$

where $t = 0, 1, 2, \ldots T$ is the number of periods.

present value determination
A procedure that discounts a future value (FV) into its present value (PV) by accounting for the opportunity cost of money.

Inverting the problem, we argue that $210 received one year from now is equivalent to having $200 today, or ($210/[1 + 0.05]). This conversion procedure is called **present value determination** because it involves **discounting** a future value (FV) into its present value (PV). Mathematically, the calculation is performed using a rearrangement of the first formula:

$$PV = FV(1/[1 + r])$$

In this form of the equation, r is called the **discount rate**, and the term $1/(1 + r)$ is called the **discount factor**. If the discounting involves more than one time period (t), the formula is written as:

discount factor
The term, $1/(1 + r)^t$, where r is the discount rate, and t is the number of periods.

$$PV = FV(1/[1 + r]^t),$$

where $(1/[1 + r]^t)$ is the discount factor. Discounting is the procedure economists use to adjust the value of environmental benefits and costs accruing in the future.

Notice that the discount rate (r) is the one variable element in present value determination. As this rate is elevated, the PV is decreased, and of course, the converse is true. Because the magnitude of the discount rate affects the conversion, its selection is critically important to benefit-cost analysis. In the context of environmental policy development, this rate can directly affect which proposals meet a given criterion and which are rejected out of hand. Because of the implications, discount rate selection is one of the more commonly debated points in the literature on present value analysis.

How *is* the discount rate selected for public policy decision making? There are many considerations, but one important position is that this rate should reflect the **social opportunity cost** of funds allocated to the provision of a public good. Why? Because monies used to support public policy initiatives are perceived as a transfer from the private sector. Therefore, it is argued that the discount rate used for public policy—called the **social discount rate**—should reflect the rate of return that *could* be realized through private spending on consumption and investment, assuming the same level of risk.

social discount rate
Discount rate used for public policy initiatives based on the social opportunity cost of funds.

Since 1992, the U.S. government has maintained a policy that its agencies use a social discount rate of 7 percent as a base-case for their benefit-cost analyses. In 2003, the U.S. Office of Management and Budget recommended using a 3 percent rate along with the 7 percent rate to illustrate the sensitivity of estimates to the discount rate employed.[1] The operative issue is whether the 7 percent base rate reflects the social opportunity cost of public expenditures. Because this rate is stated in **real** terms (i.e., net of inflationary effects), the **nominal** rate is roughly equivalent to the sum of the inflation rate and the 7 percent rate. Inflation over the last several years has been about 2.2 percent, which means that the nominal social discount rate is about 9.2 percent. If this rate is meant to capture the social opportunity cost of funds, it must be the case that society could earn a 9.2 percent return through private investment. Beyond this issue, there are also conceptual arguments about what *should* determine this rate, a question about which there is no clear consensus.

Inflation Correction

inflation correction
Adjusts for movements in the general price level over time.

Preliminary cost and benefit estimates also must be modified to account for movements in the general price level over time. This adjustment is referred to as **inflation correction**. To adjust a dollar amount in the present period for expected inflation in the next future

[1] U.S. Office of Management and Budget (September 17, 2003; October 29, 1992).

nominal value
A magnitude stated in terms of the current period.

period, the value must be converted to its **nominal value** for that period. The conversion uses a measure of price, such as the consumer price index (CPI), shown as:

$$\text{Nominal value}_{\text{period } x+1} = \text{Real value}_{\text{period } x}\,(\text{CPI}_{\text{period } x+1}/\text{CPI}_{\text{period } x}), \text{ or}$$
$$\text{Nominal value}_{\text{period } x+1} = \text{Real value}_{\text{period } x}\,(1+p)$$

where p is the rate of inflation between period x and period $x + 1$. The more generalized formula for any number of (t) periods is:

$$\text{Nominal value}_{\text{period } x+t} = \text{Real value}_{\text{period } x}\,(1+p)^t$$

real value
A magnitude adjusted for the effects of inflation.

Inverting the formula allows for the conversion of a nominal value to its **real value**, as follows:

$$\text{Real value}_{\text{period } x} = \text{Nominal value}_{\text{period } x+t}/(1+p)^t$$

deflating
Converts a nominal value into its real value.

This conversion procedure, sometimes called **deflating**, is used to assess changes over a period during which there has been inflation.[2] To illustrate, look back at the cost data presented in Table 8.3. Notice that all the annual dollar values are expressed in 2010 dollars. This is done to avoid confounding changes in control costs over time with changes in the general price level.

Summary of Deriving Time-Adjusted Benefits and Costs

present value of benefits (*PVB*)
The time-adjusted magnitude of incremental benefits associated with an environmental policy change.

present value of costs (*PVC*)
The time-adjusted magnitude of incremental costs associated with an environmental policy change.

The time-adjusted magnitudes for incremental benefits and costs are called the **present value of benefits (*PVB*)** and the **present value of costs (*PVC*)**, respectively. These are defined in real terms as:

$$PVB = \sum (b_t/[1+r_s]^t), \text{ with } b_t = B_t/(1+p)^t, \text{ and}$$
$$PVC = \sum (c_t/[1+r_s]^t), \text{ with } c_t = C_t/(1+p)^t$$

where r_s is the real social discount rate, b_t is incremental real benefits, B_t is incremental nominal benefits, c_t is incremental real costs, and C_t is incremental nominal costs. The steps to measuring PVB and PVC in real terms are shown in Table 9.1.

Example: Time-Adjusted Incremental Benefits

To illustrate how these steps might be carried out in practice, let's consider a hypothetical benefit assessment. Suppose an analyst is using the averting expenditure method (AEM) to monetize the incremental benefits of improved water quality. As part of the assessment, the analyst must estimate an individual's averting expenditures on a water-filtering system.[3] Assume the system has an initial capital cost (i.e., purchase and installation) of $200 and a useful economic life of 5 years. Further assume that annual operating costs for general maintenance and filters are quoted as $30 in nominal terms,

---

[2]Notice that the deflating formula is identical in form to the present value formula, except that the inflation rate (p) is substituted for the discount rate (r).

[3]For an actual study of this type of analysis, see Abdalla (June 1990) or Abrahams, Hubbell, and Jordan (2000).

TABLE 9.1 Deriving Time-Adjusted Benefits and Costs

Measuring Present Value of Benefits (*PVB*) in Real Terms

1. Monetize all present and future incremental benefits in nominal terms (B_t), where t refers to the appropriate time period ($t = 0, 1, 2, \ldots T$).
2. Deflate each of the B_t, converting to real dollars (b_t): $b_t = B_t/(1 + p)^t$.
3. Select the appropriate real social discount rate, (r_s).
4. Discount the b_t for each period as $b_t/(1 + r_s)^t$.
5. Sum the discounted b_t values over all t periods to find the **present value of benefits** in real dollars as $PVB = \sum(b_t/[1 + r_s]^t)$.

Measuring Present Value of Costs (*PVC*) in Real Terms

1. Monetize all present and future incremental costs in nominal terms (C_t), where t refers to the appropriate time period ($t = 0, 1, 2, \ldots T$).
2. Deflate each of the C_t, converting to real dollars (c_t): $c_t = C_t/(1 + p)^t$.
3. Maintain the same real social discount rate (r_s) used in discounting future benefits.
4. Discount the c_t for each period as $c_t/(1 + r_s)^t$.
5. Sum the discounted c_t values over all t periods to find the **present value of costs** in real dollars as $PVC = \sum(c_t/[1 + r_s]^t)$.

NOTE: The procedure is analogous for finding *nominal PVB* and *PVC*, as long as all magnitudes, including the discount rate, are properly adjusted.

due at the end of each year.[4] Part of the objective of the AEM is to determine the overall cost to the individual of purchasing, installing, and maintaining this filtration system, accounting for both the opportunity cost of money and expected inflation.

Assume for simplicity that prices are expected to rise on average by 5 percent each year. Using the conversion formula defined earlier, the real value of the operating costs at the end of the first year will be $\$30.00/(1.05)^1 = \$28.57$. At the end of the second year, the real costs will be $\$30.00/(1.05)^2 = \$27.21$, and so forth. Table 9.2 presents these calculations for each year of the system's economic life.

A further adjustment is necessary to find the present value of these future expenditures based on the opportunity cost of money. Assuming that the proper discount rate (r) is 10 percent in real terms,[5] the present value (PV) of each year's inflation-adjusted expenditures is found using the discounting formula. For example, the PV of the first year's operating costs is $\$25.97 = \$28.57(1/[1.10]^1)$. For the second year, the PV of the \$27.21 real expenditure is $\$22.49 = \$27.21(1/[1.10]^2)$, and so forth. These calculations are given along with the price adjustments in Table 9.2.

To complete the analysis, all the time-adjusted annual operating costs are added to the initial capital outlay of \$200. Notice that the \$200 is unaffected by the conversion

[4]The timing of the annual expenditures is a critical element of the discounting process. For illustrative purposes, assume that all annual costs are incurred at the *end* of each year. For more detail on this or any other discounting issue, see any public finance text, such as Rosen and Gayer (2008).

[5]Because the annual costs have been converted to their real values, the discount rate must also be expressed in real terms to adjust for the effect of inflation on the rate of return. Alternatively, one can express all the cost values in nominal terms and use the nominal discount rate.

TABLE 9.2 Assessing Future Operating Costs of a Water Filtering System

	ADJUSTMENTS FOR INFLATION (assuming an annual inflation rate (p) of 5%) Real value$_{period\ x}$ = Nominal value$_{period\ x+t}/(1+p)^t$		ADJUSTMENTS FOR OPPORTUNITY COST (assuming a real annual discount rate (r) of 10%) $PV = FV(1/[1+r]^t)$		
Year	**Expenditure in Nominal Terms**	**Future Expenditure in Real Dollars**	**Future Value (FV)**	**Discount Factor ($1/[1+r]^t$)**	**Present Value (PV)**
0	–	–	$200.00	1.0000 = $1/(1.10)^0$	$200.00
1	$30.00	$28.57 = $30/(1.05)^1$	$ 28.57	0.9090 = $1/(1.10)^1$	$ 25.97
2	$30.00	$27.21 = $30/(1.05)^2$	$ 27.21	0.8264 = $1/(1.10)^2$	$ 22.49
3	$30.00	$25.92 = $30/(1.05)^3$	$ 25.92	0.7513 = $1/(1.10)^3$	$ 19.47
4	$30.00	$24.68 = $30/(1.05)^4$	$ 24.68	0.6830 = $1/(1.10)^4$	$ 16.86
5	$30.00	$23.51 = $30/(1.05)^5$	$ 23.51	0.6209 = $1/(1.10)^5$	$ 14.60
TOTAL PRESENT VALUE IN REAL TERMS:					$299.39

procedures, since it is incurred in the present period. Overall, the *PV* of the filtering system expressed in real dollars is $299.39.

THE FINAL ANALYSIS: COMPARING ENVIRONMENTAL BENEFITS AND COSTS

The final phase of any benefit-cost analysis involves comparing the time-adjusted incremental benefits and costs and arriving at a decision based on their relative values. An interesting example of such a comparison is found in the Impact Assessment that was conducted for the Climate Change Act 2008 in the United Kingdom, which is discussed in Application 9.1.

In the broader context of environmental policy, this final phase is typically used to set a policy objective or to select a control instrument. The analysis would be carried out over a series of possible options and used to identify the "best" solution among them. Benefit-cost analysis is also used to evaluate an *existing* policy initiative, as shown in Table 9.3 for the U.S. Clean Air Act Amendments (CAAA), with estimates projected over the 1990–2020 period.

Two steps generally comprise this final phase of benefit-cost analysis, and they are:

1. Determine whether or not an option is *feasible* from a benefit-cost perspective.
2. Evaluate all feasible options based on a decision rule, and then *select* a single "best" solution.

Let's take a closer look at each of these.

benefit-cost ratio
The ratio of *PVB* to *PVC* used to determine the feasibility of a policy option if its magnitude exceeds unity.

Step One: Determining Feasibility

Feasibility Tests

To distinguish feasible options from infeasible ones, the analyst must compare the time-adjusted values of incremental benefits and incremental costs for each option under study. A common way to do this is to form a **benefit-cost ratio** of (*PVB*/*PVC*) and

APPLICATION 9.1 A Benefit-Cost Analysis of the UK's Climate Change Act 2008

In 2008, the U.K. government enacted the Climate Change Act, which establishes a legislative framework from which to establish and manage climate change initiatives. Among its provisions are:

- Establishment of binding targets on greenhouse gas (GHG) emission reductions by 2050
- Creation of a carbon budgeting system, which uses a series of carbon budgets to meet the 2050 objective
- Formation of a Committee on Climate Change to act in an advisory role to government
- Provisions to facilitate the introduction of domestic emissions trading programs

Before any such legislation can become law in the United Kingdom, an Impact Assessment (IA) must be conducted. These formal assessments include a benefit-cost analysis, which is conducted according to prescribed guidelines. In this case, the IA document offers what is characterized as a high-level discussion of the associated costs and benefits based on achieving stated goals. These goals were defined (and later revised in 2009), as medium- and long-term objectives to reduce GHG emissions by at least 34 percent by 2020 and by at least 80 percent by 2050, both relative to 1990 levels. There is no attempt, however, at estimating costs and benefits based on any specific set of policy initiatives or projected pathway to reach those goals. Rather, the estimations are presented as generalized and illustrative.

To formulate the cost estimate, the IA finds the present discounted value of the GDP reduction associated with the long-run costs of GHG mitigation. Based on the forecast used in the IA (from estimates given in the Stern Review), this GDP decline is estimated to be 1 percent (+/−3 percent), over the time period to 2050. The projected decline in GDP is then converted to a monetized value, which is found to be about £324 billion to £404 billion (or $524 billion to $653 billion) for the 43-year period up to the target year of 2050. (These values and others reported in the IA use 2008 as the base year.) The upper end of the range assumes no international trading, and the lower end assumes higher fossil fuel prices, which would reduce the quantity of energy demanded. The assessment carefully qualifies that this cost range is a partial estimate, excluding expenditures of moving to a low-carbon economy. Moreover, these costs are expected to be sensitive to a number of factors, including the policy mix that is ultimately implemented and the emissions reduction trajectory.

On the benefit side, reported values are based on the reduction in damages associated with climate change. One way to estimate the value of these damages is to use what is known as the social cost of carbon. This value represents the cost of climate change damages associated with the release of each additional ton of GHGs. In accordance with this approach, the IA reports a present discounted value of benefits for the Climate Change Act in the range of £457 billion to £1,020 billion (or $739 billion to $1,649 billion) for the period up to 2050. Again, these values are approximations, which would vary with such factors as the extent of international participation and the advance of low carbon technology. Moreover, ancillary benefits, such as greater energy security and improved health, are not included.

Although both the cost and benefit estimates are necessarily rough approximations, the IA document does use these values to project that the new legislation should yield net benefits over the 2007–2050 period in the range of £53 billion to £696 billion (or $86 billion to $1,125 billion). These results were intended to guide policy makers in evaluating the proposed legislation, which ultimately entered into law in 2008 as the Climate Change Act.

For further detail on the IA, visit **www.decc.gov.uk/assets/decc/85_20090310164124_e_@@_climatechangeactia.pdf**.

Sources: U.K. Department of Energy and Climate Change (DECC) (March 2009); U.K. Department for Environment, Food and Rural Affairs (DEFRA) (December 2, 2008; April 2008); Stern (2007).

compare the result to 1. If the ratio for a policy option exceeds 1, that option is among the feasible solutions; if not, it is rejected:

If $(PVB/PVC) > 1$ for a given option, the option is considered feasible.

present value of net benefits (*PVNB*) The differential of (*PVB* – *PVC*) used to determine the feasibility of a policy option if its magnitude exceeds zero.

By way of example, look again at Table 9.3, and notice that the benefit-cost ratio for the Clean Air Act Amendments (CAAA) is 32/1, indicating that, based on the estimates shown, these amendments represent a feasible option.

An equivalent test for feasibility is to find the **present value of net benefits (*PVNB*)**, which is $(PVB - PVC)$, and compare the result to zero. If this differential is greater than zero, the policy option is feasible; if not, it is rejected.

TABLE 9.3 Estimated Costs and Benefits of the Clean Air Act Amendments (CAAA) 1990–2020

Cost or Benefit Category	Present Value ($millions)
COSTS	
Electric Utilities	$ 53,000
Industrial Point Sources	46,510
Onroad Vehicles and Fuels	237,958
Non-Road Engines and Fuels	4,867
Other	65,547
Total Costs	$411,018
MONETIZED BENEFITS	
Avoided Mortality	$11,898,000
Avoided Morbidity	443,467
Ecological and Welfare Effects	335,304
Total Benefits	$12,980,000
Benefit-Cost Ratio	32/1

NOTES: All values are in 2010 dollars.
Estimates are generated using a 5 percent social discount rate.
Totals may not add due to rounding and discounting.

Source: U.S. EPA, Office of Air and Radiation (April 2011), Table 7-4, p. 7-8; Table 7-5, p. 7-9.

If $(PVB - PVC) > 0$ for a given option, the option is considered feasible.

What both rules communicate is that feasibility is implied if the benefits associated with a policy proposal *outweigh* the costs incurred.

How the Tests Differ

Although any policy option satisfying the $(PVB/PVC) > 1$ condition necessarily meets the requirement that $(PVB - PVC) > 0$, the equivalency ends there. Why? First, the magnitudes of the two expressions have different interpretations. The numerical value of the PVB/PVC ratio conveys the benefits of a policy option *per dollar of costs incurred*. For example, a ratio of 4.2 means that for every dollar of costs imposed on society, there are $4.20 in realized benefits. In contrast, the value of $(PVB - PVC)$ measures the dollar value of *excess benefits*, so it directly communicates the net gain to society. Second, although it might be tempting to think that either measure could be used to rank feasible projects and arrive at the same "best" solution, such a hypothesis is incorrect. In fact, of the two measures, only the benefit-cost differential can be used to establish such a ranking. Why? Because the benefit-cost ratio is not reliable for comparing feasible policy proposals.

Attempts to use the benefit-cost ratio for any kind of ranking among options will lead to ambiguous results. The problem arises because of the inherent uncertainty about whether to consider an event as an increase in costs or a reduction in benefits. Suppose, for example, that a policy option is expected to create some amount of unemployment. Should the value of that unemployment be counted as a negative benefit or a positive cost? Intuitively, it should be apparent that either approach is correct and should not affect the outcome. However, although the choice would not affect the value of the *PVNB*, it would change the benefit-cost ratio. A simple example will illustrate.

Assume that this hypothetical policy option is expected to cause unemployment valued at $2 million after discounting and adjusting for inflation. Excluding this effect, suppose also that the *PVB* is $18 million and that the *PVC* is $10 million. If the $2 million is counted as a negative benefit, the benefit-cost ratio would have a value of 1.6. If instead the $2 million is counted as a cost, the ratio would be 1.5. Try the same experiment with the *PVNB* to see that in either case the result is consistent at $6 million. What this example implies is that in the second step of evaluation, only the *PVNB* is useful in guiding the decision to select among feasible projects.

Step Two: Selecting Among Feasible Options

In practice, the decision-making process used to select among feasible options is guided by one of the following economic criteria: **allocative efficiency** or **cost-effectiveness**. Adding the dimension of time to these criteria, we arrive at the following decision rules:

To achieve **allocative efficiency:**	Maximize the present value of net benefits (*PVNB*).
To achieve **cost-effectiveness:**	Minimize the present value of costs (*PVC*) based on a preestablished benefit objective.

In the United States, the determination about which decision rule to use generally is dictated by law or by presidential executive order—an issue we will explore later in this chapter. For now, notice that both rules involve some type of optimization to guide the selection of the "best" available option.

Decision Rule: Maximize the Present Value of Net Benefits (PVNB)

maximize the present value of net benefits (*PVNB*) A decision rule to achieve allocative efficiency by selecting the policy option that yields greatest excess benefits after adjusting for time effects.

As noted previously, the **present value of net benefits (*PVNB*)** is the difference between the present value of benefits (*PVB*) and the present value of costs (*PVC*). Thus, this decision rule calls for choosing the option that gives society the highest amount of excess benefits after adjusting for time effects. The point at which total benefits exceed total costs by the greatest amount corresponds to where $MB = MC$, or where resources are efficiently allocated.

We summarize this decision rule in the following way:

To achieve allocative efficiency, maximize $PVNB = (PVB - PVC) = \Sigma(b_t/[1 + r_s]^t) - \Sigma(c_t/[1 + r_s]^t) = \Sigma([b_t - c_t]/[1 + r_s]^t)$ for all t periods ($t = 0, 1, 2, \ldots, T$) among all feasible alternatives.

Decision Rule: Minimize the Present Value of Costs (PVC)

minimize the present value of costs (*PVC*) A decision rule to achieve cost-effectiveness by selecting the least-cost policy option that achieves a preestablished objective.

The decision rule to **minimize the present value of costs (*PVC*)** guides the policymaker to select the least-cost option among those capable of achieving some preestablished objective. This is an explicit directive to set policy based on the criterion of cost-effectiveness. Of the two decision rules, this one is the more common. The reason is that the law often predetermines the level of environmental benefits to be achieved through its definition of an environmental quality objective. This means that the *PVB* is essentially fixed, leaving only *PVC* as a decision variable.

Before the cost comparison is done, the analyst has to first eliminate any options that do not achieve the requisite benefit level. Once done, the *PVC* is calculated for the remaining possibilities, and the option that incurs the lowest costs is selected. This decision rule can be summarized as follows:

> To achieve cost-effectiveness, minimize $PVC = \Sigma(c_t/[1 + r_s]^t)$ for all t periods ($t = 0, 1, 2, \ldots, T$) among all feasible alternatives that achieve a predetermined benefit level.

RESERVATIONS ABOUT BENEFIT-COST ANALYSIS

Although benefit-cost analysis is a viable risk management strategy, it is not without flaws. In fact, as a tool of environmental decision making, it has been the object of a fair amount of scrutiny. Critics of benefit-cost analysis generally point to two sources of concern. The first is the inherent problem of accurately measuring and monetizing environmental costs and benefits. The second is the failure of this approach to address potential equity problems.

Measurement Problems

Assigning a dollar value to environmental benefits and costs is a major challenge of using benefit-cost analysis. As noted in Chapter 7, estimation is particularly problematic on the benefit side. It is challenging not only to identify all the benefits of a policy proposal, but also to assign a dollar value to these gains. In fact, this measurement problem is often used to support using cost-effectiveness as an alternative criterion to allocative efficiency. On the opposite side of the analysis, capturing implicit costs poses the biggest challenge. Solving this problem continues to be the objective of ongoing research. Finally, the selection of the social discount rate, which affects the present value of both benefit and cost estimates, is the subject of much debate, even among proponents of the benefit-cost approach.

Equity Issues

In an objective assessment of benefit-cost analysis, equity concerns are not unfounded. The decision rules do not consider how the benefits and costs are distributed across segments of society. It is possible that the distribution of benefits might be skewed, so that some group of consumers or some industrial sector receives less than its fair share. Similarly, the distribution of costs might be such that some economic sector bears an inequitable share of the burden. Of course, either problem is less troublesome if the same inequity occurs on both the benefit side and the cost side, but there is no guarantee that this will be the case.

Because of the measurement problems and potential inequities associated with benefit-cost analysis, policymakers must make adjustments for any noted shortcomings. In the interim, economists are continuing to research better estimation techniques. Finding ways to improve the results is worth the effort. As a risk management strategy, benefit-cost analysis adds vital information to a very difficult undertaking. Moreover, in the United States, benefit-cost analysis is required of all major policy proposals—a decision that has a long history behind it.

U.S. GOVERNMENT SUPPORT OF BENEFIT-COST ANALYSIS[6]

Most policy analysts point to the Flood Control Act of 1936 as the first U.S. federal legislation that explicitly acknowledged benefits and costs in a public policy initiative. According to this mandate, federal funds could be allocated to water projects only if the associated benefits were found to exceed the costs. Notice that this is precisely the feasibility rule based on $(PVB - PVC) > 0$ that we discussed earlier. Since this historic first, the use of economic theory as a decision-making tool has been promoted in many public policy settings, a practice that continues into present-day rule making.

Over the last 40 years, U.S. presidents have used the power of the office to require economic analysis in federal regulatory decision making. Table 9.4 lists some of the presidential executive orders and other directives that specifically call for the consideration of benefits and/or costs in policy decisions. More information on how various U.S. regulations require the consideration of certain benefit and cost categories is available online at **http://yosemite.epa.gov/ee/epa/eed.nsf/webpages/RegulatoryEconomicAnalysisAtEPA.html**.[7] To understand the economic thinking in policy decisions, we next consider the instruments called for in the most recent of these directives, the **Regulatory Impact Analysis (RIA)** and the **Economic Analysis (EA)**.

TABLE 9.4 Recent History of U.S. Regulation Using Economic Analysis

Act/Executive Order	Year	Title of Analysis	Type of Analysis
Office of Management and Budget, October 1971 Memo	1971	Quality of Life (QOL) review	Costs, benefits
Executive Order 11821	1974	Inflation Impact Statement (IIS)	Costs, benefits, inflationary impacts
Executive Order 11949	1976	Economic Impact Statement (EIS)	Costs, benefits, economic impacts
Executive Order 12044	1978	Regulatory Analysis	Costs, economic consequences
Regulatory Flexibility Act	1980	Regulatory Flexibility Analysis	Impacts on small businesses
Executive Order 12291	1981	Regulatory Impact Analysis (RIA)	Costs, benefits, net benefits
Executive Order 12866	1993	Economic Analysis (EA)	Costs, benefits, net benefits
Executive Order 13258	2002	Economic Analysis (EA)	Costs, benefits, net benefits
Executive Order 13422	2007	Economic Analysis (EA)	Costs, benefits, net benefits
Executive Order 13563	2011	Economic Analysis (EA)	Costs, benefits, net benefits

Sources: U.S. EPA, Office of Policy, Economics, and Innovation (January 2001), pp. 7–9; U.S. EPA, Office of Policy, Planning, and Evaluation (August 1987), Table 2-1, p. 2-2.

[6]Part of this discussion is drawn from Andrews (1984), pp. 43–85; and U.S. EPA, Office of Policy, Planning, and Evaluation (August 1987), pp. 2-1–2-6, which is also available online at **http://yosemite.epa.gov/ee/epa/eed.nsf/webpages/EPAUseOfBenefitCostAnalysis.html**.

[7]At this site, click on "Table of Contents" and select "Statutory Authorities for Economic Analysis."

Executive Order 12291: Regulatory Impact Analysis (RIA)[8]

Although the framework for considering the benefits and costs of regulatory proposals had existed under previous administrations, it was not until President Reagan's first term in office that the efficiency criterion was explicitly made a part of the regulatory review process. In 1981, Reagan signed Executive Order 12291, which outlined requirements that federal agencies had to follow in proposing or reviewing any major rule or regulation, where a "major rule" referred to any regulation expected to have an annual effect of at least $100 million or otherwise adversely affect the economy. The full text of this executive order is accessible via the Internet at **www.archives.gov/federal-register/codification/executive-order/12291.html**.

Unlike prior references to economic considerations, this set of requirements outlined in the executive order explicitly called for meeting economic criteria in adopting regulatory actions. **Allocative efficiency** is referenced in two paragraphs that specifically require the maximization of net benefits. The criterion of **cost-effectiveness** is noted in another paragraph that calls for meeting a given objective on the basis of least net cost to society.

Regulatory Impact Analysis (RIA)
A requirement under Executive Order 12291 that called for information about the potential benefits and costs of a major federal regulation.

Reagan's Executive Order also required that all major regulations be accompanied by what is called a **Regulatory Impact Analysis (RIA)**. Guidelines for an RIA call for descriptions of potential benefits, potential costs, and potential net benefits, leaving little doubt about the importance of benefit-cost analysis in the regulatory review process or policy proposal assessment.

Executive Order 12866: Economic Analysis (EA)

In September 1993, President Clinton signed Executive Order 12866, which replaced Reagan's Executive Order 12291. To access the complete text of Clinton's executive order on the Internet, visit **www.archives.gov/federal-register/executive-orders/pdf/12866.pdf**. This executive order continued Reagan's commitment to economic fundamentals in policy formulation and evaluation. Among its "Principles of Regulations" that pertain directly to economic criteria are the following (emphasis added):

> When an agency determines that a regulation is the best available method of achieving the regulatory objective, it shall design its regulations in the most *cost-effective* manner to achieve the regulatory objective.
>
> Each agency shall assess both the costs and the benefits of the intended regulation and... propose or adopt regulation only upon a reasoned determination that the benefits of the intended regulation justify its costs.

Economic Analysis (EA)
A requirement under Executive Order 12866, amended by Executive Orders 13258 and 13422, and Executive Order 13563 that calls for information on the benefits and costs of a "significant regulatory action."

Similar to Reagan's executive order, Clinton's applies to all "significant regulatory actions," which includes those expected to have an annual impact on the economy of at least $100 million. For each such action, the relevant agency is required to prepare an **Economic Analysis (EA)**, which acts as the successor to the RIA. The EA follows the same logic and economic criteria outlined in an RIA. It, too, explicitly calls for a maximization of net benefits or, where otherwise required under the law, that a proposed action be the most cost-effective.

[8]For more detail concerning this executive order, see *U.S. Federal Register 46* (February 17, 1981), pp. 13193–98.

Executive Order 13422: Economic Analysis (EA)

The fundamental guidelines that support the use of economic criteria continued to be followed by President George W. Bush. This was formalized first in February 2002, with the signing of Executive Order 13258, and later in January 2007, when President Bush signed Executive Order 13422. This latter document can be accessed online at **http://edocket.access.gpo.gov/2007/pdf/07-293.pdf**. Aside from making only minor amendments to Clinton's Executive Order 12866, this executive order extends the use of economic criteria in policy design and evaluation through an Economic Analysis, as outlined previously.

Executive Order 13563: Economic Analysis (EA)

In 2011, President Obama signed Executive Order 13563, which reaffirmed and enhanced the fundamentals outlined in President Clinton's Executive Order 12866.[9] The full text is available at **www.gpo.gov/fdsys/pkg/FR-2011-01-21/pdf/2011-1385.pdf**. Like its predecessors, this order requires that a regulation be proposed or adopted only if the associated benefits justify the costs. Further, if choosing among alternative regulatory initiatives, the one that maximizes net benefits should be chosen.

To better illustrate how economic criteria actually are used in an Economic Analysis, it is helpful to examine a real-world example. One such case is the Environmental Protection Agency's (EPA) benefit-cost analysis used to support the phasedown of lead in gasoline in the 1980s.

REDUCING LEAD IN GASOLINE: A REAL-WORLD EXAMPLE

Petroleum refiners began using lead additives as an inexpensive source of octane in the 1920s. In so doing, refiners were able to reduce engine knock and increase engine performance. However, these gains were not without consequence. Coincident with the increased use of lead additives, researchers observed a link between lead exposure and adverse health effects, such as mental and cardiovascular disorders. Acting on the growing body of scientific evidence, the EPA was charged with the responsibility of achieving a reduction in the lead content of gasoline.

In 1985, the U.S. lead standard was 1.1 grams per leaded gallon (gplg). As part of its plan to reduce society's exposure to lead, the EPA proposed tightening this standard to 0.1 gplg, effective January 1, 1986. Since such a policy change came under the legal definition of a major rule, the EPA was required to conduct an RIA. This in turn meant that the agency had to examine the costs, benefits, and net benefits of approaches that could be used to meet the more stringent lead standard. The text of this RIA is available online by visiting the EPA's National Center for Environmental Economics at **http://yosemite.epa.gov/ee/epalib/ee222.nsf/vwl**, and clicking on "Lead in Gasoline."

Following the estimation of incremental benefits and costs required in the RIA, the EPA officially announced a low-lead standard of 0.1 gplg, effective January 1, 1986. The decision was supported by the economic evidence that the more stringent lead standard would be beneficial to society, as discussed in Application 9.2.

[9]In January 2009, President Obama revoked President Bush's Executive Orders 13258 and 13422 by signing Executive Order 13497.

APPLICATION 9.2 Benefit-Cost Analysis in Practice: The RIA for Reducing Lead in Gasoline

In the 1980s, the EPA conducted a Regulatory Impact Analysis (RIA) to determine the incremental benefits and costs of tightening the nation's lead standard from 1.1 grams to 0.1 gram per leaded gallon (gplg).

On the benefit side of the analysis, a key issue was to estimate the value of predicted health improvements. Of particular interest was the associated improvement in children's health based on abundant scientific evidence about the increased susceptibility of children to lead exposure. To monetize this benefit, the EPA examined the associated savings in medical care expenditures and in compensatory educational expenditures to derive an incremental benefit estimate of $600 million for 1986, the first full year of the revised lead standard. (This value and all others in the RIA report are stated in 1983 dollars.)

For adult health gains, the EPA examined the evidence linking lead exposure to elevated blood pressure in adult males. Based on this evidence, the agency measured the associated benefits of a tighter lead standard as the value of improved cardiovascular health. When monetized on a per-case base, these were reported as $220 per case of hypertension; $60,000 per heart attack; $44,000 per stroke; and $1 million per life saved. In the aggregate, the dollar value of these benefits ranged from $5.897 billion in 1986 to $4.692 billion in 1992.

Three nonhealth benefits also were assessed: the reduction in harmful emissions caused by misfueling, lower maintenance costs, and increased fuel economy. Accounting for all associated damage reductions, the EPA estimated the incremental benefits for each in 1986 to be $222 million, $914 million, and $187 million, respectively.

For the cost side of the RIA, incremental costs were based on the associated change in manufacturing costs of gasoline and other petroleum products. Using an engineering cost model, the EPA found that tightening the lead standard would generate an incremental cost of $608 million in 1986 and $441 million in 1992.

The accompanying table summarizes these results. Because the evidence for adult blood pressure benefits was not well established, the EPA chose to rely on the net benefit estimates *excluding* these values. Even using this more conservative estimate, net benefits are above $1 billion for each of the seven full years affected by the proposed ruling.

Finally, to adjust for time differences over the period, the EPA conducted a present value analysis of its findings. Using the conservative estimate that excludes blood pressure effects and selecting a *real* social discount rate of 10 percent, the present value of net benefits (*PVNB*) over the period was estimated to be $5.9 billion, which supported the proposed new lead standard.

Monetized Benefits and Costs of Reducing Lead in Gasoline (millions of 1983 dollars)

	1985	1986	1987	1988	1989	1990	1991	1992
Monetized benefits								
Children health effects	223	600	547	502	453	414	369	358
Adult blood pressure	1,724	5,897	5,675	5,447	5,187	4,966	4,682	4,692
Conventional pollutants	0	222	222	224	226	230	239	248
Vehicle maintenance	102	914	859	818	788	767	754	749
Fuel economy	35	187	170	113	134	139	172	164
TOTAL	2,084	7,821	7,474	7,105	6,788	6,517	6,216	6,211
Monetized costs								
TOTAL	96	608	558	532	504	471	444	441
Net Benefits	1,988	7,213	6,916	6,573	6,284	6,045	5,772	5,770
Net Benefits Excluding Adult Blood Pressure	264	1,316	1,241	1,125	1,096	1,079	1,090	1,079

NOTES: The estimates are reported under the assumption of partial misfueling. Sums may not equal totals shown because of rounding.

Source: U.S. EPA, Office of Policy, Planning, and Evaluation (February 1985).

CONCLUSIONS

In its most fundamental form, environmental policy is aimed at minimizing society's risk of exposure to environmental hazards. Although all of society is involved in public policy development, most of the detail of formulating, implementing, and monitoring regulatory provisions falls in the hands of public officials. By itself, this fact supports the need for analytical tools such as benefit-cost analysis to guide the decisions that define environmental policy.

Benefit-cost analysis, although not without flaws, is a useful strategic approach to environmental decision making. Its objectivity is its primary strength in helping officials evaluate the social gains and the opportunity costs of their decisions. Choosing between a policy that saves one more life in a million lives exposed and a regulation that will reduce the risk of birth defects by 10 times that amount is a tough decision by any measure. Yet these kinds of decisions are precisely what environmental policy development is all about. Every resource allocated to save a national forest is one less resource available to clean rivers and streams. Every dollar spent to clean a hazardous waste site is one less dollar available to save an endangered species. Such is the dilemma that confronts every society as it decides how to allocate scarce resources, and such is the fundamental premise of economic thought.

Guiding decision making is the purpose of benefit-cost analysis. Explaining what this analytical tool can accomplish, how it is implemented, and what it fails to achieve have been the primary objectives of this chapter. Because benefit-cost analysis is becoming a more dominant force in public policy decisions, it is critical that we comprehend how it influences environmental regulations and how, in so doing, it affects the quality of our lives.

Summary

- In benefit-cost analysis, two types of time-oriented adjustments are necessary: present value determination and inflation correction.
- To discount a future value (FV) into its present value (PV), use the conversion formula:

$$PV = FV(1/[1+r]^t),$$

where t is the number of periods, and $1/(1 + r)^t$ is the discount factor.
- To adjust a value in the present period for expected inflation in the future period, it must be converted to its nominal value for that period, using the formula:

$$\text{Nominal value}_{\text{period } x+t} = \text{Real value}_{\text{period } x}\,(1+p)^t,$$

where p is the rate of inflation.
- Time-adjusted incremental benefits and costs are referred to as the present value of benefits,

$$PVB = \Sigma(b_t/[1+r_s]^t),$$

and the present value of costs,

$$PVC = \Sigma(c_t/[1+r_s]^t),$$

respectively.
- The first step of benefit-cost analysis identifies feasible options. A policy option is feasible if

$$(PVB/PVC) > 1 \text{ or if } (PVB - PVC) > 0.$$

- The second step of benefit-cost analysis evaluates all acceptable options *relative* to one another on the basis of a decision rule.
- One decision rule is to maximize the present value of net benefits ($PVNB$), which equals $\Sigma(b_t/[1 + r_s]^t) - \Sigma(c_t/[1 + r_s]^t)$, among all feasible alternatives to achieve allocative efficiency. Another decision rule is to minimize the present value of costs (PVC), which equals $\Sigma(c_t/[1 + r_s]^t)$ among all feasible alternatives that attain a predetermined benefit level to achieve cost-effectiveness.
- Measuring and monetizing intangibles, the selection of the social discount rate, and capturing implicit costs are among the concerns of using benefit-cost analysis. Another problem is the potential for an inequitable distribution of costs and benefits.
- Over the last 40 years, U.S. presidents have required economic analysis in federal regulatory decision making. Instruments that call for benefit and cost considerations in policy decisions are the Regulatory Impact Analysis (RIA) and the Economic Analysis (EA), both of which are required by presidential executive order.
- Executive Order 12291, Executive Order 12866, which was amended by Executive Orders 13258 and 13422, and Executive Order 13563 explicitly call for the achievement of allocative efficiency and cost-effectiveness in adopting regulatory actions.

Review Questions

1. Suppose industry abatement costs rise from $850 million in 2011 to $1,000 million in 2012 in nominal terms and that the CPI is 100 in 2011 and 106 in 2012.
   a. Evaluate the change in costs over the period in real terms, first in 2011 dollars and then in 2012 dollars.
   b. Are your answers the same? Explain why or why not.
2. To examine the implications of selecting various discount rates, reconsider the water filtration system example in the text but change the discount rate from 10 percent to 5 percent.
   a. Find the present value of the system.
   b. Now compare this present value with the one calculated under the assumption of a 10 percent rate. Explain the difference intuitively.

3. Refer to the RIA for the revised lead ruling discussed in Application 9.2.
   a. Mathematically confirm that the estimate of the present value of net benefits (*PVNB*) is approximately $5.9 billion, as stated.
   b. Discuss how the *PVNB* would have changed if the EPA had used a social discount rate of 8 percent.
4. Refer back to the cost and benefit data presented in Table 9.3.
   a. Using the total benefit and cost values shown, determine if the Clean Air Act Amendments (CAAA) pass the feasibility test. Explain.
   b. Based on the information in the table, do the CAAA achieve allocative efficiency?
5. Suppose that the state of Pennsylvania is proposing an environmental policy plan aimed at reducing smog in its urban centers, particularly Pittsburg and Philadelphia. An economist estimates the present value of benefits (*PVB*) of the proposed policy to be $4.2 billion and the present value of costs (*PVC*) to be $5.6 billion. Is this proposal feasible? Explain.
6. In 1997, the EPA proposed an increase in the national air quality standards for particulate matter (PM). Because the change was considered a major rule, the agency was required to conduct an RIA and assess the potential costs and benefits of the proposed standard. The completed RIA can be accessed at **www.epa.gov/ttn/oarpg/naaqsfin/ria.html**. As part of this RIA, benefit and cost data for the *existing* standard were given and are replicated here:

Control Region	Annual Control Cost ($1990 millions)
Midwest/Northeast	380
Southeast	2
South Central	230
Rocky Mountain	210
Northwest	140
West	130
National	1,100

Source: Appendix C, Table C1: "Costs and Benefits of Achieving the Current PM10 and Ozone Standard," available at the EPA Web site provided in this question.

Benefit Category	Monetized Benefits ($1990 billions)
Mortality	
Short-term exposure	$2.950
Long-term exposure	2.860
Chronic bronchitis	2.010
Hospital admissions	
Total respiratory	0.022
Congestive heart failure	0.003
Ischemic heart disease	0.005
Upper respiratory systems	
Asthma attacks	0.001
Work loss days	0.015
Minor restricted activity days	0.057
Consumer cleaning cost savings	0.039
Visibility	0.320
Total Benefits	
Using short-term mortality	$5.400
Using long-term mortality	$5.300

Source: Appendix C, Table C1: "Costs and Benefits of Achieving the Current PM10 and Ozone Standard," available at the EPA Web site provided in this question.

   a. Examine the control cost estimates, and provide a plausible explanation of why there are incremental costs differences across regions for meeting the PM standard.
   b. Economically, can you justify the PM standard? Explain.

   c. List some of the limitations associated with benefit-cost analysis. Which of these are applicable to this particular economic analysis? Explain.

Additional Readings

Blomquist, Glenn C. "Self-Protection and Averting Behavior, Values of Statistical Lives, and Benefit Cost Analysis of Environmental Policy." *Review of Economics of the Household 2*(*1*) (March 2004), pp. 89–110.

Brent, Robert J. *Applied Cost-Benefit Analysis*. 2nd Edition: Northampton, MA: Elgar, 2007.

______. *Handbook of Research on Cost-Benefit Analysis*. Northampton, MA: Elgar, 2009.

Cropper, Maureen L. "Has Economic Research Answered the Needs of Environmental Policy?" *Journal of Environmental Economics and Management 39*(*3*) (May 2000), pp. 328–50.

Cropper, Maureen L., and Wallace E. Oates. "Environmental Economics: A Survey." *Journal of Economic Literature 30* (June 1992), pp. 675–740.

Farrow, Scott, and Michael Toman. "Using Benefit-Cost Analysis to Improve Environmental Regulations." *Environment 41*(*2*) (March 1999).

Hanley, Nick, and Edward B. Barbier. *Cost-Benefit Analysis and Environmental Policy*. Northampton, MA: Elgar, 2009.

Harrington, Winston, Lisa Heinzerling, and Richard D. Morgenstern eds. *Reforming Regulatory Impact Analysis*. Washington, DC: RFF Press, 2009.

Morgenstern, Richard D. *Economic Analyses at EPA: Assessing Regulatory Impact*. Baltimore, MD: World Resources Institute, 1997.

Noll, Roger G. "The Economic Significance of Executive Order 13422." *Yale Journal of Regulation 25*(*1*) (Winter 2008), pp. 113–24.

Pannell, David J., and Steven G. M. Schilizzi, eds. *Economics and the Future: Time and Discounting in Private and Public Decision Making*. Northampton, MA: Elgar, 2009.

Pearce, David. "Does European Union Environmental Policy Pass a Cost-Benefit Test?" *World Economics 5*(*3*) (July–September 2004), pp. 115–38.

Pearce, David William, Giles Atkinson, and Susana Mourato. *Cost-Benefit Analysis and the Environment: Recent Developments*. Paris: Organisation for Economic Co-operation and Development (OECD), 2006.

Scheraga, Joel D. *Discounting and Environmental Policy*. Burlington, VT: Ashgate, 2003.

U.S. Environmental Protection Agency, Office of the Administrator. *Guidelines for Preparing Economic Analyses*. Washington, DC: December 2010.

To access additional course materials, visit www.cengagebrain.com. At the home page, search for the ISBN of this title (shown on the back cover). This will take you to the product page where these resources can be found.

MODULE 4 The Case of Air

From a technical perspective, there has been some form of air pollution as long as the earth has supported life. Yet air pollution did not begin to be a pervasive problem until the Industrial Revolution. The combination of manufacturing processes, population growth, and motorized transportation that characterized the nineteenth century also marks the period when air quality became a real concern. As industrial growth proceeded on its course, so too did air pollution, though few realized the severity of the problem during those early days of industrial development.

In the United States, as urban smog became more apparent in cities like Los Angeles and Philadelphia, pressure began to mount for public officials to take action. Local communities and some state governments began to enact air quality laws. But it was decades before the federal government took an active role in what was fast becoming a worldwide problem.

Over time, U.S. air quality control policy has evolved into a comprehensive body of laws, with major revisions embodied in the 1990 Clean Air Act Amendments. These extensive provisions are supported by an equally sophisticated infrastructure to implement controls and monitor compliance. More recently, changes to national standards and objectives also have been proposed in the form of rules, new programs, and presidential executive orders. Among these are recent rulings and decisions that are affecting national policy on greenhouse gases. In addition, international agreements and programs have been implemented to address worldwide air pollution problems, such as global warming and ozone depletion.

On balance, it's fair to say that there has been a good deal of progress. We have a much better sense of the implications of air pollution, thanks in part to medical and scientific research. As knowledge advanced, we have become better able to find solutions, and many nations have begun substituting economic incentives for command-and-control approaches to regulating air quality. Understanding the extent of this progress and analyzing what has been accomplished require a fairly thorough examination of the facts and some sense of the evolution of policy.

To that end, this module provides a multistep approach to learning about air quality problems and policy solutions, using economics as an analytical tool. We begin in Chapter 10 with a broad-based analysis of U.S. air quality control policy and the standard-setting process that defines clean air for the nation. In Chapter 11, we study the problem of urban smog and how U.S. standards are implemented through controls aimed at mobile sources, including the new greenhouse gas emissions standards. Chapter 12 follows with an analogous presentation for stationary sources, focusing in part on how well these initiatives reduce acidic deposition. Lastly, in Chapter 13, we explore global air pollution with an overview of U.S. policy on GHG emissions, an in-depth discussion of current international policy, and various proposals aimed at ozone depletion and climate change.

CHAPTER 10

Defining Air Quality: The Standard-Setting Process

"Air pollution is not merely a nuisance and a threat to health. It is a reminder that our most celebrated technological achievements—the automobile, the jet plane, the power plant, industry in general, and indeed the modern city itself—are, in the environment, failures."

—**Barry Commoner (1917–)**

Among the most important elements of clean air policy development is determining the target level of air quality to be achieved. In simpler terms, policymakers have to decide how clean is clean air. In the United States, standards are set to define a level of air quality that will protect human health and the ecology. Many factors influence this standard-setting process, such as technological feasibility, energy requirements, and economic considerations. But more fundamentally, this process involves making a determination about which substances present a risk to society.

Identifying these contaminants is important, but it's only part of the story. Ultimately, a decision must be made about *how much* of any contaminant society can tolerate—an issue of some debate. What *is* clear is that we cannot entirely avoid the risks of air pollution without incurring unreasonable opportunity costs. To understand this assertion, consider the following.

natural pollutants Contaminants that come about through nonartificial processes in nature.

anthropogenic pollutants Contaminants associated with human activity.

Some contaminants in the atmosphere are **natural pollutants**, such as pollen, dust particles from volcanic disturbances, gases from decaying animals and plants, even salt spray from the oceans. Because these pollutants occur naturally, they are virtually beyond human control.

Other contaminants are **anthropogenic**, meaning caused by human activity. These include such substances as carbon monoxide from tailpipe exhausts and sulfur dioxide emissions from electricity generation. Although these types of pollutants are controllable and generally present a greater environmental risk than natural contaminants, they cannot be avoided completely without incurring the unrealistic opportunity cost of no industrial activity. Consequently, we must accept the reality that our air quality will not be synonymous with zero pollution. But just what level of pollution is "acceptable," and how should it be determined?

Using economics, we can identify an efficient level of air quality where the associated marginal social benefits are balanced with the marginal social costs. Economics can also be used to design market-based instruments to achieve air quality objectives. These theories can and sometimes do guide policy decisions. However, the real-world complexities of government procedures, scientific uncertainty, and political pressures tend to delay or even prohibit the realization of an economic approach.

Our goal in this chapter is to explore the realities of clean air policy development. Of particular interest is to examine how the government uses standards to define air quality and to evaluate the economic implications of using a standards-based approach. At the end of this chapter is a helpful reference list of the acronyms used in our discussion.

AN OVERVIEW OF U.S. AIR QUALITY LEGISLATION

On November 15, 1990, President George H. W. Bush signed into law extensive changes in air quality policy in the form of the 1990 Clean Air Act Amendments (CAAA). This legislative landmark received strong congressional support. Its hallmark is the integration of market-based policy instruments. The amendments came only after years of political battles, false starts, and a chronicle of events that underscores the complexity of drafting policy to deal with environmental problems.

In the Beginning

Prior to the 1950s, all air quality legislation in the United States had been enacted by state and local governments. Logically, the earliest of these legislative acts were passed during the Industrial Revolution. Even in contemporary times, state and local governments typically have taken the lead. The most notable example is California, whose often path-breaking legislation has been in direct response to the smog problem in Los Angeles. In 1961, California passed the first state-level air pollution law to control motor vehicle emissions. This legislation followed many years of struggle between California officials and the American automobile industry, the subject of Application 10.1.

At the federal level, the history of air pollution initiatives is much shorter. In fact, there were no *national* air quality laws until the Air Pollution Control Act of 1955 was passed, and there was no truly comprehensive legislation until the Clean Air Act of 1963 was enacted, nearly a century after state and local governments had begun to take action. From that point on, a series of revisions and new initiatives helped to form U.S. policy as it is now defined. Look at Table 10.1 for a synopsis of the major federal laws that map out this evolution.

Current U.S. Policy

With this chronicle of policy evolution behind it, Congress passed into law some of the most comprehensive legislation in its history—the 1990 Clean Air Act Amendments (CAAA). These amendments map out national directives for reducing the risks of air pollution. The 1990 Amendments are extensive, comprising 11 sections of new and revised statutes. Certain of the titled sections within the 1990 Amendments use market-based approaches. For example, Title IV uses tradeable allowances to control sulfur dioxide emissions, and Title VI does the same for stratospheric ozone-depleting

APPLICATION 10.1 A Look Back at California Smog and the Automobile Industry

California's severe smog problem motivated its early attempts in the 1950s to gain the support of the automobile industry in solving its dilemma. Despite evidence connecting motor vehicle emissions to smog production revealed in Dr. A. J. Haagen-Smit's 1951 study, the automobile industry denied that auto emissions were a major contributor to the problem and argued that the source of urban smog required further investigation. The result was a volley between the industry and the state government that accomplished little throughout the 1950s.

In 1959, manufacturers acknowledged that the engine's design was an important source of the emissions problem. So in 1961, automakers began to install engine ventilation devices on all new cars sold in California to control these emissions, a technology change that was eventually mandated under California law. However, this technology was not new. In fact, it had been developed in the 1930s.

Apparently weary of the lack of progress, the California government tried to force the issue. Legislation was passed calling for the installation of equipment to control vehicle exhausts as soon as it was developed either by the major automakers or by any three independent manufacturers. The automobile producers asserted that such technology simply did not exist. Yet, in June 1964, the state confirmed that independent parts manufacturers could provide the needed add-on equipment at a reasonable cost and mandated its installation on all new vehicles starting with the 1966 model year. Predictably, the automakers announced in August 1964 that they had the capability to install their own devices on 1966 models, despite the fact that in March they had claimed that 1967 was the earliest model year by which they could accommodate the change.

Not only were the major auto producers using delay tactics, but the measure of their innovative effort was unimpressive. Ford and General Motors developed only a simple air pump, while Chrysler's solution, marketed as a clean air package, involved only minor alterations to its fuel and carburetion system. It was later discovered that Chrysler's "solution" contributed significantly to nitrogen oxide emissions.

Why were there such significant delays? Part of the answer is based on economic theory. If there had been a demand for cleaner running cars, competitive firms would have recognized the advantage of being the first to satisfy that demand as well as the threat to their survival of being left behind in what is often a race to innovate. However, it is not clear that such a demand existed, and even if it had, market incentives were dampened by the market power then possessed by the Big Three—General Motors, Ford, and Chrysler.

In 1953, the three firms formally pooled their efforts by jointly investigating the pollution problem. Later, in 1955, they executed a cross-licensing agreement to share access to any patents on emission controls. In so doing, they effectively removed the incentive for any one of them to find an innovative solution. In 1969, the Department of Justice filed suit against the Big Three, charging them with collusive attempts to hinder the advancement of pollution control technology. The suit was settled by a consent decree that ended the cross-licensing agreement.

Sources: White (1982), Chapter 3; Seskin (1978); U.S. Senate, Staff of the Subcommittee on Air and Water Pollution of the Committee on Public Works (October 1973).

substances.[1] However, the underlying structure of the law is command-and-control oriented. The full text of these amendments is available online at **www.epa.gov/air/caa/**.

During George W. Bush's presidency, an initiative was put forth to amend the current legislation. Formally introduced into Congress as the Clear Skies Act, this initiative proposed a multi-pollutant approach and the use of market-based instruments to reduce power plant emissions.[2] When the bill was stalled in Congress, the Environmental Protection Agency (EPA) proceeded with a suite of actions in 2004 called the Clean Air Rules. Included were regulations aimed at mercury, tropospheric (or ground-level) ozone, fine particle emissions, releases from nonroad diesel equipment, and interstate pollution.

---

[1]Note the distinction between tropospheric, or ground-level, ozone, which is a pollutant, and stratospheric ozone, which protects the earth from ultraviolet radiation. More on these issues will be discussed in Chapters 11 and 13, respectively.

[2]Details on this proposed policy reform are available at **www.epa.gov/clearskies**.

TABLE 10.1 Retrospective of U.S. Air Quality Legislation

Legislation	Brief Summary
Air Pollution Control Act of 1955	This first federal legislation on air pollution was limited, aimed mainly at providing federal appropriations to state governments for research and training.
Clean Air Act of 1963	This act established emissions regulations for stationary sources and called for a committee to be formed with the auto industry to study the effects of motor vehicle emissions.
Motor Vehicle Air Pollution Control Act of 1965	This act authorized the Department of Health, Education, and Welfare (HEW) to set emission standards on new motor vehicles, but no deadline was set.
1965 Clean Air Act Amendments	HEW was authorized to establish the first federal standards for mobile sources.
Air Quality Control Act of 1967	Air quality control regions (AQCRs) were to be established. States were to set and implement ambient air quality standards.
1970 Clean Air Act Amendments	National Ambient Air Quality Standards (NAAQS) were established for stationary sources, and emissions limits were set for mobile sources, both to be implemented through State Implementation Plans (SIPs). New source performance standards (NSPS) were set at more stringent levels than those for existing sources.
1977 Clean Air Act Amendments	AQCRs were reclassified into attainment and nonattainment regions to protect regions that were cleaner than required by the NAAQS. These regions were termed prevention of significant deterioration (PSD) areas.
1990 Clean Air Act Amendments	Eleven titled sections outlined directives to improve air quality, including regulations to achieve the NAAQS, stronger enforcement provisions, and market-based approaches to control acid deposition and stratospheric ozone.

Sources: Portney (1990a); Wolf (1988), Chapter 2; Stern (January 1982); Mills (1978), pp. 189–94.

Although the ruling on interstate pollution, called the Clean Air Interstate Rule (CAIR), was considered to be the most important environmental initiative enacted by the Bush administration, it subsequently was challenged. As a result, in July 2008, the U.S. Court of Appeals for the D.C. Circuit vacated this rule, just as it had done for the mercury regulation earlier that year. Responding to a petition from the EPA, the Court later remanded the matter to the EPA for further rulemaking without vacating it. Hence, the CAIR would remain in effect until a new rule was issued to replace it.[3]

In July 2010, the EPA under the Obama administration proposed the new rule, known then as the Transport Rule, and renamed when finalized in July 2011 as the Cross-State Air Pollution Rule (CSAPR). Aimed at the eastern half of the United States, this rule, which replaces the CAIR, calls for significant reductions in sulfur dioxide (SO_2) and nitrogen oxide (NO_X) emissions that are transported across state borders. The targeted polluters are power plants located in 27 states. Because of the CSAPR

[3]U.S. EPA, Office of Air and Radiation (January 15, 2009); McCarthy, Parker, and Meltz (2008).

along with other directives, SO_2 emissions and NO_X emissions are projected to fall by 73 percent and 54 percent, respectively, by 2014, relative to their 2005 levels.[4] For more information on the CSAPR, visit **www.epa.gov/crossstaterule/**.

Policy revisions such as these are aimed at achieving the nation's air quality policy objectives as outlined by the CAAA. These goals are defined by national standards, which effectively identify a common air quality level for the nation. The standards are implemented through an extensive infrastructure that facilitates federal oversight. That said, the uniformity of such an approach likely impedes the achievement of allocative efficiency or cost-effectiveness. To clarify this assertion, let's begin by examining the statutory objectives of the Clean Air Act.

DEFINING THE OBJECTIVES OF AIR QUALITY CONTROL

The objectives of U.S. air quality policy today were originally defined in the first comprehensive federal law on air quality control: the Clean Air Act of 1963. Chief among these is the following:

> To protect and enhance the quality of the Nation's air resources so as to promote the public health and welfare and the productive capacity of its population.[5]

To achieve the nation's objectives, the government must understand the risks of air pollution and the abatement necessary to reduce these risks to an acceptable level. The first step is to identify the primary causes of air pollution and to isolate those contaminants deemed most harmful.

Identifying Major Air Pollutants

Think about the extraordinary responsibility of determining which pollutants are most responsible for air pollution and setting the level at which to control them. Such a decision-making process is complex given the wide range of human sensitivity to pollutants, the uncertainty about health and welfare effects—particularly over the long term—and the enormous task of assessing the effects of various combinations of pollutants.

Criteria Pollutants

criteria pollutants Substances known to be hazardous to health and welfare, characterized as harmful by criteria documents.

In the United States, official reports called **criteria documents** present scientific evidence on the properties and effects of any known or suspected pollutant. This evidence is used to identify common air pollutants known to present a risk to health and the environment. Officially, these **criteria pollutants** are identified substances known to be hazardous to health and welfare. Today, there are six identified criteria pollutants:

---

[4] U.S. EPA, Office of Air and Radiation (July 7, 2011).

[5] 42 U.S.C. §740(b)(1), July 14, 1055, as amended 1963.

- Particulate matter (PM-10 and PM-2.5),[6]
- Sulfur dioxide (SO_2),
- Carbon monoxide (CO),
- Nitrogen dioxide (NO_2),
- Tropospheric ozone (O_3), and
- Lead (Pb).

For more information on these common air pollutants, including their health and environmental effects, visit **www.epa.gov/air/urbanair**.

Hazardous Air Pollutants

hazardous air pollutants
Noncriteria pollutants that may cause or contribute to irreversible illness or increased mortality.

A second group of contaminants identified by U.S. legislation is **hazardous air pollutants**, or **air toxics**. These are noncriteria pollutants that may contribute to irreversible illness or increased mortality. What distinguishes these substances from the criteria pollutants is that the associated risk is much greater, although typically a much smaller segment of society is affected. The 1990 CAAA include a list of 189 identified hazardous air pollutants, which is to be periodically revised, as needed. Currently, the number of identified hazardous air pollutants under the EPA's control is 188. Prior to 1990, only eight such substances had been identified and placed under the EPA's control. It was precisely because of this lack of regulatory action that a more aggressive policy was enacted in the 1990 Amendments. To further investigate hazardous substances, visit the EPA's Air Toxics Web site at **www.epa.gov/air/toxicair**.

Greenhouse Gases (GHGs)

A U.S. Supreme Court decision in April 2007 found that greenhouse gases are pollutants covered by the Clean Air Act and that the EPA is authorized to regulate them.[7] **Greenhouse gases (GHGs)** are those substances collectively responsible for the absorption process that naturally warms the earth. This group of pollutants includes:

greenhouse gases (GHGs)
Gases collectively responsible for the absorption process that naturally warms the earth.

- Carbon dioxide (CO_2),
- Methane (CH_4),
- Nitrous oxide (N_2O),
- Hydrofluorocarbons (HFCs),
- Perfluorocarbons (PRCs), and
- Sulfur hexafluoride (SF_6).

In 2009, the EPA presented two key findings: (1) its Endangerment Finding—that current and projected concentrations of GHGs threaten public health and welfare; and (2) its Cause or Contribute Finding—that emissions for mobile sources contribute to GHG pollution and threaten public health and welfare.[8]

---

[6]PM-10 refers to particles less than 10 micrometers in diameter, and PM-2.5 has an analogous meaning. Standards for PM-2.5 were added by the EPA in 1997.

[7]That this decision sparked controversy and debate is evidenced in part by a bill introduced in Congress as the Energy Tax Prevention Act of 2011, which would rescind the EPA's newly granted authority to control GHGs. The bill was passed in the House in April 2011 (Tracy 2011).

[8]Greenhouse (April 3, 2007); U.S. EPA, Office of Air and Radiation, Office of Atmospheric Programs, Climate Change Division (April 14, 2011a). The official case is filed as *Massachusetts et al. v. EPA et al.*, No. 05–1120.

It has not been determined whether these pollutants will ultimately be identified within one of the existing categories, i.e., as criteria pollutants or hazardous air pollutants, and indeed, this issue is the subject of some debate and discussion.[9] What *is* known at this time is that these recent findings have led to several new rulings on GHG emissions, which will be addressed in subsequent chapters. For now, further information on GHGs is available at **www.epa.gov/climatechange/emissions/index.html**.

SETTING STANDARDS TO DEFINE AIR QUALITY

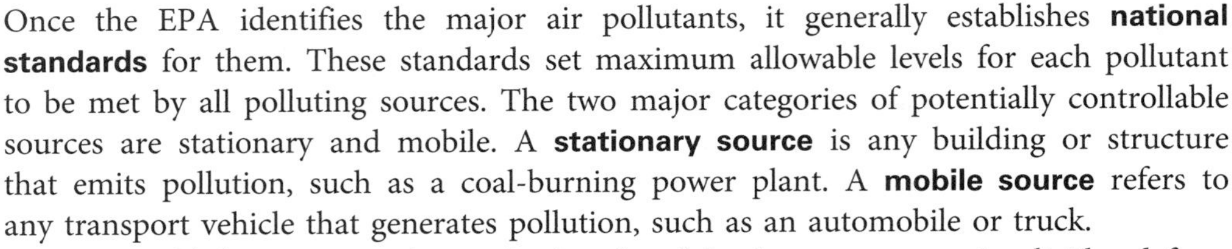

stationary source A fixed-site producer of pollution.

mobile source Any nonstationary polluting source.

Once the EPA identifies the major air pollutants, it generally establishes **national standards** for them. These standards set maximum allowable levels for each pollutant to be met by all polluting sources. The two major categories of potentially controllable sources are stationary and mobile. A **stationary source** is any building or structure that emits pollution, such as a coal-burning power plant. A **mobile source** refers to any transport vehicle that generates pollution, such as an automobile or truck.

By establishing national standards, the federal government implicitly defines air quality for the entire country—a practice not exclusive to the United States. For example, Japan, the United Kingdom, and some European countries also use a standard-setting approach to communicate an acceptable level of air quality. Although the specifics vary across countries, in almost every case, the identified air pollutants coincide with those named as criteria pollutants in the United States. See, for example, a summary of the United Kingdom's air quality standards in Application 10.2. For an overview of other nations' environmental policies, visit the United Nations Environment Programme (UNEP) at **www.unep.org**.

Standards for Criteria Air Pollutants

National Ambient Air Quality Standards (NAAQS) Maximum allowable concentrations of criteria air pollutants.

The standards for the six criteria pollutants are called **National Ambient Air Quality Standards (NAAQS)**. Within this group are two subcategories—primary and secondary NAAQS:

- **Primary NAAQS** are set to protect public health, with some margin of safety.
- **Secondary NAAQS** are intended to protect public welfare.

Originally established under the 1970 Clean Air Act Amendments, these standards have been revised from time to time. In fact, the law requires that the criteria and the NAAQS be reviewed by the EPA every five years.

Recent NAAQS Revisions

Among the most recent changes to the NAAQS is the 2010 revision to the primary standard for sulfur dioxide (SO_2). The objective is to establish better protection from short-term SO_2 exposure by setting a one-hour standard to replace both the 24-hour and the annual standards that had been in place. According to the EPA, this revision is expected to cost approximately $1.6 billion ($2010) in 2020 to fully implement, while generating health benefits between $14.1 billion ($2010) and $35.7 billion ($2010) per year.[10]

[9]The interested reader may wish to consult Meltz (2009) or Parker and McCarthy (2009).

[10]U.S. EPA, Office of Public Affairs (June 3, 2010).

APPLICATION 10.2 Defining Air Quality Standards in the United Kingdom

In the United Kingdom, as in the United States, air quality is measured relative to well-defined standards for a set of identified pollutants. Officially, these standards are concentrations of a substance considered to be acceptable based on scientific evidence of human health effects associated with exposure. These standards are used as benchmarks to establish legally binding emissions limits, which are to be attained within a stated period and not exceeded once attained. They are revised periodically as is done in the United States.

A summary of these limits for human health protection, called EU Limit Values, is shown in the following table based on the Air Quality Standards Regulations 2010 of England and the United Kingdom, which are available at **www.legislation.gov.uk/uksi/2010/1001/made/data.pdf**. These limits are defined in the 2008 directive of the European Parliament and the Council of the European Union on ambient air quality.

Notice that five of the six pollutants identified in the table are also covered by the National Ambient Air Quality Standards (NAAQS) in the United States. Moreover, although no limit values are yet in force for ozone (O_3) and small particulate matter (PM-2.5) as there are in the United States, the EU does have target values, or objectives, in place for these substances of 120 μg/m^3 and 25 μg/m^3, respectively.

Pollutant	Concentration Measured as	EU Limit Values
Benzene (C_6H_6)	Annual average	5 μg/m^3
Carbon monoxide (CO)	Maximum daily running 8-hour mean	10 mg/m^3
Lead (Pb)	Annual mean	0.5 μg/m^3
Nitrogen dioxide (NO_2)	1-hour mean	200 μg/m^3 not to be exceeded more than 18 times a year
	Annual mean	40 μg/m^3
Particles (PM_{10})	24-hour mean	50 μg/m^3 not to be exceeded more than 35 times a year
	Annual mean	40 μg/m^3
Sulfur dioxide (SO_2)	1-hour mean	350 μg/m^3 not to be exceeded more than 24 times a year
	24-hour mean	125 μg/m^3 not to be exceeded more than 3 times a year

Sources: United Kingdom (2010); European Parliament and the Council of the European Union (2008).

Another proposal for tightening the NAAQS was made in 2010 to strengthen the primary and secondary standards for ground-level ozone (O_3) to help combat urban smog. It was estimated that the new standards would generate incremental benefits within a range of \$14.1 billion (\$2010) and \$108.2 billion (\$2010) and incremental costs between \$20.6 billion (\$2010) and \$97.4 billion (\$2010), depending on the levels at which the new standards are set. Interestingly, in 2011, President Obama requested that the EPA withdraw this proposal because of concerns about the associated regulatory costs in the face of a sluggish economy. The standard is due for its five-year review and reconsideration in 2013 in accordance with the rulings of the Clean Air Act.[11]

The primary and secondary NAAQS as of 2011, which include the SO_2 revisions, are listed in Table 10.2. These are also available online at **www.epa.gov/air/criteria.html**.

[11]Solomon and Tennille (2011); U.S. EPA, Office of Public Affairs (January 7, 2010).

TABLE 10.2 National Ambient Air Quality Standards in Effect in 2011

Pollutant	Standard Value		Standard Type
Carbon monoxide (CO)			
8-hour average	9 ppm	(10 mg/m³)	Primary
1-hour average	35 ppm	(40 mg/m³)	Primary
Lead (Pb)			
Rolling 3-month average	0.15 μg/m³ (2008)		Primary and Secondary
Quarterly average	1.5 μg/m³ (1978)		Primary and Secondary
Nitrogen dioxide (NO_2)			
Annual arithmetic mean	53 ppb		Primary and Secondary
1-hour average	100 ppb		Primary
Particulate matter (PM-10)	*Particles with diameters of 10 micrometers or less*		
24-hour average	150 μg/m³		Primary and Secondary
Particulate matter (PM-2.5)	*Particles with diameters of 2.5 micrometers or less*		
Annual arithmetic mean	15 μg/m³		Primary and Secondary
24-hour average	35 μg/m³		Primary and Secondary
Ozone (O_3)			
8-hour average	0.075 ppm (2008)		Primary and Secondary
8-hour average	0.08 ppm (1997)		Primary and Secondary
1-hour average	0.12 ppm		Primary and Secondary
Sulfur dioxide (SO_2)			
Annual arithmetic mean	0.03 ppm (1971)		Primary
24-hour average	0.14 ppm (1971)		Primary
1-hour average	75 ppb (2010)		Primary
3-hour average	0.5 ppm (2010)		Secondary

NOTES: Parenthetical value is an approximately equivalent concentration; ppm = parts per million; μg/m³ = micrograms per cubic meter; ppb = parts per billion; mg/m³ = milligrams per cubic meter.
The 1978 lead standard remains in force until one year after an area is designated for the 2008 standard with some exceptions.
The 8-hour ozone standard for 1997 remains in place for implementation purposes while the EPA addresses the transition from the 1997 standard to the 2008 standard. The 1-hour ozone standard was revoked except for some areas that have continuing obligations under it.
The 1971 sulfur dioxide standards remain in force until one year after an area is designated for the 2010 standards with some exceptions.

Source: U.S. EPA, Office of Air and Radiation, Office of Air Quality Planning and Standards (November 8, 2011).

Notice that none of the standards are set at a zero level. Thus, if we accept these standards as the nation's definition of air quality, then we must conclude that acceptable air quality does not mean the absence of criteria pollutants.

National Emission Standards for Hazardous Air Pollutants (NESHAP) Standards applicable to every major source of any identified hazardous air pollutant.

Standards for Hazardous Air Pollutants

U.S. law also calls for the establishment of **National Emission Standards for Hazardous Air Pollutants (NESHAP)**. These are intended to protect public health and the environment, taking into account the costs to attain the standards, any nonair quality health and environmental impacts, and energy requirements. The NESHAP are to attain the maximum reduction for each toxic achievable, referred to as maximum achievable control technology (MACT). Where possible, this reduction should achieve a complete ban on the substance.

Recent NESHAP Revisions

Recently, a number of proposals for new NESHAP have been made, some of which have generated debate because of the cost implications for business and industry. Among these is a rule finalized in February 2011 (with reconsiderations proposed in December 2011), which is aimed at reducing mercury and other toxic emissions from about 2,400 new and about 187,000 existing, area source, industrial, commercial, and institutional boilers.[12] The motivation for this new standard is that the targeted emissions, which include mercury, other metals, and organic air toxics (such as dioxins), are known to pose serious health and environmental risks.[13] Incremental costs for this new NESHAP for area source boilers are estimated to be $493.2 million ($2010) annually. Incremental benefits for just the reduction in fine particle pollution are forecasted to be in the range of $212.7 million ($2010) to $526.7 million ($2010) in 2014, the implementation year. Among the incremental health benefits identified though not monetized by the EPA are the annual avoidance of 24–61 premature deaths, 40 nonfatal heart attacks, 17 cases of chronic bronchitis, 40 hospital and emergency room visits, and 800 cases of respiratory symptoms.[14]

Another high-profile NESHAP issued in 2011 is one directed toward coal-fired (about 1,200) and oil-fired (approximately 150) power plants. Referred to as the Toxics Rule, this NESHAP is aimed at reducing heavy metals like mercury, arsenic, and nickel, as well as acid gases, such as hydrogen chloride. These pollutants are believed capable of causing serious health problems, including cancer. The Toxics Rule replaces the court-vacated mercury regulation outlined under President Bush's Clean Air Rules, known as the Clean Air Mercury Rule. Power plants are targeted in part because they are the largest source of mercury emissions released into the air, responsible for 50 percent in the aggregate, with those facilities using coal responsible for 99 percent. Annual costs of this rule are estimated to be $11.5 billion ($2010) in 2016, the implementation year. Estimated incremental health benefits due solely to reducing exposure to fine particles are between $62.1 billion ($2010) and $147.2 billion ($2010) in 2016. Among the identified nonmonetized benefits are the avoidance of 6,800–17,000 premature deaths; 4,500 cases of chronic bronchitis; 11,000 nonfatal heart attacks; 12,200 hospital and emergency room visits; and 220,000 cases of respiratory symptoms.[15]

---

[12]Area source facilities are those emitting less than 10 tons per year. Analogous regulations were finalized for major source facilities, i.e., those emitting more than 10 tons per year, and for various incinerators.

[13]For example, lead and mercury can harm brain development in children, affecting their learning and memory. Other toxics, such as cadmium, dioxin, and formaldehyde, can cause cancer and other serious health problems in adults and children.

[14]U.S. EPA, Office of Public Affairs (December 2, 2011); U.S. EPA, Office of Air and Radiation, Office of Air Quality Planning and Standards (February 23, 2011).

[15]U.S. EPA, Office of Public Affairs (December 21, 2011); U.S. EPA, Office of Air and Radiation, Office of Air Quality Planning and Standards (March 18, 2011).

Standards for Greenhouse Gas (GHG) Pollutants[16]

As noted, the recent rulings on GHGs did not state whether these newly identified pollutants are to be considered part of one of the identified groups of pollutants covered by the Clean Air Act. Hence, no ambient standards have been established for GHGs, nor have any such standards been proposed. Again, this is an issue of some debate that will likely be assessed and analyzed for some time.

In any case, the absence of ambient standards for GHGs does not preclude the EPA from determining *how* to lower GHG emissions now that the Supreme Court has ruled that the agency has the authority to do so. In that regard, the EPA in 2010 finalized new GHG emissions standards for passenger cars and light-duty trucks for model years 2012–2016 and in 2011 proposed comparable rulings for subsequent model years. Also in 2011, the agency adopted analogous GHG standards for medium- and heavy-duty vehicles for model years 2014–2018. The EPA also is initiating plans to control GHG emissions from stationary sources, a process that includes reporting and permitting requirements in the near term with emissions standards for major GHG sources to be finalized subsequently. These new regulations will be discussed further in upcoming chapters.

IMPLEMENTING THE STANDARDS

To implement the NAAQS and NESHAP, an infrastructure has been established that involves both federal and state governments. It is defined through two components:

- State Implementation Plans (SIPs)
- Air Quality Control Regions (AQCRs)

State Implementation Plans (SIPs)

State Implementation Plan (SIP)
A procedure outlining how a state intends to implement, monitor, and enforce the NAAQS and the NESHAP.

Coordination between the two major levels of government is achieved through State Implementation Plans (SIPs). A **State Implementation Plan (SIP)** is an EPA-approved procedure of how a state intends to implement, monitor, and enforce the NAAQS and the NESHAP.[17]

The SIP system follows a federalist format by delegating certain tasks to different jurisdictions. The standard setting is assigned mainly to the federal level to standardize air quality across the country. State governments are responsible for implementing the standards and monitoring polluters, since their knowledge of the region gives them an advantage in doing so.

Air Quality Control Regions (AQCRs)

Air Quality Control Region (AQCR)
A federally designated area within which common air pollution problems are shared by several communities.

To coordinate states' responsibilities, **Air Quality Control Regions (AQCRs)** are defined within each state's jurisdiction. These are geographic areas designated by the federal government within which common air pollution problems are shared by several communities.

[16]U.S. EPA, Office of Transportation and Air Quality (August 2011; July 2011; April 2010).

[17]The NAAQS represent the *minimum* requirements to be attained by every state. However, at its discretion, a state can submit a plan to achieve more stringent standards.

TABLE 10.3 Nonattainment Areas as of 2011

Pollutant	Number of Nonattainment Areas	Population (thousands)
Carbon monoxide (CO)	0	0.0
Lead (Pb) 1997	2	4.7
Lead (Pb) 2008	16	–
Nitrogen dioxide (NO_2)	0	0.0
8-hour ozone (O_3)	44	118,188.6
Particulate matter (PM-2.5) 1997	39	88,394.4
Particulate matter (PM-2.5) 2006	32	70,176.2
Particulate matter (PM-10)	45	25,440.1
Sulfur dioxide (SO_2)	9	1,060.6

NOTES: The effective designation date is April 21, 2011.
Estimated population data for the lead 2008 nonattainment areas are not available.
Source: U.S. EPA, Office of Air and Radiation, Office of Air Quality Planning and Standards (April 21, 2011).

Currently, 247 AQCRs have been designated. These well-defined areas are monitored to determine whether or not a region is in compliance with the national standards. Today, a number of AQCRs have not met the current NAAQS for one or more of the criteria pollutants. Table 10.3 provides information on these nonattainment areas as of 2011. As these data indicate, regions have the greatest difficulty achieving the PM-2.5, PM-10, and the ozone (O_3) standards, with 32, 45, and 44 regions, respectively, classified as nonattainment for these pollutants. For updates, visit **www.epa.gov/air/oaqps/greenbk/index.html.**

Reclassification of AQCRs to Protect "Clean Air Areas"

In 1972, the Sierra Club filed suit against the EPA for failing to protect areas that were cleaner than what was required by law.[18] The environmental group argued that the existing NAAQS were aimed solely at *improving* air quality in areas that did not meet the standards. The concern was that nondegradation areas, as they are called, would be allowed to deteriorate to the existing national standards. The Sierra Club claimed that the lack of protection for nondegradation areas was a violation of the law, because one of the statutory purposes of the Clean Air Act is to *protect* as well as *enhance* the quality of the nation's air resources.

prevention of significant deterioration (PSD) area
An AQCR meeting or exceeding the NAAQS.

nonattainment area
An AQCR not in compliance with the NAAQS.

Ultimately, the group won the suit, and in 1974, a new program was established to protect these already clean areas. The new "clean air areas" program redefined the structure of federal air pollution control and played a significant role in directing future policy decisions. Regions meeting or exceeding national standards were redesignated as **prevention of significant deterioration (PSD) areas**. Those that did not were identified as **nonattainment areas**. Once determined, every PSD area was to be designated as Class I, II, or III, a progressive classification based on the maximum concentration of criteria pollutants allowed. Class I areas, the most stringently controlled, include wilderness areas and national parks.

[18]See *Sierra Club v. Ruckelshaus*, 344 F. Supp. 253 (D.D.C. 1972).

1990 Reclassification of Nonattainment AQCRs

In response to the nation's persistent problem of urban air pollution, the 1990 Amendments reclassify all nonattainment areas for the pollutants most responsible—ozone, carbon monoxide, and particulate matter—into new progressive categories based on existing pollutant concentrations. For example, ozone nonattainment areas are reclassified into five groups: marginal, moderate, serious, severe, and extreme. These new categories not only identify the severity of pollution but also provide some justification to set more stringent regulations in those areas with higher pollution levels.

Monitoring Air Quality Across Regions

Determining the compliance status of each region and the assessment of the nation's air quality depends upon a systematic measurement of the six criteria pollutants. This is accomplished either by estimating the emissions level of each pollutant or by actually measuring the ambient concentration of each in some volume of air.

Estimating Pollutant Emissions Levels

In the United States, best available engineering methods are used to derive annual emissions estimates for over 450 source categories, which include almost all anthropogenic sources. These are then aggregated to determine regional and national emissions trends. Figure 10.1 shows emissions estimates for the criteria pollutants in 2009 compared to 1990.

FIGURE 10.1 Changes in U.S. Emissions of Criteria Pollutants, 1990–2009

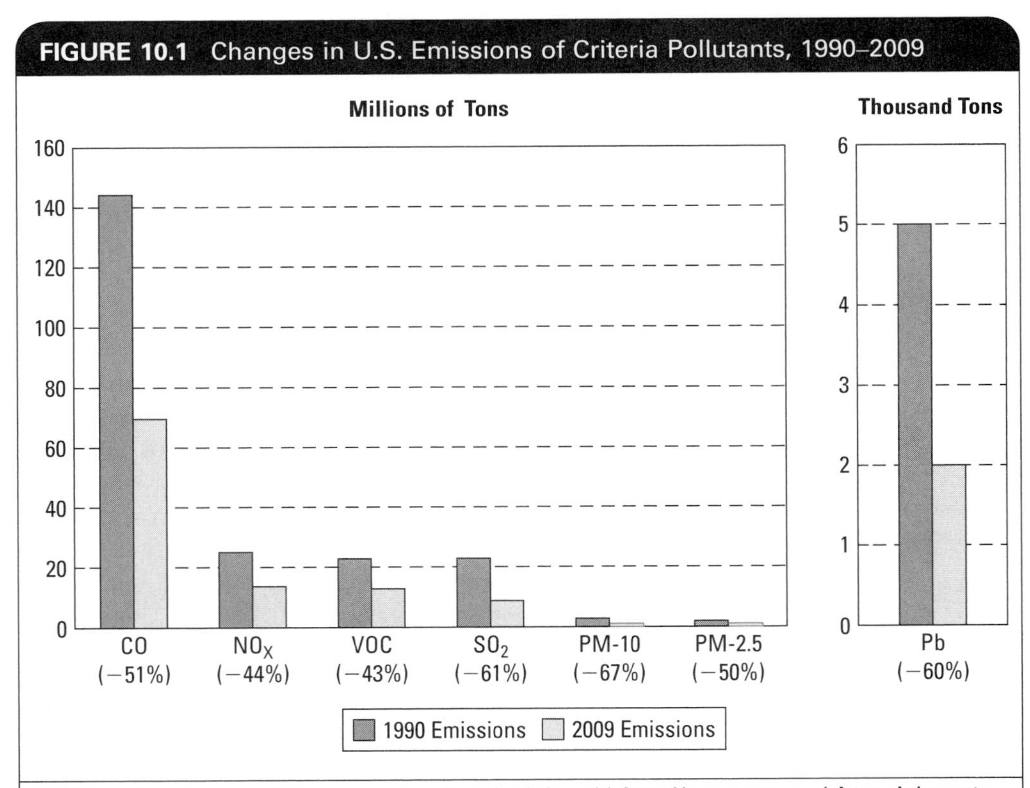

NOTE: For Internet access to EPA reports on air quality and emissions, visit **http://www.epa.gov/airtrends/reports.html**.

Source: U.S. EPA, Office of Air and Radiation (February 10, 2011).

These estimates show that emissions reductions have been achieved for all criteria pollutants over the long run. The most substantial decline is for lead, which is the result of a nationwide phase-out program. After setting a new standard for leaded gasoline in the mid 1980s, the EPA progressively tightened the standard until a complete ban was achieved by 1996. This phasedown was accomplished by a program of **tradeable credits** issued to refineries for achieving lead reductions greater than those required by law.

Measuring Pollutant Concentrations

Pollutant concentration levels are actually measured at air-monitoring station sites located throughout the country. Most of these sites are in urban regions characterized by relatively high pollutant concentrations and population exposure. All sites report their data to the EPA by means of an air-monitoring network. A wealth of information on air pollution monitoring in the United States is available at **www.epa.gov/oar/oaqps/montring.html**.

ANALYSIS OF U.S. AIR QUALITY POLICY

If we consider the evolution of U.S. air quality legislation, it is clear that this policy has become more comprehensive over time. But is it reasonable to assess accomplishments by the number of provisions that define air quality legislation or by the implementation and monitoring systems that have been established? These may be indicators that the nation is trying to strengthen policy, but do they suggest that policy is any more effective?

It seems simple enough to argue that the bottom line in assessing air quality policy is whether or not the air is any cleaner. However, such an argument is an oversimplification of what is a much more complex problem. Look back at the trend data in Figure 10.1, which suggest that air quality has improved. Although this is likely true to a point, other factors must be considered. Among these are the accuracy of the emissions estimates and the determination of what part of any observed improvement is actually attributable to effective policy.

So how *do* we assess air quality policy? To help us answer this question, we generally rely on policy evaluation criteria. Recall from Chapter 1 that these include an equity criterion—environmental justice—and economic criteria—allocative efficiency and cost-effectiveness.

Environmental Justice: An Equity Criterion

Although equity considerations are not the mainstay of economic analysis, they have taken on increasing importance in U.S. environmental policy making. In 1992, the Office of Environmental Justice was established, and in 1993, environmental justice became one of the EPA's seven guiding principles. The commitment was further underscored by President Clinton in 1994 when he issued Executive Order 12898, which directs all federal agencies to make environmental justice part of their missions.[19]

[19] This executive order can be accessed online at **www.archives.gov/federal-register/executive-orders/pdf/12898.pdf**.

APPLICATION 10.3 *Clean School Bus USA*: Air Quality and Environmental Justice

When assessing air quality policy using an equity criterion, a number of questions are relevant. Does the policy correct for any inequities across population groups? Does its implementation affect all segments of society in the same way? Although differences in air quality or exposure to environmental risk might seem to be strictly regional in orientation, it turns out that they sometimes translate to inequities across income, ethnic, and racial population groups. Why? Because a high proportion of minorities and low-income groups live in urban centers, where the ambient air is dirtier than in suburban and rural areas. Recognition of this potential inequity has prompted a national movement toward achieving **environmental justice**.

In keeping with this goal, the Office of Air and Radiation, along with all offices within the EPA, has made a formal commitment to integrate environmental justice into its programs and initiatives. Among these is a program called *Clean School Bus USA*. Launched in 2003, the program's chief objectives are:

- To decrease unnecessary school bus idling
- To replace diesel-powered school buses with cleaner-running vehicles designed to meet current pollution abatement standards
- To retrofit existing vehicles with pollution-reducing equipment and use cleaner diesel fuels

The expected environmental benefit of this effort is the improvement in children's health associated with reduced emissions from school buses.

That this initiative has potentially significant implications for society and for environmental justice is evidenced by a few simple facts. In the United States, some 24 million children ride a bus to school, spending 20 minutes to several hours each day on these vehicles. Buses built before 1990 pollute as much as 6 times the amount released by a bus built in 2004 and 60 times more than a contemporary bus equipped with newer, cleaner technology. Children are particularly susceptible to the health effects linked to diesel fuel emissions, which include respiratory disease and worsening of long-term conditions like asthma. Funding for the *Clean School Bus USA* program gives priority to helping children in environmental justice areas—minority or low-income communities that bear a disproportionate environmental burden.

In 2003, Congress awarded $5 million for grants to help school districts fund the upgrade of their bus fleets. Among the considerations used to decide which districts would receive a grant award was environmental justice. In fact, the EPA's Office of Air and Radiation explicitly stated that one of their objectives was health improvement in communities characterized as low income, minority, having high rates of asthma, or having exposure to a disproportionate level of diesel pollution. Subsequent congressional allocations of $5 million in grants for fiscal year 2004, $7.4 million for 2005, and $7 million in 2006 suggest that the successes of *Clean School Bus USA* program should continue into the future.

Visit **www.epa.gov/cleanschoolbus/index.htm** to learn more about this innovative program.

Sources: U.S. EPA, Office of Transportation and Air Quality (October 19, 2007); U.S. EPA, Office of Air and Radiation (2004).

During his campaign for office, President Obama made a formal commitment to strengthen the EPA's Office of Environmental Justice and to make environmental justice policies a priority within the EPA.[20] An example of how environmental justice is being integrated into air quality programs is discussed in Application 10.3. More information is available at the EPA's Environmental Justice home page at **www.epa.gov/compliance/environmentaljustice/**.

Allocative Efficiency: An Economic Criterion

Notwithstanding the importance of equity concerns, it is also important to consider whether air policy initiatives are allocatively efficient. Not that this is an easy task. Even if we could isolate what amount of air quality improvements are attributable to

[20]Obama and Biden (n.d.).

policy, there is still the challenge of converting any policy-driven air improvements to monetized values of social benefits, such as reduced health risks and increased visibility. One must also consider the costs borne by society to achieve whatever progress has been made. These costs should include not only explicit expenditures for compliance, monitoring, and enforcement but also implicit costs resulting from shifts in technology, substitution of fuels and other inputs, and changes in available goods and services.

Collectively, these are the tasks that comprise an economic analysis of environmental policy. By estimating and evaluating the costs and benefits associated with the Clean Air Act, one can determine whether our air control policy is **allocatively efficient**. To illustrate, we present a two-part benefit-cost investigation of U.S. air policy. The first part assesses the overall efficiency of the Clean Air Act. The second considers the efficiency of the standard-setting process itself.

BENEFIT-COST ANALYSIS OF THE CLEAN AIR ACT

Independent Benefit-Cost Analysis of the 1990 Amendments

Drawn from the work of other researchers, a benefit-cost analysis of the 1990 CAAA was conducted by Paul Portney (1990b), an economist, who at the time was affiliated with Resources for the Future in Washington, DC. To accomplish this task, Portney assigns estimated marginal benefits and costs to each of the three major sets of provisions of the 1990 law: those for urban air pollution (Titles II and V), those for hazardous air pollutants (Title III), and those for acidic deposition (Title IV).

Marginal Costs and Benefits

To facilitate the discussion, we present Portney's estimates of marginal benefits and costs in Table 10.4, which we assume are points lying on the true *MSB* and *MSC* curves.[21] The table lists the reported values in 1990 dollars for the three sets of provisions as well as for the aggregate of all three. Notice that most of the estimates are reported as a range of values, since there is some amount of guesswork involved.

Overall, Portney suggests that abatement linked to the 1990 Amendments should yield an *MSB* in the range of $6 billion to $25 billion annually, with most of the benefit resulting from urban air quality improvements. He also offers a rough point estimate for *MSB* of $14 billion. The comparable estimate for the *MSC* is $29 billion to $36 billion annually.[22] Again, most of the costs are attributable to urban air quality controls. A point estimate for *MSC* to correspond to the benefit side is $32 billion.

---

[21]Recall from Chapters 7 and 8 that technically these values are *incremental* as opposed to *marginal*, since the magnitudes represent a discrete change from one policy initiative to another.

[22]Portney is careful to qualify his findings because only *explicit* private costs are considered in the analysis.

TABLE 10.4 Portney's Benefit-Cost Analysis of the 1990 CAAA

	Acidic Deposition (Title IV)	Urban Air Quality (Titles II and V)	Hazardous Air Pollutants (Title III)	Aggregate Range	Aggregate Point Estimate
Annual Marginal Costs	$4	$19–$22	$6–$10	$29–$36	$32
Annual Marginal Benefits	$2–$9	$4–$12	$0–$4	$6–$25	$14

NOTE: All values are stated in billions of 1990 dollars.
Source: Based on Portney (1990b).

Benefit-Cost Comparison

Based on these estimates, one possible conclusion is that Titles II through V of the 1990 Amendments might be overregulating society, since *MSC* far outweighs *MSB*.[23] This relationship holds even if we conservatively compare the low end of the range for *MSC* ($29 billion) with the upper end of the range for *MSB* ($25 billion), giving an excess of $4 billion. On the other hand, the excess could be as much as $30 billion if we use the high end of the range for *MSC* ($36 billion) with the low end of the range for *MSB* ($6 billion).

To further illustrate this result, Figure 10.2 presents two models of these estimates. Figure 10.2(a) depicts the relationship between *MSC* and *MSB*. Notice that the hypothetical abatement level corresponding to the 1990 Amendments (A_{1990}) is located to the *right* of the efficient abatement level (A_E). It follows, therefore, that society would be better off with less regulation, which translates to a movement on the graph from A_{1990} to A_E.

The model in Figure 10.2(b) illustrates these same results using an approximation of the *TSC* and *TSB* as of 1990. To obtain the estimate for *TSC*, we add the estimated *MSC* for the 1990 law, $32.0 billion, to an estimate of *TSC* for 1981 estimated by Portney in a different study (1990a) valued at $17.2 billion in 1990 dollars. This yields a 1990 *TSC* estimate of $49.2 billion.

	TSC effective 1981:		$17.2 billion
+	*MSC* for 1990 Amendments:	+	$32.0 billion
	***TSC* effective 1990:**		**$49.2 billion**

Similarly, to derive an approximation of *TSB* from all air quality controls in effect as of 1990, the estimated *MSB* for the 1990 Amendments, $14.0 billion, is added to a previously derived estimate of *TSB*, $46.9 billion, based on a study by Freeman (1982). This yields a 1990 *TSB* estimate of $60.9 billion.

[23]There is, however, an important caveat. If, for example, Portney's (1990b) marginal cost estimates do not reflect least-cost decisions, meaning that they lie *above* the true *MSC* curve, then the appropriate conclusion is that the 1990 Amendments are not being implemented in a cost-effective manner. If this is the case, the efficiency of the abatement level being achieved by the 1990 law is indeterminate.

FIGURE 10.2 Costs and Benefits of the 1990 CAAA

According to Portney's estimates, the *MSC* is above the *MSB* at A_{1990}. Notice that A_{1990} is higher than the efficient abatement level, A_E. This implies that society would be better off with less regulation, which translates to a movement from A_{1990} to A_E.

(a)

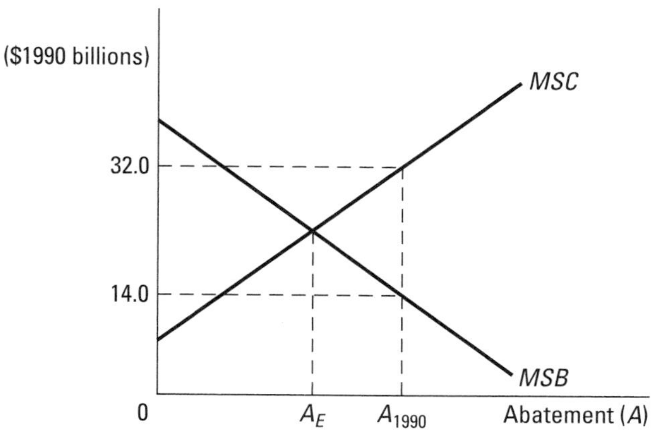

Although *TSB* exceeds *TSC* at A_{1990}, this abatement level is higher than the efficient abatement level, A_E. Just as in part (a), this model implies that the 1990 CAAA overregulate the private sector to achieve cleaner air.

(b)

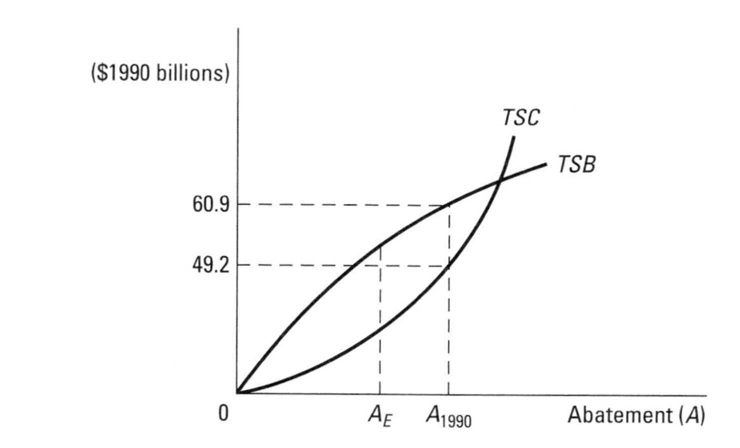

Sources: Values derived from Portney (1990a, 1990b) and Freeman (1982).

TSB effective 1978:	$46.9 billion
+ *MSB* for 1990 Amendments:	+ $14.0 billion
***TSB* effective 1990:**	**$60.9 billion**

These *TSB* and *TSC* values are shown in Figure 10.2(b) corresponding to A_{1990}. Notice that while *TSB* exceeds *TSC*, the associated abatement level, A_{1990}, is nonetheless greater than the efficient level, A_E.

The implication of these findings is that by implementing the 1990 Amendments, the U.S. government has overregulated the private sector to achieve cleaner air. Overallocation of resources to implement this legislation means that resources are being underallocated toward other uses. Many would agree with this conclusion, arguing, for example, that the allocation of government spending among environmental problems is disproportionate to the associated risks. But as we stated at the outset, benefit-cost analysis of environmental issues is not an exact science. These values are just estimates, and the scientific evidence on most, if not all, environmental problems is incomplete.

Taken in this context, Portney's research should be recognized for what it is, *and* for what it is not. Although the results have been supported by existing data and by other researchers, they are nonetheless a series of estimates and guesses, albeit educated ones. There are no hard-and-fast conclusions that can be drawn from such preliminary results. Rather, more advanced research, using new data as they become available, is necessary to make a more concrete determination about the efficiency of the 1990 Amendments.

EPA Benefit-Cost Analysis of the 1990 Amendments[24]

Under Section 812 of the 1990 CAAA, the EPA is required to prepare periodic reports to Congress on the benefits and costs of this major legislation. The first of these, a retrospective analysis, was presented in 1997 for the 20-year prior period from 1970 to 1990.[25] The subsequent report, called the first prospective analysis, assesses the 1990–2010 period and was completed in 1999. The second prospective study, for the 1990–2020 period and which updates the first, was completed in 2011, although it uses methodology that disallows comparison to prior reports. An overview of its findings is provided in Application 10.4. All three reports can be accessed on the Internet at **www.epa.gov/air/sect812/index.html.**

To shed some light on the effect of the 1990 CAAA on the nation's air quality, let's take a closer look at the first prospective analysis. To conduct its comprehensive investigation, the EPA carried out a sequence of six steps, identified as:

1. Estimate air pollution emissions in 1990, 2000, and 2010.
2. Estimate the costs of reducing emissions based on the 1990 legislation.
3. Model air quality based on the emissions estimates.
4. Quantify the associated health and environmental effects.
5. Estimate the economic value of cleaner air.
6. Aggregate the findings, and identify any uncertainties.

Based on the overall results, this analysis indicates that the total benefits of the CAAA outweigh the total costs. A brief overview of these findings follows.

Benefit-Cost Comparison

In its final report to Congress, the EPA estimates that the present value of net benefits associated with Titles I through V of the 1990 CAAA for the 1990–2010 period is $510 billion ($1990). As shown in Table 10.5 (page 226), this value arises from the estimated present value of direct benefits of $690 billion ($1990) and the comparable value for direct costs of $180 billion ($1990).

---

[24]Much of the following is drawn from U.S. EPA, Office of Air and Radiation (November 1999).
[25]U.S. EPA, Office of Air and Radiation (October 1997).

APPLICATION 10.4 Benefit-Cost Analysis of the Clean Air Act Amendments (CAAA): The Second Prospective Study

In 2011, the EPA completed its second prospective study, estimating the incremental benefits and costs of the Clean Air Act Amendments (CAAA) for the period 1990 through 2020. The baseline for comparison is the set of regulations and programs established by the 1970 CAAA and the 1977 CAAA. Two reports preceded this one: a retrospective analysis for the 1970–1990 period, which was published in 1997, and an initial prospective analysis for the 1990–2010 period, which was completed in 1999. This second prospective study expands upon the first by extending the time period to 2020 and by using better data and new modeling techniques.

In addition to its primary objective, which is to estimate the direct costs and benefits of the CAAA as a whole, this study also aims to assess economy-wide effects of the CAAA, to use as comprehensive an approach as possible, and to identify any limitations of the study and determine opportunities for improved results in future analyses.

Significant emissions sources controlled by the CAAA of 1990 are assigned to five categories: electricity generating units; nonutility industrial sources; onroad vehicles and fuel; nonroad vehicles and fuel; and area sources. Costs to reduce emissions from these sources are estimated for the 1990–2020 period with and without the controls given in the CAAA of 1990. For the final year of the study period, 2020, when the difference is highest, the estimated incremental costs are approximately $65 billion ($2006), with the largest proportion attributed to the onroad vehicles and fuel category.

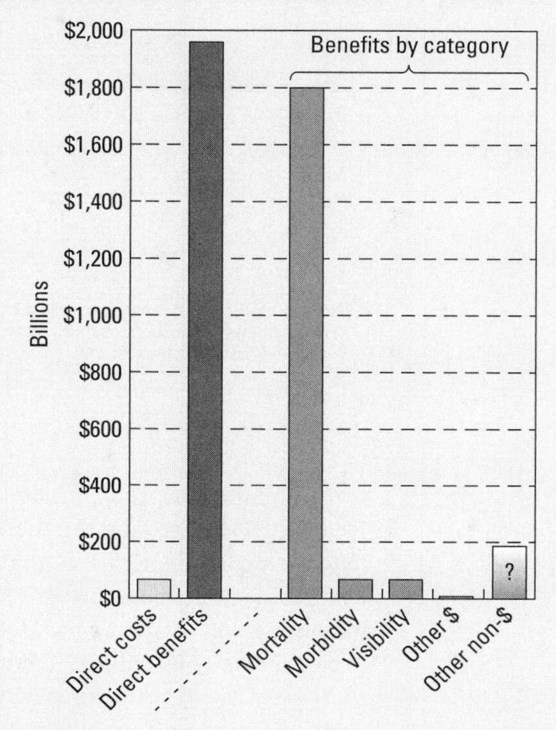

NOTE: "Other non-$" shows that some benefits cannot be monetized and are unknown.

Source: U.S. EPA, Office of Air and Radiation (March 2011), Exhibit 11, p. 19.

Estimated benefits for the study period are based upon the following categories: reducing risk of premature death from exposure to fine particle pollution; reducing risk of premature death from ozone exposure; reduction in illnesses such as acute myocardial infarction; and improved health of ecosystems such as higher farm yields and better visibility. Collectively, the value of these incremental benefits for the year 2020 is $1,951 billion ($2006). The highest estimated benefits are attributed to decreased mortality, as shown in the following figure.

Taken together, the estimated benefits significantly exceed the estimated costs, resulting in net benefits of $1,886 billion ($2006). Because of the wide margin between the two values, the EPA is confident that the CAAA of 1990 are associated with positive net benefits. A number of reasons are given in support of this assertion. Among them is the fact that the benefits linked to reduced morbidity and better visibility are more than twice the associated costs, suggesting that net benefits would be positive even without considering the benefits of reduced mortality risk. Another reason given is that all benefits could not be quantified, implying that the estimated benefits would be even larger than reported if all such gains were measurable. Lastly, the assessment of uncertainties in the study is such that the probability of negative net benefits is extremely small.

As a final note, the EPA cautions that the results of this second prospective study are not readily comparable with those presented in the first. This is due in part to the use of better data and methods in the more recent analysis, particularly for estimates related to fine particle and ozone pollution. As a result, the benefit estimates are significantly larger in this study relative to the first. Some of the difference is linked to new programs that were put in place after 1999, including the Clean Air Interstate Rule. However, the major contributing factor is the updated findings of epidemiology studies used to determine the risk of premature death linked to fine particle pollution.

For more information about this study and for online access to the full report, visit **www.epa.gov/air/sect812/prospective2.html**.

Source: U.S. EPA, Office of Air and Radiation (April 2011; March 2011).

TABLE 10.5 EPA's Benefit-Cost Analysis of the 1990 CAAA

	TITLES I THROUGH V			TITLE VI	ALL TITLES
	Annual Estimates		Present Value Estimate	Present Value Estimate	Total Present Value
	2000	2010	1990–2010	1990–2165	
Monetized Direct Costs	$19,000	$ 27,000	$180,000	$ 27,000	$210,000
Monetized Direct Benefits	$71,000	$110,000	$690,000	$530,000	$1,200,000
Net Benefits	$52,000	$ 83,000	$510,000	$500,000	$1,000,000
Benefit/Cost Ratio	4:1	4:1	4:1	20:1	6:1

NOTES: Values are expressed in millions of 1990 dollars.
Dollar estimates shown are central estimates, or mean values.
A discount rate of 5 percent was used in the calculations.

Source: U.S. EPA, Office of Air and Radiation (November 1999), p.105, Table 8-4.

Estimates for protecting the ozone layer (Title VI) are projected for a longer period, 1990–2165, to account for the lengthy process of ozone formation and for the persistence of ozone-depleting contaminants in the atmosphere. The estimated present values of direct benefits and costs for Title VI are reported to be $530 billion ($1990) and $27 billion ($1990), respectively. Hence, the estimated present value of net benefits for ozone protection is $500 billion ($1990). Adding this to the comparable estimate for Titles I through V yields a total net benefit value of $1,000 billion ($1990). More specific values also are derived for individual titled sections and for specific annual periods.

Assessing the Results

At the outset, it is important to note that since these estimates are for total benefits and costs, as opposed to marginal values, no conclusions about efficiency can be drawn. Also, because the EPA's approach and methodology are vastly different from Portney's 1990 study, it is not surprising that the resulting estimates likewise are markedly dissimilar, and in truth, the two sets of estimates are not directly comparable. For one thing, Portney's analysis is much less rigorous than the study conducted by the EPA. That said, the EPA's report identifies uncertainties inherent in their study, arising from such factors as tentative predictions about future conditions, data limitations, and available research on air pollution effects. Moreover, the quantitative results, though recognized as based on sound methods and data, are considered to be controversial on a number of fronts.

As discussed in an article by Freeman (2002), these controversies arise on both the benefit and cost sides. For example, in the benefit assessment, the EPA's study assumes that the statistical value of a human life is $4.8 million ($1990). However, this value is based on labor market studies for people younger than those most at risk for premature death due to poor air quality. On the cost side, the study assesses only direct costs, which means that changes in prices and employment are not considered.

Another article, by Krupnick and Morgenstern (2002), underscores the implications of the highly aggregate level at which the EPA's analysis is conducted. That is, while the reported net benefits are positive, the aggregation might be disguising some inefficient elements of the CAAA. Among the authors' recommendations for a future prospective analysis is to disaggregate costs and benefits to facilitate assessment of individual provisions in the nation's air quality policy.

Taken together, these assessments suggest that, until greater clarity is achieved and further research is done, the efficiency implications of U.S. air quality policy remain uncertain.

BENEFIT-COST ANALYSIS OF THE AIR QUALITY STANDARDS

One of the more important contributions of the EPA's reports to Congress and studies such as Portney's is that they motivate the need for further investigation of the Clean Air Act. To that end, we conduct a benefit-cost analysis of the standards established through this legislation, focusing on the *efficiency* implications of two specific issues:

- the **absence of cost considerations** in setting the standards
- the **uniformity of the standards**

Absence of Cost Considerations in the Standard-Setting Process

benefit-based decision rule
A guideline to improve society's well-being with no allowance for balancing with associated costs.

From an efficiency perspective, a common criticism of the NAAQS is that they are motivated solely by the anticipated benefits from protecting public health and welfare with no consideration of the economic feasibility of doing so. This use of a **benefit-based decision rule** is particularly problematic for the primary standards for health, which are to include a "margin of safety." This wording suggests that there is some pollution level that will not cause any harm to public health. But is there a pollution level that does not harm at least one individual? Probably not. So, without the balance of cost considerations, the law seems to suggest setting the primary standards for criteria pollutants at zero. Although such an extreme was not Congress's intent, the law does not provide appropriate guidelines to set the NAAQS at an efficient level.[26]

Uniformity of the National Ambient Air Quality Standards (NAAQS)

A critical observation to make about the NAAQS is that, since they are nationally based, they ignore any region-specific cost or benefit differences associated with meeting them. All nonattainment regions must meet the same uniform standards regardless of such differences as existing pollution levels, access to technology, demographics, and traffic patterns. Yet the marginal costs and benefits of reducing pollution to the same level across such highly diverse regions likely will be dissimilar. This might explain why some regions met the legislated deadlines with relative ease, whereas others continue to struggle with compliance. Thus, while the shift to federally mandated standards was intended to strengthen U.S. policy, it nonetheless contributed to the inefficiency that typically arises with a command-and-control approach.

[26]This discussion is elaborated by Burtraw and Portney (1991) and Portney (1990a).

One legislative change that allows for some region-specific differences is the recognition of prevention of significant deterioration (PSD) areas and the use of existing air quality in those regions as the relevant standard. This policy reform effectively elevates the standard for PSD areas above the NAAQS applicable to nonattainment areas. The relevant issue is whether this particular use of differentiated standards represents an efficient allocation of resources across both types of regions. This would be the case only if the efficient level of abatement in a PSD area is higher than that in a nonattainment area. Some economists also question the wisdom of assigning more resources to already clean areas in order to achieve a higher standard of air quality.[27] To investigate these issues, we use benefit-cost analysis in a simple economic model.

Benefit-Cost Analysis of Higher PSD Standards

To economically justify the relatively higher standards in PSD areas, the associated *MSC* and *MSB* curves of abatement for these regions must intersect at a higher level of abatement than the comparable intersection for nonattainment areas. For this to occur, there must be certain differences in either or both of the *MSC* and *MSB* across the two types of regions. In Figure 10.3, we analyze three possible scenarios under which the use of different standards across PSD and nonattainment areas would be efficient. The marginal cost and benefit curves for a representative PSD area are labeled as MSC_{PSD} and MSB_{PSD}, and those for a nonattainment area as MSC_{NON} and MSB_{NON}.

In Figure 10.3(a), the *MSB* functions are identical across the two types of areas, that is, $MSB_{PSD} = MSB_{NON}$ at all abatement levels (A). If this relationship holds, the efficient abatement level in the PSD area (A_{PSD}) would be higher than the efficient level in a nonattainment area (A_{NON}) only if MSC_{PSD} is *lower* than MSC_{NON}. In Figure 10.3(b), the opposite set of conditions is modeled, with the *MSC* functions drawn as identical curves. Here, efficiency holds only if MSB_{PSD} is *above* MSB_{NON}. Finally, in Figure 10.3(c), both the *MSC* curve and the *MSB* curve are allowed to be unique to each area. Under this latter assumption, one way that efficiency results is if the following three conditions hold:

1. MSB_{PSD} is *below* MSB_{NON},
2. MSC_{PSD} is *below* MSC_{NON}, and
3. the vertical distance between the *MSC* curves is sufficiently *greater* than that between the *MSB* functions to support the relative position of A_{PSD} and A_{NON}.[28]

So far, what we have illustrated is that the use of relatively higher standards for PSD areas can be efficient, but only under certain conditions. Now, we need to consider the logic of each scenario based on the expected relationship between the respective marginal costs and benefits for the two areas. For context, think of the differences between nonattainment areas and the most stringently controlled PSD areas, which are Class I. Recall that these PSD areas include national parks and wilderness regions. One reason these areas have cleaner air is that they are less populated and less industrialized

---

[27] See Portney (1990a), pp. 78–79.

[28] Other scenarios are possible under which both the *MSC* and the *MSB* curves are allowed to be unique and the A_{PSD} is higher than A_{NON}. To see this, try modeling a case different from the one shown in Figure 10.3(c).

FIGURE 10.3 Standards in PSD Areas Relative to Nonattainment Areas

In panel (a), $MSB_{PSD} = MSB_{NON}$. In this case, the efficient abatement level in the PSD area, A_{PSD}, would be higher than the efficient level in a nonattainment area, A_{NON}, only if MSC_{PSD} is *lower* than MSC_{NON}. In panel (b), $MSC_{NON} = MSC_{PSD}$. Here, efficiency holds only if MSB_{PSD} is *above* MSB_{NON}. Finally, in panel (c), both the *MSC* and *MSB* curves are unique to each area. Under this assumption, efficiency holds only under certain conditions.

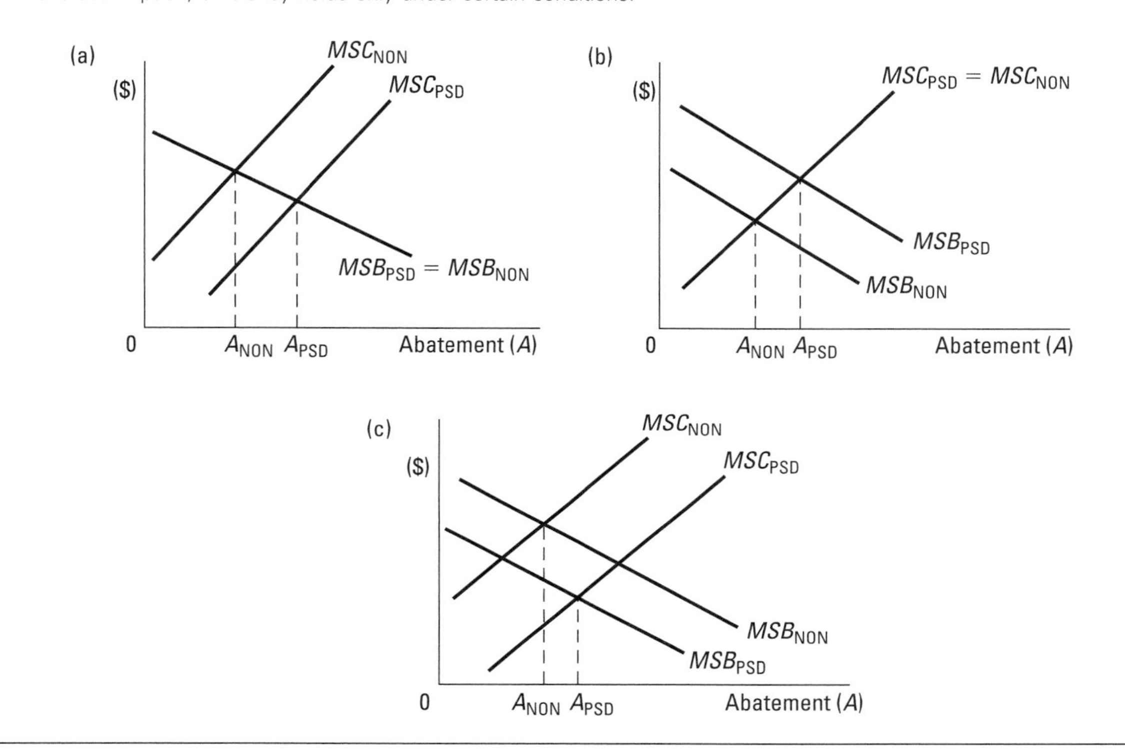

than their nonattainment counterparts. How might these conditions affect the *MSB* and *MSC*?

On the cost side, it is reasonable to expect that MSC_{PSD} is lower than MSC_{NON}, since there are fewer polluting sources to control in PSD regions. By itself, this assertion suggests that the model in Figure 10.3(b) is not a likely representation of actual conditions.

Now consider the benefit side. The primary benefits of maintaining higher air quality in PSD areas are nonhealth gains, such as enhanced recreational uses, aesthetic improvements, and protection of the ecology. In contrast, the major benefits of cleaning up the more populated nonattainment areas are improvements in human health. Since health improvements are generally valued more at the margin than nonhealth gains, MSB_{NON} is likely higher than MSB_{PSD} at all abatement levels. Based on this hypothesis, the models shown in Figures 10.3(a) and 10.3(b) appear to be inaccurate.

Taking these assumed benefit differences into account along with the expected cost differences, the model in Figure 10.3(c) appears to be more accurate than the other two. However, recall that for this model to support the efficiency criterion, several conditions have to hold. Hence, all we can conclude is that the higher standards in the cleaner PSD areas *may* be justifiable on efficiency grounds under certain economic circumstances.

Another reform that acknowledges region-specific differences was the reclassification of nonattainment areas for ozone, carbon monoxide, and particulate matter enacted within the 1990 Amendments. By imposing progressively more stringent controls on the dirtiest of these areas, these revisions may have improved efficiency. Again, this issue can be examined through benefit-cost analysis.

Benefit-Cost Analysis of Progressively Regulated Nonattainment Areas

To give our analysis context, we consider two of the new classifications for nonattainment ozone areas, marginal areas at the cleaner end of the continuum and extreme areas at the other end. Now, we can apply benefit-cost analysis to examine the efficiency of strengthening regulatory controls in extreme ozone areas.

Since the United States is imposing more stringent regulations on extreme ozone regions, the *MSC* in extreme regions will necessarily be higher than the *MSC* in marginal ozone regions at all levels of abatement. Hence, in order for the use of different control methods to make sense from an economic perspective, *MSB* at the existing abatement level in extreme regions must be greater than the *MSB* at the current abatement level in marginal regions.[29] As long as this is the case, we can argue that the higher incremental costs of more stringent regulations are at least to some extent justified by higher incremental benefits.

CONCLUSIONS

U.S. air quality policy has evolved considerably over the last several decades. Legislative initiatives and the infrastructure to implement them have become more extensive and more complex. Indeed, the 1990 CAAA are far more comprehensive than any U.S. environmental legislation enacted to date. Today, there is a massive network of agencies and interagencies working toward the objective of improving national air quality. And now, new rulings and proposals are being developed that are additive to the current regulatory structure. In some sense, such an evolution is not surprising. Air pollution is a difficult problem, and the sheer size of the United States means that air quality controls must accommodate a variety of meteorological, geographic, and economic conditions.

Despite the numerous revisions and new initiatives, U.S. legislation continues to be based primarily on a command-and-control approach, evidenced largely by the use of uniform standards to define national air quality. The problem is that such an approach is not likely to achieve an efficient solution. In fact, benefit-cost analysis suggests that clean air policy initiatives have been more costly to society than necessary. Similar conclusions follow from a qualitative evaluation of the standard-setting process. From an

[29] If the *MSB* curves are the same for both regions, this relative difference will necessarily hold, because the abatement level in extreme regions is by definition lower than that in marginal regions. If the two curves are distinct, then the *MSB* in extreme regions lies everywhere above the *MSB* in marginal regions.

economic perspective, uniform standards are a troubling element of U.S. air legislation. Likewise, the absence of cost considerations in establishing these standards defies time-tested economic theory.

Of course, there are no simple answers, nor can there be any sweeping conclusions without careful reservations. Air quality issues involve many intangibles that are difficult to quantify. Nonetheless, the analytical process guided by economic theory provides a way to better understand the motivation and economic implications of the Clean Air Act and to set more realistic expectations about achieving air quality objectives.

Summary

- There were no federal air pollution laws until 1955 when the Air Pollution Control Act was passed and no comprehensive legislation until the Clean Air Act of 1963.
- In 1990, President George H. W. Bush signed into law extensive changes in U.S. air quality control policy as the 1990 Clean Air Act Amendments (CAAA).
- An initiative to amend the CAAA was proposed by President George W. Bush in the form of the Clear Skies Act. When the proposal was stalled in Congress, the EPA established a set of actions called the Clean Air Rules, which included the Clean Air Interstate Rule (CAIR). It subsequently was challenged, and the Court remanded the matter to the EPA for further rulemaking. Under President Obama, the Cross-State Air Pollution Rule (CSAPR) was finalized as a replacement for the CAIR.
- The air pollutants that have been identified under U.S. law as causing the greatest damage to outdoor air quality are: criteria pollutants, hazardous air pollutants, and greenhouse gas (GHG) air pollutants.
- The six criteria pollutants are particulate matter, sulfur dioxide, carbon monoxide, nitrogen dioxide, tropospheric ozone, and lead.
- An initial list of 189 hazardous air pollutants is identified in the 1990 CAAA.
- The GHG air pollutants include carbon dioxide, methane, nitrous oxide, hydrofluorocarbons, perfluorocarbons, and sulfur hexafluoride.
- The National Ambient Air Quality Standards (NAAQS) state the maximum allowable concentrations of criteria air pollutants that may be emitted from stationary or mobile sources.
- Primary NAAQS are set to protect public health. Secondary NAAQS are set to protect public welfare. A recent revision to the primary standard for sulfur dioxide (SO_2) establishes better protection from short-term SO_2 exposure. A proposal for tightening the NAAQS, which is still pending, recommends strengthening the primary and secondary standards on ground-level ozone (O_3).
- National Emission Standards for Hazardous Air Pollutants (NESHAP) are established for hazardous air pollutants. Recently issued NESHAP include rules finalized in 2011 aimed at reducing mercury and other toxic emissions from area source industrial, commercial, and institutional boilers. Other new NESHAP issued in 2011 are aimed at reducing heavy metals and acid gases released from coal-fired and oil-fired power plants.
- No ambient standards have been set for GHG air pollutants, nor have any been proposed to date. However, the EPA has finalized new GHG emissions standards for light-duty vehicles and has proposed comparable rulings for medium- and heavy-duty vehicles.
- Coordination of air quality control policy between federal and state governments is formalized through the State Implementation Plan (SIP).
- There are currently 247 air quality control regions (AQCRs) in the United States. These are classified into nonattainment areas and prevention of significant deterioration (PSD) areas. In the 1990 CAAA, new classifications are established for certain of the nonattainment areas.
- Both environmental justice, an equity criterion, and allocative efficiency, an economic criterion, can be used to assess air quality policy.
- Economist Paul Portney finds that the 1990 CAAA may abate pollution beyond the efficient level.
- The EPA's first prospective benefit-cost report on the 1990 CAAA estimates that the present value of net benefits for the 1990–2010 period is $1,000 billion ($1990).
- The uniformity of the NAAQS and the absence of cost considerations suggest these standards are not set at an efficient level.

Review Questions

1. If you were responsible for setting the NAAQS for lead, what key determinants would you consider if the standard were established to meet the efficiency criterion? Be sure to itemize separately the benefits and costs associated with your decision.
2. Using the efficiency criterion, carefully analyze the problem faced by the EPA in identifying hazardous air pollutants prior to the 1990 Amendments.
3. Suppose economists estimate the following cost and benefit functions associated with the 1990 CAAA:

$$TSB = 75A - 0.4A^2 \quad MSB = 75 - 0.8A$$
$$TSC = 25A + 0.85A^2 \quad MSC = 25 + 1.7A$$

where A represents the percent of pollution abatement, and costs and benefits are measured in millions of dollars.
   a. Find the *TSB* and *TSC* associated with an abatement standard of 30. Based solely on the relative values of *TSB* and *TSC*, can you state definitively whether this abatement level is efficient? Explain.
   b. Find the efficient level of pollution abatement (A_E).
4. **a.** Briefly explain the significance of prevention of significant deterioration (PSD) areas to the setting of air quality standards.
   b. From an economic perspective, explain the paradox associated with setting higher standards in PSD areas relative to nonattainment areas. Show how an inefficient result may arise using an *MSB-MSC* model.
5. Briefly summarize Portney's overall assessment of the 1990 CAAA, and discuss the major implications of these findings for society.
6. Refer back to Table 10.4 of the marginal benefit and cost estimates for the 1990 CAAA. Conduct an individual benefit-cost analysis for each of the three major components of the amendments, namely, acidic deposition, urban air quality, and hazardous air pollutants. What do you conclude?
7. Visit the site of EPA's Green Book: Nonattainment Areas for Criteria Pollutants, which can be accessed at **www.epa.gov/oar/oaqps/greenbk/**, and use the data to investigate which AQCRs in your state have nonattainment status for each of the criteria pollutants. Summarize your findings.
8. Suppose the following are the *MSB* and *MSC* functions for PSD and nonattainment areas in Denver, Colorado:

$$\text{PSD:} \quad MSB_{PSD} = 30 - 0.2A$$
$$MSC_{PSD} = 12 + 0.1A$$
$$\text{NON:} \quad MSB_{NON} = 34 - 0.3A$$
$$MSC_{NON} = 18 + 0.2A$$

where A is the level of abatement in each region.
   a. Find the efficient abatement level for each area, i.e., A_{PSD} and A_{NON}. Based on these results, are the differentiated standards for PSD areas supported on efficiency grounds? Briefly explain why or why not.
   b. Is the relationship between MSB_{PSD} and MSB_{NON} reasonable based on the legislated distinction between PSD areas and nonattainment areas? Briefly explain.

Additional Readings

Ayres, Richard E., and Jessica L. Olson. "The New Federalism: States Take a Leading Role in Clean Air." *Natural Resources & Environment 23*(2) (Fall 2008), pp. 29–33.

Buckley, Shannon, and Myron J. Mitchell. "Improvements in Urban Air Quality: Case Studies from New York State, USA." *Water, Air, and Soil Pollution 214*(*1-4*) (January 2011), pp. 93–106.

Burtraw, Dallas, and David A. Evans. "NO_X Emissions in the United States: A Potpourri of Policies." In Winston Harrington, Richard D. Morgenstern, and Thomas Sterner, eds., *Choosing Environmental Policy*. Washington, DC: Resources for the Future, 2004.

Chay, Kenneth Y., and Michael Greenstone. "Does Air Quality Matter? Evidence from the Housing Market." *Journal of Political Economy 113(2)* (April 2005), pp. 376–422.

Clement, Douglas. "Cost v. Benefit: Clearing the Air?" *The Region*. Minneapolis: Federal Reserve Bank of Minneapolis (December 2001), pp. 19–21, 48–57.

Hall, Jane V. "Air Quality in Developing Countries." *Contemporary Economic Policy XIII(2)* (April 1995), pp. 77–85.

Hall, Jane V., and Amy L. Walton. "A Case Study in Pollution Markets: Dismal Science vs. Dismal Reality." *Contemporary Economic Policy XIV(2)* (April 1996), pp. 67–78.

Hueth, Brent, and Tigran Melkonyan. "Standards and the Regulation of Environmental Risk." *Journal of Regulatory Economics 36(3)*, (December 2009), pp. 219–46.

Hutchinson, Emma, and Peter W. Kennedy. "State Enforcement of Federal Standards: Implications for Interstate Pollution." *Resource and Energy Economics 30(3)* (August 2008), pp. 316–44.

Janofsky, Michael. "Many Counties Failing Fine-Particle Air Rules." *New York Times,* December 18, 2004.

Lee, Amanda I., and James Alm. "The Clean Air Act Amendments and Firm Investment in Pollution Abatement." *Land Economics 80(3)* (August 2004), pp. 433–47.

McKitrick, Ross. "Why Did US Air Pollution Decline After 1970?" *Empirical Economics 33(3)* (November 2007), pp. 491–514.

Palmer, Karen, Wallace E. Oates, and Paul R. Portney. "Tightening Environmental Standards: The Benefit-Cost or the No-Cost Paradigm?" *Journal of Economic Perspectives 9(1)* (Winter 1995), pp. 129–32.

Swinton, J. T. "Phase I Completed: An Empirical Assessment of the 1990 CAAA." *Environmental and Resource Economics 27(3)* (March 2004), pp. 227–46.

A Reference to Acronyms and Terms Used in Air Quality Control Policy

Environmental Economics Acronyms

MSB	Marginal social benefit
MSB_{NON}	Marginal social benefit of abatement in a nonattainment area
MSB_{PSD}	Marginal social benefit of abatement in a PSD area
MSC	Marginal social cost
MSC_{NON}	Marginal social cost of abatement in a nonattainment area
MSC_{PSD}	Marginal social cost of abatement in a PSD area
TSB	Total social benefits
TSC	Total social costs

Environmental Science Terms

C_6H_6	Benzene
CH_4	Methane
CO	Carbon monoxide
CO_2	Carbon dioxide
GHG	Greenhouse gas
HFCs	hydrofluorocarbons
$\mu g/m^3$	Micrograms per cubic meter
mg/m^3	Milligrams per cubic meter
NO_X	Nitrogen oxides
NO_2	Nitrogen dioxide
N_2O	Nitrous oxide
O_3	Ozone
Pb	Lead
PM-2.5	Particulate matter of less than 2.5 micrometers in diameter
PM-10	Particulate matter of less than 10 micrometers in diameter
ppb	Parts per billion
ppm	Parts per million
PRCs	perfluorocarbons
SF_6	Sulfur hexafluoride
SO_2	Sulfur dioxide
SO_X	Sulfur oxides
VOC	Volatile organic compound

Environmental Policy Acronyms

AQCR	Air Quality Control Region
CAAA	Clean Air Act Amendments
CAIR	Clean Air Interstate Rule
CSAPR	Cross-State Air Pollution Rule
MACT	Maximum Achievable Control Technology
NAAQS	National Ambient Air Quality Standards
NESHAP	National Emission Standards for Hazardous Air Pollutants
NSPS	New Source Performance Standards
PSD	Prevention of Significant Deterioration
SIP	State Implementation Plan

To access additional course materials, visit www.cengagebrain.com. At the home page, search for the ISBN of this title (shown on the back cover). This will take you to the product page where these resources can be found.

CHAPTER 11

Improving Air Quality: Controlling Mobile Sources

"Between a quarter and a third of Los Angeles's land area is now monopolized by the automobile and its needs—by freeways, highways, garages, gas stations, car lots, parking lots. And all of it is blanketed with anonymity and foul air."
—Alistair Cooke (1908–2004)

Because the criteria pollutants identified under the Clean Air Act are so pervasive, they are responsible for most air quality problems, even though they are less dangerous than hazardous air pollutants. The relevant point is that everyone is exposed to the risks of criteria pollutants, which explains why the Clean Air Act places such a strong emphasis on them and why Congress has legislated well-defined strategies to implement the National Ambient Air Quality Standards (NAAQS).

From a similar perspective, greenhouse gases (GHGs) are also pervasive, and this reality is part of the motivation for the development of new national initiatives to control them. As to why greenhouse gases were not identified as pollutants until recently, the explanation is complex—a combination of disagreement within the scientific community, inherent uncertainty about climate projections, economic implications of policy prescriptions, and politics. In any case, these substances, like the criteria pollutants, are now being controlled at the national level in addition to international policy efforts that have been in place for some time.[1]

Although the motivation for U.S. controls on pervasive pollutants may be well placed, it turns out that this top-down policy approach has had some adverse consequences, an assertion we will support with economic analysis in this chapter. Starting from a common vantage point, we know that the Clean Air Act sets stringent regulations on both mobile and stationary sources, because both contribute to the release of all six criteria pollutants and GHGs, though in different ways. In this chapter, we focus on mobile sources, deferring our discussion of stationary sources to Chapter 12.

Automobiles and other mobile sources are controlled through Title II of the Clean Air Act. Specific policies are often tailored to the air quality problems with which mobile sources are commonly associated. Based on available data, we know that these sources contribute over 23 percent of U.S. GHG emissions. They are also primarily responsible for

[1]More on this international effort to address greenhouse gases and climate change is discussed in Chapter 13.

carbon monoxide emissions in the United States and are major contributors to emissions linked to smog formation. In 2008, for example, mobile sources released about 80 percent of the carbon monoxide (CO) emissions and, together with industrial and other processes, were responsible for nearly 50 percent of the volatile organic compound (VOC) emissions.[2]

Because motor vehicles are the major source of emissions that cause urban smog, we begin this chapter with a discussion of urban air pollution. Once done, we begin our study of mobile source controls. An economic analysis follows, which examines the efficiency and cost-effectiveness of mobile source initiatives. A list of commonly used acronyms is provided for reference at the end of this chapter.

URBAN AIR POLLUTION

Environmentalists as well as policymakers are concerned about the air pollution that characterizes urban centers. In these metropolitan areas, the high concentration of human population, traffic, and industrial activity intensifies the concentration of criteria pollutants and hence increases the environmental risks of exposure. Because a large proportion of the population is exposed to the associated health hazards, increased abatement efforts in cities should yield a higher level of marginal benefit than in rural communities.

Measuring U.S. Urban Air Quality[3]

Air Quality Index (AQI)
An index that signifies the worst daily air quality in an urban area.

To develop a better sense of urban air quality in the United States, the Environmental Protection Agency (EPA) monitors the air in those metropolitan statistical areas (MSAs) with populations greater than 350,000. The agency reports part of its findings using the **Air Quality Index (AQI)**. The AQI is based on concentration data collected from a monitoring network that measures five of the criteria pollutants—particulate matter (PM), sulfur dioxide (SO_2), carbon monoxide (CO), ozone (O_3), and nitrogen dioxide (NO_2).[4] For each pollutant, an index value between 0 and 500 is formed from the concentration data, where a value of 100 coincides with the short-term standard for that pollutant set by the Clean Air Act. The AQI is reported as the highest of the five pollutant-specific index values for that day and signifies the worst daily air quality in an urban area over a given time period.

The descriptor words and color codes that identify the health effects associated with various ranges of the AQI are shown in Table 11.1. The AQI, its descriptors, and corresponding color codes are cited on the Internet and used by various media in the United States, such as the Weather Channel and CNN, to communicate air quality conditions to the general public. For more information on the AQI, visit **www.airnow.gov/**. In addition, this index has been adopted by other nations around the world, such as Mexico, Singapore, and Taiwan. To learn more about air quality indices around the world, visit **www.airnow.gov/index.cfm?action=topics.world**.

[2] U.S. EPA, Office of Air and Radiation, Office of Air Quality Planning and Standards (February 2010); U.S. EPA, Office of Public Affairs (December 7, 2009).

[3] Much of the following discussion is drawn from U.S. EPA, Office of Air and Radiation, Office of Air Quality Planning and Standards (February 2010; September 2003, Chapter 3), and U.S. EPA, Office of Air and Radiation (August 2009).

[4] The criteria pollutant omitted from the index is lead, because there are no short-term NAAQS for this substance.

TABLE 11.1 Air Quality Index (AQI) Categories

AQI Values	Level of Health Concern	Color
0–50	Good	Green
51–100	Moderate	Yellow
101–150	Unhealthy for sensitive groups	Orange
151–200	Unhealthy	Red
201–300	Very unhealthy	Purple
301–500	Hazardous	Maroon

Source: U.S. EPA, Office of Air and Radiation, Office of Air Quality Planning and Standards (February 2010).

To gain a sense of urban air quality in the United States, look at Table 11.2, which presents selected AQI trend data over a 10-year period. The table shows the number of days in a year each city experienced an air quality level signified by an AQI greater than 100, that is, a value not meeting the standard set by the Clean Air Act. Notice

TABLE 11.2 Number of Days With AQI Greater Than 100

Metropolitan Statistical Areas	1999	2004	2009
Atlanta	97	22	11
Baltimore	55	31	11
Boston	7	1	0
Chicago	81	28	24
Cleveland	120	62	3
Denver	11	4	4
Detroit	72	37	9
El Paso	8	5	2
Houston	74	42	18
Kansas City	13	33	41
Los Angeles	141	106	54
Miami	20	11	2
Minneapolis–St. Paul	5	11	2
New York	60	39	11
Philadelphia	99	38	7
Phoenix	52	9	6
Pittsburgh	152	94	43
Sacramento	94	56	29
St. Louis	113	34	9
San Diego	53	27	25
San Francisco	21	10	4
Seattle	6	4	8
Washington, DC	66	25	4

Source: U.S. EPA, Office of Air and Radiation (March 21, 2011).

that for almost all cities, the number of days that the AQI exceeded 100 was lower in 2009 than in 1999, and most show a consistent downward trend from 1999 through 2009. To examine trends in other cities and for updates, visit **www.epa.gov/air/airtrends/aqi_info.html**.

The data in Table 11.2 also indicate that air quality in Los Angeles is worse than most other major urban areas in the sample, a distinction that has persisted for several decades. As early as the 1940s, Los Angeles residents noticed frequent episodes of a brownish haze on the horizon—an awareness that eventually led to nationwide concern about urban air pollution. Today, we know that the brownish haze in Los Angeles and many other metropolitan areas is **urban smog**.

Urban Smog

Coined from the words *smoke* and *fog*, the word **smog** originally referred to the overall air quality first observed in London at the turn of the last century. Used in this context, smog refers mainly to the presence of particulate matter and other emissions, such as sulfur oxides, in the air.[5] Because of Los Angeles's history and its dubious distinction as America's "king of smog," many people erroneously think of this city as the only one facing a severe smog problem. It is true that the dense traffic there and the meteorological conditions make this region particularly vulnerable to smog formation. However, most major cities suffer from the effects of smog. The problem is more pervasive in highly industrialized and heavily populated parts of the world, such as Europe and Japan.

photochemical smog
Caused by pollutants that chemically react in sunlight to form new substances.

In urban areas, another type of smog can form from a chemical reaction involving several of the criteria pollutants. Scientists refer to this type of smog as **photochemical smog** to distinguish it from the more conventional "smoke and fog" variety. Photochemical smog is formed from certain air pollutants that chemically react in the presence of sunlight (thus the prefix *photo*) to form entirely new substances. Its principal component is tropospheric ozone (O_3), also referred to as ground-level ozone. Along with more than 100 different compounds, ground-level ozone is produced by a chemical reaction of nitrogen oxides (NO_X), volatile organic compounds (VOCs) (primarily hydrocarbons), and sunlight.

Mobile sources are major contributors to photochemical smog formation, since they are responsible for a large proportion of NO_X and VOC emissions, the precursors of smog. Among all transportation sources, those emitting the largest amounts of these substances are ordinary highway vehicles, particularly those fueled by gasoline. In urban centers, officials are particularly concerned with emissions from transportation sources because of traffic congestion. In China, for example, rapid economic growth and the associated increased automobile consumption have led to significant air pollution problems throughout the giant nation but particularly in its major cities. Application 11.1 discusses the air quality problems in China linked specifically to automobile emissions.

[5] Pryde (1973), p. 147.

APPLICATION 11.1 China's Automobile Demand Still a Force Even in Economic Slowdown

Over 1.3 billion people live in China, and as that nation's economy grows, so too does the population's demand for automobiles. Not unlike the experience of other nations moving toward a market economy, the associated accelerated growth comes with a heavy cost—environmental degradation—and China's air quality is no exception. Although the air pollution arises from both stationary and mobile sources, the latter has become an increasing problem as China's automobile and fuel consumption has soared. Said to be the fastest-growing automobile market in the world, second only to the United States, China's annual car sales rose more than 70 percent in 2003 and 50 percent in 2004. Such stunning growth means a rapid increase in fuel demand and dismal forecasts of worsening traffic congestion, increasing numbers of automobile accidents, and declining air quality.

Whether or not such expectations are fully realized depends in part on the current economic slowdown, which has dampened automobile demand, even in emerging markets. J. D. Power and Associates, a research and consulting firm, reduced its projections of Chinese automobile unit sales for 2008 from 5.95 million to 5.8 million. This translates to 6.7 percent annual growth in 2008, which is markedly lower than the 24 percent growth realized in the prior period. Nonetheless, global unit sales for 2008 were expected to reach 59 million, with 14 million of that coming from emerging markets in the so-called BRIC nations: Brazil, Russia, India, and China.

Just what is responsible for China's strong auto consumption, even in an economic recession? In truth, there are a number of contributing factors. One is the rapid increase in the number of automobile factories located in China. Another is rising income levels linked to economic prosperity. Still another is the degree to which automobile demand in China is unmet. Specifically, there are only about 30 automobiles for every 1,000 people of driving age in China, vastly fewer than in the United States, where the comparable number is over 900. These factors have led to a brisk rise in car sales through 2007 with apparent ongoing strength in consumption relative to other countries beyond this period.

Strong Chinese demand for automobiles along with the migration of millions of residents from rural to urban areas has one obvious outcome: China's city streets have become jammed with vehicles, which in turn means deteriorating urban air quality. According to estimates by economists at the International Monetary Fund (IMF), China will have more cars on its roads than the United States by 2030, and by 2050 will have nearly as many as the entire world does today. Adding to the problem of rising automobile consumption is poor fuel quality. Both gasoline and diesel fuel are substandard in China, causing excessive emissions on top of poor auto repair records. Pressure has been mounting for a government response despite concerns about any associated decline in economic growth. The broadest solution is to control China's energy consumption. But how?

At the most fundamental level, there is an anti-car lobby in China that is reportedly gaining momentum. There is also rising pressure on automakers to produce hybrid cars or zero-emission vehicles for China. Some manufacturers, including Volkswagen AG and Ford Motor Company, have promoted diesel-fueled vehicles, which use less fuel than their gasoline-powered counterparts. Increased fuel taxes are predicted, possibly by as much as 60 percent by 2020, according to one source. And China must toughen emissions standards on motor vehicles. In 2001, EURO I standards were adopted, and within the next few years, all new cars will have to meet the more stringent EURO IV standards.

These actions suggest a sea change for a country that has been focused far more on economic advance than environmental quality. Progress will be closely monitored by both domestic and foreign interests as China continues to seek a balance between industrial growth and environmental protection.

Sources: "A Global Love Affair" (November 15, 2008); "The Art of the Possible" (November 15, 2008); Reed and Waldmeir (October 15, 2008); Kynge (July 26, 2004); Lee (December 11, 2003); Bradsher (October 22, 2003).

CONTROLLING MOBILE SOURCES

Of all the transportation vehicles, the primary focus of U.S. policy is highway vehicles, such as passenger cars and trucks. As Figure 11.1 shows, these polluting sources emit the largest quantities among mobile sources of NO_X and VOCs, the precursors of smog, and CO.[6]

[6]More information on automobile emissions can be found at the EPA's Office of Transportation and Air Quality, accessible online at **www.epa.gov/otaq/ld-hwy.htm**.

FIGURE 11.1 Mobile Source Emissions Estimates in 2008

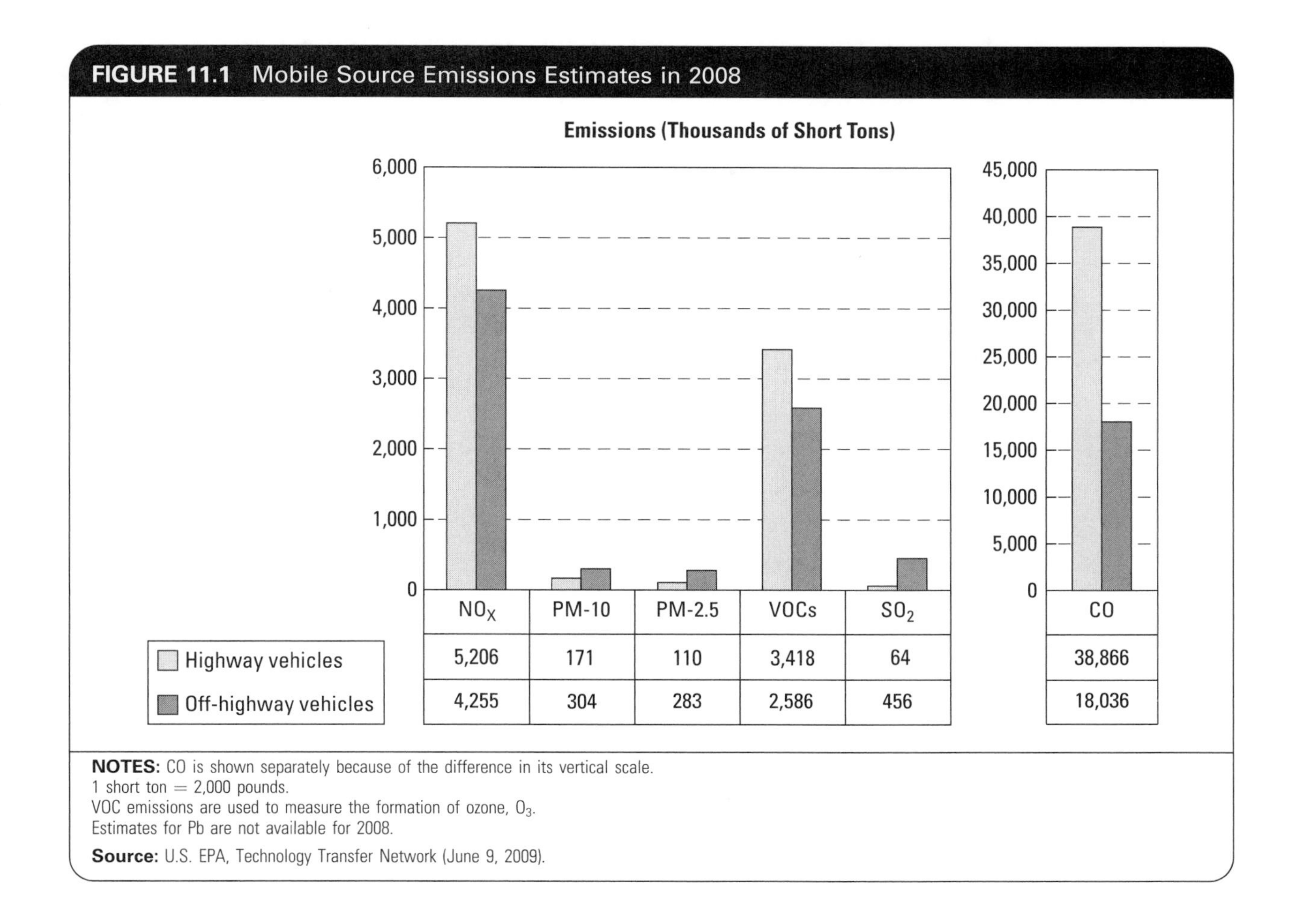

	NO_X	PM-10	PM-2.5	VOCs	SO_2	CO
Highway vehicles	5,206	171	110	3,418	64	38,866
Off-highway vehicles	4,255	304	283	2,586	456	18,036

NOTES: CO is shown separately because of the difference in its vertical scale.
1 short ton = 2,000 pounds.
VOC emissions are used to measure the formation of ozone, O_3.
Estimates for Pb are not available for 2008.

Source: U.S. EPA, Technology Transfer Network (June 9, 2009).

These data explain the policy focus on the manufacture of these vehicles and the increasing number of initiatives aimed at developing alternative fuels. To research the emissions, fuel efficiency, and environmental ranking of any make and model vehicle, visit **www.epa.gov/greenvehicles/Index.do.**

For its part, the American automobile industry has begun to take a more aggressive position toward reducing motor vehicle pollution, not only to meet government controls but also to accommodate environmentally conscious consumers. One outcome of this effort is the United States Council for Automotive Research (USCAR), accessible at **www.uscar.org**, which is aimed at investigating a variety of technologies to develop cleaner-running cars and cleaner fuels.

Such public and private initiatives were unheard of 30 years ago. The evolution of U.S. controls on motor vehicles underscores the inherent problems associated with regulation of a major industrial sector—in this case, the auto industry. Scientific evidence tying the automobile to smog formation had existed since the early 1950s, but no federal mandate to control auto emissions was legislated until 1965, and it was not until the 1970 Clean Air Act Amendments (CAAA) that emission limits appeared within the law. Automobile manufacturers often voiced strong opposition to emissions standards and fought for extensions to legislated deadlines. In some cases, compliance dates were postponed for several years.

All told, the early years of mobile source controls were marked by a series of extensions and a chronicle of delays. As the power struggle between Washington and Detroit continued, society grew increasingly impatient with the lack of progress. Washington officials knew that federal policy on motor vehicle emissions was in need of major revision.[7]

U.S. Controls on Motor Vehicles and Fuels

Attempting to correct errors of the past, Congress passed the 1990 Clean Air Act Amendments (CAAA), which strengthened U.S. controls on motor vehicle emissions and fuels through Title II. To access this titled section on the Internet, visit **www.epa.gov/air/caa/title2.html**. In addition to tougher command-and-control regulations, these rulings also incorporate incentives to encourage technological development of cleaner-running vehicles and cleaner alternative fuels. Among the more salient features of this law are:

- Emission reductions for cars and trucks
- Fuel quality controls
- Initiatives to develop clean fuel vehicles and clean alternative fuels

A brief summary of each of these initiatives follows.

Emission Reductions for Motor Vehicles[8]

The 1990 Amendments impose tougher regulations on tailpipe emissions released from all types of highway vehicles through two tiers of progressive standards. The first, Tier I, was aimed at reducing emissions of non-methane hydrocarbons (NMHC), carbon monoxide (CO), nitrogen oxides (NO_X), and particulate matter (PM), and was to be phased in between 1994 and 1998. Under this first phase, passenger cars and light trucks were subject to tighter regulation than heavier sport utility vehicles (SUVs), minivans, and pickup trucks. There also were differences based on the life of the vehicle with stricter controls for the initial 5 years or 50,000 miles of vehicle use relative to the remainder of the vehicle's useful life redefined to be 10 years or 100,000 miles. Beyond these distinctions, the emissions limits were to be implemented much as they had been in the past: uniformly applicable to every new model car.

The law also defines Tier II standards to take effect between 2004 and 2008, with tougher requirements identified for each successive model year. Unique to this second phase of controls is an interrelated system of controls on vehicles and fuels. As for the emissions limits on vehicles, these are defined to achieve further reductions in pollutant releases than those outlined by Tier I, and, perhaps more significantly, for the first time they make no distinction for vehicle size. Hence, all emissions standards are to be met both by passenger cars and heavier vehicles, such as SUVs, vans, and pickups.

Fuel Quality Controls

Just as vehicles are more stringently regulated under the 1990 law, so too are fuels and fuel additives. For example, after December 31, 1995, any fuel containing lead or lead

---

[7]To read more about this history, visit **www.epa.gov/otaq/invntory/overview/solutions/milestones.htm**.
[8]Information for this section is drawn from Ashford and Caldart (2008), Chapter 7; U.S. EPA, Office of Transportation and Air Quality (November 2005).

additives was prohibited for highway use. A similar change was called for under Tier II of the CAAA whereby refiners are required to reduce sulfur in gasoline by 90 percent. Like lead, sulfur interferes with the functioning of catalytic converters.

reformulated gasoline
Emits less hydrocarbons, carbon monoxide, and toxics than conventional gasoline.

oxygenated fuel
Has enhanced oxygen content to allow for more complete combustion.

Other provisions in the law require the use of cleaner fuels in certain of the nonattainment areas. For instance, **reformulated gasoline** must be used in some ozone nonattainment regions, and **oxygenated fuel** must be used in designated carbon monoxide nonattainment areas.

- **Reformulated gasoline** refers to formulations that emit less hydrocarbons, carbon monoxide, and toxics than conventional gasoline.[9]
- **Oxygenated fuel** contains more oxygen to allow for more complete combustion and a reduction in carbon monoxide emissions.

To encourage compliance with these new fuel laws, market-based incentives are used. Marketable credits are issued for fuels that exceed legal requirements. These credits can be used by the recipient or transferred to another individual in the same nonattainment area.

In recent years, there has been increased discussion about using ethanol as an oxygen-boosting additive to gasoline. Why? Because the once common use of methyl tertiary-butyl ether (MTBE) as an oxygenated additive has come under scrutiny for posing an environmental risk. This substance, which had been touted as a relatively inexpensive way to meet oxygenated requirements, was detected in drinking water supplies around the nation, apparently leaking from underground storage tanks. Beyond concerns that the additive is known to add an offensive taste and odor to drinking water, research also suggests that MTBE is a possible human carcinogen at high doses.

As a consequence, there have been extensive discussions at the federal level to reduce or eliminate the use of MTBE as a fuel additive, but no proposals to legislate such a change have made it through Congress. States, however, have taken a more active role, with 25 of them having enacted legislation for a partial or complete ban.[10] These actions prompted the need for substitute additives to comply with regulations on clean fuels, which lead to renewed support for ethanol blends and other alternative fuels.

clean fuel vehicle
A vehicle certified to meet stringent emission standards.

clean alternative fuels
Fuels, such as methanol and ethanol, or power sources, such as electricity, used in a clean fuel vehicle.

Clean Fuel Vehicles and Clean Alternative Fuels

Responding further to urban air quality pollution, the 1990 CAAA establish a clean fuel vehicles program. A **clean fuel vehicle** is one that has been certified to meet stringent emission standards for substances such as CO, NO_X, PM, and formaldehyde over prescribed time periods. For designated ozone and carbon monoxide nonattainment areas, states must set up programs for the adoption of clean fuel vehicles by owners or fleet operators. These programs were to be phased in starting with the 1998 model year. At this point, a fixed percentage of all new fleet vehicles bought in each area had to be clean fuel vehicles and use **clean alternative fuels**. These are fuels or power sources used in a clean fuel vehicle.

Examples of clean alternative fuels are methanol, electricity, and ethanol blends. Table 11.3 lists some of the pros and cons of using alternative fuels. Although all have

[9]To read more about reformulated gasoline, visit **www.epa.gov/otaq/fuels/gasolinefuels/rfg/index.htm**.
[10]U.S. EPA (August 2007). For information and updates about MTBE, visit the EPA's MTBE home page at **www.epa.gov/mtbe**.

TABLE 11.3 Clean Alternative Fuels

Fuel	Selected Advantages	Selected Disadvantages
Biodiesel	Reduced PM and greenhouse gas emissions; domestically produced; available in bulk; nontoxic, biodegradable; renewable; low volatility.	NO_X emissions may be higher; limited availability; more costly than conventional diesel; lower fuel economy and power.
E85 Ethanol	Reduced emissions of CO and NO_X relative to reformulated gasoline; produced domestically; renewable; higher octane rating than gasoline.	Can be used only in flexible fuel vehicles; more expensive than traditional fuels in some areas; lower energy content than gasoline; can form an explosive vapor in fuel tanks; limited but increasing availability.
Electricity	Zero tailpipe emissions; wide availability; relatively low cost for fuel and maintenance; energy efficient; low operating noise.	Higher-cost vehicles; high replacement cost of lead-acid battery; limited driving range on single battery charge.
Hydrogen	Can be produced domestically; emits no pollutants or GHGs when used in fuel cells; emits only NO_X when burned in combustion engines.	Costly to produce; limited availability; high cost and limited availability of fuel cell vehicles; low per-volume energy requiring large tank in vehicles.
Liquefied Natural Gas (methane)	Reduced smog-producing emissions; lower GHG emissions; domestically produced; lower cost per mile than diesel; cheaper than gasoline.	Higher costs of storing and dispensing; higher vehicle costs; special handling needed for storing and refueling; hydrocarbon emissions may be higher; limited availability.
Liquefied Petroleum Gas (propane)	Reduced emissions of CO and NO_X relative to reformulated gasoline; most widely available alternative fuel; higher octane rating than gasoline.	Higher-cost vehicles; higher flammability in gaseous state; lower energy content than gasoline; less available than gasoline or diesel.
Methanol	Lower CO and NO_X emissions relative to reformulated gasoline; produced domestically from renewable resources; higher octane rating than gasoline.	Slightly higher costs based on lower energy content; slightly higher vehicle cost; high distribution costs; can form an explosive vapor in fuel tanks; not commonly used.

NOTE: For more detail, visit **www.afdc.energy.gov/afdc/fuels/index.html**, the U.S. Department of Energy's Alternative Fuels and Advanced Vehicles Data Center.

Sources: U.S. Department of Energy, Office of Energy Efficiency and Renewable Energy, and U.S. EPA (February 22, 2009); U.S. Department of Energy, Office of Energy Efficiency and Renewable Energy (February 20, 2009; June 9, 2003); U.S. EPA, Office of Transportation and Air Quality (March 2002).

ethanol (E10) Known as gasohol, a blend of 10 percent ethanol and 90 percent gasoline.

ethanol (E85) Blended fuel comprising 85 percent ethanol and 15 percent gasoline.

been part of important discussions, much of the recent debate has been about ethanol blends: **ethanol (E10)**, or gasohol, comprising 10 percent ethanol and 90 percent gasoline; and **ethanol (E85)**, a blend of 85 percent ethanol and 15 percent gasoline. E10 is approved for use by all automotive manufacturers but technically is not considered an alternative fuel under some legislative acts. E85 can be used only in flexible fuel vehicles (FFVs). In addition to its environmental advantages, production of corn-based ethanol would be a boon to the ailing agricultural industry, according to its proponents. However, use of ethanol is not without debate. Some argue that it is a costly solution given that production of ethanol can be more expensive than production of traditional gasoline.[11]

[11]To learn more about the costs of ethanol production, see U.S. Department of Energy, Energy Information Administration (February 2007), which is available online at **www.eia.doe.gov/oiaf/analysispaper/biomass.html**.

Within the clean fuel vehicle provisions of the CAAA are market incentives. Credits are to be issued to fleet operators or owners who surpass the requirements in the law. They might do this by purchasing clean fuel vehicles in advance of the deadline or by buying more of these vehicles than the law requires. The awarded credits can be held, banked for future use, or traded.

Other market incentives for clean fuel vehicles have been added through more recent legislation. Specifically, under the Energy Policy Act of 2005, tax credits up to $3,400 were made available to purchasers of hybrid cars. These credits gradually were phased out once the manufacturer of the vehicle sold 60,000 hybrids. Analogous tax credits were offered for purchasers of fuel cell and alternative fuel vehicles, which varied by type of fuel used and by vehicle weight. Later, with the passage of the Emergency Economic Stabilization Act of 2008, new tax credits for plug-in hybrids were added, ranging from $2,500 to $7,500. In this case, the credits are phased out after total sales in the United States reach 200,000.[12]

Reducing Greenhouse Gas (GHG) Emissions[13]

Following years of procedural entanglement and legal debate, the U.S. Supreme Court ruled in 2007 that greenhouse gases (GHGs) fit the definition of pollutants according to the Clean Air Act. The decision meant that the EPA is authorized to regulate these gases if, in fact, the agency determines that GHGs pose a threat. In 2009, the EPA announced that this is indeed the case, a decision known as the Endangerment Finding. The agency also stated that mobile sources contribute to this threat, which is referred to as the Cause or Contribute Finding.

Consequently, in April 2010, the EPA and the National Highway Traffic Safety Administration (NHTSA) finalized a joint rule under the 1990 CAAA and the Energy Policy and Conservation Act to improve fuel efficiency and lower GHG emissions for all passenger cars and light-duty vehicles, (i.e., those with a gross weight of 8,500 pounds or less). These vehicles are responsible for almost 60 percent of all mobile source GHG emissions.

In August 2011, the two agencies announced a similar program—this one for medium- and heavy-duty vehicles under the authority of the 1990 CAAA and the Energy Independence and Security Act, respectively. The targeted vehicle class under this so-called HD National Program includes all on-road vehicles with a gross weight greater than 8,500 pounds, which are collectively responsible for about 20 percent of the GHG releases from the transportation sector in 2007.

Setting GHG Emissions Standards

As a consequence of these new programs, for the first time in U.S. history, GHG emissions standards have been established for two classes of motor vehicles by the EPA and the NHTSA. For passenger cars and light-duty trucks, the standards are aimed at model years 2012 through 2016. This class of motor vehicles will be required to meet a

[12]U.S. Department of Energy (October 21, 2008); U.S. Department of the Treasury, Internal Revenue Service (2006).

[13]Much of the following is drawn from Pew Center on Global Climate Change (2011); U.S. EPA, Office of Transportation and Air Quality (August 2011; July 2011; April 2010).

combined average emissions level of 250 grams of carbon dioxide (CO_2) per mile in model year 2016. This would be equivalent to achieving 35.5 miles per gallon (mpg), assuming that all CO_2 reductions were met through fuel efficiency gains. In addition, caps on tailpipe emissions are being set on nitrous oxide (N_2O) and methane (CH_4) at 0.10 and 0.30 grams per mile, respectively.

The comparable set of rules and standards for heavy-duty vehicles under the HD National Program targets three predefined vehicle categories for the 2014–2018 model years. According to estimates, the combined standards will reduce CO_2 emissions by some 270 million metric tons and reduce oil consumption by about 530 million barrels over the life of the affected vehicles. Also, as in the light-duty program, standards are being set on N_2O and CH_4 emissions. More detail and updates are available online at **www.epa.gov/otaq/climate/regulations.htm**.

Setting New CAFE Standards

Part of the light-duty vehicle initiative is the establishment by the NHTSA of more stringent Corporate Average Fuel Economy (CAFE) standards, which come under the Energy Independence and Security Act. These standards call for light-duty vehicles to achieve a combined average of 34.1 mpg in the 2016 model year.[14] For further information on the CAFE standards and these new rulings, visit **www.nhtsa.gov/fuel-economy**.

Next Steps

The EPA and the NHTSA have proposed a second phase of CO_2 and CAFE standards for light-duty vehicles that will be applicable to the 2017–2025 model years, becoming increasingly more stringent for each model year. The new standards are expected to achieve an industry, fleet-wide average of 163 grams per mile of CO_2, or equivalently 54.5 mpg, by model year 2025. Anticipated gains are estimated to be a reduction in GHG emissions of 2 billion metric tons and decrease in oil consumption of 4 billion barrels of oil over the life of the affected vehicles. The two agencies also are considering a second phase of GHG standards for the medium- and heavy-weight class of vehicles for the 2018 model year and beyond.[15]

ECONOMIC ANALYSIS OF MOBILE SOURCE CONTROLS

Historically, U.S. attempts to control emissions from mobile sources have been frustrated by a litany of problems and delays. In a very real sense, many of the difficulties were predictable, even avoidable. In retrospect, much of the problem was that the law failed to consider the dynamics of the marketplace and the importance of benefit-cost analysis. This failure seems to have been addressed in more recent initiatives, however, such as the clean fuel vehicle alternatives and the GHG emissions standards.

[14]This 34.1 mpg value is slightly lower than the 35.5 mpg equivalent cited previously because the NHTSA assumes that automobile manufacturers will employ certain compliance options made available within the program, which will reduce compliance costs (Pew Center on Global Climate Change, 2011).

[15]According to the EPA, these new regulations on GHG emissions from mobile sources trigger analogous rulings for stationary sources, which are currently under consideration. We will discuss these GHG controls on stationary sources in Chapter 13.

To provide structure to our evaluation of these issues, we focus our analysis on the following characteristics of mobile source emissions control policy:

- absence of benefit-cost analysis in setting standards for tailpipe emissions
- uniformity of auto emissions standards
- inherent bias against new vehicles
- implications of clean fuel and clean fuel vehicle alternatives
- market assessment of the GHG emission controls

Absence of Benefit-Cost Analysis: An Inefficient Decision Rule

After years of delay in establishing federal limits on auto emissions, it is an interesting irony that the 1970 CAAA imposed controls that by most accounts were extraordinarily stringent. In fact, most argue that the standards were not attainable and in fact were meant to be **technology-forcing** by design—that is, specifically set to compel the auto industry to find solutions. Such a strategy may have been in order, given the industry's track record up to that point. However, most have questioned the government's wisdom in setting emission reductions so far out of reach that manufacturers had a strong case to seek adjustments and postponements.

Just as important is that the implied decision rule used to establish the emissions controls was not supported by benefit-cost analysis. Rather, the standards were set to protect public health and welfare, a purely **benefit-based** objective. As shown in Figure 11.2, this is linked to abatement activity that maximizes total social benefits, which corresponds to the point where the marginal social benefit (*MSB*) curve crosses the horizontal axis

FIGURE 11.2 Inefficiency of a Benefit-Based Decision Rule

Mobile source standards are set to achieve a benefit-based objective. Such an objective calls for abatement activity that maximizes total social benefits, which corresponds to the point where the *MSB* curve crosses the horizontal axis (i.e., where $MSB = 0$), at A_0. Notice that A_0 is *higher* than the efficient level, A_E (i.e., where $MSB = MSC$), which implies overregulation.

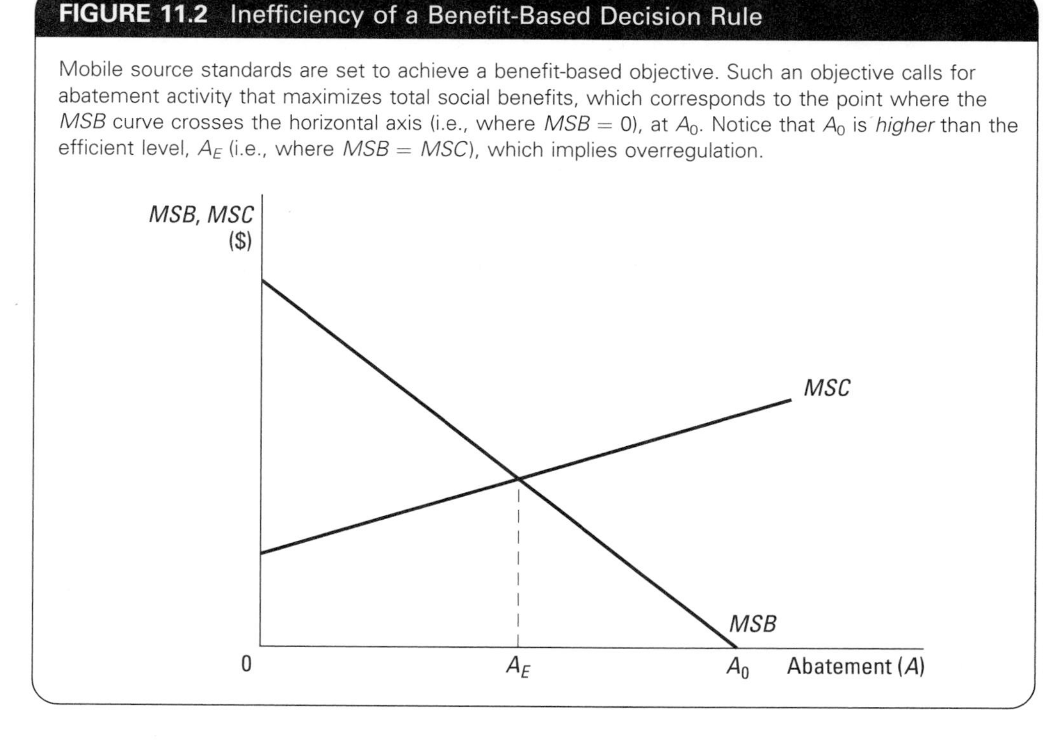

(i.e., where $MSB = 0$), at A_0. Notice that A_0 is *higher* than the efficient level A_E (where $MSB = MSC$), which implies overregulation.

As it turned out, the 1970 standards were never implemented in their original form. In fact, they were modified considerably. Worse, the haggling over the unrealistic standards between the government and the automobile industry added further to the already long delays. Had the standards been set at levels that accounted for the full extent of the associated costs, it is possible that the automobile industry would not have had sufficient support for its lobbying efforts to extend the statutory deadlines. Instead, the lack of efficient decision making left policymakers unarmed for the assault mounted by the automakers. The result? The ultimate burden was borne by society, and the ill-defined decisions of that era accomplished little more than to postpone the provision of cleaner-running vehicles to the marketplace.

Uniformity of Auto Emissions Standards

An ongoing problem with U.S. mobile source controls is that with few exceptions the emission standards are applicable across the board on every model produced and without regard to where the vehicle will be driven. For example, there is a uniform requirement in the law that calls for all cars and light-duty trucks starting with 1984 models to meet high-altitude standards, despite the fact that only a small percentage of cars sold in the United States are purchased in high-altitude areas. Ultimately, such an approach needlessly increases the costs of reducing pollution with no offsetting benefit to society.[16]

Instead of making emission controls applicable to every model produced, substantial savings could be achieved by allowing each producer to meet legislated targets based on *average* emissions of all new cars produced. The benefits based on overall emission reductions would be unchanged, but the costs of compliance would be much lower.[17] The forgone savings help to explain in part the observed growth rate of mobile source control costs. According to the EPA's first prospective analysis of the Clean Air Act, control costs of mobile sources under Title II were projected to grow to $9.1 billion ($1990) in 2010 from $7.4 billion ($1990) in 2000.[18]

Disregard for the location where a vehicle is operated overregulates vehicles sold in relatively clean areas and undercontrols those sold in more polluted areas. This is illustrated in Figure 11.3, where we logically assume that the *MSB* of abatement in dirty areas is higher than in clean areas, *ceteris paribus*.[19] Notice that if the standard, A_{ST}, is the same for both regions, the dirty area is undercontrolled, because A_{ST} is below the efficient level, A_E. The opposite holds in relatively clean areas, where A_{ST} is above the efficient level, A_E.

---

[16]Council of Economic Advisers (1982), pp. 145–46.

[17]This is discussed in the economic analysis of mobile source emissions by the Council of Economic Advisers (1982), p. 146, which asserts that such an averaging plan could save millions of dollars.

[18]U.S. EPA, Office of Air and Radiation (November 1999), Table 3.1, p. 26.

[19]It is not unreasonable in this context to assume that the *MSC* is the same for both regions, because it is referring to abatement of mobile source emissions, and abatement of a mobile source is, by definition, not specific to a geographic location.

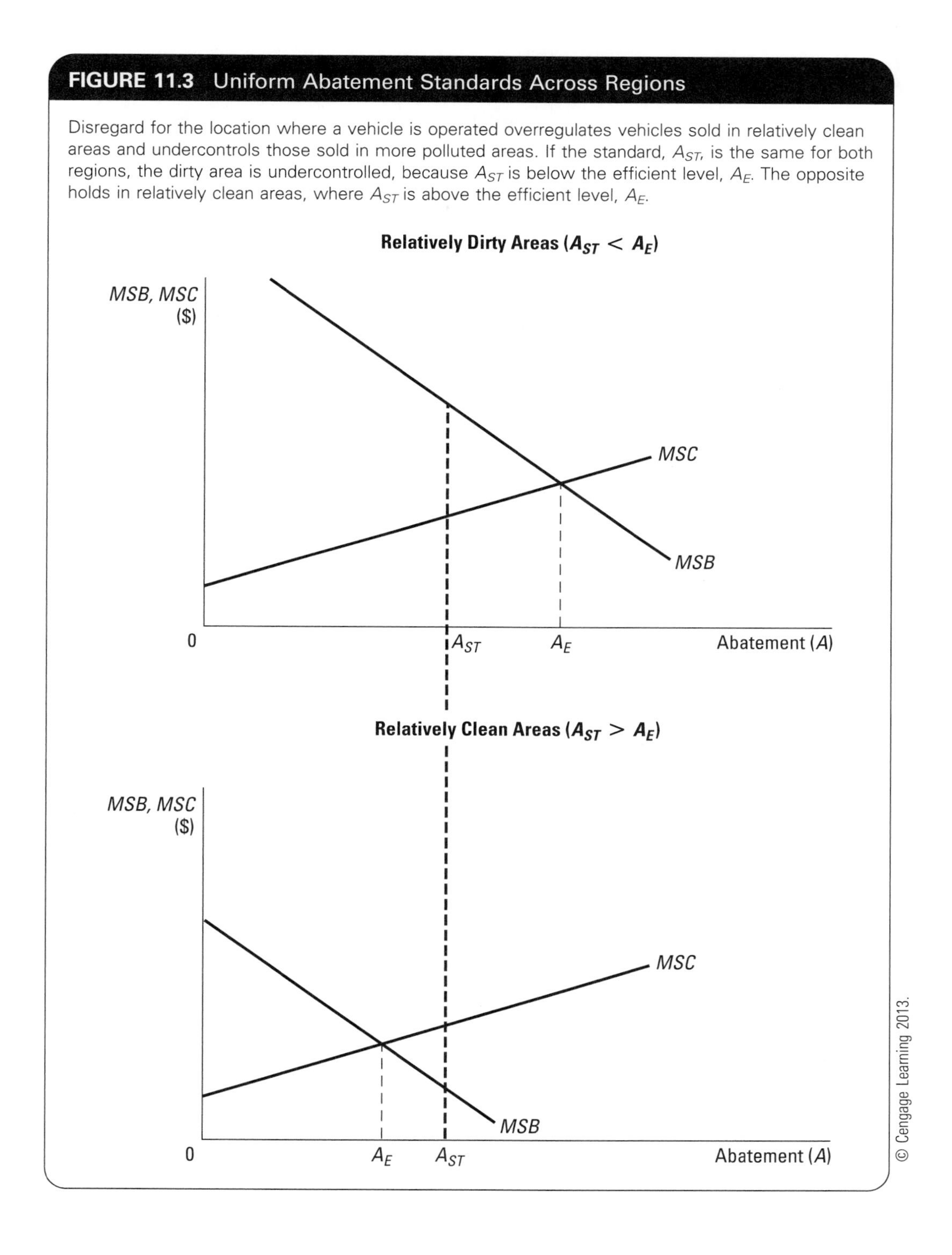
FIGURE 11.3 Uniform Abatement Standards Across Regions
Disregard for the location where a vehicle is operated overregulates vehicles sold in relatively clean areas and undercontrols those sold in more polluted areas. If the standard, A_{ST}, is the same for both regions, the dirty area is undercontrolled, because A_{ST} is below the efficient level, A_E. The opposite holds in relatively clean areas, where A_{ST} is above the efficient level, A_E.
Relatively Dirty Areas ($A_{ST} < A_E$)
MSB, MSC ($)
MSC
MSB
0
A_{ST}
A_E
Abatement (A)
Relatively Clean Areas ($A_{ST} > A_E$)
MSB, MSC ($)
MSC
MSB
0
A_E
A_{ST}
Abatement (A)
© Cengage Learning 2013.

A solution would be to have two sets of standards administered at the state level based on the degree of pollution in that area. Proponents of this more federalistic approach believe this would yield important cost savings. A study conducted by the National Academy of Sciences estimates the costs of a **uniform standard** of 0.4 grams per mile (gpm) for NO_X relative to the costs of a **two-tiered standard**, calling for a 0.4 gpm standard in seriously polluted areas (about 37 percent of motor vehicles) and a 3.1 gpm standard in all other areas. The study predicts that a $23 billion savings over the 1975–1985 period would have resulted if the two-tiered system had been used.[20]

zero-emission vehicle (ZEV)
Emitting zero tailpipe emissions and runs 98 percent cleaner than the average new model year vehicle.

partial zero-emission vehicle (PZEV)
Emitting zero evaporative emissions and runs 90 percent cleaner than the average new model year vehicle.

Current law does allow some exceptions to these uniform standards. Perhaps the most notable is that under the Clean Air Act, California has special authority to set stricter standards for vehicles than specified at the federal level subject to EPA's approval of a waiver. An important example is California's use of more stringent emission controls as part of its **low-emission vehicle (LEV)** program adopted in 1990 and amended in 1998. Some states followed California, including Maine, Massachusetts, New York, and Vermont, with more states joining the list in later years. These LEV regulations define California-specific emission standards and phase-in requirements for manufacturers.

Included in this LEV program are rulings on **zero-emission vehicles (ZEVs)** (i.e., electric or fuel cell cars) and **partial zero-emission vehicles (PZEVs)** (i.e., plug-in hybrids, hybrids, and compressed natural gas automobiles). Starting in 2005, the law called for 10 percent of every automaker's production for California to be ZEVs. However, given the emission control improvements in conventional vehicles and the increased use of PZEVs, manufacturers now have the option to meet this requirement by replacing ZEVs with PZEVs.[21] To learn more about the California LEV program, visit **www.arb.ca.gov/msprog/levprog/levprog.htm**.

These regional variances from national requirements have set off a round of debates. The auto industry points to the increased costs of complying with several sets of emissions standards. Others question the wisdom of moving to electric vehicles that have to be fueled by power plants. In some regions, such as the Northeast, electric power plants are more important contributors to smog than are mobile sources, so the net gain in air quality may not be worth the effort. Furthermore, the technology of electric car engineering is still developing, particularly in addressing the harsher conditions of cold-weather climates. An alternative to electric cars that many believe holds greater promise is the hybrid vehicle, which is discussed in Application 11.2, accompanied by an overview of the more recently introduced plug-in electric hybrid.

Inherent Bias Against New Versus Used Automobiles

One of the more troubling dilemmas in environmental law is the inherent bias caused by more stringent controls imposed on *new* polluting sources. This partiality is particularly problematic for motor vehicles, because engine performance and sticker price are at issue. As long as there are differences between the regulation of new and used vehicles, there will be price and performance impacts that can create market distortions.

To fully understand how the law can influence market decisions, consider the following scenario, which is modeled in Figure 11.4 (page 252). The automobile is a

[20]See National Academy of Sciences and National Academy of Engineering (1974).

[21]California Environmental Protection Agency, Air Resources Board (May 6, 2008; August 14, 2004; February 1999).

APPLICATION 11.2 Hybrids and Plug-In Hybrid Electrics

As their name implies, hybrid vehicles are powered by a combination of two energy sources: electricity and conventional fuels. These vehicles, first introduced to the mass market by Honda and Toyota, utilize an electric motor for start-up and low speeds and rely on a small conventional engine for cruising at normal speeds. These high-tech low-emission vehicles achieve high fuel economy, particularly for city driving, offering drivers significant savings on fuel costs and resulting in reduced emissions. The fuel cost savings rather than environmental concerns may be the strongest buying incentive. This is particularly evident in periods when gasoline prices are rising sharply. Nonetheless, society still gains the environmental benefit of lower fuel emissions, no matter what incentive triggers the demand.

Just as Japanese automakers led the industry in producing small cars during the 1970s, so too were these companies first to develop hybrid vehicles and market them on a mass scale. Today, these manufacturers are still the front-runners of the hybrid market, with Toyota as the clear leader. A case in point is Toyota's 4-cylinder Prius. Some argue that the Prius is popular because of its unusual body style, which distinguishes it clearly from conventional cars, a feature that hybrid buyers appreciate. More importantly, the Prius achieves 51 miles per gallon (mpg) in the city and 48 mpg for highway driving, far better than any of its competitors, and it has a relatively small carbon footprint.

The following table offers a comparison of the top hybrids for the 2011 model year, ranked by city mpg. For comparisons of other vehicles in 2011 or in other model years, visit **www.fueleconomy.gov**.

Make/Model	City MPG	Highway MPG	Carbon Footprint[a]
Toyota Prius	51	48	3.8
Lexus CT	43	40	4.5
Mercury Milan	41	36	4.8
Lincoln MKZ	41	36	4.8
Ford Fusion	41	36	4.8
Honda Civic	40	43	4.6
Honda Insight	40	43	4.6

[a]The **carbon footprint** shown is a measure of the vehicle's effect on climate change, quantified as the number of tons of carbon dioxide (CO_2) emitted annually.

Source: U.S. Department of Energy, Office of Energy Efficiency and Renewable Energy, and U.S. EPA (April 26, 2011).

That Toyota has enjoyed superlative positioning in the hybrid market evidenced in part by its strong market share. However, the company's litany of recalls followed by the 2011 earthquake and tsunami have diminished this strength. Nonetheless, Toyota's innovation and success in the hybrid market has not gone unnoticed by the other major manufacturers. Honda, for example, relaunched its Insight hybrid, which was initially introduced in 2000 and sold until 2006. This time around, the car is aimed squarely at Toyota's Prius with a sharply lower sticker price.

As for U.S. manufacturers, the Detroit firms have not fared well in the hybrid market relative to their Japanese counterparts. Chrysler, for example, halted production of its SUV hybrid in 2009 after only a few months of lackluster sales. Part of the problem has been the relatively high prices of domestic contenders and the overall struggle of U.S. automakers during the economic slowdown. However, these U.S. producers are participating more vigorously in the plug-in hybrid vehicle (PHEV) market. A case in point is the GM Chevy Volt, which was named North American Car of the Year, Motor Trend Magazine's Car of the Year, Green Car of the Year, and Automobile Magazine's Car of the Year in 2011.

Plug-in hybrid electric vehicles (PHEVs) are new to the market. More expensive than their predecessors, these PHEVs can be plugged into an ordinary household electric outlet to recharge the battery. Once done, the car can be operated as an electric vehicle with the gasoline engine as backup. Under the Emergency Economic Stabilization Act of 2008, tax credits for PHEVs purchased in or after 2010 are available to consumers up to $7,500. The plug-in technology adds a new dimension to the hybrid marketplace, but it is unclear how these vehicles will affect energy costs and how industry participants ultimately will fare in this evolving market environment.

Sources: U.S. Department of Energy, Office of Energy Efficiency and Renewable Energy, and U.S. EPA (April 26, 2011; April 20, 2011); Bunkley (January 13, 2009); Smith (May 2, 2008).

FIGURE 11.4 Modeling the Bias Against New Automobiles

Regulating new car emissions adds to production costs, shifting supply in the new car market to the left from S_1 to S_2 and elevating price from P_1 to P_2. Faced with higher relative prices for new cars, some consumers will purchase a used car or keep a deteriorating one, either of which is likely to be a relatively high emitter of pollutants. This would cause a rightward shift in the demand for used cars from D_1 to D_2.

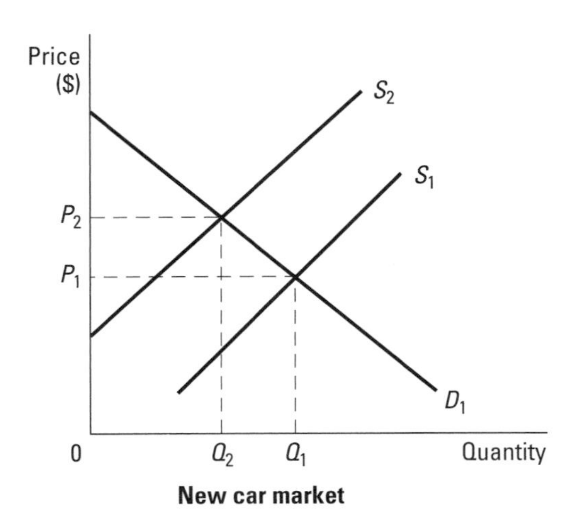

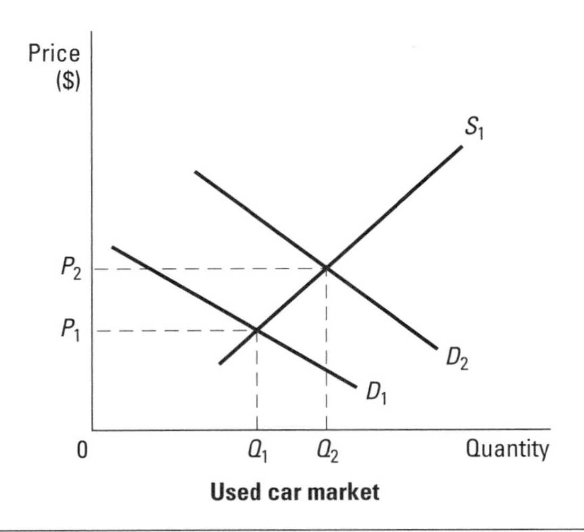

durable good, meaning its economic life extends beyond a single period. Because of this, new and used automobiles are substitutes for one another with the usual relationship between their relative prices. Regulating new car emissions adds to the production costs of those vehicles, shifting the supply curve to the left from S_1 to S_2. The result is that the higher costs are passed on, at least in part, to consumers, elevating price from P_1 to P_2.

Faced with higher relative prices for new cars, some consumers will purchase a used car or opt to keep a deteriorating one, either of which is likely to be a relatively high emitter of pollutants. As illustrated in Figure 11.4, we then would observe a shift rightward in the demand for used cars from D_1 to D_2. Because the regulation encourages consumers to substitute in favor of used cars, it effectively extends the economic life of cars already on the road. There is also the issue that emission control devices negatively affect a car's acceleration and gas mileage. Thus, the legislation confounds the signaling mechanism of price by diminishing the performance of a relatively higher priced vehicle.

Some policy changes address this dilemma. For example, the 1990 Amendments extend the economic life of the automobile over which emission standards are mandated by increasing the regulatory period from its former 5 years or 50,000 miles to 10 years or 100,000 miles. Although the standards for the second half of the period are less stringent, the extension to a longer time period helps to reduce the bias against future new vehicles, because most cars on the road ultimately will be subject to at least some degree of emissions control.

An alternative idea is to use some type of Voluntary Accelerated Vehicle Retirement (VAVR) program. These incentive-based plans allow polluters to earn emissions credits in return for buying older vintage cars and scrapping them to get these high emitters off the road. A contemporary example of a VAVR program is the recent Car Allowance Rebate System (CARS) launched under President Obama's administration, which is discussed in Application 11.3.

APPLICATION 11.3 The Car Allowance Rebate System (CARS): Cash for Clunkers Revisited

In June 2009, President Obama signed into law the Consumer Assistance to Recycle and Save Act of 2009, which in turn launched a short-term stimulus program known as the Car Allowance Rebate System (CARS). The economic purpose of this initiative was to shift consumers' future expenditures into the present, but there was an expected environmental gain as well. The environmental objective was to encourage consumers who met certain criteria to trade in their older vehicles for a credit voucher that would be used toward the purchase of a newer, and hence more fuel-efficient, vehicle. Such incentive-based plans are more generally known as Voluntary Accelerated Vehicles Retirement (VAVR) programs or less formally as "Cash for Clunkers." The origin of VAVRs dates back to the early 1990s when the United States proposed the use of these market-based programs to encourage the scrapping of older-model, high-emitting automobiles.

Administered by the National Highway Traffic Safety Administration (NHTSA), transactions under the CARS program were permitted only for a brief period, from July 1, 2009, through November 1, 2009. However, demand for the program was so high that transactions had to be halted at the end of August even after tripling the amount of allocated funding to $3 billion.

In order to participate in the program, consumers had to satisfy specific conditions of eligibility. Among these was that the traded-in vehicle had to be older than the 2008 model year with originally certified fuel economy lower than 18 mpg. Another condition was that the replacement vehicle had to have a better mileage rating than the trade-in vehicle. The vouchers had a redemption value of $3,500 ($2008) or $4,500 ($2008), whereby the higher-valued voucher was reserved for mileage improvements of 10 mpg for cars and 5 mpg for light trucks.

Ultimately, over 675,000 vehicles were traded in and replaced by more fuel-efficient vehicles, using vouchers with an average price of $4,209. As to the stimulus objective of CARS, the NHTSA estimated that the program added between $3.8 billion and $6.8 billion to gross domestic product (GDP) and created or saved more than 60,000 jobs. As for the environmental implications, the NHTSA estimates that over the subsequent 25-year period, fuel consumption would decline by approximately 824 million gallons, and GHG emissions would be reduced by 9 million metric tons. To read an economic analysis of the CARS program, visit **www.whitehouse.gov/administration/eop/cea/CarAllowanceRebateSystem/**.

In California, use of such buy-back programs has been part of the state's effort to mitigate its persistent smog problem for some time. Identified in California's State Implementation Plan (SIP), its VAVR program outlined a statewide effort for the 1999–2010 period to retire and destroy as many as 75,000 cars and light trucks each year during the period. The stated goal was to realize a reduction of 25 tons per day of ozone-forming emissions by 2010. To a large extent, California's objective has been rooted in the reality that older cars, while representing a small proportion of vehicles on the road, are responsible for a disproportionate amount of mobile source emissions.

Today, California continues to offer VAVR programs at both the state and local levels. The updated state plan, which is run by the Bureau of Automotive Repair, is called the Consumer Assistance Program (CAP). It offers eligible consumers $1,000 for their unwanted vehicles that have either passed or failed their most recent Smog Check Test. Local programs are available only where districts elect to offer and administer them. For more information on California's VAVR programs, visit the state's Air Resources Board at **www.arb.ca.gov/msprog/avrp/avrp.htm**.

Sources: California Environmental Protection Agency, Air Resources Board (January 7, 2011); U.S. Department of Transportation, National Highway Traffic Safety Administration (NHTSA) (December 2009); Council of Economic Advisers (September 10, 2009); Dixon and Garber (2001).

Implications of Clean Fuel and Clean Fuel Vehicle Alternatives

Clean fuel provisions in the 1990 Amendments are an extension of technology-based standards, which were formerly directed almost entirely toward engines and emission control systems. Furthermore, the use of more advanced fuel compositions is required only in the dirtier regions of the country. Targeting these areas makes good sense from an economic perspective. The development and use of alternative fuels generates relatively high marginal costs, which aligns with the higher expected marginal benefits that will accrue in the designated nonattainment regions. Hence, this selective legislation follows the fundamental decision rule of benefit-cost analysis, which may approach an efficient solution.

Available tax credits for clean fuels and clean fuel vehicles also may facilitate the achievement of efficiency. By way of example, consider the tax credits made available to purchasers of plug-in hybrids. These tax credits act as a subsidy (s) to promote consumption of these vehicles relative to their conventionally powered counterparts, because doing so contributes to improved air quality. Cleaner air is an external benefit of consumption. Hence, we can model the market for plug-in hybrids by explicitly showing the marginal external benefit (*MEB*) of consumption, which is additive to the marginal private benefit (*MPB*), as shown in Figure 11.5.

In the absence of the tax credit, the *MEB* is ignored by private markets, resulting in the competitive equilibrium output, Q_C, and price, P_C, both of which are lower than

FIGURE 11.5 Tax Incentives on Plug-In Hybrids

The positive consumption externality associated with the use of a plug-in hybrid vehicle gives rise to a positive marginal external benefit (*MEB*), making the marginal social benefit (*MSB*) higher than the marginal private benefit (*MPB*). In the absence of third-party intervention, equilibrium occurs at Q_C and P_C, which is a lower output and price than the efficient solution, Q_E and P_E. The government-sponsored tax credits effectively elevate the *MPB* vertically by the dollar value of the credit to $MPB_{TAX\ CREDIT}$. If the credit equals the *MEB* at Q_E, the efficient solution is achieved, as shown in the model below.

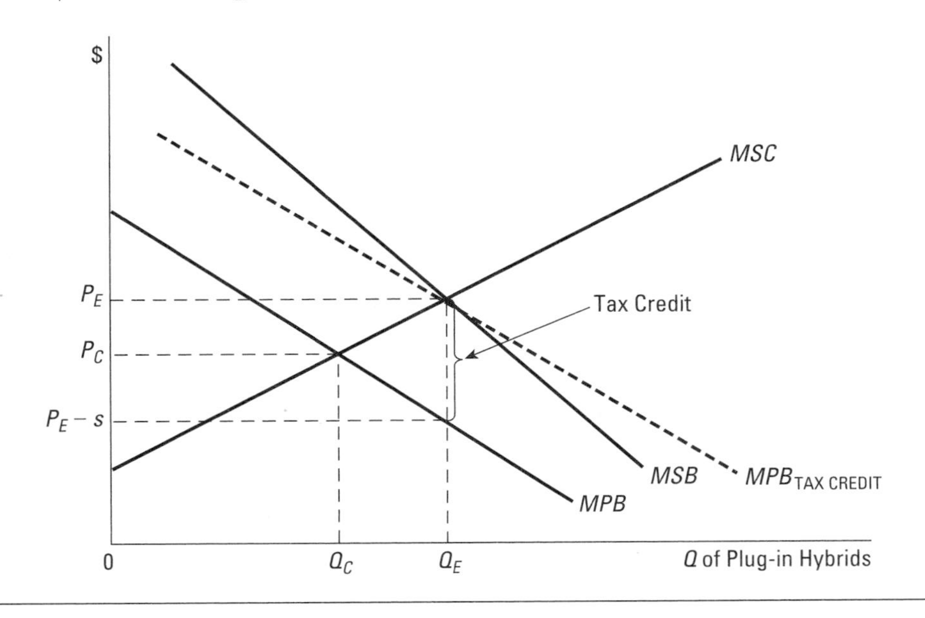

their respective efficient levels. This outcome means that resources are underallocated to the production of plug-in hybrids, and these vehicles are undervalued by the market because the external benefit is not captured by market price. The tax credit, however, effectively shifts up the *MPB* by the amount of the credit (or subsidy), and in so doing, increases equilibrium output and price. If the credit is equal to the *MEB* at the efficient output level, Q_E, the efficient solution results at Q_E and P_E, where P_E is the price received by the seller, and $(P_E - s)$ is the effective price paid by the consumer. Even if the credit is less than the *MEB*, the resulting output level would move closer to the efficient result than if the credit were not allowed, i.e., at some point between Q_C and Q_E with an associated price between P_C and P_E.

Market Assessment of the GHG Emission Controls[22]

The two national programs aimed at controlling GHGs from mobile sources are environmentally motivated by concerns about climate change and the associated implications. But what about the economics of these programs?

As it turns out, economic theory underlies and supports the feasibility of these programs and the implementation of each. First, the two joint rules finalized by the EPA and NHTSA—one for passenger cars and light-duty trucks, and the other for medium- and heavy-duty vehicles—are supported by benefit-cost analysis. Specifically, the projected benefits and costs of the proposed regulation economically justify the initiative. Second, inherent in the proposed implementation of these programs are incentive-based instruments, which are rooted in economic theory. Let's briefly consider each of these important issues.

Estimated Benefits and Costs of the GHG Vehicle Programs

For the light-duty vehicle program, overall incremental costs are estimated to be less than $52 billion, and comparable benefits are estimated at $240 billion. These monetized benefits include such gains as fuel savings, enhanced energy security, and particulate matter (PM-2.5) improvements. Nonmonetized benefits include such improvements as the effects of avoided non-CO_2-GHGs, such as methane (CH_4) and nitrous oxide (N_2O), and reduced exposure to ozone and hazardous air pollutants. According to EPA projections, this precedent-setting program is expected to save 1.8 billion barrels of oil and lower GHG emissions in the United States by 960 million metric tons over the life of vehicles sold over the relevant 2012–2016 period.

For the HD National Program, i.e., the medium- and heavy-duty vehicle program, costs to affected industries are estimated to be $8 billion. The associated benefits are estimated to be over $57 billion, including $50 billion in fuel cost savings to owners of affected vehicles. Taken together, expected net benefits to society are $49 billion.

Market-Based Elements of the GHG Vehicle Programs

Integrated into the two national programs are several market-based components. To provide flexibility to manufacturers, each program includes a system of averaging, banking, and trading (ABT) of CO_2 credits. For the light-duty program, the ABT system is

[22]The following is drawn from U.S. EPA, Office of Transportation and Air Quality (August 2011; July 2011; April 2010).

based on a manufacturer's fleet average performance. Trades will be allowed both among different manufacturers and among vehicles produced by a single firm. Within the HD National Program, there is an engine and a vehicle ABT program, both of which allow for emission and fuel consumption credits that can be averaged, banked, or traded within predefined vehicle categories.

Both programs also offer other credit options to provide added compliance flexibility. For the light-duty vehicle initiative, these options include credits for improving vehicle air conditioning performance, so-called early credits for achieving GHG emissions reductions prior to model year 2012, and advanced technology credits for commercializing innovations that reduce GHGs or improve fuel economy, such as electric vehicles, plug-in hybrid electric vehicles, and fuel cell vehicles. Within the HD National Program, identified options include early credits for making improvements before the target model year, and credit programs for advanced technologies, such as hybrid powertrains, and for innovations that lower CO_2 emissions and fuel consumption. In addition, engine, pick-up, and van manufacturers have an added option to use CO_2 credits to offset emissions of N_2O and CH_4 in excess of emission standards based on their respective global warming potential (GWP) measures.

CONCLUSIONS

After decades of delays and false starts, U.S. air quality policy aimed at motor vehicles has evolved into a fairly comprehensive set of provisions. The 1990 Clean Air Act Amendments strengthen federal regulations on tailpipe emissions for cars and trucks, particularly with the two tiers of standards, which are phased in over time. The amendments also establish rulings that effectively extend the economic life of a motor vehicle. In addition, there are initiatives to develop clean fuel vehicles. Recently, the EPA and the NHTSA launched national programs to reduce GHGs by setting new emission standards on motor vehicles and to improve fuel efficiency through revised CAFE standards with further changes in the planning phase. Beyond the controls aimed directly at vehicles, the 1990 Amendments also set policy on fuels, including provisions to mandate the use of cleaner fuels, such as requirements to use reformulated gasoline in certain ozone nonattainment areas.

Most mobile source initiatives are command-and-control in orientation, though a few provisions are market-based, such as credits for purchasing more clean fuel vehicles than required by law. These incentives should enhance compliance and help the nation achieve its overall air quality objectives. That said, economic analysis of mobile source controls suggests the need for further revision. One example is the absence of benefit-cost analysis in setting standards for tailpipe emissions. Another is the regulatory bias against new cars inadvertently caused by the nation's policy on mobile source emissions.

Taken together, these observations remind us that legislating air quality is an evolving process and that there are no perfect solutions. On net, there has been measurable progress in controlling the emissions of motor vehicles in the United States, particularly given the delays in launching policy initiatives at the federal level. Yet there is more to be done, not only in policy development but also in technological innovation, to develop more fuel-efficient engines, better emissions diagnostics, and cleaner fuels.

Summary

- In metropolitan areas, several factors increase the concentration of criteria pollutants, which elevates the risk of exposure. In the United States, the EPA gathers data in major urban centers and reports part of its findings using an Air Quality Index (AQI).
- In urban areas, photochemical smog can form from a chemical reaction involving several criteria pollutants. The principal component of photochemical smog is ground-level, or tropospheric, ozone.
- Mobile sources, particularly highway vehicles, contribute to photochemical smog formation, since they emit a large proportion of NO_X and VOCs, the precursors of smog.
- The 1990 Clean Air Act Amendments (CAAA) strengthen federal control over motor vehicle emissions and fuels. This legislation also incorporates market-based incentives to encourage development of cleaner-running vehicles and alternative fuels.
- In 2010, the EPA and the National Highway Traffic Safety Administration (NHTSA) finalized a joint rule to establish new emissions standards and more stringent Corporate Average Fuel Economy (CAFE) standards for passenger cars and light-duty vehicles for model years 2012 through 2016. In 2011, the two agencies established an analogous set of rules and standards for medium- and heavy-duty vehicles for the 2014–2018 model years, known as the HD National Program.
- In 2011, the EPA and the NHTSA proposed a second phase of GHG emissions standards for light-duty vehicles applicable to the 2017–2025 model years. The two agencies are also considering a second phase of GHG emissions standards for the medium- and heavy-weight class of vehicles for the 2018 model year and beyond.
- Stringent controls on mobile sources established by the 1970 CAAA may have been technology-forcing. Moreover, the implied decision rule used to establish the controls was solely benefit-based, which suggests overregulation.
- The uniformity of national emission standards for automobiles inflates the costs of reducing pollution with no added benefit to society.
- The use of different controls on new versus used motor vehicles creates market distortions. Tougher regulations on new vehicles can bias buying decisions in favor of higher emitting used cars.
- The use of reformulated fuels and oxygenated fuels is being called for in the nation's dirtiest regions, where they can yield the most benefit to society. Available tax credits for clean fuels and clean fuel vehicles may facilitate the achievement of efficiency in these markets.
- Benefit-cost analysis and economic incentives play a role in the new national programs aimed at controlling GHG emissions from mobile vehicles. For both the light-duty vehicle initiative and the HD National Program, projections indicate that the associated benefits of the new rulings are expected to far outweigh the costs. Second, a number of market-based options are available in each program to add flexibility in how manufacturers achieve compliance.

Review Questions

1. In the 1990 Clean Air Act Amendments, Congress and the EPA rely on the automobile industry to develop a cleaner automobile. At the same time, the government imposes a relatively minor federal tax on gasoline.

   a. Do you see any problem with the implicit signals the federal government is sending to American auto manufacturers and to American car drivers through these policies? Briefly discuss.

   b. Formulate a hypothetical economic policy to motivate automobile manufacturers to advance the technology of cleaner motor vehicles.

2. With few exceptions, automobile emissions standards are imposed uniformly under the Clean Air Act, which leads to an inefficient outcome. Suppose a uniform abatement standard (A_{ST}) has been set for PM-10 at 50 units. Also assume that the following are the *MSC* and *MSB* functions for

New Jersey (NJ) and Montana (MT), where each is measured in millions of dollars. (The *MSC* is the same for both states because mobile source abatement is not location-specific.)

$$MSB_{NJ} = 54 - 0.2A$$

$$MSB_{MT} = 32 - 0.3A$$

$$MSC_{NJ} = MSC_{MT} = 20 + 0.2A$$

a. Evaluate the *MSB* and *MSC* for each region at the uniform abatement standard. What do the relative values imply about the over- or under-regulation of each region?

b. Support your result in part (a) by comparing the uniform standard to the efficient abatement level for each state.

3. Refer to the discussion of a two-tiered system of automobile emissions standards studied by the National Academy of Sciences. Use a graph to model this system, and illustrate how the cost savings are achieved.

4. New source bias may exist for mobile sources. Briefly discuss why this bias leads to a solution that is *not* cost-effective. What policies would you implement to eliminate this bias?

5. In the United States, a buyer of a new electric car is eligible for a one-time federal income tax credit of up to $4,000. Show the effect of this tax credit graphically, assuming the $4,000 credit is a Pigouvian subsidy. Label carefully.

Additional Readings

Abrams, Burton A., and George R. Parsons. "Is CARS a Clunker?" *The Economists' Voice* 6(8) (August 2009).

Beresteanu, Arie, and Shanjun Li. "Gasoline Prices, Government Support, and the Demand for Hybrid Vehicles in the United States." *International Economic Review* 52(1) (February 24, 2011), pp. 161–82.

Brown, Jennifer, Justine Hastings, Erin T. Mansur, and Sofia B. Villas-Boas. "Reformulating Competition? Gasoline Content Regulation and Wholesale Gasoline Prices." *Journal of Environmental Economics and Management* 55(1) (January 2008), pp. 1–19.

Carlsson, Fredrik, and Olof Johansson-Stenman. "Costs and Benefits of Electric Vehicles." *Journal of Transport Economics and Policy* 37(1) (January 2003), pp. 1–28.

Chakravorty, Ujjayant, Céline Nauges, and Alban Thomas. "Clean Air Regulation and Heterogeneity in U.S. Gasoline Prices." *Journal of Environmental Economics and Management* 55(1) (January 2008), pp. 106–22.

Condliffe, Simon, and O. Ashton Morgan. "The Effects of Air Quality Regulations on the Location Decisions of Pollution-Intensive Manufacturing Plants." *Journal of Regulatory Economics* 36(1) (August 2009), pp. 83–93.

Cummings, Ronald G., and Mary Beth Walker. "Measuring the Effectiveness of Voluntary Emission Reduction Programmes." *Applied Economics* 32(13) (October 2000), pp. 1719–26.

Fullerton, D., and S. E. West. "Can Taxes on Cars and on Gasoline Mimic an Unavailable Tax on Emissions?" *Journal of Environmental Economics and Management* 43(1) (January 2002), pp. 135–57.

Hubbard, Thomas N. "Using Inspection and Maintenance Programs to Regulate Vehicle Emissions." *Contemporary Economic Policy* XV(2) (April 1997), pp. 52–62.

Klier, Thomas, and Joshua Linn. "New Vehicle Characteristics and the Cost of the Corporate Average Fuel Economy Standard." Washington, DC: Resources for the Future Discussion Paper 10-50, December 2010.

McConnell, Virginia D., and Tom Turrentine. "Should Hybrid Vehicles Be Subsidized?" *Backgrounder* (July 2010), pp. 1–72.

Mikler, John. *Greening the Car Industry: Varieties of Capitalism and Climate Change.* Northampton, MA: Elgar, 2009.

Muller, Joann, and Andy Stone. "Jump Start." *Forbes* (April 7, 2008), p. 68.

Newell, Richard G., and Kristian Rogers. "Leaded Gasoline in the United States: The Breakthrough of Permit Trading." In Winston Harrington, Richard D. Morgenstern, and Thomas Sterner, eds., *Choosing Environmental Policy.* Washington, DC: Resources for the Future, 2004.

Rask, Kevin N. "Ethanol Subsidies and the Highway Trust Fund." *Journal of Transport Economics and Policy* 38(1) (January 2004), pp. 29–43.

Rusco, Frank W., and W. David Walls. "Vehicular Emissions and Control Policies in Hong Kong." *Contemporary Economic Policy XIII(1)* (January 1995), pp. 50–61.

Sangkapichai, Maria, and Jean-Daniel Saphores. "Why Are Californians Interested in Hybrid Cars?" *Journal of Environmental Planning and Management 52(1)* (January 2009), pp. 79–96.

Sevigny, Maureen. *Taxing Automobile Emissions for Pollution Control.* Northampton, MA: Elgar, 1998.

Shirouzu, Norihiko, and Jeffrey Ball. "Revolution Under the Hood." *Wall Street Journal,* May 12, 2004, p. B1.

Wise-Sullivan, Johanna L. "The Limited Power of States to Regulate Nonroad Mobile Sources Under the Clean Air Act." *Boston College Environmental Affairs Law Review 34(1)* (2007), pp. 207–39.

A Reference to Acronyms and Terms Used in Mobile Source Control Policy

Environmental Economics Acronyms

MEB	Marginal external benefit
MPB	Marginal private benefit
MSB	Marginal social benefit
MSC	Marginal social cost

Environmental Science Terms

CH_4	Methane
CO	Carbon monoxide
CO_2	Carbon dioxide
gpm	Grams per mile
GHG	Greenhouse gas
GWP	Global warming potential
mpg	Miles per gallon
MTBE	Methyl tertiary-butyl ether
N_2O	Nitrous oxide
NMHC	Non-methane hydrocarbons
NO_X	Nitrogen oxides
NO_2	Nitrogen dioxide
O_3	Ozone
Pb	Lead
PM	Particulate matter
PM-2.5	Particulate matter less than 2.5 micrometers in diameter
PM-10	Particulate matter less than 10 micrometers in diameter
SO_2	Sulfur dioxide
VOC	Volatile organic compound

Environmental Policy Acronyms

ABT	Averaging, banking, and trading
AQI	Air Quality Index
CAAA	Clean Air Act Amendments
CAFE	Corporate Average Fuel Economy
CARS	Car Allowance Rebate System
E10	Fuel with 10 percent ethanol and 90 percent gasoline
E85	Fuel with 85 percent ethanol and 15 percent gasoline
FFV	Flexible fuel vehicle
LEV	Low-emission vehicle
NAAQS	National Ambient Air Quality Standards
NHTSA	National Highway Traffic Safety Administration
PHEV	Plug-in hybrid vehicle
PZEV	Partial zero-emission vehicle
SIP	State implementation plan
USCAR	United States Council for Automotive Research
VAVR	Voluntary Accelerated Vehicle Retirement
ZEV	Zero-emission vehicle

To access additional course materials, visit www.cengagebrain.com. At the home page, search for the ISBN of this title (shown on the back cover). This will take you to the product page where these resources can be found.

CHAPTER 12

Improving Air Quality: Controlling Stationary Sources

"Canada is a country whose main exports are hockey players and cold fronts. Our main imports are baseball players and acid rain."

—Pierre Elliott Trudeau (1919–2000)

In this chapter, we continue our discussion of national air quality policy aimed at the most pervasive pollutants, particularly criteria pollutants, but here we focus our attention on rulings that control stationary sources.[1] The motivation for these rulings is the same as it is for mobile sources. Because criteria pollutants affect all of society, the Clean Air Act imposes stringent regulations on *all* major sources of these emissions. We also provide some insight into U.S. policy to control greenhouse gas (GHG) emissions from stationary sources, which is analogous to our discussion in Chapter 11, and which follows from the 2009 Endangerment Finding by the Environmental Protection Agency (EPA). Unlike the GHG controls on mobile sources, however, those on stationary sources are still in the planning phase.

Although the motivation to control pervasive pollutants is common to both mobile and stationary sources, there are important differences in policy design and implementation. Some of these distinctions are a function of the fixed location of stationary sources, which makes certain policy instruments easier to implement and monitor. Others are related to the pollutant being controlled. For example, mobile sources contribute relatively more to the level of carbon monoxide (CO) emissions, whereas stationary sources are bigger emitters of sulfur oxides (SO_X). Consequently, some control instruments are uniquely motivated by specific emissions and their associated environmental risks.

To this latter point, an important context for our analysis in this chapter is an environmental problem tied primarily to stationary sources—acidic deposition, more commonly known as acid rain. Of particular interest is an economic evaluation of the market-based program aimed specifically at this problem, which involves emissions trading.

[1]Refer back to Chapter 10 for policies controlling hazardous emissions from stationary sources and, in particular, the new standards aimed at industrial boilers and incinerators as well as the proposed Toxics Rule directed toward power plants.

To motivate our discussion, we start with an overview of acidic deposition. Once done, we begin our study of stationary source controls as defined under the 1990 Clean Air Act Amendments (CAAA). In particular, we examine the technology-based emissions limits and trading policies that characterize national air quality policy. We also offer a brief overview of recent policy aimed at controlling GHG emissions from stationary sources, deferring a more thorough exposition of GHG policy and climate change to Chapter 13. Lastly, we conduct an economic analysis of key aspects of this policy. A list of acronyms and terms used in this context is provided at the end of this chapter.

ACIDIC DEPOSITION

Acid rain, or more accurately **acidic deposition**, has become a common subject of political and media attention. Research on the problem has intensified, and both public officials and private citizens have grown increasingly aware of the implications. As scientific knowledge has grown, so too has the motivation to enact stronger legislation. Effectiveness of policy initiatives is limited by the fact that acid rain is a *regional* air pollution problem, meaning that the polluting source is often hundreds of miles from where the effects are felt. Consequently, some federal intervention is necessary to control polluters.

Understanding Acidic Deposition

acidic deposition Arises when sulfuric and nitric acids mix with other airborne particles and fall to the earth as dry or wet deposits.

Acidic deposition occurs when the production of sulfuric and nitric acids mix with other airborne particles and fall to the earth as dry deposition or as fog, snow, or rain, thus the descriptive phrase "acid rain." These acidic compounds result from the chemical reaction of sulfur dioxide (SO_2) and nitrogen oxide (NO_X) emissions with water vapor and oxidants (e.g., oxygen [O_2] or ozone [O_3]) in the earth's atmosphere, as illustrated in Figure 12.1. Table 12.1 gives an overview of the causes, sources, and effects of this regional problem.

Of the two responsible pollutants, the more significant is SO_2. It is formed when the sulfur found naturally in coal or oil is released during combustion and reacts with the oxygen in the atmosphere. Primary generators of SO_2 emissions are fossil-fueled electric power plants, refineries, pulp and paper mills, and any sources that burn sulfur-containing coal or oil. According to recent estimates, about 66 percent of SO_2 emissions and 18 percent of the NO_X emissions in the United States are generated by electric power plants.[2]

The problem is pervasive, as nations everywhere struggle with the problem of acidic deposition. In China, for example, rapid economic growth and the associated demand for electricity led to the construction of numerous coal-fired power plants, which in turn generated acidic emissions not only for China but for neighboring countries as well. However, new technology in electricity generation holds promise for mitigating these damaging emissions, as reported in Application 12.1.

[2]U.S. EPA, Technology Transfer Network (June 9, 2009).

FIGURE 12.1 Acid Rain Formation

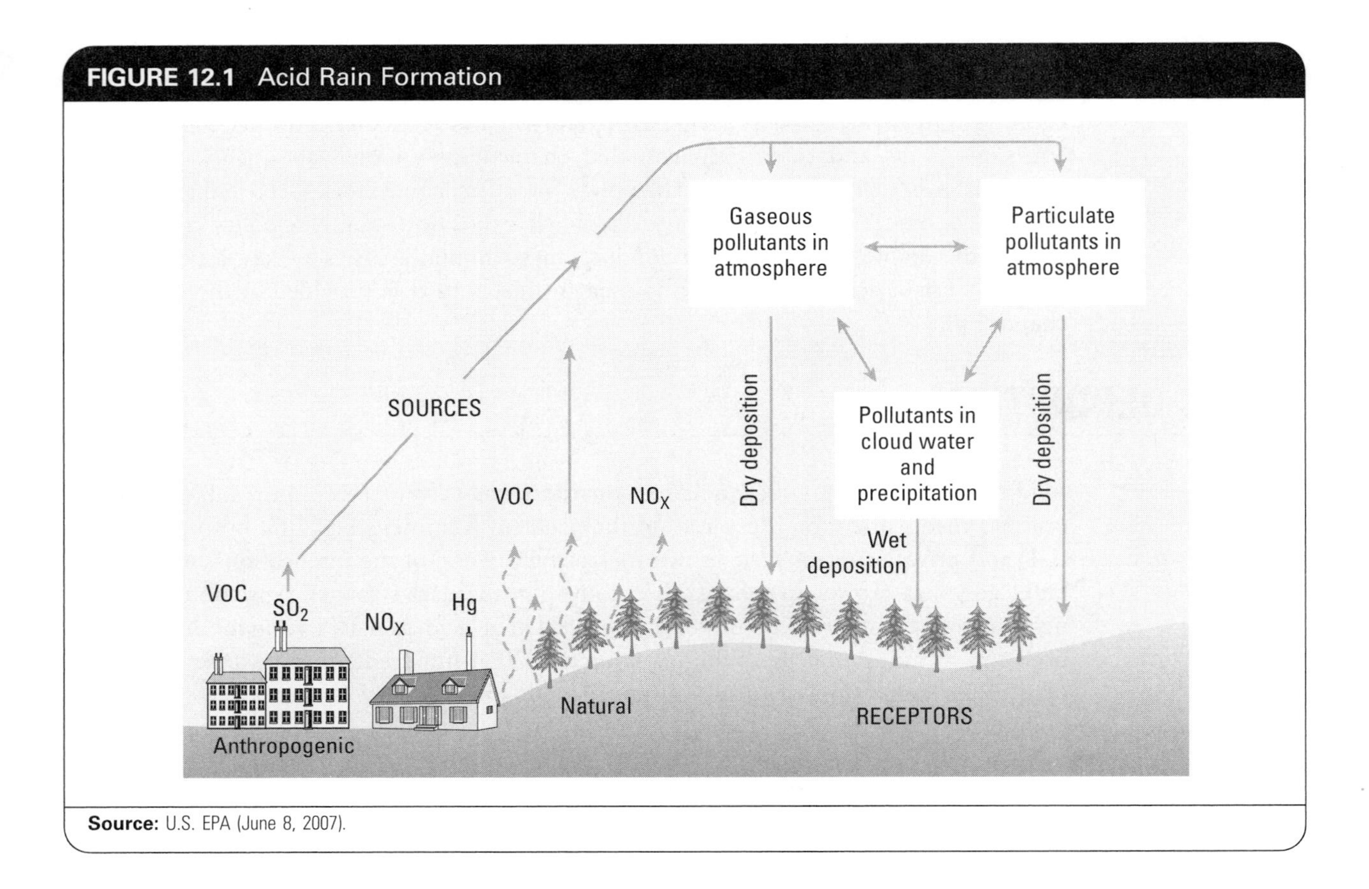

Source: U.S. EPA (June 8, 2007).

TABLE 12.1 The Problem of Acidic Deposition

Causes	Anthropogenic Sources	Major Effects
SO_2	Fossil-fuel burning primarily by electric power plants	**Ecological and forestry effects:** Acidification of surface waters; localized forest damage.
NO_X	Fossil-fuel burning primarily by electric power plants and motor vehicles	**Health effects:** Direct effects include respiratory and cardiovascular problems; some indirect effects possible from increased lead or methyl mercury exposure under extreme conditions.
		Aesthetic effects: Visibility impairment.
		Property effects: Damage to buildings, statues, and monuments.

NOTE: For more information on acid rain and its effects, visit **www.epa.gov/acidrain/**.

APPLICATION 12.1 China's Power Plants and Acid Rain

As China's economic growth has accelerated, so too has its demand for energy to support its burgeoning manufacturing industries. Between 2000 and 2004, China's electricity generation and consumption grew by 60 percent. Because China possesses the third largest coal reserves in the world—about 126.2 billion short tons—it comes as no surprise that its energy producers rely on coal for fuel. In fact, coal accounts for nearly 70 percent of China's primary energy consumption. In 2004, China used 2.1 billion short tons of coal, which represented about one-third of the world's total consumption. A similar amount was consumed in 2006. Interestingly, China is at once the largest producer and consumer of coal in the world.

On a fundamental level, this enormous dependence on coal poses a serious environmental threat because the combustion of coal, along with other fossil fuels, generates sulfur dioxide (SO_2) emissions that cause acid rain. In China, this problem is of greater concern because its coal is high-sulfur, brown coal. Another contributing factor is that almost 50 percent of China's coal-burning power plants built between 2001 and 2005 are small, dirty, and inefficient. Some were built at the local level and often without oversight or approval at the national level. The result? China has become the largest SO_2 polluter in the world.

Not only is China at risk of exposure to SO_2 emissions and the associated problem of acid rain, but so too are other countries. Reportedly, acid rain linked to China's emissions poses a threat to neighboring countries, such as South Korea, and to nations as far away as Canada and Europe.

Not unaware of the increasing environmental threat of acid rain, Chinese officials are taking steps to address the problem. One response has been to require that existing power plants mitigate the SO_2 before releasing their emissions. Another is to modernize electricity generation. Today, over 80 percent of China's electricity is generated by five state-owned holding companies, one of which, Huaneng China Group, recently opened a high-tech 1,000-megawatt plant near the city of Wenzhou. The facility's new technology generates electricity with 17 percent less coal usage than average power plants in China. According to plant officials, the new facility is capable of reducing SO_2 emissions to nearly zero. Together, the five holding companies plan to build another ten of these new technology power plants.

Interestingly, in 2007, China's Ministry of Environmental Protection reported a 4.7 percent decline in SO_2 emissions, which may have been attributable, at least in part, to these changes. Furthering these efforts, China set an explicit objective to achieve a 10 percent reduction in SO_2 emissions. To achieve this goal, officials announced plans to use a market-based approach, trading SO_2 emissions through China's Tianjin Climate Exchange. If successful, this approach, along with the infrastructure changes that are underway, should help mitigate the acid rain problem for China and neighboring nations.

Sources: "China to Kick Off Emissions Trading" (November 24, 2008); Bradsher (June 6, 2008); Ansfield (January 15, 2007); MacBean (2007); U.S. Department of Energy, Energy Information Administration (August 2006).

CONTROLLING STATIONARY SOURCES

Stationary sources contribute to the emissions levels of all criteria pollutants. This source group covers a lot of ground. Included are electric power plants, chemical plants, steel mills, and even residential furnaces. Obviously, certain categories are more important to control policy than others, as the data in Figure 12.2 attest. These data show emissions of each criteria pollutant released by the two major stationary source categories: fuel combustion and industrial processes.

Notice in Figure 12.2 how fuel combustion sources, which include fossil-fueled electric power plants, are major contributors of sulfur dioxide (SO_2) and nitrogen oxide (NO_X) emissions, the primary causes of acid rain. Furthermore, because NO_X is a precursor of photochemical smog, as discussed in Chapter 11, this means that fuel combustion sources also contribute significantly to urban smog. Similarly, note that industrial

FIGURE 12.2 Stationary Source Emissions Estimates in 2008

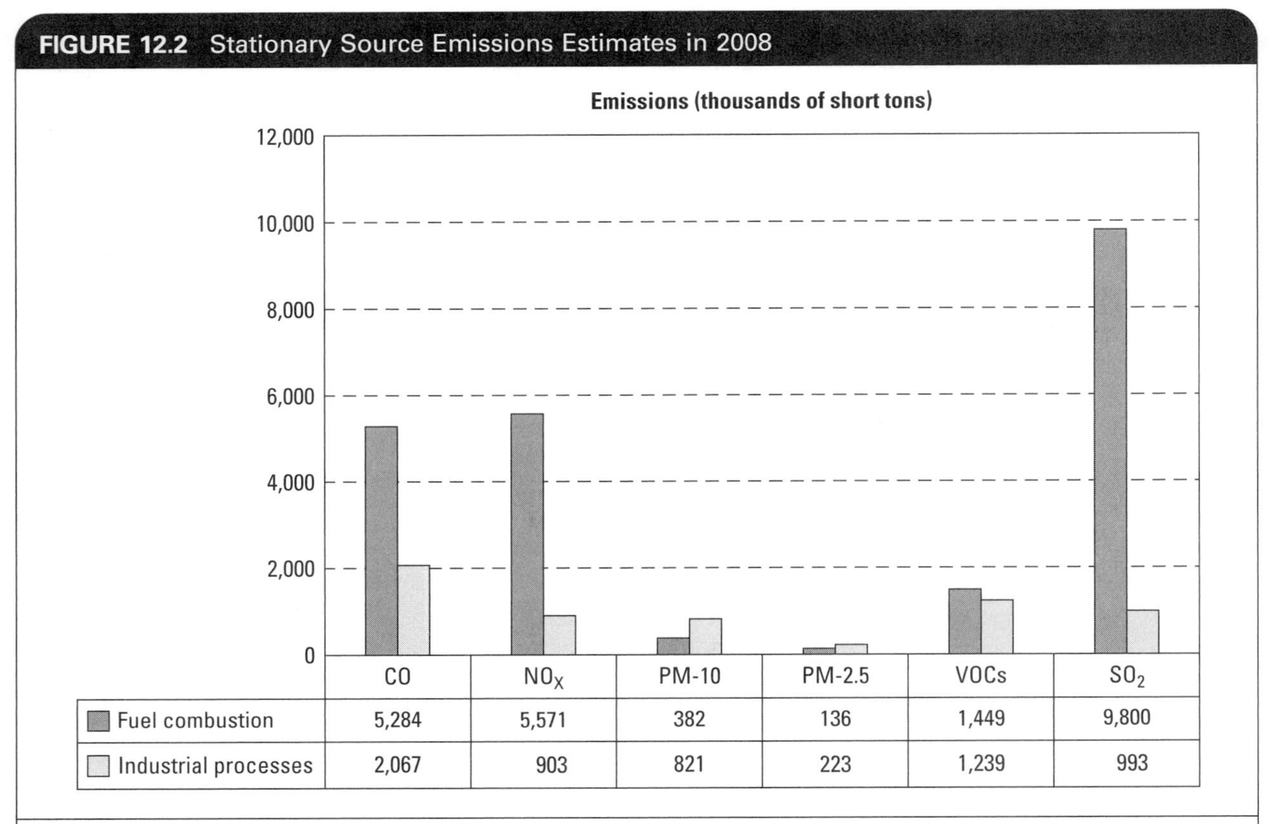

	CO	NO_X	PM-10	PM-2.5	VOCs	SO_2
Fuel combustion	5,284	5,571	382	136	1,449	9,800
Industrial processes	2,067	903	821	223	1,239	993

NOTES: 1 short ton = 2,000 pounds.
VOC emissions are used to measure the formation of ozone, O_3.
Estimates for lead (Pb) are not available.
Source: U.S. EPA, Technology Transfer Network (June 9, 2009).

sources (e.g., petroleum refineries and chemical manufacturers) are responsible for much of the volatile organic compound (VOC) emissions, making these stationary sources also important targets for policies aimed at reducing urban smog.

Emissions of all stationary sources are regulated mainly by uniform **technology-based standards**, which is indicative of a command-and-control approach. Over time, the EPA has introduced various **emissions trading programs** to existing policy. We organize our study of stationary source controls around these two very different types of policy instruments. As we proceed, note that both the standards and the trading programs are implemented in accordance with two key features of stationary sources:

- facility age (new or existing), and
- facility location (in a prevention of significant deterioration [PSD] area or a nonattainment area).

Technology-Based Standards

Age-specific control differences were established in the 1970 CAAA. The rulings require that facilities constructed or significantly modified after 1970 must meet more stringent emissions limits than those already in existence. The reasoning is that new or modified sources presumably can integrate the technology needed to meet stringent standards more easily than their older counterparts. Existing facilities would likely face high retrofitting costs, which could hurt local economies. Furthermore, they eventually would deteriorate to the point of needing to be modified or replaced. Hence, over time, the tougher standards would have to be met by *all* stationary sources.

New Source Performance Standards (NSPS) Technology-based emissions limits for new stationary sources.

Emissions limits applicable to new and modified stationary sources are called **New Source Performance Standards (NSPS)** and defined by the EPA. When states set technology-based standards for PSD and nonattainment areas (to be discussed subsequently), the NSPS act as a minimum. According to the Clean Air Act, the NSPS are technology-based in accordance with the "best system of emission reduction." This phrase suggests that the NSPS were meant to be stringent, but in this case, the stringency is tempered somewhat by allowing for costs and energy requirements to be considered as well.[3]

Because technological factors are the basis for these limits, the NSPS are allowed to vary across industry categories. However, each set of industry standards is to be applied *uniformly* across all firms within a given category. Because of the long-term importance of these more stringent controls, Congress established a **dual-control approach**, placing the responsibility for controlling new or modified stationary sources with the EPA and assigning the comparable duties for existing stationary sources to state governments.

PSD Versus Nonattainment Areas

With the official introduction of PSD areas in 1977, stationary source controls had to be revised to accommodate the change. Because the air quality standard in PSD regions is higher, controls for sources located in these areas had to be strengthened. However, new facilities in PSD areas would face tougher standards than existing ones. According to the new policy, any new or modified source in a PSD area has to meet emissions limits based on the **best available control technology (BACT)**, a standard that aligns with the *maximum* degree of pollution reduction available. Existing sources in these areas can use a more lenient technology called the **best available retrofit technology (BART)**. See Table 12.2 for a summary of these differences.

Standards for nonattainment areas are relatively less stringent. But just like their PSD counterparts, sources face different standards based on their age. As shown in Table 12.2, existing sources must use **reasonably available control technology (RACT)**, the least stringent of all the technology-based standards. New or modified sources in these areas must comply with the **lowest achievable emission rate (LAER)**, roughly defined as the most stringent emissions limit achieved in practice.[4] More on these various technologies is available at **http://cfpub.epa.gov/RBLC/**.

[3]In practice, this cost consideration is meant to imply only that the technology be affordable and *not* that it be used as part of a formal benefit-cost analysis in setting the standards (Portney 1990a, p. 38).

[4]Furthermore, although the new source standards for PSD and nonattainment areas (i.e., BACT and LAER, respectively) are supposed to be more stringent than the NSPS, in practice, the NSPS are generally accepted for either (Portney 1990a, pp. 37–38).

TABLE 12.2 Technology-Based Standards

	New or Modified Sources	Existing Sources
PSD Area	**Best Available Control Technology (BACT):** BACT refers to an emissions limit based on the maximum degree of pollution reduction that the permitting authority deems achievable on a case-by-case basis.	**Best Available Retrofit Technology (BART):** BART refers to an emissions standard applicable only to emissions that might negatively affect visibility.
Nonattainment Area	**Lowest Achievable Emission Rate (LAER):** LAER refers to an emissions rate that reflects the most stringent emission limitation outlined in a given State Implementation Plan (SIP) for the same type of source.	**Reasonably Available Control Technology (RACT):** RACT refers to an emissions standard based on the use of technology deemed practically available.

Sorting It Out

Although the collection of technology-based standards can be confusing, the key points to remember are:

- Emissions limits in PSD areas are more stringent than those in nonattainment areas.
- Emissions limits for new sources are more stringent than those for existing sources.

Reducing Greenhouse Gas (GHG) Emissions[5]

When the EPA issued its findings about greenhouse gases (GHGs) in 2009 (i.e., its Endangerment Finding and its Cause or Contribute Finding, as described in Chapters 10 and 11), it subsequently announced new GHG emissions standards for light-duty mobile sources for the 2012–2016 model years, taking effect in January 2011. As it turns out, these new GHG controls on mobile sources triggered analogous regulatory requirements for stationary sources, according to rulings in the Clean Air Act. These latter regulations would be implemented through GHG permitting programs and newly established GHG emissions standards.

GHG Permitting Requirements

Under the new ruling, all new or modified major stationary sources of GHGs would be subject to the New Source Review (NSR) requirements. These requirements call for sources to obtain permits prior to starting any new construction.[6] In turn, the permits define the requisite parameters for the planned construction, including the emissions limits to be met and the emissions control technology to be used. The GHG permitting requirements took effect in January 2011 because this was the date the aforementioned motor vehicle GHG regulations became operative.

[5]Much of the following is drawn from U.S. EPA, Office of Air and Radiation (December 23, 2010); U.S. EPA, Office of Public Affairs (December 23, 2010).

[6]To facilitate the transition, the EPA issued its GHG Tailoring Rule, which raises the emissions threshold that defines major sources subject to these regulations as a way to confine the new laws to the largest sources.

GHG Emissions Standards

In addition to the NSR permitting requirements, the EPA is establishing emissions standards for two major source groups: fossil fuel power plants and petroleum refineries. These two groups of stationary sources collectively account for about 40 percent of all GHG releases in the United States. The EPA announced its intent to propose the New Source Performance Standards (NSPS) for the designated source categories during 2011. Following the proposals, final standards are to be issued in 2012.

Emissions Trading

bubble policy
Allows a plant to measure emissions as an average of all emission points from that plant.

When the EPA launched its **bubble policy** in 1979, it was originally conceived as an option for states to use in implementing the national standards for existing sources. This instrument allows a plant to measure emissions, not from each of its release points within the plant, but as an *average* of all such points, as if the emissions were captured within a bubble. This offers firms flexibility in deciding *how* they can achieve the requisite emissions limits at least cost.

emissions banking
Accumulating emission reduction credits through a banking program.

In 1980, the EPA introduced another market-based program, called **emissions banking**. Designed to complement the bubble concept, this plan allows a stationary source to accumulate emission reduction credits if it abates more than required by law. These credits are then deposited through the banking program for future use, establishing the foundation for an emissions market.

PSD Versus Nonattainment Areas

netting
Matching any emissions increase due to a modification with a reduction from another point within that same source.

To facilitate control in PSD areas, a trading program known as **netting** was developed for use by modified sources. A type of bubble policy, netting allows emissions trading to take place among release points within a facility for the same type of pollutant. Its distinguishing feature is that any added emissions associated with a plant modification must be *exactly matched* by a reduction from somewhere else within that same plant. Hence, netting ensures that any construction project will have no detrimental effect on the air quality in a PSD area.

In some sense, implementing pollution controls in nonattainment areas must be more strategic. Obviously, there is concern about the environmental impact of allowing new facilities into an area already not meeting the national ambient standards. Yet, preventing new construction could potentially worsen air quality, because new plants likely operate more cleanly than older ones.

offset plan
Uses emissions trading to allow releases from a new or modified source to be more than countered by reductions achieved by existing sources.

In an attempt to achieve the best of both worlds, the 1977 Amendments introduced the **offset plan** for nonattainment areas. Implemented through a permit procedure, an offset plan ensures that emissions from a new or modified source are *more than countered* by the reductions achieved by existing sources. Unlike the bubble policy or netting, the offset plan involves trades between existing and new sources rather than within the same facility or plant complex. To facilitate the plan, an **emissions bank** of accumulated emission reduction credits was set up where facilities could deposit and access offsets as needed.

Although these early trading programs were intended to be subordinate to the standards, they nonetheless represented a shift toward integrating economic mechanisms into what had been almost exclusively a command-and-control approach. Title IV of the 1990 CAAA continues this trend by establishing another type of trading program aimed squarely at the problem of acid rain.

Acid Rain Program (ARP) and New Rules[7]

Because SO_2 and NO_X are the primary substances that contribute to acid rain, Title IV of the 1990 CAAA launched a direct attack on each of these criteria pollutants through the nation's **Acid Rain Program (ARP)**. This two-phase initiative (the first phase for the 1995–1999 period, and the second for 2000–2009) established a cap-and-trade allowance program for SO_2 emissions and a reduction plan for NO_X emissions. The reduction plan for NO_X under the ARP was to be achieved through performance standards set by the EPA. The goal was to decrease annual NO_X emissions to 2 million tons below what was forecasted for 2000 without the ARP. This objective was met in 2000 and in every subsequent year.[8] The text of Title IV is accessible online at **www.epa.gov/air/caa/title4.html**.

Setting the SO_2 Caps

For SO_2, the law established a permanent annual cap of 9.5 million tons for electric power plants (the major sources of SO_2) starting in 2000. This cap was tightened to 8.95 million tons for 2010. This was the first time any national limit had been imposed on a single pollutant. Once fully implemented, these limits are defined to cut SO_2 emissions in half relative to 1980. As shown in Figure 12.3, annual emissions were 10.6 million tons in 2003. Accumulated annual reductions through 2003 were 38 percent of the 1980 benchmark. By 2007, annual SO_2 emissions for the first time fell below the 8.95 million ton cap to 8.9 million tons, a full three years before the statutory deadline. Through 2009, all ARP sources had reduced annual SO_2 emissions by 67 percent relative to 1980 levels and by 64 percent compared to 1990 levels, leaving 12.3 million allowances banked.

Establishing the SO_2 Emissions Market

tradeable SO_2 emission allowances
Permits allowing the release of SO_2 that can be held or sold through a transfer program.

The success of the Acid Rain Program (ARP) has been credited mainly to the innovative market approach of using a cap-and-trade process. Specifically, to achieve the SO_2 emissions cap, the EPA issued **tradeable SO_2 emission allowances**, or rights to pollute, to stationary sources. Each allowance permits the release of 1 ton of SO_2. Because no emissions are permitted without an allowance, the aggregate number of allowances issued by the EPA effectively sets the national limit. If any polluting source releases SO_2 emissions greater than the allowances it holds, that source must surrender to the EPA enough allowances to cover the excess releases and must pay a penalty of just over $3,575 ($2010) per ton.

Once the allowances are distributed, they can be used by the recipient, sold to other sources through a transfer program, or banked for future use. This establishes a true market through which allowances can be bought and sold. Referring again to Figure 12.3, notice that there are banked allowances for each year accumulated from the prior period. These arise if actual emissions are lower than available allowances.

---

[7]Much of the following is discussed in U.S. EPA, Office of Air and Radiation, Office of Atmospheric Programs, Clean Air Markets Division (December 2010; January 2009; September 2004).

[8]In 2002, for example, a reduction of over 3 million tons below forecasted 2000 levels without Title IV was achieved, and in 2007, the analogous reduction was 4.8 million tons (U.S. EPA, Office of Air and Radiation, Office of Atmospheric Programs, Clean Air Markets Division, January 2009, p. 14).

FIGURE 12.3 SO_2 Emissions and Allowance Banking: 1995–2009

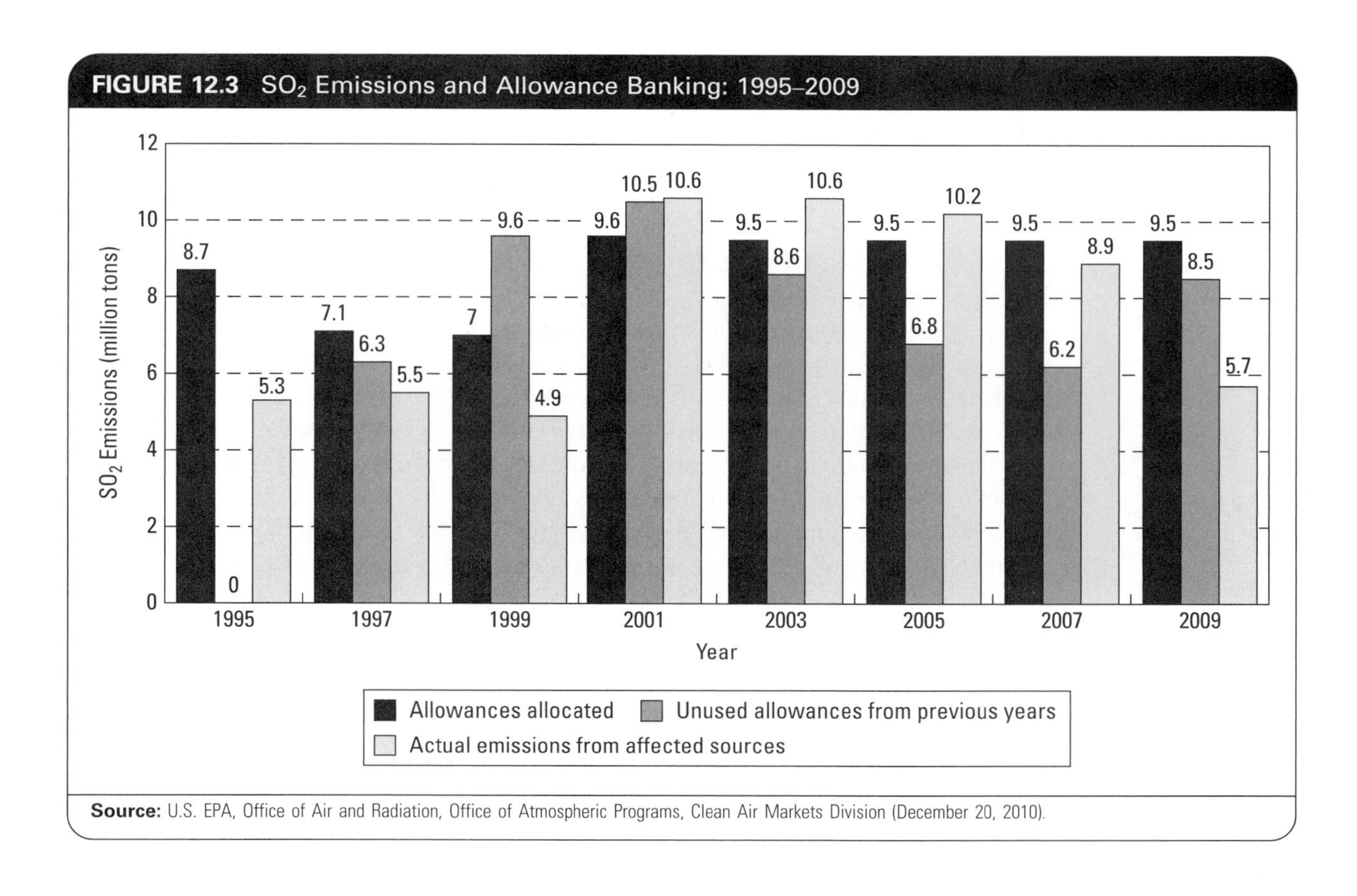

Source: U.S. EPA, Office of Air and Radiation, Office of Atmospheric Programs, Clean Air Markets Division (December 20, 2010).

For example, in 2009, 9.5 million one-ton allowances were allocated for use in addition to 8.5 million that were not used from 2008, and since only 5.7 million tons were actually released, 12.3 million tons could be banked for future periods.

From an historical perspective, it is interesting to note that the first allowance trade announced in May 1992 took place between the Tennessee Valley Authority (TVA) and Wisconsin Power and Light Company, one of the cleanest utilities in the country. TVA bought 10,000 allowances at a value of $2.5 million to $3 million. The market incentive for this precedent-setting trade was clear, once the facts were known. To meet the new federal standards *without* trading, the TVA would have had to invest in abatement technology, estimated to cost between $750 million and $850 million.

An extension of this market is accomplished through a bank of allowances available for direct sale by the EPA through an annual auction. Because this program depends greatly on external trades, the EPA originally placed the Chicago Board of Trade (CBOT) in charge of running the auctions. The expectation was that the CBOT could facilitate the exchange of allowances and possibly reduce the transaction costs of matching up buyers and sellers. Starting in 2006, the CBOT stopped participating, and the EPA now administers the auctions itself.

Interestingly, some private firms anticipated the profit opportunities of the newly created pollution market and competed with the CBOT and the EPA as allowance brokers. One example is CantorCO2e, a subsidiary of Cantor Fitzgerald, which offers full brokerage and consulting services for the SO_2 market. In fact, in the March 2008 auction, the company purchased 6,000 allowances totaling $2.4 million.[9]

Virtually anyone can enter the market for SO_2 allowances, including private individuals and environmental groups. More information on this allowance market, including specific trades, is available at **www.epa.gov/airmarkets/trading/factsheet.html**.

Clean Air Interstate Rule (CAIR) and the Cross-State Air Pollution Rule (CSAPR)

In 2005, the EPA issued the **Clean Air Interstate Rule (CAIR)**, which involved declining emissions caps and three trading programs, including an annual SO_2 trading program launched in 2010.[10] Allowances issued under the ARP were used in the CAIR, which had broader requirements and hence affected more sources. This new rule was intended to facilitate attainment of the nation's ozone standard for PM-2.5, for which SO_2 is a precursor. It was part of a suite of actions known as the Clean Air Rules issued under President Bush. A summary of this rule is available at **www.epa.gov/CAIR**.

Interestingly, however, the CAIR was challenged, and in 2008, the D.C. Circuit Court ruled to vacate it. However, the Court later granted an EPA petition to refer the case without vacating the rule. So, the CAIR was remanded back to the EPA for further rulemaking and would remain in force until a new rule was issued to replace it. In 2010, that replacement was proposed as the **Transport Rule**, which, when finalized in July 2011, became known as the **Cross-State Air Pollution Rule (CSAPR)**. As noted in Chapter 10, this rule calls for power plants to achieve major reductions in SO_2 and NO_X emissions, which can drift across state borders.[11] More on this rule is offered in Application 12.2.

Ozone Control: NO_X Trading Programs and New Rules[12]

In addition to causing acid rain, NO_X contributes to the formation of ground-level ozone, or smog. That NO_X is easily transported across state borders led to a series of interstate collaborations to control the problem. The first one was established under the 1990 CAAA and was comprised of northeast and mid-Atlantic states. The multistate alliance, called the Ozone Transport Commission (OTC), developed the **OTC NO_X Budget Program** in 1994, which was a cap-and-trade program not unlike the allowance program for SO_2. This program was the first of its kind formed by an interstate group and the first broad-based use of cap-and-trade for a problem other than acid rain. More information on the use of cap-and-trade programs is available at **www.epa.gov/captrade**.

In 1998, concerns about ozone nonattainment and interstate transport of NO_X prompted the EPA to call for NO_X emissions limits across an area comprising most OTC states plus certain southeastern and midwestern states. Known as the **NO_X SIP Call**, this

---

[9]U.S. EPA, Office of Air and Radiation, Office of Atmospheric Programs, Clean Air Markets Division (August 25, 2008); Taylor (August 24, 1993).

[10]The other two trading programs address NO_X emissions and will be discussed subsequently in the section on NO_X trading.

[11]U.S. EPA, Office of Air and Radiation (July 7, 2011).

[12]Much of the following is drawn from U.S. EPA, Office of Air and Radiation, Office of Atmospheric Programs, Clean Air Markets Division (December 2010; May 28, 2009; December 2008; August 2004).

APPLICATION 12.2 The Cross-State Air Pollution Rule (CSAPR): Successor to the CAIR

Proposed in July 2010 by the EPA as the Transport Rule, the Cross-State Air Pollution Rule (CSAPR) is aimed squarely at power plant emissions of SO_2 and NO_X that drift across state lines. These pollutants react in the atmosphere to form fine particulate matter (PM-2.5) and tropospheric (ground-level) ozone (O_3), both of which are capable of traveling long distances, impeding states' ability to meet the National Ambient Air Quality Standards (NAAQS). Once the CSAPR takes effect, it will replace the Clean Air Interstate Rule (CAIR), which was vacated in July 2008 by the U.S. Court of Appeals for the D.C. Circuit.

More comprehensive than its predecessor, this rule involves 27 states, and a subsequent proposal, if finalized, will bring the total to 28. Federal Implementation Plans for each affected state will outline the requisite steps to reduce emissions that significantly contribute to nonattainment or impede the maintenance of the NAAQS in other states. In turn, each state may elect to devise its own state plan to replace the federal plan and may select which sources are to be controlled.

Significant reductions in SO_2 and NO_X are anticipated. According to projections, the Cross-State Air Pollution Rule along with other directives should achieve a decline in SO_2 emissions by 73 percent and in NO_X emissions by 54 percent by 2014 relative to 2005 levels.

According to the Regulatory Impact Analysis (RIA) for this rule, the estimated benefits far outweigh the estimated costs. Specifically, annual social benefits in 2014 are in the range of $126.2 billion ($2010) to $294.5 billion ($2010), (using a 3 percent discount rate), and annual social costs are estimated to be $ 0.84 billion ($2010). Most of the incremental benefits are due to an anticipated reduction of between 13,000 and 34,000 premature deaths from exposure to O_3 and PM-2.5. Other identified health benefits include the avoidance of 15,000 nonfatal heart attacks, 19,000 cases of acute bronchitis, and 400,000 cases of aggravated asthma attacks. Although not all welfare benefits are quantifiable, these gains include improvements in visibility, enhanced crop yields, and improved commercial fishing, and they are believed to be substantial.

For further information on the CSAPR, visit **www.epa.gov/crossstaterule/**.

Sources: U.S. EPA, Office of Air and Radiation (July 7, 2011; June 2011).

initiative required affected states to submit revised State Implementation Plans (SIPs) aimed at achieving these limits during ozone season (May 1 to September 30), starting in 2003. The EPA recommended the use of a cap-and-trade program and established the **NO_X Budget Trading Program (NBP)** in 2003 as successor to the OTC seminal program. Eleven more states joined the effort in 2004 (and one was subsequently removed from the requirements), bringing the total to 20 plus the District of Columbia. All chose to participate in the NBP to meet their NO_X emissions budgets, or caps. Incidentally, CantorCO2e was the primary brokerage firm for trades under this program.

Clean Air Interstate Rule (CAIR) and the Cross-State Air Pollution Rule (CSAPR)

As noted in our discussion of SO_2 trading, the Clean Air Interstate Rule (CAIR) issued in 2005 included three trading programs: the annual SO_2 trading program previously discussed and two trading programs for NO_X—one for the NO_X ozone season, which actually replaced the NBP, and an annual NO_X program. In fact, the intended purpose of the CAIR was to build upon the existing trading programs for both NO_X and SO_2. Both NO_X programs began in 2009, using declining emissions caps.

The CAIR is to be replaced by the **Cross-State Air Pollution Rule (CSAPR)**. Like its predecessor, the CSAPR allows power plants the option of trading emissions allowances both within and between participating states in order to meet pollution limits set

FIGURE 12.4 States Affected by the Cross-State Air Pollution Rule

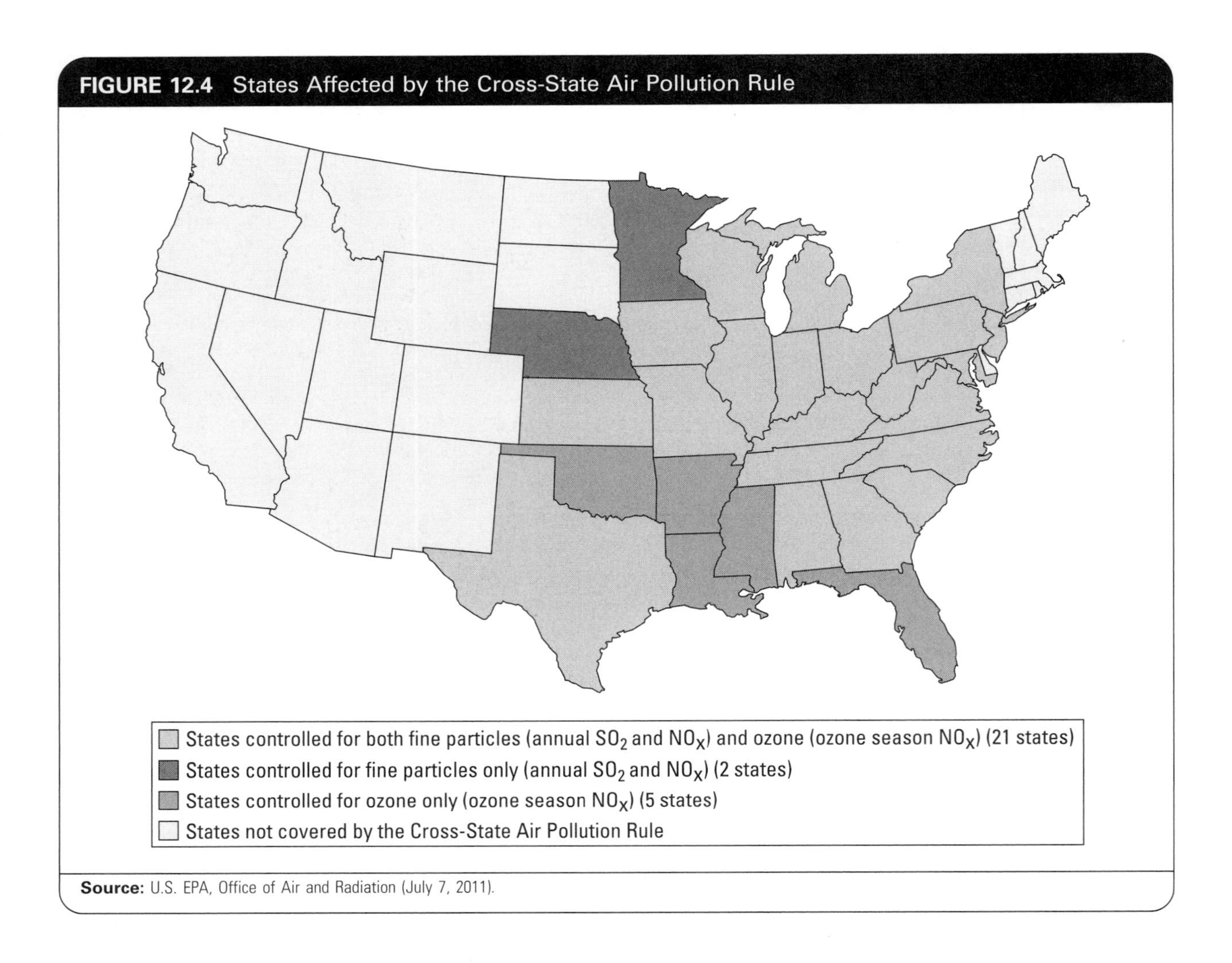

Source: U.S. EPA, Office of Air and Radiation (July 7, 2011).

by the EPA.[13] Refer to Figure 12.4 for a map of the states affected by the CSAPR. More information about this new rule is available at **www.epa.gov/crossstaterule/**.

State-Level Trading Plans

Among the first state-level trading programs was the Regional Clean Air Incentives Market (RECLAIM) program developed in 1994 by California's South Coast Air Quality Management District (SCAQMD). The basic premise of RECLAIM was the same as the national SO_2 allowance trading system: pricing pollution. However, instead of using allowances, SCAQMD polluters were required to hold RECLAIM trading credits (RTC), where each RTC was equivalent to 1 pound of emissions. The result was a regional trading market for sulfur oxide (SO_X) and nitrogen oxide (NO_X) emissions. Compared to a command-and-control approach, the RTC market allows polluters to achieve air quality

[13]U.S. EPA, Office of Air and Radiation (July 7, 2011).

objectives in a cost-effective manner. Compliance cost savings for this program have been estimated at $58 million per year. For further information about RECLAIM, visit **www.aqmd.gov/RECLAIM/**. Other states that have developed emissions trading programs include Colorado, Illinois, Massachusetts, Michigan, New Jersey, Pennsylvania, Texas, and Washington.[14]

ECONOMIC ANALYSIS OF STATIONARY SOURCE CONTROLS

Whether or not the gains of stationary source controls justify the costs can be determined through a **benefit-cost analysis**. An interesting study was done by Chestnut and Mills (2005), who estimate the overall benefits and costs of the Acid Rain Program (ARP). According to their findings, estimated annual benefits for 2010, the year by which the ARP was fully implemented, are $122 billion ($2000), and estimated annual costs are $3 billion ($2000). Interestingly, most of the benefits are attributable to reduced mortality linked to reduced emissions of small particulate matter (PM-2.5). In any case, the overall results of this study suggest that the ARP is feasible from a benefit-cost perspective, generating nearly $120 billion in net benefits each year.

We can also use benefit-cost analysis to structure an investigation of four aspects of U.S. national policy, which are:

- Relative cost of using command-and-control instruments
- Uniform technology-based NSPS
- Dual-control approach and the new source bias
- Economics of market-based trading programs

Relative Cost of Using Command-and-Control Instruments

Because of the absence of benefit-cost balancing in the standard-setting process, it would only be by chance that the technology-based standards would correspond to those that maximize net benefits to society. Despite the failure of these regulations to meet the efficiency criterion, it is still possible to minimize the cost of meeting them, efficient or not. Unfortunately, the lack of flexibility in a standards-based approach adds significantly to society's costs and offers no incentive to low-cost abaters to clean up beyond the statutory level.

There is a growing body of research that attempts to measure the extent of cost inefficiency associated with command-and-control policy. These studies use computer simulations to model existing conditions and compare the cost of implementing command-and-control instruments with that of the least-cost alternative. Table 12.3 summarizes a sample of those studies examining stationary source control costs for meeting the National Ambient Air Quality Standards (NAAQS). Each study quantifies the comparison as the ratio of the cost of using a command-and-control instrument *relative* to the least-cost method, which is shown in the last column of the table. Despite the wide range of values for the cost ratio, in nearly every case, the magnitude is significantly greater than 1.

At the high end of the range, Spofford (1984) finds that the costs of using a uniform percentage reduction exceed the least-cost alternative by a factor of 22. This implies that

[14]U.S. EPA, Office of Policy, Economics, and Innovation (January 2001), pp. 93–96.

TABLE 12.3 Cost-Effectiveness and Air Quality Regulation

Study and Year	Command-and-Control Approach	Pollutant Controlled	Geographic Area	Ratio of Command-and-Control Cost to Least Cost
Atkinson and Lewis (1974)	SIP regulations	Particulate matter	St. Louis	6.0
Roach et al. (1981)	SIP regulations	Sulfur dioxide	Four Corners: Utah, Colorado, Arizona, and New Mexico	4.25
Hahn and Noll (1982)	California emissions standards	Sulfates	Los Angeles	1.07
McGartland (1984)	SIP regulations	Particulate matter	Baltimore	4.18
Spofford (1984)	Uniform percentage reduction	Sulfur dioxide Particulate matter	Lower Delaware	1.78 22.0
Maloney and Yandle (1984)	Uniform percentage reduction	Hydrocarbons	All domestic DuPont plants	4.15
Krupnick (1986)	Proposed RACT	Nitrogen dioxide	Baltimore	5.9
Oates et al. (1989)	Equal proportional treatment	Particulate matter	Baltimore	4.0
ICF Resources International (1989)	Uniform emission limit	Sulfur dioxide	United States	5.0
SCAQMD (1992)	Best Available Control Technology	Reactive organic gases and NO_2	Southern California	1.5 in 1994 1.3 in 1997
Krupnick et al. (2000)	SIP call provisions	Nitrogen oxides	Eastern United States	1.83 (utilities) 2.00 (all sources)

Sources: U.S. EPA, Office of Policy, Economics, and Innovation (January 2001), p. 25, Table 3-2; Tietenberg (1985); and the original sources cited in the table.

the costs of using a command-and-control approach are 2,200 percent of what they would be if the policy allowed for equal marginal abatement costs across polluters. At the other end of the spectrum are the findings of Hahn and Noll (1982). Their ratio of 1.07 suggests that the command-and-control approach is not that far off from the least-cost method. In this case, however, the researchers suggest that their unusual finding may be the result of stringent controls imposed by the California control authority with the specific intent of achieving cost-effective standards.

Despite the wide range of values for the cost ratio, in nearly every case, the magnitude is significantly greater than 1. It is difficult to dispute such a consistent finding that points to the excess costs of using command-and-control instruments to implement the NAAQS. These results suggest that in most instances, society should realize important cost savings from a shift to more flexible policy instruments with no reduction in air quality benefits.

Uniform Technology-Based NSPS

Implicitly, there are two potential problems with the new source performance standards (NSPS). First, they are implemented **uniformly** across all firms in a given category. Second, because the standards are **technology-based**, firms have no flexibility in selecting *how* to meet the national emission limits. Both characteristics suggest that cost-effectiveness is not being achieved.

The intent of using uniform emission standards was to prevent regional differences in regulations from affecting new firms' decisions about where to locate, which creates a market bias. Nonetheless, just as is the case for the uniform emissions limits for new cars, this particular command-and-control approach generally disallows the achievement of a cost-effective outcome. Why? Because the cost-effective solution calls for pollution abatement up to the point where all firms face the same marginal abatement cost (*MAC*), *not* to the point where all firms abate at the same level.

In Figure 12.5, a model of two hypothetical stationary sources, A and B, is shown with the cost-effectiveness solution at A_0, where MAC_A intersects MAC_B. (Source A's allocation is measured left to right up to A_0, and source B's allocation is measured right to left.) Total abatement costs, assuming no fixed costs, are shown as the shaded area under the two *MAC* curves up to the cost-effective abatement allocation. If a uniform standard were used instead, the abatement allocation would force each firm to abate equally to the point labeled A_1. Total costs again would be measured as the area under the two *MAC* curves up to that solution point. Notice that total costs under the uniform standard approach are higher by the triangular area labeled XYZ.

Using technology-based NSPS instead of performance-based standards can also produce a result that is not cost-effective. By dictating the type of technology to be used across the board, firms are prevented from using cheaper alternative methods. The result is a waste of economic resources.

FIGURE 12.5 Cost-Ineffectiveness of the NSPS

The cost-effective solution occurs at A_0, where MAC_A intersects MAC_B. Total costs of abatement, assuming no fixed costs, are the shaded area under the two *MAC* curves up to the cost-effective abatement allocation. If a uniform standard is used, the abatement allocation across the two firms would force each to abate equally to A_1. Total costs under the uniform standard approach is higher by the triangular area, XYZ.

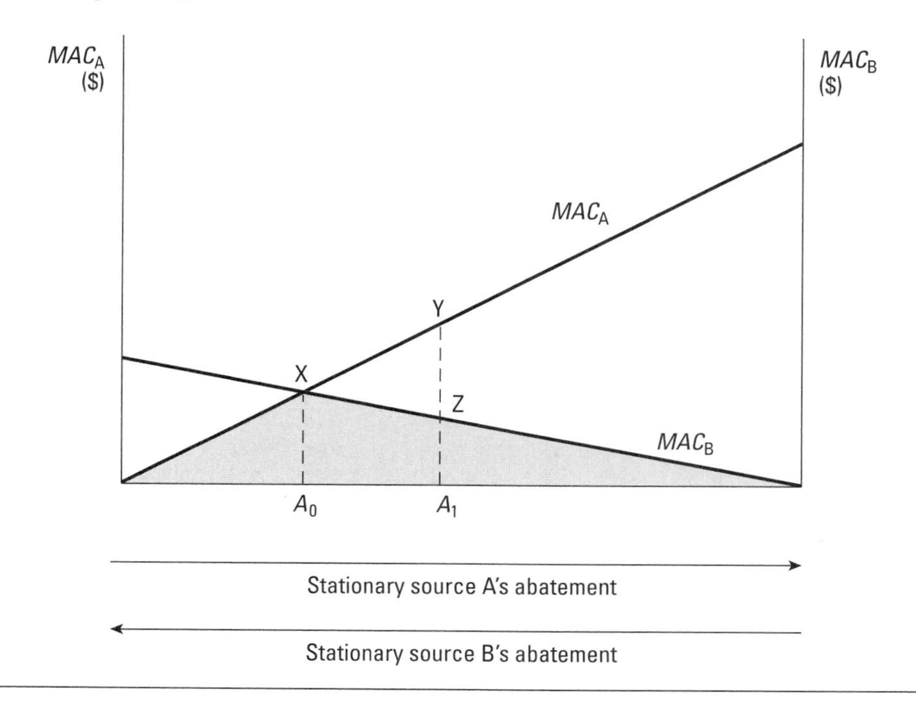

Dual-Control Approach and the New Source Bias

The dual-control approach gives states more direct supervision over those firms on which their economies have come to depend. Yet, despite its apparent logic, a dual-control system can generate market distortions. As long as there are two sets of standards, firms have both the incentive and the opportunity to avoid the more stringent, and hence the more costly, of the two. The logical expectation is that state-determined emissions limits will be more lenient, suggesting that firm decision making will be biased against new construction, a true market distortion.

To understand this latter hypothesis, consider that states have an incentive to attract business as a way to create jobs and to maintain a healthy tax base to support fiscal spending. They therefore should avoid setting stringent emissions limits that may encourage firms to relocate where regulations are more lenient.[15] In turn, firms recognize that they can avoid the federally established standards for new or modified sources simply by maintaining their existing facilities, which would be subject to less stringent state limits. The irony then is that the law sets up a disincentive for businesses to build new facilities that would likely operate more cleanly.[16] This process was the subject of litigation and considerable debate that began in 1999 and was not finalized until 2005, as discussed in Application 12.3.

It is also the case that this dual-control approach may disallow a cost-effective solution. Look at Figure 12.6, which models the marginal abatement cost (*MAC*) functions

FIGURE 12.6 The Dual-Control Approach

A_{EX1} represents the more lenient abatement level of a state-determined standard on existing sources, and A_{N1} represents the more stringent controls imposed by the NSPS. The result is not cost-effective, because the MAC_{N1} level corresponding to A_{N1} is higher than the MAC_{EX1} level corresponding to A_{EX1}. To achieve cost-effectiveness, the state could increase the required abatement for existing sources to A_{EX2} so that MAC_{EX2} equals MAC_{N1}.

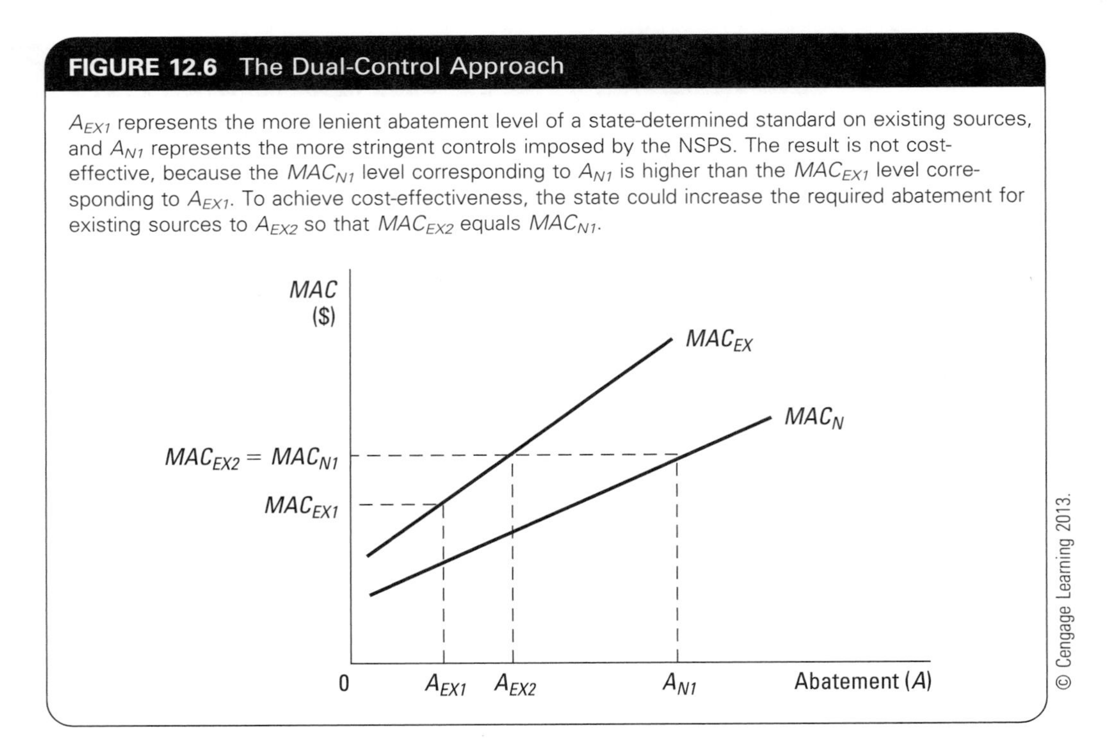

[15]In practice, there are limits as to how permissive state laws can be. Under the State Implementation Plan (SIP), each state's standards must be approved by the EPA, so they must conform to the EPA's range of reasonableness.

[16]For more on this argument, see Gruenspecht and Stavins (January 26, 2002).

APPLICATION 12.3 The New Source Review Standards Come Under Fire

In 1999, the Justice Department and several states filed suits against 13 electric utilities for violating the Clean Air Act at more than 51 of their plants across 12 states. The claim was that these utilities had failed to obtain a New Source Review (NSR) permit to undertake "significant modifications" to their facilities. According to the Clean Air Act, new or significantly modified stationary sources are subject to more stringent standards than their existing counterparts and are required by law to install more advanced technology for pollution control.

Predictably, there was an intense, negative response from the utilities. Among their objections was that the effective litmus test, "significant modification," was not well defined in the law but, in any case, should not be interpreted to include ordinary maintenance. Perhaps even more important was their assertion that the NSR program is destructive. As written, it discouraged power companies from updating or improving their facilities, because doing so would be met with tougher and more costly abatement regulations. Such an outcome is environmentally perverse, since newer power plants generally emit less pollution, and is economically damaging if it interferes with meeting rising energy demands.

A lengthy debate ensued, involving the EPA, government officials, private industry, and environmental groups, and it was not at all clear how the gridlock would be resolved. Some utilities settled or reached tentative settlements with the EPA, but others sought a decision at a higher level.

In May 2001, the Bush administration's energy task force recommended that the EPA review the NSR requirements to ascertain how these provisions affected energy efficiency and environmental quality. After the review was completed, the EPA concluded that the determination of which plant changes constituted a significant modification had been too inflexible. The EPA also announced that it would recommend legislative or administrative changes in the NSR program to the White House.

Following a series of revisions to the ruling in 2002 and 2003, a final NSR rule was issued in October 2003. In this rule, the EPA explicitly identifies the types of equipment replacements that will be exempt from NSR requirements. This component of the ruling is called the "equipment replacement provision (ERP)." Not satisfied with this ERP component, several parties filed suit against the EPA in November 2003, and the U.S. Court of Appeals was asked to grant a stay against this section. It did so that December, and as a result, the issue had to be reconsidered by the EPA. On July 1, 2004, the EPA announced its decision, concluding that no changes to the final NSR rule were necessary. Consequently, after nearly six years, the final NSR rule became effective on June 10, 2005.

For additional information and updates about the EPA's NSR permitting process, visit **www.epa.gov/nsr/actions.html**.

Sources: *U.S. Federal Register* 70 (June 10, 2005), pp. 33838–33850; *U.S. Federal Register* 69 (July 1, 2004), pp. 40274–40285; *U.S. Federal Register 68* (November 7, 2003), pp. 63021–63029; Pianin and Mintz (August 8, 2001); Cohn, Carey, and Palmer (June 11, 2001).

for new and existing sources: MAC_{EX} for an existing source and MAC_N for a new source. The MAC_N curve is lower than the MAC_{EX} curve to illustrate the more advanced control technology available to a more modern facility. Now, consider the cost implications of two distinct abatement levels: A_{EX1}, a more lenient level to represent the state-determined standard on existing sources, and A_{N1}, a more stringent control level based on the federal NSPS. Notice that MAC_{N1} corresponding to A_{N1} is higher than MAC_{EX1} corresponding to A_{EX1}. Thus, the inconsistency in controls on stationary sources has generated a solution that is *not* cost-effective.

Theoretically, cost-effectiveness could be achieved by altering one set of emission standards. For example, if the state official increases the standard for existing sources to A_{EX2}, the corresponding MAC_{EX2} would equal MAC_{N1}. (Alternatively, the federal authority could reduce the standard for new sources until the corresponding marginal abatement cost equals MAC_{EX1}.) In so doing, the bias against new construction would be eliminated. In practice, this outcome would be difficult to realize, however, because officials would need abatement cost data for every polluting source in its jurisdiction.

Economics of Market-Based Trading Programs

Although the legislative effort in the United States up through the 1977 Amendments was primarily a command-and-control approach, national policy subsequently began to use more market-based instruments to control stationary sources. Perhaps the best example of this trend was the initiation of emissions trading in the 1970s implemented through offsets, netting, bubbles, and banking.

Analysis of Early Emissions Trading Programs

To illustrate how the market process operates through these instruments, reconsider the motivations operating in the offset plan. If Firm A wishes to build a new plant in a nonattainment area, it might provide the offsetting emissions through one of its own plants in the region. Alternatively, it could negotiate to have another firm in the region, Firm B, provide the emission reductions in its place. Firm B has an incentive to provide this service if its costs of doing so are less than what Firm A is willing to pay for it. Firm A will contract out the emission reductions if it is cheaper than performing the task itself. In the end, those firms controlling most of the pollution in a region will be the ones that can do so more cheaply. Ultimately, all sources will control pollution to the point where the marginal abatement cost of doing so is equal across firms—a cost-effective solution.

Overall, studies suggest that trading can be lucrative, with cost savings ranging from \$135 million for state-approved bubbles to between \$0.5 billion and \$12 billion for netting transactions. The overwhelming majority of trades are internal (i.e., between sources within the same plant), which is explained by the inherently higher transaction costs associated with trades that are external (i.e., between sources in different facilities).[17] In theory, banking should help to reduce these costs and foster external trades, but the data indicate otherwise. It seems that some other mechanism is needed to bring firms together so that the potential savings can be realized. Such an alternative is evident in the 1990 trading program aimed at acid rain.

Analysis of SO_2 Allowance Trading Plans

Let's consider the cost-effectiveness of the allowance trading established originally by Title IV of the 1990 Amendments and extended through the CAIR and, more recently, its successor rule, the CSAPR. Polluting sources that can reduce SO_2 emissions relatively cheaply will do so and sell off their excess allowances. Each has an incentive to do so as long as its respective *MAC* is below the market price of an allowance (*P*), as shown in Figure 12.7. In the model shown, the polluter would abate A_0 units of SO_2. Excess allowances are an asset to these firms, who effectively become **suppliers** in the SO_2 allowance market. These can be used internally to offset emissions at another source within the firm, banked for future use, or sold on the open market.

On the other side of the market are buyers of allowances, or **demanders**. These are primarily firms unable to match their emissions level to the number of allowances held. For these polluters, abatement is relatively costly, so they will pay for the right to release excess emissions as long as the market price of an allowance (*P*) is below their

[17]Hahn and Hester (1989); Hahn (Spring 1989).

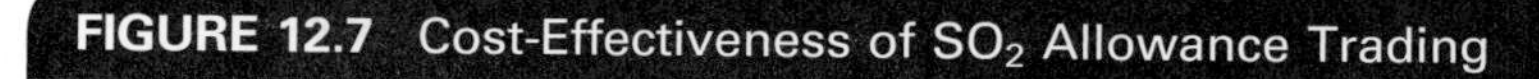

FIGURE 12.7 Cost-Effectiveness of SO_2 Allowance Trading

When the allowance price equals *P*, the polluter would abate A_0 units of SO_2. This is so because at all abatement levels less than A_0, the firm's *MAC* is lower than *P*, so it has an incentive to abate. At all abatement levels above A_0, the firm's *MAC* is higher than *P*, so the polluter would be better off buying allowances rather than abating further.

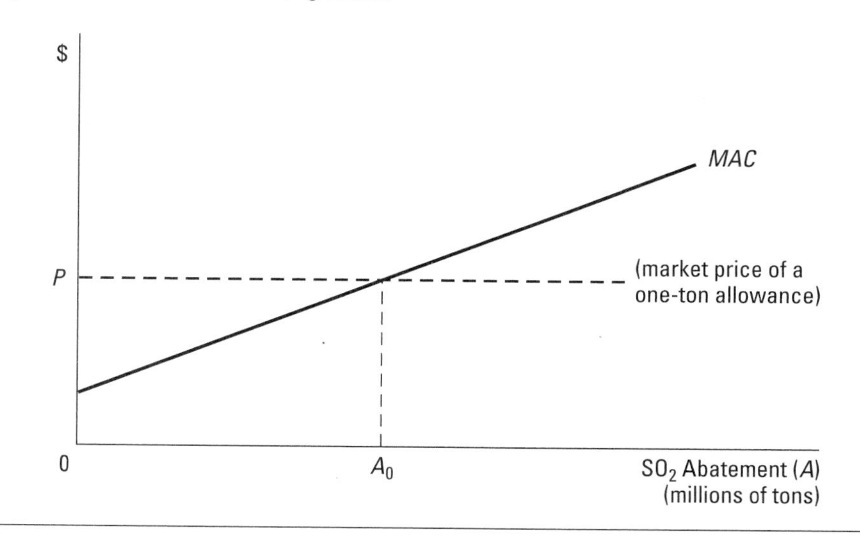

respective *MAC* levels. Other potential buyers include brokers and environmental organizations, each of which has different motivations. Brokers are motivated by profit, expecting to resell the allowances for a lucrative rate of return. Environmentalists want to buy allowances and remove them from the market, effectively increasing the abatement standard.

On net, buyers and sellers trade for allowances, and emissions should reach the level where *MAC*s are equal across firms. Creating an emissions market and allowing the usual incentives to operate freely can accomplish abatement in a cost-effective manner with no compromise to social benefits. Notice that the national emissions limit is achieved, because the number of allowances issued determines the maximum amount of SO_2 emissions that may be released. Hence, all the health and ecological benefits of pollution abatement still accrue to society, but the associated costs are lower.

Using the results of a recent annual spot auction for SO_2 allowances under the Acid Rain Program (ARP) held in March 2011, we can gain a better perspective of how this market operates. On the supply side, the EPA offered 125,000 allowances to be sold to the highest bidders, and none were privately offered.[18] On the demand side, 30 bids were submitted for 550,131 allowances. The highest per unit bid was $66.67, and the

[18]The 125,000 allowances are sold in the spot auction for use in the same year unless banked for future use. There are also 125,000 allowances sold in an advance auction for use in the seventh year after the year of sale, unless banked for future use.

lowest was $0.06. At the auction's close, all 125,000 allowances had been sold to six individual demanders who bid between $2 and $66.67. Data on the 2011 annual allowance auction and all others are available online at **www.epa.gov/airmarkets/trading/auction.html**.

Of the successful demanders, the majority of allowances were purchased by electric utilities to assist them in meeting the SO_2 abatement standard. However, allowances also were purchased by several private organizations. The Acid Rain Retirement Fund was one such organization, which successfully purchased 68 allowances. As the name suggests, this entity's intent was to withhold its 68 allowances from the market, thereby effectively removing 68 tons of SO_2 from the atmosphere. Also of interest were successful bids that came from colleges or universities, including Bates College and the University of Tampa. Actions from these private institutions signify their interest in removing allowances from the market, which in turn eliminates SO_2 emissions from the atmosphere.[19]

It is generally agreed upon that the Acid Rain Program (ARP) has been a success in large part because firms have achieved emission reduction targets at substantially lower costs than the projected estimates. Specifically, costs of the SO_2 trading program have been estimated to be between $1 billion and $2 billion per year as of 2010 when nearly full implementation was achieved. This is 25 percent of initially projected costs. And NO_X costs are expected to be no higher than $1 billion per year. These estimates are particularly important when considered relative to the associated benefits, estimated at more than $70 billion per year, higher than any federal program in the last 10 years. Moreover, the compliance rate has been consistent and remarkably high, at more than 99 percent.[20]

Allowance prices, which reflect the marginal cost of compliance, are much lower than originally anticipated. The EPA had predicted that SO_2 allowance prices would be about $750 per ton, but these forecasts proved to be inaccurate. Prior to the transition from the CAIR to the new Cross-State Air Pollution Rule (CSAPR), these prices have ranged from about $68 per ton to $883 per ton, with only two years in which average prices were above $445. See Figure 12.8, which shows the change in SO_2 allowance prices from 1993 to 2011. During 2009, prices dropped significantly, specifically by 67 percent, with further decline into 2010.[21] According to Burtraw and Szambelan (2009) at Resources for the Future, the sharp decline in 2009 is attributable to inordinately low natural gas prices at that time and to the uncertainty associated with the vacatur of the CAIR. Whether prices will rebound when the CSAPR takes effect remains to be seen.[22]

---

[19]U.S. EPA, Office of Air and Radiation, Office of Atmospheric Programs, Clean Air Markets Division (March 29, 2011).

[20]U.S. EPA, Office of Air and Radiation, Office of Atmospheric Programs, Clean Air Markets Division (January 2009; September 2004).

[21]U.S. EPA, Office of Air and Radiation, Office of Atmospheric Programs, Clean Air Markets Division (April 19, 2011; September 2010); U.S. EPA, Office of Policy, Economics, and Innovation (January 2001), pp. 80–81.

[22]A readable article that analyzes the relatively low allowance prices and other aspects of the SO_2 allowance trading program is Schmalensee *et al.* (Summer 1998). Related issues are discussed in the accompanying article by Stavins (Summer 1998).

FIGURE 12.8 Average Auction Prices for SO_2 Allowances: 1993–2011

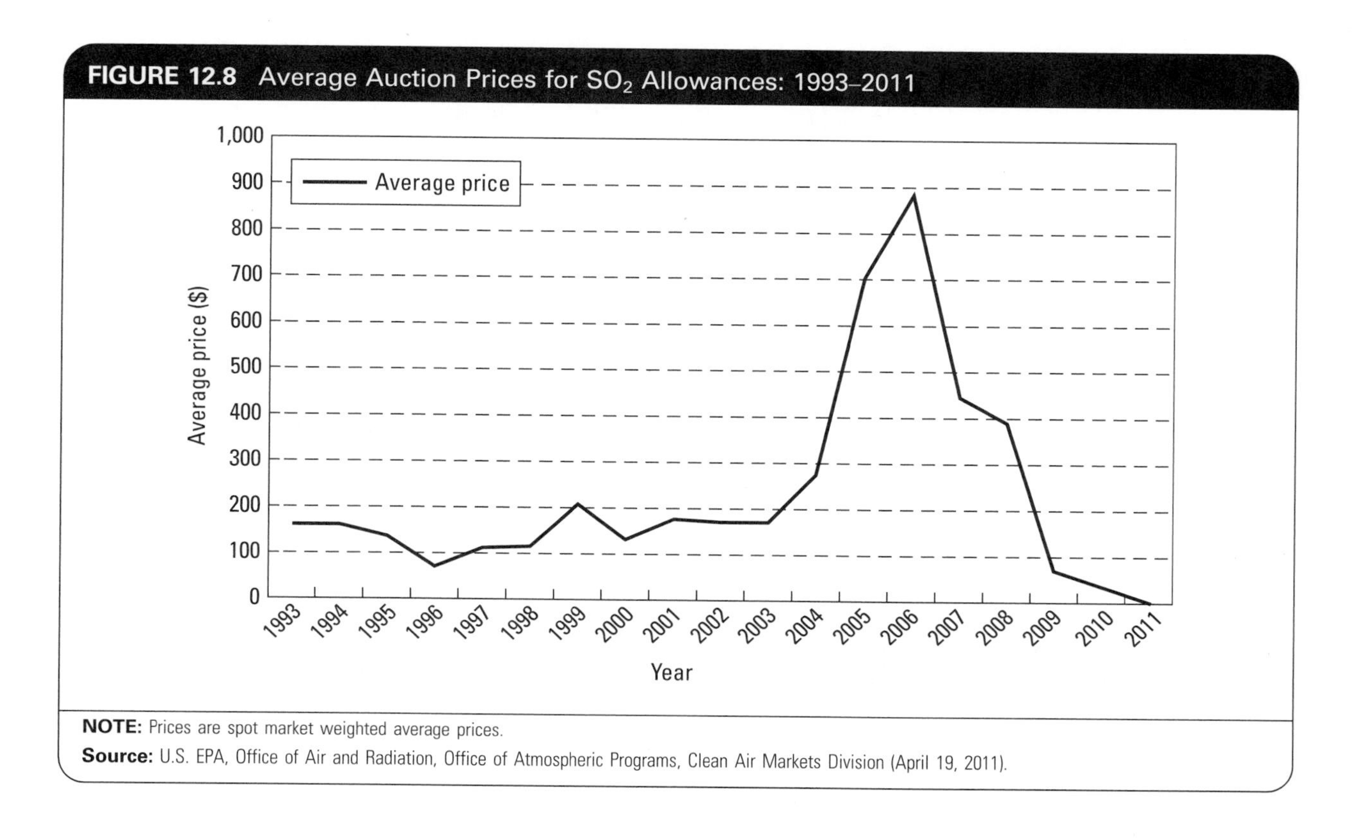

NOTE: Prices are spot market weighted average prices.
Source: U.S. EPA, Office of Air and Radiation, Office of Atmospheric Programs, Clean Air Markets Division (April 19, 2011).

Part of the impetus for developing a national market for SO_2 allowances was the expectation of significant cost savings linked to trading activity. Initially, the EPA estimated these savings to be in the range of $0.7 billion to $1 billion per year.[23] As the program evolved, a number of researchers have addressed this important issue. For example, Ellerman et al. (2000) estimates that the cost of compliance with Title IV of the Clean Air Act is $2.5 billion less annually because of allowance trading.

Analysis of NO_X Trading Programs

Like the SO_2 allowance trading program, the NO_X Budget Trading Program (NBP) and the added trading programs under the CAIR and the CSAPR were developed to give states an opportunity to reduce NO_X emissions in a cost-effective manner. That is, through the cap-and-trade plans, polluting sources able to reduce NO_X emissions relatively cheaply could do so and sell any excess allowances to firms that could not. The incentive for trading exists as long as sellers receive a price greater than their *MAC* and as long as buyers pay a price lower than their *MAC*.

At issue is to consider whether the trading programs have been successful and have achieved the goal of providing a cost-effective solution to affected states. A few observations are noteworthy. First, all states subject to the NO_X SIP Call elected to participate in the NBP, validating the cost-effectiveness incentive offered by cap-and-trade plans.

[23] *U.S. Federal Register 56* (December 3, 1991), p. 63097.

Second, annual NO_X releases from all ARP sources fell by 4.1 million tons between 1995 and 2009. Moreover, the 2009 emissions level exhibited the largest decline year over year since the trading program began, an accomplishment linked to the two added trading programs under the CAIR. Although some of the decline is attributable to falling demand for electricity, NO_X trading also played a role, indicated by the decrease in the emission rate for all ARP sources between 2008 and 2009 from 0.22 to 0.16 pounds per million British thermal units (mmBtus). Third, allowance prices under the NBP and its successor, the ozone season trading program under the CAIR, show a general downward trend, with a sharp decline during 2006, as shown in Figure 12.9. This overall trend may suggest lower control costs as trading continues over time.[24]

A caveat is in order, however. As indicated in the figure, allowance pricing exhibits volatility at times, which can be due to exogenous factors, such as energy prices and regulatory changes and hence may not reflect or predict the success of allowance trading. A case in point is the instability of NO_X allowance prices under the CAIR annual trading program, which is identified separately in Figure 12.9. As pointed out by Burtraw and Szambelan (2009), trading activity and the prices of both NO_X and SO_2 allowances

FIGURE 12.9 NO_X Allowance Daily Average Price: January 2005–December 2009

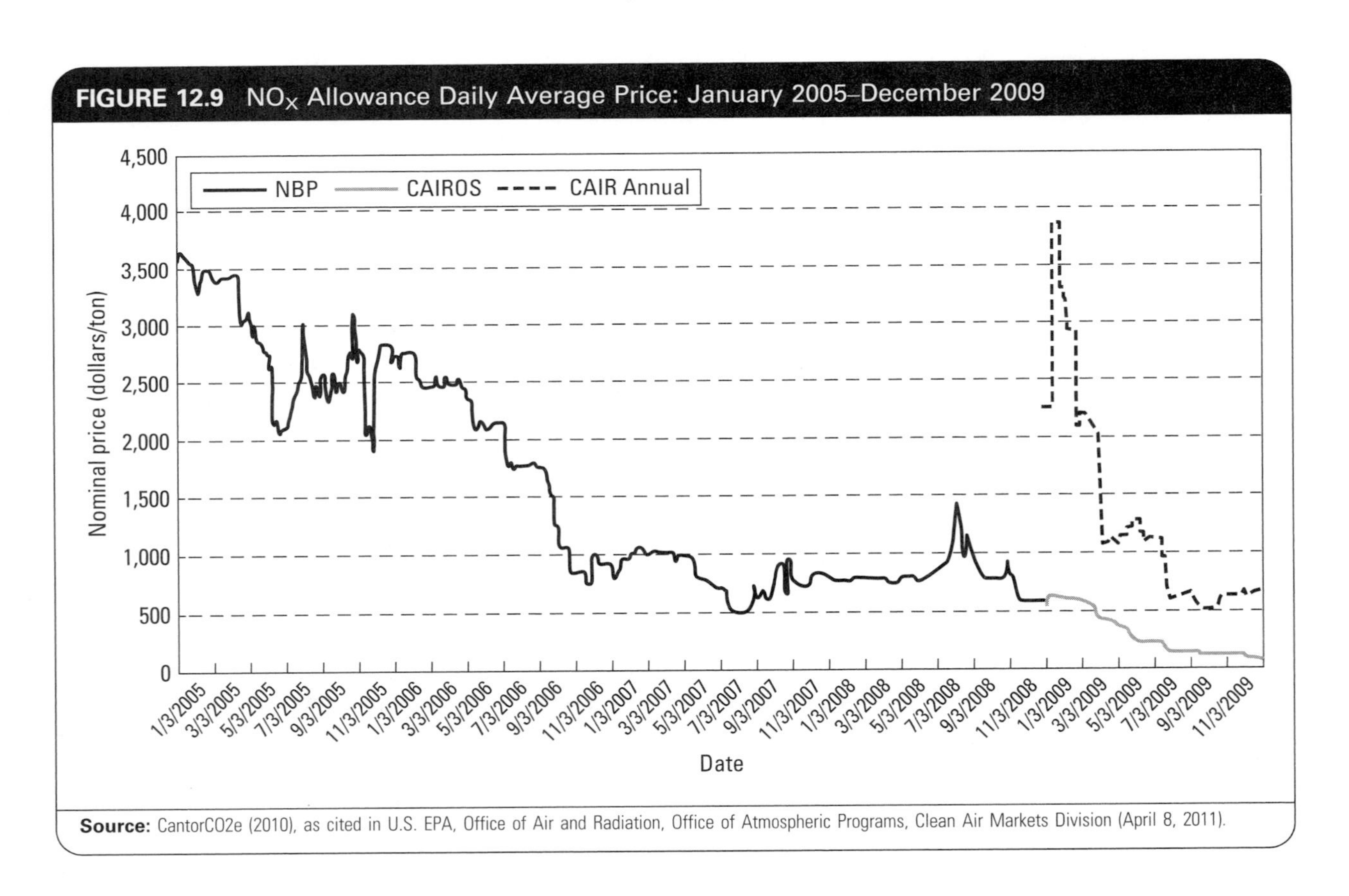

Source: CantorCO2e (2010), as cited in U.S. EPA, Office of Air and Radiation, Office of Atmospheric Programs, Clean Air Markets Division (April 8, 2011).

[24]U.S. EPA, Office of Air and Radiation, Office of Atmospheric Programs, Clean Air Markets Division (April 8, 2011; December 2010; August 2004).

were affected dramatically when the court decided to vacate the CAIR in 2008, dropping sharply in response and then rising again when the CAIR was reinstated. Subsequently, prices have fallen once more, which is linked again to regulatory uncertainty. Such price volatility can, of course, detract from expected cost savings.

CONCLUSIONS

Current U.S. air quality policy is ambitious and complex, implemented in part through a comprehensive set of controls on stationary sources. There is no doubt that the trend toward integrating more market-based initiatives, such as cap-and-trade programs, represents a landmark in environmental legislative history. Evidence is accumulating that market instruments are a more cost-effective means to achieve air quality than command-and-control policies. Other aspects of existing air control policy are less satisfying and continue to be debated. The dual-control approach to setting standards for stationary sources is one example of how government policy can create market distortions.

All of this means that, as a nation, we still have much to learn—about air quality and about policy instruments that work effectively to protect it. As we have come to discover, policy development and appraisal are ongoing processes. In the context of air quality controls, the simple truth is that it is extremely difficult to assess exactly what has been accomplished through all the revisions of the last several decades. In the interim, the United States must continue to make policy adjustments and to respond to the challenge of achieving both its air quality objectives *and* economic prosperity.

Summary

- Acidic deposition is caused by the reaction of SO_2 and NO_X emissions with water vapors and oxidants in the atmosphere. These reactions form sulfuric acid and nitric acid, which mix with other airborne particles and fall to the earth as dry or wet deposition.
- Emissions of stationary sources are regulated mainly by uniform technology-based standards, which is indicative of a command-and-control approach. Over time, emissions trading programs have been introduced.
- Standards applicable to new stationary sources are called New Source Performance Standards (NSPS). The EPA controls new or modified sources, and states set standards for existing facilities, known as the dual-control approach.
- In PSD areas, standards for new or modified sources must be based on the best available control technology (BACT). Existing facilities must install best available retrofit technology (BART).
- In nonattainment areas, existing sources must use, at minimum, reasonably available control technology (RACT). New or modified sources must comply with the lowest achievable emission rate (LAER).
- According to rulings in the Clean Air Act, new GHG controls on mobile sources triggered regulatory requirements for stationary sources. These new regulations would be implemented through GHG permitting programs and GHG emissions standards.
- In 1979, the EPA initiated a bubble policy for existing sources and subsequently launched an emissions banking program.
- Netting was developed for modified sources in PSD areas, and an offset plan was devised for new or modified sources in nonattainment areas.
- To control acid rain, Title IV of the 1990 Amendments established a reduction plan for NO_X emissions and an allowance program for SO_2 emissions.
- The aggregate number of SO_2 emission allowances issued by the EPA sets the national limit. These allowances may be used by the recipient or exchanged through an allowance market. Small banks of allowances are available for sale by the EPA or through annual auctions supervised by the EPA.
- The Clean Air Interstate Rule (CAIR) was issued by the EPA in 2005 and involved declining emissions caps and separate trading programs for SO_2 and NO_X. Because the CAIR was remanded back to the EPA by the courts, a new ruling called the Cross-State Air Pollution Rule (CSAPR) is to replace the CAIR. This rule calls for power plants to achieve major reductions in SO_2 and NO_X emissions, which can drift across state borders.
- In 2003, the NO_X Budget Trading Program (NBP) was established to help states achieve NO_X emissions limits during the ozone season.
- Under the CAIR, issued in 2005, two trading programs for NO_X were established. One of these was applicable to the NO_X ozone season and replaced the NBP, and the other was an annual program. The Cross-State Air Pollution Rule (CSAPR), which is to replace CAIR, allows power plants the option of trading emissions allowances both within and between participating states in order to meet pollution limits.
- Among the first state-level emissions trading programs was the Regional Clean Air Incentives Market (RECLAIM) program developed in 1994 by California. Several other states have launched similar initiatives.
- A growing body of research indicates that command-and-control policies are relatively costly.
- Because the New Source Performance Standards (NSPS) are technology-based and implemented uniformly, they disallow a cost-effective solution.

- Under the dual-control system, state-determined limits will likely be more lenient. Thus, firms can avoid meeting the NSPS by maintaining existing plants instead of building new ones.
- Emissions trading programs can achieve a cost-effective solution, because sources abate to the point where the *MAC* of doing so is equal across firms.
- There is some consensus that the Acid Rain Program (ARP) has been a success in large part because firms have achieved emission reduction targets at much lower costs than the projected estimates.
- The NO_X Budget Trading Program (NBP) and the recently added trading programs under the CAIR and the CSAPR were developed to give states an opportunity to reduce NO_X emissions in a cost-effective manner.

Review Questions

1. New source bias can exist for stationary sources. Discuss why this bias leads to a solution that is *not* cost-effective. What policies might eliminate this bias?
2. Distinguish between the technology-based emission standards RACT and LAER used to control stationary sources.
3. **a.** Carefully explain how economic theory supports (i) the bubble policy and (ii) the emission allowance program for sulfur dioxide.
   b. Other than these two programs, briefly summarize any two U.S. policies that use market incentives to control air pollution.
4. **a.** Based on the analysis of the bubble policy given in this chapter, discuss why transaction costs are so important in explaining the success or failure of the EPA's bubble program.
   b. How would you devise a program that minimizes the transaction costs of bringing polluters together so that they could effectively equalize the level of their individual *MAC*s?
5. In July 1997, the EPA announced new air quality standards for small (2.5 micrometers in diameter) particulate matter referred to as PM-2.5. Steel mills are a major source of these smaller particles and therefore must find ways to abate. To analyze the implications, consider the following hypothetical model of two steel plants, one owned by Bethlehem Steel (B) and one owned by National Steel (N), both located in Pittsburgh, Pennsylvania:

$$\text{Bethlehem: } MAC_B = 1.2A_B \qquad TAC_B = 0.6A_B^2$$

$$\text{National: } MAC_N = 0.3A_N \qquad TAC_N = 0.15A_N^2$$

   Assume that each plant emits 40 units of PM-2.5 for a total of 80 units. In order for the Pittsburgh area to meet the new standard, the EPA determines that the combined abatement for both plants must total 30 units.
   a. Assuming the new abatement standard is implemented *uniformly* across the two firms, find the total cost of abatement.
   b. Find the *cost-effective* solution and illustrate graphically, labeling all curves, intercepts, and relevant intersections. Calculate the associated cost savings.
6. In accordance with the dual-control approach, suppose New Mexico's NO_X abatement standard for existing sources is 5 units (A_E) and the EPA's standard for new sources is 12 units (A_N). If the two types of sources face the following *MAC* functions, where each *MAC* is in thousands of dollars, what economic criterion is violated? Show how you know.

$$\text{Existing: } MAC_E = 1.8A_E$$

$$\text{New: } MAC_N = 1.2A_N$$

Additional Readings

Banzhaf, H. Spencer. "Efficient Emission Fees in the U.S. Electricity Sector." *Resource and Energy Economics 26(3)* (September 2004), pp. 317–41.

Burtraw, Dallas. "Markets for Clean Air: The U.S. Acid Rain Program." *Regional Science and Urban Economics 32(1)* (January 2002), pp. 139–44.

Burtraw, Dallas, Karen Palmer, Ranjit Bharvirkar, and Anthony Paul. "Cost-Effective Reduction of NO_X Emissions From Electricity Generation." *Journal of Air & Waste Management 51* (October 2001), pp. 1476–89.

Busse, Meghan R., and Nathaniel O. Keohane. "Market Effects of Environmental Regulation: Coal, Railroads, and the 1990 Clean Air Act." *The Rand Journal of Economics 38(4)* (Winter 2007), pp. 1159–80.

Carlson, Curtis, Dallas Burtraw, Maureen Cropper, and Karen L. Palmer. "Sulfur-Dioxide Control by Electric Utilities: What Are the Gains From Trade?" *Journal of Political Economy 108(6)* (December 2000), pp. 1292–1326.

Ellerman, A. D. "Designing a Tradable Permit System to Control SO_2 Emissions in China: Principles and Practice." *Energy Journal 23* (April 2002), pp. 1–26.

Farrell, Alex, Roger Raufer, and Kimberly Killmer. "The NO_X Budget: Market-Based Control of Tropospheric Ozone in the Northeastern United States." *Resource and Energy Economics 21* (May 1999), pp. 103–24.

Gangadharan, Lata. "Analysis of Prices in Tradable Emission Markets: An Empirical Study of the Regional Clean Air Incentives Market in Los Angeles." *Applied Economics 36(14)* (August 2004), pp. 1569–82.

Greenstone, Michael. "Did the Clean Air Act Cause the Remarkable Decline in Sulfur Dioxide Concentrations?" *Journal of Environmental Economics and Management 47(3)* (May 2004), pp. 585–611.

Hahn, Robert W., and Gordon L. Hester. "Where Did All the Markets Go? An Analysis of EPA's Emissions Trading Program." *Yale Journal on Regulation 6(1)* (1989), pp. 109–53.

Klaassen, Ger, and Andries Nentjes. "Creating Markets for Air Pollution in Europe and the USA." *Environmental and Resource Economics 10* (September 1997), pp. 125–46.

Kruger, Joseph, and Melanie Dean. "Looking Back on SO_2 Trading: What's Good for the Environment Is Good for the Market." *Public Utilities Fortnightly* (August 1997), pp. 30–37.

List, John A., Daniel L. Millimet, and Warren McHone. "The Unintended Disincentive in the Clean Air Act." In *Advances in Economic Analysis and Policy 4(2)*, Berkeley, CA: Berkeley Electronic Press, 2004.

Markandya, Anil, Andrea Bigano, and Roberto Porchia., eds. *The Social Cost of Electricity: Scenarios and Policy Implications.* Northhampton, MA: Elgar, 2011.

Muller, Nicholas Z., and Robert Mendelsohn. "Efficient Pollution Regulation: Getting the Prices Right." *American Economic Review, 99(5)* (December 2009), pp. 1714–39.

______ "Weighing the Value of a Ton of Pollution." *Regulation 33(2)* (Summer 2010), pp. 20–24.

Muller, R. Andrew, and Stuart Mestelman. "What Have We Learned From Emissions Trading Experiments?" *Managerial and Decision Economics 19* (June–August 1998), pp. 225–38.

Newell, Richard G., and Robert N. Stavins. "Cost Heterogeneity and the Potential Savings From Market-Based Policies." *Journal of Regulatory Economics 23* (January 2003), pp. 43–59.

Schennach, Susanne M. "The Economics of Pollution Permit Banking in the Context of Title IV of the 1990 Clean Air Act Amendments." *Journal of Environmental Economics and Management 40* (November 2000), pp. 189–210.

Schwartz, Peter M. "Multipollutant Efficiency Standards for Electricity Production." *Contemporary Economic Policy 23(3)* (July 2005), pp. 341–57.

Selin, N. E. "Mercury Rising: Is Global Action Needed to Protect Human Health and the Environment?" *Environment 47(1)* (January/February 2005), pp. 22–37.

A Reference to Acronyms and Terms Used in Stationary Source Control Policy

Environmental Economics Acronyms

MAC	Marginal abatement cost
MAC_{EX}	Marginal abatement cost for an existing stationary source
MAC_N	Marginal abatement cost for a new stationary source
TAC	Total abatement cost

Environmental Science Terms

CO	Carbon monoxide
GHG	Greenhouse gas
Hg	Mercury
mmBtu	millions of British thermal units
NO_x	Nitrogen oxides
NO_2	Nitrogen dioxide
O_2	Oxygen
O_3	Ozone
Pb	Lead
PM-2.5	Particulate matter less than 2.5 micrometers in diameter
PM-10	Particulate matter less than 10 micrometers in diameter
SO_2	Sulfur dioxide
SO_x	Sulfur oxides
VOC	Volatile organic compound

Environmental Policy Acronyms

ARP	Acid Rain Program
BACT	Best available control technology
BART	Best available retrofit technology
CAAA	Clean Air Act Amendments
CAIR	Clean Air Interstate Rule
CAIROS	CAIR ozone season
CBOT	Chicago Board of Trade
CSAPR	Cross-State Air Pollution Rule
ERP	Equipment replacement provision
LAER	Lowest achievable emission rate
NAAQS	National Ambient Air Quality Standards
NBP	NO_X Budget Trading Program
NSPS	New Source Performance Standards
NSR	New Source Review
OTC	Ozone Transport Commission
PSD	Prevention of significant deterioration
RACT	Reasonably available control technology
RECLAIM	Regional Clean Air Incentives Market
RIA	Regulatory Impact Analysis
RTC	RECLAIM trading credits
SCAQMD	South Coast Air Quality Management District
SIP	State Implementation Plan

To access additional course materials, visit www.cengagebrain.com. At the home page, search for the ISBN of this title (shown on the back cover). This will take you to the product page where these resources can be found.

CHAPTER 13

Global Air Quality: Policies for Ozone Depletion and Climate Change

"Many people today assume mistakenly that the Earth is so big that we humans cannot possibly have any major impact on the way our planet's ecological system operates. That may have been true at one time, but it is not the case anymore."

—Al Gore (1948–) (from *An Inconvenient Truth*, 2006)

While most air pollutants produce localized effects, others have more far-reaching implications. Such is the case for contaminants that alter atmospheric conditions, posing a risk that is geographically without bound and generating a free-ridership problem that crosses national boundaries. Of course, the effects can vary by degree across different locations. In any case, since the associated damage is widespread and since the source cannot be linked to a specific site or region, this air quality problem is termed **global air pollution**. Controlling global air pollution is a unique policy challenge, because solutions must be developed not only through domestic initiatives but also through international treaties and programs.

In this chapter, we investigate the principal issues associated with global air pollution by studying **ozone depletion** and **climate change**. In each case, we consider theories about the causes and sources of the atmospheric disturbance and the available evidence to support these theories. Using this as a foundation, we then explore policy responses that have been set in motion in the United States and other nations, along with proposals for alternatives. Ultimately, our objective is to economically evaluate the effectiveness of these policies, given what we know about the origin of the problem and the associated risks. As in the previous three chapters, there is a reference list of acronyms and terms at the end of this chapter.

OZONE DEPLETION

ozone layer
Ozone present in the stratosphere that protects the earth from ultraviolet radiation.

Starting in the 1950s, scientists began measuring the earth's **ozone layer** in the stratosphere, which is the atmospheric layer lying between 7 and 25 miles above the earth's surface. The effort was motivated by more than scientific curiosity. Stratospheric ozone protects the earth from ultraviolet radiation.

Some variability in the depth of the ozone layer was assumed normal, including an observed thinning above Antarctica during the Southern Hemisphere spring. This would generally fill back in by November each year. However, in the early 1980s, scientists became concerned when this thinning was found to be increasing in size and persisting into December. In 1985, an "ozone hole" the size of North America was discovered over Antarctica. It was then that world attention was drawn in earnest to the problem of **ozone depletion** and the pollutants responsible for the damage.[1]

ozone depletion
Thinning of the ozone layer, originally observed as an ozone hole over Antarctica.

While all the implications are not known with certainty, there are some consequences of increased ultraviolet radiation about which there is agreement. Scientists tell us that rising levels of ultraviolet radiation can alter delicate ecosystems, diminish human immune systems, and increase the risk of skin cancer. The National Academy of Sciences estimates that 10,000 more cases of skin cancer per year would result for every 1 percent decline in stratospheric ozone. A good resource on the science of ozone depletion is available at **www.epa.gov/ozone/science**.

Searching for the Causes of Ozone Depletion

Scientists debate about the principal cause of the ozone hole, which extends approximately 9 million square miles over the Antarctic.[2] Figure 13.1 presents an image from the National Aeronautics and Space Administration (NASA) that shows the ozone thinning over Antarctica during the 2010 ozone season. Although no one theory has been able to fully explain the extent of ozone depletion, scientists agree that the presence of **chlorofluorocarbons (CFCs)** in the atmosphere is the most likely explanation—a theory originally advanced in 1974 by F. Sherwood Rowland and Mario Molina, two University of California researchers. The pair won the Nobel Prize in chemistry for this theory in 1995.

chlorofluorocarbons (CFCs)
A family of chemicals believed to contribute to ozone depletion.

CFCs are a family of chemicals that were commonly used in refrigeration, air conditioning, packaging, and insulation and as aerosol propellants. They are sometimes referred to by their trade names, Freon and Styrofoam. In fact, the energy crisis in the 1970s was responsible for an even greater use of CFCs as foaming agents in the production of home insulation. The rise of the fast-food industry was also a contributing factor to intensified CFC use, because polymer foams were utilized to produce disposable cups and food containers. These long-lived compounds are not destroyed in the lower atmosphere and therefore are able to drift up into the stratosphere, where their chlorine components destroy ozone. In addition, because of their long atmospheric lifetimes, CFCs released today affect the ozone layer for decades to come.

From an economic perspective, production of goods like Styrofoam cups is associated with a **negative externality**. As discussed in Chapter 3, a negative externality arises when there is an external effect that generates costs to a third party. Graphically, this

---

[1]For the published report, see Farman, Gardiner, and Shanklin (1985).

[2]Research shows that ozone depletion is truly a global issue, occurring over latitudes that include Asia, parts of Africa, Australia, Europe, North America, and South America (U.S. EPA, Office of Air and Radiation, Office of Atmospheric Programs, Stratospheric Protection Division, August 19, 2010a).

FIGURE 13.1 Ozone Hole over Antarctica

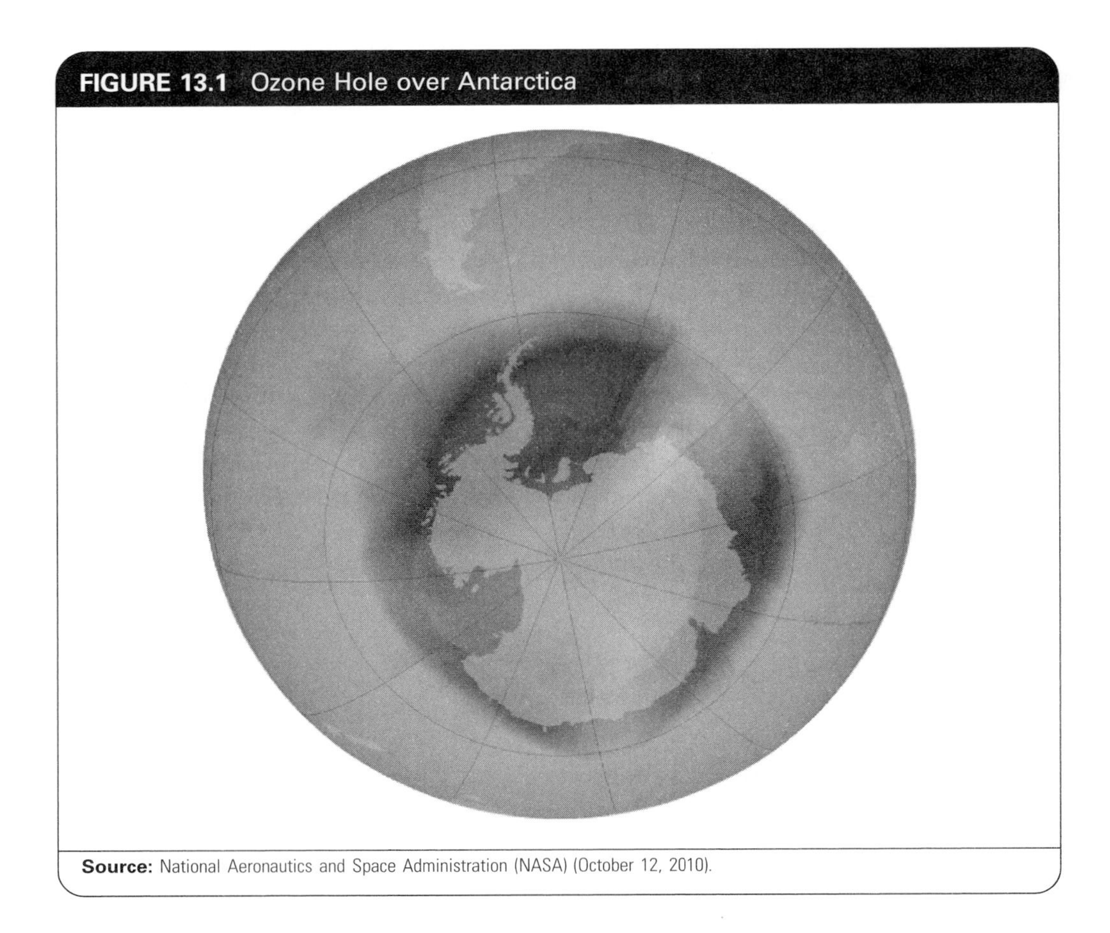

Source: National Aeronautics and Space Administration (NASA) (October 12, 2010).

externality would be modeled as a marginal external cost (*MEC*), which causes the marginal social cost (*MSC*) to be higher than the marginal private cost (*MPC*). Therefore, the *MEC* is represented as the vertical distance between *MSC* and *MPC* at each output level. This relationship is shown in Figure 13.2 for the Styrofoam cup market. Notice that the competitive equilibrium output (Q_C) is higher than the efficient equilibrium (Q_E), indicating that too much of the good is produced in the absence of third-party intervention.

Another major group of ozone depleters are **halons**, which have long atmospheric lifetimes. Before government controls, these substances were becoming increasingly important in the production of fire extinguishants. Their use is not as widespread as CFCs, but halons are known to have a higher potency for ozone depletion than their chlorine-containing counterparts.

Despite the lack of hard evidence at the time, the United States opted to ban the use of CFCs in most aerosol sprays in 1978, and other countries followed suit. However, other uses of these ozone depleters were not controlled, and little effort was aimed at finding substitutes. As a result, domestic and international CFC use continued to grow. Until governments intervened in the late 1980s, there was little question that a stronger policy position was needed to control ozone-depleting substances.

FIGURE 13.2 A Production Negative Externality in the Market for Styrofoam Cups

The negative externality of Styrofoam cup production is modeled by a marginal external cost (*MEC*) curve. The *MEC* causes the marginal social cost (*MSC*) to be higher than the marginal private cost (*MPC*) at each output level. As a result, too much of the good is produced, and the competitive equilibrium output (Q_C) is higher than the efficient equilibrium (Q_E).

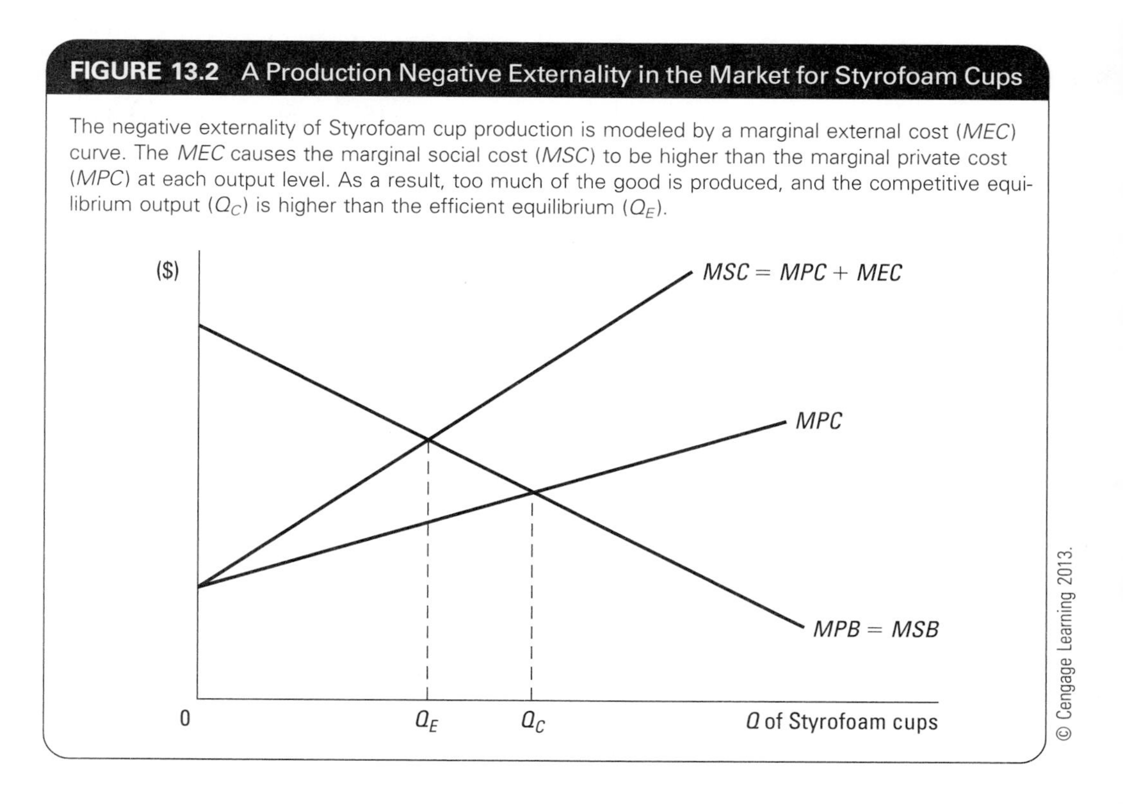

CONTROLLING OZONE DEPLETION

As a global air pollution problem, ozone depletion cannot be controlled without an integrated international effort. More formally, think of this environmental problem as an externality with transboundary implications. Domestically, the 1990 Clean Air Act Amendments (CAAA) call for the president to enter into international agreements that encourage joint research on ozone depletion and to establish regulations consistent with those in the United States. Although not without political implications, a number of international agreements and multilateral treaties have been executed or are on the negotiating table. A brief summary of the most significant of these follows.

International Agreements to Control Ozone Depletion[3]

Montreal Protocol

In 1987, 24 countries as well as the European Community Commission signed the Montreal Protocol on Substances That Deplete the Ozone Layer. Among the signatories were the major producers of CFCs. This landmark agreement called for a 50 percent reduction in CFC consumption and production, a target that was to be achieved gradually by 2000. To achieve this objective, each party to the protocol was responsible for

[3]General information in the subsequent discussion is drawn from U.S. EPA, Office of Air and Radiation, Office of Atmospheric Programs, Stratospheric Protection Division (August 19, 2010b).

designing and implementing an effective control program in accordance with the agreed-upon deadlines. Current ratification status of the Montreal Protocol and subsequent amendments is available at **http://montreal-protocol.org/new_site/en/treaty_ratification_status.php**. As of 2012, 197 nations have ratified this accord.[4]

Amendments to the Protocol

In 1990, 59 countries executed the London Amendment to the protocol. This amendment, which strengthened the worldwide commitment to protecting the ozone layer, was in direct response to reports that ozone depletion might be more severe than originally believed. The new agreement outlined a full phaseout plan for CFCs and halons, and added controls for other ozone-depleting substances, such as carbon tetrachloride and methyl chloroform. At subsequent conferences, controls for other substances were added, including additional CFCs, hydrochlorofluorocarbons (HCFCs), hydrobromofluorocarbons (HBFCs), methyl bromide, and bromochloromethane (BCM). In addition, various phaseout deadlines were advanced and then formalized. The Protocol does include special provisions for developing nations, allowing them an additional 10–12 years to phase out ozone depleters.[5] Table 13.1 gives a summary of how these phaseout agreements are defined for both developed and developing nations.

International Allowance Trading

A market approach was integrated as part of the international effort to protect the ozone layer. Specifically, production and consumption allowances were issued to the protocol participants, and transfers were permitted under certain guidelines. To ensure that the phaseouts were achieved, trading was conditioned upon revision of each country's aggregate production limits to levels lower than what would have occurred without the transfers.

Multilateral Fund

Ongoing negotiations are aimed at encouraging more nations to ratify the protocol amendments. Some countries have been hesitant because of the high costs of converting production technology to eliminate the use of ozone-depleting substances. This is particularly problematic for developing nations. In response to these concerns, an Interim Multilateral Fund of $160 million was established in 1990 to help developing countries transition toward the requisite CFC replacement technologies. In 1992, the fund became permanent, and it is replenished every three years. The replenishment amount for 2009–2011 was $400 million with a budget for the period of $490 million. As of 2011, 45 industrialized nations, including Countries with Economies in Transition, or CEIT countries, have contributed over $2.76 billion to this fund.[6] More information is available at **www.multilateralfund.org**.

---

[4]Note that Title VI of the CAAA provides the legislative guidelines for U.S. compliance with the Montreal Protocol.

[5]Changes at these later conferences became known as the 1990 London Amendment, the 1992 Copenhagen Amendment, the 1995 Vienna Convention, the 1997 Montreal Amendment, the 1999 Beijing Amendment, and the 2007 Montreal Adjustment.

[6]United Nations, Secretariat of the Multilateral Fund for the Implementation of the Montreal Protocol (September 17, 2011).

TABLE 13.1 Montreal Protocol Phaseout Schedules for Production and Consumption of Ozone Depleters

	Developed Nations	Developing Nations
CFCs	Phase out by 1996	Freeze at average 1995–1997 levels by July 1, 1999; reduce by 50 percent by 2005, by 85 percent by 2007; phase out by 2010
Carbon tetrachloride	Phase out by 1996	Freeze at average 1995–1997 levels by July 1, 1999; reduce by 50 percent by 2005, by 85 percent by 2007; phase out by 2010
Halons	Phase out by 1994	Freeze at average 1995–1997 levels by July 1, 1999; reduce by 50 percent by 2005, by 85 percent by 2007; phase out by 2010
HCFCs	Reduce by 35 percent by 2004, by 75 percent by 2010, by 90 percent by 2015; phase out by 2020 (allowing 0.5 percent for servicing reasons between 2020 and 2030)	Freeze at average 2009–2010 levels by 2013; reduce by 10 percent by 2015, by 35 percent by 2020, by 67.5 percent by 2025; phase out by 2030 (allowing 2.5 percent annual average for servicing reasons between 2030 and 2040)
Methyl bromide	Reduce by 25 percent by 1999, by 50 percent by 2001, by 70 percent by 2003; phase out by 2005	Freeze at average 1995–1998 levels by 2002; reduce by 20 percent by 2005; phase out by 2015
Methyl chloroform	Phase out by 1996	Freeze at average 1998–2000 levels by 2003; reduce by 30 percent by 2005, by 70 percent by 2010; phase out by 2015
HBFCs	Phase out by 1996	Phase out by 1996
BCM	Phase out by 2002	Phase out by 2002

For more detail on the Montreal Protocol or any of the subsequent amendments, visit the Ozone Secretariat at the United Nations Web site at **http://ozone.unep.org/new_site/en/index.php**.

NOTE: Phaseouts shown include changes through the 2007 Montreal Adjustment.

Source: United Nations Environment Programme (September 2008).

U.S. Policy to Control Ozone Depletion

The 1990 CAAA significantly strengthened U.S. policy on ozone-depleting substances through Title VI (accessible at **www.epa.gov/oar/caa/title6.html**). These provisions must comply with the nation's commitment to the Montreal Protocol. Under these rulings, Congress charged the Environmental Protection Agency (EPA) with the responsibility of identifying ozone-depleting substances. Each substance is assigned a numerical value that signifies its **ozone depletion potential (ODP)** relative to chlorofluorocarbon-11 (CFC-11). The agency also must distinguish between Class I and Class II substances, where Class I refers to those having a greater potential for damage. Then, for each substance class, phaseout schedules were outlined.

ozone depletion potential (ODP)
A numerical score that signifies a substance's potential for destroying stratospheric ozone relative to CFC-11.

With the exception of HCFCs, which are to be phased out by 2020, all ozone-depleting substances were phased out of production in or prior to 2005. However, because these substances have extended lifetimes, they affect the environment long after they have been produced. By way of illustration, a subset of ozone-depleting substances is listed in Table 13.2 with their ODP values and their atmospheric lifetimes.

TABLE 13.2 ODP Values of Selected Ozone-Depleting Substances

Substance	Atmospheric Lifetime in Years	ODP
Class I Substances		
CFC-11	45	1.0
CFC-12	100	1.0
CFC-113	85	0.8
CFC-114	300	1.0
CFC-115	1,700	0.6
Halon 1211	16	3.0
Halon 1301	65	10.0
Halon 2402	20	6.0
Carbon tetrachloride	26	1.1
Methyl chloroform	5	0.1
Class II Substances		
HCFC-22	12.0	0.055
HCFC-123	1.3	0.02
HCFC-124	5.8	0.022
HCFC-141b	9.3	0.11
HCFC-142b	17.9	0.065

NOTES: The ODP is a score signifying a substance's potential for destroying the ozone layer relative to CFC-11. The listed ODP values are those defined by the Montreal Protocol.

Source: U.S. EPA, Office of Air and Radiation, Office of Atmospheric Programs, Stratospheric Protection Division (August 19, 2010c).

Recognizing the industry's dependence on CFCs, the 1990 Amendments also include provisions to establish a mandatory national recycling program for Class I and II substances, with the intent that recycled refrigerants could be used as substitutes for virgin materials. Related provisions call for federal programs and research aimed at finding safe alternatives to identified ozone depleters. In addition, two policy instruments are legislated that explicitly use market incentives to eliminate ozone-depleting substances: an **excise tax** and a **marketable allowance system**.

excise tax on ozone depleters
An escalating tax on the production of ozone-depleting substances.

product charge
A fee added to the price of a pollution-generating product based on its quantity or some attribute responsible for pollution.

Excise Tax on Ozone Depleters

One market-based instrument used to control ozone depletion and achieve the phaseout deadlines is an escalating **excise tax** on the production of ozone-depleting substances. Enacted by Congress in 1990, the tax rate per pound is a base dollar amount multiplied by the chemical's ODP, where the base amount increases with each successive year in the phaseout schedule. The tax rate was initially set at $1.37 per pound, and by 1995, it had increased to $5.35 per pound. Starting in 1996, the tax was to increase by $0.45 per pound each year, bringing it to $13 per pound in 2012. Although the phaseout deadline has passed, the tax is still applicable to imported recycled CFCs.

From an economic perspective, the excise tax acts as a **product charge** on the ozone-depleting substance. Since production of ozone-depleters generates a negative externality, the tax can internalize this externality by elevating the producer's marginal

FIGURE 13.3 Excise Tax on Ozone-Depleting Substances

Since production of ozone-depleters generates a negative externality, an excise tax can internalize this externality by shifting up the producer's marginal private cost (*MPC*) curve. If the excise tax is set equal to the *MEC* at the efficient output level, Q_E, an efficient allocation of resources is achieved.

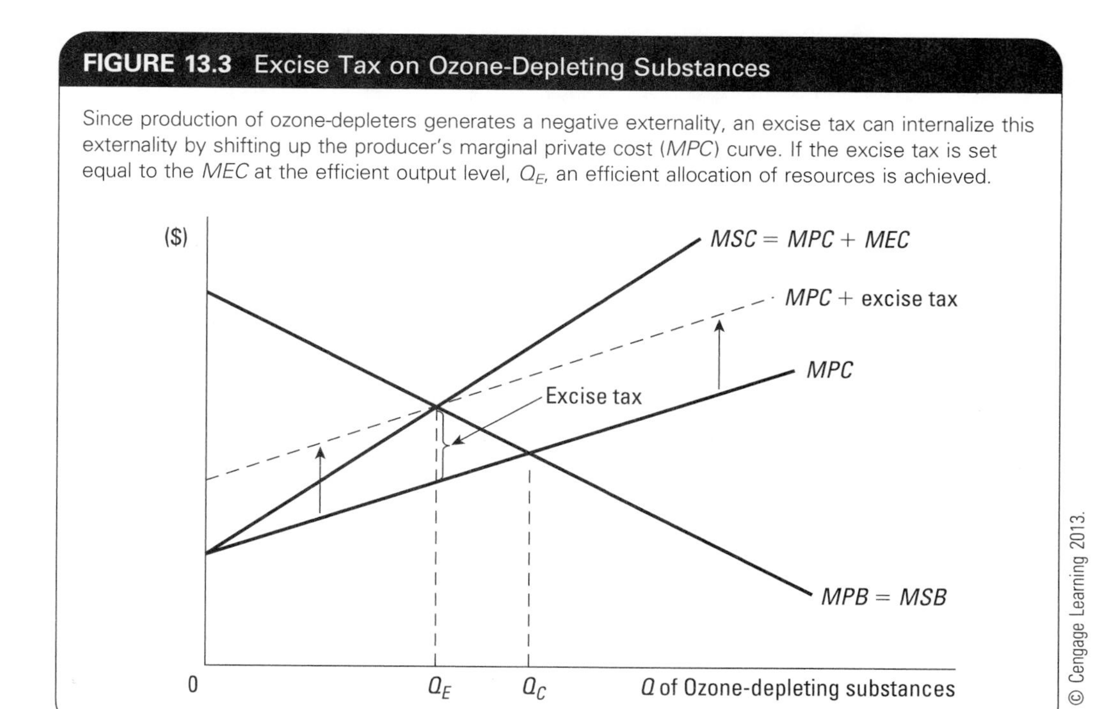

private cost (*MPC*). If, in fact, the excise tax is set equal to the *MEC* at the efficient output level, Q_E, an efficient allocation of resources is achieved. This is illustrated in Figure 13.3. Notice also that because the tax elevates the effective price of CFCs, it motivates a reduction in quantity demanded along the *MPB* curve. According to Cook (1996), consumption of ozone-depleting substances (expressed in CFC-11 equivalents) decreased from 318,000 metric tons in 1989 to 200,000 metric tons in 1990, the year the tax was put into effect.

allowance market for ozone-depleting chemicals Allows firms to produce or import ozone depleters if they hold an appropriate number of tradeable allowances.

Allowance Market for Ozone-Depleting Chemicals

The EPA established an **allowance market** to facilitate the phaseout of HCFCs. Implementation essentially follows the allowance program put in place prior to 1996 to control CFCs and certain other ozone-depleting substances. Firms will be allowed to produce or import these substances only if they hold an appropriate number of allowances. Each allowance will authorize a one-time release of one kilogram of an HCFC based on its ODP. The number of available allowances eventually will be brought to zero to meet the phaseout deadlines. For more information on this program, visit **www.epa.gov/ozone/title6/phaseout/index.html**.

ECONOMIC ANALYSIS OF OZONE DEPLETION POLICY

To understand the implications of ozone depletion policy, we can use economics to analyze key initiatives, such as the phaseout plan and the allowance trading, as well as to examine the influence of these initiatives on the market for CFCs and the market for

CFC substitutes. To begin, we review the Regulatory Impact Analysis (RIA) that was done to evaluate the phaseout plan.

Regulatory Impact Analysis (RIA) of the Phaseout

As part of its Regulatory Impact Analysis (RIA), the EPA conducted a benefit-cost study of the phaseout plan. Given the long life of ozone depleters, the agency considered the regulatory implications over a long time period, out to 2075. The agency's benefit assessment assigned a value to the damages that would be prevented by controlling these substances. These included health effects associated with increased exposure to ultraviolet radiation and nonhealth effects, such as reduced crop yields and rising sea levels. In total, the EPA estimated that accumulated damages would be approximately $6.5 trillion by 2075.[7]

On the cost side, a value had to be assigned to all anticipated market disruptions that would arise from a phaseout plan. Some 84 distinct-use categories for CFCs were analyzed—the two largest being mobile air conditioning and refrigeration. All told, the EPA's estimate of control costs associated with a phaseout plan was $27 billion through 2075.

Although the costs of the phaseout plan are significant, they pale in comparison to the dollar value of damages that would result if the United States took no action at all. Consequently, U.S. regulations to control ozone depleters were announced in August 1988, less than one year after the signing of the Montreal Protocol.[8]

Assessing Cost-Effectiveness

Following the 1978 ban of CFCs in all nonessential aerosols, the EPA began investigating the feasibility of further controls. In an EPA-commissioned study conducted by the Rand Corporation, three alternative control approaches were analyzed: a technology-based command-and-control approach, a fixed emission charge, and a tradeable emission permit system.[9] Each approach was modeled to achieve a given level of reductions over a 10-year period so that the accumulated costs of each plan could be compared. The study showed that the estimated costs for each approach were as follows:

- Technology-based command-and-control approach: $185.3 million
- Fixed emission charges: $107.8 million
- Tradeable emissions permit system: $ 94.7 million

These estimates support the expectation that allowance trading would approach a cost-effective solution. At the same time, trading should act as an incentive for the development of substitutes by firms that could do so at least cost. Hence, the phaseout in the CFC market had implications for other markets, in particular, the market for CFC substitutes, which can be analyzed using supply and demand and price movements.[10]

---

[7]Cogan (1988), p. 88.

[8]See *U.S. Federal Register* 53, 30598.

[9]Palmer et al. (1980).

[10]Of interest is the EPA's report on the supply and demand of CFC-12 in 1999 (U.S. EPA, Stratospheric Protection Division, Office of Air and Radiation, June 9, 1999).

Price Changes

Because U.S. policy was implemented through market-oriented instruments, its progress could be observed through price changes. Prices of CFCs and other ozone-depleting chemicals signaled the impact of the phaseout and the underlying market adjustments. As the phaseout plan advanced, availability of CFCs declined. In addition, an excise tax was levied on production, as discussed previously. These events are illustrated in Figure 13.4(a) by the shift leftward of the supply of CFCs, which in turn elevated their price. Manufacturers of CFC-dependent products faced higher production costs as a result and passed on at least some of this cost increase to consumers. Thus, buyers of commodities such as refrigerators and auto air-conditioning units paid higher prices over time.[11]

Also, as CFC prices increased, the demand for CFC substitutes rose, as shown in Figure 13.4(b). One potential consequence of these events is the evolution of a black market for CFCs. This developed in large part because the phaseout dates established by industrialized nations preceded those set by developing nations. The excise tax also may have contributed to this problem.

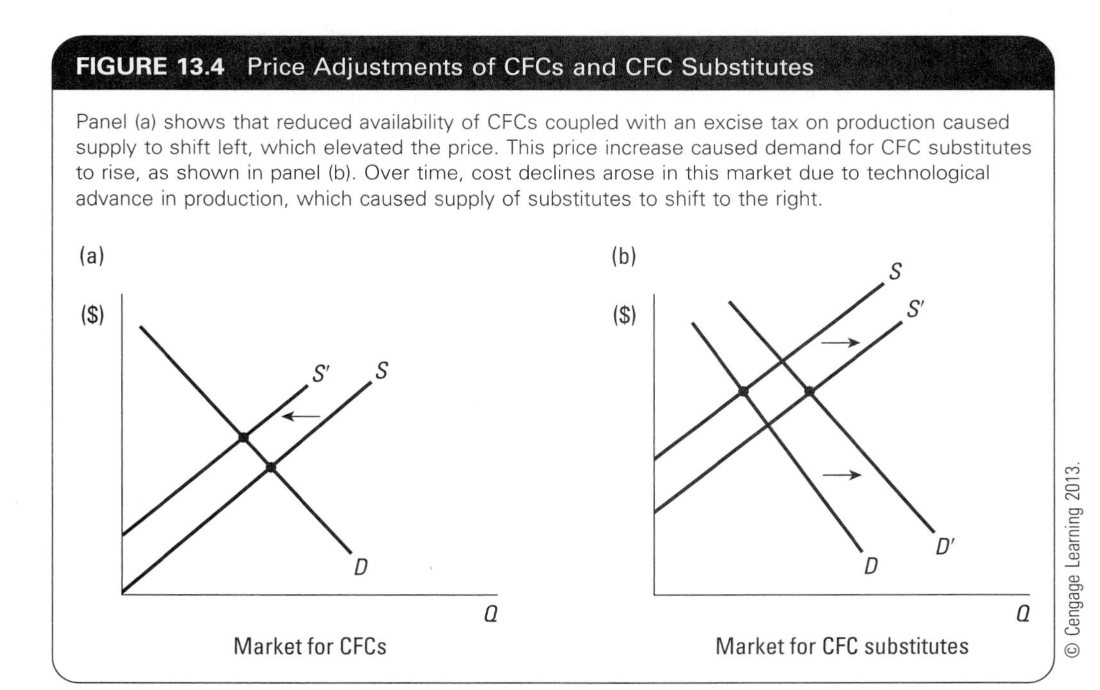

FIGURE 13.4 Price Adjustments of CFCs and CFC Substitutes

Panel (a) shows that reduced availability of CFCs coupled with an excise tax on production caused supply to shift left, which elevated the price. This price increase caused demand for CFC substitutes to rise, as shown in panel (b). Over time, cost declines arose in this market due to technological advance in production, which caused supply of substitutes to shift to the right.

[11]Of course, the ultimate change in price depended on the elasticity of demand for these products, which in turn depended on the availability of substitutes. The more substitutes there are for a product, the more elastic the demand is for that product. Because there are few good substitutes for refrigerators, we would expect demand to be price inelastic. Hence, much of the cost increase from the phaseout plan likely was passed on to refrigerator buyers.

Incentives and Disincentives to Develop CFC Substitutes

Because costs and prices were allowed to move naturally, the usual incentives encouraged a market adjustment to the observed industry declines and price changes. Theoretically, two opposing reactions were possible.

One possibility is that firms would perceive a profit advantage in developing ozone-friendly substitutes, since prices of these substances were relatively high at the outset. For example, the 1987 price of CFC-12, which was commonly used in automobile air-conditioning units, was about $0.50 per pound, whereas a substitute, HFC-134a, was $3 per pound.[12] However, this relative price difference would diminish over time as technology-driven cost declines would shift the supply of HFC-134a to the right, as shown in Figure 13.4(b), and as CFC prices continue to rise. Indeed, in 1999, the price of CFC-12 ranged between $25 and $30 per pound, while the price of HFC-134a remained stable, at about $3 per pound.[13] Such price stability likely reflects proportionate increases in supply and demand, as shown in Figure 13.4(b).

A second possibility is that the relatively few firms holding allowances possessed some measure of market power and price control. These firms would have enjoyed above-normal profits and would therefore not have been motivated to find alternatives to ozone-depleting substances. One solution would have been to transfer any excess profit to the government, which in turn could redistribute the windfall. Such a safeguard was implemented when Congress approved the escalating excise tax on ozone-depleting chemicals. A redistribution of income was achieved because fiscal spending was funded in part by the tax revenues collected from CFC producers.

The 1990 Amendments also called for a national recycling program for CFCs used in refrigeration and air conditioners. Consider the economics of this approach. By making recycled substances available in the market as a substitute, firms reduce their demand for virgin compounds needed to produce ozone-depleting products. Moreover, firms can use recycled materials beyond the phaseout deadlines, thus avoiding costly retrofitting until new substitute products are available.

Overall, the use of the tradeable allowance plan along with the excise tax, the recycling program, and the safe alternatives policy achieved the phaseout objectives in a more cost-effective manner. Such a control program was less disruptive than an immediate ban on production, which would have affected virtually every segment of society with no time to make proper adjustments.

Proposal to Phase Down Certain CFC Substitutes[14]

In 2009 a proposal was submitted by the Federated States of Micronesia and Mauritius to phasedown production and consumption of HFCs as an amendment to the Montreal Protocol. This proposal was subsequently supported by eight other countries. In 2010 and again in 2011, another proposal was made to phasedown HFCs, this time by Canada, Mexico, and the United States. A phasedown of HFCs, rather than a complete

[12]Putnam, Hayes, and Bartlett, Inc. (1987).

[13]Learner (2001).

[14]The following is drawn from U.S. EPA, Office of Air and Radiation, Office of Atmospheric Programs, Stratospheric Protection Division (May 20, 2011); Environmental Investigation Agency (November 2009).

phaseout, is being suggested because substitutes are not available for HFCs in all applications. For the status of these proposals, visit **http://ozone.unep.org/highlights.shtml**.

Why are countries looking to phase out a substitute for ozone-depleting substances? Because HFCs, while not damaging to the ozone layer, are greenhouse gases that contribute to climate change. This reality is of concern because of the rising use of HFCs as substitutes for ozone depleters and because of developing nations' increasing demand for HFC-dependent refrigeration and air conditioning. However, any phasedown of HFCs must not derail the final scheduled phaseout of HCFCs by 2020. (Refer back to Table 13.1 on page 293.)

As of 2012, the HFC phasedown proposals are still pending. If these or other similar proposals are accepted by the parties to the Montreal Protocol, the supply of HFCs would shift to the left, countering the influence of supply increases linked to technology advances, as noted previously. At the same time, the HFC phasedown should motivate the development of HFC alternatives, which in turn should influence the HFC market. To review a list of substitutes for ozone-depleting substances used in various industrial sectors, visit the EPA's Significant New Alternative Policy (SNAP) Program at **www.epa.gov/ozone/snap**.

CLIMATE CHANGE

climate change
A major alteration in a climate measure such as temperature, wind, or precipitation that is prolonged.

Climate change refers to a major alteration in a climate measure such as temperature, wind, or precipitation that is prolonged, i.e., lasts decades or longer. This type of response might be associated with a natural phenomenon like variances in sun intensity or alterations to oceanic circulation. It might also be linked to human activities, such as deforestation or the burning of fossil fuels.[15] To learn more, visit the EPA's site on climate change, **www.epa.gov/climatechange/**.

greenhouse gases (GHGs)
Gases collectively responsible for the absorption process that naturally warms the earth.

A source of controversy is the predicted climate response to the increasing production of what are termed **greenhouse gases (GHGs)**. On the agenda of the 1992 Rio Summit, the issue of accumulating greenhouse gases and the associated predictions of global temperature changes is one that continues to be discussed and debated. In fact, during 1997, the international community formulated the Kyoto Protocol, which continues the climate change initiative first discussed in Rio. The scientific community is not in complete agreement about climate change and its implications, although accumulating information and data have helped improved awareness and understanding. In any case, because of the associated complexity and uncertainty, national and international policy responses to climate change issues have been somewhat tentative.

Understanding the Potential Problem of Global Warming

global warming
Increased temperature of the earth's surface caused by accumulating GHGs that absorb the sun's radiation.

The phenomenon of **global warming** is based on the following scientific facts. Sunlight, passing through the atmosphere, hits the earth's surface and is radiated back into the atmosphere, where it is absorbed by naturally present gases such as carbon dioxide (CO_2). This absorption process heats the atmosphere and warms the earth's surface.

[15] U.S. EPA, Office of Air and Radiation, Office of Atmospheric Programs, Climate Change Division (April 14, 2011b).

This is somewhat like a greenhouse that allows sunlight through the glass but prevents the heated air from escaping back outside, thus the phrase "**greenhouse effect**." See Figure 13.5 for an illustration. This natural phenomenon is responsible for the existence of life on earth as we know it. In fact, without the so-called greenhouse gases, the earth's temperature would be some 30° to 40° Celsius cooler. Global warming falls under the broader heading of climate change, although the two phrases are often used interchangeably.

If this warming is a natural occurrence, what, then, is the problem? The issue is that GHG emissions have increased considerably over time, particularly CO_2. This trend is believed to be linked mainly to increasing human activity, such as combustion of fossil fuels (i.e., oil, coal, and natural gas) and deforestation. Because GHGs affect the earth's temperature, a significant disruption to their natural levels would generate climate changes. A landmark study conducted by the National Academy of Sciences in 1979 predicted that a doubling of CO_2 would generate a rise in the earth's temperature of 1.5° to 4.5° Celsius (or 2° to 8° Fahrenheit).[16] Climate changes in turn would alter agricultural regions, weather conditions, and sea levels. For more on the science of global warming, visit **http://epa.gov/climatechange/science/index.html**.

FIGURE 13.5 The Greenhouse Effect

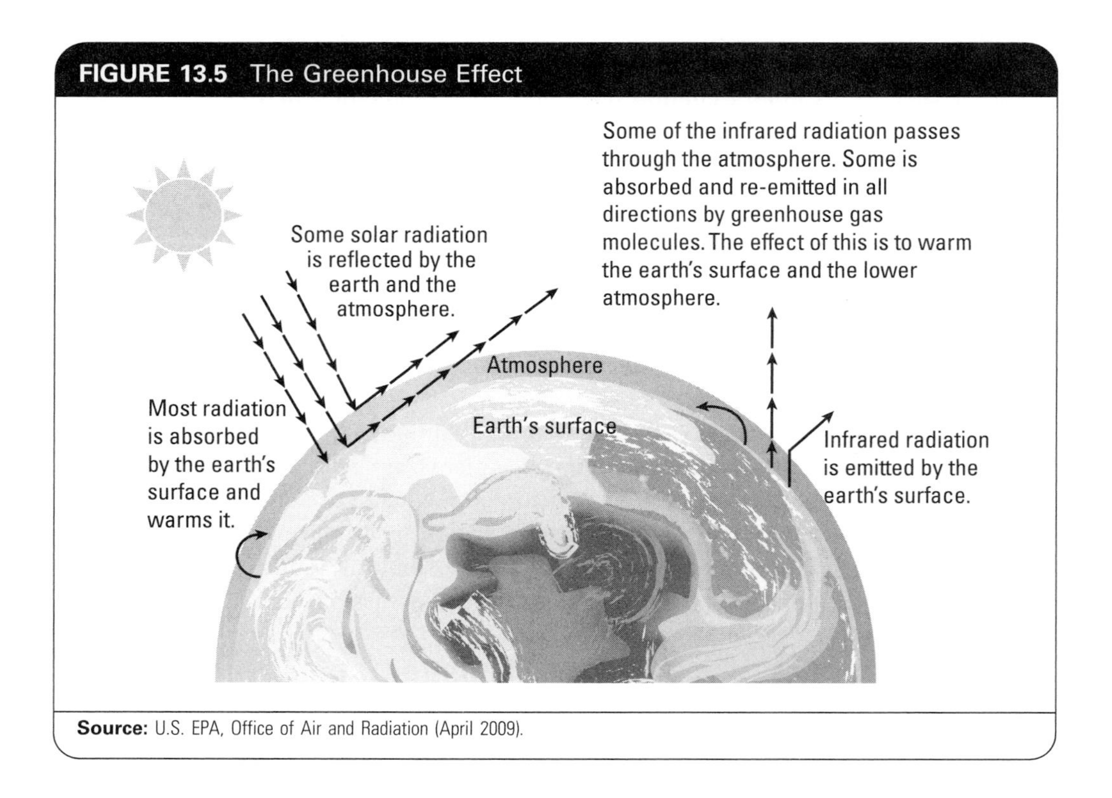
Some solar radiation is reflected by the earth and the atmosphere.
Some of the infrared radiation passes through the atmosphere. Some is absorbed and re-emitted in all directions by greenhouse gas molecules. The effect of this is to warm the earth's surface and the lower atmosphere.
Atmosphere
Earth's surface
Most radiation is absorbed by the earth's surface and warms it.
Infrared radiation is emitted by the earth's surface.

Source: U.S. EPA, Office of Air and Radiation (April 2009).

[16]Similar estimates have been found by more current analyses, such as those conducted by the Intergovernmental Panel on Climate Change (IPCC). See, for example, IPCC (2001).

Although a number of GHGs are responsible for this warming phenomenon, the primary ones are carbon dioxide (CO_2), methane (CH_4), and nitrous oxide (N_2O). Figure 13.6 gives estimates of their proportionate contribution to global warming. The relative contributions differ not only because these gases exist in varying amounts, but also because they have dissimilar capacities to absorb heat. The ability of a GHG to trap heat in the atmosphere is measured relative to CO_2 by its **global warming potential (GWP)**. Selected GWPs are shown in Table 13.3. For example, the GWP for methane is 21, which means that one unit of methane has 21 times the capacity for heat absorption as one unit of CO_2.

global warming potential (GWP) Measures the heat-absorbing capacity of a GHG relative to CO_2 over some time period.

Despite growing consensus that rising amounts of CO_2 will change the earth's climate, no one knows with certainty the timing or the extent of the outcome, in part because there are numerous factors to consider. Not the least of these is the influence of feedback effects that can either lessen or intensify the warming phenomenon.

FIGURE 13.6 U.S. Greenhouse Gas Emissions by Gas, 2009

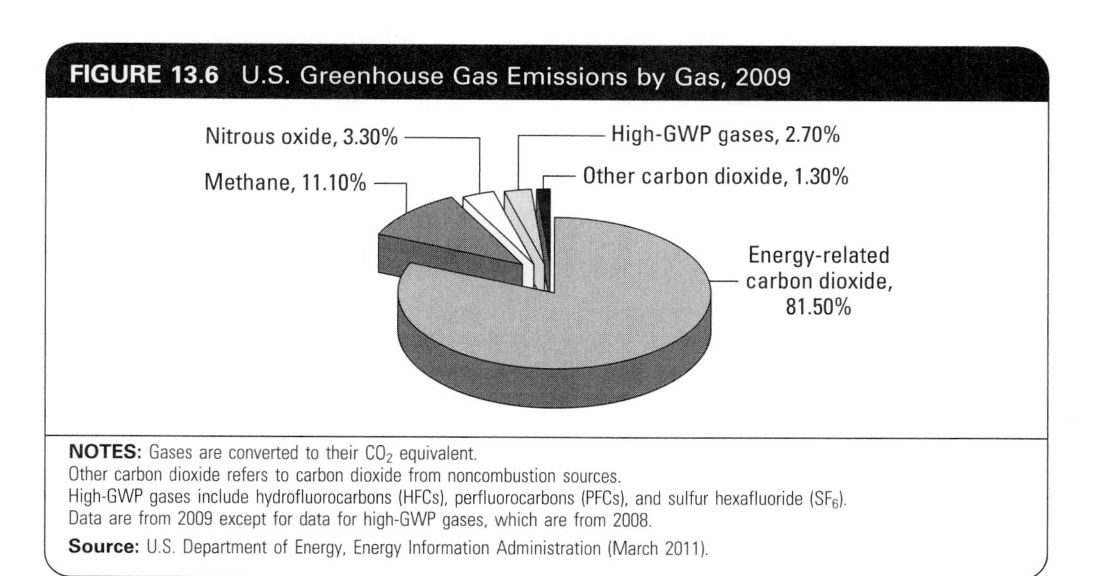

NOTES: Gases are converted to their CO_2 equivalent.
Other carbon dioxide refers to carbon dioxide from noncombustion sources.
High-GWP gases include hydrofluorocarbons (HFCs), perfluorocarbons (PFCs), and sulfur hexafluoride (SF_6).
Data are from 2009 except for data for high-GWP gases, which are from 2008.
Source: U.S. Department of Energy, Energy Information Administration (March 2011).

TABLE 13.3 GWP Values for Selected GHGs

Gas	GWP
Carbon dioxide (CO_2)	1
Methane (CH_4)	21
Nitrous oxide (N_2O)	310
Hydrofluorocarbon-125 (HFC-125)	2,800
Tetrofluoromethane (CF_4)	6,500
Perfluoroethane (C_2F_6)	9,200
Hydrofluorocarbon-23 (HFC-23)	11,700
Sulfur hexafluoride (SF_6)	23,900

Source: IPCC (2007b), Table 2.14.

carbon sinks
Natural absorbers of CO_2, such as forests and oceans.

For example, volcanic dust acts to filter the sun's warming rays, which would counter some of the influence of accumulating GHGs. Similarly, oceans and forests, which act as **carbon sinks**, are major absorbers of CO_2, although higher temperatures tend to diminish this capacity. Because of factors such as these and the overall inherent uncertainty, scientific research is ongoing in an attempt to settle at least some of the controversy.

Meanwhile, as society has become more aware of climate change issues, more attention is being drawn to carbon emissions and the influence of various products and activities on carbon levels in the atmosphere. Measuring this effect is referred to as **carbon footprinting**. Application 13.1 discusses this trend and how some firms are assessing and communicating the carbon footprints of their products.

Predicting the Potential Effects

National Assessment Synthesis Team

Because the science of global warming is itself the subject of some debate, it is not surprising that there has been a symmetric lack of substantive information on what the eventual outcome of accumulated GHGs might be. Nonetheless, through the use of computer simulation models, researchers have been able to assimilate some information about the expected implications. A comprehensive study was conducted as part of the U.S. Global Change Research Program, and the findings were published in a 2001 report by the National Assessment Synthesis Team. Estimates include predictions about forest ranges, biodiversity, the earth's sea levels, agricultural productivity, water and air quality, and health risks. A subsequent report was published in 2009, which integrates findings from the U.S. Global Change Research Program with research results from other nations around the world.[17]

Rising sea levels are considered to be one of the more probable outcomes of global warming, arising from thermal expansion of the earth's waters and melting of glaciers. Estimated U.S. changes cited in the report range from an increase of 13 centimeters to as much as 95 centimeters (or 5 to 37 inches) by 2100. According to one study cited in the 2009 report, a 3-foot rise in sea level would inundate nearly two-thirds of the marshlands and swamps in the coastal regions of the contiguous United States.[18] Other changes include substantial losses of U.S. coastal wetlands, regional flooding, and beach erosion.

On the plus side, there is some evidence to support the possibility of a beneficial fertilization effect from increased levels of CO_2, since it is a necessary component of photosynthesis. Furthermore, certain parts of the world such as Canada, Russia, and Northern Europe would profit from the northward shift of viable agricultural land that

[17]See National Assessment Synthesis Team, U.S. Global Change Research Program (2001), the source of much of this section. More than 20 syntheses and assessments on different topics were called for over a 4-year period, according to the 10-year strategic plan for the Climate Change Science Program (CCSP) launched in 2002 (U.S. Climate Change Science Program, July 2003). For the 2009 report, from which some information was drawn for this section, see Karl, Melillo, and Peterson (2009).
[18]Park, Trehan, Mausel, and Howe (1989).

APPLICATION 13.1 What Is a Carbon Footprint? Some Firms Are Finding Out

In recent years, society has become increasingly aware of climate change and global warming—perhaps even more so when the Nobel Peace Prize was awarded jointly to the Intergovernmental Panel on Climate Change (IPCC) and former U.S. Vice President Al Gore in 2007. Concerns about the potential risks of climate change prompted a grassroots interest in reducing emissions of GHGs, including CO_2. Along the way, the notion of a **carbon footprint** evolved as a way to communicate the contribution to carbon releases generated by an individual, a household, a product, or an activity.

As this trend has emerged, both public and private sectors have begun to respond. For example, many private entities, organizations, and government agencies offer an interactive carbon footprint calculator online. A few examples are:

- EPA's Household Emissions Calculator: **www.epa.gov/climatechange/emissions/ind_calculator.html**
- BP Energy Calculator: **www.bp.com/iframe.do?categoryId=9036038&contentId=7066737**
- Berkeley Institute of the Environment Carbon Footprint Calculator: **http://coolclimate.berkeley.edu/uscalc**

Most of these online tools estimate a household's GHG annual emissions based on user-provided information such as place of residence, vehicles driven, fuels used, and recycling practices. At face value, not all footprint calculators employ the same calculating method or use the same information, so the estimates are strictly that—estimates—as opposed to exact measures. Nonetheless, the intent is to educate society about energy use and to promote energy conservation and environmentalism.

Predictably, the automobile industry is a market in which carbon footprinting has become commonplace. As a point of reference, the EPA estimates that the average automobile is responsible for about 5.5 metric tons of carbon dioxide equivalent emissions per year, or about 55 metric tons over the 10-year life of the vehicle. Toyota's well-known hybrid, the Prius, has a carbon footprint of 44 metric tons, markedly less than the average, giving it a competitive edge in attracting environmentally conscious consumers. Annual carbon footprint estimates for other vehicles can be found at the U.S. Department of Energy's fuel economy Web site, **www.fueleconomy.gov**.

Other less-obvious products are also becoming the object of carbon footprinting. Examples include milk, fruit smoothies, shoes, detergent, and beer. Specifically, the New Belgium Beer Company finds that the carbon footprint of a six-pack of its Fat Tire Amber Ale is about 7 pounds, most of it attributable to refrigeration at the retail level. And a fruit smoothie made by a British firm, Innocent, bears a footprint of 294 grams, which even accounts for emissions linked to producing pesticides and fertilizers for the orchards.

Some firms have taken the idea to a higher level. The New Hampshire-based outdoor shoe manufacturer and retailer Timberland Company has assigned some 60 pairs of its shoes a numerical carbon rating from 0 to 10, which is marked inside the shoe. It announced plans to do the same for its entire product line by 2010. Zero represents less than 2.5 kilograms of carbon emissions, and 10 represents 100 kilograms, which is about equal to the emissions from an automobile driven over 200 miles. According to Timberland, the assigned value captures both a carbon footprint and the amount of chemicals used in production. With similar zeal, U.K. retailer Tesco PLC intends to give carbon labels to all the products it sells, ranging from food items to electronics. Its laundry detergent, for example, now bears a carbon footprint ranging from 1.3 to 1.9 pounds per load, depending on whether it is in liquid, powder, or solid form.

At this point, there is no standard method for computing a carbon footprint, a reality that has drawn criticism, particularly from climate change researchers who are keenly aware of the inherent complexities. Whether standardization comes to pass or whether the carbon footprint is more hype than good information remains to be seen. But for now, most agree that it is calling attention to GHG generation and raising awareness, which should be a positive to society.

Sources: U.S. EPA, Office of Transportation and Air Quality (February 13, 2009); Ball (October 6, 2008); Green and Capell (March 17, 2008).

may occur with a warming trend. Conversely, other areas, such as the southern United States, may suffer losses should this shift occur. Therefore, because some regions would gain while others lose, most global warming models predict that the net economic effect on agriculture would be relatively minor.

To read more about the specific forecasts of the National Assessment Synthesis Team, visit **www.globalchange.gov/what-we-do/assessment**.

Intergovernmental Panel on Climate Change (IPCC)

In 2007, the Intergovernmental Panel on Climate Change (IPCC) completed its Fourth Assessment Report (AR4) on the earth's climate.[19] The Fifth Assessment Report (AR5), comprising three Working Group Reports and a Synthesis Report, is in progress and should be completed between 2013 and 2014. The IPCC was established in 1988 by the World Meteorological Association and United Nations Environment Programme. The AR4, the fourth in a series of reports published every five or six years, is a compilation of in-depth research by three working groups: Working Group I on "The Physical Science Basis," Working Group II on "Impacts, Adaptation and Vulnerability," and Working Group III on "Mitigation of Climate Change."

The findings of this assessment are comprehensive, complex, and highly detailed, but at a high level, the chief conclusions of the IPCC are as follows:

- Global warming is unequivocal, based on observed increases in air and ocean temperatures, snow and ice melting, and rising sea levels.
- GHGs associated with human activity have risen by 70 percent between 1970 and 2004, and carbon dioxide (CO_2) increased by 80 percent during that period. Much of the observed rise in global average temperatures since the middle of the twentieth century is very likely linked to the observed rise in anthropogenic GHG concentrations.
- Based on current climate change policies and practices, GHG emissions are projected to grow over the next 30 years, and the effects on the climate are very likely to be greater than what was observed during the twentieth century.
- A broad range of adaptation options are available, but more are needed to lessen vulnerability to climate change. Also, there are barriers, costs, and limits that are not completely understood.
- Reasons for concern identified in the prior IPCC report remain important to defining vulnerabilities, and some are assessed as stronger than originally believed.

Detail on any of these findings as well as the full reports from the three working groups are available at the IPCC's Web site, **www.ipcc.ch/**.

Taken together, scientific predictions about the effects of global warming are not conclusive. Given the complexity of the science as well as the long-range nature of the forecasts, it is understandable that there is not complete agreement about which of the conjectured events may occur, the degree of impact, and the timing of any associated outcome. However, scientists are achieving greater clarity, and the research continues. In the interim, policymakers must decide how to respond to rising CO_2 levels and other GHGs, even in the face of uncertainty about the implications.

POLICY RESPONSE TO CLIMATE CHANGE

Setting policy in response to climate change is a difficult problem for two reasons. First, as discussed, global warming and other climate variations are complex issues, and there is some degree of uncertainty associated with them. Second, because both the source of

[19]See IPCC (2008), from which this section is drawn, and IPCC (November 2010) for information about AR5.

the problem and the predicted effects are global in scope, any effective policy solution relies on international agreement. Many nations contribute significantly to the aggregate level of GHG emissions. The top 12 national emitters of the primary GHG, CO_2, are shown in Figure 13.7.

International Response

U.N. Framework Convention on Climate Change (UNFCCC)

At the 1992 Rio Summit, global climate change was an important agenda item for the many national representatives who gathered at the 12-day worldwide conference. Among the major agreements produced at the summit was the U.N. Framework Convention on Climate Change (UNFCCC), which deals with global warming and other air quality issues. Among its major provisions are:

- Countries must implement national strategies to limit GHG emissions with the objective of reducing emissions to their 1990 levels by 2000.
- Differences in political and economic conditions among nations are to be accommodated by avoiding uniform emission targets and timetables for only one GHG.
- Signatories are encouraged to recognize climate change in the formulation of economic, social, and environmental policies.
- Industrialized nations will assist developing countries in obtaining data and in limiting emissions.

The UNFCCC was to become effective following ratification by 50 nations. In October 1992, the United States became the first industrialized country, and the fourth

FIGURE 13.7 Top 12 National Emitters of Carbon Dioxide in 2009

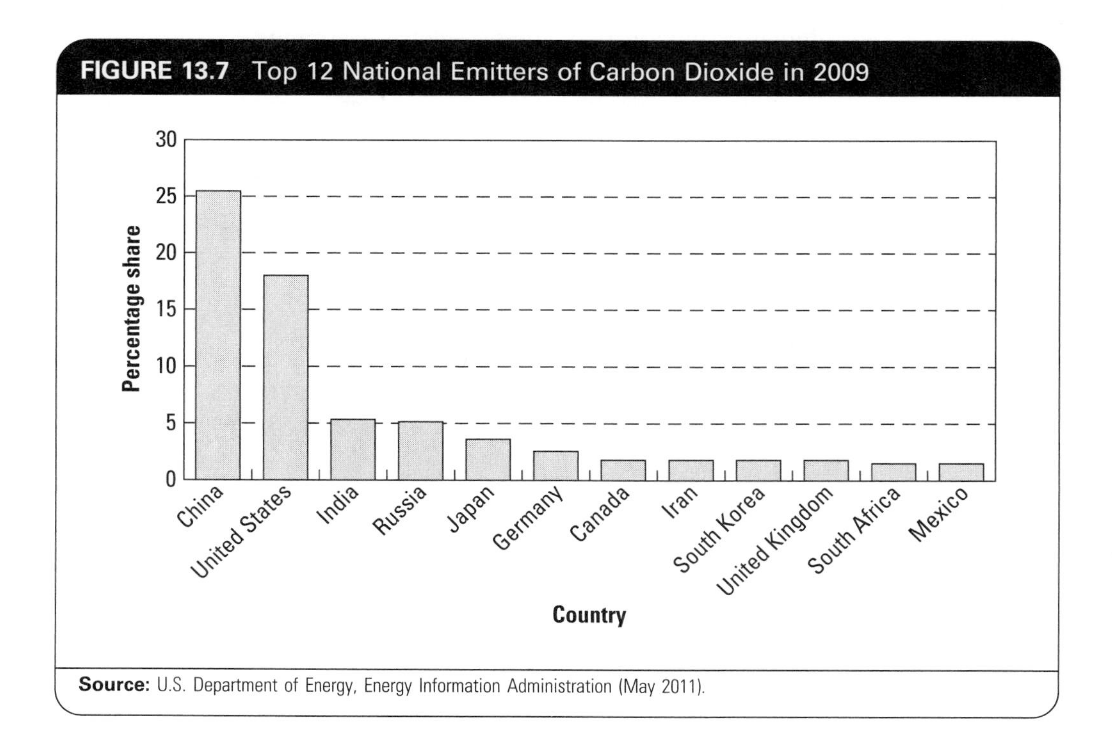

Source: U.S. Department of Energy, Energy Information Administration (May 2011).

overall, to do so, following a unanimous vote by the U.S. Senate.[20] By the close of 1993, the requisite number of nations had ratified the treaty, and it became legally binding in March 1994.

Kyoto Protocol to the UNFCCC

In December 1997, a Conference of the Parties (COP) was held in Kyoto, Japan. The goal was to reach an agreement, or protocol, that would address the issue of GHG emissions beyond 2000. A key outcome was the establishment of emissions limits for developed nations. These were to become effective once 55 nations ratified the protocol as long as these nations included developed countries responsible for at least 55 percent of CO_2 emissions for 1990. Achieving these limits was to be accomplished in part through emissions trading.

Further details of the international accord, including its market-based approach, were discussed during a 1998 meeting in Buenos Aires. While the meeting was in progress, the United States signed the protocol, though the signing was more symbolic than substantive. The protocol was not submitted to the Senate for ratification, because there was strong opposition to the treaty among senators. In addition to apprehensions about limited scientific evidence of global warming, the Senate also was concerned that the stated emissions limits would negatively affect industry and the American economy. Furthermore, the United States would not ratify the treaty until developing countries made a commitment to meet binding emissions targets along with developed nations.

Discussions, often contentious, continued at a series of COPs. Before the 2001 conference, President Bush announced his opposition to the Kyoto Protocol and took the United States out of the agreement in March 2001, a controversial decision discussed in Application 13.2. At the same time, the Bush administration promised to deliver a domestic initiative to combat global warming. Moreover, the United States would remain involved in negotiations and funding for the UNFCCC.

Ultimately, in July 2001, 178 nations reached an agreement—without the United States. The accord called for 38 industrialized countries to cut their GHG emissions to 5.2 percent below 1990 levels by 2012, leaving developing countries without any emissions requirements. Emissions targets were to be achieved during the so-called commitment phase from 2008 to 2012, using several market-based instruments, referred to as **flexible mechanisms**. Chief among these is a cap-and-trade system of GHG allowances for participating developed nations. These countries also could earn credits for carbon-absorbing forestry practices and for implementing emissions-reducing projects in other nations. But the agreement still had to be ratified.

Many believed that ratification of the accord was unlikely without the United States because of the requirement that participating developed nations represent at least 55 percent of 1990 CO_2 emissions. The United States alone was responsible for 36 percent of 1990 CO_2 emissions. Essentially, this meant that Russia, the second largest CO_2 emitter in the world, held the deciding vote. In 2004, Russia ratified the protocol, bringing the number of participants to 141, and in February 2005, the much-debated Kyoto Protocol

[20]Council on Environmental Quality (January 1993), pp. 142–43.

APPLICATION 13.2 Why the United States Is Not Participating in the Kyoto Protocol

In November 2001, the parties to the United Nations Framework Convention on Climate Change (UNFCCC) met in Marrakech, Morocco, to finalize the language and operational elements of the Kyoto Protocol. In attendance were 171 national governments and 4,500 participants. The protocol would become legally binding once ratified by at least 55 industrialized nations that account for over 55 percent of 1990 GHG emissions. Ratification was achieved in 2004 following Russia's signing of the accord, which brought the total number of participating nations to 141. Interestingly, the United States was not one of them. Why? Because in March 2001, President Bush declared the Kyoto Protocol to be "... fatally flawed in fundamental ways" and proceeded to take the United States out of the accord.

The Bush administration's position was based on the initial findings of a cabinet-level climate-change working group established by the president. In its analysis of the Kyoto Protocol, the group expressed concern that the agreement would not impose emissions cuts on developing countries, even though the net emissions from those nations exceed those of industrialized countries. Another reservation cited was that the treaty's emissions targets were based on political negotiations rather than on science.

There were also economic concerns. The protocol's emissions reduction target for the United States had been set at 7 percent from 1990 levels for each year in the 2008–2012 period. However, this target ignored emissions growth between 1990 and 2012, which, if considered, would translate to greater than a 30 percent reduction for the period. Achieving this objective would have had serious economic consequences, according to the analysis. In fact, most models predicted that U.S. GDP would have declined by 1 to 2 percent as a result.

Beyond the White House report, independent economists had used models to assess and quantify the protocol's implications for the U.S. economy. For example, Richard Schmalensee, an economist and dean emeritus at MIT's Sloan School of Management, forecasted that the United States would have had to shut down all its coal-fired facilities by 2012 just to meet the Kyoto targets halfway. In another study, analysts at MIT's Joint Program on the Science and Policy of Global Change estimated implementation costs of the Kyoto treaty on a per-household basis. Their predictions ranged from a high of $1,000 to a low of $140, with the latter assuming efficient implementation of emissions trading.

In lieu of participating in the Kyoto Protocol, the United States opted to pursue independent actions, including researching climate change and implementing domestic and international strategies aimed at reducing GHG emissions. On the research front, the Climate Change Research Initiative (CCRI) was created to facilitate the integration of scientific understanding of climate change, including areas of uncertainty, into policy decisions. Strategic initiatives included developing government-business partnerships to promote source reduction and recycling, promoting energy efficiency in government buildings, developing cleaner technologies for electricity generation and transmission, and supporting the development of fuel-efficient vehicles and renewable resources such as solar energy and hydropower.

Sources: Hilsenrath (August 7, 2001); Revkin (July 24, 2001); Fialka and Winestock (July 16, 2001); U.S. Government, White House (June 2001).

entered into force.[21] As of 2012, over 190 nations have ratified the protocol. Its key features, including the flexible mechanisms, are described in Table 13.4.

United Nations Climate Change Conferences[22]

Mexico was the host nation for the 16th Conference of the Parties (COP) held in Cancún in November–December 2010. The parties had a full agenda, elements of which were brought forward from the COP15 held in Copenhagen in 2009 and the agreements collectively known as the **Copenhagen Accord**. Included in that accord were pledges by individual

[21]The ratification status of the Kyoto Protocol can be viewed online at **http://unfccc.int/kyoto_protocol/status_of_ratification/items/2613.php**.

[22]The following is drawn from UNFCCC (May 23, 2011; March 15, 2011; January 2009); Aguilar (2011); Ball (December 3, 2007).

TABLE 13.4 Major Elements of the Kyoto Protocol to the UNFCCC

- Developed nations must reduce overall greenhouse gas (GHG) emissions by at least 5.2 percent below 1990 levels in the first commitment period, 2008–2012, by meeting individual emissions targets assigned to each country. Negotiations for subsequent commitment periods will follow.
- Emissions targets cover the primary GHGs: carbon dioxide (CO_2), methane (CH_4), nitrous oxide (N_2O), hydrofluorocarbons (HFCs), perfluorocarbons (PFCs), and sulfur hexafluoride (SF_6).
- Emissions targets are to be achieved by international emissions trading of GHG allowances.
- Carbon-absorbing activities, such as reforestation, can be used to offset emissions targets. Any GHGs removed through such activities earn credits called **removal units (RMUs)**.
- Under the *Clean Development Mechanism* (*CDM*), a developed nation can enter into emissions-reducing projects in a developing nation and use the resulting **certified emissions reductions (CERs)** to meet its own emissions targets.
- Using *Joint Implementation*, a developed nation can implement an emissions-reducing project in another developed nation and use the resulting **emissions reduction units (ERUs)** against its own emissions target.
- Participants failing to meet their emissions targets must resolve the difference in the second commitment period plus a 30 percent penalty.

NOTE: The full text of the Kyoto Protocol is available online at **http://unfccc.int/resource/docs/convkp/kpeng.pdf**.

member nations to reduce GHG emissions, but, unlike Kyoto's, these were not binding. The overall consensus was that the COP15 did not meet expectations.

The lack of progress in Copenhagen was problematic because attendees understood that Kyoto's first commitment phase would officially expire in 2012, and hence a post-2012 strategy was needed. Most understood that replacing or extending the Kyoto Protocol would not be easy, given the large number of countries involved and their differences in terms of their economic development, access to economic resources, governance structure, and positions on climate change.

In Mexico at the COP16, the member nations met again, and they secured another set of decisions, this time called the **Cancún Agreements**. Among these was a long-term goal to reduce GHG emissions to assure that the rise in average global temperature would be less than 2° Celsius above preindustrial levels. The member nations also agreed to establish the Green Climate Fund, originally announced in Copenhagen, to support developing nations' projects, policies, and other activities. A Technology Mechanism also was launched to support innovations that would facilitate GHG mitigation and to transfer needed technologies to developing countries.

Notably absent from the Cancún Agreements was the establishment of binding GHG reduction commitments to replace those stated in the soon-to-expire Kyoto Protocol. Achieving this goal is likely to be difficult given that neither China nor the United States had been subject to Kyoto's emissions caps—China, because it is a developing nation, and the United States, because it did not ratify the treaty. Yet the two nations are the largest emitters of GHGs, jointly responsible for over 40 percent of the world's releases. (Refer back to Figure 13.7 on page 305.) Consequently, industries within participating developed countries must face the potential disadvantage of facing emissions limits that their competitors within two world powers do not.

These and other issues formed the agenda for the COP17, which took place in Durban, South Africa, at the end of 2011. Following lengthy negotiations, the parties reached an agreement, the **Durban Platform**, to extend the Kyoto Protocol and to establish a successor agreement by 2015, which must take effect by 2020. Unlike the original Kyoto pact, the new protocol is expected to be applicable to both developed and developing nations. The participating nations also formally implemented the Green Climate Fund, which had been discussed at both the Copenhagen and Cancún meetings, to support the efforts of developing nations.[23] For updates and further information on the COP17 and other UNFCCC meetings, visit **http://unfccc.int/2860.php**.

European Union (EU) Response

To align with its commitment under the Kyoto Protocol, the European Union (EU) launched its own GHG trading program in 2005 for member states. This system, called the **European Union Emissions Trading Scheme (EU ETS)**, is a true cap-and-trade program designed to help the EU achieve its emissions reduction commitment under the Kyoto Protocol of 8 percent below 1990 levels for the 2008–2012 period. As such, it substantiates the claim that Europe is the only world power to establish and use a carbon market. The EU ETS has been implemented in two trading phases, with the second trading phase aligned with the first commitment phase of the Kyoto accord. Exchanges located all over Europe facilitate trade among the member countries, and allowance prices are market-determined. Allowances are not printed but instead are exchanged through electronic registries.

In the first phase of trading (i.e., between 2005 and 2007), too many emission allowances were allocated relative to actual emissions released throughout the EU. The result is rooted in fundamental economics. The excess supply of allowances drove down the price, in this case to less than €1 per ton of CO_2 emissions in 2007. Implementing mid-course corrections, the European Commission placed stringent limits on allowances during the second trading period (i.e., between 2008 and 2012), targeting the number to 6 percent below the emission releases observed during the first trading period. As a consequence, allowance prices held to between €19 ($26.77) and €29 ($40.86) in 2008.[24] Figure 13.8 illustrates the EU ETS allowance price fluctuations over the 2005–2009 period. According to the IPCC, prices ranging between $20 and $80 per ton of CO_2-equivalent should achieve stabilization of these emissions at about 550 parts per million (ppm) by 2100.[25]

As a complement to the EU ETS program, the European Union has proposed a **carbon tax** for all EU members. A carbon tax is a per-unit charge based on the carbon content of fuel. Introduced in 2011 by the European Commissioner, Algirdas Šemeta, this EU-wide market-based policy would require all member states to set a tax on CO_2 at a minimum rate of €20 ($29) per metric ton of CO_2 released from the combustion of such fuels as gasoline, natural gas, and coal. The proposal suggests a start date of 2013. Officials believe that the carbon tax is necessary in order to achieve the emissions reduction target set for 2020, which is 20 percent below 1990 levels. If enacted, the new

[23]Center for Climate and Energy Solutions (C2ES) (2011); Whiteman (2011).

[24]European Environment Agency (2008).

[25]IPCC (2008), p. 59. CO_2-equivalent represents the emissions of a GHG adjusted by its GWP to be in terms of CO_2.

FIGURE 13.8 EU ETS Allowance Prices: 2005–2009

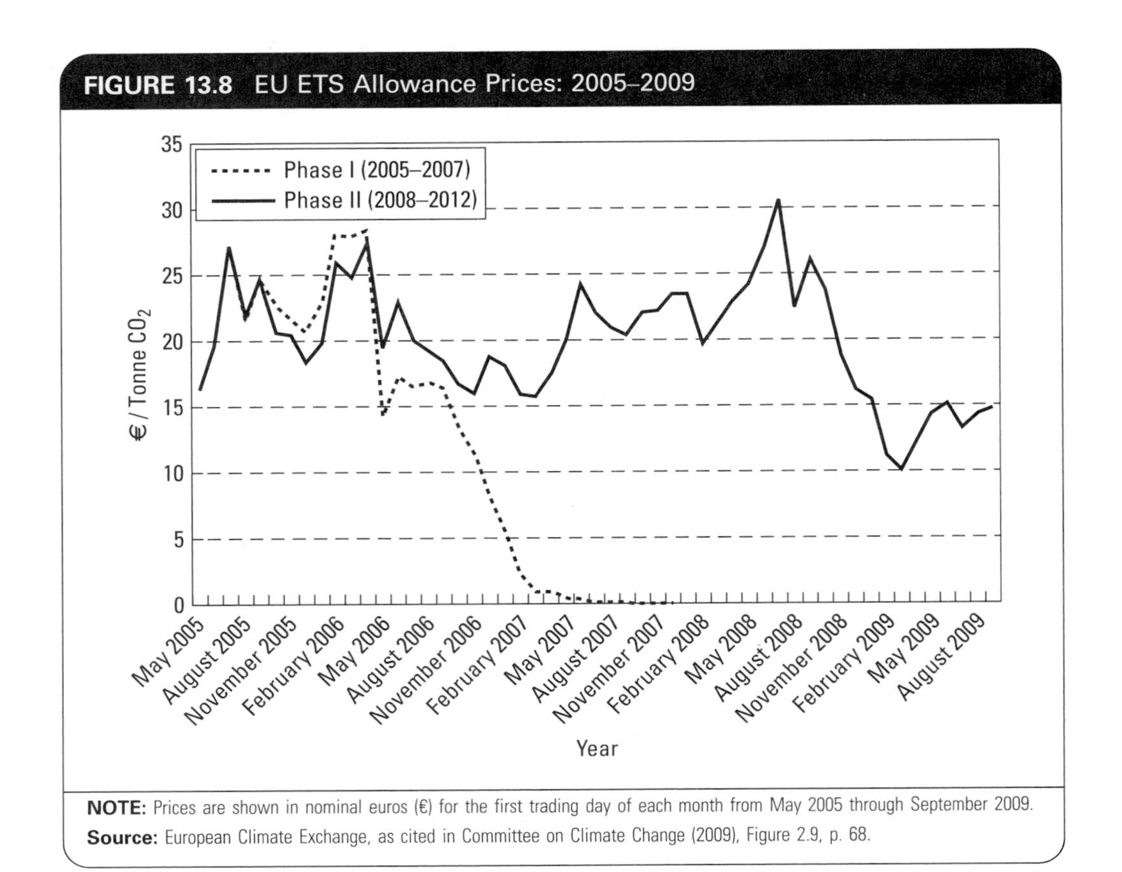

NOTE: Prices are shown in nominal euros (€) for the first trading day of each month from May 2005 through September 2009.
Source: European Climate Exchange, as cited in Committee on Climate Change (2009), Figure 2.9, p. 68.

tax would not be applicable to power plants and other facilities that are actively trading carbon within the EU ETS.[26]

Some member nations already have carbon or other energy taxes in place, including Finland, the first country to implement a carbon tax; Belgium; Denmark; Italy; Norway; and Sweden.[27] Nonetheless, opposition by some countries is anticipated, particularly those that wish to maintain national control over taxation decisions. Moreover, this matter has been debated for years with no resolution. To read more about the proposal, visit the European Commission Taxation and Customs Union at **http://ec.europa.eu/taxation_customs/taxation/excise_duties/energy_products/legislation/index_en.htm**.

U.S. National Response

After removing the United States from the Kyoto agreement in 2001, President Bush called for a cabinet-level review of the nation's policy on climate change, and a climate change working group was formed. The group in turn requested a report from the National Academy of Sciences (NAS) on what was known about the science of climate

[26]Rankin (2011); Torello (2011).

[27]U.S. EPA, Office of Policy, Economics, and Innovation (November 2004).

change. Initial findings were presented in a report that included an analysis of the Kyoto Protocol, proposals to advance needed technologies, plans to establish international partnerships, and an assessment of existing domestic strategies, including those promoting voluntarism through incentives. An example is the Energy Star® program discussed in Application 13.3, which the working group recommended expanding in scope to improve energy efficiency.

Both the working group's report and the NAS report were released in June 2001. President Bush took the reports under advisement and formulated his climate change plan to act as the U.S. alternative to the Kyoto Protocol.[28] According to this plan, the overall objective for the nation was to reduce **greenhouse gas (GHG) intensity** by 18 percent by 2012, where GHG intensity refers to the ratio of GHG emissions to economic output.

greenhouse gas (GHG) intensity The ratio of GHG emissions to economic output.

The Obama administration has taken a different position toward climate change than the previous administration. On an international scale, President Obama committed the United States to a 17 percent reduction in GHG emissions from 2005 levels by 2020 as part of the Copenhagen Accord and the subsequent Cancún Agreements, although again this pledge is nonbinding. Domestically, the commitment to decrease

APPLICATION 13.3 The Energy Star® Program

As a way to promote energy efficiency through collaboration between the public and private sectors, the EPA launched the Energy Star® Program in 1992. The initiative is a voluntary labeling program that identifies energy-efficient products and allows them to bear the Energy Star® logo. This logo acts as an explicit market signal that communicates to prospective consumers the potential cost savings and the emissions reduction associated with more efficient energy use. In 1996, Energy Star® teamed up with the U.S. Department of Energy, with each entity assuming responsibility for certain product categories.

The now-familiar logo first appeared on qualifying computers and monitors. By 1995, it had been made available to other types of office equipment, such as fax machines and copiers, as well as to heating and cooling products for private residences. Today, the label appears on over 60 product categories, including lighting, major appliances, windows, consumer electronics, even new homes and commercial buildings. Over 8,400 contractors have built nearly 1.2 million Energy Star® qualified new homes.

To communicate information about energy-efficient products, Energy Star® has set up partnerships with more than 20,000 private and public organizations. These alliances also are used to provide technical assistance aimed at helping consumers and organizations make energy-efficient decisions. The program touts that these efforts generate cost savings of over $18 billion a year on utility bills along with important energy and air quality gains. In 2010, the avoided GHG emissions associated with the program were reportedly equivalent to taking 33 million cars off America's highways.

In addition to the explicit cost savings and the environmental gains of partnering with Energy Star®, participating businesses might observe increased demand for their energy-saving products, which translates directly to revenue gains. Increased demand can arise directly from the positive image linked to the logo and from national Energy Star® performance awards that receive some measure of media attention. Visit the Energy Star® Web site at **www.energystar.gov** for more information.

Sources: U.S. EPA and U.S. Department of Energy (2011; 2005); U.S. EPA, Office of Air and Radiation (October 2008).

[28]This plan was presented as the *Global Climate Change Policy Book*. See U.S. Government, White House (February 2002).

GHG releases by 17 percent has been included in various bills submitted through Congress, some of which called for a cap-and-trade approach to achieve this objective, but none became law.[29] Formal reports on U.S. climate change progress and activities are prepared periodically in accordance with the articles of the UNFCCC and are available from the U.S. Department of State.[30]

Despite the failure to legislate a national GHG emissions limit or launch a cap-and-trade program, other regulations for both mobile and stationary sources have been put in place, or are under development, to address climate change. Before examining these, let's begin with a brief retrospective on how these rulings came about.

Court Rulings and EPA Findings on Controlling GHGs

A contentious Supreme Court decision in April 2007 during the Bush administration found that the EPA is authorized to regulate GHGs, including CO_2, under the Clean Air Act, provided that the agency determined that GHGs posed a threat. The court's 5-to-4 decision was diametrically opposed to the Bush administration's long-held position that the Clean Air Act Amendments (CAAA) provide no authority to regulate CO_2 and other GHGs. In any case, the Court's ruling placed the onus on the EPA to respond, either with new regulations or with scientific evidence that a regulatory response is unfounded.[31]

In April 2009, the EPA announced two critical findings: (1) the Endangerment Finding: that six GHGs—carbon dioxide (CO_2), methane (CH_4), nitrous oxide (N_2O), hydrofluorocarbons (HFCs), perfluorocarbons (PFCs), and sulfur hexafluoride (SF_6)—pose a threat to public health and welfare of both current and future generations; and (2) the Cause or Contribute Finding: that CO_2, CH_4, N_2O, and HFC emissions from motor vehicles add to concentrations of these GHGs in the atmosphere and therefore contribute to climate change.[32] For more information on this pair of landmark decisions, visit **www.epa.gov/climatechange/endangerment.html**.

California's Role in National GHG Regulations

The Court's decision and the EPA's findings on GHGs were linked directly to efforts by California to implement regulations the state had passed previously to control CO_2 emissions from motor vehicles. Automakers had opposed this state law in part because CO_2 had not been an identified pollutant under the CAAA. Once the Supreme Court ruled that the EPA has authority to control these emissions, California's position was strengthened. Under the CAAA, California is permitted to

[29]Among these bills was the commonly referenced Waxman-Markey Bill, or the American Clean Energy and Security Act, which was passed in the House in June 2009. See Eilperin (2010) and U.S. Department of State (2010).

[30]The most recent report, *U.S. Climate Action Report—2010: Fifth National Communication of the United States of America*, can be accessed online at **http://unfccc.int/resource/docs/natc/usa_nc5.pdf**.

[31]Greenhouse (April 3, 2007). The official case is filed as *Massachusetts et al. v. EPA et al.*, No. 05-1120.

[32]U.S. EPA, Office of Air and Radiation, Office of Atmospheric Programs, Climate Change Division (April 17, 2009).

adopt tougher pollution standards than those set at the federal level (and other states can follow suit if they wish), but it must receive a waiver from the EPA to do so. It had requested the waiver in December 2005 so that the CO_2 emission controls could take effect with the 2009 model year. Despite the Court's ruling on the EPA's power to control GHGs, the EPA blocked California's progress, denying the waiver in early 2008. California requested a reconsideration of the decision, and once President Obama took office, he urged the EPA to do so. In June 2009, the agency did grant the waiver.[33] For more information, visit **www.epa.gov/otaq/climate/ca-waiver.htm**.

Neither the granting of California's waiver request nor the two EPA findings on GHGs imposed any new national regulations on industry. However, as outlined in Chapters 11 and 12, respectively, these outcomes did set the stage for new regulations on mobile sources and stationary sources.

U.S. GHG Controls on Mobile Sources[34]

In 2010, the EPA and the National Highway Traffic Safety Administration (NHSTA) finalized a first phase of precedence-setting GHG emissions standards and Corporate Average Fuel Economy (CAFE) standards for passenger cars and light-duty vehicles for model years 2012–2016, which took effect in January 2011. The new emissions standards require this class of vehicles, which is responsible for about 60 percent of the GHGs released by the transportation sector, to meet a combined average emissions level of 250 grams of carbon dioxide (CO_2) per mile in model year 2016. This emissions level is equivalent to achieving 35.5 miles per gallon (mpg), if all CO_2 reductions had to be met through fuel efficiency gains. Complementing this initiative are new CAFE standards, calling for light-duty vehicles to achieve a combined average of 34.1 mpg in the 2016 model year.

A second phase of standards for this class of vehicles was proposed by the EPA and the NHTSA in 2011. These are to be applicable to the 2017–2025 model years and will become increasingly stringent over time. Once in place, the projected estimated gains are a reduction in GHGs of 2 billion metric tons and 4 billion barrels of oil over the lifetimes of the affected vehicles.

Also in 2011, the two agencies announced an analogous initiative, called the HD National Program, for medium- and heavy-duty vehicles, which are responsible for 20 percent of GHGs released by the transportation sector. Standards for CO_2 emissions and fuel consumption applicable to the 2014–2018 model years of this vehicle class are expected to save approximately 530 million barrels of oil and reduce 270 metric tons of CO_2 emissions over the life of the affected vehicles. Also under consideration is a second phase of rules for this class of vehicles, starting with the 2018 model year.

---

[33]U.S. EPA, Office of Transportation and Air Quality (February 17, 2009); McCarthy (July 28, 2008); "Air Pressure; Global Warming" (April 7, 2007).

[34]The following is drawn from Pew Center on Global Climate Change (2011); U.S. EPA, Office of Transportation and Air Quality (August 2011; July 2011; April 2010). Also, refer back to Chapter 11.

U.S. GHG Controls on Stationary Sources[35]

Under rulings within the Clean Air Act, the new GHG controls on mobile sources triggered analogous regulations for stationary sources. Hence, permitting requirements and emissions standards have been proposed to control GHG releases from stationary sources. The permits are to be issued under the New Source Review (NSR) requirements and are to dovetail with the new mobile source controls. The new emissions standards are aimed at fossil-fuel power plants and petroleum refineries, which are responsible for about 40 percent of all U.S. GHG releases. Officials expect to issue these new standards in 2012.

U.S. Regional Response

In the United States, several regional initiatives have evolved that are aimed at mitigating climate change. Operating on a multistate level can generate scale economies, avoid redundancies, and even create synergies, all of which can lead to better results than a single state can accomplish on its own. According to a published report by the Pew Center on Global Climate Change (January 2011), five regional climate initiatives have been formed—Regional Greenhouse Gas Initiative (RGGI), Transportation and Climate Initiative (TCI), Western Climate Initiative (WCI), Midwest Energy Security and Climate Stewardship Platform, and Midwestern Greenhouse Gas Reduction Accord. Some of these involve Canadian provinces, and a few have established a cap-and-trade program to reduce GHG emissions.[36]

To provide some insight as to how these programs operate, we offer an overview of two that have cap-and-trade components: the RGGI, which is the seminal regional program, and the WCI, which is a collaborative of not only several states but also four Canadian provinces.

Regional Greenhouse Gas Initiative (RGGI)[37]

Developed in the Northeast and Mid-Atlantic regions of the United States, the **Regional Greenhouse Gas Initiative (RGGI)** involves nine states (Connecticut, Delaware, Massachusetts, Maryland, Maine, New Hampshire, New York, Rhode Island, and Vermont), committed to reducing GHG emissions. To implement its objective, all states participate in a mandatory cap-and-trade GHG program, which was the first of its kind in the nation. RGGI participants set a multistate cap on CO_2 emissions released by power plants and agree to lower the cap over time until it is 10 percent below its initial level by 2018. To meet the emissions cap, the participating states sell tradeable allowances at quarterly auctions and use the proceeds to invest in low-carbon, clean energy technologies, such as solar and wind power. Offsets are also provided for emissions reduction activities or carbon sequestration projects external to the electricity industry. At the March 2011 auction, nearly 42 million CO_2 allowances were sold at a market clearing price of $1.89 per ton. For further information and updates, visit RGGI's Web page at **www.rggi.org/home**.

---

[35]This section relies mainly on U.S. EPA, Office of Air and Radiation (December 23, 2010); U.S. EPA, Office of Public Affairs (December 23, 2010). The interested reader should also refer to Chapter 12.

[36]Pew Center on Global Climate Change (January 2011).

[37]Drawn from RGGI (April 16, 2012; May 30, 2011; April 1, 2009).

Western Climate Initiative (WCI)[38]

Formed in 2006, the Western Climate Initiative (WCI) currently comprises seven U.S. states (Arizona, California, Montana, New Mexico, Oregon, Utah, and Washington), and four Canadian provinces (British Columbia, Manitoba, Ontario, and Quebec). The overall objective of this collaborative is to achieve a 15 percent reduction of GHG emissions relative to 2005 levels by 2020. Once the WCI is fully implemented, it will address 90 percent of GHG releases in the region. A cap-and-trade program, initially planned for launch in 2012, is one of the means by which this goal is to be achieved. Like the RGGI, offset credits are also allowed for projects in forestry and agriculture.

ECONOMIC ANALYSIS OF CLIMATE CHANGE POLICIES

Ideally, environmental policy should achieve an efficient allocation of resources, where net benefits are maximized, or equivalently, where the marginal social benefit from implementing policy is exactly offset by the marginal social cost. Although assessing benefits is always the more difficult process, it is particularly so for global warming because of the gray areas that weaken scientific predictions. Hence, before we can analyze specific policy proposals, we first need to consider the dilemma of estimating the potential benefits from *any* initiative designed to control climate change.

Estimating the Benefits

To illustrate how benefit assessment is at the root of the global warming policy dilemma, let's consider the findings of several research efforts.

A report prepared by the OECD Environment Committee estimates expected benefits from controlling global warming in the United States. The values, presented as estimated damages associated with climate change, are based on two global warming scenarios.[39] The first assumes the conventional prediction of a temperature rise of 2.5° Celsius, while the second presents estimates based on a temperature rise of 10° Celsius over a very long term. According to the findings, the benefits of controlling global warming based on conventional predictions would be $61.6 billion ($1990), or approximately 1.1 percent of GDP, and as high as $338.6 billion ($1990), or at least 6 percent of GDP, over the very long run of 250–300 years.

Economist Wilfred Beckerman (1990) argues that not only are the damages from global warming difficult to assess, but even if the most dire predictions are correct, most are not sufficient to warrant the high costs of avoidance. Beckerman cites a 1988 EPA report that estimates the net effect of global warming on U.S. agriculture to be within a range of a net gain of $10 billion and a net loss of $10 billion. The net gain is actually possible because some parts of the nation would enjoy longer growing seasons, enhanced by increased precipitation and the fertilization of increased CO_2 concentrations. At the other extreme, even if the maximum estimated loss of $10 billion is incurred, this is only about 0.2 percent of U.S. GDP.

---

[38]Drawn from Pew Center on Global Climate Change (January 2011); Western Climate Initiative (WCI) (May 2010).

[39]Cline (1992).

More recent studies conducted by Mendelsohn and Neumann (1999) and Nordhaus and Boyer (2000) present benefit estimates that align more closely with those discussed by Beckerman. Specifically, Mendelsohn and Neumann (1999) estimate that the net benefit to the United States from climate change control would be 0.1 percent of GDP, while Nordhaus and Boyer (2000) estimate the comparable value at approximately −0.5 percent of GDP, or a net cost to society. In a more recent assessment, Sir Nicholas Stern (2007) presents much different estimates. Specifically, Stern argues that without a policy response, the costs of climate change, and hence the benefits of responding, are likely to be 5 percent of global GDP per year and possibly as high as 20 percent per year, but that the costs of responding to the problem by controlling GHGs can be limited to about 1 percent of GDP each year.[40]

Although these studies share a common perspective—that benefit assessment is critical to developing climate change policy, the implications are different. The OECD study suggests that *long-term* benefits might be more relevant to assessing global warming initiatives. The other studies, which use a *shorter-term* approach, determine much lower benefit estimates that may not outweigh the associated costs. What this means is that if time is explicitly considered, policy development might take a very different direction. In any case, these analyses suggest the need for more research and help to explain the challenge policymakers face in deciding how to respond to global warming concerns.

Recognizing this challenge, economists strongly promote market-based policies designed to consider the benefits and costs of government controls. Although there are many types of instruments that use market forces to operate, all of them can be motivated by modeling the cause of global warming, rising GHG emissions, as a market failure.

Economic Model of the Market Failure

To simplify our analysis and give it context, we focus on the release of CO_2—the most prevalent GHG, arising mainly from the production of electricity. Figure 13.9 illustrates a hypothetical market for electricity generation, which consists of two groups of electric power plants: one using fossil fuels that generate CO_2 emissions and the other using alternative fuel technologies such as solar, wind, or nuclear power.

Figure 13.9(a) depicts the marginal private cost of fossil fuel users (MPC_F) and the associated marginal social cost (MSC_F). By definition, MSC_F is the vertical sum of the MPC_F and the marginal external cost (MEC_F) of electricity production, where the MEC_F is capturing the **negative production externality**. This means that MEC_F is implicitly shown as the vertical distance between MSC_F and MPC_F and represents the cost of health and property damages associated with CO_2 emissions. Figure 13.9(b) shows the marginal social cost of alternative fuel users (MSC_A). For simplicity, it is assumed that this segment of the market does not generate any negative externalities. Thus, MSC_A is exactly equal to MPC_A.

---

[40]The Stern study is available online at **http://webarchive.nationalarchives.gov.uk/+/http://www.hm-treasury.gov.uk/sternreview_index.htm**. The study sparked debate in part because Stern's estimates were markedly different from those presented by other researchers and because certain of his assumptions are questionable. For more on this debate, see Nordhaus (2007) and Weitzman (2007).

FIGURE 13.9 Modeling the Negative Externality of GHG Emissions

The competitive equilibrium at Q_C and P_C is determined by market demand (D) and *private* market supply (S_P). At P_C, fossil fuel utilities supply Q_{F1}, and alternative fuel users supply less at Q_{A1}. The efficient equilibrium at Q_E and P_E is determined by D and the *social* market supply (S_S). P_E is higher than P_C because demanders pay the full cost of their consumption activity. Fossil fuel-based production has declined from Q_{F1} to Q_{F2}, reducing CO_2 emissions, and alternative fuel-based production has risen from Q_{A1} to Q_{A2}.

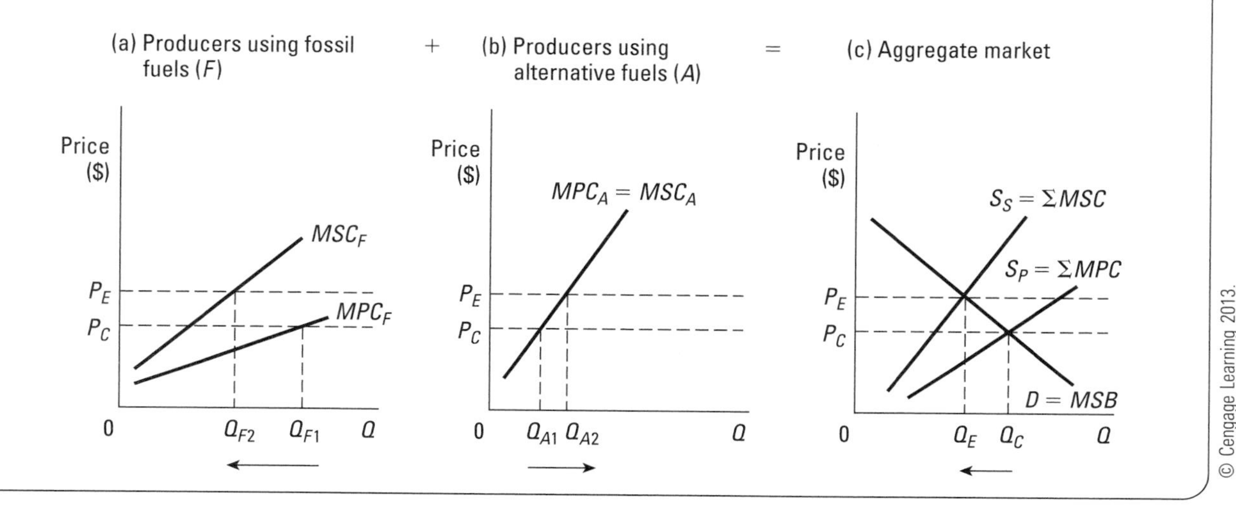

Figure 13.9(c) presents the aggregate market for electricity, shown with a demand curve (D) measuring the marginal social benefit of electricity usage (MSB) and two distinct supply curves, S_P and S_S. The S_P is the private market supply of electricity, found as the horizontal sum of MPC_F and MPC_A, and the S_S is the social market supply, found as the horizontal sum of MSC_F and MSC_A.

In the absence of CO_2 emission controls, equilibrium is determined by the intersection of market demand (D) and *private* market supply (S_P). The competitive equilibrium quantity and price are Q_C and P_C, respectively. At P_C, fossil fuel utilities are willing to supply Q_{F1}, whereas the alternative fuel users supply a much smaller amount, Q_{A1}.[41] Notice that fossil fuel users can supply most of the electricity to the market at price P_C because their MPC_F is relatively low. Without policy controls, these utilities do not consider the external costs of their CO_2 emissions. This results in private market incentives allocating too many resources to their production processes and too few resources to alternative fuel users. Consequently, P_C is sending a false signal about how to efficiently allocate productive inputs, and the market fails.

To correct the market failure, the external cost of fossil fuel emissions must be brought into the market transaction. As shown in Figure 13.9(c), this means that market price must be determined by market demand (D) and the *social* market supply (S_S),

[41] Although the market segments shown in Figure 13.9 are hypothetical, note that as of November 2008, about 71 percent of electricity generated in the United States was produced by burning carbon-based fuels (U.S. Department of Energy, Energy Information Administration, 2009).

which includes both external and private costs. Notice that the efficient equilibrium price (P_E) is higher than what the private market determines, because electricity demanders are now paying the full cost of their consumption activity. That is, the per-unit price (P_E) paid for electricity is exactly equal to the marginal social cost incurred to produce it.[42]

At this higher price level, the quantity of electricity has been reduced to its efficient level (Q_E). This new price level corrects the mix of fossil and alternative fuels used to produce electricity. At P_E, fossil fuel-based production appropriately has declined from Q_{F1} to Q_{F2}, effectively reducing the emissions of CO_2. At the same time, the higher price provides an incentive to alternative fuel users to supply more electricity, shown as the increase from Q_{A1} to Q_{A2}.

renewable energy certificate (REC) A market-based instrument that conveys rights to the nonpower attributes of renewable electricity generation.

A real-world example of how the electricity market is responding to the negative externality associated with fossil-fuel-based production is the use of **renewable energy certificates (RECs)**. These RECs are market-based instruments that convey rights to the nonpower attributes of renewable electricity generation, which include its environmental and social benefits. Their use, which has become more common over time, encourages power generation using cleaner fuels. An overview of RECs and their influence on renewable energy markets is provided in Application 13.4.

Evaluating Market-Based Policy Instruments[43]

Anthropogenic emissions of GHGs are a **negative externality** because the effects of these gases are not captured within the market transaction and therefore are borne by society. To correct the problem, policy instruments must internalize the externality so that the market participants absorb the cost of the damages. Two types of market-based controls that can accomplish this are **pollution charges** and **tradeable allowance systems**.

Pollution Charges

pollution charge A fee that varies with the amount of pollutants released.

The use of some type of **pollution charge** to reduce CO_2 emissions has received a fair amount of attention in both domestic and international policy discussions. However, the specific form of the charge has been the subject of debate. In general, three types of product charges commonly are proposed as possible candidates: a gasoline tax, a Btu tax, and a carbon tax.

- A **gasoline tax** is a per-unit tax levied on each gallon of gasoline consumed.
- A **Btu tax** is a per-unit charge based on the energy or heat content of fuel measured in British thermal units (Btus).
- A **carbon tax** is a per-unit charge based on the carbon content of fuel.

[42]Notice that Q_E still includes some fossil fuel-based production, meaning that CO_2 emissions are not totally eliminated. However, they are reduced to the level that society believes is acceptable, based on the trade-off between the *MSB* of electricity usage and the *MSC*.

[43]For an interesting overview of market-based policies on global warming, see Parker (June 20, 2002), which is a Congressional Research Service report accessible online at **www.cnie.org/nle/crsreports/climate/clim-5.pdf**.

APPLICATION 13.4 Renewable Energy Certificates (RECs): Currency for Renewable Energy Markets

Renewable energy certificates (RECs) are market-based instruments that motivate the generation of electricity from renewable (i.e., nondepleting) sources, such as wind, solar, moving water, biomass, or geothermal. Each REC represents the displacement of one megawatt-hour of power generated by conventional fossil fuels. In so doing, GHG emissions are reduced. According to EPA estimates, every REC purchased eliminates over 1,000 pounds of CO_2. So the operative question is: How do RECs work in the marketplace?

At the point of generation, renewable energy has two components: (1) the electricity itself; and (2) the environmental and other nonpower attributes of that electricity. The REC is used to transfer the latter component to the buyer, where each certificate represents 1,000 kilowatt-hours, or 1 megawatt-hour, of renewable electricity generation.

RECs are tradeable instruments—bought and sold in the market for renewable electricity generation. They can be sold by utilities, REC marketers, and other sellers, and they can be purchased by virtually any buyer of electric power, many of whom do so voluntarily. When buyers purchase electricity from their utility, they typically do not know what technology or resources were used to produce that power. But RECs provide that information. Each certificate communicates relevant facts about the electricity with which it is associated. These include the type of renewable resource used, the age and location of the renewable generator, the date the REC was created, and the associated GHG emissions, if any.

Interestingly, the renewable electricity and its REC can be sold as a bundle or separately. If sold separately, the REC still conveys to the purchaser the benefit of avoided GHG emissions linked to conventional generation of 1 megawatt-hour of electric power. This option is important for buyers whose utility does not offer the option of purchasing renewable energy. However, the renewable energy itself that is sold unbundled from its REC is no longer considered "renewable." That way, two parties cannot lay claim to the benefits of cleaner power generation.

To assure buyers that the power is in fact generated from renewable resources, the RECs are monitored and verified. Unlike the electrons that are fed to the utility grid, which cannot be distinguished by the generation resource, the RECs are identifiable, can be tracked independently of the power generated, and can be linked to a specific buyer who has the right to claim the benefits of the renewable energy. In practice, each REC is assigned a tracking number, which remains with that certificate until the final owner makes the environmental claim, at which point the REC is considered "retired" within the system. To avoid double-counting, once the purchaser makes that claim, the RECs cannot be resold.

RECs are important to markets for renewable energy and green power.* They are being accepted as evidence of compliance with Renewable Portfolio Standards defined by some states. These standards identify a requisite percentage of a utility's output that must be generated using renewable resources.

On the other side of the market, RECs provide a means for purchasers to reduce their carbon footprint. Doing so might be driven by public relations, brand imaging, or environmental objectives of stakeholders within an organization. By way of example, in 2010, Intel Corporation purchased 1.4 billion kilowatt-hours of RECs to cover over half of its U.S. power needs. Also in 2010, Whole Foods Market purchased 815 million kilowatt-hours of RECs, and TD Bank bought over 240 million. Carnegie Mellon University buys enough kilowatt-hours of green power, nearly 87 million, to meet 75 percent of the university's power requirements.

For further information on green power and renewable energy markets, visit the U.S. Department of Energy's Green Power Web site at **apps3.eere.energy.gov/greenpower/markets/index.shtml**.

*Renewable energy is not the same as green power. According to the EPA, green power is a subset of renewable energy that produces zero anthropogenic GHG emissions, using resources and technology linked to the highest environmental benefits. Sources of renewable energy may have a small effect on the environment.

Sources: U.S. EPA, Office of Air and Radiation, Office of Atmospheric Programs, Climate Change Division (October 25, 2010; October 2010; July 2008); U.S. Department of Energy, Office of Energy Efficiency and Renewable Energy, U.S. EPA Green Power Partnership, World Resources Institute, and Center for Resource Solutions (2010).

corrective tax
A tax aimed at rectifying a market failure and improving resource allocation.

Unlike taxes on income or consumption that generate market distortions, these charges are referred to as **corrective taxes**, because they are aimed at internalizing a negative externality and hence at correcting a market failure. By design, these taxes should *reduce* market inefficiency. In addition, all three taxes are revenue generating, an attribute that is sometimes used to promote this type of policy, particularly if there are national budget deficits. Beyond these common characteristics, these taxes differ in terms of their applicability, ease of implementation, and overall effectiveness in achieving environmental objectives.

Because gasoline is a carbon-based fuel, its combustion produces CO_2 as a byproduct. By levying a tax on gasoline, its effective market price will increase, discouraging consumption and encouraging the use of cleaner alternative fuels. As fewer gallons of gasoline are burned, less CO_2 is emitted. In the United States, gasoline taxes already are being collected by state and federal governments, making an increase in the tax rate to bring about a reduction in carbon emissions relatively easy to implement.

The major drawback is that the gasoline tax targets only polluting sources using gasoline, which are relatively minor CO_2 emitters, and ignores more significant sources that burn other fossil fuels such as oil and coal. Furthermore, it imposes a disproportionate burden on some segments of the economy, such as rural communities that lack good public transportation and certain industries like interstate trucking. This explains why the broader-based carbon tax or the Btu tax is often proposed as a superior alternative.

Although the Btu tax and the carbon tax each use a slightly different tax base, the general purpose of each is the same—to encourage fuel switching and conservation by elevating fuel prices. Of the two, the carbon tax is more specific, because it targets only carbon-based fuels. In fact, it is considered the more relevant form of taxation to mitigate CO_2 emissions, because the carbon content of fuel and carbon emissions are generally proportional to one another. Effectively, the carbon tax changes *relative* fuel prices and theoretically could elevate the price of fossil fuel by the *MEC* of the environmental damage caused by its combustion. In the context of Figure 13.9(a) (page 317), the tax should equal the vertical distance between the MPC_F and the MSC_F for fossil fuel users measured at the efficient output level, Q_{F2}. Such a unit charge would successfully internalize the external costs associated with the burning of fossil fuels.

Carbon taxes are becoming more prevalent on a global scale. As noted previously, in 2011 the EU proposed a new carbon tax to be applicable to all EU members. Some EU countries already use carbon or other energy taxes, including Belgium, Finland, Denmark, Italy, Norway, and Sweden. Beyond the EU-wide proposal, the United Kingdom is replacing its Climate Change Levy on energy with a tax on CO_2 of £4.94 ($8.02) per ton, starting in 2013. Tax rates are to be linked to allowance prices in the EU Emissions Trading Scheme (EU ETS) and are expected to rise to as much as £9.86 ($16.01) in 2015.[44]

Two Canadian provinces also have established carbon taxes. Quebec set its low-rate carbon tax at CAD$3.50 ($3.20) per metric ton of CO_2 in 2007, making it the first province or state in North America to do so. British Columbia followed, enacting its own carbon tax in 2008 of CAD$10 ($9.55) per metric ton of CO_2, which rises by CAD$5 ($4.77) each year up to CAD$30 ($28.64) in 2012.[45]

The EPA has studied the carbon tax as part of its investigation of alternative control options. In fact, according to an EPA report, a tax of $5 per ton of carbon would reduce CO_2 emissions by approximately 1 to 4 percent, with estimated revenues of $7 billion to $10 billion per year. By increasing the tax to $25 per ton, emissions would decline by some

---

[44]Airlie and Carr (2011); Rankin (2011).

[45]Carbon Tax Center (2011); Sumner, Bird, and Dobos (2011).

8 to 17 percent, and revenues would be between \$38 billion and \$50 billion per year.[46] The first carbon tax enacted in the United States was one implemented at the city level, in Boulder, Colorado. Taking effect in 2007, this tax was set initially at a rate of \$7 per ton of carbon, was increased to about \$12 to \$13 per ton in 2009, and is scheduled to expire in 2013. The revenues were earmarked to fund the city's climate action plan.[47]

Tradeable Allowance Systems

tradeable allowance system for GHGs
Establishes a market for GHG permits where each allows the release of some amount of GHGs.

Another market instrument used to control global warming is a **tradeable allowance system for GHGs**. Within the Kyoto Protocol, the international GHG allowance market is the primary means by which developed nations are to achieve their emissions targets. This market operates in much the same manner as national and international trading of CFC permits used to combat ozone depletion. Under Kyoto's provisions, nations also can earn credits for engaging in carbon-absorbing activities like reforestation and for initiating carbon-reducing projects in other nations. As discussed, the EU established its own emissions trading program, known as the **European Union Emissions Trading Scheme (EU ETS)** to help the EU meet its commitment under the protocol to reduce GHG emissions 8 percent below 1990 levels.

As economic theory suggests, the trading of GHG permits can lead to a cost-effective solution. Countries able to reduce emissions below the amount initially allowed by agreement can sell their excess permits to the highest bidding country, whereas those that could not achieve the needed reductions would buy the permits. If the system operates efficiently, the price of the allowance should be the dollar value of the *MEC* associated with the emissions. Again, as shown in Figure 13.9(a) (page 317), this price should equal the vertical distance between MSC_F and MPC_F.

According to predictions, the EU ETS, which accounts for over 40 percent of total GHG emissions for the EU, should bring about "… significant emission reductions between 2008 and 2012." Moreover, across 10 policies implemented in member states, the largest predicted declines in GHG emissions are linked to renewable energy initiatives and the EU ETS. As a consequence, the EU as a group is expected to meet and even exceed its Kyoto commitment, based on the use of existing domestic policies along with the adoption of additional measures under discussion and Kyoto mechanisms as planned by member states.[48]

Numerous papers and reports have been written assessing the EU ETS, and more undoubtedly will follow. Not only are the specific environmental outcomes important, but from a policy perspective, so too is the observed operation of such a complex and large trading platform. Consider that the EU ETS controls 12,000 polluting sources, compared to 3,000 under the U.S. Acid Rain Program (ARP). And the value of allowances issued annually by the ARP of \$2.25 billion pales in comparison to the comparable value issued by the EU ETS of \$37 billion.[49] In a very real sense, the progress of this large trading program could influence similar efforts in other countries, including the

[46]U.S. EPA, Office of Policy, Planning, and Evaluation (March 1991), pp. 3–5.
[47]Carbon Tax Center (2011); Sumner, Bird, and Dobos (2011).
[48]European Environment Agency (2008). In this context, the EU refers to the EU-15.
[49]Pew Center on Global Climate Change (2005).

proposed use of a national cap-and-trade GHG program in the United States. For more detail on the EU ETS, visit **http://ec.europa.eu/clima/policies/ets/index_en.htm**.

CONCLUSIONS

Formulating sound policy in response to global air pollution is a major undertaking. Scientific knowledge about atmospheric disturbances and the associated implications is still limited, particularly for global warming. As long as the extent of environmental risk is unknown, the benefits of corrective policy initiatives are likewise indeterminate. Consequently, public officials are faced with the challenge of justifying the social costs of policy controls with reliable information about the comparable benefits.

Even when the knowledge base is stronger, such as for ozone depletion, policy development is still complicated by the global nature of the problem. A successful resolution depends critically on international commitment, supported in turn by domestic initiatives. On this front, there have been several achievements. Examples include the Montreal Protocol and its subsequent amendments to control ozone depletion and the ratification of the Kyoto Protocol to limit GHG emissions, along with the Durban Platform, which extends the protocol. Nonetheless, there are still important issues to be worked out, including the participation of developing nations in a successor agreement to the Kyoto pact. These countries lack the financial resources and the technology to innovate around the causes of global air pollution. Yet their cooperation is critical, given the expected rate of industrial and economic growth in these nations and the associated ramifications for the global environment.

Despite the difficulties, there has been progress in recognizing the relevant issues, acknowledging the unknowns, and investigating alternative solutions. Furthermore, scientific research is ongoing in the hope of reaching a general consensus about the implications of global air pollution. That policy development has been tentative, particularly in responding to the risks of climate change, is recognition of how important a balancing of benefits and costs is to that process.

Summary

- Ozone depletion refers to damage to the stratospheric ozone layer caused by certain pollutants. Scientists agree that the presence of chlorofluorocarbons (CFCs) in the atmosphere is the most likely explanation for ozone depletion.
- The Montreal Protocol is an international agreement aimed at ozone depletion. This agreement and its amendments, which include provisions for allowance trading, call for a phaseout of ozone-depleting substances. A Multilateral Fund was established to provide support for developing countries transitioning toward CFC substitutes.
- Title VI of the 1990 Clean Air Act Amendments is dedicated to protecting the ozone layer. Included are two market-based instruments: an excise tax and a marketable allowance system.
- According to an EPA-commissioned study, a tradeable permit system achieves ozone-depletion reductions in the most cost-effective manner.
- Price changes for ozone-depleting chemicals signaled the effect of the phaseout and the underlying market adjustments. The excise tax helped to counter the accumulation of excess profits to those firms holding the limited number of allowances.
- Proposals have been submitted to phase down production and consumption of HFCs because these are greenhouse gases (GHGs) that contribute to climate change.
- Climate change refers to a major alteration in a climate measure such as temperature, wind, or precipitation that is prolonged, lasting decades or longer.
- Global warming is caused by the absorption of sunlight by greenhouse gases (GHGs) such as carbon dioxide (CO_2). Accumulating CO_2 arises from fossil fuel combustion and widespread deforestation.
- Significant disruption to the natural levels of GHGs is expected to cause climate changes that would alter agricultural regions, weather conditions, and sea levels. The timing and magnitude of these effects are uncertain.
- One agreement generated at the Rio Summit was the U.N. Framework Convention on Climate Change (UNFCCC), which became legally binding in March 1994.
- In 1997, a Conference of the Parties (COP) was held in Kyoto, Japan. The resulting Kyoto Protocol set emissions targets for developed nations. The accord would become effective with ratification by 55 nations that included developed countries responsible for at least 55 percent of 1990 CO_2 emissions.
- In March 2001, President Bush took the United States out of the Kyoto accord.
- In February 2005, the Kyoto Protocol entered into force. Emissions targets would be achieved using GHG allowance trading, credits for carbon-absorbing forestry practices, and credits for emissions-reducing projects in other nations.
- The 16th Conference of the Parties (COP16) was held in Cancún, Mexico, in November–December 2010. The parties to the protocol reached some decisions in Mexico, collectively known as the Cancún Agreements, but no new agreement was reached to replace the Kyoto Protocol, whose first commitment phase expires in 2012.
- Following lengthy negotiations at the COP17 in Durbin, South Africa, the parties reached an agreement to extend the Kyoto Protocol and to establish a successor agreement by 2015. This new protocol is expected to be applicable to both developed and developing nations. The COP17 participants also implemented the Green Climate Fund to support the efforts of developing nations.
- To align with its commitment under the Kyoto Protocol, the EU launched its own GHG allowance trading program for member states, called the European Union GHG Emissions Trading Scheme (EU ETS), which is a true cap-and-trade program.
- As a complement to the EU ETS program, the EU in 2011 proposed a carbon tax for all EU members at a minimum rate of €20 per metric ton of CO_2 released from fossil fuel combustion. Some EU nations already have carbon or other energy taxes in place.
- During his administration, President Bush formulated a climate change plan aimed at reducing GHG intensity by 18 percent by 2012. The plan included energy tax credits, emission reduction credits, and joint research with other nations.

- The Obama administration committed the nation to a 17 percent reduction in GHG emissions from 2005 levels by 2020 as part of the Copenhagen Accord and the subsequent Cancún Agreements, although the pledge is nonbinding.
- A Supreme Court decision in 2007 found that the EPA is authorized to regulate GHGs, including CO_2, released. In response, the EPA announced in 2009 its Endangerment Finding—that six GHGs pose a threat to public health and welfare, and its Cause or Contribute Finding—that GHGs from motor vehicles contribute to climate change.
- The Supreme Court's decision and the EPA's findings on GHGs were linked to California's effort to control CO_2 from motor vehicles and its request for a waiver to adopt more stringent controls than those set at the federal level. In 2009, urged by President Obama, the EPA granted the waiver.
- In 2010, the EPA and the NHTSA finalized GHG emissions standards on light-duty vehicles for model years 2012 through 2016. Complementary CAFE standards also were announced that call for a combined average of 34.1 mpg in the 2016 model year. A second phase of standards for this vehicle class was announced in 2011 to be applicable to the 2017–2025 model years.
- An analogous set of rulings, comprising the HD National Program, was announced by the EPA and the NHTSA in 2011 to be applicable to medium- and heavy-duty vehicles for the 2014–2018 model years. Under consideration is a second phase of rules for this class of vehicles, starting with the 2018 model year.
- Triggered by the mobile source GHG regulations, new permitting requirements and GHG emissions standards have been proposed for stationary sources.
- At the regional level, nine U.S. states participate in the Regional Greenhouse Gas Initiative (RGGI), which uses a mandatory cap-and-trade program to reduce GHG emissions, lowering the cap until it is 10 percent lower than its initial level by 2018. Another regional effort is the Western Climate Initiative (WCI), involving seven U.S. states, which also uses a cap-and-trade program to achieve a GHG reduction of 15 percent relative to 2005 by 2020.
- A number of economic researchers have estimated the benefits associated with controlling global warming. A major difference across these studies is the time period used to assess benefits.
- Accumulating GHG emissions can be modeled as a negative externality. To internalize the external costs, pollution charges or tradeable allowances can be used.
- Three types of product charges commonly proposed to reduce GHG emissions are: a gasoline tax, a Btu tax, and a carbon tax, all of which are corrective taxes.
- Carbon taxes are becoming more prevalent worldwide. In 2011, the EU proposed a new carbon tax that would be applicable to all member states. The U.K. is replacing its Climate Change Levy on energy with a tax on CO_2 of £4.94 ($8.02) per ton, starting in 2013.
- An international system of tradeable GHG allowances is a market-based instrument being used by developed countries to achieve emissions targets set by the Kyoto Protocol. The EU established its own emissions trading program, known as the European Union Emissions Trading Scheme (EU ETS). As economic theory suggests, trading of GHG permits can lead to a cost-effective solution. Moreover, across 10 policies implemented in member states, the largest predicted declines in GHG emissions are linked to renewable energy initiatives and the EU ETS.

Review Questions

1. To meet its commitments under the Montreal Protocol, the United States implemented certain market-based policy instruments, including an excise tax on ozone depleters. Consider the following market for CFC-12 before the excise tax is imposed, where P is price per pound.

$$\text{Demand}: Q = 18.40 - 0.5P$$
$$\text{Supply}: Q = 10.00 + 2.5P$$

Now assume that a 60 cent excise tax is used, which shifts the supply curve to $Q' = 8.50 + 2.5P$.

   a. Find the equilibrium price before and after the tax is implemented.

b. What do you conclude about who bears the primary burden of the tax—buyers or sellers? Briefly explain the economic sense of this outcome in this particular context.

2. Other than financial assistance, how might industrialized countries help developing countries to control ozone depletion?
3. Suppose that two major manufacturers of commercial refrigerants, Firm J and Firm K, face the following marginal abatement costs (*MAC*) for HCFCs.

$$MAC_J = 1.2A_J \quad MAC_K = 1.8A_K$$

In the aggregate, the two firms do not hold enough HCFC allowances to cover their production activity and must abate a combined level of 20 units. Should each firm abate 10 units each? Explain economically, and support with calculations.

4. Consider the distributional effects of agricultural productivity due to global warming. Discuss some of the ramifications this outcome would have on regional economies, national economies, and world trade.
5. In the market for electricity generation, suppose that the following are the estimated marginal cost functions:

$$MPC = 40 + 0.2Q \quad MSC = 40 + 0.3Q$$

where *MPC* and *MSC* are in dollars per kilowatt hour, and *Q* is in thousands of kilowatt hours.

Based on these functions, is it likely that the firms generating electricity are using fossil fuel or some alternative clean fuel? Explain.

6. a. Why is it that a carbon tax is preferred to either a Btu tax or a gasoline tax when the objective is to reduce carbon dioxide (CO_2) emissions?
   b. Instead of enacting a carbon tax, assume that Congress decides to provide tax incentives to non–carbon-based energy sources, such as solar and wind power. Would this instrument be cost-effective in reducing CO_2 emissions?
   c. Now suppose that the government chooses to initiate tax incentives (e.g., a tax credit) for these energy alternatives along with the carbon tax. Would this be a more socially optimal solution? Explain briefly.
7. Suppose the marginal benefits and costs per gallon of gasoline in the United States are modeled as follows to illustrate the negative externality of gasoline combustion:

$$MSB = 12.80 - 0.42Q \quad MPB = 12.80 - 0.4Q$$

$$MSC = MPC = 1.25 + 0.02Q$$

where Q is millions of gallons.

   a. State the equation that represents the market externality. Give the economic interpretation of this equation, using its specific numerical value(s).
   b. Find the efficient equilibrium, P_E and Q_E, for this market. (Do not round off.)
   c. Find the dollar value of a per-unit gasoline tax that would achieve the efficient solution, and calculate the tax revenues generated to the government as a result.

Additional Readings

Barker, Terry, and Paul Ekins. "The Costs of Kyoto for the U.S. Economy." *Energy Journal 25*(3) (2004), pp. 53–72.

Benz, Eva, Andreas Löschel, Andreas, and Bodo Sturm. "Auctioning of CO_2 Emission Allowances in Phase 3 of the EU Emissions Trading Scheme." *Climate Policy 10*(6) (2010), pp. 705–18.

Beron, Kurt J., James C. Murdock, and Wim P. M. Vijverberg. "Why Cooperate? Public Goods, Economic Power, and the Montreal Protocol." *The Review of Economics and Statistics 85*(2) (May 2003), pp. 286–97.

Burguet, Roberto, and Jaume Sempere. "Trade of Permits for Greenhouse Gas Emissions: Bilateral Trade Need Not Be the Answer." *Environmental and Resource Economics 46*(4) (August 2010), pp. 495–509.

Ellerman, A. Denny, and Paul L. Joskow. *The European Union's Emissions Trading System in Perspective.* Prepared for the Pew Center on Global Climate Change (May 2008).

Ellerman, A. Denny, Frank J. Convery, and Christian de Perthuis. *Pricing Carbon: The European Union Emissions Trading Scheme.* Cambridge, UK: Cambridge University Press, 2010.

Faure, Michael, Joyeeta Gupta, and Andries Nentjes. *Climate Change and the Kyoto Protocol.* Northampton, MA: Elgar, 2003.

Fell, Harrison, and Richard D. Morgenstern. "Alternative Approaches to Cost Containment in a Cap-and-Trade System." *Environmental and Resource Economics 47(2)* (October 2010), pp. 275–98.

Griffin, James M. *Global Climate Change: The Science, Economics and Politics.* Northampton, MA: Elgar, 2003.

Hart, Craig. *Climate Change and the Private Sector.* New York: Routledge, 2009.

Heal, Geoffrey, and Bengt Kristrom. "Uncertainty and Climate Change." *Environmental and Resource Economics 22(1-2)* (June 2002), pp. 3–39.

Justus, John R., and Susan R. Fletcher. "Global Climate Change." *CRS Issue Brief for Congress.* Washington, DC: Committee for the National Institute for the Environment (October 29, 2004).

Kopp, Raymond J. "How to Evaluate Domestic Climate Policy Options: When Cap and Trade Is Not on the Agenda." *Resources (176)* (Fall 2010), pp. 14–18.

Kruger, J. A., and W. A. Pizer. "Greenhouse Gas Trading in Europe: The New Grand Policy Experiment." *Environment 46(8)* (October 2004), pp. 8–23.

McKibbin, Warwick J., and Peter J. Wilcoxen. "The Role of Economics in Climate Change Policy." *Journal of Economic Perspectives 16(2)* (Spring 2002), pp. 107–29.

Metcalf, Gilbert E. "Market-based Policy Options to Control U.S. Greenhouse Gas Emissions." *Journal of Economic Perspectives 23(2)* (Spring 2009), pp. 5–27.

Miller, Clark, and Paul N. Edwards, eds. *Changing the Atmosphere.* Cambridge, MA: MIT Press, 2001.

Muller, Frank. "Mitigating Climate Change: The Case for Energy Taxes." *Environment 38(2)* (March 1996), pp. 13–20, 36–43.

Naess, Tom. "The Effectiveness of the EU's Ozone Policy." *International Environmental Agreements: Politics, Law, and Economics 4(1)* (2004), pp. 47–63.

National Academy of Sciences. *Understanding and Responding to Climate Change: Highlights of National Academies Reports; 2008 Edition.* Washington DC: National Academies Press, 2008.

Nordhaus, William D. "Global Warming Economics." *Science 294(5545)* (November 9, 2001), pp. 1283–84.

Nordhaus, William. "Economics: Critical Assumptions in the Stern Review on Climate Change." *Science 317(5835)* (July 13, 2007), pp. 201–02.

Parry, Ian W. H., and Roberton C. Williams, III. "Is a Carbon Tax the Only Good Climate Policy? Options to Cut CO_2 Emissions." *Resources 176* (Fall 2010), pp. 38–41.

Pearce, David. "The Social Cost of Carbon and Its Policy Implications." *Oxford Review of Economic Policy 19(3)* (August 2003), pp. 362–84.

Pinkse, Jonatan, and Ans Kolk. *International Business and Global Climate Change.* NY: Routledge, 2008.

Schmalensee, Richard, Thomas M. Stoker, and Ruth A. Judson. "World Carbon Dioxide Emissions: 1950–2050." *The Review of Economics and Statistics LXXX* (February 1998), pp. 15–27.

Shogren, Jason F., and Michael A. Toman. "Climate Change Policy." In Paul R. Portney and Robert N. Stavins, eds., *Public Policies for Environmental Protection.* Washington, DC: Resources for the Future, 2000.

Sorrell, S. "Carbon Trading in the Policy Mix." *Oxford Review of Economic Policy 19(3)* (November 2003), pp. 420–37.

Springer, Urs. "The Market for Tradable GHG Permits Under the Kyoto Protocol: A Survey of Model Studies." *Energy Economics 25(5)* (September 2003), pp. 527–52.

Stavins, Robert N. "Forging a More Effective Global Climate Treaty." *Environment 46(10)* (December 2004), pp. 22–31.

Stern, Nicholas. "The Economics of Climate Change." *American Economic Review 98(2)* (May 2008), pp. 1–37.

Tietenberg, Tom. "Cap-and-Trade: The Evolution of an Economic Idea." *Agricultural and Resource Economics Review 39(3)* (October 2010), pp. 359–67.

Toman, Michael A., and Brent Sohngen, eds. *Climate Change.* Burlington, VT: Ashgate, 2004.

Van Kooten, G. Cornelius. *Climate Change Economics: Why International Accords Fail.* Northampton, MA: Elgar, 2004.

A Reference to Acronyms and Terms in Global Air Quality Control Policy

Environmental Economics Acronyms

MEB	Marginal external benefit
MEC	Marginal external cost
MPB	Marginal private benefit
MPC	Marginal private cost
MSB	Marginal social benefit
MSC	Marginal social cost
S_P	Private market supply
S_S	Social market supply

Environmental Science Terms

BCM	Bromochloromethane
Btu	British thermal unit
C_2F_6	Perfluoroethane
CF_4	Tetrofluoromethane
CFCs	Chlorofluorocarbons
CFC-11	Chlorofluorocarbon-11
CFC-12	Chlorofluorocarbon-12
CH_4	Methane
CO_2	Carbon dioxide
GHG	Greenhouse gas
GWP	Global warming potential
HBFCs	Hydrobromofluorocarbons
HCFCs	Hydrochlorofluorocarbons
HFCs	Hydrofluorocarbons
N_2O	Nitrous oxide
ODP	Ozone depletion potential
PFCs	Perfluorocarbons
SF_6	Sulfur hexafluoride

Environmental Policy Acronyms

ARP	Acid Rain Program
CAAA	Clean Air Act Amendments
CAFE	Corporate Average Fuel Economy
CCRI	Climate Change Research Initiative
CCSP	Climate Change Science Program
CDM	Clean Development Mechanism
CER	Certified Emissions Reduction
COP	Conference of the Parties
CEITs	Countries with Economies in Transition
ERU	Emissions reduction unit
EU ETS	European Union Emissions Trading Scheme
IPCC	Intergovernmental Panel on Climate Change
NAS	National Academy of Sciences
NHTSA	National Highway Traffic Safety Administration
NSR	New Source Review
REC	Renewable Energy Certificate
RGGI	Regional Greenhouse Gas Initiative
RIA	Regulatory Impact Analysis
RMU	Removal Unit
SNAP	Significant New Alternative Policy
TCI	Transportation and Climate Initiative
UNFCCC	U.N. Framework Convention on Climate Change
WCI	Western Climate Initiative

To access additional course materials, visit www.cengagebrain.com. At the home page, search for the ISBN of this title (shown on the back cover). This will take you to the product page where these resources can be found.

MODULE 5 The Case of Water

Water is so much a part of what we are and where we live that it is generally taken for granted. Satellite photographs of earth are convincing evidence of the predominance of water on our planet, covering over 70 percent of its surface. Over half of the 3-mile-deep outer layer of the earth is water.[1] Yet ironically, the abundance of water on earth disguises the fact that *usable* water is scarce, and as a result, efforts to conserve and protect this natural resource have been less than adequate.

Although most of the earth is covered by water, 97.5 percent is seawater and hence unusable for drinking or crop irrigation. According to accumulated scientific research, the estimate of fresh water on the planet is less than one-thirtieth of the earth's entire supply, and very little of this amount is contained in lakes and streams. Most of our freshwater supply—some 77 percent, according to estimates—is trapped in ice and snow. Another 22 percent lies beneath the earth's surface, and most of this (about two-thirds) is not accessible, absent prohibitive costs.[2] Finally, of what fresh water is available and accessible, much of it has been damaged by pollution.

The dependence of life on this scarce resource demands that society understand the risks of water pollution and take appropriate action to minimize those risks. In the United States, water quality control policy has taken decades to develop and, by most accounts, still has much to accomplish. What *is* clear is that water quality laws have not been consistently effective. Moreover, national policy objectives generally are not efficient, and many control instruments are not cost-effective.

In this module, we undertake a collection of tasks: to understand how water resources are threatened by contamination, to examine policy initiatives aimed at the problem, and to analyze the outcome using economic criteria. We begin our study in Chapter 14, which provides a broad overview of the Clean Water Act along with recent initiatives. Chapter 15 focuses on the primary control instruments used to achieve clean water goals. In Chapter 16, we shift our attention to the Safe Drinking Water Act and the standards used to protect human health.

---

[1]Lyklema and van Hylckama (1988).

[2]Raven, Berg, and Hassenzahl (2008); United Nations (1978), as cited by White (1988).

CHAPTER 14

Defining Water Quality: The Standard-Setting Process

"Water, water, everywhere, nor any drop to drink."

—Samuel Taylor Coleridge (1772–1834)

Water is a classic example of a natural resource characterized by the absence of property rights. Unless government intervenes, water supplies likely will be overused and contaminated. Effective water policy depends on a careful appraisal of existing water quality conditions, the setting of appropriate objectives, and the design of effective instruments to bring the two together. As we have learned in previous chapters, these tasks depend on good risk assessment and sound risk management practices.

For water policy formulation, however, many problems have impeded risk analysis. National objectives have not always been properly motivated by benefit-cost analysis. Furthermore, some policy instruments have been guided more by public and political pressures than by sound environmental management practices. As a consequence, resources have been misallocated, and society has had unrealistic expectations about what federal policy can accomplish.

Beyond these procedural difficulties are the natural complexities of water resources, which present a further challenge to policymakers. Water supplies have different chemical, biological, and ecological attributes. They also serve different functions, an important factor in policy decisions. Consider how controls on drinking water have to be much more stringent than those applied to waters used only for navigation. Adding to the challenge is a complex, natural cycle that links water resources, land, and the atmosphere together. This interdependence suggests the need for an integrated approach that acknowledges the transmedia implications. Taken together, these issues add up to an ambitious policy agenda, which we begin to investigate in this chapter.

As a preface to our analysis of policy, we present a general overview of water resources and the pollution sources that threaten them. We then focus on policy goals under the Clean Water Act and how standards are used to define water quality. A two-part policy analysis follows. The first is an evaluation of the standard-setting process. The second is a benefit-cost analysis of U.S. water quality initiatives. At the end of this chapter is a list of acronyms commonly used in discussing water quality policy.

UNDERSTANDING WATER RESOURCES

Water Resources and Their Interdependence

Most of us are aware that water is a significant component of the earth's surface and its underlying geological layers. Indeed, policy specifically addresses two major categories of water resources: **surface water** and **groundwater**.

surface water
Bodies of water open to the earth's atmosphere.

groundwater
Fresh water beneath the earth's surface, generally in aquifers.

- **Surface water** refers to all water bodies open to the atmosphere, such as rivers, lakes, oceans, and streams, and also springs, wells, or other collectors directly influenced by surface water.
- **Groundwater** refers to the fresh water located beneath the earth's surface, generally in what are called aquifers, which are underground geological formations that supply wells and springs.

Setting policy to protect and maintain these resources is a major undertaking because these water supplies are so vast and because they are remarkably heterogeneous.

Surface waters represent a highly diverse group of water bodies that support distinct ecological systems, serve different uses, and often face dissimilar sources of pollution. Because of these differences, U.S. policymakers categorize surface waters into the following groups: **open ocean waters**, **ocean coastal waters**, **bays and estuaries**, **inland waters**, and **wetlands**. Although national initiatives are aimed at protecting all these waters, policy instruments and programs often are aimed specifically at one of these categories.

Groundwater is an enormous resource—by volume over 50 times the annual flow of the earth's surface waters. It is of critical importance, because groundwater is available at virtually every point on earth and because it is the only reliable source for many arid and semi-arid regions of the world. Moreover, in the United States, some 40 percent of water used for irrigation is sourced from groundwater supplies.[1]

hydrologic cycle
The natural movement of water from the atmosphere to the surface, beneath the ground, and back into the atmosphere.

Adding to the complexity of water supplies is the interdependence of all water resources, a phenomenon explained by the **hydrologic cycle**. This cycle models the natural movement of water from the earth's atmosphere to the surface and beneath the ground and back into the atmosphere through a series of natural processes. This model is depicted in Figure 14.1, which shows how the earth's water supply is continuously in motion to replenish itself.

Notice that the hydrologic cycle illustrates the natural process that joins together the waters on the earth's surface, those below the ground, and the water vapor in the atmosphere. It also communicates how the earth's atmosphere and its water supplies are interdependent. Pollutants that damage a river or stream can easily be carried to nearby groundwater supplies, and contaminants in the atmosphere can damage surface waters through precipitation.[2] Therefore, although surface water and groundwater are, in a practical sense, distinct water supplies, potential threats to one can have implications for the other.

A comprehensive online resource about water science and water quality issues is available at the U.S. Geological Survey Web site, **http://water.usgs.gov/**.

[1]For more detail, see Raven, Berg, and Hassenzahl (2008), Chapter 14; Heath (1988).

[2]Recall from Chapter 12 that sulfur dioxide (SO_2) and nitrogen oxide (NO_x) emissions contribute to acid rain, which in turn damages lakes and other surface waters.

FIGURE 14.1 Hydrologic Cycle

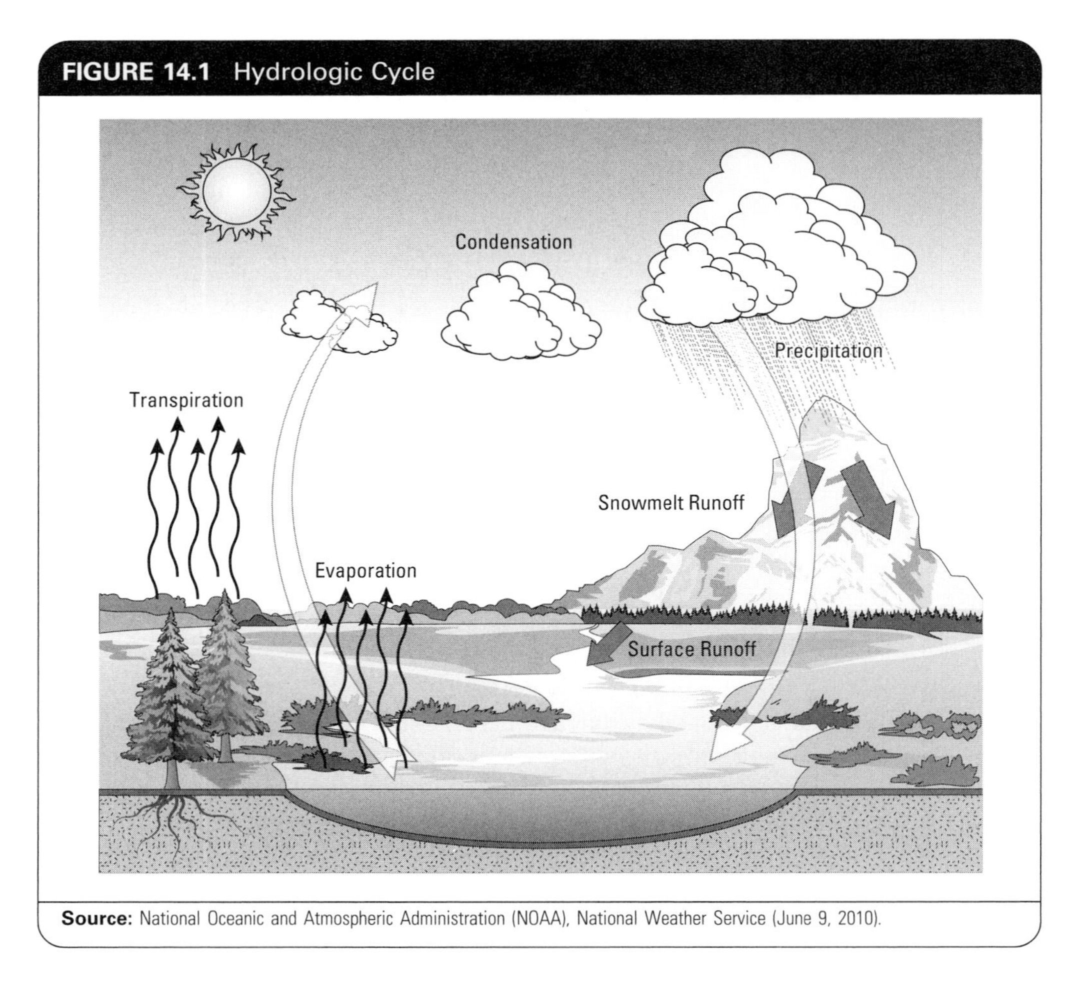

Source: National Oceanic and Atmospheric Administration (NOAA), National Weather Service (June 9, 2010).

Targeting Water Quality Policy

Both surface water and groundwater resources are vulnerable to contamination from a variety of sources, some more obvious than others. Conventionally, these are characterized as **point sources** and **nonpoint sources**.

point source
Any single identifiable source from which pollutants are released.

nonpoint source
A source that cannot be identified accurately and degrades the environment in a diffuse, indirect way over a broad area.

- A **point source** is any single identifiable source from which pollutants are released, such as a factory, smokestack, pipe, or ship.
- A **nonpoint source** is one that cannot be identified accurately and degrades the environment in a diffuse, indirect way over a relatively broad area.

As a group, point sources cover a lot of ground, so policy instruments often distinguish among the following categories:

- **Publicly owned treatment works (POTWs)**, which treat wastewaters flowing through sewer systems
- **Industrial facilities**, such as factories, mills, or other physical plants
- **Combined sewer systems**, which carry sewage and storm water runoff to waste treatment plants

Nonpoint sources by definition are difficult to identify and hence difficult to control. Examples are land runoff (such as from farms, urban areas, and construction sites), septic systems, landfills, spills, and atmospheric deposition. As an example, Application 14.1 discusses how the Chesapeake Bay has been polluted by both nonpoint and point sources and explains what is being done to save this important natural resource.

APPLICATION 14.1 Saving the Chesapeake Bay: A Renewed Effort

A recognized national treasure, the Chesapeake Bay is the largest estuary in North America with a watershed extending over 64,000 square miles, and it has been deteriorating for decades. Providing a habitat to more than 3,700 plant and animal species, this well-known estuary has suffered serious stress from a barrage of point and nonpoint polluting sources. Among them were sewage treatment plants, acid rain, urban and rural runoff, and industrial dischargers. The vulnerability of this majestic water body is explained in part by the large number of major rivers that flow into the bay from Maryland, Virginia, West Virginia, Pennsylvania, and Delaware. Furthermore, its geographic location is such that it does not benefit from the cleansing forces of the open ocean. Despite efforts to reverse this trend, this popular bay continues to show signs of decline.

The bay receives an enormous amount of nutrients and suspended solids. Most originate from municipal treatment plants and agricultural runoff. The major effects are declines in submerged vegetation and severe eutrophication, which is the gradual aging process during which a water body evolves into a marsh and eventually disappears. Because the eutrophication is extreme, hypoxia (a serious depletion of dissolved oxygen) is common in the bay. This condition has been blamed for fish kills that are both ecologically and commercially devastating. Other contaminants, such as heavy metals and organic substances, are linked to industrial and municipal pipelines as well as to urban and rural runoff.

One of the more dramatic consequences of the Chesapeake's contamination is the decline in its productivity. Commercial fishing from this large estuary had always been prolific. But in the last few decades, oyster production has fallen by 99 percent. Severe reductions also have been observed in such species as striped bass, white perch, and the blueback herring.

In 1983, a seven-year study reported on the major causes and sources of the deterioration. Later that year, the Chesapeake Bay Agreement was signed by the EPA, Maryland, Virginia, Pennsylvania, the District of Columbia, and the Chesapeake Bay Commission. This historic document authorized the Restoration and Protection Plan, the oldest estuarine program in the United States. The plan was devised as a cooperative agreement to clean up the bay and restore its productivity. In 1987, the program was made a part of the Clean Water Act, and a new agreement was drafted and signed. Then, in June 2000, the Chesapeake 2000 agreement was executed to formally extend the restoration plan through 2010.

Although these efforts did not go unrewarded, progress has been uneven. In 2007, the stock of fish and shellfish was 52 percent of objective levels, which was higher than the 48 percent achieved in 2006. However, water quality in the bay was assessed at 21 percent of its overall goal, which was lower than the 23 percent achieved in the prior year. Oysters and blue crabs were still at risk, and in many parts of the basin, water clarity continued to be a problem.

Again in 2009, water quality was reported to be extremely poor, meeting only 24 percent of the goals established by the Chesapeake Bay Commission. More than half of the streams in the watershed were rated poor or very poor (according to the biological integrity index). Poor water quality in turn threatens the habitats of fish and wildlife. For example, oyster populations are in serious decline, at less than one percent of historic levels, and these same declines have been observed for other key species of fish and shellfish.

In response to this inadequate progress, a presidential executive order was issued in 2009 that established a Federal Leadership Committee (FLC) and charged it with the responsibility of developing a strategy to protect and restore the Chesapeake Bay. The completed strategy, announced in May 2010, is structured around four key priorities: restore clean water, recover habitat, sustain fish and wildlife, and conserve land and increase public access. The FLC is also required to publish an annual Chesapeake Bay Action Plan in which funding from the federal budget would be used to achieve stated goals. For fiscal year 2011, the budgeted amount was $491 million, and the first annual 2011 Chesapeake Bay Action Plan was completed in September 2010. As required, the plan outlines how the funds are to be allocated to achieve the committee's stated objectives. With more infrastructure, strengthened commitment, and greater accountability inherent in the new strategy, optimism is growing for the restoration of this important natural resource.

The full strategy and the 2011 Action Plan are available at the Chesapeake Bay Executive Order Web site at **http://executiveorder.chesapeakebay.net/**, and more detail on the Chesapeake Bay Program itself can be found at **www.chesapeakebay.net**.

Sources: Federal Leadership Committee (FLC) for the Chesapeake Bay (September 2010; May 2010); Chesapeake Bay Program (March 2008; October 16, 2002; June 28, 2000); U.S. EPA, Office of Water (December 1995), pp. 330–31; U.S. Congress, Office of Technology Assessment (1987), pp. 21–22, 100–108, 157–64.

OVERVIEW OF U.S. WATER QUALITY LEGISLATION

Similar to the development of U.S. air policy, government controls on water quality date back to the late 1800s. Over time, the legislation has evolved into a fairly complex set of laws to protect not only human health but also the natural condition of water resources and the aquatic life they support. A summary of this legislative evolution is given in Table 14.1.

TABLE 14.1 Evolution of U.S. Water Quality Laws

Legislative Act	Major Provisions
Rivers and Harbor Act of 1899	Prohibited the discharge of refuse into navigable waters without a permit.
Federal Water Pollution Control Act (FWPCA) 1948	Called for federal government to research, conduct contamination surveys, and provide loans to municipalities for POTWs.
FWPCA Amendments of 1956	Authorized states to set water quality criteria; established a federal grant program to subsidize POTW construction up to 55 percent of costs.
Water Quality Act of 1965	Required states to set ambient standards, devise implementation plans, and set up monitoring and enforcement programs.
The Marine Protection, Research, and Sanctuaries Act of 1972	Regulates ocean dumping; acts as the domestic instrument to implement rulings under the London Convention 1972.
Coastal Zone Management Act of 1972	Establishes rulings to preserve and manage the nation's coastal waters; helps states to accept their coastal zone responsibilities through management programs aimed at proper use of land and water resources.
FWPCA of 1972	Shifted primary responsibility to federal government; established the first national goals for water quality; authorized EPA to set technology-based effluent limitations; increased federal cost share of POTWs to 75 percent.
Clean Water Act (CWA) of 1977	Postponed deadlines for meeting the effluent limitations; established treatment standards for POTWs; strengthened controls on toxics.
Water Quality Act of 1987	Replaced POTW grants with federal subsidies for a state program for construction loans; required states to set up programs for nonpoint sources.
Ocean Dumping Ban Act of 1988	Amends the Marine Protection, Research, and Sanctuaries Act; ends dumping of sewage sludge and industrial waste; regulates garbage barges; prohibits disposal of medical wastes.
Oil Pollution Act (OPA) of 1990	Establishes a National Contingency Plan to remove oil discharges and a response system to execute it; strengthens licensing standards; provides for improved construction of oceangoing vessels.
Coastal Zone Act Reauthorization Amendments (CZARA) Section 6217 of 1990	Requires states and territories with approved Coastal Zone Management Programs to develop Coastal Nonpoint Pollution Control Programs.
Beaches Environmental Assessment and Coastal Health (BEACH) Act of 2000	Provides grant funding for beach water quality monitoring and notification programs for coastal and Great Lakes recreational beach waters.
Great Lakes Legacy Act of 2002	Amends the Clean Water Act to provide federal funding to help address contamination in the Great Lakes basin.

Sources: Federal Facilities Environmental Stewardship and Compliance Assistance Center (June 6, 2011); U.S. EPA, Office of Water (March 14, 2011; December 3, 2010; February 20, 2003); Adler, Landman, and Cameron (1993), pp. 5–10; Freeman (1990), pp. 97–108; Freeman (1978), pp. 45–53.

In the Beginning[3]

U.S. water quality legislation has a long history. However, much like air quality laws, the role of the federal government in water quality legislation was limited for a long time, apparently for the same reason. There was a reluctance to bring federal intervention into what had been perceived as states' responsibilities. Consequently, there was no federal legislation dealing with water pollution per se until 1948, and even then, federal responsibility was limited. Despite legislative revisions in the 1950s and 1960s, little progress was achieved. Part of the problem was that state governments had primary responsibility for implementing policy but were ill prepared to do so. They also lacked an incentive to enforce stringent laws that would discourage industry from locating in their jurisdictions.

Responding to the failure of state-level controls and the continuing decline in national water quality, Congress passed the Federal Water Pollution Control Act (FWPCA) of 1972, considered a legislative milestone. In fact, much of what guides today's policy originated with this act. Primary responsibility for water quality shifted to the federal level, national goals were defined, and new technology-based effluent limitations were called for as the primary policy instrument. It soon became apparent, however, that this new law was overly ambitious, falling far short of its objectives. Making much needed midcourse corrections, the Clean Water Act (CWA) of 1977 extended compliance deadlines and strengthened the law on toxic water pollutants.

Current U.S. Policy

Water Quality Act of 1987

When the Clean Water Act subsequently came up for reauthorization, Congress enacted revisions and more deadline extensions—these in the form of the Water Quality Act of 1987, the law governing national water policy today.[4] Among other important changes, this act called for states to establish programs aimed specifically at nonpoint polluting sources. It also replaced a federal grant program for POTWs with a state loan program.

Since 1987, no formal reauthorization has been passed by Congress, though it has been on the agenda in some congressional sessions. In fact, in 1995, the House actually passed a reauthorization bill, but the Clinton administration opposed it, and it never became law. No analogous bills to reauthorize the Clean Water Act have been introduced since that time. And so the Water Quality Act of 1987 remains the overarching legislation governing water issues in the United States.[5]

Other Water Quality Legislation

Aside from reauthorization bills, Congress has passed other legislation on specific water pollution problems or particular water bodies. Referring back to Table 14.1, the Ocean Dumping Ban Act was passed in 1988 to halt dumping of sewage sludge and industrial waste in seawaters. In 1990, the Oil Pollution Act (OPA) of 1990 established the

---

[3]Much of the following is drawn from Adler, Landman, and Cameron (1993), pp. 5–10; Freeman (1990), pp. 97–108; Freeman (1978), pp. 45–53.

[4]Although this act officially changed the title of U.S. legislation, we will follow convention and refer to this legislation as the Clean Water Act (CWA).

[5]For more detail, consult Copeland (2010).

National Contingency Plan to respond to oil spills in the aftermath of the *Exxon Valdez* accident in 1989. Its rulings, along with those of other laws, are guiding the reimbursement of claims related to the *Deepwater Horizon* blowout of 2010. In 2000, the Beaches Environmental Assessment and Coastal Health (BEACH) Act was enacted as an amendment to the CWA to provide grants for monitoring and notification programs for coastal and Great Lakes recreational beaches. And in 2002, Congress passed the Great Lakes Legacy Act, formally amending the CWA, to provide funding for addressing contaminated sediments in the Great Lakes basin.

Executive Orders and Initiatives

Under the Obama administration, no major water quality legislation has been enacted to date. However, President Obama has signed several executive orders that are focused on water quality. These are:

- **Executive Order 13508: *Chesapeake Bay Protection and Restoration,* May 12, 2009.**
  Establishes a Federal Leadership Committee to oversee programs and activities aimed at protecting and restoring the Chesapeake Bay.
- **Executive Order 13543: *National Commission on the BP Deepwater Horizon Oil Spill and Offshore Drilling,* May 22, 2010.**
  Establishes this national commission to examine the oil spill, to develop options for prevention and mitigation, including changes to legislation and industry practices, and to submit a final report.
- **Executive Order 13547: *Stewardship of the Ocean, Our Coasts, and the Great Lakes,* July 19, 2010.**
  Establishes a National Ocean Council to oversee the protection and restoration of coastal, ocean, and Great Lakes ecosystems.
- **Executive Order 13554: *Gulf Coast Ecosystem Restoration Task Force,* October 5, 2010.**
  Establishes a task force to coordinate the Gulf Coast restoration and to develop a strategy that proposes an agenda with goals and milestones, recommendations for needed new programs, and the identification of monitoring, scientific, and research assessments needed.

These executive orders can be accessed online at **www.whitehouse.gov/briefing-room/presidential-actions/executive-orders**.

Efforts are also underway to restore the coastal regions of Louisiana and Mississippi, which were devastated by Hurricane Katrina. A working group led by the Council on Environmental Quality and the Office of Management and Budget was formed in 2009, which devised a strategic plan to protect and restore the ecosystems of that region.[6]

On a broader scale, in 2011 President Obama announced a *Clean Water Framework*, which comprises a national commitment to clean water. Included is an explicit reference to the restoration of significant water bodies, such as the Chesapeake Bay, the Great Lakes,

---

[6]Council on Environmental Quality (June 2, 2011). The strategic plan is referred to as the *Roadmap for Restoring Ecosystem Resiliency and Sustainability* and is available at **www.whitehouse.gov/administration/eop/ceq/initiatives/gulfcoast/roadmap**.

the Gulf of Mexico, and the Everglades. Another element of this framework is the provision of new federal guidance drafted by the Environmental Protection Agency (EPA) and the U.S. Army Corps of Engineers to clarify those waters protected by the Clean Water Act.[7]

POLICY OBJECTIVES

In the United States, as in many industrialized nations, many laws are aimed at protecting water quality, some directed solely at ocean waters, some at drinking water, and some at specific water bodies.[8] Of these, none is more comprehensive than the Clean Water Act (accessible online at **www.epa.gov/regulations/laws/cwa.html**). To understand its far-reaching intent, let's examine the three major goals that it defines for the nation: the **zero discharge goal**, the **fishable-swimmable goal**, and the **no toxics in toxic amounts goal**.

Zero Discharge Goal

zero discharge goal Calls for the elimination of all polluting effluents into navigable waters.

The **zero discharge goal** called for the elimination of all polluting effluents into navigable waters by 1985. The objective was ambitious. It was motivated by the ineffectiveness of states' control efforts and by accumulating evidence about the deterioration of U.S. waters. For example, in 1969, the buildup of oil and industrial wastes in Ohio's Cuyahoga River caused it to literally catch fire. Two years later, a report stemming from Ralph Nader's investigation of U.S. water quality, entitled *Water Wasteland*, cited incident after incident of damage to U.S. waters caused by pollution.[9]

Fishable-Swimmable Goal

fishable-swimmable goal Requires that surface waters be capable of supporting recreational activities and the propagation of fish and wildlife.

The **fishable-swimmable goal** was written as an interim objective to be met until the zero discharge goal could be achieved. According to this objective, by 1983, surface waters were to be capable of supporting recreational activities and the propagation of fish and wildlife. In so doing, this goal established a baseline level of water quality across all states and set a fixed deadline by which the baseline would be achieved.

No Toxics in Toxic Amounts Goal

no toxics in toxic amounts goal Prohibits the release of toxic substances in toxic amounts into all water resources.

The **no toxics in toxic amounts goal** prohibits the release of toxic substances in toxic amounts into all water resources. Singling out toxic pollutants responded to the observation of an increasing number of water bodies contaminated by dangerous chemicals. Throughout the 1960s and 1970s, chemical use had been on the rise. The discovery of new synthetic chemicals during that period added uncertainty to the potential threat.

Taken together, these three goals were intended to be the guiding principles for achieving and maintaining water quality throughout the nation. In truth, none of them were met by the stated deadlines, nor have they since been achieved, although they are still the ultimate targets of policy initiatives.

---

[7]U.S. EPA, Office of Public Affairs (April 27, 2011). The framework, entitled *Clean Water: Foundation of Healthy Communities and a Healthy Environment*, is available online at **www.whitehouse.gov/sites/default/files/microsites/ceq/clean_water_framework.pdf**.

[8]In Chapter 16, we will analyze policies aimed at protecting drinking water.

[9]Zwick and Benstock (1971), as cited in Adler, Landman, and Cameron (1993), pp. 5–6.

Identifying Water Pollutants

Water can be polluted at virtually any point in the hydrologic cycle and by many different contaminants. The contamination can arise either deliberately through illegal waste disposal or accidentally such as when oil tankers spill their cargo into the ocean, an international problem discussed in Application 14.2.

APPLICATION 14.2 Oil Tanker Spills Around the Globe: The Ecological and Economic Effects

In December of 2007, the tanker *Hebei Spirit* was hit by a crane barge in the waters off the coast of South Korea, spilling 10,500 tons of crude oil into the Yellow Sea. In 2002, the tanker *Prestige* broke in half after suffering damage to its hull off the coast of northern Spain, spilling some 63,000 tons of heavy fuel oil into the Atlantic Ocean, which ultimately damaged the northern coast of Spain and the Atlantic coast of France. In January 1993, the second of two major oil spills within five weeks of one another made world headlines. The first involved a Greek oil tanker *Aegean Sea* that ran aground in stormy seas off the coast of Spain in December 1992, dumping 74,000 tons of oil into the port of La Coruna. Damage to marine life resulted in extensive bans on fishing and shellfish harvesting. The second incident involved the tanker *Braer*, which encountered engine failure in the North Sea, ran aground, and lost its cargo of 84,700 tons of crude oil.

The largest tanker spill in history involved 287,000 tons of oil when two tankers collided in 1979 off the coast of Tobago. In the United States, the most damaging tanker spill occurred in March 1989 when the *Exxon Valdez* ran aground in Alaskan waters, and 37,000 tons of oil poured into Prince William Sound. The following table lists the five largest oil spills since 1967.

According to data gathered by the International Tanker Owners Pollution Federation Ltd. (ITOPF), the number of large oil spills, i.e., those over 700 tons, caused mainly by groundings and collisions, has declined significantly since 1970. In fact, in the 1970s, there were 253 large oil spills, which dwarf the 33 that occurred between 2000 and 2010. The improvement is assumed attributable to enhanced safety measures and tougher enforcement. Nonetheless, in the aftermath of these accidents, society must contend with the environmental damage and the economic losses.

Birds, fish, sea otters, and other marine life have been killed or found coated with oil, struggling to survive. Coastlines are contaminated as the crude oil makes its way to shore. Some of the losses have market implications, such as regional economic losses in tourism and the destruction of commercially valued fish and shellfish. Even more difficult to assess are damages based on the existence value of the ecology—the ocean itself, the marine life it supports, and natural coastlines.

No one can accurately determine the numbers of birds, fish, and sea mammals destroyed as a result of these accidents nor the loss of both recreational and commercial use of valuable resources. Even if this were possible, there is also the daunting problem of assigning a dollar value to these losses of ecological quality and biological life. Nonetheless, economists and other analysts attempt to assess and monetize these damages along with other market effects. In the *Exxon Valdez* case, between 3,500 and 5,500 sea otters were destroyed, and more than 30,000 sea birds were killed. Including these losses, dollar damages attributable to the *Valdez* incident have been estimated in the billions of dollars.

For more information on oil spills, visit **www.itopf.com/**.

Spill Size (tons)	Ship Name	Location	Year
287,000	***Atlantic Empress***	Off Tobago, West Indies	1979
260,000	***ABT Summer***	Off Angola	1991
252,000	***Castillo de Bellver***	Off Saldanha Bay, South Africa	1983
223,000	***Amoco Cadiz***	Off Brittany, France	1978
144,000	***Haven***	Genoa, Italy	1991

Source: Based on International Tanker Owners Pollution Federation Ltd. (ITOPF) (June 5, 2011).

Sources: International Tanker Owners Pollution Federation Ltd. (ITOPF) (June 9, 2011; June 5, 2011; February 28, 2008); Raven, Berg, and Hassenzahl (2008), pp. 249–50.

toxic pollutant
A contaminant that, upon exposure, will cause death, disease, abnormalities, or physiological malfunctions.

conventional pollutant
An identified pollutant that is well understood by scientists.

nonconventional pollutant
A default category for pollutants not identified as toxic or conventional.

Many pollutants are responsible for degrading the earth's water supplies, ranging from excess plant nutrients to synthetic toxics. Three categories of pollutants are relevant to national law: **toxic pollutants**, **conventional pollutants**, and **nonconventional pollutants**.

- A **toxic pollutant** is a contaminant that, upon exposure, will cause death, disease, abnormalities, or physiological malfunctions.
- A **conventional pollutant** is one well understood by scientists that can take various forms, such as organic waste, sediment, bacteria, nutrients, oil, or heat. Included are suspended solids, biological oxygen demanding (BOD) substances,[10] fecal coliform, and pH.
- A **nonconventional pollutant** is the default category for pollutants not identified as either toxic or conventional.

These categories play a role in how the **technology-based effluent limits** are implemented, an issue discussed in Chapter 15. As to their motivation, these classifications came about to allow for tougher controls on toxic contaminants.

SETTING STANDARDS TO DEFINE WATER QUALITY

receiving water quality standards
State-established standards defined by use designation and water quality criteria.

As originally required under the Water Quality Act of 1965, surface water quality is defined by **receiving water quality standards**. These state-established standards have two components:

- **Use designation** for the water body
- **Water quality criteria** to sustain the designated uses

The **use designation** identifies the intended purposes of a water body, such as for irrigation or shellfishing. The **water quality criteria** give the biological and chemical water attributes necessary to sustain or achieve these designated uses, including the maximum concentration of pollutants allowed. More information about U.S. water quality standards is available via the Internet at **http://water.epa.gov/scitech/swguidance/standards/about_index.cfm**.

Use Designation[11]

States are authorized to decide the designated beneficial uses for all intrastate water bodies, subject to EPA approval. Among the uses they should consider are public water supplies, propagation of fish and wildlife, recreational activities, and agricultural purposes. At a minimum, designated uses must be sufficient to support swimming and some fishing in order to be consistent with the national **fishable-swimmable goal**.

[10] This refers to the amount of oxygen needed by microorganisms to decompose organic compounds.
[11] The following is drawn from 40 CFR 131.10, *Designation of Uses*; U.S. EPA, Office of Water (March 1994).

Use-Support Status

use-support status A classification based on a water body's present condition relative to what is needed to maintain its designated uses.

Periodically, state authorities must determine the **use-support status** of a water body by assessing its present condition and comparing it with what is needed to maintain its designated uses. One of three classifications is assigned to characterize use-support status of surface waters. As described in Table 14.2, these classifications are **good**, **threatened**, and **impaired**. States' findings on use-support status are submitted to the EPA, analyzed, and reported to Congress as part of a biennial National Water Quality Inventory required by the CWA. The status of U.S. waters as of the most recent national inventory is summarized in Table 14.3. These biennial reports are accessible online at **http://water.epa.gov/lawsregs/guidance/cwa/305b/index.cfm**.

TABLE 14.2 Classifications of Use Support for U.S. Surface Water Quality

Use-Support Level	Definition
Good	Fully supporting designated uses
Threatened	Fully supporting designated uses, but where declining water quality is indicated
Impaired	Not supporting one or more designated uses

Source: U.S. EPA, Office of Water, Office of Wetlands, Oceans, and Watersheds (May 25, 2011).

TABLE 14.3 Use-Support Status by Water Category

	Assessed (%)	Good (%)	Threatened (%)	Impaired (%)	Leading Source	Leading Pollutant
Rivers, Streams (Miles)	26.5	49.7	<1	49.6	Agriculture	Pathogens
Lakes, Reservoirs, Ponds (Acres)	42.2	33.7	<1	66.0	Atmospheric deposition	Mercury
Bays and Estuaries (Square Miles)	21.0	36.3	<1	63.7	Municipal discharges/ sewage	Organic enrichment/ oxygen depletion
Coastal Shorelines (Miles)	4.0	62.2	–	37.8	Municipal discharges/ sewage	Mercury
Oceans and Near-Coastal (Square Miles)	10.9	17.7	–	82.3	Unknown	Mercury
Wetlands (Acres)	1.9	63.6	<1	36.4	Agriculture	Organic enrichment/ oxygen depletion
Great Lakes Shorelines (Miles)	22.8	6.3	–	93.8	Legacy/historical pollutants	Polychlorinated biphenyls (PCBs)
Great Lakes Open Water (Square Miles)	93.7	0.1	–	99.9	Atmospheric deposition	Polychlorinated biphenyls (PCBs)

NOTE: Percentages are based on the proportion of each category assessed by states.

Source: U.S. EPA, Office of Water, Office of Wetlands, Oceans, and Watersheds (May 25, 2011).

Water Quality Criteria[12]

To ensure that designated uses are achieved and maintained, states must either establish **water quality criteria** subject to EPA approval or adopt criteria set by the EPA itself. The EPA is responsible for developing criteria that reflect the most current scientific knowledge on all identifiable effects of pollution on health, aquatic life, and welfare. These criteria are pollutant specific and may be expressed as concentrations of pollutants allowed in water (numeric criteria), in qualitative statements (narrative criteria), or as comments about the overall condition of an aquatic system (biocriteria). To examine specific water quality criteria, visit **http://water.epa.gov/scitech/swguidance/standards/current/index.cfm**.

ANALYSIS OF RECEIVING WATER QUALITY STANDARDS

The **receiving water quality standards** are a critical element of U.S. water quality policy because they are linked directly to the nation's objectives, particularly the fishable-swimmable goal. However, the process of establishing these standards has been problematic, as has their reliance on the effluent limitations for implementation.

Absence of Benefit-Cost Analysis

When states were called upon to establish receiving water quality standards, the law allowed them to set different standards for the portion of any interstate water body within their jurisdictions. It was implied that they could use benefit-cost analysis in setting these standards, but they were not required by law to do so.[13] Of course, the use designation had to be consistent with national goals, indicating at minimum a water quality level to support fishing and swimming—a goal that was solely benefit-based with no cost consideration. Hence, there is no assurance that pollution abatement will be set at an efficient level. Even in states' biennial reports on use-support status, benefit-cost estimates are not provided for waters achieving fishable-swimmable use designation, even though the law calls for it.

That benefit-cost assessment is important to water quality policy is illustrated by the ongoing investigation and study of the *Deepwater Horizon* blowout in 2010. As pointed out in Application 14.3, although costs have been estimated, it is difficult to identify, much less quantify, the associated benefits.

Lack of Consistency With the Technology-Based Effluent Limitations

Another problem with the standard-setting procedure relates to the use of federally mandated effluent limitations to achieve the states' receiving water quality standards.

[12]The following discussion is drawn from 40 CFR 131.11, *Criteria*; CWA, Sec. 303; U.S. EPA, Office of Water Regulations and Standards (September 1988), pp. 8–11.

[13]Freeman (2000), p. 171.

APPLICATION 14.3 Abating the *Deepwater Horizon* Blowout: Benefits and Costs

On April 20, 2010, a massive explosion occurred on the offshore drilling rig, the *Deepwater Horizon*, located in the Macondo Prospect, which is owned and operated by British Petroleum (BP). The explosion damaged the wellhead, and the gush of oil continued for three months before the wellhead was capped. More than 170 million gallons of oil spilled into the Gulf of Mexico, making it the largest marine oil spill in American history. Once the well was capped, the complex task of assessing and abating the damages could begin in earnest, a task that will continue for years.

In theory, abatement of pollution should continue as long as the marginal benefits of doing so outweigh the marginal costs, achieving efficiency when the two are equal. In practice, this equality can be difficult to realize, particularly in a context as complex as this one.

By all accounts, the explicit costs of abating the oil spill damage are significant. According to a government report of response efforts, resources dedicated to the cleanup have included 9,700 vessels at the peak; 6,500 government and commercial vessels; 48,200 responders; and 127 surveillance aircraft. Coast Guard assets used in the response have included 7,000 personnel; 60 vessels; and 22 aircraft. As to the dollar value of the cleanup, the federal government sent 11 bills to BP by mid-2011, totaling nearly $711.7 million. BP estimates that its total costs from the spill ultimately will total $41 billion.

On the other side of the ledger, damages from the massive oil spill, and hence the benefits of abatement, will take decades to fully determine. Quantitative values have not been assigned, but officials have qualitatively identified at least some of the damages, which fall into three major categories: natural resources, economics, and human health.

Natural resources damages, or benefits of abatement, have been characterized as extensive, despite some consensus that the effects are not as severe as originally believed. Part of the assessment challenge is that there were no comprehensive data on the region's environmental status prior to the accident, and hence no benchmark for determining incremental effects. Moreover, knowledge of deepwater marine life is limited, adding to the difficulty. What *is* clear is that the region is ecologically rich and productive, making environmental threats costly.

Reports of environmental damage have included observations of lifeless deepwater corals, despoiled marshlands, and wildlife covered in oil. To provide substance to this latter observation, consider the following statistics, provided by the U.S. Fish and Wildlife Service (USFWS) and the National Oceanic and Atmospheric Administrations (NOAA). As of April 2011, over 8,200 birds were collected; almost 4,400 of which were visibly oiled; and 6,147 of which were dead. Nearly 1,150 sea turtles were collected; of these, 613 were dead and 474 were visibly oiled. Lastly, 170 mammals were collected, 157 of which were dead, and 12 were visibly oiled.

Scientists also are concerned about impairment to marine life, but particularly the so-called keystone species, which play a role in maintaining the health of aquatic habitats. Oysters are one such species, which contribute to the ecosystem by filtering impurities from seawater. Adding to concerns is the potential for harmful effects from deepwater plumes. Officials say that long-term monitoring is necessary to fully identify these damages.

Economic effects are nontrivial, particularly for the seafood and tourism industries, which are important to the Gulf region. According to U.S. Census Bureau data, tourism and commercial fishing account for more than $40 billion in economic activity in the five Gulf states. Other markets directly affected include the oil and gas industries, which were shutdown for months during the government-imposed moratorium on deepwater drilling. In response to President Obama's appeal, BP established a $20 billion fund to help compensate for these particular economic losses.

Human losses linked to the oil spill are among the most difficult to assess. At the explosion site, 11 people lost their lives; 17 more were injured; and countless others were traumatized by the horrific fires, dangerous rescue efforts, and ongoing investigations. Beyond these immediate effects are physical and emotional health risks to residents concerned about the future of their livelihoods and their communities. Although difficult to quantify, most agree that the human consequences linked to the incident are significant and, like other damages to the region, may be long-term in duration.

For more detail on the costs and benefits of cleanup efforts, visit the National Commission on the BP Deepwater Horizon Oil Spill and Offshore Drilling at **www.oilspillcommission.gov**.

Sources: RestoretheGulf.gov (June 2, 2011; May 16, 2011); U.S. Department of the Interior (DOI), Fish and Wildlife Service (FWS) (April 20, 2011); "Loss at BP as Gulf Oil Spill Costs Hit $41 BN" (February 2, 2011); National Commission on the BP Deepwater Horizon Oil Spill and Offshore Drilling (January 2011).

The source of contention is that the link between the standards and the effluent limitations is blurred. Why? Because each is motivated differently:

- The standards are motivated by *water usage.*
- The effluent limits are motivated by *technology.*

Because the effluent limits are technology-based, they are motivated by what is feasible rather than by benefit-cost analysis or environmental criteria. The intent was to avoid the problem that states had in the past of trying to estimate the pollution reduction needed to achieve a predetermined water quality level. However, it is precisely this relationship that is needed to map the effluent limits to the standards.

Another drawback is that the technological limits are applied uniformly within defined groups of point sources, so they do not account for varying conditions or use designations across water bodies. So, even if all polluters met the effluent limits, there is no guarantee that water quality as the states have defined it would be achieved. In fact, it is exactly because of this possibility that the law requires states to identify those waters for which the effluent limitations are insufficient. Officially, these waters are to be labeled "water quality limited" and placed in a priority ranking. More stringent controls must then be established, called **total maximum daily loads (TMDLs)**, which specify the maximum amount of pollution that can be received without violating the standards.[14] These additional rulings are necessary because the policy instrument (the effluent limitation) is not properly linked to the objective (the water quality standard).

total maximum daily loads (TMDLs) Maximum amount of pollution a water body can receive without violating the standards.

BENEFIT-COST ANALYSIS OF THE CLEAN WATER ACT

Water quality is a goal about which there is little debate, at least from a qualitative perspective. The issue is not a zero-one option of whether or not this objective is worth pursuing but rather a determination of the extent to which water contamination should be controlled. As is always the case in environmental policy, the question is: How clean is clean? Once this fundamental question has been addressed, a secondary issue is to evaluate how this goal, however defined, is to be achieved. Both issues are complex, but they can be addressed with considerable objectivity using benefit-cost analysis and the efficiency criterion. Relying on the careful work of well-respected environmental economists, we can examine the evidence and draw some qualitative, albeit guarded, conclusions.[15]

Benefit Estimates

Assessing benefits is the most difficult task in benefit-cost analysis of environmental policy. Nonetheless, there have been advances in the methods used to estimate environmental benefits. One in particular is the use of the **contingent valuation method (CVM)**, touted for its ability to capture existence value as well as user value of environmental resources.[16]

[14]See CWA, Sec. 303(d) (1) (A); U.S. GAO (January 1989), pp. 2–3.

[15]To review an EPA benefit-cost analysis of water policy, see U.S. EPA, Office of Water (1994).

[16]To review the CVM, refer to the discussion in Chapter 7.

An important study conducted by Carson and Mitchell (1993) uses the CVM approach to estimate the value of water quality improvements associated with U.S. regulations. The researchers surveyed individuals across the United States in 1983, asking them to assign a dollar value to the minimum levels of water quality specified in the post-1972 Clean Water Act, that is, for boatable, fishable, and swimmable surface waters.[17]

As indicated in Table 14.4, usable responses helped to form a preliminary nationwide estimate of the total social benefits (*TSB*) of water quality at $29.2 billion ($1990) per year. (All values are expressed in 1990 dollars.) Carson and Mitchell then combine this estimate with elements of an earlier study by Freeman (1982) that were not captured by their survey and arrive at an adjusted value of $39.1 billion.[18] A final modification is made to adjust for changes in the number of households, the general price level, real income, and attitudes about water pollution up to 1990. Carson and Mitchell's final estimate of *TSB* is $46.7 billion per year as of 1990. This magnitude represents the value of improving water quality from a baseline of nonboatable to swimmable water quality.

TABLE 14.4 Benefit Estimates of Water Quality Improvements

Survey Results of Contingent Valuation Study	
Water Quality	**Estimated Willingness to Pay ($1990) (average per person per year)**
Boatable	$106
Fishable	80
Swimmable	89
Total	$275
Preliminary Estimate of Annual Economic Benefits Based on Survey Results ($1990)	
Range	$24 billion–$45 billion
Point Estimate	$29.2 billion
Adjusted Estimate of Annual Economic Benefits of Clean Water ($1990)	
Adding benefits of commercial usage and marine recreational activities from Freeman (1982)	$39.1 billion
Scaling for increases in real income and changes in attitude about water pollution from 1983 to 1990	$46.7 billion

NOTE: Benefit estimates represent the value of achieving swimmable water quality from a baseline of nonboatable quality.
Source: Drawn from Carson and Mitchell (July 1993).

[17]A later study, conducted by Bingham et al. (1998), uses Carson and Mitchell's results to estimate selected benefits for those rivers predicted to actually achieve the swimmable water quality level.

[18]Their estimate is considerably higher than what Freeman found, which is $20.1 billion in 1990 dollars when adjusted for inflation and number of households. Carson and Mitchell explain that this difference is due to the broader measure of water quality improvement they use in their analysis. Freeman measures the benefit of achieving the effluent limitations, whereas Carson and Mitchell measure the benefit of achieving the swimmable goal. Their estimate might also be higher because the CVM captures the existence value of clean water.

TABLE 14.5 Cost Estimates of Water Quality and Drinking Water Control

	COST ESTIMATES ($1990)				
Program	1980	1987	1990	1995	2000
Water quality[a]	$27.2	$41.0	$46.3	$57.5	$68.7
Drinking water	2.4	3.7	4.3	6.4	7.9
Total[b]	$29.5	$44.7	$50.6	$63.8	$76.4

NOTE: Data for 1990 and beyond are projections.

[a]Water quality costs are those pursuant to the Clean Water Act as amended in 1987 and the Marine Protection, Sanctuaries, and Research Act of 1972.

[b]Some totals do not agree with components shown due to rounding.

Source: U.S. EPA, Office of Policy, Planning, and Evaluation (December 1990), pp. 2–2 to 2–3, Table 2-1.

Cost Estimates

For comparison to Carson and Mitchell's (1993) benefit assessment, we can consider cost estimates from a number of government sources. One comprehensive source is the EPA, whose estimates of annualized costs of water quality control are shown in Table 14.5. Notice that the EPA data show a projected annual cost in 1990 of $50.6 billion. Carson and Mitchell cite the most recently available cost data from the Department of Commerce, which indicate that 1988 annual costs of U.S. water quality control were approximately $37.3 billion expressed in 1990 dollars.[19] To be conservative, we use both these sources and assume that annualized total social costs (*TSC*) for 1990 are somewhere in the range of $37.3 billion to $50.6 billion, or about $44.0 billion as a midpoint estimate.[20]

Benefit-Cost Comparison

Comparing the *TSB* and *TSC* estimates (using 1990 dollars), we observe that the *TSB* of $46.7 billion is slightly higher than the comparable *TSC* estimate of $44.0 billion.

- Point estimate of *TSB*: **$46.7 billion ($1990)**
- Point estimate of *TSC*: **$44.0 billion ($1990)**

However, Carson and Mitchell (1993) suggest that costs are likely to rise above benefits in the future because of the spending needed to bring all surface waters up to the swimmable level. In any case, this comparison of magnitudes does not indicate whether the associated abatement level is efficient.

As Figure 14.2 illustrates, the efficient abatement level occurs where net benefits are maximized at point A_E, the point where *TSB* exceeds *TSC* by the greatest distance. Given the available estimates for 1990, we can conclude only that the associated abatement level lies somewhere between 0 and A_1 where *TSB* is above *TSC*. However, there is no reason to assume that this level coincides with A_E.

[19]Bratton and Rutledge (1990), pp. 32–38, as cited in Carson and Mitchell (July 1993).

[20]Note that although these estimates are comprehensive and include costs to all economic sectors, they do not capture the true social costs of water quality control, because implicit costs are not included.

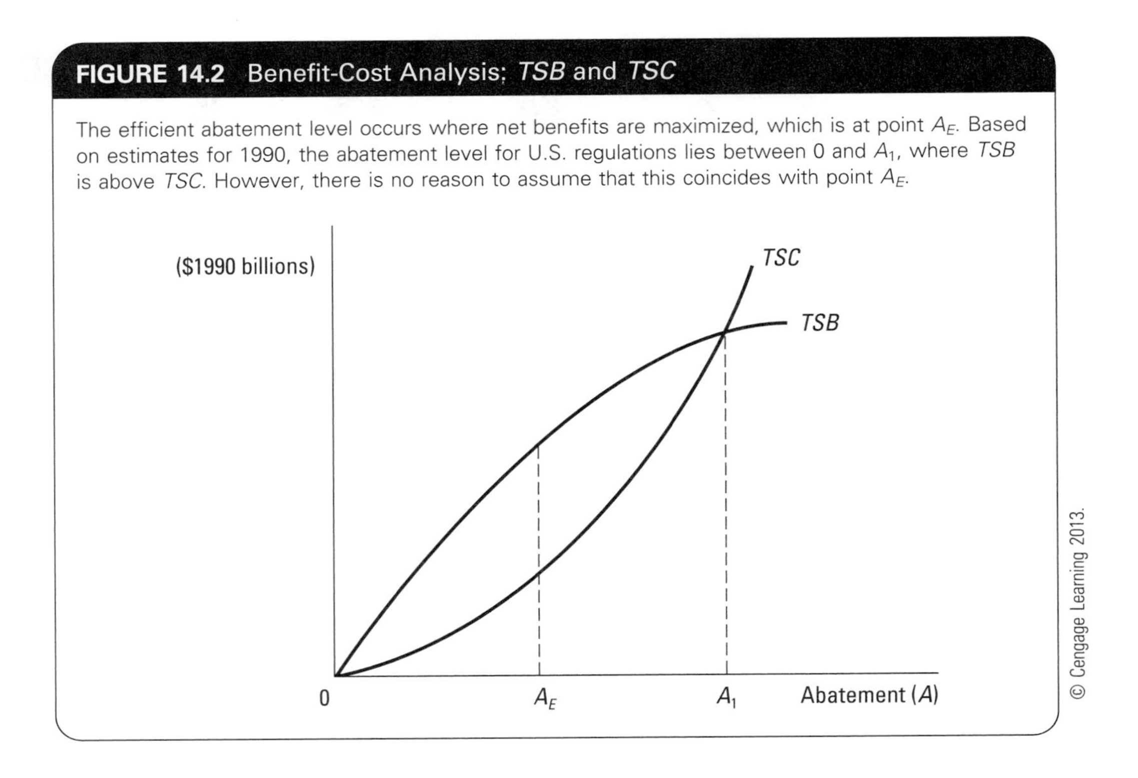

FIGURE 14.2 Benefit-Cost Analysis: *TSB* and *TSC*

The efficient abatement level occurs where net benefits are maximized, which is at point A_E. Based on estimates for 1990, the abatement level for U.S. regulations lies between 0 and A_1, where *TSB* is above *TSC*. However, there is no reason to assume that this coincides with point A_E.

To determine whether or not this is the case, we need to compare marginal social cost (*MSC*) and marginal social benefit (*MSB*) as of 1990. Although these magnitudes are not directly available, a rough approximation is possible by comparing *incremental* costs and *incremental* benefits over a relevant time period as long as the magnitudes are measured in constant dollars. In this context, the appropriate comparison is the incremental costs and benefits between 1985 and 1990, using Freeman's (1982) earlier estimates for 1985 and Carson and Mitchell's estimates for 1990. After converting all estimates to 1990 dollars, these incremental values are as follows:

	Incremental costs for 1985–1990:	
	1990 control costs (EPA and Dept. of Commerce):	$44.0 billion ($1990)
–	1985 control costs (Freeman):	34.6 billion ($1990)
	Incremental costs	**$9.4 billion ($1990)**

	Incremental benefits for 1985–1990:	
	1990 benefits (Carson and Mitchell):	$46.7 billion ($1990)
–	1985 benefits (Freeman):	20.1 billion ($1990)[21]
	Incremental benefits	**$26.6 billion ($1990)**

[21] As noted previously, this magnitude is the value of Freeman's estimate after being adjusted by Carson and Mitchell for number of households and inflation.

These calculations communicate an interesting result. Because the incremental benefits are greater than the incremental costs, there is reason to believe that the post-1972 revisions to U.S. water legislation have not achieved an efficient abatement level. This result is illustrated in Figure 14.3, where incremental costs are shown using the conventional marginal social cost (*MSC*) curve with an analogous interpretation on the benefit side. Notice that the 1990 abatement level (A_{1990}) is *lower* than the efficient level (A_E) where the *MSC* curve intersects the *MSB* curve. This suggests that the current legislation may underregulate polluting sources, and hence more stringent controls may be justified.

This assessment is interesting, but it is important to point out that it is based on crude calculations. For one thing, as Carson and Mitchell assert, most U.S. waters are at a level above the nonboatable quality baseline assumed in the benefit estimate. Hence, costs to achieve swimmable quality are lower than they would be if all water bodies were actually at the baseline level.

Furthermore, the calculations implicitly assume that the values between 1985 and 1990 are directly comparable, which is not the case on the benefit side. Because the 1990 benefit assessment is determined by the CVM, it is a more comprehensive estimate than Freeman's because it includes some measure of existence value. Consequently, Freeman's estimate of $20.1 billion likely undervalues benefits for the 1985 period, which in turn means that incremental benefits likely are biased upward. It can be argued, however, that existence values in 1985 would not be large enough to change the *qualitative* result. Also, Freeman (2000) points out that the two benefit estimates use different baselines of water quality, making the findings not directly comparable.

FIGURE 14.3 Benefit-Cost Analysis: *MSB* and *MSC*

Incremental costs for 1985 to 1990 are shown using the conventional *MSC* curve with an analogous interpretation on the benefit side. Notice that the 1990 abatement level (A_{1990}) is *lower* than the efficient level (A_E), where *MSC* and *MSB* intersect.

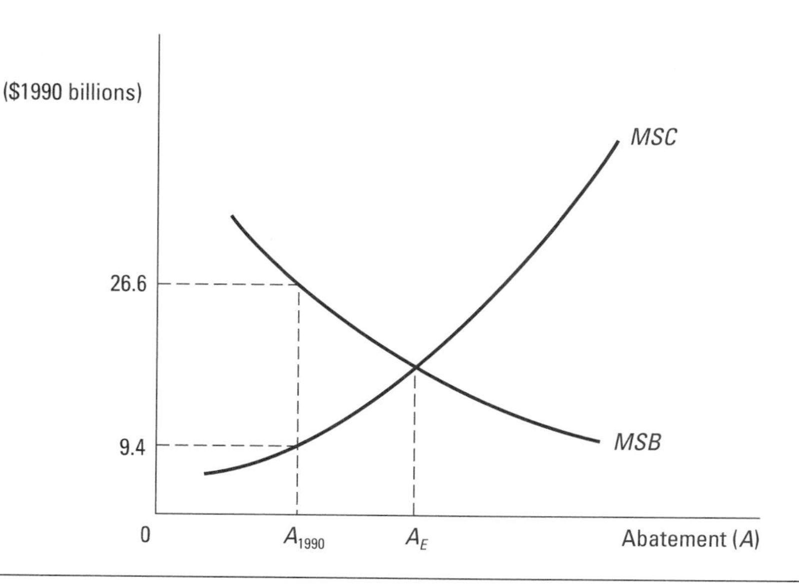

In any case, until more accurate data are available, the conclusion that current U.S. regulations underallocate resources to water quality control is a tentative one at best.

CONCLUSIONS

The water on earth has some natural capacity to replenish itself through the hydrologic cycle. Likewise, water bodies have some ability to assimilate certain of the pollutants they receive. However, there is a limit to nature's restorative powers. Furthermore, the hydrologic cycle that contributes to the restoration process is also the mechanism that can spread contamination. Water pollution is pervasive, and virtually every form of human activity and many natural processes contribute to the problem.

Although water pollution is widespread, society has come to recognize the problem and has begun to take action. In the United States, water quality legislation has been expanded and strengthened over the last several decades. However, most argue that whatever progress has been realized, it has happened at too slow a pace. There have been missed deadlines both in establishing effluent limitations and in setting the receiving water quality standards. Even without the delays, economic analyses suggest that the result may not be an efficient solution. If this is the case, valuable resources are not being employed in their best possible use, and society is paying for the misallocation.

Beyond individual programs, initiatives, and executive orders, an important motivation in moving national water quality policy forward will be for Congress to consider how much progress has been made and at what measure of costs. Once done, Congress has to determine how much of any realized improvement can realistically be attributed to existing policy. Finally, constructive revisions must be proposed, and feasible options formulated. In Chapter 15, we will examine the control instruments currently used to achieve national objectives—in a very real sense, a textbook exposition of what policymakers are now undertaking as they continue to develop water quality policy for the nation.

Summary

- The earth's water supply comes from two major sources: surface water and groundwater.
- Surface water and groundwater are linked by the hydrologic cycle, a model that explains the movement of water from the earth's atmosphere to the surface and back into the atmosphere.
- The two primary source categories of water pollution are point sources and nonpoint sources.
- Federal legislation on water pollution began with the Water Pollution Control Act of 1948. The legislation was revised in the 1950s and 1960s, but little progress was achieved.
- In 1972, Congress enacted the Federal Water Pollution Control Act (FWPCA), which shifted primary responsibility from states to the federal government. Further amendments were accomplished through the Clean Water Act (CWA) of 1977 and the Water Quality Act of 1987.
- Aside from CWA reauthorization bills, Congress has passed other legislation on specific water pollution problems or particular water bodies, such as the Ocean Dumping Ban Act, the Oil Pollution Act (OPA) of 1990, and the Great Lakes Legacy Act of 2002.
- Under the Obama administration, no major water quality legislation has been enacted to date. However, several executive orders have been signed that are focused on water quality. President Obama also announced in 2011 a *Clean Water Framework*, which comprises a national commitment to clean water.
- The three most important objectives of the Clean Water Act are the zero discharge goal, the fishable-swimmable goal, and the no toxics in toxic amounts goal.
- U.S. law distinguishes among three categories of pollutants: toxic pollutants, conventional pollutants, and nonconventional pollutants.
- Surface water quality is defined in the Clean Water Act by receiving water quality standards. These standards assign a use designation for each water body and identify water quality criteria necessary to sustain designated uses.
- State authorities determine the use-support status of a water body by assessing its present condition and comparing it with what is needed to maintain its designated uses.
- One problem with the receiving water quality standards is that states are not required by law to use a benefit-cost evaluation to establish these standards. Also, the link between receiving water quality standards and effluent limits is unclear because each is motivated differently—one by usage and the other by technology.
- A study by Carson and Mitchell, which uses the contingent valuation method to estimate benefits, found that as of 1990, the total social benefits of the Clean Water Act exceeded the total social costs.
- Over the 1985–1990 period, estimates indicate that the incremental benefits of water control policies exceeded incremental costs, suggesting that existing legislation may be underregulating polluting sources.

Review Questions

1. Use the concept of the hydrologic cycle to explain how contamination of surface waters can also cause degradation of groundwater.
2. Refer to Application 14.1 on the Chesapeake Bay. Discuss both the natural and man-made conditions that are most responsible for the degradation of this water body.
3. Nutrients and pesticides are prevalent causes of water pollution. From an economic perspective, analyze why more has not been done to control agricultural runoff, the major source of these contaminants.
4. Identify two significant trends that characterize the evolution of U.S. water quality policy up through the 1987 Water Quality Act.
5. Using economic analysis, comment on the following statement: "*The Clean Water Act's zero discharge goal was doomed from the beginning.*"

6. **a.** Discuss the major difficulties associated with estimating the benefits of water pollution abatement.
   b. Propose a policy approach that would promote states' use of benefit-cost analysis in setting and evaluating standards.
7. Assume that the following benefit and cost equations have been estimated for the abatement of an industrial solvent:

   $MSB = 15 - 0.2A \qquad MSC = 7 + 0.3A$

   where A is the percentage of abatement, and the dollar values are in millions.

   Now suppose that rulings under the Clean Water Act call for an abatement standard of 25 percent. Assess the efficiency implications of this standard.
8. Reconsider the study done by Carson and Mitchell (1993). Cite any two reasons why their benefit estimate might be inflated. Be specific.
9. Consider the following benefit and cost relationships for mercury abatement (A):

   $MSB = 30 - 0.3A \qquad MSC = 16 + 0.2A$

   $TSB = 30A - 0.15A^2 \qquad TSC = 16A + 0.1A^2$

   where A is the percentage of mercury abatement and dollar values are in millions.
   a. Suppose the mercury abatement level was set at 20 percent for 2003. Are net benefits positive or negative?
   b. On the basis of the efficiency criterion, should controls on mercury be tightened or relaxed? Support your response with specific calculations.
10. Visit the Web site at **www.epa.gov/waters/305b/index.html**, and select a state in the nation. Write a one-page summary of the surface water and groundwater quality in that state. Provide some policy suggestions to improve water quality levels. Be specific.

Additional Readings

Andreen, William L. "Water Quality Today—Has the Clean Water Act Been a Success?" *Alabama Law Review 55(3)* (Spring 2004) pp. 537–93.

Boyd, James W. "How Do You Put a Price on Marine Oil Pollution Damages?" *Resources, No. 175,* (Summer 2010), pp. 21–23.

Cohen, Mark A. "Deepwater Drilling: Recommendations for a Safer Future." *Resources, No. 177,* (Winter/Spring 2011), pp. 23–26.

Easter, K. William, and Mary E. Renwick. *Economics of Water Resources.* Burlington, VT: Ashgate, 2004.

Egan, Kevin J., Catherine L. Kling, and John A. Downing. "Valuing Water Quality as a Function of Water Quality Measures." *American Journal of Agricultural Economics 91(1)* (February 2009), pp. 106–22.

Freudenburg, William R., and Robert Gramling. *Blowout in the Gulf: The BP Oil Spill Disaster and the Future of Energy in America.* Cambridge, MA: MIT Press, 2010.

Gordon, Frederick. "Water Issues at the U.S.–Mexican Border." *Environment 46(4)* (May 2004), pp. 42–43.

Hanley, Nick, Sergio Colombo, Dugald Tinch, Andrew Black, and Ashar Aftab. "Estimating the Benefits of Water Quality Improvements Under the Water Framework Directive: Are Benefits Transferable?" *European Review of Agricultural Economics 33(3)* (September 2006), pp. 391–413.

Hansen, Kristiana, Richard Howitt, and Jeffrey Williams. "Valuing Risks: Options in California Water Markets." *American Journal of Agricultural Economics 90(5)* (December 2008), pp. 1336–42.

Harrington, Winston. "Industrial Water Pollution in the United States: Direct Regulation or Market Incentive?" In Winston Harrington, Richard D. Morgenstern, and Thomas Sterner, eds. *Choosing Environmental Policy.* Washington, DC: Resources for the Future, 2004.

Holland, Daniel S., James N. Sanchirico, Robert J. Johnston, and Deepak Joglekar. *Economic Analysis for Ecosystem-Based Management: Applications to Marine and Coastal Environments*. Washington, DC: RFF Press, February 2010.

Jacobs, J. W., and J. L. Wescoat. "Managing River Resources: Lessons From Glen Canyon Dam." *Environment 44(2)* (2002), pp. 8–19.

Kahl, Jeffrey S., John L. Stoddard, Richard Haeuber, Steven G. Paulsen, Rona Birnbaum, Frank A. Deviney, James R. Webb, *et al.* "Have U.S. Surface Waters Responded to the 1990 Clean Air Act Amendments?" *Environmental Science & Technology 38(24)* (December 15, 2004), pp. 484A–490A.

Morgan, Cynthia, and Nicole Owens. "Benefits of Water Quality Policies: The Chesapeake Bay." *Ecological Economics 39(2)* (November 2001), pp. 271–84.

Nelson, Rebecca. "Water Pollution in China: How Can Business Influence for Good?" *Asian Business & Management 7(4)* (December 2008), pp. 489–509.

Olmstead, Sheila M. "The Economics of Water Quality." *Review of Environmental Economics and Policy 4(1)* (Winter 2010), pp. 44–62.

Parris, Thomas M. "Is It Safe to Swim Here?" *Environment 46(4)* (May 2004), p. 3.

Renzetti, Steven, ed. *The Economics of Industrial Water Use*. Northampton, MA: Elgar, 2002.

Swihart, Tom. *Florida's Water*. Washington, DC: RFF Press, May 2011.

A Reference to Acronyms and Terms in Water Quality Control Policy

Environmental Economics Acronyms

CVM	Contingent valuation method
MSB	Marginal social benefit
MSC	Marginal social cost
TSB	Total social benefits
TSC	Total social costs

Environmental Science Terms

BOD	Biological oxygen demand
NO_x	Nitrogen oxide
PCBs	Polychlorinated biphenyls
SO_2	Sulfur dioxide

Environmental Policy Acronyms

BEACH	Beaches Environmental Assessment and Coastal Health Act
CWA	Clean Water Act
CZARA	Coastal Zone Act Reauthorization Amendments
FLC	Federal Leadership Committee
FWPCA	Federal Water Pollution Control Act
NOAA	National Oceanic and Atmospheric Administration
OPA	Oil Pollution Act of 1990
POTWs	Publicly owned treatment works
TMDLs	Total maximum daily loads
USFWS	U.S. Fish and Wildlife Services

To access additional course materials, visit www.cengagebrain.com. At the home page, search for the ISBN of this title (shown on the back cover). This will take you to the product page where these resources can be found.

CHAPTER

15

Improving Water Quality: Controlling Point and Nonpoint Sources

"Water is the driving force of all nature."
—Leonardo da Vinci (1452–1519)

Few would debate that the nation's water quality has measurably improved over time. Yet, some water bodies continue to deteriorate, and most are still threatened by contamination—particularly the pollution from nonpoint sources. Furthermore, several key policy instruments have come under fire. A common criticism is an overreliance on uniform technology-based limits and a lack of economic incentives to achieve national objectives. An investigation of these and other issues forms the agenda for this chapter.

Our analysis will cover the following policy instruments established by the Clean Water Act: effluent limitations and permits to control point sources, funding programs for publicly owned treatment works, and nonpoint source initiatives. In each case, we present an overview of the control approach followed by an analytical assessment. Where appropriate, the evaluation will be based on the economic criteria of efficiency and cost-effectiveness. We conclude this chapter with a discussion of two important trends in water quality policy: the use of market-based policy instruments and the watershed management approach. A list of acronyms and terms is provided at the end of the chapter for reference.

CONTROLLING POINT SOURCES

When evidence began to mount in the 1960s and early 1970s that U.S. waters were deteriorating, Congress realized that federal legislation was in need of a major overhaul. The result was a command-and-control policy dominated by uniform pollution standards—a response not unlike the initial reaction to America's air quality problems.

technology-based effluent limitations Standards to control discharges from point sources based primarily on technological capability.

Technology-Based Effluent Limitations

The chief control instruments for achieving U.S. water quality objectives are **technology-based effluent limitations**, so named because the most important consideration in setting these limits is technological capability. Following a command-and-control approach, these standards limit the pollution that may be released into surface waters by point sources. However, polluters are allowed to choose the method by which the limit is achieved, making these limitations more accurately termed **performance-based standards**. As technology advances, these standards are to be revised. Over time, the intent is for effluents to be completely eliminated to meet the zero discharge goal.

All point sources are subject to effluent limitations, which differ for various groups, such as publicly owned treatment works (POTWs), indirect industrial polluters that release effluents to POTWs, and direct industrial dischargers that release pollution directly to surface waters. However, within any designated group, standards are applied uniformly. For direct industrial dischargers, the standards are industry-specific and vary by the age of the facility and type of contaminant released. An overview of these standards is provided in Table 15.1.

Notice the distinction between **new** and **existing** industrial facilities, analogous to what exists for air quality standards, whereby new sources must meet more stringent limits than existing sources.

- **New sources** must meet standards based on the best available demonstrated control technology (BADCT).
- **Existing sources** must meet two sets of standards: those based on best conventional control technology (BCT) for conventional pollutants and those based on best available technology economically achievable (BAT) for nonconventional and toxic pollutants.

An important difference between the BCT and BAT standards is that the determinants of the BCT standards include a consideration for the relationship between the associated benefits and costs.

National Pollutant Discharge Elimination System (NPDES)
A permit system to control effluent releases from direct industrial dischargers and POTWs.

Permit System

Once determined, pollution limits for direct industrial dischargers and for POTWs are communicated through a permitting system, called the **National Pollutant Discharge Elimination System (NPDES)**. In simplest terms, this system prohibits any discharges into navigable waters without an NPDES permit. The permits state precisely what the effluent limitations are as well as the monitoring and reporting requirements.

TABLE 15.1 Technology-Based Effluent Limitations

Source Type	Pollutant Type	Standard
New Direct Industrial Dischargers	Toxic, conventional, nonconventional	Best available demonstrated control technology (BADCT)
Existing Direct Industrial Dischargers	Conventional	Best conventional control technology (BCT)
Existing Direct Industrial Dischargers	Toxic, nonconventional	Best available technology economically achievable (BAT)

NOTE: More information about these effluent limitations is available online at **http://cfpub.epa.gov/npdes/techbasedpermitting/effguide.cfm?program_id=15**.

Though the Environmental Protection Agency (EPA) bears the overall responsibility for this program, states may administer NPDES permits subject to certain federal requirements. Currently, 46 states and 1 territory have assumed this responsibility.[1] More information on the NPDES is available at **http://cfpub.epa.gov/npdes/index.cfm**.

ANALYSIS OF THE EFFLUENT LIMITATIONS

The command-and-control approach to U.S. water quality policy has been the subject of some debate. Much of the criticism has been lodged against the technology-based effluent limits, which have been blamed for the lack of progress toward achieving national objectives. Among the recognized problems are the following:

- Imprecise statutory definitions
- Meeting the zero discharge goal
- Lack of an efficiency criterion
- Cost-ineffective decision making

Imprecise Statutory Definitions

A fundamental problem with the effluent limitations is that they are not aligned with the nation's objectives. The standards are based on what is technologically feasible, as opposed to what is necessary to achieve water quality. In fact, the potential inadequacy of these standards is implied by the statutes themselves. As noted in Chapter 14, waters for which the effluent limits are insufficient are considered to be "water quality limited" and require the setting of more stringent controls, called **total maximum daily loads (TMDLs)**. These are to be met if the desired level of water quality is not being achieved, even if a polluting source is already satisfying the technology-based limits.

total maximum daily loads (TMDLs) Maximum amount of pollution a water body can receive without violating the standards.

While this may seem reasonable, it nonetheless speaks to the deficiency of the effluent limits. The fact that modifications might be needed calls into question the characterization of the technology-based limitations as "best." Moreover, phrases such as "best practicable," "best available," and "best conventional" are not exact terms. Without more careful guidelines and objective decision rules, officials are left with the task of trying to infer what these terms mean.[2]

Meeting the Zero Discharge Goal

In retrospect, the zero discharge goal was overly ambitious. Furthermore, its call for the complete elimination of water-polluting effluents is likely an inefficient objective, since it is inherently benefit-based. Lastly, although the Clean Water Act (CWA) requires the

[1]U.S. EPA, Office of Water, Office of Wastewater Management (April 14, 2003).

[2]Freeman (1990), p. 107.

EPA to advance the standards successively toward a zero limit as new technology becomes available, the EPA's track record in this regard has been less than satisfactory. In only a few instances (among them onshore oil and gas wells) has this limit been imposed.

Lack of an Efficiency Criterion

The CWA allows for many factors to be considered in setting the effluent standards beyond technological feasibility, including economic consequences. However, the provisions do little more than list the relevant factors, offering no guidance as to how they are to be used in decision making. This is problematic, because many determinants are itemized in the law, and they cross over from engineering and scientific criteria to economic considerations.

In the definition of the BCT standards for conventional pollutants, there is a troubling lack of precision in the reference to benefits and costs. According to law, officials are to consider the "reasonableness of the relationship between the costs of attaining a reduction in effluents and the effluent reduction benefits derived" (CWA, Section 304b–4-B). This is a far cry from setting abatement levels at the point where marginal benefits and marginal costs are equal. And absent from the BAT standards is any reference to economic benefits. Only the cost of achieving the required effluent reduction is among the list of determinants. As long as the law does not mandate that these standards be set to maximize net benefits, **efficient** abatement levels likely will not be achieved.

Cost-Ineffective Decision Making

In instances where the law prevents the use of the efficiency criterion, a "second-best" economic solution is to select cost-effective policy instruments to achieve an objective. **Cost-effectiveness** requires that abatement levels be set to achieve equal marginal abatement cost (*MAC*) levels across all polluters. However, the *uniformity* of the effluent limits likely prevents such an outcome.

Although the limitations are industry specific, the associated effluent reductions must be achieved by all polluting sources within each industry group, regardless of firm-level differences in resource availability or technological expertise. If all dischargers achieve the same standard, the only way their *MAC*s would be equal is if these polluting sources were identical. Because this clearly is not the case, the standards impose higher costs to society than is necessary.

Just how significant is the problem? Unfortunately, no research investigation provides a comprehensive answer to this question. However, some economic studies have estimated the cost implications of command-and-control instruments for specific water bodies. Table 15.2 summarizes some of these findings. Each analysis determines a ratio of the cost of implementing a command-and-control approach to that of the least-cost market-based method. In each case, the ratio is significantly greater than 1, meaning that the command-and-control approach is more costly than using economic incentives.

TABLE 15.2 Cost-Effectiveness of Market-Based Controls: Some Quantitative Studies

Investigators and Year	Command-and-Control Approach	Geographic Area	Ratio of Command-and-Control Cost to Least-Cost
Johnson (1967)	Equal proportional treatment	Delaware Estuary	3.13 at 2 mg/l 1.62 at 3 mg/l 1.43 at 4 mg/l
O'Neil (1980)	Equal proportional treatment	Lower Fox River, Wisconsin	2.29 at 2 mg/l 1.71 at 4 mg/l 1.45 at 6.2 mg/l
Eheart, Brill, and Lyon (1983)	Equal proportional treatment	Willamette River, Oregon	1.12 at 4.8 mg/l 1.19 at 7.5 mg/l
		Delaware Estuary in Pennsylvania, Delaware, and New Jersey	3.00 at 3 mg/l 2.92 at 3.6 mg/l
		Upper Hudson River, New York	1.54 at 5.1 mg/l 1.62 at 5.9 mg/l
		Mohawk River, New York	1.22 at 6.8 mg/l
Faeth (2000)	Equal treatment	Minnesota River Valley	2.7 at 1 ppm/l
		Rock River, Wisconsin	1.74 at 1 mg/l
		Saginaw Bay, Michigan	5.9 at 1 mg/l
Farrow, Schultz, Celikkol, and Van Houtven (2005)	Equal proportional removal	Upper Ohio River Basin	1.33 (linear cost) 2.58 (quadratic cost)

NOTES: mg/l = milligrams per liter (of some pollutant).
ppm/l = parts per million per liter (of some pollutant).
Study by Farrow *et al.* (2005) holds social damages constant rather than pollution level.

Sources: U.S. EPA, Office of Policy, Economics, and Innovation, Office of the Administrator (January 2001), p. 26, Table 3-3; Tietenberg (1985); original sources cited in the table.

THE POTW FUNDING PROGRAM

An important and sometimes controversial aspect of the Clean Water Act (CWA) is the federal funding authorized by Congress to support the construction of POTWs, which are potentially significant sources of water contamination. Among the pollutants released by these facilities are pesticides, heavy metals, viruses, and bacteria. Unless municipal wastewater is properly treated, the associated pollution threatens ground water and surface water drinking supplies, aquatic life, recreational opportunities, and the overall health and stability of ecosystems. Responding to the potential risks, the law requires POTWs to satisfy technology-based secondary treatment standards and calls for federal funding to support this mandate.

federal grant program
Provided major funding from the federal government for a share of the construction costs of POTWs.

Federal Grant Program

Prior to 1987, one titled section of the CWA was devoted to waste treatment management. The voluminous and highly detailed set of provisions outlined a **federal grant program** for POTW construction at the local level. Before its demise, the program allocated over $60 billion in federal monies to this effort.[3] The federal cost share was set at a maximum of 75 percent until 1984, when it was reduced to 55 percent.

Clean Water State Revolving Fund (CWSRF) Program[4]

Clean Water State Revolving Fund (CWSRF) program Establishes state lending programs to support POTW construction and other projects.

As part of the 1987 CWA reauthorization, capitalization grants were authorized to establish a loan program for water pollution projects. States must provide 20 percent in matching funds to support the effort. This **Clean Water State Revolving Fund (CWSRF) program**, which replaced the federal grant program, is to provide loans for POTW construction as well as for other environmental projects. All 50 states and Puerto Rico have established CWSRF programs. As of 2009, $74 billion in cumulative assistance has been made available to fund nearly 24,700 projects, as illustrated in Figure 15.1.

More recently, the American Recovery and Reinvestment Act of 2009 provided an additional $4 billion to the CWSRF program for 2009. The funding is distributed by the EPA through Recovery Act grants to states and Puerto Rico. Grant recipients are required to allocate 20 percent of their funding to four green projects: green infrastructure, water efficiency improvements, energy efficiency improvements, and environmentally innovative activities. For more details on the CWSRF program, visit **http://water.epa.gov/grants_funding/cwf/cwsrf_index.cfm**, and for further information on the Recovery Act funding of this program, visit **http://water.epa.gov/aboutow/eparecovery/index.cfm**.

FIGURE 15.1 Annual and Cumulative Funding for POTW Construction

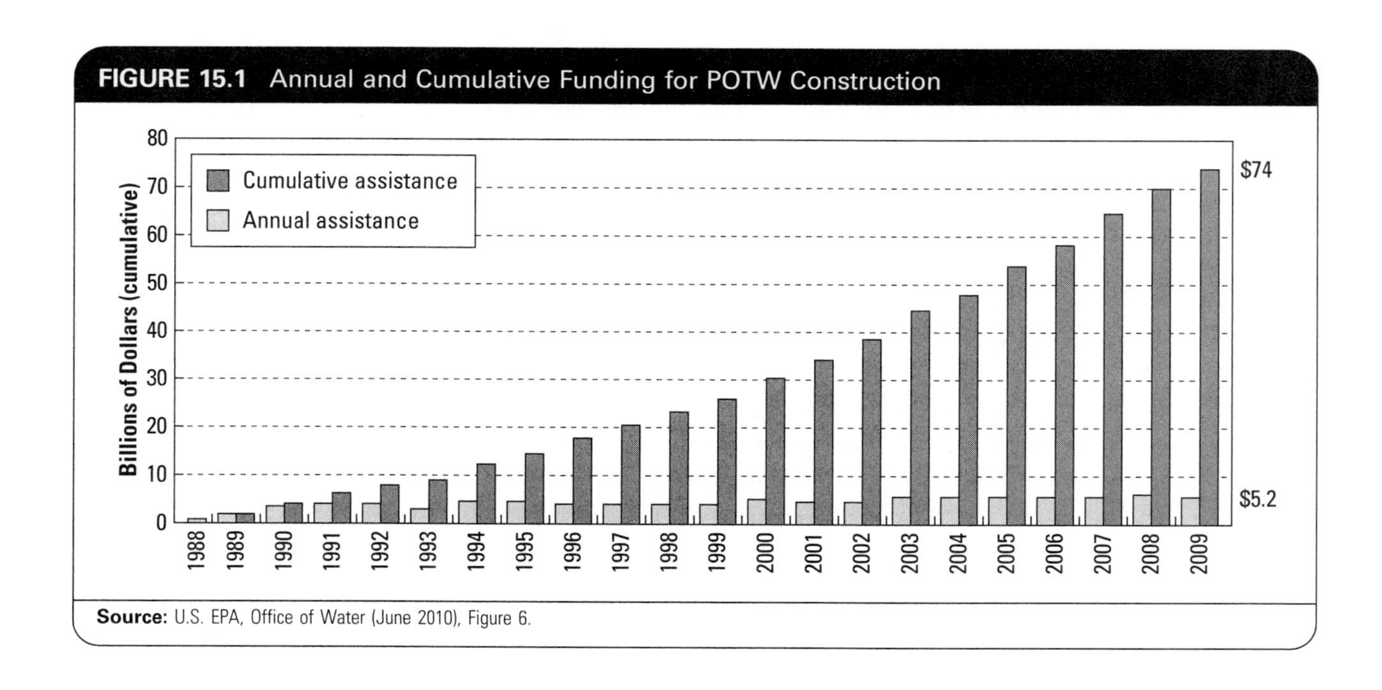

Source: U.S. EPA, Office of Water (June 2010), Figure 6.

[3]U.S. EPA, Office of Water (June 28, 2002).
[4]Drawn from U.S. EPA, Office of Water (June 2010), pp. 2–5.

ANALYSIS OF THE POTW FUNDING PROGRAM

Most understand the motivation behind the federal subsidy of POTW construction, given the potential health and ecological threat of inadequate waste treatment. Yet the issue is not whether the intent of the program is well founded but rather whether these subsidies are an effective policy instrument.

Identifying Accomplishments Attributable to Federal Subsidies

To argue that federal subsidies have been effective because municipal waste treatment has progressed measurably is falsely motivated. Such an argument implicitly assumes that the federal program is fully responsible for any observed improvement, when in fact, the evidence suggests otherwise.

It *is* true that an increasing proportion of the population is served by facilities using at least secondary treatment. Between 1968 and 1996, this proportion nearly doubled from 85.9 million to 164.8 million, and the number of people whose service comprised less than secondary treatment fell sharply during that period, from 54.6 million to 17.2 million.[5] Yet, research shows that most of the federal grant monies only *displaced* local funds that would have been allocated to POTW construction had the national program not been in place.[6] Because only a portion of the federal monies was incremental to what would have been spent at the local level, only a fraction of the improved water quality can be linked to the federal aid program.

Inefficiencies in the Grant Program

That the federal grant program achieved something less than a dollar-for-dollar improvement is an important realization, but it's only part of the story. It turns out that the well-intended plan was plagued by inefficiencies. Most, if not all, of these were due to one important void: a lack of incentives.

Fundamental to the program's design was the share of costs to be absorbed by the federal government. This share was originally set at a 75 percent maximum. Such a large proportion shifted most of the spending away from local governments, leaving them little incentive to minimize costs in the building of treatment facilities. In fact, with the addition of special aid from federal and state governments, some municipalities had to raise as little as 5 percent of construction costs. Consequently, some POTWs are larger and more elaborate than necessary, motivated by a desire to attract industry and promote growth. Such excess capacity is a waste of economic resources and has been blamed for the poor operating performance observed in some facilities.[7]

Responding to such inefficiencies, the U.S. government made several changes to the program. Aggregate grant awards were cut significantly, and starting in 1985, the federal cost share was lowered to 55 percent. To reduce excess capacity, tighter restrictions on grant authorizations were instituted, such as limiting funds to servicing the needs of

[5]U.S. EPA, Office of Water, Office of Wastewater Management (October 30, 2007).

[6]Jondrow and Levy (1984).

[7]U.S. Congress, Congressional Budget Office (CBO) (1985a), pp. ix, 12.

existing rather than projected population levels.[8] Finally, the basic premise of the program was altered from outright construction aid to funding for state-administered revolving loans. The overall effect was significant.

As more of the cost burden is shifted to local governments, municipalities must become more self-sufficient in developing revenue sources to fund wastewater treatment. This in turn should motivate more cost-effective decision making. Furthermore, concern for higher sewer fees may prompt local residents to take a more active role in ensuring that local officials carefully consider costs. According to a statistical analysis done by the U.S. Congressional Budget Office (CBO), increases in the lifetime local cost share of up to 50 percent to 60 percent led to efficiency improvements measured as declines in lifetime unit costs.[9]

Equity Implications

Beyond the efficiency issues, the POTW program is also associated with equity imbalances. Some municipalities had not yet been funded when the grant program was eliminated, placing a relatively higher cost burden on those communities. The inequity was greater for smaller, rural communities that were unable to take advantage of scale economies. The shift to the Clean Water State Revolving Fund (CWSRF) program may have provided some measure of offset, however, because state-managed loans can be tailored through interest rates or grace periods to accommodate lower income or wealth levels of certain local communities.[10]

In truth, according to the most recently available data, small communities are receiving the lion's share of assistance agreements, although the dollar value is proportionately small. Specifically, as shown in Figure 15.2, communities with populations

FIGURE 15.2 Communities Funded by CWSRFs in 2009 by Population Size

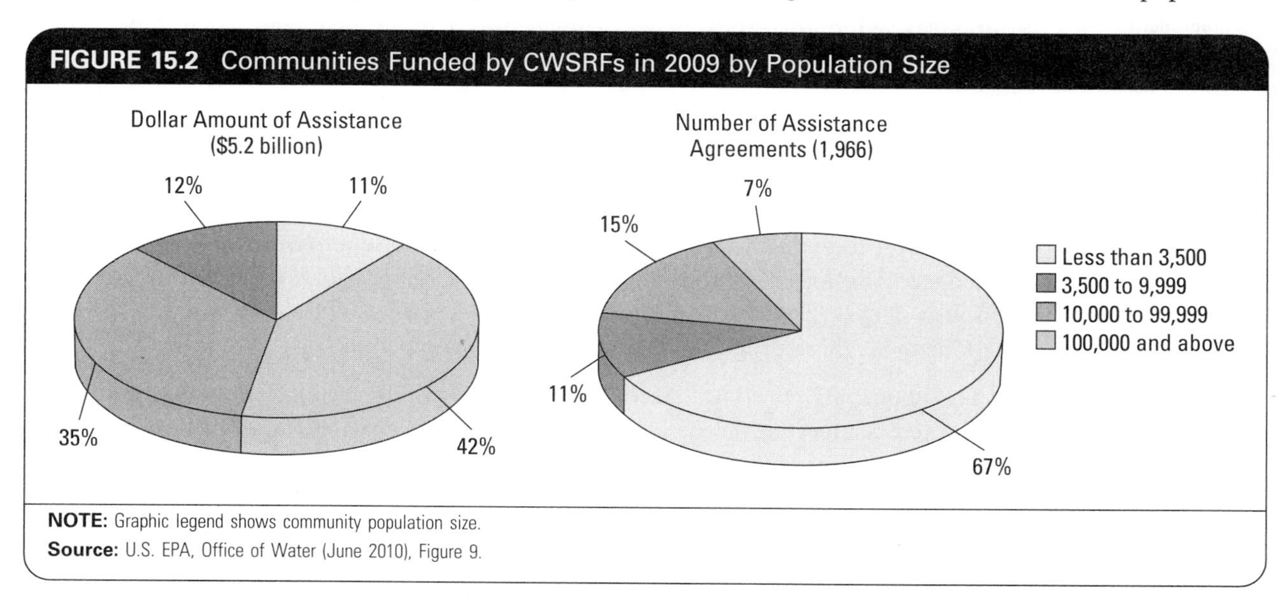

NOTE: Graphic legend shows community population size.
Source: U.S. EPA, Office of Water (June 2010), Figure 9.

[8]See U.S. Congress, CBO (1985a), p. 3. To learn more about the inherent incentive toward excess capacity, see Freeman (1978), pp. 63–64.

[9]For more details on this study, consult U.S. Congress, CBO (1985a), Chapter 2.

[10]U.S. Congress, CBO (1985a), pp. 54, 59–60.

below 10,000 received 78 percent of the number of CWSRF assistance agreements in 2009. However, in terms of the dollar value of that assistance, the proportion is 23 percent of the total $5.2 billion, or about $1.2 billion. In addition, these smaller communities were awarded more than half of the funding available through the American Recovery and Reinvestment Act as of mid-2009. In contrast, large communities with population of 100,000 people or more received 7 percent of the available assistance agreements, but this relatively small share represented 42 percent of available funding, or about $2.2 billion.[11]

CONTROLLING NONPOINT SOURCES[12]

In 1987, the federal government added a policy goal to the Clean Water Act that called for the development of programs to control nonpoint polluting sources. This addition reflected a growing awareness that these diffuse sources are major contributors to surface water and groundwater pollution. In fact, the EPA asserts that nonpoint source pollution is the greatest source of water quality impairment in the United States. A case in point is the problem of mercury contamination caused mainly by atmospheric deposition, a significant nonpoint source. Read Application 15.1 for insight into this important contemporary problem.

Nonpoint Source Management Program

Nonpoint Source Management Program
A three-stage, state-implemented plan aimed at nonpoint source pollution.

Best Management Practices (BMP)
Strategies other than effluent limitations to reduce pollution from nonpoint sources.

Officially enacted as Section 319 of the Water Quality Act of 1987, the **Nonpoint Source Management Program** was launched as a three-stage plan to be implemented by states with federal approval and financial assistance. According to the plan, states must:

1. Prepare reports in which they identify waters that cannot achieve water quality standards without action taken toward nonpoint sources and identify the sources responsible.
2. Develop programs in which they designate **best management practices (BMP)** to reduce pollution from every identified category, subcategory, or individual nonpoint source.
3. Implement these programs over a multiyear time period.

To support states' efforts, Section 319 federal grants are available for a portion of the total costs incurred. From 2000 through 2009, annual federal appropriations have averaged approximately $216.3 million.[13] To learn more about the Nonpoint Source Management Program (Section 319), visit **www.epa.gov/owow_keep/NPS/cwact.html**.

[11]U.S. EPA, Office of Water (June 2010).

[12]The following discussion is drawn from the CWA, Sec. 319. Nonpoint Source Management Programs.

[13]U.S. EPA, Office of Water, Office of Wetlands, Oceans, and Watersheds (November 23, 2009).

APPLICATION 15.1 Mercury Contamination and Fish Advisories

In the United States, mercury is among the chief causes of water quality impairment. As rising levels of mercury have been detected, concerns have mounted about the observed changes and the inherent risks of exposure. Mercury is classified as a persistent, bioaccumulative, toxic pollutant. This means it is highly toxic and persists in the environment for long periods of time, which indicates that it can accumulate in the food chain.

Although mercury is naturally present, the increasing levels discovered in water resources are linked to anthropogenic (human-induced) sources. Chief among these are power plants, which are responsible for 50 percent of mercury emissions, with coal-fired facilities responsible for most of that amount. This reality motivates in part the Toxics Rule issued in 2011 (discussed in more detail in Chapter 10). This rule establishes a new National Emission Standard for Hazardous Air Pollutants (NESHAP) aimed at reducing emissions of mercury and other heavy metals released from coal-fired and oil-fired power plants.

As mercury moves from the atmosphere to land and water resources, it undergoes a chemical transformation showing up as methylmercury, an organic form that accumulates in fish tissue. It therefore *indirectly* contaminates water resources through atmospheric deposition and runoff, which are nonpoint polluting sources. This cycle continues as mercury evaporates back into the atmosphere and is redeposited. Because the cycle of contamination is complex, so too are solutions.

Part of the dilemma is that the sources of atmospheric deposition of mercury are not geographically confined to a local or even regional area. Scientists are learning from satellite monitoring and other advanced technologies that mercury can move across continents. This explains why mercury contamination is an increasing problem in the United States despite a decline in U.S. mercury emissions. Some scientists estimate that as much as 30 percent of mercury seeping into U.S. soil and water resources comes from other nations, particularly China, where over 2,000 coal-burning power plants are subject to relatively lenient environmental controls.

States monitor chemical contamination of surface waters by testing fish tissue. Once done, they issue fish consumption advisories as needed, which are available at **http://water.epa.gov/scitech/swguidance/fishshellfish/fishadvisories/**. These advisories are not regulations; rather they represent recommendations to the public to limit or avoid consumption of certain fish or wildlife from certain water-body categories (e.g., lakes), or from specific water bodies. According to the most recently available National Listing of Fish Advisories, nearly 4,250 advisories are in effect across all 50 states, the District of Columbia, 2 U.S. territories, and 5 Native American tribes. Of this total number of advisories, 80 percent have been issued at least in part because of mercury contamination.

Populations most at risk are young children, infants, and fetuses, because mercury is harmful to developing nervous systems. Hence, the EPA and the Food and Drug Administration (FDA) make specific recommendations for young children, women who are pregnant or might become pregnant, and nursing mothers. In addition to limiting weekly consumption of fish and shellfish that are lower in mercury, like shrimp, salmon, and catfish, the advisory also recommends that these groups avoid consuming predator fish that are higher in the food chain, including shark, swordfish, and king mackerel.

Sources: U.S. EPA, Office of Air and Radiation, Office of Air Quality Planning and Standards (March 18, 2011); U.S. EPA, Office of Water, Office of Science and Technology (September 2009; July 2007; August 2004); Pottinger, Stecklow, and Fialka (December 17, 2004).

ANALYSIS OF NONPOINT SOURCE CONTROLS

Until the law called attention to nonpoint sources in 1987, policymakers had focused their energies on point sources, the more visible and obvious sources of contamination. Once the need for revision was recognized, legislators placed the onus on state governments to develop nonpoint source pollution programs. Because of the importance of this issue, we need to consider the reasonableness of this decision. What are the pros and cons of state-level control of nonpoint pollution problems? We also need to assess the role of federal government in this effort.[14]

Delegating Control to the States: The Pros

One factor in support of delegating nonpoint source control to states is the variability of nonpoint source pollution. Not only are these sources difficult to identify and isolate, but also the extent of the associated damage is unpredictable. A major source of the problem is land runoff from farms, city streets, mines, construction sites, etc. Because runoff is influenced by precipitation, the resulting contamination is affected by many exogenous factors, such as weather, geological patterns, and soil conditions. Not only are these factors uncontrollable, but they also vary considerably from location to location. Logically, the use of broad-based, uniform controls is likely to be ineffective and difficult to implement.

Another relevant factor is that nonpoint pollution is affected by land use practices, such as agriculture, mining, forestry, and urban development, all of which historically have been controlled by local governments. It therefore becomes a politically sensitive issue if federal policy dictates how a state or local community should use its land resources.

Delegating Control to the States: The Cons

Although there is logic to controlling nonpoint pollution close to the source, state governments often lack the necessary information to carry out their responsibilities fully. Data on the extent of water contamination are inadequate, in large part because of ineffective monitoring systems. Most state assessment data are collected from monitoring systems designed to detect point source pollution. Therefore, they do not give an accurate picture of the damage from nonpoint sources. Inadequate monitoring data not only hinders environmental assessment but also impedes the evaluation of states' management programs and their designation of best management practices (BMP).[15]

There is also the potential problem of inconsistent pollution controls when state governments are in charge of policy implementation. Such inconsistencies are problematic because water contamination in one state can flow downstream into another state's jurisdiction. Anticipating this possibility, Congress provided for intervention by the EPA to arrange for interstate management conferences if a state's water body is not meeting standards because of nonpoint sources in another jurisdiction.

Assessing the Federal Role

Because of the historical emphasis on point source pollution, federal funds are needed to fill information voids about nonpoint source pollution and to improve monitoring technologies aimed at this more complex problem. Attending to these needs requires substantial resources. Data on federal expenditures illustrate the national emphasis on point sources, as shown in Table 15.3. In addition to absolute differences in federal funding, notice also that the proportion of spending dedicated to nonpoint sources

[14]Much of the following discussion is drawn from U.S. GAO (October 1990), Chapter 2.
[15]U.S. GAO (July 1991), pp. 21–22.

TABLE 15.3 Expenditures on Point and Nonpoint Source Controls

	YEAR				
Program	1972	1980	1987	1995	2000
Point source	$16,997 (93.8%)	$44,002 (97.2%)	$66,932 (97.7%)	$94,106 (98.1%)	$112,618 (98.3%)
Nonpoint source	$1,128 (6.2%)	$1,287 (2.8%)	$1,551 (2.3%)	$1,776 (1.9%)	$1,908 (1.7%)
Water quality total[a]	$18,125 (100%)	$45,289 (100%)	$68,483 (100%)	$95,883 (100%)	$114,526 (100%)

NOTES: All values are in $2010 millions.
Relative proportions given in parentheses.
[a]Total excludes expenditures on drinking water controls.
Source: U.S. EPA, Office of Policy, Planning, and Evaluation (December 1990), p. 3-3, Table 3-3.

shows a decline from 6.2 percent in 1972 to 1.7 percent in 2000. Moreover, the funding allocation across the two sources is not supported by relative risk analysis. One study showed that, although health risks from point and nonpoint source pollution are comparable, nonpoint sources pose a much greater risk to ecosystems.[16]

Another issue needing federal intervention is the potential conflict between water quality objectives and other regulations. This is particularly important for policies or programs that support industries such as agriculture and forestry, which are major contributors to nonpoint source pollution. A case in point is any program-based incentive for farmers to specialize in certain crops. The problem is that repeated plantings of the same crop year after year deplete the soil and make plantings more vulnerable to pests. Consequently, farmers might rely more on agrichemicals, such as fertilizers and pesticides, both of which contribute to agricultural runoff—the leading source of pollution in U.S. rivers, streams, and lakes.[17]

MARKET-BASED SOLUTIONS

The lack of progress in achieving national water quality goals has prompted public officials and policymakers to look for alternatives. Following the lead of several European countries, the United States has begun to consider the potential cost-savings of using market-based instruments in water policy prescriptions. In fact, local governments have been experimenting with incentive-based policy tools for some time. The transition is still in process, and not every program has met expectations. Nonetheless, to understand the underlying economics as well as the practical challenges of implementation, we consider two types of market-based instruments used to manage water resources: pollution charges and tradeable effluent permit markets.

[16]U.S. EPA (August 1989).

[17]Specifically, agricultural runoff contributes 48 percent of the water quality problems in impaired rivers and streams and 41 percent in impaired lakes (U.S. EPA, Office of Water, August 2002, pp. 15, 22).

Pollution Charges

As discussed in Chapter 5, the underlying premise of a **pollution charge** is to internalize the negative externality of environmental damage by pricing the pollution-generating activity. This follows the "polluter-pays principle." In this context, we consider two types of pollution charges: an effluent fee for point sources and a product charge for nonpoint sources.

Effluent Fees for Point Sources

volume-based effluent fee
Based on the quantity of pollution discharged.

pollutant-based effluent fee
Based on the degree of harm associated with the contaminant being released.

An effluent fee is a charge based on the release of pollution. The fee can be based on the volume or type of effluent released. A **volume-based effluent fee** is imposed on a per-unit basis so that polluters pay higher amounts for larger quantities of discharges. If the intent is to discourage the release of more damaging effluents, such as those containing toxics, a **pollutant-based effluent fee** can be used. In this case, the fee is higher for discharges containing more harmful substances.

To determine the potential for cost savings, assume that the government sets a per-unit effluent fee (*MEF*) for releases of a conventional water pollutant. Each polluting source would abate up to the point where its marginal abatement cost (*MAC*) equaled the fee. This is shown in Figure 15.3 for two hypothetical firms, each facing different *MAC* curves but the same *MEF*. Firm 1 faces MAC_1 and abates A_1 units of effluent, and Firm 2 faces MAC_2 and abates A_2 units. Although the abatement amounts are different, the associated *MAC* level for each firm is the same, which means that the effluent fee yields a **cost-effective** solution.

Recognize, however, that the combined abatement level achieved by both firms would not be **efficient** unless the marginal social benefit (*MSB*) of abatement were equal to the associated marginal social cost (*MSC*) of abatement. Identifying the *MSB* and the *MSC* in

FIGURE 15.3 Cost-Effectiveness of a Per-Unit Effluent Fee

Each firm faces a different *MAC* curve but the same marginal effluent fee (*MEF*). Firm 1 faces MAC_1 and abates A_1 units of effluent, and Firm 2 faces MAC_2 and abates A_2 units. Although the individual abatement levels are different, the associated *MAC* level incurred by each firm is the same, which means the effluent charge yields a cost-effective solution.

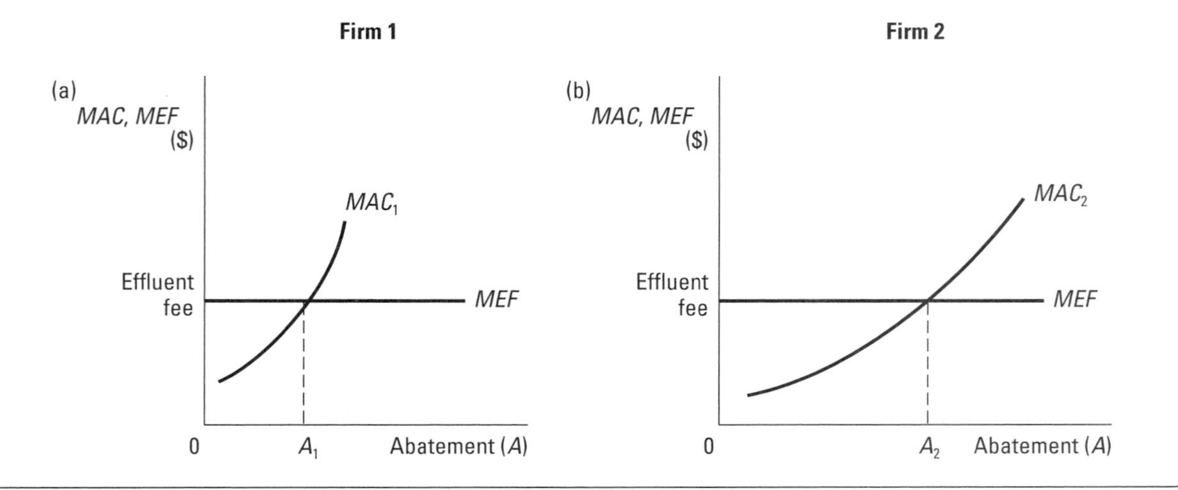

practice is difficult, and the costs of collecting the necessary data would likely be prohibitive. Even if these tasks were somehow accomplished, the solution would be efficient only in the aggregate. Unless the marginal benefits and costs at each pollution site were identical, the fee would not yield a water body–specific efficient outcome.

To illustrate, look at the model in Figure 15.4, where it is assumed for simplicity that the *MSC* is the same for two regions but that there are different *MSB* curves: MSB_{low} for a low population area and MSB_{high} for a high population area. If the effluent fee is set at *MEF*, then each region will abate A_0 units. Notice, however, that A_0 is *above* the efficient level for the sparsely populated region (A_{low}) and *below* the efficient level for the densely populated region (A_{high}). This in turn means that if the objective were to achieve efficiency at the regional level, officials would have to set a unique effluent fee for every region corresponding to the abatement level where the respective *MSB* and *MSC* curves are equal.

In practice, there are opportunities for states to institute such individualized programs, and in fact, some have begun to do so. Officials in Florida, Connecticut, and New York, for example, have instituted volume-based fees, whereas California, Indiana, Louisiana, and others have set fees that vary with volume and toxicity.[18] A similar approach is being used by some POTWs. These initiatives are extensions of the fees implemented through the NPDES. Normally, those permit fees are independent of the type or amount of effluent being discharged, which translates to an *MEF* of zero, offering no incentive for polluting sources to abate.

FIGURE 15.4 Inefficiency of a National Per-Unit Effluent Fee

The *MSC* curve is the same for two regions, but there are different *MSB* curves: MSB_{low} in a low population area and MSB_{high} in a high population area. If the per-unit effluent fee is set at *MEF*, each region will abate A_0 units. Notice that A_0 is above the efficient level for the sparsely populated region (A_{low}) and below the efficient level for the densely populated region (A_{high}).

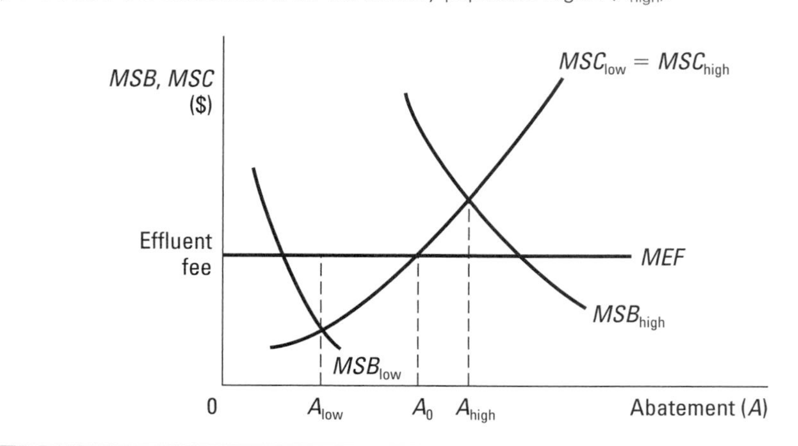

[18]Duhl (December 1993), p. 10, as cited in U.S. EPA, Office of Policy, Economics, and Innovation (January 2001), p. 36, Table 4-2.

Internationally, both developed and developing countries have instituted effluent charges, including France, Germany, the Netherlands, Malaysia, and China.[19] Application 15.2 discusses Germany's experience with this market-based instrument.

product charge
A fee added to the price of a pollution-generating product based on its quantity or some attribute responsible for pollution.

Product Charges for Nonpoint Sources

Since the pollutant release of each nonpoint source is indeterminate, an effluent fee cannot be implemented fairly, at least not in the usual sense, making an indirect approach a better alternative. A commonly cited example is a **product charge** implemented as a fertilizer tax. Because agricultural runoff is a major contributor of nonpoint source pollution, such a market instrument appears to have merit. Imposing a tax on fertilizer causes its effective price to rise, which in turn should reduce quantity demanded.

APPLICATION 15.2 Germany's Effluent Charge System

Like most advanced countries, Germany experienced substantial economic growth and industrialization during the 1960s. By the early 1970s, this surge of development had taken its toll on Germany's natural environment, particularly its water resources. In some regions, the contamination was so severe that it was impeding customary uses of some water bodies.

Germany's first line of defense was to create the Cabinet Committee for Environmental Protection, which was established in 1971. By design, this body was responsible for coordinating all environmental activities at the federal level of government. In one of its first official documents, the committee advocated the use of market-based instruments. Among its recommendations was the implementation of an effluent charge to help restore Germany's water quality. As proposed, this charge was to be levied on every discharger in an amount equal to the incremental damages caused by its effluents—precisely the kind of solution economists endorse.

Response to the committee's recommendation for an effluent charge was mixed. Support came from some international organizations, such as the Organisation for Economic Co-operation and Development (OECD) and the 1972 World Environmental Conference sponsored by the United Nations. However, the Länder (i.e., all the states of the Federal Republic of Germany) opposed the shift to an economic approach. Instead, they supported a more moderate transition with an integration of market-based instruments into the country's existing command-and-control structure. There was also strong opposition from most of Germany's industrial sector. Ultimately, a revised initiative was imposed, which integrated an effluent charge system into Germany's existing command-and-control regulatory framework.

In September 1976, the German government passed the Effluent Charge Law, which combined a discharge fee with a permit structure similar to the U.S. system. Implemented in 1981, the new law requires each German state to levy an effluent charge on all direct dischargers of such pollutants as settleable solids, mercury, and cadmium. One of the unique elements of this landmark legislation is a market-based incentive that reduces a polluter's charge liability if it complies with federally mandated minimum standards. For example, if a polluter uses technology-based standards for both hazardous and nonhazardous pollutants, its charge is reduced by 75 percent. Studies indicate that this incentive has had its intended qualitative effect, at least in some German cities and towns. Some municipalities and industrial dischargers claim that this aspect of the new law was the primary motivation for their increased investment in waste treatment.

From a financial perspective, the effluent charge adds to government revenues. There is also anecdotal evidence that this instrument can provide significant cost savings. Germany's Council of Experts on Environmental Questions claims that an effluent charge can achieve a given level of water quality for about 33 percent less money than if uniform standards were used. Finally, there are indications that the effluent charge may stimulate innovation in pollution abatement. According to the OECD, Germany's clean water technology market has grown considerably and is now the largest segment of its environmental protection market nationwide.

Sources: U.S. EPA, Office of Policy, Economics, and Innovation (November 2004); OECD (1989); Brown and Johnson (October 1984).

[19] U.S. EPA, Office of Policy, Economics, and Innovation (November 2004), Chapter 2.

In theory, the optimal tax is one that covers the marginal external cost (*MEC*) associated with agricultural production that uses fertilizers. In practice, however, the operative issue is whether the demand response to the elevated effective price is sufficient to measurably reduce the runoff problem.

In the United States, at least 46 states impose a fertilizer tax, but the rates apparently are too low to have much of an effect on consumption. Tax rates range from less than $1.00 to $4.00 per ton—a very small proportion of the product's price, which is anywhere from $150 to $200 per ton. With taxes levied at no higher than a 2.5 percent rate (and usually much lower), the decline in consumption has been negligible.[20] Even if fertilizer tax rates were elevated, the effect on quantity demanded might not be sufficient to significantly mitigate the associated pollution problem. Because fertilizer is an important input in crop production, its demand is likely to be price inelastic.

Nonetheless, some European countries have used fertilizer taxes that seem to have measurable incentive effects. For example, Austria formerly used a relatively low charge on fertilizers and reported a significant effect on quantity demanded. Sweden imposed two different fertilizer charges at 30 percent to 35 percent of selling price and also found that the charges affected use significantly.[21]

Tradeable Effluent Permit Markets

tradeable effluent permit market
The exchange of rights to pollute among water-polluting sources.

An alternative economic instrument that can be used to improve water quality is a **tradeable effluent permit market**. This market can be established using credits or allowances. Effluent reduction credits are issued if a polluter discharges less than what is permitted by law. Effluent allowances are issued up front and give the bearer the right to release pollution in the future. In either case, the permits are marketable and can be bought and sold as needed. Such a market could be established for point sources, nonpoint sources, and even for point-to-nonpoint source trades, which can be set up within a bubble. In simple terms, a bubble allows an overall abatement objective to be established for a geographical area, and then each polluting source is issued permits based on that objective.

The advantage to these trading systems, as discussed in Chapter 5, is that cost savings can be realized as long as polluters face different abatement costs to control the same pollutant. In such a case, low-cost abaters would be willing to sell excess permits at any price greater than their marginal abatement cost (*MAC*), and high-cost abaters would buy them as long as the price is less than their *MAC*. Market forces should establish a price such that all firms abate to the point where the levels of their *MAC*s are equal—a **cost-effective** solution.

Is there any evidence that bears out this theory? It turns out that a number of effluent trading programs have been established—most in the United States, with a few in Australia and Canada. A survey conducted in 2008 by World Resources Institute identified 57 trading programs, 51 of which are in the United States. Of the total, 26 are active, 21 are being considered or under development, and 10 have become inactive. An outline of selected water body–specific programs is shown in Table 15.4. In addition,

[20]U.S. EPA, Office of Policy, Economics, and Innovation (January 2001), p. 51.

[21]U.S. EPA, Office of Policy, Economics, and Innovation (November 2004), p. 20.

TABLE 15.4 Selected Tradeable Effluent Markets

UNITED STATES			
State	**Water Body**	**Pollutant**	**Type of Trading**
California	Lower San Joaquin River Watershed	Selenium	Nonpoint to nonpoint
Colorado	Chatfield Reservoir	Phosphorus	Point to point/nonpoint
Colorado	Cherry Creek Reservoir Watershed	Phosphorus	Point to point/nonpoint
Colorado	Lake Dillon Reservoir	Phosphorus	Point to point/nonpoint
Connecticut	Long Island Sound Watershed	Nitrogen	Point to point
Delaware	Delaware Inland Bays	Nutrients	Point to nonpoint
Minnesota	Lower Minnesota River Watershed	Phosphorus	Point to point
Nevada	Las Vegas Wash	Ammonia, phosphorus	Point to point
New Jersey	Passaic River Watershed	Heavy metals	Point to point
North Carolina	Catawba Watershed	Phosphorus	Point to point
North Carolina	Neuse River Basin	Nitrogen	Point to point/nonpoint
North Carolina	Tar-Pamlico Basin	Nutrient	Point to point/nonpoint
Wisconsin	Red Cedar River Watershed	Phosphorus	Point to point/nonpoint
INTERNATIONAL			
Country	**Water Body**	**Pollutant**	**Type of Trading**
New South Wales, Australia	Hunter River Salinity	not reported (nr)	Point to point
New South Wales, Australia	South Creek	nr	Point to point
Southeastern Australia	Murray-Darling Basin	nr	nr
Ontario, Canada	South Nation River Watershed	nr	point to nonpoint

Sources: U.S. EPA, Office of Water, Office of Wetlands, Oceans, and Watersheds (June 2, 2011); Selman, Greenhalgh, Branosky, Jones, and Guiling (2009).

several states have statewide trading programs, including Colorado, Idaho, Michigan, and Pennsylvania, and others have analogous programs under development, such as Minnesota and West Virginia.[22]

Unlike air quality trading schemes, however, early effluent trading programs have had limited success. One of the first point-to-point programs was designed in the 1980s to clean up the lower Fox River in Wisconsin. Despite high expectations, only one trade materialized. There was a similar outcome for a plan established at the Dillon Reservoir in Colorado, possibly the most frequently cited point-to-nonpoint trading plan.

Although early programs have yielded limited trading activity, the results seem to be linked to external factors or conditions imposed on market activity. For example, in the Fox River case, officials established stringent rules on such issues as which firms could participate in trades and when traded rights would expire. It was also the case that program participants still had to meet the technology-based effluent limits, which likely dampened their interest in trading. In the Dillon Reservoir case, wastewater treatment

[22]U.S. EPA, Office of Water, Office of Wetlands, Oceans, and Watersheds (June 2, 2011); Selman, Greenhalgh, Branosky, Jones, and Guiling (2009).

facilities had improved their abatement technologies, which logically would diminish their motivation to trade for effluent permits. Moreover, that region's economy had been slowing, which meant that wastewater discharges, and hence the need for effluent trading, had lessened.[23] More on these early programs is accessible online at **http://water.epa.gov/type/watersheds/hotlink.cfm**.

What we can take from these cases is twofold. First, lessons can be learned from these early experiences of integrating market instruments into water quality policy, both in the United States and abroad. Second, the transition from an exclusively command-and-control approach to one that integrates incentives is not seamless. In fact, most argue that despite limited success at the local level, market instruments continue to hold promise. However, the complexity of certain of these instruments, such as trading programs, requires sound management and good oversight. As we next examine the watershed management approach, we will see some of these observations carried through to a national level.

WATERSHED MANAGEMENT

Coincident with the integration of market solutions has been a general movement toward a more comprehensive approach to water quality policy. Public officials and their constituents recognize that existing regulations have been inefficient and costly. Moreover, while there has been measurable success in controlling point source pollution, the same cannot be said for nonpoint sources. Responding to these issues, policymakers are redefining how water resources should be managed and how policy should be implemented, endorsing a **watershed approach** to water quality control. By some measures, this transition may represent a true sea change in how the United States manages and protects its water resources.

A Watershed Approach[24]

watershed
A hydrologically defined land area that drains into a particular water body.

watershed approach
A comprehensive framework used to coordinate the management of water resources.

A **watershed** refers to the land areas that drain into a water body, as illustrated in Figure 15.5. It is defined by surface hydrologic features, so it naturally identifies the appropriate basis for water quality policy. Watersheds vary greatly in size, from a few acres to an area larger than several states. By way of example, the Chesapeake Bay watershed spans six states plus the District of Columbia.[25]

Following logically from the definition of a watershed, a **watershed approach** is a comprehensive framework used to coordinate the management and preservation of each hydrologically defined area. This approach builds on the rulings established for point and nonpoint sources. However, instead of addressing specific water resources or individual polluting sources, this holistic approach targets the environmental quality of

[23]U.S. EPA, Office of Policy, Economics, and Innovation (January 2001), pp. 101–02.

[24]Much of the following is drawn from U.S. EPA, Office of Water, Office of Wetlands, Oceans, and Watersheds (September 12, 2008; December 11, 2002).

[25]U.S. EPA, Office of Environmental Information and the Office of Research and Development (June 2003), pp. 2–3.

FIGURE 15.5 A Watershed

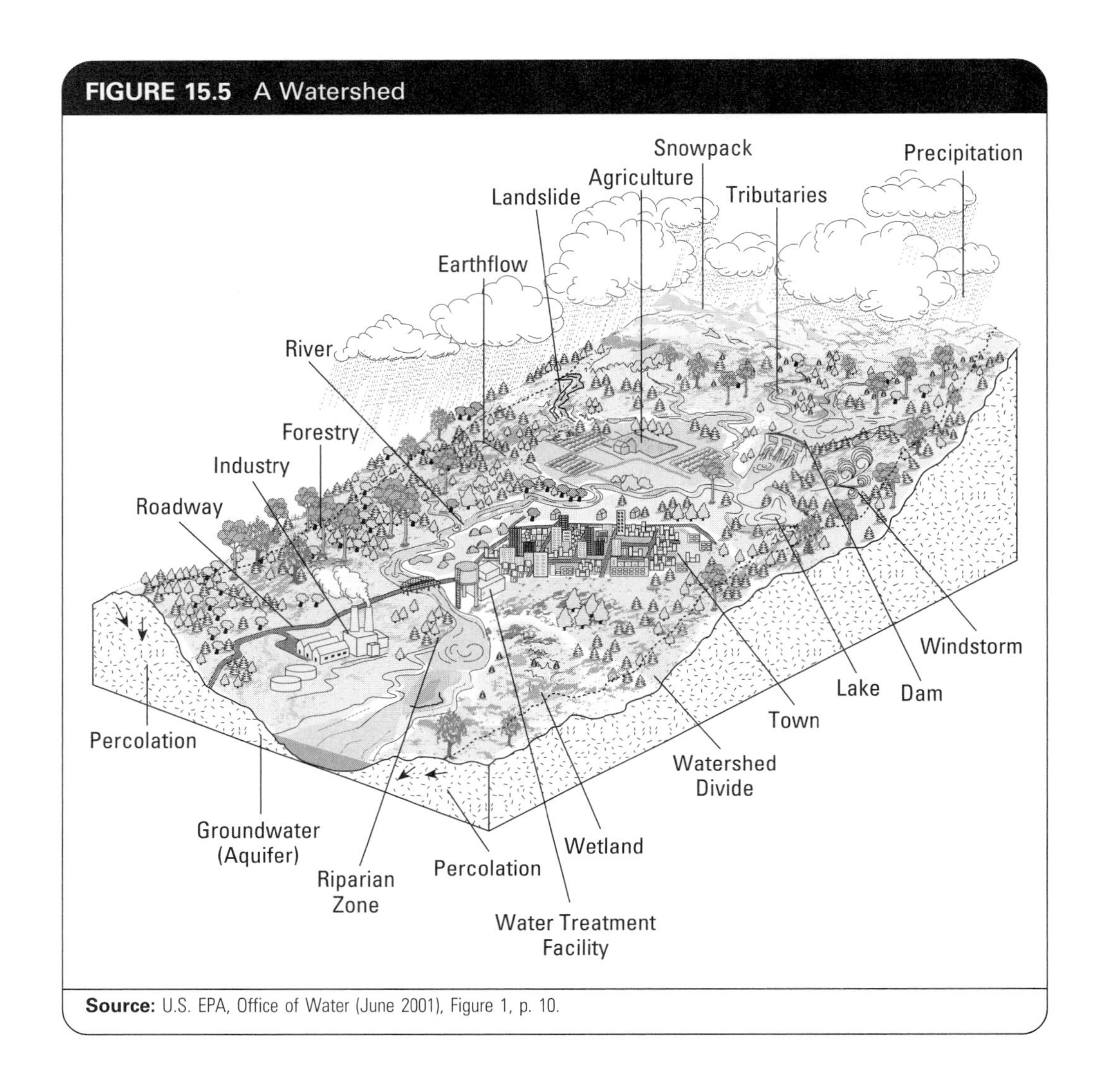

Source: U.S. EPA, Office of Water (June 2001), Figure 1, p. 10.

an entire area. In so doing, policymakers gain a better sense of overall conditions and can better identify factors that negatively affect those conditions.

The driving force is to integrate policy initiatives, using pollution prevention where possible, and to coordinate tasks and resource use among all public and private stakeholders associated with the watershed. Through such collaborative efforts, the expectation is more efficient allocation of resources and elimination of redundancies, both of which should yield more effective policy and cost savings. More information is available at **http://water.epa.gov/type/watersheds**.

Watershed Funding

In the United States, some federal grants have been earmarked to support watershed protection projects. A complete listing of these is available at **http://cfpub.epa.gov/fedfund/list.cfm**. Some are aimed at a specific water body, such as the Chesapeake Bay Small Watersheds Grant Program, and some are broad in scope, such as the Clean

Water State Revolving Fund (CWSRF) discussed previously. One program aimed exclusively at community-based watershed projects is the Targeted Watershed Grants Program, which recently funded a competitive urban watershed program. Launched in 2002 as the Watershed Initiative, this program provided funding to governmental or nonprofit, nongovernmental entities for projects using watershed-based strategies to achieve environmental goals. Since 2003, 61 organizations received more than $50 million in grant monies through this program.[26]

Key Policy Instruments

Watershed-Based NPDES Permits

Although watershed management does not displace existing policy instruments, it does amend certain initiatives to better achieve its objectives. One example is the use of **watershed-based NPDES permits**. This policy instrument provides for the issuance of permits to multiple point sources within a watershed. Although there are a number of ways to accomplish this, the most common is to reissue individual permits based on the watershed area, setting effluent limits that support watershed goals. Other options include issuing general permits to a group of sources within a watershed, such as all POTWs, or issuing an individual permit to a group of point sources.[27] More information is available at **http://cfpub1.epa.gov/npdes/wqbasedpermitting/wspermitting.cfm**.

watershed-based NPDES permit
Allows for permitting of multiple point sources within a watershed.

Water Quality Trading

In 2003, the EPA issued a final rule outlining the nation's Water Quality Trading Policy, which explicitly states that all trading activity should occur within a watershed. The ruling explicitly supports trades involving nutrients (phosphorus and nitrogen) and sediments, with trades for other pollutants to be considered on a case-by-case basis. A good example is the creation of environmental markets to accelerate the restoration of the Chesapeake Bay, an initiative discussed in Application 15.3.

The EPA's endorsement of trading is supported with economic arguments, citing compliance cost savings, scale economies, innovation incentives, and greater efficiency. Under the general guidelines, polluting sources must align trading programs with the core rulings of the Clean Water Act, including NPDES requirements. Units of trade must be clearly defined and may be effluent reduction credits. Various options are offered to encourage trades that involve nonpoint sources.[28] For further insight on this market-based program, visit **http://water.epa.gov/type/watersheds/trading.cfm**.

It is too soon to predict the long-term success of this new trading policy. However, the EPA asserts that this initiative will contribute to the achievement of national water quality goals. Moreover, the agency has touted the potential cost savings of market-based policies for some time. In this case, the agency estimates that the

[26]U.S. EPA, Office of Water, Office of Wetlands, Oceans, and Watersheds (June 22, 2011); U.S. EPA, Office of Water (June 2, 2011).
[27]U.S. EPA, Office of Water, Office of Wastewater Management (May 2003).
[28]U.S. EPA, Office of Water (January 13, 2003).

APPLICATION 15.3 Creating Environmental Markets in the Chesapeake Bay Watershed

Insufficient progress in restoring the Chesapeake Bay prompted President Obama to issue Executive Order 13508, which established a Federal Leadership Committee (FLC) to develop a new, coordinated strategy to protect and restore the 64,000 square mile watershed. As a result, the *Strategy for Protecting and Restoring the Chesapeake Bay Watershed* was announced in 2010, which establishes measureable goals, priorities, and outcomes to be achieved through 2025, with two-year milestones along the way to improve accountability.

At the outset, federal agencies were asked to identify goals for the Chesapeake Bay, which resulted in the establishment of four essential priorities: restore clean water; recover habitat; sustain fish and wildlife; and conserve land and increase public access. Supporting strategies also were identified, including an incentive-based plan to develop **environmental markets**.

Guidance for these markets is to be defined by the EPA along with states in the Chesapeake Bay area. The U.S. Department of Agriculture (USDA), together with the EPA and other federal agencies, will lead the Chesapeake Bay Environmental Markets Team (CB EMT), which will establish an infrastructure for the market environment. This charge is particularly important because the new market's success depends in part upon clearly defined procedures as well as metrics to measure environmental outcomes.

In the first two years, environmental markets are to be established for water quality, using credits, or offsets, for nutrient and sediment reductions defined by the Chesapeake Bay's total maximum daily load (TMDL). Entities that must reduce their environmental impact on the watershed become demanders of credits in an amount that is equivalent or greater than that impact. Suppliers of these credits, or offsets, are expected to be primarily private landholders, who have large holdings as well as the capacity to find cost-effective ways to achieve environmental improvements. In subsequent years, other market options will be explored for such contexts as wetlands, wildlife habitats, and carbon.

With 2011 funding of $0.37 million from the president's budget for 2011, the CB EMT began its work on the following tasks:

- Develop drivers of demand and high-quality credits
- Create a viable marketplace in which to exchange the credits
- Devise consistent methods to quantify and verify performance
- Define protocols that identify important baselines
- Establish a platform for registering, tracking, and reporting measurable benefits
- Develop cost-effective systems that will bring together demanders and suppliers

Beyond efforts at the federal level of government, the World Resources Institute (WRI) also is contributing to the long-term restoration effort. WRI received grant funding in 2010 to devise a platform that will support water quality trading among farming participants in the Chesapeake Bay Watershed. This platform is expected to comprise a credit registry, a profit calculator, and other monitoring tools.

Although the development of Chesapeake Bay environmental markets is still in its infancy, expectations for success are high. Executive Order 13508 clearly identifies environmental markets as an innovative means to accelerate the restoration of the Chesapeake Bay Watershed. The clear expectation is that devising markets for water quality, wetlands, carbon sequestration, and wildlife habitats can contribute measurably to the restoration and protection of this highly valuable and ecologically rich natural resource.

To follow the progress of this collaborative effort, visit **http://executiveorder.chesapeakebay.net**.

Sources: U.S. Department of Agriculture, Office of the Chief Economist, Office of Environmental Markets (2011; August 2010); Federal Leadership Committee (FLC) for the Chesapeake Bay (September 2010; May 2010).

compliance cost savings from effluent trading markets could be quite significant, within a range of $658 million to $7.5 billion. Of this amount, trading among point sources is expected to save $8.4 million to $1.9 billion, with point-to-nonpoint source trades estimated to save between $611 million and $5.6 billion.[29]

[29]U.S. EPA (n.d.), as cited in U.S. EPA, Office of Policy, Economics, and Innovation (January 2001), p. 106.

CONCLUSIONS

An ongoing source of debate is the nation's dependence on a command-and-control approach to achieving water quality and the methods it uses to guide critical decisions. Since 1972, U.S. water control policy has been rooted in the use of technology-based effluent limitations. Delays in setting these limits have been documented, and these delays in turn have slowed states' efforts to achieve receiving water quality standards. An efficient abatement level is not likely, given the absence of any mandate to use benefit-cost analysis in defining these standards. Furthermore, because the effluent standards are applied uniformly within major groups of polluters, a cost-effective solution is also unlikely.

Another costly policy decision has been the federal funding of POTW construction. Studies show that these funds have served primarily to displace rather than supplement local spending. In addition, by removing much of the cost burden from municipalities, local officials had little incentive to make cost-conscious decisions in building waste treatment facilities.

Beyond this dependence on command-and-control instruments, there are gaps in the overall policy approach. Nonpoint polluting sources, such as agricultural and urban runoff, are major contributors to water contamination, yet the United States has only begun to address this highly complex issue. At issue is whether or not there are sufficient resources allocated to improving monitoring technologies and to advancing the knowledge base in nonpoint source pollution.

Economic instruments such as effluent fees and trading programs may be viable alternatives to the nation's current standards-based approach. Moreover, the coincident adoption of a watershed approach may move the nation closer to its water quality objectives and do so in a way that can yield significant cost savings.

Summary

- Point sources are subject to technology-based effluent limits that vary by type of polluting source, age of the facility, and type of contaminant released.
- Effluent limits are communicated through the National Pollutant Discharge Elimination System (NPDES).
- Effluent limitations for new industrial point sources are more stringent than those for existing sources.
- The effluent limitations are not aligned with U.S. objectives, because they are based on what is technologically feasible rather than on what is needed to achieve water quality.
- The zero discharge limit was overly ambitious, and the EPA has imposed it in only a few instances.
- The Clean Water Act does not require standards to be set to maximize net benefits, which prevents an efficient solution. Also, the uniformity of the standards within identified groups disallows a cost-effective outcome.
- In 1987, the publicly owned treatment works (POTWs) grant program was replaced by the Clean Water State Revolving Fund (CWSRF) program.
- A criticism of the POTW funding program was that most of the federal monies only displaced local funding. Also, because the program shifted most of the expenditures away from local governments, they had little incentive to minimize costs.
- Under the Nonpoint Source Management Program, states must develop programs aimed at nonpoint sources, designating best management practices (BMP) to reduce pollution.
- Factors supporting states' responsibility for nonpoint sources are the location-specific nature of nonpoint source pollution and conventional land use practices. On the opposite side are information deficiencies and the potential problem of inconsistent control efforts.
- Far less federal funding has been allocated to nonpoint source controls than to point source controls.
- Market-based approaches, including pollution charges and effluent trading, are becoming integrated into water quality policy in some European nations and in the United States.
- An effluent charge can be based on either the volume or the type of effluent released.
- For nonpoint sources, product charges can be levied on a commodity whose usage adds to a known runoff problem.
- A tradeable effluent permit market is an alternative economic instrument that can be used to improve water quality. It can be established using credits or allowances.
- A watershed approach is a framework that coordinates water resource management efforts. Federal funding for watershed projects is being offered through a competitive grant program.
- Effluent trading within a watershed is encouraged by the nation's 2003 Water Quality Trading Policy.

Review Questions

1. **a.** Evaluate the use of technological attainability as the primary determinant of the effluent limitations.
   b. Economically analyze the use of *uniform* technology-based effluent limitations.
2. Discuss any incentive or disincentive implications of the federal assistance programs for POTW construction.
3. As an alternative to standards, one policy proposal is the use of permit trading among point sources of water pollution. Give the major reason why this is advantageous (a) from an economic perspective and (b) from an environmental perspective.
4. Suppose that two major dischargers (Firm 1 and Firm 2) of phosphorus face the following marginal abatement cost (MAC) functions:

$$MAC_1 = 3 + 0.75A_1 \quad MAC_2 = 2 + 0.5A_2$$

where A is units of phosphorus, and MAC is in millions of dollars.

Now, consider a government initiative to limit phosphorus emissions by establishing a marginal effluent fee (*MEF*) of $12 per unit of phosphorus.

a. Given this *MEF*, determine the abatement level for each firm.

b. Is this a cost-effective solution? Why or why not?

5. Working as part of an EPA team, your assignment is to give a clear, objective presentation of a market-based approach to reduce nonpoint source pollution. Include in your discussion the theoretical issues and any practical concerns that must be addressed before implementing your proposal.

6. a. To help fight the problem of nonpoint source pollution associated with agricultural runoff, your state is contemplating charging an annual fee of $500 to every seller of pesticides. If this fee is to achieve an efficient solution, state specifically according to externality theory (a) what the $500 fee must represent and (b) in which market.

b. Illustrate graphically, labeling where and how the fee is imposed.

7. Assume for simplicity that there are two identified point sources discharging chemical wastes into a local water body. Currently, each source releases 30 units of effluent, for a total of 60 units. To improve water quality, suppose that the government sets an aggregate abatement standard of 30 units. The two polluters' abatement cost functions are

Point source 1: $TAC_1 = 10 + A_1^2$
$MAC_1 = 2A_1$

Point source 2: $TAC_2 = 20 + 2A_2^2$
$MAC_2 = 4A_2$

a. Suppose the government allocates the abatement responsibility equally across the two point sources so that each must abate 15 units of effluent. Graphically illustrate this policy, and explain why this abatement allocation does not yield a cost-effective solution. Support your answer numerically.

b. What cost condition is required for the government's abatement allocation to be cost-effective?

c. Suppose that instead of using an abatement standard, the government institutes an effluent fee of $40 per unit of pollution. How many units of pollution would each point source abate? Is the $40 fee a cost-effective strategy for meeting the 30-unit abatement standard? Explain.

Additional Readings

Austin, Susan A. "Designing a Nonpoint Source Selenium Load Trading Program." *The Harvard Environmental Law Review* 25(2) (2001), pp. 339–403.

Balascio, Carmine C., and William C. Lucas. "A Survey of Storm-Water Management Water Quality Regulations in Four Mid-Atlantic States." *Journal of Environmental Management 90*(*1*) (January 2009), pp. 1–7.

Bunch, Beverly S. "Clean Water State Revolving Fund Program: Analysis of Variations in State Practices." *International Journal of Public Administration 31*(2) (January 2008), pp. 117–36.

Fisher, Douglas. *The Law and Governance of Water Resources*. Northampton, MA: Elgar, 2010.

Harrington, Winston. "Industrial Water Pollution in the Netherlands: A Fee-Based Approach." In Winston Harrington, Richard D. Morgenstern, and Thomas Sterner, eds. *Choosing Environmental Policy*. Washington, DC: Resources for the Future, 2004.

Heberling, Matthew T., Jorge H. García, and Hale W. Thurston. "Does Encouraging the Use of Wetlands in Water Quality Trading Programs Make Economic Sense?" *Ecological Economics 69*(*10*) (August 2010), pp. 1988–94.

Hellegers, Petra, and Ekko van Ierland. "Policy Instruments for Groundwater Management in the Netherlands." *Environmental and Resource Economics 26*(*1*) (September 2003), pp. 163–72.

Hung, Ming-Feng, and Daigee Shaw. "A Trading-Ratio System for Water Pollution Discharge Permits." *Journal of Environmental Economics and Management 49(1)* (January 2005), pp. 83–102.

Jou, Jyh B. "Environment, Asset Characteristics, and Optimal Effluent Fees." *Environmental and Resource Economics 20(1)* (September 2001), pp. 27–39.

Kampas, Athanasios, and Ben White. "Selecting Permit Allocation Rules for Agricultural Pollution Control: A Bargaining Solution." *Ecological Economics 47(2–3)* (December 2003), pp. 135–47.

Kramer, Daniel Boyd, Stephen Polasky, Anthony Starfield, Brian Palik, Lynn Westphal, Stephanie Snyder, Pamela Jakes, Rachel Hudson, and Eric Gustafson. "A Comparison of Alternative Strategies for Cost-Effective Water Quality Management in Lakes." *Environmental Management 38(3)* (September 2006), pp. 411–25.

Lankoski, Jussi, Erik Lichtenberg, and Markku Ollikainen. "Point/Nonpoint Effluent Trading with Spatial Heterogeneity." *American Journal of Agricultural Economics 90(4)* (November 2008), pp. 1044–58.

Morgan, Cynthia. L., Jay S. Coggins, and Vernon R. Eidman. "Tradable Permits for Controlling Nitrates in Groundwater at the Farm Level: A Conceptual Model." *Journal of Agricultural and Applied Economics 32(2)* (August 2000), pp. 249–58.

Ning, Shu-Kuang, and Ni-Bin Chang. "Watershed-Based Point Sources Permitting Strategy and Dynamic Permit-Trading Analysis." *Journal of Environmental Management 84(4)* (September 2007), pp. 427–46.

Ribaudo, M. O., and C. J. Nickerson. "Agriculture and Water Quality Trading: Exploring the Possibilities." *Journal of Soil and Water Conservation 64(1)* (January/February 2009), pp. 1–7.

Sabatier, Paul A., Will Focht, Mark Lubell, Zev Trachtenberg, Arnold Vedlitz, and Marty Matlock. *Swimming Upstream: Collaborative Approaches to Watershed Management*. Cambridge, MA: MIT Press, 2005.

Shaw, W. Douglas. *Water Resource Economics and Policy: An Introduction*. Northampton, MA: Elgar, 2005.

Shortle, James S., and Richard D. Horan. "The Economics of Water Quality Trading." *International Review of Environmental and Resource Economics 2(2)* (October 2008), pp. 101–33.

Tisdell, John, and Daniel Clowes. "The Problem of Uncertain Nonpoint Pollution Credit Production in Point and Nonpoint Emission Trading Markets." *Environmental Economics and Policy Studies 9(1)* (2008), pp. 25–42.

Van Schoik, R., C. Brown, E. Lelea, and A. Conner. "Barriers and Bridges: Managing Water in the U.S.–Mexican Border Region." *Environment 46(1)* (January/February 2004), pp. 26–41.

A Reference to Acronyms and Terms in Water Quality Control Policy

Environmental Economics Acronyms

MAC	Marginal abatement cost
MEC	Marginal external cost
MEF	Marginal effluent fee
MSB	Marginal social benefit of abatement
MSC	Marginal social cost of abatement
TAC	Total abatement cost

Environmental Science Terms

mg/l	Milligrams per liter
ppm/l	Parts per million per liter

Environmental Policy Acronyms

BADCT	Best available demonstrated control technology
BAT	Best available technology economically achievable
BCT	Best conventional control technology
BMP	Best management practices
CB EMT	Chesapeake Bay Environmental Markets Team
CWA	Clean Water Act
FDA	Food and Drug Administration
FLC	Federal Leadership Committee
CWSRF	Clean Water State Revolving Fund
NESHAP	National Emission Standards for Hazardous Air Pollutants
NPDES	National Pollutant Discharge Elimination System
POTWs	Publicly owned treatment works
TMDLs	Total maximum daily loads
USDA	U.S. Department of Agriculture

To access additional course materials, visit www.cengagebrain.com. At the home page, search for the ISBN of this title (shown on the back cover). This will take you to the product page where these resources can be found.

CHAPTER 16

Protecting Safe Drinking Water

"Water is the only drink for a wise man."

—Henry David Thoreau (1817–1862)

In 1993, 400,000 residents of Milwaukee, Wisconsin, became ill from a waterborne disease that was transmitted through the city's drinking water. Ultimately, more than 40 people lost their lives. Beyond the human suffering, an estimated $37 million in wages and productivity was lost. How could a city's water supply become so polluted? What went wrong? There are several theories—human error, aging facilities, illegal discharges, and monitoring failures among them. Another is that the disease outbreak was a result of poorly defined federal controls. The claim is that regulations diverted attention and resources to lower-priority problems, which in this case caused officials to miss the warning signs of the contamination.[1]

Whatever the reason, the Milwaukee incident sparked national concern about the potential for similar problems across America. It also triggered skepticism and criticism of U.S. drinking water policy. What regulations are in place to protect drinking water supplies? How effective is this part of U.S. water quality control policy?

Drinking water supplies depend on ground and surface water resources, both of which are protected by the Clean Water Act. However, this act requires a level of water quality to support aquatic life and recreational uses—a level that calls for far less stringent standards than those necessary to support safe drinking. Furthermore, although the Clean Water Act is comprehensive, its protection of groundwater resources is limited. Groundwater is the source of drinking water for about 80 percent of water systems in the United States.[2] Hence, regulations beyond those in the Clean Water Act are necessary to minimize the risks of contaminated drinking water.

In this chapter, we focus on the nation's drinking water policy, which is governed by the Safe Drinking Water Act. We begin with an overview of drinking water legislation, followed by a discussion of the standards used to define drinking water quality. Once done, we examine the efficiency implications of the standard-setting process, and we study an actual Economic Analysis (EA) conducted for a proposed tightening of the

[1] Smith (Summer 1994).

[2] Despite this statistic, most individuals (about 66 percent) use a system that draws from surface waters because large urban areas commonly use surface water, while small rural communities rely on groundwater sources. U.S. EPA, Office of Water, Office of Ground Water and Drinking Water (September 20, 2007).

arsenic standard. We conclude with the economics of pricing water supplies. A list of acronyms and terms is provided at the end of this chapter.

OVERVIEW OF U.S. DRINKING WATER LEGISLATION

In the Beginning[3]

The first U.S. law dealing with drinking water dates back to the 1800s, as shown in Table 16.1. Legislation calling for drinking water standards came nearly 20 years later. Initially these standards were to be set by the Public Health Service and targeted only contaminants capable of spreading communicable waterborne diseases. It was not until 1974 that the first Safe Drinking Water Act (SDWA) was passed. It was aimed at protecting drinking water from *any* contaminant that could threaten human health or welfare—not just bacteria responsible for communicable disease. At the same time, the responsibility of protecting drinking water shifted from the Public Health Service to the Environmental Protection Agency (EPA).

The 1986 Amendments expanded federal controls on drinking water and corrected some of the failings of the original SDWA. A significant provision was the so-called

TABLE 16.1 Evolution of U.S. Drinking Water Legislation

Legislation	Major Provisions
Interstate Quarantine Act of 1893	Authorized the surgeon general to issue regulations that prevent the spread of disease; led to the first U.S. water regulation passed in 1912 prohibiting the use of a common drinking cup on interstate carriers.
Public Health Service Act of 1912	Provided for the U.S. Public Health Service to set bacteriological standards for drinking water.
Revisions to Public Health Standards in 1925, 1942, 1946, 1962	Strengthened existing standards; required more stringent testing procedures; established maximum allowable concentrations for certain substances.
Safe Drinking Water Act of 1974	Authorized the EPA to establish drinking water standards; controlled underground injection activities and protected sole-source aquifers.
Safe Drinking Water Amendments of 1986	Accelerated procedures for the standard-setting process; provided greater protection of groundwater sources of drinking water; banned the use of lead in public drinking water systems.
Safe Drinking Water Amendments of 1996	Integrates risk assessment and benefit-cost analysis into standard-setting procedures; authorizes a \$1-billion-per-year Drinking Water State Revolving Fund; promotes pollution prevention through source water protection and better management.

Sources: U.S. Congress (August 6, 1996); Dzurik (1990); Larson (1989); Wolf (1988), Chapter 4, pp. 133–37; U.S. Congress, Office of Technology Assessment (OTA) (October 1984), Chapter 3, pp. 64–75; McDermott (1973).

[3]Drawn from Dzurik (1990); Larson (1989); McDermott (1973).

lead ban, which prohibited all future use of lead pipe and solder in public drinking water systems. Other revisions attempted to provide better protection of underground drinking water supplies. Although the SDWA of 1986 addressed some important concerns, further changes were needed. Numerous proposals for reform were discussed as part of the most recent reauthorization of this act. Ultimately, Congress passed the SDWA Amendments of 1996, which is the law currently in force.

Current U.S. Policy

Drinking Water State Revolving Fund (DWSRF) Authorizes $1 billion per year to finance infrastructure improvements.

It was President Clinton who signed into law the Safe Drinking Water Act (SDWA) Amendments of 1996. An important element of this legislation was the establishment of a **Drinking Water State Revolving Fund (DWSRF)**. Recognizing the success of the Clean Water State Revolving Fund (CWSRF) for publicly owned treatment works (POTWs), Congress initiated this similar plan for drinking water. The fund initially authorized $1 billion per year to finance infrastructure improvements. Appropriations for subsequent years may vary. Through 2009, the DWSRF has provided $16.2 billion in low-interest loans to support drinking water projects.[4] To learn more about the DWSRF, visit **http://water.epa.gov/grants_funding/dwsrf/index.cfm**.

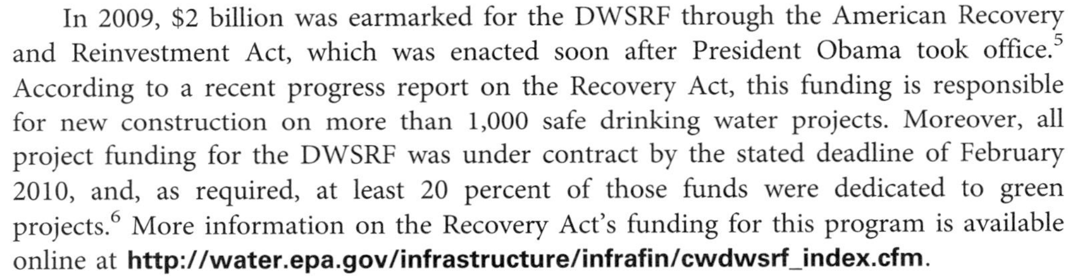

In 2009, $2 billion was earmarked for the DWSRF through the American Recovery and Reinvestment Act, which was enacted soon after President Obama took office.[5] According to a recent progress report on the Recovery Act, this funding is responsible for new construction on more than 1,000 safe drinking water projects. Moreover, all project funding for the DWSRF was under contract by the stated deadline of February 2010, and, as required, at least 20 percent of those funds were dedicated to green projects.[6] More information on the Recovery Act's funding for this program is available online at **http://water.epa.gov/infrastructure/infrafin/cwdwsrf_index.cfm**.

The 1996 Amendments also reformed the provisions for standard setting, integrating an economic decision rule into the process. Specifically, the new ruling requires that drinking water standards be based on sound risk assessment and **benefit-cost analysis**. In fact, the law specifically states [emphasis added]:

> ... in considering the appropriate level of regulation for contaminants in drinking water, *risk assessment,* based on sound and objective science, and *benefit-cost analysis* are important analytical tools for improving the efficiency and effectiveness of drinking water regulations to protect human health.[7]

Other new rulings promote pollution prevention through better management and source water protection rather than relying solely on remediation. For more details on the 1996 Amendments, visit **http://water.epa.gov/lawsregs/guidance/sdwa/laws_statutes.cfm**.

Under the Obama administration, the EPA formulated what it calls its **Drinking Water Strategy (DWS)**, which is aimed at identifying ways to strengthen protection

---

[4]U.S. EPA, Office of Water (November 2010).

[5]U.S. EPA (March 12, 2009).

[6]U.S. EPA (May 4, 2011).

[7]Pub. L. 104-182, § 3.7, 110 Stat. 1613 (codified at 42 U.S.C. § 300f).

from drinking water contaminants under existing legislation. Four goals are articulated under this strategy:

- "Address contaminants as groups rather than one at a time so that enhancement of drinking water protection can be achieved cost-effectively.
- Foster development of new drinking water technologies to address health risks posed by a broad array of contaminants.
- Use the authority of multiple statutes to help protect drinking water.
- Partner with states to develop shared access to all public water systems (PWS) monitoring data."

Thus far, the EPA has moved forward on each objective, including the identification of the group, carcinogenic volatile organic compounds.[8] For further information and updates on the progress of the DWS, visit **http://water.epa.gov/lawsregs/rulesregs/sdwa/dwstrategy/index.cfm**.

Relating the SDWA to Other Laws

It is important to have some perspective about how the Safe Drinking Water Act (SDWA) relates to other U.S. water quality laws. First, compared to the Clean Water Act (CWA), which has a comprehensive span of control, the SDWA is more focused. Its aim is to define, monitor, and enforce whatever standards are needed to ensure that water drawn from the tap is safe for human consumption. Also, note the phrase "drawn from the tap," communicating that bottled water is not regulated through the SDWA. Ensuring the potability of bottled water, a growing global market, is discussed in Application 16.1.

Second, it is important to realize that the goals of the SDWA are not independent of the CWA, which controls *all* water resources, including those used for drinking. Since the rulings of the CWA are not stringent enough to ensure the safety of drinking water, the SDWA imposes tougher standards needed to ensure potability. However, it does rely implicitly on the CWA to achieve a baseline level of water quality and to control the effluents of polluting sources.

Third, the groundwater protection in the law is indicative of the fragmentary approach to controlling groundwater contamination. There are many statutes implemented by different government agencies aimed at this important natural resource—arguably, an ineffective approach with no unified oversight.[9] This outcome evolved from the false belief that groundwater was naturally protected from contamination because of its location within layers of soil and rock. As it turns out, every state in the nation has discovered some contaminants in its groundwater at some point in time. Public concern prompted a call for more aggressive legislative controls, and existing laws were amended in an attempt to address the problem. However, no single legislative act has ever been passed whose main objective is to protect groundwater resources.[10]

---

[8]U.S. EPA, Office of Water (May 4, 2011a).

[9]Among the laws with some provisions to protect groundwater are the Comprehensive Environmental Response, Compensation, and Liability Act; the Federal Insecticide, Fungicide, and Rodenticide Act; the Resource Conservation and Recovery Act; and the Toxic Substances Control Act.

[10]U.S. EPA, Office of Pesticides and Toxic Substances (October 1991), pp. 1–2; U.S. Congress, OTA (October 1984), vol. 1, pp. 3, 63, 73–75.

APPLICATION 16.1 Worldwide Demand for Bottled Water: Is It Safer Than Tap Water?

Sales of bottled water have grown into a thriving multibillion-dollar market. In the United States, per capita annual consumption is 27.6 gallons, with 2009 sales reaching $10.6 billion. Although the United States is the largest consumer of bottled water worldwide, totaling 8.5 billion gallons in 2009, other nations also have a strong demand presence in this growing industry. For example, Mexico is the second largest buyer of bottled water, with aggregate purchases in 2009 of 6.9 billion gallons, and China and Brazil are next in the ranking, responsible for 5.7 billion and 4.3 billion gallons, respectively, in the same year. Worldwide, total consumption in 2009 was 53.5 billion gallons.

To a large extent, these impressive statistics reflect consumers' uncertainty about the quality of public drinking water. Some are concerned about aesthetic issues and others about more serious threats, like lead contamination. Whatever the cause, the response of many consumers is to substitute bottled water for tap water. Implicit in households' decisions to buy bottled water is the assumption that this substitute commodity is of higher quality than tap water. For this assumption to be valid, it must be the case that bottled water is subject to more stringent standards than is public drinking water or that suppliers of bottled water are closer to compliance with some set of universal standards than their public counterparts. Is either of these scenarios correct? To answer this question, we need to compare the regulations affecting private and public drinking water supplies.

To begin, the two types of water supplies generally are controlled by different entities. In the United States, for example, public drinking water supplies are regulated by the Safe Drinking Water Act through standards administered by the EPA and state authorities. Bottled water quality, however, falls under the jurisdiction of the Food and Drug Administration (FDA). The reason is that bottled water is considered a food and, as such, is part of the FDA's charge. This is similar to what is done in the United Kingdom under its Food Safety Act.

To ensure consumer safety, the FDA requires that bottled water products be produced in compliance with FDA Good Manufacturing Practices. All products must be clean and safe for human consumption and must be processed and distributed under sanitary conditions. Furthermore, according to a 1978 agreement, the FDA must adopt the EPA's public drinking water standards for bottled water. The FDA also has its own standards, dealing with aesthetics and health concerns.

However, according to a report by the Natural Resources Defense Council (1999) (available at **www.nrdc.org/water/drinking/bw/bwinx.asp**), there are regulatory gaps. For example, 60–70 percent of bottled water sold in the United States is exempt from FDA regulations because FDA standards are not applicable to waters packaged and sold *intra*state. And although almost 40 states regulate bottled water, they have little or no resources available for enforcement. Moreover, certain water products, such as those identified as "carbonated water" and "seltzer," are exempt from FDA controls regardless of where they are sold. Also of interest is the fact that FDA's standards are more lenient than some international regulations, such as the European Union's (EU) bottled mineral water standards and the standards for some chemicals set by the World Health Organization.

What does all this mean to the consumer? It is true that regulations are in place to protect the quality of bottled water, and there is consensus that bottled water is safe for human consumption. However, there is no reason to assume that it is safer to drink than ordinary tap water. Beyond this issue, there is also debate about whether bottled water is being unfairly or deceptively marketed. For example, it has been alleged that some bottled water is nothing more than ordinary tap water. Yet, its price can be as much as 10,000 times higher than the price of public water supplies. Further allegations have been made about misleading information on the characteristics or sources of bottled water products. To address these problems, the FDA established labeling requirements for bottled water that provide consistency to such designations as "mineral," "distilled," or "sterile" and that require the identification of bottled water drawn from municipal supplies. Even tougher rulings exist in the EU.

For more information about the bottled water industry, visit the Web site of the International Bottled Water Association at **www.bottledwater.org**.

Sources: Rodwan (April/May 2010); British Soft Drinks Association (March 24, 2009); International Bottled Water Association (IBWA) (2008); Natural Resources Defense Council, Erik D. Olson (April 1999); U.S. EPA (March 1991); U.S. EPA, Office of Drinking Water (n.d.).

Pollutants Controlled Under the SDWA

The SDWA controls all types of contaminants that may threaten human health. Its rulings refer to "contaminants" as any physical, biological, or radiological substances in water. The 1986 Amendments list 83 contaminants for which drinking water standards are to be set, with new contaminants to be drawn from a list of **priority contaminants** devised by an advisory group, published, and updated regularly.

priority contaminants
Pollutants for which drinking water standards are to be established based on specific criteria.

Among the criteria for selecting **priority contaminants** are the following:

- The contaminant must be known or expected to occur in a public water system.
- The contaminant may have an expected adverse effect on human health.

Consideration of which substances are identified on the priority list shall include but not be limited to contaminants identified in the Comprehensive Environmental Response, Compensation, and Liability Act (also known as Superfund), and pesticides registered under the Federal Insecticide, Fungicide, and Rodenticide Act.

The 1996 Amendments call for **risk assessment** and **benefit-cost analysis** to govern which contaminants are to be regulated and the standard-setting process itself. In addition to developing new rules and guidance for priority contaminants, a new infrastructure was established for future decisions, called the **National Contaminant Occurrence Database (NCOD)**. This is a collection of data on both regulated and unregulated contaminants that may occur in public water systems. The purpose of the NCOD is to facilitate identification and selection of contaminants to be controlled in the future. For more information, visit **http://water.epa.gov/scitech/datait/databases/drink/ncod/databases-index.cfm**.

SETTING STANDARDS TO DEFINE SAFE DRINKING WATER

The most important directive of the SDWA is to set standards that define drinking water quality for the nation. Under the law, there are two types of standards—**primary standards**, which protect human health, and **secondary standards**, which protect public welfare. The distinction is an important one, both in how each is defined and in how each is implemented.

Establishing National Primary Drinking Water Regulations (NPDWRs)

National Primary Drinking Water Regulations (NPDWRs)
Health standards for public drinking water supplies that are implemented uniformly.

The emphasis of the SDWA is clearly on the primary, or health, standards for drinking water. These are applicable only to public water systems and are to be implemented *uniformly* throughout the country. More formally, these standards are called **National Primary Drinking Water Regulations (NPDWRs)**. Thus far, NPDWRs have been announced for 69 organic and inorganic chemicals, 7 microorganisms, 7 disinfectants and disinfection byproducts, and 4 radionuclides.[11]

All standards are to be reviewed at least once every five years and must be amended whenever enhanced health protection is possible through changes in technology,

---

[11]U.S. EPA, Office of Water, Office of Ground Water and Drinking Water (May 2009).

FIGURE 16.1 Community Water Systems (CWSs) Meeting the Standards

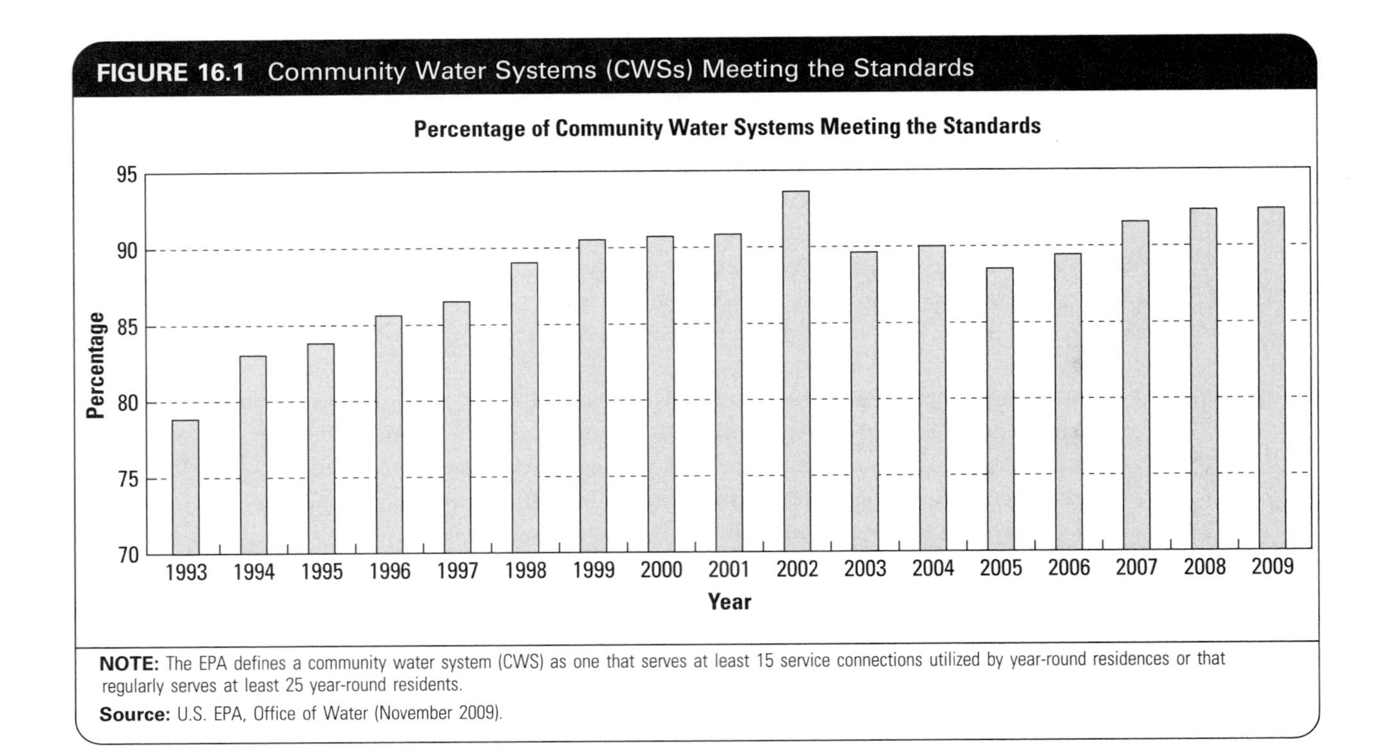

NOTE: The EPA defines a community water system (CWS) as one that serves at least 15 service connections utilized by year-round residences or that regularly serves at least 25 year-round residents.

Source: U.S. EPA, Office of Water (November 2009).

treatment, or other means. The proportion of community water systems (CWSs) meeting these standards has generally risen over time, as shown in Figure 16.1. However, testing of water on commercial aircraft in 2004 showed that a rising proportion of aircraft sampled by the EPA failed to meet the national standards for total coliforms—an issue discussed in Application 16.2.

Each NPDWR comprises three parts:

- **Maximum Contaminant Level Goal (MCLG)**
- **Maximum Contaminant Level (MCL)**
- **Best Available Technology (BAT)** for public water supply treatment

The MCLG

maximum contaminant level goal (MCLG) Component of an NPDWR that defines the level of a pollutant at which no adverse health effects occur, allowing for a margin of safety.

One of the key elements of a primary drinking water regulation is the **maximum contaminant level goal (MCLG)**. This defines the level of a pollutant at which no known or expected adverse health effects occur, allowing for an adequate margin of safety. Of particular relevance is the evidence on carcinogenicity of water contaminants. The MCLG for known or probable carcinogens is zero, meaning that no amount of such a contaminant is allowed in public drinking water. For any substance that is not a carcinogen, the MCLG is set according to the established **reference dose (RfD)** for that contaminant. An RfD is an estimate of the amount of a pollutant to which humans can be exposed over a lifetime without harm. A key point to remember is that the MCLG is *not* an enforceable standard. Instead, it serves as a target, or an objective, toward which the primary standard is to be aimed.

APPLICATION 16.2 The Aircraft Drinking Water Rule (ADWR): Ensuring Safe Drinking Water on Aircraft

The Safe Drinking Water Act protects water supplies from the tap, and that includes tap water on airliners. Drinking water on aircraft is governed jointly by the Food and Drug Administration (FDA), the Federal Aviation Administration (FAA), and the EPA. As part of its enforcement responsibilities, the EPA tested water supplies from a random sample of passenger aircraft in 2004—once in the summer and again in the fall. The results showed that these water supplies did not meet all national standards *and* that the problem was worsening. Let's examine the findings more closely.

In the first series of random tests, onboard water supplies from 158 aircraft were examined during August and September of 2004. These water supplies are those held in aircraft tanks, which are used in lavatories and in the galleys. What the agency determined was that 20 aircraft, or 12.7 percent, tested positively for total coliform bacteria. Of these 20 aircraft, two, or 1.3 percent, also tested positively for *E. coli*, a type of fecal coliform bacteria. The finding of either total coliform or *E. coli* in water supplies suggests that other disease-causing organisms may be present.

A second round of testing took place in November and December, and the results were not encouraging. This time, 169 aircraft were randomly tested, and 29 of these, or 17.2 percent, tested positive for total coliform, although none showed the presence of *E. coli*. Taken together with the first series of tests, the EPA discovered that 15 percent of the 327 planes tested in 2004 were found to be in violation of the total coliform standard.

The EPA reported these findings to the media and to the general public. The intent was to facilitate more informed decision making by air travelers, such as requesting bottled beverages and avoiding beverages made with tap water. In addition, the EPA sought and reached agreement with 12 major carriers, including American Airlines, Continental Airlines, and Southwest Airlines, to implement monitoring and disinfecting protocols. By 2005, agreements had been reached with 24 U.S. carriers to employ new water testing and disinfection protocols. The EPA also accelerated its review of existing guidance and regulations for airline water supplies. Priority issues were identified as sound preventive actions, proper monitoring activities, and maintenance procedures, such as disinfection and flushing of aircraft water systems.

Ultimately, the EPA developed and issued the **Aircraft Drinking Water Rule (ADWR)** in 2009. Its purpose is to assure that the drinking water provided to aircraft passengers and crew is safe. This rule refers only to aircraft within U.S. jurisdictions. However, the EPA supported the World Health Organization's effort to devise international guidelines for aircraft drinking water. The primary elements of the ADWR are:

- Developing and implementing operations, maintenance, and coliform sampling plans
- Routine disinfection and flushing according to manufacturer guidelines and routine monitoring
- Self-inspection of water systems on aircraft every five years
- Taking corrective action in response to sample results or other conditions that may pose a risk to public health
- Notification to the public of any possible public health risk
- Compliance audits by the EPA to be conducted periodically

Further information on the ADWR and other information on aircraft drinking water supplies are available online at **http://water.epa.gov/lawsregs/rulesregs/sdwa/airlinewater/index.cfm**.

Sources: U.S. EPA, Office of Water (May 29, 2011; May 4, 2011b; October 2009); U.S. EPA (January 19, 2005).

maximum contaminant level (MCL)
Component of an NPDWR that states the highest permissible contaminant level delivered to a public system.

The MCL

Once the MCLG is established, the primary standard is set. The primary standard gives the **maximum contaminant level (MCL)** allowed in drinking water. It is to be set as close to the MCLG as is feasible, where feasibility is defined through the best available technology (BAT). More formally, an MCL is the highest permissible level of a contaminant in water delivered to any user of a public system. It is expressed as an **action level** measured in milligrams per liter (mg/L). Unlike the MCLGs, the MCLs are federally enforceable. Moreover, the law requires a published determination as to whether the benefits of the MCL are justified by the cost. Table 16.2 gives the MCLs for selected contaminants and their associated health effects. A complete list is accessible online at **http://water.epa.gov/drink/contaminants/index.cfm#primary**.

TABLE 16.2 National Primary Drinking Water Standards for Selected Contaminants

Contaminant	MCL[a]	Health Effects
Organic Chemicals		
Atrazine	0.003	Reproductive and cardiovascular problems
Benzene	0.005	Anemia; increased cancer risk
Carbon tetrachloride	0.005	Liver problems; increased cancer risk
Chlordane	0.002	Liver or nervous system problems; increased cancer risk
Heptachlor	0.0004	Liver damage; increased cancer risk
Styrene	0.1	Liver, kidney, and circulatory problems
Vinyl chloride	0.002	Increased cancer risk
Inorganic Chemicals		
Arsenic	0.010	Skin damage; circulatory system problems; may have increased cancer risk
Asbestos	7 MFL[b]	Increased risk of developing benign intestinal polyps
Cadmium	0.005	Kidney damage
Fluoride	4.0	Bone disease; children may get mottled teeth
Lead	TT[c]Action level = 0.015	Delays in physical or mental development in infants and children; children may have slight deficits in attention span and learning abilities; kidney problems and high blood pressure in adults
Mercury (inorganic)	0.002	Kidney damage
Radionuclides		
Beta particle and photon emitter	4 mrem/yr[d]	Increased cancer risk
Radium 226/228	5 pCi/L[e]	Increased cancer risk
Microorganisms		
Legionella	TT[c]	Legionnaire's disease (a type of pneumonia)
Turbidity	TT[c]	Higher levels of disease-causing microorganisms, e.g., viruses, parasites, some bacteria
Viruses (enteric)	TT[c]	Gastrointestinal illness
Total Coliforms (including fecal coliform and *E. coli*)	5 percent[f]	Not a health threat in itself; used to indicate whether other potentially harmful bacteria may be present

NOTES:
a: MCLs are measured in milligrams per liter (mg/L) unless otherwise noted.
b: MFL = million fibers per liter, with fiber length > 10 microns.
c: TT means treatment technique requirement in effect.
d: mrem/yr = millirems or 1/1000 rem per year.
e: pCi/L = picocuries per liter.
f: More than 5.0 percent of samples are total coliform-positive in a month.

Source: U.S. EPA, Office of Water, Office of Ground Water and Drinking Water (May 2009).

Best Available Technology (BAT)

best available technology (BAT)
Treatment technology that makes attainment of the MCL feasible, accounting for cost considerations.

The law requires that each national primary drinking water regulation identify the treatment technology that makes attainment of the MCL feasible. This is characterized as the **best available technology (BAT)** observed under field conditions, taking into account the cost considerations. Since this technology is not a requirement per se, the primary water quality regulations are more accurately characterized as **performance-based standards**.

Establishing National Secondary Drinking Water Regulations (NSDWRs)

secondary maximum contaminant levels (SMCLs)
National standards for drinking water that serve as guidelines to protect public welfare.

Protection of public welfare is the statutory objective of secondary drinking water standards. More to the point, these standards deal with contaminants that so impair aesthetics and other non-health-threatening characteristics like odor and taste that a substantial number of individuals may be forced to discontinue use of the public water system. Referred to as **secondary maximum contaminant levels (SMCLs)**, these standards serve as guidelines to protect public welfare and are *not* enforceable by the federal government. Furthermore, unlike the primary standards, the secondary standards are not uniform, since they may vary with geographic or other conditions. The current list of SMCLs is available online at **http://water.epa.gov/drink/contaminants/index.cfm#sec**.

ECONOMIC ANALYSIS OF SAFE DRINKING WATER POLICY

The fundamental basis of the safe drinking water policy is the standard-setting process. In the United States, both costs and benefits guide this important process, thanks to the 1996 Amendments. Prior to these amendments, standard setting was governed solely by benefits. Let's examine the implications of this significant policy reform.

Standard Setting and Efficiency

The statutory goals of the Safe Drinking Water Act (SDWA), that is, the maximum contaminant level goals (MCLGs), are defined as the level of a pollutant at which no adverse human health effects occur with a "margin of safety." Allowing for an adequate margin of safety seems to suggest a contaminant level where no individual would be harmed. Given the variability of human sensitivities, this implies that at least some MCLGs must be set at zero. Notice that there is no mention of feasibility or cost considerations in setting these goals. They are purely benefit-based.

But what about the maximum contaminant levels (MCLs), the enforceable standards? Once the 1996 Amendments were passed, the MCLs had to be set in accordance with benefit-cost analysis. But prior to these revisions, these standards, like the MCLGs, were solely benefit-based. There was no legal requirement that the marginal benefits of achieving the MCL be balanced with the expected marginal costs. And given the language of the MCLG, there was the potential for overregulation of some contaminants.

Consider the graph in Figure 16.2, which models the total social health benefits (*TSB*) and total social costs (*TSC*) of abating some hypothetical drinking water contaminant. If the MCL is set to maximize social health benefits, the resulting abatement level is A_1, corresponding to the point where the *TSB* curve reaches its highest level. In this case,

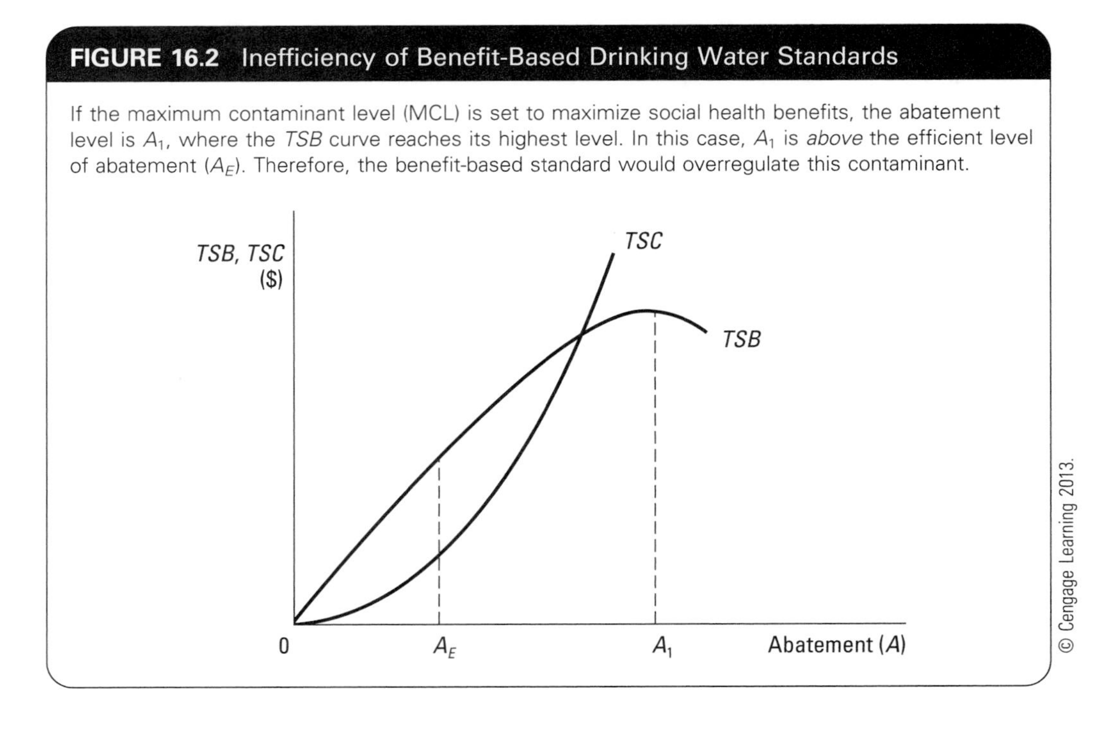

FIGURE 16.2 Inefficiency of Benefit-Based Drinking Water Standards

If the maximum contaminant level (MCL) is set to maximize social health benefits, the abatement level is A_1, where the *TSB* curve reaches its highest level. In this case, A_1 is *above* the efficient level of abatement (A_E). Therefore, the benefit-based standard would overregulate this contaminant.

Economic Analysis (EA)
A requirement under Executive Order 12866, amended by Executive Orders 13258 and 13422, and Executive Order 13563 that calls for information on the benefits and costs of a "significant regulatory action."

A_1 is *above* the efficient level of abatement (A_E). Hence, a benefit-based standard would overregulate this contaminant, imposing unnecessarily high costs on society.

The 1996 Amendments specify precisely how the MCLs are to be determined. At the time a new NPDWR is proposed, the EPA must publish a determination of whether the benefits of any MCL justify the costs. Effectively, this means that the EPA must conduct an **Economic Analysis (EA)** for any proposed NPDWR. The requisite benefit-cost analysis is to be based on specific health risk reduction and cost analysis used in the risk assessment for any proposed MCL. This premise is rooted in economic theory and recommended in President Obama's Executive Order 13563, which reaffirmed and amended President Clinton's Executive Order 12866, both of which followed the initial set of rulings in President Reagan's Executive Order 12291.

To learn more about the EPA's use of economic considerations in setting drinking water standards, visit **http://water.epa.gov/lawsregs/rulesregs/regulatingcontaminants/economicanalysis.cfm**. For now, consider the actual EA used to support a revision to the arsenic standard, which was proposed by President Clinton and later accepted by the Bush administration.

An Economic Analysis (EA) for an NPDWR[12]

For some 60 years, the MCL for arsenic had been set at 50 parts per billion (ppb). Arsenic has long been associated with various health effects, but more recent scientific

[12]This discussion is drawn from Seelye (November 1, 2001); U.S. EPA, Office of Water, Office of Ground Water and Drinking Water (December 2000). The full EA report is available online at **www.epa.gov/safewater/arsenic/pdfs/econ_analysis.pdf**.

research suggests that the risks may be greater than originally believed. Specifically, some studies indicate that high doses of arsenic in drinking water may lead to several types of fatal cancers. Visit **http://water.epa.gov/lawsregs/rulesregs/sdwa/arsenic/index.cfm** for more information about arsenic in drinking water. Citing such health concerns, in 1996, Congress directed the EPA to review the arsenic standard and issue a revision by January 2001. In May 2000, the EPA proposed a change in the MCL from 50 ppb to 5 ppb. Objections from industry and municipal water authorities prompted the EPA to change the proposed standard to 10 ppb, a decision made in January 2001 by the Clinton administration, just before President Bush took office.

Concerned that the decision to set the arsenic standard at 10 ppb may have been rushed, the Bush administration initially rescinded the Clinton standard. Then EPA administrator Christine Todd Whitman called for a comprehensive review of the scientific and economic evidence used by the Clinton administration in its decision. Although there were concerns about the underlying methodology and the benefit and cost estimates that had been cited in the **Economic Analysis (EA).** Whitman ultimately decided there was sufficient evidence to warrant the new arsenic standard. So, after much debate and political scrutiny, the MCL of 10 ppb originally proposed under President Clinton ultimately was accepted by the Bush administration.

Beyond the political and scientific complexities involved in this decision, the EA for the new standard proved to be particularly significant. This analysis was necessary because the change in the standard was expected to have a substantial impact on the regulated community—in excess of $100 million per year. A brief overview of the benefit and cost estimates cited in the EA follows, with a summary of the findings given in Table 16.3. All values are reported in 2010 dollars.

Incremental Benefits of the New Arsenic Standard

Citing numerous epidemiological studies, the EA for the new arsenic standard indicates that its major incremental health benefit is the number of lung and bladder cancer cases avoided. For the reduction in fatal cases, the value of a statistical life of $7.9 million is used to monetize these health benefits, based on current EPA guidance. Avoided nonfatal cancer cases are monetized using a willingness-to-pay value of $794,690. Other health gains, such as reductions in skin, kidney, and liver cancers, are addressed in the analysis but not monetized. Collectively, the estimated health benefits for the 10-ppb standard for arsenic are reported to be between $182.7 million and $258.9 million per year.

Incremental Costs of the New Arsenic Standard

In assessing incremental costs of the new standard, the EA identifies annual expenditures associated with compliance, monitoring, implementation, and enforcement, as shown in Table 16.3. Treatment procedures account for the highest proportion of estimated costs, at $262.6 million per year. In sum, the estimated incremental costs of complying with and enforcing the new arsenic standard are reported to be $269.1 million per year.

Net Benefits of the New Arsenic Standard

According to the overall findings of the EA, the annualized net benefits associated with the more stringent arsenic standard are between −$86.4 million and −$10.3 million per year. Based on these estimates, the Bush administration's concern that the cost of

TABLE 16.3 Economic Analysis (EA) for the New Arsenic Standard

	INCREMENTAL BENEFITS	
Health Benefits		
Descriptive assessment: Avoided cases of bladder and lung cancer. Avoided premature deaths were valued at $7.9 million per statistical life. Avoided nonfatal cancer cases were monetized using a willingness-to-pay value of $794,690.		
Monetized annual value:		$182.7–$258.9 million per year
Additional benefits (not monetized): Avoided cases of skin, liver, kidney, and prostate cancer; avoided cardiovascular and pulmonary effects; and psychological effects of knowing that the drinking water is safer to consume.		
ANNUALIZED BENEFITS		**$182.7–$258.9 million per year**
	INCREMENTAL COSTS	
Systems Costs (assuming a 7% discount rate)		
Descriptive assessment: Treatment costs; monitoring and administrative expenses.		
Monetized annual value:		
	Treatment costs:	$262.6 million per year
	Monitoring/administrative costs:	$ 5.0 million per year
State Costs		
Monetized annual value:		$ 1.5 million per year
ANNUALIZED COSTS		**$269.1 million per year**
	NET BENEFITS	
ANNUALIZED NET BENEFITS		**–$86.4 million to –$10.3 million per year**

NOTE: All values are in 2010 dollars.

Source: U.S. EPA, Office of Water, Office of Ground Water and Drinking Water (December 2000).

complying with the 10-ppb standard might not have been justified by the benefits may have been warranted. However, since a substantial number of potential benefits were not monetized in the EA, the EPA decided that the actual benefits would likely be higher than the estimates. Hence, the agency argued that the true incremental benefits would justify the incremental costs and announced the new standard in 2001. The more stringent standard of 10 ppb would take effect in February 2002, and all drinking water systems had to achieve compliance by January 2006.[13]

ECONOMIC PRINCIPLES IN PRICING WATER SUPPLIES

At all levels of government, there are significant costs associated with implementing the Safe Drinking Water Act (SDWA). In fact, for 2002, the social costs of U.S. drinking water regulations were estimated to be between $3,788.5 million and $4,643.2 million ($2010).[14]

[13]U.S. EPA, Office of Water, Office of Ground Water and Drinking Water (September 13, 2006).

[14]U.S. EPA, Office of the Chief Financial Officer (September 30, 2003), Table 4, p. 181.

Not only might these costs be inflated because of inefficient decision making, but in many communities, these costs are not properly reflected in the pricing of water supplies.

Basic economic theory argues that resources are misallocated if the price of a good is not equal to the associated marginal social cost (*MSC*) of production. In this context, society's marginal cost of *current* water usage includes the explicit costs incurred by water supply facilities plus the opportunity costs of forgone *future* consumption. All too often, water is priced at a fixed rate that is independent of use and therefore independent of rising marginal costs. The result is overconsumption. If present consumption is higher than its allocatively efficient level, then future supplies will be adversely affected. The 1996 Amendments recognize the importance of efficient water use and explicitly require the EPA to establish guidelines to encourage water conservation. To learn more about the nation's water efficiency program and guidelines for water conservation plans, visit **http://www.epa.gov/watersense/our_water/water_use_today.html**.

An International Comparison

An international comparison of water consumption patterns reveals two interesting findings. First, U.S. per capita consumption is higher than it is in most other nations. For example, according to recent data, the average American consumes about 1,690 cubic meters of water per year. Contrast this with the same statistic for Austria, which is reported at about 470 cubic meters, or Denmark at 130 cubic meters, or France, estimated at 530 cubic meters.[15] Second, water prices vary considerably around the world, as shown in Figure 16.3, which presents data for capital cities in selected Organisation for Economic Co-operation and Development (OECD) countries. Notice how cities in European countries generally face higher prices for water than urban centers in Canada, Japan, and the United States. This observation coupled with the fact that European consumption rates are generally lower than their North American counterparts supports what is predicted by the Law of Demand—that higher prices are associated with lower quantities consumed, *ceteris paribus*.

Officials in China are keenly aware of this reality as they grapple with that country's severe water shortages. These shortages, which affect about 65 percent of China's cities, are linked to the nation's rapid economic growth, its polluted water supplies, and its low water prices. In response, the Chinese government is launching a $60 billion water transfer project, establishing rationing programs, *and* raising prices sharply. Ultimately, water prices are expected to nearly double, marking the first time China has raised these prices in over 55 years.[16]

What all of this conveys is that water use is price sensitive—an important observation given that some locations in the United States employ pricing policies that actually encourage inefficient water use. By examining these pricing practices in more detail, we can identify the inherent weaknesses and consider some economically sound alternatives.

flat fee pricing scheme
Pricing water supplies such that the fee is independent of water use.

Pricing Practices of U.S. Water Utilities

Table 16.4 provides data on the actual pricing practices of water utilities based on a 2006 survey of community water systems (CWSs). Notice that a considerable percentage uses a **flat fee pricing scheme**, which refers to pricing that is independent of water

[15]Water consumption data are for 2007 or latest available year. See OECD (2010).

[16]Stein (October 4, 2004); Wonacott (June 14, 2004).

FIGURE 16.3 International Water Prices

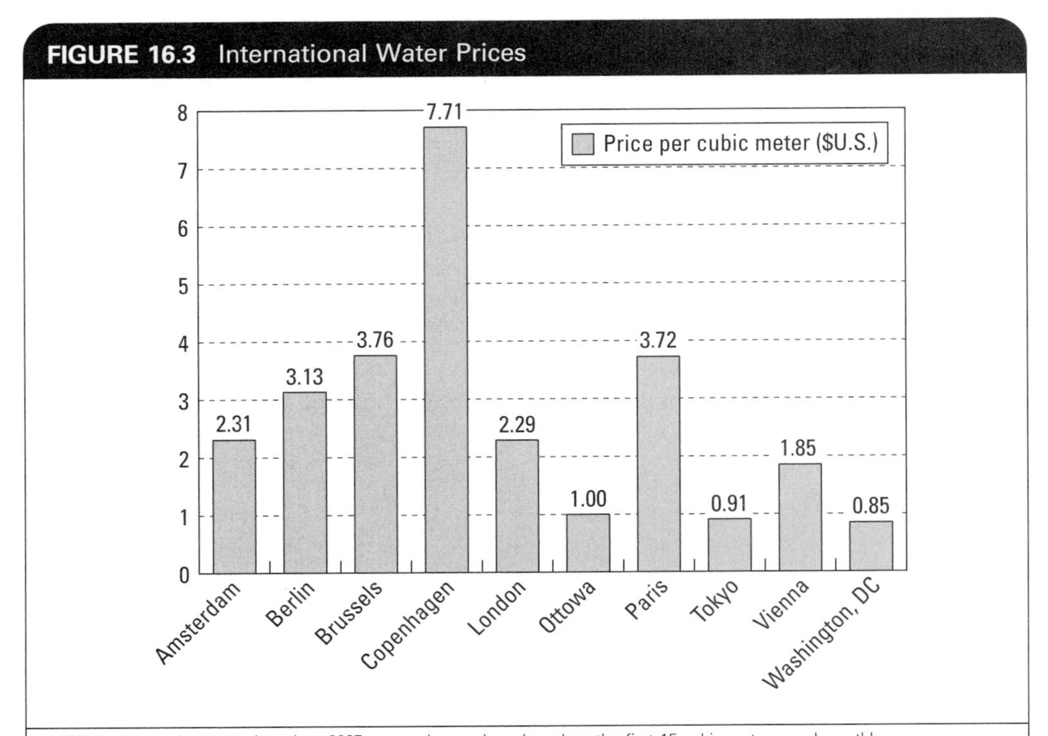

NOTE: Values shown are based on 2007 survey data and are based on the first 15 cubic meters used monthly.

Source: Global Water Intelligence and Organisation for Economic Co-operation and Development (GWI and OECD) (2007), as cited in Gleick (2009), Data Table 15.

TABLE 16.4 Percentage Use of Residential Water Pricing Structures

Pricing	Percentage of Water Systems
Metered Charges	
Uniform rate	36.4
Declining block rate	10.3
Increasing block rate	13.2
Seasonal rate	0.3
Unmetered Charges	
Separate flat fee for water	17.4
Annual connection fee	1.1
Combined flat fee for water and other services	15.7
Other	6.8

NOTES: Values are based on a 2006 survey.
Percentages do not sum to 100 because some systems use more than one rate structure.

Source: U.S. EPA, Office of Water (May 2009), Table 71.

use. Such an approach would be efficient only if the associated marginal costs were zero. Over one-third employ a **uniform rate (or flat rate) pricing structure**, which charges more for higher usage, but at a constant rate. This structure would be efficient only if marginal costs were constant and equal to the uniform rate being charged. Neither of these pricing structures reflects the rising marginal social costs (*MSC*) of water provision, and thus they act as a disincentive for consumers to economize on water usage.

uniform rate (or flat rate) pricing structure
Pricing water supplies to charge more for higher water usage at a constant rate.

The other major pricing methods represented in the table are the **declining block** and **increasing block** pricing structures. Each of these allows for changes in unit prices for different blocks of consumption levels, but obviously each works in the opposite direction of the other. Both pricing structures are illustrated graphically in Figure 16.4.

FIGURE 16.4 Alternative Pricing Structures of Water

(a) In an attempt to recover fixed costs, utilities use declining block pricing to encourage higher consumption. In so doing, they can exploit available scale economies and incur lower average costs.

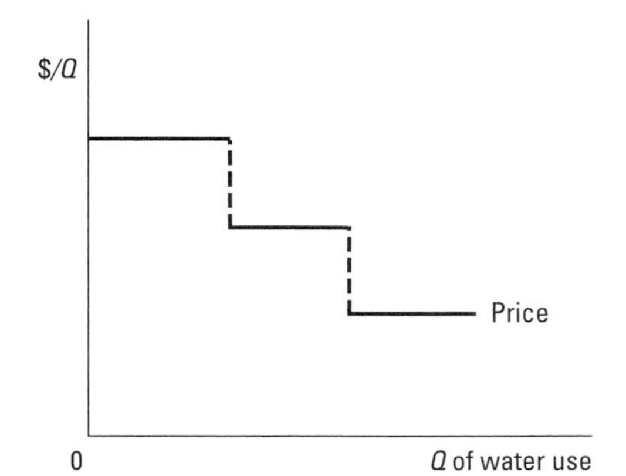

(b) Increasing block pricing provides an incentive for more conservative usage. As each additional block of water is used, higher marginal costs are considered along with the marginal benefits of consumption.

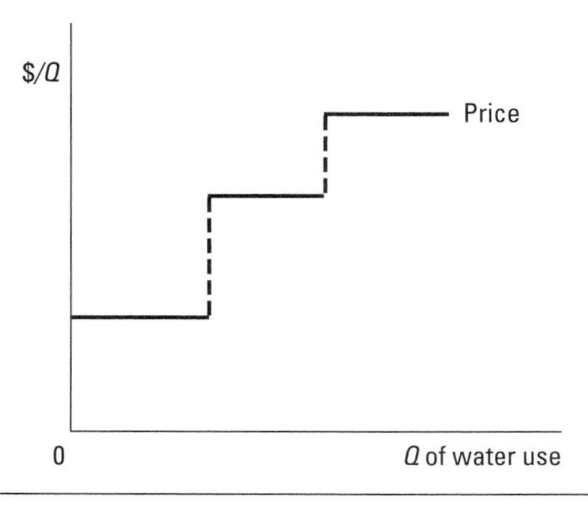

Since they are motivated differently, we need to examine them more closely and consider how each affects water usage.

Declining Block Pricing Structure

declining block pricing structure
A pricing structure in which the per-unit price of different blocks of water declines as usage increases.

Referring back to the data in Table 16.4 (page 392), the **declining block pricing structure** is used by about 10 percent of the surveyed water systems. Its usage appears to have its basis in the typical utility's cost structure. Throughout the United States, most water utilities tend to have high fixed costs and relatively low variable costs.[17] In an attempt to recover these high fixed costs, some utilities encourage higher consumption levels with declining block pricing, as shown in Figure 16.4(a). In so doing, they can exploit available scale economies and incur lower average costs.

From the utility's perspective, the declining block price structure has merit. But from society's point of view, the pricing scheme is inefficient. Although this inefficiency arises for several reasons, the major culprit is that suppliers are incorrectly using average cost pricing, as opposed to a marginal cost pricing framework.

Increasing Block Pricing Structure

increasing block pricing structure
A pricing structure in which the per-unit price of different blocks of water increases as water use increases.

Looking again at Table 16.4, notice that one of the less common pricing practices is **increasing block pricing**. As illustrated in Figure 16.4(b), the customer pays a higher unit price as water consumption increases—a practice that provides an economic incentive for more conservative water usage.

Notice how benefit-cost analysis is implicitly reflected in this approach. As each additional block of water is used, higher marginal costs are considered along with the marginal benefits of consumption. Of course, unless the increasing price blocks accurately reflect society's marginal cost, the result will not be efficient. Also, there are implementation costs to be considered in instituting this pricing structure, since metering is necessary to track usage. This factor explains why some smaller utilities favor a uniform rate system despite its inherent inefficiencies.

Based on this fundamental analysis, a reasonable argument is that government authorities should attempt to structure water pricing to reflect the associated marginal costs to society. A further consideration is to find ways to make such a pricing scheme financially feasible for all types of water utilities, including smaller facilities. Although this would be a challenging undertaking, it could make a measurable difference in both current and future water quality.

CONCLUSIONS

For over 35 years, the Safe Drinking Water Act (SDWA) has been the legislation responsible for protecting human health from contaminated tap water. Like most environmental laws, this one has been the subject of debate and criticism. Confidence in national drinking water policy has weakened in recent years. Right or wrong, most

[17]U.S. EPA, Office of Policy, Planning, and Evaluation (March 1991), pp. 4–7.

Americans had been fairly complacent about the quality of tap water. However, reports such as the 1993 Milwaukee incident have made both private citizens and public officials question the effectiveness of U.S. drinking water policy.

The 1996 Amendments address some of these concerns as well as shortcomings in the law. A major set of revisions calls for integrating benefit-cost analysis into the setting of drinking water standards. This is a major shift away from the strict command-and-control approach that characterized the pre-1996 legislation. There is also a Drinking Water State Revolving Fund (DWSRF) aimed at helping states and local communities improve the infrastructure needed to ensure clean drinking water supplies, and the Recovery Act of 2009 allocates funding to this long-term effort. Other important reforms include the integration of preventive programs to protect drinking water supplies and the ongoing efforts to ensure safe drinking water on aircraft.

Because water resources are so diverse and at the same time interconnected, policy formulation is not straightforward, nor is the revision process. Both the Clean Water Act (CWA) and the Safe Drinking Water Act (SDWA) continue to be rooted strongly in the command-and-control approach with standards providing the legislative muscle. But there has been some integration of market-based incentives into this policy structure and gradual changes to launch initiatives that *prevent* water contamination and lessen the dependence on end-of-pipe treatment. These are nontrivial revisions, as are the problems confronting any nation in its development of sound water quality policy.

Summary

- Early laws aimed at protecting the nation's drinking water made the Public Health Service responsible for setting standards, which targeted only contaminants capable of spreading communicable waterborne diseases.
- The Safe Drinking Water Act (SDWA) of 1974 authorized the EPA to set standards for any contaminant that could threaten human health or welfare. The 1986 Amendments corrected some of the failings of this law and expanded federal controls.
- The legislation currently in force is the 1996 Amendments. Among the chief revisions are the integration of benefit-cost analysis into standard setting, a new Drinking Water State Revolving Fund (DWSRF), and efforts to encourage pollution prevention. The EPA recently formulated its Drinking Water Strategy (DWS), which identifies ways to strengthen protection from drinking water contaminants.
- The Safe Drinking Water Act controls all types of contaminants (physical, biological, or radiological substances) that may threaten human health.
- National Primary Drinking Water Regulations (NPDWRs) are federal standards aimed at protecting human health. Each regulation comprises a maximum contaminant level goal (MCLG) and a maximum contaminant level (MCL) and specifies the best available technology (BAT) for public water supply treatment.
- The MCLG defines the level of a pollutant at which no adverse health effects occur, allowing for an adequate margin of safety. It is not enforceable.
- The MCL is the highest permissible level of a contaminant in water delivered to a public system and is federally enforceable. Current law requires a published determination of whether an MCL is justified by benefit-cost analysis.
- The BAT represents the feasible treatment technology capable of meeting the standard, taking account of cost considerations.
- National Secondary Drinking Water Regulations (NSDWRs) establish secondary maximum contaminant levels (SMCLs) for pollutants that impair aesthetics and other characteristics like odor and taste. These are nonenforceable federal guidelines aimed at protecting public welfare.
- Prior to the 1996 Amendments, the law did not call for a balancing of benefits and costs in setting the maximum contaminant levels (MCLs). They were benefit-based and therefore could lead to overregulation and unnecessarily high costs.
- The 1996 Amendments specify that when a new NPDWR is proposed, the EPA must publish a determination identifying whether or not the benefits of the MCL justify the costs. This means that the EPA must conduct an Economic Analysis (EA) for any proposed NPDWR.
- Cities in European countries generally face higher prices for water than urban centers in the United States and Canada. And European nations have lower consumption rates than their North American counterparts, as predicted by the Law of Demand.
- Many U.S. communities use improper pricing practices for water supplies, such as a flat fee, uniform rate, or declining block structure, none of which account for rising marginal costs. The result is overconsumption of water resources.

Review Questions

1. Consider the following benefit and cost functions for the abatement level (A) necessary to achieve the maximum contaminant level (MCL) for some identified drinking water pollutant.

$$TSB = 45A - 0.25A^2 \qquad TSC = 24A + 0.10A^2$$

$$MSB = 45 - 0.5A \qquad MSC = 24 + 0.2A$$

where all benefit and cost values are measured in millions of dollars, and A is measured in percent.

   a. Using the pre-1996 benefit-based criterion for drinking water standards, find the level of abatement that will maximize total social benefits (TSB).

b. Based on the benefit-cost criterion in the 1996 SDWA, find the maximum level of abatement that will ensure positive net benefits.

c. What level of abatement will maximize the net social benefits?

2. a. Explain the practical difference between the MCLG and the MCL.

b. Draw the relationship of marginal social benefit and marginal social cost for the MCLG for lead, assuming it is set at an efficient level. Intuitively explain your model.

3. Suppose that a proposal is under review to strengthen the NPDWS for benzene, and the associated abatement level would generate a marginal social benefit (*MSB*) that is greater than the marginal social cost (*MSC*) by $125,000.

a. On efficiency grounds, should the proposed standard be more stringent or more lenient, and why?

b. Specifically, what does the $125,000 represent?

4. Support or refute Congress's decision to make the secondary drinking water standards *not* enforceable by the federal government.

5. According to the EPA, uncontrolled disposal of pesticide residues and containers contributes significantly to groundwater contamination and hence threatens drinking water supplies. Design and evaluate a deposit/refund system for pesticide containers.

6. Determine the water pricing structure that is used in your hometown. Using the criterion of allocative efficiency, defend the current practice, or propose an alternative pricing structure.

7. Suppose the market for household drinking water in San Francisco is modeled as follows:

$$S = MSC = 10 + 0.2Q$$

$$D = MSB = 40 - 0.4Q$$

where Q is millions of gallons per day and *MSC* and *MSB* are in cents per gallon.

a. Assuming the city officials use a flat fee pricing system for drinking water, find the equilibrium amount of water that households in San Francisco would use.

b. Suppose instead that officials set the price of water to reflect the true marginal social costs of production. Show that such an approach promotes water conservation by determining the equilibrium quantity, and find the price that must be set to achieve this outcome.

Additional Readings

Baird, Gregory M. "Water Affordability: Who's Going to Pick Up the Check?" *American Water Works Association Journal 102(12)* (December 2010), pp. 16–23.

Bergstron, John C., Kevin J. Boyle, and Gregory L. Poe, eds. *The Economic Value of Water Quality*. Northampton, MA: Elgar, 2001.

Dalhuisen, Jasper M., Raymond J. G. M. Florax, Henri L. de Groot, and Peter Nijkamp. "Price and Income Elasticities of Residential Water Demand: A Meta-Analysis." *Land Economics 79* (May 2003), pp. 292–309.

Gaudin, S. "Effect of Price Information on Residential Water Demand." *Applied Economics 8(4)* (March 2006), pp. 383–93.

Innes, Robert, and Dennis Cory. "The Economics of Safe Drinking Water." *Land Economics 77(1)* (February 2001), pp. 94–117.

Levin, R. "Lead in Drinking Water." In R. D. Morgenstern, ed. *Economic Analyses at EPA: Assessing Regulatory Impact*. Washington, DC: Resources for the Future, 1997, pp. 205–32.

MacLeish, William H. "Water, Water, Everywhere, How Many Drops to Drink?" *World Monitor 13(12)* (December 1990), pp. 54–58.

Mullin, Megan. *Governing the Tap: Special District Governance and the New Local Politics of Water*. Cambridge, MA: MIT Press, 2009.

Musolesi, Antonio, and Mario Nosvelli. "Dynamics of Residential Water Consumption in a Panel of Italian Municipalities." *Applied Economics Letters 14(4–6)* (May 2007), pp. 441–44.

Newman, Alan. "A Blueprint for Water Quality." *Environmental Science and Technology 27(2)* (1993), pp. 223–25.

Roberson, J. Alan. "What's Next After 40 Years of Drinking Water Regulations?" *Environmental Science & Technology 45(1)* (January 2011), pp. 154–60.

Rosado, Marcia A., Maria A. Cunha-E-Sá, Maria M. Dulca-Soares, and Luis C. Nunes. "Combining Averting Behavior and Contingent Valuation Data: An Application to Drinking Water Treatment in Brazil." *Environment and Development Economics 11(6)* (December 2006), pp. 729–46.

Um, M. J., S. J. Kwak, and T. Y. Kim. "Estimating Willingness to Pay for Improved Drinking Water Quality Using Averting Behavior Method With Perception Measure." *Environmental and Resource Economics 21(3)* (November 2002), pp. 285–300.

Worthington, Andrew C., and Mark Hoffman. "An Empirical Survey of Residential Water Demand Modelling." *Journal of Economic Surveys 22(5)* (December 2008), pp. 842–71.

Yoskowitz, David W. "Spot Market for Water Along the Texas Rio Grande: Opportunities for Water Management." *Natural Resources Journal 39* (Spring 1999), pp. 345–55.

A Reference to Acronyms and Terms in Drinking Water Control Policy

Environmental Economics Acronyms

MSB	Marginal social benefit
MSC	Marginal social cost
TSB	Total social benefits
TSC	Total social costs

Environmental Science Terms

MFL	Million fibers per liter
mg/L	Milligrams per liter
mrem/yr	Millirems per year
pCi/L	Picocuries per liter
ppb	Parts per billion
RfD	Reference dose

Environmental Policy Acronyms

ADWR	Aircraft Drinking Water Rule
BAT	Best available technology
CWA	Clean Water Act
CWSRF	Clean Water State Revolving Fund
CWS	Community water system
DWSRF	Drinking Water State Revolving Fund
DWS	Drinking Water Strategy
EA	Economic Analysis
FDA	Food and Drug Administration
MCL	Maximum contaminant level
MCLG	Maximum contaminant level goal
NCOD	National Contaminant Occurrence Database
NPDWRs	National Primary Drinking Water Regulations
NSDWRs	National Secondary Drinking Water Regulations
POTWs	Publicly owned treatment works
PWS	Public water system
RIA	Regulatory Impact Analysis
SDWA	Safe Drinking Water Act
SMCL	Secondary maximum contaminant level
TT	Treatment technique

To access additional course materials, visit www.cengagebrain.com. At the home page, search for the ISBN of this title (shown on the back cover). This will take you to the product page where these resources can be found.

MODULE 6 The Case of Solid Wastes and Toxic Substances

In a stunning decision in 1978, the state of New York urged 1,000 families to leave their homes and ordered an emergency evacuation of 240 others—all residents of a Niagara Falls community known as Love Canal. Constructed on a site that 30 years earlier had been a chemical dumping ground, the ill-fated Love Canal was eventually declared a disaster area by President Carter. In 1982, residents of Times Beach, Missouri, learned that their groundwater and soil contained dangerously high levels of dioxins—the result of contaminated road oil used in the town 11 years earlier. Times Beach was completely evacuated. These events and countless others alerted society to the potential risks of solid waste pollution and exposure to toxic chemicals.

Although progress has been made since these dramatic events unfolded, the extent of hazardous waste problems is not completely known, and efficient solutions can be just as elusive. Not only are current waste generation and disposal practices at issue, but also damages caused by waste mismanagement of the past. In a broad sense, there are two interrelated issues to be addressed—excess waste generation and the use of toxic substances that eventually enter the waste stream.

Less severe, but nonetheless cause for concern, is the growth rate of municipal trash generation—millions of tons of bottles, cans, food scraps, and the like. In many communities, residents and public officials are still searching for cost-effective ways to collect and dispose of the accumulating heap.

Policy development aimed at solid waste pollution got a late start in the United States, and most argue that current legislation is inadequate. Why the lack of initiative? For one thing, population centers had remained fairly compact for a long time and seemingly were able to manage the wastes being generated. Managing solid wastes was viewed as a local responsibility and not one requiring national action. Furthermore, problems associated with the treatment and disposal of chemical wastes did not accelerate until the 1970s.

Regardless of the logic of these explanations, there is one root cause that underlies all of them. Society failed to recognize the significance of solid waste as an environmental and health risk. The result? Policymakers are playing catch-up with a problem that, at least until recently, has been advancing in magnitude and severity.

In this module, we assess this environmental and social dilemma and analyze government solutions. In Chapter 17, we study hazardous waste pollution—the risks it poses to society and the legislation designed to minimize those risks. In Chapter 18, we examine the challenges of managing nonhazardous wastes and associated policy efforts. Our focus shifts in Chapter 19, where we analyze laws aimed at controlling pesticides and other toxic substances *before* they are introduced into commerce and enter the waste stream.

CHAPTER 17

Managing Hazardous Solid Waste and Waste Sites

"In the last few decades, entire new categories of waste have come to plague and menace the American scene."

—Lyndon Baines Johnson (1908–1973)

An official order to evacuate homes and businesses following the discovery of a hazardous waste leak is a chilling reminder of the potential risks of solid waste pollution. All too often, there are accounts of strange odors emanating from basements, tainted water supplies, or health symptoms for which doctors have no diagnosis—reports that are eventually linked to a waste accident or cover-up from years past. Once-thriving communities in the United States, like Love Canal, New York, and Times Beach, Missouri, became ghost towns for a time because of the toxic effects of hazardous waste. Why did the problem reach such a level before something was done?

By the time federal policies were formulated in the United States, the nation was already suffering from the ill effects of mismanaged hazardous wastes that had been accumulating for decades. At the same time, the growth rate of the waste stream was rising due to urban development, industrialization, population growth, and increasing chemical usage. Additive to this was a glaring lack of information about the magnitude of the problem. In fact, some argue that we still do not have a complete picture of the extent and severity of solid waste pollution in this country.

What we *do* know is that the damage extends to all environmental media, which means that the effects are widespread, difficult to control, and costly to society. Accepting this characterization, it should not be surprising to learn that several major legislative acts are aimed at the dilemma. While some deal with the damage from past mismanagement, others attempt to control the present solid waste stream.

Investigating this multidimensional control approach is the objective of this chapter. Our aim is to explain the policy position of the United States in dealing with hazardous waste. The operative issue is whether or not existing policy is effective and, if not, whether there are viable alternatives that should be implemented. We begin with an overview of the hazardous waste problem. We then focus on assessing policy initiatives aimed at controlling hazardous wastes, which concludes with a careful analysis. Once done, we move on to examine and analyze the policy aimed at managing uncontrolled hazardous waste sites. A list of acronyms is provided for reference at the end of this chapter.

CHARACTERIZING THE HAZARDOUS WASTE PROBLEM[1]

hazardous solid wastes Unwanted materials or refuse posing a substantial threat to health or the ecology.

waste stream A series of events starting with waste generation and including transportation, storage, treatment, and disposal of solid wastes.

Hazardous solid wastes refer to any unwanted materials capable of posing a substantial threat to health or the environment. Because a diverse range of substances fall into this broad category, it is useful to consider the hazardous constituents of wastes as belonging to subgroups like pesticides and synthetic chemicals, as shown in Table 17.1. Despite their toxicity, these substances serve important functions in productive activity. Often, the properties that make them useful in production are precisely those that make them dangerous to society once they enter the waste stream. The **waste stream** refers to the series of events that starts with waste generation and continues through the **transportation**, **storage**, **treatment**, and **disposal** of these materials. Environmental and health risks arise both because of excess generation and improper waste management. For more information about hazardous wastes, visit **www.epa.gov/epawaste/hazard/index.htm**.

TABLE 17.1 Types of Hazardous Substances

Hazardous Substance	Description
Acids and Bases	Acids have a low pH, between 0 and 6, and bases have a high pH, between 8 and 14. Highly acidic or basic substances are corrosive, which means that discarded acids and bases can corrode the containers in which they are stored and leach into land and nearby waterways.
Heavy Metals	Heavy metals are highly toxic and do not readily break down in the human body. Examples include lead, arsenic, cadmium, barium, copper, and mercury.
Reactives	Reactives, which include explosives and flammables, behave violently when combined with air or water. Common examples are petroleum or natural gas by-products.
Synthetic Organic Chemicals	Synthetic organic chemicals are man-made substances that are hydrocarbon based, some created from relatively new processes like chemical splicing and molecular engineering. Two important subcategories are solvents and pesticides.
Solvents	Solvents are liquids that can dissolve or disperse one or more substances. Examples include ethylene, benzene, pyridine, and acetone.
Pesticides	Pesticides are used to prevent, destroy, or repel any unwanted form of plant or animal life. Some are highly resistant, which allows them to persist in the environment for a long time without the need for frequent reapplication. Some are nearly insoluble in water, making their use as insecticides on crops resistant to rainfall. Many are acutely toxic.

Sources: Drawn from Phifer and McTigue (1988), Chapter 3; Epstein, Brown, and Pope (1982), pp. 14–26.

[1]Following the literature and government data reporting, we have omitted from this discussion any reference to medical wastes and radioactive wastes.

Magnitude and Source of the Problem

Extent of the Problem

Hazardous waste generation across the United States is estimated at about 35.3 million tons per year, which translates to some 0.12 tons per person annually. Nations everywhere must deal with the risks of accumulating hazardous wastes. In Spain, for example, the reported annual level of hazardous wastes is 3.6 million tons, or about 0.08 tons per person; and in Italy, the comparable estimate is 6.7 million tons, or 0.11 tons per person. Developing countries also are confronting the problem as they evolve into more industrialized economies. In Bulgaria, for example, hazardous waste production is estimated at about 13.0 million tons per year, or 1.71 tons per person.[2]

All environmental media—the atmosphere, groundwater and surface waters, and soil—are susceptible to hazardous waste contamination. Ocean pollution, soil contamination, disease, fish kills, and livestock loss are among the potential damages. The health and ecological effects can be severe and long term, particularly from exposure to persistent pollutants like polychlorinated biphenyls (PCBs) and dichloro-diphenyl-trichloroethane (DDT) that bioaccumulate in the environment.

Water contamination and soil contamination are the most prevalent type of damage linked to hazardous waste sites. Contamination of drinking water can occur when rainwater absorbs pollutants as it runs through disposal sites. The resulting leachate can pollute groundwaters and surface waters—damage that is costly and sometimes impossible to rectify. Surrounding soil can be polluted in the same way. Food crops grown in such contaminated soil absorb pollutants that are later ingested by humans and animals. Particularly vulnerable are households living near hazardous waste sites. One of the most often cited instances of an entire community affected by this contamination is Love Canal, New York. A brief account of this case is given in Application 17.1.

Sources of Hazardous Waste

Just who is responsible for generating all these toxic waste materials? It turns out that every sector of the economy—households, industry, government, and institutions—contributes to the hazardous waste stream. For example, U.S. households illegally dispose of used motor oil each year that is literally equivalent to 30 times as much as the *Valdez* spilled into Alaskan waters in 1989.[3] While none of these should be ignored, there is no question that most hazardous wastes are generated by industry. In the United States, the majority of industrial hazardous wastes comes from large-quantity generators, those producing over 2,200 pounds per month.

---

[2]U.S. Census Bureau, Population Division (February 2011), Table 1; U.S. EPA, Office of Solid Waste and Emergency Response (November 2010); European Commission, Eurostat (July 19, 2011; July 12, 2011a). U.S. data are for 2009, and all other country data are for 2008. Note that international comparisons cannot be made with precision, given differences in definitions and estimation methods.

[3]City of Santa Monica (August 6, 2008).

APPLICATION 17.1 The Hazardous Waste Site Called Love Canal

In the late 1880s, an entrepreneur named William T. Love began excavating a canal that would connect the upper and lower ends of the Niagara River in upper New York State. Envisioned as the future site of a major industrial complex, the Love Canal project was designed to make use of the enormous supply of hydroelectric power that would be provided naturally by the 280-foot drop between the two ends of the river. Before the plan was finished, alternating current (AC) was developed, and the once-futuristic Love Canal project lost its appeal. Love's visionary complex was never finished, but one segment of the partially constructed canal remained. It simply filled with water and, for a time, was used for swimming.

In the 1940s, Hooker Chemical and Plastics Corporation, through agreement with the canal's owner, began using the site to dump chemical wastes. Ultimately buying the canal to use as a dump site, Hooker discarded over 21,000 tons of chemicals into the Love Canal between 1942 and 1952. Among the toxic wastes were benzene derivatives, dioxins, and trichlorophenol (TCP) contaminated with a highly potent carcinogenic substance called tetrachlorodibenzo-p-dioxin (TCDD). According to one source, TCDD is so lethal that less than three ounces could kill all the residents of New York City (Epstein, Brown, and Pope, 1982, p. 93).

Some years later, the postwar construction boom that characterized America in the 1950s prompted the city of Niagara Falls to condemn the properties surrounding the canal to make room for a new school and a residential area. In 1953, Hooker signed the canal over to the city for $1 in exchange for a release from any liability for damage associated with the dump site. Despite the danger, the canal was filled in, and construction proceeded. The school opened in 1955, and hundreds of homes were built.

Buyers of the new homes were unaware of the chemical dump site. Soon after moving in, they began to complain about fumes coming from the former dump site and chemical burns their children got from playing in nearby fields. Lawns and gardens refused to grow, and pools of black liquid began to surface in backyards. In the mid-1970s, the same black sludge started to seep into basements after heavy rains. Finally, in 1977, after numerous complaints, the city called in a consultant. The findings confirmed the residents' worst fears—toxic chemicals had leached from the canal and contaminated the groundwater and surface water.

One resident, Lois Gibbs, took action, suspecting that her son's history of health problems might have been caused by the chemical leaks. What followed was a complex and bitter struggle for Gibbs and the other Love Canal residents. Despite mounting evidence, the state's response to the problem was less than adequate. Gibbs and the other residents formed the Love Canal Homeowners Association Inc. and took matters into their own hands. Every level of government eventually became involved along with public agencies, health officials, researchers, and the media. In 1979, the Department of Justice filed a series of lawsuits against Hooker Chemical, the city of Niagara Falls, the city's board of education, and the Niagara County Health Department. In May 1980, President Carter declared Love Canal a disaster area, and that summer, $15 million in grants and loans was offered to New York to purchase new homes for the relocated families.

In September of 1983, Love Canal was added to the National Priorities List (NPL) under the U.S. Superfund program, an action that officially designated the site as among those posing the greatest risk to human health and the ecology. It took years of cleanup work, testing, and monitoring before the Love Canal site was officially removed from the NPL in September 2004. Today, the community has been successfully redeveloped to accommodate residential housing, commercial offices, and light industrial operations.

To learn more about the Love Canal and its extensive history, visit the EPA's site at **www.epa.gov/history/topics/lovecanal/**.

Sources: U.S. EPA, Office of Solid Waste and Emergency Response (April 2008); Griffin (January/February 1988); Epstein, Brown, and Pope (1982), Chapter 5; Gibbs (1982).

OVERVIEW OF U.S. HAZARDOUS WASTE POLICY

In the United States, the early development of solid waste policy was anything but aggressive. Most argue that this segment of environmental legislative history should be characterized as having been reactive—even passive at times. This lack of spirited attention to solid waste problems by federal officials was matched by the same absence in the private sector. The environmentalism of the 1960s was preoccupied with fouled rivers and streams, urban smog, and threatened wildlife, paying little attention to the accumulating trash and the risks of chemical wastes.[4]

It is often suggested that the U.S. government did virtually nothing in the way of controlling solid wastes until after the Love Canal saga made national news in the late 1970s. However, at face value, this suggestion is erroneous. The primary piece of federal legislation on solid wastes, the Resource Conservation and Recovery Act (RCRA), actually had been passed by Congress two years prior to that infamous event.[5] The point is that the United States did have a national policy agenda on solid waste in the 1970s, but the legislation had passed with little scrutiny and was not fully responsive to the magnitude of the problem. Hence, while Love Canal was not the catalyst for introducing federal solid waste legislation, this event and others like it motivated public officials and private citizens to give the matter more attention. Several federal laws and amendments speak to this evolution, as shown in Table 17.2.

In the Beginning

"cradle-to-grave" management system
A command-and-control approach to regulating hazardous solid wastes through every stage of the waste stream.

Somewhat remarkably, no federal policy on waste control existed in the United States until the Solid Waste Disposal Act (SWDA) was passed in 1965. However, this law had limited strength until it was amended in 1970 by the Resource Recovery Act. Even then, the emphasis remained on nonhazardous waste and land disposal.

More revisions came in the mid-1970s, these in the form of the Resource Conservation and Recovery Act (RCRA). The hallmark of RCRA was a distinct set of regulations that established a **"cradle-to-grave" management system** for controlling hazardous

TABLE 17.2 U.S. Legislation on Hazardous Solid Wastes

Legislative Act	Major Provisions
Solid Waste Disposal Act (SWDA) of 1965	Provided financial assistance to state and local governments for solid waste disposal programs; initiated a national research plan aimed at finding better disposal methods.
Resource Recovery Act of 1970	Encouraged recycling and technological controls to reduce waste at the generation point; continued to promote land disposal in local communities and to emphasize nonhazardous waste.
Resource Conservation and Recovery Act (RCRA) of 1976	Represented the first official position on hazardous waste control; established a "cradle-to-grave" management system for hazardous waste; delegated the administration of nonhazardous waste mainly to states.
Hazardous and Solid Waste Amendments of 1984	Reauthorized RCRA and broadened federal control; shifted emphasis from land disposal to waste reduction and toward improved treatment technologies for hazardous wastes; elevated standards for hazardous waste facilities.

Sources: Wolf (1988), Chapters 6 and 7; U.S. EPA, Office of Solid Waste and Emergency Response (October 1985); Hazardous and Solid Waste Amendments of 1984; Resource Conservation and Recovery Act of 1976.

[4]One important exception is Rachel Carson's *Silent Spring* (1962), which warned of the risks of pesticide chemicals.

[5]Landy, Roberts, and Thomas (1990), p. 140.

wastes. The metaphor refers to a policy strategy for controlling, managing, and tracking these wastes through every stage of the waste stream, as shown in Figure 17.1. A separate set of provisions delegated the administration of nonhazardous waste primarily to state governments.[6]

FIGURE 17.1 "Cradle-to-Grave" Management System Under RCRA

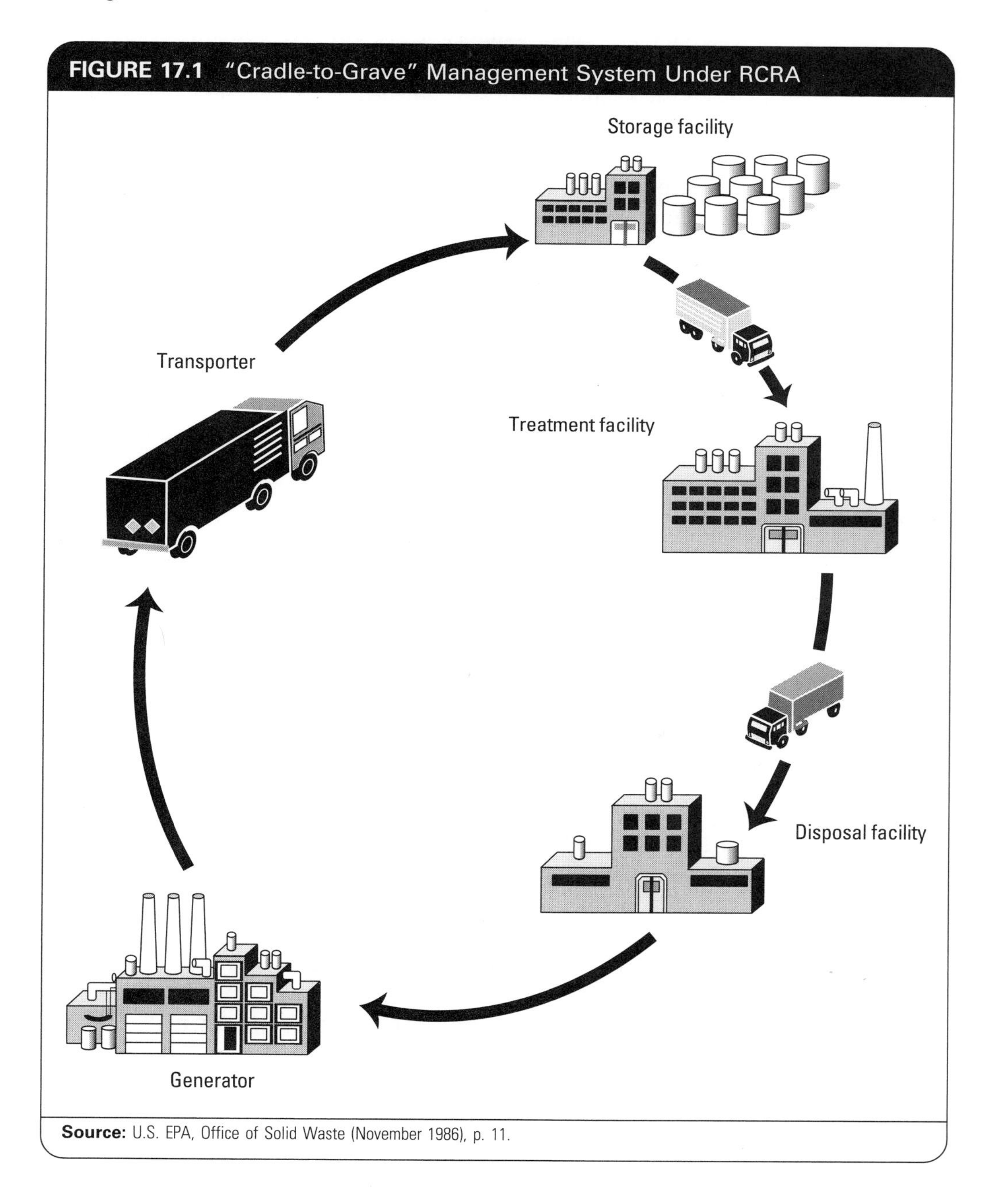

Source: U.S. EPA, Office of Solid Waste (November 1986), p. 11.

[6]The provisions applicable to hazardous waste are given in Subtitle C of RCRA. In Chapter 18, we will investigate and analyze the rulings in Subtitle D of RCRA aimed at nonhazardous municipal solid wastes.

Current U.S. Policy

Congress reauthorized and strengthened RCRA through the Hazardous and Solid Waste Amendments of 1984, which governs U.S. policy today. (Despite the change in title, we will follow conventional practice and use the acronym RCRA to refer to this legislation.) The new provisions reinforced the evolving policy shift away from land disposal and toward waste reduction efforts and better treatment technologies. Standards for hazardous waste facilities were strengthened with new safety requirements. Control was broadened by adding two groups of facilities to those regulated under RCRA—owners and operators of certain underground storage tanks and all small-quantity waste generators. Stringent rulings were imposed on underground storage tanks that house potentially dangerous substances.

CONTROLLING HAZARDOUS WASTES: RCRA

To minimize the risks of hazardous waste pollution, efforts must be made to reduce the quantity and toxicity of the waste stream. Most agree that two general approaches are needed:

source reduction Preventive strategies to reduce the quantity of any contaminant released to the environment at the point of generation.

waste management Control strategies to reduce the quantity and toxicity of hazardous wastes at every stage of the waste stream.

- **Source reduction**, which is a preventive strategy aimed at the generation stage
- **Waste management**, which refers to strategies to control those wastes that cannot be eliminated

While RCRA's objectives make a general reference to preventive strategies, there is little doubt that waste management is the primary emphasis. A command-and-control framework characterizes the implementation plan, just as is true for most clean air and water initiatives. Primary responsibility for controlling hazardous wastes is assigned to the federal government, mainly through the Environmental Protection Agency (EPA). However, Congress provided little guidance as to *how* the agency should design and implement its policy position effected under RCRA. This absence of legislative muscle is often blamed for what became long delays in launching the program and in deflecting opposition from those most affected by the new regulations.

Just as critical to the policy challenge was the uncertainty about the extent of hazardous waste pollution. Absent the obvious sources that had become national news, no one—not even major federal agencies—had a complete picture of the problem. The EPA's response was to devise a command-and-control approach intended to give structure to what must have seemed little more than ordered chaos. For more information on RCRA, visit **www.epa.gov/epawaste/laws-regs/index.htm**.

"Cradle-to-Grave" Management Approach[7]

The RCRA program emerged as the **"cradle-to-grave" management system**—a multipronged command-and-control approach to regulating hazardous waste. Its four major components are:

- **Identification** of hazardous wastes
- A **national manifest system** for tracking and monitoring the movement of wastes

[7]Much of the following discussion is drawn from U.S. EPA, Office of Solid Waste (November 1986).

- A **permit system** for **treatment, storage, and disposal facilities (TSDFs)**
- Development of **standards** for TSDFs

characteristic wastes
Hazardous wastes exhibiting certain characteristics that imply a substantial risk.

Identification of Hazardous Wastes[8]

A fundamental element of the management system is a procedure for identifying wastes considered to be hazardous and therefore subject to federal regulations. The rules are highly detailed, since an error of omission would allow a dangerous material to escape federal control. Essentially, a waste is considered hazardous under the law if it falls into one of two defined categories:

listed wastes
Hazardous wastes preidentified by government as having met specific criteria.

- **Characteristic wastes**: Those with characteristics or attributes that imply a substantial risk. In the United States, these attributes are ignitability, corrosivity, reactivity, and toxicity.
- **Listed wastes**: Those preidentified by the EPA as having met certain criteria, such as the presence of toxic or carcinogenic constituents.

For more information, visit **www.epa.gov/epawaste/hazard/wastetypes/index.htm**. The proportions of hazardous waste in each category based on 2009 data are shown in Figure 17.2.

National Manifest System

manifest
A document used to identify hazardous waste materials and all parties responsible for its movement from generation to disposal.

If a generator chooses to transfer any hazardous wastes offsite for treatment, storage, or disposal, that movement is tracked by the **national manifest system**. Once the wastes are ready for transport, the generator must prepare a document, called a **manifest**, that identifies the hazardous material and all the parties responsible for its movement. The document remains with the shipment from generation through to its final disposal. The objective is to ensure that dangerous waste materials are accounted for through every phase of the waste stream, reducing the opportunity for illegal dumping.

FIGURE 17.2 Characteristic and Listed Wastes in 2009

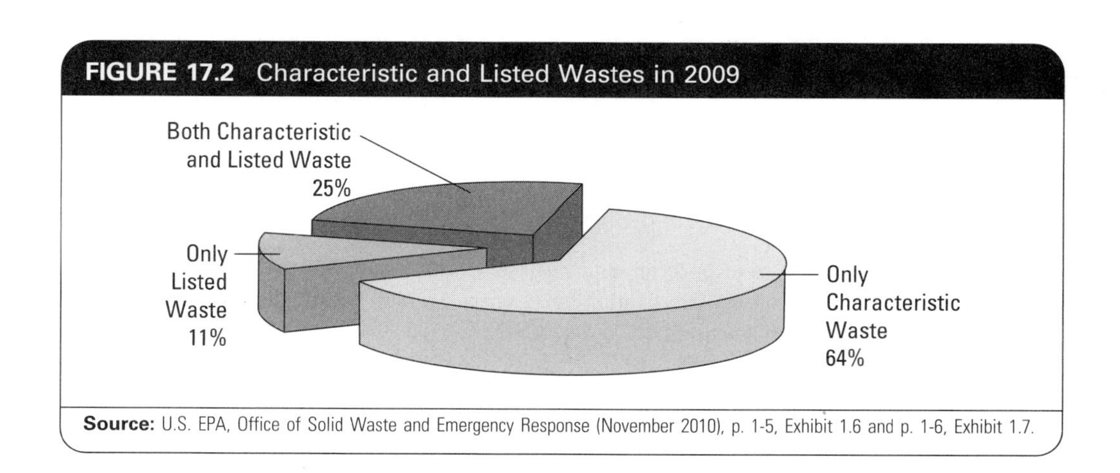

Source: U.S. EPA, Office of Solid Waste and Emergency Response (November 2010), p. 1-5, Exhibit 1.6 and p. 1-6, Exhibit 1.7.

[8]See 40 CFR 261, Identification and Listing of Hazardous Waste.

permitting system
A control approach that authorizes the activities of TSDFs according to predefined standards.

Permit System

While the tracking system monitors the location of hazardous waste, a **permitting system** controls its management at **treatment, storage, and disposal facilities (TSDFs)**. Every TSDF must obtain a permit to operate. The permitting process is stringently controlled, since its purpose is to ensure that TSDFs meet federal standards, many of which are aimed at protecting groundwater from waste contamination. More on the hazardous waste permitting process is available online at **www.epa.gov/epawaste/hazard/tsd/permitting.htm**.

Standards

Two sets of standards control the waste management practices of TSDFs. **General regulatory standards** apply to *all* types of TSDFs and control generic functions like inspections, emergency plans, and participation in the manifest program. There are also **technical regulatory standards**, which outline procedures and equipment requirements for specific types of waste facilities. For example, storage facilities must follow careful procedures to avoid leakages into the environment, and disposal sites must use technologies such as cover systems, liners, and leakage-detection systems.

Although these four components of the "cradle-to-grave" system provide controls over the hazardous waste stream, the program is inadequate in one important respect—it does not indicate nor even suggest a hierarchy of preferred methods for managing these wastes. Prior to subsequent revision, this omission meant that virtually nothing was being done to slow the overuse of land at the disposal stage.[9] From a broader perspective, it also meant that national policy was not supporting preventive strategies to reduce the size of the hazardous waste stream.

Moving Toward Pollution Prevention

Responding to these policy inadequacies, Congress passed the RCRA Amendments of 1984. Unlike their predecessor, these amendments outline in detail how hazardous wastes are to be controlled and explicitly set forth restrictions on land disposal. In so doing, the law prioritizes hazardous waste treatment technologies, which is intended to provide an incentive for source reduction and hence prevention of hazardous waste pollution.

Several references in the 1984 Amendments speak to the policy shift away from land disposal and toward preventive solutions. The law's general provisions were revised to include the following mandate:

> ... reliance on land disposal should be minimized or eliminated, and land disposal, particularly landfill and surface impoundment, should be the least favored method for managing hazardous wastes.[10]

Other rulings specifically prohibit land disposal of untreated hazardous wastes except under certain conditions. These controls are intended to encourage facilities to find

[9]For more details, see Fortuna and Lennett (1987), Chapter 9.
[10]42 U.S.C. § 6901(b)(7).

alternatives that prevent pollution. To reinforce the emphasis on prevention, the 1984 law added the following to RCRA's statement of objectives:

> The Congress hereby declares it to be the national policy of the United States, that, wherever feasible, the generation of hazardous waste is to be reduced or eliminated as expeditiously as possible.[11]

More about pollution prevention will be discussed in subsequent chapters. Information on prevention is also available on the Internet at **www.epa.gov/p2/**.

ANALYSIS OF U.S. HAZARDOUS WASTE POLICY

The nation's "cradle-to-grave" management system is a classic example of a command-and-control policy approach. As such, it lacks incentives, which decreases the likelihood of an efficient or cost-effective outcome. To illustrate this assertion, we consider the following aspects of the RCRA program:

- Risk-based uniform rules of identification
- Benefit-based uniform standards
- Failures of the manifest system
- Market implications of the land restrictions

Risk-Based Uniform Rules of Identification

Central to RCRA's control approach is the identification of hazardous wastes. The law requires that hazardous waste be identified according to risk-based characteristics and criteria. There is no provision for balancing risk with the benefits provided by these materials before they enter the waste stream. That being the case, all wastes satisfying these criteria are regulated with the same stringency regardless of their origin. Consequently, a waste material associated with a product that is vital in use is treated under the law in exactly the same way as one generated from an output of relatively low value to society.

Another problem is that the identifying criteria are applied uniformly. No qualifiers are in place to allow for different degrees of toxicity. Instead, the procedure sets up a zero-one decision-making process. If a waste falls into one of the predefined categories, it is hazardous; if not, it is nonhazardous. Also, the law does not allow for determining how much of a substance poses a hazard to society. Yet the risk of exposure to the same quantity of different hazardous substances can vary greatly.[12] Ignoring this variation is **allocatively inefficient**. If the risks across hazardous wastes are different, then the marginal benefits of abating these wastes are different as well. If marginal abatement costs are the same, society's welfare could be improved by allocating more resources to abating relatively high-risk substances and fewer to those posing a lower risk. Yet there is no provision in the law for such flexibility.

[11]42 U.S.C. § 6902(b).

[12]For more details, see U.S. Congress, OTA (1983), Chapter 6.

Benefit-Based Uniform Standards

RCRA controls hazardous waste management primarily through standards imposed on the TSDFs. According to the law, these standards must be set to protect health and the environment. Such a mandate defines the standard-setting process as benefit-based with no consideration for economic costs. As long as this is the case, these standards likely will be **allocatively inefficient**.

Another criticism of the standards-based approach is that with few exceptions, it offers states no flexibility in how they administer RCRA's hazardous waste program. While state officials can submit their own plans, these are approved only if they are equivalent to the national program or consistent with those applicable in other states.[13] There is virtually no provision for site-specific differences—either in the facilities themselves or in the types of wastes they handle. The lack of consideration for such differences means that the standards will likely be **cost-ineffective**.

Failures of the Manifest System

One of the more prominent elements of the "cradle-to-grave" approach is the manifest system. Congress's intent in establishing this elaborate tracking procedure was to deter illegal disposal. The system is a definitive command-and-control instrument, with no incentives, no pricing mechanisms, and no options. It is a no-nonsense response to a tough problem, and in that sense, the approach is understandable, albeit inefficient. Why? Because its objective is solely benefit-based—to halt the damage caused by mismanaged hazardous wastes.

Beyond the efficiency implications, a more fundamental issue is the effectiveness of the manifest plan. Despite its elaborate design, the tracking system has not been a major factor in controlling hazardous wastes because of its limited scope. For all its complexity, the tracking system applies only to wastes that are transported off-site by the generator. According to estimates, these represent only 4 to 5 percent of the nation's hazardous waste stream because most large firms manage wastes on their own premises. It is smaller businesses, particularly those in crowded urban areas, that rely on other facilities to treat and dispose of their wastes.[14]

Finally, the high compliance costs associated with the tracking program set up the potential for a perverse outcome. When the United States instituted the manifest system, it elevated the costs of legal disposal, making both source reduction and illegal disposal relatively cheaper and hence more attractive options from a profitability perspective. The key question is: Which of these two cheaper options will firms pursue? Without evidence, there is no definitive answer. Therefore, it would be naive to dismiss the possibility that this well-intended program might promote the very practice it sought to deter—illegal hazardous waste disposal.

In the interim, the EPA has taken steps to streamline the manifest system by standardizing the manifest form. This effort is expected to save the nation between $12 million and $20 million each year. Related to this change is a proposal to use an

---

[13]RCRA, Sec. 3006(b).

[14]Wirth and Heinz (May 1991), p. 47; U.S. EPA, Office of Solid Waste (November 1986), p. 9.

electronic manifest form, or an "e-manifest," although several issues have to be resolved before this innovation can be implemented.[15]

From nearly every perspective, RCRA's track record during its early years was less than impressive. There were major delays in implementation, more than a few policy reversals, several potential loopholes, and many questions about its effectiveness. These observations were largely responsible for Congress's enactment of the hard-hitting 1984 Amendments. Attempting to solve a litany of problems, these amendments continue to rely upon command-and-control policy instruments. There is still a glaring absence of economic incentives in the law, though the land restrictions may have been an exception.

Market Implications of the Land Restrictions

In the late 1970s, it became apparent that existing hazardous waste policy was doing nothing to halt the overuse of land disposal. One reason why landfilling had become so prevalent was that it was cheaper than other disposal options. This differential is due in part to more significant scale economies for landfilling relative to alternative practices, particularly waste incineration. For example, landfills face unit costs of $70 to $260 per metric ton, while incineration facilities incur much higher unit costs of $1,169 to $2,104 per ton of solid materials.[16]

In response, policymakers imposed a landfilling restriction as part of the 1984 RCRA Amendments. This command-and-control approach would elevate the costs of landfilling, which in turn should lead to higher prices. Just as in our analysis of the manifest system, we need to ask how generators are likely to respond. As profit maximizers, they would consider the relative costs of all available options. They could reduce the amount of waste generated or they could seek an alternative waste management practice, which in this case does not necessarily mean illegal disposal. Waste generators could use a treatment method like incineration to reduce the amount of waste disposed. Although the options are complex, we can gain some insight by modeling two *legal* options facing the waste generator:

- Substituting an alternative treatment like incineration
- Using a preventive strategy to achieve source reduction

Market Equilibrium Before the Land Restrictions

The two relevant markets are shown in Figure 17.3—land disposal in panel (a) and incineration in panel (b). The marginal private benefit (*MPB*) curves represent the decisions of waste generators, which are identical to marginal social benefit (*MSB*) functions, assuming no consumption externalities. On the supply side, marginal private cost (*MPC*) curves denote the decision making of the respective waste service facilities. The *MPC* curves are distinct from the marginal social cost (*MSC*) curves to indicate that **negative externalities** are associated with production in each market, with each

---

[15]U.S. EPA, Office of Solid Waste and Emergency Response (February 2005).

[16]Peretz and Solomon (April 1995), and a survey by Environmental Information Ltd. (1990) in Krukowski (1995), as cited in Sigman (2000), p. 225, Table 7-3.

FIGURE 17.3 Land Restrictions Under RCRA's 1984 Amendments

The higher costs of the 1984 land restrictions shift up the *MPC* and *MSC* curves for land disposal, decreasing the private market solution from L_0 to L_1. External costs decline from area *abc* to area *def.* If the waste generator uses less landfilling because it uses source reduction, the net result is this external cost reduction. But if the generator does not reduce its waste level, then the decrease in land disposal must have been achieved by using some alternative practice, such as incineration. Increased use of incineration shifts rightward the *MSB* curve in that market, increasing output from I_0 to I_1 and raising external costs from area *ghi* to area *gjk*. Unless the external cost decline in the landfilling market outweighs the increase in the incineration market, the land restrictions would achieve no net reduction in external costs.

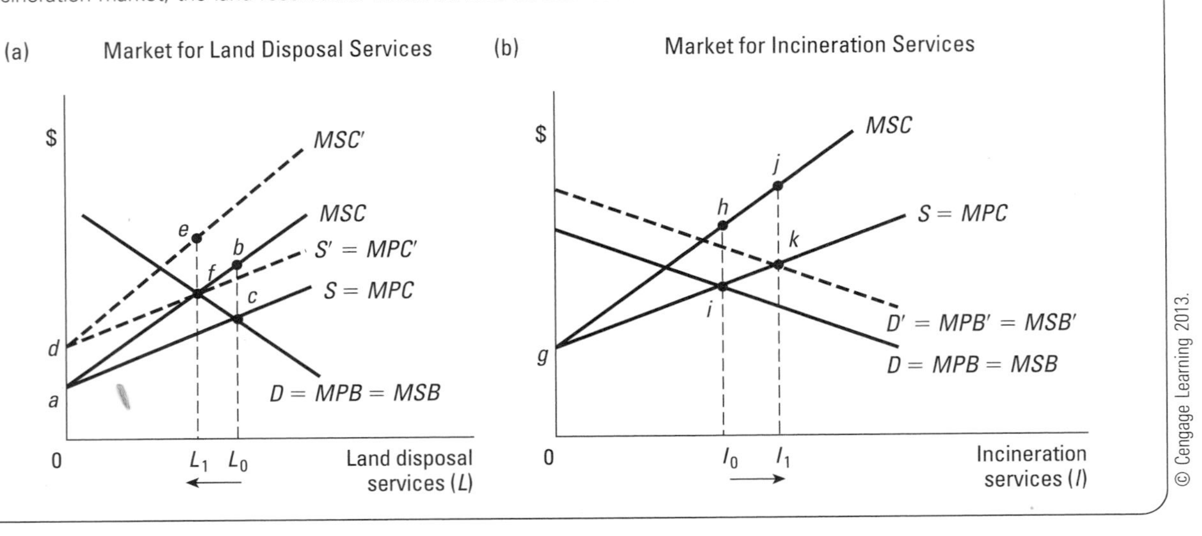

marginal external cost (*MEC*) implied by the vertical distance between each pair of *MPC* and *MSC* curves.

Prior to the imposition of the land restrictions, private equilibrium in each market is determined by the intersection of the *MPC* and *MPB* curves. In the landfill market, the equilibrium output level is L_0. Therefore, the external costs associated with L_0 is the area between the *MPC* and *MSC* curves up to that point, or triangular area *abc*. In the incinerator market, the initial equilibrium output level is I_0 and area *ghi* represents the analogous measure of external costs.

Market Equilibrium After the Land Restrictions

The effect of the land restrictions is higher operating costs for land disposal, shifting up the *MPC* and *MSC*, as shown in Figure 17.3(a). The result is a decline in landfill use from L_0 to L_1. Because equilibrium output is lower, so too are the associated external costs, which decrease to area *def.*[17] Hence, if waste generators use less landfilling because

[17]Notice that the land restrictions do not eliminate the market inefficiency, since there is nothing in these rulings that internalizes the external costs.

they pursue source reduction, the 1984 rulings should lower the external costs associated with hazardous wastes, *ceteris paribus*. However, if generators do *not* reduce their level of wastes, then the decrease in land disposal must have been achieved by using some alternative treatment. Assuming this alternative is incineration, we need to consider how the changes in that market affect the net result.

Increased use of incineration is shown in Figure 17.3(b) as a rightward shift of the *MSB* curve to *MSB′*, which increases equilibrium output from I_0 to I_1. Notice that at this higher equilibrium level, external costs have increased to area *gjk*. What can we conclude from this observation? Unless the decline in external costs in the landfilling market outweighs the increase in the incineration market, the land restrictions would achieve no net reduction in the external costs borne by society.

In sum, because waste generators have a number of options from which to choose, it is difficult to predict the net effect of the land restrictions. What we can say is that a reduction in external costs is possible as long as generators substitute land disposal with preventive strategies or an alternative waste treatment that adds less external costs than the reduction in those costs achieved from less landfilling. Hence, at least one aspect of U.S. hazardous waste policy—the land restrictions—may provide an economic incentive through price to help reduce hazardous waste generation. Absent this single element of the law, U.S. policy relies on a command-and-control approach that fails to achieve either efficiency or cost-effectiveness. This is a disconcerting observation, particularly since other nations and some state governments have used market instruments successfully to help control hazardous waste pollution.

Market-Based Policy

Market instruments use the price mechanism to make polluters confront the full costs of their actions and respond accordingly. Since there are significant risks associated with hazardous waste pollution, the external costs can be enormous. Hence, if price were somehow made to reflect these costs, polluters—or, in this context, hazardous waste generators—would have a powerful incentive to engage in source reduction activities.

waste-end charge
A fee implemented at the time of disposal based on the quantity of waste generated.

A customary market-based approach to hazardous waste control policy is the use of a **pollution charge**, which can be implemented as a **waste-end charge**. A waste-end charge is so named because it is a fee put in place at the time of disposal based on the quantity of waste generated. To achieve efficiency, such a charge would be set equal to the *MSC* of hazardous waste services at the efficient equilibrium to cover both the *MPC* of the waste service facility and the *MEC* from the associated pollution. This market-based solution is illustrated in Figure 17.4, where the efficient per-unit fee is labeled P_E.

In the United States, similar economic approaches have been employed by state governments in designing and implementing their hazardous waste programs, and the same is true for certain international venues. Application 17.2 (page 415) discusses some examples. Despite the international evidence and the experience of state governments, federal hazardous waste policy in the United States has employed no analogous market instruments to date.

FIGURE 17.4 A Waste-End Charge for Hazardous Wastes

To achieve efficiency, the waste-end charge, P_E, should be set equal to the *MSC* at the efficient equilibrium, Q_E, so that it covers both the *MPC* of the waste service facility and the *MEC* from the associated pollution.

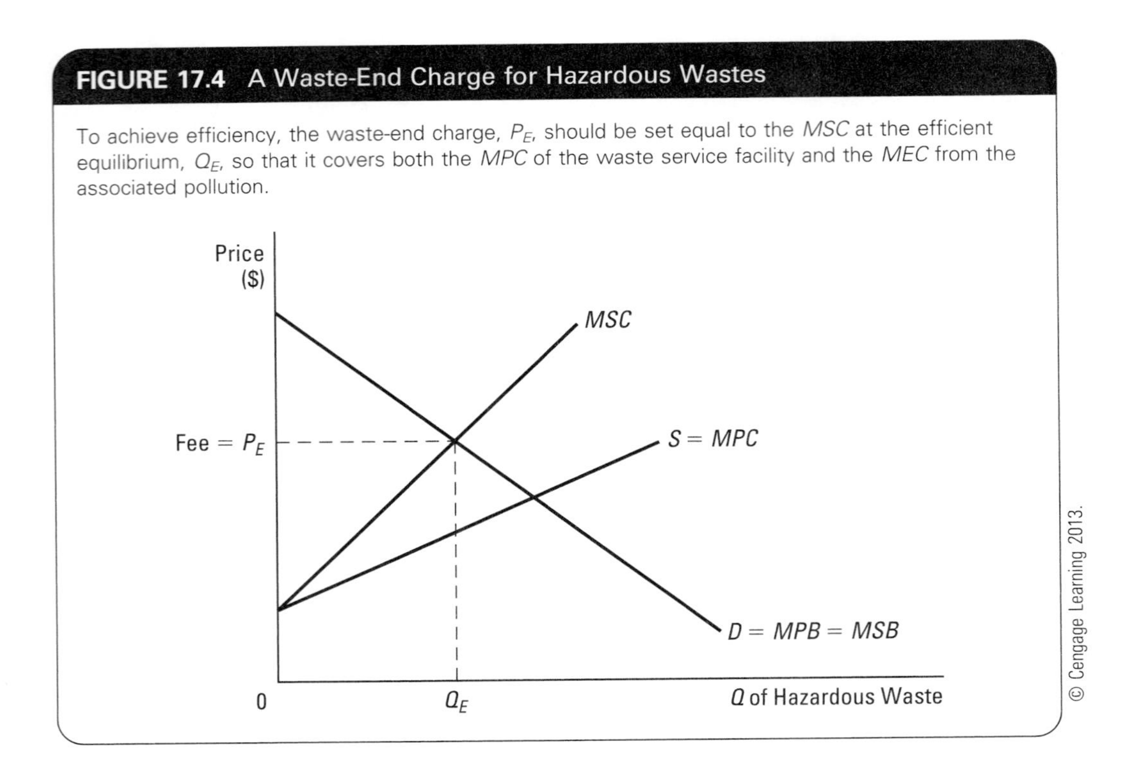

OVERVIEW OF U.S. HAZARDOUS WASTE SITE POLICY

Beyond RCRA's controls on *current* solid waste management, guidelines and financial support were needed to address the damages caused by *past* contamination and to clean up abandoned and uncontrolled hazardous waste sites. On the heels of the Love Canal disaster, the Comprehensive Environmental Response, Compensation, and Liability Act (CERCLA), or Superfund, as it is more commonly known, was passed in 1980. This marked the beginning of U.S. legislative development aimed at managing the nation's hazardous waste sites, as shown in Table 17.3 (page 416). There is also a comprehensive Web site on the Superfund program at **http://epa.gov/superfund/index.htm**.

In the Beginning

CERCLIS
A national inventory of hazardous waste site data.

Under CERCLA, the federal government was authorized to clean up contaminated sites and to recover damages from parties identified as responsible. Past and present owners of hazardous waste site facilities were required to notify the EPA of conditions at these sites by 1981. This baseline information along with data collected on an ongoing basis would comprise a national inventory called the **Comprehensive Environmental Response, Compensation, and Liability Information System (CERCLIS)**. This system is used to identify those sites posing the greatest threat to human health and the ecology—a classification referred to as the National Priorities List (NPL). To finance the cleanup of priority sites and to pay for damages, the first "Superfund" of $1.6 billion was established.

APPLICATION 17.2 Using the Market to Control Hazardous Wastes: An International Approach

Although the feedstock tax imposed under CERCLA may appear to have been motivated by the incentives of a product charge, its true intent was to provide a reliable source of funds for the nation's remediation program. Beyond this revenue-raising initiative, the use of a tax to reduce hazardous waste generation in the United States remains exclusive to state governments. While some states had introduced waste levies before the Superfund tax was put in effect, most imposed their own version of a pollution charge after the feedstock tax became law. Typically, this instrument was implemented as a **waste-end charge**.

According to a survey by the U.S. Army Corp of Engineers, 35 states have some form of direct treatment or disposal fee on hazardous wastes. Technically, such a waste-end charge is imposed to encourage waste reduction. However, this motivation is not universally apparent. In fact, revenue raising appears to be the impetus in some instances. Many states use the revenues to finance monitoring costs or cleanup efforts. Some add the waste tax revenues to their general funds, forcing environmental needs to compete for these funds along with all other state programs.

The characteristics of these market instruments vary considerably from state to state. Some states impose the tax directly on the generator at the first stage of the waste stream. Others levy the charge at the postgeneration phase by taxing the facility operator, such as a storage or treatment facility. Revenue motivations aside, taxing the generator offers a stronger incentive for waste reduction than does imposing the charge on a facility engaged in treatment, storage, or disposal. Some levy higher taxes on landfilling than on incineration, which should discourage land disposal.

The rate of the tax also differs widely. For example, the rate of tax imposed by Vermont is $112 per ton, while in Delaware, the rate is $21 per ton. More refined data show even more state-specific differences, such as varying rate schedules based on waste characteristics. For example, in California, the land disposal fee for most hazardous waste is $45.13 per ton, but for extremely hazardous waste, the fee is $223.44 per ton. If these differences properly reflect local economic and environmental conditions, the use of a hazardous waste tax at the state level may be far superior to any uniform charge imposed at the national level.

Internationally, Australia, Austria, Belgium, and Finland are among the nations that charge a fee on hazardous wastes. In each case, the waste-end charge elevates the effective price of waste management services. If the charge is high enough, it should *discourage* waste generation and *encourage* source reduction. Moreover, the revenues generated by these fees can be used to clean up the damages of hazardous waste disposal. A case in point is Austria, which uses the revenues from its waste-end charge, equivalent to $19 per metric ton, to fund the remediation of contaminated land.

Sources: U.S. EPA, Office of Policy, Economics, and Innovation (November 2004), p. 19; U.S. EPA, Office of Policy, Economics, and Innovation (January 2001); Hoerner (1998); U.S. Army Corp of Engineers (1998); Reese (1985); U.S. Congress, CBO (1985b), p. 82; U.S. Congress, CBO, Office of Technology Assessment (September 10, 1984).

As it turned out, CERCLA was a national failure on two counts. First, the $1.6 billion Superfund was used to remediate only eight of the thousands of contaminated sites across the nation. Second, the CERCLIS inventory did not assess the full magnitude of the problem, as was its purpose.[18] These failures motivated the revision and reauthorization of CERCLA through the Superfund Amendments and Reauthorization Act (SARA), passed in 1986—the law governing Superfund sites today. Visit **www.epa.gov/superfund/policy/cercla.htm** for an overview of CERCLA and SARA.

Current U.S. Policy

feedstock taxes
Taxes levied on raw materials used as productive inputs.

SARA raised the Superfund to $8.5 billion. This was to be financed primarily from **feedstock taxes**, including a petroleum tax, a tax on raw chemicals, and a corporate environmental income tax. Prior to their expiration at the end of 1995, these taxes

[18]Wolf (1988), p. 227; U.S. General Accounting Office (GAO) (December 1987), pp. 2–3.

TABLE 17.3 U.S. Legislation on Hazardous Waste Sites

Legislative Act	Major Provisions
Comprehensive Environmental Response, Compensation, and Liability Act (CERCLA) of 1980 (known as Superfund)	Called for a national inventory of hazardous waste site data from which to identify sites posing the greatest threat; placed these on the National Priorities List (NPL); designated a $1.6 billion fund to abate NPL sites and pay for damages.
Superfund Amendments and Reauthorization Act (SARA) of 1986	Reauthorized CERCLA; increased the fund to $8.5 billion; promoted permanent cleanup technologies.
Small Business Liability Relief and Brownfields Revitalization Act	Exempts certain parties from the liability associated with Superfund; authorizes grant funding up to $200 million annually for the assessment and abatement of brownfield sites.

Sources: Brownfields Act of 2001; Wolf (1988), Chapters 6 and 7; Superfund Amendments and Reauthorization Act of 1986; U.S. EPA, Office of Solid Waste and Emergency Response (October 1985); Comprehensive Environmental Response, Compensation, and Liability Act of 1980.

raised some $1.45 billion per year. Although there have been attempts to reauthorize the taxes, none have yet materialized. However, as part of the U.S. national budget for fiscal year 2012, there is a proposal to reinstate these after December 31, 2011, and before January 1, 2022. Meanwhile, even without the tax, Congress continues to appropriate funding for Superfund site cleanup at about $1.2 billion each year.[19]

In addition to annual funding approved by Congress, $600 million was appropriated for Superfund as part of the American Recovery and Reinvestment Act of 2009 enacted under the Obama administration. According to a recent performance report for this new legislation, among the goals for the funding allocated to Superfund is to accelerate abatement of NPL sites and, in so doing, create new jobs and generate economic and environmental benefits. These benefits are expected to arise from Superfund remedial actions through treating or removing contamination, improving drinking water quality to meet federal or state standards, providing substitute drinking water supplies, and reducing damages to wildlife habitats and ecosystems. As of March 2011, 51 sites and 61 projects have received funding through the Recovery Act allocation.[20]

In part, the Superfund tax has not been reinstated because Congress has failed to reauthorize CERCLA, despite efforts to do so on a number of occasions.[21] With so many unresolved issues, Superfund continues to be a source of controversy and political debate. In an effort to squelch some of the controversy, President Bush signed a bill in January 2002 that amended CERCLA and addressed certain Superfund-related concerns. This legislation, officially called the Small Business Liability Relief and Brownfields Revitalization Act, but commonly known as the Brownfields Act, addresses two issues dealing with liability and funding. First, it outlines specific exemptions from the extensive liability associated with Superfund (to be discussed later in this chapter), including

[19]U.S. Department of the Treasury (February 2011); U.S. GAO (July 18, 2008); Ramseur, Reisch, and McCarthy (February 4, 2008).

[20]U.S. EPA (May 4, 2011).

[21]To read more about attempts by Congress to reinstate the Superfund tax and reauthorize CERCLA, see Ramseur, Reisch, and McCarthy (February 4, 2008).

households and small businesses that dispose only municipal solid waste at Superfund sites. Second, it authorizes grant funding of up to $200 million annually for the assessment and abatement of brownfield sites.

brownfield site Real property where redevelopment or expansion is complicated by the presence or potential presence of environmental contamination.

Brownfield sites are abandoned or underutilized properties that are less contaminated than Superfund sites, but their redevelopment is complicated by the presence or potential presence of contamination. In support of abating these sites, $100 million was earmarked for the evaluation and cleanup of brownfield sites as part of the American Recovery and Reinvestment Act of 2009. Among the specific activities supported by this funding are: assessments to identify contaminants, direct abatement of brownfield sites, community involvement undertakings, and training for abatement and other environmental tasks. As a consequence of this financial support, 832 brownfield assessments have been initiated with 703 completed, and 36 brownfield cleanups have been started with 23 completed, as of the end of March 2011.[22]

Attention to brownfields is said to be coincident with the view that it might be time to bring the Superfund program to a close, assuming it has dealt with the most seriously contaminated sites. This type of sentiment along with concerns about the program's costs and associated litigation is partly responsible for congressional failures to reauthorize CERCLA. To learn more about brownfield sites, visit **www.epa.gov/swerosps/bf/basic_info.htm**.

Lastly, the EPA launched a new program in 2010 called the **Integrated Cleanup Initiative (ICI)**, which is aimed at integrating abatement procedures and improving cleanup across all types of contaminated sites, including Superfund, brownfields, RCRA corrective actions, federal facilities, and underground storage tank sites. The ICI is a three-year strategy developed through a collaborative effort between the EPA's Office of Solid Waste and Emergency Response and its Office of Enforcement and Compliance Assurance. The goal of this program is to identify ways to accelerate the cleanup process and to improve communication with local communities about abatement decisions.

As part of this new initiative, starting in 2011, the EPA is to report on a new metric for Superfund NPL cleanup progress, known as Remedial Action Project Completions. This new measure is intended to provide a means to quantify site cleanup progress and to improve accountability. Another important element of the ICI is a specific emphasis on enforcement aimed at ensuring that responsible parties abate contaminated sites, which in turn frees up resources for those sites where no responsible parties are identified.[23] More information on the ICI is available online at **www.epa.gov/oswer/integratedcleanup.htm**.

MANAGING UNCONTROLLED HAZARDOUS WASTE SITES: SUPERFUND

By most accounts, CERCLA, or Superfund, is considered a landmark legislative act because its intent is solely remedial rather than preventive. Despite this precedent-setting intent and a $1.6 billion fund, the 1980 law failed to execute on its commitments. The 1986 Amendments attempted to address these failures, but many problems

[22]U.S. EPA (May 4, 2011); U.S. EPA (September 17, 2004).

[23]U.S. EPA, Office of Solid Waste and Emergency Response (May 9, 2011).

remain. To understand why, we need to investigate the basic elements of this national policy.

Response and Cleanup

Superfund gives the federal government broad authority to respond whenever there is a release or the threat of a release of a hazardous substance, which it identifies as one that:[24]

- has been so designated under the Water Pollution Control Act, the Solid Waste Disposal Act, the Federal Water Pollution Control Act, the Clean Air Act, or the Toxic Substances Control Act, or
- may present substantial danger to health, welfare, or the environment when released.

Superfund cleanup process A series of steps to implement the appropriate response to threats posed by the release of a hazardous substance.

Notification of any release exceeding amounts specified in the law (1 pound unless stated otherwise) is to be made to the National Response Center (NRC). Failure to do so is punishable by a fine or imprisonment.

To implement these response actions, a comprehensive set of regulations was established, called the National Contingency Plan (NCP), which outlines procedures to put Superfund into action. The resulting steps in the **Superfund cleanup process** are summarized in Figure 17.5. More information is available online at **www.epa.gov/superfund/cleanup/index.htm.** By executive order, response activities are delegated to the EPA and must be carried out through removal actions or remedial actions.

removal actions Official responses to a hazardous substance release aimed at restoring immediate control.

Removal Actions

Removal actions are aimed at restoring immediate control to a release site. An example is removal of a leaking drum at a waste site or providing drinking water if supplies are deemed unsafe. These activities are undertaken only as short-term measures.[25]

remedial actions Official responses to a hazardous substance release aimed at achieving a more permanent solution.

Remedial Actions

Remedial actions are long-term responses aimed at finding a more permanent solution to a release site. Following the official identification of a problem, a complete assessment is made, and an inspection is ordered. The findings are evaluated using a **Hazard Ranking System (HRS)**, which assigns a numerical score to a site based on its inherent risks. If the site receives a score of 28.50 or higher out of a 100-point maximum, it is placed on the **National Priorities List (NPL)**.[26] To view a listing of NPL sites by state, visit **www.epa.gov/superfund/sites/query/queryhtm/nplfin.htm**. Only if a site is listed on the NPL can remedial actions be undertaken by the federal government. The rest are relegated to state and local officials for an appropriate response.

National Priorities List (NPL) A classification of hazardous waste sites posing the greatest threat to health and the ecology.

Once a site is placed on the NPL, a feasibility study is done, which involves a complete examination of the site and an appraisal of available cleanup options. According to

[24]CERCLA, Sec. 101(14) and Sec. 102(a).

[25]U.S. EPA, Office of Solid Waste and Emergency Response (April 2008).

[26]U.S. EPA, Office of Solid Waste and Emergency Response (April 2008), p. 4.

FIGURE 17.5 Superfund Cleanup Process

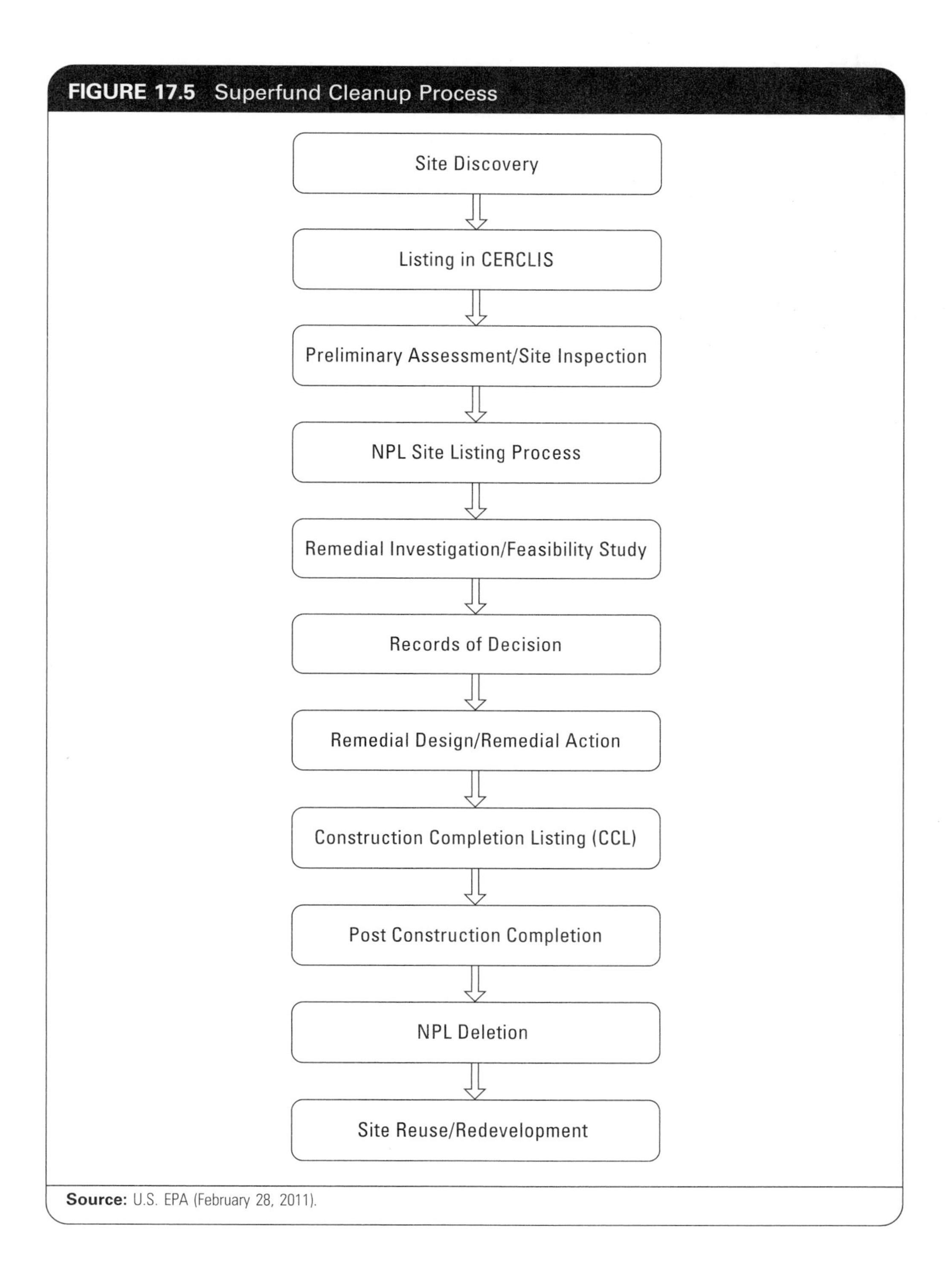

Source: U.S. EPA (February 28, 2011).

a highly regarded study conducted by researchers at Resources for the Future, this phase can take from 3.5 to 5 years. The action plan selected and support for the decision are documented and used in efforts to recover funding from parties responsible for the release. Design work follows, which takes another 2 to 2.5 years. Finally, the actual cleanup begins, which can take 3 to 4 years more. On average, the total cleanup cost for most sites is $12 million and can take over 11 years on average to complete. So-called mega sites (those with removal and remedial costs totaling over $50 million) have an average cleanup cost of $140 million.[27]

Compensation, Liability, and Enforcement

potentially responsible parties (PRPs)
Any current or former owner or operator of a hazardous waste facility and all those involved in the disposal, treatment, or transport of hazardous substances to a contaminated site.

Superfund gives the EPA authority to force those parties responsible for a hazardous substance release to correct the problem and pay for the damage. So-called **potentially responsible parties (PRPs)** include any existing or former owner or operator of a hazardous waste facility as well as those involved in the disposal, treatment, or transport of hazardous substances to a contaminated site. Exceptions have been made under the Brownfields Act, as previously discussed. The extent of the financial accountability is similarly extensive—to include all costs of removal or remedial action and any damages to natural resources or human health.[28] The intent is to force guilty parties to **internalize the externality** of whatever market transaction contributed to the release.

Emergency Planning

Toxics Release Inventory (TRI)
A national database that gives information about hazardous substances released into the environment.

Title III of the Superfund Amendments and Reauthorization Act of 1986 is a freestanding piece of legislation known as the Emergency Planning and Community Right-to-Know Act. Its purpose is to inform citizens about the existence and potential release of hazardous substances and to provide a planning system for emergencies. A related objective is to keep the public informed about the production of hazardous substances. To accomplish this goal, this law includes reporting requirements from firms that manufacture, process, or use certain chemicals above a mandated threshold. These data are used to compile a national database called the **Toxics Release Inventory (TRI)**, described online at **www.epa.gov/tri/**. Currently, information is collected on over 650 chemicals and chemical categories.[29]

ANALYSIS OF SUPERFUND

In concept, Superfund is the most unusual of all U.S. environmental laws. Unlike RCRA and other environmental legislation like the Clean Air Act, Superfund is not a regulatory program. Rather, its aim is to clean up the damage at uncontrolled or abandoned hazardous waste sites, using liability standards to identify potentially responsible parties and make them pay for their actions. While the overall intent is beyond reproach, the implementation has produced mixed results.

[27]Probst, Konisky, Hersh, Batz, and Walker (2001), Chapters 3 and 5. Cost estimates in the study are expressed in 1999 dollars.
[28]CERCLA, Sec. 107(a).
[29]U.S. EPA (December 16, 2010).

Assessing Superfund's Performance

Superfund's Remedial Program has moved at a snail's pace by most accounts, as shown in Figure 17.6. As of 2010, there were 1,628 sites on the NPL, net of additions and deletions over time. Of these, 1,098 have been placed on the **Construction Completion List (CCL)**,[30] which means that any physical construction work is finished and the site qualifies for deletion. However, as of mid-2011, only 349 NPL sites have been officially deleted. In fact, according to EPA estimates, one in four people in the United States lives within three miles of a hazardous waste site.[31] Further, it is argued that even the small number of removed sites is an inflated measure of Superfund's progress, since the EPA has not in all cases properly addressed the potential of recurring problems.[32]

If **benefit-cost analysis** guided the decision making, the high costs of cleanup would be used to determine how far remedial actions should go in cleaning up a site.

FIGURE 17.6 Number and Percentage of NPL Sites Placed on the Construction Completion List: 2000—2010

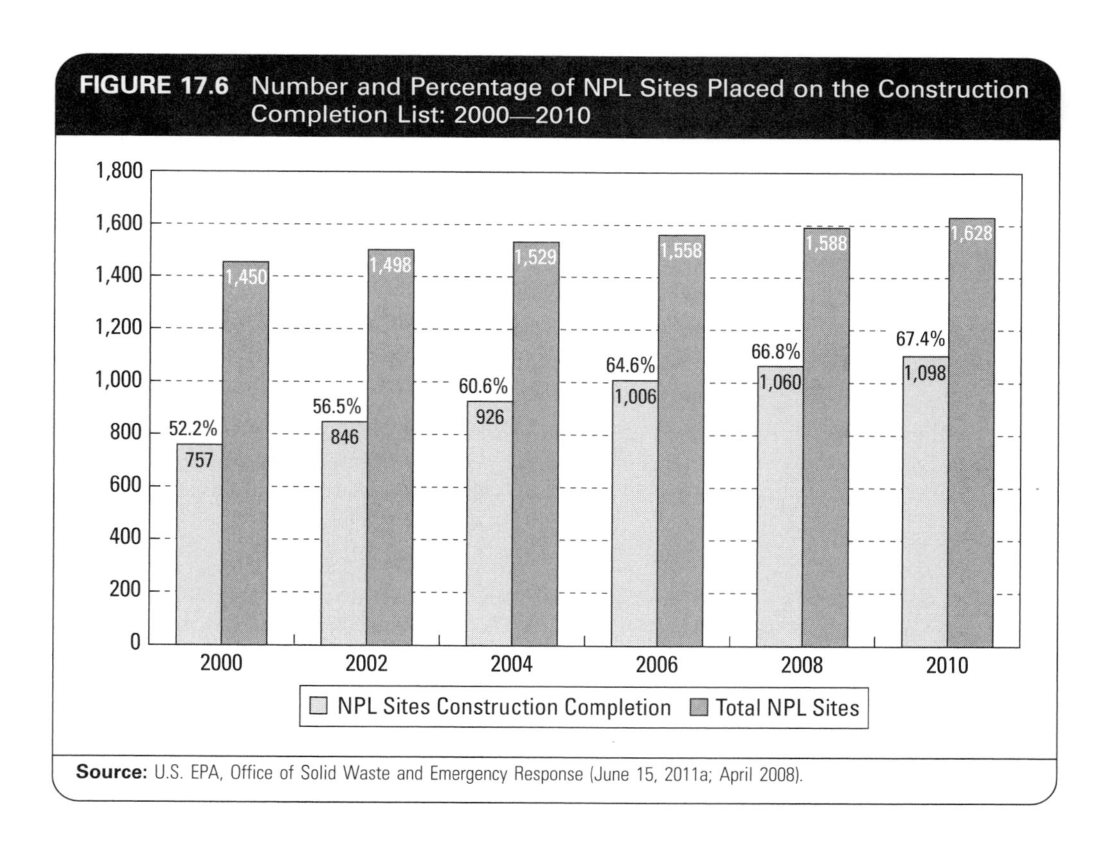

Source: U.S. EPA, Office of Solid Waste and Emergency Response (June 15, 2011a; April 2008).

---

[30]The CCL was established by the EPA in 1993 to help simplify site categorization, but it has no legal significance. See **www.epa.gov/superfund/cleanup/ccl.htm** for more information and to view the CCL by state.

[31]U.S. EPA, Office of Solid Waste and Emergency Response (June 15, 2011b); U.S. GAO (July 18, 2008).

[32]See U.S. Congress, OTA (1985).

Opponents of such an approach argue that the associated risks of a hazardous release should be reduced to zero or close to it, implying that a benefit-based standard ought to be employed. Both arguments are firmly entrenched in the issue of "how clean is clean." Until this is resolved, significant amounts of resources may be wasted. This very issue explains some of the controversy surrounding the EPA's ordering of General Electric Company to dredge PCBs from the Hudson River, a matter discussed in Application 17.3.

APPLICATION 17.3 GE Dredges the Hudson River: Will the Benefits Justify the Costs?

In one of the more controversial decisions of its history, the EPA in 2001 ordered General Electric Company (GE) to dredge polychlorinated biphenyls (PCBs) from a 40-mile length of New York's Hudson River. After much planning and debate, the first phase of dredging began in May 2009 and was completed in November 2009. The second phase was started in June 2011 and is expected to take between five and seven years to complete. GE did indeed release 1.3 million pounds of PCBs into the river. The huge conglomerate did so legally, starting in 1940 up until PCB use was banned in 1976. Nonetheless, the company caused the environmental damage and now is being ordered to clean up the mess. So why the controversy?

From a benefit-cost perspective, the multiyear dredging project makes sense only if the marginal costs of abatement are balanced by the associated marginal benefits. Superfund law requires GE to pay for the Hudson River abatement. According to recent reports, cleanup costs have already reached $830 million since 1990, and GE estimates that the entire project may end up costing as much as $1.4 billion. The expected benefits are the reduced health and ecological damages from PCB exposure, including the effect on the commercial fishing industry. The problem is that there are a number of issues surrounding the dredging order that are in dispute, and any one of these could upset the benefit-cost balance needed to achieve efficiency.

One major issue is the possibility that dredging up the PCBs and "resuspending" them may cause further environmental harm to the Hudson River and its ecosystems. Instead, goes the argument, the river should be left alone, since any risk of exposure diminishes as layers of silt naturally form on top of the chemicals. In response, the EPA argues that the dredging is no more of a threat to resuspending the PCBs than is the natural churning of the river.

A second source of controversy is that the dredging order is intended to reduce the carcinogenic risk of PCB exposure—an expected benefit of abatement. However, a recent study casts doubt on this expectation. Congress's original decision to ban PCBs followed the results of a study by Dr. Renate Kimbrough, who was then with the Centers for Disease Control and Prevention. Her findings indicated that rats that had ingested large amounts of PCBs developed liver cancer. However, more than 20 years later, in a study funded in part by GE, this same scientist along with two colleagues report that no association could be found between PCB exposure and death by cancer or any other disease. The scientists studied 7,035 people over an average of 31 years who worked in the very two GE plants that released the PCBs into the Hudson River. Their findings were published in the *Journal of Occupational and Environmental Medicine* in March 1999. If these new findings are accurate, the benefits of abating PCBs in the Hudson would be greatly diminished.

Lastly, local residents expressed concern that the massive dredging effort with its unsightly equipment and truckloads of oily sludge may wreak havoc on property values, quality of life, and local tax revenues. These added costs, although difficult to quantify, would likely be nontrivial. In a valid benefit-cost analysis, these should be estimated and added to the abatement costs.

Taken together, these concerns suggest that even the explicit costs of the Hudson dredging, let alone the implicit costs, may not be justified by the expected benefits. If so, allocative efficiency will not be achieved, suggesting that resources used for this massive project might be better utilized in some alternative effort.

For further detail on the Hudson River cleanup project, visit **www.epa.gov/hudson**.

Sources: U.S. EPA (June 15, 2011); Barrett and Glader (December 18, 2010); Fallon (March 13, 2009); Cohen (January 7, 2002; December 12, 2001); Murray (December 5, 2001); U.S. EPA, Region 2 (December 6, 2000).

What's Wrong With Superfund?

Superfund's lack of success cannot be blamed on any single factor. There are many reasons that explain its generally disappointing track record, all of which tend to fall into two major categories. The first is the lack of information about the extent of the problem, and the second is linked to the absence of incentives to advance the remediation process.

Information Problems[33]

A major obstacle in addressing the nation's hazardous waste site problem was that the EPA had to start at ground zero, with very little data on the extent of the problem. As information first began to accumulate, the agency learned that the problem was far worse than it had anticipated. There also has been a contention that all the significant problems have not been identified. Beyond these data problems, there is also a lack of knowledge about the most effective technologies to accomplish the complex cleanup task. The many phases of remedial action are time intensive and costly. Hence, as the EPA continues its cleanup efforts, it is pursuing cost recovery settlements from responsible parties to help defray Superfund's enormous expense.

Lack of Incentives

Like the provisions of RCRA, the Superfund legislation is distinguished by the conspicuous absence of market mechanisms. Even the feedstock taxes used to fund the $8.5 billion Superfund were aimed mainly at raising revenues rather than encouraging decisions to improve the environment. Congress's decision to tax oil and certain raw chemicals was motivated by the need for a substantial and fairly reliable tax base from which to collect much-needed revenues.[34] Despite the statutory label of "environmental taxes," these levies offered little motivation to diminish hazardous waste pollution.

Other incentive issues arise in how liability is established under Superfund. Motivated by the knowledge that the $8.5 billion fund would not be nearly enough to pay for the massive cleanup, Congress and the courts use whatever legislative and judicial muscle is necessary to impose the "polluter-pays principle." While the intent was to achieve swift action against guilty parties and advance the remediation process, the plan backfires because it sets up an unexpected and perverse incentive.

strict liability The legal standard that identifies individuals as responsible for damages even if negligence is not proven.

The problem is rooted in how the law identifies a potentially responsible party (PRP) and in how the liability is defined. In establishing legal responsibility, the courts use the concepts of strict liability and joint and several liability.[35] **Strict liability** means

[33]Much of the following discussion is drawn from U.S. GAO (December 1987), p. 14, Chapter 2, Table 2.1; U.S. EPA (March 1987), cited on p. 22 of this GAO report.

[34]SARA added another revenue source in 1986, calling for a new tax based on a mandated percentage of corporate income. U.S. EPA, Office of Policy, Planning, and Evaluation (July 1992), p. 3-5; CERCLA, Title II; SARA, pt. 7, Sec. 59A.

[35]Interestingly, these legal concepts are not written into Superfund's provisions. They were initially in the bill submitted before Congress, but were removed as part of an agreement between the House of Representatives and the Senate before signing it into law. However, the courts took the initiative and reinstated these notions based on what they perceived to be Congress's intent (Mazmanian and Morell, 1992, p. 36).

joint and several liability
The legal standard that identifies a single party as responsible for all damages even if that party's contribution to the damages is relatively small.

that individuals can be held liable even if negligence is not proven. **Joint and several liability** means that a single party found liable for damages can be held responsible for *all* associated costs even if that party's contribution to the damages is relatively small.[36] The use of such tough legal standards was intended to save the government the complex and time-intensive process of proving negligence and then determining what proportion of costs each PRP must pay. There was also the expectation that the use of joint and several liability would encourage an identified PRP to name all other responsible parties at a given site.

Unfortunately, these expectations proved to be off the mark. Recognizing the potential to be held accountable for *all* costs at a named waste site, few PRPs were willing to come forward with information. Furthermore, because joint and several liability can assign the entire costs of cleanup to a single party, there was a strong financial incentive for a liable party to use delay tactics or to incur the relatively lower costs of litigation to fight the charges.[37] As a consequence, the well-intended legal standards have led to long delays and a diversion of resources from cleanup to litigation proceedings.

CONCLUSIONS

Most would agree that U.S. hazardous waste policy has strengthened considerably over the past few decades. As the evolution continues, we observe public officials beginning to integrate preventive initiatives into the policy solution. By addressing the generation of hazardous waste *before* it becomes a problem, the nation can avoid some of the associated health and ecological risks, not to mention the high costs of correcting the degradation. Yet in spite of these favorable assessments and the progress that has been made, there are still questions about the efficiency and cost-effectiveness of current policy.

With 20/20 hindsight, the federal government should have acted sooner to respond to hazardous waste risks. But such an assessment assumes full information—a luxury the United States clearly did not have. Confronted with tough problems, a marked lack of data, and an angry constituency, Congress sought policy instruments that it thought would be direct and uncompromising. Given this motivation, it is not surprising that the major laws on hazardous solid waste are couched within a command-and-control framework.

Understanding this motivation, however, does not change the fact that such a policy approach came at some measure of sacrifice—a loss of efficiency and cost-effectiveness. Exacerbating the problem, many of Congress's intentions were not carried out as planned. Not only did most of the initiatives lack incentive mechanisms, but many were set in motion without a full understanding of the extent of the problem. As such, they were destined for failure. Such was the fate of the Superfund legislation, a unique body of law that continues to be a source of contention among government officials, industry, and private citizens.

[36]There are exemptions to this liability, including those outlined in the Brownfields Act.
[37]Mazmanian and Morell (1992), pp. 36–37.

Midcourse corrections might save the nation considerable expense in executing its policy missions and could restore the level of environmental risk to an acceptable level more quickly. This suggests that Congress should decide a new course of action to reauthorize the Superfund law. If nothing else, policymakers seem to have recognized the need for reform and the importance of using risk management strategies in all environmental initiatives—even in the absence of extensive information. Such an awareness may be significant to what appears to be a critical phase in the evolution of U.S. hazardous waste policy.

Summary

- Hazardous solid wastes are any unwanted materials or refuse capable of posing a substantial threat to health or the environment.
- In the United States, hazardous waste generation is about 35.3 million tons per year, or about 0.12 tons per person annually. All environmental media are vulnerable to hazardous waste contamination.
- There was no federal policy on waste control until the Solid Waste Disposal Act (SWDA) was passed in 1965. This legislation had limited strength until it was amended by the Resource Recovery Act in 1970 and the Resource Conservation and Recovery Act (RCRA) in 1976.
- Congress reauthorized and strengthened RCRA through the Hazardous and Solid Waste Amendments of 1984, which governs U.S. policy today.
- RCRA uses a command-and-control approach known as the "cradle-to-grave" management system. Its four components are: the identification and listing of hazardous waste; a manifest system; a permit system for treatment, storage, and disposal facilities (TSDFs); and standards.
- The RCRA Amendments suggest a policy shift away from land disposal and toward more preventive solutions.
- The criteria used to identify hazardous waste are risk based and applied uniformly, resulting in allocative inefficiency. Standards applicable to TSDFs are benefit-based with no consideration for economic costs and applied uniformly.
- Economic considerations play no role in implementation of the manifest system. It also suffers from high compliance costs, which may encourage illegal disposal.
- The land restrictions should lead to higher prices for landfilling, which should encourage source reduction or the use of alternative waste management practices. A net decline in external costs is possible but not ensured.
- A customary market-based approach to hazardous waste control policy is the use of a waste-end charge, which is being used internationally and at the state level in the United States.
- The Comprehensive Environmental Response, Compensation, and Liability Act (CERCLA), or Superfund, was passed in 1980. Its objective was to identify and clean up the worst inactive hazardous waste sites and recover damages. It was revised in 1986 through the Superfund Amendments and Reauthorization Act (SARA).
- The Brownfields Act amends CERCLA by providing certain exemptions to Superfund liability. It also provides grant funding for abating brownfield sites.
- In 2010, the EPA launched its Integrated Cleanup Initiative (ICI), which is aimed at integrating abatement procedures and improving cleanup across all types of contaminated sites.
- Under CERCLA, the federal government can undertake a response action whenever there is a hazardous substance release. The National Contingency Plan (NCP) outlines procedures to implement removal actions or remedial actions.
- As part of a remedial action, a site is assigned a numerical score based on its inherent risks. If the site receives a score of 28.50 or higher, it is placed on the National Priorities List (NPL).
- The EPA has the authority to force potentially responsible parties (PRPs) to pay for the damages caused by a hazardous substance release.
- The purpose of Title III of SARA is to inform citizens about potential hazardous substance releases and to provide an emergency planning system. One of its requirements calls for reports on chemical releases, which are used to form the Toxics Release Inventory (TRI).
- Superfund's lack of success is due mainly to lack of information and the absence of incentives. The use of strict liability and joint and several liability has led to delays and the diversion of resources from cleanup to litigation proceedings.

Review Questions

1. Assume that the uniform standard on TSDFs is associated with the following benefit and cost relationships:

$$TSB = 20Q - 0.1Q^2 \qquad TSC = 8Q + 0.2Q^2$$
$$MSB = 20 - 0.2Q \qquad MSC = 8 + 0.4Q$$

where dollar values are in thousands of dollars, and Q is the number of inspection procedures.

   a. Because the standards on TSDFs are benefit-based, find the Q that maximizes TSB, and evaluate TSC and TSB at that level.

   b. Show that an efficient standard would lower TSC and increase total net benefits.

2. Consider the various provisions in RCRA that discourage land-based waste disposal. Identify and explain one aspect of the law that is command-and-control in approach and one that is incentive-based.

3. Suppose that prior to the land restrictions of 1984, the market for land disposal was modeled as follows:

$$D = MPB = 80 - 2L \qquad S = MPC = 20 + 2L$$

where L is thousands of tons of landfilled hazardous waste, and the dollar values are per ton.

When the 1984 restrictions decreased the supply of land disposal services, assume that the marginal private costs (MPC) of production decreased to $MPC' = 32 + 2L$.

   a. Based on this model, quantify the decrease in land disposal accomplished by the restrictions.

   b. Graphically show the qualitative effect on total external costs in the land disposal market.

4. Propose an alternative method of identifying hazardous waste that is more efficient than the one mandated under RCRA without compromising the objective of risk reduction. Support your proposal with a well-defined risk management strategy.

5. Suppose the following equations have been estimated for the market for hazardous waste services:

$$MSB = 30 - 2.5Q \qquad MPC = 1.1 + 0.9Q$$
$$MSC = 1.5 + 1.25Q$$

where Q is thousands of tons per week, and the dollar values are per ton. Assume there are no consumption externalities in this market.

   a. State the equation that represents the negative externality in this market.

   b. Algebraically solve for the competitive and socially optimal price and quantity combinations. Quantify the extent of the resource misallocation associated with the competitive solution.

   c. Find the dollar value of a waste-end charge that would restore efficiency to this market.

6. Use benefit-cost analysis to qualitatively evaluate the Superfund remedial action program.

7. Recommend an incentive-based reform that would improve states' identification of hazardous waste sites.

8. **a.** An ongoing debate about the Superfund program is the determination of the optimal abatement level, characterized as the "how clean is clean" problem. Choose one of the risk management strategies, and propose how it might be used to resolve this issue.

   b. In your view, why has there been no movement to initiate such a proposal?

Additional Readings

Carruthers, John I., and David E Clark. "Valuing Environmental Quality: A Space-Based Strategy." *Journal of Regional Science 50*(4) (October 2010), pp. 801–32.

Chang, Howard F., and Hilary Sigman. "The Effect of Joint and Several Liability Under Superfund on Brownfields." *International Review of Law and Economics 27*(4) (December 2007), pp. 363–84.

Collins, Flannary P. "The Small Business Liability Relief and Brownfields Revitalization Act: A Critique." *Duke Environmental Law & Policy Forum 13*(2) (Spring 2003), pp. 303–29.

Ferrey, Steven. "Converting Brownfield Environmental Negatives Into Energy Positives." *Boston College Environmental Affairs Law Review 34*(3) (2007), pp. 417–79.

Fullerton, Don, and Seng-Su Tsang. "Should Environmental Costs Be Paid by the Polluter or Beneficiary? The Case of CERCLA and Superfund." *Public Economics Review* (June 1996), pp. 85–117.

Gamper-Rabindran, Shanti, and Christopher Timmins. "Hazardous Waste Cleanup, Neighborhood Gentrification, and Environmental Justice: Evidence From Restricted Access Census Block Data." *The American Economic Review 101*(3) (May 2011), pp. 620–25.

Greenstone, Michael, and Justin Gallagher. "Does Hazardous Waste Matter? Evidence From the Housing Market and the Superfund Program." *The Quarterly Journal of Economics 123*(3) (August 2008), pp. 951–1003.

Hamilton, James, and W. Kip Viscusi. "How Costly Is 'Clean'? An Analysis of the Benefits and Costs of Superfund Site Remediations." *Journal of Policy Analysis and Management 18*(1) (Winter 1999), pp. 2–27.

Harper, Richard K., and Stephen C. Adams. "CERCLA and Deep Pockets: Market Responses to the Superfund Program." *Contemporary Economic Issues XIV*(1) (January 1996), pp. 107–15.

Hibiki, Akira, and Shunsuke Managi. "Does the Housing Market Respond to Information Disclosure? Effects of Toxicity Indices in Japan." *Journal of Environmental Management 92*(1) (January 2011), pp. 165–71.

Khanna, Madhu, and Wilma Rose Q. Anton. "Corporate Environmental Management: Regulatory and Market-Based Incentives." *Land Economics 78*(4) (November 2002), pp. 539–59.

Kiel, Katherine A., and Michael Williams. "The Impact of Superfund Sites on Local Property Values: Are All Sites the Same?" *Journal of Urban Economics 61*(1) (January 2007), pp. 170–92.

Levinson, Arik. "State Taxes and Interstate Hazardous Waste Shipments." *American Economic Review 89*(3) (June 1999), pp. 666–77.

McCluskey, Jill J., and Gordon C. Rausser. "Hazardous Waste Sites and Housing Appreciation Rates." *Journal of Environmental Economics and Management 45*(2) (March 2003), pp. 166–76.

Schoenbaum, Miriam. "Environmental Contamination, Brownfields Policy, and Economic Redevelopment in an Industrial Area of Baltimore, Maryland." *Land Economics 78*(1) (February 2002), pp. 72–87.

Sigman, Hilary. "Taxing Hazardous Waste: The U.S. Experience." *Public Finance and Management 3*(1) (March 2003), pp. 12–33.

Stafford, Sarah L. "The Impact of Environmental Regulations on the Location of Firms in the Hazardous Waste Management Industry." *Land Economics 76*(4) (November 2000), pp. 569–81.

______. "The Effect of Punishment on Firm Compliance With Hazardous Waste Regulations." *Journal of Environmental Economics and Management 44*(2) (September 2002), pp. 290–308.

Viscusi, W. Kip, and James T. Hamilton. "Are Risk Regulators Rational? Evidence From Hazardous Waste Decisions." *American Economic Review 89*(4) (September 1999), pp. 1010–27.

A Reference to Acronyms and Terms in Hazardous Waste Control Policy

Environmental Economics Acronyms

MEC	Marginal external cost
MPB	Marginal private benefit
MPC	Marginal private cost
MSB	Marginal social benefit
MSC	Marginal social cost

Environmental Science Acronyms

DDT	Dichloro-diphenyl-trichloroethane
PCBs	Polychlorinated biphenyls
TCDD	Tetrachlorodibenzo-p-dioxin
TCP	Trichlorophenol

Environmental Policy Acronyms

CCL	Construction Completions List
CERCLA	Comprehensive Environmental Response, Compensation, and Liability Act
CERCLIS	Comprehensive Environmental Response, Compensation, and Liability Information System
HRS	Hazard Ranking System
ICI	Integrated Cleanup Initiative
NCP	National Contingency Plan
NPL	National Priorities List
NRC	National Response Center
PRP	Potentially responsible party
RCRA	Resource Conservation and Recovery Act
SARA	Superfund Amendments and Reauthorization Act
SWDA	Solid Waste Disposal Act
TRI	Toxics Release Inventory
TSDFs	Treatment, storage, and disposal facilities

To access additional course materials, visit www.cengagebrain.com. At the home page, search for the ISBN of this title (shown on the back cover). This will take you to the product page where these resources can be found.

CHAPTER 18

Managing Municipal Solid Waste

"This country must make every effort to stem the rising tide of garbage and industrial waste through a more aggressive use of waste minimization and recycling practices."

—George H. W. Bush (1924–)

Everywhere in the world, local communities must deal with the accumulation of what most of us offhandedly call trash. **Municipal solid waste (MSW)**, as it is more formally termed, is the collection of aluminum cans, plastic bottles, food scraps, lawn clippings, and consumer electronics that characterizes everyday living. How could such a mundane matter as everyday trash become such a serious issue? Mainly because society has seen it as exactly that—a routine part of living that merits no particular attention. As long as the unsightly pile of old newspapers and garbage is removed from the curbside, most citizens give the matter little thought—that is, until some public official proposes a new landfill site on the next block. The typical reaction is a negative one, dubbed the "not in my backyard," or "NIMBY," syndrome. This response coupled with the growth trend in municipal waste generation and an aging disposal system has left many communities with a difficult problem.

As a component of the nonhazardous waste stream, municipal solid waste poses no *direct* threat to human, animal, or plant life. Nonetheless, there are risks if too much is generated or if it is improperly managed. Unsanitary conditions at landfill sites can contaminate water and soil with disease-spreading bacteria. A more serious risk is the release of toxic substances into the environment. Toxic contamination can arise from natural decomposition processes or the presence of hazardous substances mixed in with municipal refuse. There is also the risk of atmospheric pollution caused by gases released from waste decomposition or from the incomplete combustion of incinerated wastes.

Given the potential damage, public policies are needed to control waste management practices, to find ways to recover and reuse waste materials, and to develop new technologies. In the United States, much of the responsibility for these policies is delegated to state and local governments, with some measure of federal oversight. Such an approach allows for more flexibility in devising waste management programs than is the case for hazardous waste controls. In some communities, innovative policies have been developed to deal with MSW, including the use of market instruments to achieve environmental goals.

In this chapter, we consider the risks of MSW pollution and analyze the policy response to these risks. To understand the motivation of government programs and regulations, we begin by characterizing the generation and composition of this waste stream. Once done, we outline the delegation of responsibilities between federal and state governments given by the Resource Conservation and Recovery Act (RCRA). We then present a model of MSW services, which prepares the way for an economic analysis of market instruments used to manage the MSW stream. A list of commonly used acronyms and terms can be found at the end of this chapter.

CHARACTERIZING MUNICIPAL SOLID WASTE

Observing a Trend

municipal solid waste (MSW) Nonhazardous wastes disposed of by local communities.

Many U.S. communities have been observing an increase in the size of the **municipal solid waste (MSW)** stream for some time, a phenomenon only partly linked to population growth. As shown in Table 18.1, per capita generation rates are much higher in 2010 than in 1980 despite a gradual decline after 2000. Notwithstanding this slight reversal in trend, most argue that the characterization of Americans as a "disposable society" may be well placed. In-built product obsolescence, persuasive advertising that encourages excessive consumption, and reliance on prepared, heavily packaged food products are all contributing factors. Despite the apparent growth trend in MSW generation, the development of better waste management practices has lagged behind. Recent data and other information on U.S. MSW are available at **www.epa.gov/epawaste/basic-solid.htm**.

For years, the United States and most European countries have depended primarily on landfills to dispose of MSW, a practice that seemed a reasonable solution, at least for a time. In the 1980s, reports of a "capacity crisis" warned that many municipal landfills were running out of space and others were being closed for improper waste management practices. More recently, counterarguments have emerged, stating that the so-called capacity crisis was exaggerated, occurring in only certain locations. These reports

TABLE 18.1 Trend Data on U.S. MSW Generation

	1980	1990	2000	2010
Total MSW generated by weight (millions of tons)	151.6	208.3	242.5	249.9
Total population (thousands)	227,726	250,181	282,418	310,109
Per capita MSW (pounds per day)	3.65	4.56	4.70	4.42
Gross Domestic Product (GDP) per capita (chained $2005)	$25,640	$32,112	$39,750	$42,723

Sources: U.S. EPA, Office of Resource Conservation and Recovery (November 2011), Table 1; Council of Economic Advisers (February 2011), Table B-31.

point out that the rash of dump closings has involved mainly small sites and that increased recycling and a greater use of waste incineration have contributed to what is now a surplus of landfill space in some areas.

A closer look helps to explain the disparate views. Recent Environmental Protection Agency (EPA) data do show an ongoing dependence on landfilling in the United States, with 54.3 percent of MSW landfilled in 2010. It is also the case that the number of landfills in the nation has been declining. For example, in 1988, there were nearly 8,000 landfills in the United States, and by 2010, the number had fallen sharply to 1,908. However, since the size of the average landfill has been increasing, overall capacity has remained fairly stable. The EPA asserts that capacity is not a problem in the United States but that regional dislocations do occur on occasion.[1]

Composition of MSW in the United States[2]

To get a sense of what comprises MSW in the United States, let's examine its composition both by the types of products being discarded and by the kinds of materials entering the waste stream.

Product Groups

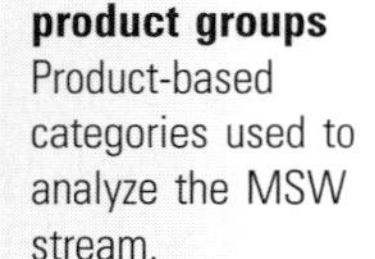
product groups
Product-based categories used to analyze the MSW stream.

The designated **product groups** are durable goods (e.g., major and small appliances, furniture, and tires), nondurable goods (e.g., newspapers, magazines, clothing, and trash bags), containers and packaging, and other wastes, such as yard trimmings and food scraps. Figure 18.1(a) gives the relative proportions by weight of these product groups based on the 249.9 million tons of MSW generated in 2010. Notice that packaging and containers account for about 30.3 percent of this total, or 75.6 million tons.

Within the durable goods product group, one area of interest is the disposal of consumer electronics, which includes smartphones, computer equipment, and televisions. Rising sales of these products along with their relatively short economic lives have contributed to their growth in the waste stream. In 2010, 0.65 million tons of the 2.4 million tons of generated electronics waste were recovered, which is proportionately lower than a number of other products in the MSW stream. More information on electronics recycling is available online at the National Center for Electronics Recycling at **www.electronicsrecycling.org/public/**. Tracking this segment of MSW is relevant not only because of its rising volume, but also because some electronics include toxic materials such as lead. Hence, efforts to recover, reuse, and remanufacture these products have become increasingly important. Further information on the growth of electronics consumption and waste generation is discussed in Application 18.1 (page 434).

[1]U.S. EPA, Office of Resource Conservation and Recovery (November 2011), Table 28; U.S. EPA, Office of Solid Waste and Emergency Response (November 2011), p. 2.

[2]Much of the following is drawn from U.S. EPA, Office of Resource Conservation and Recovery (November 2011).

FIGURE 18.1 Proportion of Products and Materials Generated in U.S. MSW in 2010

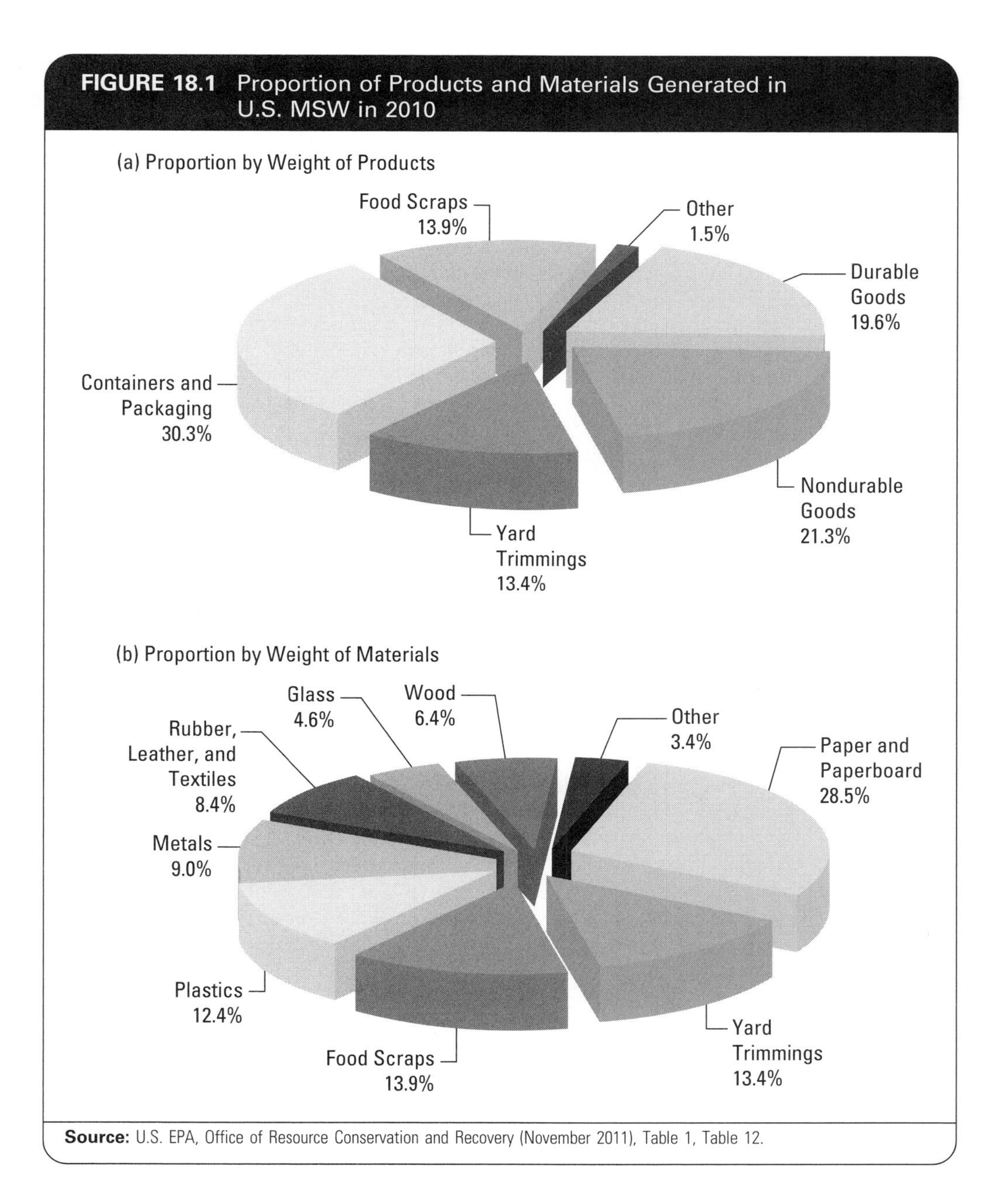

Source: U.S. EPA, Office of Resource Conservation and Recovery (November 2011), Table 1, Table 12.

Materials Groups

materials groups Materials-based categories used to analyze the MSW stream.

The **materials groups** identified in MSW are paper and paperboard, yard trimmings, glass, metals, plastics, textiles, rubber and leather, wood, and other miscellaneous wastes. Proportions by weight for these categories are shown in Figure 18.1(b). According to these data, paper and paperboard represent the largest proportion by weight of generated materials in 2010, accounting for 28.5 percent of the total.

Composition data by materials convey information about how manufacturing decisions ultimately affect waste accumulation. Table 18.2 (page 435) gives trend data on

APPLICATION 18.1 Consumer Electronics and the Growth of e-Waste

Consumer electronics have become an integral part of contemporary life, helping to define modern culture—how we communicate, how we conduct business, and how we live. Smartphones, laptop computers, digital copiers, monitors, and flat-panel televisions comprise only some of the goods commonly classified as electrical and electronic equipment (EEE). Although this rapidly growing product category has added to the quality of life, it also has contributed to MSW accumulation once these goods reach the end of their useful lives and become electronics waste, or e-waste as it is sometimes known. An added concern is that certain materials found in e-waste such as lead and mercury pose a risk to human health and the environment.

To give context to these assertions, let's consider some recent data from the EPA's assessment of MSW in the United States. In 2010, consumer electronics accounted for 2.4 million tons in the MSW stream. Although this tonnage represents only 1 percent of the total MSW generated, it is 28 percent higher than in 2000, the earliest year for which estimates are currently available. In 2000, only 10 percent of the 1.9 million tons generated was recovered for recycling. By way of comparison, in 2010, recovery was higher, at 26.6 percent, but still lower than some other components of durable goods waste. Taken together, this recovery rate is concerning, given the rising growth rate of e-waste, which in turn reflects the rapid growth rate of electronics consumption.

To gain a better sense of the size and growth of EEE consumption, discards, and recovery, the EPA launched a study of selected electronics products used by residential and commercial/institutional buyers. In its analysis, which was updated in 2011, the agency addressed the following product categories: personal computers, computer displays, keyboards and mice, hard-copy devices, televisions, and mobile devices, sold from 1980 through 2009, which sum to an estimated 6.1 billion units. The identified product groups were chosen in part because they are common targets of recycling and stewardship initiatives.

Among the EPA's findings were the following estimated values and observations. In 2009, an estimated 438 million consumer electronics were sold, which is twice the amount sold in 1997. Much of this increase was attributed to a nine-fold rise in sales of mobile devices, such as cell phones, pagers, and smartphones. In addition to new sales, another 5 million tons of electronics were estimated to have been in storage in 2009. Approximately 2.37 million tons of used products entered end-of-life management in 2009, which is over 120 percent of the e-waste in 1999. This addition brought the total number of electronics at their end-of-life stage to 3.6 billion units, or 26.8 million tons, with another 2.44 million tons projected for 2010.

Of the electronics products that were ready for end-of-life management in 2009, 25 percent by weight were collected for recycling. This percentage has risen steadily from 2006 when the recycling rate was estimated at 22 percent and was projected to increase to 27 percent in 2010. Among the electronics product groups identified, the recycling rate for computers was the highest in 2009 at 38 percent, while the rate for mobile devices, keyboards, and mice was the lowest at 8 percent.

So what is being done to improve recovery and reuse of e-waste? Unless the e-waste is identified as hazardous under the Resource Conservation and Recovery Act (RCRA), there is no federal legislation that controls e-waste in the United States. However, many states do have mandatory programs in place. These state-level rulings can be explored online at **www.ecyclingresource.org/ContentPage.aspx?Pageid=28&ParentID=0** at the National Electronics Recycling Infrastructure Clearinghouse. In addition, there are many private and government voluntary programs designed to encourage recycling, reuse, and product stewardship for electronics, including the *National Strategy for Electronics Stewardship* announced in 2011 to promote markets for recycled electronics. Learn more about this collaborative effort between government and private industry at **www.epa.gov/epawaste/conserve/materials/ecycling/taskforce/index.htm**.

Sources: U.S. EPA, Office of Resource Conservation and Recovery (November 2011, Table 12, Table 13; August 17, 2011; May 2011).

the materials content of the nation's MSW for the 1980–2010 period. Notice that the generation of paper and paperboard waste has grown fairly steadily over time, rising from 55.2 million tons in 1980 to 71.3 million tons in 2010. Paper wastes are procyclical, meaning that the amount varies directly with the economic business cycle. Interestingly, however, paper wastes have declined slightly relative to total MSW, decreasing from 36.4 percent of the total in 1980 to 28.5 percent in 2010.

TABLE 18.2 Trend Data on U.S. MSW Generation by Materials

	Millions of Tons[a]				Percentage of Total[a]			
	1980	1990	2000	2010	1980	1990	2000	2010
Materials in Products								
Paper and paperboard	55.2	72.7	87.7	71.3	36.4	34.9	36.2	28.5
Glass	15.1	13.1	12.8	11.5	10.0	6.3	5.3	4.6
Metals	15.5	16.6	18.9	22.4	10.2	7.9	7.8	9.0
Plastics	6.8	17.1	25.5	31.0	4.5	8.2	10.5	12.4
Rubber and leather	4.2	5.8	6.7	7.8	2.8	2.8	2.8	3.1
Textiles	2.5	5.8	9.5	13.1	1.7	2.8	3.9	5.3
Wood	7.0	12.2	13.6	15.9	4.6	5.9	5.6	6.4
Other[b]	2.5	3.2	4.0	4.8	1.7	1.5	1.6	1.9
Subtotal	108.9	146.5	178.7	177.9	71.8	70.3	73.7	71.2
Other Wastes								
Food scraps	13.0	23.9	29.8	34.8	8.6	11.5	12.3	13.9
Yard trimmings	27.5	35.0	30.5	33.4	18.1	16.8	12.6	13.4
Miscellaneous inorganic wastes	2.3	2.9	3.5	3.8	1.5	1.4	1.4	1.5
Subtotal	42.8	61.8	63.8	72.0	28.2	29.7	26.3	28.8
Total MSW Generated by Weight	151.6	208.3	242.5	249.9				

NOTES:
[a]The sum of any given column may not add to the total due to rounding.
[b]The category "Other" includes electrolytes in batteries and materials in disposable diapers.
Source: U.S. EPA, Office of Resource Conservation and Recovery (November 2011), Table 1.

One of the fastest growing segments of MSW materials in the United States is plastics, which has risen steadily from 6.8 million tons in 1980 to 31.0 million tons in 2010. This growth rate is just as impressive when viewed on a proportionate basis by weight, from 4.5 percent of total MSW generated in 1980 to 12.4 percent in 2010. These data explain why there has been a surge of interest in recycling plastics, an issue discussed in Application 18.2.

Recovery rates for plastics and other components of MSW are shown in Figure 18.2 (page 437). To learn more about recycling, visit **www.epa.gov/epawaste/conserve/rrr/recycle.htm**.

International Comparisons

Generation

Proper management of the MSW stream is a universal objective. Table 18.3 (page 437) shows annual MSW generation (measured in kilograms per capita) and gross domestic product (GDP) for various nations around the world. What is apparent from these data is that Americans are among the highest MSW generators in the world. Notice how the U.S. data compare with other industrialized nations, such as France and Italy. For example, Italy's per capita rate of 545 kilograms pales in comparison to the comparable statistic for the United States of 745 kilograms.

APPLICATION 18.2 The Facts on Recycling Plastics

Over the last several decades, plastic use in the United States has risen by over 10 percent per year, which translates to a substantial growth rate of plastic wastes. In 1960, plastic wastes totaled 390,000 tons, which grew to over 31 million tons in 2010. Relative to the entire MSW stream in 2010, plastic wastes represent 12.4 percent of total generation by weight, but by volume, they account for a much higher proportion. Containers and packaging are by far the largest proportion of plastic wastes, representing 13.7 million tons in the United States in 2010, or 44.1 percent of the plastics waste stream.

Beyond the sheer quantity of plastic wastes, these materials also can endanger the environment. According to the EPA, most of the wastes collected during harbor surveys and beach cleanups are plastics. In addition to the aesthetic degradation of such littering, disposal of plastics in surface waters threatens virtually all forms of marine life. Plastics also contain such additives as colorants, stabilizers, and plasticizers, some of which include toxics like cadmium and lead. Reportedly, 28 percent of the cadmium and 2 percent of the lead found in MSW arise from plastics.

Exacerbating the problem, the recovery rate for plastic wastes in the United States has been poor. In 1990, only 0.37 million tons, or 2.2 percent, of all plastic wastes generated were recovered. In 2010, the figure rose to 2.4 million tons, or roughly 7.6 percent of the total, but this is still a low rate compared to other materials. For example, the recovery rate for paper and paperboard in 2010 was 62.5 percent, and for glass, 27.1 percent. Why is the overall recovery rate for plastics so poor?

The bottom line is that recycled plastics have to compete with virgin materials. And for any recycled product to be competitive, all three steps in the recycling process—collection, separation of materials, and manufacture of new products—must be executed efficiently, which is not straightforward.

One of the key issues in producing recycled plastic is the resin content. Plastics are made from a variety of different types of resins. Currently, there appear to be sizable and lucrative markets for products made from single resins but not for commodities produced from mixed plastics, which command a lower market value. The problem is, even though many products are made of only a single resin, all of these end up together in the waste stream. To achieve a homogeneous collection of a particular resin, different kinds of plastics have to be identified and separated after collection—a costly step in the recycling process. An important problem is that plastic wastes often are not easily identified, even by experts.

Hence, in order for plastic waste recovery rates to improve, cost-effective methods must be developed to more easily identify different types of plastics and to more readily separate them into batches of single resins. Until these technologies are developed, recycling will continue to focus on easily recognizable plastic wastes that accumulate in large amounts. This explains why soft drink bottles and milk containers account for the greatest majority of recycled plastic. It also explains why many packaging manufacturers label their containers with a recycling symbol and a code number that identifies the resins that were used.

For further information on plastics and plastics recycling, visit **http://plastics.americanchemistry.com/Sustainability-Recycling**, at the Web site of the American Chemistry Council, Inc.

Sources: U.S. EPA, Office of Resource Conservation and Recovery (November 2011), Table 1, Table 2, Table 18; Resource Integration Systems Ltd. and Waste Matters Consulting (Portland, Oregon) (December 1990), pp. 7–9; U.S. EPA, Office of Solid Waste and Emergency Response (February 1990).

Some of the international differences in MSW generation are attributable to the amount of packaging used by producers. In 2009, the average American generated 466 pounds of packaging waste. Citizens of certain European Union (EU) member states are said to have produced far less than that amount. For example, residents of Germany generate approximately 408 pounds per person, Italy about 309 pounds, and Spain approximately 255.[3] These variations also may reflect cultural preferences,

[3]European Commission, Eurostat (July 19, 2011; July 12, 2011b); U.S. EPA, Office of Solid Waste (December 2010), Table ES-3, p. 3; Table 12, p. 72. Data are for the most recent comparable periods available: U.S. values are for 2009; European values are for 2008.

FIGURE 18.2 Recovery Rates (%) for Components of U.S. MSW in 2010

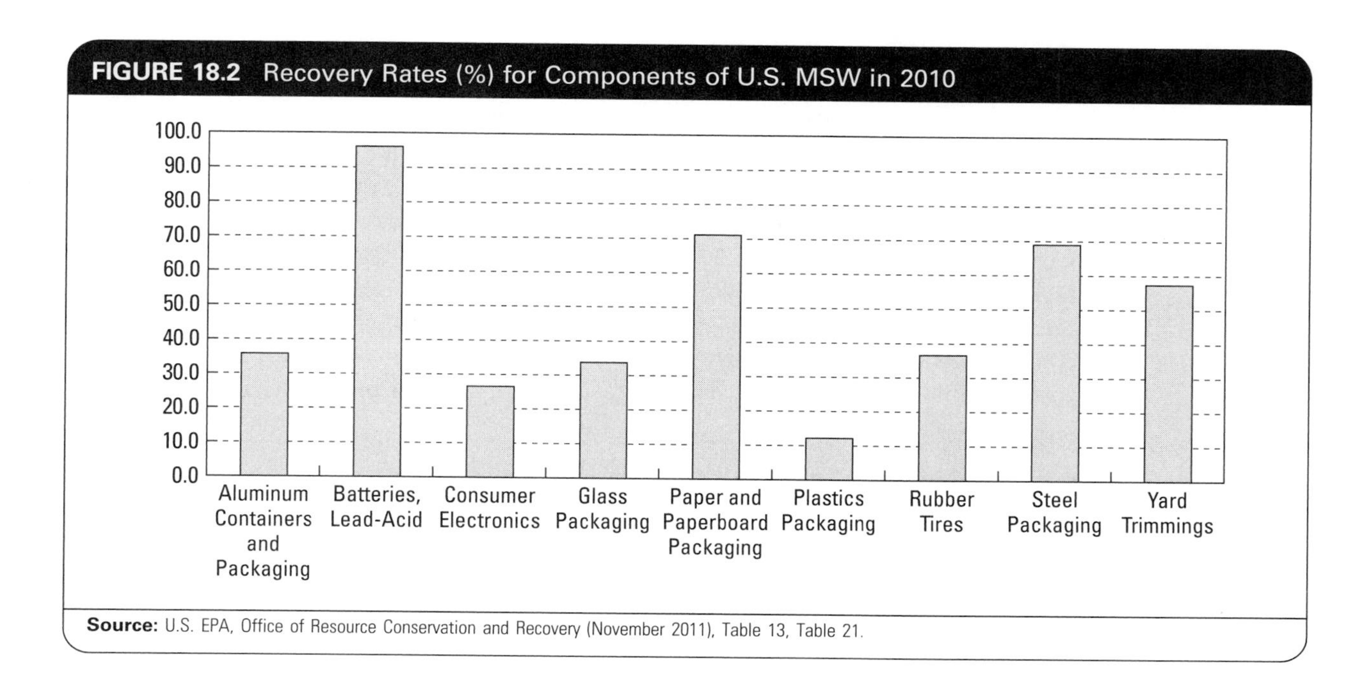

Source: U.S. EPA, Office of Resource Conservation and Recovery (November 2011), Table 13, Table 21.

TABLE 18.3 Per Capita MSW Generation for Selected Countries

Country	Waste (kilograms per capita)	GDP ($ per capita)
United States	745	43,250
Netherlands	623	38,035
Germany	581	33,663
Spain	575	27,747
United Kingdom	546	34,356
Italy	545	28,245
France	543	30,624
Sweden	515	33,744
Norway	490	49,416

NOTES: Waste and GDP figures shown are for 2008.
GDP values are in 2005 U.S. dollars.

Sources: European Commission, Eurostat (January 17, 2011), Table 4; U.S. Census Bureau (2011), Table 373, p. 229, Table 1348, p. 847.

environmental awareness, economic conditions, or government regulations. Of course, statistical comparisons across nations are not totally reliable. Estimation methods and even the definitions used for MSW are often different. Nonetheless, even assuming some measure of inaccuracy, the United States is clearly at the high end of worldwide MSW generation rates.

Recycling Rates

There is also considerable variability in recycling rates across nations. In the United States, the national MSW recycling rate for 2009 was 33.8 percent. Many European countries have practiced recycling for much longer than the United States, and their national averages reflect their experience. For example, in 2009, Germany's 48 percent recycling rate surpassed all EU members, as it has in the past. Belgium and Sweden are next in the ranking, recycling 36 percent each. Other EU members exhibiting relatively high recycling rates include Slovenia and Denmark with 34 percent, and Ireland and the Netherlands each with 32 percent.[4] Technically, however, direct international data comparisons should be adjusted for political, cultural, economic, and demographic distinctions that affect the recyclability rates of manufactured products and the effectiveness of recycling programs.

POLICY RESPONSE

As mentioned in Chapter 17, there is a set of provisions within the Resource Conservation and Recovery Act (RCRA) that applies to nonhazardous waste, including MSW.[5] For the most part, the responsibility of nonhazardous waste management is assigned to states with supervision and support provided by the federal government. This delegation of authority implies that policymakers view MSW controls as a responsibility best undertaken close to the source. Not every community faces the same problems, and even among those that do, the degree of environmental risk can vary considerably across localities. Factors such as town population, age of a landfill, and the proximity of groundwater and surface water sources to landfills are among the conditions affecting the outcome.

Federal Responsibilities[6]

Based on RCRA's provisions for nonhazardous wastes, the federal government must provide financial and technical assistance to states in designing and implementing their waste management plans. It also must encourage states to conserve resources and assist them in finding ways to maximize the use of recoverable resources. The EPA is responsible for establishing minimum criteria for sanitary landfills and other land disposal sites. Facilities not meeting the criteria are considered open dumps and must be closed or upgraded in accordance with specific rules.

In September 1991, new regulations were issued to establish tougher standards for land disposal sites starting in 1993. These new rules are effective throughout the useful life of a facility plus 30 years after its closure. In addition to the risk reduction provided by these tougher rules, the revisions are expected to encourage source reduction and recycling on a national scale. New regulations also were issued for waste-to-energy incineration facilities. These regulations establish controls on various air emissions and ensure proper combustion conditions. The rulings are of special importance, since there are about 86 municipal

[4]European Commission, Eurostat (March 8, 2011); U.S. EPA, Office of Solid Waste (December 2010).
[5]These provisions are given in Subtitle D of RCRA.
[6]Drawn from U.S. EPA, Office of Solid Waste and Emergency Response (August 1995), Chapter 9.

waste-to-energy facilities with an aggregate capacity of more than 96,150 tons per day currently in operation in the United States.[7] Prior to this time, the EPA had established guidelines, but no regulatory controls per se, for MSW incineration facilities.

Independent of RCRA and other federal legislation, President George H. W. Bush signed Executive Order 12780 in 1991, which required all federal agencies to step up their recycling efforts and support the development of markets for recycled products. With a similar intent, President Clinton issued Executive Order 13101 in September 1998, which called for all printing and writing paper used by federal agencies to contain at least 30 percent recovered paper. Then in 2007, President George W. Bush signed Executive Order 13423, which called for federal agencies to use sustainable environmental practices when purchasing goods and services, including acquisitions of products with recycled content, and to maintain cost-effective waste prevention and recycling programs.

More recently, President Obama expanded upon Bush's Executive Order 13423 by signing his own Executive Order 13514, aimed at establishing sustainability goals for federal agencies through improvements in their environmental, energy, and economic performances. Among the order's specific goals is to achieve a 50 percent recycling and waste diversion objective by 2015.

While these responses show initiative, they fall short of what some other nations are doing. For example, Germany and subsequently the EU have passed packaging ordinances that are far more aggressive in their approaches toward recycling, as Application 18.3 explains.

State Responsibilities

According to RCRA, states are to develop their own waste management plans, but these must meet certain federal requirements. For example, state plans must include provisions to close or upgrade existing open dumps and to prohibit the establishment of new ones. They also must require that nonhazardous wastes either be used for resource recovery or be properly disposed of, such as in a sanitary landfill.

RCRA also requires states to establish whatever regulatory powers they need to comply with the law. Typically, states pass their own legislation to meet federal requirements and to fulfill their responsibilities as outlined in their state plans. Some have used incentive-based approaches such as providing grant monies for their cities and towns to set up waste management programs. Others have passed laws calling for local governments to set up recycling plans, and some mandate specific recycling rates for local communities, such as Virginia and New Jersey.[8]

integrated waste management system
An EPA initiative that promotes source reduction, recycling, combustion, and land disposal, in that order.

Because states are responsible for their own nonhazardous waste plans, they have the flexibility to develop cost-effective programs. Many states have developed bona fide plans, following the EPA's **integrated waste management system**. This system promotes using a combination of techniques and programs aimed at source reduction, recycling, combustion, and land disposal—in that order. For details on any given state's program, visit **www.epa.gov/epawaste/wyl/stateprograms.htm**.

[7]Energy Recovery Council (ERC) (December 2010), as cited in U.S. EPA, Office of Resource Conservation and Recovery (November 2011), Table 27.

[8]Virginia Department of Environmental Quality (July 7, 2011); New Jersey Department of Environmental Protection (October 12, 2006).

APPLICATION 18.3 The Green Dot Program in Europe

A commonly cited example of a national recycling law is Germany's Packaging Ordinance of 1991. According to some, this 1991 law, along with its amendments, might be the most ambitious recycling program in the world. The progressive mandate set initial recycling targets of 72 percent for glass, tinplate, and aluminum, and 64 percent for cardboard, paper, plastic, and composites. As of 2010, the targets are 60 percent for plastic and aluminum; 70 percent for paper, cardboard, and tin; and 75 percent for glass.

To achieve these ambitious objectives, Germany's ordinance stipulates that industry is to be responsible for the collection and recycling of all its packaging. If businesses' efforts fail to meet the statutory targets, the German government will institute costly deposits on essentially all packaging.

Responding to these tough requirements, firms collaborated to form a private, not-for-profit company, called Duales System Deutschland AG (DSD), meaning a dual system, to provide collection and recycling services to consumers. Through its extensive disposal, sorting, and recycling network, DSD ensures that used packaging will be collected and sent to a proper recycling facility.

Of course, such an undertaking doesn't come cheap. DSD's start-up costs were estimated to be in the neighborhood of $10 billion, with operating expenses of $1 billion per year. To finance these fixed and variable costs, DSD sells licenses to participating companies for the right to place a green dot on their packages, a symbol that guarantees that the packaging will be collected, sorted, and recycled by DSD's network of partners. Estimates show that more than 18,500 licenses have been sold in Germany. Following the polluter-pays principle, the price of a green dot varies with the amount of packaging on each product and its recyclability.

As of 2010, reported recycling statistics indicate a successful outcome, with recycling rates easily exceeding the objectives of Germany's packaging ordinance. For example, recycling rates are 112 percent for paper and cardboard; 117 percent for aluminum, plastic, and tin; and 143 percent for glass. Rates exceed the 100 percent mark because some packaging waste without the green dot ends up in the Green Dot system.

Because of this recycling success, DSD provides an added environmental benefit of energy savings associated with using recycled materials. Estimates suggest that over 49 billion megajoules of primary energy were saved in 2010, which is equivalent to reducing emissions of carbon dioxide (CO_2), a major greenhouse gas, by 1.3 million tons. For more details and updates on Germany's Green Dot program, visit **www.gruener-punkt.de**.

Given the success of DSD's Green Dot program, it is not surprising that the program has grown internationally. In response to the EU's 1994 Packaging and Packaging Waste Directive, DSD founded Packaging Recovery Organisation Europe (PRO EUROPE) in 1995. PRO EUROPE initially was developed to foster international cooperation and to avoid trade barriers among countries implementing the EU's packaging directive. Over time, it has become an umbrella organization that oversees the granting of Green Dot licenses to other countries, as a way to facilitate these nations' meeting their disposal and recycling requirements under the EU's directive.

There are now 33 countries operating under the umbrella of PRO EUROPE. This participation has led to the issuance of more than 170,000 Green Dot licenses and the use of the green dot on over 460 billion pieces of packaging. As with the Germany program, the Green Dot ensures that packaging waste will be collected, sorted, and properly recycled. For more information on PRO EUROPE's recent activities, visit **www.pro-e.org**.

Sources: Duales System Deutschland AG (August 11, 2011; July 1, 2008; April 4, 2005; 2004/2005); PRO EUROPE (August 11, 2011; October 2006); Shea (July/August 1992); Cairncross (March–April 1992).

To better understand how states are responding to solid waste management issues, we need to develop a general market model of MSW services. This provides the analytical tool with which to assess the cost-effectiveness and efficiency of various state programs.

THE MARKET FOR MSW MANAGEMENT SERVICES

In the market for MSW services, the relevant commodity is a combination of several distinct activities—the collection, transportation, and disposal of MSW. Based on this output definition, we model the market for MSW services in Figure 18.3, using a market demand (D), or marginal private benefit (MPB) curve, and market supply (S), or marginal private cost (MPC) curve. Together, these curves determine the competitive equilibrium price (P_C) for MSW services and the equilibrium quantity (Q_C).

Supply of MSW Services

The supply side of the MSW services market represents the production decisions of private firms under contract with cities and towns or those of municipalities that provide these services directly to the community. In this market, the costs of production reflect the expenses of operating a fleet of disposal trucks, managing an approved landfill or incinerator, and labor. Under the usual assumptions about production and diminishing returns, the MPC curve exhibits a positive slope. Among the nonprice factors affecting supply in this market are land availability and government regulation.[9]

tipping fees
Prices charged for disposing of wastes in a facility such as a landfill.

Consider, for example, the impact of Congress's decision to tighten federal controls on landfill disposal. These tougher standards led to the closing of several hundred landfills that were unable to meet the law's new requirements. This outcome along with the NIMBY syndrome limited the availability of landfill space in some locations. As a result, landfill prices, or **tipping fees** as they are called, have risen. As Figure 18.3 shows, this outcome is modeled as an upward shift of the MPC curve, which causes an increase in price to P_C' and a decline in equilibrium quantity to Q_C'.

FIGURE 18.3 Modeling the MSW Market

The market demand (or MPB curve) and market supply (or MPC curve) determine the competitive equilibrium price (P_C) and quantity (Q_C). Congress's decision to tighten federal controls on landfill disposal shifts MPC upward to MPC', elevating price to P_C' and decreasing quantity to Q_C'. The decrease in quantity of MSW services may mean that generators are using a source reduction strategy, recycling more of their wastes, or engaging in illegal disposal.

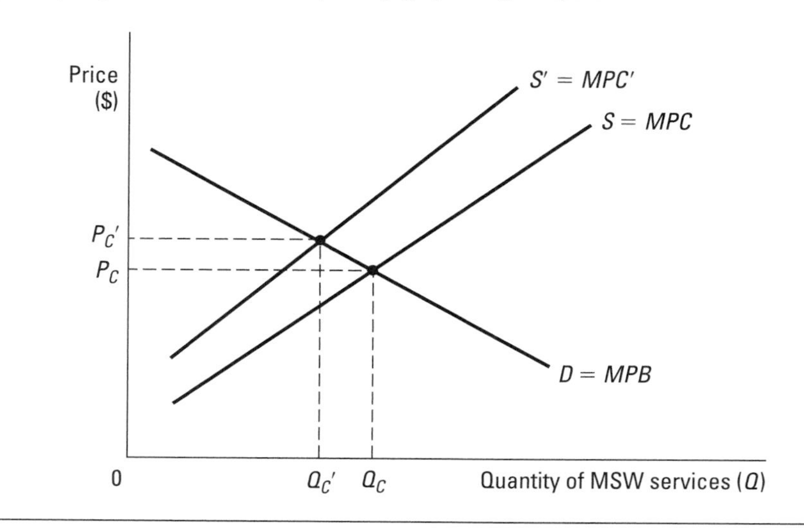

[9]See, for example, Callan and Thomas (November 2001).

Demand for MSW Services

The demand side of this market represents the purchasing decisions of MSW generators. In this context, the quantity response to changes in price takes on an important meaning about how wastes are managed. To understand this, reconsider the decline in quantity from Q_C to Q_C' due to the regulation-induced price increase described in Figure 18.3. How might generators change their behavior to achieve this reduction in quantity? One possibility is that they use a source reduction strategy and produce less trash. Another is that they generate the same amount of trash but demand less services because they recycle or reuse. Finally, they might maintain the same generation rate *and* the same recycling rate but engage in illegal disposal to avoid the higher price of MSW services.

Which of these options is chosen depends on their availability to the generator and the prices of these options relative to the price of MSW services. Recognizing the natural market response of demanders to higher-priced MSW services, a local community can encourage recycling by offering a cost-effective program to its residents. An example of such a program is the provision of curbside instead of drop-off recycling services. As of 2010, there were 9,066 reported curbside recycling programs in the United States, serving 71 percent of the population.[10] In such cases, the opportunity cost of recycling to the waste generator is reduced.[11] In the absence of such a program, some generators may be motivated to illegally dispose of their wastes.

The demand, or *MPB*, for municipal waste services also responds to certain nonprice changes. For example, more affluent individuals tend to generate larger amounts of trash, since they purchase more products and replace them more frequently. Thus, demand for MSW services likely would shift to the right as the income of a community rises, holding all else constant. Another nonprice determinant of demand is tastes and preferences. As generators become more environmentally responsible, we would expect their demand for these services to decline as they adjust their purchases toward products with less packaging. In sum, waste generators in each community likely face a uniquely shaped demand curve that responds somewhat predictably to both price and nonprice changes.

If actual MSW markets behaved in accordance with this model *and* if there were no externalities, we could conclude that MSW markets achieve an efficient solution where $MPC = MPB$. However, it turns out that these conditions typically are violated in actual MSW markets. The resulting resource misallocation is an important issue that merits further investigation.

Resource Misallocation

There are two distinct problems that arise in most MSW services markets, and both are associated with the supply side of the market:

- Pricing of MSW services does not properly reflect the rising *MPC* associated with increases in production levels.
- Production of MSW services gives rise to negative externalities.

---

[10] U.S. EPA, Office of Resource Conservation and Recovery (November 2011), Table 25, reporting from miscellaneous sources, including U.S. Census Bureau, *Biocycle* preliminary State of Garbage data, and data from various state Web sites.

[11] See, for example, Callan and Thomas (Fall 1997).

fixed fee or flat fee pricing system
Pricing MSW services independent of the quantity of waste generated.

Flat Fee Pricing of MSW Services

In most communities, suppliers charge a **fixed fee** per household or commercial establishment for MSW services. Since the fee is the same regardless of the quantity of waste generated at each location, this type of pricing scheme is known as a **flat fee pricing system**. Notice that the waste generator is charged nothing for any additional containers of trash beyond the first one, which means that the price does not reflect rising *MPC*. Demanders effectively pay a marginal price of zero and hence have no incentive to reduce wastes. More formally, the market price under this scenario is being determined *as if* the *MPC* is zero.

This phenomenon is illustrated in Figure 18.4, where the *MPC* curve is shown coincident with the horizontal axis. Notice that under such a scenario, the market equilibrates where $MPC = MPB = 0$, at Q_0. Compared to the competitive equilibrium, too many resources are being allocated to MSW services. In communities where MSW services are publicly provided, the price mechanism is further dampened because the fee is not explicitly communicated but rather collected through property taxes. But whether these services are publicly or privately provided, the use of a fixed fee has serious efficiency implications.

Negative Externalities

Even if waste generators are charged a fee equal to the *MPC*, there may still be a resource misallocation if production of MSW services gives rise to negative externalities.

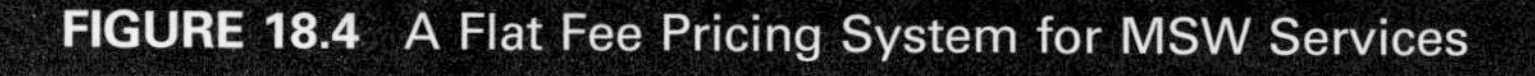

FIGURE 18.4 A Flat Fee Pricing System for MSW Services

In a flat fee pricing system, the price does not reflect rising *MPC*. Demanders are effectively paying a marginal price of zero and hence have no incentive to reduce wastes. More formally, the market price is being determined as if the *MPC* is zero or coincident with the horizontal axis. Under this scenario, the market equilibrates where $MPC = MPB = 0$, at Q_0. Compared to the competitive equilibrium, too many resources are allocated to MSW services.

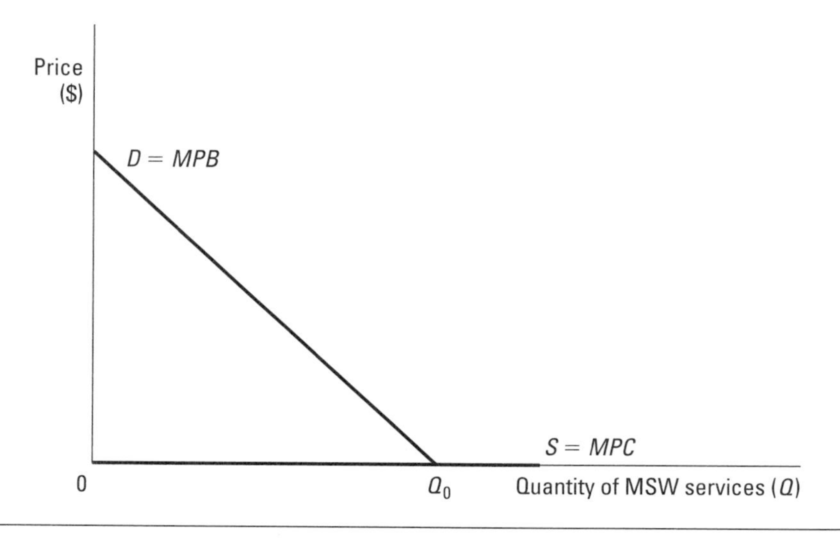

Such an outcome is not uncommon in waste management markets. In the MSW services market, external costs may be due to groundwater contamination, air pollution from incineration, or impairment of aesthetics.

As we have illustrated in other contexts, such a production externality is captured by a marginal external cost (*MEC*) curve, which must be added to the *MPC* to identify the marginal social cost (*MSC*) of producing the good. Assuming no externalities on the demand side, the marginal social benefit (*MSB*) of consumption is the same as the *MPB* function. The efficient solution is determined where the *MSC* equals the *MSB*, as illustrated in Figure 18.5. (To avoid confounding the effect of the externality with that of charging a fixed fee, we show the *MPC* as a positively sloped curve.) Notice that the efficient output level is Q_E, which is *lower* than the private market outcome, Q_C.

To summarize, we observe that private MSW markets typically do not yield an efficient outcome. The use of fixed fees distorts the signaling mechanism of price, effectively removing the incentive to economize on trash generation. The private market's disregard for external costs leads to a further distortion. Singularly or in combination, these factors contribute to an overallocation of resources to waste services and an overproduction of municipal trash. Is there any solution to these problems? Actually, a number of pricing schemes can be used to correct the inefficiency.

FIGURE 18.5 The Negative Externality in MSW Service Markets

The production externality in MSW services is captured by a marginal external cost (*MEC*) curve, which is added to the *MPC* to identify the marginal social cost (*MSC*). The efficient solution is determined where *MSC* equals *MSB*. Notice that the efficient output level, Q_E, is *lower* than the private market outcome, Q_C.

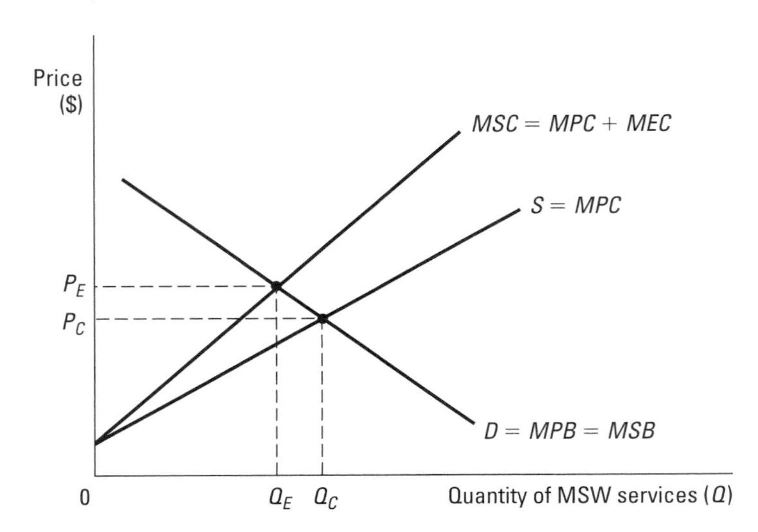

MARKET APPROACHES TO MSW POLICY[12]

In recent years, some communities have begun to institute market-based policies aimed at reducing the problems associated with MSW. Of interest are three approaches that directly exploit the signaling mechanism of price. These are:

- **Back-end or waste-end charges**
- **Front-end or retail disposal charges**
- **Deposit/refund systems**

Back-End or Waste-End Charges

back-end or waste-end charge
A fee implemented at the time of disposal based on the quantity of waste generated.

As discussed in Chapter 17, a **waste-end charge** is so named because it is implemented at the time of disposal based on the quantity of waste generated. Because such a charge varies with the quantity of waste, it avoids the market distortion caused by a flat fee pricing scheme. To achieve efficiency, the waste-end charge must be set to cover the *MSC* at the efficient equilibrium. This solution is illustrated in Figure 18.6, with the appropriate per-unit fee shown as P_E. Notice that this fee is a *price per unit of waste* to be paid by all generators, which is very different from charging a *price per household*, as is done under a flat fee system.

flat rate pricing
A unit pricing scheme that charges the same price for each additional unit of waste.

variable rate pricing
A unit pricing scheme that charges a different price for each additional unit of waste.

In practice, programs that use waste-end charges are called **unit pricing schemes**, also known as **pay-as-you-throw (PAYT) programs**, to indicate that prices for MSW services are charged on a per-unit-of-waste basis. The units may be measured either by weight or by volume generated, though the latter is more commonly used at the present time. The same price can be charged for each unit, called **flat rate pricing**, or the price can vary, called **variable rate pricing**. To learn more about these pricing systems being used in various communities, visit **www.epa.gov/epawaste/conserve/tools/payt/top13.htm**.

Perhaps the best known example of a community having implemented unit pricing is the city of Seattle, Washington. In fact, Seattle's program has come to be the prototype upon which other communities have modeled their own MSW unit pricing plans. The program is implemented by having each MSW generator register with the city for the number and volume of trash containers it expects to use in a week. Based on the registration, a fee is calculated using a schedule of volume-based prices.[13]

By moving from a fixed fee schedule to a unit pricing program, two important market incentives are restored. First, unit pricing communicates to generators that increased waste services are associated with a nonzero marginal cost of production, ideally the marginal social cost. Effectively, this provides an incentive to economize on waste generation by using a source reduction strategy. Secondly, by elevating the price of collection and disposal, the *relative* price of alternative approaches declines. Given that recycling

[12]Much of the following discussion is drawn from U.S. EPA, Office of Policy, Economics, and Innovation (January 2001), Chapters 4 and 5; Wirth and Heinz (May 1991), pp. 48–65.
[13]Seattle Solid Waste Utility (1988), as cited in Wirth and Heinz (May 1991), pp. 50–51.

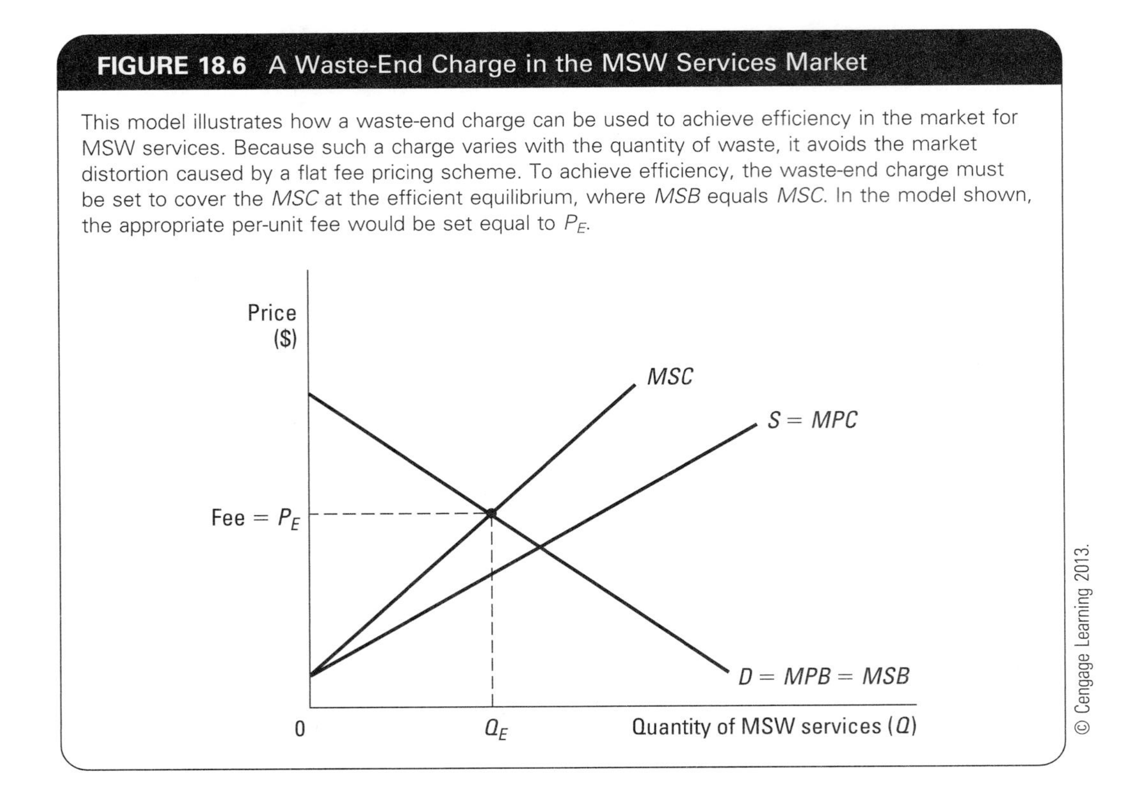

FIGURE 18.6 A Waste-End Charge in the MSW Services Market

This model illustrates how a waste-end charge can be used to achieve efficiency in the market for MSW services. Because such a charge varies with the quantity of waste, it avoids the market distortion caused by a flat fee pricing scheme. To achieve efficiency, the waste-end charge must be set to cover the *MSC* at the efficient equilibrium, where *MSB* equals *MSC*. In the model shown, the appropriate per-unit fee would be set equal to P_E.

and disposal are substitute activities, the change in relative prices may encourage more recycling, which means less disposal.

Beyond theoretical arguments, there is a growing body of empirical evidence suggesting that unit pricing is a viable policy instrument. For example, Robin Jenkins (1993) conducted a rather extensive statistical study on unit pricing. The results show that the effect of switching from a fixed fee to a volume-based fee of 50 cents per container is a decrease in waste per person per day of 0.2 pounds. This translates to an estimated decline of 3,650 tons per year for a community of 100,000 people, or 18,250 tons annually for a community of 500,000 people. Another study by Miranda and Aldy (1996) of communities in California and the Midwest found that disposal decreased by 6 percent to 50 percent following the launch of a variable rate pricing scheme, where the reductions were positively related to the price level charged.[14]

bag-and-tag approach
A unit pricing scheme implemented by selling tags to be applied to waste receptacles of various sizes.

As the evidence accumulates, more and more communities have followed Seattle's lead and have moved to unit pricing programs. Current estimates indicate that over 7,000 communities across the United States are using this form of MSW pricing.[15] In fact, some local authorities have enhanced the Seattle prototype to improve its efficiency by instituting more flexible PAYT programs whereby generators purchase stickers for receptacles of various sizes. Application 18.4 discusses this evolution in PAYT programs, including the so-called **bag-and-tag approach**, which has met with some success.

[14] See also Callan and Thomas (Spring 2006; December 1999).

[15] Skumatz and Freeman (2006).

APPLICATION 18.4 Pay-As-You-Throw (PAYT) Pricing Programs

Although the city of Seattle, Washington, has received national acclaim for its innovative unit pricing waste program, the market-based plan has not been without problems. One in particular is that per-can pricing schemes are based on residents' forecasts of their waste generation rate, which may or may not be accurate. As a result, some generators end up paying for disposal services on a full can of trash, even if it is half empty or not used at all. This potential inefficiency has prompted some communities to design modified versions of the Seattle prototype.

One such modification is called a **bag-and-tag system**. This type of PAYT program is implemented by providing collection and disposal services only to trash placed in specially designated containers sold by the municipality. Program participants purchase in advance a supply of these trash bags, typically available in various sizes and priced accordingly. In each collection period, residents use only as many of the prepaid trash bags as they need. A variation of the system operates from the same premise except that stickers or tags are sold by the municipality, and these are then affixed to residents' own trash containers.

High Bridge, New Jersey, was among the first communities to institute a tag pricing system, charging residents an annual fee for 52 stickers. The charge per household is intended to cover the program's administrative costs plus the disposal cost of one container per week for each household. Restrictions on maximum weight and volume are imposed on a per-sticker basis. To provide greater flexibility to residents, additional stickers can be purchased from the town or traded between low-use and high-use residents. In the first 10 months of the program's operation, High Bridge officials reported a 24 percent decrease in waste tonnage.

At about the same time as High Bridge got its plan under way, the town of Perkasie, Pennsylvania, began selling town-issued trash bags to implement its unit pricing program. Perkasie uses a differential charge based on bag size. A 20-pound trash bag commands a price of $0.80, while the larger 40-pound container is sold for $1.50. Like the High Bridge plan, residents pay for services only on the quantity of waste generated in a given period and not on a predetermined and likely inaccurate forecast. A 40 percent decline in waste tonnage was observed after Perkasie's first full year of operation.

Several years later, officials in San Jose, California, launched a **hybrid variable rate program** called a cart/sticker system. Local residents subscribe to designated cart sizes, as in the Seattle plan, and pay specified fees for the carts they select. However, if more disposal services are needed, households can use 32-gallon bags, as long as each bag bears a designated sticker that can be purchased for $3.50 each. This combination method adds needed flexibility to the per-can subscription method, yet still adheres to the pay-as-you-throw (PAYT) principle.

According to one report, the number of communities with PAYT programs has grown significantly—up by nearly 70 percent in the last 10 years. As of 2006, there were over 7,000 communities in the United States that use some form of these market-based programs, commonly instituted as a per-can, bag-and-tag, or hybrid system. Community administrators contemplating a PAYT program have to access their specific waste management needs and implement the approach best suited to achieving their objectives.

For information, data, and case studies on PAYT programs, visit **www.epa.gov/epawaste/conserve/tools/payt/index.htm**.

Sources: Skumatz and Freeman (December 2006); U.S. EPA, Office of Policy, Economics, and Innovation (January 2001), pp. 41–46; Wirth and Heinz (May 1991), pp. 49–53; U.S. EPA, Office of Policy, Planning, and Evaluation (March 1991), pp. 2.7–2.12; Tregarthen (September/October 1989).

Front-End or Retail Disposal Charges

front-end or retail disposal charge
A fee levied on a product at the point of sale designed to encourage source reduction.

An alternative pricing scheme aimed at the MSW problem is the use of a **front-end or retail disposal charge**. In contrast to a waste-end charge imposed on *wastes* at the point of *disposal*, a retail disposal charge is levied on *products* at the point of *sale*. Since such a charge is instituted at the pregeneration stage, its objective is to encourage pollution prevention through source reduction. This includes motivating manufacturers to seek out product designs and packaging that are more environmentally responsible.

From a practical perspective, local conditions tend to dictate when a retail disposal charge might be more appropriate than a waste-end charge. For example, a waste-end pricing scheme should be coordinated with a recycling program to deter illegal disposal. Hence, if a community has not instituted such a plan, a better option might be to use a front-end charge. A similar argument applies to communities where there is a predominance of multi-unit residences, making waste-end fees difficult to implement. Finally, a front-end charge can be used to complement a waste-end charge system to discourage the use of products that yield large amounts of waste or wastes that pose a particular threat to health or the environment. Examples include products like motor oil, batteries, and automobile tires.

To better understand the environmental and economic consequences of retail disposal charges, consider the market for household batteries. Batteries are made with heavy metals like mercury or cadmium, so their disposal poses an environmental risk to society. Battery consumers do not consider the external effects, so the *MPB* effectively overstates the true marginal benefits of consumption. This means that the *MSB* curve is actually *below* the *MPB* curve, as shown in Figure 18.7.[16] In the absence of government intervention, equilibrium output is determined where *MPB* equals *MPC*, or Q_C. However, the efficient output level is lower, determined by the intersection of the *MSC* and

FIGURE 18.7 A Retail Disposal Charge in the Household Batteries Market

Because of the negative externality of battery consumption, the *MSB* curve is *below* the *MPB* curve. In the absence of government intervention, equilibrium output is Q_C, where *MPB* intersects *MPC*. The efficient output level is lower, at the intersection of *MSC* and *MSB*, or Q_E. To correct the inefficiency, government can impose a retail disposal charge equal to the difference between *MSB* and *MPB* at Q_E, or equivalently $-MEB$. We can model the charge as a shift up of the *MSC* by distance *xy*, resulting in an effective price of P_R.

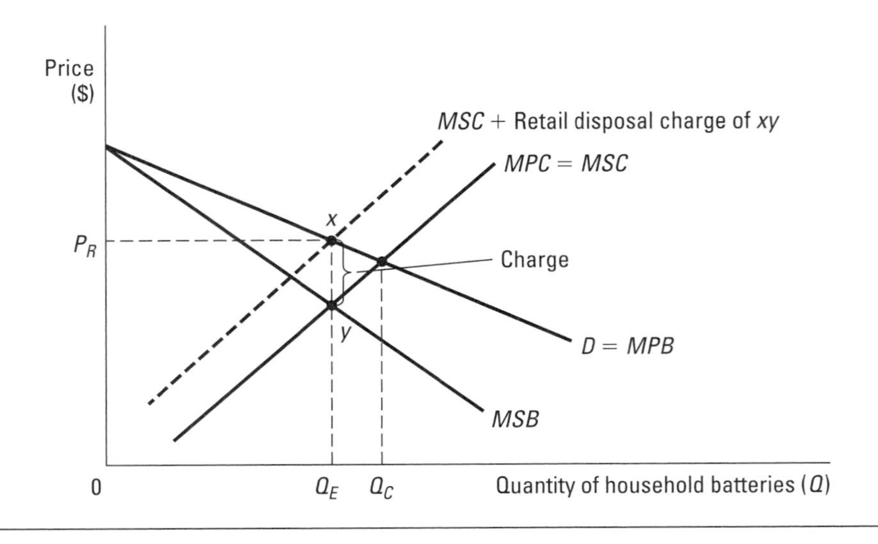

[16]The externality is not modeled as an *MEC* added to the *MPC*, because in this case, it is not associated with the production side of the market. Rather, the externality arises from consumption, or the demand side. The negative *MEB*, when added to the *MPB*, illustrates that the true benefits to society, *MSB*, are *reduced* by the amount of the environmental damage. For a more rigorous analysis of this type of externality, see Tresch (2002), Chapter 6.

MSB curves, or Q_E. (For simplicity, we assume that no production externalities exist, so $MSC = MPC$.)

To correct the inefficiency, government can impose a retail disposal charge on batteries at the point of sale equal to the vertical difference between the *MSB* and the *MPB* at Q_E, or equivalently $-MEB$. In Figure 18.7, we model the imposition of the charge as a shift up of the *MSC* by distance *xy*, resulting in an effective price to the consumer of P_R.

Retail disposal charges are used both internationally and domestically. For example, Belgium, Denmark, Hungary, Italy, Portugal, and Sweden are among the nations that have imposed these charges on batteries; and Denmark, Norway, and Sweden have used them for pesticides. Belgium, Ireland, Italy, and South Africa impose these charges on plastic bags; and Canada, Finland, Hungary, and Portugal have applied these to tires.[17] While part of the objective is to diminish the damage caused by disposal of these products, these charges also are used to generate revenue, which may help to fund cleanup efforts. Domestically, retail disposal charges are imposed on a multitude of products—many by state governments. Table 18.4 lists some common applications. Most of the target products and materials are those expected to generate high external costs. For example, Rhode Island charges $0.10 per gallon of antifreeze. Similarly, of the 35 states with tire charges, Arkansas imposes a $2.00 charge on auto and light truck tire sales and a $3.00 charge on large truck tires.

Deposit/Refund Systems

deposit/refund system
A market instrument that imposes an up-front charge to pay for potential damages and refunds it for returning a product for proper disposal or recycling.

An alternative market instrument that can mitigate MSW pollution is a **deposit/refund system**, which is actually a two-part pricing scheme. It imposes an up-front charge for potential damages caused by improper disposal, and it allows for a refund of that charge at the end of the product cycle if the consumer takes proper action to avoid those damages. Typically, the consumer must return the product or its container for recycling or safe disposal. The formal model of this system was presented in detail in Chapter 5, so it is not repeated here. More relevant to this discussion is the conceptual relationship between this instrument and the other two pricing schemes. Notice that the deposit is

TABLE 18.4 Examples of U.S. Retail Disposal Taxes

Governing Entity	Waste	Amount of Tax
35 states	Automobile and truck tires	$0.25 to $5.00 each
46 states	Fertilizers	generally less than $1.00 per ton
Federal government	Tires	$0.15 to $0.50 per pound[a]
North Carolina	White goods	$3.00 per item
Rhode Island	Lubricating oil	$0.05 per quart
	Antifreeze	$0.10 per gallon
	Organic solvents	$0.25 per gallon

NOTES: [a]An additional fixed fee is associated with higher-weight tires.
Source: Rohlwing (2005); U.S. EPA, Office of Policy, Economics, and Innovation (January 2001), pp. 50–52.

[17]OECD/EEA (2011); U.S. EPA, Office of Policy, Economics, and Innovation (November 2004), p. 20.

imposed at the time of purchase exactly like a retail disposal charge. The refund attempts to reduce disposal and encourage recycling much like a waste-end charge.

This innovative economic instrument has gained acceptance both domestically and internationally. In the United States, deposit/refund systems are aimed primarily at beverage containers and lead-acid batteries, as Table 18.5 indicates. Similar applications are quite common in other countries as well. Among the nations that use deposit/refund systems for beverage containers are Australia, Canada, Denmark, Finland, Iceland, Italy, the Netherlands, Norway, Sweden, and Turkey. Other countries, including Denmark, Poland, and Mexico, use a deposit/refund system for batteries. Denmark, Finland, Norway, and Sweden use this instrument for automobile bodies. In Denmark and Norway, new car buyers receive a larger refund than the deposit as an added incentive to properly return the used vehicle.[18]

TABLE 18.5 Selected State Deposit/Refund Programs

State	Product	Amount of Deposit
Arizona	Batteries	$15.00 maximum
Arkansas	Batteries	$10.00
California	Beverage	$0.05 for < 24 oz.
		$0.10 for > 24 oz.
Colorado	Batteries	$10.00 minimum
Connecticut	Batteries	$5.00
	Beverage	$0.05
Hawaii	Beverage	$0.05
Idaho	Batteries	$10.00
Iowa	Beverage	$0.05
Maine	Batteries	$10.00
	Beverage	$0.05–$0.15
Massachusetts	Beverage	$0.05
Michigan	Beverage	$0.10
Minnesota	Batteries	$10.00 minimum
New York	Beverage	$0.05
	Batteries	$5.00
Oregon	Beverage	$0.02–$0.05
South Carolina	Batteries	$5.00
Vermont	Beverage	$0.05–$0.15
Washington	Batteries	$5.00 minimum

Source: Container Recycling Institute (2011; 2009); Battery Council International (August 14, 2011); U.S. EPA, Office of Policy, Economics, and Innovation (January 2001), Chapter 5, pp. 57–66, including Table 5.1.

[18]OECD/EEA (2011); U.S. EPA, Office of Policy, Economics, and Innovation (November 2004), pp. 23–26.

CONCLUSIONS

Municipal solid wastes—the simple by-products of production and consumption—have become a local public issue in communities everywhere. Today, most agree that former claims of a landfill crisis were exaggerated. Nonetheless, the world was listening and, in the absence of 20/20 hindsight, responded to what it heard. Private citizens began to change their behavior, avoiding products with excess packaging and participating in local recycling programs, and they turned to government for broader solutions.

In retrospect, many argue that people responded to false signals—a crisis that did not exist and unfounded predictions of doom and gloom. This assertion is probably true, at least to a point. However, it is wrong to conclude that there were no solid waste issues to resolve and no environmental risks to manage. Open dumps had become a health hazard, and most of society had been ignorant of the potential problems of excess waste generation and poor waste management practices. Having said this, a more relevant question is whether the response itself—both by private citizens and by government, however motivated—was and is appropriate.

In the United States, Congress passed laws to restrict the use of land for waste disposal and to replace the nation's open dumps with modern sanitary landfills. It then delegated the responsibility for devising local waste management programs to the states, encouraging plans that promoted waste reduction and recycling. The land restrictions and the landfill rulings added costs to local waste management, and some states responded with market instruments that reflected these costs.

What was the result of these policy moves? At this stage, the most accurate assessment is that these efforts have yielded mixed results. Some waste generators are pursuing source reduction activities, while others are illegally disposing of their wastes. Some states and local governments have developed successful cost-effective recycling centers, while others are struggling with expensive curbside programs that have placed a heavy tax burden on their residents. Add to this the series of debates about the viability of markets for recycled materials. Netting it out, these observations imply that municipal waste policies are still evolving, and public officials at all levels of government likely should consider some needed reforms.

Summary

- The composition of municipal solid waste (MSW) can be characterized both by the types of products being discarded and the kinds of materials entering the waste stream. The fastest growing segment of MSW in the United States is plastics.
- Internationally, the major industrialized nations are among the highest MSW generators. Some of the international differences are attributable to the amount of packaging used by producers, cultural preferences, environmental awareness, economic conditions, or government regulations.
- One set of provisions in the Resource Conservation and Recovery Act (RCRA) is concerned with managing nonhazardous wastes, including the MSW stream. The responsibility of nonhazardous waste management is assigned to states with supervision provided by the federal government.
- The federal government must provide financial and technical assistance to states in designing and implementing their waste management plans. The EPA must establish minimum criteria for sanitary landfills and other land disposal sites.
- Under RCRA, states must develop their own waste management plans, but these must meet certain requirements to receive federal approval. They also must establish any regulatory powers needed to comply with RCRA.
- The EPA encourages state authorities to use an integrated waste management system, which promotes source reduction, recycling, combustion, and land disposal, in that order.
- Resources are misallocated in private MSW services markets that use a flat fee pricing system, which charges a fixed fee per household. The inefficiency occurs because the constant price does not properly reflect rising marginal private cost (*MPC*) and because production of MSW services gives rise to a negative externality.
- A waste-end charge is implemented at disposal based on the quantity of waste generated. In practice, programs using waste-end charges are referred to as unit pricing schemes to indicate that prices for MSW services are charged on a per-unit-of-waste basis.
- Retail disposal charges are levied on the product at the point of sale. Since this charge is imposed at the pregeneration stage, its objective is to encourage pollution prevention through source reduction.
- A deposit/refund system imposes an up-front charge for potential damages caused by improper disposal and allows for a refund of that charge at the end of the product cycle if the consumer takes proper action to avoid those damages.

Review Questions

1. **a.** Summarize a specific municipal solid waste problem in your hometown or one that recently has been reported in the media. Use the specific stages of the waste stream in your discussion.

   b. What policies have been proposed or implemented to address the problem you describe? Analyze these policies from an economic perspective.
2. Using what you have learned about market-based incentives, suggest a policy initiative that would discourage land disposal in the United States.
3. Using the information in Application 18.3, compare the Green Dot Program to U.S. national initiatives on recycling.

4. Diagram a general model of MSW management services. Show the simultaneous effect of the federal restrictions on landfills *and* rising consumer awareness of the benefits of "green packaging." Assuming a private competitive market, predict the effect on the price and quantity of MSW services.

5. Consider the following hypothetical model of the market for municipal solid waste (MSW) services in the town of Stonington, Connecticut:

$$MPB = 30 - 2.5Q \qquad MEB = 0$$
$$MPC = 6 + 0.5Q \qquad MEC = 0.2Q$$

where Q is the number of trash containers per household per month.

a. Assume that Stonington uses a flat fee pricing system for trash services. Under this scenario, find the flat fee equilibrium quantity (Q_{FF}).

b. Compare the flat fee equilibrium to the competitive equilibrium (Q_C) and the efficient equilibrium (Q_E).

6. Consider the following model of a hypothetical market for MSW management services:

$$MPB = 25 - 2Q$$
$$MPC = 4 + Q$$
$$MEC = 0.5Q$$

where Q is the number of trash containers per household per month.

a. Quantitatively determine the effect of the resource misallocation due to:

(i) the presence of the negative externality, and

(ii) the use of a flat fee pricing system in the presence of a negative externality.

b. Support your answer to part (a) with a graphical model.

c. Determine the dollar value of a waste-end charge that would restore efficiency to this market. Explain your answer intuitively.

7. a. Contrast the Seattle per-can pricing program with Perkasie's bag-and-tag approach, both from an environmental and an economic perspective.

b. Why might weight-based unit pricing for MSW management be more advantageous than volume-based programs?

Additional Readings

Beatty, Timothy K. M., Peter Berck, and Jay P. Shimshack. "Curbside Recycling in the Presence of Alternatives." *Economic Inquiry 45*(4) (October 2007), pp. 739–56.

Bel, Germà, and Xavier Fageda. "Empirical Analysis of Solid Management Waste Costs: Some Evidence From Galicia, Spain" *Resources, Conservation and Recycling 54*(3) (January 2010), pp. 187–93.

Brown, Kelly M., Ronald Cummings, Janusz R. Mrozek, and Peter Terrebonne. "Scrap Tire Disposal: Three Principles for Policy Choice." *Natural Resources Journal 41*(1) (Winter 2001), pp. 9–22.

Dijkgraaf, E., and R. H. Gradus. "Cost Savings in Unit-Based Pricing of Household Waste." *Resource and Energy Economics 26*(4) (December 2004), pp. 353–71.

Fullerton, Don, and Thomas C. Kinnaman, eds. *The Economics of Household Garbage and Recycling Behavior*. Northampton, MA: Elgar, 2002.

Isely, Paul, and Aaron Lowen. "Price and Substitution in Residential Solid Waste." *Contemporary Economic Policy 25*(*3*) (July 2007), pp. 433–44.

Jenkins, Robin R., Salvador A. Martinez, Karen Palmer, and Michael J. Podolsky. "The Determinants of Household Recycling: A Material-Specific Analysis of Recycling Program Features and Unit Pricing." *Journal of Environmental Economics and Management 45*(*2*) (March 2003), pp. 294–318.

Judge, R., and A. Becker. "Motivating Recycling: A Marginal Cost Analysis." *Contemporary Policy Issues 11*(*3*) (July 1993), pp. 58–68.

Kim, Geum-Soo, Young-Jae Chang, and David Kelleher. "Unit Pricing of Municipal Solid Waste and Illegal Dumping: An Empirical Analysis of Korean Experience." *Environmental Economics and Policy Studies 9*(*3*) (2008), pp. 167–77.

Kinnaman, Thomas C., ed. *The Economics of Residential Solid Waste Management*. Burlington, VT: Ashgate, September 2003.

Kinnaman, Thomas C. "Policy Watch: Examining the Justification for Residential Recycling." *Journal of Economic Perspectives 20*(*4*) (Fall 2006), pp. 219–32.

_______ "Optimal Solid Waste Tax Policy With Centralized Recycling." *National Tax Journal 63*(*2*) (June 2010), pp. 237–52.

Miranda, Marie Lynn, and Joseph E. Aldy. "Unit Pricing of Residential Municipal Solid Waste: Lessons From Nine Case Study Communities." *Journal of Environmental Management 52*(*1*) (January 1998), pp. 79–93.

Miranda, Marie Lynn, Jess W. Everett, Daniel Blume, and Barbeau A. Roy Jr. "Market-Based Incentives and Residential Municipal Solid Waste." *Journal of Policy Analysis and Management 13*(*4*) (Fall 1994), pp. 681–98.

Palmer, Karen, and Margaret Walls. "Optimal Policies for Solid Waste Disposal and Recycling: Taxes, Subsidies, and Standards." *Journal of Public Economics 65*(*2*) (August 1997), pp. 193–205.

Pellow, David Naguib. *Garbage Wars: The Struggle for Environmental Justice in Chicago*. Cambridge, MA: MIT Press, 2002.

Pickin, Joe. "Unit Pricing of Household Garbage in Melbourne: Improving Welfare, Reducing Garbage, or Neither?" *Waste Management & Research 26*(*6*) (December 2008), pp. 508–14.

Porter, Richard C. *The Economics of Waste*. Washington, DC: Resources for the Future, 2002.

Powell, Jane C., ed. *Waste Management and Planning*. Northampton, MA: Edward Elgar Publishing, Inc., 2001.

Repetto, Robert, Roger C. Dower, Robin Jenkins, and Jacqueline Geoghegan. *Green Fees: How a Tax Shift Can Work for the Environment and the Economy*. Washington, DC: World Resources Institute, 1992.

Ueta, K., and H. Koizumi. "Reducing Household Waste: Japan Learns From Germany." *Environment 43*(*9*) (November 2001), pp. 20–32.

Yang, Hai-Lan, and Robert Innes. "Economic Incentives and Residential Waste Management in Taiwan: An Empirical Investigation." *Environmental and Resource Economics 37*(*3*) (July 2007), pp. 489–520.

Zhang, Dong Qing, Soon Keat Tan, and Richard M. Gersberg. "Municipal Solid Waste Management in China: Status, Problems, and Challenges. *Journal of Environmental Management 91*(*8*) (August 2010), pp. 1623–33.

A Reference to Acronyms and Terms in Municipal Solid Waste Policy

Environmental Economics Acronyms

MEB	Marginal external benefit
MEC	Marginal external cost
MPB	Marginal private benefit
MPC	Marginal private cost
MSB	Marginal social benefit
MSC	Marginal social cost

Environmental Policy Acronyms

DSD	Duales System Deutschland AG
EEE	Electrical and electronics equipment
MSW	Municipal solid waste
NIMBY	"Not in my backyard"
PAYT	Pay-as-you-throw
PRO EUROPE	Packaging Recovery Organisation Europe
RCRA	Resource Conservation and Recovery Act

To access additional course materials, visit www.cengagebrain.com. At the home page, search for the ISBN of this title (shown on the back cover). This will take you to the product page where these resources can be found.

CHAPTER

19

Controlling Pesticides and Toxic Chemicals

"As crude a weapon as the cave man's club, the chemical barrage has been hurled against the fabric of life."

—Rachel Carson (1907–1964)

In the United States, toxic substances are extensively regulated. Air toxics, for example, are controlled under the Clean Air Act. The discharge of hazardous pollutants into waterways is governed by the Clean Water Act, and the Resource Conservation and Recovery Act regulates the hazardous solid waste stream. Notice how all these laws deal with the *release* of toxics into the environment. In this chapter, we study legislation aimed at controlling pesticides and other chemicals *before* they become residuals and enter the waste stream—when they are formulated and produced, which is essentially a preventive approach.

More than 84,000 chemical substances are produced in this country, with new ones introduced every year. Over 1 billion tons of pesticide products are used annually with expenditures on these substances totaling $12.5 billion in 2007.[1] While this increasing reliance on chemicals has added to the variety, durability, and usefulness of products available to society, there also has been a cost. Certain of these chemicals, such as polychlorinated biphenyls (PCBs) and the dichloro-diphenyl-trichloroethane (DDT), pose a threat to human health and ecosystems. And in some cases, the risks are not fully understood until long after the damage has been done.

Recognizing the potential risks of chemical usage and the impact of toxic residuals in the waste stream, Congress passed the Federal Insecticide, Fungicide, and Rodenticide Act (FIFRA) and the Toxic Substances Control Act (TSCA). These two laws, each from a different vantage point, control the production, distribution, and consumption of pesticides and other chemicals. By instituting controls *before* these substances are introduced into commerce, those that pose an unacceptable risk can be restricted in use or even banned. This approach is preemptive because it eliminates the need to treat and dispose of some amount of hazardous residuals.

[1]U.S. EPA, Office of Chemical Safety and Pollution Prevention, Office of Pollution, Prevention, and Toxics (August 5, 2011); U.S. EPA, Office of Chemical Safety and Pollution Prevention, Office of Pesticide Programs (February 16, 2011a); U.S. EPA, Office of Chemical Safety and Pollution Prevention, Office of Pesticide Programs, Biological and Economic Analysis Division (February 2011).

In this chapter, our objectives are to understand the intent of FIFRA and TSCA, to analyze their implementation, and to evaluate their effectiveness using economic criteria. To organize our investigation, we begin with an overview of FIFRA and an analysis of its overall approach. An analogous treatment is used to study TSCA. Once done, we model the underlying externality and discuss a market-based solution. A list of acronyms and terms is provided at the end of this chapter.

OVERVIEW OF U.S. PESTICIDE CONTROLS

Legislating Pesticide Use[2]

Originally passed in 1947, the Federal Insecticide, Fungicide, and Rodenticide Act (FIFRA) and its subsequent amendments are the primary laws governing current U.S. policy on pesticides. The Environmental Protection Agency (EPA) has responsibility for FIFRA, which imposes stringent controls on the storage, transportation, and disposal of pesticides, including labeling regulations on both the pesticide and its container. The rulings also control the use, disposal, refill, and reuse of pesticide containers.

The chief regulatory instrument of FIFRA is the **registration** of new pesticides and the reevaluation and **reregistration** of those already on the market. Over time, legislative reform has led to improved registration procedures and a system of one-time reregistration fees and annual maintenance fees to help pay for the reregistration process. Further revisions were made through the Pesticide Registration Improvement Act of 2003 (PRIA 1), which became effective in March 2004. Its primary goals are to extend the collection of maintenance fees, to promote quicker decisions for reduced-risk pesticides, and to create a new registration fee system. Subsequently, the Pesticide Registration Improvement Renewal Act (PRIA 2) was enacted in 2007 to revise and reauthorize this fee system, which had been scheduled to expire in 2008.

Legislating Pesticide Residues[3]

In August 1996, President Clinton signed into law the Food Quality Protection Act (FQPA) of 1996. This act amends FIFRA as well as the Federal Food, Drug, and Cosmetic Act (FFDCA), another piece of legislation that regulates pesticides—this one by establishing tolerances (i.e., maximum levels allowed) for pesticide residues in food. To read more about the FQPA, visit **www.epa.gov/pesticides/regulating/laws/fqpa/**.

The major thrust of the Food Quality Protection Act (FQPA) is to provide a more consistent set of controls by establishing a single health-based standard for all pesticides in all foods. It also gives special attention to pesticide risks faced by infants and children. By imposing one standard, the FQPA eliminates inconsistency problems posed by one of

[2]The following is drawn from U.S. EPA, Office of Chemical Safety and Pollution Prevention, Office of Pesticide Programs (February 16, 2011b); Esworthy (August 11, 2010); U.S. EPA, Office of Pesticides and Toxic Substances (December 1988); Wolf (1988), p. 154. To access the text of FIFRA online, visit **www.law.cornell.edu/uscode/7/ch6.html**.

[3]The following is drawn from Schierow (February 1, 2010).

the provisions of the FFDCA, known as the Delaney Clause, which is discussed in Application 19.1. The standard requires that all tolerances be safe, where safety in the law refers to "… a reasonable certainty that no harm will result from aggregate exposure."[4]

APPLICATION 19.1 Solving the Delaney Clause Dilemma

The Delaney Clause was perhaps the most controversial piece of environmental legislation ever passed in the United States. Named after its sponsor, former U.S. Representative James Joseph Delaney of New York, the 1958 law effectively banned the use of any food additive found to cause cancer—no matter how negligible the degree of risk. How could this ruling cause a controversy? In large part, the problem arose from inconsistencies in the law.

The Delaney Clause was included in Section 409, the food additives amendment of the Federal Food, Drug, and Cosmetic Act (FFDCA). This section governs processed food and sets tolerances for food additives, which includes pesticide residues, using a risk-based analysis. Pesticides in raw food, however, are governed by a different part of the FFDCA—Section 408, the pesticide residue amendment—and tolerances under this section are set using risk-benefit analysis, a different risk management strategy.

Sections 408 and 409 are linked by Section 402 of the FFDCA, which includes an exemption known as the "flow-through provision." This provision states that processed foods would be considered safe if the remaining residues were not in excess of what had been allowed for raw foods under Section 408, despite the Delaney Clause language. Put another way, setting a tolerance under Section 409 would become necessary only if the pesticide concentrated during processing *and* if the resulting concentration caused the residue to exceed the approved tolerance level for raw foods.

The last piece of the puzzle is that the EPA had adopted a "coordination policy" in setting pesticide tolerances. Specifically, the agency would not approve a pesticide for raw foods under Section 408 if doing so would violate Section 409 should that raw product became processed at a later point. The EPA did this because it could not discern which raw foods would be processed prior to consumption. Despite its logic, this policy extended the control of the Delaney Clause to more pesticides than it otherwise would have.

Taken together, the law as written, coupled with the EPA's coordination policy, created a number of nontrivial inconsistencies. A critical consequence was that pesticides with negligible carcinogenic risks would be banned if they concentrated during processing, yet pesticides with higher carcinogenic risks could be used on raw foods if they did not concentrate during processing, because they would not require a Section 409 ruling. Another was that a noncarcinogenic pesticide that caused serious health problems, such as birth defects, would be subject only to the more lenient risk-benefit standard of Section 408.

In response to this dilemma, which became known as the Delaney Paradox, the EPA asked the National Research Council (NRC) to review its regulatory options. The NRC did so and, in a 1987 report, recommended that consistent standards be used for all pesticide residues in food, whether raw or processed, and that tolerances be allowed for those carcinogenic pesticides posing a negligible risk. The EPA decided to follow this recommendation, but its decision was challenged in court as a violation of the Delaney Clause. In 1992, a federal appeals court ruled against the EPA's decision, and with that, the Delaney Paradox remained in place.

For several years, various constituencies, including the pesticide industry, the food industry, and environmental groups, voiced dissatisfaction with the tolerance-setting process. But no tangible solution was found until the Food Quality Protection Act (FQPA) was enacted in 1996. The FQPA did not repeal the Delaney Clause, but it did solve the Delaney Paradox by amending both FIFRA and FFDCA. The 1996 law establishes one safety standard under an amended Section 408 for pesticide residues on both raw and processed food. Further, it defines *safety* to include *all* risks—not just carcinogenic risks. And so, after decades of debate and controversy, the legal inconsistencies and the Delaney Paradox seem to have been resolved.

Sources: Schierow (February 1, 2010; September 11, 1996); Council on Environmental Quality (1997), pp. 94–96; Vogt (1995); Kimm (January–March 1993).

[4]Pub. L. 104-170, Sec. 405(b)(2)(A)(ii), 110 Stat. 1489 (1996).

CONTROLLING PESTICIDE USE

Registration of New Pesticides

pesticide registration
Formal listing of a pesticide with the EPA, based on a risk-benefit analysis, before it can be sold or distributed.

A major objective of FIFRA and its amendments is to accomplish proper **registration** of any pesticide before it can be distributed or sold. Currently, there are over 20,000 pesticide products registered for use in the United States.[5] Registration is subject to EPA approval based upon a **risk-benefit analysis**. The EPA must ensure that the benefits of using the substance outweigh the associated health and ecological risks. Hence, the law implicitly identifies those pesticides associated with unreasonable adverse effects by denying their registration.

risk-benefit analysis
An assessment of risks of a hazard along with the benefits to society of not regulating that hazard.

Based on the law, registration of a pesticide is denied if any of the following hold true:

- Its composition does not warrant its proposed claims.
- Its labeling does not comply with the law.
- It will not perform its intended function without causing unreasonable adverse effects on the environment.
- It cannot be used in widespread practice without generating unreasonable adverse environmental effects.

pesticide tolerances
Legal limits on the amount of pesticide residue allowed on raw or processed foods.

At the time of registration, the EPA sets legal limits, or **tolerances**, on the amount of a pesticide that may remain as a residue on raw or processed food without causing an unacceptable health risk. In effect, the registration acts as a license for use, and the tolerances specify the conditions under which that use is approved. Tolerances for most foods are enforced by the Food and Drug Administration (FDA) according to the provisions of the Federal Food, Drug, and Cosmetic Act (FFDCA). For online information on pesticide registration, visit **www.epa.gov/pesticides/regulating/registering/**.

Reregistration of Existing Pesticides[6]

pesticide reregistration
A formal reevaluation of a previously licensed pesticide already on the market.

Among the more important elements of amendments to FIFRA and the FQPA of 1996 are the rulings dealing with **reregistration** of previously licensed pesticides. This is a reevaluation process aimed at ensuring that formerly licensed pesticides meet current regulatory standards. In the final phase of the process, the EPA issues a **Reregistration Eligibility Decision (RED)** document for each pesticide, which gives the full results of the agency's risk assessment.

As a result of this reevaluation process, many unsafe pesticides have been restricted in use or banned. Examples include agricultural uses of chlordane and virtually all uses of DDT and ethylene dibromide (EDB). As harmful chemicals like these are restricted or eliminated, industries and consumers must find alternative, and presumably less toxic,

---

[5]U.S. EPA, Office of Chemical Safety and Pollution Prevention, Office of Pesticide Programs (February 16, 2011c).

[6]Much of the following is drawn from U.S. EPA, Office of Chemical Safety and Pollution Prevention, Office of Pesticide Programs (February 16, 2011a; February 16, 2011b); U.S. EPA, Office of Pesticides and Toxic Substances (June 1991).

pesticides. Consequently, some amount of hazardous pollution should be prevented instead of having to be remediated after the fact.

Important changes to the reregistration program came with the enactment of the FQPA of 1996. Under this law, the EPA was required to use reregistration as a means to reassess all established tolerances within 10 years to make certain that the new safety standard established by the FQPA is met. This reassessment was completed in September 2007. The FQPA further requires that the EPA reassess every pesticide registration on a 15-year cycle. To learn more about the reregistration process, visit **www.epa.gov/oppsrrd1/reregistration/index.htm**.

New Policy Direction

In 1993, a report by the National Academy of Sciences (NAS) triggered what appeared to be a major shift in U.S. pesticide policy and perhaps a resolution to some critical issues. What the NAS found was that children were not being sufficiently protected from pesticide risks by U.S. policy. At the core of this assessment is that the smaller size of children and their diets make them more vulnerable to pesticide risks. Almost immediately, the Clinton administration announced its intention to develop a new pesticide plan—a joint effort by the EPA, the FDA, and the Department of Agriculture.

Integrated Pest Management (IPM) A combination of methods that encourage more selective pesticide use and greater reliance on natural deterrents.

Ultimately launched as the **Pesticide Environmental Stewardship Program (PESP)**, the plan's major objective is to achieve **pollution prevention** by *reducing* pesticide use—not just *regulating* it. This voluntary program (discussed online at **www.epa.gov/pesp**) establishes partnerships with pesticide users to implement preventive strategies. Among the program's goals is the promotion of **Integrated Pest Management (IPM)**—a collection of methods that fosters more selective use of pesticides and greater reliance on natural deterrents. In support of this objective is an incentive-based proposal to encourage chemical manufacturers to develop alternative products.[7]

ANALYSIS OF FIFRA

Based on available estimates, the costs of pesticide programs are nontrivial and have risen over time due in part to increasingly tougher laws on registration and reregistration. Nonfederal explicit costs to comply with current pesticide regulations in 2002 are shown in Table 19.1. Notice that annual net costs are nearly $460 million ($2010), with most of these expenses, or $371.5 million ($2010), attributable to registrations and reregistrations. At issue is whether these costs are economically justified.

The fact is, controlling pesticides poses a tough problem for policymakers and for society as a whole. Pesticides can contribute positively to agriculture and other industries. Yet exposure to these substances can harm the ecology and cause serious health problems like cancer, birth defects, and neurological impairments. Both sets of factors need to be considered in a risk-benefit analysis. Hence, the $460 million cost to comply with the nation's pesticide laws may be warranted if it facilitates government's ability to properly manage whatever risks are justified by the associated social benefits.

[7]U.S. EPA, Office of Chemical Safety and Pollution Prevention, Office of Pesticide Programs (May 9, 2011); "White House to Seek Reduced Use of Pesticides" (September 21, 1993).

TABLE 19.1 Annualized Nonfederal Costs of Pesticide Controls

Entity	Annual Cost (millions)
Registrants (manufacturers)	
New registrations	$286.4
Reregistrations	85.1
State agencies	4.1
Users, agricultural	99.0
TOTAL COSTS	$474.6
Government subsidies (for research and testing)	(15.8)
NET COSTS	$458.8

NOTE: Values shown are for 2002 stated in millions of 2010 dollars.
Source: U.S. EPA, Office of the Chief Financial Officer (September 30, 2003), Table 8, p. 198.

Risk-Benefit Analysis

Under FIFRA, no pesticide may be distributed or sold unless it has been registered with the EPA, and registration is denied for any substance associated with "*unreasonable adverse effects on the environment*." Thus, the statutory definition of this phrase implicitly identifies the standard used to evaluate pesticides.

> The term "unreasonable adverse effects on the environment" means any unreasonable risk to man or the environment, taking into account the economic, social, and environmental costs and benefits of the use of any pesticide.[8]

Notice that, unlike most U.S. environmental laws, FIFRA explicitly calls for a consideration of the costs and benefits of pesticide use in determining unreasonable risk. This clearly suggests the need for **risk-benefit analysis**. That is, the risks of using a pesticide measured in terms of health and ecological effects should be weighed against the associated benefits, such as increased crop yields. It turns out, however, that risks are the dominant factor in the EPA's registration decisions.

For **new pesticides**, risks are evaluated from data on health and environmental effects submitted by manufacturers as part of the registration application. This means that the onus of proving that risks are not "unreasonable" lies with the producer. If risks are found to be negligible, the EPA generally assumes that private benefits exist based on the manufacturer's willingness to absorb the high cost of registration. If risks are found to be greater than negligible, the manufacturer typically has to formulate a risk reduction strategy or show that the benefits exceed the risks.[9] When the EPA conducts a special review of an **existing pesticide**, a formal benefit analysis is done. In these instances, benefits are measured by determining biological gains, such as changes in agricultural yields, and then these effects are monetized.

[8]FIFRA, Sec. 2(bb).

[9]Generally, manufacturers try to reduce the risks or withdraw the registration application. U.S. GAO (March 1991), pp. 3 and 9.

Problems in Risk Assessment

At least in principle, the nation's approach to pesticide control appears to have merit. So why is U.S. pesticide policy the subject of intense debate? Probably the best answer is that there is much uncertainty about the associated risks, despite the widespread exposure. Many scientists question the EPA's methods of assessing pesticide risks. Some argue that the risks are underestimated, citing the agency's lack of attention to inert ingredients or the cumulative effects from using multiple pesticides in combination. Others assert that the opposite is true, claiming that findings from animal bioassays overstate the potential harm to humans. Part of the difficulty may lie in the difference between actual risk and perceived risk, an issue addressed in a landmark investigation of risk ranking.[10]

To conduct this study, an EPA-designated task force defined the universe of environmental issues and assessed four types of risks: cancer risks, noncancer risks, ecological effects, and welfare effects. The objective was to determine the relative risk of each problem within each category and establish a priority ranking that could be used to guide national policy. These findings also were compared to an ordinal ranking of the public's perception of environmental risks based on a survey conducted by the Roper Organization.

What the study found is of interest on a number of fronts. First, the two sets of rankings were found to be inconsistent. Table 19.2 presents these rankings for chemical and pesticide risks. Second, the task force noted that the federal government's environmental priorities do not align consistently with the expert risk ranking—an observation that seems to be explained by the public's risk perceptions. Third, the task force identified environmental hazards of high risk where the EPA's control efforts are low. Among these are runoff, leaching, and air deposition associated with the use of pesticides and

TABLE 19.2 Risk Rankings

	EXPERT RISK RANKING				PUBLIC RISK RANKING
Problem	Cancer Risk	Noncancer Health Risk	Ecological Effects	Welfare Effects	Overall
Accidental releases of toxics	4	High	Medium	Low	High
Pesticide residues	1	High	High	Minor	Moderate
Other pesticide risks	2	Medium	High	Medium	Moderate
New toxic chemicals	2	Not ranked	Not ranked	Minor	Not ranked
Consumer product exposure	1	High	Not ranked	Minor	Low
Worker chemical exposure	1	High	Not ranked	Minor	Moderate

NOTE: Cancer rankings are on a scale of 1 through 5, with 1 being the highest relative risk and 5 being a "not assessed" or "no risk" rank.
Source: U.S. EPA, Office of Policy, Planning, and Evaluation (February 1987).

[10]U.S. EPA, Office of Policy, Planning, and Evaluation (February 1987). The conclusions of a follow-up study are reported in U.S. EPA, Science Advisory Board (September 1990).

agricultural chemicals. Thus, based on relative risk analysis, it appears that too few resources are being used to control pesticides.

Problems in Benefit Assessment

In a risk-benefit analysis, the benefits are the gains associated with using the hazard that generates the risk. In the context of pesticide use, the primary incremental benefits are reduced plant damage and increased crop yields. For a given agricultural crop, these changes can be modeled as an increase in supply (*S*), as shown in Figure 19.1, which raises equilibrium quantity (*Q*) and lowers equilibrium price (*P*). Since both producers and consumers are affected, the incremental benefits can be estimated by the resulting change in consumer and producer surpluses. This is shown in Figure 19.1 as area XYZ.[11]

While this approach to benefit measurement is theoretically sound, it is difficult to apply in practice. As the figure illustrates, an estimate of incremental benefits from pesticide use would require precise information on the demand, pre-pesticide supply, and post-pesticide supply for every agricultural market in which a given pesticide is used. Of course, such data are not readily available, particularly the supply changes, which would be confounded by any other factors that affect agricultural productivity, such as weather, soil conditions, and technology. Furthermore, there are secondary benefits that would be virtually impossible to estimate on a broad scale, such as improved health or worker productivity associated with increased food supplies.

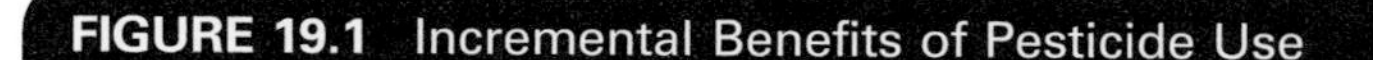

FIGURE 19.1 Incremental Benefits of Pesticide Use

For a given agricultural crop, the primary incremental benefits of pesticide use are reduced plant damage and higher yields, which can be modeled as an increase in supply. This in turn raises equilibrium quantity and lowers equilibrium price. The incremental benefits can be estimated by the resulting change in consumer and producer surpluses, which is shown as area XYZ.

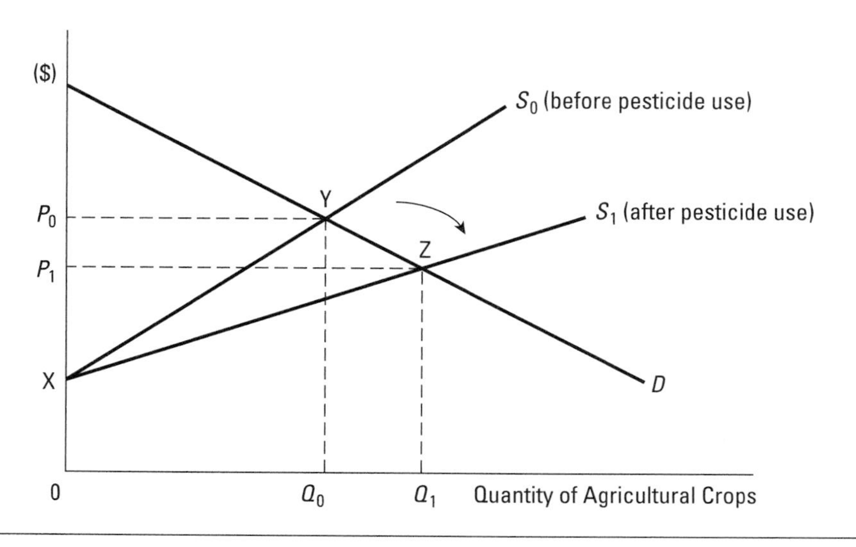

[11] Refer back to Chapter 7 to review this model, as needed.

OVERVIEW OF U.S. LEGISLATION ON TOXIC CHEMICALS

Chemicals are an integral part of modern society and represent a major industry in the U.S. economy. Used properly, most contribute positively to the quality of life. Yet exposure to chemicals also can pose significant risks. Unfortunately, some of the health risks of chemical exposure were not known until after these substances had been introduced into commerce and had become widely used. An important example is the chemical group polychlorinated biphenyls (PCBs), man-made substances formerly used in adhesives and tapes, in oil-based paints, and in cable insulation. The toxicity of PCBs, which had been manufactured since 1929, was not known until after these substances had leaked into the environment and began to accumulate in the tissues of animals and fish, thereby entering the food chain. The discovery led to a ban in 1976 on their manufacture, processing, distribution, and use in the United States within three years, except in completely enclosed electrical equipment.[12]

Policy Response to Chemical Risks[13]

Responding to the problem of chemical risks, Congress enacted the Toxic Substances Control Act (TSCA) in 1976, which governs U.S. policy today. One of the act's primary objectives is to identify and control chemical substances that present a risk to health or the environment *before* they are introduced into commerce. By confronting the risks of chemicals at the premanufacture stage, the government has an opportunity to prevent pollution rather than having to correct problems after the fact.

TSCA inventory A database of all chemicals commercially produced or processed in the United States.

Since little was known at the time about most chemical substances in use, Congress also authorized the compilation of an inventory of all chemicals commercially produced or processed in the United States between 1975 and 1979. Using data collected from manufacturers and importers, the government published the first **TSCA inventory** in 1979 and a second version in 1982, which contained information on about 62,000 chemicals. Today, over 84,000 chemicals are listed in the TSCA inventory as those present in U.S. commerce, with between 500 and 1,000 added each year. More information on the inventory, including how to access it, is available at **www.epa.gov/oppt/existingchemicals/pubs/tscainventory/index.html**.

TSCA gives the EPA authority to gather information on chemical risks from producers, to require testing on existing chemicals, and to review most newly introduced chemicals before they are manufactured and made available for use on a broad scale. Depending on its findings, the EPA takes appropriate action. The regulatory response can range from requiring warnings during production and distribution to banning the chemical's manufacture. Risk controls may be imposed at any stage in the chemical's

---

[12]U.S. EPA, Office of Solid Waste and Emergency Response (December 29, 2010).

[13]Much of the following is drawn from U.S. EPA, Office of Chemical Safety and Pollution Prevention, Office of Pollution, Prevention, and Toxics (August 5, 2011); U.S. EPA, Office of Toxic Substances (June 1987), pp. 2–8. For online access to the full text of TSCA, visit **www.law.cornell.edu/uscode/15/ch53.html**.

life cycle, including its manufacture, processing, distribution in commerce, usage, and disposal.

Despite the well-established intent and procedural detail of TSCA, the legislation is considered to be in need of updating. Moreover, some have raised questions about what is the actual number of chemicals in commerce at a given point in time. In response, the EPA recently announced a set of six principles to help guide the reauthorization and strengthening of TSCA. Identified in Table 19.3, these so-called **Essential Principles for Reform of Chemicals Management Legislation** are intended to facilitate targeting chemicals of concern and regulating both new and existing substances in a timely manner.[14]

Complementing the Essential Principles, the EPA is also launching a more comprehensive approach to its existing chemical management program. This effort is aimed in part at questions raised by the American Chemistry Council (ACC) and others about the actual number of chemicals that remain "in commerce," as reported by the TSCA Inventory. Under the new approach, the EPA plans to undertake the following initiatives:

- Take new regulatory action on selected chemicals of concern, including lead and mercury.
- Develop action plans to label, restrict, or ban selected chemicals.

TABLE 19.3 Essential Principles for Reform of Chemicals Management Legislation

Principle Number 1

Chemicals should be reviewed against safety standards that are based on sound science and reflect risk-based criteria protective of human health and the environment.

Principle Number 2

Manufacturers should provide EPA with the necessary information to conclude that new and existing chemicals are safe and do not endanger public health or the environment.

Principle Number 3

Risk management decisions should take into account sensitive subpopulations, cost, availability of substitutes, and other relevant considerations.

Principle Number 4

Manufacturers and EPA should assess and act on priority chemicals, both existing and new, in a timely manner.

Principle Number 5

Green chemistry should be encouraged, and provisions assuring transparency and public access to information should be strengthened.

Principle Number 6

EPA should be given a sustained source of funding for implementation.

Source: U.S. EPA, Office of Chemical Safety and Pollution Prevention, Office of Pollution, Prevention, and Toxics (April 28, 2010).

[14]U.S. EPA, Office of Chemical Safety and Pollution Prevention, Office of Pollution, Prevention, and Toxics (April 28, 2010).

- Strengthen requirements for information on certain chemical substances
- Improve public access to information about chemicals
- Identify priority chemicals to be reviewed for potential action under TSCA.[15]

Coordination With Other Agencies and Initiatives

To avoid the inefficiency and potential inconsistency of dual controls, TSCA acknowledges the mandates of other laws dealing with toxic substances and the agencies that implement them. Some of these are listed in Table 19.4. Under the law, the EPA must coordinate its control activities with those of other agencies, such as the Food and Drug Administration (FDA), the U.S. Department of Agriculture (USDA), and the Occupational Safety and Health Administration (OSHA). As a chemical risk is discovered, the EPA must consider whether other agencies have investigated that risk, and if so, it must determine if existing laws adequately address the problem. In fact, the EPA must use other laws it administers to reduce risk, such as the Clean Water Act and the Resource Conservation and Recovery Act (RCRA), *before* implementing TSCA's

TABLE 19.4 Selected Laws Controlling Substances Exempted Under TSCA

Legislation	Agency	Substance or Control
Atomic Energy Act	Nuclear Regulatory Commission	Nuclear waste disposal, nuclear energy production
Federal Insecticide, Fungicide, and Rodenticide Act	Environmental Protection Agency	Pesticides
Federal Food, Drug, and Cosmetic Act	Food and Drug Administration	Foods, food additives, drug additives, cosmetics
Safe Drinking Water Act	Environmental Protection Agency	Controls contaminants in drinking water supplies
Occupational Safety and Health Act	Occupational Safety and Health Administration	Controls hazards found in the workplace
Clean Air Act	Environmental Protection Agency	Controls the emissions of hazardous air pollutants
Clean Water Act	Environmental Protection Agency	Controls the discharge of hazardous pollutants into surface waters
Hazardous Materials Transportation Act	Department of Transportation	Regulates the transport of hazardous materials
Marine Protection, Research, and Sanctuaries Act	Environmental Protection Agency	Regulates waste disposal at sea
Resource Conservation and Recovery Act	Environmental Protection Agency	Regulates hazardous waste generation, storage, transport, treatment, and disposal
Comprehensive Environmental Response, Compensation, and Liability Act	Environmental Protection Agency	Provides for cleanup of inactive and abandoned hazardous waste sites

Source: U.S. EPA, Office of Solid Waste (November 1986).

[15]U.S. EPA, Office of Chemical Safety and Pollution Prevention, Office of Pollution, Prevention, and Toxics (August 18, 2011); Schierow (July 28, 2009).

provisions. Finally, eight product categories are exempted from TSCA, most of which are regulated under other laws.[16]

CONTROLLING CHEMICAL USE

New Chemicals

new chemical Any substance not listed in the TSCA inventory of existing chemicals.

premanufacture notice (PMN) Official notification to the EPA by a chemical producer about its intent to produce or import a new chemical.

Unlike the nation's pesticide law, TSCA does not use a registration procedure to control toxic chemicals. Instead, it requires manufacturers to notify the government at least 90 days before they intend to produce or import any **new chemical**, which is one that is not listed in the TSCA inventory. The notification is made to the EPA via a **premanufacture notice (PMN)**. The PMN provides information about the chemical's characteristics, the expected exposure to workers, its intended use, and the health and ecological effects.

Upon receipt of the PMN, the EPA has a 90-day review period to evaluate the risks and respond. The risks are assessed relative to the benefits in use, following a **risk-benefit analysis**. If the risks are too high or, in the absence of full information, are expected to be too high, the EPA may restrict usage of the chemical permanently or, if more data are needed for a complete risk analysis, temporarily. To accommodate these procedures, the EPA launched the **New Chemicals Program** to determine conditions, up to and including a total production ban, to be imposed on a new chemical prior to being entered into commerce. More on this program is available at **www.epa.gov/oppt/newchems/index.htm**.[17]

Over 36,000 PMNs have been submitted to the EPA since 1979, the year the TSCA inventory was established. Of this total, the EPA has restricted or prohibited the production or use of more than 1,400 chemicals pending further information. Over 300 chemicals have been controlled through voluntary agreements with producers, and more than 1,800 have been withdrawn or suspended from the review process by manufacturers in response to EPA concerns. In the aggregate, about 10 percent of the chemicals submitted for review since 1979 have been suspended, restricted, or withdrawn from production or use.[18]

Existing Chemicals

existing chemical A substance listed in the TSCA inventory.

Other provisions of TSCA deal with evaluating the risks of some 84,000 **existing chemicals**, which are those in the TSCA inventory. The law requires manufacturers to notify the EPA if any chemical is found to present a substantial risk to human health or the environment. When this occurs, the EPA's Office of Toxic Substances evaluates the information and takes action, ranging from labeling requirements to outright bans.

[16]See U.S. EPA, Office of Toxic Substances (June 1987). The eight product groups are: tobacco, nuclear material, firearms and ammunition, food, food additives, drugs, cosmetics, and pesticides.

[17]U.S. EPA, Office of Pollution Prevention (October 1991), p. 142; U.S. EPA, Office of Toxic Substances (June 1987), pp. 3–4; TSCA, Sec. 4(b)(2)(A).

[18]U.S. EPA, Office of Chemical Safety and Pollution Prevention, Office of Pollution, Prevention, and Toxics (September 1, 2010).

One of the more significant examples of such an action was the banning of certain asbestos-containing products in the late 1980s.[19]

New Policy Direction[20]

Although TSCA's premanufacture rulings for new chemicals have been effective in encouraging substitution of safer chemicals for toxic ones, there has been less success in fostering the same activity for existing chemicals. Recognizing the weaknesses in certain of TSCA's provisions, the United States is taking steps to develop chemical programs aimed at **pollution prevention** to complement existing policy.

Green Chemistry Program
Promotes the development of innovative chemical technologies to achieve pollution prevention.

One such effort is the **Green Chemistry Program**. This initiative, with domestic and international partners in industry, academia, research centers, national laboratories, and others, promotes the development and application of innovative chemical technologies to achieve pollution prevention by minimizing or even eliminating the production of hazardous substances. To accomplish its objectives, the program actively supports research in environmentally benign chemistry along with other activities such as public conferences and meetings. For more information and updates, visit **www.epa.gov/greenchemistry/**.

Among the program's voluntary partners are Dow Chemical Company, the Green and Sustainable Chemistry Network in Japan, the American Chemistry Council (ACC), the National Science Foundation, and the University of Massachusetts, Lowell, as shown in Figure 19.2. To illustrate what some of these partners are doing to promote green chemistry, Application 19.2 (page 470) discusses the Responsible Care® Program, which was developed by the chemical industry and now has been adopted by chemical associations in 53 nations.

Extended Product Responsibility (EPR)
A commitment by all participants in the product cycle to reduce any life cycle environmental effects of products.

Another preventive approach being used to reduce chemical risk is **Extended Product Responsibility (EPR)**, also known as **product stewardship**. This refers to committed responsibility by all participants in the product cycle—manufacturers, retailers, consumers, and disposers—to find ways to reduce a product's effect on the environment. Though its objectives are comprehensive, an important element of this strategic framework is to reduce the toxicity of inputs used in production. Further information is available online at **www.epa.gov/epawaste/partnerships/stewardship/index.htm**.

ANALYSIS OF TSCA

Regulatory costs associated with toxic chemicals have shown positive growth since TSCA was passed in 1976. To provide a sense of their magnitude, Table 19.5 (page 470) presents selected compliance costs for 2002. Just as is the case for pesticide control costs, it is reasonable to consider whether these costs are economically justified. Society can benefit from chemical use in a myriad of ways, but chemical exposure also can pose significant

[19]*U.S. Federal Register* 54, 29460-513 (July 12, 1989).

[20]This discussion is drawn mainly from U.S. EPA, Office of Chemical Safety and Pollution Prevention, Office of Pollution, Prevention, and Toxics (August 10, 2011a); U.S. EPA, Office of Solid Waste and Emergency Response (December 1998; August 2, 2011).

FIGURE 19.2 Selected Partners in the Green Chemistry Program

Industry	Goodrich Corporation The Dow Chemical Company Dow Corning Corporation Eastman Kodak Company
International Organizations	Centre for Green Chemistry (Australia) Organisation for Economic Co-operation and Development (OECD) Green & Sustainable Chemistry Network (Japan) Royal Society of Chemistry
Academia	Center for Green Chemistry and Green Engineering at Yale University of Massachusetts, Lowell The University of Alabama
Environmental Groups	Environmental Defense Fund (EDF)
Scientific Organizations	American Chemical Society International Union of Pure and Applied Chemistry National Research Council National Science Foundation
Trade Associations	American Chemistry Council (ACC) American Petroleum Institute Society of the Plastics Industry
National Laboratories	Los Alamos National Laboratory National Renewable Energy Laboratory
Research Centers	Center for Process Analysis and Control Emission Reduction Research Center National Environmental Technology Institute Toxics Use Reduction Institute
Government	Environmental Council of the States National Institute of Standards and Technology U.S. Department of Energy

Source: U.S. EPA, Office of Chemical Safety and Pollution Prevention, Office of Pollution, Prevention, and Toxics (August 10, 2011a).

risks to health and the ecology. Both sides of the coin are important to a risk-benefit analysis of TSCA rulings. Hence, compliance costs to chemical manufacturers make economic sense if they facilitate the government's management of those risks that are justified by associated benefits.

APPLICATION 19.2 The Responsible Care® Program: A Global Initiative

As a major contributor to the nation's hazardous waste stream, the chemical industry developed a program called Responsible Care® to signal the industry's commitment to environment, health, and safety. The objectives of the Responsible Care® Program are simple but powerful: to protect the health and safety of the workforce and to protect the environment from contamination.

Originally launched in 1985 by the Canadian Chemical Producers Association, the initiative has now been adopted by chemical associations in 53 countries. For example, in 1988, the U.S. industry's trade alliance, the American Chemistry Council (ACC) (then called the Chemical Manufacturers Association), started its Responsible Care® Program as a comprehensive chemical management plan to be adopted by its membership. Great Britain's counterpart to the ACC, the Chemical Industries Association, also joined the ranks of trade associations working to prevent pollution. Its Responsible Care® Program came on line in March 1989—the first such plan in Europe. Soon thereafter, chemical associations in other European countries followed suit. In 2007, Russia became the 53rd nation to join the global initiative.

To coordinate these worldwide efforts, the International Council of Chemical Associations (ICCA), through Responsible Care®, serves as the program's guardian. Its role is one of oversight, ensuring proper implementation and appropriate focus on key concerns. Although the ICCA provides leadership, each country's chemical association is responsible for developing its own program guidelines. For example, the U.S. plan ushered in a series of 10 guiding principles to help members reduce waste generation and dangerous releases into the environment. Among these are commitments to produce safe chemicals, to give the environment and resource conservation critical consideration for all new and existing products, and to provide information on health and environmental risks to all major stakeholders. The complete listing of the 10 principles is available online at **http://responsiblecare.americanchemistry.com/Responsible-Care-Program-Elements/Guiding-Principles**.

In 2006, the Responsible Care Global Charter was launched. Comprised of nine elements that deal with such issues as sustainable development; measures of environmental, health, and safety performance; and the adoption of Responsible Care® core principles, this charter evolved over two decades based on observations and recommendations from stakeholders around the globe. It is expected to chart the course and priorities of Responsible Care® well into the future.

Since the inception of the Responsible Care® Program, the chemical industry has achieved enhanced environmental, health, and safety performance, and its public image has improved. In fact, as of 2011, the Responsible Care® Program reports the following among the environmental gains achieved by participating companies:

- More than a 16 percent reduction in greenhouse gas intensity since 1992
- A 75 percent decrease in hazardous releases to air, water, and land since 1988
- A 10 percent rise in energy efficiency relative to 1992

Nonetheless, to gain public approval over the long run, the industry must remain committed to the program's guiding principles. To learn more about the Responsible Care® initiative and its progress, visit **www.icca-chem.org/en/Home/Responsible-care/**.

Sources: American Chemistry Council (July 2011; April 14, 2009a; April 14, 2009b); Responsible Care (April 14, 2009); International Council of Chemical Associations, Responsible Care Leadership Group (2007); U.S. EPA, Office of Pollution Prevention (October 1991), pp. 40–41; Chemical Manufacturers Association (September 1990), pp. 9–15.

TABLE 19.5 Costs of Chemical Controls to Manufacturers and Importers

Cost Category	Annual Cost (millions)
New chemicals (PMNs and exemption notices)	\$37.6
Existing chemicals (reporting requirements)	\$6.0–\$9.3[a]
Laboratory testing and administrative costs	\$45.1

NOTES: [a]These annualized costs are based on a 7% discount rate.
Values shown are for 2002 stated in millions of 2010 dollars.

Source: U.S. EPA, Office of the Chief Financial Officer (September 30, 2003), pp. 196–97.

Risk-Benefit Analysis

TSCA's statutory objectives are to obtain data on chemical risks and to regulate chemicals posing an unreasonable risk to human health or the environment. Interestingly, "unreasonable risk" is not explicitly defined in the law. However, a House of Representatives report states that the determination of "unreasonable risk" involves:[21]

> ... balancing the probability that harm will occur and the magnitude and severity of that harm against the effect of proposed regulatory action on the availability to society of the benefits of the substance or mixture.

Notice how this determination supports the use of **risk-benefit analysis** in the approval process—just as is the case for pesticides. Like FIFRA, TSCA empowers the EPA to ban or restrict the use of toxic chemicals that do not pass the risk-benefit test. Such actions should encourage the substitution of alternative, less dangerous substances, which in turn lessens the toxicity of the hazardous waste stream.

A major difference between TSCA and FIFRA is the process used to review new substances. While FIFRA calls for extensive test data as part of a complex registration process, TSCA requires only a 90-day advance notice of intent to produce a new chemical. Testing is done only upon formal request by the EPA. It has been argued that this regulatory difference properly reflects the relative magnitude of risks between toxic substances and pesticides. Pesticides generally pose a greater risk since they are biologically active. Hence, the more stringent controls and higher costs of introducing a new pesticide into commerce may be justified by the inherently greater risk potential.[22]

Problems in Risk Assessment

On the risk side of the equation, exposure to chemicals can lead to serious health and ecological consequences, and in some cases, the risks are not discovered until after a chemical has been introduced into commerce. According to one article, of the tens of thousands of chemicals registered with the EPA, health studies have been conducted on only about 10,000, and of these, only about 1,000 have been studied for acute effects.[23] There is an even greater void in formal assessments of ecological risks.

Furthermore, just as is the case for pesticides, risk perception of chemical exposure is not always aligned with actual risk. Look back at Table 19.2 (page 462), which compares expert risk rankings with that of the general public. As we noted previously, the federal government's environmental priorities do not always align with how experts assess relative risk. Indeed, the task force reported that several chemical exposure hazards, including consumer and worker exposure to chemicals, are characterized as being in a "high risk/low EPA effort category." This suggests that, just as is the case for pesticides, too few resources may be being used to control chemicals. The implication is that government

[21]Dominguez (1977), p. 5.9.

[22]Shapiro (1990), pp. 213–14.

[23]Stranahan (February/March 1990).

should intensify its efforts under TSCA, where there is greater opportunity for risk reduction.

Problems in Benefit Assessment

Chemicals are used in the production of innumerable goods and services, from household lightbulbs to sophisticated medical equipment. Hence, they add to social benefits when they improve product quality, enhance productivity, facilitate health care advances, and contribute to economic growth. New chemical formulations can even improve the environment if they are substituted for more toxic counterparts in production. So what, then, is the problem?

Estimating social benefits is a difficult task, even in a narrow context. But such estimation is particularly daunting when the benefits accrue to society in various ways and across multiple markets. As it turns out, no comprehensive benefit analysis has been undertaken for the major U.S. programs on toxic chemicals. Why has so significant an issue escaped the rigors of formal analysis? Part of the explanation is that national policy on toxic chemicals is not as well established as air and water quality controls, in part because the United States got a relatively late start in initiating this regulatory policy.

Beyond the timing issues, it is also true that certain of the expected benefits are difficult to measure because of the sheer magnitude of the relevant variables. There are over 84,000 registered chemicals, and hundreds of premanufacture notices (PMNs) are filed each year for new formulations. These numbers help to explain the uncertainty in even defining the extent of social benefits, much less assessing the progress achieved by policy.

Bias Against New Chemical Introductions[24]

A common observation about TSCA's effectiveness is that the EPA has been slow to develop data on existing chemicals, despite evidence that such information is critical. Ironically, much of the delay seems to be caused by how TSCA is written, imposing command-and-control procedures that are time intensive and possibly unwarranted. For example, the EPA literally must write a new regulation each time it wishes to test a chemical substance. Another contributing factor is the agency's current use of a single chemical review process rather than one aimed at a group of similar substances.

Overall, the analysis of existing chemicals is complex and bogged down with time-intensive procedures—markedly different from the relatively straightforward rules in place for reviewing new substances. The result? Not only do these regulations delay the development of an important database, but they also create a bias against the introduction of new chemicals. The government is much more efficient in its testing of new substances than it is for those already on the market. Consequently, chemical producers wishing to avoid such procedures and the associated costs can continue to sell existing, and possibly more dangerous, substances. Hence, certain of TSCA's rulings appear to generate the perverse outcome of deterring the development of new, safer chemicals.

[24]This discussion is drawn from Shapiro (1990), pp. 223–24, 232–36.

ECONOMIC ANALYSIS AND MARKET-BASED POLICY

As we have observed, FIFRA and TSCA attempt to control the production and use of potentially harmful substances *before* they are released into the environment. This is accomplished somewhat differently in each case, but the objective is the same—to minimize the risks associated with using pesticides and other chemicals. By restricting or even banning the use of these substances, fewer resources are allocated to production, and equilibrium quantity decreases. However, there is nothing in either law that motivates an efficient solution. To illustrate, let's begin by modeling the source of inefficiency in these markets and then discuss a market-based instrument that could achieve an efficient outcome.

Modeling the Negative Externality

In any pesticide or toxic chemical market, the source of the inefficiency arises from the presence of a negative externality. Note, however, that this negative externality is linked to use, or consumption, and not production. Think about the application of agricultural pesticides to crops, for example. It is this use, and not the manufacture of the substance, that poses an environmental risk to society.[25]

This negative externality is modeled in Figure 19.3 in the market for some chemical product, which might be a pesticide or some other toxic chemical. Supply (S) is represented by the marginal private cost (MPC), which also equals the marginal social cost (MSC), assuming no production externality. Demand (D) is shown by the marginal private benefit (MPB) curve. The consumption externality is modeled as a negative marginal external benefit (MEB), which causes the marginal social benefit (MSB) to lie *below* the MPB. In the absence of government intervention, this negative MEB is ignored. Hence, the competitive equilibrium output arises where MPB equals MPC, or Q_C, which is higher than the efficient output level, Q_E, determined by the intersection of MSC and MSB. The consequence is inefficiency, which in this case is an overallocation of resources to chemical production and a market failure shown by a price that fails to capture the external damage.

Market-Based Solution: Product Charge

To internalize the consumption externality in the pesticide or chemical market, a **product charge** could be used. This could be implemented as a unit tax (t) on the chemical. To achieve efficiency, the tax should be set equal to $-MEB$ at Q_E, as shown in Figure 19.3. The S curve would shift up by the amount of the tax, raising the effective product price to $(P_E + t)$.

Product charges are not used at the federal level in the United States for pesticides or chemicals. The fees charged for pesticide registration under FIFRA are not incentive-based but, rather, driven by revenue needs to fund the pesticide program. However,

[25]Of course, production of virtually any product gives rise to some pollution. However, in these markets, the major environmental risk is linked to consumption. So for simplicity, we will assume no production externality.

FIGURE 19.3 Negative Externality in a Chemical Market

The consumption externality of chemical use is modeled as a negative *MEB*, which causes the *MSB* to lie *below* the *MPB*. In the absence of government intervention, equilibrium arises where *MPB* equals *MPC*, or Q_C, which is higher than the efficient output level, Q_E. To internalize the externality, a product charge could be implemented as a unit tax (t). To achieve efficiency, this tax should be set equal to $-MEB$ at Q_E. The *S* curve would shift up by the amount of the tax, raising the effective product price to ($P_E + t$).

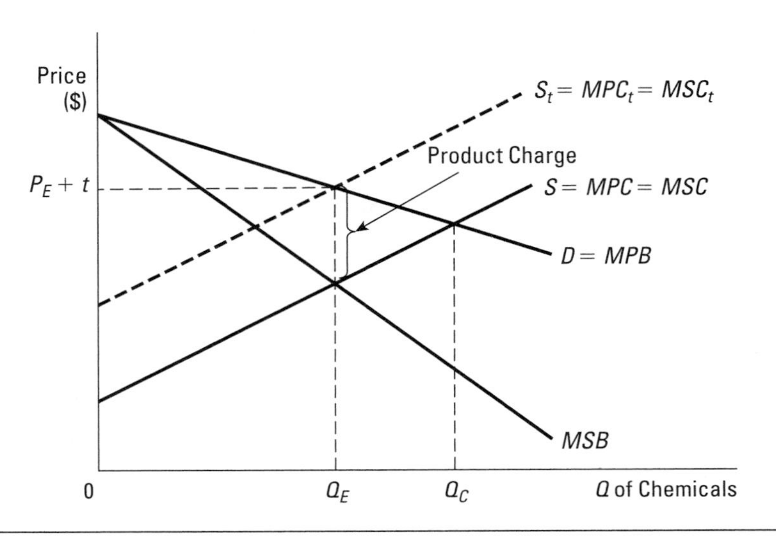

bona fide product charges are used by some state governments to discourage chemical use, and the same is true for other countries. For example, Norway has imposed a product change of \$7.50 per kilo on trichloroethylene (TCE) and perchloroethylene (PER), which are solvents. Belgium, Denmark, Norway, and Sweden use fees to control pesticide use. Norway's charge is implemented as a tax levied at 13 percent of purchase price, and Denmark's charge is set at 20 percent at the retail level. Based on an empirical estimate of demand elasticity, Denmark's tax should reduce pesticide use by 9 percent.[26]

CONCLUSIONS

Recognizing the potential risks of chemical exposure and the enormous costs of remediation efforts, the government has developed legislation aimed at controlling hazardous substances *before* they become part of the waste stream. Although many legislative acts contribute to this effort, two of the more significant are the Federal Insecticide, Fungicide, and Rodenticide Act (FIFRA) and the Toxic Substances Control Act (TSCA). Both of these are designed to prevent the production and use of substances posing an

[26]OECD (2006) Table 2.1, p. 27; U.S. EPA, Office of Policy, Economics, and Innovation (November 2004), pp. 20–21; (January 2001), p. 50.

unacceptable risk to society and to monitor the risks of those that have been introduced to the marketplace.

Although many pesticides and toxic substances are associated with health and ecological effects, both can, and do, contribute to society's well-being. Hence, it is appropriate that decision making under FIFRA and TSCA be guided by risk-benefit analysis. There are, however, practical problems that have hindered the realization of what this decision rule can achieve. For example, registration decisions for new pesticides are based on risk data submitted by manufacturers, but benefits are assumed to justify those risks based on the registrant's willingness to pay the registration fee. There are similar flaws in the implementation of TSCA, particularly those applicable to chemicals already on the market.

New programs have been developed to complement these laws and to integrate incentives to encourage the production and use of safer substitutes. This new direction fosters pollution prevention to displace some of the reliance on treatment and cleanup. According to this preventive approach, society needs to rethink how market decisions affect the size and toxicity of the waste stream. Firms must adjust how their products are designed, manufactured, and packaged, and households must modify consumption decisions. Pollution prevention is an important theme in the ongoing development of environmental policy—one of several issues discussed in the concluding module of this text.

Summary

- Current U.S. policy on pesticides is governed by the Federal Insecticide, Fungicide, and Rodenticide Act (FIFRA) and subsequent amendments. The chief regulatory instrument of FIFRA is the registration of new pesticides and the reregistration of those already on the market.
- In 1996, the Food Quality Protection Act (FQPA) was enacted, which amended FIFRA as well as the Federal Food, Drug, and Cosmetic Act (FFDCA), another law that regulates pesticides.
- A major objective of FIFRA is to register all pesticides before they are distributed or sold. Registration is based upon a risk-benefit analysis. At the time of registration, the EPA must set tolerances on the amount of a pesticide that may remain as a residue on food without causing an unacceptable health risk.
- Reregistration is a reevaluation process aimed at ensuring that formerly licensed pesticides meet current regulatory In standards. the final phase, the EPA issues a Reregistration Eligibility Decision (RED) document for each pesticide.
- The Pesticide Environmental Stewardship Program (PESP) is a voluntary program that establishes partnerships with pesticide users to implement preventive strategies. It promotes Integrated Pest Management (IPM), which fosters more selective use of pesticides and greater reliance on natural deterrents.
- For new pesticides, risks are evaluated from data submitted by manufacturers as part of the registration application. When the EPA conducts a special review of an existing pesticide, a formal benefit analysis is done.
- There is uncertainty about the risks associated with pesticide use. Part of the difficulty may lie in the difference between actual risk and perceived risk, which may cause an underallocation of resources for pesticide control.
- The primary benefits of pesticide use are reduced plant damage and increased crop yields, which are difficult to quantify on a broad scale.
- In 1976, Congress enacted the Toxic Substances Control Act (TSCA), which governs U.S. policy today. Its chief objective is to control chemicals that pose a risk before they are introduced into commerce. TSCA also monitors and regulates chemicals already on the market.
- Recently, the EPA announced six Essential Principles for Reform of Chemicals Management Legislation to help guide the reauthorization and strengthening of TSCA. The EPA is also launching a more comprehensive approach to its existing chemical management program.
- TSCA requires chemical producers to notify the government before they intend to produce or import any new chemical, using a premanufacture notice (PMN). For existing chemicals, TSCA requires manufacturers to notify the EPA if any chemical is found to present a substantial risk to human health or the environment.
- The Green Chemistry Program promotes the design of chemical products and processes that minimize or eliminate the production of hazardous substances. Another preventive program is called Extended Product Responsibility (EPR), which calls for all participants in the product cycle to find ways to reduce a product's effect on the environment.
- Resources may be underallocated to controlling chemical hazards because of differences between actual and perceived risk that may influence government environmental priorities.
- Estimating the benefits of chemical use is difficult because these benefits accrue to society in various ways and across multiple markets. No comprehensive benefit analysis has been undertaken for the major U.S. programs on toxic chemicals.
- Current regulations create a bias against the introduction of new chemicals because the government is more efficient in testing new substances than for those already on the market.
- In any pesticide or toxic chemical market, inefficiency arises from the presence of a negative consumption externality. To internalize this externality, a product charge could be used. To achieve efficiency, the charge should be set equal to $-MEB$ at the efficient output level.

Review Questions

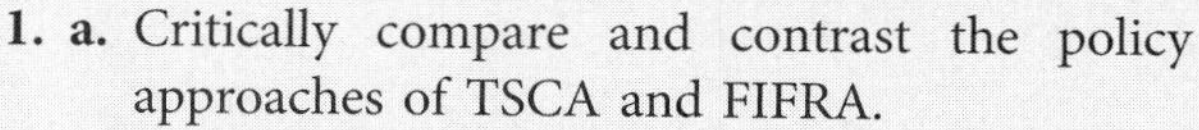

1. a. Critically compare and contrast the policy approaches of TSCA and FIFRA.
   b. In your view, which of these is more effective in preventing pollution? Explain.
2. Suggest two market-based instruments that would support the Pesticide Environmental Stewardship Program. Explain.
3. Suppose that a state government is contemplating a sales tax on pesticides. Assume that the following equations model the pesticide market before the sales tax is enacted, where P is measured in price per pound, and Q is in thousands of pounds.

$$D = MPB = 30 - 4Q$$
$$S = MPC = -2.50 + Q$$

   Now assume that an efficient sales tax of 50 cents per pound is enacted, which shifts the supply curve to $S' = MPC' = -2 + Q$
   a. Find the equilibrium price and quantity before and after the tax is implemented. Who pays the majority of the tax, suppliers or demanders? Explain why this makes economic sense.
   b. How much revenue does the sales tax generate for the state? Suggest some environmentally motivated uses for these revenues.
4. In 1976, the United States banned the manufacture of PCBs. Propose an alternative policy that would effectively reduce society's exposure to these cancer-causing substances. Support your proposal by discussing two of its strong points. Realistically, why might opponents argue against your idea?

5. Visit the Web site of the American Chemistry Council at **www.americanchemistry.com**.
   a. Summarize the economic benefits associated with any three chemical uses. Be as specific as possible.
   b. Now discuss how TSCA and the Green Chemistry Program help to reduce the risks associated with these uses.
6. Assume that a chemical company introduces a new compound, the production of which generates a negative externality. The environmental regulatory authority has estimated the following marginal costs and benefits for this new market:

$$MSB = 75 - 0.3Q \quad MPC = 25 + 0.2Q$$
$$MEB = 0 \quad MEC = 0.5Q$$

   where Q is annual output in thousands of pounds, and P is price per pound measured in dollars.
   a. For this new market, find the competitive equilibrium, Q_C and P_C, and the efficient equilibrium, Q_E and P_E.
   b. Now suppose that the regulatory authority wants to internalize the externality by imposing a product charge on the new compound. Calculate the dollar value of a product charge that would achieve the efficient solution. Graphically illustrate this product charge and the competitive and efficient solutions.

Additional Readings

Cropper, Maureen L., William N. Evans, Stephen J. Berard, Maria M. Ducla-Soares, and Paul R. Portney. "The Determinants of Pesticide Regulation: A Statistical Analysis of EPA Decision Making." *Journal of Political Economy 100(1)* (February 1992), pp. 175–97.

Delmas, Magali, and Ivan Montiel. "The Diffusion of Voluntary International Management Standards: Responsible Care, ISO 9000, and ISO 14001 in the Chemical Industry." *Policy Studies Journal 36(1)* (February 2008), pp. 65–94.

Galt, Ryan E. "Beyond the Circle of Poison: Significant Shifts in the Global Pesticide Complex, 1976–2008." *Global Environmental Change 18(4)* (October 2008), pp. 786–99.

Geiser, Kenneth. *Materials Matter: Toward a Sustainable Materials Policy*. Cambridge, MA: MIT Press, 2001.

Karlsson, Sylvia I. "Agricultural Pesticides in Developing Countries." *Environment 46(4)* (May 2004), pp. 22–42.

Selin, Noelle Eckley. "Mercury Rising." *Environment 47(1)* (January/February 2005), pp. 22–36.

Sterner, Thomas. "Trichloroethylene in Europe: Ban Versus Tax." In Winston Harrington, Richard D. Morgenstern, and Thomas Sterner, eds. *Choosing Environmental Policy: Comparing Instruments and Outcomes in the United States and Europe.* Washington, DC: Resources for the Future, 2004.

Thornton, Joe. *Pandora's Poison: Chlorine, Health, and a New Environmental Strategy.* Cambridge, MA: MIT Press, 2000.

Travisi, Chiara Maria, Peter Nijkamp, and Gabriella Vindigni. "Pesticide Risk Valuation in Empirical Economics: A Comparative Approach." *Ecological Economics 56(4)* (April 2006), pp. 455–74.

van Leeuwen, C. J., and T Vermeire, eds. *Risk Assessment of Chemicals: An Introduction.* 2nd ed., Dordrecht, The Netherlands: Springer, 2007.

Vogel, Sarah. "Why the Toxic Substances Control Act Needs an Overhaul, and How to Strengthen Oversight of Chemicals in the Interim." *Health Affairs 30(5)* (May 2011), pp. 898–905.

Wilson, James D. "Resolving the 'Delaney Paradox.'" *Resources 123* (Fall 1996), pp. 14–17.

A Reference to Acronyms and Terms in Toxic Substances Control Policy

Environmental Economics Acronyms

MEB	Marginal external benefit
MPB	Marginal private benefit
MPC	Marginal private cost
MSB	Marginal social benefit
MSC	Marginal social cost

Environmental Science Acronyms

DDT	Dichloro-diphenyl-trichloroethane
EDB	Ethylene dibromide
PCBs	Polychlorinated biphenyls
PER	Perchloroethylene
TCE	Trichloroethylene

Environmental Policy Acronyms

ACC	American Chemistry Council
EPR	Extended Product Responsibility
FDA	Food and Drug Administration
FFDCA	Federal Food, Drug, and Cosmetic Act
FIFRA	Federal Insecticide, Fungicide, and Rodenticide Act
FQPA	Food Quality Protection Act
ICCA	International Council of Chemical Associations
IPM	Integrated Pest Management
NAS	National Academy of Sciences
NRC	National Research Council
OSHA	Occupational Safety and Health Administration
PESP	Pesticide Environmental Stewardship Program
PMN	Premanufacture notice
PRIA 1	Pesticide Registration Improvement Act of 2003
PRIA 2	Pesticide Registration Improvement Renewal Act
RCRA	Resource Conservation and Recovery Act
RED	Reregistration Eligibility Decision
TSCA	Toxic Substances Control Act
USDA	U.S. Department of Agriculture

To access additional course materials, visit www.cengagebrain.com. At the home page, search for the ISBN of this title (shown on the back cover). This will take you to the product page where these resources can be found.

MODULE 7 Global Environmental Management

With time and education, society has become more mindful of environmental damage and its consequences and has recognized that economic advance and industrialization are largely responsible for the pollution problems it confronts. At the outset, such a realization seemed to suggest an unacceptable trade-off between competing social objectives—economic prosperity and environmental quality. Nonetheless, most saw the need for policy initiatives that would improve environmental quality and reduce the risk of exposure to hazards like urban smog, polluted waterways, and abandoned hazardous waste sites.

The thrust of the initial policy response was to use end-of-pipe pollution controls. Although this approach has achieved some success, public officials have been concerned about the high costs, the unattained policy goals, and the conflict between economic growth and environmental protection. In response, incentives are displacing some command-and-control instruments, integrated programs are replacing some pollutant- and media-specific initiatives, and prevention is superseding treatment as a waste management option. Taken together, these trends speak to a global transition in policy development toward broader, more long-term solutions to environmental damage.

We have reason to be optimistic about this new direction of environmental policy. The wisdom of experience coupled with scientific and technological advances can bring about significant change. Perhaps more importantly, environmental objectives have broadened to consider the future along with the present and to accommodate global interests along with national and local needs. Such is the fundamental premise of **sustainable development**—a goal that integrates economic prosperity with environmental preservation, as a legacy to future generations.

In this final module, we examine environmental management from a global perspective, with an eye toward the objective of sustainable development. Although much has yet to be defined, several themes are beginning to emerge. Among these are efforts to advance environmental literacy, to redefine policy toward prevention, to reconcile trade objectives with environmental goals, and to develop domestic and international agreements that facilitate common objectives.

In Chapter 20, we present an overview of sustainable development and the implications of population growth on the environment. We also examine the intent of international agreements to control transboundary pollution, and the interaction of international trade and environmental issues. In Chapter 21, we deal with implementation of these efforts, discussing evolving themes that are becoming long-run environmental approaches—industrial ecology and pollution prevention. Mindful of what has been accomplished and what remains to be done, we explore these themes that characterize what many view as an important transition in global environmental management.

CHAPTER 20

Sustainable Development: International Environmental Agreements and International Trade

"Far from being a burden, sustainable development is an exceptional opportunity—economically, to build markets and create jobs; socially, to bring people in from the margins; and politically, to give every man and woman a voice, and a choice, in deciding their own future."
—Kofi Annan (1938–)

Direct regulation of economic activity has characterized much of the evolution of environmental policy in most nations around the world. While this command-and-control approach has contained the problem, most recognize that reliance on end-of-pipe controls does not adequately address the long-term implications of environmental damage. Furthermore, this approach does not accommodate the broader objective of **sustainable development**—achieving environmental quality *and* economic prosperity, which is particularly difficult for third world or developing nations.

The comprehensive goal of sustainable development calls for fundamental changes in how society makes market decisions. The challenge is to achieve economic prosperity but alter market activity so that natural resources and the environment are protected. Effecting changes of this magnitude calls for a different policy approach than one that relies on rules and limits—control instruments that often run counter to the polluter's market incentive. If society is to sustain a long-term commitment to preserving the earth, there has to be a motivation to do so beyond the avoidance of penalties for regulatory noncompliance.

Logically, the motivation should be consistent with economic incentives. The premise is that economic growth and environmental quality can be reinforcing rather than competing objectives. Perceptions must be changed to recognize that resource conservation and pollution abatement can enhance private as well as social interests. Communication must be improved—both within and among nations—to share information about technologies and processes that can protect the environment without diminishing profitability. If successful, cooperation should displace what is sometimes an adversarial position between the public and private sectors of society, and there should be less reliance on costly monitoring and enforcement procedures.

In this chapter, we explore this transitional phase in environmental policy, focusing on objectives and international agreements. We begin by discussing the concept of sustainable development and explore its motivation by studying the implications of economic growth on environmental quality. We then examine selected international conferences and agreements to determine how these facilitate achieving the dual objectives of economic growth and the realization of environmental quality. Finally, we assess international trade issues and examine how trade objectives can conflict with environmental goals.

SUSTAINABLE DEVELOPMENT AS A GLOBAL OBJECTIVE

Redefining Environmental Objectives

environmental quality
A reduction in anthropogenic contamination to a level that is "acceptable" to society.

Over the last several decades of environmental policy development, the clear focus has been to achieve **environmental quality** by reducing anthropogenic pollution to a level that is "acceptable" to society. Policymakers have struggled with the tough issue of how clean is clean and the challenge of devising cost-effective instruments to achieve whatever quality level has been set. Through revisions, political debate, and a growing social consciousness, there has been progress and some needed adjustments in policy development. Yet there also has been an increasing awareness of the need for change that is more fundamental than periodic revisions of legislative provisions—a redefinition of environmental policy objectives.

Moving Toward Sustainable Development

sustainable development
Management of the earth's resources such that their long-term quality and abundance are ensured for future generations.

A consensus is forming that both public and private decision making should be driven by a broader goal that is global in scope and dynamic in perspective. One such goal is **sustainable development**—managing the earth's resources such that their long-term quality and abundance are ensured for future generations, which is discussed at length at **www.epa.gov/sustainability**. This objective, posed as a pressing issue at the 1992 Rio Summit, is also referred to as **intergenerational equity**. In any case, what this goal makes clear is that environmental policy must consider the long run. Just as it is foolhardy to pursue economic growth without regard for the environmental implications, so too is it irrational to pursue environmental objectives that ignore the future economic consequences.

Sustainable Development in Practice

To be fair, not all the commentary on sustainable development or sustainable growth is entirely favorable. One common criticism, expressed by Nobel Laureate Robert Solow (1991) and others, is that the concept itself is vague and hence not a good notion to guide policy. Furthermore, and more importantly, the idea of each generation leaving the earth as they found it for future generations is not feasible and, in some instances, not even desirable. Consider that such a premise, followed literally, would mean that society should engage in no construction activity and should use no depletable resources, such as copper or oil. Notwithstanding a rational recognition of scarce resources, such inactivity does not make sense.

As an alternative, Solow (1991) argues that sustainability ought to be considered as:

> ... *an obligation to conduct ourselves so that we leave to the future the option or the capacity to be as well off as we are.* (p. 132)

Even at that, there is still the difficulty of attempting to forecast both the preferences and the technological capability of future generations, which is no easy task by any measure. Solow also asserts, though somewhat controversially, that sustainability need not mean preservation of a particular species or tract of land. Rather, substitutability ought to be allowed in applying the notion of intergenerational equity, or distributional equity over time, to policy prescriptions. Lastly, he argues that if policy is to be guided by this idea of intergenerational equity, to be consistent, it must also consider *intragenerational* equity, with clear effort expended toward reducing poverty today.

Note that the practicality of Solow's ideas does not dispel the importance of recognizing the obligation to future generations. Nor does it minimize the challenge of doing so in the face of population growth. This reality is, of course, more of an issue for developing nations, which means we need to consider the ramifications of growth a bit more carefully.

Implications of Growth on the Environment

Sustainable development is based on the premise that economic growth and environmental quality must be reconciled. Statistical estimates on worldwide population and income growth help to explain why this is important.

What the Data Imply

According to one source, per capita income levels have to grow by at least 2 percent per year to reduce world poverty and close the gap between the rich and the poor. This communicates the importance of achieving economic growth. Furthermore, world population is growing at about 1.7 percent each year, a rate that is expected to decline only very slowly. Recognizing that economic growth has an effect on the environment, these data imply that the associated environmental impact per unit of income must decline at a rate between 3.5 and 4 percent per year to avoid further pollution and natural resource depletion.[1]

To generalize this assertion, we can characterize the relationship among population growth, income growth, and the environment at a point in time as follows:[2]

$$\begin{matrix}\text{Environmental} \\ \text{Impact}\end{matrix} = \begin{matrix}\text{Income} \\ \text{per capita}\end{matrix} \times \begin{matrix}\text{Environmental impact} \\ \text{per unit of income}\end{matrix} \times \text{Population}$$

Developing countries face more urgent conditions, struggling to advance economically to accommodate a rapidly rising population and at the same time confronted with the environmental contamination that these pursuits have exacerbated. For example, available projections estimate annual population growth through 2050 for the fastest growing nations as: 3.19 percent for Niger, 3.09 percent for Yemen, 3.03 percent for

---

[1]Nitze (April–June 1993).

[2]This relationship is just an identity, which is more apparent when the equation is rewritten using ratios, as follows:

$$\text{Environmental Impact} = \frac{\text{Income}}{\text{Population}} \times \frac{\text{Environmental Impact}}{\text{Income}} \times \text{Population.}$$

Somalia, and 2.96 percent for Uganda.[3] As population grows, so too does the demand for goods and services. A case in point is the expected increase in energy consumption for China and India, which together accounted for some 21 percent of energy use worldwide in 2008, and is expected to more than double between 2008 and 2035, reaching 31 percent of world energy use in 2035.[4] If an economy's productive capacity cannot accommodate the population growth, shortages arise and resources are misused in an attempt to compensate for the imbalance.[5]

Environmental Kuznets Curve

environmental Kuznets curve
Models an inverted ∪-shaped relationship between economic growth and environmental degradation.

The implications of the relationship between economic growth and environmental quality have motivated numerous studies. Some of these examine the possibility of a technical relationship or pattern between economic development (commonly measured as income per capita) and environmental degradation. Known as the **environmental Kuznets curve**, this pattern has been hypothesized as an inverted ∪ shape, as shown in Figure 20.1.

Such a model suggests that early stages of industrialization are associated with rising levels of pollution, when growth is a greater priority than natural resource protection and environmental controls are lenient or virtually nonexistent. It further implies that more advanced development is linked to a shift in focus in the opposite direction with increasing

FIGURE 20.1 The Environmental Kuznets Curve

The environmental Kuznets curve models a theorized relationship between economic development and environmental degradation that can be depicted as an inverted ∪ shape. This model suggests that early stages of industrialization are associated with rising levels of pollution, and that more advanced development is linked to an increasing concern for environmental quality and an associated strengthening of environmental regulation.

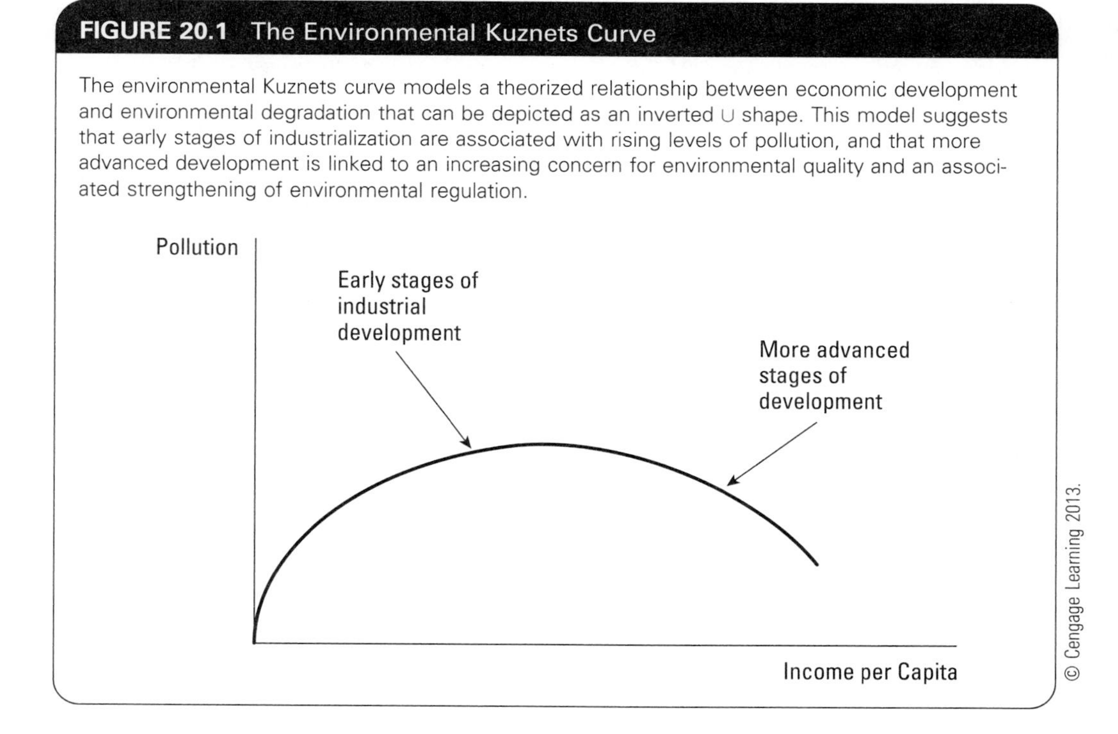

[3]United Nations, Department of Economics and Social Affairs, Population Division (2004), Table 6.
[4]U.S. Department of Energy, Energy Information Administration (September 2011), pp. 9–10.
[5]For a classic article illuminating the environmental implications of population growth, see Hardin (1968).

concern for environmental quality, essentially an income effect, and a concomitant strengthening of environmental regulation. The implications of this theory are far-reaching and have motivated debate among empirical researchers, as Application 20.1 explains.[6] That said,

APPLICATION 20.1 The Environmental Kuznets Curve

An interesting hypothesis in the environmental economics literature is that the relationship between pollution and economic development might look like an inverted U. Known as the **environmental Kuznets curve (EKC)**, this relationship implies that as a society begins to develop, it focuses more on growth than on the environment. Put another way, there is a trade-off between growth and environmental quality in the early stages of a society's development. As a society becomes more advanced, it recognizes the importance of preserving the environment and has the resources to begin implementing environmental policy. Thus, at some point, the trade-off ceases to exist. Beyond that point, environmental quality may actually improve with further growth.

Not surprisingly, the EKC has prompted debate among environmental economists. Even among those who accept the basic notion of a trade-off between growth and environmental quality, there is uncertainty about the extent of the trade-off and the point at which the trade-off ends. Simply put, exactly where is the turning point of the inverted U, and what is the shape of the curve up to and beyond that point? Such questions are part of the motivation for empirical research in this area.

As the findings have begun to accumulate, a divergence of opinions has become apparent. A recent article by Dasgupta et al. (2002) presents an overview of the literature, which suggests three different views about the shape of the EKC. Each is shown in the accompanying figure, identified as **conventional**, **pessimistic**, and **optimistic**.

The **conventional** view depicts the standard inverted U shape. In general, empirical results show that the turning point occurs between \$5,000 and \$8,000 income per capita. For example, Grossman and Krueger (1995) estimate this point as falling between \$4,000 and \$6,000 for air pollution measured by sulfur dioxide (SO_2) and particulate matter (PM). Yet, according to Stern (1998), who reviewed the literature, such findings seem to arise only for certain pollutants.

Less prevalent is the **pessimistic** view, which suggests that beyond a critical income level, the best a nation can do is maintain some existing level of pollution. In fact, some findings suggest that the EKC may not ever flatten out but continue to increase with economic growth.

At the other extreme is the **optimistic** view, which implies that the trade-off is not as severe (i.e., the curve is flatter), it does not persist (i.e., the turning point occurs at lower levels of income), and the extent of environmental damage is lower at each level of development (i.e., the curve is everywhere below its conventional counterpart). This suggests that the EKC might be shifting downward. Dasgupta et al. (2002) indicate that this depiction is the most likely. Possible causes include increasing effectiveness of environmental regulation, improvements in abatement technology, and greater public awareness of pollution and its effects.

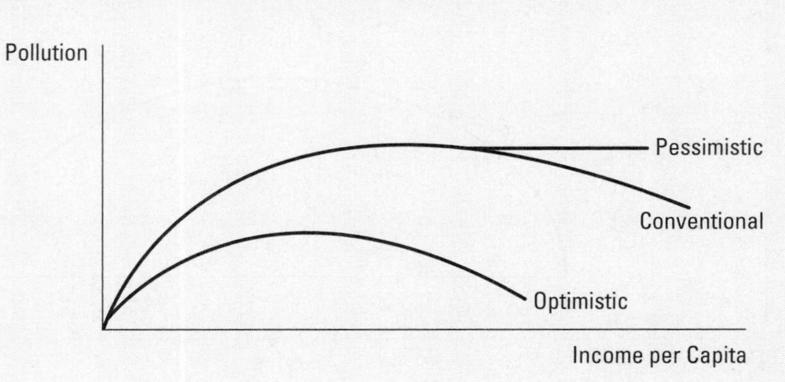

Sources: Dasgupta et al. (Winter 2002); Stern (May 1998); Grossman and Krueger (May 1995).

[6]Also, see Harbaugh, Levinson, and Wilson (August 2002).

a consensus seems to be forming that the trade-off between economic growth and environmental quality may not be as severe as once feared. This in turn suggests that progress in reconciling these goals may be realized.

GLOBAL FRAMEWORK FOR SUSTAINABLE DEVELOPMENT

Sustainable development is intended to be a global objective, the benefits of which should accrue to all segments of society and to all nations. Because of its intent and its pervasive implications, it calls for a collaborative effort from all stakeholders. Fundamental to such an effort is communication on a global scale, which can be facilitated at conferences that set an agenda and establish a framework for achieving growth and improving the environment.

United Nations Conference on Environment and Development (UNCED)[7]

More commonly called the Rio Summit, the U.N. Conference on Environment and Development (UNCED) was a 12-day worldwide forum held in Rio de Janeiro, Brazil, in June 1992. Thousands of delegates from over 170 nations attended the event to discuss issues and concerns dealing with sustainable development. Among the major documents produced from the Rio Summit are *Agenda 21* and the *Rio Declaration*. A brief description of each follows.

Agenda 21

This 40-chapter document is a voluntary action plan, outlining the course for worldwide progress toward sustainable development. Online access to the text of this document is available at **www.un.org/esa/dsd/agenda21/**. Major issues covered within the 900 pages of *Agenda 21* include:

- Financing for developing countries
- Conservation and sustainable development for forests
- Prevention and minimization of hazardous and solid wastes
- Risk assessment and management of toxic chemicals

Perhaps the most important achievement of the Rio Summit, *Agenda 21* is the result of 2½ years of negotiations to reach an international consensus.

Rio Declaration

The *Rio Declaration* outlines 27 principles to act as guidelines for achieving global environmental quality and economic development.[8] Among these principles is a requirement that environmental protection be an integral part of development and a call for a reduction in unsustainable production and consumption. Reportedly, the declaration is a compromise from what was originally anticipated. The preparatory meetings were

---

[7]The following discussion is drawn from Sessions (April–June 1993), p. 12; Parson, Haas, and Levy (October 1992); Reilly (September/October 1992).

[8]The full text of the *Rio Declaration* is available online at **www.un.org/documents/ga/conf151/aconf15126-1annex1.htm**.

intense, often immersed in seeking a balance among the views of developing and industrialized countries. So fragile was this agreement that it was adopted in Rio without further negotiation, for fear that further discussion would jeopardize what was left of the intended Earth Charter.

World Summit on Sustainable Development (WSSD): Johannesburg 2002[9]

In August–September 2002, 10 years after the Rio Summit, more than 20,000 participants converged in Johannesburg, South Africa, at the World Summit on Sustainable Development (WSSD). Heads of state, national delegates, and representatives from industry, trade unions, and the scientific community attended this major event. The objective was to renew worldwide interest in sustainable development and to assess progress achieved since the 1992 Rio Summit.

Among the summit's reported accomplishments were the following:

- Adoption of a declaration and implementation plan in confirmation of the commitment to fully implement *Agenda 21*
- Strengthening of the notion of sustainable development and the relationships among resource use, poverty, and the environment
- Establishment of over 300 partnership initiatives to complement government actions aimed at achieving sustainable development

By the conclusion of the worldwide meeting, participating nations made commitments to a variety of efforts in what are called the five WEHAB initiatives: water and sanitation, energy, health, agriculture, and biodiversity. For example, the European Union announced a new program aimed mainly at achieving water and sanitation objectives in Africa and Central Asia, and the United States committed $90 million to support programs in sustainable agriculture.

While seemingly encouraging, these accomplishments were not without critics. Reportedly, heads of state in attendance at the worldwide meeting believe that the implementation plan is too weak. Moreover, some of the action plans apparently will be difficult to monitor, erring on the side of generality and lacking specific timetables. If accurate, the summit's objectives may not be met, and the associated initiatives may be disappointing.

Rio+20—United Nations Conference on Sustainable Development

To recognize the twentieth anniversary of the Rio Summit, officials organized the Rio+20—United Nations Conference on Sustainable Development (UNCSD), which was held in Brazil in 2012. The overarching goal of this conference was to reinforce political commitment to sustainable development, to examine progress to date, and to face new challenges. To help achieve its objectives, two themes were defined to focus discussions at Rio+20:

- To examine greening markets and a green economy within a sustainable development context

[9]Drawn from United Nations (January 2003); Speth (January/February 2003).

- To address opportunities to strengthen the institutional framework for sustainable development

For updates and further information on the Rio+20 summit, visit **www.uncsd2012.org/rio20/index.php?menu=14.**

INTERNATIONAL AGREEMENTS TO CONTROL TRANSBOUNDARY POLLUTION

international externality
A spillover effect associated with production or consumption that extends to a third party in another nation.

Some pollution problems are transboundary, such as acidic deposition, ozone depletion, global warming, and some surface water pollution. This means that contamination in one nation can travel beyond its borders—a type of **international externality**. For example, air pollution associated with electricity generation in Canada near the U.S. border imposes external costs that extend beyond Canada's borders to Americans who live near the border but are not part of the market transaction. This transboundary air pollution is modeled in Figure 20.2.

Notice in the figure that the marginal private benefits (MPB_{Canada}), assumed equal to the marginal social benefits (MSB_{Canada}), accrue only to Canadian consumers. The production externality, however, which is captured by the marginal external cost ($MEC_{Canada\ and\ United\ States}$), accrues to both Canadian and American society. Just as in

FIGURE 20.2 International Externality of Transboundary Pollution

In the case of transboundary pollution from Canada to the United States, the marginal private benefits (MPB_{Canada}), assumed equal to the marginal social benefits (MSB_{Canada}), accrue only to Canadian consumers. The production externality, however, which is captured by the marginal external cost ($MEC_{Canada\ and\ United\ States}$), accrues to both Canadian and American society. The outcome is overallocation of Canadian resources at the competitive output level, Q_C, which is greater than the efficient output level, Q_E.

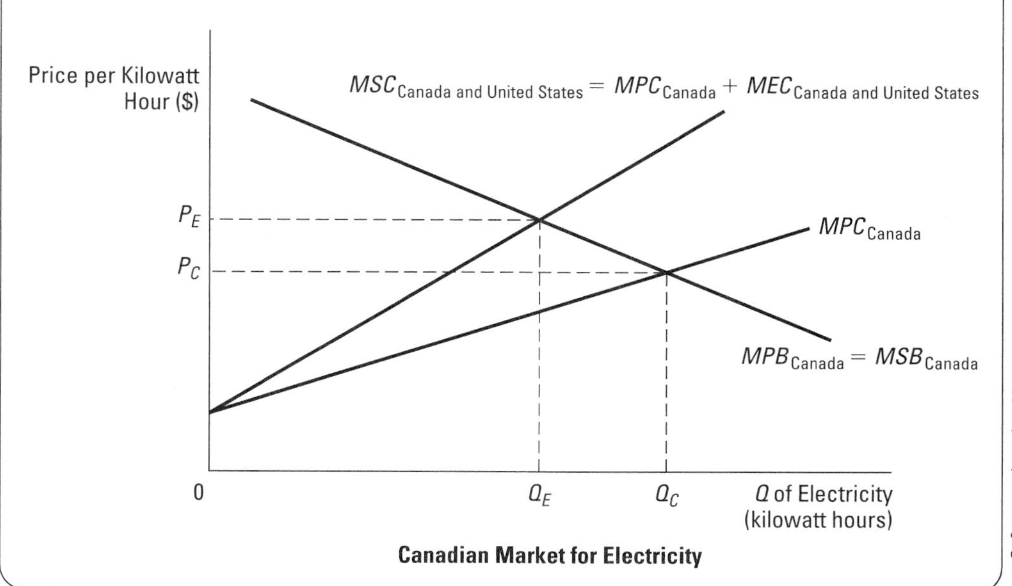

more conventional externality models, the outcome is overallocation of Canadian resources at the competitive output level, Q_C, which is greater than the efficient output level, Q_E. The difference here is that solutions are more complex. Specifically, formal treaties must be negotiated and agreed to by all affected countries to achieve a resolve. Certain of these agreements we have discussed in other contexts, while others are introduced here for the first time.

Montreal Protocol and Amendments

Originally signed by 24 countries in September 1987, the Montreal Protocol on Substances that Deplete the Ozone Layer is an important example of international cooperation aimed at environmental protection. Strengthened by a series of amendments, this treaty is aimed at phasing out **chlorofluorocarbons (CFCs)** and other ozone-depleting substances. A summary of the phaseout deadlines, as they evolved over time, is presented in Table 20.1. As of 2012, 197 nations have ratified the Montreal Protocol.[10]

chlorofluorocarbons (CFCs)
A family of chemicals believed to contribute to ozone depletion.

Recognizing the special problems of developing nations and the importance of gaining their support, industrialized countries set up a $160 million Interim Multilateral

TABLE 20.1 Montreal Protocol Phaseout Schedules for Production and Consumption of Ozone Depleters

	Developed Nations	Developing Nations
CFCs	Phase out by 1996	Freeze at average 1995–1997 levels by July 1, 1999; reduce by 50 percent by 2005, by 85 percent by 2007; phase out by 2010
Carbon tetrachloride	Phase out by 1996	Freeze at average 1995–1997 levels by July 1, 1999; reduce by 50 percent by 2005, by 85 percent by 2007; phase out by 2010
Halons	Phase out by 1994	Freeze at average 1995–1997 levels by July 1, 1999; reduce by 50 percent by 2005, by 85 percent by 2007; phase out by 2010
HCFCs	Reduce by 35 percent by 2004, by 75 percent by 2010, by 90 percent by 2015; phase out by 2020 (allowing 0.5 percent for servicing reasons between 2020 and 2030)	Freeze at average 2009–2010 levels by 2013; reduce by 10 percent by 2015, by 35 percent by 2020, 67.5 percent by 2025; phase out by 2030 (allowing 2.5 percent annual average for servicing reasons between 2030 and 2040)
Methyl bromide	Reduce by 25 percent by 1999, by 50 percent by 2001, by 70 percent by 2003; phase out by 2005	Freeze at average 1995–1998 levels by 2002; reduce by 20 percent by 2005; phase out by 2015
Methyl chloroform	Phase out by 1996	Freeze at average 1998–2000 levels by 2003; reduce by 30 percent by 2005, by 70 percent by 2010; phase out by 2015
HBFCs	Phase out by 1996	Phase out by 1996
BCM	Phase out by 2002	Phase out by 2002

For more detail on the Montreal Protocol or any of the subsequent amendments, visit the Ozone Secretariat at the United Nations Web site at **http://ozone.unep.org/new_site/en/index.php**.

NOTE: Phaseouts include changes through the 2007 Montreal Adjustment.

Source: United Nations Environment Programme (September 2008).

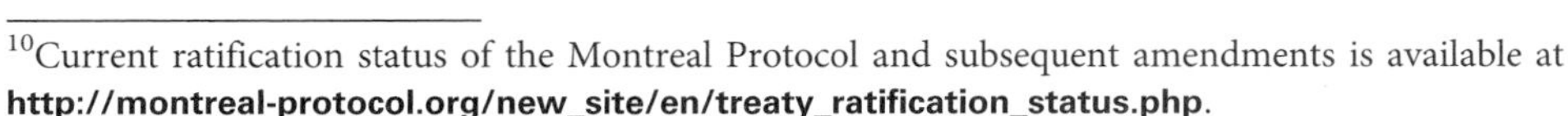

[10]Current ratification status of the Montreal Protocol and subsequent amendments is available at **http://montreal-protocol.org/new_site/en/treaty_ratification_status.php**.

Fund to assist them in transitioning toward CFC substitute technologies. Developing countries also were given a 10-year grace period to meet the targets of the agreement. The **Multilateral Fund** became permanent in 1992 and is replenished every three years. The replenishment amount for 2009–2011 is $400 million, and the budget for this period is $490 million.[11]

Important accomplishments for the Montreal Protocol were the addition of China to the group of signatories in 1991 and India in 1992. Both nations have high economic growth rates and very large populations, which means that their commitments to the protocol are critical. Specifically, China's 2000 population was estimated at 1.275 billion and is projected to grow to 1.395 billion by 2050. For India, the 2000 population was estimated at 1.017 billion, with a forecast of 1.531 billion by 2050.[12]

U.N. Framework Convention on Climate Change (UNFCCC)

greenhouse gases (GHGs)
Gases collectively responsible for the absorption process that naturally warms the earth.

Now adopted by 195 nations, including the United States, the U.N. Framework Convention on Climate Change (UNFCCC) established a baseline for a global cooperative response to climate change. A key provision of the UNFCCC, and one that prompted much contention, was a commitment by each signatory to implement a national strategy to limit the release of **greenhouse gases (GHGs)**. Each strategy was to support the common objective of reducing emissions to 1990 levels by 2000. In March 1994, the treaty became legally binding, following its ratification by the requisite number of 50 countries in 1993.[13]

Kyoto Protocol to the UNFCCC

At a Conference of the Parties (COP) held in Japan in December 1997, participants reached an agreement, or protocol, about GHG emissions beyond 2000. Specifically, GHG emission targets were set for developed nations. These were to become effective when 55 nations ratified the protocol, as long as these nations included developed countries responsible for at least 55 percent of CO_2 emissions for 1990. Achieving these limits was to be accomplished in part through emissions trading.

Although the United States signed the protocol in November 1998 during the fourth COP in Buenos Aires, its action was merely symbolic. President Bush did not submit the protocol to the Senate for ratification, because of growing opposition to the treaty on Capitol Hill. A major concern was that the mandated emission limits would significantly hurt American industry and hence the economy. Moreover, the United States was holding firm to its commitment not to ratify the Kyoto Protocol until developing nations adopted binding emissions limits along with their industrialized counterparts.

Before the seventh session of the COP convened in Morocco, the president took the United States out of the Kyoto treaty in March 2001.[14] The U.S. decision sparked

[11]As of 2011, 45 industrialized nations, including Countries with Economies in Transition, or CEIT countries, have contributed over $2.77 billion to this fund. See United Nations, Secretariat of the Multilateral Fund for the Implementation of the Montreal Protocol (September 17, 2011), available at **www.multilateralfund.org/default.aspx**.

[12]United Nations, Department of Economic and Social Affairs, Population Division (2004).

[13]UNFCCC (September 18, 2011a; September 18, 2011b).

[14]Refer back to Application 13.2 for more detail on the U.S. decision to withdraw.

concerns about the future of the treaty, particularly since the United States is responsible for about 25 percent of worldwide GHG emissions. Nonetheless, 178 countries reached an international climate change accord later in 2001. Ratification was achieved in 2004 following Russia's signing of the accord, which brought the total number of participating nations to 141. The Kyoto Protocol called for 38 developed nations to reduce GHG emissions to 5.2 percent below their 1990 levels by 2012, the official expiration of the first commitment phase, with no targets imposed on developing countries. The emissions targets specified by the Kyoto treaty were to be achieved using several market-based instruments, called **flexible mechanisms**. Among these is a trading system of GHG allowances for participating developed nations. Additional information on the key provisions of the Kyoto Protocol, including the flexible mechanisms, is presented in Table 20.2.

At the 2011 COP in Durban, South Africa, nearly 200 nations agreed to extend the Kyoto Protocol and to meet at a later date to develop a successor agreement by 2015. This new agreement is to involve *all* nations that are major GHG emitters—both developed and developing countries.[15] To learn more about the UNFCCC, the Kyoto Protocol, and subsequent COPs, visit **http://unfccc.int/2860.php**.

London Convention 1972 (LC72) and the 1996 Protocol

Ocean dumping of certain wastes, including radioactive wastes, is prohibited by the London Convention. Originally decreed in November 1972, the LC72 has been ratified

TABLE 20.2 Key Elements of the Kyoto Protocol to the UNFCCC

- Developed nations must reduce overall greenhouse gas (GHG) emissions by at least 5.2 percent below 1990 levels in the first commitment period, 2008–2012, by meeting individual emissions targets assigned to each country. Negotiations for subsequent commitment periods will follow.
- Emissions targets cover the primary GHGs: carbon dioxide (CO_2); methane (CH_4); nitrous oxide (N_2O); hydrofluorocarbons (HFCs); perfluorocarbons (PFCs); and sulfur hexafluoride (SF_6).
- Emissions targets are to be achieved by international emissions trading of GHG allowances.
- Carbon-absorbing activities, such as reforestation, can be used to offset emissions targets. Any GHGs removed through such activities earn credits called removal units (RMUs).
- Under the *Clean Development Mechanism* (*CDM*), a developed nation can enter into emissions-reducing projects in a developing nation and use the resulting certified emissions reductions (CERs) to meet its own emissions targets.
- Using *Joint Implementation*, a developed nation can implement an emissions-reducing project in another developed nation and use the resulting emissions reduction units (ERUs) against its own emissions target.
- Participants failing to meet their emissions targets must resolve the difference in the second commitment period plus pay a 30 percent penalty.

NOTE: The full text of the Kyoto Protocol is available online at **http://unfccc.int/resource/docs/convkp/kpeng.pdf**.

[15]Center for Climate and Energy Solutions (C2ES) (2011); Whiteman (2011).

by 87 nations, including the United States. In fact, all nations that have recently engaged in ocean dumping of radioactive wastes are contracting parties to the LC72. This treaty has been amended by the 1996 Protocol, which was ratified by the requisite number of nations in 2006. As such, it will eventually replace the LC72. A discussion of the key changes effected by the 1996 Protocol is the subject of Application 20.2.

acid rain
Arises when sulfuric and nitric acids mix with other airborne particles and fall to the earth as precipitation.

United States–Canada Air Quality Agreement[16]

In March 1991, the United States–Canada Air Quality Agreement was finalized to combat the problem of **acid rain** and visibility impairment. Along with this agreement, the two nations negotiated the Acid Rain Annex, which set specific goals for reducing

APPLICATION 20.2 An International Treaty on Ocean Dumping

Among the first global conventions to protect the ocean from anthropogenic pollution, the London Convention of 1972 (LC72), (formerly known as the London Dumping Convention) is an international agreement to control the practice of dumping wastes at sea. In force since 1975, it is the only such convention to which the United States is a signatory nation. As of 2011, there are 87 parties to the LC72, and these include the major industrialized countries that generate most of the wastes being dumped into the sea.

The LC72 controls ocean dumping through various means. It bans the ocean disposal of so-called black-listed substances like mercury and cadmium. The disposal of "gray-listed," or less damaging, substances is controlled through special permits, and the dumping of any other substances is managed through a general permitting system.

Starting in the early 1990s, the parties to the convention began a detailed review of the LC72, a process that had been encouraged by *Agenda 21*. As this review proceeded, a number of important amendments became legally binding. These included a ban on the disposal of all radioactive wastes; the phaseout of all industrial waste dumping by January 1, 1996; and a ban on incineration at sea of industrial wastes and sewage sludge.

By 1996, the review of the convention rules was finished, and the 1996 Protocol to the LC72 was adopted. This accord, which explicitly acknowledges international agreements executed at the Rio Summit, has been ratified by the requisite signatories and entered into force in 2006. For a time, the LC72 and the 1996 Protocol will govern in parallel, but eventually, the 1996 Protocol will replace the LC72. As of 2011, there are 40 parties to the protocol. To review this list of countries as well as those nations that ratified the original London Convention, visit **www.imo.org/OurWork/Environment/SpecialProgrammesAndInitiatives/Pages/London-Convention-and-Protocol.aspx**.

Not surprisingly, the protocol is a more stringent agreement than its predecessor. Among the more significant revisions under the new rulings is the use of a "reverse list" approach. This means that instead of listing those materials banned from dumping as under the LC72, the 1996 Protocol prohibits all dumping unless specifically identified as permitted. In addition, all waste incineration at sea is prohibited under Article 5 of the protocol, as is the export of wastes intended for ocean dumping or incineration. As participating nations implement the protocol's rulings, the onus is upon them to find alternatives to waste disposal at sea that will allow for industrial advance without transferring the environmental damage from the ocean to other natural resources.

In November 2006, the parties to the protocol adopted the 2006 Amendments to the 1996 Protocol, which address carbon sequestration. Specifically, these amendments state that storage of carbon dioxide (CO_2) under the seabed will be permitted after February 2007, the date the amendments entered into force. This agreement acts as a basis from which to guide and regulate carbon capture and storage in sub-seabed geological formations.

Sources: International Maritime Organization, Office for the London Convention and Protocol (2011a; 2011b; 2003a; 2003b); U.S. Congress, OTA (1987), pp. 73, 149.

[16]The following is drawn from U.S. EPA, Office of Air and Radiation (October 2010).

sulfur dioxide (SO_2) and nitrogen oxides (NO_X). The bilateral provisions also include directives to establish a forum for addressing other transboundary air quality issues. Among these are rulings for each nation to do research on common air pollution problems and stipulations to facilitate a regular information flow about monitoring efforts, effects of air pollutants, and control methodologies.

The U.S. commitment to these concerns is confirmed by the 1990 Clean Air Act Amendments, which refer specifically to Canada's acid rain program. The statutes call for regular reporting to Congress on the SO_2 and NO_X emissions of all Canadian provinces participating in Canada's control program. Access to recent progress reports as well as to the text of the agreement is available online at **www.epa.gov/airmarkt/progsregs/usca/index.htm**.

According to a recent progress report, both Canada and the United States report important advances relative to the Acid Rain Annex. In 2008, Canada's SO_2 emissions were 1.7 million metric tons, which is 47 percent below 1990 levels and more than 63 percent lower than the nation's 1980 emissions. Similarly, SO_2 emissions for the United States in 2008 were 51 percent lower than 1990 levels. For the U.S. electricity industry, SO_2 emissions in 2009 were 5.7 million tons, which was a significant drop from the 7.6 million tons released in 2008 and much lower than the 2010 cap of 8.95 million tons. Reportedly, part of these emissions reductions were attributable to lower electricity demand in 2009 along with more advanced pollution control technology.

Regarding NO_X emissions, Canada met its reduction target to lower emissions from major sources, including power plants, by more than 100,000 metric tons below its forecast of 970,000. As for the United States, it too surpassed its objectives, reducing its annual NO_X emissions by 2 million tons below projected levels for 2000 without the Acid Rain Annex. Most of the emissions reductions have been attributed to the Acid Rain Annex, but other initiatives, such as the Clean Air Interstate Rule (CAIR) and the NO_X ozone season and annual programs, also played a significant role.

In 2000, Canada and the United States also agreed to an Ozone Annex, whereby the two countries committed to reducing transboundary ozone-forming pollutants, NO_X and volatile organic compounds (VOCs), within predefined geographical areas collectively known as the Pollutant Emission Management Area (PEMA). Thus far, early reports are positive. Canada has met its commitment to cap NO_X emissions within the Ontario and Quebec provinces of the PEMA. It also has implemented new regulations to align its emissions standards for vehicles and engines with those of the United States. For its part, the United States is implementing existing regulations in the affected regions to achieve needed VOC and NO_X emissions reductions. These include the NO_X SIP Call, the CAIR NO_X ozone season program in the PEMA states, the New Source Performance Standards (NSPS), and regulations on vehicles and fuel quality.

Lastly, the two nations have begun discussions to negotiate another air quality agreement, this one aimed at particulate matter. Negotiating sessions already have taken place to move toward adding this new Particulate Matter (PM) Annex. Both countries expect that this next step can be realized once each has advanced its own domestic initiatives to control the emissions of PM and its precursors.

U.S.–Mexico Border 2012 Program and U.S.–Mexico Border 2020 Program[17]

In 1992, the United States and Mexico initiated an Integrated Border Environmental Plan (IBEP). The 2,000-mile border region between the two countries has been a source of concern because of the concentration of the now infamous *maquiladora* factories positioned there for access to American markets. Effluent releases into the Rio Grande River and air emissions far exceed U.S. standards. The plan, supported by $208 million from Congress and $160 million from the Mexican government, was aimed at improving water quality, monitoring air pollution, tracking hazardous waste, developing enforcement cooperation, and promoting pollution prevention.

Today, there is an ongoing binational effort aimed at some of these same border issues. Launched in 2002 as a 10-year plan, the Border 2012 Program (accessible at

www.epa.gov/usmexicoborder) extends the efforts of its predecessor plans—the IBEP and the Border XXI Program. Its stated mission is to protect human health and the environment in the Mexico–United States border region in keeping with the principles of sustainable development. Program objectives are to be implemented using a bottom-up approach, relying on local decision making that can best identify environmental issues at the border. Border 2012 is motivated toward achieving six key goals focused on pollution reduction, environmental health, and environmental stewardship. The program's hallmark is its reliance on partnership and collaboration, which is led by 10 border states, 26 U.S. tribes, many binational institutions, and participation of border communities.

According to a recent progress report, numerous projects have been undertaken aimed at program objectives: to reduce water, air, or land contamination; to improve environmental health; or to improve environmental compliance, performance, and stewardship. For example, in 2010, funding of $149,000 from Border 2012 was used to retrofit 39 school buses in Arizona border counties with devices that reduce particulate matter from tailpipe emissions. Another effort that was funded with just over $15,700 was aimed at local recycling of 17 tons of electronic wastes in Nogales, Sonara. A related initiative, launched with $84,000 from Border 2012, studied existing and potential electronic waste streams in Mexico's industrialized northern border region. The investigation showed that this border region generates nearly 19 percent of the country's overall e-waste, a finding that will support efforts to define a national policy to manage e-waste in the future.

In 2011, a draft framework was released for the U.S.–Mexico Border 2020 Program. An eight-year plan (2013–2020) to succeed the Border 2012 Program, the 2020 Program is aimed at protecting health and the environment in the border regions, using new underlying principles, an added emphasis on chemical safety, and the use of action plans to achieve objectives. Building upon Border 2012, five strategies are proposed in the draft framework aimed at the following: climate change, disadvantaged and underserved communities, children's health, environmental education, and strengthening tribal, state, federal, and international partnerships.

Notwithstanding these efforts, the region still faces difficult challenges, which are exacerbated by rising population and economic activity. Projections suggest that the

[17]Drawn from U.S. EPA (September 5, 2011; Fall 2010; 2007; May 5, 2003); Council on Environmental Quality (January 1993), pp. 52–53.

population of the border region, now at more than 14 million people, will increase by 4.6 million by 2020 and by another 9.3 million by 2030. At the same time, economic development has advanced, due in part to the *maquiladora* program and the North American Free Trade Agreement (NAFTA). Together, these realities have strained natural resource supplies and negatively affected the environment and human health. Hence, the Border 2020 Program must be strongly committed to its objectives in order to realize measurable improvement in the border region.

INTERNATIONAL TRADE AND ENVIRONMENTAL PROTECTION

International trade negotiations have always been the subject of political and economic debate. Although there are known economic advantages to trade, exchange between nations is nonetheless a complex undertaking. Trade negotiations are rarely, if ever, totally distinct from political objectives and national defense issues. The gains from trade can easily become clouded by protectionist attitudes, which are fueled by differences in product safety regulations, labor laws, and nationalism. In recent years, international trade discussions have triggered controversy of another sort—the potential conflict between the associated gains from trade and a decline in environmental quality.

Overview of the Controversy: Free Trade Versus Protectionism

free trade The unencumbered exchange of goods and services among nations.

Proponents of **free trade** argue that nations should trade with one another because there are tangible gains to be realized. Though not intended to be exhaustive, the following list identifies some of the chief benefits of international trade:

- Greater consumer choice
- Higher worldwide output
- Efficiency gains from specialization
- International political stability from forming trading partnerships
- More competition and lower prices in a global marketplace

International trade models can illustrate these gains and sort out which nations should export which products, but these are beyond the scope and intent of this discussion. What *is* relevant is to ask why there are strong opponents to trade, given that these gains from trade exist.

protectionism Fostering trade barriers, such as tariffs or quotas, to protect a domestic economy from foreign competition.

On the opposite side of the coin are those who support **protectionism**, arguing that nations are better off without engaging in international trade. Several arguments are commonly offered to support this position, such as concerns that trade can threaten national security, create unfair competition, lose jobs to nations with cheap labor, and limit the growth of infant industries. Though seemingly plausible, there are strong counterarguments to each, which are adopted by most economists. Nonetheless, protectionist arguments are used to justify trade barriers like quotas and tariffs.

Though the protectionist view is not new, concerns about environmental quality now have been added to its arsenal of arguments. In part, these concerns arise from the disparity of environmental standards among trading nations—a disparity that is particularly striking between developing nations and their more advanced trading partners.

To better understand this issue, we need to explore exactly how environmental policies can tangibly influence trading arrangements among nations.

International Trade and Environmental Objectives

A common concern about international trade is that lenient labor laws and relatively low wage rates in less developed nations can adversely affect employment in more advanced countries. Similar apprehensions have arisen about differences in environmental regulations between trading partners. The relevant issue is that production costs are lower in countries that have more lenient environmental standards. This cost differential gives producers in these nations a competitive advantage over those in countries with more stringent controls, sometimes called the **pollution haven effect.**[18] Predictably, such disparate regulations are more commonly found between developing and advanced trading partners.

pollution haven effect
Changes in trade patterns caused by cost differences among nations due to varying environmental regulations.

Another source of controversy is the quality and desirability of imports produced in nations with lax regulations on such issues as toxic chemical use, fuel efficiency, and coal consumption. Beyond cost considerations, these imports would be associated with environmental externalities that can extend beyond national borders. For example, importing goods produced with high-sulfur coal has negative implications for global air quality, which identifies the market failure as an **international externality**. At issue is whether demand for such products should be implicitly augmented as a result of an international agreement.

As a counter to this concern, trade advocates argue that the economic gains from trade will help poorer nations afford the costly cleanup of what is in many cases severe environmental pollution. In the case of transboundary pollution, the benefits from this response may accrue to bordering countries or those located downstream along a common river. An improved economy in all likelihood will also provide the financial support to design and implement more comprehensive environmental policy, as suggested by the **environmental Kuznets curve**. These arguments are consistent with the premise of sustainable development—that economic prosperity, in this context enhanced through trade, can bring about improvements in environmental quality.

These and other issues were part of the lengthy and often contentious negotiations associated with the North American Free Trade Agreement (NAFTA), the latest round of the General Agreement on Tariffs and Trade (GATT), and the World Trade Organization (WTO) agreements.

International Trade Agreements and the Environment

North American Free Trade Agreement (NAFTA)[19]

Following a difficult series of negotiations, many of which centered on environmental issues, the NAFTA was reached by the United States, Mexico, and Canada in 1992 and

---

[18]This effect is distinct from the "pollution haven hypothesis," which deals with shifts in industry locations. Empirical evidence on these theories is discussed in Copeland and Taylor (March 2004).

[19]This discussion is drawn from Export.gov (April 27, 2011); U.S. Department of Commerce, Census Bureau, Foreign Trade Division (July 2011); U.S. Department of Commerce, International Trade Administration (Summer 2008); North American Commission for Environmental Cooperation (April 24, 2005).

approved by Congress in 1993. Today, the free trade area that was created by this agreement has a population of about 454 million people and, as of 2010, produced more than $17.2 trillion in goods and services.

Among the NAFTA provisions aimed at economic advance and improving environmental quality are the following:

- Explicit language asserting the signatories' commitment to sustainable development
- Agreement to implement NAFTA in accordance with the aim of environmental protection and not to lower health, safety, or environmental standards to attract investment
- Consensus to aim for congruence of each country's respective environmental regulations, while preserving each nation's right to select a level of environmental quality that it deems appropriate
- Agreement that NAFTA dispute settlement panels will solicit environmental experts for advice on factual issues as needed

The full text of NAFTA is available online from the NAFTA secretariat at **www.nafta-sec-alena.org/en/view.aspx**.

As to how NAFTA actually affected the flow of trade, consider the following summary statistics. Since NAFTA's inception in 1994, total trade among Canada, Mexico, and the United States increased from $297 billion to $920 billion in 2010. During this same time frame, U.S. exports to Mexico and Canada rose 191 percent, from $142 billion to $412.6 billion. Similarly, Canadian and Mexican exports to the United States increased from $151 billion to $507.6 billion, representing a 236 percent increase.

To address the environmental impact of the increased trade, the North American Commission for Environmental Cooperation (CEC) was formed among the three trading nations in 1993. This international organization complements NAFTA and is aimed at protecting and improving the natural environment. The CEC has engaged in a number of environmental efforts aimed at such issues as biodiversity, pollutant reporting, chemical management, and trade-environmental issues. To learn more about the commission's programs and projects, visit **www.cec.org**.

General Agreement on Tariffs and Trade (GATT)[20]

Originally executed in 1947, GATT was a major international treaty on foreign trade. Its avowed purpose was to reduce tariffs and other trade barriers. The 107 signatories, mostly developing countries, met periodically for negotiations, called rounds, the last being the Uruguay Round. This round began in 1986 and was signed in December 1993. The associated negotiations, much like those preceding the signing of NAFTA, considered environmental protection issues. However, most environmentalists say that the long-awaited agreement did not respond adequately to their concerns.

Environmentalists' opposition to GATT was stronger than it was to NAFTA. Reportedly, the intense debate was fueled by a 1991 GATT panel ruling that the United States could not block imports of tuna from Mexico harvested under conditions yielding an unacceptable incidental kill rate of dolphins. However, the arguments that ultimately

[20]This discussion is drawn from Cough (April–June 1993); "The Greening of Protectionism" (February 27, 1993).

pitted open traders against environmentalists during the GATT round ran broader and deeper than this single incident.

Fundamentally, environmentalists are wary of any agreement that promotes economic growth and hence increases the potential for ecological damage. Furthermore, they are concerned about how GATT rulings that deter trade restrictions run counter to environmental aims. A few examples will illustrate. A nation cannot use countervailing duties on imports from a country whose environmental regulations are lower, and therefore less costly to producers, than its own. An import cannot be restricted based solely on the exporter's use of a pollution-generating input or production method. Relatively high environmental standards in one country can be viewed as restrictions on free trade.

Based on the results of the Uruguay Round, countries are required to use the least trade restrictive measures to achieve environmental goals, and in most cases, international standards should be employed rather than national ones. Economically, the concern is that such a universal set of guidelines may become a ceiling instead of a floor. Nations might recognize the incentive to keep standards and enforcement relatively low as a way to keep compliance costs down and attract foreign investment. In any case, once this pattern emerges, other countries may follow suit to maintain export competitiveness.

Another important outcome of the Uruguay Round was the formation of the World Trade Organization (WTO), which is the successor to GATT. However, the multilateral trading system that was developed and established under GATT continues under the WTO.

World Trade Organization (WTO)[21]

Formed in 1995 as the successor to GATT, the World Trade Organization (WTO) is an international organization aimed at facilitating trade between nations. Comprising over 150 member nations and customs territories, the organization performs many functions, such as:

- Administering trade agreements
- Facilitating trade negotiations and resolving trade disputes[22]
- Overseeing national trade policies
- Helping developing nations in trade policy matters

Aggregate data suggest that reduced trade barriers due in part to the WTO and its predecessor have markedly improved trade flows. For example, over the 1994–1998 period, U.S. exports increased by 36 percent, with more than 1.3 million new jobs created as a consequence. More information is available at the WTO's Web site, **www.wto.org**.

The WTO generally makes decisions by consensus among the entire membership, over 75 percent of which are developing or least-developed nations. Its high-level, decision-making body is the Ministerial Conference, which meets at least once every two years. The multilateral trading system established under GATT is referred to as the

---

[21]This discussion is drawn from World Trade Organization (2011; 2008a; 2008b; April 25, 2005).

[22]To read about some of the dispute settlements, visit **www.wto.org/english/tratop_e/envir_e/edis00_e.htm**.

set of WTO agreements (comprising some 30,000 pages), which are essentially rules of trade established through negotiation among its members. Over 97 percent of all world trade is accounted for by the WTO. Its 2011 annual budget was about 196 million SF (Swiss francs), or $215.8 million.

That the WTO has a role to play in environmental matters was initially established at the Rio Summit. The *Rio Declaration* specifically asserts that a multilateral trading system can make an important contribution to protect the environment and achieve sustainable development. Indeed, called for by the 1994 Ministerial Decision on Trade and Environment at the end of the Uruguay Round, the WTO established a Committee on Trade and Environment (CTE). The committee's directive is:

- To identify the relationship between trade measures and environmental measures to foster sustainable development
- To recommend any necessary changes to the multilateral trading system

The WTO's commitment to the environment was confirmed at the Doha Ministerial Conference held in 2001. At this meeting, an agreement was reached to begin a new round of negotiations on issues dealing with trade and the environment. A special session of the CTE was created to participate in this process and to focus on three issues: the influence of environmental measures on market access, intellectual property rights, and environmental labeling requirements.

The WTO is careful to assert that its purpose is not to act as an environmental protection agency. Rather, its competency regarding the environment is limited to trade matters and to the effect of environmental policy on trade flows. To accomplish its goals, the organization fosters coordination and cooperation among nations to address environmental issues and supports the identification of market access opportunities that will aid developing countries in achieving sustainable development.

CONCLUSIONS

While some argue that the realization was long overdue, society has begun to recognize the importance of achieving a balance between economic growth and the preservation of natural resources. Although population growth in advanced countries is stable or even declining, the same cannot be said of developing nations. Based on current trends, world population is estimated to rise to 8.9 billion by 2050, an increase of about 2.8 billion.[23] Attempts to increase production to provide for this growth will place inordinate stress on the ecology and the earth's stock of resources.

Acknowledging and understanding the relevant issues is an important first step, but the real challenge is in establishing appropriate goals and in taking responsible action to achieve them. Critical to meeting this challenge is a true understanding of the interdependence between economic activity and nature. Recognizing this connection is a precondition for effective policy development and informed decision making, both of which are essential elements of global environmental management.

[23]United Nations, Department of Economics and Social Affairs, Population Division (2004).

If sustainable development is to be achieved, there must be a cooperative and educated effort from industry, private citizens, and public officials at all levels of government and around the world. As this process unfolds, certain strategies are evolving that may be effective in realizing long-term environmental objectives. In Chapter 21, we examine these strategies to learn how the United States and other countries are attempting to implement the common goal of sustainable growth and development.

Summary

- A consensus is forming that decision making should be driven by a goal that is global in scope and dynamic in perspective. One such goal is sustainable development: managing the earth's resources such that their long-term quality and abundance are ensured for future generations.
- Sustainable development is based on the premise that economic growth and environmental quality must be reconciled.
- The environmental Kuznets curve suggests that there is a technical relationship or pattern between economic development and environmental degradation that graphs as an inverted ∪ shape. This in turn implies that early stages of industrialization are associated with rising levels of pollution, and more advanced development is linked to an increasing concern for environmental quality.
- The Rio Summit in 1992 dealt with the objective of sustainable development. Among the major documents produced at the summit were *Agenda 21* and the *Rio Declaration*. *Agenda 21* outlines a course for worldwide progress toward sustainable development. The principles of the *Rio Declaration* include a requirement that environmental protection be an integral part of development.
- In August–September 2002, 10 years after the Rio Summit, over 20,000 participants converged in Johannesburg, South Africa, at the World Summit on Sustainable Development (WSSD). To commemorate the twentieth anniversary of the Rio Summit, officials organized the Rio+20—United Nations Conference on Sustainable Development (UNCSD) to be held in Brazil in 2012.
- When pollution problems are transboundary, the contamination can generate an international externality. In such cases, formal treaties must be negotiated among all affected countries.
- Strengthened by a series of amendments, the Montreal Protocol is aimed at phasing out chlorofluorocarbons (CFCs) and other ozone-depleting substances.
- The U.N. Framework Convention on Climate Change (UNFCCC), which became legally binding in 1994, established a baseline for a global cooperative response to climate change.
- At the third Conference of the Parties (COP), held in Japan in December 1997, the Kyoto Protocol was formulated. Ratification was achieved in 2004, without the participation of the United States, following Russia's signing of the accord.
- The Kyoto Protocol called for developed nations to reduce greenhouse gas (GHG) emissions to 5.2 percent below their 1990 levels by 2012, with no targets imposed on developing countries. The emissions targets were to be achieved using several market-based instruments, called flexible mechanisms. Among these is a trading system of GHG allowances for participating developed nations.
- At the 2011 COP in Durban, South Africa, the parties agreed to extend the Kyoto Protocol and to develop a successor agreement by 2015. The new agreement is to involve all major GHG emitters—both developed and developing countries.
- Ocean dumping of certain wastes is prohibited by the London Convention 1972 (LC72). The LC72 has been ratified by 85 nations, including the United States. The 1996 Protocol amends the LC72 and entered into force in 2006.
- The United States–Canada Air Quality Agreement was finalized in March 1991 to combat acid rain and visibility impairment. Under this accord, each country must commit to national emissions caps on sulfur dioxide (SO_2) and nitrogen oxides (NO_X).
- In 1992, the United States and Mexico initiated an Integrated Border Environmental Plan (IBEP) for the border region. Extending these efforts is the Border 2012 Program, launched in 2002 as a 10-year plan. In 2011, a draft framework was released for the U.S.–Mexico Border 2020 Program, which is an eight-year plan to succeed the Border 2012 Program. The 2020 Program is aimed at protecting health and the environment in the border regions, using new underlying principles, an added emphasis on chemical safety, and the use of action plans to achieve objectives.

- A common concern about international trade is that lenient labor laws and environmental standards along with relatively low wage rates can give international competitors an unfair advantage, which may create a pollution haven effect. Another source of controversy is the quality of imports produced in nations with lax regulations on toxic chemical use and fuel efficiency, which may generate an international externality.
- Following difficult negotiations, many of which centered on environmental issues, NAFTA was reached by the United States, Mexico, and Canada in 1992.
- Originally executed in 1947, the General Agreement on Tariffs and Trade (GATT) was a major international treaty on foreign trade, whose purpose was to reduce tariffs and other trade barriers. Environmentalists' opposition to GATT was stronger than it was to NAFTA.
- Formed in 1995 as the successor to GATT, the World Trade Organization (WTO) is an international organization aimed at facilitating trade. Called for by the 1994 Ministerial Decision on Trade and Environment, the WTO established a Committee on Trade and Environment (CTE). The WTO's commitment to the environment was confirmed at the Doha Ministerial Conference held in 2001.

Review Questions

1. Critically discuss the following statement:

   Without a well-enforced command-and-control regulatory structure, society will not take the necessary steps toward a sustainable future.

2. Summarize in your own words Robert Solow's view of using substitutability in applying intergenerational equity to environmental policy. Do you agree with this view? Why or why not?
3. Assume that the following equation quantifies the relationship between a country's level of air pollution emissions (E) and its per capita income (I):

   $$E = 5(I) - 0.5(I)^2$$

   where E is measured in thousands of tons of sulfur dioxide (SO_2) emissions, and I is measured in thousands of dollars.

   a. Graph the equation, and explain how it supports the theory of an environmental Kuznets curve (EKC).

   b. If a country's per capita income (I) increases from \$2,500 to \$3,500, what is the associated change in emissions (E)? What do these changes suggest about where the country is operating on its EKC?

   c. Solve for the level of per capita income that maximizes the EKC. What is the significance of this income level? (Hint: The EKC reaches a maximum when its slope is zero.)

4. Explain the dilemma that advanced nations might have faced as they debated the ratification of the Kyoto Protocol, once the United States withdrew from the agreement.
5. In your view, does NAFTA advance or hinder the achievement of sustainable development? Explain.
6. **a.** Critically evaluate the provision resulting from the Uruguay Round that requires that the least trade-restrictive measures be used to achieve environmental goals. Include in your answer both the environmental and economic implications.

   b. Support or refute the use of international environmental standards among trading partners.

7. Suppose that under the United States–Canada Air Quality Agreement, both countries agree to a combined 30 percent SO_2 abatement standard. Further assume that the following SO_2 abatement cost functions have been estimated by each country:

   $$TAC_{US} = 500 + 1.5(A_{US})^2$$
   $$MAC_{US} = 3\ A_{US}$$
   $$TAC_{CAN} = 1000 + 3(A_{CAN})^2$$
   $$MAC_{CAN} = 6\ A_{CAN}$$

   where A_{US} and A_{CAN} represent the percent of SO_2 abatement achieved by the United States and Canada, respectively, and each TAC and MAC is measured in millions of dollars.

a. If the countries implement a uniform abatement standard, find the resulting values of *TAC* and *MAC* for each nation. Based on these values, is there an economic incentive for the two nations to participate in an emissions trading program? Explain.

b. Assuming a trading program is enacted, find the cost savings associated with a cost-effective abatement solution?

c. What must be the price of a tradeable permit to achieve the cost-effective abatement allocation?

Additional Readings

Abaza, Hussein, and Andrea Baranzini, eds. *Implementing Sustainable Development*. Northampton, MA: Elgar, 2002.

Agyeman, Julian, Robert D. Bullard, and Bob Evans. *Just Sustainabilities: Development in an Unequal World*. Cambridge, MA: MIT Press, 2003.

Aslanidis, Nektarios, and Susana Iranzo. "Environment and Development: Is There a Kuznets Curve for CO_2 Emissions?" *Applied Economics 41(6)* (March 2009), pp. 803–10.

Bithas, Kostas. "Sustainability and Externalities: Is the Internalization of Externalities a Sufficient Condition for Sustainability?" *Ecological Economics 70(10)* (August 2011), pp. 1703–06.

Bohringer, Christoph. "Climate Politics From Kyoto to Bonn: From Little to Nothing?" *Energy Journal 23(2)* (2002), pp. 51–72.

Chambers, P. E., and R. A. Jensen. "Transboundary Air Pollution, Environmental Aid, and Political Uncertainty." *Journal of Environmental Economics and Management 43(1)* (January 2002), pp. 93–112.

Commoner, Barry. "Economic Growth and Environmental Quality: How to Have Both." *Social Policy* (Summer 1985), pp. 18–26.

Dean, Judith, ed. *International Trade and the Environment*. Burlington, VT: Ashgate Publishing, 2001.

Deere, Carolyn L., and Daniel C. Esty, eds. *Greening the Americas: NAFTA's Lessons for Hemispheric Trade*. Cambridge, MA: MIT Press, 2002.

Fernandez. L. "Trade's Dynamic Solutions to Transboundary Pollution." *Journal of Environmental Economics and Management 43(3)* (May 2002), pp. 386–411.

Goldenberg, José, and Robert N. Stavins. "Beyond Kyoto: A Second Commitment Period." *Environment 47(3)* (April 2005), pp. 38–41.

He, Jie. "China's Industrial SO_2 Emissions and Its Economic Determinants: EKC's Reduced vs. Structural Model and the Role of International Trade." *Environment and Development Economics 14(2)* (April 2009), pp. 227–63.

Horn, Henrik. "The Burden of Proof in Trade Disputes and the Environment." *Journal of Environmental Economics and Management 62(1)* (July 2011), pp. 15–29.

Kates, Robert W., Thomas M. Parris, and Anthony A. Leiserowitz. "What Is Sustainable Development?" *Environment 47(3)* (April 2005), pp. 8–22.

Lesser, Jonathan A., and Richard O. Zerbe, Jr. "What Can Economic Analysis Contribute to the Sustainability Debate?" *Contemporary Economic Policy XIII(3)* (July 1995), pp. 88–100.

Levinson, Arik, and M. Scott Taylor. "Unmasking the Pollution Haven Effect." *International Economic Review 49(1)* (February 2008), pp. 223–54.

Lieb, Christoph Martin. "The Environmental Kuznets Curve and Flow Versus Stock Pollution: The Neglect of Future Damages." *Environmental and Resource Economics 29(4)* (December 2004), pp. 483–507.

Lipford, Jody W., and Bruce Yandle. "NAFTA, Environmental Kuznets Curves, and Mexico's Progress." *Global Economy Journal 10(4) Article 4* (December 2010), pp. 1–18.

Menz, Fredric C. "Transborder Emissions Trading Between Canada and the United States." *Natural Resources Journal 35* (Fall 1995), pp. 803–19.

Obasi, Godwin O. P. "Embracing Sustainability Science: The Challenges for Africa." *Environment 44(4)* (May 2002), pp. 8–19.

Pezzey, John C. V., and Michael A. Toman, eds. *The Economics of Sustainability*. Burlington, VT: Ashgate, 2002.

Sampson, Gary P. "The Environmentalist Paradox: The World Trade Organization Challenges." *Harvard International Review 23(4)* (Winter 2002), pp. 56–62.

Sampson, Gary, and John Whalley, eds. *The WTO, Trade and the Environment.* Northampton, MA: Elgar, 2005.

Smulders, Sjak, Lucas Bretschger, and Hannes Egli. "Economic Growth and the Diffusion of Clean Technologies: Explaining Environmental Kuznets Curves." *Environmental & Resource Economics 49(1)* (May 2011), pp. 79–99.

Speth, James Gustave. "Perspectives on the Johannesburg Summit." *Environment 45(1)* (January/February 2003), pp. 24–29.

Wettestad, Jorgen. "Clearing the Air: Europe Tackles Transboundary Pollution." *Environment 44(2)* (March 2002), pp. 32–40.

To access additional course materials, visit www.cengagebrain.com. At the home page, search for the ISBN of this title (shown on the back cover). This will take you to the product page where these resources can be found.

CHAPTER 21

Sustainable Approaches: Industrial Ecology and Pollution Prevention

"Intellectuals solve problems, geniuses prevent them."
—**Albert Einstein (1879–1955)**

Few would debate that sustainable development is an ambitious pursuit. Fundamental to its achievement are substantive changes in how market activity is undertaken within the circular flow and in how environmental policy is drafted. The key is to modify behavior so that economic growth and environmental protection can become reinforcing rather than competing goals. Although the needed changes will take time to orchestrate, several initiatives being developed around the world are moving public and private efforts in this direction.

Acting as a multidisciplinary framework for certain of these initiatives is the concept of **industrial ecology**. This framework fosters a systems approach to environmental management and recognizes the interaction of ecological systems with industrial systems. The goal is to minimize the impact of economic activity on the environment. In so doing, industrial ecology integrates economic objectives with environmental goals—precisely the motivation of sustainable development. Consequently, it advocates environmental strategies that are different from the traditional command-and-control approach. In particular, industrial ecology promotes the development of industrial ecosystems in which residuals or wastes from one manufacturing process are used as inputs in the production of another good.

A related and more familiar approach to achieving sustainable development is **pollution prevention**. As the name implies, pollution prevention refers to initiatives that reduce or eliminate wastes rather than deal with them at the end of a product cycle. These types of initiatives avoid unnecessary and costly abatement and are considered necessary to achieving sustainable growth and development. Though their underlying aims are different, pollution prevention and industrial ecology share a common view that end-of-pipe policy controls are not sufficient for achieving long-term goals.

These approaches and associated programs that make them operable are the focus of this chapter. We begin with an overview of industrial ecology. To provide a relevant context, we reintroduce the materials balance model presented originally in Chapter 1. This model clearly explains the relationship between industrial activity and nature, which is at the core of industrial ecology. We then study pollution prevention—what it means in practice and how it is being supported internationally. Lastly, we explore a number of voluntary programs currently used in the United States and in other nations to operationalize these approaches and move toward sustainable development.

INDUSTRIAL ECOLOGY

As a society, we are beginning to understand that end-of-pipe pollution controls are inadequate for achieving long-run environmental objectives. It is simply not enough to clean up the effects of pollution after the damage has been done. What this means is that we have to use a broader perspective when assessing the environmental effects of consumption and production and that firms must adopt sustainable business practices.

What Is Industrial Ecology?[1]

industrial ecology A multidisciplinary systems approach to the flow of materials and energy between industrial processes and the environment.

An increasingly pervasive notion in environmental policy development is that the entire life cycle of a product, including all the materials and energy flows, ought to be considered in efforts to improve the environment. This conviction underlies an emerging discipline known as **industrial ecology**. Though there is no standardized definition, industrial ecology can be thought of as a multidisciplinary systems approach to the flow of materials and energy between industrial processes and the environment. It is considered a systems approach because it supports the integration of ecological systems and industrial systems. Moreover, it promotes conceptualizing the production process itself as an industrial ecosystem, which is something we will discuss later in this chapter.

In any case, the primary purpose of industrial ecology is to promote the use of recycled wastes from one industrial process as inputs in another. It also supports optimal materials flows, or dematerialization of output, which means efficient use of materials and energy in production. Notice that this approach to environmental protection is centered on industry and is implemented over an interdependent group of firms. For further study, visit the Smart Communities Network at the U.S. Department of Energy at **www.smartcommunities.ncat.org/business/indeco.shtml** or Yale University's International Society for Industrial Ecology at **www.is4ie.org/**.

Based on its conceptual arguments, industrial ecology is intimately linked with sustainable development. In fact, some argue that industrial ecology is the means by which a society promotes sustainable development or, at the very least, that the achievement of sustainable development is its ultimate purpose. In any case, since this field deals directly with the flow of materials in an economic system, it is appropriate to examine the materials balance model in this context.

materials balance model Positions the circular flow within a larger schematic to show the connections between economic decision making and the natural environment.

Revisiting the Materials Balance Model

The dynamic relationship between economic activity and nature can be better understood by positioning the circular flow model within a larger framework to generate the **materials balance model**. Originally introduced in Chapter 1, this model illustrates the

[1]This discussion is drawn from Erkman (1997); Garner and Keoleian (November 1995).

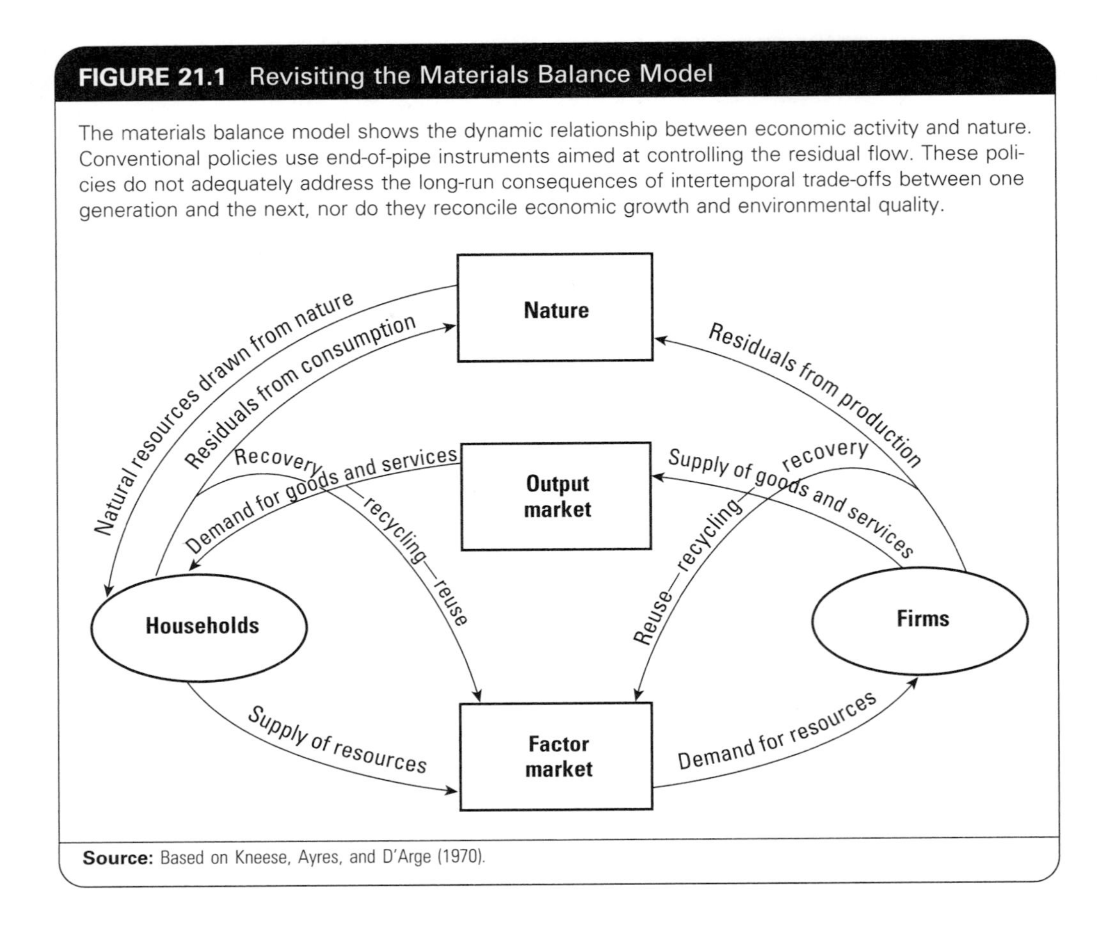

FIGURE 21.1 Revisiting the Materials Balance Model

The materials balance model shows the dynamic relationship between economic activity and nature. Conventional policies use end-of-pipe instruments aimed at controlling the residual flow. These policies do not adequately address the long-run consequences of intertemporal trade-offs between one generation and the next, nor do they reconcile economic growth and environmental quality.

Source: Based on Kneese, Ayres, and D'Arge (1970).

link between economic activity and the natural environment—a relationship critical to achieving sustainable development. As shown in Figure 21.1, this relationship is defined by the flow of resources from nature to an economic system and the return flow of **residuals** from economic activity back to the environment.

residual
The amount of a pollutant remaining in the environment after a natural or technological process has occurred.

Throughout our study of environmental economics, the primary focus has been on the flow of residuals and the potential damages associated with their release to nature. Indeed, the conventional policy approach uses end-of-pipe, command-and-control instruments aimed at limiting the amount and toxicity of residuals released to the environment *after* they are generated. Implicit in this approach is a perception that the flow of materials through the economy is linear and open.

Linear or Open Materials Flow: Cradle to Grave

linear or open flow of materials
Assumes that materials run in one direction, entering an economic system as inputs and leaving as wastes or residuals.

As the graphic in Figure 21.2 illustrates, a **linear, or open, flow of materials** assumes that materials run in one direction, entering an economic system as inputs and exiting as wastes or residuals. This so-called **cradle-to-grave flow** emphasizes use, waste generation, and disposal, which coincides with policies aimed at abating contaminating residuals only at the end of the flow. For example, conventional clean air policy efforts control harmful residuals released into the atmosphere, such as sulfur dioxide and particulate matter. Although such policy efforts can and do improve environmental quality in the short run,

FIGURE 21.2 Conventional Linear Materials Flow

The aim of performance- or technology-based standards is to control the amount of polluting residuals released to the environment. Implicit in this approach is a perception that the flow of materials through the economy is linear and open. This view assumes that materials run in one direction, entering an economic system as inputs and leaving as wastes or residuals.

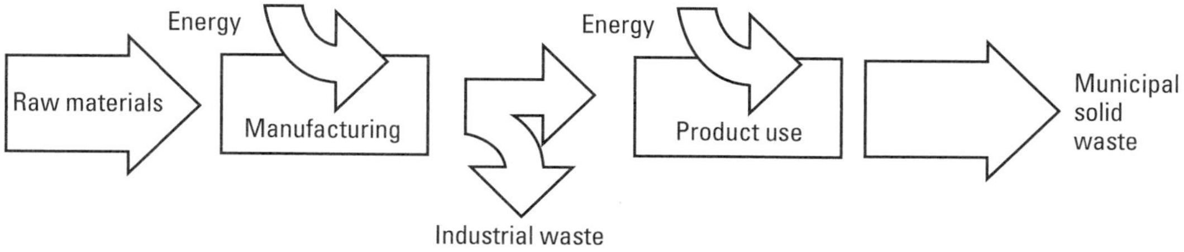

Energy
Raw materials
Manufacturing
Energy
Product use
Municipal solid waste
Industrial waste

Source: U.S. Congress, OTA (October 1992), as cited in Gibbons (September/October 1992), p. 31.

they do not adequately address the long-run consequences. To understand why, think about the dynamics of the materials balance model in the context of policy initiatives.

First, nature's capacity to convert matter and energy is limited. Although residuals can be converted to other forms that can flow back into productive use, this process is not without bound. Hence, policy decisions affecting resource use and environmental damage have implications for future generations. Look again at Figure 21.1. Government programs to promote recovery and reuse of residuals shown by the inner flows in the figure have helped delay their release but not eliminate them.

Second, end-of-pipe policy controls take the form of abatement or remediation after the damage has been done. Such methods draw resources away from other productive activities—resources that ultimately are added to the residual flow. Think about the use of raw materials and energy needed to clean up a hazardous waste site or to dredge a polluted waterway. These resources, once expended, go back into the residual flow.

Third, command-and-control initiatives generally operate at cross purposes with private market incentives and the broader social goal of economic growth. This observation points to the potential for intertemporal trade-offs between one generation and the next. The pursuit of economic development today can so harm the natural environment as to leave future generations unable to continue that progress. In the context of the materials balance model, we know that population growth and economic development increase the flow of resources into market activity and the return flow of residuals back to nature. These changes can adversely affect the well-being of future generations. On the other hand, restricting the pursuit of economic gains to protect the environment deprives today's society as well as future generations of a higher standard of living.

Cyclical or Closed Materials Flow: Cradle to Cradle

Part of the solution is to recognize that the flow of materials through an economy need not be linear. Instead, product design, manufacturing processes, and energy use can be

cyclical or closed flow of materials Assumes that materials run in a circular pattern in a closed system, allowing residuals to be returned to the production process.

modified to achieve a **cyclical, or closed, flow of materials**, as shown in Figure 21.3. Notice that this model, considered a **cradle-to-cradle flow**, suggests a much broader approach to environmental protection than controlling residuals *after* they have been generated. Waste prevention, energy and resource efficiency, and design for reuse or recycling are among the options suggested by this paradigm. In sum, the key implication of a cyclical system is that economic activity can be altered *throughout* the cycle of production and consumption to achieve a reduction in the associated environmental impacts.

Life Cycle Assessment (LCA)

life cycle assessment (LCA) Examines the environmental impact of a product or process by evaluating all its stages from raw materials extraction to disposal.

To put the notion of a cyclical materials flow into practice, a **life cycle assessment (LCA)** can be undertaken. This is an analytical tool that examines the environmental impact of a product by evaluating all stages of a product or process from raw materials extraction to final disposal. Once done, improvements can be made to lessen environmental risk.

In general, a life cycle assessment has four key components:[2]

- **Goal Definition and Scoping**: Describes the product or process to be assessed, the context for the assessment, and the environmental effects to be reviewed.
- **Inventory Analysis**: Identifies and quantifies resource usage and environmental releases.

FIGURE 21.3 A Closed System of Materials Flow

Instead of controlling residuals *after* they have been generated, this model illustrates how economic activity can be altered *throughout* the cycle of production and consumption to reduce the associated environmental effect. Product design, manufacturing processes, and energy use can be modified to achieve a cyclical or closed flow of materials.

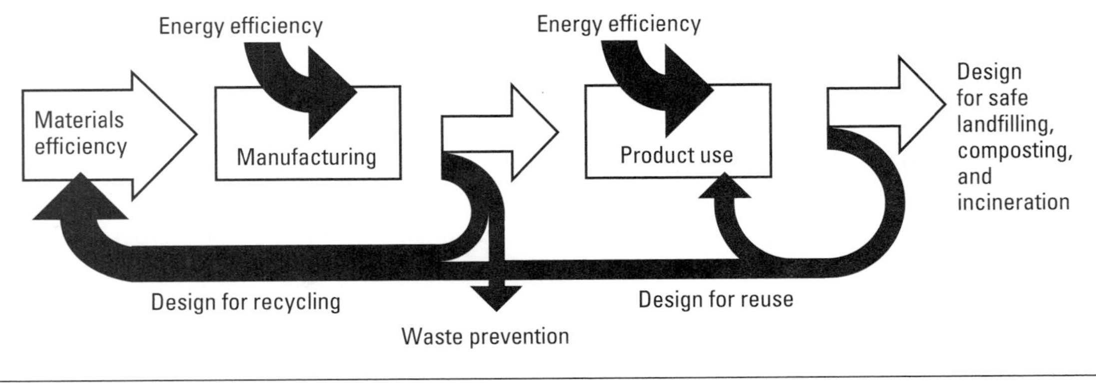

Source: U.S. Congress, OTA (October 1992), as cited in Gibbons (September/October 1992), p. 31.

[2]U.S. EPA, Office of Research and Assessment, and Science Applications International Corporation (May 2006).

- **Impact Analysis**: Assesses the human and ecological effects based on the inventory analysis.
- **Interpretation**: Evaluates the results of the inventory and impact analyses and selects the preferred product or process.

More information is available at the National Risk Management Research Laboratory's Life Cycle Assessment (LCA) Web site at **www.epa.gov/nrmrl/std/lca/lca.html**.

That life cycle assessments are becoming increasingly important is evidenced by the development of databases and software to support the effort on a wide scale.[3] In addition, the International Organization for Standardization (ISO) in Geneva, Switzerland, is working to formalize the methods used in this process on a global scale. In fact, a major component of the **ISO 14000 standards** for environmental management addresses life cycle assessment. These are voluntary international standards aimed at providing countries with a common approach to environmental issues, as discussed in Application 21.1.

ISO 14000 standards
Voluntary international standards for environmental management.

Industrial Ecosystems

The real-world outcome of implementing a closed system is the formation of an **industrial ecosystem** in which residuals from one or more manufacturing processes are reused as inputs for others.[4] In its most fundamental form, wastes from one process, such as emissions or wastewater, are collected and then redirected to an entirely different process to produce another good or service. The transfers generally take place through a collaborative effort among different companies. Though seemingly esoteric, such eco-industrial parks, as they are also known, are becoming more commonplace and are operating or being developed in Asia, Europe, and the United States. Perhaps the most well-known example is the Kalundborg ecosystem in Denmark, discussed in Application 21.2 (page 511).

industrial ecosystem
A closed system of manufacturing whereby the wastes of one process are reused as inputs in another.

In Choctaw County, Mississippi, the notion of creating an industrial ecosystem became a reality during the construction of an industrial park. The idea emerged from one element of the project—a power plant that uses lignite (a low grade of coal) for fuel. Developers began to consider the environmental effects of the plant to the surrounding community. As they did, they identified manufacturing processes that could use some of the waste products created by the power facility. The result is what officials there are calling an ecoplex. Clay from the lignite mining process to fuel the power plant will be used by a brick manufacturer. The fly ash from the combustion of lignite will be used to make wallboard or to manufacture cement. And any residual heat will be used by a greenhouse nursery, which in turn will exchange its wastes with a local fish farm.[5]

[3]U.S. EPA, Science Advisory Board (April 4, 2002).
[4]See Frosch (December 1995) and Frosch and Gallopoulos (September 1989) for more details.
[5]U.S. EPA, Office of Solid Waste and Emergency Response (March 2000), p. 11.

APPLICATION 21.1 ISO 14000 International Standards on Environmental Management

With increasing numbers of companies competing in global markets, there is a growing need to develop cooperative international partnerships as well as formal agreements on trade, economic, and political issues. Disparate regulations and practices across nations can become barriers to economic activity if not properly addressed. A case in point is the variability in environmental standards across countries, which often has been a source of debate during trade negotiations. In response, the International Organization for Standardization (ISO), a private international network involving more than 160 nations, developed voluntary standards for environmental management. These are commonly known as the **ISO 14000** series. (Note that ISO is not an acronym for the International Organization for Standardization. Rather, ISO is from the Greek *isos*, meaning equal.)

Founded in Geneva, Switzerland, ISO has developed over 18,600 standards for many technical specifications in engineering and business, including information technology, aircraft, and quality management. Each series of standards is aimed at providing consistency and homogeneity of products and technology, based on international consensus. The expected outcome is greater clarity in global market activity, which in turn should facilitate international business and trade flows.

Development of the ISO 14000 series in environmental management emerged from a consultation process begun in 1991, involving 20 nations, 11 international organizations, and more than 100 experts in environmental issues. This effort combined with ISO's commitment to sustainable development led to the formation of a technical committee on environmental management standards. The work yielded the new ISO 14000 series, which can be used by firms as well as cities, townships, municipalities, and counties.

Key areas addressed by the ISO 14000 series are:

- **Environmental Management Systems (EMS)**: Specifies requirements under the ISO 14001 EMS Standard for establishing an environmental management system aimed at producing a high-quality product with minimal environmental impact.
- **Environmental Performance Evaluations**: Gives guidance on establishing a baseline and monitoring performance.
- **Environmental Auditing**: Sets procedures and criteria for an EMS audit.
- **Life Cycle Assessment (LCA)**: Provides methodology, principles, and a framework for conducting an LCA study.
- **Environmental Labeling**: Offers guidance on terminology and procedures for declaring the environmental aspects of a product.

Achieving certification to the ISO 14000 standards is intended to benefit the individual firm as well as the international business community at large. According to the International Organization for Standardization, firms complying with ISO 14000 can expect to realize lower waste management costs, energy savings, an improved public image, and a model to help them monitor their progress in environmental performance. Moreover, an environmental management system (EMS) that meets the requirements of these ISO standards should allow an organization to control its effect on the environment, improve its environmental performance, and define an approach to setting environmental goals and meeting them. For more information on the ISO 14000 series, visit **www.iso.org/iso/home.html**.

Sources: International Organization for Standardization (October 2, 2011; April 2011; June 4, 2002; 1998); U.S. EPA, Office of Water, Office of Wastewater Management (December 17, 2007); EnviroSense (December 20, 1995).

POLLUTION PREVENTION

There is an old adage that says, "An ounce of prevention is worth a pound of cure." This time-honored advice is now being used to help achieve sustainable development. Recent policy initiatives in nations around the world have begun to emphasize preventive measures that eliminate the release of contaminating residuals and hence avoid having to deal with the associated adverse effects after the fact.

APPLICATION 21.2 Kalundborg Symbiosis in the Twenty-First Century

The generation and disposal of production wastes have become a universal concern and, as a consequence, have motivated the development of **industrial ecosystems**. This strategic approach to industrial manufacturing is based on the premise that the conventional input–output process can be transformed into a closed system, such that the residuals from one production process can be used as raw inputs for another. Generally, the link between processes occurs among several different entities, suggesting that this application of "closing the loop" calls for close coordination among the industrial participants.

The concept of an industrial ecosystem is in keeping with the broader concept of sustainable development and reconciling economic growth and environmental preservation. As such, it is directed toward two major objectives. The first is to employ energy and raw materials optimally to ensure that each productive input achieves maximum efficiency. The second is to facilitate waste reduction by minimizing the generation of nonreusable wastes that occur as part of the manufacturing function.

Consider the experience of eight firms and a municipality cooperating within a closed manufacturing system in Denmark known as the Kalundborg Symbiosis. Currently, there are nine participants:

- Kalundborg municipality
- Novo Nordisk, the largest producer of insulin in the world
- Novozymes, the world's largest manufacturer of enzymes
- Gyproc, a French gypsum board producer
- Dong Energy, which owns the largest power plant in Denmark
- RGS90, a soil and waste remediation firm
- Statoil, which owns the largest oil refinery in Denmark
- Kara/Novoren, a waste treatment facility
- Kalundborg Forsyning A/S, a water and heating supplier and wastewater disposal company

Just as the concept of an industrial ecosystem suggests, these participants link their manufacturing processes together to save resources, energy, and the environment. Just how does this cooperative relationship work?

Three waste by-products are produced by the coal-fired electrical power plant: surplus steam, fly ash, and surplus heat. In a conventional manufacturing environment, these wastes would simply be disposed of. However, as a result of establishing a closed-loop manufacturing environment, the surplus steam is sold to the local biotechnology plant and to the oil refinery for use in their respective manufacturing processes. This arrangement helps to conserve water supplies, which must be pumped from a lake seven miles away. Similarly, the surplus heat is sold by the utility to the city of Kalundborg for its heating needs. Using this surplus heat allows the city to effectively replace small oil-fired heating units for approximately 4,500 households. The fly ash associated with the utility's pollution control equipment is actually a composition of limestone. This is used by the building industry. Gypsum generated from the same process is sold to the wallboard manufacturer. Annually, 200,000 tonnes (or equivalently, nearly 220,500 tons) of gypsum are produced, which can be used in place of natural gypsum. Sludge that is generated as a residual from the local water treatment plant is used as a nutrient by the soil remediation company.

As a consequence of this industrial ecosystem, water demand in the region has declined by 25 percent, and waste energy at the power plant has been lowered by 80 percent. This innovative and successful system in Kalundborg, which began in 1961, has drawn worldwide attention, and now other industrial ecosystems are operating or are under development. For more detail about the Kalundborg system, visit the Web site of the Symbiosis Institute in Denmark at **www.symbiosis.dk/en**.

Sources: Kalundborg Symbiosis Institute (2011a; 2011b; 2002); Girard (2008); Business for Social Responsibility (BSR) Staff (December 2003); Smith and Woodruff (May 11, 1992); Schmidheiny (1992), pp. 104–06; Frosch and Gallopoulos (September 1989).

pollution prevention (P2)
A long-term strategy aimed at reducing the amount or toxicity of residuals released to nature.

What Is Pollution Prevention?

Pollution prevention, commonly referred to as **P2**, is a long-term approach aimed at reducing either the amount or the toxicity of residuals released to the environment. Operationally, it refers to practices that reduce or eliminate wastes or residuals at their source. Based on this definition, pollution prevention promotes a shift from end-of-pipe controls to front-end reduction strategies. Effectively, this implies that residual generation is a strategic variable and not a given to be dealt with after the fact. In the context of the materials balance model, preventive strategies change how economic activity is undertaken so that less contaminating residuals are released to nature. To further investigate P2, visit **www.epa.gov/p2/**.

In the United States, the policy shift toward preventing pollution rather than controlling it represented a complete reorientation of how the country would deal with environmental degradation. Having responded to highly visible and serious pollution problems in the 1970s with standards and other regulations, policy makers recognized the need for preventive approaches in the 1980s, as they observed more pervasive and indirect pollution sources.[6]

Comparing Pollution Prevention to Industrial Ecology

What pollution prevention and industrial ecology have in common is their support for a sustainable solution, using something other than end-of-pipe controls. In that sense, both views adopt the **cyclical flow of materials**, and both employ **life cycle assessment (LCA)** as an important tool to guide solutions. For these reasons, there is a real temptation to position P2 under the broad umbrella of industrial ecology. In fact, some research papers do exactly this, or even use the two concepts interchangeably. However, although comparisons are made difficult by the lack of standardized definitions for each, there are some clear differences between the two concepts.

In a very illuminating and readable paper on the subject, Oldenburg and Geiser (1997) offer the following observations. First and probably foremost, P2 promotes risk reduction through minimizing or eliminating wastes, while industrial ecology argues in favor of using wastes as inputs in other production processes. In this same regard, P2 solutions are aimed mainly at the single firm (though other sectors participate), while industrial ecology is aimed at a network of businesses. This also means, from a benefit-cost perspective, that the potential gains of industrial ecology have to be shared. Second, and not entirely unrelated to the first, recycling is not viewed as a preventive solution, while it essentially *is* the solution in industrial ecology. Third, P2 proposals generally assume some direction or oversight by government, while proponents of industrial ecosystems tend not to mention government intervention. And fourth, P2 considers efficiency as a potential tool to achieve its objectives, while industrial ecology tends to view efficiency of resource use and materials flows as an end in itself.

P2 Objectives and Strategies

source reduction Preventive strategies to reduce the quantity of any contaminant released to the environment at the point of generation.

While pollution prevention promotes the reduction of residuals or wastes generated by all segments of society, its emphasis has been on the industrial sector—the major source of hazardous pollution. In that regard, there are two major preventive objectives for industry. One of these is **source reduction**, which refers to preventive strategies aimed at reducing the quantity of any hazardous substance, pollutant, or contaminant released into the environment at the point of generation. The other objective is

[6]U.S. EPA, Office of Chemical Safety and Pollution Prevention, Office of Pollution Prevention and Toxics (June 6, 2011).

toxic chemical use substitution
The use of less harmful chemicals in place of more hazardous substances.

toxic chemical use substitution, which is the practice of using less harmful chemicals in place of more hazardous substances.[7]

In the United States, these major objectives are implemented in part through the Environmental Protection Agency's (EPA) **Pollution Prevention (P2) Program**.[8] In fact, in its five-year (2010–2014) strategic plan, the overall mission of the program is given as follows:

> ... to prevent pollution at the source, promote the use of greener substances, and conserve natural resources, which are critical steps towards achieving a sustainable society.

As part of the five-year strategic plan, five goals are identified for the program with specific measurements to monitor progress.[9] These are stated below:

1. Reduce the generation of greenhouse gas (GHG) emissions to mitigate climate change.
   Measure: Millions of metric tons of carbon dioxide equivalent reduced
2. Reduce the manufacture and use of hazardous materials to improve human and ecological health.
   Measure: Pounds of hazardous materials reduced
3. Reduce the use of water and conserve other natural resources to protect ecosystems.
   Measure: Gallons of water reduced
4. Create business efficiencies that derive economic benefits and improve environmental performance.
   Measure: Dollars saved
5. Institutionalize and integrate pollution prevention practices through government services, policies, and initiatives.
   Measure: Adoption of P2 approaches

Interestingly, the plan also identifies strategies to achieve these goals that are tailored to five sector areas in the economy: chemicals and manufacturing; hospitality; electronics; building and construction; and municipalities and institutions. These market sectors were chosen because of their effect on the environment and the expectation of significant environmental gains from using pollution prevention and source reduction initiatives.

Domestic and International Legislation

United States P2 Legislation

Beginning with the National Environmental Policy Act (NEPA) of 1969, the United States has integrated pollution prevention into federal environmental legislation. One of the express purposes of NEPA is to encourage any efforts aimed at preventing environmental damage.

---

[7]U.S. EPA, Office of Pollution Prevention (October 1991), p. 7.

[8]The EPA's P2 program is authorized by the Pollution Prevention Act of 1990, which is discussed subsequently.

[9]The following is from U.S. EPA, Office of Chemical Safety and Pollution Prevention, Office of Pollution Prevention and Toxics (February 2010).

To make pollution prevention an integral part of U.S. policy, Congress passed the Pollution Prevention Act of 1990 (accessible online at **www.epa.gov/oppt/p2home/pubs/p2policy/act1990.htm**). Its provisions outline the priorities for national environmental policy as follows:

> The Congress hereby declares it to be the national policy of the United States that
>
> - pollution should be prevented or reduced at the source whenever feasible;
> - pollution that cannot be prevented should be recycled in an environmentally safe manner, whenever feasible;
> - pollution that cannot be prevented or recycled should be treated in an environmentally safe manner whenever feasible; and
> - disposal or other release into the environment should be employed only as a last resort and should be conducted in an environmentally safe manner.[10]

Notice that this legislation explicitly ranks pollution prevention as the primary objective, with management of residuals as a secondary goal to be achieved through recycling, treatment, and disposal—in that order. This hierarchy is depicted in Figure 21.4. The reference in the law to recycling as a secondary objective has been the source of some controversy, since some have inferred from this that recycling is a solution supported by pollution prevention. However, since recycling only postpones waste disposal, it is not a true preventive solution.

P2 objectives also are supported by other U.S. policies. For example, source reduction is explicitly listed as the first priority in the EPA's integrated waste management system that guides states' solid waste programs. Also, the Federal Insecticide, Fungicide, and Rodenticide Act (FIFRA) promotes toxic chemical use substitution by empowering the EPA to cancel or restrict the use of any pesticide found to pose an unreasonable risk. And similar rulings are given in the Toxic Substances Control Act (TSCA) for non-pesticide chemicals. Lastly, Executive Order 13514 signed by President Obama in 2009

FIGURE 21.4 Pollution Prevention Hierarchy

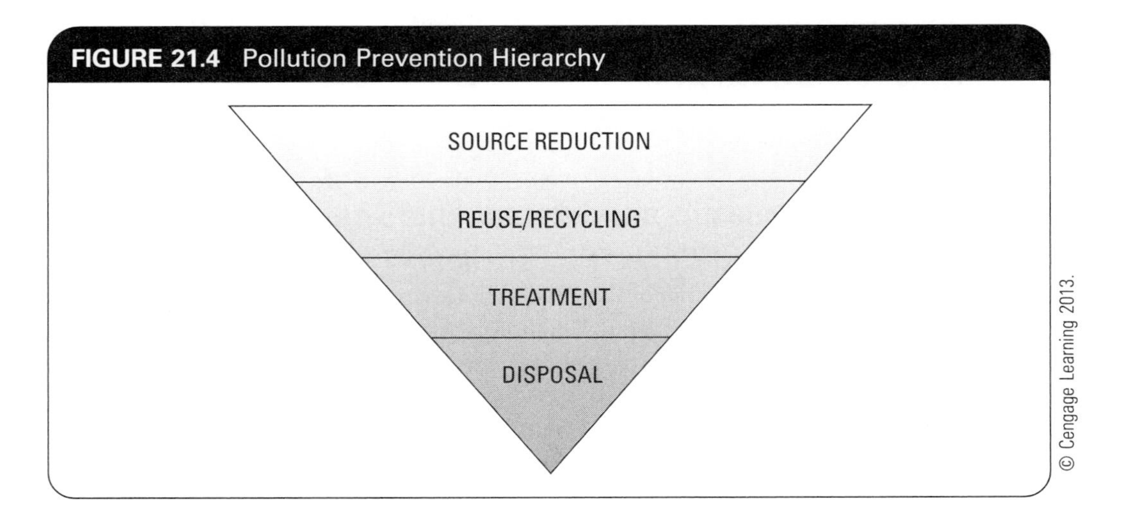

[10] Pollution Prevention Act of 1990, Sec. 6602(b).

establishes sustainability objectives for all federal agencies. Among these goals are to improve water efficiency by 26 percent by 2020, to achieve 50 percent waste diversion and recycling by 2015, and to decrease by 30 percent petroleum use in vehicle fleets by 2020.[11]

International P2 Legislation

National laws promoting pollution prevention are becoming more prevalent in countries around the world. A few examples will illustrate.

The European Union (EU) has developed a set of permitting rules called the Integrated Pollution Prevention and Control (IPPC) Directive 2008/1/EC, which calls for the use of pollution prevention by new and existing agricultural and industrial sources. This set of rules, which replaces the IPPC Directive of 1996, is implemented by outlining mandatory environmental conditions that must be met in order for a permit to be issued. Among these is to use appropriate pollution prevention, to prevent large-scale pollution, and to efficiently use energy.[12]

In Canada, the Canadian Environmental Protection Act (CEPA) of 1999 is considered to be a major legislative act and one of the primary ways in which the Canadian government achieves pollution prevention and sustainable development. CEPA's preamble specifically states:

> Whereas the Government of Canada is committed to implementing pollution prevention as a national goal and as the priority approach to environmental protection ...

The law also gives the Minister of the Environment additional authority to move toward sustainable development and to require pollution prevention plans from certain facilities. Also authorized in the law is the establishment of a national pollution prevention clearinghouse. In addition to sustainable development and pollution prevention, the guiding principles of CEPA 1999 include virtual elimination of persistent and bioaccumulative substances, use of an ecosystem approach, and the polluter-pays principle.[13]

Australia developed a National Strategy for Ecologically Sustainable Development (ESD), which outlines a framework for the promotion of ESD throughout the country. Included is a discussion of waste minimization and management in which the stated challenge specifically refers to avoiding hazardous waste generation. Australia also has adopted the preventive strategy of **cleaner production**, as defined by the United Nations Environment Programme (UNEP). Included in Australia's rationale for adopting cleaner production is what it terms a preventive approach, in which it argues:[14]

cleaner production
A preventive strategy applied to products and processes to improve efficiency and reduce risk.

> It is cheaper and more effective to prevent environmental damage than attempt to manage or fix it. Prevention requires going upstream in the production process to prevent the source of the problem instead of attempting damage control downstream.

---

[11]For more information on this executive order, visit **www.whitehouse.gov/administration/eop/ceq/sustainability**.

[12]European Commission (2011); Europa (2010).

[13]Environment Canada (October 2005; December 2004); CEPA (1999).

[14]Australia and New Zealand Environment and Conservation Council (March 1999); Australia, Ecologically Sustainable Development Steering Committee (December 1992).

Similarly, China has launched a number of initiatives aimed at making cleaner production among the nation's key strategies to achieve sustainable development. Chief among these initiatives is its Cleaner Production Promotion Law, which was approved in 2002 and took effect on January 1, 2003. In addition to supporting cleaner production, this legislation promotes higher utilization rate of resources, reducing and avoiding pollution generation, and sustainable development.[15]

Lastly, on a more aggregate level, a recent United Nations report asserts that decreases in SO_2 emissions in Europe and North America are due in part to pollution prevention strategies, such as fuel switching and coal washing.[16] Such outcomes likely reflect an observed shift in overall approach to environmental policy. For example, environmental policy across the EU moved from an end-of-pipe control approach in the 1980s to a more integrated approach in the 1990s, comprising both control and pollution prevention initiatives, and ultimately to a focus on prevention today.[17]

Corporate P2 and Sustainability Initiatives

Recognizing the cost implications of environmental laws and observing the trend toward pollution prevention, some forward-thinking firms have responded in a progressive manner rather than simply complying with new mandates. Although many business firms, from small proprietorships to Fortune 100 companies, are actively engaged in environmentalism and pollution prevention, a few stand out as being true pioneers of integrating environmental concerns into their corporate objectives.

3M Company[18]

Distinguished as one of the first U.S. companies to adopt a preventive approach, 3M Company launched its Pollution Prevention Pays (3P) Program over 35 years ago. The innovative plan relies on voluntary participation from its employees to find opportunities to reduce pollution while maintaining or improving financial performance. Over the course of the company's 3P Program, thousands of separate projects have been launched around the world—some 8,600 since 1975. According to corporate officials, the payoffs from these efforts have been substantial. Since the inception of the 3P Program in 1975 through 2011, over 3 billion pounds of pollution have been prevented, and cost savings attributable to the company's preventive initiatives total nearly $1.4 billion.

Today, the company pursues its sustainability objectives using a series of strategies, including an Environmental Health and Safety (EHS) Management System adopted in 2001, which uses a holistic approach to identifying problems and solutions, and Life Cycle Management (LCM), which requires a review of all new and existing 3M products. The firm also requires ISO 14001 certification for all its manufacturing facilities that operate transnationally.

---

[15]National Development and Reform Commission (2011); Cleaner Production Promotion Law (2002).

[16]United Nations Economic Commission for Europe (UNECE) (1979–2005), as cited in United Nations Environment Programme (2007), p. 73.

[17]United Nations Environment Programme (2007), p. 227.

[18]The following is drawn from 3M Company (October 16, 2011; 2011).

Not surprisingly, 3M has been recognized for its environmental and sustainability achievements as the recipient of numerous awards. Among these are the National Wildlife Federation's Environmental Achievement Award in 1989, the U.S. EPA's Stratospheric Ozone Protection Award in 1991, the U.S. EPA Climate Protection Award in 2005, and the 2005 Most Valuable Pollution Prevention Award from the National Pollution Prevention Roundtable.

In 2011, 3M set specific objectives to reduce its environmental footprint. These include reducing its volatile air emissions by 15 percent relative to 2010, reducing its waste generation by 10 percent compared to 2010, and improving its energy efficiency by 25 percent from 2005, by 2015. To learn more about 3M's sustainability programs, visit **http://solutions.3m.com/wps/portal/3M/en_US/3M-Sustainability/Global/**.

McDonald's Corporation[19]

One of the hallmarks of McDonald's commitment to pollution prevention is its long-term relationship with the Environmental Defense Fund (EDF). Formed in 1990, this was the first alliance between a Fortune 500 firm and an environmental group. In that sense, this was an innovative partnership that would set the stage for similar efforts to follow. The new collaborative formed a Waste Reduction Task Force, which in turn developed a Waste Reduction Action Plan (WRAP) that eventually grew to over 40 initiatives to reduce waste and recycle. In addition, the task force formulated a corporate environmental policy and integrated waste reduction efforts throughout McDonald's organization.

Although most strategies implemented recycling efforts through the McRecycle USA initiative, the task force also devised pollution prevention plans. Most were aimed at reducing McDonald's product packaging, accomplished in large part by reducing the thickness of cups, napkins, trashcan liners, and carry-out bags. As a result of these efforts, McDonald's decreased its packaging wastes by 150,000 tons over the 1990–1999 period.

Today, McDonald's continues its decades-long commitment to the environment and other socially responsible pursuits. One way this commitment is implemented is through its Eco-Filter packaging scorecard used to assess packaging changes, based on factors such as weight and recycled content. Another is through its management of a system of suppliers, which are held to the same high standards for product quality, safety, sustainability, and efficiency that are met by McDonald's. Taken together, the entire supply chain along with McDonald's are dedicated to what are called the "*Three E's of Responsibility: Ethical Responsibility, Environmental Responsibility, and Economic Responsibility*." For example, as part of the company's Sustainable Fisheries Program, McDonald's has established standards that govern the wild-caught fish it purchases, halting purchases from suppliers not meeting those standards until appropriate changes are made.

Furthering its efforts to be environmentally responsible, the firm is dedicated to reducing energy and water use and to generating less waste. Presently, about 80 percent of the packaging used in its nine largest markets is manufactured from renewable raw materials, i.e., paper or wood-fiber, and almost 30 percent of its packaging comes from recycled materials. It also is collaborating with its packaging supply chain to encourage

[19]Drawn from McDonald's Corporation (2011; 2010; June 18, 2002; May 1999).

sustainable forestry standards. For 2013, McDonald's has set sustainable goals that include using more energy efficient equipment in its restaurants, increasing energy awareness throughout the company, and employing its Supplier Environmental Scorecard to decrease environmental impacts of its supply chain. For more information on McDonald's sustainability efforts, visit **http://www.aboutmcdonalds.com/mcd/sustainability.html**.

BMW Group[20]

Recognized worldwide for its well-engineered automobiles, BMW Group also has earned a reputation as an environmentally conscious manufacturer. The Germany-based firm asserts that it is dedicated to the environment and that sustainability is integrated into every stage of its production. Part of this dedication is manifested in its "Design for Recycling" approach to vehicle construction, which facilitates recovery at the end of a vehicle's life and avoids pollution linked to waste disposal. (Refer back to Application 1.1 in Chapter 1 for more detail.) As further evidence of its commitment to pollution prevention, BMW signed the Cleaner Production Declaration of the United Nations Environment Programme in 2001.

In 2000, BMW Group adopted what it calls the *BMW Efficient Dynamics* strategy, which comprises innovations and technologies aimed at improving the vehicle's performance *and* reducing its fuel consumption. Among these technologies are brake energy regeneration, air flap control, and BMW's "Auto Start Stop Function," which automatically shuts off the car's engine if the vehicle has a manual transmission, is in neutral, and is not in motion. One outcome of this strategy was a reduction in CO_2 emissions from its fleet of automobiles in Europe by 30 percent between 1995 and 2010, forecasting further reductions by 25 percent or more by 2020 relative to 2008. Across its European fleet, BMW models now generate 148 grams of CO_2 per kilometer on average. The objective is to reduce these emissions to 140 grams of CO_2 or less by 2015.

In the next step in this *Efficient Dynamics* strategy, BMW is beginning to electrify drive trains by manufacturing hybrids, such as the BMW ActiveHybrid X6, and by producing electric vehicles, such as the BMW i3, starting in 2013. Based on BMW's forecasts, between 5 percent and 15 percent of all its vehicles will be fully or partially powered by electricity by 2020.

In 2010, for every vehicle produced, the automaker achieved a 9.6 percent decline in volatile organic compounds (VOCs), a 5.1 percent reduction in waste for disposal, and a 9.8 percent decline in water consumption compared to the previous year. BMW's overarching goal is to reduce the environmental impact of its production around the world, asserting that its resource use will have declined by 30 percent between 2006 and 2012. Ultimately, the company is dedicated to achieving zero-emissions production of its vehicles. For more on BMW's efforts to prevent pollution and promote sustainability, visit **www.bmwgroup.com**, and click on the "Responsibility" tab.

Following the model of McDonald's, 3M, BMW Group, and other corporate leaders in environmentalism, many companies have implemented their own pollution

[20]Drawn from BMW Group (2010), Chapters 2, 3.

prevention programs with encouraging results. By way of example, consider the selected corporate P2 initiatives given in Table 21.1.[21]

At a more aggregate level, a number of trade associations also have developed P2 programs for their members. The American Petroleum Institute (API), for example, has

TABLE 21.1 Corporate Pollution Prevention (P2) Initiatives

Company	Major Results and Objectives
Apple, Inc.	In 2010, avoided 27.5 million kilograms of CO_2-equivalent emissions due to renewable energy programs. Achieved a worldwide recycling rate of more than 70 percent in 2010, which is its goal through 2015. Offers take-back and recycling programs in 95 percent of the nations where products are sold, resulting in nearly 85,000 metric tons of equipment diverted from landfills since 1994. Manufactures all products to exceed Energy Star® requirements and to be free of arsenic, brominated flame retardants (BFRs), mercury, phthalates, and polyvinyl chloride (PVC).
The Boeing Company	Since 2002, reduced its CO_2 emissions by 28 percent, energy use by 30 percent, water intake by 41 percent, and hazardous waste generation by 44 percent. For the 2007–2012 period, set targets to increase the solid waste diversion from landfills by 75 percent and to reduce by 25 percent its greenhouse gas emissions, energy consumption, water intake, and energy usage.
Ford Motor Company	Reduced CO_2 emissions by 10.5 percent and 8.1 percent from U.S. and European new vehicles, respectively, in 2010 relative to the 2006 model year. As of 2010, offers four models in North America that achieve 40 miles per gallon or better. Since 2000, has saved 10.5 billion gallons of water across its production facilities worldwide. In 2010, has reduced its worldwide facility energy consumption per vehicle by over 13 percent from 2008 levels. Is committed to reducing CO_2 emissions in its facilities by 30 percent on a per-vehicle basis by 2025. Anticipates that electrified vehicles will comprise 10–25 percent of its worldwide sales by 2020.
General Electric Company	Launched its *Ecomagination* initiative in 2005, a business strategy dedicated to sustainability. As of 2010, met or surpassed every goal, including investing $5 billion in clean-tech R & D, reducing greenhouse gas releases by 22 percent, decreasing water usage by 30 percent, and saving $130 million from energy efficiency. Over the next five years, plans to increase spending on clean-tech R & D to $10 billion, reduce greenhouse gas releases and water use by 25 percent, and lower energy intensity by 50 percent.
Johnson & Johnson	By 2010, had reduced water usage by 9.5 percent from 2005 levels. Reduced CO_2 emissions by 23 percent from 1990 to 2010. Between 2005 and 2010, decreased offsite disposal of nonhazardous waste by 12 percent and hazardous waste by 25 percent. Is committed to eliminating polyvinyl chloride (PVC) in its packaging. Has a stated goal to set up an electronics take-back program in all regions to eventually accept 100 percent of electronic wastes for remanufacturing or reuse.
The Walt Disney Company	In 2010, reduced greenhouse gas emissions by 4.6 percent, lowered electricity usage by 6.6 percent, and increased its waste diversion rate from 46 percent to 60 percent since 2006. Since 2009, invested $15.5 million in offset projects to support reforestation and forest conservation. Set overall citizenship objectives to minimize water consumption, send no waste to landfills, and reach zero net direct greenhouse gas emissions.

NOTE: Some reported performance results and targets are revenue adjusted.

Sources: Apple, Inc. (2011a; 2011b); The Boeing Company (2011); Ford Motor Company (2010/2011); General Electric Company (2010); Johnson & Johnson (2011); The Walt Disney Company (2011).

[21]To examine a ranking of firms based on public perception of corporate social responsibility initiatives, which was created by the Boston College Center for Corporate Citizenship and the Reputation Institute, visit **www.bcccc.net/index.cfm?fuseaction=page.viewPage&pageID=2202&nodeID=1**, or read the report on their 2010 ranking at **www.bcccc.net/pdf/CSRIReport2010.pdf**.

devised a set of 11 environmental principles for petroleum facilities, one of which specifically sets a goal for each member to reduce emissions and waste generation and another calling for members to conserve natural resources through efficient energy use. To learn more, visit the API's *Environment, Health & Safety* page at **http://api-ec.api.org/environment-health-and-safety.aspx**.[22]

Using Economic Analysis to Implement Pollution Prevention

Firms must determine when preventive approaches should be employed, to what extent, and through which strategy. Economic criteria and their associated decision rules are just as applicable to guiding the use of preventive strategies as they are to any other policy instrument.

Cost-Effectiveness Criterion

Whether or not a firm adopts pollution prevention as part of its business strategy will depend to a large extent on the associated cost relative to other options, such as the use of treatment or abatement technologies. The goal is to select the least-cost option to achieve **cost-effectiveness** and enhance profit. Similarly, if the firm determines that pollution prevention *is* a cost-effective option, it then must select among the available strategies to implement it using the cost-effectiveness criterion as a guide. If, for example, the firm can access nontoxic raw materials more cheaply than converting its production processes, it will pursue the less costly strategy.

Given the importance of relative costs in this decision-making process, government can help promote pollution prevention through various means. For example, it could encourage the development of nontoxic chemical substitutes and cost-effective production changes through grants or subsidies. It might also develop ways to enhance the communication of effective pollution prevention strategies to industry. Another approach is for government to assist firms in making accurate cost assessments about pollution prevention strategies. This is precisely the motivation of **environmental accounting**. To learn more, visit **http://unstats.un.org/unsd/envaccounting/ceea/default.asp** at the United Nations Statistics Division.

Efficiency Criterion

It is also the case that **benefit-cost analysis** can help determine the extent to which a preventive strategy should be employed. Qualitatively, we can identify the associated benefits to the firm as abatement cost savings, the avoidance of regulatory penalties for noncompliance, and the gain in revenues associated with presenting an environmentally responsible image to consumers. On the cost side, the firm has to assess the search costs of identifying available options, the administrative costs in analyzing these options, engineering design expenses, and the expenditures to implement the strategy, such as retooling or capital investment. To achieve the **allocatively efficient** level at which a pollutant is eliminated, the decision rule is the same as in other contexts. The firm should find the point where the marginal cost of doing so is exactly offset by the marginal benefit.[23]

---

[22] American Petroleum Institute (March 7, 2011).

[23] For an economic analysis of pollution prevention, see Helfand (October 1994).

STRATEGIC INITIATIVES AND PROGRAMS

Sustainable development and approaches like pollution prevention and industrial ecology call for significant changes in how consumers and firms make decisions and, in particular, how firms design, produce, and market their products. If businesses perceive these changes as counter to their self-interest, the changes will be met with resistance at best. To foster a positive evolution in thinking and business planning, governments are working with industry and environmental groups to share information, to foster technological development, and to find solutions that satisfy both environmental and business objectives.

Although the perspective of sustainable development is global, initiatives promoting sustainability have to begin within national borders. Just as is the case for international agreements, these domestic initiatives require cooperation, but of a different sort. Often, the collaborative effort is launched by government but includes stakeholders from other economic sectors, and typically such efforts are based on voluntary partnerships. In the United States, such arrangements are generally launched by the EPA. An overview of some of these programs with links to their Web pages is given in Table 21.2. More complete information is available at **www.epa.gov/partners/**.

Of course, sustainable development programs are not exclusive to the United States. Three important initiatives that are emerging in many nations are **Extended Product**

TABLE 21.2 Selected Voluntary Partnerships

Program Name	Year Launched	Objective
Adopt Your Watershed **http://water.epa.gov/action/adopt/index.cfm**	1994	To facilitate citizen involvement in protecting and restoring local watersheds.
Design for the Environment (DfE) **www.epa.gov/dfe**	1992	To reduce risk and prevent pollution by helping firms integrate environmental considerations into product and process design.
Energy Star **www.energystar.gov/**	1994	To promote energy-efficient products and practices to consumers and firms.
Green Chemistry or Sustainable Chemistry **www.epa.gov/greenchemistry/**	1992	To support chemical technologies that reduce or eliminate generation of hazardous substances.
Green Power Partnership **www.epa.gov/greenpower/basic/index.htm**	Not available	To encourage the purchase of green power to reduce the effects of electricity generation.
PestWise Pesticide Environmental Stewardship Program (PESP) **www.epa.gov/pestwise/pesp/**	1994	To promote pesticide risk reduction and go beyond regulations to a higher level of pesticide stewardship.
Product Stewardship/Extended Product Responsibility **www.epa.gov/epawaste/partnerships/stewardship/index.htm**	Not available	To promote identifying and reducing any life cycle environmental impacts of products.
WasteWise **www.epa.gov/epawaste/partnerships/wastewise/index.htm**	1994	To promote cost savings and efficiency through source reduction and recycling.

Source: U.S. EPA (August 2, 2011); U.S. EPA, Office of Policy, Economics, and Innovation (January 2001), Table 10.3; information from the programs' respective Web pages.

Responsibility (EPR), **Design for the Environment (DfE)**, and the **Green Chemistry Program**. A brief discussion of these efforts, which are interrelated, follows.

Extended Product Responsibility (EPR)[24]

Extended Product Responsibility (EPR)
A commitment by all participants in the product cycle to reduce any life cycle environmental effects of products.

Extended Product Responsibility (EPR) refers to efforts aimed at identifying and reducing any life cycle environmental effects of products. Also known as **product stewardship**, EPR is closely linked to sustainable development and is preventive in orientation. The underlying premise of EPR is that *all* participants in the product chain—designers, manufacturers, retailers, consumers, recyclers, remanufacturers, and disposers—are responsible for a product's effect on the environment. This is a more extensive approach than **extended *producer* responsibility** adopted by some Asian countries, Europe, and certain Canadian provinces, which places essentially all the onus of product responsibility on the firm.

Under extended product responsibility (EPR), all players in a product cycle are expected to participate. Consumers, for example, should make responsible market decisions when choosing a product. They can vote with their dollars for companies that use less packaging or less energy. Of course, manufacturers typically have the largest role to play, and under EPR guidelines, they are expected to make production decisions with the environment in mind by addressing the following:

- Raw materials selection: reducing the amount and toxicity of raw materials used
- Production impacts: seeking opportunities to reduce waste and energy consumption
- Product use: designing products to use fewer resources and less waste
- Products at end of life: offering take-back programs whereby used products are collected for recycling or **remanufacturing**.

remanufacturing
Collection, disassembly, reconditioning, and reselling of the same product.

Some countries have federal laws to require certain of these EPR approaches, such as Germany's Packaging Ordinance of 1991 (discussed in Chapter 18), which requires businesses to collect and recycle their own packaging. In the United States, EPR is essentially a voluntary program, though some state programs mandate EPR-like efforts, such as requiring manufacturers to take back certain types of used batteries. But even in the absence of federal laws, some firms are participating in EPR because it is in their best interests to do so. For example, firms engage in remanufacturing because it is lucrative as well as being good for the environment, as discussed in Application 21.3.

Design for the Environment (DfE)
Promotes the use of environmental considerations along with cost and performance in product design and development.

Design for the Environment (DfE)[25]

Based on the engineering notion of "design for X," where X might be any desirable product attribute such as durability, **Design for the Environment (DfE)** is an initiative launched by the EPA in 1992 that promotes the use of environmental considerations along with cost and performance in product design and development. The DfE program in the United States is voluntary and hence relies on partnerships formed between the

[24]Drawn from U.S. EPA, Office of Solid Waste and Emergency Response (August 2, 2011; December 1998).

[25]This discussion is drawn from U.S. EPA, Office of Pollution Prevention and Toxics (August 18, 2011).

APPLICATION 21.3 An Update on Remanufacturing: A Lucrative Approach to Pollution Prevention

In an effort to be environmentally responsible, improve corporate image, and save costs, some firms are participating in a practice known as **remanufacturing**. Combined with a take-back program, remanufacturing is a process whereby a used product is collected, disassembled (sometimes called demanufacturing), reconditioned to "as good as new" working condition, reassembled, and resold.

How is this different from recycling? According to remanufacturing industry expert Robert T. Lund of Boston University, remanufacturing attempts to capture the value-added of energy, labor, and manufacturing when the good was originally produced. Recycling, on the other hand, captures only the recoverable raw materials in the product, which represent a much smaller proportion of the good's value. Energy savings are also substantial. Lund estimates that remanufacturing saves 120 trillion Btus of energy each year on a worldwide scale, which translates to about $500 million in energy savings.

Though not all products are good candidates for this process, there are many remanufactured goods on the market today, including automotive parts, toner cartridges, photocopiers, personal computers, tires, telephones, and medical equipment. In the United States, the largest single remanufacturer is the U.S. Department of Defense, which refurbishes weaponry and military equipment on a regular basis to cut costs.

Both small and large companies also participate in this process. One example is Xerox Corporation, which began its remanufacturing program in the 1990s by encouraging its customers to return used electronic equipment. Its product take-back and recycling program continues into the present. The company's well-established remanufacturing process involves converting returned equipment into products or parts that meet the same performance and quality standards of new goods. Xerox also accepts spent cartridges and toner containers through its Green World Alliance program. In 2009, more than 2.2 million of these imaging supplies were returned. These supplies are cleaned, inspected, and then recycled or remanufactured. Because of these take-back programs, over 45,000 metric tons were diverted from landfills in 2009.

With a similar motivation, Caterpillar, Inc. uses remanufacturing to keep costs down and, in so doing, maintains lower prices for its customers. This heavy equipment and engine manufacturer has been offering remanufacturing of its own products for over 30 years. The firm views the practice as both economically and environmentally sound. In fact, despite the economic downturn in 2001, company officials recognized that remanufacturing revenues were holding up. So gradually, the firm began to acquire remanufacturing operations worldwide. Eventually in 2006, the company launched its own remanufacturing division—Caterpillar Remanufacturing Services, or Cat® Reman. Today, Caterpillar's technicians are engaged in disassembling, repairing, and reassembling parts of trucks, semitrailers, and bulldozers. In a recent sustainability report, Caterpillar reports that it recycles more than 2 million end-of-life parts every year, which comprise 3.4 million pounds of wood materials, over 130 million pounds of iron, and 200,000 pounds of cardboard. Overall the Cat® Reman program has achieved a return rate of higher than 93 percent.

Remanufacturing has grown into a substantial market, estimated to be a $100-billion-per-year industry worldwide. This estimate is not surprising, given the estimated cost savings associated with this process, which underscores how both economic gains and environmental benefits are powerful motivations for continued growth in remanufacturing. More information about remanufacturing is available at the Web site of The Remanufacturing Institute (TRI), **www.reman.org**.

Sources: Caterpillar, Inc. (2010; April 23, 2009); Xerox Corporation (2010); Brat (July 5, 2006); The Remanufacturing Institute (June 19, 2002), citing Lund in Klein and Miller (1993); Ginsburg (April 16, 2001); U.S. EPA, Office of Solid Waste and Emergency Response (May 1997).

EPA and industry. This initiative also has a **Safer Product Labeling Program** in which a DfE logo is used on products that have been screened for safety by a DfE scientific team. As of 2011, the program is responsible for identifying the safety of more than 2,500 products and for eliminating 700 million pounds of potentially dangerous chemicals. In time, the EPA will require manufacturers to disclose all ingredients to consumers on any DfE-labeled product.

The DfE program directly supports a **cyclical flow of materials**, inherently recognizing that all stages of the product life cycle are important to environmental protection. Look back at Figure 21.3 (page 508), and note the flows labeled "Design for recycling" and "Design for reuse." These flows are important elements of a DfE program and have been adopted by a number of major corporations like 3M, Dell Computer, Hewlett-Packard, and SC Johnson.

By way of example, Hewlett-Packard launched its DfE program in 1992. Its primary DfE goals are energy efficiency, materials innovation, and design for recyclability. These objectives are implemented through various internal controls. Among these are: assigning environmental stewards to every product design team; using more recycled materials in product packaging; and using fewer kinds of materials, including standardizing on the plastic resins used.[26]

Another representative case is SC Johnson, maker of well-known household products like Pledge® and Windex®, which became a partner in the DfE program and is the first consumer packaged goods firm to have done so. The 125-year-old company identifies eight consumer products on its Web page that are currently part of the DfE program. As further evidence of SC Johnson's commitment to DfE principles, the company was identified as a "Champion" in the DfE's Safer Detergents Stewardship Initiative for its efforts to use safer surfactants, which are wetting agents that facilitate cleaning products.[27]

Green Chemistry Program[28]

Green Chemistry Program
Promotes the development of innovative chemical technologies to achieve pollution prevention.

Another continuing trend in environmental protection is the development of **Green Chemistry** programs, also known as benign chemistry or sustainable chemistry programs. The **Green Chemistry Program** in the United States promotes the research, development, and application of innovative chemical technologies to achieve pollution prevention in ways that are both scientifically grounded and cost-effective. Currently, this program is partnered with a broad-based group of entities. Among these are scientific organizations, such as the American Chemical Society, firms, including Monsanto and Dow Corning Corporation, trade associations like the American Petroleum Institute, and academia, including the University of Massachusetts at Lowell.

Like DfE initiatives, green chemistry recognizes the importance of a product's life cycle in pollution prevention. By finding alternatives to hazardous chemicals that are safer, health and ecological risks are reduced, production processes are safer, and final products pose less of a threat at the end of their economic lives. To realize their objectives, green chemistry programs support research, education, conferences, and international activities in environmentally benign chemistry.

In 1998, the Advisory Group of Risk Management for the Organisation for Economic Co-operation and Development (OECD) gave its support for a new sustainable chemistry project for OECD member countries. Among the nations electing to coordinate the effort were Germany, Italy, Japan, Mexico, and the United States.

[26]Hewlett-Packard Company (2011).

[27]S. C. Johnson & Son, Inc. (2011).

[28]This discussion is drawn from U.S. EPA, Office of Chemical Safety and Pollution Prevention, Office of Pollution Prevention and Toxics (August 10, 2011a; August 10, 2011b).

Ultimately, a steering group for the OECD Sustainable Chemistry Initiative was formed, consisting of 40 representatives from 10 nations. The group developed a detailed work plan to implement what it believes to be the highest priority activities. These include supporting the research and development of innovative sustainable chemistry technologies, disseminating information related to sustainable chemistry, and integrating principles of sustainable chemistry into chemistry education.

Disseminating Information and Technology on a Global Scale

Strategic advances and related environmental technologies are important to achieving economic growth while preserving the natural environment. However, they are generally developed and implemented within industrialized nations and hence need to be communicated globally, particularly to third world countries. In order for this to occur, there must be an infrastructure for sharing knowledge, and there must be an awareness of environmental issues.

Technology Transfer

technology transfer The advancement and application of technologies and strategies on a global scale.

It is often argued that **technology transfer** is critical to consistent progress toward sustainable development. This concept refers to the advancement and application of technologies and management strategies throughout the world. Environmental technologies cover a broad range of products and services. Included are the so-called **dark green technologies** that deal with the control, abatement, monitoring, and remediation of pollution. There are also more indirect but nonetheless important advances, dubbed **light green technologies**. These refer to strategies or production changes that benefit the environment, even though that is not their primary intent. An example is the use of electronic mail in place of hard-copy memos, a change that reduces waste generation.[29] Effecting technology transfer relies on a number of interdependent factors—among them, research, physical capital investment, communication, financial resources, and education. Learn more at the Technology Transfer Network at **http://epa.gov/ttn**.

Environmental Literacy

environmental literacy Awareness of the risks of pollution and natural resource depletion.

In order for knowledge and technology to be effectively disseminated, people everywhere need to be made aware of environmental risks and the importance of responding to those risks in a responsible way. The promotion of environmental education across the globe has grown over time and was an important theme at the 1992 Earth Summit in Rio. *Agenda 21* specifically references the importance of education, public awareness, and training in helping to implement the global agenda.[30] The aim is to advance **environmental literacy** across all regions of the world. Without this awareness, society will not understand the need for change, will tend not to support it, and may be unwilling to participate in the process. For further information on promoting environmental education, visit **www.epa.gov/enviroed/index.html**.

[29]"What Is Environmental Technology?" (Fall 1994).

[30]Parson, Haas, and Levy (October 1992).

CONCLUSIONS

Over the past two decades, many nations have made measurable progress in moving closer to what have become universal environmental objectives, specifically, environmental quality, biodiversity, and sustainable development. With greater emphasis on pollution prevention, product life cycle, and environmental literacy, many believe the progress can continue over the long run. That said, there is still much more to be done. Aggregate statistics disguise those regions where environmental pollution is particularly severe. There are entire nations in Asia, Africa, South America, and Central America that face extremely poor air quality, contaminated drinking water, and serious degradation of their land.

The need for setting a global environmental agenda and implementing it is more than apparent. Identifying an appropriate course of action is not easy, and it takes time. This collective effort calls for cooperation from every market sector and from every nation, a process that is now under way. As this important transition evolves, society must move forward to restore what can be repaired, to launch initiatives to prevent further degradation, and to educate people everywhere about the importance and fragility of the natural environment. This commitment was advanced at the 1992 Earth Summit in Rio and is articulated in the preamble of *Agenda 21*:

> Humanity stands at a defining moment in history. We are confronted with a perpetuation of disparities between and within nations, a worsening of poverty, hunger, ill health, and illiteracy, and the continuing deterioration of the ecosystems on which we depend for our well-being. However, integration of environment and development concerns and greater attention to them will lead to the fulfillment of basic needs, improved living standards for all, better protected and managed ecosystems and a safer, more prosperous future. No nation can achieve this on its own; but together we can—in a global partnership for sustainable development.[31]

[31]United Nations, Division for Sustainable Development, Chapter 1, paragraph 1.1 (December 17, 2004).

Summary

- Industrial ecology is a systems approach to the flow of materials and energy between industrial processes and the environment. Its main purpose is to promote the use of recycled wastes from one industrial process as inputs in another.
- A linear flow of materials assumes that materials run in one direction, entering an economic system as inputs and leaving as residuals. This cradle-to-grave flow emphasizes use, waste generation, and disposal. Most national policy focuses on abating residuals at the end of the flow.
- Product design, manufacturing processes, and energy use can be modified to achieve a cyclical flow of materials, or a cradle-to-cradle flow. A life cycle assessment (LCA) examines the environmental impacts of a good at all product stages. A major component of the new ISO 14000 standards for environmental management addresses life cycle assessment.
- The real-world outcome of implementing a closed system is the formation of an industrial ecosystem whereby residuals from one or more manufacturing processes are reused as inputs for others. The most well-known example is the ecosystem in Kalundborg, Denmark.
- Pollution prevention (P2) is a long-term approach aimed at reducing the amount or toxicity of residuals released to the environment. These major objectives are implemented in part through the EPA's Pollution Prevention (P2) Program.
- National laws promoting pollution prevention are becoming more prevalent. In the United States, Congress passed the Pollution Prevention Act of 1990. The European Union (EU) has developed the Integrated Pollution Prevention and Control (IPPC) Directive 2008/1/EC for its members. The Canadian Environmental Protection Act (CEPA) of 1999 is one of the primary ways that Canada achieves pollution prevention and sustainable development. China's Cleaner Production Promotion Law, which took effect in 2003, is part of China's strategy to achieve sustainable development.
- The economic criteria of cost-effectiveness and allocative efficiency, along with their associated decision rules, can be used to guide a firm's use of preventive strategies.
- To support sustainable development and the use of preventive approaches, various initiatives are emerging in many countries, including Extended Product Responsibility (EPR), Design for the Environment (DfE), and the Green Chemistry Program.
- Extended Product Responsibility (EPR), sometimes known as Product Stewardship, refers to efforts aimed at identifying and reducing any life cycle environmental effects of products. Design for the Environment (DfE) advances the use of environmental considerations along with cost and performance in product design and development. The Green Chemistry Program promotes the research, development, and application of innovative chemical technologies to achieve pollution prevention in ways that are scientifically grounded and cost-effective.
- Critical to consistent progress toward sustainable development is technology transfer, which refers to the advancement and application of technologies and management strategies throughout the world.
- Environmental literacy, achieved through communication and education, is part of an effective strategy to preserve and protect the earth's resources.

Review Questions

1. Assume you are a city planner working on a new industrial park and contemplating the use of an industrial ecosystem. Discuss the major advantages and disadvantages of an industrial ecosystem that you would consider in making your decision.
2. Choose a product that negatively affects the environment, assuming a linear flow of materials.

Then, use a cyclical materials flow approach, and conduct a hypothetical life cycle assessment (LCA), pointing out at least two preventive initiatives that would reduce environmental risk.

3. **a.** Identify the economic incentives that motivate private firms to engage in pollution prevention activities.
   b. How might the government devise policy initiatives to exploit these natural incentives?
4. Visit the Web site of Environmental Defense Fund, and review the list of its recent corporate partnerships at **www.edf.org/approach/partnerships/corporate**. Select one, and summarize the cooperative efforts between the Environmental Defense Fund and a private firm. Identify some of the potential environmental and economic benefits associated with that partnership.
5. Extended Product Responsibility (EPR) assumes that all participants in the product cycle play a role in finding ways to reduce environmental risk. Identify the specific role played by the average consumer in this effort.
6. Environmental technology is argued to be an important element in society's effort to achieve sustainable development.
   a. Choose a specific market-based instrument that likely would encourage the advance of dark green technologies *and* improve U.S. exports of these goods and services. Explain using economic analytical tools.
   b. Now propose a different market-based instrument that would foster the use of more light green technologies domestically and internationally.

7. Consider a corporation's decision to undertake a pollution prevention program as part of its production process. Briefly describe how this undertaking might affect the market for the company's product, and use a supply and demand model to illustrate the potential changes and the ultimate effect on equilibrium price and quantity.

Additional Readings

Alberini, A., and K. Segerson. "Assessing Voluntary Programs to Improve Environmental Quality." *Environmental and Resource Economics 22*(*1-2*) (June 2002), pp. 157–84.

Anastas, Paul T., and Joseph J. Breen. "Design for the Environment and Green Chemistry: The Heart and Soul of Industrial Ecology." *Journal of Cleaner Production 5*(*1-2*) (1997), pp. 97–102.

Ayres, Robert U., and Leslie W. Ayres, eds. *A Handbook of Industrial Ecology*. Northampton, MA: Elgar, 2002.

Banzhaf, Spencer. "Accounting for the Environment." *Resources 151* (Summer 2003), pp. 6–10.

Callan, Scott J., and Janet M. Thomas. "Corporate Financial Performance and Corporate Social Performance: An Update and Reinvestigation." *Corporate Social Responsibility and Environmental Management 16*(*2*) (March/April 2009), pp. 61–78.

DeSimone, Livio D., and Frank Popoff. *Eco-Efficiency: The Business Link to Sustainable Development*. Cambridge, MA: MIT Press, 1997.

Ehrenfeld, John, and Nicholas Gertler. "Industrial Ecology in Practice: The Evolution of Interdependence at Kalundborg." *Journal of Industrial Ecology 1*(*1*) (Winter 1997), pp. 67–79.

Geiser, Kenneth. *Materials Matter: Toward a Sustainable Materials Policy*. Cambridge, MA: MIT Press, 2001.

Geng, Yong, Pan Zhang, Raymond P Côté, and Tsuyoshi Fujita. "Assessment of the National Eco-Industrial Park Standard for Promoting Industrial Symbiosis in China." *Journal of Industrial Ecology 13*(*1*) (February 2009), pp. 15–26.

Hay, Bruce L., Robert N. Stavins, and Richard H. K. Vietor. *Environmental Protection and the Social Responsibility of Firms: Perspectives From Law,*

Economics, and Business. Washington, DC: Resources for the Future, 2005.

Hecht, Joy E. *National Environmental Accounting: Bridging the Gap Between Ecology and Economy*. Washington, DC: Resources for the Future, January 2005.

Jaffe, Adam B., Richard G. Newell, and Robert N. Stavins. "Environmental Policy and Technological Change." *Environmental and Resource Economics 22*(*1-2*) (June 2002), pp. 41–70.

Jain, Subhash C., and Ben L. Kedia, eds. *Enhancing Global Competitiveness Through Sustainable Environmental Stewardship*. Northampton, MA: Elgar, 2011.

Köhn, Jörg, John Gowdy, and Jan van der Straaten, eds. *Sustainability in Action*. Northampton, MA: Elgar, 2001.

Lesourd, Jean-Baptiste, and Steven G. M. Schilizzi. *The Environment in Corporate Management: New Directions and Economic Insights*. Northampton, MA: Elgar, 2002.

Lyon, Thomas P., and John W. Maxwell. *Corporate Environmentalism and Public Policy*. New York: Cambridge University Press, December 2004.

O'Brien, Mary. *Making Better Environmental Decisions: An Alternative to Risk Assessment*. Cambridge, MA: MIT Press, 2000.

Preston, Lynelle. "Sustainability at Hewlett-Packard: From Theory to Practice." *California Management Review 43*(*3*) (Spring 2001), pp. 26–37.

Quaddus, M. A., and M. A. B. Siddique, eds. *Handbook of Corporate Sustainability*. Northampton, MA: Elgar, 2011.

Rosa, Eugene A., Andreas Diekmann, Thomas Dietz, and Carol C. Jaeger, eds. *Human Footprints on the Global Environment: Threats to Sustainability*. Cambridge, MA: MIT Press, 2009.

Sharma, Sanjay, and Mark Starik. *Research in Corporate Sustainability*. Northampton, MA: Elgar, 2003.

van den Bergh, Jeroen C. J. M., and Marco A. Janssen, eds. *Economics of Industrial Ecology*. Cambridge, MA: MIT Press, 2005.

Vidovic, Martina, and Neha Khanna. "Can Voluntary Pollution Prevention Programs Fulfill Their Promises? Further Evidence From the EPA's 33/50 Program." *Journal of Environmental Economics and Management 53*(*2*) (March 2007), pp. 180–95.

To access additional course materials, visit www.cengagebrain.com. At the home page, search for the ISBN of this title (shown on the back cover). This will take you to the product page where these resources can be found.

References

Abdalla, Charles (1990). "Measuring Economic Losses From Ground Water Contamination: An Investigation of Household Avoidance Cost." *Water Resources Bulletin 26(3)* (June), pp. 451–63.

Abrahams, Nii Adote, Bryan J. Hubbell, and Jeffrey L. Jordan (2000). "Joint Production and Averting Expenditure Measures of Willingness to Pay: Do Water Expenditures Really Measure Avoidance Costs?" *American Journal of Agricultural Economics 82(2)* (May), pp. 427–37.

Ackley, Kate (2011). "Everybody Has a Rule They Hate." *Roll Call* (February 8). Available at **www.rollcall.com/issues/56_79/-203188-1.html**.

Adler, Jerry, and Mary Hager (1992). "How Much Is a Species Worth?" *National Wildlife* (April/May), pp. 4–14.

Adler, Robert W., Jessica C. Landman, and Diane M. Cameron (1993). *The Clean Water Act: 20 Years Later*. Washington, DC: Island Press.

"A Global Love Affair" (2008). *The Economist* (November 15), pp. 3–4.

Aguilar, Soledad (2011). "UNFCCC/Kyoto Protocol: Outcomes of the Cancún Conference." *Environmental Policy and Law 41(1)* (February), pp. 10–14.

"Air Pressure; Global Warming" (2007). *The Economist* (April 7), p. 28.

Airlie, Catherine, and Mathew Carr (2011). "U.K. Fixes CO_2 Levy at 4.94 Pounds a Ton for Utilities in 2013." *Bloomberg.com* (March 23). Available at **www.bloomberg.com/news/2011-03-23/britain-sets-minimum-emission-costs-at-16-pound-a-ton-starting-in-2013.html**.

Allen, Scott (1992). "Boston Harbor's Waters Have Started to Heal: Cleanup Helping to Shed 'Dirtiest' Label." *Boston Globe* (September 6), p. 1.

——— (1993). "Rights to Pollute Given Up." *Boston Globe* (March 20), p. 35.

Alliance of Automobile Manufacturers (2004, September 23). *New Auto Fuel Economy Regulation: Nothing in It for Californians*. Washington, DC: Alliance of Automobile Manufacturers.

American Chemistry Council (2009a, April 14). "Responsible Care® Toolkit: Guiding Principles." Available at **www.americanchemistry.com/s_rctoolkit/doc.asp?CID=1779&DID=6684** (accessed April 14, 2009).

——— (2009b, April 14). "Responsible Care® Toolkit: History of Responsible Care." Available at **www.americanchemistry.com/s_rctoolkit/sec.asp?CID=1787&DID=6615** (accessed April 14, 2009).

——— (2011, July). *Responsible Care® Fact Sheet.* Available at **http://responsiblecare.americanchemistry.com/Performance-Results/Responsible-Care-Fact-Sheet.pdf**.

American Petroleum Institute (2011, March 7). "Environmental, Health, and Safety: Environmental Principles." Available at **http://api-ec.api.org/aboutapi/principles/index.cfm** (last updated March 7, 2011).

Andrews, Richard N. L. (1984). "Economics and Environmental Decisions, Past and Present." In V. Kerry Smith, ed., *Environmental Policy Under Reagan's Executive Order: The Role of Benefit-Cost Analysis*. Chapel Hill, NC: University of North Carolina Press, pp. 43–85.

Ansfield, Jonathan (2007). "The Coal Trap; Beijing Battles for Control of a Runaway Industry That Both Powers China, and Threatens Its Future." *Newsweek [International Edition]* (January 15).

Apple, Inc. (2011a). *Apple and the Environment: The Story Behind Apple's Environmental Footprint*. Cupertino, CA: Apple, Inc. Available at **www.apple.com/environment/**.

——— (2011b). *Facilities Report: 2011 Environmental Update.* Cupertino, CA: Apple, Inc. Available at **http://images.apple.com/environment/reports/docs/Apple_Facilities_Report_2011.pdf**.

Arndt, Michael (2001). "Maybe It's Not So Hard Being Eco-Friendly." *BusinessWeek* (April 16), p. 128F.

"The Art of the Possible" (2008). *The Economist* (November 15), pp. 18–20.

Ashford, Nicholas A., and Charles C. Caldart (2008). *Environmental Law, Policy, and Economics: Reclaiming the Environmental Agenda*. Cambridge, MA: MIT Press.

"Asia: Don't Drink the Water and Don't Breathe the Air; Environmental Protection in China" (2008). *The Economist* (January 26), p. 65.

Atkinson, Scott E., and Donald H. Lewis (1974). "A Cost-Effectiveness Analysis of Alternative Air Quality Control Strategies." *Journal of Environmental Economics and Management 1*, pp. 237–50.

Austin, Duncan, Niki Rosinski, Amanda Sauer, and Colin Le Duc (2003). *Changing Drivers: The Impact of Climate Change on Competitiveness and Value Creation in the Automotive Industry*. Washington, DC: World Resources, Inc. and Sustainable Asset Management.

Australia, Ecologically Sustainable Development Steering Committee (1992, December). *National Strategy for Ecologically Sustainable Development (NSESD)*. Australia: Australia, Ecologically Sustainable Development Steering Committee.

Australia and New Zealand Environment and Conservation Council (1999, March). *Toward Sustainability: Achieving Cleaner Production in Australia*. Australia: Environment Australia.

Ball, Jeffrey (2007). "Gas Leak: Kyoto's Caps on Emissions Hit Snag in the Marketplace; U.N. Mulls How to Fix Pollution-Credit System." *Wall Street Journal* (December 3), p. A1.

——— (2008). "Environment (A Special Report); Six Products, Six Carbon Footprints: Everybody's Talking About It." *Wall Street Journal* (October 6), p. R1.

Banerjee, Neela (2011). "EPA Chief Rebukes GOP Assault on Clean Air Act; Republicans Have Drafted a Bill That Would Limit Regulation of Greenhouse Gases." *Los Angeles Times* (February 10), p. A12.

Banzhaf, Spencer (2003). "Accounting for the Environment." *Resource 151* (Summer), pp. 6–10.

Barrett, Devlin, and Paul Glader (2010). "EPA Presses GE on Cleanup of River." *Wall Street Journal* (December 18), p. A3.

Bartik, Timothy (1988). "Evaluating the Benefits of Non-marginal Reductions in Pollution Using Information on Defensive Expenditures." *Journal of Environmental Economics and Management 15,* pp. 111–27.

Battery Council International (2011, August 14). "Summary of U.S. State Lead-Acid Battery Laws." Available at **www.batterycouncil.org/LeadAcidBatteries/BatteryRecycling/StateRecyclingLaws/tabid/120/Default.aspx** (accessed August 14, 2011).

Baumol, W. J., and W. E. Oates (1975). *The Theory of Environmental Policy.* Englewood Cliffs, NJ: Prentice Hall, Inc.

Beckerman, Wilfred (1990). "Global Warming: A Sceptical Economic Assessment." In Dieter Helm, ed., *Economic Policy Towards the Environment.* Cambridge, MA: Blackwell Publishers, pp. 52–85.

Biello, David (2010). "Slick Solution: How Microbes Will Clean Up the Deepwater Horizon Oil Spill." *Scientific American* (May 25). Available at **www.scientificamerican.com/article.cfm?id=how-microbes-clean-up-oil-spills**.

Biers, Dan (1994). "China Creates Environmental Nightmare in Industrial Rush." *Brockton Enterprise* (June 16), p. 11.

Bingham, Tayler H., Timothy R. Bondelid, Brooks M. Depro, Ruth C. Figueroa, A. Brett Hauber, Suzanne J. Unger, and George L. Van Houtven (1998). *A Benefits Assessment of Water Pollution Control Programs Since 1972.* Revised Draft Report to the U.S. Environmental Protection Agency. Research Triangle Park, NC: Research Triangle Institute.

Block, Debbie Galante (1993). "CD Jewel Box Only, or Alternatives Too?" *Tape-Disc Business* (June), p. 12.

BMW Group (2001). *Sustainable Value Report 2001/2002.* Munich, Germany: BMW Group.

——— (2003). *Sustainable Value Report 2003/2004.* Munich, Germany: BMW Group.

——— (2007). *Sustainable Value Report 2007/2008.* Munich, Germany: BMW Group (September).

——— (2009). *Sustainability by Design: Taking Responsible Action.* Munich, Germany: BMW Group (September).

——— (2010). *Sustainable Value Report 2010.* Munich, Germany: BMW Group.

The Boeing Company (2011). *2011 Summary Environment Report.* Chicago, IL: The Boeing Company. Available at **www.boeing.com/aboutus/environment/environment_report_11/pdf/2011_Env_Report_Summary_English.pdf**.

Bowker, James, and John R. Stoll (1988). "Use of Dichotomous Choice Nonmarket Methods to Value the Whooping Crane Resource." *American Journal of Agricultural Economics 70(2)* (May), pp. 372–81.

Bradsher, Keith (2003). "China's Boom Adds to Global Warming Problem." *New York Times* (October 22).

——— (2008). "China Reports Declines in 3 Major Pollutants, Reversing Trend." *New York Times* (June 6), p. A12.

Brat, Ilan (2006). "Caterpillar Gets Bugs Out of Old Equipment; Growing Remanufacturing Division Is Central to Earnings-Stabilization Plan." *Wall Street Journal* (July 5), p. A16.

Bratton, D., and G. L. Rutledge (1990). "Pollution Abatement and Control Expenditures, 1985–1988." *Survey of Current Business 70,* pp. 32–38.

Bresnahan, Brian W., Mark Dickie, and Shelby Gerking (1997). "Averting Behavior and Urban Air Pollution." *Land Economics 73(3)* (August), pp. 340–57.

Bristol-Myers Squibb Company (2004). "The Taxol® (Paclitaxel) Story." Available at **www.taxol.com/timeli.html** (accessed October 19, 2004).

British Soft Drinks Association (2009, March 24). "Bottled Water Information; Legislation and Regulation." London. Available at **www.bottledwaterinformation.co.uk/default.asp?section=5** (accessed March 24, 2009).

Brookshire, D. S., and T. D. Crocker (1981). "The Advantages of Contingent Valuation Methods for Benefit Cost Analysis." *Public Choice 36* (February), pp. 235–52.

Brown, Gardner M., Jr., and Ralph W. Johnson (1984). "Pollution Control by Effluent Charges: It Works in the Federal Republic of Germany, Why Not in the U.S." *Natural Resources Journal 24* (October), pp. 929–66.

Bunkley, Nick (2009). "Honda Unveils a Cheaper Hybrid Challenger to Toyota's Top-Selling Prius." *New York Times* (January 13), p. B.3.

Burtraw, Dallas, and Paul Portney (1991). "Environmental Policy in the U.S." In Dieter Helm, ed., *Economic Policy Towards the Environment.* Cambridge, MA: Blackwell Publishers, pp. 289–320.

Burtraw, Dallas, and Sarah Jo Szambelan (2009). *U.S. Emissions Trading Markets for SO_2 and NO_X.* Discussion Paper DP 09-40 (October). Washington, DC: Resources for the Future.

Business for Social Responsibility (BSR) Staff (2003). *Waste Reduction and Resource Productivity.* San Francisco, CA: December. Available at **www.bsr.org/CSRResources/IssueBriefDetail.cfm?DocumentID=49813**.

Cairncross, Frances (1992). "How Europe's Companies Reposition to Recycle." *Harvard Business Review 70(2)* (March–April), pp. 34–45.

California Environmental Protection Agency, Air Resources Board (1999, February). *Fact Sheet: LEV II—Amendments to California's Low-Emission Vehicle Regulations.* Sacramento, CA: California Environmental Protection Agency, Air Resources Board.

——— (2004, August 14). *The California Low-Emission Vehicle Regulations*. Sacramento, CA: California Environmental Protection Agency, Air Resources Board.

——— (2004, September 24). *ARB Approves Greenhouse Gas Rule*. News Release. Sacramento, CA: California Environmental Protection Agency, Air Resources Board.

——— (2008, May 6). *The Zero Emission Vehicle Program—2008: Fact Sheet*. Sacramento, CA: California Environmental Protection Agency, Air Resources Board.

——— (2011, January 7). *Voluntary Accelerated Vehicle Retirement (VAVR) Program*. Sacramento: CA: California Environmental Protection Agency, Air Resources Board.

Callan, Scott J., and Janet M. Thomas (1997). "The Impact of State and Local Policies on the Recycling Effort." *Eastern Economic Journal 23(4)* (Fall), pp. 411–23.

——— (1999). "Adopting a Unit Pricing System for Municipal Solid Waste: Policy and Socioeconomic Determinants." *Environmental and Resource Economics 14(4)* (December), pp. 503–18.

——— (2001). "Economies of Scale and Scope: A Cost Analysis of Municipal Solid Waste Services." *Land Economics 77(4)* (November), pp. 548–60.

——— (2006). "Analyzing Demand for Disposal and Recycling Services: A Systems Approach." *Eastern Economic Journal 32(2)* (Spring), pp. 221–40.

CantorCO2e (2010). *CantorCO2e's Market Price Indicator*. New York: CantorCO2e.

Carbon Tax Center (2011). "Where Carbon Is Taxed: Country-by-Country Review." New York: Carbon Tax Center. Available at **www.carbontax.org/progress/where-carbon-is-taxed/** (last updated April 11, 2011).

Carroll, Joseph (2007). "Americans Assess What They Can Do to Reduce Global Warming." *Gallup News Service* (April 24). Available at **www.gallup.com/poll/27298/Americans-Assess-What-They-Can-Reduce-Global-Warming.aspx** (accessed January 9, 2009).

Carson, Rachel (1962). *Silent Spring*. Boston, MA: Houghton Mifflin.

Carson, Richard T., and Robert Cameron Mitchell (1988). *The Value of Clean Water: The Public's Willingness to Pay for Boatable, Fishable, and Swimmable Quality Water*. Discussion Paper 88-13. La Jolla, CA: University of California at San Diego.

——— (1993). "The Value of Clean Water: The Public's Willingness to Pay for Boatable, Fishable, and Swimmable Water." *Water Resources Research 29(7)* (July), pp. 2445–54.

Caterpillar, Inc. (2009). *Remanufacturing*. Peoria, IL: Caterpillar, Inc. Available at **www.cat.com/cda/layout?m=94424&x=7** (accessed April 23, 2009).

——— (2010). *Caterpillar 2010 Sustainability Report: Powering Change*. Peoria, IL: Caterpillar, Inc. Available at **www.caterpillar.com/cda/files/2838620/7/2010SustainabilityReport.pdf**.

Center for Climate and Energy Solutions (C2ES) (2011). *Outcomes of the U.N. Climate Change Conference in Durban, South Africa*. Arlington, VA: C2ES (December). Available at **www.c2es.org/docUploads/COP17_Summary.pdf**.

Chemical Manufacturers Association (CMA) (1990, September). *Improving Performance in the Chemical Industry*. Washington, DC: CMA.

Chesapeake Bay Program (2000, June 28). *New Chesapeake Bay Agreement Signed*. Press Release. Annapolis, MD: Chesapeake Bay Program.

——— (2002, October 16). *Chesapeake 2000 and the Bay: Where Are We and Where Are We Going?* Fact Sheet. Annapolis, MD: Chesapeake Bay Program.

——— (2008, March). *Chesapeake Health & Restoration Assessment 2007: A Report to the Citizens of the Bay Region*. Annapolis, MD: Chesapeake Bay Program.

Chestnut, Lauraine G., and David M. Mills (2005). "A Fresh Look at the Benefits and Costs of the U.S. Acid Rain Program." *Journal of Environmental Management 77(3)* (November), pp. 252–66.

"China to Kick Off Emissions Trading" (2008, November 24). *Power Finance & Risk*. London.

City of Santa Monica (2008). "Oil Recycling and Re-Refined FAQs." Available at **www.smgov.net/Departments/OSE/categories/content.aspx?id=4148** (last updated August 6, 2008).

Clawson, Marion, and Jack L. Knetch (1966). *Economics of Outdoor Recreation*. Washington, DC: Resources for the Future.

Cline, William R. (1992). *Global Warming: The Benefits of Emission Abatement*. Paris: OECD.

Coase, Ronald H. (1960). "The Problem of Social Cost." *Journal of Law and Economics 3* (October), pp. 1–44.

Cogan, Douglas G. (1988). *Stones in a Glass House: CFCs and Ozone Depletion*. Washington, DC: Investor Responsibility Research Center, Inc.

Cohen, Bonner R. (2001). "EPA Will Destroy Hudson River to Save It." *Wall Street Journal* (December 12), p. A18.

——— (2002). "EPA's Dredging Scheme Will Wreak Havoc on the Hudson River." *Human Events 58(1)* (January 7), p. 10.

Cohn, Laura, John Carey, and Ann Therese Palmer (2001). "Is the EPA Sandbagging Business?" *Business Week* (June 11), p. 136.

Committee on Climate Change (2009). *Meeting Carbon Budgets: The Need for a Step Change; Progress Report to Parliament*. London: Committee on Climate Change (October).

Container Recycling Institute (2009). "Bottle Bill Resource Guide." Available at **www.bottlebill.org/legislation/usa/allstates.htm** (last updated June 29, 2009).

——— (2011). "Bottle Bill Resource Guide: Delaware." Available at **www.bottlebill.org/legislation/usa/delaware.htm** (last updated February 4, 2011).

Cook, Elizabeth (1996). *Making a Milestone in Ozone Protection*. Washington, DC: World Resources Institute.

Copeland, Brian R., and M. Scott Taylor (2004). "Trade, Growth, and the Environment." *Journal of Economic Literature 42(1)* (March), pp. 7–71.

Copeland, Claudia (2010). *Water Quality Issues in the 111th Congress: Oversight and Implementation*. CRS Report for Congress. Washington, DC: Congressional Research Service (July 13).

Cough, Paul (1993). "Trade-Environment Tensions: Options Exist for Reconciling Trade and Environment." *EPA Journal 19(2)* (April–June), pp. 28–30.

Council of Economic Advisers (1982). *Economic Report of the President*. Washington, DC: U.S. Government Printing Office.

——— (2009, September 10). *Economic Analysis of the Car Allowance Rebate System ("Cash for Clunkers")*. Washington, DC: Council of Economic Advisers.

——— (2011, February). *Economic Report of the President*. Washington, DC: U.S. Government Printing Office.

Council on Environmental Quality (1993, January). *Environmental Quality, 23rd Annual Report*. Washington, DC: U.S. Government Printing Office.

——— (1997). *Environmental Quality, 25th Anniversary Report*. Washington, DC: U.S. Government Printing Office.

——— (2011, June 2). "Gulf Coast Ecosystem Restoration." Washington, DC: Council on Environmental Quality. Available at **www.whitehouse.gov/administration/eop/ceq/initiatives/gulfcoast** (accessed June 2, 2011).

Cox, Meg (1991). "Music Firms Try Out 'Green' CD Boxes." *Wall Street Journal* (July 25), p. B1.

Cropper, Maureen L., and Wallace E. Oates (1992). "Environmental Economics: A Survey." *Journal of Economic Literature 30* (June), pp. 675–740.

Dahl, T. E. (2006). *Status and Trends of Wetlands in the Conterminous United States 1998 to 2004*. Washington, DC: U.S. Department of the Interior; Fish and Wildlife Service.

Daley, Beth (2001). "Harbor Cleanup Falls Short at Beaches: Sewage Discharge Fouling Boston's Cleanup Success." *Boston Globe* (September 10), p. A1.

Dasgupta, Susmita, Benoit Laplante, Hua Wang, and David Wheeler (2002). "Confronting the Environmental Kuznets Curve." *Journal of Economic Perspectives 16(1)* (Winter), pp. 147–68.

Deutsch, Claudia H. (2001). "Together at Last: Cutting Pollution and Making Money." *New York Times* (September 9), p. BU1.

Dixon, Lloyd, and Steven Garber (2001). *Fighting Air Pollution in Southern California by Scrapping Old Vehicles*. Santa Monica, CA: Rand Institute for Civil Justice.

Dolin, Eric Jay (1992). "Boston Harbor's Murky Political Waters." *Environment 34(6)* (July/August), pp. 7–11, 26–33.

Dominguez, George, ed. (1977). *Guidebook: Toxic Substances Control Act*. Cleveland, OH: CRC Press, Inc.

Doneski, D. (1985). "Cleaning Up Boston Harbor: Fact or Fiction?" *Boston College Environmental Affairs Law Review 12* (Spring), p. 567.

Duales System Deutschland AG (2004/2005). "Europe Goes Green Dot." Frankfurt, Germany: Duales System Deutschland AG 2004/2005. Available at **www.pro-e.org/brochures/europe_goes.pdf**.

——— (2005, April 4). "The Green Dot." Available at **www.gruener-punkt.de/THE_GREEN_DOT.50+B6Jkw9MQ__.0.html** (accessed April 4, 2005).

——— (2008, July 1). *Press Information: Recycling Rates on Record Level.* Available at **www.gruener-punkt.de/en/info-for-journalists/research/press-information/pressemitteilungen-artikel/recycling-rates-on-record-level.html**.

——— (2011, August 11). "Mass Flow Verification and Environmental Performance Balance." Available at **www.gruener-punkt.de/en/corporate/sustainability/environmental-balance.html** (accessed August 11, 2011).

Duhl, Joshua (1993, December). *Effluent Fees: Present Practice and Future Potential.* Discussion Paper #075. Washington, DC: American Petroleum Institute.

Dunlap, Riley E. (2010). "At 40, Environmental Movement Endures, With Less Consensus." *Gallup News Service* (April 22). Available at **www.gallup.com/poll/127487/Environmental-Movement-Endures-Less-Consensus.aspx** (accessed January 17, 2011).

Dunlap, Riley E., and Lydia Saad (2001). "Only One in Four Americans Are Anxious About the Environment: Most Favor Moderate Approach to Environmental Protection." *Gallup News Service* (April 16). Available at **www.gallup.com/poll/1801/Only-One-Four-Americans-Anxious-About-Environment.aspx**.

Dzurik, Andrew A. (1990). *Water Resources Planning*. Savage, MD: Rowman & Littlefield.

Eheart, Wayland, E., Downey Brill Jr., and Randolph M. Lyon (1983). "Transferable Discharge Permits for Control of BOD: An Overview." In Erhard F. Joeres and Martin H. David, eds., *Buying a Better Environment: Cost-Effective Regulation Through Permit Trading*. Madison, WI: University of Wisconsin Press.

Eilperin, Juliet (2010). "U.S. Pledges 17 Percent Emissions Reduction by 2020." *Washington Post* (January 29, 2010). Available at **www.washingtonpost.com/wp-dyn/content/article/2010/01/28/AR2010012803632.html**.

Ellerman, A. Denny, Paul L. Joskow, Richard Schmalensee, Juan Pablo Montero, and Elizabeth Bailey (2000). *Markets for Clean Air: The U.S. Acid Rain Program*. Cambridge, UK: Cambridge University Press.

Energy Recovery Council (ERC) (2010, December). *The 2010 ERC Directory of Waste-to-Energy Plants*. Washington, DC: Energy Recovery Council.

Environmental Information Ltd. (1990). *Industrial and Hazardous Waste Management Firms*. Minneapolis, MN: Environmental Information Ltd.

Environmental Investigation Agency (2009, November). *A Global HFC Phase-Out: Essential Action for the Montreal Protocol.* Washington, DC: Environmental Investigation Agency.

Environmental Law Institute (2002). *An Analysis of State Superfund Programs: 50-State Study, 2001 Update.* Washington, DC: Environmental Law Institute (November).

Environment Canada (2004, December). *A Guide to Understanding the Canadian Environmental Protection Act, 1999.* Canada: Environment Canada. Available at **www.ec.gc.ca/lcpe-cepa/E00B5BD8-13BC-4FBF-9B74-1013AD5FFC05/Guide04_e.pdf**.

——— (2005, October). *CEPA 1999 at a Glance.* Canada: Environment Canada. Available at **www.ec.gc.ca/lcpe-cepa/EAEDF851-FB57-4525-8A9D-C728038F0B37/glance.pdf**.

EnviroSense (1995). "Partners for the Environment: EPA Standards Network." Available at **http://es.epa.gov/partners/iso/iso.html** (last updated December 20, 1995).

Epstein, Samuel S., Lester O. Brown, and Carl Pope (1982). *Hazardous Waste in America.* San Francisco, CA: Sierra Club Books.

Erkman, S. (1997). "Industrial Ecology: An Historical View." *Journal of Cleaner Production 5(1–2)*, pp. 1–10.

Esworthy, Robert (2010, August 11). *Pesticide Registration and Tolerance Fees: An Overview.* CRS Report for Congress. Washington, DC: Congressional Research Service. Available at **http://ncseonline.org/nle/crsreports/10Sep/RL32218.pdf**.

Europa, Summaries of EU Legislation (2010). "Integrated Pollution Prevention and Control (IPPC Directive)." Available at **http://europa.eu/legislation_summaries/environment/waste_management/l28045_en.htm** (last updated October 13, 2010).

European Commission (2008). *Attitudes of European Citizens Towards the Environment—Summary.* Special Eurobarometer 295 (March). Available at **http://ec.europa.eu/environment/archives/barometer/pdf/summary2008_environment_en.pdf**.

——— (2011). "The IPPC Directive." Available at **http://ec.europa.eu/environment/air/pollutants/stationary/ippc/summary.htm** (last updated June 21, 2011).

European Commission, Eurostat (2011, January 17). "Waste Statistics–Statistics Explained." Available at **http://epp.eurostat.ec.europa.eu/statistics_explained/index.php/Waste_statistics#Waste_generated_by_households** (accessed January 17, 2011).

——— (2011, March 8). "Environment in the EU27: Recycling Accounted for a Quarter of Total Municipal Waste Treated in 2009." Available at **http://epp.eurostat.ec.europa.eu/cache/ITY_PUBLIC/8-08032011-AP/EN/8-08032011-AP-EN.PDF**.

——— (2011a, July 12). "Generation of Waste by Waste Category (hazardous, non-hazardous), 2008." Available at **http://epp.eurostat.ec.europa.eu/tgm/table.do?tab=table&init=1&language=en&pcode=ten00109**.

——— (2011b, July 12). "Packaging Waste." Available at **http://appsso.eurostat.ec.europa.eu/nui/show.do?dataset=env_waspac&lang=en**.

——— (2011, July 19). "Total Population." Available at **http://epp.eurostat.ec.europa.eu/tgm/table.do?tab=table&init=1&language=en&pcode=tps00001**.

European Environment Agency (2006). *Air Quality and Ancillary Benefits of Climate Change Policies.* Copenhagen, Denmark: European Environment Agency.

——— (2008). *Greenhouse Gas Emission Trends and Projections in Europe 2008: Tracking Progress Towards Kyoto Targets: Executive Summary.* Copenhagen, Denmark: European Environment Agency.

——— (2008, March). "Impacts of Europe's Changing Climate." *EEA Briefing.* Copenhagen, Denmark: European Environment Agency.

European Parliament and the Council of the European Union (2008). *Directive 2008/50/EC of the European Parliament and of the Council of 21 May 2008 on Ambient Air Quality and Cleaner Air for Europe. Official Journal L 152* (June 11), pp. 0001–0044.

Export.gov (2011, April 27). "The North American Free Trade Agreement (NAFTA): General Information." Available at **http://export.gov/FTA/nafta/index.asp**.

Faeth, Paul (2000). *Fertile Ground: Nutrient Trading's Potential to Cost-Effectively Improve Water Quality.* Washington, DC: World Resources Institute.

Fallon, Scott (2009). "PCB Dredging Set for May." *The Record* (March 13), p. L3.

Farman, J. C., B. G. Gardiner, and J. D. Shanklin (1985). "Large Losses of Total Ozone in Antarctica Reveal Seasonal ClO_X/NO_X Interaction." *Nature 315*, pp. 207–10.

Farrow, R. Scott, Martin T. Schultz, Pinar Celikkol, and George L. Van Houtven (2005). "Pollution Trading in Water Quality Limited Areas: Use of Benefits Assessment and Cost-Effective Trading Ratios." *Land Economics 81(2)*, pp. 191–205.

Federal Facilities Environmental Stewardship and Compliance Assistance Center (2011, June 6). *Coastal Zone Management Act of 1972.* Available at **www.fedcenter.gov/Articles/index.cfm?id=2443** (accessed June 6, 2011).

Federal Leadership Committee (FLC) for the Chesapeake Bay (2010, May). *Executive Order 13508: Strategy for Protecting and Restoring the Chesapeake Bay Watershed: Executive Summary.* Available at **http://executiveorder.chesapeakebay.net/file.axd?file=2010%2f5%2fChesapeake+EO+Strategy+Executive+Summary.pdf**.

——— (2010, September). *Fiscal Year 2011 Action Plan; Executive Order 13508; Strategy for Protecting and Restoring the Chesapeake Bay Watershed.* Available at **http://executiveorder.chesapeakebay.net/file.axd?file=2010%2f9%2fChesapeake+EO+Action+Plan+FY2011.pdf**.

Fialka, John J., and Geoff Winestock (2001). "Future of Kyoto Protocol Minus U.S. Is Uncertain." *Wall Street Journal* (July 16), pp. A2, A8.

Fong, Mei, and Sky Canaves (2008). "World News: Factories on China's South Coast Lose Their Edge; Thousands Close as Increased Costs Alter the Equation." *Wall Street Journal* (February 22), p. A9.

Ford Motor Company (2010/2011). *Blueprint for Sustainability: Driving Change.* Available at **http://corporate.ford.com/doc/sr10-summary.pdf**.

Fortuna, Richard C., and David J. Lennett (1987). *Hazardous Waste Regulation: The New Era.* New York: McGraw-Hill.

Freeman, A. Myrick III (1978). "Air and Water Pollution Policy." In Paul R. Portney, ed., *Current Issues in U.S. Environmental Policy.* Baltimore, MD: Johns Hopkins University Press for Resources for the Future, pp. 12–67.

——— (1982). *Air and Water Pollution Control: A Benefit-Cost Assessment.* New York: Wiley.

——— (1990). "Water Pollution Policy." In Paul R. Portney, ed., *Public Policies for Environmental Protection.* Washington, DC: Resources for the Future, pp. 97–149.

——— (2000). "Water Pollution Policy." In Paul R. Portney and Robert N. Stavins, eds., *Public Policies for Environmental Protection.* 2nd Edition. Washington, DC: Resources for the Future, pp. 169–213.

——— (2002). "Environmental Policy Since Earth Day I: What Have We Gained?" *Journal of Economic Perspectives 16(1)* (Winter), pp. 125–46.

Frosch, Robert A. (1995). "Industrial Ecology: Adapting Technology for a Sustainable World." *Environment 37(10)* (December), pp. 16–24, 34–37.

Frosch, Robert A., and Nicholas E. Gallopoulos (1989). "Strategies for Manufacturing." *Scientific American 261(3)* (September), pp. 144–52.

Garner, Andy, and Gregory A. Keoleian (1995, November). *Industrial Ecology: An Introduction.* Ann Arbor, MI: National Pollution Prevention Center for Higher Education.

General Electric Company (2010). *Solutions for the World's Toughest Challenges: Ecomagination 2010 Annual Report.* Fairfield, CT: General Electric Company. Available at **http://files.gecompany.com/ecomagination/progress/GE_ecomagination_2010AnnualReport.pdf**.

Gibbons, John (1992). "Moving Beyond the 'Tech Fix.'" *EPA Journal 18(4)* (September/October), pp. 29–31.

Gibbs, Lois Marie (1982). *The Love Canal: My Story.* Albany, NY: State University of New York Press.

Ginsburg, Janet (2001). "Once Is Not Enough." *Business Week* (April 16), pp. 128B–128D.

Girard, Francesco Fusco (2008). "Industrial Ecology Approach for GHG Reduction in Industry." *Sixth International Conference on Ethics and Environmental Policies: Ethics and Climate Change. Scenarios for Justice and Sustainability* (October 23–25), pp. 1–6. Available at **www.webethics.net/padova2008/papers/27.pdf**.

Gleick, Peter H., Heather Cooley, Michael J. Cohen, Mari Morikawa, Jason Morrison, and Meena Palaniappan (2009). *The World's Water 2008–2009: The Biennial Report on Freshwater Resources.* Washington, DC: Island Press.

Global Water Intelligence, and Organisation for Economic Cooperation and Development (GWI and OECD) (2007). "GWI/OECD Water Tariff Survey." *Global Water Intelligence 8(10)* (October).

The Goldman Environmental Prize (2008). "Rosa Hilda Ramos." *Islands and Island Nations 2008.* Available at **www.goldmanprize.org/2008/islands** (accessed January 14, 2009).

Green, Heather, and Kerry Capell (2008). "Carbon Confusions; To Help Shoppers Make Green Choices, Companies Are Slapping Carbon Labels on Products." *Business Week* (March 17), p. 52.

Greenhouse, Linda (2007). "Justices Say EPA Has Power to Act on Harmful Gases." *New York Times* (April 3), p. A1.

"The Greening of Protectionism" (1993). *The Economist* (February 27), pp. 25–28.

Greenpeace China (2010). *Poisoning the Pearl: An Investigation Into Industrial Water Pollution in the Pearl River Delta.* 2nd Edition. Beijing: Greenpeace (January). Available at **www.greenpeace.org/raw/content/eastasia/press/reports/pearl-river-report-2.pdf**.

Griffin, Melanie L. (1988). "The Legacy of Love Canal." *Sierra* (January/February), pp. 26–28.

Groothuis, Peter A., Jana D. Groothuis, and John C. Whitehead (2007). "The Willingness to Pay to Remove Billboards and Improve Scenic Amenities." *Journal of Environmental Management 85(4)* (December), pp. 1094–1100.

Grossman, Gene M., and Alan B. Krueger (1995). "Economic Growth and the Environment." *Quarterly Journal of Economics 110(2)* (May), pp. 353–78.

Gruenspecht, Howard, and Robert Stavins (2002). "A Level Field on Pollution at Power Plants." *Boston Globe* (January 26), p. A15.

Grunwald, Michael (2010). "The BP Spill: Has the Damage Been Exaggerated?" *Time 176(6)* (August 9).

Haas, Peter M., Marc A. Levy, and Edward A. Parson (1992). "Appraising the Earth Summit: How Should We Judge UNCED's Success?" *Environment 34(8)* (October), pp. 6–11, 26–33.

Hahn, Robert W. (1989). "Economic Prescriptions for Environmental Problems: How the Patient Followed the Doctor's Orders." *Journal of Economic Perspectives 3(2)* (Spring), pp. 95–114.

Hahn, Robert W., and Gordon L. Hester (1989). "Marketable Permits: Lessons for Theory and Practice." *Ecology Law Quarterly 16,* pp. 361–406.

Hahn, Robert W., and Roger Noll (1982). "Designing a Market for Tradable Emissions Permits." In Wesley Magat, ed., *Reform of Environmental Regulation.* Cambridge, MA: Ballinger, pp. 119–46.

Hakim, Danny (2004). "Catching Up to the Cost of Global Warming." *New York Times* (July 25), p. 5.

——— (2004). "California's New Emissions Standards Ignite a Debate Over Price; Peering at the Sticker on a Cleaner Car." *New York Times* (November 2), p. 1.

Harbaugh, William T., Arik Levinson, and David Molloy Wilson (2002). "Reexamining the Empirical Evidence for an Environmental Kuznets Curve." *Review of Economics and Statistics 84*(3) (August), pp. 541–51.

Hardin, Garrett (1968). "The Tragedy of the Commons." *Science 162,* pp. 1243–48.

Hazilla, Michael, and Raymond J. Kopp (1990). "Social Cost of Environmental Quality Regulations: A General Equilibrium Analysis." *Journal of Political Economy 98*(4), pp. 853–73.

Heath, Ralph C. (1988). "Ground Water." In David H. Speidel, Lon C. Ruedisili, and Allen F. Agnew, eds., *Perspectives on Water: Uses and Abuses.* New York: Oxford University Press.

Helfand, Gloria E. (1994). "Pollution Prevention as Public Policy: An Assessment." *Contemporary Economic Policy 12* (October), pp. 104–13.

Hewlett-Packard Company (2011). *Product Design for the Environment.* Palo Alto, CA: Hewlett-Packard Company. Available at **http://www8.hp.com/us/en/hp-information/environment/design-for-environment.html** (accessed October 19, 2011).

Hilsenrath, Jon E. (2001). "Environmental Economists Debate Merit of U.S.'s Kyoto Withdrawal." *Wall Street Journal* (interactive edition) (August 7).

Hoerner, Andrew J. (1998). *Harnessing the Tax Code for Environmental Protection: A Survey of State Initiatives.* Washington, DC: World Resources Institute.

Hof, Robert D. (1990). "The Tiniest Toxic Avengers." *Business Week* (June 4), pp. 96, 98.

ICF Resources International (1989). *Economic, Environmental, and Coal Market Impacts of SO_2 Emissions Trading Under Alternative Acid Rain Control Proposals.* Report Prepared for the Regulatory Innovations Staff, Office of Policy, Planning and Evaluation, U.S. Environmental Protection Agency.

Inhofe, James (2010). "Obama's EPA a Growing Menace to U.S. Economy." *Human Events 66*(34) (October 4), p. 1.

International Bottled Water Association (IBWA) (2008). "Beverage Marketing's 2008 Market Report Findings: The 2007 Stats." Available at **www.bottledwater.org/public/statistics_main.htm**.

International Council of Chemical Associations (ICCA), Responsible Care Leadership Group (2007). *Status Report 2007.* Arlington, VA: ICCA.

International Energy Agency (2011). "End-User Petroleum Product Prices and Average Crude Oil Import Costs: January 2011." (February 4), p. 3. Available at **www.iea.org/stats/surveys/mps.pdf**.

"International: How to Cool the World; Climate Change" (2007). *The Economist* (May 12), p. 68.

International Maritime Organization, Office for the London Convention and Protocol (2003a). "A Brief Description of the London Convention 1972 and the 1996 Protocol." Available at **www.londonconvention.org/London_Convention.htm** (last revised December 16, 2003).

——— (2003b). "Frequently Asked Questions." Available at **www.londonconvention.org/FAQ.htm** (last revised December 16, 2003).

——— (2011a). "London Convention and Protocol." Available at **www.imo.org/OurWork/Environment/SpecialProgrammesAndInitiatives/Pages/London-Convention-and-Protocol.aspx**.

——— (2011b). "1996 Protocol to the Convention on the Prevention of Marine Pollution by Dumping of Wastes and Other Matter, 1972." Available at **www.imo.org/ourwork/environment/pollutionprevention/pages/1996-protocol-to-the-convention-on-the-prevention-of-marine-pollution-by-dumping-of-wastes-and-other-matter,-1972.aspx**.

International Organization for Standardization (1998). *ISO 14000: Meet the Whole Family.* Geneva, Switzerland: ISO.

——— (2002). "ISO Online." Available at **www.iso.org** (accessed June 4, 2002).

——— (2011, April). *ISO in Brief; International Standards for a Sustainable World.* Available at **www.iso.org/iso/isoinbrief_2011.pdf**.

——— (2011, October 2). "ISO 14000 Essentials." Available at **www.iso.org/iso/iso_14000_essentials** (accessed October 2, 2011).

International Tanker Owners Pollution Federation Ltd. (ITOPF) (2008, February 28). *The Environmental Impact of the Hebei Spirit Oil Spill, Taean, South Korea (7th December 2007).* London: ITOPF. Available at **www.itopf.com/news-and-events/documents/HEBEISPIRIT-EnvironmentalImpact.pdf**.

——— (2011, June 5). *Statistics.* London: ITOPF. Available at **www.itopf.com/information-services/data-and-statistics/statistics/** (accessed June 5, 2011).

——— (2011, June 9). *Case Histories: Aegean Sea (Spain, 1992), Atlantic Empress (West Indies, 1979), Braer (UK, 1993), Exxon Valdez (United States, 1989), Prestige (Spain, 2002).* London: ITOPF. Available at **www.itopf.com/information-services/data-and-statistics/case-histories/** (accessed June 9, 2011).

IPCC (Intergovernmental Panel on Climate Change) (2001). *Climate Change 2001: The Scientific Basis. Contribution of Working Group I to the Third Assessment Report of the Intergovernmental Panel on Climate Change (IPCC).* Cambridge, UK: Cambridge University Press.

——— (2007a). "Summary for Policymakers." In M. L. Parry, O. F. Canziani, J. P. Palutikof, P. J. Van der Linden, and C. E. Hanson, eds., *Climate Change 2007: Impacts, Adaptation and Vulnerability. Contribution of Working Group II to the Fourth Assessment Report of the*

Intergovernmental Panel on Climate Change. Cambridge, UK: Cambridge University Press, pp. 7–22.

——— (2007b). *Climate Change 2007: The Physical Science Basis. Contribution of Working Group I to the Fourth Assessment Report of the Intergovernmental Panel on Climate Change*. S. Solomon, D. Qin, M. Manning, Z. Chen, M. Marquis, K. B. Averyt, M. Tignor, and H. L. Miller, eds., Cambridge, UK: Cambridge University Press.

——— (2008). *Climate Change 2007: Synthesis Report*. Geneva, Switzerland: IPCC.

——— (2010, November). *The IPCC's Fifth Assessment Report (AR5)-IPCC Leaflet*. Geneva, Switzerland: IPCC.

"Japanese Environmentalist Says Gas Too Cheap in U.S." (1992). *Brockton Enterprise* (February 2).

Jenkins, Robin (1993). *The Economics of Solid Waste Reduction: The Impact of User Fees*. Brookfield, VT: Edward Elgar Publishing.

Johnson & Johnson (2011). *Our Responsibility: 2010 Sustainability Report; Summary*. New Brunswick, NJ: Johnson & Johnson. Available at **www.jnj.com/wps/wcm/connect/f9f1148046e763e7b0b4bae02a8d6552/110421_FINAL_J%26J_2010_ResponsibilityReport.pdf?MOD=AJPERES**.

Johnson, Edwin L. (1967). "A Study in the Economics of Water Quality Management." *Water Resources Research 3*, pp. 291–305.

Jondrow, James, and Robert A. Levy (1984). "The Displacement of Local Spending for Pollution Control by Federal Construction Grants." *American Economic Review 74(2)* (May), pp. 174–78.

Kalundborg Symbiosis Institute (2002). *Industrial Symbiosis: Trading Byproducts*. Kalundborg, Denmark: Kalundborg Symbiosis Institute. Available at **www.symbiosis.dk/index.htm** (accessed June 6, 2002).

——— (2011a). *Kalundborg Symbiosis: Evolution*. Kalundborg, Denmark: Kalundborg Symbiosis Institute. Available at **www.symbiosis.dk/en/evolution** (accessed October 2, 2011).

——— (2011b). *Kalundborg Symbiosis: Partners*. Kalundborg, Denmark: Kalundborg Symbiosis Institute. Available at **www.symbiosis.dk/en/partnere** (accessed October 2, 2011).

Karl, Thomas R., Jerry M. Melillo, and Thomas C. Peterson, eds. (2009). *Global Climate Change Impacts in the United States*. New York: Cambridge University Press. Available at **www.globalchange.gov/what-we-do/assessment/nca-reports/global-climate-change-impacts-in-the-us-2009**.

Kiel, Katherine A. (1995). "Measuring the Impact of the Discovery and Cleaning of Identified Hazardous Waste Sites on House Values." *Land Economics 71(4)* (November), pp. 428–35.

Kiel, Katherine, and Jeffrey Zabel (2001). "Estimating the Economic Benefits of Cleaning Up Superfund Sites: The Case of Woburn, Massachusetts." *Journal of Real Estate Finance and Economics 22(2–3)* (March–May), pp. 163–84.

Kiel, Katherine A., and Michael Williams (2007). "The Impact of Superfund Sites on Local Property Values: Are All Sites the Same?" *Journal of Urban Economics 61(1)* (January), pp. 170–92.

Kimm, Victor J. (1993). "The Delaney Clause Dilemma." *EPA Journal 19(1)* (January–March), pp. 39–41.

Kingsley-Jones, Max (2008). "From Skies to Scrap." *Flight International 173(5140)* (May 27–June 2), pp. 38–40.

Klein, Janice A., and Jeffrey G. Miller, eds. (1993). *The American Edge: Leveraging Manufacturing's Hidden Assets*. New York: McGraw-Hill.

Kneese, Allen V., Robert U. Ayres, and Ralph C. D'Arge (1970). *Economics and the Environment: A Materials Balance Approach*. Washington, DC: Resources for the Future.

Knepper, Mike (1993). "Recycling: Investment in the Future." *BMW Magazine* (January), pp. 66–69.

Kohlhase, Janet E. (1991). "The Impact of Toxic Waste Sites on Housing Values." *Journal of Urban Economics 30*, pp. 1–26.

Kopp, Raymond J., and Alan J. Krupnick (1987). "Agricultural Policy and the Benefits of Ozone Control." *American Journal of Agricultural Economics 69(5)* (December), pp. 956–62.

Krukowski, John (1995). "Survey: Thermal Prices End Free-Fall?" *Pollution Engineering 7(2)*.

Krupnick, Alan (1986). "Costs of Alternative Policies for the Control of Nitrogen Dioxide in Baltimore." *Journal of Environmental Economics and Management 13(2)* (June), pp. 189–97.

Krupnick, Alan, and Richard Morgenstern (2002). "The Future of Benefit-Cost Analyses of the Clean Air Act." *Annual Review of Public Health 23*, pp. 427–48.

Krupnick, Alan, Virginia McConnell, Matt Cannon, Terrell Stoessell, and Michael Batz (2000). *Cost-Effective NO_X Control in the Eastern United States*. Discussion Paper 00-18. Washington, DC: Resources for the Future.

Krutilla, John, V. (1967). "Conservation Reconsidered." *American Economic Review 57(4)* (September), pp. 777–86.

Kynge, James (2004). "Modern China Is Facing an Ecological Crisis." *Financial Times* (July 26).

Lancaster, Kelvin J. (1966). "A New Approach to Consumer Theory." *Journal of Political Economy 74(2)* (April), pp. 132–57.

Landy, Marc K., Marc J. Roberts, and Stephen R. Thomas (1990). *The Environmental Protection Agency: Asking the Wrong Questions*. New York: Oxford University Press.

Larson, Charles D. (1989). "Historical Development of the National Primary Drinking Water Regulations." In Edward J. Calabrese, Charles E. Gilbert, and Harris Pastides, eds., *Safe Drinking Water Act: Amendments, Regulations, and Standards*. Chelsea, MI: Lewis Publishers, Inc., pp. 3–15.

Lave, Lester B. (1982). "Methods of Risk Assessment." In Lester B. Lave, ed., *Quantitative Risk Assessment in*

Regulation. Washington, DC: The Brookings Institution, pp. 23–54.

Learner, Ivan (2001). "Fluorocarbon Industry Adapting to Regulations." *Chemical Market Reporter* (September 10), pp. 10, 14.

Lee, Jane Lanhee (2003). "China Senses Need for Cleaner Fuel." *New York Times* (December 11), p. C16.

"Loss at BP as Gulf Oil Spill Costs Hit $41 BN" (2011). *The Herald* (February 2), p. 25.

Lyklema, J., and T. E. A. van Hylckama (1988). "Water Something Peculiar." In David H. Speidel, Lon C. Ruedisili, and Allen F. Agnew, eds., *Perspectives on Water: Uses and Abuses*. New York: Oxford University Press.

MacBean, Alasdair (2007). "China's Environment: Problems and Policies." *The World Economy 30(2)* (February), pp. 292–307.

Maloney, Michael T., and Bruce Yandle (1984). "Estimation of the Cost of Air Pollution Control Regulation." *Journal of Environmental Economics and Management 11(3)* (September), pp. 244–63.

Manly, Lorne (1992). "It Doesn't Pay to Go Green When Consumers Are Seeing Red." *Adweek* (March 23), pp. 32–33.

Marshall, Alfred (1930 [1890]). *Principles of Economics*. 8th Edition. London: Macmillan.

Mazmanian, Daniel, and David Morell (1992). *Beyond Superfailure: America's Toxics Policy for the 1990s*. Boulder, CO: Westview Press.

McCarthy, James E. (2008, July 28). *Clean Air Act Issues in the 110th Congress*. CRS Report for Congress. Washington, DC: Congressional Research Service.

McCarthy, James E., Larry B. Parker, and Robert Meltz (2008). *Clean Air After the CAIR Decision: Back to Square One?* CRS Report for Congress. Washington, DC: Congressional Research Service (July 22).

McDermott, H. (1973). "Federal Drinking Water Standards: Past, Present, and Future." *Water Well Journal 27(12)*, pp. 29–35.

McDonald's Corp. (1999, May). *McDonald's Waste Reduction Action Plan (WRAP): Status Report*. Oakbrook, IL: McDonald's Corp.

——— (2002, June 18). *Tenth Anniversary of EDF/McDonald's Alliance: Fact Sheet 1990–1999*. Available at **www.mcdonalds.com/countries/usa/community/environ/info/decade/decade.htm** (accessed June 18, 2002).

——— (2010). *Worldwide Corporate Social Responsibility 2010 Report*. Available at **www.aboutmcdonalds.com/etc/medialib/csr/docs.Par.32488.File.dat/mcd063_2010%20PDFreport_v9.pdf**.

——— (2011). "McDonald's Corporate Responsibility: Values in Practice; Green Packaging Design—EcoFilter." Available at **www.aboutmcdonalds.com/mcd/csr/about/sustainable_supply/resource_conservation/sustainable_packaging.html** (accessed September 30, 2011).

McGartland, Albert M. (1984). "Marketable Permit Systems for Air Pollution Control: An Empirical Study." Ph.D. dissertation, University of Maryland.

Meltz, Robert (2009). *Legal Consequences of EPA's Endangerment Finding for New Motor Vehicle Greenhouse Gas Emissions*. CRS Report for Congress. Washington, DC: Congressional Research Service (December 15).

Mendelsohn, Robert, and James Neumann, eds. (1999). *The Impact of Climate Change on the United States Economy*. Cambridge, UK: Press Syndicate of the University of Cambridge.

Michaels, Daniel (2007). "Boeing and Airbus Compete to Destroy What They Built; How to Recycle a Jetliner Is Topic of Trash Talk; Snipping off the Wings." *Wall Street Journal* (June 1), p. A1.

Mills, Edwin S. (1978). *The Economics of Environmental Quality*. New York: W. W. Norton.

Miranda, Marie Lynn, and Joseph E. Aldy (1996). *Unit Pricing of Residential Municipal Solid Waste: Lessons From Nine Case Study Communities*. Report EE-0306 Prepared for the Office of Policy, Planning, and Evaluation, U.S. Environmental Protection Agency (March).

Mitchell, Robert Cameron, and Richard T. Carson (1989). *Using Surveys to Value Public Goods: The Contingent Valuation Method*. Washington, DC: Resources for the Future.

Mullen, John K., and Fredric C. Menz (1985). "The Effect of Acidification Damages on the Economic Value of the Adirondack Fishery to New York Anglers." *American Journal of Agricultural Economics 67(1)* (February), pp. 112–19.

Murray, Matt (2001). "EPA Orders Dredging of PCBs From the Upper Hudson River." *Wall Street Journal* (interactive edition) (December 5).

MWRA (Massachusetts Water Resources Authority) (2009, January 30). "CSO Control in South Boston." Available at **www.mwra.state.ma.us/cso/projects/southboston.htm** (last updated January 30, 2009).

——— (2009, June 3). "The Boston Harbor Project: An Environmental Success Story." Available at **www.mwra.state.ma.us/01news/2008/bhpenvironentalsuccess/bhpenvsuccess.htm** (last updated June 3, 2009).

National Academy of Sciences and National Academy of Engineering (1974). *Air Quality and Automobile Emission Control, v. 4, The Costs and Benefits of Automobile Emission Control*. Washington, DC: U.S. Government Printing Office.

National Acid Precipitation Assessment Program (NAPAP) (1991, November). *1990 Integrated Assessment Report*. Washington, DC: NAPAP Office of the Director.

National Aeronautics and Space Administration (NASA) (2010, October 12). *Ozone Hole Maps: August 2008*. Washington, DC: NASA.

National Assessment Synthesis Team, U.S. Global Change Research Program (2001). *Climate Change Impacts on the United States: The Potential Consequences of Climate Variability and Change*. Report for the U.S. Global Change Research Program. Cambridge, UK: Cambridge University Press.

National Commission on the BP Deepwater Horizon Oil Spill and Offshore Drilling (2011, January). *Deep Water: The Gulf Oil Disaster and the Future of Offshore Drilling; Report to the President*. Washington, DC: National Commission on the BP Deepwater Horizon Oil Spill and Offshore Drilling.

National Development and Reform Commission (2011). "Cleaner Production in China—Environmental Legislation." Available at **www.chinacp.com/EN/PolicyDetail.aspx?id=38** (accessed October 15, 2011).

National Oceanic and Atmospheric Administration (NOAA), National Weather Service (2010, June 9). "Hydrologic Cycle." Available at **http://forecast.weather.gov/jetstream/atmos/hydro.htm** (last updated June 9, 2010).

National Research Council (1983). *Risk Assessment in the Federal Government: Managing the Process*. Washington, DC: National Academy Press.

National Research Council of the National Academies (2008). *Science and Decisions: Advancing Risk Assessment*. Washington, DC: National Academies Press.

Natural Resources Defense Council, Erik D. Olson (1999, April). *Bottled Water: Pure Drink or Pure Hype?* Available at **www.nrdc.org/water/drinking/bw/bwinx.asp**.

"New Carbon-Neutral Packaging Is Music to Your Ears" (2011). *U.S. Newswire* (February 1).

Newcomb, Peter (1991). "Ban the Box." *Forbes* (May 13), p. 70.

New Jersey Department of Environmental Protection (2006). "New Jersey Statewide Mandatory Source Separation and Recycling Act." Available at **www.state.nj.us/dep/dshw/recycling/recy_act_link.htm** (last updated October 12, 2006).

Nitze, William A. (1993). "Stopping the Waste: Technology Itself Is Not the Problem." *EPA Journal 19*(2) (April–June), pp. 31–33.

Noah, Timothy (1993). "EPA Declares 'Passive' Smoke a Human Carcinogen." *Wall Street Journal* (January 6), p. B1.

Nordhaus, William D. (1999). "The Future of Environmental and Augmented National Accounts: An Overview." *Survey of Current Business* (November), pp. 45–49.

——— (2007). "A Review of the Stern Review on Climate Change." *Journal of Economic Literature 45*(3) (September), pp. 686–702.

Nordhaus, William D., and Joseph Boyer (2000). *Warming the World: Economic Models of Global Warming*. Cambridge, MA: MIT Press.

North American Commission for Environmental Cooperation (2005). "Our Programs and Projects." Available at **www.cec.org/programs_projects/** (accessed April 24, 2005).

Nussbaum, Bruce, and John Templeman (1990). "Built to Last—Until It's Time to Take It Apart." *Business Week* (September 17), p. 102.

Oasis Disc Manufacturing (2011). "Introducing the Zero Carbon Footprint Digipak." Available at **www.oasiscd.com/** (accessed February 2, 2011).

Oates, Wallace E., Paul R. Portney, and Albert M. McGartland (1989). "The Net Benefits of Incentive-Based Regulation: A Case Study of Environmental Standard-Setting." *American Economic Review 79*(5), pp. 1233–42.

Obama, Barack, and Joe Biden (n.d.). "Barack Obama and Joe Biden: Promoting a Healthy Environment." Available at **www.barackobama.com/pdf/issues/EnvironmentFactSheet.pdf**.

OECD (Organisation for Economic Co-operation and Development) (1989). *Economic Instruments for Environmental Protection*. Paris, France: OECD.

——— (2006). *The Political Economy of Environmentally Related Taxes*. Paris, France: OECD.

——— (2010). *OECD Factbook 2010: Economic, Environmental and Social Statistics*. Paris, France: OECD Publishing.

OECD/EEA (Organisation for Economic Co-operation and Development and European Environmental Agency) (2011). *OECD/EEA Database on Instruments Used for Environmental Policy and Natural Resources Management*. Retrieved from **http://www2.oecd.org/ecoinst/queries/index.htm** (last updated March 1, 2011).

Oldenburg, Kirsten U., and Kenneth Geiser (1997). "Pollution Prevention and/or Industrial Ecology?" *Journal of Cleaner Production 5*(1–2), pp. 103–108.

O'Neil, William B. (1980). "Pollution Permits and Markets for Water Quality." Ph.D. dissertation, University of Wisconsin, Madison.

Palmer, A. R., W. E. Mooz, T. H. Quinn, and K. A. Wolf (1980). *Economic Implications of Regulating Chlorofluorocarbon Emissions From Nonaerosol Applications*. Report R-2524-EPA, Prepared by the Rand Corporation for the U.S. Environmental Protection Agency (June).

Park, R. A., M. S. Trehan, P. W. Mausel, and R. C. Howe (1989). *The Effects of Sea Level Rise on U.S. Coastal Wetlands*. Washington, DC: U.S. Environmental Protection Agency.

Parker, Larry (2002, June 20). *Global Climate Change: Market-Based Strategies to Reduce Greenhouse Gases*. CRS Issue Brief for Congress. Washington, DC: Committee for the National Institute for the Environment.

Parker, Larry, and James E. McCarthy (2009). *Climate Change: Potential Regulation of Stationary Greenhouse Gas Sources Under the Clean Air Act*. CRS Report for Congress. Washington, DC: Congressional Research Service (December 10).

Parson, Edward A., Peter M. Haas, and Marc A. Levy (1992). "A Summary of the Major Documents Signed at the Earth Summit and the Global Forum." *Environment 34*(8) (October), pp. 12–15, 34–36.

Peretz, Jean H., and Jeffrey Solomon (1995). "Hazardous Waste Landfill Costs on Decline, Survey Says." *Environmental Solutions* (April), pp. 21–24.

Pew Center on Global Climate Change (2005). *The European Union Emissions Trading Scheme (EU-ETS): Insights and Opportunities.* Arlington, VA: Pew Center on Global Climate Change.

——— (2011). *Federal Vehicle Standards.* Arlington, VA: Pew Center on Global Climate Change. Available at **www.pewclimate.org/federal/executive/vehicle-standards** (accessed October 28, 2011).

——— (2011, January). *Climate Change 101: Understanding and Responding to Global Climate Change.* Arlington, VA: Pew Center on Global Climate Change.

Phifer, Russell W., and William R. McTigue Jr. (1988). *Handbook of Hazardous Waste Management.* Chelsea, MI: Lewis Publishers, Inc.

Pianin, Eric, and John Mintz (2001). "EPA Seeks to Narrow Pollution Initiative." *Washington Post* (August 8), p. A01.

Portney, Paul R. (1990a). "Air Pollution Policy." In Paul R. Portney, ed., *Public Policies for Environmental Protection.* Washington, DC: Resources for the Future, pp. 27–96.

——— (1990b). "Economics and the Clean Air Act." *Journal of Economic Perspectives 4(4)* (Fall), pp. 173–81.

Potter, Ned (2011). "New Planet: An Earth-Lie World, 600 Light-Years Away?" *ABC News.* Available at **http://abcnews.go.com/Technology/planet-found-600-light-years-nasa-kepler-telescope/story?id=15094976**.

Pottinger, Matt, Steve Stecklow, and John J. Fialka (2004). "A Hidden Cost of China's Growth: Mercury Migration." *Wall Street Journal* (December 17), p. A1.

Power, Stephen (2009). "U.S. News: Obama's EPA Move Likely to Spur Fight." *Wall Street Journal* (January 27), p. A3.

Probst, Katherine N., David M. Konisky, Robert Hersh, Michael B. Batz, and Katherine D. Walker (2001). *Superfund's Future: What Will It Cost?* Washington, DC: Resources for the Future (July).

PRO EUROPE (2006, October). *Europe Goes Green Dot: Uniformity in Diversity.* Brussels, Belgium: PRO EUROPE.

——— (2011). "List of Frequently Asked Questions." Available at **http://pro-e.org/Frequently_Asked_Questions.html** (accessed August 11, 2011).

Protzman, Ferdinand (1993). "Germany's Push to Expand the Scope of Recycling." *New York Times* (July 4), p. 38.

Pryde, Lucy T. (1973). *Environmental Chemistry: An Introduction.* Menlo Park, CA: Cummings Publishing Company.

Putnam, Hayes, and Bartlett, Inc. (1987). *Economic Implications of Potential Chlorofluorocarbons Restrictions: Final Report.* Prepared for the Alliance for Responsible CFC Policy, Washington, DC (December 2).

"Quietly, EPA Drops Some Tobacco Research" (1993). *Boston Globe* (January 7), p. 3.

Radnofsky, Louise (2011). "U.S. News: Business Groups' Target: EPA." *Wall Street Journal* (February 7), p. A4.

Ramseur, Jonathan L., Mark Reisch, and James E. McCarthy (2008, February 4). *Superfund Taxes or General Revenues: Future Funding Issues for the Superfund Program.* CRS Report for Congress. Washington, DC: Congressional Research Service.

Rankin, Jennifer (2011). "EU Carbon Tax Back on Commission's Agenda." *EuropeanVoice.com* (April 7). Available at **www.europeanvoice.com/article/imported/eu-carbon-tax-back-on-commission-s-agenda/70743.aspx**.

Raven, Peter H., Linda R. Berg, and David M. Hassenzahl (2008). *Environment.* Hoboken, NJ: John Wiley & Sons, Inc.

Reed, John, and Patti Waldmeir (2008). "China's Car Sales Boom Runs Out of Steam." *Financial Times* (October 15), p. 21.

Reese, Craig E. (1985). "State Taxation of Hazardous Materials." *Oil and Gas Tax Quarterly 33,* pp. 502–26.

Reifenberg, Anne, and Allanna Sullivan (1996). "Rising Gasoline Prices: Everyone Else's Fault." *Wall Street Journal* (May 1), pp. B1, B8.

Reilly, William K. (1992). "The Road From Rio." *EPA Journal 18(4)* (September/October), pp. 11–13.

The Remanufacturing Institute (2002). "Frequently Asked Questions." Available at **www.reman.org/frfaqust.htm** (accessed June 19, 2002).

Repetto, Robert (1992). "Accounting for Environmental Assets." *Scientific American* (June), pp. 94–100.

——— (1992). "Earth in the Balance Sheet: Incorporating Natural Resources in National Income Accounts." *Environment 34(7)* (September), pp. 12–20, 43–45.

Resource Integration Systems, Ltd., and Waste Matters Consulting (Portland, Oregon) (1990, December). *Decisionmaker's Guide to Recycling Plastics.* Prepared for the Oregon Department of Environmental Quality Solid Waste Reduction and Recycling Section, and the U.S. Environmental Protection Agency, Region X, Solid Waste Program.

Responsible Care® (2009). "Who We Are." Available at **www.responsiblecare.org/page.asp?p=6406&l=1** (accessed April 14, 2009).

RestoretheGulf.gov (2011, May 16). *Oil Spill Cost and Reimbursement Fact Sheet.* Available at **www.restorethegulf.gov/release/2011/05/16/oil-spill-cost-and-reimbursement-fact-sheet**.

——— (2011, June 2). "Operations and Ongoing Response." Available at **www.restorethegulf.gov/release/2011/06/02/operations-and-ongoing-response-june-2-2011**.

Revkin, Andrew C. (2001). "178 Nations Reach Climate Accord: U.S. Only Looks On." *New York Times* (July 24), p. A1.

RGGI (Regional Greenhouse Gas Initiative) (2009, April 1). "About RGGI." Available at **www.rggi.org/home** (accessed April 1, 2009).

——— (2011, May 30). "Auction Results." Available at **www.rggi.org/home** (accessed May 30, 2011).

——— (2012, April 16). "Welcome." Available at **www.rggi.org/home** (accessed April 16, 2012).

Roach, Fred, Charles Kolstad, Allen V. Kneese, Richard Tovin, and Michael Williams (1981). "Alternative Air Quality Policy Options in the Four Corners Region." *Southwest Review 1(2)*, pp. 44–45.

Roberts, Dexter (2003). "The Greening of China." *Business Week* (October 27), p. 53.

Rodwan, John G., Jr. (2010, April/May). "Challenging Circumstances Persist: Future Growth Anticipated." (Citing Beverage Marketing Corporation.) *Bottled Water Reporter* (April/May), pp. 10–16. Available at **www.bottledwater.org/files/2009BWstats.pdf**.

Rohlwing, Kevin (2005). "Tire Fees and Recycling: How Does Your State Compare." *Today's Tire Industry* (July/August), pp. 4–8.

Rosen, Harvey S., and Ted Gayer (2008). *Public Finance*. 8th Edition. Boston, MA: Irwin-McGraw-Hill.

Ross, Karl (1993). "Some Foul Air in Puerto Rico." *Boston Globe* (January 7), p. 12.

Ruff, Larry (1970). "The Economic Common Sense of Pollution." *The Public Interest 19* (Spring), pp. 69–85.

Samuelson, Paul A. (1954). "The Pure Theory of Public Expenditure." *Review of Economics and Statistics 36*, pp. 387–89.

——— (1955). "Diagrammatic Exposition of a Theory of Public Expenditure." *Review of Economics and Statistics 37*, pp. 350–56.

——— (1958). "Aspects of Public Expenditure Theory." *Review of Economics and Statistics 40*, pp. 332–38.

SCAQMD (South Coast Air Quality Management District) (1992). *Regional Clean Air Incentives Market*. Diamond Bar, CA: SCAQMD.

Scheuplein, Robert (1993). "Uncertainty and the 'Flavors' of Risk." *EPA Journal 19(1)* (January–March), pp. 16–17.

Schierow, Linda-Jo (1996, September 11). *Pesticide Legislation: Food Quality Protection Act of 1996*. Washington, DC: Congressional Research Service.

——— (2009, July 28). *The Toxic Substances Control Act (TSCA): Implementation and New Challenges*. CRS Report for Congress. Washington, DC: Congressional Research Service. Available at **http://portal.acs.org/preview/fileFetch/C/CNBP_023041/pdf/CNBP_023041.pdf**.

——— (2010, February 1). *Pesticide Law: A Summary of the Statutes*. CRS Report for Congress. Washington, DC: Congressional Research Service. Available at **www.nationalaglawcenter.org/assets/crs/RL31921.pdf**.

Schlesinger, Jacob M. (1998). "Secondhand-Smoke Study Ruled Invalid." *Wall Street Journal* (July 20), pp. A3, A8.

Schmalensee, Richard, Paul L. Joskow, A. Denny Ellerman, Juan Pablo Montero, and Elizabeth M. Bailey (1998). "An Interim Evaluation of Sulfur Dioxide Emissions Trading." *Journal of Economic Perspectives 12(3)* (Summer), pp. 53–68.

Schmidheiny, Stephan (1992). *Changing Course: A Global Business Perspective on Development and the Environment*. Cambridge, MA: MIT Press.

Schulze, William D., and David S. Brookshire (1983). "The Economic Benefits of Preserving Visibility in the National Parklands of the Southwest." *Natural Resources Journal 23(1)* (January), pp. 763–72.

S. C. Johnson & Son, Inc. (2011). *Helpful Partnerships: Design for the Environment; Safe Detergents Stewardship Initiative*. Racine, WI: S. C. Johnson & Son, Inc. Available at **http://scjohnson.com/en/commitment/focus-on/greener-products/partnerships.aspx** (accessed October 19, 2011).

Seattle Solid Waste Utility (1988). *Seattle Solid Waste Utility Rate Sheet and Customer Reply Card*. Seattle, WA: Seattle Solid Waste Utility.

Seelye, Katharine Q. (2001). "EPA to Adopt Clinton Arsenic Standard." *New York Times* (November 1), p. A18.

Selman, Mindy, Suzie Greenhalgh, Evan Branosky, Cy Jones, and Jenny Guiling (2009). "Water Quality Trading Programs: An International Overview." *WRI Issue Brief (1)* (March), pp. 1–16.

Seskin, Eugene P. (1978). "Automobile Air Pollution Policy." In Paul R. Portney, ed., *Current Issues in U.S. Environmental Policy*. Baltimore, MD: Johns Hopkins University Press, pp. 68–104.

Sessions, Kathy (1993). "Products of the Earth Summit." *EPA Journal 19(2)* (April–June), p. 12.

Shalal-Esa, Andrea (1993). "Tobacco Industry Sues EPA Over Secondhand Smoke Report." *Boston Globe* (June 23), p. 41.

Shapiro, Michael (1990). "Toxic Substances Policy." In Paul R. Portney, ed., *Public Policies for Environmental Protection*. Washington, DC: Resources for the Future, pp. 195–241.

Shea, Cynthia Pollock (1992). "Getting Serious in Germany." *EPA Journal 18(3)* (July/August), pp. 50–52.

Sigman, Hilary (2000). "Hazardous Waste and Toxic Substance Policies." In Paul R. Portney and Robert N. Stavins, eds., *Public Policies for Environmental Protection*. 2nd Edition. Washington, DC: Resources for the Future, pp. 215–59.

Sim, Glenys, and Feiwen Rong (2010). "China Is Set to Lose 2% of GDP Cleaning Up Pollution." *Bloomberg Businessweek* (September 17).

Skumatz, Lisa A., and David J. Freeman (2006, December). *Pay as You Throw (PAYT) in the U.S.: 2006 Update and Analyses*. Prepared for U.S. EPA and SERA, by Skumatz Economic Research Associates, Inc., Superior, CO. Available at **www.epa.gov/epawaste/conserve/tools/payt/pdf/sera06.pdf**.

Smith, Adam (1937 [1776]). *An Inquiry Into the Nature and Causes of the Wealth of Nations*. New York: Random House.

Smith, Emily T., and David Woodruff (1992). "The Next Trick for Business: Taking a Cue From Nature." *Business Week* (May 11), pp. 74–75.

Smith, Rebecca (2008). "Utilities, Plug-In Cars: Near Collision? Electric Firms Say Daytime Charges May Raise Costs." *Wall Street Journal* (May 2), p. B.1.

Smith, Velma (1994). "Disaster in Milwaukee." *EPA Journal 20(1–2)* (Summer), pp. 16–18.

Smith, V. Kerry, and John V. Krutilla, eds. (1982). *Explorations in Natural Resource Economics*. Baltimore, MD: John Hopkins University Press.

Smith, V. Kerry, and William H. Desvousges (1985). "The Generalized Travel Cost Model and Water Quality Benefits: A Reconsideration." *Southern Economic Journal 52* (October), pp. 371–81.

——— (1986). *Measuring Water Quality Benefits*. Norwell, MA: Kluwer-Nijhoff.

Smith, V. Kerry, William H. Desvousges, and Matthew P. McGivney (1983). "Estimating Water Quality Benefits: An Econometric Analysis." *Southern Economic Journal 50(2)* (October), pp. 422–37.

Solomon, Deborah, and Tracy Tennille (2011). "U.S. News: Obama Asks EPA to Pull Ozone Rule—President Reverses Agency's Tighter Air-Quality Standards That Republicans and Businesses Said Would Stunt Growth." *Wall Street Journal* (September 3), p. 5.

Solow, Robert M. (1991). "Sustainability: An Economist's Perspective." Paper Presented at the Eighteenth J. Seward Johnson Lecture to the Marine Policy Center, Woods Hole Oceanographic Institution, at Woods Hole, Massachusetts (June 14). In Robert Stavins, ed., *Economics of the Environment*. New York: W. W. Norton, pp. 131–38.

"Some Statutory Mandates on Risk" (1993). *EPA Journal 19(1)* (January–March), p. 15.

Spencer, Jane (2007). "Ravaged Rivers: China Pays Steep Price as Textile Exports Boom; Suppliers to U.S. Stores Accused of Dumping Dyes to Slash Their Costs." *Wall Street Journal* (August 22), p. A1.

——— (2007). "China Shifts Pollution Fight; New Rules Target Export Industry With Stiff Penalties." *Wall Street Journal* (November 1), p. A15.

Speth, James Gustave (2003). "Perspectives on the Johannesburg Summit." *Environment 45(1)* (January/February), pp. 24–30.

Spofford, Walter O., Jr. (1984). *Efficiency Properties of Alternative Source Control Policies for Meeting Ambient Air Quality Standards: An Empirical Application to the Lower Delaware Valley*. Discussion Paper D-118. Washington, DC: Resources for the Future (February).

Stavins, Robert N. (1998). "What Can We Learn From the Grand Policy Experiment? Lessons From SO_2 Allowance Trading." *Journal of Economic Perspectives 12(3)* (Summer), pp. 68–88.

Stedman, S., and T. E. Dahl (2008). *Status and Trends of Wetlands in the Coastal Watersheds of the Eastern United States 1998 to 2004*. Washington, DC: National Oceanic and Atmospheric Administration, National Marine Fisheries Service and U.S. Department of the Interior, Fish and Wildlife Service.

Stein, Nicholas (2004). "Water, Water." *Fortune 150(7)* (October 4), pp. 115–21.

Stern, Arthur C. (1982). "History of Air Pollution Legislation in the United States." *Journal of the Air Pollution Control Association 32(1)* (January), pp. 44–61.

Stern, D. I. (1998). "Progress on the Environmental Kuznets Curve?" *Environment and Development Economics 3(2)* (May), pp. 175–98.

Stern, Nicholas (2007). *The Economics of Climate Change: The Stern Review*. New York: Cambridge University Press.

Stiglitz, Joseph E. (1988). *Economics of the Public Sector*. New York: W. W. Norton.

Stranahan, Susan Q. (1990). "It's Enough to Make You Sick." *National Wildlife* (February/March), pp. 8–15.

Sumner, Jenny, Lori Bird, and Hillary Dobos (2011). "Carbon Taxes: A Review of Experience and Policy Design Considerations." *Climate Policy 11(2)* (March), pp. 922–43.

Tanner, James (1992). "Carbon Tax to Limit Use of Fossil Fuels Becomes Embroiled in Global Politics." *Wall Street Journal* (June 9), p. A2.

Taylor, Jeffrey (1992). "Auction of Rights to Pollute Fetches About $21 Million." *Wall Street Journal* (March 31), p. A6.

——— (1993). "CBOT Plan for Pollution-Rights Market Is Encountering Plenty of Competition." *Wall Street Journal* (August 24), pp. C1, C6.

Taylor, Jeffrey, and Dave Kansas (1992). "Environmentalists Vie for Right to Pollute." *Wall Street Journal* (March 26), p. C1.

Taylor, Jeffrey, and Rose Gutfeld (1992). "CBOT Selected to Run Auction for Polluters." *Wall Street Journal* (September 25), p. C1.

3M Company (2011). *2011 Sustainability Progress; Inventing Tomorrow, Sustaining Our Future*. St. Paul, MN: 3M Company.

——— (2011, October 16). *Sustainability at 3M: Timeline*. St. Paul, MN: 3M Company. Available at **http://solutions.3m.com/wps/portal/3M/en_US/3M-Sustainability/Global/VisionHistory/Timeline/** (accessed October 16, 2011).

Tietenberg, T. H. (1985). *Emissions Trading: An Exercise in Reforming Pollution Policy*. Washington, DC: Resources for the Future.

Tietenberg, Tom, and Lynne Lewis (2009). *Environmental and Natural Resource Economics*. 8th Edition. Boston, MA: Pearson Addison Wesley.

"Tobacco Firms Fail in Appeal of Ruling on Second Hand Smoke" (2002). *Wall Street Journal* (December 12), p. B6.

Torello, Alessandro (2011). "EU to Propose Carbon Tax on Fuels." *Wall Street Journal (online)* (April 8).

Tracy, Ryan (2011). "House Votes to Stop EPA From Regulating Greenhouse Gases." *Wall Street Journal (online)* (April 7).

Tregarthen, Timothy (1989). "Garbage by the Bag: Perkasie Acts on Solid Waste." *The Margin* (September/October), p. 17.

Tresch, Richard W. (2002). *Public Finance: A Normative Theory*. 2nd Edition. Boston, MA: Academic Press.

Um, Mi-Jung, Seung-Jun Kwak, and Tai Yoo Kim (2002). "Estimating Willingness to Pay for Improved Drinking Water Quality Using Averting Behavior Method With Perception Measure." *Environmental and Resource Economics 21(3)* (March), pp. 287–302.

U.K. Department for Environment, Food and Rural Affairs (DEFRA) (2008, April). *Climate Change Bill Final Impact Assessment.* London: DEFRA.

——— (2008, December 2). "Climate Change Act 2008—Key Documents." Available at **www.defra.gov.uk/environment/climatechange/uk/legislation/docs.htm** (last modified December 2, 2008).

U.K. Department of Energy and Climate Change (DECC) (2009, March). *Climate Change Act 2008 Impact Assessment.* London: DECC.

UNFCCC (United Nations Framework Convention on Climate Change) (2009, January). *Fact Sheet: Copenhagen—COP 15/CMP 5.* Bonn, Germany: UNFCCC.

——— (2011, March 15). *Report of the Conference of the Parties on Its Sixteenth Session, Held in Cancun From 29 November to 10 December 2010: Decision 1/CP.16.* Bonn, Germany: UNFCCC.

——— (2011, May 23). *The Cancun Agreements: Key Steps of the United National Climate Change Conference.* Bonn, Germany: UNFCCC. Available at **http://cancun.unfccc.int/** (accessed May 23, 2011).

——— (2011a, September 18). "Essential Background: The United Nations Framework Convention on Climate Change." Bonn, Germany: UNFCCC. Available at **http://unfccc.int/essential_background/convention/items/2627.php** (accessed September 18, 2011).

——— (2011b, September 18). *Fact Sheet: An Introduction to the United Nations Framework Convention on Climate Change (UNFCCC) and Its Kyoto Protocol.* Bonn, Germany: UNFCCC. Available at **http://unfccc.int/press/fact_sheets/items/4978.php** (accessed September 18, 2011).

United Kingdom (2010). *Statutory Instruments: 2010 No. 1001; Environmental Protection: The Air Quality Standards Regulations 2010.*

United Nations (1978). *Water Development and Management: Proceedings of the United Nations Water Conference, 4 vols.* Oxford: Pergamon Press for the United Nations.

——— (2003, January). *The Road From Johannesburg: What Was Achieved and the Way Forward.* New York: United Nations. Available at **www.un.org/esa/sustdev/media/Brochure.PDF**.

United Nations, Department of Economic and Social Affairs, Population Division (2004). *World Population to 2300.* New York: United Nations. Available at **www.un.org/esa/population/publications/longrange2/WorldPop2300final.pdf**.

United Nations, Division for Sustainable Development (2004, December 17). *Agenda 21.* Available at **www.un.org/esa/sustdev/agenda21text.htm** (last updated December 17, 2004).

United Nations Economic Commission for Europe (UNECE) (1979–2005). *The Convention on Long-Range Transboundary Air Pollution.* Geneva, Switzerland: UNECE. Available at **http://unece.org/env/lrtap/lrtap_h1.htm** (accessed April 14, 2007).

United Nations Environment Programme (2007). *Global Environment Outlook: Environment for Development GEO-4.* Valletta, Malta: Progress Press Ltd. Available at **www.unep.org/geo/geo4.asp**.

——— (2008, September). *Backgrounder: Basic Facts and Data on the Science and Politics of Ozone Protection.* New York: United Nations. Available at **http://ozone.unep.org/Events/ozone_day_2008/press_backgrounder.pdf**.

United Nations, European Commission, International Monetary Fund, Organization for Economic Co-operation and Development (OECD) and World Bank (2003). *Studies in Methods, Handbook of National Accounting: Integrated Environmental and Economic Accounting 2003.* New York: United Nations. Available at **http://unstats.un.org/unsd/envAccounting/seea2003.pdf**.

United Nations, Secretariat of the Multilateral Fund for the Implementation of the Montreal Protocol (2011, September 17). "Multilateral Fund for the Implementation of the Montreal Protocol." Available at **www.multilateralfund.org/default.aspx** (accessed September 17, 2011).

United Nations, Statistics Division (2011). "Environmental-Economic Accounting." Available at **http://unstats.un.org/unsd/envaccounting/default.asp** (accessed January 18, 2011).

U.S. Army Corp of Engineers (1998). *Report on Treatment, Storage & Disposal Facilities (TSDF) for Hazardous, Toxic, and Radioactive Waste.* Washington, DC: U.S. Army Corp of Engineers. Available at **www.environmental.usace.army.mil/library/pubs/tsdf/tsdf.html**.

U.S. Census Bureau (2008). *Pollution Abatement Costs and Expenditures: 2005.* Washington, DC: U.S. Government Printing Office (April).

——— (2011). *Statistical Abstract of the United States: 2011.* 130th Edition. Washington, DC: U.S. Government Printing Office. Available at **www.census.gov/compendia/statab**.

U.S. Census Bureau, Population Division (2011, February). "Preliminary Annual Estimates of the Resident Population for the United States, Regions, States, and Puerto Rico: April 1, 2000 to July 1, 2010." Available at **www.census.gov/popest/research/eval-estimates/eval-est2010.html**.

U.S. Climate Change Science Program (CCSP) (2003, July). *The U.S. Climate Change Science Program: Vision for the Program and Highlights of the Scientific Strategic Plan.* Washington, DC: CCSP.

U.S. Congress (1996). *Safe Drinking Water Act Amendments,* 104th Congress, Public Law 104-182, approved August 6.

U.S. Congress, CBO (Congressional Budget Office) (1984). "From Office of Technology Assessment, Statement of

Joel S. Hirschhorn for the Hearing Record." *Senate Committee on Environment and Public Works 98(2)* (September 10).
——— (1985a). *Efficient Investments in Wastewater Treatment Plants.* Washington, DC: U.S. Government Printing Office.
——— (1985b). *Hazardous Waste Management: Recent Changes and Policy Alternatives.* Washington, DC: U.S. Government Printing Office.
U.S. Congress, OTA (Office of Technology Assessment) (1983). *Technologies and Management Strategies for Hazardous Waste Control.* Washington, DC: U.S. Government Printing Office.
——— (1984, October). *Protecting the Nation's Groundwater From Contamination, Volume 1.* Washington, DC: U.S. Government Printing Office.
——— (1985). *Superfund Strategy.* Washington, DC: U.S. Government Printing Office.
——— (1987). *Wastes in Marine Environments.* Washington, DC: U.S. Government Printing Office.
——— (1992, October). *Green Products by Design: Choices for a Cleaner Environment.* Washington, DC: U.S. Government Printing Office.
U.S. Department of Agriculture, Office of the Chief Economist, Office of Environmental Markets (2010, August). *Chesapeake Bay Environmental Markets Team Charter.* Available at **www.usda.gov/oce/environmental_markets/files/FY10_11_CB_EMT_Charter_final.pdf.**
——— (2011). "Environmental Markets in the Chesapeake Bay." Available at **www.usda.gov/oce/environmental_markets/chesapeake.htm** (last modified April 28, 2011).
U.S. Department of Commerce, Bureau of Economic Analysis (2011, February 25). *Current-Dollar and "Real" Gross Domestic Product.* National Economic Accounts. Washington, DC: Department of Commerce.
U.S. Department of Commerce, Census Bureau, Foreign Trade Division (2011, July). "FTA Partners—Total Goods." Available at **www.trade.gov/mas/ian/.**
U.S. Department of Commerce, International Trade Administration (2008, Summer). *Top U.S. Export Markets: Free Trade Agreement and Country Fact Sheets.* Washington, DC: Department of Commerce. Available at **http://trade.gov/media/publications/pdf/tm_091208.pdf.**
U.S. Department of Commerce, International Trade Administration, U.S. Commercial Service (2008, October 3). "Pearl River Delta, Asia Now." Available at **www.buyusa.gov/hongkong/en/pearlriverdelta.html** (last updated October 3, 2008).
U.S. Department of Energy (2008, October 21). "Consumer Energy Tax Incentives: What the American Recovery and Reinvestment Act Means to You." Available at **www.energy.gov/taxbreaks.htm** (last reviewed October 21, 2008).
U.S. Department of Energy, Energy Information Administration (2006, August). *Country Analysis Briefs: China.* Washington, DC: U.S. Department of Energy, Energy Information Administration.
——— (2007, February). "Biofuels in the U.S. Transportation Sector." *Annual Energy Outlook 2007.* Washington, DC: U.S. Department of Energy, Energy Information Administration.
——— (2009). *Electric Power Monthly: February 2009.* Washington, DC: U.S. Department of Energy, Energy Information Administration (February).
——— (2011, March). *Emissions of Greenhouse Gases in the United States 2009.* Washington, DC: Department of Energy, Energy Information Administration.
——— (2011, May). *International Energy Statistics.* Washington, DC: Department of Energy, Energy Information Administration. Available at **www.eia.gov/cfapps/ipdbproject/IEDIndex3.cfm?tid=90&pid=44&aid=8** (accessed May 23, 2011).
——— (2011, September). *International Energy Outlook 2011.* Washington, DC: U.S. Department of Energy. Available at **http://205.254.135.24/forecasts/ieo/pdf/0484(2011).pdf.**
U.S. Department of Energy, Office of Energy Efficiency and Renewable Energy (2003, June 9). "Fuel Comparison Chart." Available at **www.eere.energy.gov/afdc/progs/fuel_compare.cgi** (last updated June 9, 2003).
——— (2009, February 20). "Alternative and Advanced Fuels." Available at **http://afdc.energy.gov/afdc/fuels/index.html** (last updated February 20, 2009).
U.S. Department of Energy, Office of Energy Efficiency and Renewable Energy, and U.S. Environmental Protection Agency (2009, February 22). "Alternative Fuels." Available at **www.fueleconomy.gov/feg/tax_hybrid.shtml** (accessed February 22, 2009).
——— (2011, April 20). "Fuel Efficient Vehicle: Tax Incentive Information Center." Available at **www.fueleconomy.gov/feg/taxcenter.shtml** (accessed April 20, 2011).
——— (2011, April 26). "Fuel Economy Guide: 2011 Hybrid Vehicles; Compare Side-by-Side." Available at **www.fueleconomy.gov/feg/hybrid_sbs.shtml** (accessed April 26, 2011).
U.S. Department of Energy, Office of Energy Efficiency and Renewable Energy, U.S. EPA Green Power Partnership, World Resources Institute, and Center for Resource Solutions (2010). *Guide to Purchasing Green Power: Renewable Electricity, Renewable Energy Certificates, and On-Site Renewable Generation* (March). Available at **http://www1.eere.energy.gov/femp/pdfs/purchase_green_power.pdf.**
U.S. Department of Justice (2004, June 23). "U.S. and PREPA Reach Agreement for More Pollution Reductions at Four Power Plants." Available at **www.usdoj.gov/opa/pr/2004/June/04_enrd_433.htm.**
U.S. DOI (Department of the Interior), Fish and Wildlife Service (FWS) (2011, April 20). *Deepwater Horizon Response Consolidated Fish and Wildlife Collection Report.* Washington, DC: U.S. DOI and FWS. Available at

www.fws.gov/home/dhoilspill/pdfs/ConsolidatedWildlifeTable042011.pdf.

——— (2012, January 6). *Species Reports: Environmental Conservation Online System*. Washington, DC: U.S. DOI and FWS. Available at **http://ecos.fws.gov/tess_public/pub/boxScore.jsp** (accessed January 6, 2012).

U.S. Department of State (2010). *U.S. Climate Action Report 2010*. Washington, DC: Global Publishing Services (June).

U.S. Department of Transportation, National Highway Traffic Safety Administration (NHTSA) (2009, December). *Consumer Assistance to Recycle and Save Act of 2009: Report to Congress*. Washington, DC: U.S. Department of Transportation, NHTSA. Available at **www.cars.gov/files/official-information/CARS-Report-to-Congress.pdf**.

U.S. Department of the Treasury (2011, February). *General Explanations of the Administration's Fiscal Year 2012 Revenue Proposals*. Washington, DC: U.S. Department of the Treasury. Available at **www.treasury.gov/resource-center/tax-policy/Documents/Final%20Greenbook%20Feb%202012.pdf**.

U.S. Department of the Treasury, Internal Revenue Service (2006). "Highlights of the Energy Policy Act of 2005 for Individuals." (January). Accessed at **www.irs.gov/newsroom/article/0,,id=153397,00.html**.

U.S. EPA (Environmental Protection Agency) (n.d.). *Effluent Trading in Watersheds*. Reinvention Activity Fact Sheets. Washington, DC: U.S. EPA.

——— (1987, March). *Site Discovery Methods*. EPA Contract 68-01-6888. Washington, DC: U.S. EPA.

——— (1989, August). *Comparing Risks and Setting Environmental Priorities*. Washington, DC: U.S. EPA.

——— (1991, March). *Bottled Water Fact Sheet*. Washington, DC: U.S. EPA.

——— (2001, September 18). *September 18: Whitman Details Ongoing Agency Efforts to Monitor Disaster Sites, Contribute to Cleanup Efforts*. Washington, DC: U.S. EPA. Available at **www.epa.gov/wtc/stories/headline_091801.htm**.

——— (2002a, September 24). "Benchmarks, Standards and Guidelines Established to Protect Public Health." Available at **www.epa.gov/wtc/benchmarks.htm** (accessed September 24, 2002).

——— (2002b, September 24). "Daily Environmental Monitoring Summary." Available at **www.epa.gov/wtc/data_summary.htm** (accessed September 24, 2002).

——— (2003, May 5). *Border 2012: U.S.–Mexico Environmental Program*. Washington, DC: U.S. EPA. Available at **www.epa.gov/usmexicoborder/docs/FrameworkDocument2012-eng.pdf**.

——— (2004, September 17). "Summary of the Small Business Liability Relief and Brownfields Revitalization Act." Available at **www.epa.gov/brownfields/html-doc/2869sum.htm** (last updated September 17, 2004).

——— (2005, January 19). *EPA Announces New Aircraft Drinking Water Quality Data*. Washington, DC: U.S. EPA. Available at **www.epa.gov/safewater/airlinewater/newreleases.html**.

——— (2007). *U.S.–Mexico Environmental Program: Border 2012: Implementation and Mid-Term Report: 2007*. Washington, DC: U.S. EPA. Available at **www.epa.gov/usmexicoborder/docs/implementation_2007_eng.pdf**.

——— (2007, June 8). *What Is Acid Rain?* Washington, DC: U.S. EPA.

——— (2007, August). *State Actions Banning MTBE (Statewide)*. Washington, DC: U.S. EPA.

——— (2008, November 19). "Lower Manhattan Test and Clean Program." Available at **www.epa.gov/wtc/testandclean/index.html** (last updated November 19, 2008).

——— (2009, March 12). "Implementation of the 2009 Recovery Act: Clean Water and Drinking Water State Revolving Funds." Available at **www.epa.gov/recovery/** (last updated March 12, 2009).

——— (2010, Fall). *U.S.–Mexico Environmental Program: Border 2012; Program Highlights*. Washington, DC: U.S. EPA. Available at **www.epa.gov/usmexicoborder/docs/reports/Border2012ProgramHighlights.pdf**.

——— (2010, December 16). *2009 Toxics Release Inventory: National Analysis Overview*. Washington, DC: U.S. EPA. Available at **www.epa.gov/tri/tridata/tri09/nationalanalysis/overview/2009TRINAOverviewfinal.pdf**.

——— (2011, February 28). "Superfund: Cleanup Process." Available at **www.epa.gov/superfund/cleanup/index.htm** (last updated February 28, 2011).

——— (2011, May 4). *American Recovery and Reinvestment Act: Quarterly Performance Report; Quarter II, Cumulative Results as of March 30, 2011*. Washington, DC: U.S. EPA. Available at **www.epa.gov/recovery/pdfs/2011_Q2_Perf_Rpt.pdf**.

——— (2011, June 15). *Hudson River PCBs*. Washington, DC: U.S. EPA. Available at **www.epa.gov/hudson/** (last updated June 15, 2011).

——— (2011, August 2). "Partnership Programs: List of Programs." Available at **www.epa.gov/partners/programs/index.htm** (last updated August 2, 2011).

——— (2011, September 5). *Border 2020: U.S.–Mexico Environmental Program; Draft Border 2020 Document*. Available at **www.epa.gov/border2012/docs/2020/border2020-draft-framework.pdf**.

U.S. EPA, Indoor Environments Division (2009, January). *A Citizen's Guide to Radon: The Guide to Protecting Yourself and Your Family From Radon*. Washington, DC: U.S. EPA, Indoor Environments Division.

U.S. EPA, Office of the Administrator (2000, September). *Guidelines for Preparing Economic Analyses*. Washington, DC: U.S. EPA, Office of the Administrator.

U.S. EPA, Office of Air and Radiation (1992, December 1). *Respiratory Health Effects of Passive Smoking: Lung Cancer and Other Disorders*. Washington, DC: U.S. EPA, Office of Air and Radiation.

——— (1997, October). *The Benefits and Costs of the Clean Air Act, 1970 to 1990.* Washington, DC: U.S. EPA, Office of Air and Radiation.

——— (1999, November). *The Benefits and Costs of the Clean Air Act, 1990 to 2010.* Washington, DC: U.S. EPA, Office of Air and Radiation.

——— (2003, June). *EPA Assessment of Risks for Radon in Homes.* Washington, DC: U.S. EPA, Office of Air and Radiation.

——— (2004). *FY 2003 Environmental Justice Action Plan—Progress Report.* Washington, DC: U.S. EPA, Office of Air and Radiation.

——— (2008, October). *Energy Star® and Other Climate Protection Partnerships: 2007 Annual Report.* Washington, DC: U.S. EPA, Office of Air and Radiation.

——— (2009, January 15). "Clean Air Interstate Rule." Available at **www.epa.gov/air/interstateairquality/** (last updated January 15, 2009).

——— (2009, April). *Frequently Asked Questions About Global Warming and Climate Change: Back to Basics.* Washington, DC: U.S. EPA, Office of Air and Radiation. Available at **www.epa.gov/climatechange/downloads/Climate_Basics.pdf**.

——— (2009, August). *Air Quality Index: A Guide to Air Quality and Your Health.* Washington, DC: U.S. EPA, Office of Air and Radiation.

——— (2010, October). *United States—Canada Air Quality Agreement: Progress Report 2010.* Washington, DC: U.S. EPA, Office of Air and Radiation.

——— (2010, December 23). *Clean Air Permitting for Greenhouse Gases With Fact Sheet: Clean Air Act Permitting for Greenhouse Gases: Guidance and Technical Information.* Washington, DC: U.S. EPA, Office of Air and Radiation. Available at **www.epa.gov/nsr/ghgpermitting.html** (last updated December 23, 2010).

——— (2011, February 10). *Air Quality Trends.* Washington, DC: U.S. EPA, Office of Air and Radiation. Available at **www.epa.gov/air/airtrends/aqtrends.html** (last updated February 10, 2011).

——— (2011, March). *The Benefits and Costs of the Clean Air Act From 1990 to 2020: Summary Report.* Washington, DC: U.S. EPA, Office of Air and Radiation. Available at **www.epa.gov/air/sect812/feb11/summaryreport.pdf**.

——— (2011, March 21). "Air Quality Index Information: Number of Days With Air Quality Index Values Greater Than 100 at Trend Sites, 1990–2009, and All Sites in 2009." Available at **www.epa.gov/airtrends/aqi_info.html** (last updated March 21, 2011).

——— (2011, April). *The Benefits and Costs of the Clean Air Act From 1990 to 2020: Final Report.* Washington, DC: U.S. EPA, Office of Air and Radiation. Available at **www.epa.gov/air/sect812/feb11/fullreport.pdf**.

——— (2011, June). *Regulatory Impact Analysis for the Federal Implementation Plans to Reduce Interstate Transport of Fine Particulate Matter and Ozone in 27 States; Correction of SIP Approvals for 22 States.* Washington, DC: U.S. EPA, Office of Air and Radiation. Available at **www.epa.gov/airtransport/pdfs/FinalRIA.pdf**.

——— (2011, July 7). *Cross-State Air Pollution Rule (CSAPR): Basic Information.* Washington, DC: U.S. EPA, Office of Air and Radiation. Available at **www.epa.gov/crossstaterule/basic.html**.

U.S. EPA, Office of Air and Radiation, Office of Air Quality Planning and Standards (2003, September). *National Air Quality and Emissions Trends Report: 2003 Special Studies Edition.* Research Triangle Park, NC: U.S. EPA.

——— (2010, February). *Our Nation's Air: Status and Trends Through 2008.* Research Triangle Park, NC: U.S. EPA.

——— (2011, February 23). *Emissions Standards for Boilers and Process Heaters and Commercial/Industrial Solid Waste Incinerators, Including Fact Sheet: Final Air Toxics Standards for Industrial, Commercial, and Institutional Boilers at Area Source Facilities.* Washington, DC: U.S. EPA.

——— (2011, March 18). *Reducing Toxic Emissions From Power Plants, including Fact Sheet: Proposed Mercury and Air Toxics Standards.* Washington, DC: U.S. EPA.

——— (2011, April 21). "The Green Book: Nonattainment Areas for Criteria Pollutants." Available at **www.epa.gov/oaqps001/greenbk/** (last updated April 21, 2011).

——— (2011, November 8). "National Ambient Air Quality Standards (NAAQS)."Available at **www.epa.gov/air/criteria.html** (last updated November 8, 2011).

U.S. EPA, Office of Air and Radiation, Office of Atmospheric Programs, Clean Air Markets Division (2004, August). *NO_X Budget Trading Program: 2003 Progress and Compliance Report.* Washington, DC: U.S. EPA.

——— (2004, September). *Acid Rain Program: 2003 Progress Report.* Washington, DC: U.S. EPA.

——— (2008, August 25). "2008 EPA Allowance Auction Results." Available at **www.epa.gov/airmarkets/trading/2008/08summary.html** (last updated August 25, 2008).

——— (2008, December). *NO_X Budget Trading Program: Compliance and Environmental Results.* Washington, DC: U.S. EPA.

——— (2009, January). *Acid Rain and Related Programs: 2007 Progress Report.* Washington, DC: U.S. EPA.

——— (2009, May 28). *NO_X Budget Trading Program—Basic Information.* Washington, DC: U.S. EPA. Available at **www.epa.gov/airmarkt/progsregs/nox/docs/NBPbasicinfo.pdf**.

——— (2010, September). *Acid Rain and Related Programs: 2009 Emission, Compliance, and Market Analyses.* Washington, DC: U.S. EPA. Available at **www.epa.gov/airmarkets/progress/ARP09_downloads/ARP_2009_ECM_Analyses.pdf**.

——— (2010, December). *Acid Rain and Related Programs: 2009 Highlights; 15 Years of Results: 1995–2009.* Washington, DC: U.S. EPA.

——— (2010, December 20). *Acid Rain Program 2009 Progress Report—Data Access.* Washington, DC: U.S. EPA.

——— (2011, March 29). "2011 EPA Allowance Auction Results." Available at **www.epa.gov/airmarkets/trading/2011/11summary.html** (last updated March 29, 2011).

——— (2011, April 8). "Quick Facts and Trends: NO_X Allowance Daily Average Price I (Prompt Vintage, MPI), January 2005–December 2009." Available at **http://camddataandmaps.epa.gov/gdm/index.cfm?fuseaction=factstrends.top_bypriceofallowances** (last updated April 8, 2011).

——— (2011, April 19). "Annual Auction." Available at **www.epa.gov/airmarkt/trading/auction.html** (last updated April 19, 2011).

U.S. EPA, Office of Air and Radiation, Office of Atmospheric Programs, Climate Change Division (2007, December 20). "State of Knowledge." Available at **www.epa.gov/climatechange/science/stateofknowledge.html** (last updated December 20, 2007).

——— (2008, July). *EPA's Green Power Partnership: Renewable Energy Certificates.* Washington, DC: U.S. EPA. Available at **www.epa.gov/greenpower/documents/gpp_basics-recs.pdf**.

——— (2008, December 17). "International Impacts." Available at **www.epa.gov/climatechange/effects/international.html** (last updated December 17, 2008).

——— (2009, April 17). *Overview of EPA's Proposed Endangerment and Cause or Contribute Findings for Greenhouse Gases Under the Clean Air Act.* Washington, DC: U.S. EPA. Available at **www.epa.gov/climatechange/endangerment/downloads/Determination.pdf**.

——— (2010, October). *EPA's Green Power Partnership: The Environmental Value of Purchasing Renewable Energy Certificates Voluntarily.* Washington, DC: U.S. EPA. Available at **www.epa.gov/greenpower/documents/gpp_basics-recs_voluntary.pdf**.

——— (2010, October 25). *2010 Award Winners: Green Power Leadership Awards.* Washington, DC: U.S. EPA. Available at **www.epa.gov/greenpower/awards/winners.htm**.

——— (2011a, April 14). *Endangerment and Cause or Contribute Findings for Greenhouse Gases Under Section 202(a) of the Clean Air Act.* Washington, DC: U.S. EPA. Available at **www.epa.gov/climatechange/endangerment.html**.

——— (2011b, April 14) "Climate Change: Basic Information; Climate Change or Global Warming?" Available at **www.epa.gov/climatechange/basicinfo.html** (last updated April 14, 2011).

U.S. EPA, Office of Air and Radiation, Office of Atmospheric Programs, Stratospheric Protection Division (2010a, August 19). "Ozone Science: The Facts Behind the Phaseout." Available at **www.epa.gov/ozone/science/sc_fact.html** (last updated August 19, 2010).

——— (2010b, August 19). "The Montreal Protocol on Substances That Deplete the Ozone Layer." Available at **www.epa.gov/ozone/intpol/index.html** (last updated August 19, 2010).

——— (2010c, August 19) "Class I Ozone-Depleting Substances; Class II Ozone-Depleting Substances." Available at **www.epa.gov/ozone/science/ods/classone.html** and **www.epa.gov/ozone/science/ods/classtwo.html** (last updated August 19, 2010).

——— (2011, May 20). *Recent International Developments in Saving the Ozone Layer.* Washington, DC: U.S. EPA. Available at **www.epa.gov/ozone/intpol/mpagreement.html**.

U.S. EPA, Office of Chemical Safety and Pollution Prevention, Office of Pesticide Programs (2011a, February 16). "The EPA and Food Security." Available at **www.epa.gov/opp00001/factsheets/securty.htm** (last updated February 16, 2011).

——— (2011b, February 16). "Pesticide Reregistration Facts." Available at **www.epa.gov/oppsrrd1/reregistration/reregistration_facts.htm** (last updated February 16, 2011).

——— (2011c, February 16). "Laws Affecting EPA's Pesticide Programs." Available at **www.epa.gov/pesticides/factsheets/legisfac.htm** (last updated February 16, 2011).

——— (2011, May 9). "Pesticide Environmental Stewardship Program." Available at **www.epa.gov/pesp/pesp/index.html** (last updated May 9, 2011).

U.S. EPA, Office of Chemical Safety and Pollution Prevention, Office of Pesticide Programs, Biological and Economic Analysis Division (2011, February). *Pesticides Industry Sales and Usage: 2006 and 2007 Market Estimates.* Available at **www.epa.gov/opp00001/pestsales/07pestsales/market_estimates2007.pdf**.

U.S. EPA, Office of Chemical Safety and Pollution Prevention, Office of Pollution Prevention and Toxics (2010, February). *U.S. Environmental Protection Agency 2010–2014 Pollution Prevention (P2) Program Strategic Plan.* Washington, DC: U.S. EPA. Available at **www.epa.gov/p2/pubs/docs/P2StrategicPlan2010-14.pdf**.

——— (2010, April 28). "Essential Principles for Reform of Chemicals Management Legislation." Available at **www.epa.gov/oppt/existingchemicals/pubs/principles.html** (last updated April 28, 2010).

——— (2010, September 1). "Summary of Accomplishments." Available at **www.epa.gov/oppt/newchems/pubs/accomplishments.htm** (last updated September 1, 2010).

——— (2011, June 6). "Pollution Prevention (P2): Basic Information." Available at **www.epa.gov/p2/pubs/basic.htm** (last updated June 6, 2011).

——— (2011, August 5). "TSCA Chemical Substance Inventory: Basic Information." Available at **www.epa.gov/oppt/existingchemicals/pubs/tscainventory/basic.html** (last updated August 5, 2011).

——— (2011a, August 10). "Green Chemistry Program at EPA." Available at **www.epa.gov/greenchemistry/pubs/epa_gc.html#partnerships** (last updated August 10, 2011).

——— (2011b, August 10). "Green Chemistry: International Activities" Available at **www.epa.gov/greenchemistry/pubs/international_activities.html** (last updated August 10, 2011).

——— (2011, August 18). "Enhancing EPA's Chemical Management Program." Available at **www.epa.gov/opptintr/existingchemicals/pubs/enhanchems.html** (last updated August 18, 2011).

U.S. EPA, Office of the Chief Financial Officer (2003, September 30). *2003–2008 EPA Strategic Plan: Direction for the Future*. Washington, DC: U.S. EPA.

U.S. EPA, Office of Communications, Education, and Public Affairs (1992, April). *Securing Our Legacy: An EPA Progress Report 1989–1991*. Washington, DC: U.S. EPA.

U.S. EPA, Office of Drinking Water (n.d.). *Bottled Water: Helpful Facts and Information*. Washington, DC: U.S. EPA, Office of Drinking Water.

U.S. EPA, Office of Environmental Information, and the Office of Research and Development (2003, June). *EPA's Draft Report of the Environment 2003*. Washington, DC: U.S. EPA.

U.S. EPA, Office of Inspector General (2004). *Evaluation Report: EPA Needs to Consistently Implement the Intent of the Executive Order on Environmental Justice*. Report No. 2004-P-00007 (March). Available at **www.epa.gov/oig/reports/2004/20040301-2004-P-00007.pdf#search=%27EPA%20Ne**.

U.S. EPA, Office of Pesticides and Toxic Substances (1988, December). *Highlights of the 1988 Pesticide Law: The Federal Insecticide, Fungicide, and Rodenticide Act Amendments of 1988*. Washington, DC: U.S. EPA, Office of Pesticides and Toxic Substances.

——— (1991, June). *For Your Information: Pesticide Reregistration*. Washington, DC: U.S. EPA, Office of Pesticides and Toxic Substances.

——— (1991, October). *Pesticides and Ground-Water Strategy*. Washington, DC: U.S. EPA, Office of Pesticides and Toxic Substances.

U.S. EPA, Office of Policy, Economics, and Innovation (2001, January). *The United States Experience With Economic Incentives for Protecting the Environment*. Washington, DC: U.S. EPA, Office of Policy, Economics, and Innovation.

——— (2004, November). *International Experiences With Economic Incentives for Protecting the Environment*. Washington, DC: U.S. EPA, Office of Policy, Economics, and Innovation.

U.S. EPA, Office of Policy, Planning, and Evaluation (1985, February). *Costs and Benefits of Reducing Lead in Gasoline, Final Regulatory Impact Analysis*. Report EPA-230-05-85-006. Washington, DC: U.S. EPA, Office of Policy, Planning, and Evaluation.

——— (1987, February). *Unfinished Business: A Comparative Assessment of Environmental Problems*. Washington, DC: U.S. Government Printing Office.

——— (1987, August). *EPA's Use of Benefit-Cost Analysis, 1981–1986*. EPA Report 230-05-87-028. Washington, DC: U.S. EPA, Office of Policy, Planning, and Evaluation.

——— (1990, December). *Environmental Investments: The Cost of a Clean Environment—A Summary*. Washington, DC: U.S. EPA, Office of Policy, Planning, and Evaluation.

——— (1991, March). *Economic Incentives: Options for Environmental Protection*. Washington, DC: U.S. EPA, Office of Policy, Planning, and Evaluation.

——— (1992, July). *The United States Experience With Economic Incentives to Control Environmental Pollution*. Washington, DC: U.S. EPA, Office of Policy, Planning, and Evaluation.

U.S. EPA, Office of Pollution Prevention (1991, October). *Pollution Prevention 1991: Progress on Reducing Industrial Pollutants*. Washington, DC: U.S. EPA, Office of Pollution Prevention.

U.S. EPA, Office of Pollution Prevention and Toxics (2011, August 18). "DfE: Safer Product Labeling." Available at **http://epa.gov/dfe/pubs/projects/formulat/saferproductlabeling.htm**.

U.S. EPA, Office of Public Affairs (2009, December 7). *EPA: Greenhouse Gases Threaten Public Health and the Environment*. Washington, DC: U.S. EPA, Office of Public Affairs.

——— (2010, January 7). *EPA Strengthens Smog Standard*. Washington, DC: U.S. EPA, Office of Public Affairs.

——— (2010, June 3). *EPA Sets Stronger National Air Quality Standard for Sulfur Dioxide*. Washington, DC: U.S. EPA, Office of Public Affairs.

——— (2010, December 23). *EPA to Set Modest Pace for Greenhouse Gas Standards*. Washington, DC: U.S. EPA, Office of Public Affairs.

——— (2011, April 27). *Obama Administration Affirms Comprehensive Commitment to Clean Water*. Washington, DC: U.S. EPA, Office of Public Affairs.

——— (2011, December 2). *EPA Proposes Changes to Clean Air Act Standards for Boilers and Incinerators*. Washington, DC: U.S. EPA, Office of Public Affairs.

——— (2011, December 21). *EPA Issues First National Standards for Mercury Pollution From Power Plants*. Washington, DC: U.S. EPA, Office of Public Affairs.

U.S. EPA, Office of Research and Assessment, and Science Applications International Corporation (2006, May). *Life Cycle Assessment: Principles and Practice*. Washington, DC: U.S. EPA. Available at **www.epa.gov/ORD/NRMRL/lcaccess/pdfs/600r06060.pdf**.

U.S. EPA, Office of Resource Conservation and Recovery (2011, May). *Electronics Waste Management in the United States Through 2009: Full Report and Executive Summary*. Washington, DC: Prepared by ICF International.

——— (2011, August 17). *Interagency Task Force on Electronics Stewardship*. Washington, DC: U.S. EPA.

——— (2011, November). *Municipal Solid Waste Generation, Recycling, and Disposal in the United States: Tables and Figures for 2010*. Washington, DC: U.S. EPA.

U.S. EPA, Office of the Science Advisor, Risk Assessment Forum (1998). *Guidelines for Ecological Risk Assessment*.

Washington, DC: U.S. EPA, Office of the Science Advisor, Risk Assessment Forum.

U.S. EPA, Office of Solid Waste (1986, November). *Solving the Hazardous Waste Problem: EPA's RCRA Program.* Washington, DC: U.S. EPA, Office of Solid Waste.

——— (2010, December). *Municipal Solid Waste in the United States: 2009 Facts and Figures.* Washington, DC: U.S. EPA, Office of Solid Waste.

U.S. EPA, Office of Solid Waste and Emergency Response (1985, October). *The New RCRA Fact Book.* Washington, DC: U.S. EPA, Office of Solid Waste and Emergency Response.

——— (1990, February). *EPA's Report to Congress on Methods to Manage and Control Plastic Wastes.* Washington, DC: U.S. EPA, Office of Solid Waste and Emergency Response.

——— (1995, August). *Decision-Makers' Guide to Solid Waste Management, Volume II.* Washington, DC: U.S. EPA, Office of Solid Waste and Emergency Response.

——— (1997, May). *WasteWise Update: Remanufactured Products: Good as New.* Washington, DC: U.S. EPA, Office of Solid Waste and Emergency Response.

——— (1998, December). *Extended Product Responsibility: A Strategic Framework for Sustainable Products.* Washington, DC: U.S. EPA, Office of Solid Waste and Emergency Response.

——— (2000, March). *WasteWise Update: Moving Toward Sustainability.* Washington, DC: U.S. EPA, Office of Solid Waste and Emergency Response.

——— (2004, September). *Cleaning Up the Nation's Waste Sites: Markets and Technology Trends, 2004 Edition.* Washington, DC: U.S. EPA, Office of Solid Waste and Emergency Response. Available at **www.clu-in.org/download/market/2004market.pdf**.

——— (2005, February). *Hazardous Waste Manifest System Streamlined: Fact Sheet.* Washington, DC: U.S. EPA, Office of Solid Waste and Emergency Response.

——— (2008, April). *FY 2007 Superfund Annual Report.* Washington, DC: U.S. EPA, Office of Solid Waste and Emergency Response.

——— (2010, November). *The National Biennial RCRA Hazardous Waste Report: National Analysis (Based on 2009 Data).* Washington, DC: U.S. EPA, Office of Solid Waste and Emergency Response.

——— (2010, December 29). "Polychlorinated Biphenyls (PCBs): Basic Information." Available at **www.epa.gov/osw/hazard/tsd/pcbs/pubs/about.htm** (last updated December 29, 2010).

——— (2011, May 9). *Integrated Cleanup Initiative Draft Implementation Plan.* Washington, DC: U.S. EPA, Office of Solid Waste and Emergency Response. Available at **www.epa.gov/oswer/docs/ici_implementation_plan.pdf**.

——— (2011a, June 15). "Number of National Priorities List (NPL) Site Actions and Milestones by Fiscal Year." Washington, DC: U.S. EPA, Office of Solid Waste and Emergency Response. Available at **www.epa.gov/superfund/sites/query/queryhtm/nplfy.htm** (last updated June 15, 2011).

——— (2011b, June 15). "Deleted National Priorities List (NPL) Sites—By State." Available at **www.epa.gov/superfund/sites/query/queryhtm/npldel.htm** (last updated June 15, 2011).

——— (2011, August 2). "Product Stewardship: Basic Information." Available at **www.epa.gov/epawaste/partnerships/stewardship/index.htm** (last updated August 2, 2011).

——— (2011, November). *Municipal Solid Waste in the United States: 2010 Facts and Figures.* Washington, DC: U.S. EPA, Office of Solid Waste and Emergency Response.

——— (2012, March 12). "State Recycling Tax Incentives." Available at **www.epa.gov/epawaste/conserve/rrr/rmd/bizasst/rec-tax.htm** (last updated March 12, 2012).

U.S. EPA, Office of Toxic Substances (1987, June). *The Layman's Guide to the Toxic Substances Control Act.* Washington, DC: U.S. EPA, Office of Toxic Substances.

U.S. EPA, Office of Transportation and Air Quality (2002, March). *Fact Sheets on Alternative Fuels (Biodiesel, Electric Vehicles, Ethanol, Liquefied Natural Gas, Methanol, Propane).* Washington, DC: U.S. EPA, Office of Transportation and Air Quality.

——— (2005, November). *Toward a Cleaner Future.* Washington, DC: U.S. EPA, Office of Transportation and Air Quality.

——— (2007, October 19). "Clean School Bus USA: Basic Information." Available at **www.epa.gov/otaq/schoolbus/basicinfo.htm** (last updated October 19, 2007).

——— (2009, February 13). "Emission Facts: Greenhouse Gas Emissions From a Typical Passenger Vehicle." Available at **www.epa.gov/otaq/climate/420f05004.htm** (last updated February 13, 2009).

——— (2009, February 17). "California Greenhouse Gas Waiver Request." Available at **www.epa.gov/otaq/climate/ca-waivepr.htm** (last updated February 17, 2009).

——— (2010, April). *EPA and NHTSA Finalize Historic National Program to Reduce Greenhouse Gases and Improve Fuel Economy for Cars and Trucks.* Available at **www.epa.gov/otaq/climate/regulations/420f10014.pdf**.

——— (2011, July). *EPA and NHTSA, in Coordination With California, Announce Plans to Propose Greenhouse Gas and Fuel Economy Standards for Passenger Cars and Light Trucks.* Available at **www.epa.gov/otaq/climate/420f11027.pdf**.

——— (2011, August). *EPA and NHTSA Adopt First-Ever Program to Reduce Greenhouse Gas Emissions and Improve Fuel Efficiency of Medium- and Heavy-Duty Vehicles.* Available at **www.epa.gov/otaq/climate/documents/420f11031.pdf**.

U.S. EPA, Office of Water (1994). *Clean Water Act Initiative: Analysis of Costs and Benefits.* Washington, DC: U.S. EPA, Office of Water.

——— (1994, March). *National Water Quality Inventory: 1992 Report to Congress.* Washington, DC: U.S. EPA, Office of Water.

——— (1995, December). *National Water Quality Inventory: 1994 Report to Congress.* Washington, DC: U.S. EPA, Office of Water.

——— (2001, June). *Protecting and Restoring America's Watershed.* Washington, DC: U.S. EPA, Office of Water.

——— (2002, June 28). "Financial Support and Flexibility." Available at **www.epa.gov/ow=owm.html/cw finance/construction.htm** (last modified June 28, 2002).

——— (2002, August). *National Water Quality Inventory: 2000 Report.* Washington, DC: U.S. EPA, Office of Water.

——— (2003, January 13). *Final Water Quality Trading Policy.* Washington, DC: U.S. EPA, Office of Water.

——— (2003, February 20). "Clean Water Act." Available at **www.epa.gov/R5water/cwa.htm** (last updated February 20, 2003).

——— (2009, May). *2006 Community Water System Survey, Volume II: Detailed Tables and Survey Methodology.* Washington, DC: U.S. EPA, Office of Water. Available at **http://water.epa.gov/aboutow/ogwdw/cwssvr.cfm**.

——— (2009, October). *Fact Sheet: Final Aircraft Drinking Water Rule.* Washington, DC: U.S. EPA, Office of Water. Available at **http://water.epa.gov/lawsregs/rules regs/sdwa/airlinewater/upload/2009_11_2_airline water_fs_adwr_finalrulefactsheet.pdf**.

——— (2009, November). *FACTOIDS: Drinking Water and Ground Water Statistics for 2009.* Washington, DC: U.S. EPA, Office of Water.

——— (2010, June). *Clean Water State Revolving Fund Programs; 2009 Annual Report.* Washington, DC: U.S. EPA, Office of Water.

——— (2010, November). *Drinking Water State Revolving Fund 2009 Annual Report.* Washington, DC: U.S. EPA, Office of Water. Available at **http://water.epa.gov/grants_funding/dwsrf/upload/dwsrf-annual report2009nov2010.pdf**.

——— (2010, December 3). "Water: Laws and Executive Orders." Available at **http://water.epa.gov/lawregs/laws guidance/index.cfm** (last updated December 3, 2010).

——— (2011, March 14). "Coastal Zone Act Reauthorization Amendments (CZARA) Section 6217." Available at **http://water.epa.gov/polwaste/nps/czara.cfm** (last updated March 14, 2011).

——— (2011a, May 4). *Drinking Water Strategy.* Washington, DC: U.S. EPA, Office of Water. Available at **http://water.epa.gov/lawsregs/rulesregs/sdwa/dwstrategy/index.cfm**.

——— (2011b, May 4). *Aircraft Drinking Water Rule (ADWR).* Washington, DC: U.S. EPA, Office of Water. Available at **http://water.epa.gov/lawsregs/rulesregs/sdwa/airlinewater/index.cfm** (last updated May 4, 2011).

——— (2011, May 29). "Basic ADWR Information." Available at **http://water.epa.gov/lawsregs/rulesregs/sdwa//airlinewater/basic.cfm** (last updated May 29, 2011).

——— (2011, June 2). "Targeted Watersheds Grant Program." Available at **http://water.epa.gov/grants_funding/twg/initiative_index.cfm** (last updated June 2, 2011).

U.S. EPA, Office of Water, Office of Ground Water and Drinking Water (2000, December). *Arsenic in Drinking Water Rule: Economic Analysis.* Developed by Abt Associates Inc., Bethesda, Maryland. Washington, DC: U.S. EPA, Office of Water, Office of Ground Water and Drinking Water.

——— (2006, September 13). "Arsenic in Drinking Water; Arsenic Rule." Available at **www.epa.gov/safewater/arsenic/regulations.html** (last updated September 13, 2006).

——— (2007, September 20). "Ground Water & Drinking Water: Frequently Asked Questions." Available at **www.epa.gov/ogwdw/faq/faq.html** (last updated September 20, 2007).

——— (2009, May). "National Primary Drinking Water Regulations: List of Drinking Water Contaminants and Their MCLs." Available at **http://water.epa.gov/drink/contaminants/upload/mcl-2.pdf**.

U.S. EPA, Office of Water, Office of Science and Technology (2004, August). *National List of Fish Advisories.* Washington, DC: U.S. EPA, Office of Water, Office of Science and Technology.

——— (2007, July). *2005/2006 National List of Fish Advisories.* Washington, DC: U.S. EPA, Office of Water, Office of Science and Technology.

——— (2009, September). *2008 Biennial National Listing of Fish Advisories.* Available at **http://water.epa.gov/scitech/swguidance/fishshellfish/fishadvisories/upload/2009_09_16_fish_advisories_tech2008.pdf**.

U.S. EPA, Office of Water, Office of Wastewater Management (2003, April 14). "National Pollutant Discharge Elimination System (NPDES): State Program Status." Available at **http://cfpub.epa.gov/npdes/statestats.cfm** (last updated April 14, 2003).

——— (2003, May). *Watershed-Based NPDES Permitting: Rethinking Permitting As Usual.* Washington, DC: U.S. EPA, Office of Water, Office of Wastewater Management.

——— (2007, October 30). "Progress in Water Quality." Available at **www.epa.gov/owm/wquality** (last updated October 30, 2007).

——— (2007, December 17). "Voluntary Environmental Management Systems/ISO 14001: Frequently Asked Questions." Available at **www.epa.gov/owm/iso14001/isofaq.htm** (last updated December 17, 2007).

U.S. EPA, Office of Water, Office of Wetlands, Oceans, and Watersheds (2002, December 11). "The Watershed Approach." Available at **www.epa.gov/owow/water shed/wa1.html** (last updated December 11, 2002).

——— (2006, February 22). *America's Wetlands: Our Vital Link Between Land and Water.* Available at **www.epa.gov/owow/wetlands/vital/wetlands.pdf** (last updated February 22, 2006).

——— (2008, September 12). "Watershed Approach Framework." Available at **www.epa.gov/owow/watershed/framework** (last updated September 12, 2008).

——— (2009, November 23). "Clean Water Act Section 319(h) Grant Funds History." Available at **www.epa.gov/owow/nps/319hhistory.html** (last updated November 23, 2009).

——— (2011, May 25). "Watershed Assessment, Tracking, and Environmental Results: National Summary of State Information." Available at **http://iaspub.epa.gov/waters10/attains_nation_cy.control** (last updated May 25, 2011).

——— (2011, June 2). "State and Individual Trading Programs: List of All Trading Programs." Available at **http://water/epa.gov/type/watersheds/trading/tradingmap.cfm** (last updated June 2, 2011).

——— (2011, June 22). "Catalog of Federal Funding Sources for Watershed Protection." Available at **http://cfpub.epa.gov/fedfund/list.cfm** (last updated June 22, 2011).

U.S. EPA, Office of Water Regulations and Standards (1988, September). *Introduction to Water Quality Standards.* Washington, DC: U.S. EPA, Office of Water Regulations and Standards.

U.S. EPA, Region 2 (2000, December 6). *EPA Proposes Comprehensive Plan To Clean Up Hudson River PCBs.* Press Release.

U.S. EPA, Risk Assessment Portal (2010a, August 19). "Risk Assessment Portal: Basic Information." Available at **www.epa.gov/risk/basicinformation.htm** (last updated August 19, 2010).

——— (2010b, August 19). "Human Health Risk Assessment." Available at **www.epa.gov/risk_assessment/health-risk.htm** (last updated August 19, 2010).

——— (2011, January 5). "Ecological Risk Assessment." Available at **www.epa.gov/risk_assessment/ecological-risk.htm** (last updated January 5, 2011).

U.S. EPA, Science Advisory Board (1990, September). *Reducing Risk: Setting Priorities and Strategies for Environmental Protection.* Washington, DC: U.S. EPA, Science Advisory Board.

——— (2002, April 4). *Industrial Ecology: A Commentary by the EPA Science Advisory Board.* Washington, DC: U.S. EPA, Science Advisory Board.

U.S. EPA, Stratospheric Protection Division, Office of Air and Radiation (1999, June 9). *Report on the Supply and Demand of CFC-12 in the United States 1999.* Washington, DC: U.S. EPA, Stratospheric Protection Division, Office of Air and Radiation.

U.S. EPA, Technology Transfer Network (2009, June 9). "National Emissions Inventory (NEI) Air Pollutant Emissions Trends Data: 1970–2008 Average Annual Emissions, All Criteria Pollutants in MS Excel." Available at **www.epa.gov/ttn/chief/trends/index.html** (last updated June 9, 2009).

U.S. EPA and U.S. Department of Energy (2005). "History of Energy Star." Available at **www.epa.gov/index.cfm?c=about.ab_history** (accessed February 26, 2005).

——— (2011). *Energy Star® Overview of 2010 Achievements.* Washington, DC: U.S. EPA and U.S. Department of Energy. Available at **www.energystar.gov/ia/partners/publications/pubdocs/2010%20CPPD%204pgr.pdf** (accessed May 23, 2011).

U.S. GAO (General Accounting Office) (1987, December). *SUPERFUND: Extent of Nation's Potential Hazardous Waste Problem Still Unknown.* Washington, DC: U.S. GAO.

——— (1989, January). *Water Pollution: More EPA Action Needed to Improve the Quality of Heavily Polluted Waters.* Washington, DC: U.S. GAO.

——— (1990, October). *Water Pollution: Greater EPA Leadership Needed to Reduce Nonpoint Source Pollution.* Washington, DC: U.S. GAO.

——— (1991, March). *Pesticides: EPA's Use of Benefit Assessment in Regulating Pesticides.* Washington, DC: U.S. GAO.

——— (1991, July). *Water Pollution: Stronger Efforts Needed by EPA to Control Toxic Water Pollution.* Washington, DC: U.S. GAO.

——— (2008, July 18). *Superfund: Funding and Reported Costs of Enforcement and Administration Activities.* Washington, DC: U.S. GAO.

U.S. Government, White House (2001, June). "Climate Change Review: Initial Report." Available at **www.whitehouse.gov/news/releases/2001/06/climatechange.pdf**.

——— (2002, February). *Global Climate Change Policy Book.* Washington, DC: U.S. Government, White House.

U.S. Office of Management and Budget (1992, October 29). *Guidelines and Discount Rates for Benefit-Cost Analysis of Federal Programs.* Circular A-94, Revised. Washington, DC: U.S. Office of Management and Budget.

——— (2003, September 17). *Regulatory Analysis.* Circular A-4. Washington, DC: U.S. Office of Management and Budget.

U.S. Senate, Staff of the Subcommittee on Air and Water Pollution of the Committee on Public Works (1973, October). *The Impact of Auto Emission Standards.* Washington, DC: U.S. Government Printing Office.

Vaughan, William J., and Clifford S. Russell (1982). "Valuing a Fishing Day: An Application of a Systematic Varying Parameter Model." *Land Economics 58(4)* (November), pp. 450–63.

Vaupel, James W. (1978). "Truth or Consequences: Some Roles for Scientists and Analysts in Environmental Decisionmaking." In Wesley A. Magat, ed., *Reform of Environmental Regulation.* Cambridge, MA: Ballinger, pp. 71–92.

Virginia Department of Environmental Quality (2011). "Mandatory Recycling Rates for Localities." Available at **www.deq.state.va.us/recycle/mandatory.html** (last updated July 7, 2011).

Vogan, Christine R. (1996, September). "Pollution Abatement and Control Expenditures, 1972–94." In *Survey of Current*

Business. Washington, DC: Department of Commerce, Bureau of Economic Analysis, pp. 48–67.

Vogt, Donna U. (1995). *The Delaney Clause Effects on Pesticide Policy*. CRS Report for Congress. Washington, DC: Congressional Research Service (July 13).

The Walt Disney Company (2011). *2010 Corporate Citizenship Report: Executive Summary*. Burbank, CA: The Walt Disney Company. Available at **http://corporate.disney.go.com/citizenship2010/downloads/**.

Weitzman, Martin L. (2007). "A Review of the Stern Review on Climate Change." *Journal of Economic Literature 45*(3) (September), pp. 703–24.

Western Climate Initiative (WCI) (2010, May). "Clean Energy: Creating Jobs, Protecting the Environment." Available at **www.westernclimateinitiative.org/component/remository/general/WCI-Brochure-(May-2010)/**.

"What Is Environmental Technology?" (1994). *EPA Journal 20*(3–4) (Fall), p. 8.

White, Gilbert W. (1988). "Water Resource Adequacy: Illusion and Reality." In David H. Speidel, Lon C. Ruedisili, and Allen F. Agnew, eds., *Perspectives on Water: Uses and Abuses*. New York: Oxford University Press, pp. 12–20.

"White House to Seek Reduced Use of Pesticides" (1993). *Boston Globe* (September 21), p. 7.

White, Lawrence J. (1982). *The Regulation of Air Pollutant Emissions From Motor Vehicles*. Washington, DC: American Enterprise Institute for Public Policy Research.

Whiteman, Hilary (2011). "Canada: First Out of a Sinking Kyoto Ship?" *CNN.com* (December 13). Available at **www.cnn.com/2011/12/13/world/americas/canada-kyoto-protocol-explainer/index.html?hpt=hp_c1**.

Wirth, Timothy E., and John Heinz (1991, May). *Project 88: Round II Incentives for Action—Designing Market-Based Environmental Strategies*. New York: Carnegie Corporation.

Wolf, Sidney M. (1988). *Pollution Law Handbook: A Guide to Federal Environmental Laws*. New York: Quorom Books.

Wonacott, Peter (2004). "To Save Water, China Lifts Price." *Wall Street Journal* (June 14), p. A13.

World Bank (2010). *World Development Indicators 2010*. Washington, DC: World Bank. Available at **http://data.worldbank.org/sites/default/files/wdi-final.pdf** (accessed January 17, 2011).

World Trade Organization (WTO) (2005). *Trade and the Environment at the WTO: Background Document*. Geneva, Switzerland: WTO. Available at **www.wto.org/english/tratop_e/envir_e/envir_backgrnd_e/contents_e.htm** (accessed April 25, 2005).

——— (2008a). *10 Benefits of the WTO Trading System*. Geneva, Switzerland: WTO. Available at **www.wto.org/english/res_e/doload_e/10b_e.pdf**.

——— (2008b). *The World Trade Organization in Brief*. Geneva, Switzerland: WTO. Available at **www.wto.org/english/thewto_e/whatis_e/inbrief_e/inbr00_e.htm**.

——— (2011). "The WTO." Available at **www.wto.org/english/thewto_e/thewto_e.htm**.

Xerox Corporation (2010). *Nurturing a Greener World Through Sustainable Innovation and Development: Our 2010 Environment, Health and Safety Report*. Norwalk, CT: Xerox Corporation. Available at **www.xerox.com/corporate-citizenship-2010/Environment_Health_Safety_Report_2010.pdf**.

Zwick, David, and Marcy Benstock (1971). *Water Wasteland: Ralph Nader's Study Group Report on Water Pollution*. New York: Grossman.

Glossary

A

abatement equipment subsidy A payment aimed at lowering the cost of abatement technology.

"acceptable" risk Amount of risk determined to be tolerable for society.

acidic deposition Arises when sulfuric and nitric acids mix with other airborne particles and fall to the earth as dry or wet deposits.

acid rain Arises when sulfuric and nitric acids mix with other airborne particles and fall to the earth as precipitation.

Air Quality Control Region (AQCR) A federally designated area within which common air pollution problems are shared by several communities.

Air Quality Index (AQI) An index that signifies the worst daily air quality in an urban area.

allocative efficiency Requires that resources be appropriated such that the additional benefits to society are equal to the additional costs.

allocatively efficient standards Standards set such that the associated marginal social cost (*MSC*) of abatement equals the marginal social benefit (*MSB*) of abatement.

allowance market for ozone-depleting chemicals Allows firms to produce or import ozone depleters if they hold an appropriate number of tradeable allowances.

ambient standard A standard that designates the quality of the environment to be achieved, typically expressed as a maximum allowable pollutant concentration.

analysis phase Identifies information to predict ecological responses to environmental hazards under various exposure conditions.

anthropogenic pollutants Contaminants associated with human activity.

averting expenditure method (AEM) Estimates benefits as the change in spending on goods that are *substitutes* for a cleaner environment.

B

back-end charge A fee implemented at the time of disposal based on the quantity of waste generated.

bag-and-tag approach A unit pricing scheme implemented by selling tags to be applied to waste receptacles of various sizes.

behavioral linkage approach Estimates benefits using observations of behavior in actual markets or survey responses about hypothetical markets.

benefit-based decision rule A guideline to improve society's well-being with no allowance for balancing with associated costs.

benefit-based standard A standard set to improve society's well-being with no consideration for the associated costs.

benefit-cost analysis A strategy that compares the *MSB* of a risk reduction policy to the associated *MSC*.

benefit-cost ratio The ratio of *PVB* to *PVC* used to determine the feasibility of a policy option if its magnitude exceeds unity.

best available technology (BAT) Treatment technology that makes attainment of the MCL feasible, accounting for cost considerations.

best management practices (BMP) Strategies other than effluent limitations to reduce pollution from nonpoint sources.

biodiversity The variety of distinct species, their genetic variability, and the variety of ecosystems they inhabit.

brownfield site Real property where redevelopment or expansion is complicated by the presence or potential presence of environmental contamination.

bubble policy Allows a plant to measure its emissions as an average of all emission points from that plant.

C

capital costs Fixed expenditures for plant, equipment, construction in progress, and production process changes associated with abatement.

carbon sinks Natural absorbers of CO_2, such as forests and oceans.

CERCLIS A national inventory of hazardous waste site data.

characteristic wastes Hazardous wastes exhibiting certain characteristics that imply a substantial risk.

chlorofluorocarbons (CFCs) A family of chemicals believed to contribute to ozone depletion.

circular flow model Illustrates the real and monetary flows of economic activity through the factor market and the output market.

clean alternative fuels Fuels, such as methanol or ethanol, or power sources, such as electricity, used in a clean fuel vehicle.

clean fuel vehicle A vehicle certified to meet stringent emission standards.

Clean Water State Revolving Fund (CWSRF) program Establishes state lending programs to support POTW construction and other projects.

cleaner production A preventive strategy applied to products and processes to improve efficiency and reduce risk.

climate change A major alteration in a climate measure such as temperature, wind, or precipitation that is prolonged.

closed flow of materials Assumes that materials run in a circular pattern in a closed system, allowing residuals to be returned to the production process.

Coase Theorem Assignment of property rights, even in the presence of externalities, will allow bargaining such that an efficient solution can be obtained.

command-and-control approach A policy that directly regulates polluters through the use of rules or standards.

common property resources Those resources for which property rights are shared.

comparative risk analysis An evaluation of relative risk.

competitive equilibrium The point where marginal private benefit (*MPB*) equals marginal private cost (*MPC*), or where marginal profit $(M\pi) = 0$.

consumer surplus Net benefit to buyers estimated by the excess of marginal benefit (*MB*) of consumption over market price (*P*), aggregated over all units purchased.

contingent valuation method (CVM) Uses surveys to elicit responses about WTP for environmental quality based on hypothetical market conditions.

conventional pollutant An identified pollutant that is well understood by scientists.

corrective tax A tax aimed at rectifying a market failure and improving resource allocation.

cost-effective abatement criterion Allocation of abatement across polluting sources such that the *MAC*s for each source are equal.

cost-effectiveness Requires that the least amount of resources be used to achieve an objective.

"cradle-to-grave" management system A command-and-control approach to regulating hazardous solid wastes through every stage of the waste stream.

criteria pollutants Substances known to be hazardous to health and welfare, characterized as harmful by criteria documents.

cyclical flow of materials Assumes that materials run in a circular pattern in a closed system, allowing residuals to be returned to the production process.

D

damage function method Models the relationship between a contaminant and its observed effects to estimate damage reductions arising from policy.

deadweight loss to society The net loss of consumer and producer surplus due to an allocatively inefficient market event.

declining block pricing structure A pricing structure in which the per-unit price of different blocks of water declines as usage increases.

deflating Converts a nominal value into its real value.

demand The quantities of a good the consumer is willing and able to purchase at a set of prices during some time period, *c.p.*

***de minimis* risk** A negligible level of risk such that reducing it further would not justify the associated costs.

deposit/refund system A market instrument that imposes an up-front charge to pay for potential damages and refunds it for returning a product for proper disposal or recycling.

Design for the Environment (DfE) Promotes the use of environmental considerations along with cost and performance in product design and development.

direct user value Benefit derived from directly consuming services provided by an environmental good.

discount factor The term $1/(1+r)^t$, where r is the discount rate, and t is the number of periods.

dose-response relationship A quantitative relationship between doses of a contaminant and the corresponding reactions.

Drinking Water State Revolving Fund (DWSRF) Authorizes $1 billion per year to finance infrastructure improvements.

E

Economic Analysis (EA) A requirement under Executive Order 12866, amended by Executive Orders 13258 and 13422, and Executive Order 13563 that calls for information on the benefits and costs of a "significant regulatory action."

efficient equilibrium The point where marginal social benefit (*MSB*) equals marginal social cost (*MSC*), or where marginal profit $(M\pi) =$ marginal external cost (*MEC*).

emission or effluent charge A fee imposed directly on the actual discharge of pollution.

emissions banking Accumulating emission reduction credits through a banking program.

engineering approach Estimates abatement expenditures based on least-cost available technology.

environmental economics A field of study concerned with the flow of residuals from economic activity back to nature.

environmental justice Fairness of the environmental risk burden across segments of society or geographical regions.

environmental Kuznets curve Models an inverted U-shaped relationship between economic growth and environmental degradation.

environmental literacy Awareness of the risks of pollution and natural resource depletion.

environmental quality A reduction in anthropogenic contamination to a level that is "acceptable" to society.

environmental risk Involuntary risk of exposure to an environmental hazard.

equilibrium price and quantity The market-clearing price (P_E) associated with the equilibrium quantity (Q_E), where $Q_D = Q_S$.

ethanol (E10) Known as gasohol, a blend of 10 percent ethanol and 90 percent gasoline.

ethanol (E85) Blended fuel comprising 85 percent ethanol and 15 percent gasoline.

excise tax on ozone depleters An escalating tax on the production of ozone-depleting substances.

existence value Benefit received from the continuance of an environmental good.

existing chemical A substance listed in the TSCA inventory.

explicit costs Administrative, monitoring, and enforcement expenses paid by the public sector plus compliance costs incurred by all sectors.

exposure Pathways between the source of the damage and the affected population or resource.

exposure assessment Measures the magnitude, frequency, and duration of exposure, pathways and routes, and any sensitivities.

Extended Product Responsibility (EPR) A commitment by all participants in the product cycle to reduce any life cycle environmental effects of products.

externality A spillover effect associated with production or consumption that extends to a third party outside the market.

F

federal grant program Provided major funding from the federal government for a share of the construction costs of POTWs.

feedstock taxes Taxes levied on raw materials used as productive inputs.

first law of thermodynamics Matter and energy can neither be created nor destroyed.

fishable-swimmable goal Requires that surface waters be capable of supporting recreational activities and the propagation of fish and wildlife.

fixed fee or flat fee pricing system Pricing MSW services independent of the quantity of waste generated.

flat fee pricing scheme Pricing water supplies such that the fee is independent of water use.

flat rate pricing A unit pricing scheme that charges the same price for each additional unit of waste.

free trade The unencumbered exchange of goods and services among nations.

free-ridership Recognition by a rational consumer that the benefits of consumption are accessible without paying for them.

front-end charge A fee levied on a product at the point of sale designed to encourage source reduction.

G

global pollution Environmental effects that are widespread with global implications.

global warming Increased temperature of the earth's surface caused by accumulating GHGs that absorb the sun's radiation.

global warming potential (GWP) Measures the heat-absorbing capacity of a GHG relative to CO_2 over some time period.

Green Chemistry Program Promotes the development of innovative chemical technologies to achieve pollution prevention.

greenhouse gases (GHGs) Gases collectively responsible for the absorption process that naturally warms the earth.

greenhouse gas (GHG) intensity The ratio of GHG emissions to economic output.

groundwater Fresh water beneath the earth's surface, generally in aquifers.

H

hazard Source of the environmental damage.

hazard identification Scientific analysis to determine whether a causal relationship exists between a pollutant and any adverse effects.

hazardous air pollutants Noncriteria pollutants that may cause or contribute to irreversible illness or increased mortality.

hazardous solid wastes Unwanted materials or refuse posing a substantial threat to health or the ecology.

hedonic price method (HPM) Uses the estimated hedonic price of an environmental attribute to value a policy-driven improvement.

hydrologic cycle The natural movement of water from the atmosphere to the surface, beneath the ground, and back into the atmosphere.

I

implicit costs The value of any nonmonetary effects that negatively influence society's well-being.

increasing block pricing structure A pricing structure in which the per-unit price of different blocks of water increases as water use increases.

incremental benefits The reduction in health, ecological, and property damages associated with an environmental policy initiative.

incremental costs The change in costs arising from an environmental policy initiative.

indirect user value Benefit derived from indirect consumption of an environmental good.

industrial ecology A multidisciplinary systems approach to the flow of materials and energy between industrial processes and the environment.

industrial ecosystem A closed system of manufacturing whereby the wastes of one process are reused as inputs in another.

inflation correction Adjusts for movements in the general price level over time.

Integrated Pest Management (IPM) A combination of methods that encourage more selective pesticide use and greater reliance on natural deterrents.

integrated waste management system An EPA initiative that promotes source reduction, recycling, combustion, and land disposal, in that order.

international externality A spillover effect associated with production or consumption that extends to a third party in another nation.

involuntary risk Risk beyond one's control and not the result of a willful decision.

ISO 14000 standards Voluntary international standards for environmental management.

J

joint and several liability The legal standard that identifies a single party as responsible for all damages even if that party's contribution to the damages is relatively small.

L

Law of Demand There is an inverse relationship between price and quantity demanded of a good, *c.p.*

Law of Supply There is a direct relationship between price and quantity supplied of a good, *c.p.*

life cycle assessment (LCA) Examines the environmental impact of a product or process by evaluating all its stages from raw materials extraction to disposal.

linear flow of materials Assumes that materials run in one direction, entering an economic system as inputs and leaving as wastes or residuals.

listed wastes Hazardous wastes preidentified by government as having met specific criteria.

local pollution Environmental damage that does not extend far from the polluting source.

M

management strategies Methods that address existing environmental problems and attempt to reduce the damage from the residual flow.

manifest A document used to identify hazardous waste materials and all parties responsible for its movement from generation to disposal.

marginal abatement cost (*MAC*) The change in costs associated with increasing abatement, using the least-cost method.

marginal cost of enforcement (*MCE*) Added costs incurred by government associated with monitoring and enforcing abatement activities.

marginal social benefit (*MSB*) The sum of marginal private benefit (*MPB*) and marginal external benefit (*MEB*).

marginal social benefit (*MSB*) of abatement A measure of the additional gains accruing to society as pollution abatement increases.

marginal social cost (*MSC*) The sum of marginal private cost (*MPC*) and marginal external cost (*MEC*).

marginal social cost (*MSC*) of abatement The sum of all polluters' marginal abatement costs plus government's marginal cost of monitoring and enforcing these activities.

market The interaction between consumers and producers to exchange a well-defined commodity.

market approach An incentive-based policy that encourages conservation practices or pollution-reduction strategies.

market demand for a private good The decisions of all consumers willing and able to purchase a good, derived by *horizontally* summing individual demands.

market demand for a public good The aggregate demand of all consumers in the market, derived by *vertically* summing their individual demands.

market failure The result of an inefficient market condition.

market supply of a private good The combined decisions of all producers in a given industry, derived by *horizontally* summing individual supplies.

market-level marginal abatement cost (MAC_{mkt}) The horizontal sum of all polluters' *MAC* functions.

materials balance model Positions the circular flow within a larger schematic to show the connections between economic decision making and the natural environment.

materials groups Materials-based categories used to analyze the MSW stream.

maximize the present value of net benefits (*PVNB*) A decision rule to achieve allocative efficiency by selecting the policy option that yields greatest excess benefits after adjusting for time effects.

maximum contaminant level (MCL) Component of an NPDWR that states the highest permissible contaminant level delivered to a public system.

maximum contaminant level goal (MCLG) Component of an NPDWR that defines the level of a pollutant at which no adverse health effects occur, allowing for a margin of safety.

minimize the present value of costs (*PVC*) A decision rule to achieve cost-effectiveness by selecting the least-cost policy option that achieves a preestablished objective.

mobile source Any nonstationary polluting source.

municipal solid waste (MSW) Nonhazardous wastes disposed of by local communities.

N

National Ambient Air Quality Standards (NAAQS) Maximum allowable concentrations of criteria air pollutants.

National Emission Standards for Hazardous Air Pollutants (NESHAP) Standards applicable to every major source of any identified hazardous air pollutant.

National Pollutant Discharge Elimination System (NPDES) A permit system to control effluent releases from direct industrial dischargers and POTWs.

National Primary Drinking Water Regulations (NPDWRs) Health standards for public drinking water supplies that are implemented uniformly.

National Priorities List (NPL) A classification of hazardous waste sites posing the greatest threat to health and the ecology.

natural pollutants Contaminants that come about through nonartificial processes in nature.

natural resource economics A field of study concerned with the flow of resources from nature to economic activity.

negative externality An external effect that generates costs to a third party.

netting Matching any emissions increase due to a modification with a reduction from another point within that same source.

new chemical Any substance not listed in the TSCA inventory of existing chemicals.

New Source Performance Standards (NSPS) Technology-based emissions limits for new stationary sources.

no toxics in toxic amounts goal Prohibits the release of toxic substances in toxic amounts into all water resources.

nominal value A magnitude stated in terms of the current period.

nonattainment area An AQCR not in compliance with the NAAQS.

nonconventional pollutant A default category for pollutants not identified as toxic or conventional.

nonexcludability The characteristic that makes it impossible to prevent others from sharing in the benefits of consumption.

nonpoint source A source that cannot be identified accurately and degrades the environment in a diffuse, indirect way over a broad area.

Nonpoint Source Management Program A three-stage, state-implemented plan aimed at nonpoint source pollution.

nonrevelation of preferences An outcome that arises when a rational consumer does not volunteer a willingness to pay because of the lack of a market incentive to do so.

nonrivalness The characteristic of indivisible benefits of consumption such that one person's consumption does not preclude that of another.

O

offset plan Uses emissions trading to allow releases from a new or modified source to be more than countered by reductions achieved by existing sources.

open flow of materials Assumes that materials run in one direction, entering an economic system as inputs and leaving as wastes or residuals.

operating costs Variable expenditures incurred in the operation and maintenance of abatement processes.

oxygenated fuel Has enhanced oxygen content to allow for more complete combustion.

ozone depletion Thinning of the ozone layer, originally observed as an ozone hole over Antarctica.

ozone depletion potential (ODP) A numerical score that signifies a substance's potential for destroying stratospheric ozone relative to CFC-11.

ozone layer Ozone present in the stratosphere that protects the earth from ultraviolet radiation.

P

partial zero-emission vehicle (PZEV) Emitting zero evaporative emissions and runs 90 percent cleaner than the average new model year vehicle.

performance-based standard A standard that specifies a pollution limit to be achieved but does not stipulate the technology.

permitting system A control approach that authorizes the activities of TSDFs according to predefined standards.

per-unit subsidy on pollution reduction A payment for every unit of pollution removed below some predetermined level.

pesticide registration Formal listing of a pesticide with the EPA, based on a risk-benefit analysis, before it can be sold or distributed.

pesticide reregistration A formal reevaluation of a previously licensed pesticide already on the market.

pesticide tolerances Legal limits on the amount of pesticide residue allowed on raw or processed foods.

photochemical smog Caused by pollutants that chemically react in sunlight to form new substances.

physical linkage approach Estimates benefits based on a technical relationship between an environmental resource and the user of that resource.

Pigouvian subsidy A per-unit payment on a good whose consumption generates a positive externality such that the payment equals the *MEB* at Q_E.

Pigouvian tax A unit charge on a good whose production generates a negative externality such that the charge equals the *MEC* at Q_E.

point source Any single identifiable source from which pollutants are released.

pollutant-based effluent fee Based on the degree of harm associated with the contaminant being released.

pollution The presence of matter or energy whose nature, location, or quantity has undesired effects on the environment.

pollution allowances Tradeable permits that indicate the maximum level of pollution that may be released.

pollution charge A fee that varies with the amount of pollutants released.

pollution credits Tradeable permits issued for emitting below an established standard.

pollution haven effect Changes in trade patterns caused by cost differences among nations due to varying environmental regulations.

pollution permit trading system A market instrument that establishes a market for rights to pollute by issuing tradeable pollution credits or allowances.

pollution prevention (P2) A long-term strategy aimed at reducing the amount or toxicity of residuals released to nature.

positive externality An external effect that generates benefits to a third party.

potentially responsible parties (PRPs) Any current or former owner or operator of a hazardous waste facility and all those involved in the disposal, treatment, or transport of hazardous substances to a contaminated site.

premanufacture notice (PMN) Official notification to the EPA by a chemical producer about its intent to produce or import a new chemical.

present value determination A procedure that discounts a future value (*FV*) into its present value (*PV*) by accounting for the opportunity cost of money.

present value of benefits (*PVB*) The time-adjusted magnitude of incremental benefits associated with an environmental policy change.

present value of costs (*PVC*) The time-adjusted magnitude of incremental costs associated with an environmental policy change.

present value of net benefits (*PVNB*) The differential of ($PVB - PVC$) used to determine the feasibility of a policy option if its magnitude exceeds zero.

prevention of significant deterioration (PSD) area An AQCR meeting or exceeding the NAAQS.

priority contaminants Pollutants for which drinking water standards are to be established based on specific criteria.

private good A commodity that has two characteristics—rivalry in consumption and excludability.

problem formulation Identifies the ecological entity that is potentially at risk.

producer surplus Net gain to sellers of a good estimated by the excess of market price (P) over marginal cost (MC), aggregated over all units sold.

product charge A fee added to the price of a pollution-generating product based on its quantity or some attribute responsible for pollution.

product groups Product-based categories used to analyze the MSW stream.

profit maximization Achieved at the output level where $MR = MC$ or where $M\pi = 0$.

property rights The set of valid claims to a good or resource that permits its use and the transfer of its ownership through sale.

protectionism Fostering trade barriers, such as tariffs or quotas, to protect a domestic economy from foreign competition.

public good A commodity that is nonrival in consumption and yields benefits that are nonexcludable.

R

real value A magnitude adjusted for the effects of inflation.

receiving water quality standards State-established standards defined by use designation and water quality criteria.

reformulated gasoline Emits less hydrocarbons, carbon monoxide, and toxics than conventional gasoline.

regional pollution Degradation that extends well beyond the polluting source.

Regulatory Impact Analysis (RIA) A requirement under Executive Order 12291 that called for information about the potential benefits and costs of a major federal regulation.

remanufacturing Collection, disassembly, reconditioning, and reselling of the same product.

remedial actions Official responses to a hazardous substance release aimed at achieving a more permanent solution.

removal actions Official responses to a hazardous substance release aimed at restoring immediate control.

renewable energy certificate (REC) A market-based instrument that conveys rights to the nonpower attributes of renewable electricity generation.

residual The amount of a pollutant remaining in the environment after a natural or technological process has occurred.

retail disposal charge A fee levied on a product at the point of sale designed to encourage source reduction.

risk The chance of something bad happening.

risk assessment Qualitative and quantitative evaluation of the risk posed to health or the ecology by an environmental hazard.

risk characterization Description of expected risk, how the risk was assessed, and areas needing policy decisions.

risk management The decision-making process of evaluating and choosing from alternative responses to environmental risk.

risk-benefit analysis An assessment of risks of a hazard along with the benefits to society of not regulating that hazard.

S

second law of thermodynamics Nature's capacity to convert matter and energy is not without bound.

secondary maximum contaminant levels (SMCLs) National standards for drinking water that serve as guidelines to protect public welfare.

shortage Excess demand of a commodity, equal to $(Q_D - Q_S)$, that arises if price is *below* its equilibrium level.

social costs Expenditures needed to compensate society for resources used so that its utility level is maintained.

social discount rate Discount rate used for public policy initiatives based on the social opportunity cost of funds.

society's welfare The sum of consumer surplus and producer surplus.

source reduction Preventive strategies to reduce the quantity of any contaminant released to the environment at the point of generation.

State Implementation Plan (SIP) A procedure outlining how a state intends to implement, monitor, and enforce the NAAQS and the NESHAP.

stationary source A fixed-site producer of pollution.

stewardship Sense of obligation to preserve the environment for future generations.

strict liability The legal standard that identifies individuals as responsible for damages even if negligence is not proven.

Superfund cleanup process A series of steps to implement the appropriate response to threats posed by the release of a hazardous substance.

supply The quantities of a good the producer is willing and able to bring to market at a given set of prices during some time period, *c.p.*

surface water Bodies of water open to the earth's atmosphere.

surplus Excess supply of a commodity, equal to $(Q_S - Q_D)$, that arises if price is *above* its equilibrium level.

survey approach Polls a sample of firms and public facilities to obtain estimated abatement expenditures.

sustainable development Management of the earth's resources such that their long-term quality and abundance are ensured for future generations.

T

technical efficiency Production decisions that generate maximum output, given some stock of resources.

technology transfer The advancement and application of technologies and strategies on a global scale.

technology-based effluent limitations Standards to control discharges from point sources based primarily on technological capability.

technology-based standard A standard that designates the equipment or method to be used to achieve some abatement level.

threshold The level of exposure to a hazard up to which no response exists.

tipping fees Prices charged for disposing of wastes in a facility such as a landfill.

total maximum daily loads (TMDLs) Maximum amount of pollution a water body can receive without violating the standards.

total profit Total profit (π) = Total revenue (TR) − Total costs (TC).

toxic chemical use substitution The use of less harmful chemicals in place of more hazardous substances.

toxic pollutant A contaminant that, upon exposure, will cause death, disease, abnormalities, or physiological malfunctions.

Toxics Release Inventory (TRI) A national database that gives information about hazardous substances released into the environment.

tradeable allowance system for GHGs Establishes a market for GHG permits where each allows the release of some amount of GHGs.

tradeable effluent permit market The exchange of rights to pollute among water-polluting sources.

tradeable SO_2 emission allowances Permits allowing the release of SO_2 that can be held or sold through a transfer program.

travel cost method (TCM) Values benefits by using the *complementary* relationship between the quality of a natural resource and its recreational use value.

TSCA inventory A database of all chemicals commercially produced or processed in the United States.

U

uniform rate (or flat rate) pricing structure Pricing water supplies to charge more for higher water usage at a constant rate.

user value Benefit derived from physical use of or access to an environmental good.

use-support status A classification based on a water body's present condition relative to what is needed to maintain its designated uses.

V

variable rate pricing A unit pricing scheme that charges a different price for each additional unit of waste.

vicarious consumption Utility associated with knowing that others derive benefits from an environmental good.

volume-based effluent fee Based on the quantity of pollution discharged.

voluntary risk Risk that is deliberately assumed at an individual level.

W

waste management Control strategies to reduce the quantity and toxicity of hazardous wastes at every stage of the waste stream.

waste stream A series of events starting with waste generation and including transportation, storage, treatment, and disposal of solid wastes.

waste-end charge A fee implemented at the time of disposal based on the quantity of waste generated.

watershed A hydrologically defined land area that drains into a particular water body.

watershed approach A comprehensive framework used to coordinate the management of water resources.

watershed-based NPDES permit Allows for permitting of multiple point sources within a watershed.

Z

zero discharge goal Calls for the elimination of all polluting effluents into navigable waters.

zero-emission vehicle (ZEV) Emitting zero tailpipe emissions and runs 98 percent cleaner than the average new model year vehicle.

Index

Note: *Italicized* page references indicate a table or figure reference.

B

E

F

G

H

N

O

P

Q

R

S

T

U

V

W

X

Y

Z

SOURCES IN SEMIOTICS SERIES,

LOGIC, PHILOSOPHY, and HISTORY

A Study
in the Philosophy of History
Based on the Work of R. G. Collingwood

by

ANTHONY F. RUSSELL

with a Foreword by

Brooke Williams

UNIVERSITY PRESS OF AMERICA

LANHAM • NEW YORK • LONDON

University Press of America,™ Inc.

4720 Boston Way
Lanham, MD 20706

3 Henrietta Street
London WC2E 8LU England

Printed in the United States of America

This book is published as a volume in the
SOURCES IN SEMIOTICS series
under the General Editorship of
John Deely and Brooke Williams

It is co-published by arrangement with the
Semiotic Society of America

ISBN (Perfect): 0-8191-3779-0
ISBN (Cloth): 0-8191-3778-2

All University Press of America books are produced on acid-free
paper which exceeds the minimum standards set by the National
Historical Publications and Records Commission.

To the Memory of

A. M. WOODBURY

who lit the light for this work

and to

RAYMOND and YVONNE FRENETTE

who kept it burning

ACKNOWLEDGMENTS

The author wishes to thank John Deely and Brooke Williams as the Editors of the Sources in Semiotics series for their encouragement and help. He is indebted also to Ms. Felicia Kruse who bore the brunt of the final copyediting, proofreading and indexing of the work, and made many clarifying and constructive suggestions.

Acknowledgment is made, finally, to the following publishers and learned societies for permission to make citations from the published works of the various authors referred to throughout this study:

The British Academy, London, for citation from I. A. Richmond, "An Appreciation of R. G. Collingwood as an Archaeologist," in *The Proceedings of the British Academy*, XXIX (1943).

Desclée de Brouwer, Paris, for citations from J.-H. Nicolas, *Dieu connu comme inconnu*, 1966.

Dover Publications, Inc., New York, for citations from the Elizabeth S. Haldane and G. R. T. Ross translation of *The Philosophical Works of Descartes*, 1955.

Indiana University Press, Bloomington, Indiana, for citation from Louis O. Mink, *Mind, History, and Dialectic*, 1969.

Macmillan Publishing Co., Inc., New York, for citations from L.-M. Régis, *Epistemology*, 1959. © The Macmillan Company 1959.

Martinus Nijhoff, The Hague, The Netherlands, for citation from Herbert Spiegelberg, *The Phenomenological Movement*, 3rd edition, 1982.

Oxford University Press, Oxford, England, for excerpts from the following works of R. G. Collingwood: *An Autobiography*, 1939; *An Essay on Metaphysics*, 1940; *An Essay on Philosophical Method*, 1933; *The Idea of History*, 1946; *Speculum Mentis*, 1924. Permission was also granted to cite from Alan Donagan, *The Later Philosophy of R. G. Collingwood*, 1962.

Presses de l'Université Grégorienne, Rome, Italy, for citations from Joseph de Finance, *Essai sur l'agir humain*, 1962.

The Review of Metaphysics, The Catholic University of America, Washington, D.C., for citations from John Wild, "Being, Meaning and the World," vol. XVIII (1965) and from Otis Lee, "Dialectic and Negation," vol. I (1947).

Routledge & Kegan Paul, Ltd., London, for pages based on and containing citations from Hans Vaihinger, *The Philosophy of "As-If,"* trans. C. K. Ogden, 1924.

Charles Scribner's Sons, New York, for citation from James H. Breasted, *A History of Egypt*, 1959. Copyright 1905, 1909 Charles Scribner's Sons; copyright renewed 1933, 1937.

The University of Chicago Press, Chicago, for citations from Michael Polanyi, *Personal Knowledge*, 1962. © Copyright 1958, 1962 by Michael Polanyi.

The University of North Carolina Press, Chapel Hill, North Carolina, for citations from *The Nature of Historical Thinking* by Robert Stover. Copyright 1967 by The University of North Carolina Press. Reprinted by permission of the publisher.

Yale University Press, New Haven, Connecticut, for citations from *The Logic of the Humanities* by Ernst Cassirer, trans. Clarence Smith Howe, 1960. Copyright © 1960, 1961 by Clarence Smith Howe.

TABLE OF CONTENTS

Part II: EVALUATION

FOREWORD

Whatever the cautiousness among historians concerning analysis by philosophers on the nature of history, R. G. Collingwood stands out as a philosopher whose influence on historians has been profound, and whose philosophy of history, whatever the criticisms, remains highly respected after the test of time. Granted that he himself was not only a philosopher but an historian too, what seems more significant in explaining his lasting impact is his view of the common ground on which these distinct disciplines stand in his call for a *rapprochement* between history and philosophy.

Anthony F. Russell goes a long way toward elucidating and evaluating the nature of Collingwood's proposed *rapprochement* between history and philosophy. At the heart of the matter is his distinction between the abstract thinking of traditional propositional logic whose modes of inference are induction and deduction, which are characteristic of science, and concrete thinking whose proper mode of reasoning is the logic of question and answer, which is characteristic of history.

Russell explains that the latter logic is underdeveloped because the possibility of such a logic was a non-question in the pre-historically-minded past during which the realists criticized by Collingwood refused to admit the reality of becoming, with the result that they reduced becoming propositions into being propositions. The very arising of the question at hand is itself but a piece of a whole historical fabric in which historical thinking as such has come to penetrate the contemporary Western mind. That is, it is only in relation to this historical becoming that the possibility of *rapprochement* logic is revealed in the first place precisely as a question.

What is revolutionary is that this *rapprochement* has the foundational potential to unify knowledge in its diversity by means of the interrogatory knowledge proper to historical thought which is grounded in "the concrete universal." Because of an awareness of the historical context in which even the development of logic itself takes place, it can now be seen that scientific, artistic and religious pursuits which have been alienated from one another since the

Renaissance, because of their distinct modes of thinking, need no longer be separated in an artificial way which cuts them off from human thought in its integral wholeness, wherein abstract and concrete, speculative and practical thinking, find a common ground. This is not to deny the proper logic of each, but to admit the richness of human experience and consciousness which is at once concrete and universal. This is the direction in which the unexploited logic of history points. Collingwood's thought on this point is pivotal insofar as he has marked out the precise intersection of philosophy and history which integrates rather than divides knowledge.

Russell establishes not only the existence of this unexploited logic, but sheds a sympathetic yet critical light on its nature, structure, and modes of operation. He contrasts it with a lifeless propositional logic by showing that what is quintessential about the logic of history is that it detects rather than dismisses the rich web of relations in human culture wherein the part can be understood in relation to the whole and the whole in relation to the part. The concrete universal cannot be observed from without as if the whole and the part were independent entities whose relations to one another are only external. The logic of history penetrates to internal relations by way of the historian's participation in the inward reality of human thought as distinguished from its outward signs and action. It is for this reason, Russell points out, that attempts to separate logic from life in order to make historical thinking conform to the abstract mold of science always turn out to be aberrant. This point might have been more effectively developed by addressing in some depth the school of analytic philosophers who, following the lead of Carl Hempel, oppose Collingwood's claim.

Russell's comprehensive synthesis of Collingwood's position on the logic proper to history shows conclusively and precisely how historical thinking is concrete thinking. Specifically, human products, such as culture, or war, or a city, or a Gothic cathedral, are "a web of factors" which are each external in appearance, but "intrinsically and dynamically integrated by internal relations resulting from human thoughtful and purposeful activity." Because the whole and the part of such human entities are not mutually exclusive, or related merely by external relations, but are, rather, interrelated, the question arises: "How do all these things fit together?" The reader of this work will revel in the fascination of the concrete as Russell details how this method of inquiry is the same as that of crime detection in its concern for the following: the facts; the integration of facts (their "going together"); their manifestation as facts to the detective or historian, which itself is one of the facts affecting the nature of all the other facts; the questioning mind, whose activity makes the fact evident as a fact.

What makes this method of inquiry indeed a "method" is the inner logic of the concrete universal, the logic of question and answer. By this logic the questioning mind seeks its answers from within the concrete events themselves through the universality of thought as such, which can exist beyond local and temporal existence and which is therefore the prerequisite of all inference. This delineation of method goes further than Sherlock Holmes' "fact-grubbing" in that Collingwood emphasizes the need to use "the little gray cells" by asking a dominating question which will suggest what to look for and to select as fact in the first place. One begins, that is, not by collecting facts but by asking questions, thus turning the facts into evidence, as nothing is evidence except in relation to some definite question.

Russell holds that Collingwood's claim that such an underdeveloped logic exists is indeed to be admitted, though Russell posits some sound reservations, which stem from the depth of his knowledge of Western philosophy, and which point the way to further development of the important issues he raises. The contribution of this book lies in the appraisal of Collingwood's position from the standpoint of the criticisms offered both for and against it. Russell's own position is well-argued and clearly presented. The scholarship, moreover, is both accurate and comprehensive on Collingwood (as of the date completed), and the writing, even on many obscure points, is clear. The long synthesis of Collingwood's position, and of others', is especially helpful to the non-specialist in either Collingwoodian or other schools of thought considered here in relation to Collingwood in new ways. The book as a whole is structurally as well as conceptually integrated, yet not quite as self-evidently as it might have been. If the relations between the whole and the part had been made more explicit rather than remaining implicit, the result would have been a major reinterpretation of Collingwood's position in addition to the already comprehensive setting of it within philosophical tradition.

There is therefore a further contribution of this book which lies between its pages in potential, and which, I will venture to say, when realized, will be far-reaching. Russell seems to be on to something which the existing state of scholarship on Collingwood has not yet, to my knowledge, made explicit, but which, if grasped, would integrate intrinsically the whole of this work as well as lend greater clarity and significance to Collingwood's analysis of the concrete universal. There seems to be, in other words, a "missing link," which is all the more significant because it is so elusive. Russell himself might have shortened his lengthy volume had he found it. But in his conclusion he raises a promising question regarding the concrete universal: ''But, if this linguistic unity in diversity is the logical principle, the one in the many, at the basis of history and of research, what can be said regarding its principles?''

The historical climate today is conducive to taking on this question as a ''dominating'' one in its own right if seen in the context of today's rising semiotic consciousness. Semiotics has now historically come of age as an intellectual movement because, as the philosopher John Deely points out with a sense of history in his path-mapping book, *Introducing Semiotic: Its History and Doctrine,* the inherent interdisciplinary nature of semiotics can establish a *rapprochement* between specialized disciplines, now become insular, which have resulted from the rise of modern science. Collingwood's call for a *rapprochement* between philosophy and history, and the path of contemporary semiotics, converge. This convergence lies at what I will call an ''intersection'' between philosophy and history, precisely at the point at which *semiotic* doctrine can clarify the *relation* between what is abstract and what is concrete. To take one without the other severs the whole which participates in the part, or the part which participates in the whole. To do that rends asunder Collingwood's concrete universal.

For the first time the relation of logic to semiotics has surfaced from layers of history to be perceived in the light of day. Such an emergence well illustrates Collingwood's view of history as the present, for John Locke's proposal for an analysis of signs under the name of ''semiotic'' is now seen as a breakthrough which, precisely because it failed to be seen as such in the past, remained in the realm of non-being until the present recreated it as ''history.''

In the final pages of Locke's *Essay* of 1690, ''semiotic'' means quite precisely the foundational doctrine of signs as these underlie and structure the whole of our experience and consciousness. A semiotic logic, he suggests, could bring into existence ''a different sort of logic and critic than we have been acquainted with heretofore,'' and one which puts Collingwood's concrete universal in a new perspective, thereby clarifying what Russell's work tends toward — what I call his ''missing link.''

I will single out only one aspect of semiotic which shows what I think Russell implicitly grasps, but would couch more effectively in a semiotic context, thereby rendering an inherent ''unity in diversity'' to his entire endeavor. Semiotic transcends as well as integrates the traditional division of being into what is independent of the mind and what is dependent on it, because semiotic brings both orders into relation precisely through signification or the sign. Although Locke continues to follow Aristotle's division of knowledge into speculative and practical, Locke's breakthrough is now seen to be that he kept these branches distinct, yes, but not separate (see Deely's work). By way of an (arbitrary) analogy, he did not sever the connection between the two hemispheres of the brain; on the contrary, he shows how the two communicate. His "new logic"—semiotic—proposes a systematic study of the means whereby any knowledge is acquired, developed, and shared in community. Semiotic thus integrates logically what is already integrated in consciousness. In so doing, semiotic opens the door to studying the orders of external and internal relations, or mind-independent and mind-dependent relations, or being and non-being, not as if their boundaries were staked out in advance, but as constantly shifting because of the interaction *between* the orders as the mind participates in what it observes through the function of signs. Semiosis thus structures experience so flexibly that the line between fact and fiction is, as it were, drawn free-hand, as any historian knows who has observed fiction harden into historical fact.

Semiotic thus offers a third term in place of the tired debate between realism and idealism. Set in this light, Collingwood's proposed logic of history takes on its proper hue. Collingwood was emphatically not a realist, but it has only been for want of a more precise term and owing to a false dichotomy that he has "therefore" been called an idealist. What he really offers is a third term between realism and idealism, which is, namely, "semiotic" without the name. Russell's work, while a significant contribution in its present terms to existing Collingwood scholarship, might also pave the way to a new generation of Collingwood scholarship developed within an explicitly semiotic perspective.

In order to open the door to further discussion, I will illustrate the relation of Collingwood's logic of history to semiotic by interpreting semiotically his view of the way in which the past exists in the present as "history." What the historian's mind does is to appropriate its object as something there in itself as thought, yet not there in the immediate context of past thought. What the mediacy of the logic of history does, now understood to be logic as a semiotic art, is to enable the historian to see sign systems in the past in the perspective of present sign systems, insofar as all thought is through signs. The relations observed between sign systems, as well as within sign systems, are dependent upon the mind of the individual historian as observer, who, through his own semiosis, participates in the selection of what facts are significant, that is, what facts count as evidence.

To take a classic example of the mind-dependent nature of history, we may question whether Charlemagne was really crowned Holy Roman Emperor by the Pope. Some observers say he was, and some say he was not. The papal account has it that the Pope "with his own hand crowned" Charlemagne, signifying thereby that this fact was evidence that papal authority made him Emperor. The Royal Annals, in contrast, report the Frankish point of view that the Pope "put the crown on his head," signifying thereby that the same fact was evidence that Charlemagne did not owe his crown to papal authority. What was "signified" by the external, or physical, act of the coronation has been a question subject to endless controversy precisely because of its mind-dependent status both at that time and after that time. For that matter the skeptic has well pointed out that the "Holy Roman Empire" was neither holy, Roman, nor an empire.

Although Collingwood's philosophy of history is idealist in so far as he emphasizes the mind-dependent aspect of history, his position is founded on the absolute presupposition that past events existed in themselves independently of the observing mind re-enacting them through semiosis. Even though the historian as observer cannot get outside his own mind-set, he nonetheless observes some aspects of mind-independent reality. In Collingwood's words he does observe "the *real* tree in different aspects." The actual and the ideal do converge the closer the historian gets to "a real knowledge of the infinite world of fact." But because the infinite world of fact cannot be observed in all its concreteness and universality by the finite mind, this convergence is only "asymptotic." The nearer the actual comes to the ideal, the greater becomes the force generated by this very approach to prevent a still closer convergence. Thus, the greater the historian's knowledge, the more acutely he becomes aware that he will never know anything with finality, and that all his so-called knowledge is "to an unverifiable extent erroneous," precisely because it is mind-dependent even when observing real, or mind-independent, relations.

Here we come to that convergence of the scientific, artistic, and religious pursuits of man which, understood in its precise nature as semiotic, offers a subtler yet more sophisticated alternative to realism or idealism. Collingwood sees that while the historian cannot escape his own point of view, even if he changes perspectives, his problem is not that there is no real tree, but that he cannot see it from all perspectives at once in an infinity of possible perspectives. He is caught in an "egocentric predicament," which is his own semiosis. Whereas medieval man in such a predicament contemplated "the love that moved the sun and the other stars," modern man contemplates their collapse into "black holes." Neither perspective is founded directly on the actual or the ideal, but on something between the two, that is, on the mediacy of the sign through which the mind structures its experience and consciousness of the intercommunicating universe of mind-independent being.

In the perspective of semiotic, mind-independent fact and mind-dependent fiction are distinct but not separate beings. Collingwood's logic of history is semiotic in that it admits both orders of being precisely as they are interrelated. Such logic uses the methods of objective inquiry in order that the sequence of observation/inference/conclusion may be replicated, not simply to verify coherence *qua* coherence but to get beyond coherence as such to *truth* as coherence, which is the most probable account of what actually happened in the past. The object of historical inquiry is thus not the actual past, but "what the evidence proves." In such a context history is neither actual nor ideal, mind-independent fact nor mind-dependent fiction, being nor non-being, but a relation between the two orders of being established through the historian's interpretation of the sign. Historical thinking proceeds by the logic of question and answer addressed to mind-independent traces (such as documents) which are interpreted as signs of mind-dependent being, or the internal relations of thought upon which the whole human edifice of thought and action rests.

Looked at semiotically, the historian participates as observer in this concrete universal, which is perfused with signs. While he observes from the egocentric reference point of the ever-changing sign configuration of his own time and consciousness, he likewise participates in the ever-changing sign configurations of past thought. In recreating history as the present, the historian lives in both present and past sign systems at once. Because of his very familiarity with a sign system foreign to his own, he experiences in consciousness the shifting ground between mind-dependent and mind-independent "realities" as these constantly change places across time according to human design. The logic of history approaches the concrete universal (by way of the sign) from within human reality in precisely the way covering laws imposed from without cannot do because they cannot explain the concreteness of human experience as it is structured through semiosis. Herein lies the living appeal of Collingwood's philosophy of history to historians. The logic of history is a semiotic art which presupposes experience with life itself, and which would seem to explain why historians tend to mature at a later age than mathematicians.

What I have here only adumbrated is the kind of study of Collingwood's thought which would further develop his own insights in the light of the historical *prise de conscience* of semiotics and the consequent development of a doctrine of signs. His philosophy of history opens the door to a common understanding of signification viewed not only speculatively but practically as well, as signification manifests itself in human awareness across time in the relation between thought and action. The coming together of philosophy and history is fundamental to an anthroposemiotic view of man as a being in time whose transmissible history gives rise to culture. What Russell's study points toward is a semiotic overview of Collingwood's contribution to this understanding.

Brooke Williams
Stonecliffe Hall
Fall of 1983

AUTHOR'S PREFACE

This work was written some years ago, after I had become intrigued with Collingwood's idea of a hitherto undeveloped logic, a logic of questioning which would be adequate to deal with distinctively human problems and, as such, would be the logic proper to historical thought. He likened this new logic to the procedure of a detective solving a crime. My own habits of thought inclined me toward the scientific rather than the literary or historical, so that I tended to idealize scientific procedures, to think in inductive and deductive patterns, and, without realizing it, to adopt a somewhat trivializing attitude toward historical knowledge.

However, it is impossible to read Collingwood's *An Autobiography* and the Prologue to his *Speculum Mentis* without feeling the shock of the contrast he draws between the one same event, the first World War, as an unprecedented success for natural science and at the same time an unprecedented disgrace to the human intellect. He contrasts the success, on the one hand, of modern European minds in controlling almost any situation which involves physical bodies and physical forces, and, on the other hand, the inability of these minds to control situations that involve human beings and mental forces. Collingwood blames disasters such as the first World War on trying to solve human problems with the methods of natural science instead of with the instrument properly adapted to such problems, which he regards to be history and its proper procedures. He stresses that logicians of the twentieth century should give serious attention to the development of the logic of history, which is a concrete logic of question and answer, of the raising and solving of problems, which activity, according to Collingwood, is an activity identical with the mind itself, since "the mind is what it does."

The present work is the result of my research to find out what Collingwood had to say about this logic and to take a position one way or another on his contention. While I felt that I had to agree with his general position on the existence of an interrogative logic proper to history, something important seemed to remain undisclosed, something that the General Editor in her Preface refers to as 'a missing link.' Collingwood seemed to be saying something massively important, but, like Heraclitus of old, what he said seemed as much to conceal as to reveal this important aspect. The crux of the problem as I felt it seemed to turn on getting a handle onto cultural concepts and the type of understanding and reasoning that went with them. I found considerable help in this respect in the works of Ernst Cassirer, Robert Stover, John Wild and Michael Polanyi. But still something seemed to be missing. The natural world was still there in all its brilliance, illuminated by science and its abstractive, inductive and deductive ways, whereas the human world and its history, although in no way appearing to be as trivial as before, was still irritatingly elusive.

During the last four or five years I have become acquainted with, and exposed to, the fresh winds of a newly developing discipline, semiotic. This movement seems to have come to recognizable self-consciousness in the United States at a conference on intercultural and interdisciplinary communication sponsored by the U.S. Office of Education and held at the Bloomington campus of Indiana University in 1962. But its advent had long since been foretold by John Locke in the concluding pages of his *Essay Concerning Human Understanding* in 1690. There Locke divided the sciences into three kinds: Natural Science, competent to explain the natural world; Ethics (in the wide sense of *practica*), competent to deal with what is right in human action; and a third type, which he refers to as "the doctrine of signs" and which he names with the Greek word Σημιωτικὴ, which renders into English as Semiotic. According to Locke, this hitherto non-existent discipline is concerned with the way in which the human agent relates the objective world to himself by the use of signs such as words and ideas, and will "afford us another sort of logic and critic than what we have been hitherto acquainted with." Collingwood's interrogative logic of history presents us with the human mind in dynamic interaction with its environing world, its world being what that mind has made the environment to mean for itself, and that mind being what it has made of itself in making the environment meaningful to it. The Self and its world are strictly correlative and mutually implicatory. Strains develop in this communication situation, which are responded to by the mind interrogating any perturbing facts that are discordant with other facts. The life of the mind is an ongoing process of raising and solving problems, which is also a process of humanizing the world, of making its world an amenable place to live in. When Collingwood refers to questions as turning facts into evidence, making raw data into signs, he is referring to his logic of questioning as something which is thoroughly semiotic and which aims at a harmonious balance between the Self, a concrete mind as a sign-making agency, and its environing world as a meaningful whole made evident—that is, significant, sign-bearing—through the mind's questioning activity.

This work as it stands shows Collingwood's thought clearly to be a precursor to semiotic thought. Further understanding of his proposed logic, with its goal of being a service to human amelioration, will benefit enormously from the advancement of John Locke's foretold semiotic logic now in the process of historical unfolding. Collingwood's concrete universal at the basis of his interrogative logic stands out as a whole-of-meaning through the signification (sign-bearing, evidentiality) of the parts in relation to the whole and simultaneously of the whole in relation to its parts, which signifying function, or 'understandability,' derives from the mind's proper activity, the activity of questioning. Collingwood's constellation of consupponible presuppositions, including both absolute and relative presuppositions, represents a mind-set of structured subjectivity in dynamic intercommunion with its environing world, each having its signification or meaning-bearing value or evidentiality in relation to the other. Semiotic as ''the doctrine of signs,'' as Locke called it, provides us with a new tool for the interpretation of Collingwood's position; and, reciprocally, Collingwood's theory of the logic of questioning as the proper historical instrument for turning facts into significant facts, into evidence, provides semioticians with a unique sign-making system which they cannot afford to neglect. Collingwood scholars and semioticians stand to benefit from a mutual consideration of each other's work, and Locke's 'new logic and critic' might turn out to be very intimately connected with Collingwood's interrogative logic proper to history and to problem solving generally.

In view of the growing interest of the semiotic movement in the nature of history, Collingwood's contribution, seen semiotically, might well lay the foundation for that *rapprochement* between the human disciplines that is at the heart of Collingwood's thought and of the semiotic movement, and especially for the *rapprochement* between philosophy and history, to the achievement of which Collingwood devoted his life's work.

Beresford
New Brunswick, Canada
August 15, 1983

Introduction

R. G. Collingwood is perhaps the most outstanding thinker of the English speaking world in recent years to draw attention to the determining role of history in human affairs today, and to make the point that as "the chief business of seventeenth-century philosophy was to reckon with seventeenth-century natural science" so "the chief business of twentieth-century philosophy is to reckon with twentieth-century history."[1] Collingwood's emphasis on the importance of history both for harmonious management of human affairs[2] and as the activity which demands first priority attention of the philosopher today, no doubt is ultimately traceable to his own innate predisposition towards history and its method of thinking and proceeding, just as Descartes had gravitated towards the mathematical mode of thinking by a sort of innate predisposition. "Philosophy" for Collingwood "is

[1] R. G. Collingwood, *An Autobiography*, London, Oxford University Press, 1939, pp. 78-79.

[2] In his *An Autobiography* (p. 90) he attributes the contrast between the war of 1914 as "an unprecedented triumph for natural science" and at the same time "an unprecedented disgrace to the human intellect" to neglect of history, whose proper function, according to Collingwood, is to provide "more understanding of human affairs and more knowledge of how to control them" (p. 92) just as natural science has made possible "the success of modern European minds in controlling almost any situation in which the elements are physical bodies and the forces physical forces . . ."

reflective,''[3] by which he means that ''the philosophising mind never simply thinks about an object'' but ''always, while thinking about any object, thinks also about its own thought about that object,'' and is therefore ''thought of the second degree, thought about thought.''[4] Furthermore, ''philosophy is never concerned with thought by itself; it is always concerned with its relation to its object, and is therefore concerned with the object just as much as with the thought.''[5] Now, the thoughts which are the relevant consideration of the philosopher are the predominating considerations which determine the character of any given historical period. Greek philosophy is characterized by a preoccupation with mathematics, according to Collingwood.

> Now, our philosophical tradition goes back in a continuous line to sixth-century Greece, and at that time the special problem of thought was the task of laying the foundations of mathematics. Greek philosophy therefore placed mathematics in the centre of its picture, and when it discussed the theory of knowledge it understood by it first and foremost the theory of mathematical knowledge.[6]

Collingwood then distinguishes ''down to a century ago, two great constructive ages of European history,'' namely, the Middle Ages and the Renaissance periods, each having its own characteristic thought occupation which in turn gave to the philosophy of the period its characteristic form.

> In the Middle Ages the central problems of thought were concerned with theology, and the problems of philosophy therefore arose out of reflection on theology and were concerned with the relations of God and man. From the sixteenth to the nineteenth centuries the main effort of thought was concerned with laying the foundations of natural science, and philosophy took as its main theme the relation of the human mind as subject to the natural world of things around it in space as object.[7]

During these periods, people thought historically but their historical thought was simple and rudimentary. But from the eighteenth century onwards there developed a growing concern with history.

> . . . in the eighteenth century people began thinking critically about history, as they had already learnt to think critically about the

---

[3] R. G. Collingwood, *The Idea of History*, London, Oxford University Press, 1946, p. 1.

[4] *Ibid*.

[5] *Ibid*., p. 2.

[6] *Ibid*., p. 4; see also *Speculum Mentis*, London, Oxford University Press, 1924, pp. 164 et seq.

[7] R. G. Collingwood, *The Idea of History*, pp. 4-5.

> external world, because history began to be regarded as a special form of thought, not quite like mathematics or theology or science.
>
> The result of this reflection was that a theory of knowledge proceeding on the assumption that mathematics or theology or science, or all three together, could exhaust the problems of knowledge in general, was no longer satisfactory. Historical thought has an object with peculiarities of its own.[8]

Collingwood tells us that "theories of knowledge designed to account for mathematical and theological and scientific knowledge thus do not touch on the special problems of historical knowledge" and furthermore if such theories of knowledge "offer themselves as complete accounts of knowledge they actually imply that historical knowledge is impossible."[9] Collingwood regarded his life's work as "in the main an attempt to bring about a *rapprochement* between philosophy and history."[10]

As a consequence of this change of preoccupation characteristic of contemporary Western thought, of this awakening of the historical consciousness in the eighteenth century, the logical types of reasoning which were developed to facilitate mathematical and scientific reasoning are no longer adequate to deal with the mode of reasoning characteristic of historical inquiry. The deductive syllogism, according to Collingwood, was developed by the Greeks as the method of reasoning adequate to mathematics,[11] and the method of induction was developed at the time of the Renaissance, when deductive reasoning proved no longer adequate to handle the new types of problems dealt with by Renaissance science.[12] And just as the ancient syllogistic-deductive logic was found to be inadequate to deal with Renaissance experimental and observational science, so likewise neither is deductive nor inductive reasoning adequate to guide the type of investigation and inference employed by the historian in his attempts to reveal the past.[13]

Thus, according to Collingwood, just as the awakening of the scientific consciousness of Western European man in the sixteenth century demanded a new logic to supersede the syllogistic deductive logic as an adequate logical instrument to handle scientific problems, so today the awakening of the historical consciousness of Western

---

[8] *Ibid.*, pp. 4-5.
[9] *Ibid.*, p. 5.
[10] See *An Autobiography*, p. 77.
[11] See *The Idea of History*, p. 253.
[12] See *ibid.*, p. 254.
[13] See *ibid.*

man demands that logicians give attention to developing the type of inference which is proper and peculiar to historical investigation, which is neither deductive nor inductive.

> The main scientific achievement of the ancient Greeks lay in mathematics; their main work on the logic of inference was naturally, therefore, devoted to that form of inference which occurs in exact science. When at the end of the Middle Ages the modern natural sciences of observation and experiment began to take shape, a revolt against Aristotelian logic was inevitable; in particular, a revolt against the Aristotelian theory of demonstration, which could by no manner of means be made to cover the technique actually used in the new natural sciences. The text-books of logic in use today still bear the marks of this revolt in the distinction they draw between two kinds of inference, 'deductive' and 'inductive.' It was not until late in the nineteenth century that historical thought reached a stage of development comparable with that reached by natural science about the beginning of the seventeenth; but this event has not yet begun to interest those philosophers who write text-books of logic.[14]

According to Collingwood, the logic proper to historical reasoning is characteristically a logic of question and answer[15] in which every statement is meaningful only as the answer to a definite question,[16] and the mode of inference, as contrasted with inductive and deductive reasoning, is by way of an ordered series of questions and answers.[17] Collingwood contrasts this logic of question and answer with what he calls the traditional propositional logic in which propositions have meaning and truth in themselves independently of any questions to which they might be the answer,[18] and whose traditional modes of inference are induction and deduction.[19] The ultimate difference between the two logics, according to Collingwood, is the basis of the difference between abstract thinking and concrete thinking, namely, the difference between the abstract universal and the concrete universal. Propositional logic is the logic proper to abstract

---

[14] *Ibid.*, pp. 253-254. See also H.-I. Marrou, *De la connaissance historique*, Paris, Editions du Seuil, 1966, p. 85. (English translation, *The Meaning of History*, p. 89). "If we try to account for this understanding process [characteristic of history], we cannot make use of any transposition of the methods of the sciences of nature. Properly speaking, the historian does not proceed by way of deduction or induction."

[15] See *The Idea of History*, pp. 268-270; 273-275; *An Autobiography*, pp. 24-28; 124-126.

[16] See *An Autobiography*, p. 33; also *An Essay on Metaphysics*, London, Oxford University Press, 1940, pp. 23-25.

[17] See *The Idea of History*, pp. 273-275; *An Autobiography*, p. 37.

[18] See *An Autobiography*, pp. 30-39.

[19] See *Speculum Mentis*, pp. 163-169; 176-180, especially p. 179.

thought; the logic of question and answer is the logic proper to concrete or historical thought, whose proper mode of reasoning is a process of explicating the implicit through ordered questioning. "The concrete universal is the daily bread of every historian and the logic of history is the logic of the concrete universal."[20]

It was noted above that Collingwood declared his life's work to be that of bringing about a *rapprochement* between history and philosophy.[21] In the Prologue to *Speculum Mentis* he shows at great length that the scientific mentality, based as it is on abstract thinking, has resulted in the sundering of our way of life since the Renaissance into mutually exclusive and conflicting activities: artistic, religious and scientific pursuits which are alien to each other and result in a divided and inhuman way of life. Collingwood's call for a *rapprochement* between history and philosophy is also a call for a *rapprochement* between the various forms of human activity and a harmonious human life which has been impossible since the Middle Ages. History is the science of human affairs and the study of history will enable us to regulate human affairs as natural science enables us to control the world of nature. The instrument of this twofold *rapprochement* is the logic of question and answer, which may also be called *rapprochement* logic, as contrasted with inductive and deductive propositional logic whose abstractive mode divides and alienates the various human disciplines and the very mode of daily human living.

It is therefore fitting to examine Collingwood's claim regarding the existence of this unexploited logic, to determine whether this claim can be substantiated, and to determine as far as possible the nature and implications of such a logic. To this end the following work is offered as a contribution.[22]

---

[20] *Ibid.*, p. 221; see also *ibid.*, pp. 217-221; p. 244; p. 286; p. 299. Collingwood's *An Essay on Philosophical Method*, London, Oxford University Press, 1933, is also an elaboration of the logic of the concrete universal, especially Chapters 2 and 3. See below, Chapter 4, pp. 97-117.

[21] See above, p. 3.

[22] According to Louis O. Mink in his *Mind, History and Dialectic*, Bloomington, Indiana University Press, 1969, p. 131, "the logic of question and answer is not a theory of logic at all, in any ordinary sense of that term, nor is it even a theory of semantics; it is a hermeneutics and as such it supplements but does not replace formal logic." This may be true, but it seems to weaken Collingwood's very emphatic insistence on the need to develop the logic appropriate to historical inquiry as something for which inductive and deductive logic cannot supply. Alan Donagan, in his *The Later Philosophy of R. G. Collingwood*, London, Oxford University Press, 1962, p. 56, goes so far as to say that "Collingwood's forays into logic resemble Caesar's invasions of Britain: their execution was bungled, and his account of them, which unlike Caesar's was given in advance, did not accord with his actions. Although in his *Autobiography*

The method of procedure will be as follows. Firstly, Collingwood's theory of the logic of questioning as proper to history will be expounded as far as his published works permit; this will be done in chapters 1 to 6 inclusive. Then, the remaining chapters will attempt to evaluate Collingwood's theory.

As regards the first, expository, part, the chapters will be divided according to the main line headings delineated by Collingwood himself in regard to his theory of questioning. Chapter 1 will present Collingwood's call for the development of an interrogative logic proper to history and quite distinct from deductive and inductive logic, which logical systems, according to Collingwood, are appropriate to exact (mathematical) and empirical science respectively. The principal documentary source of this subject matter will be Collingwood's posthumously published "Historical Evidence," found in the Epilegomena (part 5) of *The Idea of History*, pp. 205-282.

Chapter 2 will expose Collingwood's epistemology of questioning, in which knowledge is understood in terms of a question and answer complex, and propositional meaning is to be taken in reference to the question which the proposition is intended to answer. The essentially historical meaning of propositions will appear, since every question is asked by a conscious subject historically situated and conditioned. Collingwood's *Autobiography* will be the main source of this embryonic epistemology identifying knowledge with the process of asking and answering questions.

The third chapter will elaborate on Collingwood's emphasis on the active aspect of knowledge. Questioning, inasmuch as it involves the activity of supposing, requires that the questioning mind confront the objects of its knowledge in an active, dominating way, not, as Positivists would have it, as a mere compresence with already existing

---

he proclaimed himself 'in logic . . . a revolutionary' (A.52), his revolutionary manifesto turned out to be an embarrassment, and in what he afterwards wrote about logic he more and more ignored it." David Rynin remarks in his "Donagan on Collingwood: Absolute Presuppositions, Truth and Metaphysics," *The Review of Metaphysics*, XVIII (1964-1965), p. 307: "Collingwood could, of course, be as poor a thinker as Donagan makes him out to be on the subject here under discussion [Metaphysics as historical study of absolute presuppositions], but one wonders then what point there is in writing a book about these inconsistencies and confusions. In any case, I wish to read Collingwood on the assumption that he is not an intellectual dolt, and the least I can do is to seek out, if possible, some interpretation of his thought that leaves it not actually ridiculous and obviously mistaken, to see whether what he seems to be saying may not be worth saying and even listening to." This present investigation into Collingwood's theory of the logic of questioning is undertaken in the same spirit as that expressed by David Rynin.

objects or 'hard facts.' Supposing involves consideration of the as yet non-existent, and any question involves supposed alternatives, the actualization of one or another of which will provide the answer to the question. Chapter 3 will therefore expose Collingwood's theory of questioning as supposing or "the cutting edge of the mind." *Speculum Mentis* will provide much of the material of this section. The questioning activity of the mind considered precisely as an exercise of creativity, which is the proper function of art, whose activity is characteristically that of supposal, will be deferred from Chapter 3 to an appendix at the end of the work. Its presence as an integral part of Chapter 3 would disturb the flow and continuity of thought, as its material, drawn from an analysis of Collingwood's *The Principles of Art,* is somewhat voluminous in detail and turgid in treatment.

The central core of Collingwood's logic of questioning, his theory of the concrete universal, which is 'the daily bread of every historian,' will be exposed in Chapter 4. The concrete thought of the historian, if it has logical value, must somehow have about it the character of universality. The universal import of concrete fact with which the historian trades is shown to be a universal as concrete and fully determinate as the facts which it proposes to make intelligibly meaningful. This concrete unity in diversity, and the logic of history which it founds, will be presented mainly by drawing on Collingwood's *Speculum Mentis* and his *Essay on Philosophical Method,* the former work being mainly concerned with the concrete and dialectical character of this universal, the latter with its intelligible structure as an overlap of classes in a scale of forms, which structure, being at each point internally affected by a temporal before-and-after relationship, is essentially historical in nature, so that the before part is related questioningly to the after part and the after part is related answeringly to the before part.

Chapter 5 deals with Collingwood's important point that relevant questions have their logical efficacy in presuppositions, ultimately in absolute presuppositions, which are characteristic of the thought patterns which determine and historically situate a particular culture. Collingwood advances his well known theory that the proper subject matter of metaphysics (not sufficiently recognized by traditional metaphysicians) is the study of these absolute presuppositions and of the manner in which one constellation of presuppositions gives way to another constellation and consequently to another way of viewing and questioning the world of experience. In doing so, Collingwood denies the traditional notion of metaphysics as being concerned with eternal problems and regards it as a historical investigation of particular thought-principles (that is, absolute presuppositions) held by particular peoples at a particular time and a particular place and which are subject to historical development from one period to another.

Chapter 6 examines Collingwood's notion of history, since he has insisted that a logic of questioning is at the basis of historical thought. Here, Collingwood's notion of history is presented as the account of self-developing, self-creating mind in accordance with his insistence throughout all his works that 'mind is what it does.' Mind, as self-creating activity, acts and constructs itself as an overlap of classes in a scale of forms by the process of 'raising and solving problems,' or asking and answering questions. The activity which *is* mind is therefore together radically historical and radically interrogative.

Chapter 7 begins the work of evaluating Collingwood's theory of interrogative logic proper to history. This chapter criticises and evaluates Collingwood's theory of knowledge as identical with the questioning and answering process which was exposed in Chapter 2. Collingwood's theory that propositions are meaningful only as answers to questions is evaluated partially in this chapter and more completely in Chapter 9. Chapter 8 continues the critical work and takes Collingwood to task for neglecting to recognise an already well developed logic of interrogation in the Aristotelian tradition. An attempt is made to give the main outlines of this theory of questioning as found in Aristotle's *Topics* and *Posterior Analytics* and developed by certain Aristotelian commentators. An attempt is then made to see Collingwood in relation to this Aristotelian interrogative theory.

Chapter 9 faces the crucial question as to whether Collingwood is correct or not in asserting that historical thought operates by a logic of questioning distinct from the deductive logic of strict science or the inductive logic of empirical science. An attempt is made to separate out what is acceptable in Collingwood's thought and making a genuine contribution to the philosophy of history from what is unacceptable or requiring emendation.

Having taken an affirmative position, subject to certain important qualifications, regarding Collingwood's theory of the existence of a logic of questioning proper to history, the tenth and concluding chapter will endeavour to indicate along what lines such a logic might be developed as far as documentary sources available permit such indications to be made.

PART I: EXPOSITION

CHAPTER ONE

The Need for a New Logic of Question and Answer

A. The Logic of Question and Answer in the Works of R. G. Collingwood

Perhaps the most explicit statement of the need for a new logic of question and answer, and of its particular relevance to historical thought, is to be found in the third section of part 5 of the posthumously published (1946) *The Idea of History*. This part of *The Idea of History* is an edited collection, made by T. M. Knox, of various works, some previously published, others not, such as the section headed "Historical Evidence," which, according to Knox,[1] was written in the spring of 1939 as part of a projected work, *The Principles of History*. It is in this previously unpublished fragment that we find Collingwood's expressed view of the need for a new logic, a logic of question and answer, as the appropriate mode of inference in historical research, as induction is the appropriate mode of inference in the experimental sciences and deduction the appropriate mode of inference in the exact, mathematical sciences.

Collingwood's *Autobiography* details the genesis and development of his original thinking on question and answer, and provides an

[1] See editor's preface to *The Idea of History*, Oxford University Press, 1946, p. v; see also A. Shalom, *R. G. Collingwood, philosophe et historien*, Paris, Presses Universitaires de France, 1967, p. 157.

embryonic epistemology of the activity of questioning.[2] In this work we see how this theory of questioning arose as a result of his increasing dissatisfaction with the realism of Cook Wilson and the Oxford 'realists,' and how this growing opposition to realist methods of thinking inevitably led him to develop a critical mode of thinking more and more in line with his archaeological practice of proceeding according to definite questions.

Chapters 4 through 7 of *An Essay on Metaphysics* deal explicitly with the theory of questioning, and specifically with the presuppositions from which relevant questions arise. A novel account of metaphysics as the historical study of absolute presuppositions is presented. This alignment of metaphysics with historically predetermined presuppositions to scientific questions has occasioned rival interpretations of Collingwood's thought. Shalom and Donagan follow Knox's radical conversion theory according to which, some time after 1936, Collingwood changed from thinking that philosophy was distinct from history to a train of thought which culminated in *An Essay on Metaphysics*, which, according to these commentators, ended in radical historicism, identifying philosophy with history.[3] On the other hand, Lionel Rubinoff argues convincingly for an internally coherent system throughout the whole of Collingwood's works, and explicitly refutes the radical conversion hypothesis and that Collingwood's final thought identifies philosophy with history.[4]

The works mentioned: "Historical Evidence," *An Autobiography* and *An Essay on Metaphysics*, all written during the period 1936-1939, contain Collingwood's most direct treatment of his thought on question and answer. Important references, however, are made to it in other works.

The earlier *Speculum Mentis* (1924) contains basic information for

---

[2] It is well to note that Alan Donagan, in his *The Later Philosophy of R. G. Collingwood*, London, Oxford University Press, 1962, p. 1, takes seriously T. M. Knox's warning that Collingwood's own narrative of his development is not trustworthy.

[3] See T. M. Knox's preface to Collingwood's *The Idea of History*, pp. x-xi; A. Donagan, *The Later Philosophy of R. G. Collingwood*, pp. 10-12; A. Shalom, *R. G. Collingwood, philosophe et historien*, pp. 210-215; A. Shalom, "R. G. Collingwood et la métaphysique," *Les Etudes Philosophiques*, X (1955), pp. 706-708. Although Shalom must be classed with the radical conversion theorists if this means those who maintain that Collingwood's *Essay on Metaphysics* identifies history and philosophy, nevertheless it is questionable if this classification fits Shalom in virtue of his conclusion that Collingwood's historicist conclusion is implicit in the principles of his earliest works.

[4] See Lionel Rubinoff, *Collingwood and the Reform of Metaphysics*, Toronto, University of Toronto Press, 1970, pp. 23, 213-214. Louis O. Mink, *Mind, History, and Dialectic*, 1969, also argues for systematic continuity.

the understanding and interpretation of Collingwood's thought on questioning. In the third section, on Art, there is inserted (p. 76, subsection 5) a treatment of 'Knowledge as Question and Answer' in which an intimate connection is established between the activity of supposal and both art and questioning. In addition to this special reference the work generally is a manifestation of the mind as a questioning and problem solving activity.

Collingwood's final work, *The New Leviathan*, being an ethico-political work, emphasizes the practical aspect of the activity of the mind rather than the epistemological. However, reference is made to knowledge as the asking and answering of questions in the context of appetite (pp. 74-79); brief reference is also made to question and answer in the first chapter, pp. 6-7.

An early publication (1920), *Ruskin's Philosophy*,[5] which was an address delivered August 8th, 1919, at the Ruskin Centenary Conference, Coniston, needs mentioning, for " 'Ruskin's Philosophy' must . . . be regarded as representing an important stage in the development of Collingwood's thought. Not only does it appear to support Collingwood's own claim in the *Autobiography* to have arrived at the theory of presuppositions and the logic of question and answer early in his philosophical career but it is also probably the closest record we have of the actual views of 'Truth and Contradiction.' "[6]

The work "Truth and Contradiction," referred to here by Lionel Rubinoff, concerns an early work which Collingwood tells us, in his *Autobiography* (p. 42; p. 99), he wrote in 1917 but was unable to have published owing to prevailing wartime conditions. This work, Collingwood says in the *Autobiography* (p. 42), contained a 'writing out' of his thoughts until that time on questioning activity as a philosophical alternative to the realist's correspondence-passivity theory of knowledge. Rubinoff argues convincingly[7] that " 'Truth and Contradiction' was an earlier version of what later became *An Essay on Metaphysics*" (a conclusion detrimental to the Knox-Donagan radical conversion hypothesis previously referred to).

In addition to the aforesaid works more or less directly referring to question and answer logic, *An Essay on Philosophical Method*, although

---

[5] See R. G. Collingwood, *Ruskin's Philosophy*: An Address delivered at the Ruskin Centenary Conference, August 8th, 1919, Titus Wilson and Son, Kendal, England, 1920; reprinted in Alan Donagan's edition of *Essays in the Philosophy of Art* by R. G. Collingwood, Bloomington, Indiana University Press, 1964, pp. 3-41.

[6] L. Rubinoff, *Collingwood and the Reform of Metaphysics*, p. 230.

[7] See *ibid.*, p. 15; see also p. 230.

not explicitly mentioning questioning, will be seen to be the logical theory of the concrete universal which is the basis of the dialectical logic of questioning.[8] This work proceeds by comparing the elements of traditional (i.e., genus-species abstract propositional, inductive-deductive logic) with the elements of dialectical or interrogative logic.

Having outlined the various sources in Collingwood of more or less explicit reference to questioning and the logic of questioning, we may fittingly begin with "Historical Evidence" (Section 3 of Part 5 of *The Idea of History*) in presenting Collingwood's theory that there is a distinct, as yet undeveloped logic proper to historical thought, a logic of question and answer which is distinct from what Collingwood calls the traditional propositional logic, whose familiar modes of inference are deduction and induction.

B. The Status of History as a Science

In the Epilegomena (Part V) of *The Idea of History*, Section 3, entitled "Historical Evidence,"[9] Collingwood deals with history as a unique human discipline, and shows that it operates according to a logic quite different from the traditional logic of inductive and deductive inference.

In what sense is it legitimate to call history a science? It is not a science in the sense in which the principle of intelligibility and meaning is ultimately derived from abstract universal laws and principles. In this sense history is not a science, whereas biology, physics and meteorology are sciences. But history, like biology, physics and meteorology, is an organized and systematized body of knowledge, which, Collingwood says, is a traditionally accepted meaning of the term 'science.'[10]

But every science, understood as an organized body of knowledge, is organized in its own special way. History as an organized body of knowledge is organized quite differently from the ways meteorology

---

[8] See Lionel Rubinoff, *Collingwood and the Reform of Metaphysics*, p. 160.

[9] Written during 1939 as part of a projected *Principles of History*: see Knox's preface to *The Idea of History*, p. v.

[10] See *The Idea of History*, p. 249. See also H.-I. Marrou, *The Meaning of History*, Montreal, Palm Publishers, 1966, pp. 34-35: ". . . we must say specifically that when we use the word 'science' with reference to history it is not the Greek word for knowledge—'επιστήμη—that we have in mind but rather the idea of τέχνη, an art—or technical method. In other words, as opposed to the ordinary knowledge of daily experience we have reference to an elaborate knowledge set forth in terms of a systematic and rigorous method which has proved to be productive of the *optimum* measure of truth."

and chemistry are organized. The body of knowledge which is meteorology is organized by collecting observations concerned with certain kinds of events which can be watched as they occur though they cannot be produced at will by the scientist. And meteorologists, like astronomers, will undertake costly and difficult journeys to observe the kind of event they are interested in, which cannot be trusted to inexpert witnesses. Wars and revolutions, events in which the historian is concerned, are not observed in the way meteorological or astronomical events are observed. Nor does the historian make such events happen under strictly controlled conditions as does the chemist with regard to the events which he observes and records. This is not because historians are less energetic or that such means are too costly, but simply because the facts which might be learnt through large costly expeditions or by deliberately fomenting revolutions at home would not teach historians what, as historians, they want to know.[11]

Observational and experimental sciences are alike in that they seek to find the constant or recurring features in all events of a certain kind. A meteorologist studies one cyclone but only in its relation to others so as to find out what belongs to cyclones as such. But the historian does not study, say, the Hundred Years' War or the Revolution of 1688 to reach conclusions about wars as such. This is because history is organized in a way different from the ways in which the observational and experimental sciences are ordered.

> In the organization of history, the ulterior value of what is known about the Hundred Years War is conditioned, not by its relation to what is known about other wars, but by its relation to what is known about other things that people did in the Middle Ages.[12]

History likewise differs from the exact (mathematical, deductive) sciences in its method of organization. Both agree in that they are inferential, going from a certain starting point through a process of inference to what that starting point proves. But the starting points, the intervening process and the resulting conclusions are diverse in each case. The starting points of the exact sciences are assumptions, traditionally expressed in the form 'Let ABC be a triangle, and let AB = BC.' But in history the starting points are not assumptions but facts coming under the historian's observation, such as that which is open before him in what purports to be a charter by which a certain king grants certain lands to a certain monastery. The conclusions

---

[11] See *The Idea of History*, pp. 249-250.

[12] *Ibid.*, p. 250.

in exact science are about things which are not restricted in space or time; if they are anywhere they are everywhere; if they are at any time they are at all times. But in history the conclusions are about events each of which has a place and date of its own; the exactitude with which these are known may vary, but the historian knows that there is such a definite time and place and within limits he knows what it is.[13]

> These differences in starting point and conclusion imply difference in the entire organization of the respective sciences. When a mathematician has made up his mind what the problem is which he desires to solve, the next step before him is to make assumptions which will enable him to solve it; and this involves an appeal to his powers of invention. When an historian has similarly made up his mind, his next business is to place himself in a position where he can say: 'The facts which I am now observing are the facts from which I can infer the solution of my problem.' His business is not to invent anything, it is to discover something. And the finished products, too, are differently organized. The scheme upon which the exact sciences have been traditionally arranged depends on relations of logical priority and posteriority: one proposition is placed before a second, if understanding of the first is needed in order that the second should be understood; the traditional scheme of arrangement in history is a chronological scheme, in which one event is placed before a second if it happened at an earlier time. History, then, is a science, but a science of a special kind. It is a science whose business is to study events not accessible to our observation, and to study these events inferentially, arguing to them from something else which is accessible to our observation, and which the historian calls 'evidence' for the events in which he is interested.[14]

History, then, according to Collingwood, is a science not only inasmuch as it is an organized body of knowledge, but also inasmuch as it is inferential; that is, it consists of (not merely propositions inferentially related, as does both exact and observational science, as will appear later, but) knowledges (which, as will be shown later, according to Collingwood include not merely propositions but the questions to which such propositions are answers) ordered among themselves so that one item of knowledge (a question and its answer) is related to another in such a way that one conditions the other, and itself is yet conditioned by a prior, together forming a systematic and organized whole.

According to Collingwood, this organization through inference

---

[13] See *ibid.*, pp. 250-251.

[14] *Ibid.*, pp. 251-252.

differentiates history from memory, which is not organized, not inferential at all.[15] To say that I remember writing a letter to someone last week is not an historical statement, only a statement of memory. However, if I add that my memory is not deceiving me because here is my correspondent's reply, then I am talking history because my statement about a past event is based on evidence.[16]

To say that history is inferential is to say that it is grounded in evidence which is available to oneself and to anyone else willing and able to follow the demonstration of such evidence.

> The knowledge in virtue of which a man is an historian is a knowledge of what the evidence at his disposal proves about certain events.[17]

> . . . the historian is not allowed to claim any single piece of knowledge, except where he can justify his claim by exhibiting to himself in the first place, and, secondly, to anyone else who is both able and willing to follow his demonstration, the grounds upon which it is based. This is what is meant . . . by describing history as inferential.[18]

But, different kinds of knowledge are related in different ways to the grounds upon which they are inferentially based. Collingwood criticises the conviction, which he alleges to have originated with Aristotle, that a study of the general nature of inference as such will enable one to evaluate correctly the inferential capacity of any piece of evidence solely by inspecting its form without considering the relevant subject matter.[19] He then proceeds to distinguish deductive inference and inductive inference as modes of inference differentiated by reason of the diverse subject matters to which each is appropriate.

As the main scientific achievement of the ancient Greeks was in mathematics, so also their main contribution to the logic of inference was the development of deductive demonstration in the form of syllogistic rules.[20] According to Collingwood, the chief characteristic of inference in the exact (i.e., mathematical) sciences is "a kind of logical compulsion whereby a person who makes certain assumptions

---

[15] See *ibid.*, p. 252.

[16] See *ibid.*, pp. 252-253. However, see later in *The Idea of History*, pp. 293 et seq., for identification of history with memory "but a peculiar case of memory . . . that special case of memory where the object of present thought is past thought. . . ."

[17] *The Idea of History*, p. 252.

[18] *Ibid.*

[19] *Ibid.*, p. 253. Correlate with p. 49 of *Speculum Mentis* on the logical result of classifying religion and science as species of a genus as giving to logic the status of a master-science having jurisdiction over the whole field of knowledge.

[20] See *The Idea of History*, p. 253.

is forced, simply by so doing, to make others."[21] "When a mathematician has made up his mind what the problem is which he intends to solve, the next step before him is to make assumptions which will enable him to solve it; and this involves an appeal to his powers of invention."[22] Having made his original assumptions (e.g., 'Let ABC be a triangle in which AB = BC'), if he is to go on thinking he is under a compulsion to arrive at definite conclusions resulting from that assumption. He is free either to refrain at the beginning from making such assumptions, or, having made them, to stop thinking at any time, but "what he cannot do is to make the initial assumption, to go on thinking, and to arrive at a conclusion different from that which is scientifically correct."[23]

But the development of the modern natural sciences at the close of the Middle Ages precipitated a revolt against Aristotelian syllogistic-deductive science as inappropriate to the subject matter of those sciences.[24] As the Aristotelian theory of demonstration increasingly manifested its unadaptability to the techniques of the new sciences, "by degrees, there came into existence a new logic of inference, based on analysis of the procedure used in the new natural sciences."[25]

Inductive thinking, according to Collingwood, differs from deductive in that there is no sustained compulsion from some initial assumption. The essence of the process is that a pattern is found to be made by several observations taken together, which pattern is then extrapolated indefinitely. A few points on squared paper suggest a parabola, which can then be drawn, as much as one likes, in either direction. This is the meaning of what is technically described as 'proceeding from the known to the unknown' or 'from the particular to the universal.' Collingwood insists that the extrapolating step is essentially taken without any logical compulsion; the thinker who takes such a step is free to take it or not to take it; there is nothing about the

---

[21] *Ibid.*, p. 254. The assumptions refer to Plato's 'hypotheticals' from which, as principles, the mind goes to conclusions implied by those principles; correlate with *Essay on Philosophical Method*, pp. 13-14.

[22] *The Idea of History*, p. 251. This activity of 'supposal' which dominates mathematics in principle, Collingwood identifies with imaginative activity (see *Speculum Mentis*, pp. 76-80), which activity, absolutised (abstractively) by disassociating it from the world of fact into a self enclosed world is the activity of art (*Speculum Mentis*, pp. 80-91).

[23] *The Idea of History*, p. 254; see also *Speculum Mentis*, pp. 163-169, for further elaboration of the mode of proceeding in deductive science; see also, below, Chapter 4, pp. 75-82.

[24] See *The Idea of History*, pp. 253-254.

[25] *Ibid.*, p. 254.

suggested pattern formed by his (or someone else's) observations which obliges him in that particular way, or indeed obliges him to extrapolate at all; he freely follows a suggestion, that is all.[26] Hypnotization by the prestige of Aristotelian logic has led people to see a closer resemblance between inductive and deductive thinking than really exists, and consequently between sciences of experiment and observation and exact science. In both cases there is an initial starting point and a terminal point, called the premises and the conclusion, and in both cases the premises 'prove' the conclusion. "But whereas in exact science this means that they enforce the conclusion, or make it logically obligatory, in the sciences of observation and experiment it means only that they justify it, that is, authorize anybody to think it who wishes to do so. What they provide, when they are said to 'prove' a certain conclusion, is not compulsion to embrace it, but only permission; a perfectly legitimate sense of the word 'prove' (*approuver, probare*), as there should be no need to show."[27]

Collingwood says that there is a feeling of compulsion which seemingly attaches to this permission to extrapolate freely, so that one feels not free but under obligation to do so and to do so in certain ways: "obligations which, when we inquire into their history, we find to have their roots in certain religious beliefs about nature and its creator God."[28] The motivation impelling (obligating) scientists is basically a religious one: "Take away Christian theology, and the scientist has no longer any motive for doing what inductive thought gives him permission to do. If he goes on doing it at all, that is only because he is blindly following the conventions of the professional society to

---

[26] However, if this 'extrapolation' is as free as Collingwood asserts it to be, it would seem that the astronomer Percival Lowell was fully justified in his inferring from the observed markings on the surface of the planet Mars that these markings represent canals constructed by intelligent beings to irrigate the surface of the planet, in spite of the almost universal opposition to this interpretation by equally competent observers.

[27] *The Idea of History*, p. 255.

[28] *Ibid*. In *Speculum Mentis*, pp. 160-161, Collingwood explains that the historical reason for the transition from exact science to observational science is the Christian religion. Collingwood maintains that natural laws presuppose the concept of a creator-God; see *The Idea of Nature*, pp. 8-9, 102-103; *An Essay on Metaphysics*, pp. 217-227. See also Alan Donagan's commentary on this in his *The Later Philosophy of R. G. Collingwood*, pp. 297-304. Collingwood's attitude in this matter is probably influenced by what Kant says regarding the understanding of particular empirical laws in his third Critique (see J. H. Bernard's translation of Kant's *Critique of Judgement*, London, Macmillan, 2nd revised edition 1931, p. 19) ". . . particular empirical laws . . . must be considered . . . as . . . if an Understanding (although not our Understanding) had furnished them to our cognitive faculties, so as to make possible a system of experience according to particular laws of nature."

which he belongs.''[29] The point Collingwood is here making is that any compulsion to extrapolate from observed instances is not a logical compulsion, as is the compulsion to follow through from the hypotheses in demonstration in exact sciences.[30] He seems to mean by this that in demonstrative inference the mind imposes on itself conditions according to which it must think, and after having freely submitted to such hypotheses or assumptions it is no longer free as to how it will think or to what it will conclude, whereas, on the other hand, in inductive inference the mind retains autonomy over its thinking processes and thinks not as it is forced to think but only as it is justified or authorized or 'given a clearance' to think.

With the development of historical thought in the nineteenth century another logical crisis was felt similar to the crisis felt in the sixteenth century with the rise of the observational sciences. History, as a mode of thinking with its own subject matter quite diverse from that of the exact or experimental sciences, gave rise to the felt need for the development of a mode of inference proper to itself.

Before discussing the positive characteristics of historical inference, Collingwood discusses various counterfeits of the historical method and consequently of what he calls 'bogus history.'[31]

The first type of bogus history considered by Collingwood is that which he calls 'scissors-and-paste' history, in which ''history [is] constructed by excerpting and combining the testimonies of different authorities.''[32] In this type of history the historian is at the mercy of the testimony of others; in principle he cannot go beyond what others have told him in their records; his historical account will therefore be a selection made and compiled from various previous testimonies. The historian is passive with respect to his sources. ''Scissors and paste history,'' Collingwood says, ''was the only historical method known to the later Greco-Roman world or the Middle Ages.''[33]

---

[29] *The Idea of History*, pp. 255-256.

[30] For further elaboration of Collingwood's views on inductive reasoning and experimental science, see *Speculum Mentis*, pp. 176-180. See also Chapter 4, below, pp. 75-82.

[31] See *The Idea of History*, p. 256. This method of arriving at a statement as to what something is, derived through a consideration of what it is not, is essential to Collingwood's method of procedure and is in effect an exercise in his proposed dialectical logic of question and answer, according to which questioning the criticising of imperfect forms reveals their deficiency and therefore brings to a knowledge of the more perfect form. The procedure of *Speculum Mentis* is also an example of Collingwood's dialectical, questioning procedure.

[32] *The Idea of History*, p. 257.

[33] *The Idea of History*, p. 258. ''Collingwood, who does not hide his contempt for

The second type of bogus history is what Collingwood calls 'critical history' "as it was worked out from the seventeenth century onwards, and officially acclaimed in the nineteenth as the apotheosis of the historical consciousness."[34] Collingwood says that only when the post-mediaeval reform of natural science was completed in the seventeenth century did historians begin to feel the need also to set in order their own house. As a result two new movements in historical method began: one, "a systematic examination of authorities, in order to determine their relative credibility, and in particular to establish principles according to which this determination should be carried out,"[35] the other "a movement to broaden the basis of history by making use of non-literary sources, such as coins and inscriptions and suchlike relics of antiquity which hitherto had been of interest not to historians but only to collectors of curiosities."[36]

The first of these movements resulted in documents being referred to as sources rather than authorities, indicating that they contain statements but without a commitment as to their value until a systematic enquiry into the credibility of the author had been made. 'Critical History' recognizes that it is the historian who judges. To that extent the historian is not merely passive to his documents; he has to make the value-judgment as to the veracity of the statements contained therein in terms of the trustworthiness and competence of those who made and recorded such statements.[37]

Collingwood makes two observations about critical history: first, "it is still only a form of scissors-and-paste history"[38] for, ultimately, after his evaluative work, the work of the historian will be restricted to repeating what he has critically extracted from the documents as trustworthy statements. The only improvement introduced by critical history, from this point of view, is that some form of test is made to decide whether a recorded statement is to be accepted or rejected. This having been done, the result is simply to "pass [it] as fit for the scrap-book" or to "consign [it] to the waste paper basket."[39] Thus,

---

such a conception of 'pre-fabricated historical knowledge, needing only to be gulped down and then disgorged,' calls it 'history compiled with scissors and paste.' [I.H. 246, 257.] The irony of his remark is fully deserved, for nothing could be less accurate than an analysis or enquiry that completely overlooks the actual proceedings of the historian's mind." H.-I. Marrou, *The Meaning of History*, p. 56.

[34] *The Idea of History*, p. 259.

[35] *Ibid.*, p. 258.

[36] *Ibid.*

[37] *Ibid.*, p. 259.

[38] *Ibid.*

[39] *Ibid.*

the final result of the historian's work gives us no more than previously made statements which are accepted or rejected. The point is that for all his critical activity the critical historian is still tied in principle to ready made statements of others; there are no statements in his historical account which are properly his own made on his own authority.

The second point is that critical history had already (in the nineteenth century) "in principle, been superseded by something very different."[40]

Collingwood says that many nineteenth century historians and even some eighteenth century historians recognized 'as a commonplace' that "if in some source you found a statement which for some reason could not be accepted as literally true, you must not on that account reject it as worthless. It might be a way, perhaps a well established way according to the custom of the time when it is written, of saying something which you, through ignorance of that custom, did not recognize its meaning."[41]

Collingwood credits Vico as being the first to point this out at the beginning of the eighteenth century.[42] According to Vico, "the important question about any statement contained in a source is not whether it is true or false, but *what it means*."[43] "And to ask," Collingwood continues, "*what it means* is to step right outside the world of scissors-and-paste history into a world where history is not written by copying out the testimony of the best sources, but *by coming to your own conclusions*."[44]

Thus the movement of critical history, notwithstanding that it was unable to transcend successfully the limitations of scissors-and-paste history, yet in principle transcends it inasmuch as it introduces the notion that the sources are questionable.[45]

---

[40] *Ibid.*

[41] *Ibid.*, p. 259. H.-I. Marrou (*The Meaning of History*, p. 108) provides an illustration of what Collingwood means here: "When I was a young man I wrote a whole chapter trying to show that Saint Augustine was unable to write well. This judgment merely revealed my incompetence as a young barbarian, both ignorant and presumptuous. When I was a little better informed about classical rhetoric (that subtle and refined technique of which St. Augustine was an incomparable master) I realized that what I had regarded as decadent unskilfulness was really the refinement of an art so perfectly sure of itself that it avoided quick impressions and was not reluctant to take the risk of deformed expression. With greater discernment I gradually perceived the vain pride of up-to-dateness, that pride of the barbarian who scorns whatever he does not know—and I understood a little better."

[42] See *The Idea of History*, p. 259.

[43] *Ibid.*, p. 260, emphasis added.

[44] *Ibid.* Emphasis added.

[45] See *ibid.*, pp. 260, 258, 259.

The other movement which Collingwood says gave new life to history is the archaeological movement which "was totally hostile to the principles of scissors-and-paste history, and could have arisen only when those principles were moribund."[46] Collingwood explains the transcendence of scissors and paste methods as follows:

> No very profound knowledge of coins and inscriptions is needed in order to realize that the assertions they make are by no means uniformly trustworthy, and indeed are to be judged more as propaganda than as statements of fact. Yet this gives them an historical value of their own; for propaganda, too, has a history.[47]

The next instance of 'bogus history' is that attempt found mainly in the nineteenth century, although also spilling over into the twentieth, to 'raise history to the rank of a science' modelled after the methods of the observational sciences. Reacting against the mere transshipment of ready-made information from one mind into another which characterized all forms of scissors-and-paste history, and feeling that history might be something more than this, the analogy of natural sciences came to the aid of historians and suggested what this 'something more' might be. Just as natural science since Bacon was commonly seen to begin by collecting facts, then going on to construct theories, to extrapolate the discernible patterns in the already collected facts, so the historians sought to "put together all the facts that are known to historians, look for patterns in them, and then extrapolate these patterns into a theory of universal history."[48]

Collingwood says that the acceptance of any "of these pigeonholing schemes . . . as a means for discovering historical truths not ascertainable by the interpretation of evidence" was due not to their intrinsic value but because it became an orthodoxy among what was in fact though not necessarily in name a religious community, such as was the case to some extent by Comtism and to a much greater extent by Marxism. Such historical schemes "proved to have an important magical value, as providing a focus for emotions and in consequence an incentive to action."[49]

---

[46] *Ibid.*, p. 260.
[47] *Ibid.*
[48] *Ibid.*, pp. 264-265.
[49] *Ibid.*, pp. 265-266. For Collingwood's understanding of magic as a ritual procedure to evoke certain kinds of emotive response, see his *Principles of Art*, pp. 57-77, especially pp. 65-69.

C. Positive Description of Historical Inference

"The hope that scissors-and-paste history would one day be replaced by a new kind of history that should be genuinely scientific" without attempting to mimic the observational sciences "was a well grounded hope, which," says Collingwood, "has in fact been realized."[50] This new kind of history also fulfilled the historian's hope for a kind of history which "would enable the historian to know things that his authorities could not or would not tell him":[51] that is, the hope for a history whose method would enable the historian to make genuine discoveries and so progress in historical knowledge beyond what his authorities could tell him.

Collingwood describes the method of reasoning proper to historical inference by comparing it to the procedures of crime detection. "The methods of criminal detection are not at every point identical with those of scientific history . . .";[52] ". . . in criminal detection probability is required, of a degree sufficient for the conduct of daily life, whereas in history we demand certainty. Apart from that, the parallel is complete."[53] Whereas the rules of evidence recognized in courts of law are affected by pragmatic needs so that "a jury has to content itself with something less than scientific (historical) proof, namely with that degree of assurance or belief which would satisfy it in any of the practical affairs of daily life."[54] On the other hand "the historian is under no obligation to make up his mind within any stated time. Nothing matters to him except that *his decision*, when he reaches it, *shall be right: which means, for him, that it shall follow inevitably from the evidence*."[55] Thus, the manner of inferring in crime detection and in history is exactly the same, except that in court proceedings in establishing guilt there are *ab extra* restrictions on the manner of usage of evidence and on what may be admitted as evidence.

To illustrate this distinctive mode of reasoning, Collingwood

---

[50] *The Idea of History*, p. 266.
[51] *Ibid.*
[52] *Ibid.*, p. 268.
[53] *Ibid.*, p. 270.
[54] *Ibid.*, p. 268.
[55] *Ibid.*, italics added. Charles Rescher and Carey B. Joynt have made a close analysis of "Evidence in History and in the Law," *The Journal of Philosophy*, LVI, 1959, pp. 561-578. They discuss, among other things, restrictions courts put on admissible evidence for the safeguarding of the rights of the accused; such a safeguard is not needed in the historian's handling of evidence; he looks merely to logical requirements of inference and need not have an eye to safeguarding rights.

invents "a fable" of crime detection: "Who killed John Doe?"[56] What is distinctive about the mode of reasoning is that *a conclusion is reached* as to who is the murderer *by a process of ordered questioning*. The facts and investigation are as follows: Early one Sunday morning John Doe was found lying across his desk with a dagger through his back.[57] There was a public outcry that the murderer be brought to justice; the police were hopeful of succeeding, though the only evidence they had was the little fresh green paint on the knife handle similar to that on the iron gate between the (neighboring) rector's and the murdered man's gardens.

No one entertained the possibility that a witness would enter a police station and denounce the murderer, nor that the murderer would denounce himself (the allusion is to the scissors-and-paste historian's passive dependence on witnesses' testimony presupposed). When accusers and self-denouncers did in fact present themselves, elementary common sense was enough to discredit their testimony: an elderly neighborly spinster claiming that she killed John Doe herself because he attempted to violate her; the local village poacher who said that he saw the squire's gamekeeper climbing into John Doe's study through the window.[58] Then, the rector's daughter eventually, in great agitation, made the statement that she did it herself; upon which the village constable merely rang the local inspector reminding him that the girl's fiance, Richard Roe, was a medical student, presumably knowing well where to find a man's heart, and that he had spent the night at the rectory within a stone's throw from the murdered man's house. Upon subsequent questioning, the rectory parlormaid told the Inspector that Richard Roe's shoes were very wet that morning (there had been a heavy rainstorm between midnight and 1:00 A.M. that night). Questioned, Roe admitted going out, but refused to say where and why.[59]

---

[56] See *The Idea of History*, pp. 266-268, 270-273. H.-I. Marrou (*The Meaning of History*, p. 140) is unjustly unkind to Collingwood when he refers to Collingwood's crime-detection illustration as "history of a very elementary kind, almost crude—because the event to be reconstituted (in this instance, a dagger stab) is so simple factually, easily recognizable and 'understandable' "; he forgets that the work of an example is to illustrate, not to prove anything, and that an example is a good one according as it well or ill does its work of illustrating or manifesting the point, and in doing this simplicity is essential and crudity may well be an advantage. See also Robin W. Winks, editor, *The Historian as Detective*, New York, Harper and Row, 1968, pp. 39-60.

[57] See *The Idea of History*, p. 266.

[58] See *ibid.*, p. 266.

[59] See *ibid.*, p. 270.

The questioning procedure, which resulted in the disclosure of the murderer, takes place as follows:

The village constable, (not exceptionally bright, but nevertheless) a scientific thinker (for which he need not be clever; all he needs is merely to know how to do his job, that is, to know what questions to ask),[60] starts with the rector's daughter's *statement* that she killed John Doe. He begins by using the methods of 'critical history,' torturing (that is, 'putting to the question') not her body but *her statement* that she killed John Doe. He rejects her story, because whoever did the murder requires greater strength than she has together with some knowledge of anatomy; she is known not to have attended ambulance classes.

With the rejection of her statement as untrue, the critical historian has no further interest in it; but it is precisely there that the scientific historian *begins* to be interested. He subjects the fact *that* she made the statement to tests for reactions; and he does this *by subjecting it to questions*. Why does she lie thus? Because she is shielding someone? Who? Her father? . . . or her young man? Certainly not her father (after all, the rector!); therefore her young man. Are her suspicions well founded? Possibly; he was there at the time; he is strong enough; as a medical student he knows enough anatomy.

Being not as thorough in his job as a more experienced man, the village constable stops his questioning there. In doing so, his mistake was that he lost sight of the original question 'Who killed John Doe?,' stopping short at the question 'Whom does this girl suspect?'[61] Inspector Jenkins, not necessarily a more clever man, but one who has learned his job more thoroughly, goes about the interrogation as follows: Why does the rector's daughter suspect Richard Roe? Most likely she knows that he was involved in something queer which happened at the rectory that night; he was out in the storm, which fact alone would be sufficient to arouse the girl's suspicions. Did he kill John Doe? If so, when? After the thunderstorm broke? . . . or before? Not before; his tracks go both ways in the mud of the rectory garden path; they begin a few yards from the garden door; going away from the house, so that is where he was, and the direction in which he was going when the downpour began. Did he carry mud into John Doe's study? No; none there. Did he take off his shoes before entering the study? Impossible. What position was John Doe in when he was

---

[60] See *ibid*.
[61] See *ibid*., p. 271.

stabbed? He could only have been leaning right forward, either asleep or awake; if asleep, the murderer need only step forward and plunge in the dagger; if awake, only a little more caution needed. So, in either case, no time for removing shoes before entering the study. So, absence of mud in the study lets Richard Roe out as murderer.

But, then, why did he go out into the garden? Something must have been going on out there; something queer that we do not yet know about. What could it have been? If the murderer came from the rectory (the paint on the dagger like the fresh paint on the garden fence suggests he did), and if Richard Roe saw him from his window (the murderer got to John Doe's house before the rain fell, whereas Richard was caught in it ten yards from the garden door) what would follow? He probably returned to the rectory afterwards. But no tracks; why? Because he was sufficiently familiar with the garden, and the path, to stay on the grass all the way, even in the darkness. If so, he knew the rectory very well and spent the night there.

Was it the rector himself?[62]

Why does Richard refuse to say why he went into the garden? It can only be to shield someone from trouble, and almost certainly trouble connected with the murder. Not himself, so who else? Might it be the rector? If so, how would the rector have done it? Quite easily: go out about midnight in tennis shoes and gloves; no gravel on the rectory paths to make noise while walking; reach the iron gate; does he know there is wet paint on it? Probably not, since it was painted only after tea. So he grabs it, gets paint on the glove, probably on the jacket too. Then he walks on the grass to Doe's study window. Doe is leaning forward in his chair, maybe asleep. No difficulty for a good tennis player to place his left foot inside the study window, right foot to the right, grab the nearby dagger-like instrument, left foot forward, then plunge it in.

What was John Doe doing at his desk? There was nothing on it when the murder was discovered. Does a man sit at an empty desk? There must have been something there. What was it?

What do we know about Doe from Scotland Yard records? He is a blackmailer. Was he blackmailing the rector? . . . and perhaps gloating over the letters, or whatever they were, he had? Did the rector not possibly find him asleep on top of them?

But this line of questioning is taking us away from our main concern, namely, 'Who killed John Doe?,' so let us return to it. Given

---

[62] See *ibid.*, p. 272.

that the rector is in there, in Doe's study, what does he do to get out? It is now raining heavily; he has to go back through it; more paint picked up off the gate; walking on the grass avoids bringing mud into the house. Inside, he is all soaked, and with paint on his gloves also. He wipes any paint off the door-knob and locks the door. He puts the letters (or whatever it was that he removed from Doe's study desk) and the stained gloves in the hot-water furnace; the ashes may still be in the dustbin. He puts all clothes in the bathroom cupboard (they will be dry in the morning), but the jacket will be hopelessly out of shape when dry; if it has paint on it nothing else to do but destroy it (a difficult task in a house overrun with women). If no paint on it, he could quietly give it to a poor man.[63]

But is this story true or not? Two questions have to be answered to decide: first, can the ashes of the gloves, and their metal buttons, be found? This would confirm the story; and finding a lot of writing-paper ash would confirm the blackmail aspect also. Second, where is that jacket? If there is the tiniest speck of John Doe's paint on it the case is closed.

In the rectory dustbin there were a lot of ashes, mostly from writing paper, but also including some from leather, probably from a pair of gloves. Among the ashes were also found metal buttons bearing the name of a famous Oxford glove-maker. The jacket, which the rector gave to a deserving parishioner on Monday morning, had John Doe's paint on the right cuff, and was hopelessly out of shape by a recent wetting. (The Detective-Inspector was later severely reprimanded for permitting the rector to see the direction in which his enquiries were leading him, thus giving the rector the opportunity to evade custody by suicide).[64]

The papers in front of John Doe at the time of the murder bore the address John Doe, Esq. in the rector's wife's handwriting. Doe had been blackmailing the rector on the ground that the rector's daughter was born six months after the marriage; only then did the rector know who his wife's seducer was, and who was the father of his supposed daughter.[65]

Having presented this illustration of the process of questioning by his analysis of the John Doe murder case fable, Collingwood then proceeds to emphasise two points about this process which, he

---

[63] See *The Idea of History*, pp. 272-273.
[64] See *ibid.*, p. 268.
[65] See *ibid.*, p. 267.

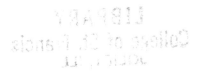

says, ''is the dominant factor in history, as it is in all scientific work.''[66]

The first point: Every step in the argument (i.e., inference; remembering that Collingwood has said that it is neither inductive nor deductive) depends on asking a question.

> The question is the charge of gas, exploded in the cylinder head, which is the motive force of every piston-stroke. But the metaphor is not adequate, because each new piston-stroke is produced not by exploding another charge of the same old mixture but by exploding a charge of a new kind. No one with any grasp of method will go on asking the same question all the time, 'Who killed John Doe?' He asks a new question every time. And it is not enough to cover the ground by having a catalogue of all the questions that have been asked, and asking every one of them sooner or later; they must be asked in the right order.[67]

Collingwood here refers to ''Descartes, one of the three great masters of the Logic of Questioning (the other two being Socrates and Bacon)''[68] as having insisted on this cardinal point of ordered in-

---

[66] *The Idea of History,* p. 273. Jacques Barzun's *The Delights of Detection* (New York, Criterion Books, 1961) presents a collection of detective stories intended to highlight the logical and intellectual element in crime detection which gives to it its popular fascination. ''The Nine-Mile Walk,'' by Harry Kemelman, pp. 199-208 of Barzun's collection, could well substitute for Collingwood's illustration above.

[67] *The Idea of History,* p. 273.

[68] *Ibid.* See also *The New Leviathan,* London, Oxford University Press, 1942, p. 2, paragraph 1.19: ''We of the twentieth century hold ourselves bound to the tradition in these matters laid down by Bacon and Descartes in the seventeenth: to speak not merely 'to the subject' but 'to the point'; to divide our subject into parts, to arrange the parts in such an order that what is said about each prepares the way for what is said in the next; and to say about each not all we know but only what need be said for the sake of that preparation.'' This commitment echoes Descartes' summary of his Regulae as this appears in his Discourse on Method, Pt. 2 (p. 92 of Vol. 1 of the Haldane-Ross edition of Descartes' *Philosophical Works,* ignoring paragraph breaks): ''. . . the four [rules] which I shall state [to be] quite sufficient, provided that I adhered to a firm and constant resolve never on any single occasion to fail in their observance. The first of these was to accept nothing as true which I did not clearly recognise to be so [the critical questioning character]: that is to say, carefully to avoid precipitation and prejudice in judgments, and to accept in them nothing more than what was presented to my mind so clearly and distinctly that I could have no occasion to doubt it. The second was to divide up each of the difficulties [read questions for difficulties for Collingwood's position] which I examined into as many parts as possible, and as seemed requisite in order that it might be resolved in the best manner possible. The third was to carry on my reflections in due order, commencing with objects that were the most simple and easy to understand, in order to rise little by little, or by degrees, to knowledge of the most complex, assuming an order, even if a fictitious one, among those which do not follow a natural sequence relatively to one another [for reflections and objects, read questioning and questions for Collingwood's position; also Descartes' allowance of a fictitious order aligns with Collingwood's right

terrogation in scientific method, which, Collingwood repeats, has so far been ignored by modern works of logic. He accuses modern logicians of conspiring "to pretend that a scientist's business is to 'make judgments', or 'assert propositions', or 'apprehend facts', and also to 'assert' or 'apprehend' the relations between them," which suggests that "they have no experience whatever of scientific thinking,

---

although false answer which allows the questioning process to get ahead]. The last was in all cases to make enumerations so complete and reviews so general that I should be certain of having omitted nothing [for Collingwood the seried questioning is seen as an ordered body of knowledge in retrospect, when all the questions have been asked and answered in their due order]." Further basis in Descartes' works for Collingwood's acknowledgement of him as an authority on questioning may be found in the last part of Rule 12 of the Regulae (p. 48 of Haldane-Ross edition, Vol. 1) in which Descartes gives the projected plan of this (uncompleted) work. The first (and completed) twelve rules concern simple propositions. "But as to 'questions' some of them can be perfectly well comprehended, even though we are ignorant of their solution; these we shall treat by themselves in the next twelve rules. Finally there are others [i.e., other questions] whose meaning is not quite clear, and these we reserve for the last twelve [rules]." Again in regard to questions (Rule 13; see p. 52, Vol. 1 of the Haldane-Ross edition of Descartes' works) Descartes says: "However, though, in every 'question' something must be unknown otherwise there is no need to raise it, we should nevertheless so define this unknown element by means of specific conditions that we shall be determined towards the investigation of one thing rather than another." Descartes then proceeds to detail how the specific conditions of the unknown element are to be defined, and to warn of the pitfalls to be avoided in doing this.

Collingwood's reference to Bacon as one of the three masters in questioning doubtless refers to the rules of induction (later perfected by J.S. Mill) of his new organon, intended to displace the Aristotelian logic as the genuine method of the new science of the Renaissance. The negative moment, the expurgation of the mind of its four idols, parallels Descartes' methodic doubt as the preliminary step to the positive constructive work. The three tables, and the fourth step, the process of exclusion, are the rules according to which nature is to be 'put to the question' for the purpose of controlling nature and its forces. Bacon was a lawyer by profession and it could be asked how far he conceived the scientific questioning of nature in terms of the lawyer's cross-examining of a recalcitrant witness. The role of separating positive and negative instances, of recognizing alternatives, is central both to the Baconian method and to Collingwood's notion of questioning (see *Speculum Mentis*, p. 78; *The New Leviathan*, p. 74, paragraph 11.12). Negative instances assume a positive value in dialectical thought such as Collingwood's. See Bernard Bosanquet, *The Essentials of Logic* (London, Macmillan, 1897), ch. 8, for the role of negation in thought, especially pp. 136 et seq., for the defining function of the negative, and for the highly positive value of the negative moment of the crucial experiment (just when x ceases, just then y begins), pp. 133-135. For a good treatment of the positive function of negation in dialectical thinking, see Otis Lee, "Dialectic and Negation," *The Review of Metaphysics*, I (1947), pp. 3-23.

The inclusion of Socrates as one of the masters of questioning needs no explaining. But it is surprising that Collingwood omits reference to Aristotle's treatment of the order of Scientific Interrogation in his *Posterior Analytics*, Bk. 2, Chapters 1 and 2 (see L.-M. Régis, *Epistemology*, New York, Macmillan, 1959, pp. 128-137, for an exposition of the Aristotelian scientific interrogation). This omission is doubly inexcusable in Collingwood's writing, not only as a philosopher but as one insisting on the basic requirement of historical accuracy in philosophy. See below, Chapter 8, pp. 318 et seq.

and wish to palm off, as an account of science, an account of their own haphazard, unsystematic consciousness."[69]

The second point to be noted about this procedure is that "these questions are not put to one man by another man, in the hope that the second man will enlighten the first man's ignorance by answering them. They are put, like all scientific questions, to the scientist by himself."[70] Collingwood says that this is Plato's designation of the Socratic idea of thought, defined as 'the dialogue of the soul with itself,' and adds that the Platonic context makes it clear that the dialogue referred to was a process of question and answer. When Socrates taught his pupils by asking them questions he was teaching them how to ask questions of themselves and showing them how the most obscure subjects can be amazingly illuminated by asking oneself intelligent questions about them instead of simply gaping at them after the fashion of modern anti-scientific epistemologists who advocate that our minds should be made 'a perfect blank' so as to 'apprehend the facts.' The allusion is to the positivistic passive approach of the mind to "ready-made facts."[71]

D. The Autonomy Proper to History as a Science: The Historian as His Own Authority

Collingwood has pointed out that 'scissors-and-paste' history, even in its most refined form as critical history, depends on the statements of others, that is, that what has not been said by someone previously can never be said by the historian. Witnesses are required, and the critical work is to apply criteria to determine whether their ready-made statements are to be accepted as trustworthy or rejected as untrustworthy. So, at its best, 'scissors-and-paste' history is merely a restatement of the statements of others; there being no room for statements arising from the historian's own initiative, that is, for statements that would be authentically his own.

On the contrary, scientific history contains no ready-made statements at all. Confronted with the ready-made statement of an authority, the scientific historian does not ask himself 'Is his statement true or false?', and consequently 'Shall I incorporate it into my historical account of this or that subject, or not?' The question he asks himself is 'What does it mean that this statement is made?' This is not the

---

[69] *The Idea of History*, pp. 273-274.
[70] *Ibid.*, p. 274.
[71] See *ibid.*

same as the question 'What does the one who made the statement mean by making it?', although this latter question will not be without relevance to the inquiry. When he asks, 'What does it mean that so-and-so makes this statement?', he is asking, 'What light is thrown on the subject in which I am interested by the fact that this person made this statement, meaning by it what he did mean?' This is equivalent to saying that the scientific historian is interested in statements that are made, not precisely as they are statements, but precisely as they are evidence. The scientific historian thus takes ready-made statements not as true or false accounts of what has happened but "as other facts which, if he knows the right questions to ask about them, may throw light on those facts."[72]

Thus, whereas the explanation of facts in the observational and experimental sciences, by the process of inductive inference, in terms of some trans-factual (universal, ideal) order of which the explained facts are more or less irrelevant instances, the explanation of facts in historical reasoning is by means of other facts which are uncovered by means of the question and answer process, which discloses these other facts precisely as they are the explanation of the facts 'under question.' History as a science does not merely chronologically catalogue facts (that is, tell that they happened and when and where they happened) but must explain them, that is, tell why they happened. Thus, the historian, in dealing with facts, deals with other facts as their explanation. And the procedure of going from facts to other facts as their explanation is the work of historical inference. History has to be able to infer if it is to merit the title of scientific, and to infer is to disclose the fact or facts which explain the fact with which we are concerned.

Thus, applying this to the John Doe murder fable, the rector's daughter makes the statement that she killed John Doe. The critical historian discards the statement when he knows that it is false. But the scientific historian begins his proper work precisely where he stops considering the statement as a true or false account of what the rector's daughter did, and begins treating it as a fact that she did make that statement (whether true or false), which fact is vouched for on his own authority, as 'ear-witness', and which he may be able to use to throw light on 'Who killed John Doe?' It is a fact useful to him because he knows what questions to ask about it, beginning with the question 'Why does she tell this story?' It is the question he puts to it

[72] *Ibid.*, p. 275.

that turns it into evidence, that is, makes it able to illuminate the point of interest. A 'scissors-and-paste' historian is interested only in the content of the statement, that is, 'what is said' in the statement. The scientific historian is interested in the fact that the statement is made, that is, he regards it as a fact in its own right calling for explanation, which explanation can be found by questioning that fact correctly.

When a historian reads, or hears a statement being made, he is simply receptive of what is said, and in no way the author of it. But when he makes the statement to himself that 'I am now reading (or hearing) a statement to such and such effect' he is himself the author of that statement, making it on his own authority. This autonomous statement is the point of departure for the scientific historian. The evidence from which the constable infers that the rector's daughter suspects Richard Roe is not her statement that she killed John Doe but his own statement that 'the rector's daughter tells me that she killed John Doe.'[73]

Collingwood says that not only does the historian get conclusions from his own autonomous statements of the fact that statements have been made but also that he can get conclusions even when no statements at all are made to him. This is because, being autonomous, he is not dependent on ready-made statements; the premisses from which he argues are his own autonomous statements, and it is not necessary that his own autonomous statements be themselves statements about other statements. Thus, the premisses from which the Detective Inspector argued to conclude to Richard Roe's innocence were all premisses which consisted of the Detective Inspector's own statements, autonomous statements, resting on no authority but his own, not one of them being a statement about statements made by anyone else. The essential points in his inferential argumentation were as follows: First, Richard Roe got his shoes muddied while going away from the rectory, second, no mud was found in John Doe's study, and third, the circumstances were such as not to have allowed him to stop to clean or to remove his shoes. Each of these three points is itself the conclusion of an inference, and the statements which made these points evident were no more statements about other people's statements than are the three points themselves.

Furthermore, the ultimate case against the rector did not depend upon the Detective Inspector's statements about statements made by other persons, but depended solely upon the presence of certain

[73] See *ibid.*, pp. 275-276.

objects in a certain dustbin, and a smear of paint on the cuff of a clerically styled jacket shrunk by wetting. All these facts, from which it was inferred that the rector killed John Doe, were vouched for, or authorized by, his own observations.[74]

Collingwood then corrects a possible misunderstanding. It is not being denied that ready-made statements are useful to the scientific historian, but only that he depends on such statements. This settles the controversy between those who maintain that history is ultimately dependent on written sources and those who maintain that it can be constructed from unwritten sources. He adds that historians schooled to regard history as a scissors-and-paste affair began timidly to see the possibility of something quite different.[75]

E. Evidence Which Is the Basis of Scientific History Is Always in Relation to a Definite Question

In scientific history, sources are valuable only inasmuch as they can be made to function as evidence. "If history means scientific history, for 'source' we must read 'evidence'."[76] But it is not easy to say exactly what evidence is. "When we try to define 'evidence' in the same spirit in which we defined 'sources,' we find it very difficult."[77] There seems to be no a priori criterion to decide whether, say, a certain book will or will not provide evidence about a given subject; indeed there is no reason why evidence should be restricted to books or written sources. Collingwood says that indexes and bibliographies of sources are records not of evidence but of previous discussion on a topic which the historian may or may not take as a starting point for his own investigations. If they do function as evidence this will derive from what the historian himself does in regard to them, or, more exactly, what, by reason of his skill in questioning them, he can make them do for him as regards illuminating a given subject.

In the John Doe fable there is only one obvious characteristic common to all the items of evidence used by the Detective-Inspector in his investigations: *they are all things which he observed by himself*. But, if we ask what kind of things they are, it is not easy to give the answer. They include such things as the existence of certain footprints in certain mud, their number, position, direction, their resemblance to prints produced by a certain pair of shoes and the absence of any

---

[74] See *ibid.*, p. 276; 280.
[75] See *ibid.*, pp. 276-280.
[76] *Ibid.*, p. 279.
[77] *Ibid.*

others, the absence of mud on the floor of a certain room, the position of a dead body, the position of a dagger in its back, the shape of the chair in which the body was sitting, and so on; a most variegated collection. Thus, it can be safely said, says Collingwood, that no one could possibly know what could or could not find a place in that list "until he had got all his questions not only formulated but answered."[78] That is, the list of items constituting the evidence is *a posteriori* or *post factum* in relation to the process of interrogation, inasmuch as it is the very process of interrogation which makes such items function as evidence; they are not evidence in themselves but only in relation to the process of inquiry, and to the manner in which that inquiry makes use of them. "In scientific history anything is evidence which is used as evidence, and no one can know what is going to be useful as evidence until he has had occasion to use it."[79]

According to 'scissors-and-paste' history, all extant statements about a given subject constitute potential evidence, and actual evidence is that selection of those statements which the historian decides, after critical inquiry as to trustworthiness, etc., to incorporate into his account. But, "in scientific history the idea of potential evidence disappears; or, if we like to put the same fact in these other words, everything in the world is potential evidence for any subject whatever."[80] The mentality of the 'scissors-and-paste' historian is to think that before one can know which facts are actually to be of service one needs to have already some inventory of facts which might possibly be of service. But one who understands the nature of scientific thinking, whether historical or not, will not regard this as a difficulty. For when a historian asks a question he already has some provisional idea of what he wants as evidence.

> . . . every time the historian asks a question, he asks it because he thinks he can answer it; that is to say, he has already in his mind a preliminary and tentative idea of the evidence he will be able to use. Not a definite idea about potential evidence, but an indefinite idea about actual evidence.[81]

Collingwood adds that "to ask questions which you see no prospect of answering is the fundamental sin in science."[82] He says further that "question and evidence, in history, are correlative. *Anything is*

[78] *Ibid.*, p. 280.
[79] *Ibid.*
[80] *Ibid.*
[81] *Ibid.*, p. 281.
[82] *Ibid.*

evidence which enables you to answer your question—the question you are asking now.''[83] Note that Collingwood is here saying that the actual criterion as to whether something is evidence or not is its ability to enable you to answer your question. A question which you think you will, now or later, have evidence for answering is a sensible question, the only type of question a scientifically competent man should ask. If you think you have the evidence here and now the question is an actual one, such as 'What position was John Doe in when he was stabbed?' But if you think you will have it sooner or later, the question is a deferred question, such as 'Who killed John Doe?'

Following through on the analogy of historical inquiry with crime detection, Collingwood has severe criticism for the Sherlock Holmes approach of 'fact grubbing,' ''the 'human blood-hound' who crawls about the floor trying to collect everything, no matter what, which might conceivably turn out to be a clue.''[84] On the other hand, he admires Monsieur Hercule Poirot who insists that ''the secret of detection was to use what . . . he called 'the little grey cells.' ''[85] Poirot means that ''you can't collect your evidence before you begin thinking . . . because thinking means asking questions (logicians please note), and nothing is evidence except in relation to some definite question.''[86]

According to Collingwood, the difference between Poirot and Holmes reflects the changed mentality in regard to the understanding of historical method in the last forty years (writing in 1939), and the revolutionary principles which overturned 'scissors-and-paste' history had by now become common property.[87]

F. Summary and Comment on the Foregoing

Collingwood has insisted on the scientific character of history, by which he means that it is an organized body of knowledge, that is, that one thing follows *inferentially* from another, that facts are connected with other facts in a serial order of explaining to explained, and that the explaining fact is inferred from the fact to be explained by a process of reasoning which consists in a seried order of questions and answers.[88]

---

[83] *Ibid*. Emphasis added.

[84] *Ibid*.

[85] *Ibid*.

[86] *Ibid*.

[87] *Ibid*., p. 282.

[88] H.-I. Marrou in *The Meaning of History*, p. 63, likewise speaks of the ''logical priority of the 'question' which the historian poses in the presence of the documents.'' He says that 'to lay hold of [the mysterious Past] the historian must encompass it

In this process of serial questioning, the mind actively confronts a subject or topic, not simply to 'gape' at it and passively receive impressions from it or merely to 'apprehend' what is there to apprehend (against Positivism), but in the way in which a crime detective, or judge and jury, confronts a crime suspect, that is, in a questioning manner, by 'putting him to the question,' and by such activity of questioning to draw from him the evidence whereby the fact is explained. Active questioning turns the statements made by others, and even physical objects, into evidence whereby a question is answered and a fact explained. This question-answer type of inference, according to Collingwood, differs from the traditionally recognized forms of induction and deduction. Deduction proceeds from a trans-factual, hypothetical principle and concludes to the implications contained in the hypothesis; the principle, the process and the conclusion are therefore only incidentally referred to fact if they are at all referred to fact. Induction begins from fact but proceeds to find the explanation of the fact in a trans-factual, universal, abstract entity, of which the fact is merely illustrative. But in historical inference, facts are inferentially explained by other equally factual things or events. The explanation of or answer to the question 'Who killed John Doe?' is the factual implementation of the rector's intention to rid himself of a troublesome blackmailer together with all the factual details of such implementation.

There seems to be no doubt that Collingwood recognizes an as yet undeveloped logic of question and answer proper to historical inquiry. Louis O. Mink says that Collingwood's "logic of question and answer is not a substitute for propositional logic but rather a theory of the process of inquiry; its locus is, so to speak, not logic but reflection on logic."[89] This statement, though difficult to deny, seems at odds with Collingwood's obvious reprimand to the textbook logicians for neglecting this special mode of inference.

---

tightly within a network of questions . . . compelling it to reveal itself frankly. . . . Logically, the process of the elaboration of history is set in motion not by the existence of documents but by an initial step, the 'posed question' inscribed in the choice, deliberation and conception of the subject."

[89] *Mind, History, and Dialectic*, Bloomington, Indiana University Press, 1969, p. 123.

Chapter Two

Collingwood's Identification of Knowledge with Question and Answer

In his *Autobiography*,[1] published in 1939, Collingwood treats extensively of his theory of questioning, and of his position that knowledge is essentially a matter of question and answer. He shows the fundamental character of the theory of questioning in his thought and gives an account of the genesis and evolution of his thought on the matter.

A. Collingwood's Question and Answer Theory of Knowledge Rooted in His Opposition to the Oxford Realists

Collingwood's initiation into philosophy was through the lectures and tutorship of his Oxford professors, E. F. Carritt, John Cook Wilson, etc., who were members of the movement known as English neo-realism.[2] He was thoroughly indoctrinated in their methods and principles, though from the start his adherence was not without reservations. His dissatisfaction with the movement was precipitated by G. E. Moore's article, "The Refutation of Idealism." Moore, in Collingwood's estimation, was the Cambridge equivalent of the Oxford

[1] R. G. Collingwood, *An Autobiography*, London, Oxford University Press, 1939. See Pierre Fruchon, "Signification de l'histoire de la philosophie selon l'Autobiographie de Collingwood," *Les Études Philosophiques*, XIII (1958), pp. 143-160, for a resumé and favorable evaluation of Collingwood's Autobiography, referred to as "un document de première importance pour celui qui s'intéresse aux problèmes posés par l'histoire, et en particulier par l'histoire des idées . . ." (p. 143).

[2] See *An Autobiography*, p. 22.

realist type. In this article, Moore in criticising Berkeley attributes to Berkeley a position which in fact, according to Collingwood, is the very position Berkeley himself opposes. All one has to do to verify this is to place the two texts, Berkeley's and Moore's versions of it, together and compare them. And he says that the same thing happened in Cook Wilson's attacks in his lectures on positions he alleges to have been Bradley's. A comparison of Cook Wilson's statement of Bradley's position with Bradley's statement itself would show that the two did not square up.

Collingwood tells us that in his early student days he preferred to distinguish and qualify rather than, as he says he would have done at a more mature stage, to break his attachment to the school convicted of such gross errors on such important matters. He argued that the 'realists' were professing philosophy and not history, and that their business as philosophical critics was to show whether a certain doctrine was sound or not and that this could be done without regard to their historical inaccuracies as to whether a certain author did or did not hold such a doctrine. The philosophical issue was thus one thing and the historical issue another.[3]

Collingwood says that he felt logically bound to remain a realist until he had satisfied himself either that the positive doctrines of the school were false, or that its critical methods were unsound. He did not answer these questions until he took his degree and commenced working as a philosophy teacher. Then it became clear to him that Cook Wilson's positive teaching could not stand up beside his (Cook Wilson's) own critical methods.

In 1914, when his academic life was interrupted by the war, Collingwood faced the three following alternatives: (1) If there was a logical connection between the positive doctrine and the critical methods this was fatal for both of them. But, if there was no such logical connection, then, either (2) the positive teaching might be mistaken and the critical methods valid, or (3) vice versa, the positive teaching might be sound but the critical methods invalid.[4]

B. Origin of Collingwood's Theory of Questioning in His Archaeological Practice

When he became a teacher of philosophy, Collingwood spent his summers studying history and archaeology. From 1913 onwards, he

---

[3] See *An Autobiography*, pp. 22-23.

[4] See *ibid.*, p. 23.

was directing his own excavations. In his archaeological work he says that he found himself conducting thought experiments in the knowledge laboratory.[5] The following is his account of his method of procedure in archaeological work:

> . . . I found myself . . . at first asking myself a quite vague question, such as: 'was there a Flavian occupation on this site?' then dividing that question into various heads, and putting the first in some such form as this: 'are these Flavian sherds and coins mere strays, or were they deposited in the period to which they belong?' and then considering all the possible ways in which light could be thrown on this new question, and putting them into practice one by one, until at last I could say, 'There was a Flavian occupation; an earth and timber fort of such and such a plan was built here in the year a plus or minus b and abandoned for such and such reasons in the year x plus or minus y. Experience soon taught me that under these laboratory conditions *one found out nothing at all except in answer to a question*; and not a vague question either, but a definite one. . . . What one learned depended not merely on what questions one was asking; so that a man who was asking questions of one kind learned one kind of thing from a piece of digging which to another man revealed something different, to a third something illusory, and to a fourth nothing at all.[6]

Collingwood says that in these reflections on his activity as an archaeologist he merely rediscovered for himself, in the practice of historical research, principles which Bacon and Descartes had stated three hundred years earlier in regard to the natural sciences. Each had clearly showed (i) that knowledge comes only by answering questions, and (ii) these questions must be the right questions, and (iii) they must be asked in the right order.[7]

C. Confrontation with the Oxford Realists: Active Questioning vs. Passive Compresence

Collingwood criticises the Oxford realists for talking as if knowing were simple intuiting or simple apprehending of some 'datum'

---

[5] See *ibid.*, pp. 23-24. E. Gilson, in his *Being and Some Philosophers*, Toronto, Pontifical Institute of Mediaeval Studies, 1949, p. 6, defends the usage of the term 'experiment' with respect to ideas. This is intimately tied up with the theory of supposal which plays a large part in Collingwood's questioning theory, as will be shown later.

[6] *An Autobiography*, pp. 24-25. Emphasis added. Note the formalizing function of the question. "The same data . . . can yield a variety of different kinds of information depending upon the questions one asks. More importantly, if one asks *no* questions, it does not seem that the data can yield any information." L. Armour, *The Concept of Truth*, Assen, Van Gorcum, 1969, p. 213.

[7] See *An Autobiography*, p. 25.

confronting the mind, to which the mind must submit itself when it knows. He also criticises Moore at Cambridge for speaking of 'the transparency' of the act of knowing, and likewise Alexander at Manchester for describing knowledge as the simple 'compresence' of two things, one of which is a mind. All these positions regarded knowledge as a simple condition or situation, devoid of diversities and complexes, 'nothing except just the act of knowing.' They conceded that there was activity in knowledge, inasmuch as a man might have to work in very complicated ways to put himself 'in the position' in which something could be 'apprehended.' But once having attained that position there was nothing more to do but simply 'to apprehend' it, or perhaps to fail to apprehend it.[8]

In opposition to this realist attitude Collingwood says:

> The questioning activity, as I called it, was not an activity of achieving compresence with, or apprehension of, something; it was not preliminary to the act of knowing; it was one half (the other half being answering the question) of an act which in its totality was knowing.[9]

From these reflections, and from his estimation of Moore's 'refutation' of Berkeley and Cook Wilson's 'refutation' of Bradley, he concluded that Oxford philosophy stood in need of a background of scholarship which would form in its students such habits of mind as would safeguard them from such falsifications of the positions of other thinkers.[10]

The first rule to be followed in reading, expounding or commenting on a text[11] is: 'never accept criticism of any author before satisfying yourself of its relevance.' Collingwood insisted that his pupils satisfy themselves, by firsthand study of what is said or written, just what the speaker's or writer's philosophy is before criticising it, or before accepting anyone else's criticism of it. In his first lectures on Aristotle's *De Anima*, he concentrated on the question, 'What is Aristotle saying and what does he mean by it?,' foregoing, however alluring it may be, the question, 'Is it true?,' and leaving for other teachers the further business of criticising its doctrine. By this approach, he intended to train his pupils in a scholarly approach to a philosophical text.

Collingwood tells us that, with the outbreak of the 1914 war, he

---

[8] See *ibid.*, pp. 25-26.
[9] *Ibid.*, p. 26.
[10] See *ibid.*, pp. 26-27.
[11] See *ibid.*, pp. 27, 74.

had not yet answered the three questions regarding the conflict between the doctrine and the method of the realist,[12] but he felt that he had made a significant step in that direction since he recognized that he had made himself proficient in a certain kind of research and found out how to use it as a laboratory testing ground for epistemological theories.[13] This perfected technique, together with what he says might be called his 'early manner' of philosophical teaching, tended towards an attack on realism as *a philosophy which erred through neglecting history*. A foreshadowing of this coming attack appeared in the form of such statements as the following:

> You must pay more attention to history. Your positive doctrines about knowledge are incompatible with what happens, according to my own experience, in historical research; and your critical methods are misused on doctrines which in historical fact were never held by those to whom you ascribe them.[14]

Collingwood remarks that the interruption of his academic life by the war served to focus his attention on the as yet confused issues so as to further clarify them; which brings us to his meditations on the Albert Memorial.[15]

D. Meditations on the Albert Memorial

Collingwood's war work with the Admiralty intelligence during the 1914-1918 war required him to walk daily across the Kensington Gardens past the Albert Memorial. He says that by degrees it began to obsess him. He likened it to Wordsworth's leechgatherer, and, for him, it took on a strange air of significance. "Everything about it was visibly mis-shapen, corrupt, crawling, verminous"; he says that "for a time I could not bear to look at it, and passed with averted eyes."[16] Recovering from this nausea, he forced himself to look at it, and "to face day by day the question: a thing so obviously, so incontrovertibly bad, why had Scott done it?"[17] He proceeded to grope for possible answers. It would be mere tautology to say that Scott was a bad architect and an evasion by a *suggestio falsi* to say that there is no accounting for tastes. The question then subdivides into other questions. "What relation was there . . . between what he had done and

---

[12] See above, pp. 38 et seq.
[13] See *An Autobiography*, p. 28.
[14] *Ibid.*, p. 28.
[15] See *ibid.*, p. 29.
[16] *Ibid.*
[17] *Ibid.*

what he had tried to do? Had he tried to produce a beautiful thing; a thing . . . which we should have thought beautiful?'' If so he had of course failed. ''But had he perhaps been trying to produce something different?'' If this is the case he may well have succeeded. Again: ''If I found the monument merely loathesome, was that perhaps my fault? Was I looking in it for qualities it did not possess, and either ignoring or despising those it did?''[18]

Collingwood says that his daily communings with the Albert Memorial, together with his work in archaeology, impressed upon him the importance of the activity of questioning in knowledge. It further resulted in his impossibility of being contented with the realist's intuitionist theory of knowledge (a simple inspection of a datum already there before the mind simply to be contemplated). He says that the effect of these considerations brought about a revolt in his mind against the current logical theories similar to the reaction against the scholastic logic produced in the minds of Bacon and Descartes when they reflected on the experience of scientific research as this began to take shape in the late sixteenth and early seventeenth centuries. Bacon's *Novum Organum* and Descartes' *Discours de la méthode* began to have a new significance for him in the light of these considerations.

> They [Bacon's and Descartes' works referred to] were the classical expressions of a principle in logic which I found it necessary to restate: the principle that a body of knowledge consists not of 'propositions', 'statements', 'judgments', or whatever name logicians use in order to designate assertive acts of thought (or what in those acts is asserted: for 'knowledge' means both the activity of knowing and what is known) but of these together with the questions they are meant to answer; and that a logic in which the answers are attended to and the questions neglected is a false logic.[19]

Collingwood then gives an account of how his thought on the relevance of questioning in the theory of knowledge developed as his meditations on the Albert Memorial continued. First, however, he interrupts to clarify terminology. He says that he proposes to use the word 'proposition' for the 'entity' which the idealistic logicians call 'judgment' and the realists call 'statement', and that it will denote a

---

[18] *Ibid.*, p. 30. I think that here is an early instance of Collingwood's later notion of history as a rethinking of the thoughts of another. Also, perhaps, there is the influence of Vico's insistence on the need to penetrate *from inside* the mentality of other cultures, and a foreshadowing of Collingwood's later doctrine regarding history as knowledge by 'insight.'

[19] *Ibid.*, pp. 30-31.

logical, not a linguistic, entity. Collingwood's biographical account of his evolving thought is as follows:

> I began by observing that you cannot find out what a man means by simply studying his spoken or written statements, even though he has spoken or written with perfect command of language and perfectly truthful intention. In order to find out his meaning you must also know what the question was (a question in his own mind, and presumed by him to be in yours) to which the thing he has said or written was meant as an answer.[20]

Collingwood stresses that in his understanding of the matter there is a strict correlation between question and answer,[21] so that a proposition is not an answer, or, at any rate, not the right answer, to any question that might have been answered otherwise. He explains by stating that "a highly detailed and particularized proposition must be an answer, not to a vague and generalized question, but to a question as detailed and particularized as itself."[22] A question such as 'Why won't my car go?' must be broken down into detailed and particularized questions of which it is really only a summary, such as: 'Is it because No. 1 plug is not sparking?' . . . or No. 2 plug?, etc. And when it can be said that No. 1 plug is functioning, this is not recording failure to answer the sustained question 'Why won't my car go?' but success in answering the question 'Is the stoppage of my car due to failure in No. 1 plug?'

This 'principle of correlativity between question and answer' results in clarity which disposes of a good deal of claptrap, such as speaking of "a savage as 'confronted by the eternal problem of obtaining food'. But what really confronts him is the problem, quite transitory like all things human, of spearing this fish, or digging up this root, or finding blackberries in this wood."[23]

Collingwood says that his next step was to apply this to the idea of contradiction. The current logic accepts that two propositions,

---

[20] *An Autobiography*, p. 31. See also Collingwood's *The Principles of Art*, pp. 265-266, for the need to know even the tone of voice in which a speaker utters a proposition in order to know its meaning.

[21] A similar statement by the historian-philosopher E. Gilson may be worth nothing (*The Christian Philosophy of Thomas Aquinas*, London, Gollancz, 1961, p. 223): "If we want the Thomistic solution of the problem of knowledge, it must come first from St. Thomas. If it is to come really from him and not from Descartes or Kant, we must not approach it with a Cartesian or Kantian question on our lips, because *philosophers' questions are one with their replies*." (Emphasis added.)

[22] *An Autobiography*, p. 32.

[23] *Ibid.*, pp. 32-33. See Alan Donagan's *The Later Philosophy of R. G. Collingwood*, pp. 60-61, for his criticism of Collingwood's strict correlation of a definite answer to a definite question.

simply as propositions, might contradict one another, so that by simply examining them as propositions one could determine whether in fact they did contradict each other or not. Collingwood opposes this position on the ground that in order to know whether two propositions are in contradiction with each other one needs to know their meaning, but one does not know the meaning of each without together knowing the question to which each is intended to be the answer.

> If you cannot tell what a proposition means unless you know what the question is it is meant to answer, you will mistake its meaning if you make a mistake about that question.[24]

A symptom of *mistaking the meaning* of a proposition is mistaking it to contradict another proposition which in fact it does not contradict.

> No two propositions, I saw, can contradict one another unless they are answers to the same question. It is therefore impossible to say of a man 'I do not know what the question is which he is trying to answer, but I can see that he is contradicting himself.'[25]

Collingwood applies his principle of the correlativity of question and answer to the idea of truth. If the meaning of a proposition is relative to the question it answers, so also is its truth. Neither (i) meaning, nor (ii) agreement and contradiction, nor (iii) truth and falsehood belong to propositions by themselves, but only inasmuch as they are answers to questions, each proposition being strictly correlative as answer to its relevant question.

Collingwood then criticises what he calls 'propositional logic,' under which classification he groups the so-called traditional logic, the idealistic logic of the eighteenth and nineteenth centuries and the symbolic logic of the nineteenth and twentieth centuries. All of these logics are in agreement in treating truth and falsehood, which, says Collingwood, is the chief concern of logic, as belonging to propositions as such. In such a doctrine the proposition is regarded as the unit of thought, the smallest 'entity' about which truth and falsity can be said, which cannot be said of its parts—subject, predicate and copula—taken separately.[26]

Collingwood attributes this position of propositional logic to the early liaison of logic and grammar, according to which the logician's

---

[24] *An Autobiography*, p. 33.

[25] *Ibid*.

[26] See *An Autobiography*, p. 34. In *Speculum Mentis* (pp. 114-116) Collingwood attributes to Religion the discovery of truth and its expression in propositions asserted to be true which are contradictorily opposed to their denial, recognized to be thereby false.

proposition was regarded as a kind of ghostly double of the grammarian's sentence, much in the same way as primitive views regarded minds as ghostly doubles of bodies. Among the various kinds of verbal expressions (of command, of wishes, etc.) grammarians recognize that special kind, the indicative sentence whose function it is to express statements. Collingwood says that "logicians have almost always tried to conceive the 'unit of thought', or that which is either true or false, as a kind of logical 'soul' whose linguistic 'body' is the indicative sentence."[27]

This correlation of the logical proposition with the grammatical indicative sentence was never entirely satisfactory. Bacon and Descartes, Plato and Kant, are examples of those who saw that the true unit of thought was not the proposition but something more complex in which the proposition functioned as answer to a question.[28]

> When Plato described thinking as a 'dialogue of the soul with itself', he meant (as we know from his own dialogues) that it was a process of question and answer, and that of these two elements the primacy belongs to the questioning activity, the Socrates within us. When Kant said that it takes a wise man to know what questions he can reasonably ask, he was in effect repudiating a merely propositional logic and demanding a logic of question and answer.[29]

Collingwood says that the one sentence expressive of one proposition, as maintained by propositional logic, does not accord with the way words are actually used in certain situations, as when expressions of thought are 'elliptical' or 'pleonastic.' Again, it is generally held that the indicative sentences in a work which is no more than a work of fiction do not express propositions. The central doctrine of propositional logic is that "there is, or ought to be, or in a well constructed and well used language would be, a one-one correspondence between propositions and indicative sentences expressing a proposition, and a proposition being defined as the unit of thought, or that which is true or false."[30]

Collingwood says that all the well known theories of truth presuppose this one-one correspondence. One school maintains that truth and falsity are qualities of propositions in themselves. Another school maintains that to call a proposition true is to assert a relation of 'cor-

---

[27] *An Autobiography*, p. 34. See Alan Donagan, *The Later Philosophy of R. G. Collingwood*, pp. 58-59, for a criticism of Collingwood's interpretation here.

[28] See *An Autobiography*, p. 34.

[29] *Ibid.*, p. 35.

[30] *Ibid.*, pp. 35-36.

respondence' or 'non-correspondence' between the proposition and something which is not a proposition, some 'state of things' or 'fact.' A third school holds that truth or falsity asserts a relationship of a proposition to other propositions with which it 'coheres' or 'fails to cohere.' According to pragmatists, to call a proposition true or false is to assert the utility or inutility of believing the proposition. Collingwood rejected all these theories because they presupposed the principle of propositional logic which he denied altogether.[31]

For a logic of propositions he wanted to substitute what he called a logic of question and answer. For Collingwood, *truth in its proper sense,* truth *as he was familiar with it,* was truth taken in the sense in which a philosophical theory or historical narrative is called true, and this "was something that belonged not to any single proposition, nor even as the coherence theorists maintained, to a complex of propositions taken together; but to a complex consisting of questions and answers."[32]

Collingwood accuses propositional logic of completely ignoring the structure of such a complex, but, he says, with a little help from Bacon, Descartes and others, he found himself able to hazard a few statements about it. He gives the following as characteristics of the structure of his interrogative logic: First, the characteristics of Context and Relevance: "Each question and each answer in a given complex had to be relevant or appropriate, had to 'belong' both to the whole and to the place it occupied in the whole"; second, the characteristic of 'arising' that accompanies a relevant question: "Each question had to 'arise'; there must be that about it whose absence we condemn when we refuse to answer a question on the ground that it 'doesn't arise' "; thirdly, the characteristic of 'rightness of the answer': "Each answer must be 'the right' answer to the question it professes to answer."[33] Collingwood insists that by 'right' (answer) he does not mean 'true.' The right answer, according to him, is the one which

---

[31] See *ibid.,* p. 36.

[32] *Ibid.,* pp. 36-37. The identifications Collingwood makes here, of truth in its proper sense with truth as he was familiar with it (and his archaeological practice is to be kept in mind here) and with philosophical and especially historical truth, are important in evaluating Collingwood's theory. As for Descartes the paradigm of knowledge was mathematics, and for Kant, Newtonian physics, so for Collingwood the paradigm of knowledge, and truth, is historical knowledge and truth (to which philosophy is assimilated, rather than to science). ". . . exclusive reliance on history as the type of all knowledge is the exact counterpart of the positivist reliance on natural science and is open to much the same objections." W. H. Walsh, "R. G. Collingwood's Philosophy of History," *Philosophy,* XXII (1947), p. 153.

[33] *An Autobiography,* p. 37.

enables you "to get ahead with the process of questioning and answering."[34] Cases are common, he says, in which the 'right' answer to a question is 'false.' As examples, he cites cases in which a thinker is following a false scent, either inadvertently, or in order to construct a *reductio ad absurdum*. Again, Polemarchus' answer to Socrates' question in the *Republic* (333B) whether in a draughts game he would prefer as partner a just man or one able to play draughts. The answer 'a man able to play draughts' is false, since it presupposes that justice and the ability to play draughts are incompatible, but it is nevertheless 'right' because it is a link, and a sound one, in the chain of questions and answers by which the falseness of the aforesaid presupposition is manifested.

Collingwood appeals to what he regards as *the ordinary meaning* of calling a proposition true as he understands it. Its characteristics are: (i) that the proposition belongs to a question-and-answer complex which as a whole is 'true' in the proper sense of the word; (ii) within this complex the proposition functions as an answer to a certain question; (iii) the question to which it is the answer is what would commonly be referred to as a sensible or intelligent question, that is, it is a question which 'arises,' in Collingwood's terminology, to which is contrasted a silly question; (iv) the proposition is the 'right' answer to that question.[35] From this it follows that you cannot tell whether a proposition is true or false until you first find out the question it was intended to answer. Furthermore, Collingwood adds, a proposition which in fact is true can always be thought to be false by anyone who goes to the trouble to think out a question to which it would have been the wrong answer, but convinces himself that it is the question which it is intended to answer.[36] And again, a proposition which is in fact significant can always be thought to be meaningless by anyone who convinces himself that it was intended to answer a question when in fact it does not either rightly or wrongly answer the question, in spite of the intention of the one who offers it to be such answer.[37]

---

[34] "*La notion fondamentale pour Collingwood est celle d'un ordre à la fois notionnel et historique selon lequel les problèmes se succèdent et se commandent.*" P. Fruchon, "Signification de l'histoire de la philosophie selon l'Autobiographie de Collingwood," *Les Etudes Philosophiques*, XIII (1958), p. 152 (emphasis in original).

[35] It is questionable whether this is what is commonly meant by 'true' as Collingwood claims that it is; it seems that '*ordinary usage*' would rather endorse the position Collingwood is opposing, namely that statements are true or false as such.

[36] See *An Autobiography*, p. 38.

[37] See *ibid.*, pp. 38-39. It is unfortunate that Collingwood does not clarify these

> Whether a given proposition is true or false, significant or meaningless, depends on what question it was meant to answer; and any one who wishes to know whether a given proposition is true or false, significant or meaningless, must find out what question it was meant to answer.[38]

E. First Consequence: The Historicity of Propositional Meanings

Collingwood draws a far reaching consequence from his reduction of propositional meaning to its function in answering a question. The consequence is that the meaningfulness of propositions can only be determined by historical methods. History, therefore, is going to acquire a commanding position in any epistemological theory. For the question 'To what question did So-and-so intend this proposition for an answer?' (to which question reduces the question 'What does So-and-so mean by this proposition?') is an historical question, and, as such, cannot be settled without recourse to historical methods. Collingwood maintains that this is not an easy question to answer, since past writers always write for their contemporaries about questions that 'everyone is asking,' and so do not go to the trouble of explicitly stating what is the question that their propositions are intended to answer. But subsequently, when the writer has become a 'classic' the question has become forgotten since the contemporaries who asked it are

---

cases with examples; his meaning is not clear. What he says seems to argue for his position in one respect and against it in another respect. If one can mistakenly think that a proposition answers a certain question when in fact it really answers another, this seems to argue that the proposition *from itself* has some *innate* meaning whereby from its own resources it can contribute to the answering of more than one question. It seems to argue that the *source of meaning* to a proposition is not *totally* borne by the question. On the other hand, differentiation in meaning as the proposition is applied to answer now one, now another question certainly argues to the derivation of meaning to the proposition from the question. What seems to follow from this, and it is generally the position which will be taken in evaluating Collingwood's theory, is that propositions are able to draw meaning from the question to which they are the answer *provided that from themselves, as propositions, they have (at least some) meaning of their own*. In other words, Collingwood seems to exaggerate the *contextual* meaning of the proposition into its *only* meaning, its meaning *in use* into its meaning *absolutely*. In regard to terms, logicians from mediaeval times distinguished the meaning a term has *in its propositional context*, which they called *the supposition of the term*, from the meaning it has *in itself*, which they called *the signification of the term* (see Jacques Maritain, *Formal Logic*, New York, Sheed and Ward, 1946, pp. 59-72). It would seem that a similar distinction should be maintained between the meaning of propositions *in themselves* and their meaning *in use*. Collingwood, like Wittgenstein in this respect, seems to reduce meaning to usage, but, further, unlike Wittgenstein, to reduce usage (and therefore meaning in usage) to usage in answering a question. See in this respect, Donagan's discussion (*The Later Philosophy of R. G. Collingwood*, pp. 60-61) on the statement, 'That is the Bodleian,' as able to function as answer to more than one question.

[38] *An Autobiography*, p. 39.

long since dead, and especially is this so in regard to a question which was 'rightly' answered, for then the writer's contemporaries would have ceased asking it and begun asking the next question which 'arose.' So that "the question asked by the original writer can only be reconstructed historically, often not without the exercise of considerable historical skill."[39]

Collingwood illustrates by comparing the work of understanding Plato's *Parmenides* with the archaeological work of distinguishing the various periods of construction in the remains at the south gate of Househeads. Just as that archaeological reconstruction requires penetration into the purpose the builders had in mind at each period, so the *Parmenides* can only be understood by understanding the questions asked at that period and understanding the statements of the *Parmenides* precisely as they are answers to those questions.[40]

It forcibly struck Collingwood at the time that "whereas no two propositions can be in themselves mutually contradictory, there are many cases in which one and the same pair of propositions are capable of being thought either that or the opposite, according as the questions they were meant to answer are reconstructed in one way or in another."[41] Collingwood accuses critics, who regard such metaphysical statements as 'The world is both one and many' as self-contradictions, of ignorance of the question which it was intended to answer by those who made it, just as a critic of the statement 'The contents of this box are both one thing and many things,' accusing the one who made it of contradiction, thinking that two incompatible answers are being given to one question, 'Are the contents of this box one x or many x's?,' whereas, in fact, there are two questions being asked: (i) 'Are the contents of this box one set of chessmen or many sets?' and (ii) 'Are the contents of this box one chessman or many chessmen?' According to Collingwood, there is no contradiction between saying that something, whether it be the world or the contents of a box, is one and saying that it is many. There would be contradiction only if it were said that something was both one x and many x's; but in the original statement nothing is said about one x or many x's; this

---

[39] See *An Autobiography*, p. 39.

[40] See *ibid.*, pp. 39-40. The analogy is very illuminating regarding Collingwood's thought. Archaeological interpretation is by the very nature of the case highly contextually conditioned, and the analogy is therefore likely to maximize this aspect at the expense of other aspects which *in a different context* might have a greater significance than from *context alone*.

[41] *An Autobiography*, p. 40.

was foisted on the statement by the critic, and the contradiction he protests about never existed in his victim's philosophy until the critic himself planted it there, like planting treasonable correspondence in his coat pockets, with the intention of obtaining a reward by denouncing him.[42]

> Thus, if a given doctrine D is criticized as self-contradictory because it is divisible into two parts E and F, where E contradicts F, the criticism is valid only if the critic has correctly reconstructed the questions to which E and F were given as answers. A critic who is aware of this condition will of course 'show his working' by stating to his readers the evidence on which he has concluded that the author criticised really did formulate his questions in such a way that E and F in his mouth were mutually contradictory. Failing that, a reader disinclined to work the problem out for himself will naturally assume the criticism to be sound or unsound according as he has found the critic to be, in a general way, a good historian or a bad one.[43]

This realization enabled Collingwood to answer the question left unanswered since 1914, namely whether the realists' critical methods were sound. His only answer could be that they were not, since their method consisted only in analysing the position being criticized into various propositions between which they detected contradictions. ''Following the rules of propositional logic, it never occurred to them that those contradictions might be the fruit of their own historical errors as to the questions which their victims had been trying to answer.''[44] And he was far from being favorably impressed by the realists' attitude towards history.

Collingwood wrote out the fruits of his meditations at considerable length and with many illustrations in a book called *Truth and Contradiction*. War conditions, however, discouraged its publication, and Collingwood tells us later in his *Autobiography* that he destroyed the manuscript of this work.[45]

F. Second Consequence: Philosophical Problems Reduced to Historical Problems to Be Solved by Historical Methods

Collingwood insisted that in any philosophical discussion it was necessary to get into the mind of the speaker to understand what was the problem or the question which was in his mind. Only then

---

[42] See *ibid.*, p. 41. No doubt the allusion is to the criticism of Berkeley by Moore and that of Bradley by Cook Wilson referred to above, pp. 37-38.

[43] *An Autobiography*, p. 42.

[44] *Ibid*.

[45] See *ibid.*, pp. 42-43; 99.

could the position he advocated be understood, for his position was his answer to his question. Before criticising an opponent's position, then, it is necessary to attempt actively to understand the opponent's position in terms of his question to which his exposition is intended to be the answer.[46]

> According to my own 'logic of question and answer,' a philosopher's doctrines are his answers to certain questions he has asked himself, and no one who does not understand what the questions are can hope to understand the doctrines. The same logic committed me to the view that any one can understand any philosopher's doctrines if he can grasp the questions which they are intended to answer. These questions need not be his own; they may belong to a thought-complex very different from any that is spontaneously going on in his mind; but this ought not to prevent him from understanding them and judging whether the persons interested in them are answering them rightly or wrongly.[47]

Collingwood found it extremely valuable to follow habitually and take part in discussions concerning subjects and methods which were not his own but other people's. For ". . . developing their positions and applying them to topics they had not dealt with, to reconstruct their problems in my own mind . . ." was "a delightful task" and "a magnificent exercise."[48] Which evoked Pritchard's exasperated comment, "I wish you'd get off the fence."

Collingwood says that this way of dealing with other people's thoughts had already become a fixed habit before he realized that it is formally deducible from his 'logic of question and answer.'[49] Thinking this way about the philosophies of others is to think about them historically, and he tells us that as early as the age of six or seven he saw that the only way to tackle any historical question was to see what the different people involved were trying to do.[50] As an example,

---

[46] See *ibid.*, pp. 53-55.

[47] *Ibid.*, p. 55.

[48] *Ibid.*, p. 57.

[49] "He had . . . supreme intellectual self-confidence which prevented him from being taken by surprise, shocked or rattled by anything that might be said. He rarely contradicted or engaged in emphatic dissent. He would not rudely dismiss a statement made to him, but would usually deal with it in the Socratic manner by asking questions often very embarrassing to the person who had raised the subject. This was not always agreeable to the rash and self opinionated." R. B. McCallum, "Robin George Collingwood 1889-1943" (accompanying obituary notice), *Proceedings of the British Academy*, XXIX, 1943, p. 467.

[50] See *An Autobiography*, p. 58. Collingwood thus acknowledges an innate predisposition to think in this way, just as Descartes was innately disposed to think mathematically.

he refers to the Battle of Trafalgar, since "naval history was a childish passion of mine, and Trafalgar my pet battle."[51]

> History did not mean knowing what events followed what. It meant getting inside other people's heads, looking at their situation through their eyes, and thinking for yourself whether the way in which they tackled it was the right way. Unless you can see the battle through the eyes of a man brought up in sailing ships armed with broadsides of short-range muzzle-loading guns, you are not even a beginner in naval history, you are right outside it. If you allow yourself to think for a moment about the tactics of Trafalgar as if the ships were driven by steam and armed with long-range breech-loading guns, you have for that moment allowed yourself to drift outside the region of history altogether.[52]

Collingwood again takes up his quarrel with the realists, for whom the problems of history are unchanging problems.[53] In thinking that Plato, Aristotle, Epicurus, the Stoics, the Schoolmen, the Cartesians, etc., were each giving different answers to the same recurring perennial question, so that the question somehow remained outside the flux of time and therefore of history, they implied that the question itself had no history, that the problem discussed did not itself change from thinker to thinker who successively took it up. He accused the realists of thinking that the problems discussed in modern ethical theory are the same problems as those discussed in Plato's *Republic* or Aristotle's *Ethics*, and that the philosopher's task is to ask himself whether, for example, Aristotle or Kant gives 'the right answer' to the problems of duty.[54] They regarded the history of philosophy as a mere cataloguing, in order and date, of the various answers given by successive philosophers to the eternal (i.e., non-historical) questions of philosophy. For them, therefore, the question 'What was

---

[51] *An Autobiography*, p. 58. H.-I. Marrou, in his *Meaning of History*, pp. 37; 103-111, insists on the need for a sympathetic connaturality between a historian and his subject of inquiry which enables the historian to appreciate his subject 'interiorly.' This is important for understanding Collingwood's theory of 'insight' into the 'inner' (i.e., 'thought') dimension of an event which he regards as proper to history. But such sympathetic connaturality is very much in opposition to Collingwood's 'Baconian attitude,' that the subject studied be treated as an unwilling witness whose testimony has to be extracted by torture in the process of scientific interrogation. It may be well to remember that Bacon was a lawyer, and, like Collingwood, may have carried over attitudes proper to his profession into other fields in an illegitimate universalizing process. Basic moral attitudes, especially sympathy and connaturality with the subject, are capital in the historian, as Marrou well points out, and as Collingwood concords with in this example he cites.

[52] *An Autobiography*, p. 58.

[53] See *An Autobiography*, p. 59.

[54] See *ibid*.

Aristotle's theory of duty?' is a historical question wholly separate from the question 'Is what Aristotle said true?,' which was a philosophical question.[55] Collingwood severely criticizes Oxford philosophical training for which "the history of philosophy was a subject without philosophical interest,"[56] inasmuch as Oxford philosophy was then under the influence of the realists.

One of the problems Collingwood had set himself to reconsider during the war and in the course of his Albert Memorial meditations was this attitude of the realists towards the history of philosophy.

> Was it really true, I asked myself, that the problems of philosophy were, even in the loosest sense of that word, eternal? Was it really true that different philosophies were different attempts to answer the same questions? I soon discovered it was not true; it was merely a vulgar error, consequent on a kind of historical myopia which, deceived by superficial resemblances, failed to detect profound differences.[57]

Collingwood compares Hobbes' treatment of the nature of the state in the *Leviathan* with Plato's in the *Republic*. They are obviously two different political theories, but are they two different theories about the same thing? If you pressed for a definition of the state by each, you would find that the differences between the two widely separated thinkers went down to essentials and beyond mere superficialities. Collingwood allows that you can call the two things the same if you so wish, but you must admit that it got *diablement changé en route*, so much so that the nature of the state in Plato's time was genuinely different from the nature of the state in Hobbes' time. Collingwood stresses that he does not merely mean the empirical nature of the state, but the very ideal nature; Plato's *Republic* is an attempt at a theory of one thing and Hobbes' *Leviathan* is an attempt at a theory of something else. "What even the best and wisest of those who are engaged in Politics are trying to do has altered."[58]

Collingwood does not deny that there is some connection between these two things called by the same name, the State, by these two thinkers, but he wants to insist that the connection is not the one maintained by the realists. He agrees that no one would dispute that what Hobbes' *Leviathan* and Plato's *Republic* are about is in some way the

---

[55] See *ibid*.

[56] *An Autobiography*, p. 60.

[57] *Ibid*., pp. 60-61. For a discussion of "Collingwood on Eternal Problems" see Errol Harris, *The Philosophical Quarterly*, 1 (1951), pp. 228-241.

[58] *An Autobiography*, pp. 61-62.

same but in another way different. The dispute is precisely about the kind of sameness and the kind of difference. Whereas for the realists the sameness is that of a universal and the difference that between two instances of that universal, Collingwood maintains that "the sameness is the sameness of a historical process and the difference is the difference between one thing which in the course of that process has turned into something else, and that other thing into which it has turned."[59] A traceable historical process relates Plato's polis to Hobbes' Absolutist state, a process by which one is turned into the other. Anyone who argues that the two political theories are contradictorily related, and that therefore one must be right and the other wrong, ignores that process and denies that difference between them which results from that process.

Thinking along these lines forced Collingwood to the conclusion that the history of political theory is not the history of different answers to the one same question, but is rather the history of a more or less changing problem whose solution was likewise changing along with the problem. Thus, the form of the polis is not one unchanging ideal of human society eternally laid up in heaven and the eternal goal of the efforts of all good statesmen of every age. It was the ideal of human society as conceived by the Greeks of Plato's time. But by Hobbes' time people had changed their minds as to what they regarded as the ideal of social organization. The ideals in each case were different.[60]

---

[59] *Ibid.*, p. 62.

[60] See *ibid.*, pp. 62-63. Notwithstanding that Collingwood presents his thinking here in his *Autobiography* as the fruits of his own labouring mind, it is difficult not to see reflected in them what Croce had written in his *Logica come scienza del concetto puro*, first published in 1905: "Every definition is the answer to a question, the solution to a problem; and there would be no need to enunciate it if we did not put questions to ourselves, and did not propose problems. Why otherwise give ourselves this inconvenience? What need would constrain us? Definition arises, as does every mental work, from a contrast, a struggle, a war which seeks a peace, from a darkness which seeks light . . . it is a question which demands an answer. Not only does the answer suppose a question, but such an answer supposes such a question. The answer must be proportioned to the question, otherwise it would not be an answer but the illusion of an answer. . . . But the question, the problem, the doubt is always individually conditioned: the doubt of the child is not the doubt of the adult; the doubt of the uncultured man is not the doubt of the cultured man; the doubt of the novice is not the doubt of the indoctrinated; the doubt of an Italian is not the doubt of a German, and the doubt of a German of 1800 is not the doubt of a German of 1900. In fact, the doubt formulated by an individual at a determinate moment is not the same as that formulated by that same individual a moment later. By way of simplification one tends [suole] to affirm that the same question has been posed as such by many men in many centuries and varying times. But in saying this, one makes a simplification which amounts to an abstraction. In reality every question is different from every other, and every definition, however constant it may sound, and however cir-

Likewise, what the Greeks meant by 'ought' in Ethics is not the same as what Kant means when he is discussing moral obligation. The ideals of personal conduct are as impermanent as the ideals of social organization and so have to be approached from the historical point of view; and even what is meant by calling them ideals is subject to similar change.[61] The realists erred in thinking 'ought to behave' has an eternal, unchanging, unique meaning about which different peoples at different times and places held differing views, some wrong, some right. But if they had merely consulted the texts on their shelves as to the meaning therein they would have been alerted to this, instead of systematically mistranslating passages in the light of their mistaken preconceptions. An acquaintance with the history of science should show anyone that when Einstein introduced discussion about relativity into philosophy, any convictions about the eternity of problems and conceptions were as unfounded as a young girl's conviction that this year's hats are the only ones any sane woman would wear. Collingwood thus became convinced that Metaphysics, rather than being a futile attempt to know what lies beyond the limits of experience, is primarily an attempt to discover what the people of a given time believe about the general nature of the world, such beliefs being the presuppositions of all their physics, and therefore of all their (relevant) questionings into detail. In addition, he recognized that Metaphysics is an attempt to discover the corresponding presuppositions of other peoples at other times and to show the historical process by which one set of presuppositions has turned into another.[62]

Collingwood recognizes that discovering what presuppositions underlie the physics of a given people at a certain time, which is the question metaphysicians have to answer, is a purely historical question. He denies that it is their business to raise the further question whether among the various beliefs so held at different times and places, this or that one is true and opposed to the others as false. Such a question, according to Collingwood, has consistently been found to be unanswerable.

---

cumscribed by certain determinate words, in reality is different from every other, because the words, even when they seem to be materially the same, are effectively different, according to the spiritual diversity of those who pronounce them, these being individuals who find themselves, therefore, in new and individual circumstances." Translated from the work cited, appearing as Vol. II, Logica, *Filosofia come scienza dello spirito*, Bari, Gius, Laterza and Figli, 1964; see pp. 133-134.

[61] See *An Autobiography*, pp. 63-65.

[62] See *An Autobiography*, p. 66. The question of Metaphysics as being concerned about the presuppositions to questions will be taken up later (in Chapter 5).

> . . . if there is anything in my 'logic of question and answer' that is not to be wondered at, for the beliefs whose history the metaphysician has to study are not answers to questions but only presuppositions to questions, and therefore the distinction between what is true and what is false does not apply to them, but only the distinction between what is presupposed and what is not presupposed.[63]

Collingwood clarifies this by distinguishing what he calls relative presuppositions from absolute presuppositions, which latter are the proper concern of the metaphysician. Some presuppositions to questions are themselves answers to prior questions. But "the beliefs which a metaphysician tries to study and codify are presuppositions of the questions asked by natural scientists, but are not answers to any questions at all," which "might be expressed by calling them 'absolute presuppositions.' "[64]

Collingwood does not deny that the statements made by metaphysicians are true or false, for their statements (about absolute presuppositions) are answers to questions about the history of these presuppositions. Metaphysics can rightly lay claim to being a science, not in the same way as a naturalistic science, but as an organized body of knowledge in which:

> on the one hand, the beliefs of a given set of people at a given time concerning the nature of the world are exhibited as a single complex contemporaneous fact . . . and on the other hand the origin of these beliefs is enquired into, and it is found that during a certain space of time they have come into existence by certain changes out of certain others.[65]

Collingwood came by degrees to recognize that no branch of philosophy was exempt from the historical conditioning of its problems and solutions; the only sense in which he would admit the eternity of an historical fact was in the sense that it had happened once and for all.

Collingwood then grappled with the realist's distinction between the 'historical' question 'What was So-and-so's theory on such and such a matter?' and the 'philosophical' question 'Was he right?' He rejected this distinction since it depended on the presupposition that there are permanent and eternal problems in history.

---

[63] *Ibid.*, p. 66.
[64] *An Autobiography*, pp. 66-67.
[65] *Ibid.*, p. 67.

> If there were a permanent problem P, we could ask 'what did Kant, or Leibniz, or Berkeley, think about P?' and if that question could be answered, we could then go on to ask 'was Kant, or Leibniz, or Berkeley, right in what he thought about P?' But what is thought to be a permanent problem P is really a number of transitory problems p_1 p_2 p_3 . . . whose individual peculiarities are blurred by the historical myopia of the person who lumps them together under one name P. It follows that we cannot fish the problem P out of the hyperuranian lucky-bag, hold it up, and say 'what did So-and-so think about this?' We have to begin, as poor devils of historians begin, from the other end. We have to study documents and interpret them. We have to say 'here is a passage of Leibniz; what is it about? what is the problem with which it deals?' Perhaps we could label that problem p_{14}.[66]

The next question is 'Does Leibnitz deal rightly or wrongly with his problem?' The answer to this question is not as easy as the realists think. If in writing the relevant passage Leibnitz happened to be so confused in his own thinking as to completely mess up the job of solving his problem he would have thereby so mixed up his own tracks that no reader would be able clearly to unravel exactly what his problem was.

> For one and the same passage states his solution and serves as evidence of what the problem was. The fact that we can identify his problem is proof that he has solved it; for we only know what the problem was by arguing back from the solution.[67]

We can only discover what tactical problem Nelson set himself at Trafalgar by studying the tactics he pursued in the battle and argue back from the solution to the problem. Even if we had the original typescript of the coded orders to his captains a few hours before the beginning of the battle, this would not tell us that he had not changed his mind at the last moment and thought out a new plan, trusting to his captains to understand what he was doing and back him up. Collingwood alleges that it is for this reason that naval historians give much consideration to Nelson's tactical plan, but not to Villeneuve's. Because Nelson won the battle, we can reconstruct what his plan was, but no one will ever know what Villeneuve's plan was because he did not succeed in carrying it out; one may hazard a guess at it, but guessing is not history.[68]

---

[66] *Ibid.*, p. 69.
[67] *Ibid.*, p. 70.
[68] *Ibid.*

Collingwood draws the far reaching conclusion that "there are not two separate sets of questions to be asked, one historical and one philosophical, about a given passage in a given historical author. There is one set only, historical."[69]

From this it follows that Plato is to be studied in exactly the same way as Thucydides. The study of Greek philosophy is a historical study just as is the study of Greek warfare. Collingwood does not reject the question 'Was Plato right to think as he did on such or such question?,' but he points out that it is to be answered in the same way as the question 'Was Phormio right to row round the Corinthian's circle?' It is just as much history that he beat the Corinthians by doing it in the way he did it. Collingwood warns against 'the ghost of Ranke' gibbering about 'what exactly happened?,' for he alleges that such an attitude ignores that victories as well as tactical manoeuvres are things that happen.

G. Collingwood's Conclusion from These Considerations

Collingwood tells us that these ideas, which were only partially worked out before his return to Oxford after the (1914-18) war, became clear to him on his return there. Rather than discuss them with his Oxford colleagues, he put them into practice with his pupils in accordance with an ancient Oxford tradition of reading, expounding and commenting on philosophical texts. He held that the best way of teaching undergraduates exactly how the subject under consideration has been revolutionized is by lecturing on the old classics, whose authors in their day may have revolutionized the subject, and showing in one's commentary how their doctrine has been modified. He observes that this was a field of activity exactly suited to him inasmuch as his inclinations led him towards detail rather than towards generalization.[70]

Collingwood insisted that his students obey two rules governing the procedure of textual analysis before going on to the further task of criticising its doctrine. The first rule was that criticism of an author was not to be accepted before satisfactorily establishing that the criticism is relevant. The second rule (from his meditations on the Albert Memorial) was to "reconstruct the problem," or to "never

[69] *Ibid.*, p. 72.

[70] See *ibid.*, p. 74. This admission alerts us to ask how far native temperament and acquired inclinations have played a determining (and perhaps unexamined) part in his judgments concerning philosophical procedure, just as Descartes' natural mathematical bent is material in assessing his verdict on the nature of philosophical method.

think you understand any statement made by a philosopher until you have decided, with the utmost possible accuracy, what the question is to which he means it for an answer.''[71]

> Thus the history of philosophy . . . was no longer a 'closed' subject. It was no longer a body of facts which a very, very learned man might know, or a very, very big book enumerate, in their completeness. It was an 'open' subject, an inexhaustible fountain of problems, old problems re-opened and new problems formulated that had not been formulated until now. Above all it was a constant warfare against the . . . dogmas of that putrefying corpse of historical thought, the 'information' to be found in text-books. For, in the history of philosophy, as in every other kind, nothing capable of being memorized, is history.[72]

[71] *Ibid.*

[72] *Ibid.*, pp. 75-76.

Chapter Three
Supposal and Questioning

A. Questioning as Supposing, the Cutting Edge of the Mind

The activity of questioning presents itself in two ways in Collingwood's earlier work, *Speculum Mentis* (1924), in which he distinguishes five forms of human experience in dynamic evolution: Art, Religion, Science, History and Philosophy. The evolutionary transformation of one life form into the next (in the order given above) is by way of a question and answer process, the lower form (e.g., Art) being related to its next higher (e.g., Religion) as a question to its answer. Each lower form as it tends to its own fulfillment tends simultaneously to its own disintegration and transformation, inasmuch as it is in dynamic disequilibrium with itself in such a way that its fulfillment is in another, higher life form to which it gives way. Each life form is a system which makes a promise it cannot fulfill; the system thus breaks down under increasing strain of internal contradiction which precipitates criticism (i.e., questioning) of the validity of that form. The answer to the question thus internally provoked is provided by transition to the next higher life form.

Thus, "art by its own dialectic gives rise, over against itself, to criticism, which is science."[1] "The point at which a child begins to

[1] *Speculum Mentis*, London, Oxford University Press, 1924, p. 101.

ask whether stories are true, and passes through the crisis of learning to disbelieve in fairies . . . indicates the emergence of religion from art.''[2] ''It [i.e., art] is pregnant with a message it cannot deliver.''[3] ''When [language is seen to express not itself but thought as its meaning] art and religion cease to exist as self-contained and autonomous forms of experience and give way to a life no longer of expression but of thought.''[4] ''This revolution in the use of language [from the metaphorical, as in art and religion, to the literal] is the birth of science.''[5] Then ''the discovery of the abstractness of science is . . . the birth of history.''[6] ''Science is a question whose answer is history.''[7] ''Science is explicit to itself as thought, but it turns out on enquiry to be identical with the questioning activity; that is, it realizes the contradiction of a type of thought which is not thought precisely because it is thought's opposite, intuition.''[8] ''. . . history solves the question which science asks but cannot answer.''[9] ''This discovery [that science presupposes history and can never go behind history] . . . implicitly resolves science into history.''[10] ''History is an unstable attitude which leads either back into science or forward into philosophy.''[11] ''The transition . . . effected from history to philosophy is . . . merely the making explicit of what was implicit already,'' namely ''that in the last resort nothing but the knower can be known.''[12]

Generally speaking, then, *Speculum Mentis* is a phenomenology of human experience[13] in which the dynamism throughout is a raising and solving of problems, a constant raising and answering of questions.

> . . . the life of the mind consists of raising and solving problems, problems in art, religion, science, commerce, politics, and so forth. The solution of these problems does not leave behind it a sediment of ascertained fact, which grows and solidifies as the mind's work goes on. Such a sediment is nothing but the externality of a half-solved problem: when the problem is fully solved the sediment

---

[2] *Ibid.*, p. 113.
[3] *Ibid.*, p. 110.
[4] *Ibid.*, p. 155.
[5] *Ibid.*, p. 157.
[6] *Ibid.*, p. 180.
[7] *Ibid.*, p. 186.
[8] *Ibid.*, p. 188.
[9] *Ibid.*, p. 193.
[10] *Ibid.*, p. 202.
[11] *Ibid.*, p. 230.
[12] *Ibid.*, pp. 245-246.
[13] See Louis O. Mink, *Mind, History, and Dialectic*, Bloomington, Indiana University Press, 1969, p. 61.

of information disappears and the mind is left at liberty to go on. Philosophy therefore, is not a prerogative kind of knowledge immune from this reabsorption into the mind's being: it is nothing but the recognition that this reabsorption is necessary and is indeed the end and crown of all knowledge, the self-recognition of the mind in its own mirror.[14]

But, in addition to this general identification of the activity of the mind with a questioning and answering process, Collingwood identifies questioning in a special way with the aesthetic activity characteristic of that life form which is the artistic, which, according to Collingwood, is the first and basic form of human experience. In the section of *Speculum Mentis*[15] entitled 'Knowledge as Question and Answer,' Collingwood makes an intimate tie-up of the mind's questioning activity with the imaginative activity characteristic of art.

What characterizes the life form of art is that "the aesthetic experience is imaginative not in the sense that its objects are fictitious but in the sense that it treats them indifferently, whether they are fictitious or real."[16] So, according to Collingwood, what is characteristic of imaginative activity is its attitude of indifference to the reality or non-reality of its objects. The objects may in fact be real, or they may be unreal, but this aspect is of no concern to the imagination as such. It is the attitude "which neither asserts reality truly or falsely, nor denies it, but merely imagines."[17] Shakespeare's Hamlet, as a work of art, may or may not conform to real characters and incidents; this argues neither for nor against its merits as a work of art. Collingwood notes that this mental attitude, which neither asserts nor denies reality, truth or falsity, has been given a great deal of attention by recent philosophers "under the name of hypothesis, intuition, supposal, representation and so forth."[18]

---

[14] *Speculum Mentis*, p. 317.

[15] Section 5 of Ch. 3, pp. 76-86.

[16] *Speculum Mentis*, p. 76.

[17] *Speculum Mentis*, pp. 60-61; p. 76. See also *The Principles of Art*, London, Oxford University Press, 1938 (Galaxy p/b. 1958), pp. 136-138.

[18] *Speculum Mentis*, p. 76. The 'philosophers' referred to no doubt include Bradley and Bosanquet. See for example Bradley's "On Floating Ideas and the Imaginary," *Mind*, LX (1906), pp. 445-472 (the same is repeated as Ch. 3 of Bradley's *Essays on Truth and Reality*, Oxford, Clarendon Press, 1914). See also Bradley's *The Principles of Logic*, London, Macmillan, 1928, Vol. I, pp. 80-90 on supposal and hypothetical judgments; B. Bosanquet, *Knowledge and Reality*, London, Swan and Sonnenschein, 1892, pp. 33-58, on supposal and hypothetical judgments; pp. 287-306 for supposal, hypothesis and experiment in inference. See also the discussion between Professor G. F. Stout and Professor R. Hoernlé; R. Hoernlé, "Professor Stout's Theory of Possibilities, Truth, and Error," *Mind*, XL (1931), pp. 273-284; G. F. Stout, "Truth and

Collingwood criticises those philosophers who err by isolating this mental attitude from the general context of human experience, thereby regarding it as a self-contained and autonomous phase of experience which would then require some sort of a bridge to make the transition to other phases, principally those of assertion and denial.[19] If the mind in fact succeeded in achieving that condition in which it neither asserts nor denies but merely intuitively represents its object to itself it could never acquire the ability, nor would it ever have a reason to pass out of that condition. It is useless to appeal to intervention of the concept to bridge this unbridgeable gulf from supposal to assertion, for this would be merely to postulate (suppose) that the problem had been solved without facing the difficulty as to how the concept would be able to intervene in such a situation. But, says Collingwood, this is a pseudo-problem, since the mind never operates in a vacuum of pure imagination or supposal. There is no bridge because there is solid ground all the way between supposal and assertion.

> Supposal and assertion are not two independent chapters in the history of mind; they are two opposite and correlative activities which form as it were the systole and diastole of knowledge itself.[20]

Falsity," *Mind*, XLI (1932), pp. 297-310. Especially noteworthy is pp. 299-302 of Professor Stout's article on questioning and supposing. The date, 1932, precludes this material being a source of Collingwood's similar doctrine. Either Stout is indebted to Collingwood or both Collingwood and Stout point to a more widespread common thinking going back at least to the date of *Speculum Mentis*. Through Stout, English philosophy is injected with the thought of Brentano, Husserl and especially Meinong on the status of objects of thought; see Passmore's *A Hundred Years of Philosophy*, London, Penguin Books, 1968, pp. 194-200, especially pp. 194-195.

[19] See *Speculum Mentis*, p. 76. The aesthetic life, according to Collingwood, is a self-contradicting attempt to live out this attitude of supposal, which breaks down, and gives way to religious experience as the contradiction becomes felt and apparent. Aesthetic activity creates a world in which it is possible to live as an artist. But life-experience intrudes many things that cannot be meaningfully included in that world. Witness the struggle many artists have lived with all their lives, namely, how to live the artistic life and make a living. The artistic world is not therefore an all inclusive world; the world horizon of meaning has to give way and become expanded if certain experiences are to be meaningfully accounted for.

[20] *Speculum Mentis*, p. 77. This, in substance, would be Collingwood's answer to the question which tormented Bradley, whether there are any Floating Ideas. "In the end and taken absolutely . . . there can be no mere idea. Reality is always before us, and every idea in some sense qualifies the real. So far as excluded it is excluded only from some limited region, and beyond that region has its world. To float in the absolute sense is impossible. Flotation means attachment to another world, a world other than that sphere which for any purpose we take here as solid ground and as fact." F. H. Bradley, "On Floating Ideas and the Imaginary," *Mind*, LX (1906), p. 451 (pp. 35-36 of *Essays on Truth and Reality*). For Collingwood, supposal and assertion are two sides of the one entity which is knowledge of reality. Ernst Cassirer, *The Logic of the Humanities*, New Haven, Yale University Press, 1960, p. 56, says that "all concrete

Collingwood criticises 'a crude empiricism' which regards knowledge to be only a matter of assertion, for it is only in retrospect, by looking back over the traversed road, that a knower identifies knowledge with assertion. Knowledge as a past, accomplished fact, such as one finds in encyclopaedias and textbooks, is indeed assertion. But first hand acquaintance with knowledge, as it is an activity of the mind, as it is something the mind does, not just something it 'apprehends,' shows that knowledge is a matter of voyage and discovery in which assertions are only answers to questions. Thus it was for such as Plato, Bacon and Kant:

> So, Plato described true knowledge as 'dialectic,' the interplay of question and answer in the soul's dialogue with itself; so Bacon pointed out once for all that the scientist's real work was to interrogate nature, to put her, if need be, to the torture as a reluctant witness; so Kant mildly remarked that the test of an intelligent man was to know what questions to ask. . . .[21]

Collingwood calls questioning "the cutting edge of knowledge" and assertion "the dead weight behind the edge that gives driving force."[22] Questioning apart from information and information apart from questioning are valueless.

> Questions undirected by positive information, random questions, cut nothing; they fall in the void and yield no knowledge. Information, when it is not ground to the keen edge of inquiry, is not knowledge but mere pedantry, the talent buried in the earth.[23]

Collingwood allows that textbooks and encyclopaedias are invaluable as a body of information which directs further questioning, but he regards them as contemptible if considered to constitute and exhaust knowledge. "Information may be the body of knowledge, but questioning is its soul."[24]

Collingwood then identifies questioning with the activity of supposal, which is the activity characteristic of the imagination and aesthetic activity.

---

spiritual events, all genuine 'history' is only the reflection . . . of this continually renewing process, this 'systole' and 'diastole', this separation and reintegration."

[21] *Speculum Mentis*, p. 77.

[22] *Speculum Mentis*, p. 78. Compare *The Idea of History*, p. 237 (on Historical Evidence): "The question is the charge of gas, exploded in the cylinder head, which is the motive force of every piston-stroke." See above, Chapter 1, p. 27.

[23] *Speculum Mentis*, p. 78. See also B. Bosanquet, *Knowledge and Reality*, p. 36: "The guide of research is the body of existing knowledge, and it is this that endows the will to observe which in ignorance must be the servant of caprice, with aims properly subordinated to the discovery of truth."

[24] *Speculum Mentis*, p. 78.

> Supposal and questioning are at bottom the same thing; or rather, supposal when seen in its proper perspective as an integral element in knowledge turns out to be questioning.[25]

This questioning aspect of knowledge, inasmuch as it is an activity other than asserting, since it is supposing, is a puzzle to empiricists, for whom something has to really exist, and therefore be an object of assertion, if it is to be an object of knowledge, for they regard knowledge as the imprint forced on our mind by a really existing object demanding our attention.[26] But, contrary to this, asking a question always means, to some extent at least, envisaging alternatives of which one only can exist.[27] Because questioning thus always to some degree contemplates the non-existent it is essentially a suspension of the activity of asserting, and this is precisely how aesthetic experience or imagination is defined.

But, whereas for aesthetic experience this suspension of assertion seems to be an end in itself, not looking forward to its own negation in the renewal of assertion, true questioning on the other hand is a suspension of assertion which looks forward to a renewal of this asserting activity in the shape of an answer.

> Art, as pure imagination, imagination without assertion, may be paradoxically defined as a question which expects no answer: that is, a supposal.[28]

---

[25] *Speculum Mentis*, p. 78. See also B. Bosanquet, *Knowledge and Reality*, p. 36: "The differentia of supposal which it shares with experiment as such is the replacement of intellectual by arbitrary choice."

[26] Applied to history, this means that for the positivist facts are 'hard facts' with an identity of their own, independent of the mind's questioning activity, 'out there' for the mind to apprehend.

[27] See also *The New Leviathan*, p. 74, par. 11.12: "Asking a question implies contemplating alternatives. A question that offers no alternatives is a bogus question"; pp. 74-75, par. 11.22 ". . . a proposition is an answer to a question and a question offers alternatives." However, Herbert Butterfield in his *Man on His Past* (Cambridge: Cambridge University Press, 1955, pp. 171-201), shows that the historical problem of the Saint Bartholomew massacre was solved finally by admitting the coexistence of the alternatives. Whether Rome was to be considered guilty or not depended on whether there was premeditation or not. Both sides were vigorously defended by opponents, till finally it was agreed that the evident premeditation was not the precipitating cause of the uprising. And, similarly, Leslie Armour (*The Concept of Truth*, Assen, Van Gorcum, 1969, p. 181) writes: "If we are considering more general historical theses [than whether or not Charles I was beheaded], we may not be so sure that there *is* a single right answer or that two answers, which conflict in important ways may not be both right."

[28] *Speculum Mentis*, p. 79. See also B. Bosanquet, *Knowledge and Reality*, p. 36: "Supposal as such excludes the affirmation of what is supposed in the sense that an act which is supposal is not *qua* supposal, also affirmation."

But just as questioning looks beyond itself to assertion, having no end in itself, so neither does it have an absolute beginning in itself. The supposals, either of questioning or of aesthetic experience, both of which are acts of imagining, cannot be themselves absolute beginnings. There is no pure act of imagining and there is no pure questioning.[29] Even the silliest or most irresponsible question prerequires some information already to hand, and therefore presupposes assertion as a condition of itself. A pure question would ask nothing; it would be no more than the empty form of questioning, a mere interrogation mark with nothing preceding it.[30] But even the empty form of questioning, signified by the question mark (?) implies that there is information (and therefore assertion) available.

> Any act must start somewhere and a pure act of imagination just because it is completely divorced from fact, would have nowhere to start from and would therefore have no reason for determining itself in one way rather than any other.[31]

Collingwood is saying here that every question is a determinate question, a fully determined act. But it would not be so determined except by the existence of available information. A question, therefore, seems, according to Collingwood's explanation here, to be a form which (previously asserted) information can be given by the imaginative activity whereby that information is (imaginatively) made to implicate (imagined) alternatives, one or other of which alternatives will be determined by assertion and thereby become the answer to the question. The form in question, whereby past information becomes a question, the form which, when applied to that information, is the 'cutting-edge' of the mind, seems to be the form of the imagination (i.e., one of the forms which the mind's activity takes; for Collingwood the imagination is not a faculty, a department of a fixed entity, but a mode of the activity which *is* the mind) with a dynamic thrust pointing towards information to be had by way of an answer, something like an arrow directed to a target. Already possessed factual information (past assertions), knowledge of the textbook or encyclo-

---

[29] See *Speculum Mentis*, p. 79.

[30] See also S. Lupasco, *Du devenir logique et de l'affectivité*, Paris, Vrin, 1935, t. 2, pp. 83-85, on the impossibility of reduction of radical doubt to pure questioning activity. See also K. Rahner, *Spirit in the World*, Montreal, Palm Publishers, 1968, pp. 57-77, for a discussion of the being of man as a questioning being. ". . . in human existence the question is that *fact* which *absolutely* refuses to be replaced by another fact, to be reduced back to another fact. . . ." But this questioning, identified with the being of man, is absolute *as a fact*, that is, only inasmuch as its own existence is *presupposed-affirmed* (i.e., asserted to be a fact) *by* and indeed *in* the very questioning itself.

[31] *Speculum Mentis*, p. 79.

paedia variety, is taken as a material by the mind and formed into a question.[32] These previous assertions, congealed (encapsulated?) as available information, are then given an imaginative non-assertive form by being given imaginative implicatory reference to possible alternatives, neither of which is presently asserted to be the real or true state of affairs, the mind being simultaneously under dynamic tension to resolve the alternative possibilities one way or another.[33] But, which of the visualized alternatives is to be asserted, and thus function (not just as an abstract proposition, but) as answer to the question, is to be decided and determined by reference to further facts to be sought precisely in order to answer the question (e.g., the clues for which the detective looks). Thus, already available factual information provokes questions, and these questions in turn call forth the ascertainment of further information (asserted fact) in order that they may be answered.

Thus, questioning and artistic activity, being all imaginative activity, is itself embedded in a matrix of concrete fact; it begins from there; it thrives there and in concrete fact asserted it results.

> Even if art is pure imagination it must spring from a soil of concrete fact; the artist must really exist in a real world, and his works of art are necessarily a kind of sublimated version of his experience as a real person, however unconscious of this fact he may be.[34]

Even though the imaginative work, as imaginative, is indifferent to reality and to truth, nevertheless it is not just *imaginatively* imagined; it is *really* imagined, and therefore by a *real* person in a *real* situation.[35] The artist, engrossed in his artistic, imaginative work *as such* knows nothing of the *reality* of that work or of the reality of the situation in which he works. Everything is *imagined* to be put into the work expressed, even his whole self and his whole world of reality. Nevertheless that very all-absorption presupposes the *reality* of that activity

---

[32] This 'available information' is nothing other than man's past. Thus history provides the mind with 'the weight' on which the mind will 'grind' the cutting edge of inquiry and thrust forward to hitherto unknown information; the future would seem to be nothing other than 'the dynamic direction' of this activity to be brought into the present (and subsequently into the past, to become itself part of history) by this dynamism of questioning. Questioning therefore is the very dynamism of history, the activity of the mind, and the mind is what it does.

[33] This is how I read Collingwood's alignment of questioning with desire in his *New Leviathan*, Chapter 11.

[34] *Speculum Mentis*, p. 79.

[35] See K. Rahner quote, p. 67 note 30 above.

and the reality of the artist situated in the world of fact quite distinct from the artistic work.

> The work of art is an imaginative cutting edge to a mind whose solid backing of factual experience may be forgotten for the moment, but is none the less very real; or rather, it is not forgotten but distilled into the work of art, present to the mind in this form and no other.[36]

That is, the artist, in the act of imagining, has imaginatively translated his whole world of lived experience into that one activity which, for him, presents to him the whole meaning of the world and of himself as something not distinguishable from that world. This seems to imply that the work of art is the concrete universal attained obscurely and confusedly (a 'felt' identity in difference, prior to mediation by thought in the concept), which, when the level of philosophical reflection is attained, will be revealed distinctly as a whole of identity in difference.

But that act of imagining would not be done if the artist did not draw on his lived experience and the concrete facts of his daily life.

> Thus the imaginative activity of art is itself supported and surrounded by a medium of fact; but the essence of the purely aesthetic frame of mind is that this medium or background is overlooked.[37]

But, if the artist, in his artistic activity, may overlook this real matrix of fact, the philosopher is not excused from doing so. A purely imaginal philosophy would be basically unsound. Whereas reality and truth and fact is ignored by the artist, it belongs to the philosopher's stock in trade. "Imagination does not exist in the free state, and itself requires a basis of fact."[38]

But the other side of the picture is that:

> This basis of fact in turn requires a basis of imagination, for no fact can be known until it has been sought by the imaginative act of questioning, and this question itself requires a further basis of fact, and so on *ad infinitum*.[39]

Thus, just as imagination is fact-based, so also fact is imagination-based; each mediates the other, and, in turn, is mediated by the other. Fact is not something 'out there' to be apprehended, as realist

---

[36] *Speculum Mentis*, p. 79.
[37] *Ibid.*, p. 80.
[38] *Ibid.*
[39] *Ibid.*

empiricists would have us believe.[40] The very facticity of a fact already involves the stabilizing work of the imagination working on the flux of sensation, since it, that is, fact, is an object of consciousness and consciousness is always of something, of some definite object, never pure, undifferentiated, indeterminate sensation.[41] The 'sense-data' of empiricists are already interpretations, not only of imagination, but even of the understanding, for the term involves their relationships to one another.[42] The objects of imagination also are interpretations of sensation, not interpretations at the level of understanding, involving conceptual and differentiation and interrelation, but imaginative interpretation. So, the work of imagination is involved in every fact.

But, conversely, the work of imagination is not only implied by, but itself implies, a preexistent basis of fact. Collingwood asks, 'Does this involve us in an infinite regress of fact dependent upon imagination and imagination dependent on fact, and so on?' His answer is 'No':

> . . . only because the two moments, question and answer, are not actually separate. Their distinction is an ideal distinction only, and the presupposition of each by the other is only a way of stating their inseparability.[43]

Collingwood then states that:

> the process of knowledge is therefore, strictly speaking, not so much an alternation of question and answer as a perpetual restatement of the question, which is identical with a perpetual revision of the answer.[44]

Anticipating the objection that this reduces all the diversity of knowledge to a bare identity in which there is only one judgment judging one truth, Collingwood replies: "Our answer—to be given in full later—will be that this identity contains all diversity within itself."[45]

---

[40] That is, there are no 'hard facts'; see G. Buchdahl, "Logic and History (An Assessment of R. G. Collingwood's Idea of History)," *Australasian Journal of Philosophy*, XXVI (1948), p. 95.

[41] See *The Principles of Art*, p. 203: "The place which Hume's ideas inhabit is the empty room of Locke, progressively furnished with what 'the busy and boundless Fancy of man' provides. *And it is imagination, not sensation, to which appeal is made when empiricists appeal to 'experience.'* " (Emphasis added.)

[42] See *Principles of Art*, p. 214.

[43] *Speculum Mentis*, p. 80.

[44] *Ibid*.

[45] *Ibid*. Collingwood's own thought has been followed here as closely as possible, for this, and his immediately preceding statements, are of the highest importance for the interpretation of his thought.

This 'identity which contains within itself all diversity' will be shown later to be the key concept of Collingwood's whole thought, especially in regard to the logic of question and answer; it is the notion of the concrete universal which Collingwood inherits from the British Idealists, especially Bradley and Bosanquet.[46]

The interpretation of Collingwood's treatment of the totality of the life of the mind in *Speculum Mentis* in the light of what he says in this particular section (pp. 76-80) of *Speculum Mentis* can be illustratively summarized by imagining the mind's activity as a moving hand on the face of a clock. If we imagine the full sweep of the hand around the face of the clock from 6 o'clock position back to 6 o'clock position through the 12 o'clock position and all the intervening positions, we may regard the beginning, 6 o'clock, position of the hand as representing experience devoid as yet of any explanation, and the terminating 6 o'clock position as representing fully explained experience, the intervening positions representing the various moments of the movement of the activity of explanation. The 12 o'clock position of the hand would represent the greatest elongation (abstraction) of explanation from the experience it purports to explain. This would represent that of abstract mathematical or exact science in which pure explicative thought is maximally independent of experienced actuality. Art, then religion, would be located somewhere on the passage from 6 o'clock to 12 o'clock; experimental science and then history would be represented somewhere along the return sweep from 12 o'clock back to 6 o'clock according as greater and greater account is taken of fact. Philosophy may be regarded as the central position on the face from which the hand radiates in its sweep around the dial; it is concerned with all respective positions. History in its culmination at the 6 o'clock position is at the same time the recognition of the clock *as a clock*, that is, the fully explicit recognition of what the overall philosophical view mani-

---

[46] This "identity" which "contains all diversity in itself" is "the absolute object [which] is individuality, for individuality is concreteness," that is, fully determined. (*Speculum Mentis*, p. 218.) It is a 'concrete universal' which, in the tradition of British Idealism, is the basis of logic and intelligibility. The interpretation of Collingwood has therefore to be made against the background of Bradley's and Bosanquet's (especially the latter's) teaching on the concrete universal. Likewise, the identification of conscious awareness with one sustained judgment is necessary to understand what Collingwood says here; see B. Bosanquet, *The Essentials of Logic*, London, Macmillan, 1897, pp. 21-41, especially p. 33. Whereas, for Bosanquet, consciousness is identified with one sustained affirmative judgment, for Collingwood consciousness is a sustained ('restatement') answer to a sustained ('restated') question. This is how I read his insistence on philosophy as being a constant return to its beginnings; see his *An Essay on Philosophical Method*, pp. 22-23, 168.

fests in regard to all particular aspects, the view of the whole as a whole in relation to all its parts.

If we, then, imagine ourselves seated somewhere at the pointing end of the moving hand we may imagine ourselves looking either in the backward direction from which the hand has travelled or in the forward direction towards which it is travelling. If we adopt the backward looking aspect what we will see has the aspect of available information, or answers to questions. But if we take the forward looking aspect what we see takes on the aspect of questioning. Thus, any point indicated by the moving hand is *both* a questioning *and* an answering position (just as at any moment the hand is *together* moving away from the 6 o'clock position *and* back towards the 6 o'clock position); it all depends on whether you regard the one same concrete actuality of movement in relation to what it has left behind or in relation to what it is approaching. An anti-clockwise sweep of the mind's eye round the dial sees everything as answers, whereas a clockwise sweep of the mind's eye around the clock face will see everything as questions.

If this helpful illustration is kept in mind, many of the texts just analysed become extremely illuminating in their relation to Collingwood's global position and to many of its details. Reconsider, for instance, Collingwood's statement[47] that supposal and assertion are not two independent chapters in the mind's history but two opposite and correlative activities forming the systole and diastole of knowledge, and, again, his statement that it is only in retrospect, by looking back over the road travelled, that knowledge comes to be identified with assertion, with accomplished past fact, such as encyclopaedia knowledge.

Collingwood's statement that "the progressive alienation of the mind from its object is in history complete"[48] would seem to argue against locating mathematics and exact science at the 12 o'clock position of the above clock-face illustration as the position of greatest elongation of abstractive distancing from experience, located at the 6 o'clock position. But what Collingwood says repeatedly (especially in *Speculum Mentis*, pp. 163-169), seems rather to qualify exact science for the occupancy of this position of abstractive extremity.

B. The Nature of Supposal as an Activity of the Imagination

As stated above[49] Collingwood aligns the activity of questioning

[47] See p. 64, this chapter, above.
[48] See *Speculum Mentis*, p. 238.
[49] This chapter, p. 63.

with the imaginative activity of supposal whose proper characteristic is that it treats objects indifferently as to whether they are real or not.

If absolutized as a distinct life form and thus lived for its own sake this imaginative activity of supposal is the same as the aesthetic activity of art. In order more fully to understand this activity of supposal so relevant to questioning it will be necessary to analyse what Collingwood says regarding the nature of art and try to disengage those aspects of his theory of art which throw light on the nature of supposal as this latter applies to the theory of questioning. This will require an analysis of Collingwood's *The Principles of Art* (1938). Since the relevant material is somewhat involved, it is considered advisable to move this whole section to an appendix, rather than disturb the continuity of thought in analysing Collingwood's theory of questioning. Having dealt with the nature of questioning as supposal or the active thrust of the mind in transforming experience into rationally meaningful experience, the next consideration, and the most basic consideration from the logical point of view, is that of the type of universal according to which experience is structured when it is the experience of a questioning mind. This is the concrete universal which is the basis of concrete and historical thinking as opposed to the abstract genus species logic based on the abstract universal proper to scientific thinking. The treatment of Collingwood's theory of art as this bears on his theory of questioning will be found in an appendix at the end of this work.

CHAPTER FOUR

Concrete Dialectical Rapprochement Question and Answer Logic *versus* Abstract Static Divisive Propositional Logic

A. The Concrete Universal vs. the Abstract Universal

According to R. G. C. Mure,[1] the doctrine of the concrete universal originates in Hegel's rejection of Kant's explanation of the moment of individual reality in human knowledge. Rejecting Kant's passive sense-intuition of an alien thing-in-itself, "Hegel sets his own interpretation on the original unity of thought and being adumbrated in Kant's unity of apperception."[2] ". . . the Notion . . . contains within itself, and itself sets forth, the Negation which determines it. Inasmuch as its significant negation is its own activity, nothing outside it determines it. And that is to say that it is the Universal—the Universal determining itself as Individual."[3] "Here . . . originates the doctrine of the concrete universal which has inspired British nineteenth century idealism and found perhaps its most consistent expression in Bosanquet's Gifford lectures."[4]

---

[1] *A Study of Hegel's Logic*, London, Oxford University Press, 1950.

[2] *Ibid.*, p. 165. See also James E. Griffiss, Jr., "The Kantian Background of Hegel's Logic," *The New Scholasticism*, XLIII (1969), pp. 509-529, especially pp. 512-513; 518-520; 522-523.

[3] R. G. C. Mure, *op. cit.*, p. 165.

[4] *Ibid.*, p. 166. Bosanquet's most significant treatment of the concrete universal is his second lecture, The Gifford Lectures for 1911 at the University of Edinburgh, published as *The Principle of Individuality and Value*, London, Macmillan, 1927, pp. 31-81 (abstract of lecture 2, pp. xix-xxi.)

Lionel Rubinoff says[5] that "Collingwood's conclusion . . . concerning the nature of historical explanation is clear and unequivocal" when he (Collingwood) asserts[6] that ". . . the concrete universal is the daily bread of every historian, and the logic of history is the logic of the concrete universal." If, as Collingwood elsewhere argued,[7] the logic of history is distinctively a logic of question and answer, it follows that this logic of question and answer is a logic of the concrete universal. This logic is quite distinct from the abstract, formal logic which characterises scientific thought.

> Formal logic, whether in its scholastic or mathematical form, is not only reflection upon thought in the special sense of scientific thought, but in the special sense of scientific reflection; and this is the fact which divides formal logic from that dialectical logic whose point of view is the point of view not of science but of philosophy, and whose object is not the abstract thought of science but the concrete thought of history and philosophy.[8]

A.i. The Abstractive Character of Science

Collingwood constantly contrasts the abstract universal, which is the basis of the logic operative in science, with the concrete universal, which is the basis of the logic operative in history and philosophy.

The abstraction characteristic of science derives from its origin in religion.

> In religion the mind is accustomed to regard the unity of God as standing in perfectly self-contained independence over against the plurality of empirical fact; the one and the many, the universal and its particulars are thrown outside one another and regarded not as correlative aspects of a single concrete whole but as independent concrete beings. This is a consequence of the imaginative nature of religion which inevitably personifies abstractions or presents ideal distinctions in the guise of concrete objects.[9]

This abstraction, which characterizes science as a hangover from religious transcendence, results in the logic of science being a logic of external relations.

---

[5] *Collingwood and the Reform of Metaphysics*, Toronto, University of Toronto Press, 1970, p. 160.

[6] *Speculum Mentis*, p. 221. See this chapter, below, pp. 89-90.

[7] See "Historical Evidence," *The Idea of History*, pp. 249-282. See also Chapter 1, above, pp. 22-34.

[8] R. G. Collingwood, *Speculum Mentis*, p. 195.

[9] *Ibid.*, p. 159.

Once we have, by that relic of religious transcendence which infects all science, divorced the world of thought from the world of sensation, the mutual externality of these two worlds reproduces itself *ad infinitum* within science itself, and gives rise to that logic of external relations which is just the mark of science's failure to free itself altogether from the domination of a religious habit of mind.[10]

Externality may thus be regarded as the *a priori* form of the objects of science, using Kant's terminology.

Abstraction, according to Collingwood, is the error (characteristic of the scientific mode of thought, subsequently resulting in its self-transcendence) of separating in thought what is not separated in reality.

To abstract is to consider separately things that are inseparable: to think of the universal, for instance, without reflecting that it is merely the universal of its particulars, and to assume that one can isolate it in thought and study it in this isolation. This assumption is an error. One cannot abstract without falsifying. To think of them as they are not and to plead that this initial severance makes no essential difference to their inner nature is only to erect falsification into a principle.[11]

Science is the affirmation of the abstract concept as real. As abstract it is apprehended by abstract thought without any aid from sensation. A scientist merely has to think; sensuous experience is unnecessary to him.[12]

Mathematics is *a priori* or deductive science par excellence;[13] it is built up by simply drawing out the implications of a merely asserted abstract concept independently of all reference to sensuous experience.[14] The only really *a priori* concept is the concept of class as such, since any difference between one class and another is empirical. Mathematics, as a pure science, is simply the theory of classificatory order as such, that is, structure in its most abstract form.[15]

The assertion of the abstract concept, by which the scientific concept comes into being, marks all science with the three inseparable

---

[10] *Ibid.*, p. 166.

[11] *Ibid.*, p. 160. See also A. J. M. Milne, *The Social Philosophy of English Idealism*, London, Allen & Unwin, 1962, Chapter 1, pp. 15-22.

[12] See *Speculum Mentis*, pp. 163-164.

[13] See *ibid.*, pp. 164-166.

[14] See *ibid.*, p. 164.

[15] See *ibid.*, p. 165. See also A. J. M. Milne, *The Social Philosophy of English Idealism*, pp. 16-17.

characters of mathematics, mechanism and materialism. These are products of the classificatory frame of mind, as corollaries of the fact that this frame of mind arbitrarily separates the universal and the particular in asserting the universal in its barren and rigid self-identity.[16] Materiality means abstractness and subjection to the formulae of mechanical determination and mathematical calculation, which subjection is imposed by the arbitrary act of abstraction which falsifies the object's nature by sundering what in fact is united.[17]

As long as science remains rigidly *a priori*, mathematics is the only possible science. But other sciences are possible if empirical elements are admitted into science, or if science is allowed to indulge in hypotheses. Thus, Euclidean geometry may be obtained either as a systematic account of actual empirical space, or as a pure deduction from the wholly *a priori* hypothesis that Euclidean space exists. In the first case, science would not be pure (as admitting data from the world of sensation), and in the second case it would not be true (being based on a hypothesis, not on concepts asserted as constituting the real world).[18]

But, the concept of reality, as a concept of pure, *a priori* science, is absolutized externality, absolutized exclusiveness, and thus is a class with no members. It turns out to be thought thinking about nothing real.

> The self-identity of the concept, in abstraction from the difference of the sensible world, collapses into nothingness and leaves us with the empty form of a thought which thinks nothing.[19]

Thus, pure deductive science collapses under the strain of this internal contradiction.

> Pure deductive science breaks down because its object turns out to be a chimera, a structure which is the structure of nothing, a law with no instances.[20]

In opposition to the barrenness of pure a priorism, the Renaissance scientists fought for the recognition of facts. They insisted that observation and experiment are the true process of scientific knowledge. They saw that the purely abstract or deductive ideal of science gives birth to mathematics and to nothing more. But they also saw that,

---

[16] See *Speculum Mentis*, p. 167; p. 293.
[17] See *ibid.*, p. 168.
[18] See *ibid.*, pp. 168-169.
[19] *Ibid.*, p. 169.
[20] *Ibid.*, p. 176.

in an atmosphere of pure deduction, mathematics cannot survive. Concrete reality can only be an element in the totality of science if the deductive element is supplemented and reinforced by an inductive or empirical element. They did not seek to supplant mathematics by empirical elements, since mathematics represents the ideal of science, but they sought to give the ideal real content by supplementing it with empirical elements derived from observation and experiment.[21]

Greek biology, astronomy, etc., were not purely mathematical, but their empirical element was implicit only. The Renaissance battle with Aristotle explicitated the implicitly empirical element in Greek thought. This had far-reaching results, both theoretical and practical.

> The result was on the theoretical side the substitution of inductive logic for deductive; on the practical side, an immense expansion and consolidation of scientific thought, which, because freed from the false ideal of pure deduction, could now for the first time realize its true nature as based on the empirical study of fact.[22]

But the emergence into explicitness of the empirical character does not make science cease to be science. Inductive logic stands over against deductive logic in an ambiguous relationship. On the one hand, observation and experiment are presupposed by *a priori* deduction.

> Inductive logic does not really supersede deductive; it stands over against it as at once a friend and an enemy. Observation and experiment do not supersede *a priori* deduction; rather they are said to underlie and be presupposed by it, every deductive process resting on a previous induction. Thus the theory arose of which Mill is the great representative, that science proceeds by an alternating rhythmical process of which induction is the first moment and deduction the second.[23]

On the other hand, induction itself is found to presuppose the very same deductive principles which it is supposedly in search of.

> This theory [of Mill, that induction and deduction are alternate moments of the scientific process] breaks down on discovering that induction itself rests on a principle, variously described as the uniformity of nature, the law of universal causation, and so forth, which, just because induction rests upon it, induction is powerless to establish. Thus the whole of induction falls to the ground because it is found to assume that of which it is ostensibly in search.
>
> This principle of uniformity, which induction assumes, is also the principle which deduction assumes, namely the existence of

[21] See *ibid.*, p. 177.
[22] *Ibid.*, p. 178.
[23] *Ibid.*, pp. 178-179.

> abstract concepts in general; for uniformity is nothing but the abstractness of the concept, its indifference to the variations of its own instances. Thus induction does not overcome the fundamental abstractness of deduction but simply reasserts it, and this fundamental assertion in which both agree is nothing but the definition of the scientific consciousness. Both deductive and inductive logic are therefore developed out of the definition of science as such.[24]

On account of its postulatory abstraction, science is caught in a vicious circle from which it cannot escape, save by denying the postulate that the universal and particular are not distinct things but merely two distinct elements of one whole. But such a recognition is the very denial of science itself. It therefore alternates between induction and deduction in an attempt to operate within its own arbitrarily self-imposed duality.

> Therefore science is bound to maintain both at once, and to do what it can towards sandwiching the deductive and inductive accounts of itself in slices as thin as possible but never actually fused together. If this fusion took place, which is the only possible solution of the problem of logic, science would cease to exist, for the distinction between induction and deduction would have vanished, and this would imply the denial of any distinction between the process from particular to universal and that from universal to particular, and this again would involve the discovery that the universal and the particular are not separate things, but only distinct elements of a whole which even in thought is not divisible. This recognition is the death of science as such, because science is the assertion of the abstract concept.[25]

Collingwood says that the realization of this led scientists themselves away from the nineteenth century conviction that science discovered the laws of nature to the present day schools of thought, some of which maintain that the object of science is the possible, not the real, and others which maintain that science is not true but useful.[26] A large number of eminent men, and apparently pragmatists maintaining the utility of science, hold the position which regards science as hypothesis.[27] This means that science asserts not what is true but what would be true if some condition were supposed. It remains in the hypothetical order, never dealing with the real order of actuality and fact. It is pure implication with nothing asserted.

---

[24] *Ibid.*, p. 179.
[25] *Ibid.*, pp. 179-180.
[26] See *ibid.*, pp. 180-181.
[27] See *ibid.*, p. 182.

> Science asserts, not the actual truth, but what would be true if something were true which is laid down as an hypothesis. It asserts, never that S is P, but that if there were an S it would be P. Its procedure therefore consists, first, in making an assumption, secondly in deducing the consequences of that assumption. Throughout this process it never makes an assertion, in the sense of a categorical judgement, at all. Its judgements are hypothetical from beginning to end. . . . Science is a pure tissue of implications, none of which are asserted as facts. The scientist may frame any hypothesis he likes; the freedom of untrammelled supposal is his; he must merely suppose something, and then work out the consequences—all equally matter of supposal, never of assertion—which it entails.[28]

Science never really grasps the individual, or historical fact, which is the concrete basis presupposed by all scientific abstraction. The Renaissance scientists explicated this implicit presupposition in discovering that science is an abstraction from the concrete reality of history. Scientists recognize fact indeed, but "a scientific fact is a fact purged of its crude and scientifically scandalous concreteness,"[29] and isolated from its historical setting by being reduced to the mere instance of a rule. The concrete fact has been turned from an individual into a particular.

> The facts with which empirical science concerns itself are facts thus de-individualized, de-factualized, and this is what distinguishes the sense in which even the most empirical science uses the word 'fact' from the sense it bears in history.[30]

Collingwood concludes that science is therefore explicitly supposal, which is identical with questioning, the cutting edge of the mind, implying behind it a body of information or assertion. The body of science is therefore not in a scientific form but in the form of history. Every scientist knows this in acknowledging that only the possession of an ordered body of facts enables him to frame the hypotheses which are the very essence of the scientific life. Collingwood's conclusion, therefore, is that:

> Science is the question whose answer is history. To ask that question implies that history is already in existence; and thus we get a process of history—science—history. But history on its first appearance is implicit; it is not known for what it is, and is indeed despised as the mere world of empirical or sensuous reality. It is

[28] *Ibid.*, p. 183.
[29] *Ibid.*, p. 186.
[30] *Ibid.*, p. 186.

> only when it has been distilled into terms of science and then restored to itself in the form of concepts or laws that it is recognized for what it really is.[31]
>
> The paradox of science may be expressed by calling it intuitive thought. Intuition is the questioning, immediate side of experience: thought is the asserting, explanatory side. Science is explicit to itself as thought, but it turns out on inquiry to be identical with the questioning activity; that is, it realizes the contradiction of a type of thought which is not thought precisely because it is thought's opposite, intuition.[32]

But, however erroneous science may be in abstracting the universal from its instances, it is a necessary stage in human development. It has the character of a utilitarian good, as the Pragmatists well recognize, first, in providing us with the material needs of life, and, second, in freeing us from subjection to our mythologizing imagination.

> As art and religion lift man above the level of the beasts, science lifts the civilized man above the level of the savage. The material utility of science, its service in feeding and clothing and sheltering us, carrying us from place to place and providing us with comforts, is the least part of its importance. Its real gift is simply the end of dreaming and the promise of a waking life. It sweeps aside with a ruthless hand all mythology, all symbols that are heavy with unrealized meanings and dark with the terrors of dreamland, and bids the mind face the world's mystery armed with nothing but its five senses and the sling of its wit.[33]

A.ii. Distinction between Understanding and Reason

Collingwood says that ". . . the revolt against science . . . is summed up in the antithesis of understanding and reason. Understanding is abstract thought, the 'faculty of concepts', thought spontaneously originating concepts or categories out of itself with no assistance from the world of fact."[34] Plato's concepts were postulated as self-subsistent entities, and the Baconian concepts were allegedly the actual structure of the world of fact, discovered embedded in that world by the mind. Collingwood says that the Platonists and Baconians err in overlooking the original act of abstraction which severed the universal from its particulars. The universal, which is the object of the understanding, is the universal torn away from its particulars by an act of abstraction and not the universal in its true actuality.

---

[31] *Ibid.*, pp. 186-187.
[32] *Ibid.*, p. 188.
[33] *Ibid.*, p. 194.
[34] *Speculum Mentis*, pp. 195-196.

But, whereas understanding is abstract thought, reason is concrete thought, whose object is a universal which deals with facts as they are, respecting the differences which are organic to that universal itself. Reason, therefore, deals with things in terms of their internal relations and their necessary mutual implication.

> Reason is concrete thought, thought which does not arbitrarily create to itself, by abstraction, any object it pleases for the sake of ease in thinking it, but sets out to study facts as they are, and to conceive a universal which is truly the universal of its own particulars. Hence reason thinks the concrete universal, not the bare self-identity of science which leaves all difference outside itself, but the identity to which difference is organic and essential. Understanding hypostatizes the concept into an object of intuition by itself, outside its own particulars; and this object is nothing real, but simply the fruit of an error. Reason finds the concept in the particulars, forming with them an inseparable unity.[35]

Reason, therefore, has as its object the concrete universal, which is the immanent structure of the world of experience, not an independent and isolated entity extrinsic to its particulars.

> The concept is not something outside the world of sensuous experience: it is the very structure or order of that world itself. The arrangement and the material arranged are only distinguished by an abstract and arbitrary distinction within the indivisible whole. The universal is only real as exemplified in the particular, the particular as informed by the universal. . . . The meaning [of a linguistic symbol] or concept or universal is not a separate object of consciousness other than the world of sense; not something seen through a veil of sense, but the structure of that veil itself. This is the point of view of concrete thought.[36]

Whereas the unity of the abstract universal is unity *in spite of* differences, a unity to which diversity is extrinsic, the unity of the concrete universal is a unity *because of* its differences, a unity to which the differences are relevant. It is a unity which demands diversity within it, a true unity of opposites, a unity in and through difference, a unity which essentially requires its differences. The concrete universal is thus a unity of universal and particular, it is the universal *of* its particulars and they are the particulars *of* their universal.

> The characteristic of reason is the unity of universal and particular, or in general the unity of opposites. The concrete reality of

[35] *Ibid.*, p. 196.

[36] *Ibid.*, p. 159. "The recognition of this logical form [the concrete universal, unity in diversity] as the true type of universality is the key to all sound philosophy." B. Bosanquet, *The Principle of Individuality and Value*, London, Macmillan, 1927, p. 39.

> any one opposite is its union with the other; not a bare indistinguishable identity but a union in which the two sides can be distinguished but not separated.[37]

The identity of the concrete universal is, therefore, not the abstract self-identity usually expressed by the formula A is A, which is identity in indistinction, but is coincident identity of opposites. Whereas the abstract principle of identity rules thought in the form of understanding, the concrete principle of identity, the principle of coincidence of opposites, rules thought in the form of reasoning.

> The identity, in the sense of an indistinguishable identity, of opposites is not the principle of reason but a disease, endemic and mortal, of understanding. From the violent and arbitrary separation of two opposites, the assertion that A shall be only A and not-A only not-A, it necessarily results that A, now falsely asserted as a self-subsistent concrete reality, generates a not-A within itself and the not-A conversely generates an A. Thus each of the terms, A and not-A, produces its own opposite, and when that has happened they cannot be any longer distinguished. This is the *coincidentia oppositorum* which always dogs the footsteps of abstract or scientific thinking. Matter and mind, affirmation and negation, good and evil, truth and falsehood, universal and particular, are no sooner resolutely separated than they turn into each other. Hold up a stick, and distinguish its top and bottom: there you have a concrete synthesis of opposites in an individual whole. Take a knife and cut it in two in the middle, into a top half and a bottom half. You have now separated the opposites. But the instant the separation is complete, the top half has its own bottom and the bottom half its own top. The top half is no longer simply a top and the bottom half no longer simply a bottom; each is at once a top and a bottom, each is indistinguishable from the other. Your opposites have now coincided. Abstract thought fends off the *coincidentia oppositorum* by deliberately ignoring the bottom of the top half and the top of the bottom half, and in defiance of facts simply (that is, abstractly and falsely) calling the one half all top and the other half all bottom. Thus 'the universal' as such is a particular universal, and 'the particular' as such is nothing but the universal of particularity; but formal logic conceals this fact and makes its very living by the pretence that it has not hypostatized (particularized) the universal.[38]

---

[37] *Speculum Mentis*, pp. 196-197. See also B. Bosanquet, *The Principle of Individuality and Value*, p. 37.

[38] *Speculum Mentis*, pp. 197-198. See also B. Bosanquet, *The Principle of Individuality and Value*, pp. 36-38; 47-50. The objection of the British Idealists, mainly Bradley and Bosanquet, to the formulation of the law of identity as A is A is that it is unsuitable for expressing any judgment. Any assertion, to be meaningful at all beyond tautology, requires that subject and predicate be in some way different. The formula A is A uses the same symbol for the predicate as it does for the subject, and, as

The antithesis or mutual externality of understanding and reason is the same as the antithesis of science and history, and is the work of the understanding and scientific thought. Reason repudiates this separation of reasoning from the understanding by repudiating the abstraction that sunders them and recognizes the concrete identity of understanding and reasoning itself. Understanding, the scientific consciousness, is committed to the error of denying its own identity with history, whereas history overcomes that error and sees itself as science transcending science, seeing itself (i.e., history) as a concrete universal of which science is one of its moments.[39]

A.iii. History as Concrete Thought

Collingwood attributes the revelation that the hypotheses or abstractions of science rest on the knowledge of fact to Descartes' discovery that all science rests upon the one indubitable certainty that I think, therefore I exist. The thought and the existence spoken of by Descartes were not abstractions. What Descartes said, and meant, was that the concrete historical fact, the fact of my actual present awareness, is the root of science. Collingwood says that Descartes was only going a step beyond Bacon when he saw, more profoundly, that before a scientist can make use of any fact he must observe it, and his observing of it is the fact that really matters in the last resort. Descartes' formula is the deepest and most fruitful expression of the discovery that science presupposes history and can never go behind history. And this discovery implicitly resolves science into history.[40]

In perception there is found the identical process of reconstruction from data which constitutes the essence of history, namely, to grasp the object as a whole in a synthesis of front and back, top and bottom, past, present, and future.[41] The historian's business is with fact; history is the affirmation of fact. The business of history is to state what happens and has happened, and that only. And the true historian is not content with stating facts, but endeavours to under-

---

such, fails to bring out the essential of any judgment which is the assertion of an identity in a diversity. See H. B. Acton, "The Theory of Concrete Universals," *Mind*, XLV (1936), p. 427. "Identity without difference (which is called abstract identity), would make judgment impossible. . . . If, then, judgment *is* possible, there must be some *other* sort of identity, which is called identity in difference or *concrete* identity." (*Ibid.*, p. 428.) "The reason why, according to idealists, there can be no difference without identity, is that difference is a relation, and no terms can be related unless they are identical in some respects." (*Ibid.*, p. 429.)

[39] See *Speculum Mentis*, p. 198.

[40] See *ibid.*, p. 202.

[41] See *ibid.*, p. 212.

stand them. He seeks to know not only what happened but why it happened.[42]

Fact is by definition concrete, and the historical fact in its full concreteness already contains its explicative causes and motives within itself; for the historian to know what happened is to know why it happened; its relations to its context of fact are not extrinsic but intrinsic to it. To know the fact is to know it in its relations to other facts. So, whereas science knows in terms of external relations, history knows in terms of internal relations, that is, relations that make the fact be the kind of fact it actually is.

> To understand the facts is to affirm them not in arbitrary isolation but in their actual relation to their context. The reason why an event happened is sought by the historian not in an abstract scientific law but in facts, and facts again. The cause of an event in history is its intrinsic relation to other events in history, and the causal nexus is not external to them but lies in their very nature. The motives of historical personages are not psychical forces brooding above the flow of historical events: they are elements in these events, or rather, they are simply these events themselves as purposed and planned by the agents.[43]

---

[42] See *ibid.*, p. 217.

[43] *Ibid.*, p. 218. It may be noted in passing the light that this throws on Collingwood's assertion that history gives 'insight' into events (see *An Autobiography*, p. 101). The scientist is external to the objects he studies, and they are external one to the other, and the abstract universal, the principle of unity and intelligibility in science, is external to its particulars. The form of externality rules the mode of knowing entirely; all that is known scientifically is known 'exteriorly' and 'in its externals.' But the historian must penetrate to the 'inside' of events, to know them not in their external relations to one another, but in their internal relations whereby they are essentially interrelated according to what they are. The historian is immanent to his facts and situates himself so that he sees events 'from the inside,' that is, according to their internal relations. But the inner side of events is thought, namely human motives and purposes. To know the thought within the event is to know together what happened and why it happened. This is to have 'insight' into the event, to know it 'from the inside' according to its internal relations to contextual events. Thought has the mode of 'insidedness' or 'interiority to self' or concreteness, whereas science and its object, nature, have the mode of exteriority, which sets up the relationships of other to other, whose unity can then only be that of mere external relatedness. So, according to Collingwood, whereas science proceeds by way of induction based on observation and experiment which provide the facts, history proceeds by way of 'insight.' According to Bernard Bosanquet, *The Principle of Individuality and Value*, pp. 74-77, the natural order, which is characterised by spatial exteriority, is subordinated to and ultimately reduced to the inwardness of the spiritual order. External relations are to be explained ultimately in terms of internal relations, or science in terms of human consciousness. ''Inwardness, when meant to be the equivalent of Individuality or the character of spirit, should be taken as a type of experience superior to externality and including it.'' (*Ibid.*, p. 74.) ''Externality can subsist only as subordinated to inwardness, but inwardness can subsist only in the conquest of externality.'' (*Ibid.*, p. 76.) When scientific facts, derived from observation and experimentation, and their uni-

Whereas art ignores reality and religion ignores thought and science ignores fact, there is no feature of experience nor attitude of mind towards its object that history ignores. The recognition of fact as fact is the recognition of whatever in any sense is real. And the historically determined fact is the absolute object whose distinctive mark is individuality, which is concreteness.[44] That is, the meaning of a fact as a fact is its implication *of* and implication *by* every other fact. The world of meaning by which a fact is mediated to know the meaning of that fact is the all inclusive world of everything that is, including that fact itself. It is a systematic whole to which that individual fact, together with every other individual fact, is essential, or relevant.

> The object as individual is the whole of what exists, and this is concretely articulated into parts each of which is again individual, and so to infinity. Within the cycle of this infinite articulation of the absolute object the historical spirit moves freely in all directions, never finding anything that is not individual and unique, never finding anything that is not, on the one hand, composed of individual and unique parts, and, on the other, itself a part of an individual

formities expressed in the different laws of nature, are systematised into the organic totality and seen as members of a world or cosmos, the external relations of the natural order are reduced to the internal relations of a conscious totality. Collingwood, contrary to Bosanquet, would say that the external relations according to which science totalises experience are ultimately to be reduced to the internal relations according to which the historical consciousness totalises by means of historical 'insight,' rather than through scientific systematisation. For Bosanquet, on the contrary, concrete totality is found in scientific systematisation when science achieves a cosmic or world view of its object (*ibid.*, pp. 35-37). But, given Collingwood's identification of historical knowledge with the inner side of events, and what he calls 'insight,' the reduction of externality to spiritual inwardness of Bosanquet becomes Collingwood's reduction of abstract, scientific knowledge to concrete historical knowledge. Thus he says (*Speculum Mentis*, p. 293) that "all externality is imaginary, for externality—a mutual outsidedness in the abstract sense of the denial of a mutual insidedness—is as such abstraction, and abstraction is always intuition or imagination." Collingwood seems to be much more logical than Bosanquet in accepting history as the science of concrete individuality and universality *par excellence*. Bosanquet's depreciation of history in this respect is found in his *The Principle of Individuality and Value*, pp. 32-33 and 78-81, where he alleges that the best examples of a concrete universal are to be found in Art, Religion and Philosophy, which "take us far beyond the spatio-temporal externality of history" (p. 80). "History is a hybrid form of experience, incapable of any considerable 'degree of being or trueness' " (pp. 78-79). Collingwood remarks (*The Idea of History*, p. 143) that "Dr. Inge . . . follows Bosanquet in conceiving the proper object of knowledge Platonically as a timeless world of pure universality." However, see below, this chapter, p. 136, n. 228, for implications within Bosanquet's same work for amendment of his allegedly static timeless notion of concrete totality in line with Collingwood's notion of factual, historical concreteness. Correlate with the treatment of 'insight' in chapter 6, below, pp. 216-221 and n. 45 of p. 221.

[44] See *Speculum Mentis*, p. 218.

and unique whole. The object, as a system of fact so organized, is objective throughout, for every part is a true microcosm, and is truly infinite.[45]

The concreteness and individuality which is finally attained by the historical consciousness, was what the earlier forms of consciousness were searching for. But each grasped it in such a way as to destroy that individuality.

The aesthetic consciousness expresses itself in a single work of art. Here it finds a cosmos which like the world of history is individual and unique, whose organization is a systematic structure of parts each displaying the same individuality, the same uniqueness, in its indispensable contribution to the whole. But it is only by an abstraction that the single work of art is a cosmos. Outside it are other works of art, each a cosmos in itself, and these do not combine into a totality which is itself a single all-embracing work of art. Thus the aesthetic consciousness, instead of systematic, is monadic; and the monads, being windowless, are not so much reflections of the whole universe as rival claimants to be the whole universe. This rivalry is fatal to all the claims alike. Art is thus even within its own limits a false form of individuality.

Religion goes a step farther and finds exactly the same individuality in God. God is the monad of monads, a cosmos whose structure is that of the absolute object. But God always stands over against a world whose very nature is to be outside him, and thus God forfeits his own absoluteness. As long as the world stands over against him, his own individuality is unattained. And because religion is conscious of this, its search for objectivity is confessed to be vain.

In science the attempt is made to bridge the gulf between God and the world. Instead of God we find the concept of law, which because it is the law of the world no longer stands outside the world. Here what was transcendent has become immanent. But in this very success, failure is revealed; for it becomes plain that the immanence is a false immanence. The world is not a world of individuals but

[45] *Ibid.*, pp. 218-219. H. B. Acton says that, according to Bradley, the abstract universal is nothing but an adjective, a shadow which is nothing divorced from its body and, as such, cannot exist. "What Bradley . . . means by a *concrete universal* . . . [is] an individual, and his reason for holding that individuals are universal is that they are entities which preserve an identity amidst differences." H. B. Acton, "The Theory of Concrete Universals," *Mind*, XLV (1936), p. 421. "Bradley's theory then is that what exist or are real are individuals, and individuals, being identities in difference, are universals." (*Ibid.*, p. 422.) "In [Idealists'] terminology, 'individuals', 'concrete identities', and 'concrete univerals', all stand for the same thing." (*Ibid.*, *Mind*, XLVI [1937] p. 1). Acton cites Bosanquet (*Principle of Individuality and Value*, p. 40) as saying that the most convenient example of a universal is not a quality such as redness, but a person, such as Julius Caesar. "Julius Caesar is a universal because he is the *same* individual, although engaged in *different* activities, such as fighting in Gaul, marching into Italy, and so on. Abstracted from his various activities he is nothing; they are the differences which are required to make him the identity he is." (*Ibid.*, p. 1.)

a world of particulars, and because the concept is indifferent to the various particulars in which it is embodied, their diversity remains meaningless and the world is to that extent a chaos. What is individual and organized as a system of individuals is not the world but only the concept, so that the failure of religion is repeated and the quest for immanence ends in transcendence, but an abstract transcendence more intolerable than the concrete transcendence of religion.

Thus the work of art, God, and the abstract concept are all attempts on the part of thought to reach the organized individuality of history. Art comes nearest to success; religion fails more openly, and science most openly of all. But this order of relative failure is due to an inverse order of seriousness in facing the problem. Art ignores the real world altogether, and constructs an arbitrary cosmos of its own; religion contents itself with a cosmos outside the world; and science alone tries to bring the concrete world into the unity, but destroys its concreteness in the attempt. But because all are agreed that the real object must be an absolute individual, their failure is wholly redeemed by the success of history, which actually achieves the idea of an object beyond which there is nothing and within which every part truly represents the whole.[46]

Collingwood then gives a very clear description of what he means by the concrete universal and of its relevance to history and to the logic of history.

This absolute whole is the concrete universal; for concrete universality is individuality, the individual being simply the unity of the universal and the particular. The absolute individual is universal in that it is what it is throughout, and every part of it is as individual as itself. On the other hand it is no mere abstraction, the abstract quality of individualness, but an individual which includes all others. It is the system of systems, the world of worlds. Everything in it is determined by its place in the whole, but this is not determinism because every part determines the whole and therefore by implication every other part: so that each part taken separately may be regarded as the crucial determinant of everything else, just as every separate link bears the whole responsibility for keeping the chain together. Everything in it is as unique as the whole, and the uniqueness of every part is based upon the uniqueness of every other. The principle of its structure is not classification, the abstract concept, but the concrete concept, which is relevance, or implication. The only reason why this notion of a concrete universal is thought puzzling or paradoxical is that our attempts at philosophical theory suffer from the obsession of regarding science as the only possible kind of knowledge. For the concrete universal is the daily bread of every

[46] *Speculum Mentis*, pp. 219-220.

historian, and the logic of history is the logic of the concrete universal.[47]

This identification of individuality with concrete universality by Collingwood was already well elaborated by Bernard Bosanquet in his *The Principle of Individuality and Value*, and Collingwood's thought is greatly illuminated by a brief consideration of what Bosanquet says.

Individuality, according to Bosanquet, means full determination, the determination whereby something not only 'is' but is fully 'what it is.' Whatever exists, exists as fully determinate; there is no meaning in 'it is' apart from 'it is what it is.' Both Plato and Hegel agree that this full determination of each thing involves somehow the unity of 'is' and 'is not,' as, for instance, what is when a man is is not the same as what is when a horse is; the 'being' of a man is the simultaneous 'not-being' of a horse. Anything which involves together 'is' and 'is not' so far fails to maintain itself, and its maintenance in being is therefore to be referred to some other, which fully maintains itself in being without together failing to do so. This is what is meant by a whole, by an individual, and especially by the primary instance of individual wholeness which one finds, according to Bosanquet, in a world or cosmos. A world or cosmos is that which contains all differences within itself, and itself is not differentiated from anything else which is not itself. That is, outside itself there is nothing; within itself is everything. It is that to which all things affected by otherness, relativity or negation in their being look for their maintenance and reason in being and in being what they are, that is, through their determinate differences.[48]

> . . . in as far as "is" affirms a certain determinate self-maintenance, and "is not" affirms a different one, or the character of otherness in general, so far to attach the two as predicates to the same point of being is to allege that in its self-maintenance it fails to maintain itself. This is so far to destroy the character of being as an expression for any positive experience. It is to posit and to annul in the same act. In so far, then, as an experience presents an appearance of this kind, a combination of "is" and "is not" (or "is other") without any distinction in the subject of affirmations, it falls short of the character of being. . . . It undoes itself, and fails to conserve itself in any actual character. In as far, on the other hand, as the appearance of hostility to itself is removed, by transforming the content of the experience in question into what is relatively a system, such as to accept both *this* and *the other* as cooperative and no

---

[47] *Ibid.*, pp. 220-221. See also B. Bosanquet, *The Principle of Individuality and Value*, lect. 2, pp. 37-41 and p. 77.

[48] B. Bosanquet, *The Principle of Individuality and Value*, pp. 44-45.

> longer conflicting members, the experience "is" in a higher degree; its self maintenance includes more reality; and is *pro tanto* less likely to be confronted with external facts beyond its power to assimilate.[49]

Dynamic self-containment and self-maintenance in being without essential reference to other, is therefore the mark of truth, individuality and wholeness. The more self-contained a whole is the truer it is, and the greater is its capacity to maintain and so account for itself and its members who are maintained through containment in it. This "appeal to the whole is not a detached or arbitrary procedure, but the same thing with the principle otherwise known as the principle of non-contradiction."[50] Doubt presupposes this whole of "all that is," for itself, that is the activity of doubting, exists as something within that whole, and is unable to advance anything against it. Any attempt to advance a contrary is already to agree that "something is" and this "something which is" is already contained in the system against which the doubter reacts. In other words, the doubter, in doubting, locates himself within a system of totality which makes that activity of doubting meaningful.[51] Truth cannot be contradicted because it leaves nothing outside itself (the whole of all that is) upon which a contradiction could be grounded.[52]

> Individuality is the ultimate completeness of that character of wholeness and non-contradiction which we first generalized under the name of logical stability.[53]

It makes no difference whether one says that the criterion of ultimate reality is wholeness, or individuality, or non-contradiction. Each is a name for "that which must stand . . . which has nothing without to set against it, and which is pure self-maintenance."[54] The tendency of our fullest experience is to ultimate individuality, or the absolute.[55]

The whole effort of rational experience is to unfold that experience so that it is the experience of a rational being, fully conscious of what its experience means, and able to give a full account to itself of its experience. It is the full manifestation of experience as an ordered totality, a passage from 'that it is' to 'what it is' made from within the resources of the experience itself, that is, by transforming the experi-

---

[49] *Ibid.*, p. 45.
[50] *Ibid.*, p. 44.
[51] *Ibid.*, pp. 43-44.
[52] *Ibid.*, p. 41.
[53] *Ibid.*, p. 68.
[54] *Ibid.*, p. 68.
[55] *Ibid.*, p. 69.

ence from being merely experienced, or felt, to an experience fully accounted for, that is, an experience explained.[56] It is what Collingwood calls a passage from the implicit to the explicit as warranted by the logical potentiality of the concrete universal, whose logical potentiality consists in relevance or implication, as was said above.[57]

When Collingwood says here that the principle of the structure of the concrete universal is relevance or implication I understand him to be referring to the constitution of fact by internal relations whereby fact is related to fact essentially or internally, or 'relevantly or implicatively.' The logical operation, then, is explication of the implicit, and unfolding of those internal relations whereby fact implicating other facts together implicates the totality of fact in which each fact has its reason and intelligible meaning. This, according to Collingwood, is a logical operation quite distinct from either deduction or induction in which there is movement from something known to something other not previously known, not previously implicated, but added and related to prior knowledge by extrinsic relations. This latter type of inference is characteristic of abstract or classificatory logic.[58]

I understand the foregoing to mean the following. Science deals with units of intelligibility which have their intelligibility *in themselves*; their relations to concrete facts of experience are by means of extrinsic superadded relations which leave those unit intelligibilities unaffected in themselves. The universals which are the objects of science are intelligible in themselves, and their relations to the facts which they make intelligible are extrinsic to them. This is the doctrine of realism which Collingwood constantly combats.[59] Positivistic history, or history proceeding according to the mode of science, regards historical facts as atomic units, intelligible absolutely in themselves, and contextualized with other facts by means of external relations. The whole, constituted by such externally related units, is something secondary to those related units themselves, the relation whereby they are related falling between them, not within them.

But the reason sees the parts as made intelligible through and in the whole. The whole and the parts are in mutual inter-implication;

---

[56] Refer to clockface illustration, Chapter 3, pp. 71-72 above.

[57] See above, this chapter, pp. 86, 89.

[58] See this chapter. p. 86 n. 43 above, and *Speculum Mentis*, p. 218.

[59] "The theory of the concrete universal is presupposed throughout the whole of Collingwood's philosophy. It is . . . the basis upon which he launched his lifelong attack on philosophical realism and the theory of the abstract universal." L. Rubinoff, *Collingwood and the Reform of Metaphysics*, p. 154.

the relationship of the part to the whole defines the part, and it is intelligible only through its relationship to every other part and, with every other part, to the whole of which it is the part.[60]

Central to the understanding of Collingwood's rapprochement logic is the notion of system.[61] Concreteness does not mean atomic unity, as it does for the Positivists, but systematic unity, the unity of a system. And central to the notion of systematic unity is the notion of a standard or criterion. Even the artist proceeds according to some standard or criterion whereby he makes choices as he proceeds to the completion of his artistic work. But this standard or criterion, which is not explicit to the artist, refers to wholeness or integrity which gives meaning to the artist's work and is the means whereby he judges throughout the execution of his work. The artist grasps the concrete universal of unity in diversity not in an intelligible, systematized concept, but at the level of expressed feeling.[62]

Likewise, the historian chooses his facts in accordance with some criterion, so that "no historian, however innocent, can proceed wholly without a system of *Quellenkritik*, because his very life as an historian is the perpetual exercise of some such criterion."[63]

The achievement of this concrete or historical point of view is together the recognition and transcendence of the abstractness of the scientific point of view.[64]

---

[60] L. Rubinoff, *op. cit.*, p. 156, writes: "The important thing about the concrete universal is that it treats differences as essential. Each 'moment' of a concrete universal expresses a different yet identical aspect of the whole, and in so far as it is different contributes essentially to the unity of the whole." H. B. Acton, "The Theory of the Concrete Universal," *Mind*, XLV, 1936, p. 431, says: "Idealists believe literally that nothing can be identical with itself unless it is different from itself." Acton says that this is what Bosanquet calls the paradox at once of reality and of inference and that it is upon this contention that the theory of concrete universals rests.

[61] See this chapter, below, pp. 138 et seq.

[62] See *Speculum Mentis*, pp. 63-66, 71, 94, and especially pp. 98-102. According to A. J. M. Milne in *The Social Philosophy of English Idealism*, the concrete universal, which is the central notion of nineteenth century Idealism, is bound up with a theory of rational activity and can only be understood in the light of that theory (p. 15). It is neither an empirical nor a linguistic theory (p. 22) but precisely "a theory of the rationality of rational activity, of what acting rationally is. It is criteriological rather than logical in the technical sense, being an attempt to give an account of the criterion or standard, namely, rationality. It starts from the assumption that rational activity is going on and that we have a working practical knowledge of it . . . as a theory its aim is to take what we already have a working knowledge of, and to try to get to know it better" (p. 22). This says in other words that the logical operation in dealing with the concrete universal is explicitating the already implicit; see this chapter, above, pp. 86-87 and pp. 91-93.

[63] *Speculum Mentis*, p. 213.

[64] See *ibid.*, p. 222.

> The doctrine of the concrete universal is thus a necessary presupposition of every genuine theory of history. It is in fact one of the criteria which historiography must employ in order to distinguish history from nature; for the difference between historical and scientific thinking (in general) is precisely the difference between thinking about the concrete and thinking about the abstract universal.[65]

Concrete thinking is contextual thinking, that is, thinking to which the context of the object is essential.

> . . . history is nothing but conceiving the object as concrete fact, fact to which its context is not irrelevant but essential.[66]

That is, everything, to the historical consciousness, is what it is precisely because of what everything else making up its context does to it, precisely through being its relevant context. This involves denial of the traditional doctrine of fixed substances implied in the dictum which states that "everything is what it is and not another thing" and in the Positivist doctrine of "hard facts."

For Collingwood, error is understood as a lapse from concreteness into abstraction, and all abstraction is dogmatism. Dogmatizing is the reason of error and it consists in failure to criticize our assumptions and consequent failure to recognize them as assumptions.[67] "Our enemy is abstraction,"[68] and "abstract knowledge is the same as error."[69]

Rejection of this synthesis of opposites, according to Collingwood, is at the heart of realism, whose logic is the logic of abstraction.[70] The modern realism (i.e., of Cook Wilson, etc.) is a confused running to and fro between two principles, the abstract concept and the concrete fact. Kant saw that these two opposites existed only by implicit assumption of the one by the other, but he erred in attributing the unity of sensation and thought to consciousness in general, which is a lapse into abstraction, being a failure on Kant's part to identify

---

[65] L. Rubinoff, *Collingwood and the Reform of Metaphysics*, p. 159.

[66] *Speculum Mentis*, p. 234.

[67] See *ibid.*, p. 288.

[68] *Ibid.*, p. 268.

[69] *Ibid.*, p. 313. For Collingwood, error plays an essential part in the mind's dynamic structuring of itself into a fully self-conscious, self consistent whole. It is error, through abstraction, that gives rise to the mind's immanent restlessness impelling it to transcend its present unsatisfactory condition and achieve self-equilibrium in a fuller and more integrated expression of itself.

[70] See *Speculum Mentis*, p. 285. See also *Speculum Mentis*, p. 310: "The synthesis of opposites . . . is the life of concrete thought."

the empirical and transcendental ego, or mind in its immediacy with mind in its ideal perfection.[71]

B. Logic of the Concrete Universal as Logic of Overlap of Classes

According to Lionel Rubinoff, "the logic of the concrete universal, referred to in *Speculum Mentis*, is no other than the logic of the overlap of classes which is systematically expounded in *An Essay on Philosophical Method* (1933)."[72] This logic, underlying the scale of forms displayed in *Speculum Mentis*, is developed in *An Essay on Philosophical Method* by way of contrasting the logic of the concrete universal, a logic of overlapping classes, with the classificatory logic of the abstract universal.

The classificatory logic of genus and species, which plays a useful and vital role within the boundaries of natural science, and is even a *conditio sine qua non* of the validity of its methods,[73] has for its object the abstract universal. Its aim is to guarantee the complete autonomy of each part of the classificatory whole, that is, each species of the genus.[74]

Provided that the logical doctrine of classification and division remains within the boundaries of the natural sciences no serious error is involved. Serious difficulties arise when this method is applied to deal with concepts belonging more properly to philosophy. The attempt to classify a song or an opera as either poetry or music considered as species of the genus art is frustrated from the start.[75]

The essential difference between the logical structure of scientific concepts and the logical structure of philosophical concepts makes impossible, without contradiction, the application of classificatory division into genus and species to philosophical thought.[76] Philosophical species are not mutually exclusive; they overlap.[77]

An Essay on Philosophical Method, in which Collingwood unfolds the logical characteristics of the concrete universal, which is the object of philosophy as the abstract universal is the object of science,[78] develops in the context of answering the question "What is Philos-

---

[71] See *ibid.*, p. 285.
[72] *Collingwood and the Reform of Metaphysics*, p. 160.
[73] *Ibid.*, p. 162.
[74] *Ibid.*, p. 161.
[75] See *ibid.*, p. 163.
[76] *Ibid.*
[77] *Ibid.*
[78] *Ibid.*, p. 161.

ophy?"[79] Collingwood shows throughout the course of this work that the answer to that question cannot be given, without destroying the very nature and distinctiveness of philosophy, by distinguishing thought, as a classificatory genus, into a number of species and sub-species and locating philosophy as an autonomous, exclusive species or sub-species of the common genus, thought. To proceed thus would be to absolutize logic "as a master science having jurisdiction over the whole field of science"[80] and falsely absolutize abstract thought and the abstract universal which is the principle of all classificatory, hence scientific, thinking.

The doctrine of classificatory genus and species gets into difficulties and soon becomes unmanageable when applied to "the species of a philosophical concept such as the life of spirit or mind,"[81] "to the classification of the sciences in general,"[82] "to the explanation of ethical behaviour [as] provided in the *New Leviathan*"[83] and to moral concepts,[84] "to the solution of the problem of evil,"[85] and generally "similar difficulties arise when we attempt to apply this doctrine to those concepts pertaining to human nature and the human condition,"[86] including history, of which the view according to Collingwood[87] "sees everything in that history as having its own *raison d'être* and coming into existence in order to serve the needs of the men whose minds have corporately created it." The principle whereby these entities are made intelligible is not the abstract universal, but a universal of another kind, a concrete universal of identity in difference in which the differentials are intimate to the common genus, and the members thereof are related, not by the external relations of mutually exclusive, self autonomous species, but by internal relations of mutual implication and dependence.

An Essay on Philosophical Method develops the contrast of the logical aspects of philosophical thought with that of scientific thought within the framework of the traditional division of logic into the consid-

---

[79] See *An Essay on Philosophical Method*, p. v and pp. 1-2.
[80] L. Rubinoff, *Collingwood and the Reform of Metaphysics*, p. 161; see also *Speculum Mentis*, p. 49.
[81] L. Rubinoff, *op. cit.*, p. 165.
[82] *Ibid.*, pp. 161, 162.
[83] *Ibid.*, p. 168.
[84] *Ibid.*, p. 164.
[85] *Ibid.*, p. 172.
[86] *Ibid.*, p. 163.
[87] *The Idea of History*, p. 77; see also L. Rubinoff, *Collingwood and the Reform of Metaphysics*, pp. 170-171.

eration of concepts, judgments and reasonings. He first contrasts the logic of concepts, showing how the logical behaviour of concepts differs in their philosophical usage when compared with their scientific usage, which latter usage has been the exclusive concern of the traditional logical treatment of concepts.

B.i. The Logic of Concepts

Collingwood begins by contrasting the exclusivist characteristic of concepts in science with their mutual implicatoriness in philosophy, and concludes by showing that philosophical concepts constitute an overlap of classes in a scale of forms, as contrasted with the isolationist classificatory nature of genus-species conceptualization which defines the scientific way of conceiving.

B.i.a. The Exclusivist Nature of the Scientific Concepts of the Exact and the Empirical Sciences

In regard to concepts, Collingwood first shows the exclusivist character of concepts as these are considered by traditional logicians in their doctrine of classification and division. A generic class is divided into species which are further subdivided into sub-species. Individual instances are then classified under the generic concept in such a way that each individual appears as a member of one specific class only.[88]

> Thus, every individual present in the generic class will be present in one and only one, of the specific classes, which are thus exclusive in relation to each other and exhaustive in relation to the generic class.[89]

In general, the concepts of the exact sciences conform to the rules of classification and division as laid down by logicians[90] and the same general conformity appears in the concepts of empirical science.[91] At every stage in the division which Natural History makes, of organisms into animals and vegetables, the animal kingdom into vertebrates and invertebrates, vertebrates into mammals, birds, reptiles and fishes, etc., there occurs a division of one concept, a logical genus, into others, its logical species, which are mutually exclusive and together exhaust the genus. Doubtful cases leave the classificatory system intact in principle.

---

[88] See *An Essay on Philosophical Method*, pp. 26-28.
[89] *Ibid.*, p. 28.
[90] *Ibid.*, p. 29.
[91] *Ibid.*, p. 30.

B.i.b. Mutual Implication of the Species of Philosophical Concepts

But this traditional theory of classification and division, though a true account of scientific concepts, whether of exact or empirical science, is found to be inapplicable without some modification to philosophical concepts. The mutual exclusiveness characteristic of the species of a generic class is not characteristic of philosophical concepts.

> The specific classes of a philosophical genus do not exclude one another, they overlap one another. This overlap is not exceptional, it is normal; and it is not negligible in extent, it may reach formidable dimensions.[92]

The predicates unity, goodness and truth are assigned to every being, and if, with Aristotle, we reject their equivocity we must say that they are concepts of a peculiar kind which defy any classificatory system.

> Just as Aristotle showed that the concept of good overlaps or transcends or diffuses itself across the divisions of the categories, so according to this traditional formula there is a similar overlap or transcendence or diffusion in the concepts of unity and reality.[93]

Some terms admit of both a philosophical and scientific usage, but they undergo a change of meaning in passing from one sphere to another. The concept of matter is an instance of this.

> The difference between the two phases of the concept is that in Newtonian physics matter is the name of a certain class of things separate from other classes of things, such as minds, and appearances like colours or sounds depending for their existence on the mind to which they appear; in materialistic metaphysics it is the name of reality as a whole, and every distinction like that between so-called matter and so-called mind is reduced to a distinction within matter itself.[94]

Similarly, mind, for the scientific psychologist, is the name for something outside other things regarded as material, but for the spiritualistic philosopher it is a name for all reality.[95]

> It appears from these instances that when a concept has a dual significance, philosophical and non-philosophical, in its non-philosophical phase it qualifies a limited part of reality, whereas in its philosophical it leaks or escapes out of these limits and invades the

[92] *An Essay on Philosophical Method,* p. 31.
[93] *Ibid.,* p. 33.
[94] *Ibid.,* p. 34.
[95] *Ibid.* Correlating with what has been noted elsewhere, scientific concepts conceive their object exclusively and in terms of external relations, philosophical concepts conceive their object inclusively and in terms of internal relations.

> neighbouring regions, tending at last to colour our thought of reality as a whole. As a non-philosophical concept it observes the rules of classification, its instances forming a class separate from other classes; as a philosophical concept it breaks these rules and the class of its instances overlaps those of its coordinate species.[96]

Collingwood then shows that the attempt to deal logically with the concept ''thought'' as a genus divided into the species judgment and inference and the attempt to deal ethically with ''good'' as a genus divided into the species the pleasant, the expedient and the right, soon breaks down. Judgments turn out to implicate inference, and inferences to implicate judgments, the useful can be pleasant, and the useful and pleasant can be right, and the right can be either useful or pleasant or both.[97]

Then again, the attempt, in philosophy, to treat logic and ethics as two distinct wholes independent of one another also breaks down. We can consider thought (in logic) and action (in ethics) as distinct in essence as much as we like, but in their concrete existent instances they are so connected that an instance of one is an instance of the other. The same mutual implication is found in metaphysics, aesthetics and the other philosophical sciences.[98]

> . . . this type of structure in which specific classes overlap, is so deeply rooted in the subject-matters with which philosophy has always been concerned, that to set one's face against it means abjuring any attempt to think seriously about matters of that kind.[99]

If philosophical concepts have a peculiar type of logical structure in which the specific classes overlap, no method can be used in philosophy which presupposes that species as such are mutually exclusive. We cannot philosophically study the specific forms of the generic concept action the way a scientist collects instances of a generic concept (say roses) and sort them into specific, mutually exclusive classes. If we thus sorted actions into those done from duty, those done from interest, those done from inclination, and then proceeded to examine each species in turn, before long we would find ourselves confronted by an overlap; certain actions done from interest would be right, certain actions done from duty would be pleasant, and so on. The mutual exclusiveness characteristic of the specific classes of a common genus would break down. The mutually exclusive external relationships of the

---

[96] *Ibid.*, p. 35.
[97] See *ibid.*, pp. 36-43.
[98] See *ibid.*, pp. 43-44.
[99] *Ibid.*, pp. 44-45.

different classes with respect to each other would surreptitiously give way to a system of internal relationships according to which instances of one kind would be found simultaneously to be instances of another kind.[100]

We can begin to see that what is forcing itself upon our attention when we come to look at things philosophically, as distinct from looking at them scientifically, is that behind the diversity recognized by science there is a more fundamental and basic unity-in-diversity which is the object of philosophical thought. Scientific thought operates through dispersal and diversification of its objects whereas philosophical thought operates by bringing all things back to their fundamental unity. The movement of philosophy is thus a movement, not of dissipation in specialization, but of *rapprochement* or return to reestablished unity and harmony, the need for which Collingwood underscored heavily in his *Speculum Mentis*.[101]

Collingwood distinguishes two fallacies committed by anyone who persists in using classificatory, scientific logic beyond the domain of inquiry into the natural world in asserting the autonomy of coordinated species.[102]

The first is the fallacy of precarious margins,[103] which falsely supposes, owing to undue commitment to classificatory logic, that

> The overlap which has already affected a certain area of the class in question can be trusted not to spread, and that beyond its limit there lies a marginal region in which the instances exhibit only one of the specific forms, uncontaminated by the presence of the other.[104]

This is precarious, Collingwood says, because, once admitting the overlap in principle, there is no reason to assume that it will stop at any given point.[105]

The second fallacy, which implicates and is mutually implicated by the first, is the fallacy of identified coincidents,[106] which states that every instance of a concept belongs to that concept in exactly the same way; the essence of each instance will then be the same (abstract)

---

[100] See *ibid.*, pp. 42-43; 46-48.

[101] Pp. 25-36. Thus, p. 36: ". . . we now recognize the nature of our disease. What is wrong with us is precisely the detachment of these forms of experience—art, religion, and the rest—from one another. . . ."

[102] See L. Rubinoff, *Collingwood and the Reform of Metaphysics*, p. 162.

[103] See *An Essay on Philosophical Method*, pp. 48-49.

[104] *Ibid.*, p. 48.

[105] *Ibid.*

[106] See *ibid.*, pp. 48-49; see also L. Rubinoff, *Collingwood and the Reform of Metaphysics*, p. 162.

characteristic, so that in the end the instances cannot really be separated.

> I propose to call the fallacy of identified coincidents . . . the false principle . . . that where there is no difference in the extension of two concepts, there is no distinction between the concepts themselves.[107]

These two fallacies are alternative applications of one single principle which is false in philosophy, however true it may be in science, namely that when a generic concept is divided into its species there is a corresponding division of its instances into mutually exclusive classes.

> I call this the fallacy of false disjunction, because it consists in the disjunctive proposition that any instance of a generic concept must fall either in one or in another of its specific classes; and this is false because since they overlap, it may fall in both.[108]

Collingwood then explains that this has a positive and a negative application.

> Applied positively, this yields the fallacy of precarious margins: namely that, since there admittedly is a distinction between two concepts, there must be a difference between their instances. Applied negatively, it yields the fallacy of identified coincidents: namely that, since the instances can admittedly not be separated, there is no distinguishing the concepts.[109]

Collingwood summarizes this into ''the first rule of philosophical method,'' which is:

> . . . to beware of false disjunctions and to assume that the specific classes of a philosophical concept are always liable to overlap, so that two or more specifically differing concepts may be exemplified in the same instances. . . . The rule may be put . . . by saying that any distinction in philosophy may be a distinction without a difference; or, alternatively, that where two philosophical concepts are distinguished Aristotle's formula [for the overlap of classes—the two concepts 'are the same thing'—the traditional way of referring to this principle is to speak of 'a distinction without a difference', that is, a distinction in the concepts without a difference in the instances] may hold good, that the two are the same thing but their being is different.[110]

Philosophy, therefore, cannot definitively classify its subject matter the way a naturalist or botanist arranges species of plants or animals.

---

[107] *An Essay on Philosophical Method,* p. 49.
[108] *Ibid.*, p. 49.
[109] *Ibid.*
[110] *Ibid.*, pp. 49-50.

If the philosopher does employ classification of the genus-species type it is only provisional and as looking forward to something more fundamental. We may indeed classify actions as those done from duty and those done from inclination, but we must be ready to recognize that instances of one may also be instances of the other.

> We may use the classification merely as a means to fixing our situation on the specific peculiarities of acting from duty as such and acting from inclination as such. The true work of philosophy will be the distinguishing of concepts like these coexisting in their instances.[111]

But this existential coexistence, on the other hand, will not be simply enumerative, for there will be logical relations between the members.

> In an empirical concept like man, there is no apparent connexion between such elements as having ten toes and having the power of speech, but in a philosophical concept there cannot be this looseness of structure and the various elements must be somehow interrelated. Hence no object of philosophical thought can be rightly conceived as a mere aggregate, whether of logically distinguished elements or of spatial or temporal parts . . . either of these would imply that the connexions between the parts are accidental whereas they must in reality be essential.[112]

This means that what from the aspect of a classificatory system is considered to be a collection of specifically distinguished elements, externally related and differentiated by differences extrinsic to their common, abstract genus, when philosophically considered as existentially realized, are seen to be internally connected, that is, connected by internal relations in the realization of a concrete universal to which they as its participants or members are intrinsic, essential and relevant, there being no understanding of the universal apart from the instances, or of the instances apart from the universal.

A consequence of operating with such intrinsically related instances is that philosophy brings along with it its starting point which is constantly revised in the process of philosophizing.[113] Philosophizing is simply the elaboration of that starting point in terms of intelligible and meaning giving factors that are intrinsic to it, the work of

---

[111] *Ibid.*, p. 51.

[112] *Ibid.*, pp. 51-52. That is, logical implications are intrinsic to the philosophical existent. As Bosanquet says in *The Principle of Individuality and Value*, p. 44, "There is no meaning in 'it is' apart from 'it is what it is.' " That is, the philosophical existent is fully determined, and this, as will be shown later, is through significant negation. See below, section D of this chapter, especially pp. 146-147.

[113] Correlate with p. 71, n. 46 of Chapter 3, above.

philosophy being to bring that starting point (lived experience) from a state of implicit intelligibility to explicit intelligibility by etching into it the relevant distinctions which will bring it from mere experienced actuality to an unfolded actuality, seen in terms of the intelligible distinctions within it whereby it is an intelligible and rational experience without in any way stepping outside the experience, but rather, working within the experience and elaborating and clarifying it from within.[114] Thus, in a sense, the starting point is never left (as science starts from experienced fact, but steps outside it, into the world of abstract, hypothetical, universals for its explanation, which explanation always remains extrinsic to the instances of the experience that it purports to explain).

> Thinking philosophically, whatever else it means, means constantly revising one's starting point in the light of one's conclusions and never allowing oneself to be controlled by any cast-iron rule whatever.[115]

B.i.c. Distinction and Opposition of Philosophical Concepts

Collingwood then asks, what is the kind of difference that exists between the species of a philosophical concept that makes such an overlap possible?[116]

Subdividing the question, by distinguishing differences into differences of degree and differences of kind, he asks whether one or the other of these kinds of difference can explain the overlap.

Mere differences of degree are rejected, since the crucial point at which any specific class ended would be the point at which another began, hence mutually exclusive of each other reckoned from that point.[117]

Mere differences of kind are also excluded as impotent to account for the overlap of classes typical of philosophical concepts. Sensation differentiated solely according to kind, as into seeing, hearing, smelling, etc., simply divides up the genus into the separate species, which are then allocated to different disciplines (Psychology, Physiology, etc.) for investigation. The species are thus diversified away from each other rather than seen as overlapping.[118]

If, then, neither differences of degree nor differences of kind taken separately can explain the overlap of philosophical classes, can it be

---

[114] Refer to clockface illustration, Chapter 3, pp. 71-72, above.
[115] *An Essay on Philosophical Method*, p. 54.
[116] See *ibid*.
[117] See *ibid*.
[118] See *ibid*., pp. 55-56.

explained by some combination of difference of degree with difference of kind?[119]

B.i.d. The Scale of Forms

In such a system of specifications, in which difference of degree is compounded with difference of kind,

> the two sets of differences are so connected that whenever the variable, increasing or decreasing, reaches certain critical points on the scale, one specific form disappears and is replaced by another. . . . A system of this kind I propose to call a scale of forms.[120]

Instances of familiar experience are a breaking strain, freezing and boiling points, the maxima and minima of taxation brackets, etc.

Distinction is next made between a non-philosophical scale of forms, in which the variable is something extraneous to the generic essence,[121] and a philosophical scale of forms, in which the variable is identical with the generic essence itself.[122]

Thus, in the case of water, the various forms, solid, liquid, gaseous, form a non-philosophical scale of forms, since heat, the element differentiated in degree, is not included in the formula H_20 which expresses the essence of water and is applicable without differentiation to the different forms of water. Each of the three forms of water embodies the generic essence equally and unchanged; each is unqualifiedly water. There is here concurrence of difference of degree (in temperature) with difference in kind in the water that leaves the waterness of the water unchanged throughout the different forms which the water takes at different temperatures.

But it is otherwise when, as for Plato, knowledge is presented as a scale of forms from nescience through opinion to knowledge, or again from conjecture, through opinion and understanding to reason, or again from poetry, through mathematics to dialectic, or when being is presented as a scale of forms extending from nothing, through half-being to true being, or pleasures are graduated through those of the body and those of the soul, or the forms of political institutions are graduated, or, again, when Aristotle

> recognizes the same type of logical structure, for example when he distinguishes the vegetable, animal, and human 'souls' as three

[119] See *ibid.*, p. 56.
[120] *Ibid.*, p. 57.
[121] See *ibid.*, p. 59.
[122] See *ibid.*, p. 60.

> forms of life arranged on a scale so that each includes its predecessor and adds to it something new.[123]

Here

> the variable [is] identified with the generic essence. The result of this identification is that every form, so far as it is low in the scale, is to that extent an imperfect or inadequate specification of the generic essence, which is realized with progressive adequacy as the scale is ascended.[124]

Before answering whether such a scale of forms, in which the variable is identified with the differentiated generic essence, can explain the overlap characteristic of philosophical concepts, Collingwood faces the difficulty that such a conception "falls to the ground, condemned as a tissue of contradiction, a logician's nightmare,"[125] which difficulty "proceeds from the assumption that the doctrines concerning the structure of the concept, which are expressed in elementary textbooks of logic, must be accepted without question and applied without modification as principle of philosophical method."[126]

Collingwood resolves this difficulty by a distinction deriving from

---

[123] *Ibid.*, pp. 58-59.

[124] *Ibid.*, p. 61. At this stage of the exposition, one familiar with the traditional moderate realist scholastic doctrine of the analogy of proper proportionality will immediately recognize that both are talking about the same thing. Yves Simon in *The Great Dialogue of Nature and Space*, New York, Magi, 1970, writing (p. 64) of "the great analogy, that did not acquire a distinct name until very late, the so-called analogy of proper proportionality . . ." says (p. 65), "considering . . . being as divided into 'being in act' and 'being in potency,' note that in the case of 'being in act' the common ground, being, is purely asserted. In the second case, in the case of being in potency, the common ground, namely, being, is asserted, indeed, but also negated. This is the secret of analogical thinking. In the case of the second analogate you have an assertion of the common ground coupled with a negation of the common ground . . . an association of assertion and negation which concerns the common ground. If the common ground were a genus, assertion and negation would concern only the differential factor . . . the association of assertion and negation concerns the differential. With analogates the association of assertion and negation concerns the common ground." Again (p. 82): "The general criterion which distinguishes a unity of analogy from univocity is that when you move from one term to the other, there is in the first an assertion of the common ground which receives a qualified negation in the other term." And (p. 84): "In the positive sciences, and especially in mathematics, we are concerned with understanding, distinguishing and relating to each other various species of objects. Consequently, univocal concepts are best suited for these areas. Philosophy, however, since it is concerned with providing an overall synthetic view of reality, has to use terms in extended and analogous senses because it is only in this way that it can deal with the relative unity of essentially diverse kinds of things." For a full discussion of analogy of proper proportionality see Y. Simon's "On Order in Analogical Sets," *The New Scholasticism*, XXXIV (1960), pp. 1-42. See also Chapter 7, below, pp. 297-299.

[125] *An Essay on Philosophical Method*, p. 62.

[126] *Ibid.*, p. 63.

Croce,[127] between opposites and distincts, and to the question "whether the idea of a scale of forms [in which lower and higher species are lesser and more perfect expressions of the generic essence] serves to

---

[127] William M. Johnston, in *The Formative Years of R. G. Collingwood,* The Hague, Nijhoff, 1967, p. 74, says that "in this work [*Saggio sullo Hegel*, 1906], Croce advances a notion which is crucial to his critique of Hegel and which will serve as the basis for the total revision of the *Logica* in its second edition of 1909. This notion is the distinction between two types of opposition: opposites and what Croce calls 'distincts.' Opposites are concepts which exclude each other, like hot and cold. Distincts are concepts, which though contrary, yet imply each other, like one and many or body and spirit. To use Collingwood's terms, the distincts are 'distinct but not separate.' Croce contends that Hegel erred in constructing his dialectic because he treated all forms of opposition as opposites. This led him to forced conclusions about the relationship of concepts which are really distincts. Hegel failed to see, says Croce, that the one and the many comprise a single whole, just as do the spirit and body. This notion of concepts which form distinct yet inseparable parts of a larger whole comprises the basic logical doctrine of Collingwood's *Speculum Mentis*" (see p. 170 for the phrase "distinct but not separate"). James Collins, in "The Role of Monistic Idealism in Croce's Esthetic," *The New Scholasticism,* XVII (1943), pp. 34-35, writes: "Instead of the Hegelian triad of Logos, Nature and Spirit, Croce substituted the theoretical and the practical as the basic divisions in his Philosophy of the Spirit. This modification was made in the light of his criticism of the Hegelian dialectic, which so emphasized the opposition between thesis and antithesis that the resultant synthesis could unite these moments only by emptying them of all significance. To overcome this untenable dualism of *opposites,* Croce proposed a conjunction of *distinct* factors or aspects of the concrete concept. In this way philosophical investigation can be limited to the manifestations of the spirit alone, banishing from the provenance of true knowledge such a pseudo-concept as 'an external universe or objective reality.' When this is opposed to the notion of internal spiritual reality, both terms of the opposition are reduced to abstract concepts which can never be synthesised. The triadic scheme must be replaced by the doctrine of the grades of the Spirit, in which both moments are concrete and need not be overcome. 'For those degrees, considered in their distinction, are the concept of the spirit in its determinations, and not the universal concept of spirit considered in its dialectic of synthesis of opposites.' [*What Is Living and What Is Dead of the Philosophy of Hegel*, pp. 92-93]. Within the theoretical sphere are found art and philosophy: the knowledge of individuals through images and the knowledge of the universal through the pure concept. The individual spirit passes constantly from one manifestation to the other, although the dissatisfaction which engenders this oscillation does not spring from either form considered as distinct." It seems, however, that Collingwood, in *An Essay on Philosophical Method,* inverts the terminology while retaining the doctrine of Croce. In *What Is Living and What Is Dead of the Philosophy of Hegel* (Douglas Ainslie's translation of Croce's *Saggio sullo Hegel*), Croce (p. 11) says: "It is impossible to confuse the two series, distincts and opposites. . . . Now, if distinction do not impede, if indeed it rather render possible the concrete unity of the philosophic concept, it does not seem possible that the same should be true of opposition. Opposition gives rise to deep fissures in the bosom of the philosophic universal and of each of its particular forms, and to irreconcilable dualisms. Instead of finding the concrete universal, the organic whole of reality which it seeks, thought seems everywhere to run against two universals, opposing and menacing each other." Clearly, for Croce opposites are the externalizing repellants and distincts the unifying concretisers, whereas for Collingwood it is just the reverse; distincts are mutually exclusive and non-philosophical concepts, whereas opposites are mutually implicatory and the proper object of concrete philosophical consideration. This is very clear when,

explain the overlap between the species of a philosophical genus," he answers: "Yes, if these species are opposites: no, if they are distincts."[128]

Actions, for example, may be divided into good actions and bad actions, and, as so divided, are divided into opposites, that is, into opposite species, which can be arranged on a scale with infinity at one end and zero at the other and the variety of intermediate forms in between, partaking of both opposites, each being more and less good than another, more and less bad than another. The extremities, absolute good and absolute bad, are pure abstractions falling outside the scale itself. But actions may be specified not only by division into opposites, good and bad, but also into distincts, such as just, generous and courageous actions.

Opposite species like good and bad would then belong to the philosophical phase of their genus and provide the appropriate subject matter for philosophical thought, whereas distinct species would belong to the non-philosophical phase of the concept.[129]

If this were the case, Collingwood says, we would have a straightforward and simple rule of philosophical method, namely:

> . . . since philosophical specification is into opposites and non-philosophical into distincts, any distinctions found in a philosophical subject-matter must be either banished from it as alien to the sphere of philosophy or else interpreted so as to appear cases of oppositions.[130]

B.i.e. Philosophical and Non-Philosophical Understanding of the Distinction between Distinction and Opposition

But this simple, straightforward solution becomes threatened when it retroacts over the principle of the solution, namely the distinction of relations into distincts and opposites.

The solution, that "philosophical specification is into opposites and non-philosophical [specification is] into distincts" so that "any distinctions found in a philosophical subject-matter must be either banished from it as alien to the sphere of philosophy or else inter-

---

following the citation above, Collingwood adds: "It is not enough to show that these distinctions contain in themselves an element or aspect of opposition; that will not save them; the element of distinction must be completely eliminated and nothing except pure opposition allowed to remain." I can offer no explanation for Collingwood's apparent reversal of term usage.

[128] *An Essay on Philosophical Method*, p. 63.

[129] See *ibid.*, p. 64.

[130] *Ibid.*, p. 65.

preted so as to appear cases of opposition''[131] immediately precipitates the question as to whether ''the relation between philosophical specification by opposites and non-philosophical specification by distincts [is] itself a case of distinction or of opposition.''[132]

The critical moment of this question is immediately apparent. *If* the dichotomy between opposites and distincts is an ultimate and radical dichotomy, that is, *if* the distinction itself is an instance of distinction and not of opposition, as it is if opposites and distincts are two mutually exclusive species of a classificatory genus, *then* non-philosophical thinking predominates in human thought, and philosophical thought becomes subordinated to non-philosophical thought as just another species of thought alongside and competing with other forms of thought, e.g., scientific and historical. On the other hand, *if* that distinction into opposites and distincts is an instance of opposition *then* ''philosophical logic as the logic of opposites has triumphed over non-philosophical logic as the logic of distincts, and with this triumph it has destroyed the distinction between itself and its opponent.''[133]

This requires, quite consistently with Collingwood's notion of philosophical inquiry, that we go back and revise our starting point in the light of the conclusion reached,[134] that the distinction between opposites and distincts has to account for *both* philosophical *and* non-philosophical thinking *without absorption of one into the other, or obliteration of one by the other.*

If, in order to avoid the absorption of either form of thought by the other, it is said that the relation between distinction and opposition is itself an instance neither of distinction nor of opposition but some third kind of relation, then not only has the dualism been rejected, since a third term intervenes, but also the question is immediately raised as to whether philosophical specification is by pure opposition or by this postulated third principle.[135]

Preliminary to answering this crucial question, Collingwood asks us to reconsider the examples of differences of degree when we say that one action is better than another and when we say that one body is hotter than another. The difference between these two is that the heat in a body is measurable, whereas the goodness of a man or of

---

[131] *Ibid.*, p. 65.
[132] *Ibid.*, p. 68.
[133] *Ibid.*, p. 68.
[134] See *ibid.*, p. 69; p. 160.
[135] See *ibid.*, pp. 68-69.

an action is not, just as it is also impossible to measure the degrees of beauty, truth, pleasantness or any other philosophical concept. If one book is estimated to be higher than another by measurement, it can be said that it is twice or three times higher, or so many inches or feet higher. But it is meaningless to say of some pleasant experience, estimated to be more pleasant than another, that it is twice or ten times or so many inches or feet more pleasant. At the most, one is using quantitative terms to express something non-quantitative but qualitative when one says that a certain person is more intelligent than another. Measurements of intelligence, or of degrees of painfulness or pleasantness of some stimulus, and even of light intensity, where numbered scales of some kind are used, simply place the members according to more or less *in an order with respect to each other*. An increase of heat *as felt* is a change *in the kind of experience,* from a faint warmth felt through a decided warmth, first pleasant, then slightly painful, then noticeably painful, then sharply painful. The addition of each *degree* is *also* the addition of a different *kind*. Calling them differences *of degree* is using the physicists' thermometer language, but it is used metaphorically to name something quite different.[136] All differences of degree among philosophical concepts are never merely differences of degree but imply concomitant differences of kind; they are at once differences of degree and differences of kind.

> . . . this is the real basis of the distinction between a philosophical scale of forms and a non-philosophical: in a non-philosophical scale there are differences of degree, and coordinated with them differences of kind; in a philosophical scale there is only one set of differences having this peculiar double character.[137]

Collingwood draws attention to the fact that, though peculiar, this type of difference is quite familiar, both in its philosophical and non-philosophical phase. We familiarly speak of degrees of kindred and affinity, degrees of punishment, structural degrees in society such as those of nobility and gentry, university degrees, degrees of grammatical comparison and so forth. Such differences fuse differences of degree with differences in kind.

Differences of degree and differences of kind, non-philosophically considered, are mutually exclusive species of a genus, but philosophically considered they must overlap in a form of difference which

---

[136] The question regarding the meaning of quantitative terms applied to qualities is considered by P. Hoenen in his *Cosmologia*, Rome, Gregorian University Press, 1936, Vol. 2, pp. 468-478.

[137] *An Essay on Philosophical Method*, p. 73.

shares the nature of both. In the case of water, there are two concurrent types of difference, one in degree of heat, the other in physical structure which gives rise to mutually exclusive specifications coordinated with the varying degrees of heat. In its non-philosophical phase, opposition is a relation subsisting between a positive term and its own mere negation or absence; the physicist understands cold as a lack of heat. But cold *as we feel it* is not a *mere lack of the feeling* of heat, but *another kind of feeling* which has its own positive character. But the feeling of cold and the feeling of hot are not merely two *distinct* feelings but also two *opposite* feelings. Physical cold is related to physical heat of which it is the negation by a relationship of *mere opposition*. But felt cold is related to felt heat as *at once opposed to it* and *distinct from it, opposition and distinction being fused into a single relation*.

> In general the kind of opposition which is found among philosophical terms is at once opposition and distinction, and subsists between terms each having a definite character of its own and yet forming together a true pair of opposites.[138]
>
> . . . Differences of degree and differences of kind, which in non-philosophical thought can be disentangled from one another, are in philosophy fused into a new type of difference uniting the characteristics of both. Distinction and opposition, which in non-philosophical thought are two mutually exclusive kinds of relation, in philosophy coalesce into one, so that what seems at first sight a mere opposition—the relation, that is, between a term and its own absence—turns out to be also a distinction between two terms, and vice versa.[139]

With this in mind, Collingwood returns to the difficulty encountered above (pp. 107-108), namely, that a specific form must either embody the generic essence completely or be completely outside such essence. The difficulty arises from forcing upon the facts of philosophical thought an interpretation in which the terms difference of degree and difference of kind bore the special meanings proper to them in a non-philosophical context. A work of art said to be more beautiful than another is *also* beautiful *in a different way*; it is not merely an excess over the other in beauty but an *excess* which has its own peculiar *kind* of beauty. The same is true of pleasure, goodness and the other concepts belonging to the sphere of philosophy. Superficially these appear to obey the traditional rules of specification modified by an overlap of classes, but closely scrutinized they reveal the characteristic *fusion of differences in degree with differences in kind*. When we dis-

---

[138] *Ibid.*, pp. 75-76.
[139] *Ibid.*, p. 76.

tinguish different kinds of goodness, as distinct from merely differentiating classes of good things, these kinds are more truly goodness than others.[140]

I understand this solution of Collingwood in the following way. The principle that relations are divisible into opposites and distincts may be taken either non-philosophically or philosophically. If it is taken non-philosophically, then philosophy itself is a species of the classificatory genus, thought, alongside other species, such as science and history. But it may also be taken, as it is taken in philosophy, as a philosophical distinction, in which case the division will be understood as a mode of opposition in which distincts and opposites overlap, and philosophical thinking is a mode of thinking which overlaps with other forms, such as science and history (as in *Speculum Mentis*). This means that, according to the philosophical understanding of the principle, distinction and opposition are not merely concurrent, and that *specification* and *graduation by degrees* are not merely together, but are together *in such a way that the graduation of the generic term is intrinsically the reason of its division into species*. Pleasure, as graduated, is itself different in definition and species according to the graduations. Knowledge, as *graduated* through conjecture, opinion and certitude is different *in kind,* and *differently defined at each grade on the scale*, whereas the definition of water remains unchanged throughout its different forms on the temperature scale.

> When attempts are made not merely to differentiate classes of good things but to distinguish kinds of goodness, it is constantly found that some of these kinds are more truly goodness than others.[141]

That is, differentiation in those cases *comes from inside the differentiated term, the common genus,* not from outside, as in the case of the classificatory division of a common genus. The difference adds *qualitatively* and internally, not merely quantitatively and externally.

> Distinction and opposition are two species of relation; and where the term relation is applied to a philosophical subject-matter it acquires the special colouring proper to philosophical concepts, that is . . . it denotes a generic concept whose specific classes overlap. In philosophical thought, therefore, distinction and opposition will necessarily combine into a peculiar type of relation which is neither mere distinction nor mere opposition, but partakes of both these characters; a relation which subsists between terms at once opposed and distinct.[142]

[140] See *ibid.*, p. 78.
[141] *Ibid.*
[142] *Ibid.*, p. 76.

The differences of degree found in philosophical concepts are therefore not of the same type as differences of degree found in non-philosophical concepts. The gradations of a scale of pleasures, and of intelligence by I.Q. tests, cannot be handled mathematically; to attempt to do so is to commit the fallacy of calculation. The fallacy of indifference is committed by attending to the degree of variation, ignoring the built-in difference of kind. Both fallacies are based on the false disjunction that a difference of degree cannot also be a difference of kind.[143]

B.i.f. Synthesis of Difference of Degree and Difference of Kind in Regard to the Scale of Forms

The fusion of distinction and opposition also modifies the corresponding idea of a scale of forms. If the variable is identical with the generic essence, the zero end forms no part of the scale, since at the zero end the generic essence would be altogether absent; the lower end of the scale is thus a unit of minimal realization of the generic essence.

The lowest member of the scale is a genuine realization of the generic essence, and as the limiting case at the lower extreme it is an opposite relatively to the other members on the scale.[144] The badness of a bad moral act is nothing other than its low degree and kind of goodness in relation to the degrees and kinds above it.[145] Even the minimal case of goodness, which in relation to higher kinds is bad, is good. There is no purely bad act that does not appear to the one who does it as somehow good.[146] Considered in itself it is good, but all changes when it is related to higher cases.[147] Pleasure is good in itself as far as it goes, but in relation to the pursuit of duty it is comparable as not merely less good but as positively evil. Thus, the lowest case in the scale of goodness compared with the next above it acquires the character of badness, and, as lowest on the scale, becomes identical with evil in general in which the abstract idea of evil finds concrete embodiment. And the same relation which obtains between the lowest member of the scale and the next above it reappears between any two adjacent forms. Each is good in itself, but bad in relation to that above it.[148]

---

[143] See *ibid.*, pp. 80-81.
[144] See *ibid.*, p. 82.
[145] See *ibid.*
[146] See *ibid.*, p. 83.
[147] See *ibid.*, p. 84.
[148] See *ibid.*, p. 84.

B.i.g. The Scale of Forms and the Overlap of Classes

Collingwood then clarifies how the scale of forms thus understood accounts for the overlap of classes which characterizes philosophical concepts.[149] The lower of any two adjacent terms is good in itself but bad in relation to its neighbour above it. The same is true of a scale of beauty and ugliness, true and false, simple and complex[150] or any such pair of polar opposites.[151]

The lower (good, or true, or simple, or beautiful) is not only such (good, etc.) *in general* but is such *in a specific way*. When compared with its above neighbour it loses not only its goodness or truth in general but its *specific kind* of goodness or truth *at that level*. And the higher term in comparison with the lower possesses not only that kind of goodness belonging to it in its own right but also retains, though modified, the kind of goodness which belonged to the lower in itself in its own right. It not only surpasses its neighbour *in degree* (of goodness, beauty or truth) but exceeds it on its *own ground*. The lower promises more than it can perform; in professing to exhibit a certain kind of goodness it fails to do so save approximately and inadequately.

> Thus, each term, which in itself is simply one specific form of goodness, has also a double relation to its neighbours: in comparison with the one below it is what that professes to be; in comparison with the one above, it professes to be what that is.[152]

The lower promises what the higher performs; the higher is the reality of which the lower is the appearance; the higher is the ideal to which the lower is the approximation; the higher is the truth of which the lower is a perversion or distortion. But these metaphors are intelligible only in the light of the presupposed relation which explains them.

> It is a purely logical relation . . . [as such] it is a synthesis of the four relations . . . difference of degree, difference of kind, relation of distinction, and relation of opposition. The higher term is a species of the same genus as the lower, but it differs in degree as a more adequate embodiment of the generic essence, as well as in kind as a specifically different embodiment; it follows from this that it must be not only distinct from it, as one specification from another; but opposed to it, as a higher specification to a lower, a relatively adequate to a relatively inadequate, a true embodiment of the generic essence to a false embodiment; as true, it possesses

[149] See *ibid.*, p. 86.
[150] See *ibid.*, p. 85.
[151] See *ibid.*, p. 86.
[152] *Ibid.*, p. 87. The relations obtaining between the forms of experience as explained in *Speculum Mentis* is a good instance of this.

not only its own specific character, but also that which its rival falsely claimed.[153]

The affirmational and negational relationship between the members on the scale, which Collingwood goes on to assert, is important for the understanding of the dynamic, dialectic nature of the genus of a scale of forms.

> The higher thus negates the lower, and at the same time reaffirms it: negates it as a false embodiment of the generic essence, and reaffirms its content, that specific form of the essence, as part and parcel of itself.[154]

This logical presentation of a scale of forms in which a higher beats the lower on its own ground is familiar in regard to a philosophical scale of forms. If justice and expediency are adjacent on a scale of moral values, expediency finds its fulfillment in going beyond mere expediency to the level of justice, so that justice is not merely justice but also expedient. This leads to the next important point in understanding the dialectical nature of the philosophical scale of forms: ''Each term in the scale . . . sums up the whole scale to that point.''[155]

Just as the zero form was eliminated from the scale as also an elimination of the generic essence, so also the infinity form is eliminated since any concretely *realized* specific form is the generic concept itself *so far as our thought has yet conceived it*, always allowing for further self-transcendence in higher forms of embodiment of the generic essence.

> All [the] lower stages in the scale are telescoped into this [presently achieved] situation. They are in fact summed up in it twice over: once falsely, in the proximate specification, which misinterprets their significance and combines them into a false unity, and once truly, in the culminating form.[156]

The denial of absolute opposites at either end of a scale of forms is extremely important in Collingwood's theory. What is found at either end of a scale is not something that is *first* posited as something in itself *and then* the intermediate members constituted by a differential combination of these opposites. What is given concretely is this or that particular member, which, if analysed, analyses into two opposite respects, one respect towards a prior term or terms, another respect towards a subsequent term or terms, much in the same way as an astrophysicist analyses starlight by spraying it out into a band, along

[153] *An Essay on Philosophical Method*, pp. 87-88.
[154] *An Essay on Philosophical Method*, p. 88.
[155] *Ibid.*, p. 89.
[156] *Ibid.*, pp. 89-90.

a graduated scale which points in two directions. What is given is starlight, and the polar opposites, red and blue colours, are the result of an analytical operation on that concrete entity analysed. This is true of a scale of forms since no member is considered as something absolutely in itself, for that would be precisely the position of classificatory species; rather each is what it is by reason of its position on the scale with respect to its neighbours in two opposite directions. This follows, as will be clear after the dynamic, dialectical consideration of a scale of forms is made, from the basically active, rather than substantive or entitative nature of the members of a scale of forms. In other words, since, for Collingwood, becoming, or identity in difference, is more basic than being, whatever is conceived as becoming can only be conceived as doubly relative, once to a prior condition from which it comes, and secondly, although simultaneously, to a posterior condition towards which it is becoming. Historical events, which are more properly processes than events (which term savours rather of classificatory thought) are thus understood as overlapping classes on a scale of forms, for any member, that is, any specific "event" dynamically considered in process, has to be understood relatively to what went before, namely its past, from which it came, and relatively to that into which it developed, namely its future. Thus, past and future are the opposites into which some "present" "spectroanalyses" when it is historically analysed and explained. Past and future are not things in themselves, as realist philosophy would say, but are aspects, relationships built into the very essence of some actuality.

Thus is explained the overlap of classes in a philosophical genus:

> The higher of any two adjacent forms overlaps the lower because it includes the positive content of the lower as a constituent element within itself. It only fails to include the lower in its entirety because there is also a negative aspect of the lower, which is rejected by the higher: the lower, in addition to asserting its own content, denies that the generic essence contains anything more, and this denial contains its falsehood.[157]

Thus also appears the role of significant negation in dialectically relating the members on a philosophical scale of forms. The limitation intrinsic to the lower member is a negation of its own intrinsic fulfillment. The negation of this negation of the lower by the higher overcomes the limitation in the fulfillment of the lower by and in the higher form. Utilitarianism rightly affirms expediency as a form of moral goodness but falsely denies that there is anything in higher forms

[157] *An Essay on Philosophical Method,* p. 90.

not explainable in terms of expediency. Duty rejects expediency inasmuch as expediency refuses to accept duty as a legitimate, let alone higher, form of goodness. But duty reaffirms expediency through subjecting it to its own principles as a constituent element of duty itself. Duty and expediency thus overlap; a dutiful action has its own kind of expediency, and an expedient action thus partakes of the nature of duty. Thus:

> The lower overlaps the higher in a different sense than that in which the higher overlaps the lower: it does not include the higher as part of itself, it adopts part of the positive content of the higher while rejecting another part.[158]

These explanations, says Collingwood, show "clearer than before what exactly this overlap is."[159] The higher takes the lower up into itself adding something; the lower rejects this added element in the higher.

> The overlap consists in this, that the lower is contained in the higher, the higher transcending the lower and adding to it something new, whereas the lower partially coincides with the higher, but differs from it in rejecting this increment.[160]

The overlap, says Collingwood, is not an overlap of extension between classes but an overlap of intension between concepts, each concept, in its own degree, being a specification of the generic essence, but embodying that generic essence more fully and adequately. That is, the differences, differentiating a higher from a lower, are "additions" obtained from within the generic essence, an intensification of that generic essence, not an addition of a difference extrinsically, from outside the generic essence. It is something like the growth of an organism from a single cell. A subsequent state arises from a prior state by division and resulting addition from within. Individuality is progressively acquired with increasing determination arising from self-constructive differentiation. And so it is with all dialectical activity; self-identity is acquired by medium of self-differentiation and simultaneous self-reintegration.[161] Such a process is explicable as an overlap of classes in a scale of forms. The same whole (generic essence) is present identically throughout; its identity is an identity in a differentiation of forms and its totality is the totality of its different forms, not juxtaposed one on top of the other from outside each other, but as

---

[158] *Ibid.*, p. 90.
[159] *Ibid.*, p. 91.
[160] *Ibid.*
[161] This point is developed at length in Chapter 9, below, pp. 398 et seq.

overlapping in a mutually implicating, self-transcending fashion, the final form gathering together within itself in a summary fashion all the previous forms that lead up to it, just as a man's life at any moment is the summation of all the self-perfecting differentiation which took place since he first appeared as a fertilized ovum. Relating this to the process of question and answer we might say that the question is the differentiation, and bifurcation into alternatives, the internal opposition of self to self, and the answer the reintegration, or reestablishment of harmony at a higher level.[162] And, if in adopting the Crocean theory of the distinction of opposites and distincts, Collingwood is also adopting the Crocean *reason* for making that distinction, namely the substitution of a two-term dialectic of opposites (which Croce however calls distincts; see n. 127, p. 106 above) for the classical three term thesis, antithesis, synthesis dialectic, we may further say that a moment of present actuality, which in relation to its past has the aspect of information available, or answers to questions, is the answer, the synthesis of opposites arising from and in past questions, *and together* the question, the antitheses, the contemplated alternatives, with respect to future answers or syntheses. The totality at any moment is present actuality. The understanding of this present actuality of lived experience consists in being able to see it as answeringly synthesising with respect to past questions and *at the same time* questioningly antithesising within itself its past (summarized in the present actuality) with respect to anticipated answers in the future. Thus, by medium of Croce's *reason* for distinguishing distincts from opposites, we can see how Collingwood can *together* identify the dialectical activity of the mind with the process of questioning and answering on the one hand, and with the historical bifurcation of present actuality into past and future respects on the other hand.[163]

B.ii. The Logic of Judgments

Having exposed the type of concept characteristic of philosophical thought, Collingwood goes on to consider judgments and then reasonings as these are found in philosophy.

B.ii.a. Abstract and Concrete Affirmation and Negation

Judgments are traditionally divided into affirmative and negative.

---

[162] See below, Chapter 9, pp. 401 et seq.

[163] This point is brought out well by the clockface illustration in Chapter 3, above, pp. 71-72.

The relation between affirmation and negation is peculiarly intimate in regard to philosophical statements. Dissent from the statement of a philosophical position is not simply to deny it but to counteract it with our own view which "need not be on the tip of our tongue: it may be something with which our mind, as Socrates would say, is pregnant, and which needs both skill and pains to bring it forth."[164] Without this feeling of its being "quick within us" we should not meddle with the discussion in question. This is a corollary of the Socratic principle that philosophy is never a transition from sheer ignorance to sheer knowledge but a progress in which we come to know better what we already knew. In philosophy every negation implies an affirmation, which rule is "the principle of concrete negation" and its neglect is "the fallacy of abstract negation."[165]

Correlative to this there is "a principle of concrete affirmation" and a corresponding "fallacy of abstract affirmation."[166] In a philosophical affirmative judgment there is a peculiar intimacy in the relation to the negative elements implicated by the affirmative. The negative elements make precise the affirmations by pointing out exactly what those affirmations, as affirmations, intend to deny. Every philosophical affirmation rejects some definite proposition regarded as erroneous. This is not necessarily the case regarding non-philosophical thought. The philosophical statement that the species of a philosophical genus overlap intends to deny the proposition that they are mutually exclusive.

> . . . where the generic concept predicated by a judgment is philosophical, specified in a scale of forms of which the judgment is intended to affirm the highest . . . its denial of all the inferior forms is summarized in one detail, namely that of the proximate form; since each summarizes the whole scale up to that point, and the denial of that involves the denial of all that it summarizes.[167]

The truth in the proximate lower form, and so in all the lower forms, is retained in the highest form, so, the contrast of the proximate form with the highest is a compendium of all the errors which the assertion of the highest form intends to deny. It is a principle of method, not merely a statement of observed fact in philosophy, that "a philosophical assertion whenever it affirms something definite also denies something definite."[168]

---

[164] *An Essay on Philosophical Method,* p. 103.
[165] *Ibid.,* p. 106.
[166] *Ibid.*
[167] *Ibid.,* p. 108.
[168] *Ibid.,* p. 108. This is the dialectical character found in all concrete thinking.

The principle of concrete affirmation may be applied to one's own, or another's, thought. Applied to one's own thought it means that if one wants to be clear about one's assertions one has to be clear about what one is denying. Applied to the comprehension of another's thought it means that one who is reading or listening to a philosopher must never be content to ask oneself what the philosopher intends to affirm, without together asking what he means to deny. If we do not know the doctrines Plato or Parmenides meant to deny in their affirmations we can be sure we do not understand what they intend to affirm.[169] This is both difficult and of great importance, since the works of the great philosophers of the past have formed their own views by criticizing their predecessors whose works may not have come down to us except insofar as we can construct them from these same criticisms. The rejected view need not have been an actually maintained position, but need only seem plausible to the one who rejects it. The controversy of rejecting it is then one within the mind of a single philosopher, what Plato called a dialogue of the soul with itself.[170]

Thus, whereas outside philosophy a judgment is either affirmative or negative, affirmation and negation being two distinct species of the classificatory genus, judgment, "in philosophy there is such a balance of the two that no properly weighed and considered judgment is more affirmative than negative, or more negative than affirmative."[171] What makes a judgment properly philosophical is "the peculiar intimacy of the relation between its affirmative and negative elements."[172] This relationship is such that where P and Q are equally definite and alternative specific answers to the question 'What is S?' P cannot be validly affirmed of S while Q is left indeterminate, nor can Q be validly denied of S while P is left indeterminate. A precise question details the alternative possibilities and a precise answer

---

[169] *Ibid.*, p. 109. The importance of this for questioning may be noted. Plato or Parmenides had certain questions in mind, which questions specified certain alternative positions on certain matters. Their position is understood as an answer which affirms one of the alternatives while rejecting the others. An understanding of the philosopher's position, therefore, means understanding the alternatives affirmed and denied. But this means understanding his statements as determinative of alternatives, that is, understanding his statements not merely as statements, but as answers to questions.

[170] In other words, the questioning of a position requires not that it be maintained as affirmed but only that it be supposed.

[171] *An Essay on Philosophical Method*, p. 110.

[172] *Ibid.*, p. 110.

affirms one of the alternatives and together denies the other alternatives.[173]

B.ii.b. Abstract and Concrete Universality

In regard to the traditional logical division of judgments into universal, particular and singular, Collingwood next asks whether the term universal may not have a special shade of meaning in the judgments of philosophy.[174]

The terms universal, particular and individual name three elements common to all judgments, which naturally overlap, rather than three species of judgment. The universal judgment 'all men are mortal' does not exclude but includes the judgment that some men are mortal and that this man (Socrates) is mortal. Even if we consider it as a universal judgment, the three elements of universality, particularity, singularity, introduce differentiations into its significance. The universal judgment, in relation to each of these three elements may be taken diversely. As a pure universal it means that man *as such* is mortal, as a universal of particulars it means that *every kind of man* is mortal, as a universal of singulars it means that *every individual man* is mortal.[175]

---

[173] See *An Essay on Philosophical Method*, p. 110. See also Collingwood's *Speculum Mentis*, p. 78, and *The New Leviathan*, p. 74, par. 11.12. Speaking of the disjunctive judgment as clearly delineating alternatives, B. Bosanquet, *Essentials of Logic*, London, Macmillan, 1897, p. 125, has the following to say: "The disjunction seems to *complete the system of judgments, including all others in itself* . . ." (emphasis added), and " . . . disjunction in itself implies a kind of individuality which is beyond mere fact and mere abstract truth, although allied to both; and all intelligible negation is under, not of, a disjunction." Again (p. 123): "The Disjunctive Judgment in its perfect form . . . is appropriate to the exposition of a content as a system . . ." And (p. 124): "The peculiar point of the Disjunctive is that it *makes negation positively significant.*" (Emphasis added.) If we are impatient at this stage and ask what all this has to do with "Who killed John Doe?" we can foresee in some way what the result will be. The answer to that question will depend upon our ability to grasp a global situation, defined by the question, which divides a horizon of relevance from that which is irrelevant (that is, has no bearing on the case); the global situation will be in the nature of a concrete universal. The work will then be, by a process of discriminate questioning, to etch in *the relevant disjunctions* so as to see how the parts (the detailed facts of the case) clearly enmesh. The investigator continually asks himself, 'How does this fact tie in with that fact in the overall picture?' An indication that further investigation (questioning) is called for is the observation that "this fact does not 'fit' with this fact or with these other facts, when we view the whole situation." Crime detection, as also historical research, is somewhat like solving a jigsaw puzzle; fitting together the pieces in the whole picture. See this chapter, below, pp. 137-138 on rational experience and pp. 151-153 on the relevance of the concrete universal to the logic of questioning.

[174] See *An Essay on Philosophical Method*, p. 68.

[175] *Ibid.*, p. 111.

But these three elements are related in different ways to the different types of universal judgment.

One kind of thought bears on the singular as its determining element. By an examination of each individual instance of an S we find it to be P, and, by necessity of thought, we reject this as mere coincidence and feel justified in thinking that all the S's are P because they are S. That is, S as such, is P. This is common sense generalization.

> Here the singular element is primary, the universal secondary. A universal judgment of this kind is called a generalization.[176]

A second type of thought begins from particulars rather than singulars, and judges that *each kind* of S is P, from which it goes on to judge that S *as such* is P. The mind here bears *on the specific difference* between the instances.

> This is the type of universal judgment which is normal in empirical science where the importance of the plurality of instances towards establishing a universal proposition lies not in their numerical difference, as in generalization proper, but in the specific differences between them.[177]

In a third type of thought the universal is of primary interest. From thinking that S *as such* is P we see that any specific kind of S is P and that each instance of S is P. "This is the type of universal judgment which obtains in exact mathematical science."[178] The assertion of the property of a triangle rests only on those features which belong to it as a triangle, and not on its individual or particular features.

In philosophical judgments these three elements are present but none of the above three types of structure are adequate to it. The framing of "a universal judgment in philosophy by generalization from instances is to commit the fallacy of identified coincidents."[179] Selection of the instances so as to avoid this results in "the fallacy of precarious margins."[180]

Empirical science looks for a generic essence as something common to its variously distinguished species as indifferently present in them all. Collingwood says that philosophers instinctively avoid this way of proceeding since they feel that the full nature of what is being

---

176 *Ibid.*, p. 112.

177 *Ibid.*

178 *Ibid.*

179 *Ibid.*, p. 113. See also above, this chapter, p. 101.

180 See *An Essay on Philosophical Method*, p. 113. See also above, this chapter, pp. 100-101.

investigated is exemplified only in the highest forms of it and whatever there is in the lower forms is only a minimum form of the generic essence, not its completion.

Nor can they proceed as in exact science from a knowledge of the generic essence in itself as implicating certain properties, attempting to know the properties of its various particulars and individuals. The difficulty is that "any statement about a generic concept which is true as applied to one of its specific forms is likely to require modification before it can be applied to any other."[181] What is true of knowledge as intellectual will need modification before it can be applied to knowledge as sensitive.

B.ii.c. Mutual Implication of Universality, Particularity and Singularity in Concrete Concepts

To avoid the above three fallacies we need not look for a fourth way of arranging the three elements but to find *a method that uses all these three methods at once,* by a process of *checking each by means of the others.* A philosophical judgment embodies all three types of structure. "Philosophy can and does generalize, or assert of the concept as such what is found in its single instances, but subject to the provision that, by itself, this is only a clue towards answering its question, and not a substantive answer."[182] And, "like empirical science, philosophy can and does argue that if different species of a concept agree in a certain respect this should be a feature belonging to the generic essence; but . . . they must be checked by arranging the species in a scale and showing that the features of the generic essence shine out more clearly as the scale reaches its culmination."[183] Furthermore, "philosophy like exact science aims at determining *a priori* the characteristics which belong of necessity to its concepts as such in their true universality."[184] But such statements in philosophy are tentative until verified by reference to facts. "[A] philosophical theory must show that what it claims as necessary in the concept is possible in every specification of the concept and actual in its instances."[185]

To assert a proposition in mathematics does not require belief in the actual existence of the subject of discourse, all that is required is to suppose it. The mathematician frames a supposition and sees what

---

[181] *An Essay on Philosophical Method,* pp. 114-115.
[182] *Ibid.,* p. 115.
[183] *Ibid.,* pp. 116-117.
[184] *Ibid.,* p. 117.
[185] *Ibid.*

follows from it. The body of mathematics is composed of this complex which logic calls the hypothetical proposition.[186]

The empirical scientist deals with facts, not abstractions, but he ignores the facts in their multiform variety and is especially concerned to find a certain framework into which his facts can be fitted. Without such a framework which enables him to group the facts around fixed points, as *loci* for his thought, there is no science. The scientist studying the facts of tuberculosis looks for what belongs in the tubercular character, treating the disease *itself* as an entity, ignoring the *peculiarities of its instances* which merely exemplify it. The empirical scientist deals with the notion of a specific disease as he deals with the notion of a specific plant. It is the specificity that interests him primarily, and the instances are of interest to him only as they embody the specific character.

The universal propositions of empirical science therefore are somewhat like mathematics in that they have a hypothetical character. Statements about tuberculosis in medical textbooks refer only to the standard case from which it does not follow that the standard case exists.[187]

> . . . the body of scientific knowledge is expressed in propositions that are logically intermediate between these two orders of categoricals, the statements of fact which are its data and the statements of fact which are its applications; and this body itself consists of hypothetical propositions.[188]

B.ii.d. Universality of Concrete Universal Not Merely Hypothetical

Philosophical thought differs both from mathematics and empirical science in that its body or substance is composed of propositions which are not merely hypothetical but in essence and fundamental intention categorical.[189] By categorical Collingwood explains that he uses the word in the logician's sense[190] and that he means that a statement is categorical which itself actualizes an instance of what is asserted.[191]

One cannot study logic or ethics without a commitment to the

---

[186] See *ibid.*, pp. 117-118.
[187] See *ibid.*, pp. 118-121.
[188] *Ibid.*, p. 121.
[189] *Ibid.*
[190] *Ibid.*, p. 122.
[191] *Ibid.*, pp. 123-134.

view that the subject matter being studied actually exists.[192] The body of geometrical science is heterogeneous with its subject matter, but in logic it is homogeneous and even identical therewith.

> The propositions of which logic consists must conform to the rules which logic lays down, so that logic is actually about itself. . . . It follows that logic cannot be in substance merely hypothetical.[193]

Geometry can afford to be indifferent to the existence of its subject matter; it suffices that it merely supposes it. But logic is not thus indifferent. By thinking logically and propounding logical propositions logic generates an actually existing subject matter to itself. There are thus no logical propositions about a merely supposed object; if logic is being exercised its very object is thereby existentialized. In saying that 'all squares have equal diagonals' there is no assertion that any squares or diagonals exist. But to say that 'all universal propositions distribute their subject' is not only to talk about universal propositions but to produce an actual instance of the thing under discussion. Logic contains as well as discusses reasonings. A logician cannot therefore deny the existence of any actual reasoning without thereby disowning himself as a logician.[194] Logic, therefore, is such that the very conception of its subject matter, namely thought, is at the same time the existential instancing of that subject matter.[195] The propositions constitutive of the subject matter of logic cannot be merely hypothetical, as likewise the subject matter of ethical thought must be conceived as something whose essence involves its existence.[196] Ethics discusses not merely ideas about action divorced from the action itself but the moral consciousness. It is therefore both normative and descriptive, describing something that is already somewhat what it ought to be.[197] This holds generally of the subject matter making up the body of any philosophical science. It consists of categorical propositions unlike that of the exact and empirical sciences whose subject

---

[192] *Ibid.*, p. 128.
[193] *Ibid.*, p. 129.
[194] *Ibid.*, p. 130.
[195] *Ibid.*, p. 131.
[196] *Ibid.*, p. 133. Just as to the religious consciousness God cannot be believed except as existing, as the Anselmian argument expresses, so the philosophical consciousness in every philosophical science, logic, ethics and metaphysics, "stands committed to maintaining that its subject-matter is no mere hypothesis, but something actually existing," that is, philosophy is "a form of thought in which essence and existence . . . are conceived as inseparable"; see *An Essay on Philosophical Method*, p. 127.
[197] See *ibid.*, p. 132.

matter consists merely of hypothetical propositions.[198] Thus, whereas in exact and empirical science categorical and hypothetical are mutually exclusive species of the genus 'proposition,' in philosophy they overlap; the propositions of philosophy being together categorical and hypothetical. Thus, the universal judgments of science can be purely hypothetical whereas those forming the body of philosophy cannot be purely hypothetical but must be categorical at the same time. But, furthermore, the togetherness of these two characteristics in philosophical propositions is mutually implicatory, each demands the other for itself to be present.[199]

> . . . in the case of philosophical judgments the overlap [is] peculiarly intimate; the categorical element is no longer something external to the hypothetical, even if necessary to it; both elements are alike of the essence of philosophy as such.[200]

Thus, whereas according to a classificatory system, hypothetical and categorical are names for two specifically distinct and mutually exclusive kinds of statement, in the concrete thought of philosophy they are two aspects of the one same kind of thought, the hypothetical aspect being the questioning aspect, the categorical aspect being the answering aspect, the ''systole and diastole'' of all knowledge, as Collingwood expresses it.[201] Question and answer, therefore, according to classificatory logic are two different species of mental activity, but according to the logic of overlap they are two forms in which the one same thing, knowledge, is realized; two modalities, as it were, that go together in the constitution of the one concrete activity which is knowledge. Knowledge, therefore, as concretely experienced, ''spectro-analyses'' in its explanation, into a questioning moment and an answering moment, neither existing as something *in itself* independently of the other, as classificatory logic would have it.

B.iii. The Logic of Reasoning

Collingwood then goes on[202] to enquire into the nature of reasoning in philosophy, which he does by comparing it with deductive and inductive reasoning.

In all inference there are three things: data from which we argue,

---

[198] *Ibid.*, p. 133.
[199] *Ibid.*, p. 134.
[200] *Ibid.*
[201] See *Speculum Mentis*, p. 77. See also Chapter 3, above, p. 64.
[202] *An Essay on Philosophical Method*, pp. 151-154.

principles according to which we argue and the conclusions to which we argue.

B.iii.a. Exact Science Ruled by Supposal

The data of exact science consist of suppositions, namely the principles or axioms of the science. The conclusions are inferred demonstratively, as following with perfect logical rigor from those principles. There are two kinds of axioms in exact science. First, there are axioms belonging to logic, which are principles according to which demonstration as such must proceed; they are necessary to the exact science but not part of it; they are thus presuppositions *of the science,* although not mere suppositions in themselves absolutely. It is the work of logic to inquire into these principles, not that of the science itself. That is, they are not answers to questions which the science itself asks, although they are answers to questions asked elsewhere, namely in logic.

The second kind of axioms form part of the science, but in a distinctive way. The older view of exact science regards these axioms as self-evident, not themselves requiring demonstration. Exact science, thus, has two kinds of certainly true propositions, the indemonstrable or special axioms and the demonstrated conclusions. Collingwood proceeds 'on the assumption'(?) that the special axioms of exact science are not *known,* but only *assumed* to be true. But, he says, the logical axioms cannot be merely assumed, since being philosophical propositions they are always categorical, for one cannot proceed in thinking merely as if the principles of thinking were true; if they were not true one would not be thinking. But since we may regard the special axioms as mere assumptions we may also regard the whole science as consisting of mere assumptions, the axioms as primary or fundamental assumptions and the conclusions as secondary or derivative assumptions.

Characteristic of exact science, then, is a certain irreversibility, according to which all logical dependence is of conclusions upon axioms, there being no reciprocal dependence of axioms on the conclusions. The results to which the axioms lead us have no effect on our attitude to these axioms; our acceptance of the conclusions is totally dominated by our commitment to the axioms. This irreversibility goes necessarily with exact science. There is only forward argument, from principles to conclusions, never backwards argument, from conclusions to principles.

B.iii.b. Philosophical Reasoning both Categorical and Rigidly Coherent

Philosophical reasoning resembles that of exact science in its demand for close and cogent reasoning. Each demands valid and sufficient reason to be given for its conclusions. But philosophical reasoning does not entirely resemble that of mathematics.

Firstly, there is no division of axioms into those of the special science and those properly belonging to logic. Logic is a branch of philosophy not separable from its other branches; one cannot investigate an ethical problem entirely setting aside logical points; these must be faced sooner or later. So, whereas other sciences can neglect their own presuppositions, philosophy cannot do so. Philosophy has only one kind of axiom, that kind which forms part of its own body.

Secondly, because its axioms are philosophical propositions, they must be categorically asserted, never mere supposals or assumptions.

This second difference suggests that these axioms are self-evident propositions forming the first principles of an irreversible deductive system of thought. Collingwood now asks whether this is true of philosophical reasoning.

If we turn to the great mathematician-philosophers of the seventeenth century, in whom we would expect to find such an attitude—for instance, Descartes, "when we forget his theory of method and turn to his practice," for "Descartes was far too good a philosopher to neglect the necessary differences between philosophical and mathematical reasoning"—we will find that "his first principle, 'I think, therefore I am', is neither a self-evident truth nor an assumption."[203] Although this assertion is a starting point for his metaphysical reasoning it is established by a proof in the passage in which he enunciates it. But it is a peculiar kind of proof, one which Kant will later call a transcendental deduction.

> In Kantian language, the principle *cogito ergo sum* is in this passage transcendentally deduced, that is, shown to be the condition on which experience as it actually exists, in this case the experience of systematic doubt, is alone possible. If I did not exist as a thinking being, I could not doubt. Even doubt of my own existence is therefore a guarantee of my existence.[204]

Here Descartes touches in practice on the difference between the mathematical and philosophical methods of inquiry, whatever his explicit theory may have been. Notwithstanding his advocacy of clear and distinct perception of the principles of the mathematical method, he recognizes that "on the contrary, nothing in metaphysics causes

---

[203] *An Essay on Philosophical Method*, pp. 155-156.
[204] *Ibid.*, pp. 156-157.

more trouble than the making the perception of its primary notions clear and distinct.''[205]

Descartes' successors interpreted his precepts in the light of his practice. When Spinoza lays down what purports to be the first definition of his Ethics, 'By *cause of itself* I understand that whose essence involves existence, and whose nature cannot be conceived except as existing,' he is not merely defining terms *more geometrico*, for ''his statement is not a definition but a theorem: a philosophical position . . . an arguable and argued position.''[206] Leibnitz's *Monadology* begins with two clauses which taken together state the conception of an inextended and indivisible substance which appears to be an axiomatic definition. Collingwood says, however, that Leibnitz, like Spinoza, knew that he ''was beginning his treatise not by defining his terms like a mathematician but by laying down a whole metaphysical system in a nutshell, and not by stating a self-evident axiom but by affirming a highly controversial theorem.''[207]

Kant, subsequently, laid it down that there are no axioms in philosophy, for its first principles required proof, but proof of a special kind. Rigid adherence to the mathematical method in philosophy would lead to a 'house of cards,' and Hegel, following Kant, saw that philosophy was in the peculiar situation of having to justify its own starting point. But if philosophy has to justify its own beginnings ''this can be done only if the arguments of philosophy, instead of having an irreversible direction from principles to conclusions, have a reversible one, the principles establishing the conclusions and the conclusions reciprocally establishing the principles.''[208]

The charge that there is a vicious circle in such a reasoning, in which there is reciprocal justification of principles by conclusions and conclusions by principles, is averted by noting a peculiar feature of philosophical thought, namely, ''the Socratic principle that philosophical reasoning leads to no conclusions which we did not in some sense

---

[205] Reply to the Second Set of Objections; see E. S. Haldane and G. R. T. Ross, eds., *Philosophical Works of Descartes*, Vol. II, N.Y., Dover, 1955, p. 49.

[206] *An Essay on Philosophical Method*, p. 158.

[207] *Ibid.*, p. 159. Anticipating what is to come for purposes of finding our direction, let it be said that Collingwood is referring to a concrete universal when he speaks of these first quasi-axioms of Leibnitz and Spinoza. Not just a *supposed* beginning is *postulated*, but *a whole concrete system* is *affirmed categorically*, within which, by a 'systole-diastole' dynamic reciprocity of question and answer, this continually sustained beginning will be rationally and systematically unfolded, in which unfolding the already implicit is explicitated.

[208] *An Essay on Philosophical Method*, p. 160.

know already."[209] It has widely been recognized that philosophy, unlike exact or empirical science, does not bring us to know something of which we were previously simply ignorant, but rather brings us to know in a new and different way something we knew in some way already.

> If the species of a philosophical genus overlap, the distinction between the known and the unknown, which in a non-philosophical subject-matter involves a difference betwen two mutually exclusive classes of truths, in a philosophical subject-matter implies that we may both know and not know the same thing: a paradox which disappears in the light of the notion of a scale of forms of knowledge, where coming to know means coming to know in a different and better way.[210]

The establishment of a proposition in philosophy, therefore, is not a transference from the unknown to the known, but from the known (in one way) to the same but known in a new way, or as Collingwood frequently puts it, a passage from the implicit to the explicit. Philosophical method is a method of explicating what is already known, but implicitly only. This is the case if we know *that* something happens, but do not know *why* it happens. It would be a better knowledge to know both that and why it happens together. In philosophical knowledge the facts are manifestations of principles and simultaneously principles render the facts understandable; the intelligible principles are manifested in the observation of facts at the same time as the observed facts are rendered understandable by the principles. The principles of understanding are implicated in the observed fact and the observation of fact implicates the understanding of its principles. This is simply repeating what was said above,[211] following Bernard Bosanquet, that individuality involves full determination, that whatever 'is' not only 'is' but also 'is what it is,' so that to say of something that 'it is' is to say something meaningless if it is taken apart from 'what it is.' As Collingwood expresses it:

> If by seeing certain facts in the light of certain principles we come to understand the facts and at the same time to have visible confirmation of the principles, this is a gain to our knowledge both of the principles and of the facts.[212]

---

[209] *Ibid.*, p. 161.

[210] *Ibid*. Philosophy thus begins 'from a question,' or in wonder, for, in questioning, we both *know* and *lack knowledge* about that which we question or wonder.

[211] See this chapter, above, pp. 90-92.

[212] *An Essay on Philosophical Method*, p. 162.

In this does philosophical knowledge differ from that of the exact sciences. The truth of mathematical conclusions is totally dependent on the proof. The conclusion of a philosophical proof is known prior to the proof; the proof merely assures us, not that it is so, but why it is so, thus enabling us to know it better.

> . . . philosophical systems in general . . . exhibit as a reasoned and ordered whole of interconnected knowledge what was already in substance known before the work of philosophizing began.[213]

It follows from this that the conclusions of philosophy can be checked by comparison with the anticipations of experience, so that the direction of the argument regarding principles and conclusions is reversible, each being able to be established by an appeal to the other. This is not a vicious circle, since 'established' here means being raised to a higher grade of knowledge. What was mere observation has become an understood observation, an explained or explicated observation. And an otherwise merely abstract principle is no longer a merely abstract principle but a factually verified principle.

The chief implication of this for the theory of inference in philosophy is as follows:

> If the substance of philosophical knowledge is known to us, however dimly and confusedly, before philosophical reasoning begins, the purpose of that reasoning can only be to present it in a new form, that is, the form of a system constructed according to certain principles. The philosopher who unfolds such a system is not spinning a web of ideas from the recesses of his own mind; he is expressing the results of his own experience and that of other people in a reasoned and orderly shape.[214]

Collingwood says that this means that instead of asking himself only one question at every step in the argument, as in exact science, namely, 'What follows from the premisses?', the philosopher has also to ask another question as well, namely 'Does what follows from the premisses agree with what we find in actual experience? Any philosophical argument that neglects the test of actual experience is defective. Thus, a philosophical argument does not hang with its whole weight on the starting-point; it is supported throughout its texture by cross-references to experience.[215]

In this it resembles the inductive procedure of empirical science.

---

[213] *Ibid.*, p. 163.
[214] *Ibid.*, p. 164.
[215] See *ibid.*

Does this mean that philosophical theory is wholly based on observation and experience, and is therefore an empirical science?[216]

B.iii.c. Extrinsic Relationship of Fact and Theory in Empirical Science

The aim of inductive reasoning is to establish universal propositions from individual facts. Its starting point is the facts as data, the universal propositions are its conclusions; the data are known empirically, by perception or by the historical record of perceptions in the past; the conclusion must already be present at the beginning in the form of an hypothesis which is to be tested by bringing it into relation with the data. At first the hypothesis is a mere possibility and the aim of the inductive procedure is to convert that possibility into a probability. In empirical science to establish means to establish as probable, more or less. The data, however, are certain from the beginning and the inductive process leaves this certainty untouched, for that rests entirely on observation and in no way on the demonstration. The work of induction is to conform to the facts; it depends on them and they in no sense depend upon it.[217]

Just as in the case of exact science, the principles of induction are of two kinds. First, logical principles which retain their own certainty throughout the inductive process and, as in exact demonstration, are logically presupposed by all induction, never being established by induction itself. The second kind of principles, those which belong properly to induction itself, such as the principle that the future will probably resemble the past, are assumptions that are necessary if there is going to be any argumentation at all. The successful conduct of the argument based on these principles in no way confirms them; their assumption is a necessary precondition of the very existence and possibility of inductive argumentation. We are conscious throughout the more or less successful progression of the induction that these assumptions are assumptions and nothing more. What increases in successful inductive inquiries is not the probability of such *principles* (as that the future will probably resemble the past) but the probability of *hypotheses* such as that fermentation is due to microorganisms. The principles never appear as conclusions even in the modified sense in which inductive thinking has conclusions.

The logical movement of inductive thought is therefore irreversible in the same sense as that of exact science. The principles on

[216] See *ibid*.
[217] See *ibid*., pp. 165-66.

> which induction rests receive in return no support from the inductive process itself. Either they are certain from beginning to end, or from the beginning to end they are mere assumptions.[218]

Exact science admits of reversibility of its data (though not of its principles), whereas inductive argument is not reversible regarding its data, since the data are facts vouched for by perception. An as yet unobserved fact inferred by induction is merely inferred, that is, established as probable and not perceived. It becomes certain only by observation subsequent to the inductive argumentation.

B.iii.d. The Rapprochement, or Mutual Implication of Fact and Theory, Principles and Conclusions, Essence and Existence, Experience and Explanation in Concrete Philosophical Reasoning

Notwithstanding the resemblance between philosophical reasoning and scientific induction in that both possess knowledge prior to the reasoning process, there are notable differences.

The initial knowledge of philosophy forms the substance, the material out of which the final, systematized knowledge is constructed. The theory of cyclones does not include in it a statement of the individual barometric observations on which it depends, whereas in philosophy the very same proposition first known to be true is reaffirmed in the conclusion with proofs in the body of a system. In empirical science the initial knowledge is individual facts; in philosophy, the initial knowledge, being homogeneous with the conclusion, must consist of universal propositions.

> The data of philosophy are thus never mere facts in the sense of individual events, individual objects, individual actions or the like; they are always universal.[219]

Again, whereas the individual facts dealt with by science are apprehended by perception, the facts of philosophy must be apprehended in another way.

> . . . the data of philosophy, if they are universal propositions, cannot be apprehended in that way [by perception]; they must be grasped by something in the nature of what we call, as distinct from perceiving, thinking.[220]

Therefore, the experience which provides the data of philosophy is the experience not of a perceiver but of a thinker, or of an intellectual perception.

---

[218] *Ibid.*, p. 167.
[219] *Ibid.*, p. 168.
[220] *Ibid.*, p. 169.

The initial knowledge of philosophy differs from that of empirical science not only in its relation to the reasoning process and in its own constitution and in the way we come to possess it, but also in what happens to such data throughout the development of thought. In adding the theory purporting to explain the facts in empirical science, the theory is another knowledge in addition to the original facts which themselves are not known in any new way. The process is one of accumulation. But in philosophy the knowledge why things are so makes a difference to the knowledge that they are so.

> [In philosophy] the new knowledge imparts a new quality to the old; in seeing why things are thus, we are not merely adding one piece of knowledge to another, we are coming to know the old better. Our knowledge is not simply accumulating, it is developing; it is improving, as well as increasing; it is widening and strengthening itself at once.[221]

There is a consequent difference in the conclusions of philosophy on the one hand and science on the other. In empirical science the result of the induction is "an hypothesis standing, somewhat nebulously, outside the facts on which it depends . . ." whereas "in philosophy, the theory that emerges from consideration of the facts is no mere hypothesis, it is the facts themselves more thoroughly understood."[222] So,

> In philosophy there is continuity between the experience and the theory; the theory is nothing but the experience itself, with its universality further insisted upon, its latent connexions and contradictions brought into the light of consciousness.[223]

In science there is a separation between the facts explained and the theory doing the explaining. But in philosophy experience as philosophically experienced is already found to be going over into theory, and theory, in explaining, remains an experience.

> . . . the separation between theory and fact . . . in philosophy is no separation at all, but only a distinction of the kind . . . between specific developments of the same genus . . . articulated in a scale of forms.[224]

Thus, in accordance with such a distinction, the pre-philosophical knowledge from which the philosophical process begins is only in a

---

[221] *Ibid.*, pp. 169-170.

[222] *Ibid.*, p. 170. The hypothetical and the assertive are mutually implicated, the answer is (implicitly) in the question, and the question is (explicitly) in the answer.

[223] *Ibid.*, pp. 170-171.

[224] *Ibid.*, p. 171.

relative sense non-philosophical; it is less philosophical than that into which it develops but is not absolutely non-philosophical. The knowledge from which philosophy begins is already full of philosophical elements, not at the zero end of the scale but at least at unity. Thus the 'conclusions' of philosophical thinking and the 'experience' on which they are based are names for any two successive stages in the scale of forms of philosophical knowledge. The rational element of any one stage on the scale of philosophical thought may be irrational relatively to the next stage above it.

One stage in the scale of forms of philosophical knowledge is related to the next higher as knowledge questioned to *explicitation* of that same knowledge by a knowledge which is *an answer* to a question provoked by the prior knowledge. The answer explicitates what is questioned, namely the prior experience. Questioning is experience seeking its order in the whole of experience; seeking to be experienced *as whole with constitutive parts.*

> . . . what is asked of the higher is not simply that it should agree with the lower, but rather that it should explain it: perpetuate its substance in a new form, related to the old somewhat as a fact plus the reasons for it is related to the bare fact.[225]

Thus, every philosophical experience becomes the object of a question which requests that the experience be explained by a theory which consists in the explication of that experience, in which explicitation the original experience persists but with a new quality. The (relatively) bare experience goes over *from within its own resources* to becoming a rational experience, to becoming the experience of a rational being who knows not merely that he has an experience but why he has it, and knows why in terms of elements provided from within the original experience itself.[226] The original experience, experienced as un-

[225] *Ibid.*, p. 172. Thus, again, whether a certain content is question or answer depends not on there being two entities having those characters as mutually exclusive species, but precisely on the attitude the mind takes to the one same content. If the mind's attitude is one of supposal (although with reference to further assertion) then that content is the content of a question. But if the mind's attitude to that content is assertive, then the content is a proposition having the character of answer to the question which demanded that assertion. What may provide the answer to previous questions may also provide the supposal which causes further questioning to arise. 'John Doe has been murdered' is *assertive* with respect to questions regarding John Doe's present whereabouts, but it is *supposed* with respect to the further question 'Who killed John Doe?'

[226] According to B. Bosanquet in *The Principle of Individuality and Value*, pp. 31-41, all experience is characterised by a tendency to universality inasmuch as from itself it tends to go beyond partiality and manifest a wholeness which both illuminates dif-

explained, as calling for rational explanation, is a question, or the experience of questioning, or a questioning experience, which, from within its own inner resources, resolves the conflicting alternatives which constitute the question, and passes over into the experience of explanation, or providing the answer to the prior question. Thus, the question and answer process forms one continuous whole analogous to a single cell which divides *from within itself* by a process of *self-differentiation* to a final well distinguished equilibrium state.

> At every stage in the scale, there is a datum or body of experience, the stage that has actually been reached; and there is a problem [question], the task of explaining this experience by constructing a theory of it, which is nothing but the same experience raised by intenser thought to a higher level of rationality. The accomplishment of this task is only the continuation of a process already begun; it was only by thinking that we reached the point at which we stand, for the experience upon which we philosophize is already a rational experience; so our reason for going on is that we already stand committed to the task. But the new and intenser thinking must be thinking of a new kind; new principles are appearing in it, and these

---

ferent contexts and receives light from them. This tendency is towards an ideal of truth, that is, an experience rationally accounted for through elimination from it of conflicting interpretations of the same facts in the same relations. General rules represent a step in the direction of the completion of knowledge as a coherent whole, but generalization pursues an identity apart from differences; its method is omission and is achieved by attending only to the common qualities of individuals while disregarding their differences. This procedure prohibits the consideration of any world structure of which the concrete individuals would be members and likewise prohibits the ascertainment of the full concrete nature of any individual itself. Whole areas of experience are thus left unaccounted for when identity is sought through exclusion of differences and individuals are taken merely as members of a class whose totality is by way of a whole of repetition. The generalisation method of totality thus rules out on principle any consideration of the differences within that common characteristic and those which constitute the individual in what he is distinctively. Bosanquet maintains that contradiction is most successfully removed from experience when experience is regarded as having that unity which is proper to a whole of parts, namely an organism, or more generally, that of a world or a cosmos. The unity implied by the terms world or cosmos is the most inclusive, and therefore the most effective, type of universal, and "a world or cosmos is a system of members, such that every member, being *ex hypothesi* distinct, nevertheless contributes to the unity of the whole in virtue of the peculiarities which constitute its distinctiveness." Bosanquet contrasts "the concrete universal . . . with the general rule as a centre of radii compared with a superficial area." He says that its test of universality does not consist in the number of subjects which share a common predicate but rather in the number of predicates which can be attached to a single subject. This, he says, "is the degree in which a systematic identity subordinates diversity to itself or, more truly, reveals itself as the spirit of communion and totality within which identity and difference are distinguishable but inseparable points of view." Thus, Bosanquet maintains, *the power of the concrete universal*, or of individuality which constitutes the true type of universality, *is its capacity to unify experience*.

> give a criterion by which the principles involved in the last step are superseded. Thus the stage last reached, regarded as a theory, is now a theory criticized and refuted; what stands firm is not its truth as theory, but the fact that we have experienced it; and in criticizing and demolishing it as a theory we are confirming and explaining it as an experience.[227]

This citation very well sums up the process of the unfolding of experience as it was presented in detail by Collingwood in his *Speculum Mentis*. Note that it is not a matter of external addition, after the manner in which empirical science would explain, but rather of internal "addition," more exactly of self-differentiating division from within; not the constructing of a system of external relations but the unfolding of a system of internal relations; not a quantitative addition but a qualitative addition, that is, not an extensification of thought coming to know more things, but an intensification of thought, by coming to know better and more fully what was already less well known. Since the different levels are different levels of actual experience, the earlier and later stages are not merely logically earlier and later but also temporally so. Time is thus intrinsic to the elaboration of the concrete universal,[228] for which reason history and philosophy are related by intrinsic relations. Whereas natural accumulation is the addition of something from outside according to spatial relations, experiential accumulation is by way of self-differentiating addition from within, according to temporal relations determinative of thought. The

[227] *An Essay on Philosophical Method*, p. 173.

[228] Notwithstanding that Collingwood accuses Bosanquet of "conceiving the proper object of knowledge Platonically as a timeless world of pure universality" (*The Idea of History*, p. 143) and that Bosanquet himself says (*The Principle of Individuality and Value*, pp. 78-79) that "History is a hybrid form of experience, incapable of any considerable degree of 'being or trueness' " and (*ibid.*, p. 80) that Philosophy and Religion "take us far beyond the spatio-temporal externality of history," nevertheless Bosanquet himself seems to provide a jumping off ground for the amendment of his position more into Collingwood's line of thinking, in which history and temporality are of the essence of individuality and concreteness. Thus, also in *The Principle of Individuality and Value*, p. 72, Bosanquet identifies individuality with spirituality and then states that memory is bound up with the spiritual being of the mind. "The Individual is one with the spiritual, and the characteristic of the spiritual in its proper nature is inwardness as opposed to externality . . . Memory is inward because its diversity is bound up with the being of the mind; you cannot take the one and leave the other. Inwardness is diversity without dissociation." Now, if memory is bound up intimately with mind and its inward respect it necessarily follows that mind and its inward respect are intimately bound up with time, since time is of the essence of memory. Again, if, as Bosanquet says (*ibid.*, p. 70) the uniqueness which marks individuality is the uniqueness of purpose, and purpose is meaningful only in the context of a forward-looking aspect into the future, then individuality, or concrete universality, essentially implicates futurity, therefore temporality, therefore a historical dimension.

seeds of this notion are found in Kant and their elaboration also in Bergson, for whom external spatial relations characterise the external world of material nature whereas internal temporal relations characterise the internal spiritual world of consciousness. The shadow of Descartes is also present, identifying the external world with its spatial dimensions, and the internal world of self-consciousness with thought. And when Collingwood, in the above citation, says that "our reason for going on is that we stand committed to the task," I interpret this to refer to the emotive dynamism already mentioned in regard to questioning[229]—the restlessness of the mind to pass on from an inadequate experience to a more adequate experience,[230] such inadequacy not merely known but also emotively felt.

This felt tension within the conscious experience of a rational agent to transform its given experience into a more fully rational, accounted for experience is well described by B. Bosanquet:

> [Thought] presses beyond the given, following the "what" beyond the limits of the "that." But it is also true that in following the "what" it tends always to return to a fuller "that." If its impulse is away from the given it is towards the whole—the world. And as constituting a world it tends to return to the full depth and roundness of experience from which its first step was to depart. In a "world," a "concrete universal," we do not lose directness and significance as we depart from primary experience; on the contrary, every detail has gained incalculably in vividness and in meaning, by reason of the intricate interpretation and interconnection, through which thought has developed its possibilities of "being."[231]

Bosanquet further says that the notion of reality, the concrete whole of individuality is essentially the notion of a spring of adjustment within the self,[232] whereby the self seeks to bring the fragmentary elements of experience into a systematized whole. Experiences such as the apprehension of beauty and those of the moral and religious consciousness are experiences in which the aspiration of all experience to be a unified whole at least partially comes into its own.[233] The main point in the contrast between the relations of abstract generalization on the one hand and concrete thinking on the other is the

---

[229] See Chapter 3, p. 67 above; see also Collingwood's *The New Leviathan*, pp. 47-48 and pp. 50-53 on appetite as feeling-discomfort pointing to a 'there-and-then' satisfying state of feeling.

[230] See this chapter, above, pp. 134-136.

[231] *The Principle of Individuality and Value*, p. 56.

[232] *Ibid.*, p. 53.

[233] *Ibid.*, pp. 56-57.

capacity of the latter to give completeness and fulfillment to experience. By pursuing the overcoming of contradiction within itself, by seeing its multiple elements as contributing by their differences to the meaning of a whole, the mind arrives at an incomparably more vital and intense experience as well as at a more logical and rational experience than that of the unexamined life of everyday perception.[234]

> Form, interdependence, significance, self-completeness are characteristics of thought at its best, so of vitality at its highest. This is the general character by which the concrete universal gives us the clue to the individual.[235]

The datum of philosophy does not therefore remain as a fixed point, it undergoes development as it undergoes analysis and finally vanishes in its old form to reappear in a new.[236]

Philosophy embodies the systematizing function of deductive thought, since every phase in its development is at least ideally a complete system connected throughout by strict logical bonds. But the system of philosophy is something more than a deductive system since its principles of systematization are always open to criticism, that is to questioning, and stand or fall according to their success or failure in explaining our experience (as in *Speculum Mentis*).

Likewise, philosophy embodies the inductive function of discerning the principles that permeate experience, making it a rational whole. But the experience on which it bases its theories is the theorizing experience of rational living. The theory of philosophy is thus an essential part of philosophy; the activity of philosophizing is a datum to philosophy; the data from which it begins are homogeneous with the theories by which, as conclusions, the data are explained.

C. The Systematic Unity of the Concrete Universal According to Collingwood

As said above,[237] the notion of system is central to the understanding of Collingwood's concrete *rapprochement* logic, for the unity in diversity in which consists the concrete universal is systematic unity as opposed to isolated, self-contained, atomic units of abstract intelligibility such as Positivistic science trades in.

---

234 *Ibid.*, p. 58.
235 *Ibid.*, p. 59.
236 Thinking is thus not a fixed *state* for Collingwood but an *activity* of becoming.
237 See this chapter, above, p. 93; see also Collingwood's *Speculum Mentis*, p. 221.

Collingwood's treatment of "The Idea of System"[238] adapts Bosanquet's tendency to totality which characterises rational experience to his own systematisation of thought as an overlap of classes in a dynamically unfolding scale of forms.

Collingwood elaborates his idea of system in the context of explaining the constructive moment in philosophical thought which follows upon a period of scepticism.[239] He says that the idea of a system implies finality, completeness, objectivity and unity.[240] But, a problem arises as to whether any of these notes can properly be applied to any type of thought, such as philosophical thought, which constantly advances through new discoveries to new points of view.[241]

Against the objection that any constantly changing branch of thought cannot have systematic finality, Collingwood replies that any advancement in knowledge requires that the student take stock from time to time as to where he stands at the point at which he has arrived. But this can be done only by stating his result systematically, showing by what devious routes he has reached his present position.[242]

> The philosopher . . . like every student, must sum up his progress from time to time, and express his conclusions in a systematic form, if progress is to continue.[243]

To difficulties advanced against the possibility of systematizing the unwieldy mass of what constitutes modern philosophical knowledge Collingwood argues that systematic completion does not require an encyclopaedic summation of all available facts. An adequate survey is quite possible in order that philosophy may "deal with its own special problems in its own special way."[244] No branch of knowledge, least of all philosophy, is a mere unwieldy accumulation of ascertained facts. Any thinker, any problem solver, works within a well defined area which is in organic (or internal) relationship with the work of predecessors and contemporaries, in such a way that he "finds himself shepherded, as it were, into a particular line of study by the fact that others round him are doing work which requires this as its complement."[245] Thus, the completion proper to systematic philosophy is

---

238 See *An Essay on Philosophical Method*, p. 176-198.
239 See *ibid.*, p. 176.
240 See *ibid.*, p. 177.
241 See *ibid.*
242 See *ibid.*, p. 179.
243 *Ibid.*, p. 180.
244 *Ibid.*, p. 181.
245 *Ibid.*, p. 182.

that the investigator stands at the term of intrinsic relations with past and with contemporary thinkers so that his special problems are determined for him by the context in which he is working.

To the objection that a personal contribution, which every thinker's contribution must be, cannot be objective and therefore cannot be systematic, Collingwood replies that the contribution, as a justification of making a genuine contribution to the body of human knowledge, must have a theory of the place that his contribution has in the whole; that is, he must be able, in his result, to show the internal relationships which his personal thought bears to that of his predecessors and contemporaries. This means that he must contribute his answer as to what philosophy is. In adding to the fund of human knowledge he does not just add another item to the inventory but reshapes in his own mind the idea of philosophy as a whole.[246]

Systematic thought does not mean adherence to rigidly preestablished formulae, as if the diversity of philosophical problems required that they be kept separate and solved by their own special methods.[247] The very concept itself of philosophy is a philosophical concept, so that its specific classes overlap.[248]

This overlap of the various classes of the concept of philosophy takes place in three different ways. The so-called departments metaphysics, logic, ethics, etc., overlap; secondly, the work of an individual thinker and that of his contemporaries overlap; thirdly, the work of a contemporary thinker overlaps with that of his predecessors.

Just as the work of various individual thinkers, both past and present, does not accumulate aggregationally, by addition from outside one another, so neither do the various parts of philosophy—logic, ethics, etc.—develop in isolation one from the other and add together merely aggregationally. Progress in one is intrinsically related to, and vitally affects, what is going on in every other part. Philosophy as a whole of parts is thus a concrete universal, not an abstract, classificatory universal. So, the various parts which together make up the content of a philosophy will occur as terms on an overlapping scale of forms, each expressing distinct aspects of the one same subject

---

[246] *Ibid.*, pp. 182-184. The same might be said of the historian; that he rewrites the past as he has transformed it in his mind; it has changed inasmuch as now it is something he has come to know; it includes his knowledge of it with which subsequent historians have to reckon.

[247] See *ibid.*, p. 185.

[248] See *ibid.*, p. 188.

matter which together will permit a more adequate expression of the nature of one substance.[249]

> Each form in such a scale sums up the whole scale to that point; that is to say, each form is itself a system in which the topics and methods of the subordinate forms find a subordinate place.[250]

Philosophizing as a system involves the responsibility of showing that theories otherwise true would have to be condemned as errors if considered as distinct and autonomous philosophies.[251]

> . . . the philosopher, in constructing a system, has his place in a scale whose structure is such that every term in it sums up the whole scale to that point . . . he is always [only] at a relative end . . .[252]

Just as the various parts of philosophy stand in intrinsic relation to the whole, so also in regard to any philosophy in regard to its historical situation: ''As one form in a scale, an individual philosophy is one among many, a single moment in the history of thought, which future philosophers will have to treat as such.''[253] Philosophy, as reaffirming the efforts of past philosophers as elements within itself, summarizes the whole previous course of its history, and is thus universal as well as individual,[254] ''. . . a new philosophy which is at the same time an improved version of the old.''[255]

> In a history of this kind all the philosophies of the past are telescoped into the present, and constitute a scale of forms, never beginning and never ending, which are different both in degree and in kind, distinct from each other and opposed to each other.[256]

A genuine philosophy arises in a concrete situation in which a philosopher finds himself, and this happens by an objective necessity. The situation and problems confronted by any individual philosopher are unique to himself, so that if his thought is to become acceptable to another, it will require modification and reaffirmation as an element which the other incorporates into his own philosophy.

> That each [philosopher] must reject the thoughts of others, regarded as self-contained philosophies, and at the same time reaffirm them as elements in his own philosophy, is due not to causes

[249] See *ibid.*, p. 189.
[250] *Ibid.*, p. 190.
[251] See *ibid.*
[252] *Ibid.*, p. 191.
[253] *Ibid.*
[254] See *ibid.*, pp. 191-192.
[255] *Ibid.*, p. 195.
[256] *Ibid.*

> in taste and temperament but to the logical structure of philosophical thought.[257]

Philosophy is not something rigidly predetermined as regards method or content and is not achieved by the application of rigid formulae. It is characterized by a methodical flexibility in which the method changes from one topic to another, since form and content undergo mutually interdependent development as thought traverses its scale of forms as it gradually approximates to the ideal of a perfectly philosophical subject-matter treated by a perfectly philosophical method.[258]

> The idea of a system is nowhere finally and completely realized; but it is always tending to realize itself wherever any diversity is recognized in the subject-matter and methods of thought.[259]

The tendency of philosophical experience to take on a systematic shape is expressed in an infinite variety of ways, a new way every time a new kind of diversity is encountered demanding to be organized into a whole.[260]

Regarding the systematization of (a) the various branches of philosophy, (b) the history of philosophy, (c) the varieties of contemporary philosophies, Collingwood comments as follows.

The various departments of philosophy are only minimally systematized, but systematization is to some extent realized for these differ both in kind and in degree as various ways of dealing philosophically with various topics.

> It is only in a rough and approximate way that the conventional canon of philosophical sciences corresponds to the idea of system; but it is only in so far as it does correspond that it can claim philosophical importance; otherwise it represents a merely empirical grouping of philosophical topics.[261]

The history of philosophical thought is genuine only insofar as the events contained in it are seen to lead one to the other, that is, insofar as each philosopher derived his philosophy by studying the work of his predecessors.

A philosopher genuinely philosophizes only insofar as he does what his predecessors did in assimilating what is true and reject-

---

[257] *Ibid.*, p. 192.
[258] See *ibid.*, p. 192.
[259] *Ibid.*, p. 193.
[260] See *ibid.*
[261] *Ibid.*, p. 194. This seems to suggest that this division of philosophy into parts is more a classificatory division than an overlap of classes.

ing what is false, and thereby produces a new and improved version of the old.[262]

> In a history of this kind all the philosophies of the past are telescoped into the present, and constitute a scale of forms, never beginning and never ending, which are different both in degree and in kind, distinct from each other and opposed to each other.[263]

Finally the philosophy of a philosopher's own time can only be set out by way of the various relations that subsist in and between the various apparently conflicting and opposing prevailing views. "These various relations are not given to their terms from without, by an arbitrary act of a systematizing intellect; they really subsist in and between the terms, and to apprehend the terms without apprehending these relations is to misapprehend them."[264] If these relations are grasped, interiorly, from within, the divergent views will reveal themselves as nodal points in a system of thought as a whole which may rightly be regarded as the philosophy of the present day.

Such a system is conceivable only as a scale of forms. On such a scale, ideally, the crudest and least philosophical view can be located at the lower end which tends towards zero without actually reaching it. The second phase of such a scale is reached when the chaos of conflicting views is sorted out, by consolidation and by criticism, into more or less definite and organized groups of theories called philosophy today. The third phase is reached when these conflicting theories are seen to participate in varying degrees and kinds in a single common spirit.[265]

A philosopher proceeds in the same way in trying to think out his own philosophy. He begins with a welter of half-truths and quarter-opinions which he allocates to the lower end of the scale of forms. These are sifted, diversified and consolidated so that, eventually, well defined interrelated groupings of views are recognizable. Then he tries at a higher level to see these various positions as parts of a connected whole, adjusting not only the parts to the idea of the whole but also adjusting the idea of the whole to meet the demand of the parts, so that the idea of the whole is itself undergoing transformation as the scale of forms which constitutes the whole is traversed upwards.[266]

---

[262] See *ibid.*, pp. 194-195.
[263] *Ibid.*, p. 195.
[264] *Ibid.*, pp. 195-196.
[265] *Ibid.*, pp. 196-197.
[266] See *ibid.*, pp. 197-198.

> Such a system is only an ideal in the sense that it regulates the procedure of a philosopher trying to answer the question at issue, and cannot be expected to present itself fully formed in his answer.[267]

Collingwood thus develops more explicitly than Bosanquet the progressive characteristic of the individuality of the concrete universal, no doubt due to his predominating interest in historical thought. Collingwood's systematization of the concrete universal as a scale of forms is most likely intended to overcome the difficulties that H. H. Joachim found in developing the coherence notion of truth in line with Bosanquet's notion of the perfect whole as that of a world or cosmos.

> A theory of truth as coherence, we may say must enable us to conceive the one significant whole so as to satisfy certain requirements. We must so conceive it that it is a timeless actuality, maintaining and fulfilling itself through the setting up within itself of modes, which yet are independent; and by creating an inner otherness or duality, which yet is continuously subdued to unity.[268]

The overlap of classes in a scale of forms is a reaction to this need ''to conceive the one significant whole . . . as . . . . a timeless actuality . . .'' The significant whole, which for Collingwood is that at which rational experience aims, is not a timeless actuality but a historically unfolding actuality which has the unity of a scale of forms intrinsically related by internal relations. Collingwood's rendering of the concrete universal as an overlap of classes in a scale of forms is more faithfully in line with Bosanquet's notion of mind as the active form of totality,[269] which, in turn, is more faithful to Collingwood's notion of mind as self-actualizing activity which both presupposes and is realized in a historical, temporalized process.

D. The Dialectical Dynamism of the Concrete Universal

Collingwood's commitment to the theory of the concrete universal and the corresponding logic entailed thereby, is bound up with his rejection of substance in preference to action as the basic characteristic of all that is. And this repudiation of the category of substance[270]

---

[267] *Ibid.*, p. 197.

[268] H. H. Joachim, *The Nature of Truth*, London, Clarendon Press, 1906, p. 175.

[269] B. Bosanquet, *The Principle of Individuality and Value*, p. 59.

[270] Thus, the mind is not a substance underlying its different activities. ''Hume was . . . right to maintain that there is no such thing as 'spiritual substance', nothing that a mind is, distinct from and underlying what it does.'' *The Idea of History*, p. 222.

and static essences[271] is bound up with his rejection of the abstract universal.

> In place of a plurality of abstract essences each constituting one species of knowledge, Collingwood proposes, on the side of the subject, a scale of historical activities, art, religion, natural science, political science, economic science, history, and so on, having as their objects a scale of corresponding activities: the relation of subject and object being in each case such as to constitute a fundamental 'unity', each of which is a concrete universal, and all of which together constitute the moments of the concrete universality (i.e., activity) of mind.[272]

The concrete universal has no identity apart from the existence which it has in its members or instances which manifest it. Its essence does not in any way transcend the particularity of the existent instances. What makes the instances be what they are is precisely the membership which they have in that universal. They are connected among themselves and to the whole by essential relationships. The universality of the concrete universal is therefore not the universality of a substance sharing a common undifferentiated nature. Such an abstract substantialism is incompatible with historical knowledge, since distinction of the natural world from the historical world depends upon it.

> A concrete universal is not a universal substance whose nature is shared in common by its instances. It is, on the contrary, a kind of synthesis which embraces differences as well as similarities. The instances of a concrete universal do not belong to that universal in the sense that they all possess, in precisely the same way, the same abstract characteristics which are the essence of that universal. If this were the case, then differences would be accidental and unimportant. And since differentiation and change are precisely those features which distinguish history from nature, while there can be scientific, there cannot be historical, knowledge of abstract universals. The genuine universal, however, does not enjoy the separate existence of the abstract universal. Apart from its members, or instances (or, as Hegel preferred to call them, its 'moments'), it does not exist at all; its essence does not transcend the particularity of its instances like the essence of an abstract universal. One might almost say that the instances of a concrete universal, rather than

---

[271] Lionel Rubinoff, in *Collingwood and the Reform of Metaphysics*, p. 155, explains that the state for Collingwood is not, as is generally regarded, a substance having a fixed essence whose principles are deducible from the definition expressing that essence, but is a historically changing dynamic process. The question, for Collingwood, is not what are the attributes of the state, but what is political action.

[272] *Ibid.*, pp. 155-156.

> sharing a common abstract characteristic, possess—to borrow Wittgenstein's phrase—"family resemblances"; and it is in virtue of these "family resemblances" that the various moments of the universal are recognized to belong to the same class.[273]

Reality, or that which is at the basis of everything that is, is an activity, or process, or an identity in difference of which polar opposites are the manifestation (as red and blue are the polar opposites of a spectroanalysed white light). More basic than the character of polar opposition, by which the dynamic process of dialectic activity is manifested, is creative or significant negation which is nothing other than the activity itself as generative of its polar opposites. As explained by Collingwood previously[274] the polar opposites are not entities in themselves from which events or processes are derived by a combination thereof, rather the process or activity itself is what is basically given, and the opposites are aspects which one analyses out of this one concrete activity. But these aspects are dynamic entities, they are the opposite features that result from the process of negation which belongs to the very nature of the basic activity.[275] "At the centre of dialectic is the notion of 'determinate negation.' "[276]

'Otherness' is the first and simplest form of negation. Every other form presupposes it and implies it.[277] Otherness, or difference, is the form of negation closest to common sense, which accepts plurality as a datum. But negation is more fundamental than a plurality of entities, for it also constitutes relationships between the entities of the plurality. These relationships may be regarded as a factor intrinsic to existence, or to our interpretation of it. That is, negation can be understood in terms either of thought or of existence or both.[278] Facts, for every philosophy of opposites, include positive and negative aspects, and are understood both as regards what they are and as regards what they are not.[279]

Dialectic is basically an existential aspect of things and is used in thinking about existence, as a factor in existence. The concept of dialectic may refer either to the dialectical process or operation of 'negating,'

---

[273] *Ibid.*, p. 156.

[274] Where he clarifies the kind of opposition found in philosophical concepts, *An Essay on Philosophical Method*, pp. 68-77. See this chapter, above, pp. 107-110.

[275] See O. Lee, "Dialectic and Negation," *The Review of Metaphysics*, I (1947), p. 3.

[276] L. Rubinoff, *Collingwood and the Reform of Metaphysics*, p. 176.

[277] See Otis Lee, "Dialectic and Negation," *The Review of Metaphysics*, I (1947), p. 1.

[278] *Ibid.*, p. 3.

[279] *Ibid.*, p. 4.

or to the result of the negating activity, or to the relation intrinsic to the opposition of affirmation and negation. Dialectic regards the action or process as the most basic of these three aspects, the entity which results, and the relation (of opposition) is a derivative of it. The concept of substance, typical of preceding historical periods, put static existence before change, but absolute idealism places becoming before being and holds that the basic entity of all that is, namely spirit or mind, is an endless becoming.[280]

Collingwood explains the meaning of a dialectical series in *Speculum Mentis* in the context of the forms of experience understood as a dialectical series. "The terms of a dialectical series are not related to one another in terms of degree, but by the assertion in each term of something which in the previous term was wrongly denied."[281] "A dialectical series means a series of terms, each one of which is an erroneous description of the next."[282] Collingwood exemplifies this:

> In a dialectical series A B C, the truth is C; B is a distorted account of it, and A is a distorted account of B. Now if C represents the true nature of mind, B is a mistake—partial of course; no mistake is a mere mistake—which C makes about itself. This mistake will recoil on C's own nature, for a mind which makes mistakes about its own nature will find its conduct, which is its nature, affected by these mistakes. Its nature will not be so far altered as to coincide with the false conception, but it will be disturbed by that conception. Thus we get B, what the mind now thinks it is, and C_1, what it actually is. The next downward step will be to fail even to get the notion B coherent, to distort and degrade it into A. This still further lapse from truth in the self-knowledge of the mind C is now an error of the second degree, and therefore reacts in the second degree on the conduct and nature of the mind. It no longer therefore behaves as C_1 but as B_1, which is equivalent to C_2. In a sense we may say that it still is C, though it does not know it; but because its self-ignorance affects its conduct, to call it C is misleading, for it does not behave like C but like C_2.[283]

Collingwood elaborates this by describing in full the condition of the mind in this activity. Implicitly it is C, but it is trying to conceive itself as B, although there is no such thing as B. But, in trying to conceive itself as B it confuses itself even in doing this and results in thinking of itself as A. The successive misconceptions B and A so retroact

[280] *Ibid.*, pp. 4-6.
[281] *Speculum Mentis*, p. 208.
[282] *Ibid.*, p. 206.
[283] *Ibid.*, pp. 206-207.

over the mind's true nature, C, that it comes out in the form C_2. C_2 is its actual form (i.e., C, its true nature after retroaction of misconceptions B and A), A is its explicit form, basically or ultimately it is C. Thus, the aesthetic mind (A = pure imagination) calls itself an artist, but will be basically historical (C = historical experience), but the historical nature (C) will be so affected by the description of itself as aesthetic that it will not appear even to a dispassionate observer as pure or unqualified history. The distortion in its nature disappears when it discovers its true nature through discovering the error of its false conceptions. A mind ignorant of its true nature does not in the fullest sense possess its nature. The true nature of the mind is not something ready-made waiting to be discovered; it exists only when it is discovered, yet, previous to its discovery, it exists in a confused and distorted form (i.e., implicitly), for errors are partial truths, and the dialectical work is to bring that true nature into existence by clearing up the confusions which appear as inconsistencies experienced as a conflict by the mind between what it is and what it ought to be.[284]

Another presentation of an instance of a dialectical series is given by Collingwood in *An Autobiography* (pp. 97-99) illustrating that

> the past which a historian studies is not a dead past, but a past which in some sense is still living in the present . . . [which] I expressed . . . by saying that history is concerned not with 'events' but with 'processes' . . . [which] are things which do not begin and end but turn into one another . . .[285]

Again, in the same work,[286] as already explained above,[287] the state is not a transcendent eternal entity exemplified in Plato's *Republic* or Hobbes' *Leviathan*, but an active process which changes from what Plato describes it to be to what Hobbes describes it to be.[288] Again, as L. Rubinoff well explains,[289] in his article entitled "Political

---

[284] See R. G. Collingwood, *Speculum Mentis*, p. 207. See also B. Bosanquet, "Contradiction and Reality," *Mind*, XV (1906), p. 2: ". . . the unrest of action and cognition seems to arise from the perpetual presence of implicit contradiction in the nature of actual fact, a presence which becomes explicit on the slightest reflection and forces us to go further in the hope of faring better. It must, I infer, be admitted that fact, as given in ordinary experience, is both actual and self-contradictory." This experienced conflict, which spurs the mind to reflect and explicitate its implications, is appetite; see also Collingwood's *New Leviathan*, pp. 47-53. Felt contradiction thus drives the mind to develop itself into more perfect forms.

[285] *An Autobiography*, pp. 97-98; see Chapter 6, pp. 217-220 below.

[286] See *An Autobiography*, pp. 61-62.

[287] See Chapter 2, above, pp. 53-54.

[288] See also Chapter 6, below, pp. 217-220.

[289] See his *Collingwood and the Reform of Metaphysics*, p. 155.

Action"[290] Collingwood develops the concept of the state in terms of dialectically dynamic political action rather than in terms of a static entity entailing properties as abstract deductive logic would treat it.

Each member of such a series occurs on a scale of forms and its place is determined by its internal relations with respect to its higher and lower members. The negation whereby it negates the limitation (negation, hence negation of negation) of the lower is not a static relation thereto but a dynamic, active relation thereto. The higher comes out of the lower when the lower collapses under the strain of its built-in contradiction. This negation of the negation that occurs from within the lower member is at the same time the coming into being of the higher. The process then continues; as each member develops it gives way to something higher "beating it on its own ground" by active negation of its own limitations.

Thus, active negation is the dynamic differentiation of one form from another arranged on a dialectic scale of forms. The differences of species arrayed on a dynamic scale of forms are not static extrinsic differences, but differences intrinsic to the differentiated form, consisting in the active, dynamic, self-transcendence of that form. Thus, in a dialectic scale of forms, the 'otherness' whereby one form differs from another is a self-determining activity, an activity which is not other than the active self but of the very essence of that self. Thus, 'the other' comes from 'a self' inasmuch as that self actively distances itself from itself; it alienates itself from itself from within, by reason of the very activity which is nothing other than itself.

Hence the concrete universal, as an actively, dynamically developing entity, is an 'identity-in-difference'; it has itself actively by a self-development which results in its giving way to something other than itself, in which other is found the fulfillment of the active striving with which it is identified.

E. The Concrete Universal and the Logic of Question and Answer

Collingwood's explicit denial of the distinction between philosophy and history, as this distinction is understood by the realists, is the foundation of his logic of question and answer,[291] and, therefore, his question and answer logic is the logic of the *rapprochement* he sought, as his life's work, to bring about between philosophy and

---

[290] See *Proceedings of the Aristotelian Society*, (1928-1929), 155-176.

[291] L. Rubinoff, *Collingwood and the Reform of Metaphysics*, p. 15; see also R. G. Collingwood's *An Autobiography*, pp. 65-66.

history. And "when Collingwood . . . argues in the *Autobiography* that the history of all thought, including metaphysics, can be approached through the logic of question and answer (A, 58, 65, 67), he is implicitly declaring that the question-and-answer complexes revealed by applying the logic of question and answer to the history of thought assume a dialectical scale of forms"[292] and therefore must be regarded as a concrete universal. We have already seen that Collingwood says that "the process of knowledge . . . is not so much an alternation of question and answer as a perpetual restatement of the question with a perpetual revision of the answer,"[293] but knowledge *as a process* (as distinct from knowledge as a series of events or catalogue of facts such as abstractionist, classificatory logic would regard it, such as, for example, Positivism regards historical events when it seeks "hard facts") is rather a scale of forms consisting of a series of questions and answers arranged according to their mutual implications, or internal relations. The knowledge of "who killed John Doe?" for example, consists of a series of questions and answers, beginning with that question, ending with the answer that the rector killed John Doe, as was explained in Chapter 1, above. Inasmuch as one possesses this answer in an historically scientific way, one knows that the rector killed John Doe in terms of all the reasons—in their proper order, as they form a coherent whole—why one maintains that proposition. To know this scientifically[294] one has to know the whole series of questions and answers as synthetically integrated. That is, one has to know that whole series of questions and answers as a concrete universal, as a scale of forms, each member being related to its prior and posterior members by necessary implication. The question "who killed John Doe?" lays down an area of inquiry, thereby delineating a certain intelligible whole, which area is progressively elaborated as the inquiry proceeds. It is the focal point of attention, a whole within which some facts or events will be included as relevant, others excluded as irrelevant. The sorting of the irrelevant from the relevant by means of the questioning process is the building up of that scale of forms which constitutes a concrete universal. A scientific inquiry, therefore, consists in visualizing a certain whole from the beginning; the process will consist in seeing that whole as explicated in terms of its internal

---

[292] L. Rubinoff, *Collingwood and the Reform of Metaphysics*, p. 252.

[293] R. G. Collingwood, *Speculum Mentis*, p. 80; see also L. Rubinoff, *Collingwood and the Reform of Metaphysics*, p. 252. See also, above, Chapter 3, pp. 70-71.

[294] 'Scientific' here refers to the method of orderly procedure according to which history is called a science, as was explained in Chapter 1, above, pp. 12-13.

relations, the parts, which, with these relevant distinctions and interrelations, constitute that synthetic totality. It is the whole, first delineated by the dominating question, that gives meaning to the parts (dividing the relevant from the irrelevant) and reciprocally, the parts that give meaning to the whole. In other words, whole and part are not mutually exclusive, as if the parts were *entities in themselves* related merely by external relations, as classificatory genus-species logic would analyse a whole, but are rather themselves overlapping classes in a scale of forms, the whole being realized only in and through its parts, and other parts being what they are through constituting the whole.

Thus, crime detection consists in striving to grasp a situation, the experience of a number of facts, as a whole. The perpetually restated question that is restated in every different question the crime solver asks is 'How do all these things fit together?' It is this 'fitting together' of the 'pieces' that more and more explicates the insight into the whole picture in which everything is explained as interrelated. Those familiar with the Perry Mason type of inquiry will recognize how one well loaded question to an interrogated suspect almost precipitates its own answer so that 'everything falls into place'; so much so that for the suspect to be faced with the question immediately provokes the answer merely by his behavioral response, without waiting for him to supply an articulated verbal reply. The principle, the criterion dominating the progress of the inquiry throughout, which criterion places one always in the presence of an intelligible totality, is that human events do not happen in isolation, but are intimately connected by internal relations, so that any one event is not properly understood until its place in the context of other events is appreciated. Concrete thinking is thus essentially contextual thinking, thinking of events not as isolated units (which the term 'event' itself unfortunately may suggest) but as intertwined into the fabric of a whole which at the same time takes its meaning from that whole and conveys meaning to that whole. This grasping of a situation as a whole, in terms of its internal relations, in terms of the 'inside' of events, is what Collingwood calls 'insight.'[295] But inward reality, as distinct from outward manifestation, is spirit or thought.[296] It is through the inward human thoughts and purposes that external human events are internally connected into a systematic unity in diversity, that is, a concrete universal. By means of 'insight' into the thoughts and purposes within human events a coherent,

[295] See R. G. Collingwood, *An Autobiography*, p. 101. See Chapter 6, below, p. 221 n.45..

[296] See B. Bosanquet, *The Principle of Individuality and Value*, p. 72.

interconnected whole is constituted. Specifically human entities, such as a culture, or a city, or a war, are constituted by a web of factors, each external as far as appearance is concerned, but intrinsically and dynamically integrated by internal relations resulting from human thoughtful and purposeful activity.[297]

Similarly, history in its concrete elaboration by the historian begins with a question,[298] and seeks to elaborate a series of events in such a way that there is no longer merely a series of disconnected facts, but a process in which the succeeding events are seen to give way one to the other by internal necessity, so that a coherent whole is presented in the form of an overlap of classes in a scale of forms. This explains why past facts change in meaning as their implications are developed throughout the course of history by being reassimilated by succeeding thoughts in the pursuit of new purposes. Thus, the tombs excavated by Flinders Petrie in Egypt now have as part of *their* history the fact that they were excavated by Petrie at that particular time and for the particular motives which Petrie had in undertaking such excavations.

Thus, in crime detection and in historical inquiry, the elements of the complex knowledge, consisting of seried questions and answers, fit together to form one complete picture somewhat in the way the parts of a jigsaw puzzle fit together. What guides the jigsaw puzzle-

---

[297] According to B. Bosanquet, *The Principle of Individuality and Value*, p. 70, the uniqueness which marks individuality, the concrete universal, is what is commonly regarded as the uniqueness of purpose. A purpose which is the practical element of a logical whole is nothing but a want, or a wanted object. Correlate this with Collingwood's "On the So-called Idea of Causation," *Proceedings of the Aristotelian Society*, XXXVIII (1938), pp. 85-112, for Collingwood's understanding of the primary sense of the term 'cause' to be that of purpose. See also Collingwood's *Autobiography*, p. 101, and Ch. 6, pp. 224-226, below, for Collingwood's assertion that historical insight is concerned with purposive activity. See also his *The New Leviathan*, pp. 47-60, 67-82; especially p. 72, par. 10.5, where he writes: "Shame . . . the critical point in the process converting fear into anger, is . . . a critical point in the whole development of mind . . . The importance of anger as a bridge from the lower levels of consciousness where thought is at first merely apprehensive, capable of taking what is 'given' to it, and then merely conceptual, capable of framing abstractions from what is 'given,' to the higher levels of consciousness where thought is first 'propositional,' capable of discriminating good from evil and truth from error, and then 'rational,' capable of understanding both itself and other things, has been long expounded in many forms." Thus, the theory of the individual, of the concrete universal, is intimately tied up with the theory of the primacy of practical (i.e., purposive) knowledge. One's unique individuality is at one with one's life goals. As Aristotle said (*Nicom. Ethics*, 114a 30-35): ". . . the end appears to each man in a form answering to his character."

[298] R. G. Collingwood, *passim*. See also H.-I. Marrou, *The Meaning of History*, p. 131. See also Ch. 9, pp. 355-359 below, for a consideration of various historians' attitudes to history as proceeding according to an interrogative method.

solver is the criterion of a whole picture which results from the fitting together of all the parts. But the illustration defaults, inasmuch as the picture, the whole, is superimposed on the parts as not entirely one with the reality of the parts, since the partial picture on each section can be rubbed off. But, in regard to human events, with which history, and crime detection, deal, the thoughts and purposes which constitute the 'inside' of such events, are embodied right into the fabric of the reality of the events themselves. "The historian asks himself, 'Does this fit in with everything I know about the world of facts, the nature of the case, the liability of the informant to error and mendacity, and so forth?' "[299] This means that (a) the facts, (b) the integration of the facts, that is, their "going together," or their "identity in difference," (c) their manifestness as facts to the investigator (historian or detective) which *itself* is one of the facts which internally affects the nature of all the other facts,[300] and (d) the questioning mind,[301] whose questioning activity makes the facts evident as facts, are all mutually implicated, that is, related to each other and to the totality by internal relations. That is, the facts are not "hard facts" set over against the mind as something merely contemplated by it, but form, among themselves and with the mind that knows them, a concrete totality of mutual implication.

Finally, it may be noted that, in the question and answer complex which forms the knowledge element in crime detection and in history, we are dealing with truth as coherence. This coherence notion of truth finds perhaps its most explicit exposition in H. H. Joachim's *The Nature of Truth*,[302] with which Collingwood was familiar.[303] But, whereas for Joachim truth is a coherence of judgments, for Collingwood truth is the coherence of a complex of questions and answers.[304]

---

[299] R. G. Collingwood, *Speculum Mentis*, p. 214.
[300] See *ibid.*, pp. 218-221, p. 240.
[301] See *ibid.*, pp. 244-246. See also Chapter 1, above, pp. 32-34.
[302] Oxford, Clarendon Press, 1906.
[303] See R. G. Collingwood's *Autobiography*, p. 18.
[304] See *ibid.*, pp. 33-34.

Chapter Five

Questioning, Presupposing and Metaphysics

A consideration of Collingwood's theory of Metaphysics as the historical investigation of the absolute presuppositions of scientific enquiry is necessary in order to understand his theory of questioning, since science is an ordered body of knowledge,[1] that is, of questions and answers,[2] and what causes relevant and meaningful questions to arise are the presuppositions made by the questioner.[3] "Furthermore, the doctrine of absolute presuppositions . . . is part of a wider doctrine . . . previously discussed under the heading of the 'concrete universal,' "[4] and ". . . the theory of presuppositions first expounded in *Truth and Contradiction* was part of a general attempt to develop an alternative logic to the propositional logic of realism."[5] And, again, "Metaphysics is an application of the logic of question and answer to the organization of the history of absolute presuppositions into a scale of forms (or concrete universal) . . ."[6]

A different way of speaking in *An Essay on Metaphysics* from that of the earlier *Speculum Mentis* seems to indicate a change of doctrine

---

[1] See Chapter 1, above, pp. 12-13.
[2] See *ibid.*, pp. 21-22; also Chapter 2, above, pp. 50-52.
[3] See R. G. Collingwood, *An Essay on Metaphysics*, pp. 23-33.
[4] L. Rubinoff, *Collingwood and the Reform of Metaphysics*, p. 248.
[5] *Ibid.*
[6] *Ibid.*, p. 262.

in Collingwood regarding the nature of questioning. In *Speculum Mentis*, and even in *An Essay on Philosophical Method*, the concrete universal is explained solely in terms of experience polarized into questions and answers. But here, in *An Essay on Metaphysics*, there are three explanatory factors, namely questions, answers and presuppositions. Previously, as explicitly stated in *Speculum Mentis*,[7] questions belonged to the order of suppositions, whereas here, in *An Essay on Metaphysics*, questions seem to be some sort of intermediary between presuppositions (whose logical efficacy provokes relevant questions[8] and which are alleged to be spun out of the mind by itself)[9] and answers. Presuppositions as determinants of world view meanings, appear now to have a recognizable distinctness from experience, so that coherence is no longer to be found only within experience itself but in experience under the influence of some experience-transcending *a priori* mind-stuff.[10] In other words, the rational explanation of experience seems now no longer to be regarded in terms of relational coherence between parts and whole within that experience itself, that is, the concrete universal of the earlier works, but to be understood in terms of a subsumption of experience under some transcendent mind-determinant in a quasi-Kantian fashion.

If this is so, then the *rapprochement* which Collingwood sought to establish as his life's work has already broken down in principle from within the system itself. Before attempting to take a position on this apparent discrepancy in Collingwood's theory of questioning, his treatment of presuppositions as presented in his *Essay on Metaphysics* will be considered, after which a return will be made to the problem highlighted above.

To avoid misapprehensions due to Collingwood's manner of expressing himself in the *Essay on Metaphysics*, L. Rubinoff points out that it is necessary to read the *Essay*, and the theory of absolute presuppositions found there, as a reply to Logical Positivism, specifically to the challenge of A. J. Ayer's *Language, Truth and Logic* which appeared in 1936, in which Ayer maintains that experimental verifiability is the sole criterion of truth, apart from tautologies. Collingwood maintains against this that there is a class of statements ultimately involved in and therefore prior to the process of verification, and a condition of the possibility of verification. These statements Colling-

[7] See Chapter 3, above, pp. 65-67.

[8] See R. G. Collingwood, *An Essay on Metaphysics*, pp. 25-27.

[9] See *ibid.*, p. 197. See also this chapter, below, pp. 157-158.

[10] See below, this chapter, pp. 205 et seq.

wood calls presuppositions. These statements are neither true nor false, and their acceptance before all enquiry is a necessary condition of enquiry from which truth and falsity itself will arise.[11] But this denial of truth and falsity to presuppositions, Rubinoff warns, is due to Collingwood using a neutral language to endeavour to communicate with the Logical Positivists on their own ground. Dialectically, the truth or falsity of absolute presuppositions will be reckoned according to their place on a scale of forms of questions, answers and presuppositions.[12] Collingwood himself says, speaking of a 'false' answer which is nevertheless 'right', "it is 'right', because it constitutes a link, and a sound one, in the chain of questions and answers by which the falseness of that presupposition is made manifest."[13]

A. Presuppositions and the Logic of Question and Answer

In the *Autobiography* Collingwood dealt with statements as being meaningful only as answers to questions. Such statements, made as answers to questions, are what are usually called propositions. But now Collingwood recognizes another type of statement involved in the question-answer process. Statements which are not consequences of questions, as answers to questions, but necessary conditions of the arising of relevant questions, are called presuppositions by Collingwood. Among this latter type of statement there are relative and absolute presuppositions. Metaphysics, according to Collingwood, is the historical investigation of the latter.

Collingwood's doctrine of absolute presuppositions is tied to his own experience of knowledge as an active interrogation, as opposed to the realist doctrine of knowledge as 'compresence' with an object. He contrasts his own personal experience of the activity of knowledge with that of S. Alexander, who was reputed to be able to see the solutions to problems intuitively with no apparent effort.[14]

> . . . with a slow and feeble thinker like myself there is nothing when I am thinking hard and efficiently that is like 'compresence' with any 'object' whatsoever. There is at first a whole nest of problems all tangled up together. Then by degrees the tangle is reduced to order. Here and there in it one problem is seen to depend for its solution upon the solution of another. Given luck and great patience and strict attention to the rule of never asking two questions together but always separating them out and asking the first

[11] See L. Rubinoff, *Collingwood and the Reform of Metaphysics*, pp. 241-248.
[12] See *ibid.*, pp. 248-251.
[13] *An Autobiography*, p. 38.
[14] See R. G. Collingwood, *An Essay on Metaphysics*, pp. 172-180.

one first, I can sometimes solve the whole lot. If I tried to deal with them according to Alexander's presumption I know by experience that I should never solve any.[15]

For Collingwood, absolute presuppositions are *a priori* in the sense of not being derived from experience, and function in systematic enquiry as "catalytic agents which the mind must bring out of its own resources to the manipulation of what is called 'experience' and the conversion of it into science and civilization."[16] They are therefore self-made tools of knowledge according to which the mind thinks and systematizes experience into an intelligible experience. They therefore precede knowledge as preconditions of scientific inquiry, that is, systematic interrogation.

Collingwood says that his *Essay on Metaphysics* was written "neither to expound my own metaphysical ideas, nor to criticize the metaphysical ideas of other people; but to explain what metaphysics is, why it is necessary to the wellbeing and advancement of knowledge, and how it is to be pursued."[17]

Collingwood holds that "Metaphysics is for us the name of a science, and has been for many centuries, because for many centuries it has been found necessary to think in a systematic or orderly fashion about the subjects that Aristotle discussed in the front of treatises collectively known by that name."[18] He rejects Aristotle's understanding of metaphysics as a science of pure being[19] but accepts Aristotle's position that it is the science of first principles, which he, Collingwood, understands as the presuppositions which underlie all scientific thinking,[20] that is, orderly interrogation of a subject matter.

To Kant's question, 'Under what conditions is Metaphysics a genuine science?', Collingwood's answer is that it is a science in the sense in which any organized body of knowledge is a science.[21] It is concerned to discover what are the presuppositions which underlie scientific enquiry[22] and it is possible only on condition that it is conducted as a historical investigation into the absolute presuppositions which were made at various historical periods, and the continuity existing between them.[23]

---

15 *Ibid.*, pp. 177-178.
16 *Ibid.*, p. 197.
17 P. vii. Note departure from historical procedure.
18 See *ibid.*, p. 4.
19 See *ibid.*, pp. 11-16. See also L. Rubinoff, *Collingwood and the Reform of Metaphysics*, pp. 221-222.
20 See *An Essay on Metaphysics*, pp. 11; 20; 47.
21 See *ibid.*, p. 4.
22 See *ibid.*, pp. 34-48; esp. p. 40.
23 See *ibid.*, pp. 49-77, esp. 55; 58-61, p. 77.

L. Rubinoff argues convincingly that the substance of Collingwood's *Essay on Metaphysics*, "which is treated [in the *Autobiography*] as a part of the more general 'logic of question and answer,' " "was originally conceived during the first world war and first written out in the unpublished and subsequently destroyed 'Truth and Contradiction.' "[24] Furthermore, the substance of Collingwood's Metaphysics as the science of absolute presuppositions was foreshadowed in Collingwood's address at the Ruskin Centenary Conference in August of 1919,[25] which, Rubinoff notes, is damaging to the Knox-Donagan radical conversion theory.

Rubinoff aligns the constellation of absolute presuppositions of the later *An Essay on Metaphysics* with what is referred to in the Ruskin centenary address as the basic principles to which a man adheres, which define what is referred to as one's philosophy of life.

> . . . there are certain central principles which a man takes as fundamental and incontrovertible, which he assumes as true in all his thinking and acting. These principles form, as it were, the nucleus of his whole mental life: they are the centre from which all his activities radiate. You may think of them as a kind of ring of solid thought—something infinitely tough and hard and resistant—to which everything the man does is attached.[26]

Collingwood seems to tie up this "constellation of absolute presuppositions" of the *Essay*, or "ring of solid thought" of the Ruskin centenary address, with the basic character determination of the moral personality,[27] which concords well with his *rapprochement* of thought and action, and is reminiscent of the Chestertonian remark regarding a landlady interrogating a prospective tenant, that it is just as important that she knows his philosophy as it is to find out what his income is.

> As in his later thought, so in 1919 Collingwood regarded a man's philosophy as a function of his character. He therefore declares that

[24] L. Rubinoff, *Collingwood and the Reform of Metaphysics*, p. 223.

[25] See *ibid*. Collingwood's Ruskin Centenary address, "Ruskin's Philosophy," is republished in Alan Donagan's edition of Collingwood's *Essays in the Philosophy of Art*, Bloomington, Indiana University Press, 1964. (The essay was originally published by Titus Wilson and Son, Kendal, England, 1920.) Rubinoff, *op. cit.*, p. 225, writes: "Not only is the essay an occasion for Collingwood to bring his own presuppositions to the surface; but the entire essay may be viewed as an application and restatement of the theory of philosophy which Collingwood claims to have worked out in 'Truth and Contradiction' a few years earlier."

[26] R. G. Collingwood, "Ruskin's Philosophy," see Alan Donagan's edition of R. G. Collingwood's *Essays in the Philosophy of Art*, p. 10. See also L. Rubinoff, *Collingwood and the Reform of Metaphysics*, p. 224.

[27] See L. Rubinoff, *Collingwood and the Reform of Metaphysics*, p. 224.

a man ". . . is a great man or a little, a valuable man or a worthless, largely according as this ring is strong or weak in structure, good or bad in material" (RUP, 7). And since all of the acts and decisions which shape a man's life are suspended from this ring (one thinks immediately of Aristotle's notion of *hexis*) we may, according to Collingwood, regard a man whose ring is weak as having an unworthy character, while a man whose ring is composed of "unsound" and "untrue" principles may be regarded as "bad" and "foolish." And while Collingwood does not expressly say so, it is implied by what he does say that just as the character of an individual is bound up with the quality of his beliefs, so the character of an entire age is similarly to be regarded. This view clearly anticipates the attitude towards the dangers of first-level dogmatism in *Speculum Mentis,* the theory of the corrupt consciousness in the *Principles of Art,* and the critique of irrationalism in the *Essay.*[28]

The essentially historical character of the investigation of these primary thought determinants is already recognized in the Ruskin centenary address. Corresponding to the distinction made later in the *Autobiography* between "realism" and "the logic of question and answer" there is found in the Ruskin centenary address the distinction between the logical method "which proceeds on the assumption that every individual is an instance of some eternal and unchanging principle" and the historical method which differs from the logical method in that "where the logical mind looks for general laws, the historical mind looks for individual facts, and it explains these facts by appealing not to laws but to other facts."[29] Rubinoff says, "Presupposed here is the distinction drawn in *Speculum Mentis* between the logic of the abstract universal and the logic of the concrete universal."[30]

In the Ruskin centenary address Collingwood distinguishes the logicist ('realist' of the *Autobiography* and *Essay*) and historicist ('question and answer' of the *Autobiography* and *Essay*) attitudes to contradiction.

The old logic lays it down that of two contradictory propositions one must be false and the other true. To contradict yourself, on this

[28] *Ibid.* What Collingwood speaks of as absolute presuppositions or primary convictions seems to be what J. Maritain refers to as the primary self-determination which results from "The Immanent Dialectic of the First Act of Freedom," see his *The Range of Reason*, London, Bles, 1953, pp. 66-85. See also what Henry Veatch says in his *Two Logics*, Evanston, Northwestern University Press, 1969, pp. 242-245, regarding the logical import of character traits.

[29] See R. G. Collingwood, "Ruskin's Philosophy" in Alan Donagan's *R. G. Collingwood's Essays in the Philosophy of Art*, p. 12; p. 14. Also see L. Rubinoff, *Collingwood and the Reform of Metaphysics*, p. 225.

[30] L. Rubinoff, *op. cit.*, p. 225.

view, is a sign of mental confusion: the wise man never contradicts himself. The alternative view starts from the axiom that there are two sides to every question, and that there is right on both sides; from this, the inference is drawn that truth is many-sided and that self-contradiction may easily be a mark not of weakness but of strength—not of confusion, but of a wide and comprehensive view which embraces much more truth than the one-sided consistency of the logicians.[31]

Commenting on this passage of "Ruskin's Philosophy," Lionel Rubinoff says that the old logic, or logic of logicism, is the abstract logic of classification and division, the logic of genus and species. The new logic, or logic of historicism, is undoubtedly the basis of what Collingwood himself later describes in the *Essay on Philosophical Method* as the dialectical logic of the overlap of classes and as the logic of question and answer in the *Autobiography*. Rubinoff points out that Collingwood attributes this new historicist logic mainly to the efforts of Hegel and nineteenth century historical idealism and concludes therefore that "Ruskin's Philosophy" provides clear and unequivocal evidence that the logic of question and answer of the *Autobiography* is in fact Collingwood's development of the dialectical logic of Hegel, which conclusion, Rubinoff says, supports his claim that the logic of question and answer is part of the wider dialectical logic of the scale of forms and overlap of classes, sometimes referred to as the logic of the concrete universal.[32]

Rubinoff further concludes that the theory of truth referred to in the above passage, according to which truth is reached only through contradiction or determinate negation, presupposes the logic of question and answer of the *Autobiography* and *Speculum Mentis*, and is virtually the same position defended in *An Essay on Philosophical Method* and in Collingwood's 1935 correspondence with Gilbert Ryle, in which Collingwood defends the position that philosophical arguments must proceed by means of systematic fallacies.[33]

---

[31] R. G. Collingwood, "Ruskin's Philosophy", in Alan Donagan's *R. G. Collingwood's Essays in the Philosophy of Art*, p. 22.

[32] See L. Rubinoff, *Collingwood and the Reform of Metaphysics*, p. 226.

[33] See *ibid.*, pp. 226-227. Rubinoff further says, *op. cit.*, pp. 230-231, that: ". . . 'Ruskin's Philosophy' must therefore be regarded as representing an important stage in the development of Collingwood's thought. Not only does it appear to support Collingwood's own claim in the *Autobiography* to have arrived at the theory of presuppositions and the logic of question and answer early in his philosophical career but it is also probably the closest record we have of the actual views of 'Truth and Contradiction.' If this is so then my earlier contention that 'Truth and Contradiction' may be regarded as an early statement of the doctrines of the *Essay on Metaphysics* is given further support. At the same time, as I have tried to show, just as the *Essay*

B. Presuppositions and Questions

In his *Essay on Metaphysics* Collingwood first reiterates what he has already said in the *Autobiography* concerning the meaning of propositions as functions of answering questions, and, in addition, introduces the role played in the question and answer process by presuppositions and especially what he calls the absolute presuppositions.

Whenever anyone makes a statement of his thought in words there are many more thoughts in his mind than are expressed in his verbal statement. Among these some stand in special relationship to the stated thought; they are not just its context but also its *presuppositions*. When I say 'that is a clothes-line' I mean that it was put there for the purpose of hanging clothes. In saying that, I am presupposing that it was put there for some purpose, otherwise the question for what purpose would never have *arisen*.[34]

Presupposition means priority, not in time, but *logical*. We do not go from first thinking of purpose in general for which the clothes-line in question substitutes a determinate purpose, nor do we first

---

may be regarded as fulfilling the requirements of *Speculum Mentis*, so the latter may be regarded as a systematic exposition of some of the main points of the essay on Ruskin. It therefore follows that, contrary to Donagan's interpretation, far from experimenting with a different doctrine, Collingwood's activity from 1924 to 1936, as he himself declared in the *Autobiography*, consisted primarily in applying the new 'historicist logic' to the solution of various philosophical problems; thus the actual theory of metaphysics expounded in the *Essay* may be regarded as the formal exposition of a doctrine which had been consistently applied since 1924. 'I did not really feel,' wrote Collingwood concerning the youthful period of *Truth and Contradiction*, 'any great desire to expound the philosophical ideas I have been setting forth in these chapters [he is referring of course to the doctrine of presuppositions] whether to my colleagues or to the public.' Instead he continues, ' . . . I felt justified in turning to the more congenial task of applying them and thus testing them empirically' (A, 74). *Speculum Mentis* may therefore be regarded as an attempt to expound the presuppositions of consciousness in so far as the latter exists at the first ontological level. The *Idea of Nature*, the *Idea of History*, and the *Principles of Art* are attempts to expound the presuppositions of particular forms of experience, each of which has its own history and exists, as I have argued, on three distinct but related ontological levels. In practice, of course, Collingwood does more than merely describe these presuppositions; he criticizes them as well, by explaining how they arose in the first place, and in so doing 'vindicates' them from the absolute standpoint, which is precisely what the metaphysician of the *Essay on Metaphysics* is required to do. The metaphysician's business, he writes, is not only to identify several different constellations of presuppositions but also to find out on what occasions and by what historical processes one set has turned into another (EM, 73; A, 66). Presupposed by these claims are a special set of presuppositions which the metaphysician discovers when he reflects on his own experience as metaphysician—which is precisely what Collingwood himself is doing in the main body of *An Essay on Metaphysics*, the *Principles of Art*, and the 'Epilegomena' of the *Idea of History*.''

[34] See R. G. Collingwood, *An Essay on Metaphysics*, p. 21.

think of the clothes-line as having some purpose and then go on to ask what is that purpose which it has. We *first* think 'that is a clothes-line,' *then,* by a kind of reflective analysis, come to see that this presupposition was made however little one was aware of it at the time.[35]

> Here lies the difference between the desultory and casual thinking of our unscientific consciousness and the orderly and systematic thinking we call science. In unscientific thinking our thoughts are coagulated into knots and tangles; we fish up a thought out of our minds like an anchor foul of its own cable, hanging upside down and draped in seaweed with shellfish sticking to it, and dump the whole thing on deck quite pleased with ourselves for having got it up at all. Thinking scientifically means disentangling all this mess and reducing a knot of thoughts in which everything sticks together anyhow to a system or series of thoughts in which thinking the thoughts is at the same time thinking the connexions between them.[36]

Collingwood says that logicians have paid a good deal of attention to some kinds of thought connections, but that they have neglected others, such as the theory of presupposition. He adds that perhaps this is responsible for the unsatisfactory condition of metaphysical theory, which depends upon the theory of presupposition. Collingwood then proposes to deal summarily with the theory of presuppositions to the extent to which his present purpose deems necessary. He proceeds to expound his theory in a series of numbered propositions, after the manner of a Spinozistic *a more geometrico.*[37]

Proposition one is: 'Every statement that anybody ever makes is made in answer to a question.' He then explains that by statement he includes not only those spoken out loud to someone else but also statements made interiorly by one to himself in the course of solitary thinking. Likewise, questions include questions addressed to oneself. He appeals to familiarity with scientific procedure to back up the proposition that:

> In proportion as a man is thinking scientifically when he makes a statement, he knows that his statement is the answer to a question and knows what that question is.[38]

In our least scientific and reflective moments we scarcely know that the thoughts we have are answers to questions, let alone what the questions are. It is only by *analyzing what is meant* by saying that

---

[35] See *ibid.*, p. 22.
[36] *Ibid.*, pp. 22-23.
[37] See *ibid.*, p. 23.
[38] *Ibid.*, p. 24.

'this is a clothes-line' that I realize it to have been the answer to the question 'What is that thing for?'[39] A question is *logically* prior to its own answer and, when thinking is ordered scientifically, this logical priority is accompanied by temporal priority. But the fact that the question is formulated first in thinking scientifically does not mean that it ceases when the question begins to be answered; this is a special type of temporal priority in which the prior event or activity does not stop when the posterior begins.

> The act of asking the question begins and takes a definite shape as the asking of a determinate question before the act of answering it begins; but it continues for the whole duration of this latter.[40]

This is so because, if a question did not endure during the answering of it, the questioner would have since lost interest in his subject, and the alleged answer would not be an answer at all, but a meaningless form of words. A question does not cease to be a question by being answered, but ceases only to be an unanswered question.[41]

By way of clarifying the meaning of his terms, Collingwood defines what he means by 'proposition.' Let that which is stated (i.e., that which can be true or false) be called a proposition, and let the stating of it be called propounding it.[42] Collingwood warns that English usage refers to a question or a supposition also as a statement, and this jargon is customary among logicians.

Then Collingwood lays down his second proposition: 'Every question involves a presupposition.'[43]

Ordinarily, Collingwood explains, a question involves large numbers of presuppositions, but a distinction must be made between direct and immediate presuppositions and indirect and mediate presuppositions.

> Directly or immediately, any given question involves one presupposition and only one, namely that from which it directly and immediately 'arises.'[44]

But, this direct and immediate presupposition has in its turn other presuppositions which are indirectly presupposed by the original question. But if this immediate presupposition were not made, the question to which it is logically immediately prior could not be *logically*

---

[39] See *ibid.*
[40] *Ibid.*
[41] See *ibid.*, pp. 24-25.
[42] See *ibid.*, p. 25.
[43] See *ibid.*
[44] *Ibid.*

asked. The question 'have you stopped beating your wife?' could not be *logically* asked without the presupposition on the part of the questioner that the one questioned had been in the habit of beating his wife. In the absence of that presupposition the question 'does not arise.' It is not denied that such a question can be verbally asked but only that it can be logically asked.

To say that a question 'does not arise' is the ordinary English way of saying that it involves a presupposition which is not in fact being made. A question which 'does not arise' is a nonsensical question, not intrinsically nonsensical, but nonsensical *in relation to its context*, and specifically to its presuppositions. One asking a question which 'does not arise' is talking nonsense and inviting the one so questioned to join him in nonsense talk.[45]

Collingwood says that just as one can ask questions without knowing it, and *a fortiori* without knowing what questions one is asking, so also one can make presuppositions without knowing it, and *a fortiori* without knowing what presuppositions one is making. Thus, in asking 'What is that for?' I need not necessarily be aware that I am presupposing that it is 'for' something. Only to the extent that I think scientifically am I aware of my presuppositions. Thus, in deciphering a worn and damaged inscription, I am aware that before asking 'What does that mark mean?' I must have first assured myself that the mark is part of the inscription, that is, I must first answer the question 'Does it mean anything?' An affirmative answer to the question 'Does it mean something?' causes the further question 'What does it mean?' to arise.[46]

The causing of a question to arise is called the logical efficacy of whatever so causes it to arise. Mere assumption, or mere supposition for the sake of argumentation, is sufficient to make the question 'what does this mark mean?' arise in the mind of any epigraphist when he is not sure whether a certain mark is part of an inscription or not. The logical efficacy of the supposition that the mark means something is identical with the logical efficacy of the proposition that it means something.[47]

Collingwood then lays down that to assume is to suppose by an act of free choice,[48] which is explained by saying that anyone who

---

[45] See *ibid.*, p. 26. See also L. Rubinoff, *Collingwood and the Reform of Metaphysics*, p. 232.

[46] See R. G. Collingwood, *An Essay on Metaphysics*, pp. 26-27.

[47] See *ibid.*, p. 27.

[48] See *ibid.*

'makes an assumption' is making a supposition about which he is aware that he might, if he so chose, make another. All assumptions are suppositions, but not all suppositions are assumptions, for some are made unawares and others of which one is aware of making may be made without the awareness of the possibility that other assumptions may have been made. When we use the word 'assumption' we always imply free choice, as for example when we say 'let us assume x equals ten.' It may be used, however, with malice aforethought to insult another, as when one says to another, 'you are assuming that no one will work except for payment,' implying that only a fool would make such an assumption. It may also be used by one wishing to be insulting, as when one asks 'what do you mean by treading on my toe?' knowing well that the treader meant nothing by it.[49]

The logical efficacy of a supposition does not depend upon the truth of what is supposed, or even on its being thought to be true, but only on its being supposed.[50] Collingwood appeals first to practice in scientific reasoning, where the validity of an argument is in no way affected by making assumptions believed to be false, and then to the conduct of daily affairs, when a receipt is asked for, based on the assumption that the payee is capable of dishonesty notwithstanding the belief that he is not.[51]

Collingwood says that a presupposition is either relative or absolute (here reference is not to the act of presupposing but to that which is presupposed). A relative presupposition is one which stands relatively to one question as its presupposition and to another question as its answer. In using a tape measure, for example, I *presuppose* that the answer to the question 'what is the distance between these two points?' will be *right*. The *accuracy of the tape* is thus a presupposition of the question 'what does this distance measure?' But its affirmation is *the answer* to the question 'is this tape accurate in length?' One may use a tape measure without ever questioning its accuracy, in which case the assumption of its accuracy remains unquestioned, which might lead us to suppose that it is not a relative presupposition. But, that certain presuppositions are open to being questioned is not disproved by the fact that someone who makes them fails to see their questionability, just as the necessary consequence of a conclusion from premisses is not disproved if someone who states the premisses fails to see the consequence. "The business of logical inquiries, like that

---

[49] See *ibid.*, pp. 27-28.
[50] But see this chapter, Section G, pp. 193-205, below.
[51] R. G. Collingwood, *An Essay on Metaphysics*, pp. 28-29.

on which we are now engaged, is a study of high-grade or scientific thinking: their conclusions are not impaired by the fact that low-grade or unscientific thinking also exists.''[52]

To question a presupposition is to ask for its verification, that is, to demand that a question should be asked which will have that presupposition itself as its answer, and so become a proposition. To speak of verifying a presupposition involves supposing that it is a relative presupposition.[53]

But ''an absolute presupposition is one which stands, relatively to all questions to which it is related, as a presupposition, never as an answer.''[54]

Thus, a pathologist, if asked, 'What is the cause of event E which sometimes happens in this disease?' will reply, 'The cause of E is C.' Questioned further, he might add that this was established by So-and-so in such and such experiment, now classical. Questioned further, 'I suppose that before So-and-so found out what the cause of E was, he was quite sure it had a cause?' the answer would probably be 'Quite sure, of course.' If the further question 'Why?' were pressed, the answer would likely be 'because everything that happens has a cause.' If the further question 'how do you know that everything that happens has a cause?' is imprudently asked, the one questioned will probably blow up ''because you have put your finger on one of his absolute presuppositions, and people are apt to be ticklish in their absolute presuppositions.''[55]

But one thus questioned, if he keeps his temper so as to be able to give a civil and candid answer, would probably reply that it is something he takes for granted in his job. It is not brought into question. It is not subjected to the process of verification. It is not something that anyone has discovered, but something everyone takes for granted. Such a one is saying that it is an absolute presupposition of the science which he pursues. The illustration was set in terms of a pathologist's science, which, Collingwood says, is one science which still maintains that it is an absolute presupposition that events have causes, whereas most other sciences in the past hundred years have abandoned that as an unquestioned presupposition.[56]

Collingwood says that the idea of verification does not apply to

---

[52] *Ibid.*, p. 30.
[53] See *ibid.*
[54] *Ibid.*, p. 31.
[55] *Ibid.*, p. 31.
[56] See *ibid.*, p. 32.

absolute presuppositions however much we might like to verify them. Verification presupposes that the presupposition in question is a relative presupposition. If anyone protests that if they are not verifiable they are useless in science, the answer is that their use in science is their logical efficacy and that this logical efficacy does not depend on its being verifiable, because it does not depend on its being true. It only depends on its *being supposed*.[57]

Absolute presuppositions are not propositions because they are never answers to questions, whereas a proposition is that which is stated, and whatever is stated is stated in answer to a question.[58]

The logical efficacy of an absolute presupposition is independent of its being true, since the distinction between truth and falsehood does not apply to absolute presuppositions at all but only to propositions. This is the same as saying that absolute presuppositions are never propounded, not merely in the sense that no one ever propounds them but that they are unpropoundable, for "to be propounded is not their business; their business is to be presupposed."[59]

> The scientist's business is not to propound them but only to presuppose them. The metaphysician's business, as we shall see, is not to propound them but to propound the proposition that this or that one of them is presupposed.[60]

From this Collingwood draws the conclusion that to question an absolute presupposition is to ask a nonsense question.

> Hence any question involving the presupposition that an absolute presupposition is a proposition, such as the question 'Is it true?' 'What evidence is there for it?' 'How can it be demonstrated?' 'What right have we to presuppose it if it can't?', is a nonsense question.[61]

Collingwood adds that it is nonsense to say, as some modern logicians say, that "supposing' is one of the several attitudes that one can take towards a proposition, meaning by a proposition something that can be either true or false. Collingwood warns that this is a device for imposing on unwary readers the dogma that all presuppositions are relative, or, in other words, that there are no absolute presuppositions.[62]

---

[57] See *ibid*. But, see this chapter, section G below, regarding the truth and falsity of absolute presuppositions.

[58] R. G. Collingwood, *An Essay on Metaphysics*, p. 32.

[59] *Ibid*., p. 33.

[60] *Ibid*.

[61] *Ibid*. But see later, Section G below, regarding the question of the truth of absolute presuppositions. See L. Rubinoff, *op. cit*., pp. 234-235.

[62] See R. G. Collingwood, *An Essay on Metaphysics*, p. 33.

C. Metaphysics as the Science of Absolute Presuppositions

Collingwood says that in low grade or unscientific thinking we are hardly aware of making presuppositions, since the tangled condition of thoughts coming up from the bottom of our minds gives a deceptive appearance of 'immediacy.'[63] If I merely reflect on the thought that 'this is a clothes line' without analysing it, I think that I have been *confronted* with something which *in itself* is a clothes line, *quite apart from anyone's thoughts about it,* and so I think that I have simply 'apprehended' or 'intuited' that clothes line *for what it really is.* If I never get beyond this casual way of thinking, Collingwood says, I will always remain satisfied that knowledge is no more than the simple apprehension or intuition of (or compresence with) what confronts us as being absolutely and in itself what we intuit or apprehend it to be.[64] This attitude is what Collingwood means by realism, which is founded on nothing other than human stupidity. Any one can 'convince' himself of the truth of realism if only he will let his mind drift and think so casually and haphazardly that he is hardly thinking at all. When he succeeds in doing that, he will find that he does automatically what the realists say we always do whenever we think.[65]

Collingwood will admit the legitimacy of realism as a study of the way anchors get fouled up and covered with seabed sediment. In calm weather and neap tides, he says, such an anchor is good enough to hold the ship.[66]

But the danger of such realist thinking, according to Collingwood, is to rest satisfied with such assertions as 'That is a clothes line,' or 'What I am looking at is my hand,' or 'The bookcase is further away than the table.' The danger in accepting such statements as *right in themselves* is, in supposing that rectitude, to suppose also that this realism is achieving again, and better, what Kant and Descartes achieved in their theories of knowledge. But Kant and Descartes were concerned more with how to make anchors hang the right way and get the ship home during a storm rather than with merely considering how anchors fouled up. Collingwood accuses realists of not realizing that the best account of unscientific knowledge can never be preferable to even the worst account of scientific knowledge.[67]

Collingwood says that in this low grade thinking, at its lowest,

---

[63] See *ibid.*, p. 34.
[64] See *ibid.*
[65] See *ibid.*, p. 35.
[66] See *ibid.*
[67] See *ibid.*, p. 35.

we are quite unaware that every thought is an answer to a question, and wholly unaware that the question arises from a presupposition. Such low grade thinking will never give rise to metaphysics and theories of knowledge. One who accepts such low grade thinking as the model of thought will never understand what metaphysics is about, or why it should be.[68]

But, because man is capable not only of low grade thinking but also of high grade or scientific thinking, he has succeeded in dominating the natural forces within and around him, and thereby given himself the unique character of a self-made inhabitant of a world called civilization which he has made as an environment fit for himself to live in, the original nature, both of himself and his surroundings, serving only as the raw material of his craft.[69]

> High grade thinking means thinking energetically instead of idly: thinking hard instead of allowing your mind to drift.[70]

Collingwood says that higher types of animal are higher because instead of functioning only at a low intensity they store energy for a future occasion which they meet by an expenditure of that stored energy, which elevates their mechanical effectiveness above its average level so as to be able to overcome the obstacle. What higher animals have succeeded to do with their bodies man has learned to do with his mind; he is the only animal who has learned to think hard enough so as to be able to transform the whole structure of his life by this means. ''Everything that we call specifically human is due to man's power of thinking hard.''[71]

In addition to increase of effort, the increased effort requires to be well directed if its effectiveness is to be increased. ''High grade thinking, therefore, depends on two things: increase of mental effort, and skill in the direction of that effort.''[72]

Collingwood says that increase of mental effort brings about not only a difference of degree in the intensity of thinking but also a difference of kind in its quality. At its lowest level of intensity, one is conscious only of 'intuiting' or 'apprehending' what presents itself to the mind. To say that something presents itself to one's mind is only another way of saying that one thinks about it without noticeable effort. Collingwood says that when one becomes aware of effort

---

[68] See *ibid.*, pp. 35-36.
[69] See *ibid.*, p. 36.
[70] *Ibid.*
[71] *Ibid.*, p. 37.
[72] *Ibid.*

one at the same time becomes aware of a hunger that is no longer satisfied by what swims into one's mouth; such a one wants what is not there and will not come of itself and such a one swims about hunting for it.[73]

> This ranging of the mind in search of its prey is called *asking questions*. To ask questions, knowing that you are asking them, is *the first stage* in high-grade thinking: the first thing that distinguishes the human mind from the sea-anemone mind of the 'realist' theory of knowledge.
>
> The *second stage* is not merely to hunt one's prey but to hunt it *cunningly*. To hunt it at all is to ask questions; to hunt it cunningly is to ask questions with skill, or scientifically.[74]

Collingwood further subdistinguishes two stages in this second stage: "The first is disentangling, the second is arranging."[75]

> When a question first comes into one's mind it is generally . . . a confused mass of different questions all of which, because all must be answered before I can catch my dinner, and because I am hungry, I ask at once. But they cannot all be answered at once. Before they can be answered they must be distinguished, and the nest of questions resolved into a list of questions where each item is one question and only one.[76]

The question 'Have you stopped beating your wife?' is a stock logic text illustration of what is called the fallacy of many questions, which is a logical vice of *grammatically* confusing into one question what is *logically* many questions. Collingwood says that a skillful thinker, practised in disentangling knots, will quickly resolve it into four distinct questions:

1. Have you a wife?
2. Were you ever in the habit of beating her?
3. Do you intend to manage in the future without doing so?
4. Have you begun carrying out that intention?

After the work of distinguishing, there follows the arranging, or ordering of the questions.

> After disentangling comes arranging. The reason why questions have to be arranged is because one of them may be contingent upon a certain answer being given to another. The question whether you

---

[73] See *ibid.*, p. 37. Correlate with *The New Leviathan*, Ch. 8, on hunger.
[74] R. G. Collingwood, *An Essay on Metaphysics*, p. 38. Emphasis added.
[75] *Ibid.*, p. 38.
[76] *Ibid.*

> ever beat your wife does not arise unless an affirmative answer has been given to the question whether you ever had one.[77]

Scientific or orderly thinking is orderly in the sense that it deals with things in their logical order, putting what is presupposed before what presupposes it. The resolution of the grammatically single question 'Have you stopped beating your wife?' into four questions also shows how these four questions have to be arranged in their logical order, so that each arises when, and only when, an affirmative answer has been given to the one next before it. The thought process from question to question does not depend on *the truth* of *the answer* but only on this, that it *answers*; not upon the questioner's thinking the answers to be true but only on his accepting the answers given, or assuming them for the sake of argument. It was pointed out previously[78] that the power of causing a question to arise, or the logical efficacy of a supposition, does not depend on the truth of what is supposed but only on its being supposed.

The work of disentangling and arranging questions can also be described as the work of *detecting presuppositions*. The question whether a man has left off beating his wife yet presupposes that he has formed the intention of leaving off, which in turn presupposes that he used to beat her, which in turn presupposes that he has a wife. All these are relative presuppositions, each standing now as a presupposition to one question, now as an answer to another. Each is both presupposition and proposition.[79]

But in addition to relative presuppositions there are absolute presuppositions "and the analysis which *detects* absolute presuppositions I call metaphysics."[80]

Before a presupposition can be called relative, one has to have asked whether it is relative or absolute if one is thinking scientifically. The question 'what does this presuppose?' itself presupposes an affirmative answer to the question 'does it presuppose anything?' But to ask that question is to consider the possibility that it may be an absolute presupposition, and to claim that you could identify one if you found it.

Collingwood says that the procedure and qualifications required

---

[77] *Ibid.*, p. 39. See also L. Rubinoff, *Collingwood and the Reform of Metaphysics*, pp. 235-236.

[78] See page 166, above. See also below, Chapter 7, pp. 280 et seq. on the criticism of this dialectical attitude.

[79] See R. G. Collingwood, *An Essay on Metaphysics*, p. 40.

[80] *Ibid.*

to carry out this metaphysical analysis are in no way different from the pure and simple analysis as described above. In either case the question which is constantly being asked is 'is this presupposition relative or absolute?' The *modus operandi* is the same whichever answer is given, so that the *modus operandi* of all analysis is metaphysical, and, since it is analysis which confers the scientific character on science, science and metaphysics are intimately united and stand or fall together. The establishment of orderly thinking is together the birth of science and the birth of metaphysics; each must live together and die together.[81] This is based on the assumption that metaphysics is the science of absolute presuppositions, which Collingwood says he assumes because he finds it in Aristotle, the inventor of metaphysics (after purging him of what has been shown to be nonsensical—Collingwood is referring to Aristotle's position that metaphysics, besides being the science of first principles, that is, absolute presuppositions for Collingwood, is also the science of pure being and of first being, or Theology, which Collingwood rejects as the nonsense element).[82] Collingwood visualises an objector saying that if he (Collingwood) wants to make such an assumption it is his business but that he (the objector) is not interested until such an assertion is proven. Collingwood replies that such an objector does not know what he is talking about or asking for, for the attitude of a reasonable man is simply to examine a few metaphysical problems and see whether or not they are concerned with absolute presuppositions.

Little analytical ability is needed to convince ourselves that we take a good many things for granted without asking whether they are true or not. But most often this analysis is not pressed far enough for one to decide whether these things are so taken for granted on account of lack of due, unprejudiced enquiry, or whether they are in fact principles which neither admit nor require verification.[83]

At first glance it might seem as if the distinction between absolute and relative presuppositions is not so much a distinction between two kinds of presupposition as a distinction between two ways of presupposing, and that an ordinary intelligible and reasonably honest man could say exactly which of the two he was doing. But the position is not so easy, for "in modern Europe absolute presuppositions are unfashionable."[84] Collingwood then points out the obvious truism

[81] See *ibid.*, pp. 40-41.
[82] See *ibid.*, p. 41. See also *ibid.*, pp. 13-15 and 17-20.
[83] See *ibid.*, pp. 41-42.
[84] *Ibid.*, p. 42.

that when things are conventionally regarded with disapproval people will go to remarkable lengths to convince themselves and others that they do not engage in these things. Denying the existence of absolute presuppositions is regarded as the smart thing to do. Even those who recognise the ridiculousness of this fashion may be so affected by a kind of mass suggestion in this regard that at a critical moment, when every ounce of mental determination is required to decide whether a given presupposition is absolute or relative, they will be persuaded to decide in favour of its being relative, even if in fact it is an absolute presupposition.[85]

Furthermore, honesty, candor and veracity, even in conjunction with intelligence, can do little to facilitate answering the question, because it is not the type of question that can be settled by introspection.

> Introspection can do no more than bring into the focus of consciousness something of which we are already aware. But in our less scientific moments, when knowledge appears to us in the guise of mere apprehension, intuiting that which simply confronts us, we are not even aware that whatever we state to ourselves or others is stated in answer to a question, still less that every such question rests on presuppositions and least of all that among these presuppositions some are absolute presuppositions. In this kind of thinking, absolute presuppositions are certainly at work; but they are doing their work in darkness, the light of consciousness never falling on them. It is only by analysis that any one can ever come to know either that he is making any absolute presuppositions at all or what absolute presuppositions he is making.[86]

Collingwood then proceeds to outline the manner in which such an analysis proceeds. He says that the analysis may be conducted by one or another of two types of thought—experiment, either upon one well trained in a certain type of scientific work[87] or upon oneself, the latter being the more preferable.[88]

If the enquirer, says Collingwood, can find someone well trained in a certain type of work to which he is devoted, but not yet accustomed to metaphysics, and if the enquirer is skillful enough in questioning, and the subject questioned cooperative, at least as far as relative presuppositions are concerned, it will be found that, as soon as one of his absolute presuppositions is touched, he will react violently at

---

[85] See *ibid.*, pp. 42-43.
[86] *Ibid.*, p. 43.
[87] See *ibid.*
[88] See *ibid.*, p. 45.

any attempt to bring them within the orbit of the questioning process. This rejection is a symptom that the subject, who has cooperated in the analysis, has been brought to seeing that the presupposition which he is being asked to justify is an absolute presupposition and that the kind of work to which he is devoted depends on that presupposition being unquestioningly accepted.

> This is what . . . I called being 'ticklish in one's absolute presuppositions'; and the reader will see that this ticklishness is a sign of intellectual strength combined with a low degree of analytical skill. A man who is ticklish in that way is a man who knows 'instinctively' as they say, that absolute presuppositions do not need justification. In my own experience I have found that when natural scientists express hatred of 'metaphysics' they are usually expressing this dislike of having their absolute presuppositions touched.[89]

Collingwood regards this method of dialogue with another as somewhat precarious, owing to the delicacy of the qualifications it requires in the questioned subject, who, as soon as he understands what is happening to him, will likely lose that 'ticklishness' which gives the distinctive value to that untouchable area. It is a sort of violation of virginity in the reflective faculties. When Socrates is accused of corrupting the young men of his time, the word translated by 'corrupt' is the same as the word for seduction.

The only altogether satisfactory method, says Collingwood, is for the analyst to experiment on himself.[90] It is only in this case that the experiment is more rather than less valuable to the subject, but it requires great resolution, and the temptation to cheat is unexpectedly strong.[91]

The process, simply as analytic, is the same as analysis found ordinarily in science, but its purpose is to find out what absolute presuppositions are in fact being made on certain occasions and certain kinds of occasions. In the scientific analysis the relative presuppositions are brought to the fore and subjected to verification, the absolute presuppositions being put to the side. But "in metaphysics it is the relative presuppositions that are thrown back and the absolute presuppositions that are put into the basket; not in order to justify them, because to talk of justifying them is to talk nonsense; but in order to have them scientifically described."[92]

---

[89] *Ibid.*, p. 44.
[90] See *ibid.*, p. 45.
[91] See *ibid.*
[92] *Ibid.*

Collingwood says that Aristotle's identification of metaphysics with theology reminds us that these two opposite procedures of analysis are contemplated by people with quite different feelings with respect to them.

> You may call it superstition or what you will, but hard names make no difference to the fact that there is something uncanny about absolute presuppositions. They give people more than a touch of the feeling which Rudolf Otto called numinous terror.[93]

This terror used to be dealt with by the procedures of magic. Our time prides itself with having abolished superstitions, but in fact, there are as many as ever, the only difference is that we have lost the magical art of conquering them.

> So it is a special characteristic of modern European civilization that metaphysics is habitually frowned upon and the existence of absolute presuppositions denied. This habit is neurotic. It is an attempt to overcome a superstitious dread by denying that there is any cause for it. If this neurosis ever achieves its ostensible object, the eradication of metaphysics from the European mind, the eradication of science and civilization will be accomplished at the same time.[94]

Collingwood says that the metaphysical enterprise of attempting to find out what absolute presuppositions have been made by different persons on different occasions will give rise to the further considerations whether absolute presuppositions are made singly or in groups; how such groups were organized; whether different races or classes made different absolute presuppositions; whether some have been made *semper, unique, ab omnibus*, or all differing on different occasions for different discussions.[95]

Pseudo-metaphysics is the kind of thought which questions what are in fact absolute presuppositions on the mistaken belief that they are relative presuppositions and, as propositions, susceptible to truth or error. As, for example, if one were to ask about the absolute presupposition AP, 'Is AP true?' 'Upon what evidence is AP accepted?' 'How can AP be demonstrated?' 'What right do we have to presuppose it if it is not demonstrable?' The answers to such questions are neither true nor false but nonsense and arise from thinking that supposing is one of the attitudes we can adopt towards a proposition,

---

[93] *Ibid.*, p. 46.
[94] *Ibid.*
[95] See *ibid.*, p. 47.

as some logicians think, as if what is supposed absolutely must be either true or false.[96]

Collingwood adds a lengthy footnote[97] which contains statements and implications highly relevant to his notion of the historical character of absolute presuppositions. He faces, in this footnote, the objection that since he (Collingwood) admits that absolute presuppositions change, that he is saying that they are merely changes of fashion. To this objection Collingwood replies that a change of fashion is a superficial change symptomatic perhaps of deeper and more important changes, but itself is of little depth or importance. New superficial changes of fashion can be initiated simply because others do so, or salesmen or advertisers promote it. But absolute presuppositions are not dodges, and those who start one do not do so because they 'like' to start it. Usually people are not aware of their presuppositions and consequently are not aware of changes therein, so such a change is not a matter of choice. Nor is it a superficial or frivolous change, for:

> It is the most radical change a man can undergo, and entails the abandonment of all his most firmly established habits and standards for thought and action.
>
> Why . . . do such changes happen? . . . because the absolute presuppositions of any given society, at any given phase of its history, form a structure which is subject to 'strains' . . . of greater or less intensity, which are 'taken up' . . . in various ways, but never annihilated. If the strains are too great, the structure collapses and is replaced by another, which will be a modification of the old with the destructive strain removed; a modification not consciously devised but created by a process of unconscious thought.[98]

The last statement seems to be saying that absolute presuppositions are functions of a cultural nature, a resultant of socio-cultural forces which build up into tensions and strains until something in the nature of a socio-cultural 'earthquake' happens in the social structure and finally an equilibrium is reached.

D. Metaphysics as an Historical Science

Collingwood maintains that all metaphysical questions, that is, questions about absolute presuppositions, are historical questions,

---

[96] See *ibid.*, pp. 47-48; see also this chapter, p. 168 above.

[97] See R. G. Collingwood, *An Essay on Metaphysics*, p. 48, footnote added to Chapter 5.

[98] *Ibid.*, p. 48. This reflects the fundamental character traits referred to above in comparing the absolute presuppositions of the *Essay on Metaphysics* with "the ring of solid thought" of the earlier "Ruskin's Philosophy."

since they are about the absolute presuppositions certain people make on certain occasions. "All metaphysical questions are historical questions and all metaphysical propositions are historical propositions."[99]

> Every metaphysical question either is simply the question what absolute presuppositions were made on a certain occasion, or is capable of being resolved into a number of such questions together with a further question or further questions arising out of these. This is the central point of the present essay.[100]

The various presuppositions regarding causation made by physicists during different historical periods illustrate the point at issue. Newtonian Physics presupposes that *some* events have causes, those not due to the operation of laws. Thus, the time at which a moving body passes a definite point on its trajectory is calculable in advance according to a law. But if a colliding body changes its course, this change of direction would be accounted for not by a law, but through a cause.

The presupposition of nineteenth century physics was that *all* events have causes, which seems to originate with Kant.

> The peculiarity of the Kantian metaphysics is that it uses the notion of cause and the notion of law, one might almost say, interchangeably: it regards all laws of nature as laws according to which causes in nature operate, and all causes in nature as operating according to law.[101]

But for modern physics *nothing* happens according to causes; all happens according to laws. Cases of the impact of one body with another are no longer regarded as instances of interference rendering inoperative the Laws of Motion; they are regarded as cases of 'free' motion, that is, motion not interfered with, under peculiar geometrical conditions with the straight line of Newton's first law substituted for by some other kind of line.

Collingwood says that we might be led to think from this that there are three schools of thought in physics, the Newtonian, for which some events have causes, the Kantian, for which all events have causes, the Einsteinian, for which no events have causes, and it might seem as if the metaphysician has to tell which of these three schools is based on a false foundation and which is not. The dogmatic metaphysician, Collingwood says, will not hesitate to pronounce in favour of one or other of them as being 'self-evident.' But the conscientious meta-

---

[99] R. G. Collingwood, *An Essay on Metaphysics*, p. 49.
[100] *Ibid.*
[101] *Ibid.*, p. 50.

physician will be embarrassed, because he will see that there is no way in which that can be decided; it is a pseudometaphysical problem of the type previously discussed, 'is AP true?' The reason why that is not a metaphysical problem is that it is not an historical problem.

> The sentences [Some events have causes, All events have causes, No events have causes], express absolute presuppositions made respectively in three different schools of physical science. Each is important, to the science that makes it, because it determines the entire structure of that science by determining the questions that arise in it, and therefore determining the possible answers.[102]

So, every detail in these respective sciences depends on what absolute presuppositions each respectively makes.

> But this does not mean that it depends on these presuppositions being thought true, or that the truth of the conclusions arrived at depends on the presuppositions being in fact true. For the logical efficacy of a supposition does not depend on its being true, nor even on its being thought to be true, but only in its being supposed . . .[103]

Therefore, according to Collingwood, it is a mistake to think that the metaphysician can pronounce on the fundamental correctness of one school as against the others by investigating their absolute presuppositions from the aspect of their truth or falsity. To enquire into the truth of a presupposition is to assume that it is not an absolute but a relative presupposition. But such an enquiry into the truth of an absolute presupposition is nonsense and pseudometaphysics.[104]

Collingwood says that there are two things that can be done about absolute presuppositions. You can suppose them, which is what the ordinary scientist does, or you can investigate to find out what they are, which is what the metaphysician does. The metaphysician's job is to find out which absolute presuppositions are in fact being made; it is not his job to find out what it is to be an absolute presupposition, for this is the proper work of logic.

> When I say that this is what metaphysicians do I mean that this is what I find them doing when I read their works from Aristotle onwards.[105]

Collingwood says that the business of the ordinary scientist in his scientific work, if he is a Newtonian, is to presuppose that some events have causes, if he is a Kantian, to presuppose that all events

---

[102] *Ibid.*, p. 52.
[103] *Ibid.*
[104] See *ibid.*, pp. 53-54.
[105] *Ibid.*, p. 54.

have causes, and if he is an Einsteinian, that no events have causes, whereas the business of the metaphysician is to *find out that* Newtonian scientists presuppose that some events have causes, *that* Kantian scientists presuppose that all events have causes, and *that* Einsteinian scientists presuppose that no events have causes. These are true metaphysical propositions and their contradictories would be false metaphysical propositions.

> It will be clear that the true metaphysical propositions are true historical propositions, and the false metaphysical propositions false historical propositions. It is the proper business of a metaphysician to answer the question what absolute presuppositions are or were made by Newtonians, Kantians, Einsteinians and so forth. These are historical questions.
>
> The historical nature of the metaphysician's inquiries is at once evident when the propositions he makes are stated as they are above . . . . What makes it evident is that the wording of each statement includes the formula 'so-and-so presupposes (or presupposed) that . . . .' Since the presupposition alleged to be made is an absolute presupposition, and since the question whether it is made is not a personal one but one concerning the peculiarities of a certain phase of scientific thought, the formula would be more accurately rendered: 'in such and such a phase of scientific thought it is (or was) absolutely presupposed that . . . .' This formula I call the 'metaphysical rubric.'[106]

According to Collingwood, history has its own rubric, namely, 'the evidence at our disposal obliges us to conclude that' such and such an event happened. The rubric customarily used by legendary literature is 'the story says that . . .' or 'the story goes on to say that. . . .' But just as in these cases the reader is left to supply these rubrics for himself, since he is supposed to know the ropes, so also the metaphysical rubric is most often left to the reader to be put in for himself, on the presumption that he is intelligent enough, and accustomed enough to this kind of literature.

Another reason, Collingwood tells us, for omitting these rubrics is that the writer himself does not see their necessity.

> It is only when a man's historical consciousness has reached a certain point of maturity that he realizes how very different have been the ways in which different sets of people have thought. When a man first begins looking into absolute presuppositions it is likely that he will begin by looking into those which are made in his own time by his own countrymen, or at any rate by persons belonging to some group of which he is a member. This of course, is already

---

[106] *Ibid.*, p. 55.

> an historical enquiry. But various prejudices current at various times which I will not here enumerate have tended to deceive such enquirers into thinking that the conclusions they have reached will hold good far beyond the limits of that group and that time. They may even imagine that an absolute presupposition discovered within these limits can be more or less safely ascribed to all human beings everywhere and always. In that case, mistaking the characteristics of a certain historical milieu for characteristics of mankind at large, they will leave out the metaphysical rubric on purpose, and present a piece of purely historical research as if it were a research into the universal nature of understanding. But their mistaking it for something else does not alter the fact that it is history.[107]

Collingwood concludes, therefore, that metaphysics has to be rethought so as to take into account its own historical dimension as required by the demands of the recently developed historical consciousness. He then proceeds to outline his proposal for the reform of metaphysics as the historical science of absolute presuppositions.

E. Collingwood's Proposed Reform of Metaphysics as an Historical Science

Kant, living at the time when the Newtonian Physics was making its impression in the field of science, sought to reform metaphysics using as his model the highly successful method of Newtonian physics as the paradigm of genuine knowledge. Collingwood, impressed by the methods of history, sought to bring metaphysics into line with what he thought to be the acme of human thought, namely historical reasoning.

Collingwood says that the historical character of metaphysics has not always been evident to metaphysicians, since

> it is only within the last half-century that the aims and methods of history have defined themselves with the sort of precision that those of natural science achieved around about the year 1600. Until that happened people did not understand that history is a kind of thinking whereby cogent inferences about the past are drawn from interpretation of the evidence it has left behind.[108]

So, according to Collingwood, what obscured the historical character of metaphysics was a misunderstanding regarding the nature of history itself, which, until recently, had been regarded as a scissors-and-paste affair, in which historians use ready-made statements in documents regarded as 'authorities.' According to Collingwood, the

---

[107] *Ibid.*, p. 57.
[108] *An Essay on Metaphysics*, p. 58.

correct historical method proceeds by using documents not as 'authorities' but *as evidence*.[109] And Metaphysics, Collingwood says, is an historical science, not in the sense of scissors-and-paste history of accepting testimony, but in the sense that it

> proceeds according to a method called metaphysical analysis, by which the metaphysician discovers what absolute presuppositions have been made in a certain piece of scientific work by using the records of that work as evidence.[110]

But history today is no longer a scissors-and-paste affair in which the historian repeats statements accepted on the testimony of authorities. Instead of this

> the historian of today makes his own statements on his own authority according to what he finds the evidence in his possession to prove when he analyses it with a certain question in his mind.[111]

Collingwood says that anyone understanding the nature of present day historical thought needs no further convincing that metaphysics is an historical science; all they need to understand is that statement itself in order to see that it is true.

The dissatisfaction felt with the state of metaphysics, especially since Kant's time, "can be removed by taking seriously the proposition that metaphysics is an historical science.[112] Collingwood says that "the peculiar perplexities and obscurities that have always been felt to surround the work of the metaphysician will disappear" if "it be understood both by metaphysicians and by their critics that metaphysics is the science of absolute presuppositions," and that "the distinction between metaphysics and pseudo-metaphysics be firmly grasped," and "it be understood that the business of metaphysics is to find out what absolute presuppositions have actually been made by various persons at various times in doing various pieces of scientific thinking," and "it be understood that if a certain absolute presupposition has been made on one occasion by one person this fact makes it probable that the same presupposition has been made by other persons having in general what may be called the same cultural equipment as himself . . . but improbable that it has been made by persons whose cultural equipment was noticeably different."[113]

---

[109] See Chapter 1, above, pp. 32-34.
[110] R. G. Collingwood, *An Essay on Metaphysics*, p. 59.
[111] *Ibid*. See also Chapter 1, above, pp. 29 et seq.
[112] R. G. Collingwood, *An Essay on Metaphysics*, pp. 59-60.
[113] *Ibid.*, p. 60.

Collingwood adds, however, that probabilities are not history, for history requires proof and the only proof that someone has made a certain absolute presupposition is by analysis of the records of his thoughts.

Two results will follow if this is done in regard to the metaphysician's work. Perplexities and obscurities will disappear, and the scope of metaphysical enquiry will be enlarged.

As regards the removal of perplexities, Collingwood says that this will take place in four areas of consideration: the subject of metaphysics, its method, its form, and its effect on the mind of its readers.

Concerning the perplexities regarding the subject of metaphysics, that is, what metaphysics is about, Collingwood says that clarification is achieved here if we recognise that Aristotle both knew well enough that metaphysics was a science of absolute presuppositions and also that he is altogether responsible for the notion of metaphysics as the barren search for pure being, as if to suggest that the science of pure being and that of absolute presuppositions were one and the same.[114]

With the removal of this perplexity there is also removed, as Collingwood says, the perplexity as to how the metaphysician should train himself to do his work. The Middle Ages thought it was in logic, the seventeenth century, physics, and the nineteenth century, psychology. But having clarified that the subject matter about which metaphysics is concerned is absolute presuppositions, it follows that the metaphysician is a special kind of historian, so that his training should be first in general historical education, secondly in special attention to the history of science, and finally in concentration on such problems as the light which documents providing evidence about the history of science throw on the question as to what absolute presuppositions have been made.[115]

The removal of perplexities as to what metaphysics is about results in removal of perplexities as to how it should proceed. Collingwood maintains that, with the rejection of Aristotle's science of pure being, metaphysics no longer continues to be what it was as a result, namely a blind groping for something that in fact is not there, which in effect is to be doing something futile in regard to an inaccessible object. But, when it is clear that metaphysical problems are historical problems, it is likewise clear that its methods are historical methods. The

---

[114] See *ibid.*, p. 61.

[115] See *ibid.*, p. 62. This seems to make metaphysics a species of the genus history.

metaphysician is then a man who has to get at facts; he must be clear about the facts he wants to get at and the evidence he proposes to use to get at them. There is no excuse in the twentieth century for not knowing the methods of history.

Collingwood faces the objection that, if metaphysics is the science that investigates presuppositions, then it itself must refrain from making any itself. But he refutes this on the grounds that it is a science and, like any science, its questions arise from presuppositions, ultimately from absolute presuppositions.

> If metaphysics is a science at all it is an attempt to think systematically, that is, by answering questions intelligently disposed in order. The answer to any question presupposes whatever the question presupposes. And because all science begins with a question . . . all science begins with a presupposition. Metaphysics therefore either has presuppositions or is no science.[116]

To attempt a metaphysics devoid of presuppositions, says Collingwood, can only result in an unscientific tangle of confused thoughts which glories in its confusion. Furthermore, not only is metaphysics a science having quite definite presuppositions, but everyone knows what some of them are, for, as an historical science, it shares in the presuppositions of all history with which everyone nowadays has some acquaintance.[117]

As to form, two perplexities can be removed. The first concerns completeness, whether there is a set of problems which constitute 'the' problems of metaphysics, and the second concerns the mutual implication of the various solutions to metaphysical problems.[118]

The systematization of the problems of metaphysics is not perplexing when it is realized that "Metaphysics aware of itself as an historical science will be systematic in the sense in which all historical thought is systematic and in no other," for "the historian's work is never finished; every historical subject, like the course of historical events itself, is open at the end, and however hard you work at the end it is still open." Thus "the idea that these problems form a closed repertory, or even a repertory with the door ajar, is the purest illusion,"[119] for

> people who are said to 'make history' solve the problems they find confronting them but create others to be solved, if not by themselves,

[116] R. G. Collingwood, *An Essay on Metaphysics*, p. 63.
[117] See *ibid.*, pp. 63-64.
[118] See *ibid.*, p. 65.
[119] *Ibid.*, p. 65.

> by their survivors. People who write it, if they write well, solve problems too; but every problem solved gives rise to a new problem.[120]

But, if there is a formal structure uniting the various problems of metaphysics, so that the answer to one problem implicates the answer to subsequent problems, the question immediately arises: "Is metaphysics a 'deductive' science?"[121]

A definite 'No' in answer to this question removes many perplexities that otherwise arise from it, since "metaphysics is an historical science" and "the things which it studies, namely absolute presuppositions, are historical facts."[122] Anyone reasonably well acquainted with historical work knows that there is no such thing as an isolated historical fact, for every historical fact forms a constellation or complex with other historical facts. As a consequence, the question as to what absolute presupposition is made by such and such a piece of thinking cannot be answered by reference to a single absolute presupposition. It must be given with reference to a constellation of absolute presuppositions.[123]

This immediately raises the question regarding the logical relations subsisting between the presuppositions making up this constellation. The relationships are not those of logical entailment in the deductive sense of antecedents necessitating consequents. The different presuppositions, constituting one complex piece of thinking, are all made at once, in that one same piece of thinking. Their relationship, one to another, is a relationship of befittingness, in the same sense as when we are able to say that one item of clothing "goes with" another item.

> The constellation, complex though it is, is still a single fact. The different presuppositions composing it are all made at once, in one and the same piece of thinking. They are not like a set of carpenter's tools, of which the carpenter uses one at a time; they are like a suit of clothes, of which every part is worn simultaneously with all the rest. That is to say that, since they are all suppositions, each must be *consupponible* with all the others; that is, it must be logically possible for a person who supposes any one of them to suppose concurrently all the rest.[124]

---

[120] *Ibid.*
[121] *Ibid.*
[122] See *ibid.*, p. 66.
[123] See *ibid.*, p. 66.
[124] *Ibid.*, p. 66. L. Rubinoff, *Collingwood and the Reform of Metaphysics*, p. 237, comments: ". . . instead of pointing out that the rules governing the relations among presuppositions are derived not from propositional but from dialectical logic, Colling-

Not only is the complex constellation of absolute presuppositions a single historical fact, but any constituent absolute presupposition within the constellation is also a single historical fact which the metaphysician is able to discover in the same way that any historian discovers any historical fact, by interpretation of the evidence. No single fact is deducible, *a priori*, from the constellation of which it is a constituent, it has to be uncovered by the historical method of interpreting facts from historical evidence.

> If a given person in a given piece of thinking makes the absolute presuppositions AP_1, AP_2, AP_3, AP_4 . . . each of these is a genuinely independent presupposition which can no more be deduced from the rest than waistcoat can be deduced from trousers and coat together.[125]

Collingwood says that the awareness of metaphysics as an historical science banishes once for all the hope of making it quasi-deductive or mathematical. Thus, when a metaphysician sets himself to discuss the absolute presuppositions of the thought of his own time, the literary form in which he expresses his thought cannot be that of a continuous argument from point to point by way of quasi-mathematical demonstration, such as is found in the *Ethics* of Spinoza. It must be in the form of a *catalogue raisonné*, such as is found in the fourth book of Aristotle's *Metaphysics*, or in the *Quaestiones* of a mediaeval metaphysician.[126]

Furthermore, metaphysics aware of itself as an historical science will give up the idea of founding 'schools' or establishing doctrines.

> When I accept what Spinoza says on this subject [that Nature is the same as God] I am no more going Spinozist in a war of metaphysical sects than I am going Tacitean in a war of historical sects when I accept Tacitus's statement that Agricola conquered southern and central Scotland. What I am doing in either case is to say: 'Here is a statement as to certain facts made by a contemporary writer. The evidence at my disposal proves that it is true.'[127]

---

wood attempts, in his effort to avoid using the language of dialectical logic, to communicate the meaning of dialectical logic by means of a neutral language. He therefore declares that presuppositions are not related as parts of a deductive system but instead constitute a 'constellation.' The latter however is simply a new way of characterizing what has hitherto been described as a dialectical system. Finally, he declares, the logical relation holding among presuppositions making up a constellation is one of consupponibility. . . ."

[125] R. G. Collingwood, *An Essay on Metaphysics*, p. 67.

[126] See *ibid.*, pp. 67-68.

[127] *Ibid.*, p. 69.

Thus, to say today that 'all events have causes' would not be the assertion of a metaphysical doctrine, but a blunder, the mistake of saying that an absolute presupposition is being made when in fact it is not being made.

The discovery that metaphysics is an historical science, concerned with the absolute presuppositions of a period, greatly enlarges its scope of enquiry, according to Collingwood. Ignorance of this has led metaphysicians to confine their attention to the absolute presuppositions of the present, which, in fact, is not really a present but a relatively recent past, whereas the so-called past is a more remote past. Realisation of the historical nature of the enquiry opens it up to the study of presuppositions not only of the so-called present, our immediate past, but of the more or less remote past, and to the study of the transition effected when the presuppositions of one period gave way, by historical change, to those of a succeeding period. Metaphysics then becomes the history of the absolute presuppositions made throughout the various periods of history. This, then, is an amplification of its scope of enquiry, it not now being restricted to the (so-called) present absolute presuppositions. Aristotle's metaphysics was confined to the absolute presuppositions of the Greek science of the 4th century B.C. and that of St. Thomas was confined to those of the central Middle Ages and that of Spinoza to those of European science of the 17th century.[128]

As a result, the metaphysician, instead of having before him only one constellation of absolute presuppositions, has before him an infinite number of them, and so has many worlds to conquer. He can study the presuppositions of European science at any phase of its history for which he has evidence. He can also study the absolute presuppositions of Arabic, Hindu or Chinese science in whatsoever phase he is able to find historical evidence for them. He can study the presuppositions of primitives and prehistoric peoples. All these belong not merely to an historical background of his work but to his own proper work itself, once the historical character of his science is revealed.[129]

When he has acquired knowledge about several different constellations of absolute presuppositions the metaphysician can begin to make comparisons. This study will convince the metaphysician that there are no eternal or crucial problems. It will get him out of the

---

[128] See *ibid.*, pp. 70-71.
[129] See *ibid.*, p. 71.

parish pump provincialism of thinking that the metaphysical problems of his own generation are those which all metaphysicians everywhere worried about and that they are the only ones there are to be worried about now or in the past or the future. He will then realize that different sets of absolute presuppositions correspond not only to differences in the structure of scientific thought but also to differences in the entire fabric of civilization.[130]

The metaphysician, when he uncovers the presuppositions of this or that period, must not think that they are something static which can be comparatively studied merely by noting likenesses and differences. What is essential about phases of history is that each dynamically gives place to the succeeding, not by the violence of war from without or revolution from within, but simply because "each of them while it lives is working at turning itself into the next."[131] Thus:

> To trace the process by which one historical phase turns into the next is the business of every historian who concerns himself with that phase. The metaphysician's business therefore, when he has identified several different constellations of absolute presuppositions, is not only to study their likenesses and unlikenesses but also to find out on what occasions and by what processes one of them has turned into another.[132]

Collingwood refers to this as "the only legitimate (that is, historical) way in which he, or anybody else, can answer the question 'why did such and such people at such and such a time make such and such absolute presuppositions?' "[133] He adds that that question, like all questions in metaphysics, is either an historical question or a nonsense question.

> It is a nonsense question if the answer it expects is one which identifies the cause of the historical fact in question with something outside history, like geographical or climatic environment of that fact or the psycho-physical characteristics of the human beings concerned in it. It is a significant question if it expects an answer in the form: 'Because they or the predecessors from whom they inherited their civilization had previously made such and such a different set of absolute presuppositions, and because such and such a process or change converted the one set into the other.' If any one is dissatisfied with this kind of answer his dissatisfaction shows that the question, as he was asking it, was a nonsense question.[134]

[130] See *ibid.*, pp. 72-73.
[131] *Ibid.*, p. 73.
[132] *Ibid.*
[133] *Ibid.*
[134] *Ibid.*, pp. 73-74.

Collingwood adds that it is not enough to know that each phase in the dynamics of history is converted into the next but that it is further necessary to know the more intimate relation between one phase and the next according to which one phase, by reason of an unstable equilibrium within itself, had in itself the seeds of change, and indeed of that particular change which in fact took place. The fabric of any period is always under a strain. The world studied by history is one wherein *tout passe, tout lasse, tout casse,* and it is an important part of the historian's work to analyse the internal strains to which a given constellation of historical facts is subjected, and the means by which it takes up these strains and prevents them from breaking it to pieces. Collingwood criticises Gibbon for lack of sensitivity to the internal strains of the periods about which he writes, but lauds Hegel, because, for Hegel, historical study was primarily a study of such internal strains, thereby opening the way to Marx's brilliant feat of analysing 19th century economic society in terms of its internal strains. He criticises Spengler for historicising as if piecing together a jigsaw puzzle, with every piece placidly fitting together into the picture. Collingwood says that where there is no strain there is no history. A civilization works itself out by a dynamic logic,

> in which different and at first sight incompatible formulae somehow contrive a precarious coexistence; one dominant here, another there; the recessive formula never ceasing to operate, but functioning as a kind of minority report which, though the superficial historian may ignore it, serves to a more acute eye as evidence of tendencies actually existing which may have been dominant in the past and may be dominant in the future.[135]

The mediocre historian may in his study neglect these strains which a recessive element presents and the peculiar quality which it imparts to the whole, even though he may have sensitive enough feelings to detect it. But the man of action cannot neglect them, for his life may depend on his ability to see where they are and to judge correctly their strength. Cortez destroyed Montezuma not merely using gunpowder but using it to reinforce the strains already present and tending to break up Montezuma's power.[136]

This same characteristic of internal stress and strain will be found in any constellation of absolute presuppositions, and a metaphysician with a general grounding in history will know to look for it.

---

[135] *Ibid.*, p. 75.

[136] See *ibid.*, pp. 75-76.

> He will expect the various presuppositions he is studying to be consupponible only under pressure, the constellation being subject to certain strains and kept together by dint of a certain compromise of mutual toleration having behind it a motive like that which causes parties to unite in the face of an enemy.[137]

Collingwood adds that this is why the conception of metaphysics as a deductive science, the presenting of the constellation of absolute presuppositions as a strainless structure like a body of propositions in mathematics, is a pernicious error, because mathematical propositions are not historical propositions, whereas those of metaphysics are.[138]

> A reformed metaphysics will conceive any given constellation of absolute propositions [sic] as having in its structure not the simplicity and calm that characterize the subject matter of mathematics but the intricacy and restlessness that characterize the subject matter, say, of legal or constitutional history.[139]

Collingwood concludes by adding that the only way that metaphysics can become a science is by becoming more completely and consciously what in fact it has always been, namely an historical science. He maintains that its much needed reform can only be brought about "by its adoption of principles and methods which are now common form among historians."[140] By achieving such a reform, Collingwood feels that metaphysics will be able to withstand all the criticisms previously brought against it, which criticisms were justified by defects in its own practice.[141]

F. Presuppositions as Beliefs

In his Ruskin's Centenary address, Collingwood referred to the fundamental principles of a man's philosophy as "uncontrovertible," as something "he assumes as true in all his thinking and acting," "the centre from which all his activities radiate" and "to which everything a man does is attached."[142] In the *Essay on Metaphysics*, Collingwood identifies the absolute presuppositions with the religious beliefs of individuals and civilizations which it is the business of institutions to perpetuate.

---

[137] *Ibid.*, p. 76.
[138] See *ibid.*, pp. 76-77.
[139] *Ibid.*, p. 77.
[140] *Ibid.*
[141] See *ibid.*
[142] See "Ruskin's Philosophy" in Alan Donagan's *R. G. Collingwood's Essays in the Philosophy of Art*, p. 10.

> In any civilization it is man's religious institutions that refresh in him from time to time the will . . . to retain the presuppositions by whose aid he reduces such experience as he enjoys to such science as he can compass; and it is by dint of these same religious institutions that he transmits these same presuppositions to his children.[143]

Religion is concerned with expressing these presuppositions as basic convictions according to which we regulate our behaviour, whereas science is concerned to interpret and systematize our experience in terms of these general convictions which basically predetermine our view of the world.

> . . . if science is 'experience' interpreted in the light of our general convictions as to the nature of the world, religion is what expresses these convictions in themselves and for their own sake and hands them on from generation to generation and it does this irrespectively of whether we know by means of metaphysical analysis what these convictions are.[144]

This holding firm by personal convictions, as an act of religious faith, to our absolute presuppositions is at the same time the belief of some attribute regarding God.

> The act by which we hold such presuppositions . . . is religious faith; and God is that in which we believe by faith; therefore all our absolute presuppositions must be presuppositions in holding which we believe something about God.[145]

This, according to Collingwood, is why Aristotle's Metaphysics is alternatively called Theology. "The ordinary name for that which is the logical ground of everything else is God."[146] And Collingwood

---

[143] R. G. Collingwood, *An Essay on Metaphysics*, p. 198.

[144] *Ibid.*, p. 198. This, however, seems to involve a contradiction. In *Speculum Mentis*, Collingwood says (p. 110) that "Art asserts nothing; and truth as such is matter of assertion." P. 113: "The point at which a child begins to ask whether stories are true, and passes through the crisis of learning to disbelieve in fairies . . . indicates the emergence of religion from art." P. 114: ". . . the assertion is the transcendency of this monadism [of art], for to make any given assertion is to commit oneself to the denial of whatever contradicts it. Religion is essentially assertion, belief." P. 115: ". . . religion by its very nature is pledged to selectiveness, to a discrimination between the utterances of the spirit, to a dualism between true vision and false vision." This means that what religion asserts is asserted as true. But Collingwood has insisted that truth and falsity do not apply to absolute presuppositions. The difficulty could be averted by saying that the error of religion is to assert as a proposition what is really a presupposition. But, then, why need institutions to perpetuate error? To express them as convictions without asserting them as true seems to involve a contradiction, which, I think, is a felt difficulty throughout Collingwood's whole discussion of absolute presuppositions.

[145] R. G. Collingwood, *An Essay on Metaphysics*, p. 216.

[146] *Ibid.*, p. 10.

explains that, regarding Anselm's argument that 'God exists,' "what it proves is not that because our idea of God is an idea of *id quo maius cogitari nequit* therefore God exists, but that because our idea of God is an idea of *id quo maius cogitari nequit* we stand committed to belief in God's existence."[147] "That God exists is not a proposition, it is a presupposition,"[148] and it is a presupposition upon which, since Aristotle, natural science depends.[149] The Patristic writers, according to Collingwood, saw that the pagan world was collapsing due to its failure to keep alive its own fundamental convictions. Their remedy was a new analysis and formulation of these basic convictions codified in what they called the 'Catholic Faith.'[150]

> The presuppositions that go to make up this 'Catholic Faith' preserved for many centuries by the religious institutions of Christendom, have as a matter of historical fact been the main or fundamental presuppositions of natural science ever since. They have never been its only absolute presuppositions; there have always been others and these others have to some extent differed at different times. But from the fifth century down to the present day all these differences have played their changing parts against a background that has remained unchanged: the constellation of absolute presuppositions originally sketched by Aristotle, and described more accurately, seven or eight centuries later, by the Patristic writers under the name of the 'Catholic Faith.'[151]

Faith or belief is the commitment to fundamental convictions which will give meaning to our experience and which will constitute us as having a world view, a standpoint from which an ordered perspective can be had.[152] Experience is the consequence of acting according to our basic convictions or presuppositions.[153] Scientific experience is the result of thinking systematically according to any given set of presuppositions.[154] Metaphysics is the act of thinking systematically about what presuppositions are actually in use.[155]

Absolute presuppositions are historico-culturally conditioned *a priori* principles of thinking coherently. Kant discovered by analysis

---

[147] *Ibid.*, p. 190.
[148] *Ibid.*, p. 188.
[149] See *ibid.*, p. 206; also pp. 210-214.
[150] See *ibid.*, p. 225.
[151] *Ibid.*, p. 227.
[152] See L. Rubinoff, *Collingwood and the Reform of Metaphysics*, p. 285.
[153] See *ibid.*, p. 284; p. 285.
[154] See R. G. Collingwood, *An Essay on Metaphysics*, p. 196.
[155] See L. Rubinoff, *op. cit.*, pp. 284-285; see also R. G. Collingwood, *An Essay on Metaphysics*, p. 197.

what he thought were the permanent, immutable and eternal principles of science, conceiving mind as a fixed universal nature. But they were no more permanent than the presuppositions of Newtonian, eighteenth century physics. An analysis of physics today, or of that of the Renaissance or that of Aristotle, would give a different set.[156] The mind's essence is not that of a static substance, but is dynamic and historical so that its determining categories have their own history, and are therefore related as members of a dialectical scale of forms.

> . . . [absolute presuppositions] are a special kind of synthetic a priori. They derive from the categories of mind, and are entailed by the statement that mind exists. But since the essence of mind is historicity, each category has its own history which takes the form not of a random sequence but of a progressive or serial scale of forms.[157]

G. Truth and Falsity of Absolute Presuppositions

Collingwood has expressed in his *Essay on Metaphysics* that truth and falsity do not apply to absolute presuppositions.[158] But one of his principal commentators, Lionel Rubinoff, maintains that Collingwood is simply denying that they possess the truth or falsity of empirical verifiability. According to Rubinoff, Collingwood is arguing against A. J. Ayer, for whom truth consists in empirical verifiability, and that he (Collingwood) is using a neutral (rather than dialectic) language, and, instead of arguing in terms of the truth and falsity of absolute presuppositions as they are serially related as a scale of forms in an overlap of classes, he takes his stand on Ayer's notion of truth as equivalent to verifiability and simply denies that they are true or false, that is, subject to an empirical verification process, since, in fact, they are prerequisites of that verifiability process itself. Rubinoff maintains that Collingwood's way of expressing himself is "a misleading consequence of his obsession with the need to cure positivism of its anti-metaphysical tendencies,"[159] and, more specifically, a consequence of Collingwood's growing concern to combat "The Elimination of Metaphysics," of Ayer's *Language, Truth and Logic* (1936), whose "thesis is that, since metaphysical statements are neither tautologies

---

[156] See R. G. Collingwood, *An Essay on Metaphysics*, p. 179.
[157] L. Rubinoff, *Collingwood and the Reform of Metaphysics*, p. 286.
[158] See his *Essay on Metaphysics*, pp. 28-29 and especially pp. 32-33. See also, above, this chapter, pp. 156-157.
[159] L. Rubinoff, *Collingwood and the Reform of Metaphysics*, p. 234.

nor empirical hypotheses subject to empirical verification, they are meaningless."[160]

Since we have here a case of a major commentator interpreting Collingwood in a sense quite other than that which Collingwood's own texts would seem to imply, it is necessary to examine Rubinoff's position on this point. Rubinoff maintains that the real point of Collingwood's discussion is to demonstrate that there are different ways of evaluating and justifying truth and falsity, but that the method of verification which is the only one admitted by the positivists, together with its implied propositional logic, is inappropriate to the work of metaphysics in the same way that the methods of the natural sciences are inappropriate to history.[161]

Rubinoff argues that metaphysics, like history, is a work of the *a priori* imagination, and that since this is a critical activity it seeks to establish not only what is the case but also whether it is true. The truth appropriate to an absolute presupposition is its consupponability, without introducing strain into the constellation, with other presuppositions, and its degree of truth or falsity is to be reckoned according to the integrity it maintains on the scale of forms to which it belongs.[162] Rubinoff states his position thus:

> To show that a presupposition is indispensable is equivalent to showing that it is true only because its indispensability has survived the critique through which it has been elicited. The aim of such a critique is to show that it is consupponible with a variety of other presuppositions which together define the structure of a given form of life—in which case it would be better to speak of the "truthfulness" rather than truth of a presupposition. Truth is a property which belongs only to the whole, but a presupposition is "truthful" to the extent to which it contributes to the integrity of the whole—in the sense in which each form on a scale maintains the integrity of the whole scale to which it belongs. And since a scale may be internally consistent with itself and inconsistent with respect to other scales the very same presupposition may be "truthful" with respect to its place on the scale to which it most immediately belongs and false with respect to an entirely different scale. The purpose of metaphysics is therefore to expose both the implicit truthfulness (or falseness) of a presupposition with respect to its immediate form of life and its implicit falseness (or truthfulness) with respect to the world of knowledge as a whole.
>
> The real point of the essay, then, is not simply to deny that absolute presuppositions are truth functional; its more important purpose

[160] See *ibid.*, p. 241.
[161] See *ibid.*, p. 234.
[162] See *ibid.*, pp. 234-235.

> is to outline the method whereby absolute presuppositions can be elicited and evaluated. The "reform of metaphysics" does not lie in banishing truth from the realm of presuppositions by means of a quibble; it lies rather in the development of a methodology which respects the notion of truth as a scale of forms. And this amounts virtually to an extension of the doctrines of *An Essay on Philosophical Method* to the problem of metaphysics.[163]

Thus "instead of trying to explain the real basis of his new dialectical logic, which would have meant representing some metaphysical statements as a special type of synthetic *a priori*, he (Collingwood) chose to emphasize only those statements that report historical facts about what so-and-so absolutely presupposed."[164] But ". . . metaphysics, in addition to being a descriptive-historical science, is also an activity through which absolute presuppositions are critically evaluated according to the criteria of dialectical logic; in which case they are to be regarded as a special type of the class of synthetic *a priori* statements."[165]

Rubinoff maintains that Collingwood's doctrine of absolute presuppositions, with its implied rejection of eternal problems, is but part of the wider doctrine of the concrete universal. He rejects the historicist charges of Knox,[166] who regards Collingwood's *Essay* as a total absorption of philosophy into history without remainder. According to Rubinoff, this identification of philosophy and history, affirmed by Collingwood since his earliest work, *Religion and Philosophy*, is to be understood in terms of the *rapprochement* of the dialectical logic of the overlap of classes in a scale of forms in a concrete universal.

> . . . the rejection of the realist distinction between philosophy and history is a direct implication of the rapprochement of *Speculum Mentis*: thus committing myself to the further doctrine that so far as there is any implied identity of philosophy and history in the *Essay* it ought to be understood according to the terms of the same rapprochement. As we have seen, Collingwood himself declares in the *Autobiography* that the theory of presuppositions first expounded in "Truth and Contradiction" was part of a general attempt to develop an alternative logic to the propositional logic of realism. If, as I have already suggested, this alternative logic is the dialectical logic of *Speculum*

[163] *Ibid.*, p. 235.

[164] *Ibid.*, pp. 245-246.

[165] *Ibid.*, p. 245.

[166] See *ibid.*, pp. 238-241. Knox's charges are accepted by Alan Donagan; see Donagan's *The Later Philosophy of R. G. Collingwood*, p. 12: "Knox's demonstration that between 1936 and 1938 Collingwood radically changed his mind about the relation of philosophy to history must be the foundation of any interpretation of his later works."

> *Mentis* and *An Essay on Philosophical Method*, then the identity of philosophy and history which is implied by the theory of presuppositions must be what follows from the rapprochement between philosophy and history which it is the purpose of dialectical logic to achieve.[167]

Granted that absolute presuppositions are appropriately handled only by dialectical logic, Rubinoff says that certain things can be said about them.

Firstly, they are subject to two sorts of criteria, one dealing with meaning, the other dealing with truth.

As regards meaning, a statement is meaningful in one or another of two ways, either as the appropriate answer to a given question, in which case it is propositionally meaningful, or as logically presupposed as a *sine qua non* condition of the intelligibility of the question and answer complex as a whole.

In other words, there are some statements which are concerned with one's overall world view to which everything that is said and done is referred (either explicitly or implicitly) in regard to its meaningfulness. Absolute presuppositions, and the constellational whole which they form, constitute a horizon of meaning usually referred to as one's world view. This constellation of basic commitments is what is ultimately responsible for one's world being recognizable as a world, as a whole of meaning. The "Map of Knowledge" which Collingwood draws in his *Speculum Mentis* ends by presenting us with the external world as the mind's reflection of itself.

> The mind is not one among a number of objects of knowledge . . . it is what is really known in the ostensible knowing of any object whatever . . . [but] it can only know itself through the mediation of an external world, know that what it sees in the external world is its own reflection.[168]
>
> The true object [of the mind] . . . is the mind itself. And thus the external world is not a veil between it and its object, but a picture of itself, drawn to aid its own self vision; a picture which as it grows firmer and harder, takes surface and polish and steadiness, becomes the mirror of the mind; and all the detail in it is seen by the mind to be the reflection of its own face.[169]

But, since the mind is what it does, the mind is nothing other than this construction of a world in which it deploys itself. Absolute presuppositions thus seem to be nothing other than the mind itself

[167] L. Rubinoff, *Collingwood and the Reform of Metaphysics*, pp. 248-249.
[168] R. G. Collingwood, *Speculum Mentis*, p. 315.
[169] *Ibid.*, p. 313.

precisely as it is determinative of itself through the determination of its world which it determines in the very determination of its own view of the world. The absolute presuppositions thus seem to be the mind itself if we view the mind as possessing, from within its own resources, certain structural determinations which will function as the armature, so to speak, upon which the mind will weave its lived experience in the rational process of explanation.[170]

This conception of absolute presuppositions as world view determinants finds a surprising reechoing in John Wild's article, "Being, Meaning and the World,"[171] in which Wild maintains that the world horizon is wider than that of being,[172] that certain *meanings* of being *presuppose* that of a world view,[173] and that

> being is not necessarily found with meaning, as the major streams of Western thought, and also Heidegger, have supposed. Contrary to these teachings, they may fall apart, and they have fallen apart in the world of our time.[174]

This statement reflects Collingwood's strains in the constellation of meaning-giving absolute presuppositions, whose rupture and realignment (subsequent consupponability) results in the breakdown of a definite culture and its world view with the subsequent rise of a new culture and new world view in which a consistent whole is recognizable and more or less provides for a harmonious way of life and thinking.

---

[170] This conception of Collingwood's absolute presuppositions is not without its difficulties, however. There seems to be an internal tension within the theory between a Kantian type of dualism and a dialectical monism. If, as Rubinoff says (*Collingwood and the Reform of Metaphysics*, p. 286), Collingwood's absolute presuppositions "are a special kind of synthetic *a priori* . . . [which] derive from the categories of the mind," notwithstanding that they are admitted to be historically conditioned, their *a priori* character seems to set them apart from experience. On the other hand, identity of form and content is a fundamental tenet of dialectical thinking which does not lend itself readily to being understood in terms of an *a priori* (form) and *a posteriori* (experiential content) dichotomy. The above explanation takes its point of view from the dialectical identity of form and content and sees the mind's determinants as *the mind itself so determined* allowing for the further conception of the mind *experienced as further determinable*. It is the same mind (which is what it does) which, looked at one way is determined as regards how it thinks, looked at in another way is subject to modification and further determinability. This tension within Collingwood's thinking between a sort of Kantian dualism and idealist, dialectical monism, already noted at the beginning of this chapter, will have to be faced in its own right later in the chapter; see below, p. 205 et seq.

[171] *The Review of Metaphysics*, XVIII (1965), pp. 411-429.

[172] See *ibid*., pp. 419-421.

[173] See *ibid*., pp. 421-426.

[174] *Ibid*., p. 427.

Wild further maintains that "we are more concerned with being than with the world of meaning. So we seem to have closed ourselves off from the wonderful."[175] This seems to echo Sartre's repugnance when faced with the actuality of meaningless being.

Wild then ruminates in a self-interrogating fashion as follows:

> What do we feel when, after careful manoeuvering through the wreckage, a gleam of possible meaning breaks through the clouds and we gain a brief glimpse of the elements, the waves and the sea and the clouds all together? In this sudden dawn of a possible meaning, the disjoined becomes joined on an all encompassing ground, and the utterly alien now becomes strange.
>
> Is this not the feeling of *thauma*, or, as we say, wonder with which philosophy begins?[176]

If we read questioning for wonder and constellation of absolute presuppositions for Wild's world horizon of meaning, we seem to have Collingwood's position on absolute presuppositions as this is interpreted by Rubinoff.[177] Being, for Wild, seems to refer to experienced actuality, and meaning seems to refer to what is traditionally called essence. If I understand him rightly, for Collingwood, being refers to the totality of all that is, expressed by the formula 'identity in difference.' It is experience precisely as this is the experience of mind systematically organizing its experience into a rational totality. The organized totality in the world of actual experience is possible only through the world view constituted by the constellation of absolute presuppositions, which is logically prior to the mind's organization of experience.

The second criterion to which, according to Rubinoff, a presupposition is subject is that of truthfulness, since every presupposition is part of a question and answer complex which occupies a place on a dialectical scale of overlapping forms. The theory of truth according to which absolute presuppositions are evaluated is more accurately a dialectical history of errors, according to which truth is the product of error proceeding and developing by means of contradiction, paradox and ambiguity, and arrangeable on a scale of forms in which they are related as opposite coincidents, whereas if taken singly, as propositional logic would take them, absolutely in themselves, they would be regarded as incompatible opposites. There would be no progress if successive absolute presuppositions did not overlap on a dialec-

---

[175] *Ibid.*
[176] *Ibid.*
[177] See L. Rubinoff, *Collingwood and the Reform of Metaphysics*, pp. 249-250.

tically evolving scale of forms, for progress is possible only inasmuch as each succeeding age can reflect on the absolute presuppositions of its predecessors which formed the basis of the question-and-answer complex characteristic of their age. This means being able to answer the question as to how these absolute presuppositions came to be presupposed in the first place, which question is answerable in terms of the dynamic strains in the preceding constellation of absolute presuppositions of which the consequent absolute presuppositions are the resultant equilibrium state. To understand one's own age, therefore, requires that one become the historian of one's past.[178] History then becomes an essential ingredient in the very constitution of the described facts and not merely the distanced observation and recording of them. Metaphysics as the historical specialization in the analysis of absolute presuppositions is therefore vital to the preservation and perpetuation of culture and civilization, for the world, in the sense in which we can speak of it as a cultural determinant (as when we speak of the Greek or Roman or Renaissance world), is determined by its absolute presuppositions, or what the people of that age were basically committed to as their fundamental source of meaningful activity. The survival of a culture or civilization therefore requires institutional protection and propagation of its foundations, the absolute presuppositions which are the very catalytic agents by which the mind converts experience into science and civilization.[179]

Rubinoff further says that Collingwood's argument in the *Autobiography*, that the history of all thought, including metaphysics, can be approached through the logic of question and answer, implies that the question and answer complexes, which result from the application of the question and answer process to the history of thought, assume the intelligible pattern of a dialectical scale of forms. Rubinoff illustrates this from Collingwood's account of the rise of natural science in Greece as this is presented by Collingwood in his *The Idea of Nature* and *An Essay on Metaphysics*.

> Natural science, as practised by the Greeks, emerged from a context which was dominated by the attitude of magic. Suppose we were to compare the witch doctor's account of nature with that of the Ionian philosophers. A realist would no doubt regard each as a different answer to the same question, What is nature? What is more, he would declare that in answering this question the Greeks were right and the witch doctors wrong. According to the logic of

[178] See *ibid.*, p. 249.
[179] See *ibid.*, p. 287.

> question and answer, however, the Greeks and the witch doctors were not really competing for the same prize; if we take into account the fact that every question rests on certain presuppositions we will become aware of hitherto unnoticed differences between the questions of magic and the questions of Ionian natural philosophy. Indeed, the logic of question and answer may be regarded as an effective antidote to the historical myopia of realism. For example, when the witch doctor asks, What is nature? (i.e., Why do things happen as they do?) he is presupposing absolutely 'that all things are full of Gods' (an attitude which for the sake of convenience I shall call 'polytheistic animism'). His answer therefore takes a mythopoeic form while his actions take the form of magic. Given his presuppositions, his answer is the 'right' one in the sense that it is the only appropriate response to the question he is asking.[180]

These answers, which are mutually conflicting if considered (as the realists consider them) as propositions meaningful in themselves, and as different answers to the same question, are all right answers if understood in relation to the mentality dominated by different absolute presuppositions. Nevertheless, Rubinoff carefully points out that if the question-answer complex is placed in dialectical relation with other succeeding complexes, then the answers are wrong because, by virtue of such a complex in which the answers are meaningful (and right) in accordance with the absolute presuppositions vivifying that complex, those who are committed to such a complex are restricted in the questions they can put to their experience, and the answers which they can expect from it. But dialectical logic regards their presupposition in the first place as necessary to the manifestation of their falsity, which manifestation appears in lived experience when the explanation of that experience in accordance with those presuppositions is inadequate to the experience. The magical world view of the witch doctor sooner or later comes up against experiences which it cannot handle and which demand a new attitude superseding the superstitious attitude of magic.[181]

The strains experienced through thinking in the context of such absolute presuppositions led to the transformation of the religious consciousness from polytheism to monotheism with a corresponding change of presuppositions and a corresponding change of world view. Instead of presupposing that all things are full of gods, and adopting a magical attitude to the world, the new outlook is determined by presupposing (1) that there are natural things (*The Idea of Nature*, p.

---

[180] *Ibid.*, pp. 252-253.
[181] See *ibid.*, p. 253.

29), (2) that natural things make up one natural world (*ibid.*, p. 29), and (3) that all natural things in common are made up of a single material substance (*ibid.*, p. 30). Rubinoff says that all this is implicit in the religious belief which expresses that 'God exists.'[182]

> In *Speculum Mentis* religion was shown to have its ground in the realization that the explicitly pluralistic monadology of art implicitly presupposes the monistic principle which lies at the basis of science. In the *Essay on Metaphysics,* this phenomenological transition is exemplified in the historical transition from the aesthetic religion of polytheism to the more "religious" religion of monism, which by virtue of its monistic character contains within it the seeds of science. Thales' achievement is therefore represented as one of the implications of the transition from a polytheistic to a monotheistic religious consciousness. And, as in *Speculum Mentis,* this transition is regarded as a progress from error to truth.
>
> In declaring that the scientific world view supersedes the religious world view (which in turn has superseded the aesthetic) we are recognizing that science at the same time implicitly sums up the truth of the more primitive standpoints. From the magical world view is retained the wondering attitude of supposal and questioning; from the religious world view is retained not only the notion of unity but the attitude of faith with which the *conditio sine qua non* of all science must ultimately be embraced. Magic, science, and religion, then, are not just different answers to different questions, they are, at the same time, internally related forms on an overlapping scale.[183]

Rubinoff follows with a lengthy exposition as to how the same dialectical development took place within the history of science itself. Regarding the transition from Thales to Pythagoras he says:

> The transition from Thales to Pythagoras is a dialectical transition from one set of presuppositions to another: a transition which both exposes the absurdity of Thales' presuppositions and preserves their truth. It was, in other words, a rational (i.e., dialectical) transition in which a genuine progress through error can be detected. Indeed, it is only because the presuppositions underlying the question-and-answer complexes which make up the history of thought are so related that there can be historical progress at all.[184]

Rubinoff says that the historian in his critical function acts as a catalyst of the historical process, and brings about progress by means of his critical rejection of presuppositions and, consequently, of the complex of questions and answers whose presuppositions they are.

---

[182] See *ibid.*
[183] *Ibid.*, p. 254.
[184] *Ibid.*, pp. 255-256.

This critical rejection of presuppositions comprises two functions, first a rejection of presuppositions arrived at irrationally, that is, by a dogmatic fixation which refuses the critical examination and by prejudices. Secondly there is critical rejection of presuppositions hitherto fruitful in a previous standpoint, but no longer adequate to present problems which they themselves have precipitated. Such rejected presuppositions are false, not as failing to conform to an abstract criterion of truth but for the historical reason that they are necessary forms of error on a developing scale of truth.[185] It is in this way that past philosophical systems are accepted, rejected or modified, inasmuch as their presuppositions are rated true or false through criticism operating by dialectically ordering them on a scale of forms. Rubinoff cites Collingwood's criticism of Aristotle in his *Essay on Metaphysics* as a good example of the evaluation of absolute presuppositions as regards their truth or falsity.[186]

Rubinoff's conclusion regarding Collingwood's treatment of absolute presuppositions in his *Essay* is as follows:

> Collingwood himself declares that precisely the same considerations which led him to his theory of the history of philosophy led him also to his view of the nature of metaphysics. I have already argued that the former—which Collingwood explicitly regards as an application of the logic of question and answer (A,58)—rests upon the doctrine of the concrete universal and constitutes, therefore, a dialectical history of errors. But if the logic of question and answer functions for metaphysics as it does for the history of philosophy, then my contention, based on the foregoing examples, that metaphysics is an application of the logic of question and answer to the organization of the history of absolute presuppositions into a scale of forms (or concrete universal), is given further support.[187]

Rubinoff adds that, for Collingwood, metaphysics as the historical science of absolute presuppositions of questioning may be regarded as a dialectical history of errors, and, as such, rests on two related sets of concepts. The first is the concept of logical efficacy, according to which presuppositions are judged to be 'right' but not judged to be 'true or false.' The second is the concept of historical change, according to which absolute presuppositions are aligned on a scale of forms in which "(a) one set of presuppositions is replaced

---

[185] See *ibid.*, p. 260.
[186] See *ibid.*
[187] *Ibid.*, pp. 261-262.

by another and (b) there is exhibited in this change a gradual progress towards truth,"[188] from which they are rated true or false.

The attitude to error of realist propositional thinking and of dialectical thinking is thus quite opposite. Realist propositional logic understands error in an absolute sense, and it therefore is to be detected for the purpose of rejecting it. But in dialectical logic, error is functional. It is a catalyst of the mind in forging out the truth, and therefore a necessary means to the truth. The error of a given form on a dialectical scale of forms is systematically related to the form which, in relation to it, is seen to be true. In dialectical logic, therefore, rightness and falseness can coexist as regards any particular proposition.

> The same considerations apply to the dialectic of standpoints in *Speculum Mentis*. Each standpoint is 'right' and yet 'false'. For it is only by committing the errors from which the differentiae of the separate standpoints derive, that the abstract construction of art, religion, science, and history come into existence at all. Thus Collingwood could easily have asserted, about the forms of experience in general . . . that art looks at reality from the point of view of the imagination; religion from the point of view of faith; science from the point of view of the abstract concept; history from the point of view of the concrete fact. There is no point in asking which is the 'right' point of view. Each is the only one possible for the person who adopts it.[189]

Collingwood's metaphysical analysis, which according to Rubinoff is modelled on Kant's transcendental analytic, and thus a regressive analysis into the conditions of scientific thought, is identified with history. For Kant, the mind's determining forms belong to the permanent and unchanging structure of the human mind. For Collingwood, the absolute determinants which condition all our thinking are *a priori* structures which are themselves subjected to historical change in accordance with Collingwood's position, and the being of mind is historical, in contrast to the being of nature which is cyclical and repetitive.

> For Kant the behaviour of mind derives from a permanent and unchanging structure and the principles derived from this structure are eternal truths. For Collingwood, the *a priori* is itself a product of historical change. Transcendental analytics is therefore history. The presuppositions according to which we think and to which we are (under certain conditions) obliged to commit ourselves, are

[188] *Ibid.*, p. 264.
[189] *Ibid.*, p. 267.

likewise subject to historical change. In other words, the very *a priori* criteria according to which the objectivity of historical metaphysics is guaranteed are themselves not only historically grounded but historically relative.[190]

Rubinoff reconciles the *a priori* and transcendental character of the absolute presuppositions with their historical relativity by distinguishing the dialectical sense of these terms from that of propositional logic. They are the categories of the mind whose essence is historical, so that its categories are serially ordered according to a dynamically evolving scale of forms.

> Their meaning and truth do not depend simply upon their correspondence with objectively existing facts. To apprehend the truth of such principles is not like apprehending the truth of the axioms and definitions of geometry. They are not analytic. Nor are they purely synthetic, being wholly derived from, and verifiable only by means of, experience. They are, as I have already suggested, a special kind of synthetic a priori. They derive from the categories of mind, and are entailed by the statement that mind exists. But, since the essence of mind is historicity, each category has its own history which takes the form not of a random sequence but of a progressive or serial scale of forms.[191]

The *a priori* and transcendental character of absolute presuppositions must therefore be understood in terms of mind as a concrete universal and the existence of the mind as identical with its action.

> Thus the essential meaning of the statement, Mind exists, changes as the scale develops. The statement, ''All thought is for the sake of action,'' for example, which is an absolute presupposition of experience in general, can only be understood in terms of a concrete universal whose meaning and truth value, although it changes from age to age, is nevertheless connected in a coherent system. For this reason, the history of past thought is never final. Each generation must reconstruct the past for itself, and in so doing, reveal to itself the a priori principles which lie at the basis of experience.[192]

Becoming conscious of the logical inconsistency, that is, unconsupponability, of absolute presuppositions either results in, or is indicated by, corresponding felt strains in lived experience. The dialectic between presuppositions as implicated in lived experience and as explicitated by thought gives rise to strains which constitute the dialectical driving force of history.

---

[190] *Ibid.*, p. 286.
[191] *Ibid.*
[192] *Ibid.*, pp. 286-287.

> Becoming conscious of the principles which are implicit in experience leads in some cases to a recognition that there is a basic inconsistency between the implicit presuppositions of an experience and its explicit presuppositions. The result is that the latter are replaced by the former. Such changes are neither random nor irrational. They are, on the contrary, progressive, having the characteristics of a scale of forms. The idea of the past is therefore a scale of forms which takes shape in the course of the mind's own discovery of itself. It is an innate idea which is at once both the source of the activity of historical thinking and the product of that activity; for it is only by means of this activity itself that "we endeavour to provide this innate idea with detailed content. And this we do by using the present as evidence for its own past" (IH,247).[193]

As regards the rightness and the truthfulness of absolute presuppositions Rubinoff says finally:

> The presuppositions according to which we think and act are the product of this a priori imagination. We can never subject them to any special kind of proof, but the whole point of Collingwood's argument is to show that this is not necessary. Their meaningfulness, in other words, does not depend upon this kind of verification. The very fact that they are absolute presuppositions is proof that they are 'right'. As for their ultimate truth when judged from the absolute standpoint, there is no way of determining this except by continuing to use them until we are obliged to abandon them.[194]

H. A Problem of Consistency

By way of conclusion of the latter section on the truth and falsity of absolute presuppositions, and of this whole chapter on absolute presuppositions, a return is made to the difficulty highlighted at the beginning of the chapter. There it was remarked[195] that the presuppositions of *An Essay on Metaphysics* seemed to introduce a dichotomy into Collingwood's thought prejudical to the identity of form and content essential to dialectical unity in diversity, and therefore into the *rapprochement* which Collingwood sought as his life's work. The earlier doctrine of supposition as integral to the activity of questioning seems quite consistent with concrete unity in diversity, whose intelligibility is in terms of internal relations within the concrete whole. But the presuppositions of the later *Essay*, as *a priori* to the experience of questioning (notwithstanding that, unlike Kant's *a priori* structures of the mind, these are historically conditioned and mutable) seem to

---

[193] *Ibid.*, p. 287.
[194] *Ibid.*
[195] See this chapter, above, pp. 155-156.

introduce a rift between the experience which is to be (questioned and) explained and the explicative factors, namely these absolute presuppositions. Rather than internal relations between members of a whole we now seem to be dealing with a subsumption, in Kantian fashion, of experience under a quasi experience-transcending determinant.

One way of interpreting Collingwood's position so as to maintain consistency with earlier works is to regard the absolute presuppositions, taken in their constellational unity, as a particular conception which consciousness has of itself at any stage of its development as regards the manner in which it sums up its past within itself precisely as this self-summation gives direction (articulated in the form of relevantly arising questions) to its own self-unfolding towards its future. They would thus be in the nature of "soundings"[196] indicative of the determination of self-developing consciousness at this or that particular stage of its development. Dogmatism would regard them as static entities, whereas dialectical thought would regard them as overlapping classes in a scale of forms.

Absolute presuppositions would thus seem to be embodied in the factual information which forms the body of the question, somewhat in the way in which Stephan Strasser[197] explains that a "fact" embodies in its structure a methodic idea. Strasser's methodic idea would seem more or less to correspond to Collingwood's presuppositions, including his absolute presuppositions.

In this way, conscious experience can still be regarded as an all-embracing totality of unity in diversity, and thus still be able to be understood consistently in terms of the concrete universal as elaborated in Chapter 4, above.

But this favorable interpretation is not without its own internal strains. Why speak now of *pre*suppositions rather than suppositions as previously? An obvious reply is that they are *pre*suppositions with respect to this or that particular question, or this or that complex of questions and answers, whereas with respect to the activity of mind as a whole they are its suppositions. If we begin with this or that question, or this or that complex of questions and answers, and ask the Kantian question "What are its (or their) unconditioned conditions?" we would eventually come to its (or their) absolute presuppositions. They are absolute presuppositions, or unconditioned conditions with

---

[196] See Collingwood's *The New Leviathan*, pp. 63-64.

[197] See Strasser's *Phenomenology and the Human Sciences*, Pittsburgh, Duquesne University Press, 1963, pp. 124-125.

respect to this or to that particular question, or to this or that particular complex of questions and answers. They are thus absolute only in a restricted order and *not simply absolute*. For if one looks at them in the total context of the mind's activity, they are not absolute but relative, and it is precisely as such that they have the truth or falsity which Rubinoff accords to them in the dialectical sense. Looked at from the life of the mind globally, from the absolute aspect, that of philosophy which views precisely in terms of the concrete universal, the concrete totality of all that is, the absolute presuppositions are not absolute but relative, and this on two counts.

Firstly, these absolute presuppositions are conditioned (therefore not unconditioned) by their own retroactive feedback, inasmuch as the question and answer complex to which they effectively give rise and condition can retroact over these same absolute presuppositions so as to modify, or even displace them.

Secondly, they are conditioned (therefore not unconditioned, so not absolute) by reason of the life forms, together with their absolute presuppositions, which preceded them and which gave way to them.

If, therefore, one looks at the overall activity of the mind, rather than at the particular cluster of question and answer complexes, the absolute presuppositions are not really absolute but are just as conditioned as the question and answer complexes which they themselves condition.

It is submitted that, if one looks to the absoluteness of absolute presuppositions, rather than at their implication in this or that question and answer complex, there is really only one strictly absolute presupposition in the whole Collingwoodian enterprise, namely that of self-positing mind. This ties up with what Collingwood said in his *Essay on Philosophical Method*[198] regarding the object of philosophy as being one sustained and constantly recurring starting point, in which the object cannot be posited in thought without at the same time being posited in actuality. In this radical, unique presupposition, expressed by the dictum "Mind is what it does," Collingwood is laying down his whole philosophical system in germ, in which act of positing supposal overlaps with assertion. The positing of this starting point is implicitly the positing of all that will develop into explicitness by the mind's questioning activity which is one with what the mind is. It may be further remarked, I think, that this ever recurring starting point, ever recurring according to different forms of itself, is the one eternal

---

[198] See pp. 158-159 and 169-175. See also above, Chapter 4, pp. 132-138.

reality in Collingwood's thought, notwithstanding his aversion for eternal realities. It is eternally so that "Mind is what it does," notwithstanding that what it does is to continually transform itself.

Since history according to Collingwood is the unfolding of this self-developing mind, the manifestation of creative mind in all its multifarious variety and concretions, whose logic of manifestation is by way of question and answer, it is fitting now to pass to a consideration of Collingwood's notion of history as the story of self-developing mind.

CHAPTER SIX

History as the Science of the Mind

A. History as Knowledge of Concrete Fact

The position of *Speculum Mentis* regarding history is that "the object of history is fact as such,"[1] and that "the historical consciousness asserts concrete fact."[2] The scientific consciousness denies the concreteness of this fact in regarding facts as mere instances of an abstract scientific law.[3] Attempts to raise history to the rank of a science by extracting general laws from it end in bankruptcy.[4] In science, the general law is the end and the facts merely the means to it. Whether Newton's apple really fell or not is of little moment as long as one grasps Newton's law of gravitation. But it is just the opposite in history; the individual fact is the end and the general law is important only

---

[1] R. G. Collingwood, *Speculum Mentis*, p. 211.

[2] *Ibid.*, p. 208.

[3] See *ibid.*

[4] See Collingwood's 1930 paper "The Philosophy of History" in William Debbins' *Essays in the Philosophy of History* by R. G. Collingwood, Austin, University of Texas Press, 1965, p. 133. Probably the best known currently discussed attempt to explain history in this way is what is known as the Popper-Hempel Covering Law Theory. See William Dray, *Laws and Explanation in History*, London, Oxford University Press, 1957, Ch. 1 for a statement of this position. See also Henry Veatch, *Two Logics*, Evanston, Northwestern University Press, 1969, pp. 222-241 for a discussion of this theory from the logical aspect.

to the extent that it enables us to determine the fact.[5] The nineteenth century contributes to the philosophy of history by emphasizing the individuality of the historian's object and the generality of the scientist's.[6]

B. History as Knowledge of the Past

Collingwood says that "history is the study of the past; to be an historian is to know how things came to be what they are."[7] But since a past is something which everything has, as the explanation as to how it came to be what it is, the historical aspect of things is a universal and necessary aspect of them.[8] Science studies the nature of selected parts of the world as a whole, whereas philosophy is thinking about the world as a whole. The biological scientist studies organisms as they are special parts of the world. But when Whitehead says that the world as a whole is an organism, he is talking as a philosopher, not as a biologist. Thus, whereas science studies the particular and contingent characteristics of things, philosophy studies the universal and necessary characteristics of things.[9] There can be a philosophy of something only if that something is not merely a fragment of the world "but . . . an aspect of the world as a whole—a universal and necessary characteristic of things."[10] The justification of our ability to speak of the philosophy of art, religion, history, etc., is that art or religion or history is somehow a universal and necessary characteristic of things and not merely a particular and contingent characteristic of a certain group of things.[11] So, if we can legitimately speak of a philosophy of history, history must be not merely a trade or amusement but a universal and necessary human interest, that is, an interest in some universal and necessary aspect of the world.[12] Since the time of Plato it has been regarded that genuine knowledge "must be universal, not particular; necessary, not contingent; of eternal truths, not of transitory facts."[13] Since everything in the world has a past, "history as a study of the past is therefore a universal and ne-

---

[5] See R. G. Collingwood, "The Philosophy of History," in William Debbins' *Essays in the Philosophy of History* by R. G. Collingwood, p. 132.
[6] See *ibid.*
[7] See *ibid.*, p. 124.
[8] See *ibid.*
[9] See *ibid.*, p. 121.
[10] *Ibid.*, p. 122.
[11] See *ibid.*
[12] See *ibid.*, p. 123.
[13] *Ibid.*

cessary human interest—interesting to anybody who is interested in anything—and not the affair of a special professional group."[14]

C. Philosophy and the Method of History

From the vantage point of his fiftieth year, when he was writing his *Autobiography*, Collingwood saw his life's work as that of effecting a *rapprochement* between philosophy and history.[15] In addition to demanding that philosophers respect standards of historical thinking when they are dealing with the history of their own subject, Collingwood insists that a philosophy of history is required of them, in the sense of a special branch of philosophical inquiry which would investigate the special types of problems that are raised by historical thinking. But the realization of the need for a new branch of philosophy soon developed into a demand for a new kind of philosophy, a new way of philosophising.

From the seventeenth century onwards, the problems that since ancient Greek times have gone by the name of Physics proved readily solvable by the new methods of mathematics and experiment. Man now had the means of discovering all nature's secrets; it was no longer merely a matter of gaping at the natural world; nature merely presented riddles which man had now learnt the trick of answering.[16] Man did all the asking; putting nature to the torture until she gave him the answer to his questions.[17]

---

[14] *Ibid.*, p. 124.

[15] See R. G. Collingwood, *An Autobiography*, p. 77. See also L. Rubinoff, *Collingwood and the Reform of Metaphysics*, pp. 3-5.

[16] Descartes, whom Collingwood regards as one of the three masters on interrogation, regards the role of experience in Physics as that of providing problems which the intelligence will solve by its method of intuition and deduction. "Si l'expérience ne peut pas résoudre les questions, du moins, par elle-même elle ne trompe pas; ce qui lui permettra de jouer un rôle positif: celui de poser les problèmes que l'entendement tentera de résoudre par l'analyse. Un problème est un ensemble de propositions obscures et complexes à rendre intelligibles par l'intuition et la déduction: ces propositions renferment des natures composées qu'il faut réduire en éléments simples. S'interroger sur l'origine des problèmes, c'est se demander quelle est la source des natures composées." J.-L. Allard, *Le Mathématisme de Descartes*, Éditions de l'Université d'Ottawa, 1963, p. 80. The same author, in an unpublished paper, "The Role of Experience in the Philosophy of Descartes," p. 12, writes: "Evidently, the problems concerning the material universe arise in the human mind through sense experience," and he cites A. Gewirtz, "Experience and the Non-Mathematical in the Cartesian Method," *Journal of the History of Ideas*, 1941, p. 192: "Observation and experiment, then, by supplying the subject matter for problems, are the necessary beginnings of the method when it is applied to the investigation of nature." See also Pt. 6 of Descartes' *Discourse on Method* (pp. 120-121 of Vol. 1 of Haldane and Ross' *Philosophical Works of Descartes*), for Descartes' discussion of the relevance of observation and experiment in physics.

[17] See R. G. Collingwood, *An Autobiography*, pp. 77-78.

But Collingwood became convinced that just as "the chief business of seventeenth-century philosophy was to reckon with seventeenth-century natural science" so "the chief business of twentieth-century philosophy is to reckon with twentieth-century history."[18] Until the late nineteenth and early twentieth centuries, historical studies were much in the same state of development as were the natural sciences before Galileo. Just as from that time onwards in natural science something big happened which enormously increased the speed of its progress and enlargement of its vision, so likewise about the end of the nineteenth century something similar was happening in history, perhaps not so spectacularly or rapidly, but none the less certainly.

The writer of history till the end of the nineteenth century was restricted by scissors-and-paste methods to what the authorities said about a subject. Where there were no authorities nothing could be said. From early familiarity with archaeological procedures[19] Collingwood learned another foundation for historical method than that of scissors-and-paste, which "would teach you, not indeed everything, but a great deal, about subjects whose very existence must remain permanently unknown to historians who believed in authorities," and which methods "might be used to correct the authorities themselves, where they had been mistaken or untruthful."[20] Thus, the idea of the dependence of the historian on his authorities was exploded.

Collingwood says that the possibilities of such a new method could have been got from books written since Boucher de Perthes' gravel pits investigations became known (although Collingwood puts it on record that it was never easy for him to learn from books and newspapers; a half hour with a student at an excavation site taught him more than a beautifully illustrated handbook or a friend's article on an excavation in *The Times*).

> I had learnt by first hand experience that history is not an affair of scissors-and-paste, but is much more like Bacon's notion of science. The historian has to decide exactly what he wants to know; and if there is no authority to tell him, as in fact (one learns in time) there never is, he has to find a piece of land or something that has

---

[18] *Ibid.*, pp. 78-79.

[19] ". . . since I was three weeks old" when he says that his father "took me in a carpenter's bag . . ." *An Autobiography*, p. 80.

[20] *Ibid.* See also Allan Nevins' *The Gateway to History*, New York, Doubleday-Anchor, revised edition 1962, Ch. 5, pp. 137-155 on "The Cheating Document," and Ch. 6, pp. 159-168 on "The Dubious Document."

got the answer hidden in it, and get the answer out by fair means or foul.[21]

Collingwood says that this is as far as his philosophy of history got before he went up to Oxford where the revolution in historical method was already making headway since Sir Arthur Evans earlier in the century well illustrated the new method by unearthing and reconstructing the history of the bronze age Knossos.[22] These discoveries had a drastic effect on the rewriting of Oxford ancient history. Everything up till the first Olympiad was cut out. And Mommsen showed how the use of inscriptions could enable the historian to answer questions no one had hitherto dreamed of asking. Dragendorf classified and, with others, began to date Samian pottery. The excavator could reconstruct the history of Roman cities not mentioned by any authority, and establish events not mentioned previously in any books. Collingwood accredits Haverfield, with whom he worked in his (Collingwood's) early career, with embedding these new notions and techniques at Oxford which completely transformed the study of the Roman Empire. But although Roman history was making tremendous strides under the new methods, Greek history was still in the grip of scissors-and-paste methods.

According to a tradition going back to the seventeenth century, the methods of natural science received from philosophers the most painstaking scrutiny. The books, lectures and conversations of these philosophers gave Collingwood the impression that they did not have the least awareness of what was happening in regard to historical method. This, Collingwood thought, was particularly odd in regard to Oxford philosophers, none of whom had any training in natural science, whereas practically all of them had read 'Greats' and had undergone advanced study in ancient history. Joseph's Logic perhaps comes closest, with a chapter on the Historical Method, which, on examination, turns out to be no more than the method used in natural science. Collingwood regretted that not one of the thirty or forty professional philosophers at Oxford gave himself to the study of the method of history, since, however modest it may be in certainty, dignity and utility, history was still a respectable intellectual activity. This fact, together with the fact that obscure provinces of Roman Britain interested him, presented Collingwood himself with a challenge which demanded the invention of new methods.

---

[21] R. G. Collingwood, *An Autobiography*, p. 81.

[22] See *ibid*. See also C. W. Ceram, *Gods, Graves and Scholars*, New York, Knopf, revised edition of 1968, Ch. 8, pp. 61-72 on Evans, Crete and the Minotaur.

> Obscure subjects, by forcing us to think harder and more systematically, sharpen our wits and thus enable us to dispel the fog of prejudice and superstition in which our minds are often wrapped when we think about what is familiar to us.[23]

Collingwood was thus led to believe that by concentrating on historical methodology he might discover truths in the theory of knowledge which were hidden from the realists because of their 'conventional and second-hand' ideas about the methods of natural science. Current theories of scientific method did indeed give recognition to the dependence of scientific knowledge on historical knowledge, but the expression of this was done in such a way as to suggest that it was hoped that the reader would not notice the fact.

> No one, when he said that scientific knowledge depended on experiment, meant that a given scientific theory arose in the scientist's mind contemporaneously with the experiment (or rather, experiments) upon which it was based. He meant that a scientist, in framing a theory, made use of certain historical knowledge in his possession, as to what experiments had been tried and what their results had been. It was a commonplace, though a concealed one, that all 'scientific' knowledge in this way involves an historical element; and it was clear to me that any philosopher who offered a theory of 'scientific method', without being in a position to offer a theory of historical method, was defrauding his public by supporting his world on an elephant and hoping that nobody would ask what kept the elephant up. It was no mere question of adding a theory of historical method to the already existing theory of 'scientific' method. It was a question of making good a defect in current theories of 'scientific' method by attending to an element in 'scientific' knowledge about which there seemed to be a conspiracy of silence, namely the historical element.[24]

Collingwood foresaw the possibility of being on the threshold of a new age in which history would be as important to the world as natural science proved to be between the seventeenth and twentieth centuries. He felt that "the wise philosopher would concentrate with

---

[23] Collingwood's statement here summarises in a very succinct way what is really the whole process of concrete dialectical thinking. The continuously sustained starting point (see Chapter 4, above, pp. 128 et seq.) is known first obscurely; then, by systematic interrogation, the object is increasingly clarified, much in the same way as a landscape obscured by fog becomes clearer and clearer as the fog is dissipated. In the case of knowledge, the fog arises by reason of low grade, that is, uncritical, unquestioned, thinking, such as is found in the dogmatic mode of thinking which, according to Collingwood, characterizes realist, propositional thought. (See also this chapter, below, pp. 249 et seq.).

[24] R. G. Collingwood, *An Autobiography*, p. 87.

all his might on the problems of history, at whatever cost, and so do his share in laying the foundations of the future.''[25]

D. Socio-Cultural Need for a Sound Theory of Historical Methodology

Collingwood saw the 1914-1918 war as ''an unprecedented triumph for natural science.'' The Baconian promise of power to control the forces of nature was fulfilled, but it turned out to be a power to destroy the bodies and souls of men more rapidly than any preceding human agency. Its triumph paved the way not only for others, such as improvements in transport, sanitation, medicine and commerce, but also, and above all, for the next war.

But the war (of 1914-1918) was also an unprecedented disgrace to the human intellect. Few of the combatants wanted it; things increasingly got out of hand; fighting ended without victory on either side. The contrast was glaring between the success of the modern European mind in controlling physical forces and natural realities and its utter inability to control human situations and mental forces.

> It seemed almost as if man's power to control 'Nature' had been increasing *pari passu* with a decrease in his power to control human affairs. . . . But it was a plain fact that the gigantic increase since about 1600 in his power to control Nature had not been accompanied by a corresponding increase, or anything like it, in his power to control human situations. And it was also a plain fact that the ill effects of any failure to control a human situation were more serious now than they had ever been before, in direct proportion to the magnitude of the new powers put by natural science . . . into the hands of the evil and the good, the fool and the wise man.[26]

Collingwood insists that the need of our time is progress in our ability to control human affairs similar to that made in the control of nature since the sixteenth century Baconian revolution.

Collingwood refused to consider the possibility that Psychology, the 'Science of Mind' being newly developed by Freud and others, could perhaps put into our hands an effective means of controlling human affairs as natural science had done in regard to the natural world. His reason is that ''the mind, regarded in this way [that is, scientifically, by the methods of observation and experimentation

---

[25] *Ibid.*, pp. 87-88. One may wonder whether he might not be indulging in the kind of prediction he regards as out of bounds for the historian.

[26] R. G. Collingwood, *An Autobiography*, p. 91. See also Chapter 8, below, pp. 341-343, for a reconsideration of the problem expressed here as the problem of urgent personal concern which dominated the whole of Collingwood's thought.

appropriate to nature] ceases to be a mind at all."[27] Collingwood recognises that psychology has its traditionally recognised legitimate study as the scientific investigation of sensation, appetite and its connected emotions,[28] the science of the body in its psychological sense, that is, as identical with 'feeling.'[29] Collingwood acknowledges that Freud reached a high scientific level in dealing with problems in psychotherapy, but he also says that he reached an unprecedented low in dealing with ethics, politics, religion and social structure.[30]

Not to psychology, the 'pseudo-science of mind,'[31] but to history Collingwood turned for "the help we need in diagnosing our moral and political problems. . . . precisely because history offered us something altogether different from rules, namely insight."[32] But "that man should come to a better understanding of human affairs by studying history," that history was "the thing which in future might play a part in civilized life analogous to that of natural science in the past" could not be admitted if history were understood as a scissors-and-paste affair.[33] However, it is quite otherwise "if the historian resembles the natural scientist in asking his own questions, and insisting on an answer," and provided he is not asking "questions whose answers, however interesting, were of no practical use."[34]

Two questions have to be settled before history's status as the science relevant to human situations can be admitted. The first concerns the status of the object of history, the human past, and the

---

[27] *Ibid.*, p. 93. See also Collingwood's *The New Leviathan*, Chapter 1, pp. 5-7, paragraphs 1.7, 1.83, 1.84. In his *Essay on Metaphysics*, p. 114, Collingwood refers to ". . . the modern methods of psychology, *the science of feeling*." (Emphasis added.) Alan Donagan, in his *The Later Philosophy of R. G. Collingwood*, p. 162, comments on Collingwood's latter statement as follows: ". . . Collingwood denounced as the besetting sin of psychology [its] attempt to study the processes of thought by the methods which are appropriate only to feeling." In his "On Human Nature and Human History" in *The Idea of History*, p. 231, Collingwood says that "these irrational elements [in the mind, not as it is spirit but as it is soul, that is, as it is concerned with the body as feeling—see *The New Leviathan*, pp. 16-17, paragraphs 3.6-3.73] are the subject matter of psychology."

[28] See R. G. Collingwood, *An Autobiography*, p. 95.

[29] See R. G. Collingwood, *The New Leviathan*, pp. 15-16, pars. 3.13, 3.54, 3.6.

[30] See R. G. Collingwood, *An Autobiography*, p. 95. William James' psychological study of religion, *The Varieties of Religious Experience*, is especially singled out for attack by Collingwood. See also L. Rubinoff's editing of Collingwood's views on the philosophy of religion in the volume *Faith & Reason*, Chicago, Quadrangle Books, 1968, p. 11. See also Alan Donagan, *The Later Philosophy of R. G. Collingwood*, pp. 159-168. See also Collingwood's *An Essay on Metaphysics*, pp. 101-132.

[31] R. G. Collingwood, *An Essay on Metaphysics*, pp. 103-105.

[32] R. G. Collingwood, *An Autobiography*, p. 101.

[33] See *ibid.*, pp. 95-96.

[34] *Ibid.*, p. 96.

practicality of historical knowledge. How does knowledge of the human past enable us to control human situations, instead of letting them get out of control, as instanced by the 1914-1918 war?

Regarding the status of the object, Collingwood insists that the past which the historian questions is not a dead past but a living past, a past in some way persisting into the present.[35] The historian can answer questions about the past only on condition that he has evidence about it, which evidence, if it is something the historian can be said to 'have,' must be existing here and now in the historian's present world. An event which left no traces in the present world could not be known, since there would be no 'evidence' for it.

But to leave 'a trace' of itself in the present world, a past event would have to be something more than a material body or state of a material body if it is to function as evidence. The middle ages can be studied by a modern historian only because they are not dead, which means that not only are their writings, etc., still in existence as material objects, but also that their ways of thinking are still in existence as ways in which people still think; for example, the habit of reading and understanding Latin has survived. Indeed such ways of thinking can be resurrected from the dead after a long period of discontinuity, as, for example, the ancient Egyptian and Mesopotamian languages.[36]

Collingwood says that by about 1920 he arrived at his first principle of a philosophy of history, that "the past which an historian studies is not a dead past, but a past which in some sense is still living in the present," which he then expressed by saying that "history is concerned not with 'events' but with 'processes.' "[37]

This notion of 'process' is most important to understand Collingwood's dialectical line of thought in regard to history. Processes are not for him static identities having a well defined beginning and end, but entities which turn into one another. If we consider a process P_1 which turns into process P_2 we cannot designate any dividing line at which P_1 stops and P_2 begins. P_1 does not stop but continues on in the changed form P_2. Likewise P_2 cannot be said to have a well defined beginning; it was previously going on in the earlier form P_1.

---

[35] See *ibid.*, p. 96. See also Collingwood's *The Idea of History*, p. 282: "When a man thinks historically, he has before him certain documents or relics of the past. His business is to discover what that past was which has left these relics behind it."

[36] See R. G. Collingwood, *An Autobiography*, p. 97. See also H.-I. Marrou, *The Meaning of History*, p. 119, regarding the resurrecting of an ancient language.

[37] See R. G. Collingwood, *An Autobiography*, p. 97.

In history there are no beginnings and no endings. History books may begin and end, but the events they describe do not.[38]

Collingwood argues that if P_1 has left traces of itself in P_2 in such a way that an historian living in P_2 can interpret the evidence and see that what is now P_2 was once P_1, it follows that 'the traces' of P_1 persisting in the present are not, as it were, the corpse of a dead P_1 but the real P_1 itself living and active, although incapsulated in the other form of itself which is P_2.[39]

---

[38] See *ibid.*, p. 98. This is in sharp contrast to Aristotle's notion of an event as something 'definite,' as when Aristotle writes in the *Posterior Analytics*, Book 2, Ch. 12, 95b (see p. 173 of Richard McKeon's edition of Aristotle's *Basic Works*, N.Y., Random House, 1941) that ". . . past events are limits and atomic . . . just as points are not 'contiguous' neither are past events, since both are indivisible." H.-I. Marrou in his *The Meaning of History*, p. 48, says that because historical events do not abruptly begin or abruptly end the historian must "begin with an introduction which indicates the antecedents of the phenomenon that is studied. [The historical account] must also include an epilogue that aims to answer the question: What happened afterward? The study must not begin and end abruptly, like the movie screen that is lighted at the start of the film only to be darkened suddenly at the end."

[39] See R. G. Collingwood, *An Autobiography*, p. 98. Compare with this description the unfolding of the life-forms from one to the other in *Speculum Mentis*. See also the evolution of the meaning of the State from Plato's *Republic* to Hobbes' *Leviathan*, Chapter 2, above, pp. 53-54. What we are involved with here is not an abstract universal but a concrete universal, which was dealt with in Chapter 4, above. See also in this regard, L. Rubinoff's *Collingwood and the Reform of Metaphysics*, p. 155: ". . . the *Republic* is an account of the 'polis' of the fifth century B. C. while the *Leviathan* is an account of the absolute monarchy of the sixteenth and seventeenth centuries in England. It is true that both the 'polis' and the 'absolute monarchy' are 'states.' But the 'state' is not an eternal and unchanging substance; it is an historically changing dynamic process which recognizes profound and essential differences between one historical manifestation and another. As Collingwood puts it in *Speculum Mentis*, summing up the whole matter: 'The state is an historical, not a scientific conception—a concrete, not an abstract universal' (*Speculum Mentis*, p. 174).

"This conception of the state as a concrete universal is explored further in an article of 1929 entitled 'Political Action.' The state, Collingwood argues, is generally conceived as a substance having an essence and attributes. This is the concept of the state as an abstract universal. Following this concept, political theory conceives itself either as an attempt to deduce a priori the implications of this essence, or as an inductive inquiry about the various attitudes of sovereignty found to exist in various states. All such theories, deductive or inductive, are agreed in accepting the limitations of the category of substance and attribute ('Political Action,' p. 155). Collingwood proposes to approach political theory from a different angle. Instead of putting the central issue in the form of the question, 'What are the attributes of the state?' he proposes to put it in the form of the question 'What is political action?': 'That is to say, I propose to take my stand, not on the category of substance and attribute, but on the category of action.' "

For Collingwood, the mind is not a substance but pure activity: "The mind is what it does." In his *The Idea of History*, p. 222, he says "Hume was therefore right to maintain that there is no such thing as 'spiritual substance,' nothing that a mind is, distinct from and underlying what it does." Again, in *Religion and Philosophy*, p. 34, Collingwood says that "[the] idea of the mind as a thing distinguished from its

So, if P_1 symbolizes a certain characteristic of a certain historical period (say, the polis of Plato's *Republic* in ancient Greece) and P_2 symbolizes the corresponding although different (and, therefore, contradictory or incompatible) characteristic of a succeeding period (the absolute monarchy of Hobbes' *Leviathan* in Cromwellian England) the successor P_2 always contains a survival of P_1. P_2 is not an opaque self-contained 'atomic' unit, but is "transparent, so that P_1 shines through it and their colours combine into one."[40] Collingwood warns of the dangers of too thoroughly delineating the characteristic features of this or that period, which is "to forget that the silk of the period is in reality always a shot silk, combining in itself contradictory colors."[41]

Collingwood accuses the realists of inability to come up with a viable theory of history because they refuse to admit the reality of becoming, with the result that they inevitably reduce 'becoming'

---

own activities does not seem to be really tenable; the mind *is* what it *does*; it is not a thing that thinks, but a consciousness; not a thing that wills, but an activity." This activist, anti-substantialist notion of mind is fundamental to Collingwood's thought and is maintained throughout his works. This means that the mind's activity, therefore the mind *itself*, is essentially historical. That is, for Collingwood, the activity which is the mind is an historically, dialectically unfolding activity in all its forms, namely, political, religious, artistic, scientific, historical and philosophical. And, as Otis Lee well puts it in his "Dialectic and Negation," in *The Review of Metaphysics*, I (1947), p. 11: "for dialectic, change is sheer creativity. There is only creation and generation, with no decay or passing out of existence. . . . The present includes the past, which has not really gone; and even the future is real, though it does not yet exist, for spirit is eternal." Correlate this with what is said later, in Chapter 9, below, pp. 378-383, regarding the reality of past, present and future as modalities of human existence. It has been shown, in Chapter 4, above, that Collingwood's Logic of Question and Answer is a dialectical logic based not on the abstract universal of Substance and Accident Propositional Logic which founds Science (in the sense of Exact [mathematical] and Experimental Science) but upon the concrete universal whose differentiations are internal to its identity, or 'identity in difference.' The object of history, as also of morality and political action, is a concrete universal, a dialectical, dynamic, ongoing entity. Chapter 9, below, will take up the question of the logic of historical inquiry and of history as a concrete universal.

[40] R. G. Collingwood, *An Autobiography*, p. 98. In *The New Leviathan*, p. 65, pars. 9.5, 9.51, Collingwood calls this persistence of the past in the present the Law of Primitive Survivals, which "runs as follows: *When A is modified into B there survives in any example of B, side by side with the function B which is the modified form of A, an element of A in its primitive or unmodified state.*" Once again, this is a case of a concrete universal, or unity in diversity, or identity in difference, or overlap of classes in a scale of dialectically developing forms which lies at the basis of Collingwood's logical theory. See also Alan Donagan, *The Later Philosophy of R. G. Collingwood*, pp. 28-30, on the Law of Primitive Survivals. The blending compared to the combination of a variety of colours into one is no doubt the work of the creative imagination if one considers what Collingwood writes in "The Historical Imagination" in *The Idea of History*, pp. 231-249, together with his profound description of the way imagination operates creatively in his *Principles of Art*, pp. 142-144.

[41] R. G. Collingwood, *An Autobiography*, p. 98.

propositions into 'being' propositions. In such a reduction "the true proposition 'P_1 becomes P_2' " is made to analyse into "the complex of propositions 'P_1 is P_1', 'P_1 is not P_2', 'P_1 ends where P_2 begins', 'P_2 is P_2', and 'P_2 is not P_1', all of them either tautologous or false."[42]

Returning to the question whether history could provide a school of moral and political wisdom, Collingwood rejects the old pragmatic idea, since it was based on the scissors-and-paste conception in which the past is a dead past about which nothing more can be known except what authorities have said about it. Such knowledge is useless as a guide to action, since history never exactly repeats itself, so that any problem before me now is never sufficiently like any problem described by my authorities to justify application of their solution to my problem. Collingwood insists that it is the historian's business to reveal the less obvious features hidden from a careless eye in the present situation. "What history can bring to moral and political life is a trained eye for the situation in which one has to act,"[43] just as the trained woodsman can say 'there is a tiger in this grass' whereas an ignorant traveller will see nothing but grass and trees. But, in addition, the historian must not merely be able to reveal the special features of the situation but also provide the rules for acting in such a situation, just as it is not much use detecting a tiger in the grass unless a rifle can be provided to shoot it.[44] Collingwood felt that two things were necessary if history was to provide that guidance in action.

Firstly, if you are looking for ready-made rules for dealing with situations you should look to natural science for them, just as if you want a gun the place to go is to a gunsmith. The bankruptcy of civilization between the seventeenth and twentieth centuries was due to

---

[42] Collingwood says that he wrote the substance of these thoughts in a short book and sent a typed copy to his friend Guido de Ruggiero, but that, upon writing his *Autobiography* he destroyed his own only copy together with the manuscript of the unpublished "Truth and Contradiction." See his *Autobiography*, p. 99. See also L. Rubinoff, *Collingwood and the Reform of Metaphysics*, p. 397, for further information on the "Truth and Contradiction" manuscript. Leslie Armour, in his *The Concept of Truth*, Assen, Van Gorcum, 1969, p. 210, n. 1, states that he feels that "it seems unlikely that there was, originally, only *one* copy of *Truth and Contradiction* . . . and a copy may some day turn up."

[43] R. G. Collingwood, *An Autobiography*, p. 100.

[44] Such problems as these, which Collingwood says history is competent to diagnose and solve, are well discussed in the difficulties that arise in legal enactment and application of legal rules, in Gidon Gottlieb's *The Logic of Choice*, London, Allen and Unwin, 1968. It is especially revealing that the logical problems of legal procedure have characteristics similar to the difficulty of applying induction and deduction to historical thought. See especially Chapter 2, pp. 14-32, of Gottlieb's *The Logic of Choice*.

over-reliance on natural science and its capacity to provide ready-made rules. The result was a neglect to acquire the insight needed to decide which rules apply not in a predefined situation but in the one in which one found oneself.

> It was precisely because history offered us something altogether different from rules, namely insight, that could afford us the help we needed in diagnosing our moral and political problems.[45]

Secondly, there are situations which can be easily handled without any appeal to ready-made rules, provided you have insight into them. All you need in such cases is to see what the situation is, then you can straightway extemporize a satisfactory way of dealing with it. If you are sure that it is a tiger in the grass, and that tigers are things to be shot at, the thing to do is to take a rifle with you . . . but be sure it is not your own child which turns out to be playing in the grass.

Collingwood is not referring to actions which are spontaneous and unreflected responses to stimuli, but to the kind of action an agent performs because he knows or believes himself to be in a certain situation. And by acting according to rules Collingwood says he means

---

[45] R. G. Collingwood, *An Autobiography*, p. 101. The 'insight' which Collingwood repeatedly attributes to the historian seems to mean nothing other than 'thought knowing thought,' for 'thought' according to Collingwood is the 'inner' ('inside') dimension of an event whose physical characteristics are its 'outer' or 'external' dimension, about which the natural sciences are concerned. Whereas the natural scientist knows the outer face of the event, the historian knows its 'inner' side, because he knows the thought behind the event. 'Inner' knowledge, which is proper to the historian, is knowledge of 'the thought' behind or within a given event. See Collingwood's "The Subject-Matter of History" in *The Idea of History*, pp. 302-315, especially pp. 305-307. Granted this interpretation, Collingwood's Logic of Question and Answer, *as proper to history*, will be a means of discovering the *thought content* of an event or situation. See also H.-I. Marrou, *The Meaning of History*, pp. 103-111, for what seems to be in substance a very similar, if not the same position as Collingwood's on this point. According to Marrou, "History is an encounter with others" (p. 107), a "dialectical relationship of the Same with the Other . . . [which] presupposes the existence of a broad basis of fraternal communion between the subject and the object, between the historian and document" (p. 104). "In our encounter with Others in historical documents we must proceed as in daily life. Is it a good way to become acquainted with a new friend—or a document—if we subject either to a whole series of questions concerning our interests of the moment? Certainly not, for we must be open to the Other, quite forgetting ourselves. *We must try to perceive the inner essence as something different and other than we are* . . ." (p. 111, emphasis added). Marrou shows throughout his work that whatever obstructs or impedes sympathetic communion with the 'other,' the subject of study, proportionally impedes historical understanding of that subject. See, for example, his account of his changed attitude toward , and changed ability to comprehend, Augustine (*op. cit.*, pp. 280-281). See also Chapter 4, above, pp. 86-87 and n. 43 for correlation of 'insight' and the 'inside of events' with the internal relations constitutive of the concrete universal.

acting because the rule is known and the agent decides to apply that rule, so that he excludes cases where an agent acts according to a rule but is unaware he is so acting. Most of our actions are successful because they are standard types whose rules we know and regularly apply. But action according to rules is not the only type of action. When in a situation not recognizable as belonging to any known types there is no ready rule to guide you; you have to improvise the method of acting. Again, when the referable rules inadequately apply to a given situation, since the typical situation to which the rule adequately refers does not exactly cover the present actual situation, you are thrown back on improvisation. Thus, there are rules for fair dealing with one's tailor inasmuch as he is a tailor, but the attempt to apply the rules to deal with one's tailor who is aged 60, with a weak heart and invalid daughter, becomes so complex and unwieldy that its practicality is impossible. The complexity of required modifications to apply the multiplex of known rules forces one back to improvisation.[46]

The situation in which we have to act apart from rules because the situation is not of a recognizable type most often arises from inexperience of life and is commonest among the young, or when for some reason or other (as travel in a foreign land) we encounter the unfamiliar. But the second case, where the situation involves multiple classification and application of a complexity of rules, is found only among mature people of experience and intelligence who treat the situation seriously, so seriously that they reject as an adequate guide not only desire and self-interest but even that their conduct be right as measured by recognized rules. One who insists on guidance from prescribed rules is clinging to the low grade morality of custom and precept. Such an agent sees only the elements of the situation he already knows how to deal with but shuts his eyes to whatever in the situation would lead him to think that his ready-made rules are inadequate for the conduct of life. Collingwood regards dependence on rules of conduct as keeping action at a low potential, involving a certain blindness to the realities of the situation. To act at a higher potential the agent must see the situation in which he is to act more clearly.

Science may equip us with rules adequate to act in typical cases we already know, but history will equip us to handle concrete situations in all their actuality, provided its job is to inform us about the present insofar as that present is seen to incapsulate its past as a con-

---

[46] See R. G. Collingwood, *An Autobiography*, pp. 103-104.

stitutive of that present, which may not be obvious to an untrained eye. Then only is history of the highest relevance for practical life, but not if it is confined to mere authoritative repetition of a past regarded as dead and gone.[47] Scissors-and-paste history, with its passive dependence on the testimony of authorities regarding a dead past, is inadequate to equip man with the control over human situations analogous to the control natural science provides him with in regard to natural forces.[48]

E. Collingwood's Discovery of the New Historical Method

Collingwood says that his conviction regarding the possibility of a new method of history took a decisive step forward in 1928 during a vacation at LeMartouret when he was able to reflect over his last nine years of historical research.

The first fruit of this "long and oppressive period of gestation" was to make the distinction between history proper and pseudo-history. By pseudo-history he means the *prima facie* appearance of history taken on in the early nineteenth century by such natural sciences as geology, palaeontology and astronomy.[49] Reflection over his practice as an archaeologist enabled him to disengage the stratiographical aspect of his procedure, which resembled the methods of geology, from aspects which differed from such methods. The use of potsherds by the archaeologist for purposes of correlation and dating no doubt resembles the geologist's similar use of fossils, but the difference outweighs the superficial similarity.

> For the archaeologist, these things are not stone and clay and metal, they are building stones and potsherds and coins; debris of a building, fragments of domestic utensils, and means of exchange, all belonging to a bygone age whose purposes they reveal to him.[50]

Whereas sedimentation and fossilization are remains of natural events, potsherds, building stones and domestic utensils are remains

---

[47] See *ibid.*, pp. 105-106.

[48] See *ibid.*, p. 106. L. Rubinoff, in his *Collingwood and the Reform of Metaphysics*, p. 306, says that "while the historian cannot make scientific predictions concerning the future, he is nevertheless able to anticipate it, and his experience as an historian therefore enables him to cope more effectively with the crises of life." See appendix 2, below, pp. 488-496 for a consideration of the way a researcher conducts himself in breaking new ground where there are no laws to guide him.

[49] See R. G. Collingwood, *An Autobiography*, p. 107; see also H.-I. Marrou, *The Meaning of History*, pp. 36-37, for prehistory as including both a non-historical consideration of man's past and also a genuine historical consideration of that same past. See also Chapter 9, below, p. 428.

[50] R. G. Collingwood, *An Autobiography*, p. 108.

of human events. Only the latter are properly admissible as evidence according to properly historical method, and they can be used as evidence only insofar as the archaeologist understands what each one is for. If an object does not tell him that then he has no use for it as an archaeologist, although one more resourceful than himself may be able to detect human purposes in such an object.[51]

Collingwood is recognizing that we have the ability to diagnose consciously intended purposes as distinct from mere natural happenings in the remains which are the results of such purposes. But consciously purposeful activity is thought, or activity done with a reason, or by one knowing both that and why he is acting. When Collingwood thus says that we can recognize in certain remains traces of purposeful activity he is at the same time well on his way to his notion of history as a rethinking of past thoughts. Collingwood says that pre-nineteenth century natural scientists thought that they were detecting purposes, not human purposes but divine purposes, in natural events, but no nineteenth century scientist conceives the world of nature anthropomorphically, as acting purposefully. Purpose is not now a scientist's working category.

> You are thinking historically . . . when you say about anything, 'I see what the person who made this (wrote this, used this, designed this, etc.) was thinking.' Until you can say that, you may be trying to think historically but you are not succeeding. And there is nothing else except thought that can be the object of historical knowledge.[52]

---

[51] H.-I. Marrou in *The Meaning of History*, p. 37, writes: "[When] the . . . prehistorian studies objects that bear the marks or traces of man's voluntary action, [and] [w]hen he tries through them to gain understanding of the material or spiritual techniques (magic, religion) and to some extent of the feelings or ideas of their authors, he is really doing something that pertains to archaeology, which is a branch of history. In this sense prehistory becomes real history in the full meaning of the term. When Norbert Casteret, for example, discovered in the grotto of Montespan a small clay image representing a quadruped adorned with the skull of a bear cub, an image that was pierced by iron-tipped spears in several places, he encountered no difficulty in reconstructing the rite of 'sympathetic' magic in which prehistoric hunters had participated. Eskimos of our own time have also practiced this same rite. *We understand this type of behavior interiorly*. Direct comprehension of this kind is something quite different from that of the physicist who 'understands' the disintegration of the atom. *It is our interior knowledge of man and his potentialities that enables us to understand these prehistoric hunters*. In this sense they are quite historical." (Emphasis added.) See also Collingwood's criticism of Croce's advice to one who would want to understand the history of a blade of grass advising such a one to try to become a blade of grass in the same way that one trying to understand the history of a neolithic should try to become a neolithic; see Collingwood's *The Idea of History*, pp. 199-200.

[52] R. G. Collingwood, *An Autobiography*, p. 110. See also the whole of "The Subject-Matter of History," in *The Idea of History*, pp. 302-315.

Thus, the historian of political history does not concern himself with the pitch of a man's speech but rather with what he is trying to say, that is, the thought expressed in his words and subsequently in the minds of his hearers and finally in written records for rethinking by future minds. And military history is not just a description of marches or battles but rather of plans and counterplans, that is, of thinking about strategy and tactics, and finally of what the men in the battle ranks thought.

F. History as Rethinking Past Thoughts

Whereas the scientist is concerned with observable events, the historian is concerned with events not as such but only inasmuch as they express thoughts.[53]

Natural events are merely phenomena which the scientist contemplates from the outside, but historical events are penetrated by the historian discerning the thought within.

> To the scientist, nature is always and merely a 'phenomenon', not in the sense of being defective in reality, but in the sense of being a spectacle presented to his intelligent observation; whereas the events of history are never mere phenomena, never mere spectacles for contemplation, but things which the historian looks, not at, but through, to discern the thought within them.[54]

This means that when the historian investigates a past event he makes a distinction between the outside and the inside of the event. The outside of an event is whatever can be described in terms of bodies and their movements, that is, its externals, such as, for example, the passage of Caesar and certain men across a certain river, or the spilling of his blood on the floor of the senate house. But the inside of an event is that in it which can only be described in terms of thought, such as Caesar's defiance of Republican law, or the constitutional policy clash between himself and his assassins. But the historian is interested not merely in one of those aspects to the exclusion of the other. He is interested in actions which synthesize an outer aspect and an inner aspect. The outside of an event is the external manifestation of the thought within the agent's action.

> The historian is . . . investigating not mere events (where by a mere event I mean one which has only an outside and no inside) but actions, and an action is the unity of the outside and inside of an event. . . . His work may begin by discovering the outside of

---

[53] See R. G. Collingwood, *The Idea of History*, p. 217.
[54] *Ibid.*, p. 214.

> an event, but it can never end there; he must always remember that the event was an action, and that his main task is to think himself into this action, to discern the thought of its agent.[55]

In penetrating to the inside of events to detect the thought they express the historian is doing something the scientist cannot do. The historian's task is thus both simpler in one way and more complex than the scientist's in another way. The scientist's event is discovered when perceived, but its explanation has to be further sought in its causes. The historian's event is not available to perception, but has to be constructed (rather reconstructed). But when the thought within the event is discovered, the historian's work is done, there is no further search for causes, for when he knows what is the thought within the event he already knows why it happened.

> For science, the event is discovered by perceiving it, and the further search for its cause is conducted by assigning it to its class and determining the relation between that class and others. For history, the object to be discovered is not the mere event, but the thought expressed in it. To discover that thought is already to understand it. After the historian has ascertained the facts, there is no further process of inquiring into their causes. When he knows what happened, he already knows why it happened.[56]

Thus, the processes studied by the historian are quite different from the process of natural events studied by the scientist.

> It is now clear why historians habitually restrict the field of historical knowledge to human affairs. A natural process is a process of events, an historical process is a process of thoughts. Man is regarded as the only subject of historical process, because man is regarded as the only animal that thinks.[57]

G. The Conditions of the Possibility of Historical Knowledge as Knowledge of Past Thought

In order that past thought may be known, two conditions must be fulfilled. First, the thought must be expressed, either linguistically, or in one or another of the ways in which activity is expressive of thought, as the outstretched arm expressive of a command, or running away as expressive of defeat. Secondly, the historian must be able to think again for himself the thought whose expression he

---

[55] *Ibid.*, p. 213. This penetration to the inside of an event to know the thought therein I understand to be what Collingwood means by 'insight.' See this chapter, above, p. 221 n. 45.

[56] *Ibid.*, p. 214.

[57] *Ibid.*, p. 216.

must interpret. If for any reason he is incapable of doing this he should leave the problem alone. One who cannot think mathematically is incapable of writing a history of mathematics.[58]

These two conditions are formulated in the following propositions: the first, 'all history is the history of thought,' the second, 'historical knowledge is the re-enactment in the historian's mind of the thought whose history he is studying.'[59]

Thus, says Collingwood, for the historian to know what Nelson meant by saying 'in honor I won them, in honor I will die with them', and for the historian to know why he said that, is for the historian to know what thoughts Nelson thought in saying that, which means that the historian must think those same thoughts for himself, namely that 'this is not the time to remove the decorations of honor merely to save one's life.'

But this re-enactment, whereby the historian *qua* historian thinks the thoughts of another, involved Collingwood in the dilemma as to whether the thoughts of Nelson as rethought by the historian are Nelson's or the historian's thoughts. He adds that "no question in my study of historical method gave me so much trouble; and the answer was not complete until some years later," that is, after 1928, at LeMartouret.[60]

The answer, when he finally had it, is that "the difference is one of context."[61] That thought, as thought by Nelson, was a present thought, whereas, as (re)thought by the historian it is a past thought, not the past thought of a dead past but a past thought living in the present, which Collingwood refers to as 'incapsulated' in the present.

---

[58] R. G. Collingwood, *An Autobiography*, pp. 111-112. H.-I. Marrou well explains that the encounter with the other involved in history is achieved only by a sympathetic attunement to that other. ". . . it is always sympathy, the source of understanding, which represents the constructive phase. . . . If Denifle made a positive contribution to our growing knowledge of Luther . . . it was not because of his criticism. On the contrary, it was rather to the extent that his personal competence as a medievalist and his own experience as a Catholic belonging to a Religious Order (and as a theologian) led this Dominican of the nineteenth century to sympathize with the Augustinian of the fifteenth, in spite of himself." *The Meaning of History*, pp. 105-106. A criticism to be noted later (Chapter 7, below, pp. 271-273) against Collingwood will be in regard to his excessive emphasis (under Bacon's influence) on the critical function of history to neglect of the essential sympathetic connaturality which Marrou decisively points out as an indispensable ingredient of a true historian. Criticism presupposes sympathetic attunement, as Collingwood admits regarding the history of mathematics above.

[59] See R. G. Collingwood, *An Autobiography*, pp. 110, 112.

[60] See *ibid.*, p. 112, also p. 107. See also Collingwood's "History as a Reenactment of Past Experience" in his *The Idea of History*, pp. 282-302.

[61] R. G. Collingwood, *An Autobiography*, p. 113.

He explains that by an 'incapsulated' thought he means "a thought which, though perfectly alive, forms no part of the question-answer complex which constitutes what people call the 'real' life, the superficial, or obvious present, of the mind in question."[62] Collingwood explains that, for him as a historian thinking Nelson's thought, the question 'Shall I take off my decorations' does not arise. The questions which do arise in his present experience are: 'Shall I continue reading this book on the battle of Trafalgar?' or 'What did the Victory's deck look like to someone contemplating his chances of surviving the battle?' or 'What would I have done in Nelson's situation?'

This possibility of historical knowledge as the reenactment of past thoughts incapsulated in the present, requires that a distinction be made between a primary series of questions which constitute 'real' life, and a secondary series constitutive of a secondary life which is prevented from flowing over into the primary life by that condition of 'incapsulation.'

> No question that arises in that primary series, the series constituting my 'real' life, ever requires the answer 'in honor I won them, in honor I will die with them.' But a question arising in that primary series may act as a switch into another dimension. I plunge beneath the surface of my mind, and there live a life in which I not merely think about Nelson but am Nelson, and thus in thinking about Nelson think about myself. But this secondary life is prevented from overflowing into my primary life by being what I call incapsulated, that is, existing in a context of primary or surface knowledge which keeps it in its place and prevents it from overflowing.[63]

Thus, the knowledge I have when I know the question 'Will I remove my decorations?' together with its negative answer, exists as a thought in my mind, but the question and answer complex which is its context is not part of the question-answer complex in which I think such thoughts as 'I am a little boy in Jersey,' 'This is my father's study carpet, not the Atlantic,' 'That is the study-fender, not the Spanish coast,' which later thoughts belong to the question and answer complex which I recognize as my 'real' life.[64]

---

[62] *Ibid.*, p. 113.

[63] *Ibid.*

[64] See *ibid.*, pp. 113-114. Note that one's 'real' life is distinguished from whatever other lives one may live by a difference according to question-answer complexes (and, following the theme of the previous chapter, we may add different fundamental life commitments, beliefs or absolute presuppositions, or, again, different world-view horizons).

Collingwood therefore adds a third proposition[65] definitive of historical knowledge: 'Historical knowledge is the reenactment of a past thought incapsulated in a context of present thoughts which, by contradicting it, confine it to a plane different from theirs.''[66]

This causes the further question to arise: ''How to determine which of these planes is real?'' To which Collingwood replies: ''By watching the way in which historical problems arise.''[67] Contrary to the opinions of scissors-and-paste historians, Collingwood says that all historical problems arise out of real life. Scissors-and-paste people think that they arise because of the habit of reading which simply puts questions in their mind by way of mere curiosity without any serious bearing on the problems of human everyday living. But the kind of history which Collingwood says he is discussing is something quite different. It is a kind of history which he has practised all his life in which historical problems arose out of practical problems. He maintains that we study history in order to see more clearly into the situation in which we are called to act (the full force of which is realized if we recall that the type of problem Collingwood is basically concerned with, faced with the disasters of the world war is: ''How are we to be delivered from these present distresses?''). It is on the plane of real life that all problems arise, and the plane to which they are referred for solution is history. History, therefore, is Collingwood's answer to the question raised by contemporary situation ethics, namely, 'What am I to do confronted as I am with this real-life problem which eludes decision being made through reference to ready-made rules to hand?' It is by a grasp of the concrete situation as a concrete universal by the historical mind that I will know with certainty what is my duty in the concrete context in which my problem arises.

Self-knowledge is essentially tied to one's ability to think another's thoughts.

> In rethinking what somebody else thought, he thinks it himself. In knowing that somebody else thought it, he knows that he himself

---

[65] The first and second are stated on pp. 86 and 88 of Collingwood's *Autobiography*.

[66] See *ibid.*, p. 115. Note the determining factor of 'context.' The Concrete Universal, which is the very dynamic reality of the mind, actively constructs itself historically by contextualizing in the present thoughts enacted in a previous context. Context is not a superaddition to concrete reality but belongs to the very fabric of its being. This I understand to be of the essence of the concrete universal. The problem to which this gives rise is whether context can be absorbed into a thing's essential structure and retain any intelligible meaning as context.

[67] *Ibid.*, p. 114.

> is able to think it. And finding out what he is able to do is finding out what kind of a man he is.[68]

Thus, I am able to understand the Einsteinian theory of relativity if I can think the thoughts Einstein thought in thinking out his theory as the solution of the practical problem that drove him to think out that theory as the answer to his problem. If a man finds that he is capable of thinking the thoughts of a great many men he is thereby able to understand the thoughts of a great many different kinds of men by rethinking them. It follows that he must be a great many kinds of men, in fact, a microcosm of all that history can know. Thus, his own self-knowledge is at the same time his knowledge of the world of human affairs, or, in other words, a concrete universal.[69]

Collingwood says that this train of thought was not complete until about 1930. But, with its completion, he was able to answer the question which had troubled him since the 1914-1918 war, namely, how can we construct a science of human affairs from which men could learn to deal with human situations as skillfully as natural science had taught them to deal with situations in the world of nature? It was now clear to him that the answer was that the science in question is history.

Collingwood says that such a discovery was not possible before the nineteenth century until history began to undergo a Baconian revolution and arise from its scissors-and-paste chrysalis of the eighteenth century to become a science in the full sense of that word. Men of the eighteenth century saw the need for a science of human affairs but were unable to see that history was that science. They erroneously tried to shape a science of human affairs after the manner of a natural science, such as Hume's 'Science of Human Nature.' The nineteenth century sought it in the science of psychology, in which the mental became reduced to the psychical, with the result that the very distinction between truth and falsehood is jettisoned and the very idea of science thereby negated, psychology itself being bankrupted in the process.[70]

But, says Collingwood, the revolution in historical method which superseded scissors-and-paste history introduced a rapidly progressing form of knowledge which for the first time put man in a position "to obey the oracular precept 'know thyself,' and reap the benefits that only such obedience can confer."[71]

---

[68] *Ibid.*, pp. 114-115. See also Collingwood's *The Idea of History*, p. 10, and see also Chapter 9, below, pp. 418-419.

[69] See *ibid.*, p. 115.

[70] See *ibid.*, pp. 115-116.

[71] *Ibid.*, p. 116.

Collingwood acknowledges that the work of thinking out these ideas was laborious work "because of the method used," for "every detail arose out of reflection on actual historical research in which I had therefore to be incessantly engaged, and was tested over and over again by fresh pieces of research devised to that end."[72]

H. Historical Knowledge as Knowledge by the Mind of Its Acts which are Identical with the Mind Itself

Historical knowledge has for its proper object thought, the act of thinking itself, not merely things thought about.[73] It is therefore a re-doing of what the mind has already done in the past.

> Historical knowledge is the knowledge of what mind has done in the past, and at the same time it is the re-doing of this, the perpetuation of past acts in the present. Its object is therefore not a mere object, something outside the mind which knows it; it is an activity of thought, which can be known only insofar as the knowing mind re-enacts it and knows itself as so doing.[74]

Thought is not merely presupposed by the historical process and the knowledge of that process, it is an integral element in that process, and the knowledge of the process is not distanced from the process, but one with it. The historian knowing history is thereby making it.

> Thought is therefore not the presupposition of an historical process which is in turn the presupposition of historical knowledge. It is only in the historical process, the process of thoughts, that thought exists at all; and it is only insofar as this process is known for a process of thoughts that it is one. The self-knowledge of reason is not an accident; it belongs to its essence.[75]

In order to discover what was the thought constituting the inner dimension of some past event the historian has to make it his own thought by thinking that same thought again for himself as his own thought.[76] The history of thought, which means all history, is the re-enactment of past thought in the historian's own mind.[77] The only way a historian of philosophy can know what Plato meant when he

---

[72] *Ibid.*, p. 117.
[73] See R. G. Collingwood, *The Idea of History*, p. 305.
[74] *Ibid.*, p. 218.
[75] *Ibid.*, p. 227.
[76] See *ibid.*, p. 283 and p. 215.
[77] See *ibid.*, p. 215.

expressed himself in certain words is by thinking that thought (of Plato's) for himself, as his own (the historian's) thought. This is what is meant by 'understanding' what Plato wrote. The historian of politics or warfare will study Julius Caesar's actions by trying to discover the thoughts in Caesar's mind which determined him to do the actions of which documentary evidence remains. To do this the historian has to envisage for himself the situation in which Caesar stood by imaginatively reconstructing that situation and think for himself what Caesar thought about the situation and the possible ways of dealing with it.[78]

Historical knowledge therefore is not thinking about something purely objective, for it is thinking about the very thought the historian is presently thinking in reenacting that thought. However, it is not on that account merely subjective, for that thought is something that can be thought about by anyone with the historical capacity to think it for himself.[79]

To be able to be reenacted in an historian's mind it is not sufficient that something be merely an object of experience, it must be an experience. There is thus no history of nature.[80]

> To the historian, the activities whose history he is studying are not spectacles to be watched, but experiences to be lived through in his own mind; they are objective, or known to him, only because they are also subjective, or activities of his own.[81]

Thus, the objectivity of historical knowledge is of a peculiar kind, different from the objectivity of a scientist's study of the natural world. The object is not something there before the mind with which the mind has to get acquainted; it has to be brought into existence as an object in the very act of studying it; the knowing of it is the (re)constructing of it. This means that knowledge of the historical object is at the same time knowledge of one's own knowledge. Historical knowledge is both objective and at the same time self-knowledge.

> But, because (as I have already tried to show) it is never merely objective, it requires to be thought about in a peculiar way, a way only appropriate to itself. It cannot be set before the thinking mind as a ready-made object, discovered as something independent of that mind and studied as it is in itself, in that independence. It can never be studied 'objectively', in the sense in which 'objectively'

[78] See *ibid.*, p. 215, p. 242, p. 247.

[79] See *ibid.*, p. 292.

[80] See *ibid.*, p. 302. See Collingwood's criticism of Croce on this point in Collingwood's *The Idea of History*, pp. 197-200.

[81] R. G. Collingwood, *The Idea of History*, p. 218.

> excludes 'subjectively'. It has to be studied as it actually exists, that is to say, as an act. And because this act is subjectivity (though not mere subjectivity) or experience, it can be studied only in its own subjective being, that is, by the thinker whose activity or experience it is. This study is not mere experience or consciousness, not even mere self-consciousness; it is self-knowledge.[82]

It is not enough merely to think the thought of another to be thinking historically; one has to know what one is doing oneself when one thinks historically.

> . . . unless he knows that he is thinking historically, he is not thinking historically. Historical thinking is an activity which is a function of self-consciousness, a form of thought possible only to a mind which knows itself to be thinking in that way.[83]

This means that "historical thinking is always reflection, for reflection is thinking about the act of thinking, and . . . all historical thinking is of that kind."[84]

This means that thought is transcendent in its own nature, in the sense that it is "not wholly engaged in the flow of experience," as is feeling, "so that we constantly reinterpret our past thoughts and assimilate them to those we are thinking now."[85]

But thought does not so transcend experience as to be a mere object set over against the thought which thinks it. The peculiar objectivity of thought consists in that it is able to be re-enacted in a mind other than the mind actually presently thinking that thought, otherwise solipsism would prevail.

> Thought can never be mere object. To know someone else's activity of thinking is possible only on the assumption that this same activity can be re-enacted in one's own mind. In that sense, to know 'what someone is thinking' (or 'has thought') involves thinking it for oneself. To reject this conclusion means denying that we have any right to speak of acts of thought at all, except such as take place in our own minds, and embracing the doctrine that my mind is the only one that exists.[86]

The conclusion, that historical thinking is rethinking past thoughts, leads to a certain precision regarding the statement that history studies the past. Historical thinking is the same kind of thinking as that by which we rethink and rediscover the thought of Hammurabi or Solon

---

[82] *Ibid.*, p. 292.
[83] *Ibid.*, p. 289.
[84] *Ibid.*, p. 307.
[85] *Ibid.*, p. 296.
[86] *Ibid.*, p. 288.

in the remote past and that by which we discover the thought of a friend who writes us a letter, or a stranger who crosses the street, that is, the immediate past.[87]

Nor is it necessary that the past thought being rethought be the thought of another, for it is only by historical knowledge that I know what I thought in my own past.[88]

> It is only by historical thinking that I can discover what I thought ten years ago, by reading what I then wrote, or what I thought five minutes ago, by reflecting on an action that I then did, which surprised me when I realized what I had done. In this sense, all knowledge of mind is historical.[89]

Thus rethinking of my own past thoughts is not merely remembering it, but a reliving of a past phase of my life. Memory merely objectivizes the past, as a spectacle. In historical thought the subject re-enacts that thought in his mind.

> In thus rethinking my past thought I am not merely remembering it. I am constructing the history of a certain phase of my life: and the difference between memory and history is that whereas in memory the past is a mere spectacle, in history it is reenacted in present thought. So far as this thought is mere thought, the past is merely reenacted; so far as it is thought, the past is thought of as being reenacted, and my knowledge of myself is historical knowledge.[90]

The possibility of rethinking is based on the fact that thought, unlike feeling, can be disengaged from one context and given another. Thought in its immediacy is thought in a context of emotions and other thoughts. The thought has an identity in itself, independently of this context.

> The self-identity of the act of thinking that these two angles are equal is not only independent of such matters as that a person performing it is hungry and cold, and feels his chair hard beneath him, and is bored with his lesson; it is also independent of further thoughts, such as that the book says they are equal, or that the

[87] See *ibid.*, p. 219. See also H.-I. Marrou, *The Meaning of History*, p. 91: ". . . there is nothing unique in our understanding with regard to the past. It is definitely the same process that takes place in our understanding of other men in the present, and particularly in the understanding of articulated language." In regard to the fundamental role of language in history, and in the construction of the concrete universal proper to history, see Chapter 9, below, p. 399 et seq.

[88] See R. G. Collingwood, *The Idea of History*, p. 219.

[89] *Ibid.*, p. 219.

[90] *Ibid.*, p. 293. See also Collingwood's *Autobiography*, pp. 113-116; see also, this chapter, above, pp. 227-228.

master believes them to be equal; or even thoughts more closely relevant to the subject in hand, as that their sum, plus the angle at the vertex, is 180 degrees.[91]

A thought is neither pure immediacy nor pure mediation. If it were pure immediacy it would be inextricably involved in the flow of consciousness in which it occurs and would then never become an object thought about. If it were pure mediation it would be utterly detached from the flow of consciousness, and thus outside the world of experience, and so non-existent. But an act of thought is together something that actually happens in a context of experience and, as an act of thought, able to sustain itself.

> Every act of thought, as it actually happens, happens in a context out of which it arises and in which it lives, like any other experience, as an organic part of the thinker's life. Its relations with its context are not those of an item in a collection, but those of a special function in the total activity of an organism . . . But an act of thought, in addition to actually happening, is capable of sustaining itself and being revived or repeated without loss of its identity. . . . But it cannot repeat itself *in vacuo*, as the disembodied ghost of a past experience. However often it happens, it must always happen in some context, and the new context must be just as appropriate to it as the old. Thus, the mere fact that someone has expressed his thoughts in writing, and that we possess his works, does not enable us to understand his thoughts. In order that we may be able to do so, we must come to the reading of them prepared with an experience sufficiently like his own to make those thoughts organic to it.[92]

Since an act of thought is knowledge and not mere sensation or feeling, it is something more than immediate consciousness and so not a mere flow of consciousness.[93] "The positive peculiarity which distinguishes thought from mere consciousness is its power of recognizing the activity of the self as a single activity persisting through the diversity of its own acts."[94] This self-sustaining and self-reviving power of thought is what distinguishes it from a mere event.

---

[91] R. G. Collingwood, *The Idea of History*, p. 298.

[92] *Ibid.*, p. 300.

[93] See *ibid.*, p. 287.

[94] *Ibid.*, pp. 305-306. See H. B. Acton, "The Theory of Concrete Universals (II)," *Mind*, XLVI (1937), p. 1, for a consideration of "the most convenient example of a universal" to be "not a quality, such as redness, but a person, such as Julius Caesar" who "is a universal because he is the *same* individual, although engaged in different activities, such as fighting in Gaul, marching in Italy, and so on." Thus, thought, as distinct from mere consciousness, recognizes the self (the person) as an identity in difference, that is, a concrete universal. See above, Chapter 4, p. 88 n. 45. See also, F. H. Bradley, *The Principles of Logic*, London, Oxford University Press, Vol. 1, corrected edn. of 1928, pp. 191 et seq.

> The peculiarity of thought is that, in addition to occurring here and now in this context, it can sustain itself through a change of context and revive in a different one. This power to sustain and revive itself is what makes an act of thought more than a mere 'event' or 'situation'.[95]

To reduce thought to its immediate experience is to deny its character as thought.

> . . . to say that because the theorem, as an act of thought, exists only in its context we cannot know it except in the context in which he actually thought it, is to restrict the being of thought to its own immediacy, to reduce it to a case of merely immediate experience, and so to deny it as thought.[96]

The constituents of immediacy cannot be re-enacted, so cannot be objects of history.

> The immediate, as such, cannot be re-enacted. Consequently, those elements in experience whose being is just their immediacy (sensations, feelings, &c. as such) cannot be re-enacted; not only that, but thought itself can never be re-enacted in its immediacy.[97]

Plato's thought and my thinking that thought are two different thoughts in their immediacy but are identical in their mediation.[98] Plato's argument as an experience of his own arose out of some discussion. Without knowing that context of discussion I can read and understand his argument, rearguing it with and for myself. The process of argumentation which I go through is identically Plato's, not merely similar to it. The argument in itself, from premisses to conclusion, can be developed in Plato's, my own, or anyone else's mind; and thus it is thought in its mediation.[99]

But the thought itself considered in its immediacy, as a unique act in the unique context of the life of the individual thinker, cannot be re-enacted, and therefore cannot be the object of historical knowledge.

> The historian cannot apprehend the individual act of thought in its individuality, just as it actually happened. What he apprehends of that individual is only something that it might have shared with other acts of thought and actually has shared with his own. But this something is not an abstraction, in the sense of a common characteristic shared by different individuals and considered apart from the

[95] *Ibid.*, p. 297.
[96] *Ibid.*, pp. 298-299.
[97] *Ibid.*, p. 297.
[98] See *ibid.*, p. 301.
[99] See *ibid.*, p. 300.

individuals that share it. It is the act of thought itself, in its survival and revival at different times and in different persons: once in the historian's own life, once in the life of the person whose history he is narrating.[100]

To say, then, that history is knowledge of the individual is true only if individual is taken to include universal significance and not its exclusive uniqueness to a definite person at a definite time.

> Thus the vague phrase that history is knowledge of the individual claims for it a field at once too wide and too narrow: too wide, because the individuality of perceived objects and natural facts and immediate experiences falls outside its sphere, and most of all because even the individuality of historical events and personages, if that means their uniqueness, falls equally outside it; too narrow, because it would exclude universality, and it is just the universality of an event or character that makes it a proper and possible object of historical study, if by universality we mean something that oversteps the limits of merely local and temporal existence and possesses a significance valid for all men at all times.[101]

What distinguishes the mere memorization of a past thought from historical re-enactment of it is that whereas mere memory is present thought of past experience as such "historical knowledge is that special case of memory where the object of present thought is past thought, the gap between present and past being bridged not only by the power of present thought to think the past, but also by the power of past thought to reawaken itself in the present."[102] That is, if the past thought is presently thought of merely *as an event* it is merely remembered. But if it is thought *as a thought* it is present as an activity, that is, it is reactivating itself in the present. Then historical knowledge is had.

Thought which can survive the flow of experience is more than mere consciousness of self; it is self-consciousness, that is, a self-sustaining activity. Thinking implies some standing firm against the flow of immediate experience.

> The peculiarity of thought, then, is that it is not mere consciousness but self-consciousness. The self, as merely conscious, is a flow of consciousness, a series of immediate sensations and feelings; but as merely conscious it is not aware of itself as such a flow; it is ignorant of its own continuity through the succession of experiences.

---

[100] *Ibid.*, p. 303.

[101] See the remark at the conclusion of Chapter 5 above, pp. 207-208, that mind, which is identical with its activity, is the one eternal reality in the whole of Collingwood's philosophy.

[102] *Ibid.*, p. 294.

> The activity of becoming aware of this continuity is what is called thinking.[103]

I. Historical Thinking as a *Rapprochement* between Thought and Action, Fact and Theory, History and Philosophy

According to L. Rubinoff "the unity of thought and action is . . . a cardinal principle of Collingwood's philosophy."[104] Collingwood says that he was first motivated to effect this *rapprochement* by reason of the moral corruption which he regarded as an essential consequence of the realist attitude that theorizing leaves its subject matter intact, that what we think about our actions makes no difference to them. To this he opposes what he regards as the familiar experience of every human being, who

> . . . in his capacity as a moral, political, or economic agent . . . lives not in a world of 'hard facts' to which 'thoughts' make no difference, but in a world of 'thoughts'; [so] that if you change the moral, political and economic 'theories' generally accepted by the society in which he lives, you change the character of his world; and . . . if you change his own 'theories' you change his relation to that world; so that in either case you change the ways in which he acts.[105]

To admit this is to deny that a clear-cut distinction can be maintained between philosophy and history, that is, between the facts of one's situation, knowledge of which is historical knowledge and undoubtedly necessary for action, and philosophical thinking concerned with timeless 'universals' which is not necessary for action. The realist position collapses as soon as the distinction between 'facts' and 'theories' and so between 'history' and 'philosophy' is brought into question.'[106] Although in 1919 his *rapprochement* between history and philosophy was incomplete, he saw already that the *rapprochement* between theory and practice was one of intimate and mutual dependence, thought depending on what the thinker learned by experience in action, and action depending upon how he thought of himself and the world. He recognized that scientific, historical and philosophical thinking depended as much on the moral qualities as upon the intel-

---

[103] *Ibid.*, p. 306.

[104] L. Rubinoff, *Collingwood and the Reform of Metaphysics*, p. 306. See also R. G. Collingwood, *An Autobiography*, p. 147.

[105] R. G. Collingwood, *An Autobiography*, p. 146.

[106] See *ibid.*, pp. 147-148.

lectual qualities of the thinker, and that "moral difficulties were to be overcome not by 'moral' force alone but by clear thinking."[107]

Historical knowledge of one's past renders the mind capable of acting with respect to projects to be realized in the future.

> The past, insofar as it is appropriated into the present, is in that appropriation 'projected' into the future; a mind which has reconstructed its own past becomes a mind which is capable of behaving in new and unpredictable ways.[108]

Activity as such is both individual and universal (so, together, an object of both history and philosophy). Reflexive or deliberative activity is done having previously conceived in thought the universal plan or idea which is the criterion guiding the conscious realization of the act.

> An act is more than a mere unique individual; it is something having a universal character; and in the case of a reflective or deliberate act (an act which we not only do, but intend to do before doing it) this universal character is the plan or idea of the act which we conceive in our thought before doing the act itself, and the criterion by reference to which, when we have done it, we know that we have done what we meant to do.[109]

The universality, falsely absolutized by science, is preserved in the predicate of the individual historical judgment.

> The individual judgment of history contains within itself, in the shape of its own predicate, the universality of science; and history is shown to be, not something that falls short of scientific accuracy and rationality and demonstrativeness, but something that possesses all this and, going beyond it, finds it exemplified in an individual fact.[110]

The mental activity which is historical knowledge is not merely activity but is an activity which is together a self-discovery and a self-making, and the discovery is made in the making.

> Historical thinking is therefore not just the discovery of mind but its actualization as well. All thought exists for the sake of action and it is only through such action that the mind's nature is created at all.[111]

---

[107] *Ibid.*, p. 150.

[108] L. Rubinoff, *Collingwood and the Reform of Metaphysics*, p. 306. See also Chapter 9, below, pp. 378-383, and 404.

[109] R. G. Collingwood, *The Idea of History*, p. 309.

[110] R. G. Collingwood, "The Philosophy of History," in Collingwood's *Essays in the Philosophy of History*, ed. William Debbins, p. 136.

[111] L. Rubinoff, *Collingwood and the Reform of Metaphysics*, p. 302. See also R. G. Collingwood, *The Idea of History*, p. 226.

A man is what he knows himself to be capable of doing, and he knows this through historical knowledge of what man has done in the past. Man's being is therefore his history.

> Knowing yourself means knowing, first, what it is to be a man; secondly, knowing what it is to be the kind of man you are; and thirdly, knowing what it is to be the man *you* are and nobody else is. Knowing yourself means knowing what you can do; and since nobody knows what he can do until he tries, the only clue to what man can do is what man has done. The value of history then is that it teaches us what man has done and thus what man is.[112]

> . . . since the questioning which is the mind's knowledge of itself, and through which the pure state of *speculum speculi* is achieved, is essentially an historical activity, history is therefore the true science of mind and the true medium of human self-making.[113]

This self-making activity is an explication of the immanence of the transcendental conditions of the possibility of all being and knowing, which is what is religiously called God and philosophically called the absolute. Historical activity is therefore immanent-transcendent activity.

> God or the absolute (or however else you wish to describe the transcendental *a priori* ground of the possibility of all being and knowing) is immanent as well as transcendent. Thus the activity of finite mind consists (when it raises itself to the level of science) of an attempt to render explicit the implicit self-identity of absolute knowledge. This activity is precisely what Collingwood means by self-making.[114]

The historical process therefore is radically different from all natural processes. It is a self-making process "in which man creates for himself this or that kind of human nature by recreating in his own thought the past to which he is heir."[115] Such an inheritance is possessed not by passive reception of natural transmission but by an active self-giving process which is identical with the very thought by which it is possessed.

> There is not, first, a special kind of process, the historical process, and then a special way of knowing this, namely historical thought. The historical process is itself a process of thought, and it exists only insofar as the minds which are parts of it know themselves for parts of it. By historical thinking, the mind whose self-knowledge is history

---

[112] R. G. Collingwood, *The Idea of History*, p. 10. See also Chapter 9, below, p. 418 et seq.

[113] L. Rubinoff, *Collingwood and the Reform of Metaphysics*, p. 301.

[114] *Ibid.*, pp. 300-301.

[115] R. G. Collingwood, *The Idea of History*, p. 226.

> not only discovers within itself those powers of which historical thought reveals the possession, but actually develops those powers from a latent to an actual state, brings them into effective existence.[116]

This process of self-creativity, identified with man's historical, self-endowed nature, is not merely a barely factual self-knowledge but a critical self-knowledge. It consists in knowing also what one is to be in addition to knowing what one is.

> Self-creation, moreover, is self-knowledge; not the self-knowledge of introspection, not the examination of the self that *is*, but an examination of the self that is *to be*—which is, of course, the very criterion according to which the self makes itself.[117]

Rubinoff points out that what makes the mind's acts historical according to Collingwood is not just that they happen in time but that they become known through a re-thinking of the thought which created the situation being investigated and thereby coming to understand the situation.[118]

The self-making process of historical existence is thus the active and conscious appropriation of the past. The possibility of such appropriation, that is, the possibility of historical knowledge, makes creative self-making possible.[119]

> Since mind is what it does, and human nature, if it is a name for anything real, is only a name for human activities, this acquisition or ability to perform determinate operations is the acquisition of a determinate human nature. Thus the historical process is a process in which man creates for himself this or that kind of human nature by recreating in his own thought the past to which he is heir.[120]

Mind is not something presupposed by history, for history is the very life which the mind is, and the mind is not a mind except in as much as it together lives the historical process and in the living of it knowing itself to be doing what it is doing.[121] The mind is not a substance underlying its acts; to know its acts is to know what it itself is.

---

[116] *Ibid.*, p. 226. See also pp. 215, 228; also L. Rubinoff, *Collingwood and the Reform of Metaphysics*, pp. 303, 308.

[117] L. Rubinoff, *Collingwood and the Reform of Metaphysics*, p. 301.

[118] See *ibid*. See also R. G. Collingwood, *The Idea of History*, p. 218.

[119] See L. Rubinoff, *Collingwood and the Reform of Metaphysics*, p. 301.

[120] R. G. Collingwood, *The Idea of History*, p. 226. See also L. Rubinoff, *op. cit.*, p. 302.

[121] See R. G. Collingwood, *The Idea of History*, p. 227.

> But any study of mind is a study of its activities; if we try to think of a mind absolutely at rest, we are compelled to admit that if it existed at all (which is more than doubtful) at least we should be quite unable to study it.[122]

The difference between what we can and cannot know historically defines the power of a man's mind.

> It may thus be said that historical inquiry reveals to the historian the powers of his own mind. Since all he can know historically is thoughts that he can re-think for himself, the fact of his coming to know them shows him that his mind is able (or by the very effort of studying them has become able) to think in these ways. And conversely, whenever he finds certain historical matters unintelligible, he has discovered a limitation of his own mind; he has discovered that there are certain ways in which he is not, or not yet, able to think.[123]

How this knowledge of myself and my capabilities is achieved through historical inquiry is thus explained by Collingwood:

> The only way in which I can know my own mind is by performing some mental act or other and then considering what the act is that I have performed. If I want to know what I think about a certain subject, I try to put my ideas about it in order, on paper or otherwise; and then, having thus arranged and formulated them, I can study the result as an historical document and see what my ideas were when I did that piece of thinking: if I am dissatisfied with them, I can do it over again. If I want to know what powers my mind possesses as yet unexplored, for example, whether I can write poetry, I must try to write some and see whether it strikes me and others as being the real thing. If I want to know whether I am as good a man as I hope, or as bad as I fear, I must examine acts that I have done, and understand what they really were: or else go and do some fresh acts and then examine those. All these inquiries are historical. They proceed by studying accomplished facts, ideas that I have thought out and expressed, acts that I have done. On what I have only begun and am still doing, no judgement can as yet be passed.[124]

Historical rethinking is not merely rethinking but evaluative rethinking, that is, rethinking critically by passing a judgment upon it in terms of the new context of the historian's own mind in which that thought is rethought.

> The historian not only re-enacts past thought, he reenacts it in the context of his own knowledge and therefore, in re-enacting it,

---

[122] *Ibid.*, p. 221.
[123] *Ibid.*, p. 218.
[124] *Ibid.*, p. 219.

> criticizes it, forms his own judgment of its value, corrects whatever errors he can discern in it. This criticism of the thought whose history he traces is not something secondary to tracing the history of it. It is an indispensable condition of the historical knowledge itself. Nothing could be a completer error concerning the history of thought than to suppose that the historian as such merely ascertains 'what so-and-so thought', leaving it to some one else to decide 'whether it was true'. All thinking is critical thinking; the thought which re-enacts past thoughts, therefore, criticizes them in re-enacting them.[125]

There is a standard operative in every creative activity by which a man strives to complete his human nature, which is never something fixed. This standard is, in philosophical terms, absolute truth or absolute self-knowledge, or self-knowledge of the absolute. This standard appears to the scientific consciousness under the form of reality and to the religious consciousness under the form of God or the Holy, and to the artistic consciousness under the form of the beautiful.[126] This standard is made explicit when the self brings it into existence by questioning the grounds of its own being.[127]

Reflective acts, that is, acts we do on purpose, which are the subject matter of history,[128] are acts done with knowledge of what we are trying to do, and when done can be compared with their standard, the initial conception of the act, and evaluated according to their conformity with that standard pre-conception. Reflective activity is therefore activity we can perform by knowing in advance how to perform it.[129] Historical judgment is therefore self-judgment. "It is the historian himself who stands at the bar of judgment, and there reveals his own mind in its strength and weakness, its virtues and its vices."[130]

Furthermore, the historian does not employ a criterion other than himself. His criterion is himself understood to include everything he knows.

> The explicit criterion of mature historical thought is nothing but historical fact itself: the historian asks himself, 'Does this fit in with everything I know about the world of facts, the nature of the case, the liability of the informant to error and mendacity, and so forth?"[131]

---

[125] *Ibid.*, pp. 215-216.

[126] See L. Rubinoff, *Collingwood and the Reform of Metaphysics*, p. 301. See also R. G. Collingwood, *Speculum Mentis*, p. 66, p. 90, p. 120.

[127] See *ibid.*, p. 301.

[128] See R. G. Collingwood, *The Idea of History*, p. 309.

[129] See *ibid.*, p. 308.

[130] *Ibid.*, p. 219.

[131] R. G. Collingwood, *Speculum Mentis*, p. 214. H.-I. Marrou in *The Meaning of History*, p. 161, states that the historian brings with him a philosophy of man and

Thus, to know Plato's philosophy I must re-think it in my own mind in which I am thinking also the thoughts which provide a context in which I can judge that thought of Plato which I re-think. In becoming recontextualized in the context of other thoughts which also exist in my mind I am able to judge it in the light of those other thoughts.[132]

But in becoming my thought, without losing its objectivity, it is also not merely a self-consciousness but a self-knowledge, that is, a critical study of my own thought, not merely an awareness that I am thinking.

> Thus the act of thought in becoming subjective does not cease to be objective; it is the object of a self-knowledge which differs from mere consciousness in being self-consciousness or awareness, and differs from being mere self-consciousness in being self-knowledge: the critical study of one's own thought, not the mere awareness of that thought as one's own.[133]

J. The Three Ontological Levels of Consciousness and Their Logical Correlatives

Lionel Rubinoff, with profound insight into Collingwood's

of life. "All our ideas about man are instruments whereby we strive to recapture the human past, and are related to a philosophy of man." Again (p. 246): ". . . there is no true history independent of a philosophy of man and of life. . . . The truth of history is a function of the truth of the philosophy used by the historian." In this context, W. H. Walsh, "R. G. Collingwood's Philosophy of History," *Philosophy*, XXII (1947), p. 160, writes: ". . . we can say with Collingwood that history does give us an understanding of what human beings are and can do. But it is questionable, all the same, whether we ought to see it as the science of human nature (or all that that science can be). The judgments of history, as Collingwood rightly insists, are individual judgments; but unless I am greatly mistaken, there are presupposed in the making of them, certain universal judgments, and it is these, only part of the raw material of which is provided by history, which constitute the science in question. In this connection, the view held at one time by Dilthey is of considerable interest, that behind all the human studies (history . . . sociology, etc.) there lies a more fundamental study concerned with human nature." Theodore Litt, in "The Universal in the Structure of Historical Knowledge," in R. Klibansky's and H. J. Paton's *Philosophy and History*, p. 134, writes: ". . . in the . . . case [of the investigator of the spiritual world as contrasted with that of the investigator of the natural world], and in it *alone* what the thinking mind presupposes forms at the same time a part of what it wants to investigate. The 'spiritual world' whose nature the investigator would like to determine in universal statements includes *in itself* all those spiritual actions, and abilities which are presupposed, and that not as merely incidental motives, which can perhaps be dispensed with, but as quite fundamental and ever-present functions of spiritual being and doing." Chapter 9, below, will investigate in depth the type of understanding appropriate to history and human facts as distinct from the type of understanding in which the natural sciences trade.

[132] See R. G. Collingwood, *The Idea of History*, p. 301.

[133] *Ibid.*, p. 292.

thought, explains how the dialectical development of the mind is a movement from the implicit to the explicit and how this progressive explicitation manifests itself in three distinct ontological levels which are correlated with distinct logical attitudes and therefore highly relevant to the understanding of Collingwood's logic of questioning.

J.i. The Mind's Dialectical Development as an Explication of the Implicit

The dominating concept of Collingwood's *rapprochement* logic is that of identity in difference. In his first work, *Religion and Philosophy,* Collingwood argued for the identity between religion, philosophy and history because they are all about the same object: the real world, the totality of existence, or historical fact. *Speculum Mentis* reaffirms this same identity, but in a way which more adequately accounts for the differences between the various standpoints. Whereas *Religion and Philosophy* treated the various forms of experience in too abstract a manner, as univocal expressions of the single act of knowing, as differentials of a univocal genus, *Speculum Mentis* departs ''from the more abstractly conceived system of univocal relations which tends to characterize the standpoint of *Religion and Philosophy,*'' and ''gives way to a dialectical system according to which the various sciences form a logical hierarchy or scale of overlapping forms [which] reflects . . . the dialectical growth of consciousness itself.''[134] Collingwood acknowledges that in *Religion and Philosophy* ''the identity of the forms of experience is too 'abstract' and does not account for the important 'concrete' distinctions which exist between them.''[135]

This error is overcome by recognition and application of the principle of the distinction between the explicit and the implicit referred to above. If one is unable to give a rational account of one's own, or one's observation of another's, experience its principles are said to be implicit or ''unconscious.'' Thus, the primitive world view is made explicit, or explained, when one finds out that for the primitive mind ''all things are full of gods.'' However, before explication by scientific inquiry, this belief is already implicit in the experience in question. In terms of Collingwood's later philosophy this means that an historian or observer can give an account of a given form of experi-

---

[134] L. Rubinoff, *Collingwood and the Reform of Metaphysics*, p. 50. See also A. Shalom, *R. G. Collingwood, Philosophe et historien*, pp. 11-12, pp. 16-20, pp. 22-24.

[135] R. G. Collingwood, *Speculum Mentis*, p. 108. See also L. Rubinoff, *Collingwood and the Reform of Metaphysics*, p. 51.

ence only in terms of the absolute presuppositions which are implicit in that experience.[136]

This means that the various standpoints of art, religion, science, history and philosophy are naturally ordered in such a way that each is the explicitation of the principles or presuppositions of the one prior to it on the scale, and implicitly embodies the principles or presuppositions of the one subsequent to it. Thus, religion explicitates what is implicit in art, and history explicitates what is implicit in science, etc.[137]

> Implicitly then, all of the sciences are about the same object. In this sense they are identical. This identity, however, becomes explicit only through the logical and phenomenological development of consciousness in the course of which philosophy gradually emerges. Philosophy is therefore not just one among the variety of sciences but the ground of the unity of this variety.[138]

Rubinoff argues convincingly that Collingwood's works are instances of the distinction between implicit and explicit.[139]

For Collingwood, the answer to the question 'what is beauty?' presupposes the answer to the question 'what is mind?,' and, granted certain presuppositions about the nature of mind, such as that it is identical with its acts, and that it makes itself through its acts, that is, that it undergoes dialectical development, it follows that reality and truth are subject to the same dialectical self-making.[140] Mind is thus not something distinct from its appearances, but rather it is those appearances inasmuch as they are not univocally but dialectically related according to a scale of forms which presents one form of knowledge as it grows out of another as a result of internal strains deriving from innate inconsistences in the earlier form.[141]

J.ii. The *Ab Extra* and *Ab Intra* Moments of Philosophy's Criticism of Experience from the Absolute Standpoint

Philosophy is an absolute standpoint which views all the standpoints as interrelated by internal, dialectical relations.[142]

---

[136] See L. Rubinoff, *Collingwood and the Reform of Metaphysics*, p. 52. See also Chapter 5, above, pp. 200-202.

[137] See L. Rubinoff, *Collingwood and the Reform of Metaphysics*, p. 53. Refer also to the clockface illustration of Chapter 3, above, pp. 71-72; see also Chapter 4, above, pp. 88-90.

[138] L. Rubinoff, *Collingwood and the Reform of Metaphysics*, p. 53.

[139] See *ibid.*, p. 53.

[140] See *ibid.*, p. 54.

[141] See *ibid.*, pp. 53-54.

[142] See *ibid.*, pp. 53-54. See also above, Chapter 3, the clockface illustration, pp. 71-72.

Rubinoff explains[143] that the philosophical master-criticism of experience is effected in two ways. One way is as a form of internal criticism, in which philosophy assumes the standpoint of a particular point of view, such as religion or art, and reveals and defends the presuppositions of, and thus justifies the validity of, that standpoint. This is called *ab intra* criticism. The other form of philosophical criticism is *ab extra*, or from the absolute standpoint, that is, according to criteria that transcend every particular standpoint.[144] Dogmatism is the absolutization of any particular standpoint, and dogmatic philosophy is philosophy degraded to the status of a particular standpoint among other particular standpoints and thus conceived "as the science of criticism," rather than a standpoint that transcends, and includes within its own, all other standpoints.[145] But this dogmatic "claim to truth of an erroneous standpoint is a necessary stage in the therapeutic process which begins with the discovery of an error and ends with its correction."[146]

The critic has a double task. As historian he must disclose the presuppositions lying at the basis of any particular standpoint. As philosopher, at the same time, he must judge and reveal the error at the basis of those presuppositions, which disclosure becomes the basis of a new standpoint which will be more satisfactory in terms of its ability to explain experience.[147]

> It is the very essence of a dialectical development that each phase in it should contain the next implicitly, and it is this implicit presence of elements which are, as it were, submerged in the immediacy of a particular phase. Every phase of experience is implicit in its predecessor, and therefore it is not surprising that science should be implicit in religion; but the scientific content *in tabu* and the like is felt not as science but as religion.[148]

Rubinoff points out that dialectical criticism, like metaphysics, is both historical and the medium by which the mind makes itself. Thus,

---

[143] See his *Collingwood and the Reform of Metaphysics*, p. 60.

[144] *Ibid.*, pp. 60-61. See also p. 63: "The above distinction between *ab intra* and *ab extra* is offered in part as a more adequate rendering of a distinction, which Collingwood himself introduces in *Speculum Mentis*, between the crude but popular conception of dogmatic philosophy and the more sophisticated conception of critical philosophy (SM, 254; see also EPM, 217-20). Dogmatic philosophy, writes Collingwood, is generally regarded 'as the procedure of thought without inquiry into its own powers,' while critical philosophy is 'the investigation by thought itself of the limitations of its capacity' (SM, 254)."

[145] See L. Rubinoff, *Collingwood and the Reform of Metaphysics*, p. 61.

[146] *Ibid.*

[147] See *ibid.*, p. 62.

[148] R. G. Collingwood, *Speculum Mentis*, p. 164.

dialectical history is a dialectical drama manifesting a kind of dramatic inevitability, which is a form of creative rationality, and not to be confused with either logical deduction or metaphysical determinism. Dialectical history both recounts past errors and, in doing so, shows that each error has made its contribution to present day knowledge.[149]

J.iii. The Ontological Levels of *Ab Intra* Criticism and Their Logical Correlatives

Rubinoff shows that superimposed on the aforesaid methodological distinction of *ab intra* and *ab extra* inquiry is the distinction of the operation of the former on three distinct but related ontological levels, or "soundings" in the terminology of *The New Leviathan*.[150]

> These levels are not simply three self-contained types like the co-ordinate species of a genus, nor are they three pigeon holes into which experiences may be conveniently sorted. They are, on the contrary, a scale of dynamically changing and overlapping forms, having a natural order of their own and subject to a special set of rules. Experience, in other words, undergoes a logical and dialectical development not only from one form of experience to another (from art to religion, for example) but also, *within* each of the five major forms of experience, from one level to another.[151]

Each of these three *ontological* levels, on which each of the five forms of experience may exist, correlates with a definite *logical* attitude, and so is of great importance to understand Collingwood's dialectical logic of *rapprochement*.

The first ontological level on which any form of experience is realized is unequivocal and dogmatic regarding its interpretation of itself. Two presuppositions characterize first level dogmatism: the first is that each form is fixed and self-identical, the second is that that particular viewpoint represents an absolute viewpoint.

> At the first ontological level each of the particular forms makes an unequivocal and dogmatic interpretation of its own essential nature . . . First-level dogmatism rests on two fundamental presuppositions. The first is that each form has a fixed and given self-identical nature. The second is that the particular conception of reality implied by any given standpoint is the only true and valid one, which itself presupposes that there is only one true view of reality (SM, 41). The theoretical moment of each standpoint, in other

---

[149] See L. Rubinoff, *Collingwood and the Reform of Metaphysics*, pp. 65-66.
[150] See *ibid*., p. 66.
[151] *Ibid*.

> words, presents itself not simply as a dogmatic *Weltanschauung* but as a *Weltanschauungslehre*. Thus, for example, not only does art assert itself as pure imagination, but the qualities which define the objects of imagination are extended to the entire world of reality. It is not surprising, then, that such philosophies as exist at this level tend to assume an attitude of total indifference, if not intolerance, to any other interpretation of reality.[152]

The logical attitude characteristic of first level dogmatism is that of *equivocity*. Each standpoint recognizes itself only and allows nothing in common with any other standpoint. Two standpoints, as far as first level dogmatism is concerned, are related purely equivocally.

In second level dogmatism, which arises as a result of the dialectical interplay between *ab intra* and *ab extra* approaches, any particular standpoint relinquishes its exclusiveness, and recognizes that it is one among a number of different standpoints.

> As a consequence of the dialectical interplay between the *ab intra* and *ab extra* approaches, however, consciousness is forced to the recognition that the absolute presuppositions of the first level of criticism are inconsistent. Once it is realized that each standpoint, as defined at the first level, rests upon a false distinction, consciousness is obliged to redefine the nature of each standpoint and start anew. Each standpoint will now conceive of itself as only one of a number of equally valid standpoints and proceed to define itself accordingly. Thus art, for example, which at the first level posits itself as the only fundamental mode of existence, co-exists at the second level with other forms whose validity and claims to knowledge it no longer denies. Likewise, each of the other forms conceives of itself as one of a number of equally valid and true standpoints, which arise as a result of the various ways in which mind views what is essentially the same object. At the same time each form recognizes that it has a history which to some extent exemplifies the influence which the other forms have had upon it.[153]

Comparing this second with the subsequent, third ontological level of consciousness, there are two differences, namely, submission to the subject-object dichotomy, and acquiescence in abstract, propositional logic, in which a particular form of experience regards itself as one species among other species, united as diverse species of one abstract, univocal genus. *Genus-species, propositional logic is therefore correlated with Rubinoff's second ontological level of consciousness.*

---

[152] *Ibid.*, pp. 67-68.
[153] *Ibid.*, p. 68.

> But the philosophy or reflective-theoretical moment which seeks to justify this point of view is characterized by two peculiarities which set it apart from the third level. In the first place, at the second level, consciousness has not yet overcome the subject-object distinction—the negation of which is the differentia of absolute philosophy. It therefore continues to posit the object of each form as an independently existing entity. In the second place, second-level consciousness treats each form as though it were the species of a genus, and any overlap which exists is therefore treated as a mere overlap of extension between classes—that is, the forms are externally related through the simple mediation of the universal genus.[154]

Self-transcendence as a result of dialectical interplay between *ab intra* and *ab extra* consciousness results in *the third ontological level of experience* in which *rapprochement* between the forms is accomplished, and the subject-object dichotomy, characteristic of realism and propositional logic, is transcended. This is the truly philosophical level in which each form constituting the life of the mind is seen to be, not the species of a genus, but an overlap of intension between concepts in dialectical, internal relationship one with the others.

> . . . again, as a result of the dialectical interplay between the *ab intra* and the *ab extra*, consciousness transcends itself. The relation between subject and object becomes one of dialectical identity while the relation among the forms becomes one of rapprochement. The result is the discovery of a new and higher level of existence in which philosophy treats each form not as the species of a genus, but as a member of a scale of forms whose overlap is an overlap of intension between concepts or categories rather than an overlap of extension between classes. As Collingwood himself puts it in *An Essay on Philosophical Method*, each category in its degree specifies its "generic" essence, but each embodies it more adequately than the one below. Or, to put it another way, according to the logic of the overlap of classes the generic essence is immanent in or identical with the variable element.[155]

Rubinoff cites Collingwood in describing this principle of *rapprochement* philosophy.

> . . . our five forms of experience are not five abstractly self-identical types of event which, by their recurrence in a fixed or changing order, constitute human experience; but types whose recurrence perpetually modifies them, so that they shade off into one another and give rise to new determinations at every turn.[156]

---

[154] *Ibid*.

[155] *Ibid*., pp. 68-69. See also R. G. Collingwood, *An Essay on Philosophical Method*, pp. 92-103.

[156] R. G. Collingwood, *Speculum Mentis*, p. 86.

It is only at this third level, says Rubinoff,[157] that for the first time there is introduced the possibility of a genuine philosophy of history, mind, and nature, etc. And it is only at this level that are transcended both the prejudice and intolerance of vicious (first level) dogmatism, and the pseudo liberal-humanism of "toleration," for genuine community of intersubjectivity.[158]

The logical attitude to which this third ontological level corresponds is that of analogical thinking. Thus, says Rubinoff,[159] for Aristotle, being *qua* being is to be understood as being *qua* modes of being and the categories are consequently to be understood as the various modes through which being expresses itself. Substance, Aristotle's highest category, is itself a scale of forms which leads to the notion of a highest substance. Being thus is not a genus, not an abstract, univocal class concept.

> Since being is not a genus, the categories are not species each of which expresses the genus in precisely the same way; nor is substance a genus, dispersing itself into a plurality of substances. On the contrary the categories are a scale of forms each of which expresses the essence of being more adequately than the one below.[160]

The basic problem which Aristotle faces is the problem of relating the 'manyness' of being to its 'unity.' He solves it by speaking of the many as unified through relation, each in its own way, to one central point. The term being not only extensively includes the group of beings as a whole, as does a univocal, abstract class concept, but also refers intensively to what each of the members of that group have in common.[161]

Collingwood's solution to Aristotle's problem, following Hegel, is, according to Rubinoff, only through the realization that being or reality is not some permanent and eternal substance, but rather a constantly changing activity which, as Hegel said, is nothing other than mind knowing itself. Thus, Collingwood locates the object of knowledge (being) within the mind. Being remains, as it was for Aristotle, a plurality of activities related (each in its own way) to the highest being. But the highest being is not, as it was for Aristotle, a separately existing, eternal substance, but, on the contrary, an immanent-

---

157 *Collingwood and the Reform of Metaphysics*, p. 69.
158 See *ibid.*, p. 69.
159 See *ibid.*, p. 188.
160 *Ibid*.
161 See *ibid.*, p. 189.

transcendent activity which Collingwood calls absolute mind,[162] whose absolute standpoint, having as its object all other standpoints together with its own, is the object of philosophy.

In his earlier *Ruskin's Philosophy* Collingwood contrasts the logicist method (the classificatory method of abstract science) with the synthesizing historicist method in terms of what he calls the technique of "analogy," which he regards as the only appropriate one for studying the relations between the various activities of mind.[163]

> This analogical method of reasoning . . . is a weapon of immense power, clearing the ground of unnecessary argument and accompanying a vast amount of varied work with the least possible waste of energy.[164]

According to Rubinoff, Collingwood's method of analogy contrasts both with the univocal attitude towards the mind's activities, which attitude believes that there is no distinction between the faculties, and with the equivocal attitude towards them, which attitude believes, on the contrary, that the mind's activities are a set of equivocals proceeding from different faculties.[165]

This distinction, implicit in Collingwood's works, and brought to explicitness by Rubinoff, is necessary to achieve *rapprochement* between Collingwood's earlier and later writings, which otherwise give the appearance of insoluble antinomies.

> This threefold distinction—which is crucial not only for the purpose of reconstructing and understanding Collingwood's thought but also for establishing a continuity between his early and later writings—is not one which is explicitly formulated by Collingwood himself. It is justified, however, not only by implication but by certain distinctions which Collingwood does make in various places for the purpose of elucidating his subject matter.[166]

K. The Archaeology of Roman Britain and the Development of Collingwood's Logic of Questioning

Collingwood's philosophical development was intimately tied to

---

[162] In the clockface illustration of Chapter 3, above, pp. 71-72, the absolute viewpoint of philosophy is represented by the view of one situated at the centre of radiation of the hands of the clock, and able to recognise the clock *as a clock*, together with all its hand positions at once.

[163] See L. Rubinoff, *Collingwood and the Reform of Metaphysics*, p. 228.

[164] R. G. Collingwood, "Ruskin's Philosophy," in Collingwood's *Essays in the Philosophy of Art*, ed. A. Donagan, p. 32.

[165] See L. Rubinoff, *Collingwood and the Reform of Metaphysics*, p. 228. See also R. G. Collingwood, "Ruskin's Philosophy," in Collingwood's *Essays in the Philosophy of Art*, ed. A. Donagan, pp. 30-31.

[166] L. Rubinoff, *Collingwood and the Reform of Metaphysics*, p. 69.

his constant engagement in historical studies,[167] and, conversely, his philosophical convictions greatly influenced his methods of procedure in his historical researches, and, in the opinion of colleagues, not always to the best advantage.[168] He specialized in archaeology, specifically the archaeology of Roman Britain, and after the war of 1914-1918 he was the only surviving pupil of his master Haverfield and the only Oxford trained Roman Britain specialist. In 1921 he wrote, in two days, a short elementary book on *Roman Britain*, about which he said that

> it served to lay down once for all my general attitude towards problems, and, even more important, my general conception (partly due to Haverfield, but partly different from his) of what the problems were; it gave me a first opportunity of finding out, more clearly than was possible within the limits of a short article, how my conception of historical research was developing . . .[169]

An enlarged edition of this work appeared ten years later (1931), which had to be revised again in 1934, the same year he wrote the British section in Tenney Frank's *Economic Survey of Ancient Rome*. In 1935 he wrote the sections on prehistoric Britain for the *Oxford History of England*. The invitation to write these large scale volumes gave him the welcome and timely opportunity to exchange the field laboratory for study, arrangement and publication of what he had hitherto learned as a result of all this archaeological and historical work regarding the philosophy of history. He says that a full length book on Roman Britain "would serve to display in a concrete form the principles of historical thinking as I now understood them."[170] Long practice in archaeology taught Collingwood that the most important condition of success is that the person responsible for the digging should know exactly why he is doing it. The central principle of Collingwood's logic of question and answer as applicable to archaeology is that the director of the digging "must first decide what he wants to find out, and then decide what kind of digging will show it to him."[171] Collingwood

---

[167] R. G. Collingwood, *An Autobiography*, p. 120.

[168] See I. A. Richmond, "An Appreciation of R. G. Collingwood as an Archaeologist", (accompanying obituary notice), *Proceedings of the British Academy*, XXIX (1943), p. 476: "Here . . . emerged in rather crude form a tendency, which marked and sometimes marred his work, to drive the evidence hard and to build upon it a series of conclusions whose very artistry disguised the inherent weakness of foundation."

[169] R. G. Collingwood, *An Autobiography*, pp. 120-121.

[170] *Ibid.*, p. 121.

[171] *Ibid.*, p. 122. It was the implementation of this principle which drew upon Collingwood the criticism of I. A. Richmond, as noted immediately above (n. 168). On pp. 478-479, *art. cit.*, Richmond comments as follows: ". . . Collingwood's powers of analysis and appreciation outshone his ability in field work. His attitude to exca-

criticises early archaeology in which digging was done blindly with no definite question formulated whose answer was being sought. He severely criticized the choosing of excavation sites for purposes other than that it contains the solution to a burning problem, such as choices made which would attract financial contributions to finance digging operations. Historians themselves needed to be made aware of the fecundity of the Baconian revolution in their own field.

> If historical studies were to pass through a Baconian revolution—the revolution which converts a blind and random study into one where definite questions are asked and definite answers insisted upon—the first thing to be done was to preach that revolution among historians themselves.[172]

Some, but very little, progress in this regard had been made when Collingwood began his study of Roman Britain, but Haverfield, under whose guidance Collingwood was trained, and his colleagues had fully implemented the Baconian method of systematic interrogation in their diggings with great success at comparatively little cost.

> Haverfield and his colleagues of the Cumberland Excavation Committee in the eighteen nineties had been *consciously and completely Baconian in their methods*. They never dug a trench without knowing exactly what information they were looking for; they knew *both* that this information was the next thing they needed for the progress of their study, and also that this trench would give it them. That is why they could settle highly intricate and abstruse problems at a cost of never more, and often much less, than thirty or forty pounds a year.[173]

Collingwood criticises excavations done according to the principles of General Pitt-Rivers during the last quarter of the nineteenth century.

---

vation was profoundly influenced by the selective method of excavation, introduced and perfected by F. G. Simpson on Hadrian's Wall: and the first lesson which he drew from these methods, that excavations should be conducted with specific problems in mind upon sites likely to provide an answer, was salutary and useful. But Collingwood's corollary, that to pose a problem permitted its answer to be predicted, was a product of the study rather than the field. For there are problems thus soluble, particularly those purely philosophical problems which it was Collingwood's daily task to consider. But to the field-worker excavation, no matter how carefully planned in advance, is always a plunge in the dark: and, while problems under consideration form a more regular pattern on Hadrian's Wall than on most Roman monuments, even there no excavation has ever followed a preconceived course: the expected is always accompanied, and often overshadowed, by the unexpected: and while Collingwood's favourite *dictum*, that 'what you are not looking for you do not see,' is of wide application, an excavator's first duty is to see everything without the blinkers, imposed by prearranged concentration of vision." Refer to Chapter 7, below, pp. 308-309.

[172] R. G. Collingwood, *An Autobiography*, p. 124.

[173] *Ibid*.

He acknowledges Pitt-Rivers as "a very great archaeologist and a supreme master in the technique of excavation" but severely criticises the pre-Baconian stage of his problem-solving methods, for "he dug in order to see what he could find out. He had not applied to archaeology the famous advice of Lord Acton, 'study problems, not periods.' "[174]

In addition to this first principle which should guide historical and archaeological research, that the whole work should be the solving of a definite problem guided by definite and methodical questioning, Collingwood adds a second, to the effect that history does not study events but actions which express an agent's intention.

> Since history proper is the history of thought, there are no mere 'events' in history: what is miscalled an 'event' is really an action, and expresses some thought (intention, purpose) of its agent; the historian's business is therefore to identify this thought.[175]

Applied to archaeological work, this means that all objects must be interpreted in terms of purposes, and that the question to be asked about any object found is 'What is it for?', followed by the question 'Was it good or bad for that purpose?', that is, 'Does this object successfully or unsuccessfully embody that purpose?' Such questions, being historical questions, can be answered not by guesswork but only according to historical evidence, which means that anyone who offers an answer to those questions must be able to show that his answer is the one which the evidence demands.[176]

In an article, "The Purpose of the Roman Wall,"[177] Collingwood opposed the currently accepted vague description of the wall as a frontier defense, maintaining that the more precise explanation is that it was an elevated sentry-walk. He reached this conclusion by asking such questions as 'How did it work?' 'Was it intended to work like a town-wall from the top of which defenders could repel attacks?' The question-answer procedure involved is as follows:

> A question answered causes another question to arise. If the Wall was a sentry-walk, elevated from the ground and provided (no doubt) with a parapet to protect the sentries against sniping, the same sentry-walk must have continued down the Cumberland

---

[174] *Ibid.*, p. 125. Pitt-Rivers' methods were continued and developed by Sir Mortimer Wheeler. See Glyn Daniel, *Man Discovers His Past*, N.Y., Cromwell, 1968, p. 70, pp. 74-75. Also see Mortimer Wheeler's *Archaeology from the Earth*, London, Pelican Books, 1956.

[175] R. G. Collingwood, *An Autobiography*, pp. 127-128.

[176] See *ibid.*, p. 128.

[177] In *The Vasculum*, VIII, 1, pp. 4-9.

> coast, beyond Bowness-on-Solway, in order to keep watch on vessels moving in the estuary; for it would have been very easy for raiders to sail across and land at any unguarded point between Bowness and St. Bee's Head. But here the sentry walk need not be elevated, for sniping was not to be feared. There ought, therefore, to be a chain of towers, not connected by a wall but otherwise resembling those on the Wall, stretching down that coast. The question was, did such towers exist?[178]

The evidence with which Collingwood alleges he was able to supply that answer was provided by a ''search in old archaeological publications'' which ''showed that towers of exactly the right kind had been found; but their existence had been forgotten, as generally happens with things whose purpose is not understood.''[179]

Not merely archaeology, but every kind of history operates on the principle of getting to the thought behind the deed and not merely stopping at the event. The event is the deed in its externals, that is, as described 'from the outside,' whereas the thought is the 'inside' of the deed, the purposefulness of the activity which it is the historian's business to penetrate and disclose. Written sources testify to the invasion of Britain by Julius Caesar on two successive years. These are 'events.' Collingwood criticizes historians for scarcely asking 'Why did he do it?', which question relevantly arises given that information. Comparison of the strength of Caesar's expeditionary force with that sent over by Claudius nearly a century before settled for Collingwood that the answer to that question was that Caesar did not intend merely a punitive expedition but complete conquest of the country. Caesar's silence suggests that he intended concealment of his purpose since he failed to achieve it. Consistently with what he said earlier regarding Villeneuve's plan at the Battle of Trafalgar, that since he failed to implement it in action we had no way of knowing what it was, Collingwood admits that his answer to the question regarding Caesar's intention to send expeditionary forces to Britain may be mistaken, but ''future historians will have to reckon with the question I have raised, and either accept my answer or come up with a

---

[178] *An Autobiography*, p. 129.

[179] *Ibid*. This evidence seems to be somewhat embarrassing to the very point Collingwood is making. For the very condition upon which the question is answerable is the existence of old archaeological records, which would not have existed unless the excavators, whose work was recorded in those publications, noted not only something they sought as answers to *their own* precise questions, but, contrary to Collingwood's advice, also 'whatever may have been of interest.'

better one,''[180] since this answer is the only answer to that question which the available evidence authorises one to give.

A third principle arises from Collingwood's reflections over his years of archaeological practice, namely, that ''no historical problem should be studied without studying what I called its second-order history; that is, the history of historical thought about it,''[181] and ''just as philosophical criticism resolved itself into the history of philosophy, so historical criticism resolved itself into the history of history.''[182]

Collingwood says that archaeology especially exemplifies his historical principles, since its sources are not pre-existing narratives which could function as 'authorities,' but are what is usually referred to as 'unwritten sources,' so that here, in archaeology, ''the issue raised by the project of a Baconian revolution is unmistakable.''[183]

> When history is based on literary sources the difference between scissors-and-paste, or pre-Baconian history, where the historian merely repeats what his 'authorities' tell him, and scientific or Baconian history, where he forces his 'authorities' to answer the questions he puts to them, is not always quite clear. It becomes clear enough on occasion; for example, when he tries to get out of his 'authorities' the answer to a question which they did not expect a reader to ask (as when we try to get out of an ancient writer answers to economic and demographic questions) or when he tries to get out of them facts which they wished to conceal. On other occasions it sometimes does not leap to the eye. In archaeology, however, it is obvious. Unless the archaeologist is content merely to describe what he or someone else has found, which it is almost impossible to do without using some interpretative terms implying purpose, like 'wall', 'pottery', 'implement', 'hearth', he is practising Baconian history all the time: asking about everything he handles, 'What is this for?' and trying to see how it fitted into the context of a peculiar kind of life.[184]

Collingwood says that for this reason archaeology provides an extremely sensitive method for getting answers to questions which are not answerable by appeal to literary sources, even by the most ingenious interpretation of them. The modern historian wants to ask all kinds of questions which are basically statistical, but the sources from which he has to get his answer were written by men not statistically minded, and statistical evidence is required if you want to answer statistical questions. This kind of evidence is something the

---

[180] R. G. Collingwood, *An Autobiography*, pp. 130-131.
[181] *Ibid.*, p. 132.
[182] *Ibid.*, pp. 123-133.
[183] *Ibid.*
[184] *Ibid.*, p. 133.

archaeologist can provide when his work has reached a certain volume. In England, where Roman archaeology has unceasingly progressed ever since the seventeenth century, there is a great bulk of material from which many questions of a statistical kind can be answered, if not conclusively, at least within a reasonable margin of error.[185]

Collingwood says that, in order to get to the purpose underlying an event, the first thing is to ask oneself what can be expected as the normal development in such cases, but if the event is exceptional then exceptional conditions have to be sought.

> If you want to know why a certain kind of thing happened in a certain kind of case, you must begin by asking, 'What did you expect?' You must consider what the normal development is in cases of that kind. Only then, if the thing that happened in this case was exceptional, should you try to explain it by appeal to exceptional conditions.[186]

How the reconsideration of questions of principle rather than discovery of fresh evidence historically manifests the *rapprochement* between philosophy and history is shown clearly from the solution to a much debated problem current when Collingwood was writing the first part of *The Oxford History of England*. Why did British Celts, who adopted Roman fashions in arts and crafts and persisted in them for three centuries, revert to their old Celtic types after Roman withdrawal? How could they revert to their original Celtic tastes after three centuries of adoption of Roman tastes in these matters? Several theories prevailed; one, the tradition never really died but survived and ran concurrently with the Romanized tradition, another, some Celtic tribes escaped submission to Roman rule and retransmitted the preserved

---

[185] See *ibid.*, pp. 134-135.

[186] *Ibid.*, p. 140. But a certain inconsistency seems to be involved here, if we take into consideration what Collingwood said earlier in the *Autobiography*, p. 106, that "rules of conduct keep action at a low potential because they involved a certain blindness to the realities of the situation. If action was to be raised to a higher potential, the agent must open his eyes wider and see more clearly the situation in which he was acting." Now, here above, Collingwood tells us to organize questions in terms *of the normal* and in terms *of the exceptional* to the norm. This seems to suggest that high grade thinking presupposes low grade thinking, or thinking according to rules. For how are we to recognize the exceptional, so as to ask what are its exceptional conditions, if presupposed to this line of questioning is not had the idea of the normal, or what is in accord with rules, which Collingwood himself has stigmatized as low grade thinking? This is an indication of what will be exposed later, in Chapter 9 (see below, p. 429 et seq.) that natural law thinking, and its propositional logic, which is the principle of intelligibility in scientific thought, enjoys its own *sui generis* autonomy in relation to the dialectical unfolding of self-developing consciousness, which is meaningful in relation to the personal interests of human freedom.

culture after Roman withdrawal. A third theory was to posit a Celtic temperament which required definite conditions for its flowering, which conditions prevailed before and after but not during the Roman period.

Against the first and second opinions Collingwood argued that there was no evidence for such survival and "no historian is entitled to draw cheques in his own favour on evidence that he does not possess, however lively his hopes that it may hereafter be discovered."[187] Against the third opinion, Collingwood rejects the psychological invocation of occult entities such as Celtic temperament as a substitute for historical evidence. Collingwood concluded that difficulties in solving the problem derived probably from a misconception of the nature of the historical process. Recalling what he had discussed in his small work *Libellus de Generatione* he said that "any process involving an historical change from P_1 to P_2 leaves an unconverted residue of P_1 incapsulated within an historical state of things which superficially is altogether P_2."[188] Anticipating a rebound from the psychologists, objecting that he himself is involved in the process of introducing occult entities in speaking of incapsulation, Collingwood replies that "incapsulation is not an 'occult entity' " but "my name for such facts as this . . . that a man who changes his habits, thoughts, etc., retains in the second phase some residue of the first."[189] The desire to smoke in a man who gives up smoking survives, not in smoking, but in the form of an unsatisfied desire. If he later resumes smoking that does not prove that he never gave up smoking but perhaps that he never lost the desire, which may again be satisfied when the reasons against its satisfaction no longer prevail. Something similar happens in regard to the continuity of a tradition without implying a 'racial temperament' or 'racial unconscious' or other such 'occult entities.'

> If the members of a certain society have been in the habit of acting or thinking in certain ways, and if at a certain time they try to stop acting and thinking in those ways, and do their best to act and think in different ways, the desire to go on acting and thinking in the old way will probably persist. It will certainly persist, and persist in a lively form, if they were accustomed to think and act in those ways very effectively and found great satisfaction in doing so. The

[187] R. G. Collingwood, *An Autobiography*, p. 139.

[188] *Ibid.*, pp. 140-141.

[189] *Ibid.*, p. 141. This survival of an earlier stage in a later stage of development is what Collingwood calls the Law of Primitive Survivals in *The New Leviathan*, p. 65, par. 9.51. See also Alan Donagan, *The Later Philosophy of R. G. Collingwood*, p. 28.

tendency to revert to the old ways would in that case be very strong.[190]

Collingwood faces the objection that such a tendency would not survive into a second generation unless there were something of the kind called an occult entity, like racial temperament or inheritance of acquired characteristics, and that he is therefore talking psychology. His reply is: "I am not talking psychology, and shall not ask help from its exponents; for I regard the kind of psychology that deals with this kind of question as a sham science. I am talking history."[191] A warlike people at a certain historical crisis turns peaceful; its warlike tendencies survive in the first generation and are sternly repressed; the moral education of the children carefully points out that they must on no account indulge in the forbidden pleasures of war. The parent is asked, 'What is War?' In explaining, emphasising the moral wrongness of war, the parent makes plain that war was a great thing while it lasted and that he would love to fight his neighbors again if only he did not know that he ought not. The children thus learn not only that war is wrong but that it was something grand and pass this on to their own children.

> Thus the transmission by educational means of any moral ideal which involves the outlawry of an institution or custom, and the repression of a desire for it, entails the simultaneous transmission of that desire itself. The children of each generation are taught to want what they are taught they must not have.[192]

Where the new ways prove successful and satisfying the old ways will disappear more rapidly. But where the new ways are displayed with only a mediocre degree of success you can be sure that the discarded ways are remembered with regret and that "the tradition of their glories is being tenaciously kept alive."[193]

Applying this to the then current problem of the revival of Celtic art he was able to establish that there was in fact a connection between two well known facts hitherto not thought to be connected, namely, the badness of Romanizing British art on the one hand and the Celtic revival on the other. The applied principle also unexpectedly threw light on another puzzling fact, namely, the recognized masterpiece of the degenerate period, the famous Bath Gorgon which

---

[190] R. G. Collingwood, *An Autobiography*, p. 141.
[191] *Ibid.*, p. 143.
[192] *Ibid.*
[193] *Ibid.*

was shown to be inspired, not by the 'classical' art, but by Celtic art, and was probably not the work of Briton but of Gaul.[194]

While he recognizes that the "use of epigraphic material is a magnificent exercise for an historian just beginning to shake himself free from the scissors-and-paste mentality"[195] Collingwood says that "the epigraphic historian as such can never be wholly Baconian in spirit," for "regarded as documents, inscriptions tell you less, under critical scrutiny, than literary texts . . ."[196] He says that inscriptions themselves were of little service to his Romano-British studies[197] and "on the questions which I particularly wanted to ask it happened that inscriptions threw hardly any light."[198] Such work he regarded as building a monument to the past rather than forging a weapon for the future of historical research.[199]

L. Transition to Evaluation

The foregoing chapters have attempted to expose Collingwood's theory of the logic of question and answer as this is alleged by him to be the proper logical instrument of historical thought. It now remains to review Collingwood's position as set out above in order to evaluate it critically and try to distinguish what is acceptable from what is to be emended or rejected.

First, Collingwood's identification of the questioning and answering activity with knowledge will be critically evaluated and found to be not totally in accordance with facts, especially certain facts pertaining to scientific discovery. This will be done in Chapter 7. In this chapter we will make a preliminary evaluation of Collingwood's identification of propositional meaning with the answer to a question, but the full evaluation of this central point will be in Chapter 9.

Chapter 8 will criticize Collingwood for neglecting to give due consideration to an already well developed logic of interrogation in the Aristotelian tradition, and, in terms of the criteria of that logic, the criticism will be made that Collingwood himself proceeds by a defective order of questioning.

Chapter 9 will investigate the questioning methodology which many historians, in agreement with Collingwood, allege to be the pro-

---

[194] See *ibid.*, p. 144.
[195] *Ibid.*, p. 145.
[196] *Ibid.*, p. 146.
[197] See *ibid.*, p. 145.
[198] *Ibid.*, p. 146.
[199] See *ibid.*

cedure proper to historical thought. An important correlation will be made between the concrete universal as exposed by Collingwood and the concrete unity in diversity in human consciousness effected by linguistic communication which operates basically by a question and answer process. It will be shown, in agreement with Collingwood, that man, inasmuch as he is a being who increasingly acquires command over his self-development through free, creative, developing self-consciousness, has a history which unfolds as a process of raising and solving problems. By historical knowledge, past achievements are constantly renewed in present consciousness. By a critical, questioning activity centered on the vestiges of the human past these vestiges become portents of future progress. But Collingwood will be criticized for absorbing all human activity and all human meaning into that which is created by unfolding consciousness. It will be shown that the natural world, including man's consciousness as this is part of the natural world, in addition to being meaningful in relation to freely posited human projects, has also a meaning in its own right, and that science, inasmuch as it is concerned about this natural world, enjoys its own autonomous domain of competence without being absorbed into history. Propositional meaning will be shown to be of two kinds. One kind of propositional meaning is in relation to freely posited human projects and interests. This meaning consists precisely in the proposition's function as answering a question, which question articulates the historically determined human interest of the questioner. But it will further be shown, in opposition to Collingwood, that propositions, as saying something about the natural world, have a meaning in themselves, quite distinct from their capacity to answer questions (though not necessarily separate from that capacity).

Chapter 10, finally, will endeavour to come to grips with the logic of questioning at the basis of history in terms of trying to show how human progress in the raising and solving of problems, which basically characterizes human history, is subject to rational and logical direction. An attempt by the present author to carry Collingwood's ideas on interrogative logic some steps further is made in Appendix II.

Part II: EVALUATION

Chapter Seven

Knowledge as Question and Answer

A. Criticism of Collingwood's Attack on Realism

Collingwood's reasons for identifying knowledge with the activity of raising and answering questions are, not surprisingly, historical in origin. On his own account in his *Autobiography* this identification is his reaction to Cook Wilson's position that "knowledge makes no difference to what is known."[1] He criticizes Alexander, who recommends that metaphysicans adopt "an attitude of natural piety"[2] consequent upon Alexander's description of knowledge "as the mere 'compresence' of a mind with an object."[3] He criticizes the Positivists' attitude that "if the function of thought is to classify observed facts, there must be facts available for classification before thought can begin to operate." Thought, for them, is basically a passive submission to facts, "and once facts are available there is no need to presuppose anything. You just set to work and classify them."[4] Throughout Collingwood's works there is the overriding presupposition of the pure activity of the mind, expressed in the statement that "the mind is what

[1] R. G. Collingwood, *An Autobiography*, p. 44.
[2] R. G. Collingwood, *An Essay on Metaphysics*, p. 174.
[3] *Ibid.*, p. 177.
[4] *Ibid.*, p. 146.

it does,"[5] which is essentially tied up with the assertion that "thought is primarily practical; and only in the second place theoretical."[6] "Man's mind is made of thought,"[7] and in this activity of thinking, the questioning activity, the mind determines itself to be the kind of thing that it is. Even though "for a man about to act, the situation is his master, his oracle, his god," nevertheless "the freedom that there is in history consists in the fact that this compulsion is imposed upon the activity of human reason not by anything else, but by itself."[8]

Collingwood argues that realistic attitudes, according to which the mind in knowing would simply be submissive to an external object, which enjoys some identity in itself independently of its being known by the mind, and to which "knowledge makes no difference," spells bankruptcy in moral matters and political affairs.[9]

But Collingwood's extreme reaction to the realists' position, which maintains that "knowledge makes no difference to what is known," seems to be based on a naive and simplistic interpretation of that statement by Cook Wilson. For exactly the same reasons we could reject Collingwood's own theory of conscious reflection.

> Colour, or anger, which is no longer merely seen or felt but attended to, is still colour or anger. When we become conscious of it, it is still the very same colour and the very same anger. But the total experience of seeing or feeling it has undergone a change, and in that change what we see or feel is correspondingly changed . . . the change which Hume describes by speaking of the difference between an impression and an idea.[10]

Just as feelings attentively reflected upon remain feelings (although becoming conscious feelings), so also things known remain the things that they are, although, in knowledge, they become known. It is in this sense that knowledge makes no difference to what is known, just as becoming conscious of our feelings makes no difference to feelings inasmuch as they still remain feelings. If realism is to be rejected because it asserts that knowledge makes no difference to what is known, so likewise Collingwood's various levels of conscious reflection are to be rejected. Collingwood takes the dictum "knowledge makes no difference to what is known" as if it meant that the situa-

[5] R. G. Collingwood, *Religion and Philosophy*, p. 34; see also A. Shalom, *R. G. Collingwood, philosophe et historien*, p. 7. See also Chapter 6 above, pp. 231-238.

[6] R. G. Collingwood, *The New Leviathan*, p. 5.

[7] *Ibid*.

[8] R. G. Collingwood, *The Idea of History*, pp. 316-317.

[9] See R. G. Collingwood, *An Autobiography*, pp. 47-50.

[10] R. G. Collingwood, *The Principles of Art*, pp. 206-207.

tion of knowing leaves affairs just as they were if no knowing were done. If I know that arsenic is deadly poisonous, my knowledge is not going to make one bit of difference to the poisoning capacity of arsenic, but it will certainly make a great deal of difference to the way I behave in regard to arsenic. Thus, it will be restrained as regards its opportunities to poison, given that I know its poisoning potential.

So, knowledge both does and does not make a difference to what is known, just as the attention by which I become conscious of my feelings does and does not make a difference to those feelings. The feelings become conscious feelings and subjected to my control, but they do not thereby cease to be feelings. Arsenic still poisons just as much after I know this as it does before, but it ceases to have some opportunities of doing so. The Cook Wilson realists in asserting one aspect of knowledge are not thereby denying the other aspect, just as when Collingwood asserts that attention makes no difference to feelings (as feelings) he is not denying that consciousness makes a great deal of difference to feelings (which are transformed into domesticated, objectivized, regulated feelings). John Wild neatly expresses the matter as follows:

> The act of knowing internally relates itself to the object known. To attain such knowledge makes an important difference to the knower. But this relation is wholly external to the entity known, which suffers no real internal change from being known.[11]

In his act of knowing a knower is really related to what he knows by that act. That is, being related to the object known belongs to what it is to be an act of knowing; it is constitutive of the act in its very being an act of knowledge. But in being known, the thing known gets its relation to the act of knowledge in that act in the knower, not in its own entity. The moon is known to be a moon not in the moon itself but in the mind of an observer. The manifestness which it acquires in becoming known is not a change internal to it. A *known moon* is not a different kind of moon from a real moon. Whatever change becoming known effects, it does not affect whatever internally constitutes the moon to be a moon.[12]

B. Language Describing Sensations Presupposes Corresponding Physical Objects

Again, in *The Principles of Art*, [13] Collingwood assumes the Berke-

[11] *The Challenge of Existentialism*, Bloomington, Indiana University Press, 1955, p. 68.
[12] See Appendix 1, below, pp. 441-444.
[13] p. 169.

leyan principle 'esse est sentiri' in regard to our sensations. But when he attempts to explain the meaning of the statement 'this is my hat,' he inadvertently slips into physical-object language, describing that complex sensation not merely in terms of sensa and the relations between them, as a consistent phenomenalist should, but in terms of two physical objects, the hat and the peg in the hall on which it hangs.

> . . . instead of identifying the class of sensa he meant without referring to physical objects he located them in physical space by reference to two physical objects. No phenomenalist reduction of objects of sensation to sensa whose *esse* is conceived as *sentiri* can escape this difficulty; the class of sensa to which the objects of sensation are reduced can only be identified by reference to physical objects.[14]

Donagan explains that Collingwood confuses the inability of a sensum to be a sensum when it is not actually being sensed with its inability to be anything *at all* when it is not being sensed.

> Collingwood seems to have confounded the obvious truth that only while it is sensed does a sensum exist as a sensum with the pleasantry that when it is not being sensed it cannot exist as anything whatever. While the former is compatible with the commonsense view that the moon is your sensum (i.e. what you see) whenever you see it, the latter is not; for although the moon ceases to exist as your sensum when you stop looking at it, it does not cease to exist as the moon.[15]

Donagan does not merely repel Collingwood's anti-realist phenomenalist attack, but advances positive evidence for maintaining that the *esse* of a sensum is not reduced without remainder to its *sentiri*.[16] He says that the strongest argument for the realism of sensed objects is that every natural language speaks meaningfully of our sensations by referring them to a world of physical objects.

> Can it be shown that such things as the moon are genuinely 'sensed'? . . . The strongest argument for asserting that most of our sensa are physical objects, and *so can exist unsensed*, is that every natural language in which men speak of what they see, hear, feel, taste, and smell is a physical object language. It does not seem to be possible to think in any sustained way about more than a fraction of what you 'sense' except in terms of physical objects. The reason for believing that the greater part of what you 'sense' con-

[14] A. Donagan, *The Later Philosophy of R. G. Collingwood*, p. 35.
[15] *Ibid.*, p. 33.
[16] See *ibid*.

> sists of physical objects in a physical world is not that physical objects are data of sensation, but that the only coherent general account of what you sense that can be given is in terms of physical objects.
>
> Berkeley was unable to forestall this argument, and Collingwood fared no better.[17]

In his last work, *The New Leviathan*,[18] Collingwood asks, "Are there objects of feeling or not?" and answers, "I do not know. Nobody knows. Some have said there are, some have said there are not." He denies that an answer is possible on positive grounds, but gives a negative answer on methodological grounds. "The question is whether a theory of feeling needs objects as well as modes. The Lockean theory does, the Cartesian does not. By Occam's Razor the Cartesian theory is preferable,"[19] so that "what is felt consists simply of modes of feeling, and not of objects. . . ."[20]

But, for the reasons given immediately above, Donagan insists that feeling has not merely modes,[21] but also objects which can only be physical realities.

> Feeling has objects; and since it appears to be impossible to form any intelligible notion of what are most of the objects that anybody feels, except the common-sense one that they are parts of his body and its physical surroundings, it is reasonable to conclude that most of the objects of feeling are physical.[22]

Although this may not suffice to set realism up as an established theory of knowledge in place of Collingwood's anti-realist phenomenalism, it does point out that Collingwood's reasons against realism retroact just as unfavourably against his own theory of conscious reflection, and that the natural language used by Collingwood to express his own convictions seems to be voided of meaning in the absence of the presupposition of an objective physical world to which it refers. Granted the fundamental role which Collingwood accords to language following Hobbes,[23] and the importance he attaches

---

[17] *Ibid.*, p. 34. Emphasis added.

[18] p. 28.

[19] R. G. Collingwood, *The New Leviathan*, p. 31.

[20] A. Donagan, *The Later Philosophy of R. G. Collingwood*, p. 36.

[21] "At the level of first-order consciousness, a man will be conscious of how what is felt looks, sounds, feels, tastes and smells, and of its corresponding emotional charges. So much appears to be secure." A. Donagan, *The Later Philosophy of R. G. Collingwood*, p. 36.

[22] *Ibid.*

[23] See R. G. Collingwood, *The New Leviathan*, Ch. 6, pp. 40-46. See also Collingwood's *Principles of Art*, Ch. 11, pp. 225-269 and Alan Donagan, *The Later Philosophy of R. G. Collingwood*, pp. 40-46; 50-54.

to the ordinary usage of language,[24] this latter objection is especially serious.

C. Not All Scientific Discoveries Are Answers to Questions

Another serious objection to Collingwood's reduction of knowledge to the activity of question and answer is the fact that many genuine discoveries in science are either made without there being any deliberate searching for what was found, or something quite unexpected turns up in the search for something quite different.

Norbert Hanson[25] speaks of what he calls the 'trip-over' discovery. ''What characterizes the 'trip-over' discovery is the absence of any expectation of such a happening ever turning up at all.''[26] He discusses cases ''distinguished by their total lack of relevant anticipation—psychologically and theoretically'' in which the discoverer ''is totally unprepared for the totally unexpected—both in the psychological *and* in the theoretical sense.''[27] He cites as instances[28] first the Coelacanth, which ''turned up in 1938 when fishermen threw their nets into the sea off Madagascar in search of their next meal, and then dredged up the remarkable 'living fossil' so much discussed since''; and, secondly ''the black swan Captain Cook is reputed to have encountered in Australia . . . Cook was not searching for black swans; indeed . . . probably he would have denied the possibility of a swan being black.'' Hanson says[29] that ''these two discoveries, of the Coelacanth and of the black swan, are 'trip-over' discoveries *par excellence*. There was no reason to expect objects of this kind to materialize.'' ''It just happened that way.'' Again, ''Becquerel wasn't opening his desk drawer *looking* for an example of radioactivity, for he hadn't an inkling that there ought be such a phenomenon. He was not in a 'set of expectation' psychologically or theoretically, with respect to such a happening.''[30] Similarly ''Herschel . . . was not searching the heavens for some as yet undetected object. [His] goal was only to map the sky in detail. Uranus just 'turned up' before his telescope.''[31]

---

[24] See R. G. Collingwood, *The Principles of Art*, pp. 250-251.

[25] See his ''An Anatomy of Discovery,'' in *The Journal of Philosophy*, LXIV (1967), pp. 321-352.

[26] *Ibid.*, p. 334.

[27] *Ibid.*, p. 335. Emphasis in original.

[28] *Ibid.*

[29] *Ibid.*

[30] *Ibid.* Emphasis in original.

[31] *Ibid.*

Perhaps the best example of the 'trip-over' discoverer is Galileo, whose "initial encounters with Jupiter's satellites, with Saturn's rings, and with our moon's craters, were unprepared for both psychologically and theoretically."[32] Hanson maintains that all scientific theories prevailing at the time would have been hostile to the discovery of craters on the surface of the moon, which, "being *a celestial* object, ought not to be pock-marked and 'imperfect' as Galileo reported it to be."[33] "Almost all discoveries of comets are 'trip-over' discoveries."[34] Carl Anderson's discovery of the positive electron in 1932 was made while he was engaged in work of a quite distinct kind which at no time included the discovery of a positively charged electron in his original project's formal expectations. And "when the remarkable trace of August 2 'materialized', Anderson and his associate Neddermeyer were stunned. It required four months of study and deliberation for these men to 'see' the particle properly."[35]

'Trip-over' discoveries, "in which there is no theoretical posture at all with respect to the discovery in question," in which "no antecedent theoretical or psychological expectations furrow the frontiers of inquiry," contrast with what Hanson calls 'back-into' discoveries, in which there is not merely no pre-setting of the discoverer but rather "a strong resistance in the investigator against expecting such a discovery *ever*."[36] In these instances "all theoretical expectation is *against* coming up with this discovery . . . [but] the facts won't let him do anything *but* discover the awkward thing in question—and so much the worse for his theories."[37] Thus "the negative effect of the Michelson-Morley experiment (which suggested the nonexistence of the aether) constituted an encounter of just this variety."[38] The response which Michelson and Morley got to the question which their experiment directed towards reality, namely, how fast does the aether pass the earth as the earth travels through it, was a retroaction over the very conditions of the meaningfulness of the question (or over the presuppositions which made it a reasonable question in Collingwood's terminology).

---

[32] *Ibid.*
[33] *Ibid.*, pp. 335-336. Emphasis in original.
[34] *Ibid.*, p. 336.
[35] *Ibid.*
[36] *Ibid.*, pp. 336-337. Emphasis in original.
[37] *Ibid.*, p. 336.
[38] *Ibid.*, p. 337.

> The objective of setting up his ingeniously designed interferometer was just to check the so-called 'aetherwind'—its magnitude and precise physical effects. Michelson had no doubt that there *was* an aether and a resultant "wind." That his careful experiment generated a negative effect . . . was a theoretical disaster for classical electrodynamics. If one considers Michelson's research expectations and the theoretical framework within which he was working, the last thing he expected was a negative effect of this kind—this being tantamount to a discovery of the nonexistence of the aether.[39]

Such a 'discovery,' if one can say that Michelson 'discovered' the non-existence of an aether, can only be, Hanson says, a 'back-into' discovery, a regression, on which a previously accepted standpoint (a presupposition in Collingwood's terminology) which formed the basis of the investigation had to be relinquished. So that the answer to the question, if answer it can be called, is to nullify one of the prerequisites of the validity of the question. Consequently, "his 'negative effect' became the major observational datum in support of Einstein's theory of Special Relativity."[40]

Other instances of the 'back-into' discovery cited by Hanson are the discovery of the positron by Dirac, "who made every conceivable effort from 1928 until 1931 to 'cook-away' this awkward theoretical blemish within his otherwise spectacularly successful electron theory." After every attempt to explain the facts without recourse to 'positive electrons' he was forced finally to admit their existence, which gave rise to the question as to why they had not hitherto been observed. But within six months the hitherto unknown positrons showed their tracks in "a 'trip-over' discovery by Anderson in confirmation of a 'back-into' discovery."[41]

The inadequacy of any advance programming of scientific research in terms of predetermined questions is well stated by Michael Polanyi as follows:

> The science of today serves as a heuristic guide for its own further development. It conveys a conception about the nature of things which suggests to the enquiring mind an inexhaustible range of surmises. The experience of Columbus, who so fatefully misjudged his own discovery, is inherent to some extent in all discovery. The implications of new knowledge can never be known at its birth. For it speaks of something real, and to attribute reality to something

[39] *Ibid.*, p. 337.
[40] *Ibid.*
[41] *Ibid.*, p. 338.

is to express the belief that its presence will yet show up in an indefinite number of unpredictable ways.[42]

D. Even the Question Itself Is Not Purely an Activity of the Mind

This is not by any means a substitution of 'expect the unexpected' for Collingwood's "revolution which converts a blind and random study into one where definite questions are asked and definite answers are insisted upon."[43] Between the random study of searching "just to see what we can find out"[44] and rigid adherence to the clear-cut question formulated, there is a middle way more in accord with experience. All the while remaining "certain that you can satisfy an inquirer who asks you: What are you doing this piece of work for?"[45] one can and must maintain a constant reservation, or questioning attitude, *even with respect to the very question itself* which dominates and motivates the investigation. Not only does the answer to a question depend on the question, but the very question of the questioner has a certain dependence on the answer and has to carry with it a certain abeyance or suspension with respect to the forthcoming, hoped-for answer. This is particularly true in historical investigation, as is well explained by Marrou:

> The historian begins by posing a question to himself. . . . The question which started the whole process in motion does not maintain its original identity, but in contact with the documentary data it is continuously changing.
>
> Suddenly the historian realizes, for instance, that the question was preposterous and anachronistic ("the problem does not even arise"). He learns to formulate it in more precise terms, better adapted to the nature of the object. This is the advantage to be derived from a provisional self-effacement. Instead of an impatient interrogation constantly interrupting the witness to tell him, "Get back to the question!" the historian asks the document, "What are you? Help me to know you."[46]

E. Collingwood's Theory of Knowledge Based on Undue Absolutization of the Juridical Mode of Questioning

It is here that we must take issue with what Collingwood calls his Baconian attitude to questioning. This attitude unduly inflates the

---

[42] Michael Polanyi, *Personal Knowledge*, Chicago, University of Chicago Press, 1962, p. 311. See also Appendix 2 below, p. 484.
[43] R. G. Collingwood, *An Autobiography*, p. 124.
[44] *Ibid.*, p. 122. See also p. 125.
[45] *Ibid.*, p. 126.
[46] H.-I. Marrou, *The Meaning of History*, p. 131.

autonomy of the questioner and the dominating activity of the mind to the depreciation of the positive activity on the part of the interrogated object or person in the contribution to the answer to the question.

Bacon, whose scientific method of interrogating nature is Collingwood's model for historical methodology, "wrote like a lawyer arguing a case, rather than like a scientist solving a problem; or, as was said of him, he wrote philosophy 'like a Lord Chancellor'. Well he might: He was Lord Chancellor."[47] Since the time of Descartes, the fear of error has taken on the aspect of an obsessive neurosis in modern thought. Descartes' second *Meditation* betrays this exaggerated fear in "setting aside all that in which the least doubt can be supposed to exist, just as if I had discovered that it was absolutely false."[48] The same neurotic obsession with the danger of error, more exactly with the danger of being 'deceived,' pervades Bacon's expurgation of the mind of its various idols.[49] Bacon, the philosopher of scientific methodology, exorcising the mind of all its anticipations of nature, and putting nature to the question, is heavily under the influence of Bacon the jurist, the Lord Chancellor.[50]

Collingwood's questioning theory of knowledge and of historical methodology, based as it is on Descartes' and Bacon's radical mental hygiene, is both heir to and victim of this exaggerated fear of deception by a presumed untrustworthy witness.[51] Even the existence of cheating documents is no excuse for the exaggerated mental hygiene of excessive precaution against likely deception. The sober, balanced

---

[47] T. W. Organ, *The Art of Critical Thinking*, Boston, Houghton Mifflin Co., 1965, p. 110. It may also be noted that Vico, whose name stands at the fountainhead of the modern attempt to do justice to history and its methodology, the man whose *New Science* sought to do for historical methodology what Descartes sought to do for mathematical methodology, was a lawyer. B. Mazlish, in *The Riddle of History*, New York, Harper and Row, 1966, p. 13, says that "Vico's legal studies colored his whole thought, and it is important to emphasize here that Vico, like his contemporary, Montesquieu, was a jurist."

[48] See Vol. 1 of the E. Haldane and G. R. T. Ross edition of the *Philosophical Works of Descartes*, New York, Dover, 1955, p. 149.

[49] J. Spedding, R. L. Ellis and D. D. Heath edition of *The Works of Francis Bacon*, London, Longman and Co., 1857-1874. See Vol. 4 (1868), Aphorisms XXXVIII to LXII, pp. 53-64.

[50] Bacon, however, allowed himself a moment of juridical forgetfulness when he more humanely said that "entrance into the human kingdom based on sciences is like entrance into the kingdom of heaven in that each proceeds with the simplicity of a child." See Vol. 4 of Spedding, Ellis and Heath edition of *The Works of Francis Bacon*, Aphorism LXVIII, p. 69.

[51] H.-I. Marrou, in his *The Meaning of History*, p. 103, remarks that "methodical distrust is the form that the Cartesian principle of methodical doubt will take when applied to history."

attitude which the historian should maintain in regard to possibly deceptive documents is thus explained by A. Nevins:

> The cheating document seldom imposes (at least for any length of time) upon any but the ignorant or careless. A combination of scholarly knowledge and common sense can in ninety-nine instances out of a hundred expose it—but they must be reinforced by constant vigilance.[52]

This constant vigilance advocated by Nevins is very different from treating as false anything one may have the least suspicion to doubt, as Descartes does.

The point to be made here, in criticizing Collingwood's theory of knowledge as questioning, is that fear of error arising from the possibility of deception by an untrustworthy witness, or witnesses telling conflicting stories about the same thing,[53] is not the only reason we may have for questioning. That is, not all questioning is the questioning of a hostile witness suspected of intent to deceive.

F. Questioning Not a Mere Extortion of Information Under Duress: Questioning from Wonder and from Sympathetic Connaturality

From the beginning of human inquiry, and throughout its history, there has always been recognized a more fundamental, more congenial motivation for questioning than that designed to expose a deceiver. There is that inquiry in which Plato and Aristotle consider philosophy to have had its beginning, namely the inquiry that starts from wonder.

> . . . Wonder is the feeling of a philosopher, and philosophy begins in wonder . . .[54]

> . . . it is owing to their wonder that men both now begin and first began to philosophize; they wondered originally at the obvious difficulties, then advanced little by little and stated difficulties about the greater matters, e.g., about the phenomena of the moon and those of the sun and of the stars, and about the genesis of the universe. And a man who is puzzled and wonders thinks himself ignorant . . . therefore since they philosophized in order to escape

---

[52] A. Nevins, *Gateway to History*, New York, Doubleday, 2nd ed., 1962, p. 137.

[53] See, for example, L.-M. Régis, *Epistemology*, New York, Macmillan, 1959, pp. 7 et seq., for the problem arising due to two storytellers in us regarding the external world, the senses testifying to movement and multiplicity, the intellect testifying to the unity and self-identity of reality.

[54] Plato, *Theatetus*, see Vol. 2 of B. Jowett's translation of *The Dialogues of Plato*, New York, Random House, 1937, p. 157.

from ignorance, evidently they were pursuing science in order to know, and not for any utilitarian end.[55]

A striking instance of philosophical wonder closer to our own time is recorded by H. Spiegelberg in regard to Husserl.

> Phenomenology in general may be characterized as a philosophy which has learned to wonder again and to respect wonders for what they are in themselves, where others see only trivialities or occasions to employ the cleaning brush. But not all these wonders are of equal importance. To Husserl in particular there was one wonder which exceeded them all, ''the wonder of all wonders,'' as he called it: ''the pure ego and pure consciousness.'' [Ideen III, *Husserliana*, Vol. 75.] The wonder about this phenomenon seems to have been the focal and fundamental experience of Husserl's philosophical existence, and it became so increasingly, as his phenomenology developed. The central mystery was to Husserl not Being as such, but the fact that there is such a thing in this world as a being that is aware of its own being and of other beings. This fascination accounts for Husserl's growing emphasis on the subjective aspect of phenomenology and for its shift from the ''object'' to the subjectivity of the existing ego.[56]

At the beginning of his *Metaphysics*, Aristotle says that ''all men by nature desire to know.'' Prior to any anxiety to take precautions against likely deception, there is the questioning which arises spontaneously upon the encounter with the presence of 'the other' in lived experience.[57] Wonder is the expression of this spontaneous desire to get to know that 'other.' In wonder, there is knowledge (e.g., of an eclipse experienced to be taking place) plus ignorance (as to why the sun's light should be darkened) plus fear on account of that ignorance, not fear of being deceived, but fear simply of the unknown, since the unknown is threatening.

In the experience of wonder the mind is split within itself, into the mind knowing and the (same) mind not knowing in regard to the same object. Knowing that it does not know, that is, knowing its ignorance, it is literally diseased within itself and consequently experiences a tendency to shake off that disquieting feeling by striving to know what it does not know. This striving within itself to over-

---

[55] Aristotle, *Metaphysics*, Book I, Ch. 2, 982b 10-25, trans. W. D. Ross in *The Basic Works of Aristotle*, ed. Richard McKeon, New York, Random House, 1941, p. 692.

[56] H. Spiegelberg, *The Phenomenological Movement*, The Hague, M. Nijhoff, 1969, Vol. 1, p. 87.

[57] See M. Heidegger, *Being and Time*, translated by John Macquarrie and Edward Robinson, New York, Harper and Row, 1962, pp. 24-35; also K. Rahner, *Spirit and the World*, translated by W. Dych, Montreal, Palm Publishers, 1968, pp. 57-65.

come its own ignorance is expressed and articulated as a question. Because the mind is split into a duality of itself-knowing and itself knowing its lack of knowledge it can dialogue with itself. The strive to overcome ignorance thus takes the form of a self-interrogation.

Questioning, therefore, may arise simply from wonder, as a spontaneous response to a together known and unknown presence, and not exclusively from doubt caused by conflicting testimonies on the suspicion that one is being deceived.

L.-M. Régis contrasts the ignorance at the basis of wonder with the ignorance at the basis of doubt.

> The ignorance at the root of wonder is directly concerned with things; its object is the complex aspects of the same reality, a complexity that seems disordered and therefore unintelligible. The ignorance that is the wellspring of doubt is not concerned with things but bears directly upon the proofs of our causal knowledge of things. It has as its object the *explanation* or vindication of the truths we possess.[58]

Régis says, further, that wonder can result from our ignorance as is indicated by the infinite number of 'whys' asked by children "who wonder at everything because they know nothing." And then there is the wonder which accompanies us throughout our lives "*whenever we embark upon a field of research new to us.*"[59] Wonder is outgoing to things to be known, doubt is reflexive and concerned with the insecurity of our knowledge of things.

> Doubt is not concerned with things, or with primary evidence, or with the immediate knowledge of contingent things, but solely with *concluded* truth whose cause is not evident when it should be. Strictly speaking doubt is the contrary of science, when the latter is taken in the formal sense of a conclusion known through the causality of premisses.[60]

Skepticism is aligned with doubt but not with wonder, although skepticism is by no means identical with doubt.[61] The opposite of wonder is an insensitivity to the presence of something exceeding our knowledge of it.[62]

> The scientific quest for knowledge takes us away from the immediacy of the wonder-producing encounter. The object of scientific

---

[58] L.-M. Régis, *Epistemology*, p. 26.
[59] *Ibid.*, pp. 26-27. Emphasis added.
[60] *Ibid.*, p. 27. Emphasis in original.
[61] See *ibid.*, pp. 27-28.
[62] See Sam Keen, *Apology for Wonder*, New York, Harper and Row, 1969, pp. 26-27 and 29-40.

thought is not a presence, a thou, or a mystery, but a problem to be solved.[63]

The deficiency in Collingwood's theory of questioning should now be making its presence felt. Any philosophy which absorbs the whole of that which is into mind thinking about that which is, is going to maximize or inflate questioning arising from doubt and minimize or ignore questioning arising from wonder, or from concern with the objectively present as a genuine other. This is specially true in history, in which the primary orientation of the historian's questioning is not to torture the truth from a hostile and deceptive witness but rather to get to know the other as another self. History puts the historian 'in the presence' of 'other peoples,' with their thoughts, purposes and achievements, as objects 'to be wondered at' and whom the historian naturally wants 'to get to know.'

G. Historical Knowledge as Familiar Knowledge Attained by Friendly Rather than Extortionate Questioning

Marrou exploits the notion of sympathetic connaturality between the historian and the object of his inquiry. He explains that the basic attitude of the historian to the object of his investigation, whether a present document or past event known by means of the document, is that of a friend to a friend seeking familiarity one with the other. Speaking precisely of 'history as friendship,' Marrou says:

> Anyone who thinks too much, too soon or solely of using his friends can neither really love them nor know them; friendship and business relations are not to be confused. History furthermore assumes a centrifugal rather than an egocentric inner attitude, and an opening out towards others which demands that we mute our existentialist preoccupations.[64]

The knowledge sought by the historian is familiar knowledge, familiarity with the object of his research, the type of knowledge by which the members of a family know and desire to know one another. Marrou insists that the basic condition of historical knowledge of the past is that the historian, confronted with the document or other trace of the past event, or with the deciphered past event itself, sees in that 'other' something of 'self.' A basic connaturality, a basis for trustworthiness and familiarity, is the first condition of the possibility of historical knowledge of the past.

---

[63] *Ibid.*, p. 33.
[64] H.-I. Marrou, *The Meaning of History*, p. 224.

> If [historical] understanding is really that dialectical relationship of the *Same* with the *Other* which we have described, it presupposes the existence of a broad basis of fraternal communion between the subject and the object, between historian and document. (Let us also say, more precisely, that it is presupposed between the historian and the man who is revealed through the document as by a sign or symbol.) How can we understand unless we have that attitude of mind which makes us connatural with others? It is this that enables us to feel their passions and re-conceive their ideas in the very light in which they were experienced—in short, it permits us to commune with them. Even the word "sympathy" is insufficient in this respect. Between the historian and his object a friendship must be formed, or how else can the historian understand?[65]

This, incidentally, hits quite hard against Collingwood's notion of history as a reliving of past thoughts as distinct from past feelings. But, if one goes to Grand Pré, Nova Scotia, to an inconspicuous iron cross on which is inscribed:

> The dry bed of the creek which is in sight a few paces in the marsh is the SPOT where THE VICTIMS OF THE EXPULSION OF THE ACADIANS OF 1755 were embarked on the small boats to be rowed over to the transports lying at anchor in Minas Basin

one relives not merely the past thoughts, but the anguished feelings of those people.

Another historical incident in which the elements of feeling are relived just as much as the thoughts is related by Colin Ronan.[66] He tells how Sir William Huggins, when in 1908 found he no longer had the vitality to use his elaborate spectroscopic equipment, asked the Royal Society to see that it went to an institution which would carry on the spectrographic work he had pioneered. It was agreed, and arranged that the equipment be all removed to the Cambridge Department of Astrophysics. Howard Grubb came from Dublin to supervise the removal, and Ronan cites the following account from Grubb's notes:

> The Equatorial had been partially dismounted; all the numerous parts and attachments had been removed and were scattered over the floor, which was encumbered and littered with axes and various parts of the instrument, some of which had been already placed in packing cases; and in the midst of this litter, wrapped in a large cape and seated on a packing case, was Sir William himself, and his faithful collaboratrice who was flitting about watching the packing with keen interest and loving care. . . . Lady Huggins had asked

[65] *Ibid.*, p. 104. Emphasis in original.
[66] Colin Ronan, *Astronomers Royal*, New York, Doubleday, 1969.

> me to let her know when I was ready to close the box (which contained the large object-glass from the telescope), and when I intimated that I had it safely in the case, she took Sir William by the hand and brought him across the room to have a last look at their very old friend . . . They gazed long and sadly before I closed the lid.[67]

Ronan concludes by saying that two years later Huggins died. In this historical account one relives not merely past thought decontextualized from its associated emotion, but, if anything, the emotions rather than the thoughts.

Marrou says that when Norbert Casteret discovered the clay image of a quadruped pierced with iron-tipped spears he immediately recognized it as an instance of sympathetic magic practised by prehistoric hunters similar to that practised by Eskimos of our own time. Marrou says that we understand this type of behaviour quite differently from the way we understand the disintegration of an atom. We understand it interiorly; it is through our interior knowledge of man and his potentialities that we understand these prehistoric hunters.[68]

However, we can only understand another by reason of his resemblance to ourselves. "If the Other were completely dissimilar and totally alien it is difficult to see how any understanding would be possible at all."[69] But, paradoxically, if it is through self-knowledge that the historian knows the other, it is also through a certain self-forgetfulness, in which the Other is permitted to present its affiliation with the historian's self.

> The historian appears to us as a man who can transcend himself through epokhé, a real self-suspension in order to be outgoing in his encounter with the Other. We can give this Virtue a name: it is called sympathy.[70]

Marrou, on the other hand, is careful to point out that this connatural sympathy in "the dialectics of the Same and the Other"[71] does

---

[67] *Ibid.*, pp. 163-164.

[68] H.-I. Marrou, *The Meaning of History*, p. 37. Marrou thus agrees with Collingwood that we have an 'insight' into, or interior way of knowing, the human past, but Marrou's explanation in terms of sympathetic connaturality is quite diverse from Collingwood's overly intellectual explanation. Correlate with Collingwood's criticism of Croce's advice to try to become a blade of grass if one wants to know its history; see *The Idea of History*, pp. 199-200.

[69] H.-I. Marrou, *The Meaning of History*, p. 93.

[70] *Ibid.*, p. 103. Again, p. 267, he says, "History begins only where the historian forgets himself enough to come out of himself and go forward, openminded, toward the discovery and the encounter with the other."

[71] *Ibid.*, p. 93.

not prejudice the critical spirit of the historian,[72] but this critical part of the historian's work is always subordinate to the positive, constructive part.

> . . . it is always sympathy, the source and condition of understanding which represents the constructive phase. Criticism demolishes the provisional edifice of imperfect knowledge and suggests requirements that are useful in the latter reconstruction, but of itself it contributes little.[73]

Historical knowledge is the same type of knowledge as is found in our everyday experience when we become acquainted with another.

> We enter into possession of the human past on those psychological and metaphysical terms that in our daily life permit us to increase our knowledge of this other. . . . [H]istory comes before the bar of human reason with the same claims of credibility as all experience of the other.[74]

H. Some Presuppositions to Questions Are Genuine Insights into Reality: Intuitive Reason vs. Constructive Reason

N. Belnap, in an as yet unpublished report,[75] says that prior to Harrah's work on the logic of questioning (that is, prior to 1961) "most attempts at analysing questions concentrated on the consideration that asking a question ordinarily articulates a state of doubt."[76] But Belnap says that "Harrah's insight was that asking a question also articulates a state of information. That is, although asking a question is not the same thing as making an assertion, nevertheless, to ask a question is, partly, to make an assertion."[77] Now, when Belnap wants to illustrate this, he uses the very same example Collingwood had used in his *Essay on Metaphysics* to illustrate his (Collingwood's) theory of presuppositions, namely, the question "Have you stopped beating your wife?"[78] Belnap says that to ask Jones this question is to *assert*

---

[72] *Ibid.*, p. 104.

[73] *Ibid.*, p. 105.

[74] *Ibid.*, pp. 241-242.

[75] *An Analysis of Questions*. Preliminary Report, June 3, 1963. A document produced in connection with a research project sponsored by the independent research program of the System Development Corporation, Santa Monica, California. (Copy of this report made available by courtesy of System Development Corporation.)

[76] *Ibid.*, p. 8.

[77] *Ibid.* For David Harrah's position, that "the question and answer process is interpreted as an information-matching game," see his "Logic of Questions and Answers," in *The Philosophy of Science*, XXVIII (1961), pp. 40-46.

[78] *An Essay on Metaphysics*, pp. 25-26, 38-39. See also above, Chapter 5, p. 165.

that Jones is married and used to beat his wife. Collingwood had said[79] that to ask Jones this question is to *presuppose* that Jones is married and used to beat his wife. And for Collingwood, to *suppose* is not the same as to *assert*. Assertion belongs to propositions as answers to questions, supposal belongs to questioning as such.[80] Belnap says that "[to] ask, 'Did she wear the red hat or the green one?' is to *assert* that she wore one or the other but not (presumably) both."[81] That questions have an information-bearing content is familiar to any examiner who has to be careful to word his questions in such a way that the information content is not so great as to bear within it the answer to the question, or such information as would imply the answer.

> It is true that to ask a question is not the same as to make a statement; nevertheless, since questions have presuppositions, it is clear that in asking a question we do *also* assert something . . . and that questions do, like statements, have information-content. The student of chemistry can learn something from being asked, "What is the proportion of sodium to chlorine in common table salt?", the student of human nature from overhearing, "Who was that woman I saw you with last night?"[82]

The question we must ask here, crucial to the evaluation of Collingwood's theory of knowledge as questioning activity, is whether the presuppositions to questions are always merely supposal, or whether at least some of these presuppositions, while remaining presuppositions as far as the question is concerned, are not *also* necessarily *insights into reality*.

Francis H. Parker, in his article "Traditional Reason and Modern Reason,"[83] draws attention to "two alternative conceptions of reason: the conception of reason as intuitive and the conception of reason as constructive."[84]

The conception of reason as constructive is a modern conception dating from the time of Kant. For Kant, intuition is a prerogative of sense; understanding intuits nothing, only reflects.

> Given a stock of intuited sensory particulars as materials, the understanding can set about its proper business of constructing

[79] *Ibid.*

[80] See *Speculum Mentis*, pp. 76-80. See also Ch. 3 above, p. 66 et seq.

[81] N. Belnap, *op. cit.*, p. 8. Italics added.

[82] *Ibid.*, pp. 129-130. Emphasis in original. Belnap, unlike Collingwood, seems to equate presuppositions to assertions here.

[83] See *Philosophy Today*, VII (1963), pp. 235-244.

[84] *Ibid.*, p. 237.

orderly phenomenal objects according to its purely *a priori* blueprint.[85]

The universals which are the proper concern of the understanding are *a priori* and non-empirical. The understanding has no peculiar intuitions, no proper objects of its own.

> Since all empirical data or intuitions are sensory and therefore particular and contingent, and since necessary propositions essentially involve universals, no necessary proposition can be empirical, *a posteriori*, or factual.[86]

The conception of understanding, intelligence or reason which underlies this belief, the purely reflective or non-intuitive conception of reason, is what Parker calls 'constructive reason.' But, according to Parker, there is "an alternative conception of reason which . . . is essentially foreign to modern thought. It is peculiarly classical and mediaeval, and it originates most clearly in Aristotle."[87] This is "the conception of reason as an intuitive agency, a mode of cognition, distinct both from sensation on the one hand and from reflective discursive reason on the other."[88] Parker cites Aristotle as attributing to the thinking part of the soul a capacity for receiving the form of the object. The apprehensions of this thinking part of the soul give rise to the universal and necessary judgments which form the first principles of scientific knowledge. This apprehensive faculty Aristotle calls νοῦς, which Parker renders as 'intuitive reason.'[89] This intuitive reason has

---

[85] *Ibid.*, p. 236.

[86] *Ibid.*

[87] *Ibid.*

[88] *Ibid.*

[89] *Ibid.* See also Aristotle's *De Anima*, Bk. 3, Ch. 8, 432a 4-6; *Nicomachean Ethics*, Bk. 6, Ch. 3, 1139b 31-1140a 8. See also Marjorie Grene, *A Portrait of Aristotle*, London, Faber and Faber, 1963, pp. 241-247, for a discussion of Aristotle's *nous*. Grene understands Aristotle to be saying that the *nous* which discovers the first principles "is not the mind turned inward on itself . . . but the mind methodically and appropriately facing a suitable reality, a suitable specimen of the kind that is to be known. *Things* speak to the mind of their being-what-they-are and the mind, finding the universal *in* the perceived individual, putting its insight into words in the correct definition, thus comprehends the substance of the thing, and, as formal grasp of the formal, actual understanding of the actual, is identified with it. Knower and known are united because, being in nature suited to each other, they have never been wholly apart." Having thus clearly articulated Aristotle's position, Grene takes a characteristically modern attitude, for, to the question ". . . do things show us their being-what-they-are? Is the peculiar substance of each kind of thing there for us to grasp, vaguely in perception, step by step through induction, directly and luminously through rational and necessary intuition?" she replies emphatically in the negative and says that "the *what* of each kind of thing held the conjectures of the Aristotelian scientist firmly in check. For us, the strand of contingency is ultimate; we may transcend it, we must

its own proper objects, namely the forms, essences or characteristics of things which are given as data embedded in the particular, contingent data of sensation.

> Since . . . there is nothing outside and separate in existence from sensible spatial magnitudes, *the objects of thought are in the sensible forms*, viz. both the abstract objects and all the states and affections of sensible things.[90]

This means that the propositions which define these forms are also *a posteriori*, factual and existential.

> Consequently the assumption of rational intuition of formal structures coeval with and immersed in sensory data implies the possibility of propositions which are both essential and *a posteriori*, both necessarily and factually true.[91]

Such propositions are certified in immediate experience since they are true *in virtue of their meanings alone*. No further, subsequent experience is thus required to verify them. Such meanings, which immediately guarantee these necessary existential propositions, are factually real, and as proper objects of the intuitive reason are embedded in experienced reality.[92]

Whereas the constructive reason reveals only *our attitudes* toward things, the intuitive reason, on the other hand, is revelatory of factual *reality*, and "the replacement of intuitive reason by constructive reason might be well regarded as a fundamental theme of the rise and development of modern philosophy."[93]

Ancient philosophy is characterized by the struggle for the distinctness of reason from sensation, and thus the distinction of man from the brute. In mediaeval philosophy the struggle was of a differ-

---

transcend it, but in peril. Both the source of nature's reality and nature itself have lost their radical determinancy, and so lost also the radical intelligibility of Aristotelian substances. And so when we understand one aspect or another of the world around us, that understanding becomes our venture, not the world's. If we succeed, that is our achievement; if we fail, our 'understanding' turns out to have been our dream. Aristotelian knowledge could be impersonal because things, including minds, were thoroughly determinate. The determination *we* impose on things is in part at least of our own making, and in what part we can never precisely say." Emphasis in original. It is this constructive concept of mind that, following Parker, and in accord with Aristotle, we are taking issue with here.

[90] Aristotle, *De Anima*, Bk. 3, Ch. 8, 432a 2-5, cited from Richard McKeon's edition of *The Basic Works of Aristotle*, New York, Random House, 1941, p. 595. Emphasis added.

[91] F. H. Parker, *art. cit.*, p. 236.

[92] See *ibid.*, pp. 236-237.

[93] *Ibid.*, p. 237.

ent kind, namely, to justify the distinction of reason from revelation, and consequently to defend the distinctness of the specifically human from the purely spiritual, the angelic and the divine.[94] Man is thus between the purely animal and the purely spiritual. And it is man's reason, understood as man's natural faculty of discerning universals in and through particular sense experience,[95] which distinguishes him from either, and locates him in an intermediary position.

But the basic characteristic of early modern philosophy is a revolt against this traditional concept of reason, which could go in one or the other of two ways, either by way of a reduction of reason to sensation or of a reduction of reason to revelation.[96] These two forms of revolt against traditional reason developed concurrently. The reduction of reason to sensation meant the substitution of empirical science for rational philosophy and thus the confusion of philosophy with science. The reduction of reason to revelation meant the substitution of religious faith for philosophy and resulted in the confusion of philosophy with religion.[97] This double-phased revolt continued down into and through modern times, and, as it proceeded, it led to the development of reason as constructive.[98] The early stages of the reduction of reason to sensation occur mainly in William of Ockham and Nicholas of Autrecourt, and the early stages of the reduction of reason to revelation are found mainly in Meister Eckhart and Nicholas of Cusa. The next stage in the loss of the traditional intuitive reason and development of constructive reason is found in the intermediate conception of reason such as it is found in Descartes and the continental rationalists, for whom reason was the faculty of deducing universal truths *a priori* from innate ideas which are wholly independent of sense experience. The revolution of Kant, consequent upon the British empiricist discarding of the rationalist conception of *a priori* reason and innate ideas, gave to modern philosophy the fully developed conception of reason as constructive.[99] Parker summarizes the various conceptions as follows:

---

[94] See *ibid.*, p. 238.

[95] See *ibid.*, p. 237 and p. 238.

[96] See *ibid.*, p. 238. One immediately recalls in this context Collingwood's account of the patristic codification of "fundamental convictions which would form the presuppositions of a new way of living, a new science and a new civilization, which codification was called the 'Catholic Faith.' " See *An Essay on Metaphysics*, pp. 225-227; see also above, Chapter 5, p. 190 et seq.

[97] See F. H. Parker, *art. cit.*, p. 238.

[98] See *ibid.*

[99] See *ibid.*, pp. 238-239.

> For the tradition the intelligence officer [i.e., reason] . . . finds meaning in experience; for continental ''rationalism'' meaning is deduced from innate ideas independently of experience; for British ''empiricism'' meaning is imparted only by the subrational faculties of passion and habit; and for Kant and his voluntarist and pragmatist followers, utilizing the conception of reason as constructive, meaning is created out of the mind itself.[100]

I. Distortion of the Nature of the Abstract Universal Arising from Exaggerating the Activity of Reason into Creative Activity

From this Kantian constructive conception of reason, Hegel and his followers, including Collingwood in this regard, passed to a creative conception of reason, reason as the activity of a mind 'which is what it does,' which is nothing more or less than what from its own activity it forms itself to being. Human reason is defined and spoken of in ways previously reserved for the thought and the being of God. Aristotle's self-thinking thought and Aquinas' Pure Act are now characteristics of the human reason.[101]

The theory of the concrete universal, exposed in Chapter 4 above, has its historical roots in this Kantian evacuation of real content from the abstract universal, which then becomes a mere classificatory device of the constructive reason.

> The name 'horse' is the name of a certain class of individuals. But we also have a general idea or concept of a horse. This is an intellectual construction made by us out of attributes which we have abstracted from individuals. It might be described as a second order abstraction, being a construction out of abstractions. The name 'horse' thus applies to or denotes an aggregate of individuals. It also connotes an intellectual construction, namely an idea or concept formed by us of a type of individual.[102]

A. J. M. Milne goes on to warn that failure to understand this may lead to a wild-goose chase in a search for a supposed essence embodied in the members of a class. This is a clear rejection of the traditional notion of the comprehension of the universal as the real

---

[100] *Ibid.*, p. 239.

[101] Not only is there achieved here a radical *rapprochement* between philosophy and religion, but the knowing (questioning in Collingwood) activity of the human mind is identified with what in mediaeval theology was understood as the (not merely productive, which presupposes pre-existing matter, but) creative activity of the divine mind (which presupposes no pre-existent material).

[102] A. J. M. Milne, *The Social Philosophy of English Idealism*, London, George Allen and Unwin, 1962, p. 18.

object of intuitive reason. For Milne the connotation of the universal notion is a construction formed by the mind to typify the members of the class.

J. Nature of the Abstract Universal and Its Value in Knowledge

Disturbed by the consequences of the teaching of Bradley and Bosanquet regarding the abstract, and because he found himself unreflectingly adopting the prevailing opinion, thereby depreciating the abstract as *merely* abstract, Norman Kemp Smith submitted both the abstract universal and the concrete universal of Bradley and Bosanquet to a thorough scrutiny.[103] He criticizes the identification of the abstract universal with the class-concept.

> . . . it is by no means clear by what right Bosanquet should thus identify the abstract universal with the class-concept. In so far as the abstract universal is some one character, it is indeed applicable as a predicate to any and every concrete existence in which the character is found; and we may, if we please, describe this range of application as being a class. Our purpose in selecting the character need not, however, be classificatory; we may be investigating the character in and for itself; and it may be in the very variety and heterogeneity of its embodiments that we obtain the needful data.[104]

Smith attacks Bosanquet's contrasting of the concrete universal with the abstract universal on the ground that one signifies a world and the other a class, and whereas "it takes all sorts to make a world, a class is essentially of one sort only."[105] Smith remarks that examples of universals which stand closest to concrete particulars, such as that of the Death's Head beetle, represent, like a 1925 model Ford car, a standardized type that repeats itself without notable change in the infinite number of its instances. Here the feature of monotonous repetition is at a maximum. But, on the contrary, if we choose a class-concept higher in the biological hierarchy, which is therefore more abstract, we can no longer say that while it takes all sorts to make a world a class is of one sort only. The universal 'beetle' covers some half million different species that vary among themselves in colour and shape and a variety of characteristics. Smith argues that just as

---

[103] This analysis by Norman Kemp Smith is to be found in two articles, "The Nature of Universals" in *Mind*, XXXVI (1927), pp. 137-157, 265-280, and 393-422, and "The Fruitfulness of the Abstract" in *The Proceedings of the Aristotelian Society*, XXXVIII (1927), p. 149. See also B. Bosanquet, *The Principle of Individuality and Value*, p. 37.

[104] Norman Kemp Smith, "The Nature of Universals," *Mind*, XXXVI (1927), p. 149.

[105] See *ibid.*, p. 150. See also B. Bosanquet, *The Principle of Individuality and Value*, p. 37.

a world may or may not be a macrocosm of microcosms (traditional mechanistic physics presents a world notable for its uniformity of structure) so also a class may or may not be a type of types. Even if it were, it is not in proportion to its abstractness but rather to its closeness to the concrete that reduces it to a uniformity of simple recurrence.[106]

Smith points out that Bosanquet himself admits[107] that the abstraction or notion which is to include both men and animals has to provide for a variable animality, for animality in man is different from animality in brutes. The concept animal is therefore not formed simply by the omission of certain qualities which differentiate man from the other animals. There is something corresponding to rationality in animals, namely (animal) intelligence.[108] Accordingly, "we can neither merely omit the characters which vary in their instances nor proceed by simple inclusion of others that are strictly uniform."[109] Far from being "little more than an unfortunate necessity due to the limitations in our powers of attention," abstraction is "a methodological device" enabling us to study partial features "with the ultimate purpose of so mastering the material, part by part, that through gradual reversal of this abstractive procedure, we may finally be in a position to grapple with the experienced in all its concreteness."[110] Jacques Maritain says[111] that there is no true knowledge of unity where there is not also knowledge of distinction. Any serious attempt, he says, at metaphysical synthesis must distinguish in order to unite, especially when dealing with the complex riches of knowledge and of the mind. A reflexive and critical philosophy must above all discriminate and discern the various degrees of knowing with its organization and internal differentiation. Maritain criticizes contemporary idealism's refusal to recognize any nature or structure of the mind, regarding it as pure freedom and pure movement. It thus systematically neglects vast areas of knowledge and tends to reduce the diversities of the life of the mind to a noetic monism.

---

[106] See Norman Kemp Smith, "The Nature of Universals," *Mind*, XXXVI (1927), p. 150.

[107] See B. Bosanquet, *Logic*, Oxford, Clarendon Press, 1888, pp. 66-67 (Smith erroneously cites pp. 58-59).

[108] See John N. Deely, "Animal Intelligence and Concept-Formation," in *The Thomist*, XXXV (1971), pp. 43-93, for an interesting discussion on this point.

[109] See Norman Kemp Smith, "The Nature of the Universal," *Mind*, XXXVI (1927), p. 151.

[110] See Norman Kemp Smith, "The Fruitfulness of the Abstract," *Proceedings of the Aristotelian Society*, XXVIII (1927-1928), p. 204.

[111] See his *The Degrees of Knowledge*, London, Bles, 1959, p. ix.

Norman Kemp Smith attacks the negative attitude to abstraction which sees in it only omission.

> . . . abstraction is not merely omission. Indeed, it is questionable whether the term omission has any relevance. All that we are justified in asserting is that the mind abstracts from certain factors, not that it omits them. That, however, is a minor point. What calls for special insistence is that the process mainly at work is integrative, not separative in character.[112]

Collingwood regards abstraction as synonymous with error. ''Our enemy is abstraction''[113] and ''abstract knowledge is the same as error.''[114]

N. Lobkowicz points out[115] that, in response to the objection that ''since in reality the forms of material things do not exist in separation from particulars, our intellect must be wrong if it grasps such forms as abstract,'' Aquinas distinguishes between two meanings of abstraction, separative abstraction and abstraction through simple non-consideration. One may assert by a judgment that f does not inhere in x, thereby abstracting f from x. Thus, to say that the apple is not colored is to abstract color from the apple by denying its existence in the apple. If in fact the apple is colored, such abstraction is impossible without falsifying the whole factual situation. But, on the other hand, one may simply consider the color which in fact is in the apple without considering the apple in which it is and without considering any of the other characteristics (taste, weight, etc.) which are together with the color in the apple. There is no assertion here that the color is separated from the apple in fact, merely the non-consideration of all the characteristics other than the color which also are in fact in the apple. And ''thus abstraction involves no falsity, so long as it remains a *simplex et absoluta consideratio,* i.e., does not involve a judgment.''

Abstraction is rather a focusing of attention[116] on a special aspect

---

[112] Norman Kemp Smith, ''The Fruitfulness of the Abstract,'' in *The Proceedings of the Aristotelian Society,* XXVIII (1927-1928), p. 206.

[113] *Speculum Mentis,* p. 268. See above, Chapter 4, p. 94.

[114] *Speculum Mentis,* p. 313.

[115] See his ''Abstraction and Dialectics,'' in *The Review of Metaphysics,* XXI (1967-1968), pp. 468-469.

[116] Questioning is concerned primarily with this focusing function. ''Questions, of course, do not establish the answers—they merely establish the relevance of a range of answers. The 'right' answer must in some sense 'match' the world and this seems to commit us to the outlines of a correspondence theory as a component in the ultimate situation.'' Leslie Armour, *The Concept of Truth,* Assen, Van Gorcum, 1969, p. 211. The position to be taken here is that the question, when directed at reality, aims at a 'docking' of the mind with its corresponding real object. The 'correspondence' of

of a concrete individual fact or event; it presents the mind with a definite content which can be examined *in itself,* without confusing it with the context in which it is realized. A red object can be examined and questioned precisely inasmuch as it is red. The fact that we do not attend to the object as being steel or heavy in no way invalidates what we can know of it through knowing that it is red.

> Abstraction . . . is merely the more attentive apprehension of what has been antecedently known in some concrete setting; and it would thus seem to be simply the rendering of our initial awareness more definite and precise.[117]

Smith criticizes the presumption that there is no problem regarding the apprehension of concrete particulars which are supposed to be apprehended in ordinary experience in all their concrete particularity. Having successfully done this, it is then asked, how do we arrive at universal concepts? These are said to be the outcome of abstraction, consisting in a comparison of particulars and arriving at a class-concept by disregarding the differences.

But the uniqueness of the unique, the individuality of the individual, is just as much abstract and universal as is the redness or hardness. We cannot talk meaningfully about the individuality or uniqueness of the concrete if we do not abstract these universal characteristics as meaningful objects. "Indeed, may we not say that this is an ideal goal towards which we can only approximate?"[118] Upon reflection, it seems rather that we know the generic before the individual peculiarities.

> We start, in experience and knowledge, not with the individual, but with the vaguely generic, and advance in knowledge is to be measured quite as much by increased appreciation of differences as by capacity to generalize. The child can usually recognize trees before he can distinguish the different species of trees; and to the end all recognition, even of what is discriminated as being an indi-

---

the mind with reality here is vastly different from a naive correspondence theory, as if the mind merely ran parallel with the object *without in any way engaging in a very definite type of docking arrangement.* As will be shown later, the primary 'docking' of the mind with reality is through answer to the 'what' question, as is well explained by Aristotle (see Chapter 8, below, p. 337 et seq). But, in order that this 'docking' be successfully achieved, the questioning of reality, prefaced by the questioning as to what past thinkers have thought of reality (which is all Collingwood allows), has to be carefully and correctly enacted.

[117] Norman Kemp Smith, "The Fruitfulness of the Abstract," in *The Proceedings of the Aristotelian Society,* XXVIII (1927-1928), p. 207.

[118] See *ibid.,* pp. 207-208.

> vidual distinct from all others, is by means not of absolutely unique features, but of types.[119]

The fact that we frequently err in our identification of individuals indicates that the absolutely unique is a term of reference whose evidence consists in more than purely immediate experience.

Comparison according to similarities and dissimilarities is not itself abstraction but a prerequisite facilitating and disposing for abstraction. "The individual is, indeed, initially known only in its relations within some complex context, and this context may be temporarily left out of account."[120] For the completion of abstraction there must be not merely knowledge of like and unlike characters, but identification of what the common element is as *itself* constituting a *type* or *kind*.

> The abstracted factor is, therefore, abstracted in this sense only, that it is at once distinguished from the conflicting features that accompany it in the various instances, and at the same time identified, not *qua* existent but in type or kind, as being the same in them all.[121]

Knowing that something is a man is together a knowing of what man is and a re-cognizing *that* it is here in this individual.

> To cognize is to recognize, to recognize is to identify a recurrent type, and to identify a recurrent type is to have selected for special—i.e., abstractive—attention, in one or more instances of the type, the feature or features which go to make up the content of the type.[122]

To distinguish successfully just acts from unjust acts is possible only if we are able *infallibly* to *recognize* which acts are just and which are unjust. But, since justice is not a directly experienced quality or a sensible character, its *recognition* presupposes possession of the concept we profess to be seeking in the comparison of instances.[123]

Everything in our experience has a dual aspect: its thisness and its suchness.

> As features of the particulars they share in the 'thisness' of the particulars: they are themselves particulars. On the other hand, as making up the content of the type, they are apprehended as 'suchnesses'; and thus, in their regard, as in regard to all else, we have

---

[119] *Ibid.*, p. 208.
[120] *Ibid.*, p. 211.
[121] *Ibid.*, p. 210.
[122] *Ibid.*
[123] See *ibid.*, p. 209.

> to accept as ultimate the dual 'this-such' nature of everything experienced.[124]

Corresponding to this dual character of what we experience, there are two kinds of recognition, and consequently two kinds of presupposed identity.

> Identity, and consequently recognition, is found in two forms. There is recognition as ordinarily understood, namely, recognition of an individual as having been previously experienced. Secondly, there is the recognition which logic has mainly in view, but which is not ordinarily so called, namely, recognition of a particular as being in this or that regard of this and that kind. In other words, there is recognition of individuality, and there is recognition of type. The two forms of recognition are correlative to the two species of identity—the self-identity of a particular and the identity of a type in distinguishable particulars.[125]

Smith insists that "it is important to discriminate the two species of identity as sharply as possible."[126] He explains that identity of *type* is what is meant by the universal and that this is representable by invariable symbols and persists unvaried with variation of differences. (That is, it is a true identity-in-difference.)

> The identity of a type or kind I take as being precisely what is meant by a universal. It is that which can be asserted to be identically the same in one and all of a number of instances; and as such it is invariable. While allowing of [sic] combination with this and that other universal, it does not in any such combination admit of modification by these others. It is an identity that is found amidst differences, but which is never in any degree or fashion altered in its identity by these differences. It is, indeed, a condition of the combination of universals that together they are found to constitute a new type—that is, to constitute what is more than a mere combination of the simpler types. Otherwise we should not be conceiving a new type, but only conceiving simultaneously the previously conceived types. But into the conception of the new type the simpler universals enter without change of meaning. Thus the universal or type 'triangularity' enters into the more complex universals, 'equilateral' and 'scalene', and as a constituent of the two latter has a meaning that is uniform for both; though they are species of 'triangle', they are not species of 'triangularity'. Similarly, 'animality' has a uniform meaning as predicated of man and of the various species of brutes, and this is why they can all alike be said to be

---

[124] *Ibid.*, pp. 210-211.
[125] *Ibid.*, p. 211.
[126] *Ibid.*, p. 212.

> animal species. Universals, being thus fixed in meaning, can always be represented by symbols that are invariables.[127]

Individuals enjoy individual identity which persists through its variations. Its differences, unlike the differences of the universal type, are intrinsic to it.

> Individual things, on the other hand, are admittedly variable; each is an identity that is compatible with, and indeed can be conceived at all only in and through the diversity of its distinguishable qualities, and in and through the changes whereby it responds to changes in its environment. It is not merely, like type, an identity amidst differences, it is an identity in and through its differences. The differences enter into its innermost being; and yet do so in a manner which allows of their own variation.[128]

Although universals are never found save as embodied in changing particulars they find a quite constant, not a variable, embodiment. The individual differs from the universal in that it preserves its identity through its variations. Consequently, the universal has a definite identifiability, but the individual an indefinite identifiability. The universal can be expressed, the individual only indicated.

> . . . whereas the universal, as being a type, consists in a definitely enumerable set of constituents, each of which can be abstractly considered, and in this manner exhaustively known, the individual can never be equated with any finitely exhaustible number of known characters. . . . Every individual, though apprehensible in terms of an indefinite variety of different types, according as it is regarded in this or that aspect, is never thereby exhausted.[129]

But, since "we cannot recognize what we do not apprehend" "though identity is of these two species, identity *in and through* difference, *i.e.*, identity of a continuant, and identity *amidst* difference, *i.e.*, identity of type, the mode of their recognition is found, on closer examination, to be one and the same in both cases."[130] When the individual is recognized as identical, "what we recognize as evidencing the continuing identity" of that which is elsewhere and otherwise experienced "is the special manner in which the types . . . are combined with one another."[131]

---

[127] *Ibid.*
[128] *Ibid.*, p. 212-213.
[129] *Ibid.*, p. 213.
[130] *Ibid.*
[131] *Ibid.*, pp. 213-214.

> The special manner of their combination, that is, the special permutation of common factors, is our sole clue to the identity of the individual, as found at different times and in differing situations. In other words, identification of the individual is thus itself still a form of identification by type. It is capacity to recognize type, "suchness," that makes possible the capacity to apprehend the individual.[132]

Smith says that this is the chief way in which the abstract manifests its fruitfulness, namely, as making possible the apprehension of its counterpart, the uniquely individual. No inferences are possible either from or to the unique as unique. Such inferences

> are possible only in virtue of some identity, as defined by means of type. Accordingly, in apprehending the individual in terms of type, we bring it, notwithstanding its uniqueness, within the range of all those inferences which are relevant to the type.[133]

The individual, as such, is outside the range of apprehension and reasoning, since the universal is a prerequisite of all inference.

> . . . that apprehension of type is what can alone bring the individual within the province of reasoning. A universal is a prerequisite of all inference; and the fixity of the abstract type is what enables the concept of the type to supply this need.[134]

But it is important to note, in order to avoid charges of the unreality of the abstract universal, that this universal, which is the medium of all our inferential knowledge, can be obtained only as supplied by experience and has its value only as referred back to experience.

> If we are to arrive at knowledge of what is not itself experienced, the means of conceiving it must none the less still be supplied in experience. For just as the mind can only isolate what is actually given, so it can only synthesize what has already been apprehended.[135]

If disengagement from concreteness in particulars is necessary for knowledge of the universal, embodiment in concrete particulars is necessary for its reality. It is important not to confuse *abstract knowledge* with *knowledge of the abstract*, as is often done.[136]

---

[132] *Ibid.*, p. 214.
[133] *Ibid.*
[134] *Ibid.*, p. 215.
[135] *Ibid.*, p. 217.
[136] See J.-H. Nicolas, *Dieu connu comme inconnu*, Paris, Desclée de Brouwer, 1966, pp. 36-37. Also pp. 37-38: ". . . la distinction entre les deux manières dont une connaissance peut être dite universelle: 'Connaître quelque chose universellement s'entend

K. Abstract Universal Not Merely Classificatory but Also Penetrative of Reality

The abstract universal, therefore, has a twofold relationship: one to the mind as that which the mind immediately apprehends[137] and the other to reality, as the intelligible face which reality presents, as object, to the mind[138] Now, if one considers the aspect of relation to the mind, and the mind's requirements on account of this relation, we find that genus-species classificatory logic has priority.[139] But if we consider abstract knowledge as *a penetration of reality*, the logic not of univocal genus and species but of analogy is required.[140]

Human reason is both intelligence and reason. As intelligence it has access to being in all its intelligible clarity and absolute universality. As reason it progresses step by step in the acquisition of the most profound truths.[141]

A profound antinomy is immanent to the laws of reason which the scientist and philosopher use in studying reality. On the one hand, they are characterized by an absolute rigour by reason of which logic is a genuine science and an absolutely sure art in the work of the intellect. On the other hand, they are characterized by a suppleness, flexibility and a variety of nuances which permit them to be adapted

---

de deux manières. Ou bien il s'agit de la chose connue, et on parle alors de n'en connaître que la nature universelle. En ce sens connaître universellement est une imperfection: il connaîtrait l'homme imparfaitement celui qui saurait seulement qu'il est dans le genre *animal*. Ou bien on pense à ce par le moyen de quoi on connaît. Alors il est plus parfait de connaître quelque chose universellement: plus parfait en effet est l'intellect capable de connaître les singuliers selon ce qu'ils ont de propre dans un 'medium' universel, que l'autre qui ne le peut.' Il faut donc distinguer la *connaissance de l'universel* qui est une imperfection, et la *connaissance dans l'universel* qui au contraire est une perfection. Mais pour la pensée humaine, tout entière conditionnée par le procédé abstractif, l'universel sous ces deux formes est le fruit de l'abstraction: 'Il est accidentel à l'universel d'être abstrait des singuliers, cela vient de l'intellect qui le connaît et qui tire sa connaissance des choses. S'il est un intellect ne tirant pas sa connaissance des choses, l'universel connu par lui ne sera pas abstrait des choses.' C'est ainsi qu'il nous faut distinguer dans la connaissance humaine: une connaissance de l'abstrait, qui se termine, comme à un 'quod cognitum,' à une essence encore indéterminée, à une détermination partielle; et une connaissance abstraite qui se termine à une détermination achevée et atteint en elle et par elle le réel concret."

[137] See this chapter, above, pp. 287-288.

[138] See this chapter, above, pp. 281-283.

[139] See Jean Petrin, "Univocité et analogie dans les lois de la logique," in *Angelicum*, XXVI (1949), pp. 233-242. See also L.-B. Geiger, *Philosophie et spiritualité*, Paris, Editions du Cerf, 1963, Vol. 1, pp. 76-77.

[140] See Jean Petrin, *art cit.*, pp. 242-249. See also L.-B. Geiger, *op. cit.*, pp. 77-86.

[141] See Jean Petrin, *art. cit.*, p. 233.

increasingly to the thousand and one forms of intelligible reality.[142] The rigorous and clear value which these laws possess depends upon an ensemble of *univocal* concepts, which concepts consequently enjoy a priority *in the logical order*. On the other hand, their suppleness and infinite variety in application depends upon an ensemble of *analogical* concepts which carries with it a certain primacy *in the order of intelligibility*.[143]

L. Being as Class Concept and as Active Presence

It will be recalled that in his *Essay on Metaphysics*, Collingwood rejected metaphysics as a science of abstract being, or "ontology . . . my name for a mistake which people have made, Aristotle first and foremost . . ."[144] since "there can be no science nor even a quasi-science or pseudo-science of pure being."[145] Collingwood's reason is that "abstraction means taking out. But science investigates not what is left out but what is left in. To push abstraction to the limiting case is to take out everything; and when everything is taken out there is nothing for science to investigate."[146] But this is to overlook completely the fact that the concept 'being' as it is the object of metaphysics "is not situated rigorously on the same level as other concepts, and that it is quite irrelevant (une manière toute matérielle) to treat what it presents simply as the most general class."[147]

Because of the unique situation of the concept of being in the life of the mind, it has the unique function of presenting to our intelligence both the similarities and dissimilarities of beings.[148]

Our concepts are elaborated in two quite different but complementary climates. On the one hand we are preoccupied with analysing, defining and classifying, through noting likenesses and differences, when placed in the presence of things by our sensory equipment. This classification through differentiation and similitude is the principle of genus and species conceptualisation which is adopted essentially unchanged by the methods of science. Most certainly these concepts place us in the presence of being, but implicitly only. But "what we explicitly note and directly express by these concepts is the result of the relationships discovered between beings or between parts of beings, or

---

[142] See *ibid.*, pp. 233-234.
[143] See *ibid.*, p. 234.
[144] *An Essay on Metaphysics*, p. 17.
[145] *Ibid.*
[146] *Ibid.*, p. 14.
[147] L.-B. Geiger, *Philosophie et spiritualité*, Vol. 1, p. 73.
[148] See *ibid.*, pp. 73-74.

even their relationship with our body, by sensible knowledge under all its forms."[149] In this way we see being much in the way an explorer observes, registers and describes the particularities of a region he is investigating. He supposes that the country he is investigating exists, that he himself exists, that what he is doing is inserted in reality, but his geographical description does not attend to that aspect of his object, however fundamental it may be.[150]

> The same is true of all our universal concepts. . . . In effect it is by the work of comparison between different beings or their actions and thus by an ensemble of relative data which, in default of grasping the essence directly, we form generic and specific concepts.[151]

Such comparisons are made above all on the sensible plane. Thus it is in the structure of the sensible that we find the source of the univocal likenesses and diversities at the basis of our genus and species concepts. On this level, the likenesses of things exclude the differences, and it would be contradictory to base on the same principle both the likeness and the unlikeness between beings.[152] At this level, our concepts do not grasp essences as such, that is, as determinate ways of exercising the act of being. Beings are there, and we know that, but we do not attend to that aspect. We analyse, compare and classify beings, but not inasmuch as they are beings. We do not formally envisage the essences as essences but as quiddities. Rationalism excessively separates the work of reason elaborating quiddities from their ontological base which, in fact, it seeks to attain.[153]

But immediately when the mind focuses on the being of things as its direct object, everything changes.[154] The mind interrupts its investigations to perceive clearly the *reality* of that which exists. Until then, it merely manipulates and names in function of its manipulations in terms of genus-species relationships. As soon as it adverts to the object of its manipulations as being, as something which *is*, the mind operates in an entirely new dimension. The definition is now not merely an ensemble of relations of similarity and dissimilarity, but signifies a way of exercising the act of being, that is, of formally comprehending an essence.[155]

---

[149] *Ibid.*, p. 76.
[150] See *ibid.*
[151] *Ibid.*
[152] See *ibid.*, pp. 76-77.
[153] See *ibid.*, p. 77.
[154] And, it may be added, the mode of questioning changes.
[155] See L.-B. Geiger, *op. cit.*, p. 78.

> What I know of man, and which up to that moment was a differential knowledge based on sensible representations and rational elaboration [of the differential as manifested by the relations of sensible things] appears now in itself in its own right, in its independence with regard to my knowledge [of it].[156]

The term 'man' and the term 'being of man' both attain the one same reality; one attains it by a system of relations, the other in an absolute way, as a mode of being. In Kant's famous problem of the hundred thalers there are really two problems. On the one hand, there is the distinction made by Kant between the hundred possible thalers and the hundred actual thalers, that is, in Kant's thought, between essence and existence. On the other hand, and perhaps more subtly, there is the distinction between the knowledge of a hundred possible thalers inasmuch as their number can be placed in a series of possible numbers, that is, put in relationship with four hundred or a thousand thalers, and the knowledge of those same hundred thalers inasmuch as in some way they qualify real or possible existence, that is, inasmuch as they represent an essence of some sort.[157]

A mathematician does not inquire regarding the being-value of his object. He is satisfied to construct, define, and demonstrate, operating all the while at the interior of a world populated, like our real world, by well-determined beings, but resulting as the fruit of the constructive activity of the mind.[158] But if that same mathematician reflects so as to ask what is mathematical *being* he knows full well that he has altered his point of view. To deal with a fourth dimension as something seen in relationship to a first, second, third, or even fifth dimension is properly mathematical. But to ask if a fourth dimension is as real as a third is to ask a non-mathematical question.[159]

In passing from the world of mere intelligible consistency to that of reality one passes from the world of representation to that of presence. More exactly, one enriches the world of representation with that of presence, real or intelligible presence. Representation and presence seem to be the best terms to characterize, from the point of view of the theory of knowledge, the relations between our univocal concepts and that of being.[160]

Being, therefore, is not an abstract concept obtained by leaving out all the differences of being, as Collingwood says, allegedly follow-

---

[156] *Ibid.*
[157] See *ibid.*, p. 78.
[158] See *ibid.*, p. 79.
[159] See *ibid.*, pp. 79-80.
[160] See *ibid.*, p. 80.

ing Aristotle,[161] that is, a character which one finds in everything to the exclusion of differences. Being is indeed found in everything which is, but in a diverse manner. If it is something absolutely universal, it is because it is present in everything, but both as regards their likenesses and as regards their unlikenesses. This relative unity of being and the absolute diversity of beings does not imply a contradiction, since each belongs to two different points of view: the point of view of rational construction (de la *ratio*) on the one hand, and on the other hand, the order of penetration (de la saisie), or of presence, or of intellect (*intellectus*).[162] Intellect and reason (or intuitive and constructive reason) are not two faculties but one only. Because the penetration of penetrating reason, or intellect, is always only partial, that is, abstractive, it is also relationally constructive (of its partial insights) in order to obtain a synthetic world view. "To reason is proper to a human intellect."[163]

M. Being as a Concrete Universal of Identity in Difference

How beings are one as such and at the same time differ according to their being is the problem of analogical unity, as distinct from the univocal unity of the generic universal. Here is realized a true unity in diversity, identity in difference. What makes a horse a different kind of being from a man is itself being, the same as that which makes them the same. The intelligible grasping of things is a grasping of them in their very communication in diversity (*communion dans l'altérité*). Thus, beings can manifest to one and the same intelligence that they are, that they communicate in the act of being, and also that in their essences and proper content they are diversified as beings.[164]

What this means is that whereas the univocal genus 'animal' is differentiated into the species 'man' the difference, extrinsic to the common genus, bears the whole weight of the differentiation. And man, differing from dog as rational from irrational, differs inasmuch as there is affirmation and negation of 'rational,' that is, of the difference, not of the common genus. But, when the common term, being, is differentiated into (say) human being and vegetable being, the differentiation does not at all take place outside the common term, but takes place precisely through simultaneous affirmation and negation

---

[161] See this chapter, above, p. 294 et seq.

[162] See L.-B. Geiger, *op. cit.*, p. 82. Correlate with F. H. Parker's intuitive reason and constructive reason, this chapter, above, pp. 280-284.

[163] L.-B. Geiger, *op. cit.*, p. 82.

[164] See *ibid.*, pp. 84-85.

of the common term. Man *is* a being, and thus *affirms* being. But as his being is *not* the being of a vegetable (cabbage) he simultaneously *negates* being, the being that is distinctly that of the cabbage.

> When I understand a chimpanzee, I posit something, let us say, the characteristics of this particular species, and simultaneously I deny the characteristics of another species. But notice that the association of assertion and negation concerns the differential. With *analogates the association of assertion and negation concerns the common ground.*[165]

This is "the secret of analogical thinking," namely that

> *you have an assertion of the common ground coupled with a negation of the common ground.* . . . You have here a togetherness of the 'yes' and the 'no,' an association of assertion and negation which concerns the common ground. *If the common ground were a genus, assertion and negation would concern only the differential factor.*[166]

What immediately becomes clear at this stage is that what dialecticians such as Hegel and Collingwood are striving to do by means of auto-determination through significate negation in their philosophy of mind, in which the foundation of all that is is self-actuating, self-creating mind, actualizing itself through simultaneous affirmation and negation of itself, the proponents of the philosophy of being are striving to do by means of their theory of the analogical unity of being, as that which is simultaneously affirmed and negated in the very constitution of everything that is. Thus, the logical problem at the basis of both systems is the old Parmenidean problem, 'How can that which is basically at the root of all that is (mind for the Hegelians, being for realists) be simultaneously both affirmatively and negatively the reason why each thing is what it is?'

Whereas Hegelians, accepting the Kantian critique, which concludes that understanding is always mediate and unable to reach things in their own reality, but rejecting Kant's impasse at the antinomies of reason,[167] seek to find in reason, that is, in constructive reason, the lost intuitive contact with the foundation of all that is, and thus elaborate the doctrine of the concrete universal as concrete unity in diversity attained by the self-creative activity of self-actuating reason,

---

[165] Yves Simon, *The Great Dialogue of Nature and Space,* New York, Magi Books, 1970, p. 66. Emphasis added.

[166] *Ibid.*, p. 65. Emphasis added.

[167] See J. E. Griffiss, "The Kantian Background to Hegel's Logic," in *The New Scholasticism,* XLIII (1969), p. 517.

on the other hand, philosophers of being, such as F. H. Parker[168] and Y. Simon[169] regard the understanding as enjoying its own type of immediacy with the concrete world of experience, distinct from the immediacy of sense, so that, for them, the abstract universal is not *merely* abstract,[170] but the intelligible face of present reality[171] as being. Being as thus something present to the intelligence, or to intuitive reason, is the foundation both of its being present and of its being what it is as distinct from what it is not. This being of things, inasmuch as it is that which is universally pervasive and together the concrete determinant of the being of each thing is just as much a concrete unity in diversity as is the concrete universal of Bradley, Bosanquet and Collingwood. It is "a universal which determines its own particularization." And "in this sense the universal conceived by Aristotle was, in intention at least, concrete . . ."[172]

> The movement of thought which began with Galileo and which founded the natural sciences may be described as the insurrection of matter against [Aristotle's] imposed determination by form. It developed in conscious opposition to the doctrine of 'substantial forms', and its fundamental principle of efficient causation was nothing less than the assertion that the existence of things is determined by other principles than those which determine their kind. The revolt was successful, and destroyed for ever after the claim of the Aristotelian universal to be concrete.[173]

N. Interim Summary

The foregoing sections of this chapter have taken issue with Collingwood's position that propositions are meaningful only as answers to questions and that knowledge therefore is a matter of propositions taken in union with the questions they propose to answer. First, Collingwood's *reason* for asserting his position was questioned, namely his opposition to Cook Wilson and the Oxford Realists and their realist principle that "knowledge makes no difference to what is known." It was shown in what sense this principle is quite legitimate, and that, if this legitimate sense is denied for the reasons alleged by Collingwood, then Collingwood's own theory of conscious reflection over feelings is likewise prejudiced. Then, second, it was shown that many

---

[168] See above, this chapter, p. 280 et seq.
[169] See above, p. 298.
[170] See above, this chapter, p. 292 and note 136.
[171] See above, this chapter, p. 293 et seq.
[172] M. B. Foster, "The Concrete Universal: Cook Wilson and Bosanquet," in *Mind*, XL (1931), p. 1.
[173] *Ibid.*, p. 2.

scientific discoveries have been made which were independent of, and sometimes quite opposed to, the questions in the minds of the researchers. Third, it was shown that the very questioning activity of the mind is not solely a matter of an active domination of the subject matter by the mind, as the Baconian and juridical method of interrogating a recalcitrant witness suggests, but itself involves a respectful receptivity towards that which is questioned, whether person or document. And furthermore, the questioning involved in historical research was shown, following Marrou, to presuppose a basic sympathetic connaturality with the questioned object, in which the basic attitude of the questioner with respect to the questioned object is that found in everyday life between members of a family, that is, of familiarity, in which the other is sympathetically respected as a friend, and the interrogation of the historical object is similar to that which takes place between friends trying to get to know one another. Thus the questioning traditionally associated with the attitude of wonder was shown to be other than that associated with behaviour proceeding from suspicion and doubt, where the predominating concern of the questioner is calculated to avoid deception. Questioning is thus not merely the activity of extorting information reluctantly given, but may be, and often is, a cooperative activity in which the questioner and his interrogated object concur spontaneously and connaturally in an exercise of familiarity. Fourth, Collingwood's epistemology was questioned from the point of view that the content of the question carried by its presuppositions might not be merely presuppositions catalytically drawn out of the mind by its own activity of supposal, but might in fact be genuine intuitive penetrations into reality present to the mind. Collingwood was criticised for being implicated in a theory of reason originating from Kant, according to which theory thought is always reflexive and mediate, enjoying no intuitive immediacy with reality. Following F. H. Parker, it was shown that the Aristotelian theory of abstraction attained universals as the structures of concrete reality present to the mind and not merely as mental classificatory devices. Then, sixth, it was shown that the abstract concepts of the mind are adapted both to the mind's demands for clarity and distinction, for which the mind requires univocal concepts of the genus-species variety, which form the stock in trade tools of the scientist, and also to the transcendent fullness of present reality, for which analogical concepts are required. The concept of being was then shown not to be an empty class concept, as Collingwood would have it, but an analogical concept which is a genuine concrete universal whose differentiations are intrinsic to the common term itself. Metaphysics is thus concerned with being as an active presence of that which is to the mind, both as regards its common being and as regards its differentials in being and not merely with presuppositions which the mind actively presents to itself solely from within its own resources. Consequently, the natural world, as philosophically viewed, is therefore not merely an understanding of objects in terms of external, spatial relations, but an active presence of that which is to mind exercising its own distinctive way of being. This ontological dimension of the natural world will be of the highest importance in Chapter 9 below, in defending, against Collingwood, the meaningfulness of propositions in themselves, as expressive of penetrations into the natural world, in addition to the meaning which they have as answers to questions, in which function their meaning is in terms of human concerns.

O. Instance in which Knowledge Clearly Demands Relation to Question

But if the foregoing sections of this chapter tend to undermine Collingwood's theory that every meaningful statement is by way of an answer to a question and that knowledge is therefore an activity including both questions and their answers, there are instances which clearly favor Collingwood's position. Perhaps one of the clearest possible instances of the meaning of a proposition being missed because the correct question was not asked is to be found in Hans Vaihinger's *The Philosophy of 'As If'*,[174] in which Vaihinger records the interesting case of Berkeley, who discovered the answer to a question he had never asked, and because he had not asked the question that he should have asked when he had its answer in his hands, he failed to appreciate exactly the knowledge he had within his grasp through not seeing what he did know as precisely the answer to a crucial question.

The context is that of the use of systematic errors, especially in mathematics. This methodical procedure consists in allowing a deviation or contradiction to enter one's system, provided that a compensating correction is admitted elsewhere into the system in order to cancel out the original deviation.

> If, in fictions, thought contradicts reality, or if it even contradicts itself, and if in spite of this questionable procedure it *nevertheless* succeeds in corresponding to reality, then—and this is a necessary

[174] English translation by C. K. Ogden, London, Routledge and Kegan Paul Ltd., 2nd edition, 1925. See pp. 109-124 on the method of antithetic error and the case of Berkeley's involvement therewith.

inference—*this deviation must have been corrected and the contradiction must have been made good.*[175]

Where the correction that must be made is that of aligning mental constructs with reality, the logical procedure is called the *method of correcting deviations made arbitrarily.* Where self-contradiction is wholly within the logical construct itself, as when something is added to one side only of a mathematical equation and the correction must be made in the only way possible, namely by an equivalent, cancelling error of an opposite nature (one is reminded of the Hegelian negation of the negation), the procedure is called the *method of antithetic error.*[176]

To the method of correcting deviations between thought constructs and reality, and the method of antithetic error, which remains wholly internal to the logical system, we may add a third type of systematic correction which is confined wholly to experienced reality. As an instance, a structural problem in the engineering complex of the 200-inch Mount Palomar telescope may be cited.[177] The mirror and the observer's cage were to be located at opposite ends of a tube suspended pivotally at a gravitational centre between them. The problem was to support the 40-ton mirror assembly in any desired position in mid-air, and sixty feet away at its focal point float a six-foot house full of mirrors and instruments and carrying an observer several tons in all—and at no time let the two elements slip out of line with each other by more than 8/100 of an inch. Such rigidity seemed an impossibility from an engineering point of view. Serrurier gave a brilliant solution to the problem by first agreeing with the engineers that such rigidity *was* impossible. Of whatever strength the tube may be those great weights suspended at both ends were bound to bend it. The solution was "very well, then let them bend it. Simply design the tube," he said, "so that the mirror and the observer's cage *will move together, by the same amount and in the same direction,* and *thus in relation to each other stand still.*"[178]

To define a circle as an ellipse is a self-evident error, but if one adds that the distance between the foci is zero, the assertion immediately becomes intelligible. One error has been made good by another. It is logically nonsensical that a distance equals zero; a distance equal to zero is no distance at all, but the non-occurrence is regarded as

---

[175] H. Vaihinger, *The Philosophy of 'As If'*, p. 109. Emphasis in original.
[176] See *ibid.*, p. 109.
[177] See D. O. Woodbury, *The Glass Giant of Palomar*, London, Heinemann, 1940, pp. 226-228.
[178] D. O. Woodbury, *op. cit.*, p. 226. Emphasis added.

an occurrence in the negative sense. Two mistakes have been committed here: first, the assertion that the circle is an ellipse, which is compensated for by the second, namely, that a certain distance be allowed to equal zero, which if literally taken is the same kind of contradiction as it is to say that a circle is an ellipse.[179]

Many problems in mathematics are solved by a procedure in which a discrepancy introduced into the data at the beginning is cancelled out when the process is finished. A problem for a long time insoluble was the determination of the conditions under which a line a divided into parts x and $a - x$ shall have $x^2 (a - x)$ at a maximum.[180] Fermat solved it by an arbitrary substitution of $x + e$ for x. This transforms the original $x^2(a - x)$ into $(x + e)^2 \cdot (a - x - e)$ and *treats the latter as if equal to the former when in fact it is not*. Thus, $6^2(9 - 6) = 108$, whereas $(6 - 1)^2 (9 - 6 - 1) = 50$. The original and the transformed equations are then expanded as follows:

$$x^2(a - x) = x^2a - x^3 \quad \text{.........(1)}$$

$$(x + e)^2(a - x - e) = (x^2 + 2ex + e^2)(a - x - e) \quad \text{.........(2)}$$

$$= ax^2 + 2aex + ae^2 - x^3 - 2ex^2 - e^2x - ex^2 - 2e^2x - e^3$$

Equating (1) and (2) gives:

$$x^2a - x^3 = ax^2 + 2aex + ae^2 - x^3 - 2ex^2 - e^2x - ex^2 - 2e^2x - e^3 \quad \text{.........(3)}$$

This (i.e., equation 3) reduces to the following:

$$2aex + ae^2 = 3ex^2 + 3e^2x + e^3$$

. . . which in turn reduces to:

$$2ax + ae = 3x^2 + 3x^3 + e^2$$

But, in the next step, Fermat simply cancels out the previously introduced error by saying that $x + e$ was a *catalytic fiction* merely to facilitate the calculation. The equivalence of equations (1) and (2) is possible only if $e = 0$, and then all expressions containing e drop out, giving

$$2ax = 3x^2$$

$$2a = 3x$$

$$\frac{2a}{3} = x$$

This means that if a line a equals 12 inches, the only division in which a part x and the remaining part $a - x$ will give a maximum value for

[179] See Hans Vaihinger, *op. cit.*, pp. 112-113.
[180] See *ibid.*, pp. 113-115.

$x^2(a - x)$ will be when $x = 8$ and $a - x = 4$, and $8^2(4) = 256$. Every other division gives a smaller result than that.

Fermat's reasoning in obtaining this solution was that the fictitious quantity $x + e$ is not equal to the quantity x if e is a real value, but is equal only if $e = 0$. The whole method is based on *first assuming e to be real* and *then equating it to zero*. Fermat called this an *adaequalitas*, that is, something approximating a complete equation being treated *as if* it were a complete equation, although according to strict mathematical and logical code x can never be equal to $x + e$. Nevertheless, a correct result is obtained by interpolating the fiction $x + e$ and equating it with x. Fermat simply *deliberately made one error and cancelled it in the course of the procedure by allowing the catalytic quantity e to drop out*. The equality then ceases to be imaginary and becomes real in the end. By this antithetic operation an important and hitherto unobtainable result was achieved.

Another somewhat similar procedure is adopted to solve second order equations, such as $x^2 + px = q$, which are solved by introducing the auxiliary quantity $\left(\frac{p}{2}\right)^2$ so that $x^2 + px + \left(\frac{p}{2}\right)^2 = q$. But, *to correct the error* of adding that quantity to one side of the equation, *an antithetic operation* is immediately performed by adding the same term to the other side, giving $x^2 + px + \left(\frac{p}{2}\right)^2 = q + \left(\frac{p}{2}\right)^2$.

The equation can now be solved thus:

$$\left(x + \frac{p}{2}\right)^2 = q + \left(\frac{p}{2}\right)^2$$

$$x + \frac{p}{2} = \pm\sqrt{q + \left(\frac{p}{2}\right)^2}$$

$$x = \pm\sqrt{q + \left(\frac{p}{2}\right)^2} - \frac{p}{2}$$

Mathematical thought arrived at this conclusion by adding $\left(\frac{p}{2}\right)^2$ to both sides of the equation, the addition to the right-hand side being *to correct the error* of adding it to the left side.

The second procedure resembles that of Fermat in some respects, but whereas in the first case the methodology consists in the *introduc-*

tion of a mental fiction (an unreality), in the second case the methodology consists in the deliberate *introduction of error into the reality of the case*, requiring a *corresponding balancing counter-error*. Upon this Vaihinger makes the following comment:

> I am not claiming to have reduced the logical functions to mathematical ones, but to have shown the formal identity of scientific methods in the various fields of scientific inquiry, and in particular of methods which attempt to attain their goal by an alteration of the given facts, by an arbitrary deviation from reality. The mathematical example given merely happens to be the most transparent and shows how it is conceivable that thought should be able to progress precisely *because* of such deviations from reality. The logical function finds itself faced by the stubborn data, the material of thought, but without allowing itself to be frightened thereby it arbitrarily alters the facts, sets the ideas in motion, and then quietly changes the mistakes it first made. The formal identity of these logical devices is thus obvious.[181]

Vaihinger then discusses how Berkeley showed that the differential calculus can be similarly explained by reduction to Fermat's schema. He concluded that a double error had been made, and his objections were developed at great length in fifty sections of his neglected work, *The Analyst*.

But Vaihinger says that in manifesting the contradictions in the mathematical operations of the calculus, instead of recognizing that he had made a brilliant discovery regarding mathematical methodology, Berkeley rejected it as illogical. His conclusion was that mathematicians had no right to scoff at the incomprehensible mysteries of Christianity, since their own subject suffered from exactly the same illogicalities. Vaihinger maliciously adds that "like a good Englishman . . . the Bishop of Cloyne did not realize that he would then also have to reject the basic principles of Christian theology."[182] Vaihinger comments that here "we have a strange spectacle, and one not likely to occur again in the history of science, of a thinker *with a key to a problem in his hands—but without the problem itself*."[183]

The explanation Vaihinger offers is as follows. Irritated by mathematical 'free-thinkers' expressing themselves sceptically about the incomprehensible mysteries of Christian dogma, Berkeley tried to direct the mathematicians' attention to their own sorry plight and recognize that their fluxions were full of contradictions. In doing so,

[181] H. Vaihinger, *op. cit.*, pp. 116-117. Emphasis in original.
[182] *Ibid.*, p. 117.
[183] *Ibid.* Emphasis added.

quite by chance he discovered and demonstrated that the method proceeds by cancelling one error with another and thereby arriving at the correct result.

But, Vaihinger carefully points out, "the real problem was and still remains this:—How comes it that in mathematics a correct result can be obtained by means of contradictions such as those involved in fluxions?"[184] "The answer to the problem," Vaihinger says,[185] "is that the correct answer is obtained by the method of antithetic error."

Berkeley's effort was confined merely to showing the existence of contradictions in the operations of fluxions. He actually accomplished more than he intended, but *because he did not have the relevant question in his mind he failed to see the discovery he had made,* which discovery is seen for what it is *only when seen as an answer to that question.*

> [Berkeley] accomplished more than he intended for he also discovered the answer to a question that he had never asked. And yet this answer should have led him to the correct question. We certainly have here an unusual occurrence in the history of science.[186]

It is doubtful whether a more forceful example of Collingwood's thesis, that knowledge is by way of an answer to a question, could be found than the one recorded by Vaihinger. Had Berkeley seen the *propositional data* which he had *as an answer to a question* he would have seen what those propositional data *meant*. He knew that it was a fact that there were contradictions in the procedures of calculus (fluxions). He stopped there. Had he asked 'Why do such contradictory procedures lead to correct results?', he would have seen those contradictions for what they really are in those mathematical operations, namely, the method of antithetic error, and not merely blatant illogicalities which are all that they appeared to be to him.[187]

P. General Conclusion Regarding Collingwood's Identification of Knowledge with the Mind's Questioning Activity

What these arguments both for and against Collingwood's position manifest is that questions and their answers can exist quite inde-

---

[184] *Ibid.*, p. 118.

[185] *Ibid.*

[186] *Ibid.*

[187] Having read this incident as reported by Vaihinger, what is one to think of the following statement? "The Renaissance studies of Plato suffer from . . . the mania of trying to find in a philosopher answers to questions he had never asked." Emile Bréhier, "The Formation of Our History of Philosophy," translated by Mary Morris, in *Philosophy and History*, The Ernst Cassirer Festschrift, edited by Raymond Klibansky and H. J. Paton, New York, Harper Torchbooks, Harper and Row, 1963, p. 163.

pendently of one another. Questions may be asked which, perhaps quite by chance, may only be answered generations or perhaps centuries later.

> . . . a knowledge that is incomplete in a thousand ways may still be the answer to a certain question. Someone may raise a question a few centuries before the answer is obtained, and one day when the answer to precisely this question is obtained, we do not yet know everything, we do not know the whole about anything, but we have nevertheless obtained a true answer, a true proposition.[188]

Then again, as in the Berkeley case narrated by Hans Vaihinger above,[189] a scholar may possess the answer to a question without possessing the question to which it is an answer. Collingwood himself recognizes this. During his archaeological investigation of the Cumberland Wall as a sentry walk, Collingwood theorized that if the wall was not a defense barrier but in fact a sentry walk, "there ought . . . to be a chain of towers, not connected by a wall but otherwise resembling those on the wall, stretching down that coast."[190] To the relevantly arising question "did such towers exist?" the answer in the affirmative was revealed by "search in old archaeological publications [which] showed that towers of exactly the right kind had been found; but their existence had been forgotten, as generally happens with things whose purpose is not understood."[191] Now the crucial point here is that had the contributors to the old archaeological publications merely recorded, in strict accordance with Collingwood's proposed Baconian method of rigidly adapting findings to a preconceived questionnaire, only what answered *their own* questions, they may never have seen fit to record such information as was to prove crucial to the answering of questions raised later by Collingwood. *They* had the answer to Collingwood's question *without having the question*. But had they discarded that information as being irrelevant to their own investigations, Collingwood later would have had his question *without having the answer* to it. Again, Collingwood writes:

> In England, where Roman archaeology has gone forward incessantly in most parts of the country ever since the seventeenth century, there is a gigantic bulk of material from which many questions of this kind [concerning Graeco-Roman population statistics] can be answered, if not conclusively, at least within a reasonable margin of error.[192]

---

[188] Yves Simon, *The Great Dialogue of Nature and Space*, New York, Magi, 1970, p. 68.
[189] See this chapter, pp. 301-306.
[190] *An Autobiography*, p. 129.
[191] *Ibid.*
[192] *Ibid.*, p. 135.

Had the collectors responsible for this gigantic bulk of material simply dealt with it as precise answers to their own questions, it would not be subsequently available as a means of answering many newly arisen questions. This stock of information, in the form of statements, is valuable *as such* even before the questions arise to which it will provide the answer.[193]

The conclusion of all the foregoing, therefore, is that propositions, besides being answers to questions, have a meaning *in themselves in an absolute way*. They not only relate to the questioning mind as answers to its questions, as the satisfaction of its inquiry, but they relate also to objective reality and have meaning as so related, as signifying real-

---

[193] Leslie Armour, *The Concept of Truth*, Assen, Van Gorcum, 1969, p. 213, agrees with Collingwood's criticism of the archaeologists who "were in the habit of ransacking historical sites, piling the loot neatly, and then trying to see what 'natural kinds' the objects sorted themselves into. The main result, according to Collingwood (and he was probably right) was simply the destruction of a great deal of historical evidence. . . . The archaeologists were foolish because they failed to ask themselves such simple questions as 'How many different sorts of people lived on this site and in what order?'—a question which would have readily suggested to them the fact that the order of the material in an ancient garbage heap might well be extremely important and that not all objects which looked more or less alike necessarily belonged together." Armour points out that "if one's interest in the objects is purely aesthetic, the classification of objects might well be very different than if one's interests were both aesthetic and historical or only historical. Again, if one's interests lay in the molecular structures of the garbage, one would get a still different set of classifications." He concludes that "the same data . . . can yield a variety of different kinds of information depending on the questions one asks. [And] if one asks *no* questions, it does not seem that the data could yield any information." Armour and Collingwood are certainly correct in maintaining that the yield, in terms of intelligible meaning content, of the excavation results depends on the direction of the excavator's questions, but exclusive insistence on the role of the excavator's questions, dominated as they will be by the excavator's interests in the 'dig,' may result in the very thing which both Collingwood and Armour warn against, namely the destruction of evidence, for future investigators may well find in the same excavation yield answers to *their* questions which may be questions the original investigators never envisaged. The warning is that an excavator who approaches an excavation site with aesthetical questions in mind should also keep in mind the possibility that other investigators, now or at some future time, may be interested in the results of the excavation from the optic of historical or chemical (scientific) questions. If he is not careful to preserve or record the results of his investigations in such a way as to be available to these other researchers, he may well destroy forever the very evidence that these other investigators need to answer *their* questions provoked by their respective interests. Criticism of indiscriminate digging applies well to men like Belzoni (see C. W. Ceram, *Gods, Graves and Scholars*, New York, A. Knopf, 1968, revised ed., pp. 117-120), who ransacked Egyptian tombs with no other motive than personal glory or monetary gain. But an excavator with a minimum of scientific respectability will approach any excavation with reasonable caution so that his work will yield the maximum of information, answering not only his own particular questions, but, as far as possible, being available to answer subsequent, and perhaps hitherto unknown, questions.

ity.[194] The Berkeley case as narrated above by Hans Vaihinger clearly shows this. That is, the meaningful content of a proposition does not totally derive from the question to which it is the answer, but also from its relationship to objective reality which it is able to present to the mind and thereby answer the mind's questions addressed to reality. This means that statements made in any historical period are not restricted in their meaningfulness to the questions currently raised during that period. Collingwood's insistence on the matching of meaning and evidence exclusively to questions involved serious defects in his archaeological work, as is testified by I. A. Richmond.[195] "That excavation should be conducted with specific problems in mind upon sites likely to provide an answer, was salutary and useful,"[196] but its "corollary, that to pose a problem permitted its answers to be predicted was a product of study rather than the field,"[197] and involved Collingwood in prediction rather than discovery: ". . . at King Arthur's Round Table, near Penrith . . . he had made up his mind in advance what he was to find and found it with fatal precision."[198] The attitude of natural piety, which Collingwood regarded as a vice in Alexander,[199] but which is a virtue absolutely essential to any researcher, was completely lacking in Collingwood. "He had . . . supreme intellectual confidence which prevented him from being taken by surprise . . ."[200] This exclusive attention to knowledge as active questioning, characterized by "a power of artistic criticism and draughtmanship,"[201] stifled that sense of wonder and openness which recognizes that "the expected is always accompanied and often overshadowed by the unexpected" and that "an excavator's first duty is to see everything without the blinkers imposed by prearranged concentration of vision."[202]

---

[194] See below, Chapter 9, p. 430 et seq.

[195] See Richmond's "Appreciation of R. G. Collingwood as an Archaeologist," accompanying obituary notice in *Proceedings of the British Academy*, XXIX (1943), pp. 476-480.

[196] *Ibid.*, p. 478.

[197] *Ibid.*

[198] *Ibid.*, p. 479.

[199] R. G. Collingwood, *An Essay on Metaphysics*, p. 174.

[200] R. B. McCallum, "Robin George Collingwood," accompanying obituary notice in *Proceedings of the British Academy*, XXIX (1943), p. 467.

[201] I. A. Richmond, *art. cit.*, p. 476.

[202] *Ibid.*, pp. 478-479.

CHAPTER EIGHT
Logic and Questioning

A. The Awakening of Reason and the Birth of Logic

Ernst Cassirer tells how the ancient Egyptians held in special reverence the moon-god Thoth, "the scribe among the gods," from whose hands writing came into being and who "allows gods and men to know what is their due; for he determines the measure of things."[1] Speech and writing—the precondition of all human intercourse and community—"are valued as the origin and measure; for the capacity to fix the fleeting and changing, and thereby remove the accidental and capricious, inheres, above all, in them."[2]

Cassirer says that already in myth and religion it is detected that human culture is not self-evidently given, a mere datum to be accepted as such, but a miracle of some sort which has to be explained. This feeling generates the deeper self-awareness which not only drives man to *pose the questions* which ask for that explanation, but dares him to go further and to construct an authentic and independent method by which he can *answer* them. The Greeks first took this decisive step, and thereby revealed the new power which alone could lead to a

---

[1] Ernst Cassirer, *The Logic of the Humanities*, New Haven, Yale University Press, 1960, p. 44.

[2] *Ibid.*

science of nature and a science of human civilization. Thought awakened strives to grasp the unity that pervades the whole of reality.

> In place of the unrestricted multiplicity of the mythical attempts at explanation, which focus on first one phenomenon and then another, there emerges the representation of a thoroughgoing unity of being, which must correspond to a similar unity of explanation. This unity is accessible only to pure thought. The diverse and numerous creations of the mythmaking imagination were henceforth rejected and eradicated by the critique of thought.[3]

Thus, intellectual awakening occurred with this awareness that a radical unity pervaded the hitherto discordant multiplicity. From its own resources critical thought had to supply the answers which would replace the destroyed multiplicity of mythical explanations. This task proceeded from the admirable consistency of the Pre-Socratics to the solution given in Plato's theory of ideas and Aristotle's metaphysics, which remained decisive and exemplary for centuries. Many apparently diametrically opposed tendencies contributed to the statement of the question and solution of the problem. The unifying concept in the diversity of the labour of Greek thought is that of *logos*.

> . . . if we look at it in terms of its origin and goal, this whole colossal labor of thought, in a certain measure, yields to being comprehended within *one* fundamental concept, which Greek philosophy was the first to discover and which it worked out and perfected in all its factors. In the development of Greek thought the concept which played this role was that of *logos*.[4]

By means of this concept, Heraclitus saw the world as a totality of materials which reciprocally metamorphose into one another. The Ionians were not satisfied to know merely the "what" of things, but demanded also to know their "how" and their "why."[5] Heraclitus faced these questions with the realization that perception alone could not give a satisfying answer. The answer could be given only by thought, for thought alone permits a man to transcend subjective enclosure within his own private individuality. "He no longer pursues his 'subjective understanding,' but comprehends one that is universal and divine. A universal law [of the world] replaces . . . 'private insight.' "[6] With the recognition of this *logos*, this 'universal

---

[3] *Ibid.*, pp. 44-45.

[4] *Ibid.*, p. 45. Emphasis in original.

[5] These are three of the four questions which Aristotle later will systematize into the scientific interrogation of reality as distinct from dialectical interrogation of prevailing beliefs and convictions about reality. See later, this chapter, p. 330 et seq.

[6] E. Cassirer, *The Logic of the Humanities*, p. 46.

law of the world,' a man is first liberated from his mythical dream-world of imagination and narrow and limited world of private sense perception.

> For just this is the character of being awake and of awakened being—that individuals possess a common world, whereas in dream(s) each lives only in his own world and remains mired and imprisoned in it.[7]

Western thought found it impossible ever to deviate from this new direction imparted to it by Greek thought.

> From the time this thought first permeated the school of Greek philosophy, all knowledge of reality was bound, to a certain extent, by this basic concept of '*logos*'—hence by 'logic' in the widest sense.[8]

Logic in its widest sense, therefore, is this acceptance of thought as bounded by a radical principle of unity and coherence at the basis of everything that is, and it is in terms of this principle that all meaning and intelligibility is to be found.

This newly awakened consciousness, searching for the explicative principle of all that is, soon became commercialized by the Greek sophists, for whom "what they believed to be wisdom (because they confused wisdom and power) was but an art of seducing and persuading minds."[9] For them the intellectual life was a sport, a contest, a battle of eloquence whose sole purpose was a public victory over their adversary's thesis, whether by fair means or by foul. The audience played the primary role, as the verdict of the bystanders was the final judgment. These men were professors of the 'successful' life through the techniques of opportunism, especially in the political field.[10] But by the dialectic of history "they made Socrates possible,"[11] whose "critical spirit (critical in the philosophical sense of the word) apprehends *questions* and tests ideas according to their *intrinsic value*, independently of the judgment of the crowd."[12] What matters now is not the blows exchanged in the arena of public contestation, but what is achieved in the workshop of trial and proof within the mind itself.

---

[7] *Ibid*. One recalls Collingwood's transcendence of art and religion and emergence of the scientific consciousness, as described in *Speculum Mentis*.

[8] E. Cassirer, *op. cit*., p. 46.

[9] J. Maritain, *Moral Philosophy*, London, Bles, 1964, p. 5.

[10] See *ibid*., p. 5.

[11] See *ibid*.

[12] *Ibid*.; emphasis added.

"The passion for truth has taken the place of the passion for success."[13]

From this, the notion of science clearly emerged, the notion of authentically intellectual knowledge, established on the level of the intellect's own lights and proper demands.[14] The first and most basic demand of authentically scientific knowledge is the search for what things are, their definitions. Scientific discussion regarding what is just and what is not just all turns on our ability to answer the question of what is justice.[15] This is extremely important in evaluating Collingwood's theory of the logic of questioning, for on the one hand he rejects the definitional 'what' question as vague and meaningless, and to be substituted for by the 'which' question, since the 'what' question offers no alternatives;[16] on the other hand, he frequently sets out from a 'what' question, although, during the course of the investigation, he transforms it into a 'which' question, dealt with by 'the

[13] *Ibid.*, pp. 5-6. Aristotle will later distinguish in his *Topics* (VIII, 5, 159a 25-40) argument *for the sake of investigation and teaching* from the argument of those engaged in a competition. In the first case, there is a *common seeking of a common aim,* namely knowledge of some subject matter; in the second *each* contestant seeks to win over the *other,* not to achieve a common purpose. Bacon's scientific method of interrogating a recalcitrant and hostile witness is of the latter type.

[14] J. Maritain, *Moral Philosophy*, p. 6.

[15] See *ibid*.

[16] See Collingwood's *The New Leviathan*, p. 74, par. 11.12: "A question that offers no alternatives is a bogus question. The technique of knowing proper, or what is called scientific method, depends on replacing questions which, being vague or confused, are unanswerable, by real questions, or questions which have a precise answer. (11.13) The vague question 'What do I want?' is thus replaced by the precise or real question: 'Which do I want, a or b?' " See also Collingwood's depreciation of definition in his *Speculum Mentis*, p. 111: "Giving and collecting definitions is not philosophy but a parlour game." Louis O. Mink, in his "Collingwood's Dialectic of History," in *History and Theory*, VII (1968), p. 4, writes: He [Collingwood] did *not* ask or attempt to answer the question "What is History?" Elsewhere he specifically renounced all questions of the form 'What is X?' " Mink refers us to pars. 31.61-31.68 of Collingwood's *The New Leviathan*, where we read: "Ancient [Greek] sciences . . . defined their aims by asking questions like: 'What is Nature?' 'What is man?' 'What is Justice?' 'What is Virtue?' A question of this sort was to be answered by a definition of the thing. From this definition, which had to state the 'essence' of the thing defined, implications could be derived, each implication being the statement of some 'property.' . . . To seek for x a single essence from which all the 'properties' of x are logically deducible is to propose a science of x with an unlimited objective. That is implied in any question of the form: 'What is x?' To a question in this form, for example: 'What is Nature?' modern science answers: 'I do not know. What the essence of Nature is nobody knows, and nobody need care. When they asked that question the Greeks were asking a question too vague to be precisely answered.' " Correlate this statement by Collingwood with Marjorie Grene's comments on Aristotle's *nous;* see Chapter 7, above, pp. 281-282 n. 89. See also criticism of this position, this chapter, below, pp. 337-338 and pp. 347-349.

plain historical method.'[17] This transformation of 'what' questions (such as what is art? what is religion? what is science?) into questions about the idea of art held by certain peoples at various historical periods, about the idea of religion, about the idea of science or about the idea of nature prevailing at certain periods, is the transformation of the scientific (in the strict sense; since science in the strict sense is dominated by knowledge of what the subject matter is *in itself*—a question out of bounds in European thought since the Kantian

[17] Thus in *The New Leviathan*, p. 1, pars. 1.1. et seq.: "What is man? . . . [which] is asked because we are beginning an inquiry into civilization, and . . . (1.13) civilization is a condition of communities; so to understand what civilization is we must first understand what a community is. (1.14) A community is a condition of men . . . so to understand what a community is we must first understand what men are. (1.15) This gives us the scheme of the present book . . ." This is nothing other than the old Aristotelian order of interrogation: to know why men are civilized we must know what community life is; to know what community life is we must know what men are precisely as men, that is, what is man? That is, properties are known ultimately as necessitated by the essence of the subject. The essence, expressed by the definition, of the ultimate subject of discussion—the answer to the question 'What is it?' regarding the subject—is the point of departure for any scientific discussion. In the case of *The New Leviathan* this definition is that man as man is subject of community living and by medium thereof of civilized living. But Collingwood does not follow through with this method outlined at the beginning of *The New Leviathan*. On p. 61, par. 9.1 et seq., he says that "the account of man as mind given in this first part, the account of community to follow it in the second, and the account of a civilized community to be given in the third, are all constructed on what Locke called the 'historical plain method.' (9.11) The essence of this method is concentration upon facts. . . . (9.16) A study of mind on the historical method involves two renunciations. First it renounces with Locke all 'science of substance'. It does not ask what mind is: it asks only what mind does. (9.17) You can have your cake and eat it too by holding that mind is 'pure act', so that the question *what mind is* resolves itself without residue into the question *what mind does*; but [note well]: *whether this is defensible I shall not ask*. [Latter emphasis added.] (9.18) Secondly, it renounces all attempts to discover what mind *always and everywhere does*, and asks only *what mind has done* on certain definite occasions . . . [and] (9.21) whatever I need to know for this purpose is about the modern European mind. . . ." But this evasion of the 'what' question at one level reinvolves it at another. For in order to know what has been done by *the modern European mind* we need to know *what the modern European mind is* in order to know what is to be included and what is to be excluded from one's 'plain historical method.' A clue to the reason why Collingwood slips out of gear from one method into another, proceeding, as he does, as a dialectician, is given by what Aristotle says in the *Topics*, Bk. 8, Ch. 2, 158a 15-20: "Not every universal question can form a dialectical proposition as ordinarily understood, e.g., 'What is man?' or 'How many meanings has the good?' *For a dialectical premiss must be of a form to which it is possible to reply 'Yes' or 'No'*, whereas to the aforesaid it is not possible. *For this reason questions of this kind are not dialectical unless the questioner himself draws distinctions or divisions before expressing them*, e.g., 'Good means this or this, does it not?' For questions of this sort are easily answered by a yes or a no." The straight 'what' question is a direct interrogation of *reality*, hence a *scientific* question; but 'is it this or this?' transforms it into a question about *our knowledge* of reality, that is, a dialectical question. See later in this chapter, pp. 349-350 below.

critiques) questioning and investigation *of reality* into dialectical questioning and investigation of *what people think* or *have thought* about reality. This is the core critique against the whole Collingwoodian enterprise: the most basic scientific (in the strict sense) question is put out of bounds by exaggeration of the historical dimension in human knowledge and consequent absorption of all inquiry into dialectical inquiry. The root cause of this is a historical cause, the rupture of viable contact of the understanding intellect with concrete reality in itself by the Kantian critique, and the Hegelian alleged restoration of rational contact with reality by historical/dialectical self-creating, self-unfolding reason.[18] To argue, as Collingwood does in his *Autobiography*,[19] that 'what is the State?' is a different question for Plato and for Hobbes is to presuppose that the only knowable object regarding the State is what Hobbes or Plato or Marx or somebody *thought it to be*, and to put out of play completely the vital scientific question 'what is the State?' which means 'what *is* the State *really*?' as distinct from what anybody *thinks* it to be. If one argues, with Collingwood, that the 'what' question such as 'what is the State?' is reducible without residue to what a, b, c, d, etc., *thought* it to be, then one thereby abandons the properly scientific and philosophical enterprise which began with the discovery by the Greeks of all-pervading *logos*, and lapses into a new though perhaps more refined sophism, which masks its sophistry behind historical respectability.

Jacques Maritain warns against misunderstanding the Socratic ignorance. It is an ignorance *of which we are aware*, and one of which Socrates took great pains to make us conscious. But Maritain asks, "If the idea of science were not there, would I have the idea of my ignorance?"[20] For Socrates, ignorance is something pointing to something else. As blindness would be meaningless if sight were not a reality (trees lack sight but are not blind; the term is meaningless in their regard) so ignorance, being ignorance, would be meaningless if science were not a reality. Awareness of ignorance presupposes awareness of knowledge, of knowledge which dispels the ignorance, whose known absence constitutes that very ignorance. Maritain cites Aristotle as assuring us that Socrates strives in regard to the moral virtues to discover their definitions. "He [Socrates] is in search of *what things are*: because he applied himself to doing syllogisms, and the

---

[18] See Chapter 7 above, pp. 298-299.
[19] Pp. 61-64.
[20] J. Maritain, *Moral Philosophy*, p. 6.

principle of syllogisms is *what things are*."[21] Maritain comments as follows:

> The whole Socratic enterprise bears witness to the fact that it must in the end be possible to define the virtues. And thus what matters above all is that the ideal of a knowledge which is firm and incontestable in itself, a *science* of moral matters, is now brought out.[22]

With Socrates and Plato, the element of ignorance and consciousness thereof prevailed—as is clear from the variety of opinions on different matters, each ably defended by one school of sophistry or another—and the search for definitions remained a protracted search rather than an achievement. But with Aristotle, reaping the fruits of the constructive questioning of Socrates and Plato, a solid core of well-defined scientific knowledge becomes available. The dialectic give-and-take between opinions culminates, at least to some extent, in the insight from which true science begins, the insight into what (e.g., movement or nature) is. Definitions, as expressing insights into the constitution of what things are, do not come out of the blue, but only after perhaps years or even generations of discussion, sifting what the various schools of thought held on a question. The 'state of the question'—that is, the ordered catalogue of opinions *pro* and *con* a given thesis—is still the normal and natural lead-up to the true beginning of any scientific inquiry. Having considered all the 'points of view,' one is proximately disposed to 'see,' to have 'insight,' into 'what the thing is' which previous discussion has been about. But once this insight is secured[23] into what the subject matter is (expressed by the real definition), the opinions drop out of the picture as no longer relevant. They were approaches to the principle or principles of the subject-matter itself, which principles then dominate thought totally in the elaboration of genuine science. The clustering of the opinions pro and con a given question serves to focus the mind of the investigator on what exactly is the subject matter he is dealing with, so that he is made ready to have the insight into the definition which is going to function as medium in demonstration. The readiness to pass from the opinions to the insight into what will function as the demonstrative

---

[21] *Ibid.*, p. 6. See also Aristotle, *Metaphysics*, Bk. 13, Ch. 4, 1078b 18-25. Emphasis added.

[22] J. Maritain, *Moral Philosophy*, p. 6. Emphasis in original.

[23] By intuitive reason, as explained in the previous chapter; see above, pp. 280-284.

medium[24] is what Aristotle calls 'quick wit.' "Quick wit is a faculty of hitting upon the middle term instantaneously."[25]

B. The Aristotelian Contribution to the Logic of Questioning

It has already been noted[26] that Collingwood neglected to make any mention of Aristotle's contribution to the methodology of questioning. On the other hand, L.-M. Régis goes so far as to accredit Aristotle with the invention of two techniques of interrogation.[27] One such technique is designed for orderly and correct questioning of our own or of another's knowledge.

> One [technique] is for the purpose of discussing knowledge that is already possessed with other philosophers and with his own mind, in order to classify and set it in order, to defend it against opponents, to expose sophisms in argumentation, to dispose of adversaries whose contentions are irrational and contradictory. This first technique is called "dialectics" and is set forth in the *Topics*, as are its three functions concerning properly philosophical knowledge.[28]

The other technique is for the direct interrogation of reality, and constitutes the basis of strictly scientific and philosophical investigation.

> The other technique is the very backbone of scientific and philosophical knowledge and consists in the direct interrogation of reality by the intellect, an interrogation made up of a precise number of questions to be answered by the universe under the threat of remaining obscured by a heavy veil, hiding it from the intellect and leaving visible only its outer contours. This questionnaire is fully set forth, along with its vindication, in the second book of the *Analytics*.[29]

In another work[30] Régis notes that Aristotle uses the word *logic* only in connection with the dialectical treatise. Only in the *Topics*[31] does Aristotle use the word *Organon* (ὄργανον), whereas for the theory of demonstration or science he uses the word *analytics* (ἀναλυτικός) as in the *Prior* and *Posterior Analytics*.[32] Régis remarks that Descartes reverses this usage, understanding by the term "logic" the science

---

[24] And, as will appear later, all questioning is a search for a middle which will determine one way or another two possible alternative statements.

[25] Aristotle, *Posterior Analytics*, Bk. 1, Ch. 34, 89b 10.

[26] See Chapter 1, above, p. 27, n. 68.

[27] See his *Epistemology*, New York, Macmillan, 1959, p. 127.

[28] *Ibid.*, p. 127.

[29] *Ibid.*

[30] L.-M. Régis, *L'opinion selon Aristote*, Paris, J. Vrin, 1935, p. 208, n. 2.

[31] As in 163b 11; 105a 21; 108b 32.

[32] As in *On Interpretation*, 19b 31 and *Topics* 162a 11.

of demonstration. For Aristotle, then, the *Analytics* is the method of proceeding scientifically and must be known before the scientific work begins, not learnt during the scientific procedure.[33] In his *Metaphysics*[34] Aristotle insists upon this necessity to have already been instructed on the way to proceed in each subject-matter, for it is absurd to seek at one and the same time a science and its method.

The syllogism is the foundation stone of the whole Aristotelian logical edifice, but there are two ways in which it operates: scientifically and demonstratively in the *Analytics*, dialectically and inventively in the *Topics*.[35] In demonstration, the syllogism proceeds from established principles founded on the real definition of the subject-matter. In dialectic, the syllogism proceeds from generally held convictions about subject-matters, and one of its principal functions is to terminate in the discovery of real definitions, which then become the principles of demonstration in science, opinions dropping completely out of the picture.[36] Dialectical reasoning, operating as it does in the area of held convictions, is confined to the order of probable knowledge, or knowledge of what is generally 'approved' to be the case, as distinct from objectively certain scientific knowledge.[37]

C. Aristotelian Dialectical Interrogation of Human Knowledge

Aristotelian dialectic appears to be somewhat insignificant and second-rate if compared with that of Plato, for whom the dialectician is metaphysician and dialectic the pathway to knowledge of the really

---

[33] See L.-M. Régis, *L'opinion selon Aristote*, p. 211.

[34] Bk. 2, ch. 3, 995a 10-15: ". . . it is absurd to seek at the same time knowledge and the way of attaining knowledge; and it is not easy to get even one of the two." See R. McKeon's edition of *The Basic Works of Aristotle*, New York, Random House, 1941, p. 715.

[35] L.-M. Régis, *L'opinion selon Aristote*, p. 213.

[36] See James Hogan, "The Dialectic of Aristotle," in *Philosophical Studies*, V (1955), pp. 13-14.

[37] However, as A. Gardeil has shown, probable knowledge has its own genuine type of certitude; see his "La 'Certitude Probable' " in *Revue des Sciences Philosophiques et Théologiques*, L (1911), pp. 237-266 and pp. 441-485. This is extremely important in order to recognize that historical knowledge possesses a genuine certainty as distinct from opinion in the accepted sense of being merely conjectural. Jean Danielou, in his *Christ and Us* (English translation by Walter Roberts of *Approches du Christ*), London, A.R. Mowbray, 1961, p. 2, writes that ". . . documents are the sole means of access to the historical realities. This does not mean that historical realities are not capable of a certainty as great as that of the realities of mathematics or physics. It would be a distortion of the reason to suppose that this is the case. Historical certainties are just as absolute *in their own sphere* as mathematical certainties. But they require to be securely established [by] the rigorous use of methods appropriate to their subject matter."

real.[38] This is because, for Plato, the forms definitive of things, what we mean when we talk about what things are, are self-subsistent; to reach the defining forms is to reach ultimate reality. Thus, dialectic has metaphysical value. But for Aristotle, the defining forms are a constituent of experienced reality, and definition is through abstraction of that form from the contingency, mutability and multiplicity it has in experienced reality. Aristotle, like Plato, distinguishes knowledge from opinion, but, whereas for Plato knowledge is by way of reminiscence of previous direct vision of the subsistent forms as ultimate reality, and opinion is "a likely account of a likely thing"[39]—that is, merely a conjectural view of the fleeting appearances of reality—for Aristotle both knowledge—specifically, the certain knowledge of science—and opinion concern the reality of the mobile world of everyday experience. Scientific knowledge is knowledge of that experienced world in its necessary and determinate aspects, its inherent forms. Opinion, on the other hand, is a personal conviction based on motives other than determinate reality. A pupil, for example, may be convinced because of the weight of authority of a teacher that the three angles of a triangle equal two right angles without seeing how the reality of triangle, or what a triangle is, imposes this consequence. As soon as the pupil sees the implication of what being a triangle is he has no need of a master's authority as his motive for belief. Thus, the one same conviction may be either a scientific conviction, based on the compelling evidence of *what the thing is*, or an opinion, a personal conviction based on some other motive, the trustworthiness of an authority for instance. Unlike Plato, however, Aristotle attaches dialectic to opinion rather than to knowledge.[40] Distinguishing the Platonic tradition of seeking true knowledge in a sense-transcendent world from the earlier Socratic tradition, Aristotle regards himself as developing the earlier Socratic tradition.

> [Aristotle] sees himself as developing the Socratic method of definition by means of inductive arguments, starting from the ground of probabilities, that is, from common or generally accepted opinions and cases such as were brought forward, mainly in ethical questions, in the Socratic interrogatories. In his first reaction against Plato's objective idealism Aristotle appears to have reverted to the dialectic of Socrates which had the advantage of providing a method of

---

[38] See James Hogan, "The Dialectic of Aristotle," p. 3.

[39] See Plato, *Timaeus*, pp. 29-30. See also A. E. Taylor, *Plato: The Man and His Work*, London, Methuen, 1952, p. 441.

[40] See James Hogan, *art. cit.*, p. 3.

exploring, testing, and clarifying the currency of probable knowledge or opinion.[41]

Whereas the scientific or demonstrative syllogism is based on the principle that we can only get manifestly true conclusions from premisses that are manifestly true, the dialectical syllogism, although as syllogism, as a way of proceeding from premisses to conclusion, it is the same in all cases,[42] nevertheless as dialectically operative it does not follow this principle, since the premisses of the dialectical syllogism are at most probably and at least possibly true, being based as they are on prevailing beliefs and opinions.[43]

But a philosophy which takes an experiential basis seriously cannot ignore prevailing opinion. What may be scientifically evident to a scientist may be available to a non-scientist on a basis of trust in the competence and honesty of the scientist.[44] Even the philosopher, in regard to immediately evident principles, quite frequently is not able, in spite of every effort that he can arouse, to see unwaveringly the full force of such intrinsic self-evidence; and the more he tries the more it may just look like smoke. At such times he is *believing* on the strength of *his own authority* as having seen in the past the indubitable self-evidence of such principles. If such is true of self-evident principles, how much more so must it be of demonstrated conclusions, scientific laws and even the facts of daily experience.[45] Apart from these rare moments of clear vision in the presence of objectively evident and self-compelling reality, we live out our daily lives, scientists as well as lay people, in the world of belief and opinion, trusting to our witness and trusting in the witnessing of others.[46] Aristotle and Hume have much in common in this respect.

Aristotle is well aware that exact or proven knowledge is possible for only a limited part of human activity, that the greater part

[41] *Ibid*.

[42] See L.-M. Régis, *L'opinion selon Aristote*, p. 213.

[43] See James Hogan, *art. cit.*, p. 3. It will be shown later (see this chapter, below, pp. 352-353) that Collingwood, in assigning absolute presuppositions as basic convictions determining the questioning procedure, reduces scientific interrogation of reality directly to dialectical interrogation of what people basically think about reality.

[44] See A. Gardeil, "La 'Certitude Probable,' " *Revue des Sciences Philosophiques et Théologiques*, L (1911), p. 258.

[45] See *ibid.*, pp. 258-259.

[46] See M. Polanyi, *The Tacit Dimension*, New York, Anchor Books, Doubleday, 1967, p. 64: "The acceptance of scientific statements by laymen is based on authority, and this is true to nearly the same extent for scientists using results from branches of science other than their own. Scientists must rely heavily for their facts on the authority of fellow scientists."

of man's practical life lies within the vast debatable land of opinion. He is disposed to much the same view as David Hume, that philosophy originates in our attempts to systematise "the reflections of common life," though Aristotle would not have agreed that philosophy ends, as well as begins with them.[47]

Aristotle is distinguished among philosophers for the respectful attention with which he collects and builds upon the beliefs prevailing among mankind.[48] He reacts against Plato's intellectual absolutism which dismisses all that is not intelligible as mere appearance or illusion, and sees that the way to certainties may be through probabilities, beliefs and opinions. It is in Aristotle's dialectic that is to be found his counterpart of the modern free use of hypotheses, assumptions and postulates which may or may not prove to be tenable.[49]

An exhaustive study of Aristotle's dialectical method, mainly in the *Topics*, but interspersed throughout his scientific treatises, would require a lengthy separate treatment of its own. We are here concerned with it as an already well-developed theory of discovery through questioning which was completely ignored by Collingwood. Short of a thoroughgoing analysis, the following relevant points may be made in regard to it.

The first book programmes the treatise and indicates its relevance. Aristotle then defines what constitutes a dialectical proposition and a dialectical problem. Dialectical propositions exclude both what one holds and what is obvious to everyone, that is, what no one would doubt in his right senses.[50] A dialectical proposition is to be found between these extremes.

> A dialectical proposition consists in asking something that is held by all men, or by most men or by the philosophers, i.e. either by all, or by most, or by the most notable of these, provided it be not contrary to the general opinion. . . . Dialectical propositions also include views which are like those generally accepted, also propositions which contradict the contraries of opinions that are taken to be

[47] James Hogan, *art. cit.*, p. 3.

[48] *Ibid.*, p. 4.

[49] See *ibid.*, p. 4. Dialectic is not concerned with the truth or falsity of its premisses or conclusions, but merely with internal consistency. See G. Grote, *Aristotle*, one-volume edition edited by Alexander Bain and G. Croom Robertson, London, John Murray, 1880, p. 271. An exclusively dialectical procedure, as in Collingwood, will not require that the premisses of argumentation be true or false in themselves but only that they be held, believed or presupposed for sake of argument. Truth, then, in the last analysis, can be nothing more than internal coherence within the various propositions in the reasoning process.

[50] See Aristotle's *Topics*, Bk. 1, Ch. 10, 104a 1-10.

generally accepted, and also all opinions that are in accordance with the recognized arts.[51]

A dialectical proposition is therefore a statement ascertained from prevailing opinions and beliefs.[52] A dialectical problem is that about which an inquiry is conducted, which will lead to action or to knowledge.

> A dialectical problem is a subject of inquiry that contributes either to choice and avoidance, or to truth and knowledge, and that either by itself, or as a help to the solution of some other such problem. It must, moreover, be something on which either people hold no opinion either way, or the masses hold a contrary opinion to the philosophers, or the philosophers to the masses, or each of them among themselves.[53]

Problems also include questions in which there are reasons for holding contrary positions or for which no definite reasons are available, in other words, the doubtful.

> Problems also include questions in regard to which reasonings conflict (. . . there being convincing arguments for both views); others also in regard to which we have no argument because they are so vast, and we find it difficult to give our reasons.[54]

Aristotle says that 'practically' all dialectical problems are now called 'theses,' although, strictly, "a 'thesis' is a supposition of some eminent philosopher that conflicts with the general opinion," and "a thesis is also a problem, although a problem is not always a thesis, inasmuch as some problems are such that we have no opinion about them either way."[55]

There are two kinds of dialectical argument, induction and reasoning.[56] Reasoning is an argument in which certain things being laid down, something other than these necessarily comes about through them. Reasoning is demonstration (i.e., scientific) when the premisses from which the reasoning starts are true and primary, or are such that our knowledge of them has originally come through premisses which

---

[51] *Ibid.*, 104a 5-15.

[52] In other words, one which, in Collingwood's terms, would reduce into prevailing presuppositions. If dialectic is the only mode of reasoning, ultimately we will end up talking about the foundations of beliefs, that is, some form or another of prevailing presupposition, as Collingwood does.

[53] Aristotle, *Topics*, Bk. 1, Ch. 11, 104b 1-5, trans. W. A. Pickard-Cambridge in Vol. 1 of *The Works of Aristotle*, ed. W. D. Ross, Oxford, Clarendon Press, 1928.

[54] *Ibid.*, 104b 12-17.

[55] *Ibid.*, 104b 15-20 and 25-35.

[56] See *ibid.*, Bk. 1, Ch. 12, 105a 10-15.

are primary and true. Reasoning is dialectical if it reasons from generally accepted opinions.[57] Induction is a passage from individuals to universals.[58] Comparing the two, Aristotle says:

> Induction is the more convincing and clear: it is more readily learnt by the use of the senses, and is applicable generally to the mass of men, though Reasoning is more forcible and effective against contradictious people.[59]

Analysing the structure of an inquiry, Aristotle explains that arguments start with propositions, and the subjects on which reasonings take place are problems. But every proposition, and every problem, turns around either a genus or a peculiarity or an accident, and around differentia inasmuch as they are applicable to the genus.[60] Neither of these four elements from itself *constitutes* a proposition or a problem but every proposition and every problem *arises from* one or other of these.[61]

Aristotle explains the difference between a proposition and a problem:

> The difference between a problem and a proposition is a difference in the turn of the phrase. For if it be put in this way, ' ''An Animal that walks on two feet'' is the definition of man, is it not?' or ' ''Animal'' is the genus of man, is it not?' the result is a proposition: but if thus, 'Is ''an animal that walks on two feet'' a definition of man or no?' (or 'Is ''animal'' his genus or no?') the result is a problem.[62]

Aristotle says that it is clear from this that propositions and problems are the same in number and that a problem can be made out of every proposition if the phrasing is altered.[63]

---

[57] See *ibid.*, Bk. 1, Ch. 1, 100a 25-31. Collingwood's doctrine of presuppositions is likewise a grounding of reasoning in prevailing convictions. He does not recognize demonstrative premisses which are primary and true. As Rubinoff showed, their truth is their serial arrangement on a scale of forms; it is relative to a preceding and succeeding stage of development, not a quality of the premisses in themselves. Scientific demonstration in the Aristotelian sense does not exist for Collingwood.

[58] See Aristotle, *Topics*, Bk. 1, Ch. 12, 105a 10-15. It is difficult to be precise about induction in Aristotle's works. It is perhaps better to leave it in this wide sense in absence of special treatment in its own right.

[59] *Ibid.*, 105a 15-20.

[60] See *ibid.*, Bk. 1, Ch. 4, 101b 10-20.

[61] See *ibid.*

[62] *Ibid.*, 25-35. Both propositions and problems are interrogatives under dialectical consideration. A proposition asks a leading question giving the respondent a positive motivation to respond in one way; a problem proposes the alternatives indifferently.

[63] See Aristotle, *Topics*, Bk. 1, Ch. 4, 101b 35-102a.

Generally, all propositions turn on identity and diversity, or sameness and difference. Definitions set a subject apart, distinguishing what it is from what it is not. Widely, every predicate defines a subject in a certain way. But since there is sameness and diversity *in sameness and diversity itself,* things are definable in a number of different ways.

> The notion of Sameness or Identity occurs so often in dialectic debate, that Aristotle discriminates its three distinct senses or grades: (1) *Numero;* (2) *Specie;* (3) *Genre.* Water from the same spring is only *idem specie,* though the resemblance between two cups of water from the same spring is far greater than that between water from different sources. Even *idem numero* has different significations: sometimes there are complete synonyms, sometimes an individual is called by its *proprium,* sometimes by its peculiar temporary accident.[64]

Something may be specifically the same as another though individually other; something may be specifically different from another but generically the same. Things the same in strict definition are not only generically but also specifically the same.

Aristotle explains that all disputes, whether of propositions or problems, turn on one or other of the four ways something can be said of a subject: either definitionally, or as property, or as genus, or as accident.

> Aristotle takes the four heads—Accident, Genus, Proprium, and Definition, in the order here enumerated. The thesis of which the predicate is enunciated as accident, affirms the least, is easiest to defend, and hardest to upset. When we enunciate Genus or Proprium, we affirm, not merely that the predicate belongs to the subject (which is all that is affirmed in the case of Accident), but, also something more—that it belongs to the subject in a certain manner and relation. And when we enumerate Definition, we affirm all this and something reaching yet farther—that it declares the whole essence of the *definitum,* and is convertible therewith. Accordingly, the thesis of Definition, affirming as it does so very much, presents the most points of attack, and is by far the hardest to defend. Next in point of difficulty, for the respondent [to the questioner], comes the Proprium.[65]

---

[64] George Grote, *Aristotle,* ed. Alexander Bain and G. Croom Robertson, London, John Murray, 2nd ed., 1880, p. 277. The idealism of Bosanquet and Collingwood denies the validity of this distinction of identity in difference into these three species, identifying as it does identity in difference with concrete identity in difference or individuality. It is therefore a partial denial of the full amplitude of the meaning of identity in difference.

[65] G. Grote, *op. cit.,* pp. 284-285.

Books 2 through 7 of the *Topics* deal with the problems arising from each of these modes of predicating. Book 8 deals with the interrogation rules to be respected by the questioner and by the respondent in the various types of dialectical situations which differ according to the purpose of the discussion. In a didactic situation (and didactic is not primarily a dialectical situation, but may use dialectical methodology) the questioner-respondent relationship is that of teacher-student. The purpose of the teacher as questioner is to teach what he knows and the respondent to learn what he does not know. In a contestational situation the questioner and respondent are duelling, each seeking only victory over the other. In purely investigative argumentation the aim of both parties is neither to teach nor to conquer, but the common aim of testing the argumentative consequences of different admissions and obtaining proficiency in the chains of reasoning pro and con a certain topic.[66]

In the exercise of dialectical discourse two talkers are assumed: the respondent who sets up the thesis he undertakes to defend, and the questioner who interrogates with a view to impugn it or to involve the proponent in self-contradictory answers.[67]

> The dialectician, being engaged in debate, must shape his questions, and regulate his march as questioner, according to the concessions obtained or likely to be obtained from his respondent; who, if a question be asked having an obvious reference bearing on the thesis, will foresee the consequences of answering in the affirmative, and will refuse to grant what is asked.[68]

This is quite other than what the philosopher does, for he does not face an opponent in a strategy estimating situation.

> On the contrary, the philosopher, who pursues investigation with a view to his own satisfaction alone, is under no similar restriction. He looks at once for such premisses as conduct straight to a conclusion; and, the more obvious their bearing on the conclusion is, the more scientific will the syllogism be, and the better will he be pleased.[69]

[66] See *ibid.*, p. 354.

[67] See *ibid.*

[68] *Ibid.*

[69] G. Grote, *op. cit.*, p. 354. Collingwood, not admitting the type of interrogation which deals directly with experienced reality, transforms all interrogation into contestative interrogation with other thinkers, or the interrogation which takes place as a dialogue of the soul with itself, taking now one side, now another, and thus absorbs all philosophical and scientific interrogation into dialectical interrogation, in which there is only interrogation regarding what is *held or believed* about a given topic, and never direct asking of experienced reality: What are you? How did you get to be this way?

The rules for correct interrogation and response differ according as the purpose of the dialogue is one or another of the three enumerated above. Aristotle claims to have been the first to distinguish the third purpose and investigate its particular rules of inquiry.[70] Book 8 of the *Topics* then details rules of procedure for questioner and respondent in the different interrogation situations.[71]

What is important in this regard, in evaluating Collingwood's position on interrogative logic, is that questioning, inasmuch as it is considered to be a procedure of logical reasoning, implies the syllogism, and sometimes induction, as an integral part of that process. In a contestational situation the respondent is maintaining a thesis. He must avoid, in his responses to his interrogator, conceding anything from which the interrogator can deduce, as implied, the contrary or contradictory of his thesis.[72] The questioner must interrogate in such a way that he does not put the respondent on guard by a too direct question, whose affirmative or negative response is too blatantly in opposition to the defendant's thesis.

> . . . Aristotle administers counsel to the questioning as well as to the responding partner. You as questioner have to deal with a thesis set up by the respondent. You see at once what the syllogism is that is required to prove the contrary or contradictory of that thesis; and your business is so to shape your questions as to induce the respondent to concede the premisses necessary towards that syllogism. If you ask him at once and directly to concede these premisses, he sees your drift and answers in the negative. You must therefore begin your approaches from a greater distance. You must ask questions bearing only indirectly and remotely upon your ultimate conclusion.[73]

The exercise and practice of dialectic requires that one be accustomed to argumentation by converting the syllogisms of which the

[70] See G. Grote, *op. cit.*, pp. 354-355 and p. 362. See also Aristotle's *Topics*, Bk. 8, Ch. 5, 159a 25-37.

[71] For commentary on this see G. Grote, *op. cit.*, pp. 353-375. See also Grote's commentary on the *Sophistici Elenchi, op. cit.*, pp. 376-421, which contains a notable amount of material on orderly dialectical questioning.

[72] In real life situations, this may be a matter of life and death. An F.B.I. agent, impersonating an espionage agent, has to be very careful in any of his responses not to supply those among whom he is an impersonator of one of their own with information from which they can deduce his real identity.

[73] G. Grote, *op. cit.*, p. 355. Compare Collingwood's 'Who killed John Doe?' investigation; see above, Chapter 1, p. 26, and Collingwood's *The Idea of History*, p. 268. "The Detective-Inspector was severely blamed later on, for allowing the rector to see in what direction his inquiries were tending, and thus giving him an opportunity to take cyanide and cheat the hangman."

arguments consist, and of testing every thesis by first assuming it to be true, and then assuming it to be false, and following out the consequences on both sides. Having hunted out each train of argument, one ought to look for the counter-arguments against it. This strengthens one's position both as questioner and as respondent. The exercise is so valuable that one should go through the process with oneself if no companion is available. The different trains of argument bearing on the same thesis ought to be put side by side for comparison. A command of a large number of both affirmative and negative arguments serves well both for attack and defense.[74] If this is so, Collingwood's argument that interrogative reasoning is *specifically other than* deductive, syllogistic reasoning, cannot be sustained, as interrogation *as a form of reasoning* necessarily implicates syllogistic reasoning. An analogy with the wave-transmission of light, as explained by physicists, may be helpful. In a travelling wave front of light from, say, a point source, each point on the wave front may be considered to emit a series of secondary waves. The waves so interfere with one another that the energy is transmitted along the perpendiculars to the original wave front, but completely cancelled in every other direction in the forward hemisphere. Diffraction involves the incomplete interference of secondary wavelets at the edge of the light beam.[75] The analogy with dialectical argumentation is that every admission, concession or statement, together with the presuppositions of questions, are 'point sources' from which deductive inferences flow, and any statement made along the way could be looked at as itself a 'point source' of implications. But the different implications of the progressing statements and presuppositions as the discourse continues have an interfering effect on each other, some not consistent with others, therefore requiring that some or another admission or presupposition be rejected, or that the argument go along this or that direction.

> The business of Dialectic . . . is to find and apply the syllogizing process to any given thesis, with premisses the most probable that can be obtained bearing on the thesis . . . [The questioner's] purpose is to investigate and impart this syllogizing power—the power of questioning and cross-examining a respondent who sets up a given thesis, so as to drive him into inconsistent answers.[76]

---

[74] See Aristotle, *Topics*, Bk. 8, Ch. 14, 163b 5. See also G. Grote, *op. cit.*, pp. 372-373.

[75] George Dimitroff and James G. Baker, *Telescopes and Accessories*, Philadelphia, Blakiston Publishing Company, 1945, pp. 6-7.

[76] G. Grote, *op. cit.*, p. 417.

Another important point for the purpose of evaluating Collingwood's position is the indifference of dialectical reasoning to the truth or falsity of any of the statements involved. All that is of interest to the partners is consistency. The questioner attempts to attack the respondent's thesis by forcing from him admissions inconsistent with that thesis, and the defendant is concerned to guard himself against answering in a way inconsistent with his maintained position, all the more so if such a position is *in itself* false (such as that of a murderer in a trial trying to maintain his plea of innocence in the cross-examination).

> . . . neither the direct purpose of the debaters, nor the usual result of the debate, is to prove truth or to disprove falsehood. Such may indeed be the result occasionally; but the only certain result is that an inconsistency is exposed in the respondent's manner of defending his thesis, or that the assailant fails in his purpose of showing up such inconsistency. Whichever way the debate may turn, no certain inference can be drawn as to the thesis itself . . .[77]

If, as for Collingwood, dialectical reasoning is the paradigm form, the truth of the constituent premisses in themselves is meaningless, and the only allowable meaning of truth is coherence. And this is exactly the position of Collingwood, as also of Bosanquet. The absolutizing of dialectic and truth as correspondence necessarily go together.

Again, dialectic is an art, not a science, and one who reduces reasoning to dialectical reasoning will reduce science to art and mental activity to productive activity, as does Collingwood.

> . . . Dialectic does not prove any thing. . . . Dialectic . . . is not a definite science or body of doctrine, but, like rhetoric or medicine, a practical art or ability of dealing with the ever varying situations of the dialogue; of imagining and enunciating the question proper for attack, or the answer proper for defence, as the case may be.[78]

The last words of this quotation very aptly describe Collingwood's whole attitude to the process of knowledge. Because Collingwood had no genuine appreciation of science as distinctively a knowing function, distinct from whatever pragmatic and creative functions it may associate with it, he was forced to subordinate science and scientific reasoning, erroneously understood as merely classificatory, to the self-creative, artistic activity of the mind manifested by a dialectical ques-

---

[77] *Ibid.*, p. 271.
[78] *Ibid.*, p. 273.

tion and answer discourse. Collingwood's priority of the practical over the speculative, his theory of art as basis of all human activity, his theory of dialectical primacy of reasoning are all mutually implicatory —that is, dialectically justifiable—*provided* the premisses be accepted and not questioned as to their truth or falsity. As Collingwood himself says:

> We do not acquire absolute presuppositions by arguing; on the contrary, unless we have them already arguing is impossible to us . . . We must insist on presupposing them in all our thinking without asking why they should be thus accepted.[79]

D. Aristotelian Scientific Interrogation of Reality

But, as noted above,[80] Aristotle is credited with the elaboration of a second questioning technique, one designed for the direct interrogation of reality itself, which is properly the scientific investigation, as distinct from the interrogation of various opinions and convictions held regarding reality. Here, man dialogues with the universe itself; for its correct and successful conduct he must be equipped with the (not merely art, but) science of asking questions, "the technique of asking the questions the human mind must ask the universe if it would lay bare the latter's secrets and discover its mysteries."[81] L.-M. Régis commends Kant for modelling the inquiry of his first critique on the interrogation of the universe by man, but at the same time criticizes Kant's method, whose "questioning is of the type to which a judge subjects a witness, whereas that of Aristotle and St. Thomas corresponds to the respectful but avid questions a pupil asks his teacher."[82] The requirements of the art of philosophical dialogue, however, are imposed not by the demands of an overbearing, belligerent mind, but by the nature and complexity of the subject matter being investigated.[83]

A question is defined in terms of wonder and the desire for that knowledge of which one knows one is ignorant.[84] In every question there is involved both something known and something not known which is sought.[85] The subject of questioning is not what is evident

---

[79] *An Essay on Metaphysics*, p. 173.

[80] See this chapter, p. 318.

[81] L.-M. Régis, *Epistemology*, p. 127.

[82] *Ibid*. See also Chapter 7, above, pp. 271-273, for criticism of the excessively juridical approach in investigation.

[83] L.-M. Régis, *op. cit.*, p. 129.

[84] See Chapter 7, above, pp. 273-276.

[85] See L.-M. Régis, *op. cit.*, p. 128. See also Aquinas' *Commentary on Aristotle's Metaphysics*, Bk. 7, lect. 17, n. 1669 (*In Metaphysicam Aristotelis Commentaria*, Taurini, Librairie Marietti, 1926).

but only what needs to be made evident. What is immediately evident is excluded from questioning by the very nature of the case. Questioning, therefore, bears only on what has to be made known through mediation, that is, through demonstration.

> . . . science is knowledge acquired through demonstration. But we must acquire the knowledge by demonstration of those things which were unknown before, and we ask questions concerning those things [of which we are ignorant].[86]

Since there is no question about that which is evident, or known, and what is questionable is an admixture of something known and something not yet known, the question or problem must take its beginning from the knowledge-context of the problem.[87] According to Aristotle:

> The kinds of question we ask are as many as the kinds of things which we know. They are in fact four: (1) whether the connection of an attribute with a thing is a fact; (2) what is the reason of the connection; (3) whether a thing exists; (4) what is the nature of the thing.[88]

Aristotle shows that these four questions can be ordered into two groups of two, each based on the simplicity or complexity of the interrogated subject. The subject is considered simply when either its existence or its nature is in question. It is considered as a complexity when the existence of one or other of its attributes, or the reason for the inherence of such attribute, is in question.

Concerning the questioning of a subject considered as a complex of thing and attribute, Aristotle explains the procedure thus:

> [On the one hand] we ask whether the thing is thus or otherwise qualified—whether, e.g., the sun suffers eclipse or not—then we are asking as to the fact of a connexion. That our inquiry ceases with the discovery that the sun does suffer eclipse is an indication of this; and if we know from the start that the sun suffers eclipse, we do not inquire whether it does so or not. On the other hand, when we know the fact [of the connection] we ask the reason; as, for example, when we know that the sun is being eclipsed and that an earthquake is in progress, it is the reason of eclipse or earthquake into which we inquire.[89]

---

[86] Aquinas, *Exposition of Aristotle's Posterior Analytics*, Bk. 2, lect. 1, n. 6. See Pierre Conway's translation, Québec, La Librairie Philosophique, 1956, p. 291. See also L.-M. Régis, *op. cit.*, pp. 128-129.

[87] See L.-M. Régis, *op. cit.*, p. 129.

[88] Aristotle's *Posterior Analytics*, Bk. 2, Ch. 1, 89b 20-35. See Richard McKeon's edition of *The Basic Works of Aristotle*, New York, Random House, 1941, pp. 158-159.

[89] *Ibid.*, 89b 25-30.

Thus 'whether?' (whether the sun is eclipsed?) and 'why?' (why is the sun eclipsed?) are the two questions that can be asked of a complex. For some objects of inquiry, however, the questions asked are different. Aristotle explains:

> [On the one hand we ask] whether there is or is not a centaur or a God. (By 'is or is not' I mean 'is or is not, without further qualification'; as opposed to 'is or is not (e.g.) white'.) On the other hand, when we have ascertained the thing's existence, we inquire as to its nature, asking, for instance, 'what, then, is God?' or 'what is man?'[90]

Aristotle concludes by saying that "these . . . are the four questions we ask, and it is in answers to these questions that all our knowledge consists."[91]

Aristotle then says that since all questioning is a search for evidence, "it is clear . . . that all questions are a search for a 'middle,' "[92] that is, for a medium to demonstrate and make evident what is not already, i.e., immediately, evident. Every question is a request for a demonstration, so that "in all our inquiries we are asking either whether there is a 'middle' or, if so, what that 'middle' is."[93]

A difficulty occurs in regarding the request for definition, the *quid sit* question, as a request for a demonstrative medium and demonstration therefrom. The difficulty comes from the fact that the definition is never the conclusion of a demonstration but a principle, a prerequisite to the very possibility of demonstration.[94] And ". . . even if division does demonstrate [the definition's] formula, definition at any rate does not turn out to be a conclusion of inference."[95] Again: ". . . induction proves not what the essential nature of a thing is but that it has or has not some attribute."[96]

The search for, and manifestation of, essential nature is tied intimately to the way in which we know its existence. What a thing is presupposes answered the question whether it is, for ". . . to search for a thing's essential nature when we are unaware that it exists is to search for nothing."[97] And, furthermore, "the degree of our knowl-

---

[90] *Ibid.*, 89b 30-35.
[91] *Ibid.*, 89b 35-38.
[92] *Ibid.*, Bk. 2, Ch. 3, 90a 35.
[93] *Ibid.*, Bk. 2, Ch. 2, 90a 5-6.
[94] See *ibid.*, Bk. 2, ch. 4, 91a 10-15. The parallel in Collingwood is that we must not ask for verification of absolute presuppositions, which are the prerequired conditions of the possibility of demonstration.
[95] Aristotle, *Posterior Analytics*, Bk. 2, Ch. 5, 92a 5.
[96] *Ibid.*, Bk. 2, Ch. 7, 92a 35-92b.
[97] *Ibid.*, Bk. 2, Ch. 8, 93a 25-30.

edge of anything's essential nature is determined by the sense in which we are aware that it exists.''[98] One or other, or several, of the elements of a thing's definition may be known, inductively or perceptively, as *attributes* of that known existent without it yet being known that they are in fact elements of its definition. The darkening of the moon is an attribute of the moon being eclipsed. When we know its darkening, we know that an eclipse is occurring, but we may not know why this darkening occurs, namely, by the interposition of an obscuring screen between the sun and the moon. As soon as we ask why this happens, why the moon has this attribute, we are asking for the *defining conditions* of the attribute, namely, what is an eclipse? When we see that the reason why the moon is eclipsed is that a screen (in this case the earth) interposes between it and the sun's light, we know that an eclipse is the interposition of an opaque body between the sun and the body illuminated by it. Thus, the definition of an eclipse is manifested *in* the demonstration which demonstrates the cause why the moon is eclipsed, although it itself is not demonstrated *by* that demonstration. For ''. . . while there is no syllogism—i.e. no demonstrative syllogism— of essential nature, yet it is *through* syllogism, viz. demonstrative syllogism, that essential nature is exhibited.''[99] That is, the explicative reason why some factually experienced attribute is in a given subject is ultimately the definition of that attribute. And the definition becomes known precisely in knowing it as *the reason why* the attribute inheres in the subject. That is, we co-know its essential nature through knowing the causality of which that essential nature is an unconditioned condition. In other words, even though definitions cannot be proved *as conclusions of demonstrations*, they are manifested (and in *that* sense demonstrated) precisely in the exercise of the causality which it is the job of the demonstration to manifest. In manifesting the *causality* of that which embodies the definition, *the definition itself* is co-manifested. Therefore definitions, their goodness or badness, their completeness or incompleteness, are ultimately tied, as far as their manifestation (i.e., demonstration) goes, to the way in which we experience the realization of the defined essence.[100] It follows, therefore, that however abstractly and decontextualized the for-

---

[98] *Ibid.*

[99] *Ibid.*, 93b 15-20. Emphasis added.

[100] This intimate tie-up of definition formation with the way we experience the defined reality is brought out clearly if we consider the difficulties of trying to form the concept of a unicorn, such as are considered by Leslie Armour in his *The Concept of Truth*, pp. 91 et seq.

mation of definitions may be presented in a logic treatise, in the concrete, in the actual implementation of definition and concept-formation, how we become aware of the existence of that which is to be defined is of vital importance *and cannot be programmed in advance by any purely formalized rules.*[101] In the most general consideration, then, what the thing to be defined already means to the one seeking the definition will be the starting point for his construction of the real definition. The normal starting point is, therefore, what the name means to the one who seeks to find out what the thing is, or its real definition. The search for real definitions begins in nominal definitions, or the way people generally designate things in everyday discourse. Dialectic, therefore, will play a large part in the passage from nominal definition, in which non-defining attributes are probably confused with defining elements,[102] but its function is merely catalytic and directed to immediate confrontation with the reality in its essential attributes.

E. The Scientific Questionnaire for the Interrogation of Reality

The four questions for the scientific interrogation of reality are therefore formulated as follows.[103] First, *Quia est*? (Régis), or *Quod*? (Grote), which asks whether a fact or event is so-and-so, or whether a given subject possesses this or that attribute, or is in this or that condition (Grote), and thus deals with the existence of an accident in a substance or subject, such as: "Is God eternal?" (Régis). The second, *Propter quid*? (Régis), or *Cur*? (Grote), which asks why is it that a certain attribute exists in a certain subject (why is the moon eclipsed?), or what is the cause or reason for such fact, or such conjunction of subject and attribute (Grote), and is thus concerned with the bond between accident and substance, as "Why is God eternal?" (Régis). The third, *An sit*? (Grote and Régis), or "Does the subject

---

[101] The implication of this is that a fully formalized logic, such as is sought by those designing a logic for computer purposes, is already a long way from base point and has a long chain of presuppositions preceding its own starting point. This attempt to decontextualize meaning from lived experience is never fully successful. As Collingwood well said: ". . . technical terms . . . are invented solely to serve the purpose of a particular scientific theory; but as they begin to pass current in the scientist's speech or writing they express to him and to those who understand him the peculiar emotions which that theory yields." *Principles of Art*, p. 268. Most contemporary research on the theory of questioning is concerned with computer programming and thus is of little relevance in assessing Collingwood's theory of questioning.

[102] The second book of Aristotle's *Topics* discusses the attribution of defining elements to a subject in an accidental way.

[103] See G. Grote, *op. cit.*, p. 238. See also L.-M. Régis, *Epistemology*, p. 131.

exist?'' (Régis), which deals with the predicate of the act of existence (Régis).[104] The fourth, *Quid sit*? (Grote and Régis) asks what is the essence of the subject (Grote), and is thus concerned with the nature of the subject (Régis).

Two of these questions, therefore, bear on existence, one on the existence of the subject absolutely, the other on the existence of an attribute in the subject. The other two questions bear on the nature, one of the subject absolutely, the other of the connection of the attribute with the subject, which ultimately goes back to the nature of the attribute and the nature of the subject.

The two existential questions, the first and third above, namely *Quod* and *An* (Grote's terminology), seek to find out whether or not a middle term exists which will demonstrate that the subject exists, or that the attribute exists in the subject. The second and fourth questions, *Cur* and *Quid* (Grote's terminology), assume that there is a middle, and one then tries to find out what that middle is.[105]

> The inquiry *Cur* is in the main analogous to the inquiry *Quid*; in both cases, we aim at ascertaining what the cause or middle term is. But, in the inquiry *Cur*, what we discover is perhaps some independent fact or event, which is the cause of the event *quaesitum*; while in the inquiry *Quid*, what we seek is the real essence or definition of the substance—the fundamental, generating, immanent cause of its concomitant attributes.[106]

The problem now arises as to the order of these four questions in a scientific interrogation of reality. According to Grote[107] there are two distinct pairs of questions, *Quod* and *Cur* forming one pair, and *An sit* and *Quid sit* forming the other;[108] and obviously, says Grote, the *Cur* question presupposes the answer to the *Quod* question, and the *Quid sit* question presupposes an affirmative answer to the *An sit* question. But Grote says that Themisticus and other expositors have proposed that a more suitable arrangement would be that the *An sit*

---

[104] E. Gilson, in his *Being and Some Philosophers*, pp. 192 et seq., refuses to allow that existential, as distinct from copulative, propositions predicate existence as an attribute of the subject; what he alleges to be asserted is the absolute positing of the subject, together with all its implied predicates, in existence. Régis' formulation would perhaps be better rendered as 'the question which deals with the unqualified existence of the subject.'

[105] See G. Grote, *op. cit.*, p. 239.

[106] *Ibid.*, pp. 239-240.

[107] See *ibid.*, pp. 238-239.

[108] Corresponding to the division in the previous paragraph above into two existential and two essential questions.

and *Quid sit* couplet should precede the *Quod* and *Cur* couplet. The reason is given as follows:

> . . . the third and fourth (*An sit, Quid sit*) are simpler, and come earlier in the order of philosophical exposition, while the first and second (*Quod, Cur*) are more complicated, and cannot be expounded philosophically until after the philosophical exposition of the others.[109]

But Grote maintains that this problem is cleared up by adverting to the Aristotelian distinction between what is first in the order of cognition relatively to us (*nobis notiora*) and what is first in the order of recognition by nature (*natura notiora*). He (Grote) explains this as follows, first dealing with the order of knowledge relatively to us.

> *To us* (that is to men taken individually and in the course of actual growth) the phenomena of nature present themselves as particulars confused and complicated in every way, with attributes essential and accidental implicated together: we gradually learn first to see and compare them as particulars, next to resolve them into generalities, bundles, classes, and partially to explain the *Why* of some by means of others. Here we start from facts embodied in propositions, that include subjects clothed with their attributes.[110]

Then he deals with the order of knowledge relative to the subject matter.

> But in the *order of nature* (that is, in the order followed by those who know the *scibile* as a whole, and can experience it scientifically) that which comes first is the Universal or simple Subject abstracted from its predicates or accompaniments: we have to enquire, first, whether a given subject exists; next, if it does exist, what is its real constituent essence or definition.[111]

These two processes, therefore, correspond to the order of the acquisition of science by one who is as yet untrained and to the order of the seasoned scientist equipped with the habit of thinking scientifically, able to deal with the subject matter of the science on *its* terms rather than on terms of his own meagre knowledge.

> We thus see the reason for the order in which Aristotle has arranged the two co-ordinate pairs of *Quaesita* or Problems, conformable to the different processes pursued, on the one hand, by the common intellect, growing and untrained—on the other, by the

---

[109] G. Grote, *op. cit.*, pp. 238-239.
[110] *Ibid.*, p. 239.
[111] *Ibid.*

mature or disciplined intellect, already competent for philosophical exposition and applying itself to new *incognita*.[112]

In other words, there are two scientific questionnaires, one corresponding to science in its becoming, in its acquisition, the other corresponding to science in being, as a fully achieved habit of thought in the mature scientist.

Régis explains the questionnaire of the mature scientist as follows.[113] The first of the four questions, the *An sit* question, can only be asked when the existence of the subject in question escapes our immediate perception. I cannot ask if there is an eclipse of the moon going on if I am presently perceiving it. We can only ask this question regarding the causes or effects of what is immediately perceived to exist. The objects of astronomy and microphysics fall within the scope of this question, as well as the existence of God and of spiritual realities other than the human soul. Thus, the answer to the *An sit* question may fall within the range of sense perception, as astronomy and microphysics endeavours to answer certain questions. The question of the existence of God is answerable as falling within the range of the cause-effect relationship immanent to sensibly perceived things, as the existence of fire is known from the existence of smoke. Only dialectical questions can be asked regarding what is immediately perceived to exist, but not properly scientific or philosophical questions. Again, to ask the question as to the existence of something, the nominal definition, or what the name means, has to be presupposed. To ask meaningfully whether or not unicorns or centaurs or flying saucers exist or not, I need at least to know what I mean by using those names.

Régis then says that the second, *Quid sit*, question is the most important of all, since it is that which the human intellect naturally and spontaneously seeks to ask and answer in regard to the physical objects of daily experience. It is the most difficult to answer because what we immediately are aware of in regard to experienced realities are their accidental attributes, or perhaps more exactly their obvious attributes, without being able to detect whether they are essential or non-essential attributes. We can see straightway that snow is white, but not straightway whether its whiteness is of its essence and definition or not. As soon as we realize that things other than snow are white, we then know that whiteness does not belong to its essence, but that it is an

---

[112] *Ibid*. Compare with Michael Polanyi's account of the probation of a scientific researcher; see Appendix 2, below, p. 479 et seq.

[113] See L.-M. Régis, *Epistemology*, pp. 131-135.

accident. Régis explains that this question falls primarily upon substances and on other things through relation to substances.

> The reason substance has such exclusive rights over definition, in the proper sense, is that the definition, as its name indicates, implies strict unity (i.e., indivision in itself and division from everything else), and only substance has a sufficiently rich mode of being to possess the unity that makes it intelligible. When the question *Quid*? is asked about nonsubstantial beings, it assumes a derivative and relative meaning, which must be clearly recognized if we would avoid impasses or pseudoproblems.[114]

This essence, which definition expresses to the mind as quiddity, in answer to the *Quid sit* (what is it?) question, is not grasped by sense in perception, since sense only grasps sensible accidents (as snow being white, but not what snow is). But in perception the intellect, as the faculty of intuitive reason as explained previously,[115] abstracts the essential from the non-essential and visualizes it in its own right, thereby answering to itself what the thing is. The essence of things which we do not immediately perceive are known as causes or effects of what we do perceive. What God is, for example, will be answerable only insofar as there is some resemblance, through causal connection, between what perceived objects are and what God is. What as yet unrealized artifacts are, as, for example, flying machines before they became realized achievements, are known in relation to the purposes for which they are intended.[116]

The third question, *Quia ita est*? concerns not the absolute actuality of the subject, the subject as a substance, but its actual attributes. Concrete questions of this kind are, "Is the soul immortal?" "Is knowledge true or false?" that is, "Is truth and falsity a property of knowledge?"

The fourth question, *Propter quid ita sit*?, is, according to Régis, the most important question in the order of demonstrative knowledge, since in it consists the very soul of demonstration. Having answered the previous question to the effect *that* certain attributes are inherent in the subject under question, the question next is in regard to *the nature of the bond* between the attribute and the subject, whether it is

---

[114] *Ibid*., p. 132. What Régis calls the most important question of all, Collingwood, on account of his anti-substantialist dynamism, rejects as a vague question to be converted into clear 'which' questions. See this chapter, above, pp. 314-316 and nn. 16 and 17.

[115] See above, Chapter 7, p. 280 et seq.

[116] This latter mode of defining is extremely important in regard to historical or cultural concepts; see below, Chapter 9, pp. 359-372.

necessary or incidental. To answer this question, the reason of the inherence must be known. And this question is intimately connected with the answer to the second, *Quid*? question.

> The *propter quid* that is the object of this last question is nothing but the *quid* of the second question, not in that it constitutes the nature of the thing, but that it *causes* this nature's properties, as efficient cause, final cause, or, again, as material cause *in qua*.[117]

Having considered each of these questions in its own right, the question of their order in a philosophical questionnaire then arises. Guided by Aquinas' commentary[118] on Aristotle's *Posterior Analytics*, Régis[119] explains that, first, in regard to *Quia* and *Propter quid*, sometimes we know that something is so, but do not yet know why it is so (as, for example, we may know *that* the moon is eclipsed, as when it is darkened, but not know why it is eclipsed). At other times, however, the answers to *both* questions are *immediately* apparent to us (a favorably placed observer could see together *that* and *why* the moon is eclipsed). But it is impossible to know why something is so without knowing that it is so. Régis summarizes this as follows.

> [We can say] three things about the third and fourth question: first, that there is no question if the existence of the effect and the nature of the cause are known; second, that there can be no question about the nature of the cause as long as the existence of the effect is unknown; and, finally, that when the existence of an effect is known, the only question that can be asked is *Propter quid*?[120]

Regarding the *An sit* and *Quid sit* questions, sometimes we know that something is but not perfectly what it is. Sometimes both that a thing is and what it is are known together, but one cannot know what something is without knowing whether it is.[121] Régis summarizes this as follows:

> . . . concerning the first two questions: As long as the existence of a being is not known (*An sit*?), the problem of its nature (*Quid sit*?) is unintelligible; when the existence and the nature of a being

---

[117] L.-M. Régis, *Epistemology*, p. 134. Emphasis in original.

[118] See Pierre Conway's translation of Aquinas' *Commentary on Aristotle's Posterior Analytics*, Québec, La Librairie Philosophique, 1956, Bk. 2, lect. 7, n. 5, p. 335.

[119] See L.-M. Régis, *op. cit.*, p. 135.

[120] *Ibid*.

[121] On the dependence of the definition upon the existence of a thing, see above, this chapter, pp. 332-334.

are known simultaneously, there is no question. Finally, the question *Quid est*? presupposes that the existence of this being is known.[122]

An overall summary of the order of the four questions is presented by Régis thus:

> Therefore, the normal and complete order of the four questions is as follows: The existence of a thing must be evident before we can ask questions about its nature; then, we must have evidence of the existence of accidents or effects in order to ask questions about the cause of these effects. The immediate conclusion to be drawn from these statements is that, chronologically speaking, the first object of the philosophical questionnaire is a being whose existence is immediately known by the human intellect, i.e., material things in their concrete and physical existence. If this existential starting point is not accepted, there can be no philosophy, since the questions *Quid*? and *Propter quid*? can never be asked.[123]

F. Transition to Evaluation of Collingwood's Theory of the Logic of Questioning

This lengthy analysis of Aristotle's theory of questioning methodology has been necessary to evaluate Collingwood's position regarding an allegedly as yet undeveloped logic of questioning. It is clear that Collingwood has neglected, both as philosopher and as historian, to take into consideration Aristotle's quite considerable contribution to the theory of questioning. Aristotle's *Topics* and *Sophistical Refutations* (regarded by most commentators as the last book of the Topics[124]) contain a well-articulated theory of the dialectical interrogation of opinions and convictions held about various subject matters. The second book of the *Posterior Analytics* contains a well-articulated theory of the scientific interrogation of reality.

Furthermore, by applying the Aristotelian scientific questionnaire to Collingwood's problem of knowledge and to his subsequent identification of knowledge with the question and answer procedure characteristic of historical thought, one can regard this theory as the result of asking questions out of order, an especially relevant criticism, since it accuses his theory of being a violation of his own insistence on the primacy of orderly interrogation.

By reason of his own innate disposition and of the historical situa-

---

[122] L.-M. Régis, *op. cit.*, pp. 135-136.
[123] *Ibid.*
[124] See G. Grote, *op. cit.*, p. 262.

tion in which he worked, Collingwood was led to pose the question regarding knowledge in a peculiar way, as also were Kant and Descartes before him.

G. Collingwood's Basic Problem and the Logical Order of Questioning

How does Collingwood pose the problem regarding knowledge? What questions does he ask? and in what order does he ask them?

Collingwood's problem of ultimate concern, as has already been shown, arose out of the deep impression made on him by the 1914-1918 war as "an unprecedented triumph for natural science" and *at the same time* "an unprecedented disgrace to the human intellect."[125] The Baconian promise of knowledge as power had been fulfilled in the "power to destroy the bodies and souls of men more rapidly than had ever been done by human agency before."[126] From this culminating triumph others followed, such as improvements in transport, in surgery, medicine, psychiatry, commerce and industry "and above all . . . preparations for the next war."[127]

The war broke out not because anyone wanted it but because a situation got out of hand and continued to get more and more out of hand as it proceeded. Fighting ended not because control was regained but because one side was fought to a standstill. The situation was more out of hand than ever with the signing of the peace treaty.[128]

This admixture of outstanding success combined with abysmal failure of the human intellect is Collingwood's central problem.

> The contrast between the success of modern European minds in controlling almost any situation in which the elements are physical bodies and the forces physical forces, and their inability to control situations in which the elements are human beings and the forces mental forces, left an indelible mark on the memory of every one who was concerned with it. . . . I seemed to see the reign of natural science within no very long time, converting Europe into a wilderness of Yahoos.[129]

Collingwood's problem, therefore, begins in wonder. He wonders why it is that the same human intellect which has achieved such outstanding success in the control of natural forces is such a colossal

---

[125] See R. G. Collingwood, *An Autobiography*, p. 90. See also the introductory chapter, above, p. 1 and Chapter 6, pp. 215 et seq.

[126] R. G. Collingwood, *An Autobiography*, p. 90.

[127] *Ibid.*

[128] See *ibid.*

[129] *Ibid.*, pp. 90-91.

failure in its handling of human forces. Descartes had wondered why it is that the human intellect achieved such outstanding success in mathematics but such hopeless confusion in philosophy, which had largely degenerated into scepticism.[130] Kant wondered why it is that the same human intellect which achieved such success in natural science was so barren in metaphysics.[131]

Descartes moved from his problem thus posed to its solution by equating genuine knowledge with knowledge of the mathematical type[132] and asked what it is that mathematics has, which, unless

---

[130] See L.-M. Régis, *Epistemology*, pp. 33-36.

[131] Kant, indeed, in order to relegate logic to the status of propaedeutic of science, had said, "We do not enlarge but disfigure sciences, if we allow them to trespass upon one another's territory." (Norman Kemp Smith's translation of Immanuel Kant's *Critique of Pure Reason*, London, Macmillan, 1933, p. 18). But having said this, he proceeded to erect mathematics, and especially the newly developed Newtonian natural science, into the paradigm of true science, and to ask whether metaphysics can have what they have which makes them genuine science. "What then, is the reason why, in this field [of metaphysics], the sure road to science has not been hitherto found?" (*Ibid.*, p. 21.) "The examples of mathematics and natural science, which by a single and sudden revolution have become what they now are, seem to me sufficiently remarkable to suggest our considering what may have been the essential features in the changed point of view by which they have so greatly benefited. Their success should incline us, at least by way of experiment, to imitate their procedure, so far as the analogy which, as species of rational knowledge, they bear to metaphysics may permit." (*Ibid.*, pp. 21-22.) ". . . we must therefore make trial whether we may not have more success in the tasks of metaphysics if we suppose that objects must conform to our knowledge. This would agree better with what is desired, namely that it should be possible to have knowledge of objects *a priori*, determining something in regard to them prior to their being given. We should then be proceeding precisely on the lines of Copernicus' primary hypothesis." (*Ibid.*, p. 22.) Note the assumption of the experimental method of procedure, by hypothesis and verification. Again (*ibid.*, p. 22): ". . . we are brought to the conclusion that we can never transcend the limits of possible experience though that is precisely what this science [metaphysics] is concerned, above all else, to achieve. This situation yields, however, just the very experiment by which, indirectly, we are enabled to prove the truth of this first estimate of our *a priori* knowledge of reason, namely, that such knowledge has to do only with appearances, and must leave the thing in itself, as indeed real *per se*, but as not known by us." Later (*ibid.*, p. 55), Kant writes: "Now the proper problem of pure reason is contained in the question: How are *a priori* synthetic judgments possible?" Kant states (*ibid.*, pp. 55-56) that "in the solution of [this] problem we are at the same time deciding as to the possibility of the employment of pure reason in establishing and developing all those sciences which contain a theoretical *a priori* knowledge of objects, and have therefore to answer the questions: How is pure mathematics possible? How is pure science of nature possible? Since these sciences actually exist, it is quite proper to ask how they are possible; for that they must be possible is proved by the fact that they exist." See also L.-M. Régis, *Epistemology*, pp. 36-38.

[132] See Descartes' *Rules for the Direction of the Mind*, end of Rule 2, in Vol. 1 of the Haldane and Ross edition of *The Philosophical Works of Descartes*, Dover, 1955, p. 5: ". . . one conclusion now emerges out of these considerations, viz. not, indeed, that arithmetic and geometry are the sole sciences to be studied, but only that in our search for the direct road towards truth we should busy ourselves with no object

philosophy has it also, philosophy cannot be genuine knowledge.[133] Kant moved from his problem thus posed to asking, what it is that natural science (i.e., Newtonian physics) has, that, unless metaphysics has it, metaphysics cannot be genuine knowledge.[134] Now, how does Collingwood proceed in the development of his problem?

He cannot argue that since science is such an unprecedented success, it follows that its methods should be applied to human affairs and a science of human affairs developed analogous to physics as a science of nature, biology as a science of life, and psychology as a science of mind. He is forestalled from this equation precisely because it is the unprecedented triumph of science *as such* that is *at the same time* the reason of the unprecedented disaster in human affairs. The trouble is *in the very nature of science* as adapted for successful control over nature. If there is a genuine science of human affairs, it is not a science analogous to the sciences of physics or biology or psychology.

Let us follow his thought further in this matter. Why is science, or, if you will, the mind thinking scientifically, so disastrous in human affairs in contrast to such success in natural matters?

The acceleration of scientific progress from the mediaeval windmills and water-wheels through Galileo's discoveries to the almost incredible power and delicacy of the modern machine contrasted with human dealings of man with fellow man, which remained much as they had been in the middle ages. Whereas well-meaning babblers talked of the need for a change of heart, Collingwood felt that the trouble was obviously in the head. "What was needed was not more goodwill and human affection, but more understanding of human affairs and more knowledge of how to handle them."[135]

This progress of science went hand in hand with the departmentalization and resulting fragmentation of human life. The life forms—art, religion, etc.—took on scientific forms in which "each tended to become a specialized activity pursued by specialists for the applause of specialists."[136] Mediaeval art, religion and philosophy were perhaps mediocre as art, as religion and as philosophy, but the men who practised them "were men, whole of heart and secure in their grasp on life."[137]

---

about which we cannot attain a certitude equal to that of the demonstrations of arithmetic and geometry."

[133] See L.-M. Régis, *Epistemology*, p. 139.

[134] *Ibid.*

[135] R. G. Collingwood, *An Autobiography*, p. 92.

[136] R. G. Collingwood, *Speculum Mentis*, p. 34.

[137] *Ibid.*, p. 35.

Today, art, religion and philosophy are fully developed in their own right, but a genuine human life is impossible.

> Today we can be as artistic, we can be as philosophical, we can be as religious as we please, but we cannot even be men at all, we are wrecks and fragments of men, and we do not know where to take hold of life and how to begin looking for the happiness which we know we do not possess.[138]

It is this departmentalization that characterizes the scientific way of thinking which makes it so unsuitable for the study of human affairs. The integrity of a harmonious human existence is radically sundered by its departmentalization into specialized operations. Religion becomes one man's specialty, art another man's specialty, science another man's specialty and philosophy another man's specialty. A harmonious, well-balanced life then gives way to the competition of, say, religion with science, science with philosophy, philosophy with history, and so on.[139] The end result: the First World War, followed soon after by the Second . . . and, we may confidently add on Collingwood's behalf, followed immediately by the threat of a third.

But this departmentalized way of living is due to "the specific character of science, its abstractness."[140] And,

> to abstract is to consider separately things that are inseparable: to think of the universal, for instance, without reflecting that it is merely the universal of the particulars, and to assume that one can isolate it in thought and study it in isolation.[141]

Now, the mode of proceeding of the mind which abstracts is that of genus-species logic. But "the five forms of experience—and whatever others there may be—are not species of a genus, which may be indifferently taken in any order; they have a natural order of their own."[142] Absolutization of the scientific way of thinking results in the absolutization of formal logic and the maintenance of the five forms of life experience as coordinate species of a genus.

> There is therefore only one form of knowledge, describable in terms of logic, which is directed indifferently upon five classes of

---

[138] *Ibid.*

[139] ". . . we now recognize the nature of our disease. What is wrong with us is precisely the detachment of these forms of experience—art, religion, and the rest—from one another." *Speculum Mentis*, p. 36.

[140] *Ibid.*, p. 160.

[141] *Ibid.*

[142] *Ibid.*, p. 50.

> objects. This is pure intellectualism, and leads us to look for syllogisms in music, inductions in religion, and so forth.[143]

But this absorption of all forms of thought into logic as a supreme form of thought is *together a result* and *a rejection* of absolutized genus-species logic, or scientific method; it "precisely contradicts the thesis with which this view began, namely the independence of these various fields of thought."[144]

Science, therefore, with its abstractive mode of conceptualizing and genus-species mode of proceeding logically in induction and deduction, while quite legitimate as the proper way of controlling nature and natural forces, is inherently untenable in dealing with human affairs and the life forms of human experience. The attempt to deal with human experience and the affairs of men by the method of science and its genus-species logic breaks down under the strain of its own internal contradiction.

It is this segregational character of science which together facilitates knowledge of nature and at the same time frustrates knowledge appropriate to human affairs.

But, still following Collingwood's train of thought, why is the abstractive, separative character of scientific procedure so suitable for the investigation of nature and so unsuitable for the study of human affairs? He says that it is because the world of nature is characterized by external relations, whereas human affairs are characterized by internal relations.[145] Human affairs have an inside as well as an outside, and what is needed in regard to knowledge of human affairs is 'insight,' or knowledge of this distinctive 'inside' content which is alien to natural realities.[146] But that which exists with the mode of interiority to self is thought, so that to know the thought within the event is to have 'insight' into the event and thereby to know the event in its internal relations with other events, for thought is characterized as purposive activity, and internal connectedness is through human purposes.

> . . . all thought is for the sake of action and . . . every one who offers us a philosophy must answer the question "What shall we do to be saved from these present distresses?"[147]

[143] *Ibid.*, p. 49.
[144] *Ibid.*
[145] See Chapter 4, above, pp. 86-87.
[146] See Collingwood's *The Idea of History*, pp. 199-200.
[147] R. G. Collingwood, *Speculum Mentis*, p. 35. And, *ibid.*, p. 36: ". . . our cure can only be [the] reunion [of the forms of experience] in a complete and undivided life, our task is to seek for that life, to build up the conception of an activity which is at once art, and religion, and science, and the rest."

But is there any science or human discipline with which we are acquainted, which (1) instead of proceeding abstractly proceeds concretely; and (2) instead of classifying diverse species, externally related with respect to each other, and externally unified under an external, abstract, classificatory genus, grasps its object as an unmutilated whole, as a concrete totality in terms of mutually implicatory internal relations within the one individual whole; and (3) instead of explaining through recourse to external universal covering laws, explains *from within the events themselves*, so that to know what happened is at the same time to know why it happened?

With his question posed in this way, Collingwood has the answer ready at his fingertips. The answer is 'history.' If there is to be a science properly adapted to the knowledge and control of human affairs, it is history. Science itself then becomes an aspect of human affairs when it is seen historically. And, historically, history itself comes into being with the collapse of science as an experienced self-contradictory, autonomous life form. In the control of human affairs, then, science itself has to be dealt with otherwise than by its own methods, namely by the methods of history. The historical point of the conversion of science into history took place when Descartes saw that all science "rests upon one indubitable certainty that I think and that therefore I exist."[148] Collingwood says that Descartes meant what he said and that what he said was that the concrete historical fact, the fact of one's present actual awareness, was the root of science. Descartes saw that the scientist's act of observation preceded any natural fact and that the fact of his observing the event is the fact that really matters. The discovery of which Descartes' *cogito* is the deepest and most fruitful expression is that science presupposes history and can never go behind history, and "this discovery implicitly resolves science into history."[149]

What has been said in the expository chapters (1 through 6 above) then follows coherently from this central line of reasoning, and it stands or falls according as this line of reasoning stands or falls.

But what has happened here if we look at this line of thought *as an exercise of orderly interrogation*?

Collingwood does not *first* ask 'what is knowledge?' and *then* ask 'why does knowledge have the scientific and historical (and philosophical) modes?' His line of reasoning, as sketched above, leads him to answer the third question, *Quia ita sit*? without having asked it, by

[148] *Ibid.*, p. 202. See also above, chapter 4, p. 85 et seq.
[149] R. G. Collingwood, *Speculum Mentis*, p. 202.

equating knowledge with its historical mode. Just as Descartes, by reason of the concrete situation in which his critical problem arose, equated genuine knowledge with one of its attributes, namely the mathematical mode, so Collingwood, by reason of the concrete situation in which he posed his problem of knowledge, equated genuine knowledge with its historical mode. In effect, by presupposing as answered the third question independently of the second, each inadvertently converted the third question into the second and *equated an attribute of knowledge with its essence or definition.*[150] Genuine knowledge, for Descartes, is knowledge of the mathematical mode, and the criterion of true knowledge is the evidential character of mathematical clarity and distinction. All probable knowledge is thereby rejected as knowledge, and common sense convictions are rejected as erroneous. Genuine knowledge for Kant is natural science (which for him is identified with Newtonian physics), and whatever does not measure up to the experiential criterion of Newtonian physics is not genuine knowledge. Metaphysics, failing as it does to take account of experience, and endeavouring to attain knowledge of things as they are in themselves without recourse to experience, is an impossible ideal. Metaphysics, therefore, as a claim to reach things as they are in themselves, is not genuine knowledge. Similarly for Collingwood: Knowledge comes to be equated with one of its modes, namely historical knowledge, so that whatever does not meet the standards of historical knowledge is defective as knowledge.

In answering the third, *Quia ita sit,* question independently of the *Quid sit* question regarding knowledge, Collingwood equates an attribute of knowledge with its essence. Knowledge is had of concrete singular human events, i.e., historical happenings. But this is known not as belonging to the essence of knowledge, as if that statement were an answer to the second, *Quid sit,* question regarding knowledge, rather than an answer to the third question, 'by what modalities do we know?' or 'what properties does our knowledge have?' Answering this third question without answering the second inadvertently converts it into the second, and its answer, which enunciates a *property* or *modality* of knowledge, becomes *the definition* of knowledge *as such* and therefore the criterion which separates what will be diagnosed as true knowledge from its counterfeits. Scientific knowledge is there-

[150] We may recall that Book 2 of Aristotle's *Topics* concerns conflicting opinions which arise from confusing an attribute with the definition and essence, especially the generic essence of some subject matter.

fore not a genuine mode of knowledge but an abstraction, and therefore a falsification of what it knows since it abstracts from concreteness and singularity, which are of the essence of historical knowledge.

We have seen at the beginning of this chapter that the dominating question in philosophy is the what, *quid sit*, question, the search for definition. It was upon the ability to answer this question, to say what justice is, what nature is, what movement is, that genuine science (in the ancient sense), as *propter quid* (answer to fourth question) knowledge, depends. We know *why* the moon is eclipsed *scientifically* when we know that 'why' in terms of 'what' an eclipse is, i.e., in terms of the answer to the second, *quid sit*, question asked regarding an eclipse. To circulate within the third and fourth questions, going from experienced attributes to explanations as to why the attributes are found in a subject, without attacking head-on the definitional, *quid sit*, question is to remain within the order of empirical generalization through induction and never to allow direct knowledge of the cause, the *propter quid*, in its causality, necessitating the attribute by reason of *what it is*. Because an eclipse is the interposition of an opaque body between an illuminating body and an illuminated body, the illuminated body is necessarily darkened whenever such an opaque body thus intervenes. As was shown above, sometimes the *what* and the *why* are shown *simultaneously*, but sometimes the *attribute* is known without it being known *why* it is an attribute, because the *definition* of the attribute is not known.[151]

If one is to rise above mere inductive generalization to *properly philosophical knowledge* of the *cause convertible with the effect*, one has to rise above the merely experienced fact of attributes said of subjects, and inductive generalization therefrom, and reach knowledge of the *Quid sit*. When we know that an opaque body (the earth) is in fact between the sun and the moon we know not merely *that* the moon is darkened but also *why it must be* darkened. The *Quid sit* question is therefore the key question in any philosophical questionnaire. To short-circuit it out is to run the risk of confusing *essences* with *attributes* and, however good or complete one's inductions may be, to fail, or to explain erroneously the facts calling for explanation.

---

[151] Ernst Cassirer notes that much of the scientific research concerning whether or not animals speak is questionable, if not quite worthless, because the researchers are without any precise defining concept of speech; that is, they have not asked or answered the question 'what is speech?' See Cassirer's *The Logic of the Humanities*, pp. 133-135.

Collingwood, like Kant and Descartes, did not ask *what is knowledge* and, *in the light of the answer to that question*, go on to ask *why* it has the attributes of scientific, mathematical and historical modes. But, worse than Kant and Descartes, Collingwood, by reason of the way he approached the question of knowledge, was forced to put out of play the very question upon which all other questions concerning knowledge depends. Because of his attitude to abstraction, and to scientific knowledge as abstract knowledge, he rejected the validity of the *Quid sit* question, the properly philosophical question, transforming it, as he always insisted on doing, into a dialectical question with specified concrete alternative instances from which to answer.[152]

Because he thus bypassed the *Quid sit*, definitional question regarding knowledge, his criterion as to what constitutes genuine knowledge had to come from another source than the very essence of knowledge itself. It came from his dialectical interrogation of the concrete problem in which he was historically situated. Knowledge in the genuine sense is then knowledge of the integral, concrete whole, namely historical knowledge. Sciences, abstract knowledges, are not genuine instances of knowledge, but errors due to separating part of the concrete whole and treating it as if it were the whole, as does the artist: ". . . abstraction is always intuition or imagination."[153]

But if the scientific order of interrogation as exposed above had been followed, the first question to settle would have been *what is knowledge in itself*, before inquiring into its modalities in this or that way in which it is in fact realized. In this way, scientific knowledge would have been explained as knowledge realized in a particular way, historical knowledge would have been explained as knowledge realized in another, but equally valid, way. *What knowledge is* would be seen as realizable in its scientific form, in its mathematical form and in its historical form, each form being respected in its own right without illicit reduction of one to the other.

H. Collingwood's Key Theses as Result of His Order of Questioning

But there are other far-reaching consequences of Collingwood's equation of genuine knowledge with concrete historical knowledge.

First, historical knowledge is constructive of its object; the historian has to reconstruct the past from present traces. Historical knowl-

---

[152] See this chapter, above, pp. 314-316 and nn. 16 and 17.
[153] R. G. Collingwood, *Speculum Mentis*, p. 293.

edge is also knowledge of events as embodying human purposes. It is knowledge of action and, according to Collingwood, for the sake of action. With the designation of historical knowledge as the paradigm of genuine knowledge, knowledge becomes essentially practical and constructive, there being no genuinely autonomous speculative science in its own right.

Second, historical knowledge is by way of interrogation of documents as traces of the human past to discover the thought and purpose of which it is the trace. It is an interrogation of events and thought-expressions to discover the thought of which the event is the expression. Philosophical knowledge, inasmuch as it is genuine knowledge, conforming to the historical mode of knowing, is restricted to *what has been thought* about some particular question. The question 'what is the state?' is the same as the questions 'what did Socrates, Plato, Aristotle . . . Hobbes . . . Marx, etc., say the state is?' This means that the properly philosophical question, 'what is the state?' becomes converted into a dialectical question which always remains in the order of what someone or other considered something to be, without ever attaining the really scientific (in the ancient sense) question regarding the reality as it is in itself.[154] The properly scientific and philosophical questions about any subject matter then become stigmatized as dogmatisms with critical thought supreme, being a dialectical interrogation of the *prevailing opinions* throughout history on a given question. The knowledge of any subject matter then becomes a scale of forms in which a subsequent view is seen to be based on critical interrogation of a previous view, and itself is subsequently critically interrogated, resulting in a subsequent view, and so on. What the state (or nature or history) is becomes converted into what it is thought to be at various periods of history, and that thought is to be taken in its historical evolution through critical interrogation serially arranged in an overlap of classes in a scale of forms. All conceptions of thought are thus reduced to their historical equivalents without remainder.

Third, the universal, the unity in diversity through which all rational explanation is made, must be a concrete unity in a concrete diversity. The whole is the concrete individual whole whose parts are intelligible as members contributing to that individual whole. This results in *a contextual theory of meaning*. Every thing is what it is, not

---

[154] The scientific question, in the fullest sense of the word, asking *what* the thing is *in itself* (that is, as distinct from what someone *thinks* it to be) has been out of bounds in modern and contemporary thought since Kant's first Critique. See Chapter 7, above, p. 280 et seq.

by reason of some essence determining it *in itself absolutely,* but by reason of *its relationship to everything else in the concrete whole.*[155] This is the *absolutization of context,* which, in effect, is also its destruction *as context,* since it then becomes *converted into the essential determinant of each thing.* There is, then, in effect, in a Spinozistic way of thinking, only one substance, one being in itself, one thing which really is, namely the whole, and all other beings are *exhausted by their contextual relations in that whole,* and their whole meaning-manifestative function is to manifest themselves as parts of that absolute individual whole. True unity in diversity is therefore ultimately lost, for we are left with the unity as absolute and the multiplicity a merely *relational function* in the context of the whole.[156] A genuine unity in diversity is had only if there is admitted a diversity of substances, a diversity of absolute unities whose coexistence in the whole of the universe is without prejudice to their own intrinsic, essential and definitional unity. There are diversities of things in the universe of experience which have their definitions as exclusively their own, about which the what, *Quid sit,* question can be asked in their own right. The unity of the universe is not the unity of one substance whose parts are merely relational parts in the whole, but a unity through interaction, and consequent interpassion, of each substantial whole on every other substantial whole. That is, there is diversity in the world of experience

---

[155] See *Speculum Mentis,* pp. 218-220. See also Chapter 4, above, pp. 86-93. This absorption of context into definition, or, if you will, of progressive definition through widening of context so that the thing to be defined is presented in its relationship to everything else, is perhaps most clearly manifested in Collingwood's definition of religion and his comments thereon in *Speculum Mentis,* pp. 111-112: ". . . to assert what it imagines . . . to believe in the reality of the figments of its own imagination. . . . this is the definition of religion, so to speak, from beneath: the purely abstract or formal definition whose purpose is to give the minimum account of the lowest and most rudimentary religious consciousness. It is the mere armature on which our concrete conception of religion is to be built up, and the reader need not trouble to point out its inadequacy as a description of the higher religions. Here and elsewhere, in fact, the reader is earnestly implored to resist the vice of collecting 'definitions' of this and that and the other, as if any one but a fool imagined that he could compress a thing like art or religion or science into an epigram *which could be lifted from its context and, so lifted, continue to make sense.* Giving and collecting definitions is not philosophy but a parlour game. The writer's definition of religion (as of art and so forth) is coextensive with this entire book, and will nowhere be found in smaller compass. *Nor will it be found in its completeness there; for no book is wholly self-explanatory,* but solicits the cooperation of a reasonably thoughtful and instructed reader." (Emphasis added.)

[156] And, ultimately, with Bradley, we must confess the relational as such (whether internal or external) as mere appearance. See F. H. Bradley's *Collected Essays,* Vol. 2, Oxford, Clarendon Press, 1935, pp. 635-650 and Bradley's *Appearance and Reality,* Oxford, Clarendon Press, 9th corrected impression 1930, pp. 21-29. See also M. S. Gram, "The Reality of Relations," *The New Scholasticism,* XLIV (1970), pp. 49-68, esp. pp. 49-53.

which is *not merely relational* but at least absolute in some way. That is, each is what it is *in itself*, expressed by its definition which says what it is in answer to the *Quid sit* question, whatever else it may subsequently be through its relationships with other things and with the whole of that which is.

> L'univers n'existe comme tel qu'en tant que ses parties agissent et réagissent. Ainsi, l'*esse* de l'univers comme univers, ce n'est pas simplement l'ensemble des *esse* de ses éléments, c'est l'ensemble et de ces *esse* et des actions qui les expriment, en actualisant leur mutuelle relativité. (Car tout *esse* fini, de par sa finitude même, est relatif à un autre, à un complément, à un "remède" de cette finitude). Dès lors, l'action apparaît comme l'acte dernier de l'univers en tant qu'univers, sa forme, et, mieux encore, son exister.[157]

Fourth, the function which properly should come from the definitions, manifesting the essence of the subject matter as they do, thereby illuminating the subsequent questions, is usurped by a thought-product, the absolute presuppositions which unconditionally condition any interrogation procedure, according to Collingwood. Instead of proceeding in the light of insights into the very constitution of the reality of a given subject matter, the mind proceeds under the domination of catalytic agents which it forms out of its own absolute dynamism, which catalytic agents are no more than prevailing beliefs or convictions about the world. The mind's questioning procedure is thus locked in dialectical reasoning, always tied at its roots to prevailing convictions and forestalled *in principle* from ever breaking out of self-enclosure within prevailing viewpoints to direct confrontation with reality itself. Collingwood's theory of absolute presuppositions as some kind of commitment to a way of thinking is his counterpart to Aristotle's definition and the principles dependent on definition which result from asking the *Quid sit* question of reality itself. Having outlined nine characteristics of Collingwood's absolute presuppositions, David Rynin states the following:

> I may say at once that the above characteristics appear to me to be the very obvious properties of *definitions* themselves in at least one important sense of 'definition,' namely, the sense in which they are *decisions*, commitments, acts, and as such neither true nor false, although generally being *expressed* in declarative sentences that may be interpreted also as analytic truths.[158]

---

[157] J. De Finance, *Essai sur l'agir humain*, Rome, Presses de l'Université Grégorienne, 1962, pp. 10-11.

[158] David Rynin, "Donagan on Collingwood: Absolute Presuppositions, Truth and Metaphysics," *The Review of Metaphysics*, XVIII (1964-1965), p. 309. Emphasis in original.

Having put out of play the *Quid sit* question, the intellectual guide which should come from reality in answer to that question is now regarded as coming from some commitment or decision or belief as to the structure of lived experience, not from experience itself. When historical knowledge becomes the paradigm of knowledge, the directives of our thought processes are no longer definitions of real essences as explicative of their properties, but are personal or social or cultural historically determined beliefs or commitments. The thought-determinant is then from thought itself, as a catalytic agent formed out of its own radical, dynamic self-forming activism. All discussion is locked radically in beliefs held about this or that subject matter, never achieving insight into what the subject matter is in its own right. All discourse is then dialectic discourse, never truly scientific or philosophical discourse.

It is important to stress here that the aforesaid four-point basic criticism of the Collingwoodian system is made precisely on the score of its resulting from the criticism made regarding an incorrect ordering of questions. The whole point at issue in this chapter is Collingwood's theory of logic as ordered interrogation. Collingwood has been criticised for historical neglect in failing to take into consideration the head-on treatment Aristotle gave to questioning methodology. In the light of the scientific order of interrogation as explained by Aristotle, Collingwood's whole superstructure can be viewed as a result of neglecting, indeed positively rejecting, the centrality of the *Quid sit* question, whose answer provides the basic criterion for all else that follows in the *Quia ita sit* and *Propter quid* questions in philosophy. What is said above regarding Collingwood's identification of genuine knowledge with historical knowledge has been said solely from the point of view of its being a consequence of disordered interrogation, without considering whatever other reasons may be available for criticising Collingwood's basic position on other grounds.

But here we have been concentrating on questioning as an exercise of logic, forgetting for the time being Collingwood's assertion that questioning is properly the methodology of history. Apart from whether or not Collingwood absolutizes history (with its methodology) as the paradigm case of genuine knowledge, we can still ask whether or not he is right in regarding historical methodology as properly a logic of questioning. That is the concern of the next chapter.

Chapter Nine
Logic, Questioning and History

A. Questioning as the Method of Procedure Proper to History

The answer to the question whether history is an organized body of knowledge proceeding by a distinctive logical method of question and answer cannot ignore a consideration of what historians do.

Henri-Irénée Marrou probably comes closest to Collingwood in his direct assertion that "properly speaking, the historian does not proceed by way of deduction or induction,"[1] but "begins by posing a question to himself," which "question which started the whole process in motion does not maintain its original identity, but in contact with the documentary data it is continuously changing."[2] Again, "any knowledge the historian may acquire will obviously depend on the question or questions he chooses to investigate,"[3] and "the copiousness of historical knowledge will depend directly on the skillfulness and ingenuity with which the initial questions are posed."[4] Documentary sources are not complete in themselves "since their compilers did not have in mind (and could not have conceived as possible) all the questions which the documents themselves suggest to us, [and

[1] H.-I. Marrou, *The Meaning of History*, p. 89, also p. 124.
[2] *Ibid.*, p. 131.
[3] *Ibid.*, p. 68.
[4] *Ibid.*, p. 69.

thus] they do not provide us with the means of discovering such questions.''[5] A source of information is entitled to be called a document in the broad sense ''if the historian's mind can extract something from it to increase our knowledge of the human past considered with reference to the question that has been posed regarding it.''[6] Again, he says, ''the truth of the conclusions obtained will be the direct task of the historian's technique, and of the skill and judiciousness with which he formulates the questions and the answers.''[7] Referring to the perfected history of today, Marrou says that ''there is a history that now confronts the past with questions that are always new, more varied, more extensive or more penetrating. And there is a corresponding inquiry that is enlarged and extended in every direction.''[8]

Regarding the logic of the elaboration of historical knowledge, Marrou says that ''the fundamental logical operation in action . . . is the process of understanding,'' and ''the understanding of history apparently consists in the interpretation of meaningful symbols,'' such as inscriptions, the ashes of a home and hearth or fingerprints, by means of which immediately present remains ''we succeed in perceiving something about man in former times,'' which ''includes his activity and behaviour, his ideas and inner self, or sometimes simply his presence—anything indicating that man has passed this way.''[9] This questioning procedure by which we understand the human past is not unique but ''is definitely the same process that takes place in our understanding of other men in the present, and particularly in the understanding of articulated language,''[10] which is nothing other than what Marrou elaborates at length as a ''dialectical [questioning] relationship of the *same* with the *other*. . . .''[11]

Louis Gottschalk[12] explains that in any historical exposition or narrative ''historical facts have to be (1) selected, (2) arranged, (3) emphasized or minimized, and (4) placed in some sort of causal sequence.''[13] Selection is determined by what is considered to be relevant. The

---

[5] *Ibid.*, p. 79.
[6] *Ibid.*, p. 81.
[7] *Ibid.*, p. 136.
[8] *Ibid.*, p. 84.
[9] *Ibid.*, p. 88.
[10] *Ibid.*, pp. 91-92.
[11] *Ibid.*, p. 104. See later, E. Cassirer's incisive explanation of language as establishing a concrete unity in diversity (of persons). See this chapter, below, pp. 399-400.
[12] In his *Understanding History*, New York, Knopf, 2nd ed., 1969.
[13] *Ibid.*, p. 207.

problem, then, is the criterion of relevance about which Gottschalk suggests social means of determining, but "usually the process can be simplified by converting propositions into queries" so that "the subject of any historical investigation may be expressed in the form of an interrogative hypothesis."[14] But Gottschalk feels that "both the unifying proposition and the interrogative hypothesis can be useful only for subjects so monographic in nature that their themes can be contained in a single proposition or interrogation," which "rules out all subjects that have no narrative, descriptive or causal synthesis but are held together only by association in time, place, or persons, or only by analogy," which would include such subjects as "the history of a given region at a given time, of a collectivity of leaders of parties or movements, of schools of art and thought, and of analogous movements and institutions." However, Gottschalk seems to weaken his exception of these from the interrogative process when he adds that "for synthetic subjects like these, analysis by aspect key words nevertheless can be useful if each of the separate elements in the synthesis is regarded as a separate monograph whose theme can be stated as a proposition or an interrogative hypothesis."[15] For Gottschalk, relevance is the overriding concern of the historian in any subject and "the decision of what is relevant is largely a matter of personal judgment," and in the final analysis "the individual historian must be left to make his own selection of his data."[16] But if this is not to be purely capricious, and if his account is to have universal public interest, it seems that some definite determination by the historian, some definite question in his mind, has to be made explicit if there is to be any realistic criterion distinguishing the relevant from the irrelevant.

Wood Gray and his collaborators[17] explain historical methodology as consisting of six steps: (1) selection of a topic, (2) search for evidence, (3) note-taking on evidence, (4) critical evaluation of evidence collected, (5) arrangement of material meaningfully, and (6) interesting presentation which will command readers' attention and communicate the maximum of understanding.[18] Regarding the pursuit of evidence, these authors have the following advice to give:

> Your pursuit of evidence will have had much in common with a detective's search for clues. In the evaluation of the evidence which

[14] *Ibid.*, p. 208.

[15] *Ibid.*, p. 211. This seems to come at least close to Collingwood's method of breaking the subject matter up into a series of questions and subquestions.

[16] *Ibid.*, p. 211.

[17] In their *Historian's Handbook*, Boston, Houghton Mifflin Co., 2nd ed., 1964.

[18] See *ibid.*, p. 9.

> you uncover, you will follow rules of evidence comparable to those of a court of law. An obligation rests upon every historian as it does upon a member of a jury to render a verdict to the best of his ability solely in accordance with the evidence and with as full a sense of his responsibility. . . . You will have to weigh conflicting evidence, discount special pleading, and bear in mind the fallibility of human observation and memory. And likewise your decision will be subject to review.[19]

W. L. Lucey[20] says, "It is obvious that history is an inquiry and that the historical method of investigation easily qualifies as scientific."[21] He maintains that history cannot be an exact science since "man is a free agent and his freedom of choice is the most important causal factor in human history."[22] Lucey says that man's freedom remains the most important factor in regard to future events and is the reason why the future is not predictable,[23] yet he admits that history qualifies as a science.

> History is knowledge, systematized knowledge, arrived at by scientific procedure. If it is weak on the score of general laws and generalizations, one can only point out that the nature of its subject matter severely restricts the freedom to generalize.[24]

Marc Bloch,[25] complaining of the oversimplified presentation of the historian's method, says that "even those texts or archaeological documents which seem the clearest and the most accomodating will speak only when they are properly questioned."[26] He adds that "it is a prime necessity of a well conducted historical research to force the witnesses to speak against their will once we accept no longer to simply record the witnesses' word."[27]

It seems to be generally accepted among historians, then, that the materials functioning as documents and sources, as evidence of what happened in man's past, exercise this function of being evidence only as a result of an active, orderly interrogation by the historian. Colling-

---

[19] *Ibid.*, p. 57.

[20] In his *History: Methods and Interpretation*, Chicago, Loyola University Press, 1958.

[21] *Ibid.*, p. 10.

[22] *Ibid.*, p. 13.

[23] *Ibid.*

[24] *Ibid.* This distinction, between history as the study of free causality and natural science as the study of laws and generalizations expressive of necessary causality, will be highly relevant later in this chapter when the analyses of R. Stover and E. Cassirer are exposed. See below, pp. 360-372 and pp. 384-414.

[25] See his *The Historian's Craft*, New York, Knopf, 1963.

[26] *Ibid.*, p. 64.

[27] *Ibid.*

wood's thesis would thus seem to be substantiated by being in agreement with what the above sampling of historians say about their method of procedure.

But does this interrogative uncovering of the human past differ significantly from the scientist's investigations, which uncover the intelligible laws of the natural world? The affirmative answer to this question involves distinguishing the type of intelligibility which supplies meaning and explanation when we are dealing with man as a free being and free agent from the type of intelligibility which supplies meaning and explanation when we are dealing with the natural, unfree, deterministic world (including man to the extent that he also belongs to that world).

B. Two Ways in Which the World of Experience Is Intelligently Meaningful

To evaluate Collingwood's assertion that history proceeds by a distinctive logic of question and answer, it is necessary to consider two fundamental ways in which, according to a number of recent authors, the world of lived experience can be made intelligibly meaningful and rationally accounted for. Collingwood has insisted that "scientific history contains no ready made statements at all," and that ". . . confronted with a ready made statement about the subject he is studying the scientific historian . . . asks himself . . . 'what does this statement mean?' "[28] Collingwood further insists that "the scientific historian does not treat statements as statements but as evidence . . . as other facts which, if he knows the right questions to ask about them, may throw light on these facts."[29] The following analysis will show that one of two distinguishable types of meaning is the proper concern of the historical investigation and that this type of meaning becomes disclosed in an interrogational dialogue of an intersubjective nature, or, as Collingwood expresses it, "by rethinking past thoughts."[30] The criticism to be made against Collingwood will be that he reduces both types of meaning to this one particular type, which is the proper concern of the historian, and that in doing so, he channels all investigations ultimately into investigations of an historical nature, thereby effectively denying to science its own autonomous access to its proper type of meaning and explanation. This con-

---

[28] R. G. Collingwood, *The Idea of History*, p. 275. See also Chapter 1, above, p. 29.

[29] R. G. Collingwood, *The Idea of History*, p. 275.

[30] See Collingwood's *Autobiography*, pp. 112-116. See also his *The Idea of History*, pp. 282-302. See above, Chapter 6, pp. 225-226.

clusion will be arrived at by drawing on the insights and conclusions of recent thinkers, principally Robert Stover, Ernst Cassirer and John Wild.

Robert Stover distinguishes natural order intelligibility and a distinctive type of intelligibility arising from man's ability to disclose the world in alternative ways according to greater or lesser suitability to human and personal needs. Ernst Cassirer distinguishes nature concepts, founded in what he calls ''thing'' perception, from culture concepts, founded in what he calls ''expression'' perception. The ideal of objectivity aimed at by nature concepts excludes every personal manifestation, whereas culture concepts, and the meaningfulness of the moral, cultural and historical world built up by means of them, are intelligible only as expressions of personal attitudes consciously adopted with respect to the encountered (natural and human) world, or, in Stover's terms, expressions of care and concern. John Wild's analysis of man as a ''care-taking'' being greatly illuminates Stover's analysis of meaning deriving from the attitude of careful confrontation of the world, and in doing so throws light on the nature of what is historical being, which contributes greatly to our ability to answer questions about the object and methodology of the historian.

C. Natural Order Intelligibility and Intelligibility from the Standpoint of Living in the World

Robert Stover's[31] ''thesis is that there are essential presuppositions and requirements which define *what it is for an occurrence to be intelligible to us* and that one set of such presuppositions and requirements defines natural order intelligibility''[32] and another set defines what he calls the intelligibility of the standpoint of living in the world.[33] This latter is the standpoint of confronting the world caringly,[34] that is, ''with an attitude of concern, which poses questions about the world—questions as to the relative preferability of one or another alternative states of affairs,'' which ''is to presume that one or another alternative might be preferable.''[35]

---

[31] See his *The Nature of Historical Thinking,* Chapel Hill, University of North Carolina Press, 1967.

[32] *Ibid.*, p. 5. Emphasis in original.

[33] See *ibid.*, p. xiii and p. 147.

[34] See *ibid.*, p. 149.

[35] *Ibid*. Note the interrogative character of this attitude of concern. Correlate with Collingwood's *The New Leviathan,* Chapter 11, on questioning and desire, par. 11.39: ''. . . the first part of knowing yourself is knowing what you want''; (pars. 11.22-11.23) ''. . . a proposition is an answer to a question; and a question offers alternatives . . . so

These two ways of thinking about the world, inasmuch as they define two distinct ways of questioning, are called standpoints.

> Confrontation of the world in such a way that a distinctive question arises is what we mean by a standpoint. The queries that give rise to critically differentiated lines of questioning develop from everyday encounters with the world.[36]

But Stover warns that "at the outset of man's thinking it is not apparent that there are fundamentally distinct question-posing ways in which the world is there for us, let alone what these are," for "it is a philosophical error . . . to suppose that our everyday thinking has its own framework of intelligibility and provides the key to the intelligibility of the world"; rather, "*differentiation* of these standpoints from one another" parallels "refinement in conception of standpoints."[37] This illuminates Collingwood's position on the primacy of the questioning activity in knowledge. It is by the taking of different standpoints in confrontation with the world by questioning from such standpoints that a definite type of meaningfulness and explanation is recognizable. Meaning and explanation are not explicated in our day-to-day encounters with the world until we confront the world with a definite type of question characteristic of a definite type of standpoint. A standpoint is therefore a definite position or situation taken up by an agent or observer confronting a world of events questioned as to their meaning and explanation. Questioning, therefore, implies a situated observer or agent, and a world of events which such observer, or agent, confronts as puzzling to him in some way. Situation and world are therefore mutually implicatory terms. A situation takes place only in and with reference to an environing world; a world

---

desire asks and answers the question: 'What do I want?' which it begins by converting into: 'Which do I want, *a* or *b*?' [which] . . . are put before you as alternatives . . ." [for] (11.11-11.12) "knowing involves asking questions and answering them . . . and . . . asking a question implies contemplating alternatives."

[36] R. Stover, *op. cit.*, p. 147. However, Stover's warning is not to be understood as if he sets everyday thinking in radical opposition to historical thinking. On p. 186 he writes: "The extraordinary interesting fact remains that we find the same historian sometimes adopting the standpoint of the agent and sometimes that of natural order. This is one way in which the historian's thinking resembles our thinking in everyday life. In everyday life we are not specialists. We pass from one standpoint to the other a thousand times a day without any hesitation or sense of discontinuity. Like many an historian we use both." Thus, far from being in opposition to Marrou's position that historical knowledge resembles familiar day-to-day knowledge (see Marrou's *The Meaning of History*, pp. 89-91 and pp. 103-111; see also Chapter 7, above, pp. 276-279). Stover's position, as clarified by himself here, illuminates also Marrou's position.

[37] R. Stover, *op. cit.*, p. 147. Emphasis in original.

is what situates by environing that which is situated in some meaningful way.[38]

Stover holds that there are two fundamentally, as distinct from derivatively, distinguishable ways in which the world of events can be confronted questioningly.

> . . . we shall maintain that there are two fundamental lines of questioning, i.e., two ways in which we make occurrences intelligible, and, corresponding to them, two fundamentally distinct types of historical thinking: thinking *from the standpoint of natural order* and thinking *from the standpoint of living in the world*.[39]

Natural order thinking is based on the observed uniformity in occurrences. Stover explains this as follows:

> Roughly speaking, natural order is an orderliness manifested in occurrences: under the same circumstances the same things happen, either always or with the same frequency. To think about the world, to observe it, to form concepts, and to reason with a view to making judgments about the orderliness of happenings, is what we call thinking from the standpoint of natural order, and the objective of thinking from this standpoint is natural order intelligibility.[40]

Stover says that this standpoint ''characterises much of what we call scientific thinking,'' although scientific thinking is not to be equated with this type of thinking, since, notwithstanding that ''the quest for valid empirical generalizations, for unifying theories, for systematic knowledge, all can be carried on from this standpoint,'' nevertheless ''scientific thinking . . . is not always from the standpoint of natural order, and even when it is, specific objectives of the scientist are likely to be selective, narrower in scope than those of the standpoint itself.''[41]

Just as the scientist may proceed not exclusively in his characteristic mode of natural order thinking, so the historian need not proceed by a distinctively historical type of thinking that totally excludes natural order thinking, which latter is characteristic of scien-

---

[38] See R. G. Collingwood, *Speculum Mentis*, p. 248, for the correlativity of the self and its world. See this chapter, below, p. 376. See also this chapter, below, pp. 410-411 n. 262 for John Wild's explanation of the meaning of the term 'world.'

[39] Robert Stover, *op. cit.*, pp. xii-xiii. Emphasis in original.

[40] *Ibid.*, pp. 3-4.

[41] *Ibid.*, p. 4. Emphasis in original. For example, finances may force a researcher to restrict the focus of his inquiry. His economic concern, a factor extraneous to scientific inquiry as such, may condition the type of inquiry beyond that of properly scientific considerations. The two standpoints, though distinct, mutually condition each other in their use.

tific thinking. The causes and effects of human events and actions from the standpoint of natural order are of interest to historians. Stover cites G. M. Trevelyan[42] as conceding that at least guessing about such things is part of the historian's business.[43]

> . . . natural order intelligibility is a plausible objective of thinking about particular occurrences of the human past. . . . It is the principal objective of *some* historians on *some* occasions . . . it is among the objectives of *many* historians on *many* occasions. It constitutes one very common idea of what historical inquiry is all about: a quest for the order exhibited in world happenings.[44]

Natural order intelligibility as an ideal[45] aims at universal determinism; ". . . this idea is an important constituent of the scheme of natural order intelligibility";[46] "we postulate universal determinism presuming that, pending evidence to the contrary, *any* given occurrence is an instance of natural order."[47] It likewise aims at unrestricted universality of nomological laws; "our second requirement was that a nomological universal not be formulated in terms that entail a scope of predication restricted to a definite number of occurrences."[48] But, notwithstanding these ideals of natural order intelligibility, those seeking such intelligibility frequently settle for something less, though approximating that ideal.[49]

Natural order intelligibility is a way of thinking of which the historian can and sometimes does avail himself without precluding him from making use of other ways of thinking.[50] But Stover warns that "sometimes the initial mistake is made of attributing a characteristic to a subject matter that should be attributed to a way of thinking."[51]

---

[42] See *Clio: A Muse*, London, Longmans, Green and Co., 1930, pp. 144, 148.
[43] See R. Stover, *op. cit.*, p. 4.
[44] *Ibid.*, p. 4. Emphasis in original.
[45] See *ibid.*, Ch. 2, pp. 15-53.
[46] *Ibid.*, p. 46.
[47] *Ibid.*, p. 47. Emphasis in original.
[48] *Ibid.*, p. 12.
[49] See *ibid.*, pp. 16-40. As a deviation from natural order intelligibility, Stover notes, *ibid.*, p. 48, that "the only available example of experimentally verified *dis*order is in the field of quantum mechanics."
[50] See *ibid.*, p. 57.
[51] *Ibid.* This is a point to be borne in mind in judging the Popper-Hempel theory of historical knowledge. This warning also removes the *prima facie* contradiction which may appear between Stover's recognition of the historian's use of natural order intelligibility and Ernst Cassirer's opposing of nature concepts to cultural and historical concepts; see later in this chapter, pp. 403-414 below.

Stover correctly points out that "rational actions can be descriptively understood and thought about from the natural order standpoint,"[52] which means that "they can in principle be accounted for deterministically."[53] Barbara Mertz noted, "Sooner or later, most historians succumb to the urge to discover causes in history,"[54] and Robert Stover states that ". . . most 'causal' accounts of events to be found in historical works are a special form of natural order intelligibility."[55] The historian, in having recourse to causality as essentially connoting a confirmed regular relation, is implicitly taking a stand on the meaning of causality in history. He is thereby subscribing to a "generality" interpretation of causality, in the sense that causal accounts imply a general proposition of the sort called nomological universals.[56] The historian E. H. Carr seems to be in agreement with this, and states that the historian constantly uses generalizations to test his evidence. If, for instance, it is not clear from the available evidence whether Richard murdered the princes in the Tower, the historian will ask, consciously or unconsciously, whether rulers of the particular period habitually liquidated rivals to their throne, and his final judgment will be influenced by this generalization.[57]

The historians' recourse to deterministic causality in order to explain has its parallel in psychology, and what F. J. Braceland and M. Stock[58] say in that context is equally valid for the historian. Their position is that if we look along a series of events involving free decisions *after the fact*, we can assign the *actual* causes which operated in the series, whereas if we look along the series '*from the front end,*' so that at each stage we find ourselves confronted with alternatives, "we cannot see whether or not [an event] will occur, for while we see the

---

[52] R. Stover, *op. cit.*, p. 93.

[53] *Ibid.*

[54] Barbara Mertz, *Temples, Tombs and Hieroglyphs*, New York, Howard McCann, 1964, p. 326.

[55] Robert Stover, *op. cit.*, p. 93.

[56] See *ibid.*, p. 9. I am not sure that I agree with Stover here. Actions based on care for the world are determined by free personal decisions. The free causality of human initiative is thus an alternative mode of causality to the "generality" (Stover's term) type of causality based on the necessity implied by nomological universals. Nevertheless, I think Stover is correct in saying that when historians do have recourse to causal explanation it is to the type of causality based on nomological universals that they have in mind.

[57] See E. H. Carr, *What is History*? Harmondsworth, Penguin Books, 1964, p. 63. See also Henry Veatch, *Two Logics*, Evanston, Northwestern University Press, 1969, pp. 242-245, regarding character traits as bases for causal inferences based on nomological universals.

[58] See their *Modern Psychiatry*, New York, Doubleday, 1963, pp. 252-255.

motivations which are possible, we cannot predict which ones will be freely chosen."[59] Thus, "in psychology, a causal sequence explains what has happened, but it cannot foretell what will happen in the next similar event, and the difference lies in freedom of choice."[60] If in the following we read 'historical' for 'psychological,' we can easily see how the historian is justified in giving the deterministic natural law type of causal explanation for historical events, notwithstanding that the events are manifestations of the self-determinations of free agents:

> In effect, then, we may and should accept scientific determinism when studying psychological data, accepting the principle that every psychological phenomenon which occurs comes from a cause. Simultaneously, we can accept free will, asserting that before they have occurred, many psychological phenomena are subject to free disposition by the human agent. And this seems to solve the dilemma.[61]

The second fundamental standpoint of thinking distinguished by Stover is "the standpoint of living in the world,"[62] whose principle of meaning and intelligibility is that of care or concern in regard to how the world is disposed with reference to a situated agent's preferences.[63] "To confront the world as living in the world is to confront it caringly. How the world is matters."[64]

Caring, Stover explains, is a mental attitude of concern which questions the world as to the relative preferability of one or another recognizable state of affairs. Caring assumes that things might be otherwise than they are, and that the subject has a relation of preferability to one or another of these possible states. It may not always be apparent what is the preferable state of affairs, but the prevailing opinions about this will precipitate discussion and inquiry. The existence of recognizable alternatives as objects of preference is essential

---

[59] *Ibid.*, p. 253.

[60] *Ibid.*, p. 254. See p. 364 n. 56 above for the question regarding the identification of causality with deterministic causality. I think that the determinative activity of free agents is itself a distinctive type of causality, namely free causality. That is, I object to identifying causality with predetermination.

[61] F. J. Braceland and M. Stock, *op. cit.*, p. 254. This is a very important insight and, if it is kept in mind throughout the remainder of this chapter, it will forestall many difficulties that might otherwise arise by reason of a presupposed freedom-determinism antinomy.

[62] Robert Stover, *The Nature of Historical Thinking*, pp. 147 et seq.

[63] See *ibid.*, pp. 149-152.

[64] *Ibid.*, p. 149.

to this way of thinking. "Thinking from the standpoint of living in the world takes the form of ascertaining what is preferable."[65]

Stover distinguishes "between two sorts of judgments of preferability: evaluations and practical reasoning."[66] By evaluations he means "judgments as to the comparative merits or shortcomings of a situation or aspect of a situation cared about."[67] Thus, "we may evaluate the effects of certain events on our own human situation," and "what is evaluated can be past, present or future."[68] Probably most evaluations are in connection with practical considerations such as deliberating whether to act, what course of action should be taken and what are the consequences of choices already acted upon, although evaluations are possible without essential connection with proposed rational action.[69]

Contemplated alternatives relative to one involved in a decision-making situation are essential to this mode of thinking, and it is a fundamental note distinguishing it from natural order thinking. A question of preferability as to situations, institutions, courses of action, etc., is presupposed to evaluation and practical reasoning. Only a matter of concern can precipitate questions whose answers are evaluations and practical reasonings, culminating in decisions and action thereon. Whether or not someone in fact is concerned about such matters, there must at least be matters about which concern can be had in order that there be possible this distinctive intelligibility based on the world as lived in. There is always a personal reference of something judged as preferable to or for someone.[70]

Many of the meanings by which events are intelligible from the standpoint of the suitability or unsuitability to human concern accompany the human situation as such. Consequently, it is possible to recognize the meaning that some past or distant object or event had for those confronting it in *their* situation of time and place. It is recognizably meaningful to us simply by consulting the orientation of our own preferences and value judgments.

> *Identifying understanding* is descriptive understanding *together with an awareness* that the *kinds* of distinctively human meaning relations that characterize the subjective or objectified manifestations

---

[65] *Ibid.*

[66] *Ibid.*

[67] *Ibid.*, p. 150.

[68] *Ibid.*

[69] See *ibid.* Again note the importance of alternatives, which, according to Collingwood, are made known by and in a question. See above, Chapter 3, p. 66.

[70] See Robert Stover, *op. cit.*, pp. 151-152.

> we are understanding, are *the same kinds* of meaning relations that characterize *our* subjective experience and that are involved in meanings objectively manifested in *our* artifacts, institutions and behavior. These kinds of meaning are correlates of human standpoints—common ways of confronting the world, of asking questions about what goes on, and of striving for intelligibility.[71]

Thinking from this standpoint of living in the world is characterized by temporality, concreteness and particularity. The temporal character of our careful concern for the world is thus explained by Stover:

> The questions we ask and answer from the standpoint of living in the world vary with the temporal perspective. We do not deliberate about an action already done; we do review and criticize past deliberations and compare the actual with the foreseen consequences. On the other hand, we do not criticize ourselves for a decision not yet made nor do we compare envisioned consequences with actual consequences of a decision we have yet to execute.[72]

The concreteness and particularity characteristic of our thinking from the standpoint of care for the world is then explained as follows:

> The thinking we do from this standpoint reflects not only the particularity of situations but the multiplicity of ways in which a human being can fare well or ill, do well or badly, make things well or poorly. It reflects the diverse kinds of interrelationships we have with things human and non-human: the precarious combination of environmental conditions upon which survival depends; the physiological complexity of our bodies, which exposes us to a diversity of stimuli and gives us the ability to do countless things as well as to extend the range of possible activities by manipulating our environment; our involvement in numerous and varied cultural systems and correlated social organizations; and so on. Whence questions arise: Which of several possible actions or activities, which state of affairs, which product of our making is, in a particular situation or in general, the preferable one?[73]

Both types of meaning, that deriving from natural order and that deriving from careful concern for the world, derive their meaning in relation to human beings, so that there is a general sense in which both meanings can be said to be human. But meaning based on care for the world is human in a very special way, inasmuch as this meaning is a relationship to the preferability of human persons in human

---

[71] *Ibid.*, p. 93.

[72] *Ibid.*, p. 150. See this chapter, below, pp. 373-384, for John Wild's explanation of human temporality in this context.

[73] R. Stover, *op. cit.*, p. 150.

situations.[74] Natural order meaning is human in the sense that it is meaning *for* human beings, whereas "the personal reference essential to preferability is a reference to persons *as human beings*, as in the human situation."[75] Since natural order meaning lacks this essential relation to human preferability, the meaning of the world in terms of natural order is a neutral type of meaning when contrasted with "the other distinctively human modes of meaning: being satisfactory for someone (evaluations), and being the appropriate thing for someone to do or the appropriate way for someone to proceed (practical judgments)."[76]

Now, the historian is able to make past events intelligible in terms of their preferability with respect to the interests and concerns of persons contemporaneous with those events.

> Judgments of preferability are personal in reference. *For* persons living in the past—past relative to the present in which the historian sets about making the past intelligible—there were questions of preferability.[77]

These questions of preferability may either be questions of satisfactoriness, that is, evaluations, or questions as to the correct way to proceed in decision and action. As regards the first:

> Some questions of satisfactoriness for them (matters of evaluation) were actually posed by them. Some of these questions they answered thoughtfully, using generalization and judgment. An indefinite number of questions of satisfactoriness for them could have been posed.[78]

As regards the second type of question, questions of preferability which the historian may recognize in the past as making it intelligible, Stover has the following to say:

> Similarly, there were an indefinite number of questions for these persons as to what to do and how to proceed (practical questions), of which some were actually posed and some were answered thoughtfully by them. Even when no question was explicitly asked or answered, they engaged in activities and viewed happenings with the understanding—sometimes resulting from previous thinking—

[74] *Ibid.*

[75] *Ibid.*, p. 152.

[76] *Ibid.*

[77] *Ibid.*, p. 153. Emphasis in original. It is important to keep in mind here the clarification which was borrowed from psychology in regard to human events seen from their standpoint facing *their own* indeterminate future. See this chapter, pp. 364-365, above.

[78] R. Stover, *op. cit.*, p. 153.

> that what was being done was appropriate, or that what was happening was satisfactory (or unsatisfactory).[79]

Stover makes a further important distinction in regard to the ways in which the historian can deal with past happenings. The historian may deal with past events either by descriptive understanding or by identifying understanding. He explains descriptive understanding as follows:

> The historian can *describe* the questions asked, the thinking done (or the manner in which the questions were resolved, whatever it may have been), and the temper of mind, mood, or outlook with which activities were carried on and happenings regarded. This is *descriptive* understanding, i.e., description of meaning relations; in this case, description of distinctively human meanings concretely experienced or objectified.[80]

Such descriptive understanding, notwithstanding that it is dealing with persons acting preferentially in their particular situation, deals with such person-involved happenings from the outside, so to speak, as considered by an impartial observer.

But, contrarily, when the historian takes the viewpoint of the evaluating or deliberating agent, seeing the events from the agent's point of view, from within his personal evaluating and decision-making situation, he is using what Stover calls identifying understanding.

> . . . the historian can also *identify* with the standpoint of living in the world of those persons in the past; in so doing he regards the questions as universally human. It is this latter way of regarding the lived experience of the past which distinguishes *identifying* understanding from mere description, *descriptive* understanding.[81]

Identifying understanding presupposes descriptive understanding, though this does not mean that they are exercised as two sequentially distinct operations. The value of the distinction, according to Stover, lies in the fact that it allows for the recognition of another way of meaningfully objectifying lived experience, namely as manifesting natural order. Thus, "[a] historian's description can be auxiliary to both standpoints of thinking; it can serve two ultimate aims: identifying understanding or natural law thinking."[82]

---

[79] *Ibid*.

[80] *Ibid*. Emphasis in original.

[81] *Ibid*. Emphasis in original. It is from this point of view that a past human act appears as a free decision-making act whose own future looks to that act as its determinant.

[82] *Ibid*., pp. 153-154.

Descriptive understanding, then, considers an event simply as something which happened, with minimal, if any, penetration into the meaningful content of the happening. History written solely in this way would seem to be what is called mere chronicle, a mere statement of a sequence of happenings and their dates of occurrence with little or no regard to precisely what happened or why it happened. Natural order understanding would attempt to supply such meaning and explanation in terms of the intelligibility of what regularly happens according to some recognizable nomological universal law (as, for example, rulers levying heavy taxation are unpopular and provoke discontent). Identifying understanding, on the other hand, would seek to situate itself within the concern of the agent involved in his (the agent's) present (historian's past) situation, and to understand the situation and what happened in terms of the agent's action by reflecting the agent's interests and concerns.

Stover takes Collingwood to task for identifying (a) the *descriptive understanding* of person-oriented events and happenings with (b) *identifying understanding* when he (Collingwood) says that in history to know what happened is also to know why it happened by insight into the thoughts and motives of the agents concerned. Stover says that this identification forestalls allowing that actions initiated by purposive agents, and their effects, can be treated also from the standpoint of natural order, deterministic intelligiblity,[83] which Collingwood does not allow. Collingwood correctly recognised the distinctiveness of the volitional attitude, the standpoint of the agent whose action involves thought, and he was correct to recognize that causal generalizations or nomological universals or their approximations are not implicitly contained in identifying understanding as such. But due to his (Collingwood's) faulty conception of understanding, "he failed to grasp clearly the difference between understanding the practical reasons or reasoning of others and thinking from the standpoint of natural order."[84] Thus, Collingwood "mistakenly supposed that when the historian understands, happenings (actions, in this case) thereby become intelligible *in the same way* that deterministic accounts make them intelligible."[85] Collingwood's error was to think that when the historian understands he thereby knows the determining conditions of the events in question; knowing the thoughts behind the agent's

[83] See *ibid.*, pp. 77-79; 90-95; 104-107.
[84] *Ibid.*, p. 106.
[85] *Ibid.* Emphasis in original.

action is the same as knowing *the determinate reasons why* the action happened. This, Stover says, is to equate understanding with understanding *the determinate reasons why* an action occurred, and, therefore, in Stover's judgment, to regard understanding as *an alternative of the same type* of understanding as is found in natural order, deterministic thinking. Had Collingwood, in Stover's opinion, seen the difference between understanding which can be *merely descriptive,* and *understanding by identification* with the agent, he would have been able to allow for the possibility of understanding from the natural order standpoint making its own distinctive contribution to historical explanation.[86]

I am not sure that Collingwood's position is as easily disposed of as Stover makes it out to be here. Collingwood's identification of historical understanding with what Stover calls identifying understanding is hardly a mere slip or oversight on Collingwood's part. He has too clearly and consistently argued the impotence of thought in its scientific mode of procedure, that is, through abstract universals and general laws, to deal with the problems proper to history. I agree with Stover that the explanation of some historical event, the satisfying answer to someone's question regarding a certain historical happening, may well be of the type "tyrants do that sort of thing," or "heavy taxes always breed discontent," or "kings of that period usually were cruel." Consequently, I disagree with Collingwood's position not allowing answers of a universal and general type as answers to specific historical questions. But I likewise disagree with Stover's statement that the trouble on Collingwood's part is his inability to distinguish descriptive understanding from identifying understanding. Collingwood's *whole system of thought* leads him to *conclude* that abstract thought, as found in natural science, is erroneous, and that truth is had only in thought which achieves concreteness and totality. Only historical thought, of the type Stover calls identifying understanding, fulfils the conditions of concrete and genuine thought for Collingwood. So it is in principle, rather than as a result of a negligent oversight, that Collingwood identifies understanding in history and what Stover calls identifying understanding. From his (Collingwood's) view, Collingwood saw only too clearly that natural order intelligibility could not contribute to historical explanation. In effect, Stover here is failing to take precisely that standpoint of identifying understanding in regard to Collingwood; he does not look at Collingwood's denial of that distinction from the standpoint of the problem

[86] See *ibid.*, p. 106.

which Collingwood faced in his own mind. Furthermore, historical thought, for Collingwood, is conducted at a level of consciousness which transcends scientific thought. The historian is thinking about thought (i.e., past thought, though not necessarily thinking about the activity itself of thinking about [past] thought, which, for Collingwood, is philosophical thought),[87] whereas the scientist thinks not about thought, but about spatialized, deterministic objects. The issue with Collingwood is rather with the fundamentals of his whole system, which forbid him even to visualize the possibility of the distinction which Stover seems to regard as a *de facto* error that crept in as some sort of oversight.

But, Stover continues, had Collingwood distinguished descriptive understanding from identifying understanding, he would have seen that

> descriptive understanding places rational activity at the disposal of the historian-scientist: *what* happened was purposive thinking, the solving of a *particular* problem (an emphasis repeatedly made by Collingwood); the historian-scientist's problem then is to find out the determining conditions of the fact that the agent reached the decision that he did reach, solved his problem as he did solve it, and acted on the basis of his decision as he did act.[88]

Identifying understanding is possible, and it is able to supply an intelligible account of past events precisely because those living in the past, and ourselves in the present striving to understand that past, confront the world as fellow human beings.

> It is intelligible to us having been intelligible to them, as having been thought about and judged by them *as fellow human beings*. The kind of intelligibility that the world had for them is recognized as one of the *kinds* of intelligibility that "our" world has for us, a kind we ourselves at least sometimes pursue, the goal of one form of our questioning.[89]

D. Man as a Historical Being

Robert Stover has well distinguished pre-determined causation, characteristic of scientific explanation, from self-determining causation, characteristic of human decision-making in free choice. The

---

[87] See Collingwood's *The Idea of History*, pp. 2-3. See also p. 2 of the introductory chapter, above.

[88] R. Stover, *op. cit.*, pp. 106-107. Emphasis in original.

[89] *Ibid.*, p. 154. Emphasis in original. This is an important point to be developed later by Ernst Cassirer. The principle which gives care-concern concepts a logical value is the underlying unity of man. See this chapter, below, pp. 387-388.

basic distinction is between necessary causality and free causality, between what could have been otherwise and what could not have been otherwise, as everyday colloquialism expresses it. The reason why something happens may be due to the efficiency of preexisting actualities, or it may be due to the autonomous initiation of a free being on account of as yet unrealized but realizable foreseen eventualities that he seeks to bring into being by his free decisions. His determination of a foreseeable future is thus a reason for and explanation of the events resulting from his actions.

The recognition of man's free, self-determining causality provides the key to distinguishing the historical being proper to man from that of natural being (which man shares with natural things as part of the world of nature). This character of man as a self-responding situated being is well explained by John Wild in his *The Challenge of Existentialism.*[90]

Man's situatedness and facticity is first made aware to him by certain emotive responses which disclose much more than any amount of theorizing. Mood and feeling disclose objects that threaten me in such a way that I am made aware, all together, of the object, the feeling itself, and myself as experiencing that modality of feeling. Such a mood is then not as a thing-object, to be looked at, but rather as a way I am, which reveals my being as a being situated in the world.

This revelation discloses the naked facticity of the situation. I feel myself oppressed by the situation into which I am thrown and feel myself trying to disengage myself from that situation felt as something oppressing.[91]

To a certain extent this clarifies what Collingwood explained[92] regarding appetite as making us aware of a future as yet unrealized, harmonious state of well-being through making us presently aware of an unsatisfactoriness, or felt disquietude, in our present situation. But, I think Wild's explanation is more in accord with experience, inasmuch as he allows that these emotive moods *themselves* have revelative capacity, whereas for Collingwood the selective activity of conscious attention has to intervene to disengage the 'there-and-then' of feeling from the 'here-and-now.'[93] Collingwood seems to exaggerate

---

[90] Bloomington, Indiana University Press, 1955. See Ch. 4, pp. 86-115, on "Human Awareness and Action."

[91] See *ibid.*, p. 8.

[92] Correlate with Collingwood's *The New Leviathan*, pp. 21-26, pars. 4.37-4.89 and pp. 47-53, pars. 7.1-7.69.

[93] See *ibid.*, pp. 21-23, pars. 4.4-4.6 and p. 51, par. 7.53.

the active role of consciousness over feeling when he says, "This practical directive or selective act [of attention] 'makes', as we say, the distinctions between what we attend to and what we attend 'from' or 'repress.' "[94] To the crucial questions 'Are there objects of feeling or not?'[95] and 'Is feeling active or passive?'[96] he answers that he does not know in both cases. But to the first he gives methodological reasons for preferring that what are felt are modes rather than objects of the activity. To the second he says he cannot give even methodological reasons for answering one way or another. So without being able to answer these crucial questions from *positive* grounds, he already loads the issue heavily on the side of the activity of consciousness, overriding the passivity and contribution from its object side.

If mood and feeling awaken me to the facticity of my situation, it is understanding which "focuses on my essential possibilities and projects—that part of me which is always ahead of itself in the future," which, according to Heidegger, is the central core of my being. Understanding is thus a kind of creation, the projection of new being.[97]

The whole world is ordered and understood with reference to the ultimate projects I set for myself. To judge something is to interpret it in the light of some project and, ultimately, in the light of my ultimate concern.[98]

Our primitive insights are practical. If the hammer I am using is too big I look for another. But when I have leisure, and am freed from the demand on me to act, I can look at the hammer from a detached point of view and inquire what goes with hammers as such. From such contemplation theoretical science arises. Such activity presupposes leisurely detachment from pressing practical everyday needs and has its own distinctive mood of relaxation, detachment and concentration. Theoretical knowledge bears on something as a fixed entity of some sort, and is thus incapable of grasping human possibility.[99] Practical knowledge, grasping human possibilities, does so from a situation of dynamic involvement in which these very possibilities arise. Inspired by Heidegger, Wild explains as follows:

> Our primordial and true insight into the world is derived from our active projection of possibilities ahead of ourselves. This projec-

---

[94] *Ibid.*, pp. 22-23, par. 4.52.

[95] See *ibid.*, p. 28, par. 5.2.

[96] See *ibid.*, p. 31, par. 5.4.

[97] See John Wild, *The Challenge of Existentialism*, pp. 87-88.

[98] See *ibid.*, p. 89.

[99] So it is clear why dialectical thought gives primacy to action rather than to substantiality in explaining.

tion is the first act of understanding. This orders ourselves and all other entities into an intelligible world. Interpretation is the detailed development of this frame, and its application to concrete objects confronting us. To fit something into such a frame is to disclose its being. Such disclosure belongs to *Dasein*. He is the disclosure he makes. Some of them are true and reveal the entity itself.[100]

This remarkable quote clarifies and ties up much that has been said in the foregoing. The ''interpretation'' referred to here seems to be the same as Collingwood's ''interpretation''[101] or ''insight''[102] into a situation needed to understand what one ought to do in that situation. It is clear that the understandability of the situation as described here by Wild is through nothing other than what has been previously explained at length[103] as a concrete universal. The development of the frame of reference in which we order our projects is the building up of a concrete universal which gives meaning to all those things brought together and related among themselves according to their necessary involvement in that overall framework. The relevance of all that Stover has said above regarding man's careful concern for the world as a source of distinctive meaning is also implicated here. The

---

[100] John Wild, *op. cit.*, p. 90.

[101] See Collingwood's treatment of 'insight' into the meaning of law in his *Speculum Mentis*, p. 225, where he says that the important thing about legal rules is not their rigid formulation but their 'interpretation,' whereby they are seen to illuminate dynamically a concrete situation. I think that what is said above, together with what is said by Collingwood regarding the role and function of law, would greatly clarify many of the issues raised by Gidon Gottlieb in his investigation into legal reasoning in his *The Logic of Choice*. See, for example, Gottlieb's two cases of the shooting of an elephant, one in a U.S. Central Park zoo, the other in Burma (*op. cit.*, pp. 60-61). ''All the circumstances, the nature of the offence, the motives of the accused, were in all respects identical . . . The only difference between the two cases was that one arose in the U.S. and the other in Burma. There too, a variety of facts was brought to the attention of the court but for the same two facts: the colour of the clothes of the accused and the colour of the elephant's skin. In both instances, however, the elephant was a white elephant. The colour of the elephant was of no special interest in the U.S. . . . but it was of crucial importance in Burma, where white elephants are sacred and their slaughter sacrilegious. To omit in Burma any reference to the colour of the elephant was tantamount to concealing from the court a most important fact.'' Gottlieb then makes his point in regard to the interpretation of laws about shooting elephants: ''Facts and cows have this in common: they may be sacred in India but not in the U.S. The importance of cows and the relevance of what happens to them depends on their place and function in the community; what may be significant here may be—to enrich our metaphor—a red herring elsewhere.'' See also Appendix 2 below, p. 491, regarding judicial interpretation of law.

[102] For Collingwood's discussion of the need for 'insight' into a given situation in order to decide how to act in the concrete circumstances see his *Autobiography*, pp. 101-106. See also Chapter 6 above, pp. 220-221.

[103] See Chapter 4, above.

mutual clarification of Stover's treatment of intelligibility deriving from man's deliberated choice, Collingwood's treatment of insight into and interpretation of a situation, and the elaboration of this meaning of the concrete universal by Bosanquet and Collingwood, all seem to be nicely summed up and articulated in the above quote of Wild. The progressive determination of our projects involves the elaboration of enmeshing relations which at the same time *create* and *reveal* the meanings of all that is thus related. Understanding in this context is thoroughly pragmatized, as it is also in Sartre and Heidegger.[104]

> The meaning of an event is the way it fits into a plan of action, and all understanding presupposes an ultimate end that has been chosen. Theory is assimilated to practice. My choice of myself and my world order is at the same time my discovery of the world. Awareness and choice are really one.[105]

Wild, immediately after this, adds that "this pragmatic conception determines the existentialist attitude towards the discipline of logic,"[106] which he proceeds to consider, and his consideration is of the highest importance in enlightening the verdict to be passed on Collingwood's theory of logic, especially as it is relevant to history. Like Collingwood,[107] Wild criticises the formal or symbolic logic which prevails in many parts of the world inasmuch as "conceptual meaning has been ignored and sometimes reduced to the instrumental symbols of language," and "the syntactic structure of these symbols has been cut off from their referential functions."[108] The result is that logic becomes sundered from reality and knowledge, and becomes the plaything of technicians, each playing his own game.[109]

> According to Heidegger, the judgment is logistically dissolved into a system of rules, and becomes the object of a sort of reckoning rather than the theme for ontological interpretation.[110]

In opposition to this, Wild insists on the intentionality of logic as a human discipline. "The logic of science and of all living discourse is referential."[111] To reintroduce this classical notion, Heidegger,

---

[104] John Wild, *op. cit.*, p. 95.
[105] *Ibid.*
[106] *Ibid.*
[107] See, for instance, Collingwood's comments on symbolic logic in his *The Principles of Art*, pp. 268-269.
[108] John Wild, *op. cit.*, p. 95.
[109] *Ibid.*
[110] *Ibid.*, p. 96.
[111] *Ibid.*

according to Wild, feels the need to regard the primordial insight into existence as achieved only in terms of our active projects.[112] In Collingwood the same thing is found in his recognition of the primacy of the practical over the theoretical.[113] Collingwood is obviously doing what Wild says Heidegger and the existentialists do in this respect, namely "rescue logic from the hands of minute technicians . . . to restore it to the position it once had as the interpreter and guide of meaningful discourse."[114] Whatever may be the interest and importance of turning logic into the theory of calculating machines, such "mathematical logic has certainly not shed much light on the *Logos*."[115] Like Collingwood, Wild insists, "Instead of becoming obsessed with minor details of mathematics, it is the function of logic to inquire into the structure of human meaning as such."[116] He reechoes Collingwood's need "to study the intentional structure of living discourse and to devise a human discipline for its guidance."[117] As human awareness is primarily practical rather than theoretical and human awareness cannot be separated from human action, the revitalization of logic must be as a logic of human choices, decisions and action.[118] Since this is the stuff history is made of, any distinctive logic of history will be a logic of human action and human awareness as it arises in and through action.

According to Wild, following Heidegger, it is the feeling of dread that awakens me to that which distinguishes me as a *historical being*, namely *a being who faces possibilities to be determined and actualized through choices*.

---

[112] *Ibid*.

[113] *Passim*. See specifically his *Autobiography*, pp. 147-167, and his *New Leviathan*, p. 5, par. 1.66. Collingwood's thesis, that 'mind is what it does,' I think, seems to sum up this attitude in one sentence.

[114] J. Wild, *op. cit*., p. 97.

[115] *Ibid*. See also E. Cassirer on the meaning of *Logos*; see this chapter, below, pp. 387-389.

[116] John Wild, *op. cit*., p. 97.

[117] *Ibid*., pp. 97-98. Chapters 1 and 2, above, expose Collingwood's call for this new logic.

[118] John Wild, *op. cit*., p. 97. This means that Collingwood's (and Wild's) call for a revitalized logic must be a logic of what Stover, above, called man's 'careful concern for the world,' or the logic of human decision-making. This aspect of logic is not as neglected today as it was in Collingwood's time. See, for example, Chapters 1, 13, 14 and 15 of W. Edgar Moore's *Creative and Critical Thinking*, Boston, Houghton Mifflin Co., 1967. See also the first four chapters of Troy W. Organ's *The Art of Critical Thinking*, Boston, Houghton Mifflin, 1965. The former work stresses the human decision-making dimension of logic, the latter its interrogative aspect. See also Henry S. Leonard's *Principles of Reasoning*, New York, Dover, 1967, pp. 22-86, on the recognition of problems and the methods of their solution.

> Dread singles me out and leaves me alone to face what I still might be. It also touches the whole structure of my being in the world, but with a poignant emphasis on those real possibilities which I really am, though they be ahead of me.[119]

Wild then rejoins Stover in singling out "that care or commitment which characterizes all specifically human action." The term 'care' expresses the conative structure characteristic of human existence as concrete and dynamic.[120] Human existence does not just flow down like a stream from a past through a present to some unforeseeable future. As a historical being my past and future are together with me in any present now in my careful concerns.

> . . . my activity is always to a future which I have already projected ahead of myself by understanding and mood. . . . The past and future are not separated from each other by a punctual present. They are held together in a certain unity which can be clearly seen in the structure of care. This care is always ahead of itself and never confined to the present. . . . It is this envisaged future that directs my care.[121]

Wild distinguishes a "threefold order of human existing as the structure of care: (1) ahead of itself . . . (2) as already in the world . . . (3) as being with . . ."[122] If this is so, then we may regard this as the basic structure of a concrete universal, and so it is of the highest importance in explicating that notion. Care means more than active co-presence. One can act carelessly regarding a present object. Care is a special kind of co-presence peculiar to man. If that which is co-present is another man one does not merely act on him. This careful concern does not come and go but lies *at the very core of man's concrete being*. The care-taking agent is a centre of care (in its variety of forms, namely, of drive, urge, desire, wish or will), that is, a projective structure. Even the theoretical order is basically practical and embodies this structure when it is a question of the possibility of a truth I do not yet know and to be sought out by carefully chosen means.[123] This includes scientific research, and one may add all such investigations

---

[119] John Wild, *op. cit.*, p. 99. The feeling of dread thus seems itself to delineate man in his factual situation from the environment in which he has to make choices. Therefore, before the conscious activity of selective attention, a well defined structure is presented at the level of feeling. See criticism of Collingwood on this point, above, pp. 373-374.

[120] See John Wild, *op. cit.*, p. 100.

[121] *Ibid.*

[122] *Ibid.*, p. 101.

[123] See *ibid.*

as Collingwood's 'who killed John Doe?' "This possibility [of some as yet unknown truth] ahead of me arouses my concern[;] it takes over any factual existence and directs it to the necessary procedures and acts."[124]

The most significant differentiating factor in distinguishing the different levels of care is the weight of the futurity factor. Drives and urges are dominated by an object as present; they seem to flow out of the past, so that the image of the stream is more fittingly said of them. Desires and interests, at a higher level, are almost totally lost in what is present, with at best only a dim view of futurity. Wishes manifest a limited view of real possibilities. But "the more authentic kinds of care are manifested in those acts of purposive choice where basic possibilities are clearly grasped. They take over the whole personality in decisive commitment."[125]

This threefold structure of the all-pervasive human phenomenon of care opens the way to the understanding of the problem of time and of the connections of past, present and future as these are historical characteristics.[126]

The more spatialized and familiar conception of time is that of a succession of 'now' moments. The present is real at any given moment. The past was once a (present) now and the future will be a now when it really is. Present nows seem to flow into the past and be replenished from future nows flowing into the present. This goes on indefinitely, and all that happens, including myself, happens in this flow of time. The present can be examined as an object before my senses, the past I remember, the future I anticipate or predict.[127]

But, Wild asks, do not these characteristics belong to what is *in* time rather than to time itself? Is *time* itself divided into past, present and future? or rather "is not time itself dissolved with the dissolution of the past and future?"[128]

Following Heidegger, Wild says that "time lies at the root of the structure of human care." It "is an existential structure which pervades man's being in the world."[129]

---

[124] *Ibid*. Thus, the investigative stage, or way of discovery of some theoretical concern, is itself not merely theoretical but also practical, involving careful deliberation and decision. See below, Appendix 2, p. 488 et seq.

[125] *Ibid*., p. 102. I think that we may say that the degree of expectation of purpose here is at the same time the measure of a man's disengagement from the natural world and the constituting of himself as a free, autonomous, and thus historical, being.

[126] See *ibid*., p. 103.

[127] See *ibid*., pp. 103-104.

[128] *Ibid*., p. 105.

[129] *Ibid*., p. 106.

Time is not a sort of space which things are *in*, nor is the human person, the historical being, a thing, or a set of events *in* time. Time is rather a 'stretch' or 'intension' of man's being. His being is not *first* confined to a given moment and *then* stretched out into past and future; from its beginning to its end his being is thus characterized.[130]

> From the very beginning, his being is stretched out into possibilities ahead of himself, and a past which he must take over if he is to be with things at a factual present.[131]

What Heidegger calls the three ecstasies of time are thus integrated into the unity of man's existential being. "He is not *in* time, but rather he *is* it, and exists it. To have time for something is to have self-devotion to it. To lose time, or to waste it, is to waste the self."[132] Man's historic being is thus described:

> I am not confined to a present moment nor to a specious present. I am the future I have projected ahead of myself, and I still am the past that I have been.[133]

Wild stresses that "it is the future that must take hold of the past, and guide it in present action."[134] But this addition to my present is not an accidental addition to an already there substantial being. *At no moment am I ever all there.* Stationing myself in this projected future, from this vantage point I can direct and criticize my action. "It is from this being ahead of myself that I regard the flow of my being out of the past, through the present, towards where I am primarily, at my end in the future."[135] Note what is implied here. My life is seen by me as a meaningful whole only if I see myself *as a completed whole* from some point of goal-achievement located in the as yet unrealized future.

But the being of man is not exclusively futuristic; it is stretched over the past which he has been and bears with concern on present

---

[130] *Ibid.*, pp. 106-107. If this is so, we can see the relevance of Collingwood's explication of man's historical being as an overlap of classes in a scale of forms, each later form taking up and modifying the earlier. Rather than an ejection of his being into a dead past there is a constant retaking up of that past, an injecting of it into the present (Collingwood's reliving of past thoughts), and a directing of the whole present actuality (fertilized by its own revived past which it sums up in itself) into the as yet unrealized but realizable future. Any stage of its active development can be represented together as the fulfilment (summation) of its past to that point and as its forward thrust to the self-realization of future projected possibilities.

[131] J. Wild, *op. cit.*, p. 107.

[132] *Ibid.*

[133] *Ibid.*

[134] *Ibid.*

[135] *Ibid.*

things and actions. Hence, the present does not just come after the past nor the future after the present. These are inseparable phases of a single integrated structure whose "unity must be maintained by resolute choice that holds them tightly together with the future, not after but *before* the present."[136] Laziness and irresolution may rupture this unity; then man passes from past to future through present in the way any physical object does so, as something which just happens to it, not as something he carries himself through by decisively and freely "taking himself in hand." As *things* observed scientifically they are in time, and their past, present and future are related by external relations.[137] But as man is consciously aware of them, his past, present and future are not *things* but modalities of his (conscious) existence related by internal relations through human action in decisions and the realization of projects. In this threefold modality,

> the future is in the mode of possibility, [so that] no part of it can become actually present without altering its mode of existence. The past is already there; its factual content fixed, and not open to choice. The present moment is not yet there to be forgotten or taken over. It is still indeterminate and open to choice.[138]

My past is not something over and done with. It is that which I have been so long as I am. It is unfinished because it never was all there. The future is more than a not-yet-now that will be, which would seem to imply that *until then* it is nothing. But *even now* I am this future which stretches ahead of me in the mode of possibility. It is what determines the meaning of my existence, and from whose vantage point I interpret the past and guide the present in action. The three modes of existence temporalize themselves each with respect to the other and each in its own way.[139]

---

[136] *Ibid*. Emphasis in original. Again we have, in Wild's words, what may be taken as an excellent summary statement of Collingwood's explanation of history as a concrete universal constituted by a scale of forms in an overlap of classes.

[137] This lazy and irresolute attitude seems to correspond with Collingwood's identification of 'realist' thought with low-grade and unscientific, inertial thinking. See his *Essay on Metaphysics*, p. 34. See also above, Chapter 5, p. 169.

[138] John Wild, *op. cit.*, p. 108. Emphasis in original.

[139] *Ibid.*, pp. 108-109. Luigi Sturzo, in his *Inner Laws of Society*, trans. Barbara Barclay Carter, New York, P. J. Kenedy & Sons, 1944, p. xxv, says: "The human present . . . is not something static and definitive; it is, on the contrary, dynamic. It projects itself into the future, which presses on to become present and lose itself in the past. The dynamism towards the future in human process translates itself into finalism, while the activity that in realizing the present loses itself in the past translates itself into causality. The past is the moment of causalities, the future that of purpose, while the present is the reality of existence. Past and future do not exist in themselves, but in the present, that is, in real human existence, which is in fact a continuous succession of presents."

History is usually referred to as a succession of events in the past, the role of the past being regarded as so predominant as to define history as the study of the past. It is often thought of as a great stream flowing down from the past through the present to the future. Under the influence of naturalistic thinking the past is thought of as a fixed entity on whose sure basis we must build the future.[140]

But this image of the stream is inaccurate, since at any designated moment the *whole stream* exists with all its parts from beginning to end. But men are not *in* a stream of history; they *are* their history. Social man, like individual man, exists with these modalities of past, present and future, no single one existing without the essential implication of the others. Social man, like individual man, is always ahead of himself. But if he is not to remain lost in mere possibilities, for their realization he must appropriate his factual past and put it to work in present action directed towards the realization of future projects.[141] From the point of view of the self-constituting consciousness, therefore, the present is together an active appropriation of the past and at the same time an active construction of the as yet unrealized but realizable future.

Thus appears the concern for the historic past; it is not something dead and gone but something still with us. It is what we (who are now) have been. The past is not all gone because *it never was all there in the first place*. It never occurred as *an event* but as something ahead of itself in its (own) future.[142]

It is impossible, therefore, to understand a historic fact without understanding *its* future. Possibility is at the very heart of human history as the guiding centre of human care. So the past which is the object of human history is *a past having a human future,* a past whose fulfilment in reality is to be found in a future of its own.[143]

The stream-image is defective because its past, present and future parts are all in the same mode of being, namely 'being-there,' or as 'events.' A historically past event was not a floating futurity, it also had *its own* past and *its own* future in the light of which *that* past was reappropriated in acts of self-determining choice. Thus, the past, as

---

[140] See John Wild, *op. cit.*, pp. 110-111.

[141] See *ibid.*, p. 111. This is more or less what I understand Collingwood to mean when he says that questions, the cutting edge of the mind, determinative of future possibilities, are not the pure form of questioning, but are formed out of a preexisting, that is, past, body of fact. See his *Speculum Mentis*, pp. 78-79. See also Chapter 3, above, pp. 66-68.

[142] See J. Wild, *op. cit.*, pp. 111-112.

[143] See *ibid.*, pp. 112-113.

known historically, can disclose to us *its own* possibilities (in the light of what was future *for it*), some of these which became realities. Insight is required to resolve these possibilities now buried in the mists of ambiguity. The unity of history is not a completed fact already there waiting just to be accepted; it is a struggle to be won and to be maintained against the threat of being lost again.[144]

Thus men, their being in the world, and the world in which they are, are all historical. The things which go to make up the world were not once all there in a certain form; they have their history, so that the things we now have are different. The world itself, in which men and things are said to be, was never fixedly all there; it was always temporalizing itself by actively endowing itself with the realization of future projects based on past achievements.[145]

It is the essential possibilities open to man which give unity to history. Wild thus states the task of the professional historian:

> He is studying the being of man. His duty is to clarify such insights as we already possess, and to show how those real possibilities have been won or lost in the past. By helping us to understand more clearly those that still remain to us, he is participating in the actual process of history. This understanding is the most primordial part of history. Without the relating of real possibility to brute fact, there would be no process of history. In this sense, the science of 'history' is history itself, and the single name is not only appropriate but required.[146]

It is thus in the mind of the historian, or rather in the mind of man as he is a historian, that past and future are integrated in an intelligible present, which is thereby understandable. As both Collingwood and Marrou insist, man the historian is one with his history and cannot be intelligibly separated therefrom.

But our concern is mainly with the logic of this historical being, specifically with that logic as it is a logic of questioning. Ernst Cassirer has given considerable thought to the logic of the properly human world, including its history, and what he has to say is highly relevant to the evaluation of Collingwood's position on the interrogative logic proper to history. We now turn to consider Cassirer's account of the logical ordering of the human, cultural, moral and historical world. Here the optic is epistemo-logical, that is, the ordering of the

---

[144] See *ibid.*, p. 113.
[145] See *ibid.*, p. 114.
[146] *Ibid.*

phenomenal multiplicity into a recognizably intelligible unity of universality.

E. The *Logos* of the Physical Natural Order and of the Spiritual Moral Order

Ernst Cassirer points out that from the awakening of critical thought, two meaningful spheres of order and intelligibility confronted man. There is the world of nature, which was most likely the first to be brought from chaos to ordered cosmos by the science of astronomy. But more intimate to man is his own human world, the world of custom, culture, civilization and morality. Primitive mythology explained both orders as gifts of the gods or superior beings. Even the tools produced by his own technology, including his intellectual tools, speech and writing, were attributed to the gods as gifts to man. Language, for example, was regarded by the Egyptians as a gift of the moon-god Thoth.[147]

Speaking of the mythico-theological climate of the priesthood of Memphis on the eve of the accession of Akhnaton to the throne of Egypt, James Breasted[148] has the following to say:

> Ptah had been from the remotest ages the god of the architect and craftsman, to whom he communicated plans and designs for architectural works and the products of the industrial arts. Contemplating this god, the Memphite priest, little used as his mind was to abstractions, found a tangible channel, moving along which he gained a rational, and with certain limitations, a philosophical conception of the world. The workshop of the Memphite temple, where, under Ptah's guidance, were wrought the splendid statues, utensils and offerings for the temple, expands into a world, and Ptah, its lord, grows into the master-workman of the universal workshop. As he furnishes all designs to the architect and craftsman, so now he does the same for all men in all that they do; he becomes the supreme mind; he is mind and all things proceed from him.

With the awakening of critical thought questioning the world of experience, both natural and human, man sought to explain such experience in terms of some recognizable unity in the phenomenal multiplicity. With this need to reduce the privately experienced phenomenal multiplicity to some commonly accessible rationale as principle of unity of order, logic in its widest sense is born.

---

[147] Ernst Cassirer, *The Logic of the Humanities*, pp. 40-44.

[148] See his *A History of Egypt*, New York, Scribner's, 1959, (first pub. 1905), pp. 356-357.

> The Ionians . . . are not . . . satisfied with mere knowledge of the "what"; they also ask for the "how" and the "why". . . . Only *thought* can give us the answer: for in it, and in it alone, is man freed of the limitations of his own individuality. He no longer pursues his "subjective understanding," but comprehends one that is universal and divine. A universal law . . . replaces . . . "private insight." It is by so doing that man, according to Heraclitus, first frees himself from the mythical world of dream(s) and from the narrow and limited world of sense perception. For just this is the character of being awake and of awakened being—that individuals possess a common world, whereas in dream(s) each lives only in his own world and remains mired and imprisoned in it.[149]

This ideal of bringing the world of experience under a universal reason, according to Cassirer, was unable to live harmoniously with the Christian revelation during the Middle Ages.

> The attempt to bring the logos-concept and the Johannine gospel under *one* denominator . . . will always remain abortive. For within these two the manner of mediation between individual and universal, finite and infinite, man and God, is radically different.[150]

Descartes sought to universalize experience in a world unified by the mathematical ideal of intelligibility.

> This fundamental idea [of a unified world and unified knowledge thereof] of modern (scientific) research found its decisive philosophical legitimation in Descartes' concept of *mathesis universalis*. The cosmos of universal mathematics, the cosmos of order and measure, comprehends and exhausts all knowledge. In itself it is completely autonomous; it requires no support and could recognize none, other than that which it finds within itself. Now, for the first time, reason comprehends the whole of being in its clear and distinct ideas, and now, for the first time, it is able to penetrate and dominate this whole by means of its own powers.[151]

Cassirer says that "what is decisive in Descartes' system of metaphysics is his original conception of a single all-inclusive method of knowledge."[152] But the dualism of (spiritual and material) substances qualifies and limits the ideal of the Cartesian monistic method. The world of spirit and the world of nature retained their radical exclusive dualism. But, with Grotius, for whom there is a thorough analogy between knowledge of law and of mathematics, and Spinoza, for

---

[149] E. Cassirer, *op. cit.*, p. 46. Emphasis in original.
[150] *Ibid.*, p. 47. But see Sam Keen, *Apology for Wonder*, New York, Harper and Row, 1961, p. 17, for an opposite opinion on this.
[151] E. Cassirer, *op. cit.*, p. 49. Emphasis in original.
[152] *Ibid.*

whom God is the one substance having two manifestations, "the human spirit was to cease to be a 'state within a state'; it was to be known by the same principles, and the same lawfulness was to underlie it that underlies nature."[153] Consequently, "Spinoza creates a new form of ethics . . . by taking geometry as its pattern. . . ."[154] Radical unification of thought and methodology is now achieved.

> Here, for the first time, the sphere (of knowledge) appears to be complete; the circle of mathematical thought has been able to encompass, with the same method, the world of matter and the world of the soul, the being of nature and the being of history.[155]

But the unification of the outer, physical and inner, spiritual, moral world under the one same *mathesis universalis* was short-lived. Implicitly, Descartes' English contemporary, Thomas Hobbes, and explicitly the Italian, Vico, a century later, effected a Copernican revolution that was to subordinate the mathematical to the cultural in the paradigm position of universalizer. According to Hobbes, the meaning of the natural world is confined to the propositions which we are able to form in its regard. These propositions are intelligible to us because we are their authors. God alone knows the natural world itself, since he alone is its maker. Vico elevates this attitude into the status of a fundamental principle: Verum = Factum.[156] According to Vico, mathematics, as Descartes rightly says, gives clear and distinct knowledge. But this is so, says Vico, only because mathematical entities are the products of our own thought. Since we are their authors, we are able to have comprehensive knowledge of them. But mathematics is confined to the sphere of concept-formation, and so remains in the ideal. It is only in the works of civilization that man has knowledge of an object which is both penetrated conceptually and concretely real. These objects are not only conceivable but are also fully determinate, concretely individual and historically existential.[157]

---

[153] *Ibid.*, p. 51.

[154] *Ibid.*

[155] *Ibid.*, pp. 51-52.

[156] See Bruce Mazlish, *The Riddle of History*, New York, Harper and Row, 1966, pp. 23-28. See also Collingwood's translation of Croce's *The Philosophy of Giambattista Vico*, New York, Russell and Russell Inc., 1964, especially pp. 24-29. See also, above, Chapter 1, p. 15 et seq., for Collingwood's attitude to mathematics and deductive inference. The inferential compulsion in mathematical, deductive reasoning is a compulsion which the mathematician imposes on himself in virtue of the commitment to the original assumptions which he makes in order to solve the problem to hand.

[157] E. Cassirer, *op. cit.*, pp. 52-53.

From then onwards, thinkers such as Herder looked to man's cultural products as the proper objects of man's knowledge.

> [In] Herder's work [in] the philosophy of language, the theory of art, and the philosophy of history . . . we are able to study . . . the rise and ultimate breakthrough of a new form of knowledge, which . . . is not to be abstracted from its matter, but which becomes manifest only in the free shaping of this matter and in its spiritual domination and penetration.[158]

But, since the focal point of our inquiry is the logic of history, we must ask what is the *logos*, the one in the multiple, which enables us to deal rationally with such phenomena. The unity behind, or rather within, all such diversity, its principle of intelligibility and meaningfulness, is the unity of man. The infinite varieties of human manifestations in all their forms are unified inasmuch as they are the manifestations of man, of the human spirit.

> . . . whatever man is required to perform must spring from the condensation and unbroken unity of all his powers; all that is isolated is evil.[159]

This is the 'unity in diversity' which is the principle of unity, and therefore of intelligibility, in the human sciences and in history. It is Marrou's "same in the other which . . . presupposes a broad basis of fraternal communion between the subject and the object, between the historian and document."[160] Speaking of the Ancient Egyptians, Barbara Mertz[161] well expresses this same-in-other, this unity in diversity, as follows:

> How richly grotesque—how bizarre—was the spiritual world which these long-dead aliens envisaged. And then we come upon a single sentence, or an isolated phrase, and the mask of ceremonial vanishes to expose the familiar poignancy of man's quest for immortality, with all its uncertainty and its aching desire. "No one has returned from thence to tell us how they fare." The lament for a dead child, the demand for justice, the lover's yearning for his

[158] *Ibid.*, p. 54. Italics deleted. Here is the seed of Hegel's later dialectical unity of form and content.

[159] E. Cassirer, *op. cit.*, p. 55. This new outlook, which sees isolation of what springs from the unbroken unity of all of man's powers as evil, is most likely the source of Collingwood's notion of evil as outlined in Chapter 4, above, p. 112, in which the same act may be regarded as good or evil according as it is integrative of all that preceded it, or, contrarily, as yet still divided away from the fulfillment it will achieve at a higher level on a scale of forms.

[160] H.-I. Marrou, *The Meaning of History*, p. 104.

[161] See her *Temples, Tombs and Hieroglyphs*, New York, Howard McCann, 1964, p. 338.

> beloved—before our recognition of the universality of human emotion, time and distance shrink, the barriers of language, color, and nationality go down; we look into the mind of a man three millenia dead and call him "brother."

Poetry, "the mother tongue of the human race" according to Vico and Hamann, is the living expression of man's concrete unity.

> It is in poetry that he seeks to realize and make live again that original unity which, in the beginnings of historical language and myth, had fashioned history and poetry into a genuine totality, into an unpartitioned whole.[162]

But the penetration of our knowledge into the human products of language, art, myth and history is impeded so long as physical objects remain the only goal of inquiry. As long as the ideals of natural science monopolize the methods of inquiry "this question is not only unsolved, but cannot even be stated in its true and full sense."[163] Cassirer insists that we must come to recognize that universal natural science and the physical cosmos are only an instance of a much more general inquiry which is now more and more supplanting the pan-mathematical ideal which has dominated philosophical thought since Descartes.[164] Cassirer concludes that the idea of a cosmos and universal order is not restricted to the mathematical and physico-astronomical world. The lawfulness of natural phenomena, of the world of matter, is not the exclusive type of order and intelligibility. Wherever unity in diversity is recognized there is order and intelligibility.

> We encounter the idea of a cosmos, the idea of a thoroughgoing order wherever a determinate and unified lawful structure becomes apparent within multiplicity and diversity. This "holding good" of such lawful structure—this, in the fullest sense of the term, is the most general manifestation of what we mean by "objectivity."[165]

But what is meant by a 'cosmos' in this wider and all-embracing sense? "A 'cosmos', an objective order and determination," Cassirer informs us, "is present wherever individual persons relate themselves to and participate in a common world."[166]

---

[162] E. Cassirer, *op. cit.*, p. 55. What is said here about poetry seems to be much the same as Collingwood says about art generally, that it is an expression of man as a whole, not of one or several of his faculties. The central role accorded here to language agrees with Collingwood's regarding linguistic activity as the basic form of art.

[163] *Ibid.*, p. 57.

[164] See *ibid.*

[165] *Ibid.*

[166] *Ibid.*, pp. 57-58.

Now, where, first, do we find this breaking free of the bonds of isolated individuality so that a shared, public world is clearly observed? Cassirer says that this is found in the phenomenon of speech. The word is an *intended meaning* which exists only as *constructed within a whole of communication* in which the word passes from one person to another, *fusing both into one living dialogue*. A dialogue, therefore, is *a living unity in diversity* which exists as a form established by the very activity itself of language and speech.[167]

> . . . understanding through linguistic communication becomes for Herder, as it had been for Heraclitus, the genuine and typical expression of our understanding of the world. *Logos* creates the bond between the individual and the whole; it gives assurance to the individual that, instead of being enclosed within the waywardness of his own ego . . . he is capable of a universal existence. . . .[168]

The rationality of physical science is a derivative of the more fundamental "rationality invested in language and expressed in its concepts," from which "the path points to the rationality of science."[169] Language is not from itself able to generate science, yet it is an indispensable stage on the way thereto. Language "constitutes the only medium in which the knowledge of things can arise and progressively develop."[170]

The fundamental unity in diversity in human affairs is established by the act of naming. This harks back to Hobbes, as Collingwood notes. Following Hobbes, Collingwood says: "From being an indispensable means to the diffusion of knowledge, language has become the precondition and foundation of knowledge, so far as knowledge is science."[171] Thus, "man becomes capable of reason because he has *invented* speech."[172] Cassirer says that the act of naming is the indispensable first step and prerequired condition of the determinative act which constitutes the scientific task. He insists, consequently, that a theory of language is a necessary component in the development of epistemology.[173]

---

[167] *Ibid.*, p. 58. In Bosanquet's terminology, it is an "active form of totality." See above, Chapter 4, p. 144.

[168] E. Cassirer, *op. cit.*, p. 58.

[169] *Ibid.*

[170] *Ibid.*

[171] R. G. Collingwood, *The New Leviathan*, p. 44, par. 6.47.

[172] *Ibid.*, p. 45, par. 6.53.

[173] See E. Cassirer, *The Logic of the Humanities*, p. 58. Compare Collingwood, *The Principles of Art*, pp. 264-266. A. Brunner, in his *La Connaissance humaine*, Paris, Aubier, (1943?), in regard to "Le problème de la connaissance; comment se pose-t-il?" (p. 7),

Cassirer, like Collingwood,[174] insists on the creative character of language. It is never a simple copy, registering only the contents and relations presented to us immediately by sensation. As creative, language is a self-actualizing activity of the mind.

> . . . language is a determinate and fundamental tendency of the mind's activity—an ensemble of psychological and intellectual acts; and it is in these acts that a new aspect of reality—the actuality of things—first discloses itself to us.[175]

Cassirer's profound insights find striking confirmation in Luis Alonso Schökel's consideration of language in the context of Biblical interpretation. Schökel attacks the naive realism of a positivistic attitude to language in underscoring the formative role that language has on our experience of reality.

> . . . any explanation in terms of language necessarily involves a transformation of reality. Reality and our experience of it are stylized and expressed through the medium of language. Thus reality and experience are not *de*formed, but rather *con*formed through language. Language is a power which teaches us to understand and to order the world as well as to shape our own experiences. The great writers, while enriching the instrumentality of language, have likewise enriched our capacity for ordered perception. For the most part, therefore, the naive realism of language can no longer be maintained.[176]

How the external, natural world is humanized in perception and given a distinctive meaning due to being humanly perceived is thus explained:

> The external world confronts man and longs to be discovered. When man perceives this world, he humanizes it. This does not mean that man deforms or falsifies the world, but that their mutual relationship is actuated in a moment of plenitude, since knowing is

---

after examining various traditional positions regarding "le point de départ de la critériologie" (p. 21, pp. 7-17), says: "Nous prendrons notre point de départ dans un phénomène concret, réel, connu de tous les hommes, et nécessairement impliqué dans la recherche même qui nous occupe. . . . Le phénomène qui nous paraît fournir, et seul fournir un point de départ, est le *dialogue*. La première raison de ce choix est précisément le fait que le langage est l'instrument indispensable de la vie intellectuelle et spirituelle de l'homme. Toute philosophie se constitue par l'intermédiaire du langage. . . . En outre, le langage est un fait tellement évident que celui qui voudrait le nier, ne pourrait le faire qu'en l'employant, en s'infligeant par conséquent un démenti par son affirmation même" (pp. 22-23).

[174] See Collingwood's *The Principles of Art*, pp. 128-131; p. 275.

[175] E. Cassirer, *op. cit.*, p. 59.

[176] Luis Alonso Schökel, *Understanding Biblical Research*, Montreal, Palm Publishers, 1963, p. 94.

> the perfection of man and being known is the perfection of the external universe. The dawn which a man contemplates in a moment of sorrow is not the same dawn which is seen by a cow when it bellows toward the sky. Nor is the moon of which the romantic poet sings the same as the moon which prompts the dogs to bark during the night. Thus the world is humanized by our living encounters with it, and all the more so according to the intensity which we bring to this experience.[177]

This humanization of the world, already begun at the level of perception, is continued at the level of linguistic expression into the interpretative sphere.

> These living experiences, however, possess a quality of totality or completeness and are presented to us as a formless or only half-formed continuum. It is the specific function of language to convert this continuum into discrete parts by applying a system of vocabulary, morphology, and syntax. In this way the experience is given a new and human form. This is the second stage of transformation or interpretation. It does not deform the things of the external universe, but rather gives them a human dimension.[178]

The social, interpersonal dimension essential to language contributes to the humanization and meaningful interpretation of human experience.

> Further, the linguistic form by which I order my experience also has a social function, namely to communicate my interior life to other men, and thus to bring about a strictly personal encounter which is altogether superior to the mere knowledge of objects. This social and interpersonal orientation of language is an additional interpretative factor, one which raises language to all the grandeur of the human.[179]

Schökel notes the seriousness with which the Biblical account of man's origin treats man's activity of naming the things about him. It has something of the quality of God's creative activity exercised simply by ''calling'' (naming) a thing into being.

> In the Garden of Paradise, man alone was able to give names to the animals. By doing this, he brought the human order into existence, an order over which he was lord and master. In this he resembled God, who went about giving names to His creatures, thus calling them into existence. Just as man is created in the image and

[177] *Ibid*. Emphasis in original. Schökel here, and Cassirer later (see this chapter, below, pp. 407-408) note that the stabilizing of the Heraclitean flux is already at work in perception.

[178] *Ibid*., p. 95.

[179] *Ibid*.

> likeness of God, so he creates the world of language according to his own image and likeness.[180]

The past comes down to us only as transformed by and into language but in such a way that the result is faithful to objective reality.

> All language, in fact, consists in a transformation. The occurrences of the past come down to us transformed into linguistic terms, possessing a new, intermediate reality which faithfully, though inadequately, represents the objective reality. One must have this fundamental aspect of language as his point of departure if he is to go on in his study to the forms of literary language.[181]

Even language itself, in metaphor, is subjected to its own transformative, meaning-giving, creative activity.

> In literary language there is also another type of common transformation which can affect a word or group of words—the metaphor, the image, and the allegory. Not only is reality or experience transformed through linguistic expression, but even language itself can undergo a transformation.[182]

Cassirer says that language-awareness brings with it an awakening to a life of 'meanings.' These meanings, as distinct from mere impulsive reactions, are something repeatable and recurring, not restricted to the momentary here and now but remaining "one-and-the-same in countless life-moments and in the appropriation and use of countless other persons,"[183] that is, constituting a true universality and principle of intelligibility.

These two fundamentally diverse world-views, that of the natural world and that of the moral and cultural world, with their corresponding diversities of meaning and explanation, have a common root in primary perception. To understand more clearly the nature of these two different spheres of meaning it is necessary to examine the structure of primary perception inasmuch as this concretely embodies within itself a double rationale which founds and gives rise to these two different ways of meaningfully interpreting experience. The analysis is furthermore decisive for the understanding of the nature of cultural phenomena, since it articulates very clearly the nature of cultural entities and how they arise through the creative activity of expression-perception, of which language is the primary instance. The

---

[180] *Ibid.*, pp. 95-96.
[181] *Ibid.*, p. 96.
[182] *Ibid.*, p. 97.
[183] E. Cassirer, *op. cit.*, p. 60.

analysis greatly illuminates Collingwood's position on the primacy of art and language in properly human activity. The diversification and complexity built up by personal, creative, expressive activity will be seen to realize the characteristics of what in Chapter 4, above, was said of the concrete universal.

F. Expression-Perception and Thing-Perception

The universality whereby particular phenomena are meaningful and intelligible is not to be considered in isolation from its particulars. "Concepts without percepts are empty," and "the logical order must be tied throughout to the perceptual order."[184] Consequently, "analysis of the form of concepts as such is not capable of bringing complete clarity to what specifically distinguishes the humanities from natural science."[185] Cassirer resorts to a pre-logical phenomenology of perception to contribute to the solution of this problem.[186]

Primitive perception is 'two-faced,' fusing together two irreducibly distinct factors. Every perceptive act is that of a perceiving subject and that of a perceived object.[187] Consequently, the world encountered in perceptual experience is a "thing" or "it" world on the one hand, and an "I" or a "you," that is, a person-world, on the other hand. An "it" is something totally unlike "ourselves," totally alien, totally "the other," in no way "a self." A "you" is another that is "not totally other," for it is perceived as "another self."[188]

These two opposite characteristics, already apparent in primary perception, provide the foundation for two opposite directions in which the percipient subject can go in meaningfully elaborating and explicitating that original perceptual experience. If elaborated meaning is taken from the subjective side of perception then we are dealing with expression-perception or perception of expression. Myth making is explanation exclusively in terms of expression perception. Thus, thunder and lightning as "meaning" that this or that god is angry is an explanation of this phenomenon in terms of personal expression-factors. The thunder and lightning are "expressions" of a personal disposition, that of anger. Theoretical explanation, on the other hand, such as is characteristic of science (and of philosophy inasmuch as it is like science) explains in terms of "objective," imper-

---

[184] *Ibid.*, pp. 64-65.
[185] *Ibid.*, p. 93.
[186] See *ibid.*, p. 93 and p. 118.
[187] See *ibid.*, p. 93.
[188] *Ibid.*

sonal, "thing" factors; it seeks not only "to replace mythical explanations but to battle with and attack the whole mythical interpretation of reality at its root . . . which is nothing other than the perception of expression."[189] The mythical world view gives primacy to expression-perception over "thing" perception. The scientific world view does just the opposite. The mythical world view recognizes no fixed world of cause, which, however, is the very goal of all theoretical knowledge. Every feature of the experienced world can metamorphose at any time. In true Heraclitean fashion, anything can give way to anything else. The world, both natural and human, is the expression of an emotion of the moment, the love or anger of a man or a god. The phenomenal world is thus a "face" which is the bearer of emotive expressions, and the emotion expressed is the explanation of the phenomenon.

> For [myth], the world can assume a new face at any moment; for it is the emotion of the moment which determines this facial expression. In love and hate, in hope and fear, in joy and terror the features of reality are transformed. Any one of these emotions can give rise to a God-of-the-moment in a new mythical configuration.[190]

The whole world of events, the natural as well as the human, is thus reduced, in meaning and explanation, to a god, a superman, or a man. Anthropomorphism interprets the happenings of nature as expressions of will, of emotions, of loves and hates, such as we have subjective and personal experience of them in ourselves. For myth, then, the phenomenal world is a face, or series of faces, which manifests the personal dispositions of their bearer.[191]

Both science and philosophy collaborate in an active "endeavour to stop up the source from which myth forever takes its nourishment, by disputing every claim of the perception of expression."[192] In opposing the mythical approach science proposes that all expression qualities in which we recognize such characteristics as the trustworthy, the friendly or the fearful, be replaced by pure sense qualities, such as colors, tones and suchlike. Even these are further reduced as secondary properties based on more primary properties whose essential feature is mathematical, that is, pure quantitative determination.

---

[189] *Ibid.*, p. 94.

[190] *Ibid.*, p. 95.

[191] It is a "face," not a "mask," for it *expressively reveals* the subject and the subjective state, whereas a mask hides and functions more as a "thing" interposed between the bearer and external observer.

[192] E. Cassirer, *op. cit.*, p. 95.

This quantitative feature is all that remains of reality when the ideal of objectivity as sought by physics is achieved.[193]

Within the framework of this ideal of physics, not only proofs for the existence of other selves but also the question of their existence is evacuated of all meaning. The very question itself is mythical and so not worthy of serious consideration.[194]

The result is that the only language recognized to be subjectively meaningful is that of physics, whose meaning would be universally translatable (into numerical pointer-readings—recall Eddington's scientific table)[195] and would therefore be exactly the same unvariably for all (observing) subjects. The very meaning of language itself would be reduced to its physical characteristics, to phonetics and the physiology of sound. All 'meaningful expression,' as a manifestation or revelation of the psyche, the self, revealing its concerns, its loves and fears, would be put out of play.[196] The scientific study of the domains of art, religion and all other cultural domains would fare likewise. They could not be dealt with as expressions of meaning. The holy could not be distinguished from the profane, nor the beautiful from the ugly, nor the morally approvable from the morally reprobatory.

The illegitimacy of this physicalist and mathematicist reduction is that it renders impotent before it is asked the very vital question "what do these activities (i.e., linguistic, artistic, religious activities) mean?" for to ask this question is to ask what mental representation or thought or feeling these phenomena express and manifest.[197]

Cassirer concedes that physicalism does make one important contribution to the problem of understanding cultural phenomena, namely that of highlighting the important truth "that knowledge of physical things constitutes the foundation and substratum for every other construction of this kind."[198] In the cultural world, the world of expression-meaning, "there is no purely 'ideal entity.' . . . The ideal exists only insofar as it presents itself sensuously and materially in some manner and embodies itself in this presentation."[199] Religious, linguis-

---

[193] See *ibid*. Correlate with Collingwood's explanation of deductive science in his *Speculum Mentis*, pp. 164-169. See also, above, Chapter 4, pp. 76-82.

[194] See E. Cassirer, *op. cit.*, pp. 95-96.

[195] See Sir A. S. Eddington, *The Nature of the Physical World*, London, J. M. Dent & Sons, 1935, pp. 5-6.

[196] See E. Cassirer, *op. cit.*, p. 96.

[197] See *ibid.*, pp. 96-97.

[198] *Ibid.*, p. 98.

[199] *Ibid.*

tic and artistic meaning can be grasped only inasmuch as their embodiment in tokens, memorials, etc., constitutes the locus in which we come to know their meaning.

The cultural and historical object, a parchment, a stone monument, etc., is located in space and time and it comes to be and passes away just as does any physical object. But this physical object is recognized to have a *new function* other than any which is describable in terms of mere physical determinants. Something else besides the physical object emerges, and "what emerges is a 'meaning' which is not absorbed by what is merely physical, but is embodied upon and within it."[200]

This 'meaning'-factor which arises is, according to Cassirer, what is common to that area and to its content which we recognize as "cultural."[201] Cassirer illustrates this by pointing out that a Raphaelic painting of Plato and Aristotle includes the physical reality, such as colors, etc., *through* which we see the *objective presentation*, Plato and Aristotle in conversation. But, in addition to this objective presentation, there is also the *subjective presentation of Raphael himself*. Inasmuch as that painting is "a work" of Raphael it is an "expression" of himself, an expressive "face" whereby he is placed in our presence. Any genuine "work," Cassirer says, embodies all these three characteristics, namely, a certain physical thereness, an object-presentation, and the evidence of a unique personality. Such a "work" is more than merely a "result," and belongs not only to the natural world but also to the world of culture.[202]

Because of its ideal of clarity and fear of becoming lost in obscurity, Positivism does not admit those three dimensions. Cassirer agrees with Positivism in that when we compare expression-perception with thing-perception, expression-perception does seem to have an inherent extraordinary difficulty and inconceivability. But he points out that this difficulty and inconceivability does not exist for the naive world

---

[200] *Ibid.*

[201] See *ibid.*

[202] See *ibid.*, p. 99 and pp. 146-158. H.-I. Marrou, in his *The Meaning of History*, p. 115, speaking of the Christian gospels, says that they "are not direct testimony concerning the life of Christ, but rather a primary document of incomparable value regarding the primitive Christian community. We can only approach Jesus through the image of himself that was formed by his disciples . . ." This means, in terms of what Cassirer has explained above, that the Christian gospels, as 'works' or 'expressions' of the first disciples of Christ, manifest immediately to us the beliefs of these first Christians, as a facial expression immediately manifests joy or sorrow. From these immediately manifested convictions we are led, mediately, to the person and teachings of Christ.

view.[203] Expression-perception is the noetic daily bread of everyday living, of our knowledge of the familiar world. No question of proofs for the existence of other selves even suggests itself and no theoretical arguments can shake it in its conviction. But the scandal Kant sought to remove "of being forced to accept the existence of external things merely on faith . . . becomes all the more violent when we [reflexively] turn from the question of external things to that of other subjects."[204] But Cassirer maintains that the testimony of that reflection cannot be regarded as the sole authority in the area of the really basic problems. Recourse must be had to sources of knowledge of a different and more original kind. The process of substantiation must be grounded on something itself not deductively demonstrable, but able only to be pointed out. Even Descartes' *cogito,* according to Cassirer, is no formal argument but a purely intuitive cognition.[205]

Phenomenological analysis does not substantiate the assumption that the entities immediately accessible to knowledge are determinate data, with the consequence that the sensuously given is alone capable of immediate experience, all else, especially spiritual existence, remaining uncertain. Neither observation of function nor of genesis justifies granting to sense perception primacy over expression-perception. The genetic account of the development of consciousness, both in the individual and in the species, shows that the supposed starting points of all knowledge of reality, the so-called sense data, are rather late products, and are themselves results of lengthy processes of abstraction whereby they are extracted from the whole of human experience. From the functional point of view, unbiased psychological observation testifies to the fact that the first experiences of the child are experiences of expression. The perception of things and thing qualities comes only later.[206]

It is to language primarily that Cassirer attributes the transition from expression-perception to thing-perception since he maintains that capacity for objective representation grows in direct ratio as we begin to give, and progress in giving, linguistic expression to our experience of the world. It is by this linguistic expression that our experience of the world ceases to be merely a matter of passively received impressions.[207]

---

[203] See E. Cassirer, *op. cit.,* p. 99.
[204] *Ibid.,* p. 100.
[205] See *ibid.,* p. 101.
[206] See *ibid.,* pp. 101-102.
[207] See *ibid.,* p. 102. Correlate with Collingwood's *The New Leviathan,* p. 42, par.

The theoretical world view, that of philosophy and science, did not begin by regarding the world as the sum total of merely physical things: "this is a *terminus ad quem,* not a *terminus a quo*—an end, not a beginning."[208] For Plato and Aristotle, the soul is the principle of motion without which the world would come to a standstill. Descartes for the first time gives us the notion of a strictly mathematical and mechanical universe. But this view is terminal, not initial, the product of abstraction. Here the innate tendency of theoretical science and philosophy to suppress the personal factor has been fully achieved.[209]

> Not only does it [natural science] seek increasingly to suppress all that is "personal" it strives toward a conception of the world from which the "personal" has been eliminated [sic]. It achieves its true aim only by disregarding the world of self and other.[210]

Astronomy was probably the first to achieve this goal, and recent philosophy has gone further in the same direction, demanding "the elimination of 'occult' psychic qualities, not only from astronomy and physics, but from all natural events."[211] Even in biology the dominance of mechanism has ended with life itself banished not only from the organism but from organic nature. The organism is explained by the laws of mechanics, of pressure and impact without residue or qualification.[212]

But the true starting point for the understanding of cultural, as distinct from natural, phenomena, Cassirer maintains, is *the functional meaningfulness of "the self" as this is immediately manifest in the activity of language.* "I" and "you" are never *given* as *absolutes in themselves* who would communicate by some process of action at a distance. ". . . the I and the you exist only in so far as they exist 'for each other,' only in so far as they stand in the functional relation of being reciprocally conditioned."[213] Strict naturalism attempts to explain the forms of language, art and religion as a simple sum-total of individual effects. For it, language is something agreed upon by encountering individuals. But there would be no agreement possible except in an already exist-

6.26: "To name the feeling awakens . . . consciousness of it," and p. 43, par. 6.36: "The man who names his feeling thereby immediately becomes conscious of it."

[208] See E. Cassirer, *op. cit.,* pp. 102-103.
[209] See *ibid.,* p. 103.
[210] *Ibid.,* pp. 103-104.
[211] *Ibid.,* p. 104.
[212] See *ibid.*
[213] *Ibid.,* p. 107.

ing medium of speech. Society, togetherness of individuals, could not arise from a contract, since a contract is the *product* of individuals *already united* in their agreement to execute such a contract. Metaphysical theories, on the other hand, would trace language and society back to supra-personal powers such as a world-spirit. But this is to abandon scientific explanation and fall back into myth, in which the cultural world is a kind of overworld working itself out in the physical world and in the lives of human beings. A critical philosophy of culture must steer a midway course, avoiding both of these extremes.[214] The true starting point is thus indicated:

> . . . if we no longer begin with the ''I'' and the ''you'' as two substantially separate entities, but locate them in the meeting point of that reciprocal *transaction*, which consummates itself in speech or any other culture-form, our perplexities dissipate. In the beginning is the act; always in the function of speech, in artistic creation, in the process of thought and inquiry there is expressed a specific *activity*. And only in this activity do the ''I'' and the ''you'' exist with the possibility of simultaneously distinguishing themselves from each other. They exist both within and next to each other as they preserve this unity within speech, thought and all manner of artistic expression.[215]

For Cassirer, then, the ''I'' and the ''you'' do not exist as autonomous, atomic units externally related to each other by the external mediation of language. They exist, as ''I'' and as ''you'', only inasmuch as they communicate in linguistic discourse, that is, inasmuch as linguistic discourse relates them internally and essentially. Language, that is, language operative in discourse, is not some third thing falling between the two communicants, but, together with them, constitutes that concrete totality in which they are communicants. The linguistic, dialogical situation is therefore a concrete whole in which the (multiple) participants are (no longer multiple, but) a concrete unity. So, language *constitutes* a concrete unity in diversity, a unity which demands the diversity of participants, and simultaneously a diversity which demands that unity. The unity is meaningless without the diversity, and conversely, the diversity is meaningless without the unity. That is, we have here that concrete unity in diversity which is constitutive of the concrete universal, as was explained in Chapter 4, above.[216]

---

[214] See *ibid.*, p. 108.

[215] *Ibid.*, p. 109. Emphasis in original.

[216] The point is perhaps well made by saying that not only is language something *in us*, but it is equally true to say that we are *in our language*. ''We *live within* the words

However, a difficulty presents itself at this point regarding Collingwood's understanding of the unity in diversity in linguistic discourse. On the one hand he seems to insist on the primacy of individual autonomy and on the socially communicative function of language as being something secondary. This seems implied in the following:

> In its most elementary form, language is not addressed to any audience. A child's first utterances are so completely unaddressed that one cannot even describe them as addressed to the world at large or to itself. The distinction between speaking to oneself, speaking to the world at large, and speaking to a particular person or group, is a later differentiation introduced into an original act which was simply the act of speaking.[217]

On the other hand, in a succeeding passage, he seems to speak exactly in accordance with Cassirer's position. Thus, he says that "the child's discovery of itself as a person is also its discovery of itself as a member of a world of persons," and "as persons, they construct a new set of relations between themselves, arising out of their consciousness of themselves and one another; these are linguistic relations."[218] He then insists that my discovery of myself as a person is the discovery that I can speak, and in speaking I both speak and hear myself speaking, and "since *the discovery of myself as a person is also the discovery of other persons around me*, it is the discovery of speakers and hearers other than myself."[219]

Cassirer says that although it is recognizable that there is "expression of emotion" even in the animal world, this is passive and mere spontaneous reaction to environmental stimuli. But when we come to the sphere of human existence and human culture something new is encountered, for all culture-forms are *active expression-forms*, not mere reactions, like blushing, frowning, or doubling of the fist. They are genuine actions as distinct from mere events which play themselves out in our bodies. They are energies through the exercise of which the cultural world of language, art and religion form themselves. In addition to their sheer physical presence, of which we are informed

---

of language, the shapes of poetry and plastic art. . . ." Ernst Cassirer, *op. cit.*, p. 143. Emphasis in original. See also this chapter, below, p. 409.

[217] R. G. Collingwood, *The Principles of Art*, p. 247.

[218] *Ibid.*, p. 248.

[219] *Ibid.* Emphasis added. See also J. de Finance, *Esai sur l'agir humain*, Rome, Presses de l'Université Grégorienne, 1962, pp. 177-178, for language as primarily directed towards another and only secondarily directed towards oneself.

by sense experience, there is something symbolic, that is, something 'meant.'[220]

Just as there is passive expression in animal behaviour, so also there is passive speech which does not go beyond mere habit. True speech is never purely imitative but is also productive. It is only within this latent energy which becomes manifest in this productive function that the other energy called "thinking" is displayed.[221]

> The true relationship between "I" and "you" has its being in the act of sharing a common universe of discourse. It is continual daily engagement in this common language-world that gives rise to the "I-you" relationship.[222]

The world of the "I" is never given as a finished, enduring, substantial entity, with words doing no more than transporting this givenness from one subject to another, the mere building of bridges between inner worlds of different subjects. The true relation is quite otherwise. Cassirer explains as follows:

> In speech and art the individuals not only share what they already possess; it is only by virtue of this sharing process in speech and art that individuals have attained what they possess.[223]

This is observable in any living and meaningful conversation in which "it is never simply a question of imparting information, but of statement and response."[224] Thus:

> In question and answer "I" and "you" must be distinguished, not only that they may understand each other, but even if each is ever to know himself. Here both factors are in continual interplay. The thought of one partner is kindled by that of another. And by virtue of this interaction each constructs for himself a "shared world" of meaning *within the medium of language*. Wherever this medium

---

[220] See E. Cassirer, *The Logic of the Humanities*, p. 110. This seems more or less the distinction Collingwood makes (*Principles of Art*, pp. 228-238) between psychic expression and linguistic expression. See Appendix 1, below, pp. 467-469. See also Alan Donagan, *The Later Philosophy of R. G. Collingwood*, p. 114.

[221] See E. Cassirer, *op. cit.*, p. 112.

[222] *Ibid*. This citation and what immediately follows show that Cassirer is thinking in terms of what in Ch. 4 above was explained as a concrete universal. Whole and member-parts have their identity and meaning only in virtue of mutual implication and internal relation; whole mediates member-part, and simultaneously member-part mediates whole. See this chapter, above, p. 399.

[223] E. Cassirer, *op. cit.*, p. 113.

[224] *Ibid*. Here Cassirer joins company with Collingwood's fundamental thesis of question and answer as the motor drive of knowledge as an active, self-determining evolution.

fails us our own grasp becomes unsure and dubious. All thought must prove itself in language.[225]

Language is therefore *not merely an externalization* of ourselves but, like art, also a *pathway to the realization* of ourselves. Consciousness and knowledge of ourselves is first achieved by its mediation. Even in a monologue, a "conversation of the soul with itself," the predominating function is *division* of self. In a dialogue it is *reintegration* that is central.

> . . . the "conversation of the soul with itself" is only possible by virtue of the fact that in the process the soul in a sense undergoes a division within itself. It must undertake both the function of speaking and that of hearing, of questioning and answering.[226]

This double function of division and reintegration appears also in artistic activity, in which "continual self-renewing flow of communication . . . the artist feels himself isolated and thrown back within the limits of his own ego."[227] The artist "continues to create, for he knows that it is only by doing so that he can discover and gain possession of his own self."[228] The same duplexity appears in religion. The believer wants to "spurn the world and break all ties, all social realities binding man to man . . . [so that he] knows only himself and God and does not want to know anything else." But, at the same time, he must prove his belief in making it public. "He must communicate his belief to others, he must fill them with his own religious passion and fervor, in order to be certain of his belief."[229] Cassirer adds that "the life of the spirit consists in this very act of severing what is whole in order that what has been severed may be even more securely united."[230]

---

[225] *Ibid*. This coincides with Collingwood's anti-substantialism and his understanding of spirit as active self-evolution. Not substance but action, indeed autoactivity, is the basic category of reality. It is only in the linguistic interaction situation that the selves are recognized as selves.

[226] E. Cassirer, *op. cit*., p. 114. This statement is fully in accord with Collingwood's position. In a monologue, the soul opposes itself to itself as a speaker and one spoken to, as questioner and respondent. In the monologue, dialectical opposition and subsequent reintegration takes place within the same mind by its dynamic self-differentiating, self-perfecting activity. In a dialogue the opposition of speaker and one spoken to takes place within the all-embracing whole established by the very act of dialoguing itself. In both cases there is the building up of a concrete universal in terms of which the situation is intelligible.

[227] *Ibid*., p. 115.

[228] *Ibid*. This is the same dialectical self-creative function of artistic expression as recognized by Collingwood.

[229] *Ibid*.

[230] *Ibid*., p. 116.

Having seen the universality by which distinctively human phenomena are intelligible, we can now proceed to consider how, following Cassirer, we can distinguish cultural and historical concepts from nature concepts.

G. Distinction of Culture Concepts from Nature Concepts

The foregoing has shown that cultural and specifically human concepts are distinguished from scientific nature-concepts as "function" concepts from "thing" concepts. That is, the concept whereby we understand some cultural phenomenon is not the concept of a completed "thing" or characteristic thereof, but is the conception of an ongoing, self-realizing activity (however static and 'thing'-like the physical conveyance—monument, parchment, etc.—may appear).[231] Thus, it is not by pointing out specific and distinguishing features that we understand the world of culture and the world of history (e.g., features that might serve to identify Baroque art or Renaissance man). When we pass from the animal to the human world all determinations undergo a change of *function*. The response of a man, precisely as he is a man, is the autonomous response of a self-initiating, free being, whereas the response of an animal (e.g., the cry of an animal in response to a painful stimulus) is always naturally determined and more or less predictable. Within certain limits (man's freedom is not absolute, but conditioned, yet is nevertheless a genuine freedom), indeed by means of them, man can form a breadth and self-sufficiency of movement accessible to him alone.[232]

> The various symbolic [meaning-manifesting] forms—myth, language, art, and science [as a cultural form]—constitute the indispensable precondition for this process. They are the true media—which man himself has created—by virtue of which he has been able to separate himself from the world, and in this very separation, to bind himself all the closer to it. This feature of mediation characterizes all human knowledge. It is also distinctive and typical of all human action.[233]

---

[231] It is clear why dialectical thinking, which trades in this sort of concept (i.e., concrete universals), substitutes action as the basic category in place of substance. Self-creating activity is the root of meaning for this type of thought, not things (substances) having certain fixed properties, including that of acting in definite ways.

[232] E. Cassirer, *op. cit.*, pp. 73-74.

[233] *Ibid.*, p. 74. Note the dialectical separation and reintegration which is of the essence of these dynamic forms. This active self-separation and self-recovery is the means whereby the self acquires determination, which in this case means increasing self-identity.

Even plants and animals build themselves up by receiving stimuli from their environment and in a way answer (respond in their own way) to such stimuli. But their answer or response is fixed and unified, following always from the same conditions, that is, marked by deterministic response rather than by free self-initiating response. But this is radically altered the moment we find ourselves in the sphere of human actions. The change in the mode of behaviour is obvious in the use of tools. To invent a tool, attention has to be lifted, at least momentarily, from total immersion in the impulse and the necessity of the moment. The tool-user has to foresee some as yet unrealized but realizable goal for the attainment of which he fashions the means and orders them to the realization of that preconceived goal. Man, as *homo faber,* thus lives in the future. The future is already present for man acting humanly. Not that it is a present actuality, but that it is intentionally present in the agent's decision and consequent action to bring it into fully actual realization.[234]

> The purpose which the tool serves involves within itself a definite pre-vision. The impulse does not derive solely from the force of the present; instead, it belongs also to the future, which must be anticipated in some fashion in order to become effective in this way. This "pre-presentation" of the future characterizes all human action.[235]

Cassirer says that the theoretical sphere resembles the practical in this regard. All theoretical concepts bear the character of instruments, tools fashioned for the solution of specific tasks.[236]

By tools in the practical sphere and concepts in the theoretical sphere man is able to go ahead of his experience in an anticipatory way. Instead of mere subjection to the environing world and total immersion in the facticity of brute experience, he dominates it and subordinates it, to some extent at least, to his own freely chosen goals. He is not merely situated by his situation but is able to face his situation and actively take a stand with respect to it. To know a situation *as a situation,* that is, to know what it means to be situated, is already to be liberated in some way from the situation; as we often say: "He

---

[234] E. Cassirer, *op. cit.,* pp. 74-75. Here we are dealing with what Robert Stover calls intelligibility deriving from careful concern for the world. See this chapter, above, pp. 360-372.

[235] *Ibid.,* p. 75. Concretely realizable alternatives, the objects of practical decisions and preference (Stover), presuppose the realm of the future as meaningfully recognizable, whatever problems may be raised regarding its reality-status.

[236] Cassirer here reechoes Collingwood's primacy of the practical.

has taken the situation in hand." In the choosing and realizing of self-appointed goals a man is master of himself and of his destiny.

All cultural phenomena are, thus, expressions of personal decisions,[237] not explicable merely in terms of a deterministic, depersonalized, scientifically objectivized world. Thus, for example, the civilization of ancient Egypt is to be appreciated meaningfully inasmuch as its tombs, temples and hieroglyphs are expressions of (individual and collective) personal, self-set goals, and realizations thereof. These material entities (parchments, stone monuments, embalmed corpses, etc.) are "faces"[238] by which the Egyptians' self-achieving dialectical activity is expressed and made manifest. Thus, for example, the concern to secure personal immortality is embodied in all their extant works. These monuments, etc., are intelligible only in reference not merely to a future, but even to the future of a life after death. Remove this urgent concern to secure immortality in an afterlife from one's stock-in-trade of meaning-conveying factors, and then there would remain no way of understanding and intelligibly explaining these stone monuments and parchments other than that of the impersonal analysis of the chemist or physicist in terms of chemical substances, atoms, weights and measures, etc.

But what logical value can we find in the concepts and meanings of the cultural and historical world? The logical structure of a science is understandable when we are clear about the way in which it subsumes the particulars under the universal.[239] Cassirer objects to the attitude which opposes the universal concepts of a natural science to the individual concepts of historical science, since the logical function of *any* concept is to achieve a unity of the manifold, a relation between individual and universal. To set either factor in isolation from the other is to destroy the synthesis aimed at by every concept. The universal must always accomodate itself to the particular, but the way in which this accomodation is achieved is what distinguishes one science from another.[240] Thus, the manner in which unity in diversity is achieved differs as "we compare the system of mathematical concepts with the empirical concepts of nature and the latter with historical concepts."[241]

---

[237] All that was said earlier in this chapter regarding expression-perception together with what was said in regard to careful concern for the world has to be kept in mind here and in all that follows.

[238] But see this chapter, below, pp. 415-416, for the consideration of these external physical media also as masks which hide the expression which they bear.

[239] E. Cassirer, *op. cit.*, p. 135.

[240] *Ibid.*

[241] *Ibid.*, pp. 135-136.

The simplest form of this relationship of universal to particular is the expression of the universal as the concept of law from which particular instances may be deduced. All concepts of empirical, natural science strive in some way for this ideal.[242]

But the form and style concepts characteristic of the humanities do not accomodate particulars to a universal in this way. No one historical individual can be singled out as an instance of the concept "man of the Renaissance."[243]

" 'Subsumption' can never be taken here in the same sense in which we subsume a body given here and now, that is, a piece of metal, under the concept 'gold' after finding that it fulfills all the conditions of gold known to us."[244] Cassirer explains the difference as follows:

> When we characterize Leonardo da Vinci . . . Michaelangelo and Cesare Borgia as "men of the Renaissance," we do not mean to say that there is to be found in them a definite and inherently fixed distinguishing trait in which they all agree. We perceive them to be not only completely different, but even opposed. What we are asserting of them is that in spite of this opposition, perhaps just because of it, they stand to each other in a specific ideal connection: each in his own way is contributing to the making of what we call the "spirit" of the Renaissance or the civilization of the Renaissance.[245]

Cassirer says that the universalizing unity which is striving to be expressed here is that of unity of *direction*, not unity of *actualization*.[246]

> The particular individuals belong together, not because they are alike or resemble each other, but because they are *cooperating in a common task*, which, in contrast to the Middle Ages, we perceive to be new and to be the distinctive 'meaning' of the Renaissance.[247]

Cassirer remarks that "all . . . concepts of style in the humanities reduce, when analyzed more precisely, to such conceptions of meaning."[248] Thus, the artistic style of a period can only be determined if

---

[242] *Ibid.*, p. 136.

[243] *Ibid.*

[244] *Ibid.*, p. 139.

[245] *Ibid.* Compare this with Collingwood's explanation of the individual and the whole in the constitution of the concrete universal; see his *Speculum Mentis*, pp. 218-221. See also Chapter 4, above, pp. 86-91.

[246] See E. Cassirer, *op. cit.*, p. 139.

[247] *Ibid.*, pp. 139-140. Emphasis in original.

[248] *Ibid.* Style concepts which characterize a definite historical period, then, in R. Stover's terminology (see this chapter, above, pp. 360-372) are universal in terms of the unity of a common concern of some sort. The common concern of all is reflected

we gather into a unity all its divergent and often patently disparate artistic expressions recognized to be expressions or manifestations of a specific "artistic will." Such expressions *characterize* but do not *determine*, for the particulars comprehended by that concept cannot be deduced from it, since there is as much dependence of it upon them as of them upon it.[249] But there is true conceptualization here, not just intuitive description, since there is true unity in diversity. Thus, "we are dealing with a distinctive manner and direction of characterization, with a logico-intellectual activity which is *sui generis*."[250]

This distinction, between culture concepts of form and style and naturalistic, 'thing' concepts, manifests the same opposition which was encountered in the preceding analysis of perception. "It is the logical translation of a definite *opposition in orientation*, which as such is not encountered solely in the domain of concepts, but whose roots run deep into the subsoil of perception."[251] Cassirer explains this further as follows:

> Here conception expresses "discursively" what perception apprehends in the form of a purely "intuitive" knowledge. . . . The "reality" which we apprehend in perception and direct intuition

---

in each individual, but each in his own way. The whole of that society is in each member in virtue of that common goal, and each member is in that whole as contributing to the achievement of the coordinated aim. The same can be said, for instance, of the concept of a collective experience such as the First World War. Thus, Luigi Sturzo, *The Inner Laws of Society*, New York, P. J. Kenedy & Sons, 1944, pp. xviii-xix, writes: "When we speak of collective experiences, we do not do so as a datum of experience specifically distinct from individual experience, but as an experience that is made *in similar form at the same time, and for the same reason*, by a group of persons in some way bound *together*. Thus, that of the war of 1914 was a collective experience for whole peoples, on a wider and more general scale than had ever been known. But, it was made up of innumerable personal experiences, each of which was in *no wise reducible to that of another, although taken as a whole there were the same outcomes for determined groups*. Individual experience is relative, limited, circumscribed; so are collective experiences, so is the succession of experiences or process. By this very fact, it can take on the character now of progress, now of regression, according to the points from which it is seen." (Emphasis added.) Thus, if a bereaved mother wonders why her son should have died in the First World War, the answer would be to know what that war was precisely as he was related to that global event, and it to him, by intrinsic, essential relationship. His death is understandable through the intelligibility of a concrete unity (of totality) in diversity (of parts essentially diverse and essentially related to the whole), that is, in terms of the intelligibility of a concrete universal constituted ultimately by one common (care-concern) purpose diversely realized in many individuals.

[249] I understand this to be the mutual implication that exists between form and content in dialectical thought and concrete universality.

[250] E. Cassirer, *op. cit.*, p. 140.

[251] *Ibid*. In other words these concepts have to be taken in concrete continuity with their corresponding originating perceptual acts in order that this distinction between the concepts themselves be recognizable.

> presents itself to us as a whole in which there are no abrupt separations. And yet it is both "one and twofold," for, on the one hand, we apprehend it as thing-like and, on the other hand, as "personal."[252]

Science and philosophy, by theoretical knowledge, seek the property-constants and law-constants as two essential features of the physical world whereby the Heraclitean flux of becoming is stabilized and the enduring distinctions discerned and an orderly cosmos recognized. This tendency to "fixing" is already at work in perception.[253] Science is distinguished from perception in that it requires strict determination, whereas perception is merely estimative. This scientific determination involves that the "essences" of things be elaborated entirely in numerical concepts and physical and chemical constants which characterize every class of objects as things. These constants are then united through fixed functional relationships and expressed by equations which state the dependence of one magnitude on another. This is the pure way to "objective" reality by which the common (conceptual) world of things is constitutively realized.[254]

But this rigid objectivization of reality has its price. The resultant "thing-world" is inherently soulless; every reference to the "personal" experience of the ego is obliterated. Human culture is quite meaningless in such a deviscerated world.[255]

But the natural world, or cosmos, built up by scientific "thing-oriented" objectivity, is at the same time an *inter-subjective* world, a world *shared in common*, that is, a world in which a multiplicity of subjects *participate*. But likewise, the cultural world, the unity of many subjects as subjects, as expressing, jointly and severally, cares and concerns regarding the lived world, is an *intersubjective* world, a world in which I *participate* with other selves. But this participation, whereby I participate with them in the cultural world, is a different sort of participation from the participation whereby I, together with others, participate in the physical world of nature.[256] Cassirer explains this contrast as follows:

> Instead of relating themselves to the selfsame spacio-temporal cosmos of things, they find and relate themselves within the medium

[252] *Ibid.*, p. 141. See also above, this chapter, p. 361 for Stover's statement that our everyday thinking does not distinguish the different standpoints.

[253] See this chapter, above, p. 391 n. 177.

[254] E. Cassirer, *op. cit.*, p. 142.

[255] See *ibid.*

[256] See *ibid.*, pp. 142-143.

> of the various worlds [artistic, linguistic, religious, etc.] of form out of which culture comes into being.[257]

This seems to be the same distinction which some authors make between a "we" communion and an "I-thou" communion.

> In an I-thou communion, we are, as it were, in front of the other person and we look *at each other*. In the "we communion," on the contrary, we look together with the other person *at some object*.[258]

The commonly shared relationship *to some third*, an "it," is what constitutes participation in the natural world. The commonly shared relationship *to each other*, distinctive of the various cultural interactivities, is what constitutes participation in the cultural world.

This active relating of each of the selves to each of the other selves in the formation of the different cultural forms takes place already at the perceptual level, that is, at the level of expression-perception. But the expression-perception by which this is achieved is not the passive experience of expression, or perception as mere feeling (which is somewhat analogous to "impression" in the knowledge of object-things), but rather "this genuine 'synthesis' is first realized in the active exchange which we encounter directly in every verbal (act of imparting) 'information.' "[259]

Cassirer says that "the constancy required for this [and hence for universality proper to the cultural area] is not that of properties or laws, but of meanings."[260] He clarifies this as follows:

> The more culture develops, and the more the particular areas into which it unfolds, the more richly and multifariously does this world of meanings shape itself. We *live within* the words of language, the shapes of poetry and plastic art, the structure of music, the framework of religious representation and religious beliefs.[261]

The familiar notion of the concrete universal is once again appearing. The parts are mediated by the whole, and simultaneously the whole is mediated by the parts. It is in this active, self-constituting, dialectical mediation that meanings characteristic of the human, cultural, spiritual world are created and recognized. Each member building up himself is together building up his world.

---

[257] *Ibid.*, p. 143.

[258] Dietrich von Hildebrand, *Man and Woman*, Chicago, Franciscan Herald Press, 1965, p. 16. Emphasis added.

[259] Ernst Cassirer, *op. cit.*, p. 143.

[260] *Ibid.*

[261] *Ibid.* See also above, this chapter, pp. 399-400, n. 216

What Cassirer says here both illuminates and is illuminated by what Collingwood says in this regard in his *Speculum Mentis*.[262]

---

[262] P. 248. John Wild's explicitation of the meaning of the term 'world' is also quite illuminating here. See his "Being, Meaning and the World," *The Review of Metaphysics*, Vol. XVIII (1965). On pp. 413-414 he says that "world . . . is a relatively recent term of Germanic origin with a temporal connotation. It is probably derived from *wer* (man) and *ald* (age), age of man. . . . Various beings in the world are not instances of the universal concept, 'world,' nor are they portions of world in the sense of a collection. . . . They are in the world rather as in a region of care, as we say that a doctor is *in* medicine, a soldier *in* the army, or that a kettle is *in* the kitchen, and a book *in* the study. . . .

"[The term] world is used derivatively for any far-ranging region of care, as the world of finance and the world of medicine. But in its strict sense, as *the* world, the ultimate horizon of meaning and care, it includes all such regions, and indeed, all beings, even those beyond the range of our knowledge but which we are concerned to know. In this sense *the* world, like being, is all-inclusive, but in a different way. It includes them not as a concept 'includes' its instances, but as beings are included in a field of care. And when used in this way, *the* world is not a timeless, abstract, universal term, but a spatio-temporal, concrete . . . singular term. However, it is not a single thing nor any collection of things. It is rather the horizon of real meaning within which any such thing or collection must occur, if it is really to be. Hence if we consider actual usage, we find that the notion of world is quite distinct from that of being, and prior to it."

Again (pp. 417-418) he says: "I am not in the world [as a drop of water contained in the jar]. I am in it rather as in a field of care, as we say that a student is in medicine, or a salesman is in business. Similarly, I am in the world as in the widest horizon of my concern, as dwelling in it, and caring for it in a certain way. . . . All the beings I know, and even those I know that I do not know, are in the world. . . . Even beings of which I am totally ignorant are the objects of such care, namely the care to know them and find them out.

"This care has always been active in human history and is, in fact, its moving spring. There are now vast and conspicuous regions of industry and learning which are devoted exclusively to this care, like the research laboratories of a great university. Those engaged in such enterprises are trying to reveal things which are as yet unknown. . . . [T]hese things are certainly *in the world*, though not as yet in any explicit version. This brings out the fact that every being in the world has some actual, or at least some potential meaning. In this horizon, beings are gathered together and separated in such ways as to bring out their real sense. This horizon, therefore, includes both the beings about which we know something, and those about which we know nothing except that we do not know.

"[W]e may use the term *being* for things with only a minimum of meaning, just as we use the term *meaning* for imaginative schemes with only a minimum of being. *The* world is a union of the two, the field of beings actually gathered together in such a way as to make sense and to appear as what they really are.

"Beings, as such, do not lie around any natural center. They simply *are*. There are no regions in being, and no distinct places for different things. . . . *The* world horizon may have no human center. But the beings in my version of the world lie around me, and those in the human world lie around man. And this centeredness is not arbitrary, for the germs of world meaning are found in his versions of the world. It is through him that beings are brought into meaningful relations with one another and allowed to appear as what they really are. . . ."

If this is so, the term 'world' itself is to be understood as belonging to Stover's second type of intelligibility based on care and concern. The term 'care' already seems

We know each other only within the media of these cultural forms which constitute meaningful worlds. This (basically linguistic) medium operates and builds up the various cultural forms (art, religion, etc.) at a pre-scientific level beginning at the level of perception, as was explained above. We understand one another in speech without first acquiring technical knowledge of linguistics and grammar, just as we eat before acquiring a technical knowledge of dietetics.[263] But this spontaneous, naive, pre-reflexive knowledge of each other soon reaches its limits and, due to strains that develop within it, demands that the work of science come in to clarify and explain it.[264]

When natural science, in its penetration into the depths of cosmic space, raises itself to the knowledge of universal laws which do not differentiate between the near and the distant, it becomes the master of the distant. Thus, when physical science universalizes to the extent of the universal law of gravitation its scope is the whole of cosmic space.[265] It is thus *extensively* universalizing.

But this form of universality is out of bounds to the science of culture, which cannot renounce anthropomorphism and anthropocentrism. Its subject matter is not the world as such but a particular

---

to implicate the term 'world,' and the very term 'meaning' itself would seem to be meaningless without reference to some center as agent of concern, within whose very concern, manifested by questioning, meanings arise, and a world of meaning is constituted.

This notion of world has many similarities to the concrete universal discussed in Chapter 4 above, in which, it will be recalled, B. Bosanquet says that a 'world' is the best illustration of what is meant by a universal. But this separation of being and meaning by Wild seems inadmissible if we take the position that to say that a thing is a being is merely to say *that* it is and in no way to say *what* it is. In other words, the essential, specifying determinant of the being of a thing, for Wild, seems extraneous to its being. But in Chapter 7, above, pp. 297-299, the position was taken that the being of a thing refers not only to its active presence but also to its determinate mode of presence. That is, *being is differentiated from within being itself*, not by extrinsic superaddition of differences from without. Being has to account not only for the fact of the existence of each thing but also for its determinate way of existing. If this is so, then being as determinative of the being of each thing is also determinative of its meaning. And this meaning is ascertained by the abstractive capacity of intuitive reason (as explained in Chapter 7, pp. 280-284 above) and expressed by the real definition which says what the thing is, namely its quiddity. If this distinction between essence, or what the thing is, and its existence is inflated into a separation, then being has to get meaning from some source other than its own determinations; from, say, presuppositions which determine the attitude of a conscious being, and therefrom the being itself, inasmuch as being is referred to the designs of a conscious and free being for its meaning. In true Kantian fashion, then, we do not know things as they have meanings *in themselves*, but only inasmuch as they have meanings *for us*.

[263] See Collingwood, *The Principles of Art*, pp. 226-228, on this point.

[264] See E. Cassirer, *op. cit.*, pp. 143-144.

[265] See *ibid.*, p. 144.

region of it. But it strives to penetrate fully and completely this sphere defined by the limits of the human world. That is, it universalizes *intensively*, qualitatively and penetratingly, through progressively increasing determinate manifestations or expressions of the same abiding human reality. Cassirer further explains this type of universality:

> Its goal is not the universality of laws; but neither is it the individuality of facts and phenomena. In contrast to both, it sets up an ideal of knowledge of its own. What it seeks to realize is the *totality* of the *forms* in which human life is realized. These forms are endlessly divergent and, yet, they are not without unified structure. For in the end it is ''one and the same'' human nature which meets us again and again in a thousand manifestations and in a thousand masks in the development of culture.[266]

This universalization of the multiplicity of cultural forms is based on the identity of the same human reality of which its various cultural forms are a manifestation and expression in an infinite variety of ways.[267] This identity, that is, this 'remaining itself to and for itself' in its progressive self-determinations, is not made known by watching, weighing and measuring, nor by some sort of psychological introspection or induction, but is ''demonstrated,'' that is, manifested or expressed, only within the activity which is its very actualization and by which it is realized. A culture is meaningfully accessible to us only if we are actively involved in it. The translation of the Egyptian Rosetta Stone involved an active, vital penetration into the linguistic world of the Egyptians, that is, a forming of the translator as in some way one with the scribe, by setting up a dialogue with him, a union with him by medium of language. Egyptian culture and history thus ''come alive'' again in that unified, living dialogue into which the translator entered (and entered precisely by effecting it into existence) in the exercise of his translating activity. As Marrou nicely puts it, ''if I knew Arabic as well as Massignon, the history of Arabia would also become a part of my past.''[268] A little further ahead in the same place, he asks what it matters if some group of men as foreign as possible to me are not related to me by blood relationship, for ''*if I write their history I*

---

[266] *Ibid.*, p. 144. Collingwood's *Speculum Mentis* can be considered as his attempt to realize the goal of this type of universality.

[267] See this chapter, above, pp. 387-388, for the citation from Barbara Mertz, in which she well expresses this human solidarity with ourselves which we experience when we manage to penetrate to the meaning of the cultural remains of the ancient Egyptian civilization, recognizing those remains as expressive of the distinctively human.

[268] H.-I. Marrou, *The Meaning of History*, p. 222.

too am descended from them henceforth'' (emphasis in original), and ''if I have shown that I am capable of understanding their past, it is because I feel the family relationship.'' The linguistic breakthrough that took place in deciphering the Ancient Egyptian hieroglyphs immediately set up a dynamic, dialogical unity, a shared world, in which the expressions (surviving in and through the present parchment or monument) of the thoughts and concerns of the perhaps four thousand year old, long since dead scribe, come alive again in that unified world constituted by language. The living, linguistic form is able actively to straddle that four thousand years of elapsed time and bring together the historian-archaeologist and the ancient scribe in one living, active dialogical unity. That linguistic form, expressed by the scribe (in stone or parchment) and possessed by the translator, is a true (formal) unity in diversity (of individuals). Language as the active unifying form operative here is not language considered analytically, as a set of grammatical rules and parts of speech such as a linguist would regard it, but language *as a living, exercised activity in which expression is exercised and communication achieved*. This seems to be equivalent to language in the sense of literature, a living whole, which precedes its analytic dismemberment into alphabet, dictionary meanings and grammatical rules. This also seems to correspond with what Collingwood says in his *Essay on Philosophical Method*,[269] where he treats ''philosophy as a kind of literature,''[270] and says that ''the language of philosophy is therefore, as every careful reader of the great philosophers already knows, a literary language and not a technical,''[271] so that ''the reader of philosophical literature . . . is reading a language not a symbolism.''[272] Collingwood explains this as follows:

> The duty of the philosopher as a writer is therefore to avoid the technical vocabulary proper to science, and to choose his words according to the rules of literature. His terminology must have that expressiveness, that flexibility, that dependence upon context, which are the hall-marks of a literary use of words as opposed to a technical use of symbols.[273]

Language as literature has always been regarded as the vehicle of the expression of the spirit, that is, the living insights and concerns

---

[269] Pp. 199-220.
[270] *Ibid.*, p. 201.
[271] *Ibid.*, pp. 206-207.
[272] *Ibid.*, p. 207.
[273] *Ibid.*

of a people which mark them with cultural and historical distinctiveness. However heavily the Egyptologist may depend on such analyses as Gardiner's *Egyptian Grammar* or Champollion's techniques of decipherment of hieroglyphs, he is not yet functioning properly as a historian until he penetrates beyond this hieroglyphic symbolism to make the language hidden therein function as literature, as the vehicle of the expression of human concerns.

Man, therefore, as a linguistic being, *is* a unity in diversity. Dialogue, i.e., linguistic activity, is possible only within the human world and then only inasmuch as this is a world of multiplicity and diversity. Once again we may use the example of the process of cell division borrowed from the biological world to illustrate the linguistic, dialogical question and answer activity. One single cell by its own internal dynamism is able to oppose itself to itself, thus dividing itself within itself so as to repossess itself with greater and greater differentiation and self-determination, so that, finally, a fully integrated, living organism results from that initial single cell. Likewise, when a man enters into dialogue with another, the dialogue situation is, as it were, a single human ''culture cell'' which, by repeated position and opposition, by question and answer, finally ends up (although every such ending is a new beginning) in the manifestation of a more fully determined, more fully perfected humanity. But the bearer of this fulfilling determination is not abstract, objectively defined humanity, but humanity as concretely realized in the individuals who participate in it. But they are the bearers of such self-autonomous realization only by medium of their interaction one with the others, which interaction, as expressed and manifested, is nothing other than a (dynamic, active) cultural form (as the form of language, the form of art, the form of religion, the form of science, etc.).

H. History as an Inquiry Proceeding According to a Logic of Question and Answer

Robert Stover has shown[274] the difference between the meaning and intelligibility of concepts based on careful concern for the world and of concepts based on the uniformity of behaviour observed in the natural world. John Wild elaborated[275] on the nature of care, from which the meaningfulness of careful concern concepts is derived, and, in particular, clearly showed the distinctive meaning of the temporal

---

[274] See above, this chapter, pp. 360-372.
[275] See above, this chapter, pp. 373-384.

dimensions of past, present and future in the human care-taking situation, thereby giving us an insight into historical time proper to human duration as distinct from the receptacle concept of physical time *in which* physical things are located. Ernst Cassirer explained[276] at length the logical dimension, or properly conceptual value, of concepts based on careful concern for the world. He showed that the distinction in conceptualization, that is, in universalization, is not independent of the perceptual roots from which it arises and to which it continually refers. Expression-perception both precipitates and effects linguistic intercommunion in a dialogue situation, which dialogue situation is of the nature of a concrete universal, involving, as it does, a genuine unity in diversity in which the diversity is essential to the unity and vice versa. All cultural concepts, as concepts, that is, as logically significant, as unifying a multiplicity in thought, reduce back to this *sui generis* type of universality.[277] Whatever logic will be elaborated in the human disciplines will therefore have to be adapted in its methodology to this type of universal.

It is clear from what has been said in this chapter regarding this type of universal that it is not an abstract, stabilized object like the universal law of gravitation in physics, which enjoys a recognizability independently of its mobile instances. On the contrary, it is a concrete, active, ongoing, self-developing process or ''function'' (Cassirer's term), which he (Cassirer) explicitly refers to[278] as a process of dialogue involving question and answer. And, inasmuch as it has been shown[279] that the historian in his proper function as a historian penetrates the world of the period or the subject or person or institution he is studying, forming with those of that time and place a concrete living unity through dialogue, his method is one of question and answer. This substantiates Collingwood's thesis that the logic of history is one of question and answer. In addition, it gives a deeper insight, first, by the medium of Cassirer, into the precise nature of the concrete universal logically operative, and second, by the medium of Robert Stover and John Wild, into the properly human character of historical meaning, based, as it is, on the conceptual meaning deriving from man's free, self-creative activity and consequent careful concern for the world. Inasmuch as documents, monuments and such traces of the human past as artifacts, etc., can be made to function, by

---

[276] See above, this chapter, pp. 384-414.
[277] See this chapter, above, pp. 406-407.
[278] See this chapter, above, pp. 401-402.
[279] See this chapter, above, pp. 398-399 and pp. 413-414.

the use of historical techniques, as bearers of the expression of the interests and concerns of men of the past, Collingwood's position that history is a rethinking of past thought is correct, whatever criticism may be made of that position on other grounds. Furthermore, it is by a questioning technique that the historian is able to make these physical objects transcend their brute physicality (as stone, paper, etc.) and function expressively, as media conveying expression meaning. In an immediate conversation of an everyday nature with one's contemporaries, the facial configurations, voice sounds, hand gestures (all of which Collingwood calls language in its widest sense)[280] are the bearers of the expression of some sort of concern. In historical investigation a physical medium intervenes as bearer of such expressions of concern. Such a physically interposed medium functions not only as an expressive "face," but, precisely as physical, it functions also as a mask hiding the expressive manifestation. Work has to be done on the physical object to remove the concealment function and allow full play to the manifestation function. The making of the physical object manifest the expression that it bears as something transcending its brute physicality is the constituting of the physical object into evidence of the past human event. In crime detective jargon a physical object which does not yet manifest expressively the human interest of which it is a relic, yet suggests that, properly interpreted and understood, it can be made to so function, is called a clue. A clue is turned into evidence by interrogation, either of itself or of others functioning as witnesses.

So to the question, "Does history proceed by a distinctive logic of question and answer, as Collingwood maintains?" the answer, by way of summary of all that has been said so far in this chapter, would seem to be simply "yes." But a simple "yes" would be an oversimplified reply. Certain reservations and qualifications must still be considered before a definitive answer is given.[281]

It has already been mentioned[282] that Collingwood's questioning approach to his own archaeological researches was not always successful and fruitful and on certain occasions drew upon him the criticism of his colleagues. In explaining the questioning procedure in history[283]

---

[280] See his *The Principles of Art*, p. 236, and *The New Leviathan*, p. 41, par. 6.1. See also Alan Donagan, *The Later Philosophy of R. G. Collingwood*, pp. 41-43.

[281] See this chapter, below, pp. 430-431, for a precise statement of the qualifications to be taken into account.

[282] See Ch. 7, above, pp. 307-309.

[283] See Ch. 7, above, p. 271.

Collingwood exaggerates the dominance of the questioner and the activity of questioning, as if the asking of the question is the sole active influence in the production of evidence. This is quite in accord with Collingwood's absolutizing of the activity of human consciousness, which activity, for him, is basically a questioning process.[284] More soberly, H.-I. Marrou, who agrees with Collingwood that the method of history is a questioning one, insists that this questioning procedure in history, as in an everyday conversation, is an attempt to get to know another as another self. The fact that "the other" is "a self" that one is trying to get to know must not be allowed to override the fact that this "self" is genuinely also "another." The ever present reduction to consciousness in Collingwood results in the correlative reduction of the "self" to the self-conscious self, culminating in absolute thought as the absolute self. There is then "only one judgment judging one truth."[285] In such a context all dialogue becomes "a dialogue of the soul with itself," and any genuine otherness of another self is distorted and finally obliterated.[286] But once one recognizes that "the self" one is in dialogue with and trying to get to know is truly "another," one will also recognize its right, as another, to participate actively in making its own distinctive contribution to the questioner's answers.[287] The questioner has to be at least passive enough to allow the other to reply in his own right; otherwise the dialogue becomes a self-sterilizing monologue.

The issue being taken with Collingwood here is that in a postulatory,[288] dogmatic way, he absolutizes the activity of human consciousness and identifies the questioning procedure with this absolutized, self-creating activity, leaving no room to recognize the contribution to the answering of questions from sources beyond the knower's questioning activity. This has already been criticized above[289] in dealing with Collingwood's opposition to the realist's principle that "knowing makes no difference to what is known."[290]

In this absolutization of the questioning activity of the mind in

---

[284] See *Speculum Mentis*, p. 317.

[285] See *ibid.*, p. 80.

[286] See *ibid.*, p. 245, and pp. 248-249.

[287] See, for example, Marrou's explanation of the development of the dialogue of the historian with the document as explained on pp. 131 et seq., of his *The Meaning of History*.

[288] See A. Shalom, *R. G. Collingwood: Philosophe et Historien*, Paris, Presses Universitaires de France, 1967, pp. 516-517.

[289] See Chapter 7, above, p. 271.

[290] See Collingwood's *Autobiography*, pp. 25-28.

knowledge, all meaning becomes absorbed into what Robert Stover[291] has called care-for-the-world meaning. All meaning is then taken through relation to a consciousness capable of selecting from proposed contemplated alternatives a situation of well-being which the conscious subject can effectively bring about. As a consequence of this, even man's knowledge of what it is for him to be a man becomes reduced to his knowledge of his history. Consequently, what he can do is the measure of what he is, and what he can do is determined by what he has done, that is, by his past, his history. The achievements of the past create foreseeable possibilities realizable in the future. V2 rockets as achievements of Second World War science open the possibility to future realizable (and presently being realized) space travel. The alternatives for future realization are determined by "care-concern" questioning of what has been presently actualized. Man's being, taking it historically, is thus a careful and concerned questioning of past achievement with a view to alternative possibilities for the future (contained in the question according to Collingwood),[292] the answer being the determination of the alternatives to one or another according to the estimated greater well-being offered by one or the other of such alternatives.[293] This procedure of questioning the past with the view to its answer as the self-determination of future well-being from specified alternative possibilities is the activity of taking one's life experience in hand, the assuming of authentic control over one's life,[294] and the giving of meaning to one's life. But to know one's possibilities is to know one's past; that is, as Collingwood expresses it, the measure of what man can do is what man has done as revealed to him by his history.

> It is . . . of importance to man that he should know himself: where knowing himself means knowing not his merely personal peculiarities, the things that distinguish him from other men, but his nature as man. Knowing yourself means knowing, first, what it is to be a man; secondly, knowing what it is to be the kind of man you are; and thirdly, knowing what it is to be the man *you* are and

---

[291] See this chapter, above, pp. 360-372.

[292] See his *The New Leviathan*, p. 75, par. 11.12. See also *Speculum Mentis*, p. 78.

[293] This answers the question as to how a question contains its answer. A question contains its answer as one of a number of possible alternatives. As long as the alternatives are in the state of multiple possibility the question is an unanswered question. The answer is the determination of the multiple in possibility to one in actuality. On this point, see the illuminating article of G. F. Stout, "Truth and Falsity," *Mind*, XLI (1932), pp. 297-310. See especially pp. 301-302.

[294] For an interesting elaboration of this point see Robert Johann's *Building up the Human*, New York, Herder & Herder, 1968, especially pp. 37-53.

nobody else is. Knowing yourself means knowing what you can do, and since nobody knows what he can do until he tries, the only clue to what man can do is what man has done. The value of history, then, is that it teaches us what man has done and thus what man is.[295]

Thus, if man is only what he does, which is what man is inasmuch as man is mind,[296] for mind is what it does,[297] then history is knowledge in the plenary sense of the word, and subordinates all other forms (science, religion, etc.) to itself as imperfect moments of its plenary self. In taking this position, Collingwood is clearly in the Vician 'Verum = Factum' tradition, by medium of Croce.[298] When, as noted above,[299] Vico overturned the supremacy of Cartesian rationalism by reducing mathematical clarity and distinction to its alleged source in man's productive activity, he made what man can do the measure of what man can know, and located the source of speculative knowledge in practical knowledge, which now assumes the place of primacy, as Collingwood repeatedly insists.[300] Mathematical entities are clear and distinct because we made them and we know what we make. Once that position is taken, it inevitably follows that the order of the natural world known by science is known precisely because the human productive consciousness made that order. Theoretical science is practical in its roots,[301] and the order previously accorded to the natural world becomes more and more to be regarded as an order made by men in their practical knowledge and concern in regard to it. Collingwood's philosophy of nature is a clear instance of this.[302] His *The Idea*

---

[295] R. G. Collingwood, *The Idea of History*, p. 10. See also his *Autobiography*, p. 116.
[296] See *The New Leviathan*, pp. 5-7.
[297] See *The Idea of History*, pp. 221-222.
[298] See this chapter, above, p. 386.
[299] *Ibid.*
[300] See the opening lines of the Prologue of *Speculum Mentis*, p. 15: "All thought exists for the sake of action." See also *The New Leviathan*, p. 5, par. 1.66, and *An Autobiography*, pp. 147-167.
[301] It is worth noting here that the French philosopher Maurice Blondel elaborated a philosophy on the principle that the speculative order is practically based. The speculative principle of contradiction in logic is presented as having its roots in a more fundamental contradiction encountered in the practical exigencies of lived experience. See his *L'Action: Essai d'une critique de la vie et d'une science de la pratique*, Paris, Presses Universitaires de France, 1950 (originally published 1893), especially pp. 470-474. See also John McNeill's *The Blondelian Synthesis*, Leiden, E. Brill, 1966, pp. 200-236, and J. E. Somerville's *Total Commitment: Blondel's L'Action*, Washington, Corpus Books, 1968, especially pp. 335-352.
[302] John Wild seems to some extent at least to be under this Vician influence. See above, this chapter, for his way of expressing the primacy of the practical over the speculative. However, this is not necessarily his position, but may be only his statement of the position of contemporary existentialism. But as Cassirer well pointed out

of Nature is reduced to a history of the ideas men have elaborated about the world of nature. It is a history *of ideas*, not an account *of natural things*.[303]

But, just as Cassirer pointed out[304] that absolutization of the object-pole of perception leads to absolutization of the "thing-ist," scientific, naturalistic type of explanation in which persons and personal concerns have no place, might it not also be the case that absolutization of the subjective pole and inflation of the personal concern type of meaning will blind one to the correct status of the natural order? If the perceptive act, following Cassirer, is to be taken as the focal point which radiates in two opposite meaningful directions, and therefore the phenomenological point of departure in assessing the validity of these two systems of meaning, the natural and the personal, it is of the utmost importance that this phenomenological investigation itself be conducted with its correct guide-questions in mind. That is, taking seriously Collingwood's advice in this matter regarding the importance of correct questioning in any investigation, we must ask whether the phenomenological description alone of the lived experience of perception is the absolute arbiter regarding the structure of perception as this is the source of two opposite systems of meaning, or whether the outcome of our phenomenological description is not perhaps in some way pre-determined by the questions with which we approach the subject asking it to give an account of itself. We are approaching the perceptive act with a specified purpose in the course of a specified investigation. The dominating guide-question with which we are approaching perception is "How is it that this fontal act of lived experiential immediacy is simultaneously the source of impersonal, thing, natural meanings and of personal, care and concern meanings?" We are approaching perception with that precise question in mind, precipitated by our experience of the antinomy of two systems of meaning, which question perception has to answer, and the phenomenological analysis has to be tailored accordingly. If we were approaching it with another question, say the question, "How does perception appear as an act of consciousness?" or the question "How does perception appear when it is the perception of a scientist

(see above, this chapter, p. 404), in order to invent a tool, attention has to be lifted, at least momentarily, from total immersion in the impulse and necessity of the moment. So the practical depends *to that extent at least* on the speculative.

[303] This is pointed out by R. C. Cragg in his Ph.D. dissertation, *Collingwood's Logic of Question and Answer*, University of Toronto, 1948, p. 279.

[304] See this chapter above, pp. 395-396 and p. 408.

involved in behavioral observation?'' its reply would correspondingly differ. We are therefore not starting from a mere primitive, brute experience simply to be described, but from a question applied to an already distinctively recognizable type of activity which has been experienced to precipitate two opposite kinds of result. The phenomenological analysis, if it is to be meaningfully revelative, must itself take account of its initiating guide questions, which have their origin in the object of the inquirer's concern. The phenomenological inquiry, like any inquiry, is a search for meaning, and therefore presupposes the conditions under which the meaning-searching question arises, and under which a meaningful answer will be recognizable when found. The context giving rise to the phenomenological analysis of perception here is our experience of perception as the source of two opposite systems of meaning and of explaining, and our question, resulting from that context, concerns the structure of perception insofar as it has the virtuality of originating those two meaning and explaining systems.

This being granted, might it not be the case that two worlds, each enjoying its own proper autonomy, are communicating in some way in the act of perception? Might it not be that in that *sui generis* type of act, the spiritual, subjective, interior world of consciousness is achieving some sort of unity with the physical, objective, exterior world of things? It is not merely a question of correspondence but of identity, an identity which remains consistent with a diversity of another order. Might it not be that the conscious subject, which is also a thing belonging to the natural world, precisely in that act of perception overcomes its physical duality with another physical object to enjoy an identification therewith, respecting all the while their natural, physical duality? If identity in difference is admissible in other contexts, why not here also? This at least is indicated, if not solidly established, by the fact that perception has the potential to ''spectro-analyse,'' so to speak, and be elaborated in these two opposite ways, and has in fact been so elaborated. Might not the scientific way of explaining in terms of physical natures and things (including in this explanation the subject, not precisely and distinctively as a conscious subject but as something also found in the world of nature) be an autonomous and valid way of thinking alongside and distinct from, though not necessarily separate from, explanations of a cultural and historical sort in terms of human conscious concerns?

Robert Stover opts for a ''radical pluralism''[305] in answer to this

---

[305] See his *The Nature of Historical Thinking*, p. x.

question. Speaking of the two ways he had distinguished of making the world intelligible, he writes:

> Broadly speaking, the whole issue of universal determinism lies outside of the scheme of intelligibility of thinking from the standpoint of living in the world. . . . This allows for the compatibility of the two schemes of intelligibility: the presumption of determinism being integral to the natural order scheme of thinking; indifference to the issue of universal determinism being an essential characteristic of the scheme of evaluation. But compatibility, we must remind ourselves, is not equivalent to interrelatedness.[306]

What Stover calls ''ordinary history'' reflects this pluralism. Whereas ''some historians will focus their efforts on giving natural order accounts on the plausible assumption that particular human events can be accounted for deterministically'' and ''others will concentrate on making judgments from the standpoint of living in the world,'' ordinary history ''reflects the ambiguity of judgment characteristic of man's ordinary, everyday, unspecialized thinking,'' and ''whether the ordinary historian recognizes it or not, the vision of the world that moulds his work mirrors completely the pluralism painstakingly discerned and articulated through critical reflection.''[307]

The criticism directed against Collingwood here is that which is directed generally against those in the Vician tradition, namely that this line of thinking, dominated by the 'Verum = Factum' ideal, absorbs the natural into the historico-cultural, so that the world of nature ceases to be available to knowledge in its own right, and science as knowledge of nature is reduced to a mere subordinate member on a scale of forms, subordinated as error to history which supersedes it, allegedly by overcoming its falsifying abstraction. The world of nature is then nothing more than the ideas which the scientific interests of a particular period of history have articulated about it. The unfolding of the natural world by science is then nothing more than a stage in man's unfolding consciousness of himself, in which ultimately the idea of nature gives way to the idea of history.[308] Inasmuch as history both proceeds according to, and has as its object, human concerns and responses to human problems, it is a questioning and answer procedure. But if man is regarded solely as an active, self-determining agent—and this is precisely what history questioningly investigates according to Collingwood—then history exercises an overriding

---

[306] *Ibid.*, p. 171.
[307] *Ibid.*, pp. 270-271.
[308] See R. G. Collingwood, *The Idea of Nature*, p. 177.

autonomy, subordinating all else, including science, to itself.[309] Inasmuch as every science has its history of achievements, science can be regarded as a series of solutions to problems of personal concern to the particular scientists at particular periods of human development, and this series can be displayed meaningfully as an overlap of classes in a scale of forms.[310] But this is not an absolute and overriding subordination of science to its history, for it must allow for the equally valid subordination of the history of science to the science of which it is the history, inasmuch as scientific knowledge not only satisfies a question of pressing concern but also aims at the natural world as something enjoying some status independently of the concerned investigator and his questions. This seems to be clear from the obvious fact that there are books on, say, astronomy, which, notwith-

---

[309] If genus-species propositional thinking gives logic of the scientific type, that is, induction and deduction, an undue absoluteness (see *Speculum Mentis*, p. 49), might it not be equally so that absolutized 'overlap' logic accords to history an undue supremacy over science?

[310] Joseph de Finance, in his *Essai sur l'agir humain*, Rome, Presses de l'Université Grégorienne, 1962, pp. 176-177, very well illustrates how the meaning of both a human artifact, such as a watch, and a natural entity in its cultural signification, such as, for example, the sun, can be understood as a scale of forms. Regarding the latter illustration he writes: "Considérons, par exemple, le soleil. La connaisance scientifique que j'en ai ne fut possible que moyennant d'innombrables observations et calculs que d'autres ont exécutés au cours des âges et sur lesquels présentement je m'appuie. Aujourd'hui où les enfants, dès le premier éveil de leur raison, apprennent que la terre tourne, le système de Ptolémée nous paraît saugrenu et il en est de même pour d'autres théories, à présent périmées, mais considérées en leur temps comme vérités acquises: par exemple, la doctrine des quatre éléments. Mais si la rotation de la terre et maintes autres notions de ce genre se présentent maintenant comme familières et 'allant de soi,' c'est parce que d'autres les ont, au prix de beaucoup d'efforts et de luttes, apprivoisées, en quelque sorte, pour nous, comme furent domestiqués pour nous le chien, le boeuf et le cheval. Pour toute connaissance, si vulgaire soit-elle, nous sommes, pour la plus grande part, débiteurs des autres. Ces savoirs élémentaires, ces techniques de la vie quotidienne condensent le progrès de nombreux millénaires. Lire, écrire, compter nous semblent des opérations toute simples: tout cela cependant, comme les humbles et si longtemps immuables techniques agricoles et artisanales, a dû être conquis de haute lutte et représente capitalisé l'acquis d'innombrables générations.

"Mais ce n'est pas tout. L'idée du soleil ne s'épuise ni dans la représentation vulgaire, ni dans celle que la science en fournit. Elle comporte un savoir confus de ce que le soleil est pour l'homme: pour le paysan, pour le promeneur, pour ceux qu'il écrase de sa chaleur et ceux qui saluent sa lumière comme une bénédiction, pour le poète et pour l'artiste . . . L'idée complète du soleil inclut le mythe de la caverne et les comparaisons des *Ennéades*; elle inclut les cultes solaires: Mithra, le *sol invictus* etc. Ainsi, le monde où nous nous mouvons se présente à nous comme un monde humanisé, socialisé. Notre rencontre avec lui n'est pas une rencontre solitaire: les choses nous apparaissent enveloppées de signification humaine, de sorte que nous ne pouvons ni les penser ni les nommer sans participer du même coup aux expériences, aux intérêts, aux peines et aux joies de l'humanité."

standing any amount of historical information, are properly scientific treatises dealing with the natural, astronomical world itself. There are also books on astronomy which are quite clearly histories of astronomy, notwithstanding that they contain a large amount of properly scientific information about the natural astronomical world.

The very fact of this dualism, both in our perception of objects and in our explanations (scientific and historical), is a sign that the two directions in which the perceptive act points (on the one hand to a perceiving subject, whose perceptions both arise from and result in personal cares and concerns, and, on the other hand, to a natural world objectively present) point also to two foundations essential to the perceptive act, the foundation of a natural world in which things, including percipient subjects, are naturally distinct, and another foundation, that of the world of conscious subjects, which, as such, and manifestly in the perceptive act, are able to transcend their physical duality in the unity of the perceptual, knowing act. As Aristotle says, "The activity of the sensible object and that of the percipient sense is one and the same activity, and yet the distinction between their being remains."[311] Again, if concrete unity in diversity is an acceptable notion in other contexts, might we not also have here in perception a true unity (of the activity which is together a perceiving and a being perceived) in diversity (of man as a natural being and object-thing as likewise another natural being)?

It may be objected that this reintroduces the bogey of substantialism so odious to dialectical thought. The answer is that substantialism is rejected by that mode of thought on account of its own undue absolutization of self-unfolding activity, which absolutization, in fact, is made in virtue of the equally odious process of abstraction. For self-creative activity is not only abstracted away from man, the agent of such activity, but it is then treated as if it were all that there were in man the agent. An abstraction is thus made and subsequently ignored; precisely the situation in which abstraction involves error.[312] But, opposed to this position, what man *is* is seen to be more basic to man that what he *does* and is presupposed to what he does. Indeed, his actions are unfoldings and manifestations of what he is. If there is freedom *in his doings* there is presupposed freedom *in his being*. Whatever *determines* me to be a man at the same time constitutes me

---

[311] *De Anima*, Bk. 3, Ch. 2, 425b 25-30. The translation by J. A. Smith in *The Basic Works of Aristotle*, ed. Richard McKeon, New York, Random House, 1941, was used.

[312] See Chapter 7, above, p. 287.

open to self-determination as regards whether I become an engineer, a carpenter, an accountant, etc. If this open possibility were not *already a fact* of man's *being*, there would be no self-creative *activity* whereby man could take himself in hand and make something of himself. The moral order of self-determination is thus built on a more fundamental order of being in which the *conditions of the very possibility* of acting freely and morally are *already* constituted.

Now, if we begin with conscious activity as the overriding absolute, then the concept of man as *Dasein*, man in his state of ''thrownness,'' is perhaps the closest we can get to admitting the natural dimension of man's being. But, if we examine closely the implications of this notion, man as he finds himself, man in his state of ''thrownness,'' are we not led to what it already supposes, namely man's natural being as something already constituted and ''there'' previous to critical, conscious reflection? Man ''finds'' himself ''there'' *as already there waiting to be so found*. That is, he is not merely an autonomous consciousness but also a natural being. Collingwood's reticence regarding the existence of objects of feeling and regarding the activity or passivity of feeling[313] has at least the suggestion of dishonesty about it; one almost feels that the prejudice against admitting a 'thing' world, which feeling comes up against and experiences a passivity thereto, stifles the asking of the question even before any attempt to answer it can be made. Dr. Johnson's response to Berkeley, by simply kicking the stone, seems much more honest and convincing in this regard.

But is man in fact absolutized activity, or rather, do we not find men as both acting subjects and things who exercise their activity? Cassirer has insisted[314] that the ''I'' and the ''you'' perceived in dialogue are *constituted* by the dialogical activity, as poles thereof. But does the dialogue constitute the dialecticians *simply and absolutely*, or does it not rather *presuppose them in a primordial way* such that if there is no one already there to dialogue there is no dialogue? It is not denied that the self-constructed personalities of the dialecticians are constructed only in the dialectical communication when it is asserted that these self-constructing personalities, and their self-constructing communication, presuppose their constitution as natural members ''of a natural world.''[315] It is submitted that the ''I'' and the ''you'' who

[313] See Collingwood's *The New Leviathan*, p. 28 and p. 31.

[314] See this chapter, above, pp. 398-399.

[315] This statement is not to be taken as if to imply that they are merely material entities, but that they are members of a world found to exist as preceding any human activity. See this chapter, below, p. 426 et seq.

communicate in dialogue are possibly not *manifested to themselves* except in and through the mediation of that communicating activity, but to admit that is not the same as to say that what that activity *manifests* is nothing more than the *parties as constituted by that activity*. It is submitted that what that activity manifests to the involved parties is precisely their distinct autonomy *as things* and *as subjects*, which distinctness they enjoy as a very precondition of that unifying dialogical activity. Cassirer in this respect, like Collingwood, is in the 'Verum = Factum' Vician line of thinking, so that, since a man knows only what he purposefully makes, the self he knows through self-knowledge is nothing other than the self he makes, and the making of the self is in dialogical communication. But, discarding the commitment one makes as soon as one accepts the Vician principle 'Verum = Factum' as a presupposition, what is the true disclosure of the dialogical interaction situation? Does it not disclose two individual agents, two in what they *are* even if brought into a *communicating unity* by what they *do*? Just as the activity of sexual intercourse is *one act done presupposing a duality of agents in the unity of that action*, so also does the activity of dialogue presuppose a dimension of being in which the dialoguing agents are not one but two. Notwithstanding that they are one *in their dialoguing activity* they are two in what they *are*, that is, as things which are, prior to and as a precondition of their unifying activity. Their individual duality may not be *disclosed* to them prior to their activity in dialogue, but they have *to be* and *to be as two* and *to be as things-able-to-dialogue* (unlike being a stone) before that self-revelational activity takes place. Simply, this is to assert that, besides being dialogical, self-constituting beings, they are also natural beings. Prior to any of his actions—of consciousness, of artistic production, or whatever one will—and indeed as a determining condition of the existence of any of these distinctively human activities, man is found to have a well-defined structure in his being and in his mode of acting. To say this is to say, in Aristotelian terms, that man exists 'by nature.' Distinguishing those things which exist 'by nature' from those which exist otherwise than by nature, Aristotle says that what characterizes them is that "each of them has within itself a principle of motion and of stationariness."[316] Thus a bed and a coat, as such, are not natural, since, as such, they exist only as a result of human production.[317]

---

[316] See his *Physics*, Bk. 2, Ch. 1, 192b 10-15. The translation of R. P. Hardie and R. K. Gaye in *The Basic Works of Aristotle*, ed. Richard McKeon, was used.
[317] *Ibid.*, 192b 10-20.

Furthermore, as such, that is, as artistic products as distinct from beings 'from nature,' they have no innate principle within them, "no innate impulse to change."[318] Beings which are distinctively natural have a way of acting which has its reason within the acting being itself. Thus "man is born from man, but not bed from bed. This is why people say that the figure [of a bed] is not the nature of a bed, but the wood is—if the bed sprouted not a bed but wood would come up."[319] Beds do not exist in the natural world but only in the human, cultural world, resulting from human transformation. But wood does so exist, and in the manner in which it is found prior to any human tampering, it has very definite inbuilt principles of acting. In trees, in which wood is found 'in nature,' there is the capacity to grow and generate its own kind. Likewise in regard to man. There is a well-defined region of his being which did not result from his own self-development. He is already, in what he *innately* is, distinguished from the non-human world. That some things exist 'by nature' "it would be absurd to try to prove; for it is obvious that there are many things of this kind, and to prove what is obvious by what is not is the mark of a man who is unable to distinguish what is self-evident from what is not."[320] It is likewise obvious that man bears the marks of existing 'by nature' as well as existing 'by his own ingenuity.' That is, besides being a member of the cultural world, the world which *results from* man's free disposition of things, he also exists as a member of the natural world, the world which *preexists all human cultivation and development*. Man is 'by nature' intelligent and 'by nature' free. His intellectual and free activities come from inbuilt principles which enable him to so act, which principles *precede* him in an absolute way, as in no way *resulting from* his disposition of himself, but utterly presupposed to the very possibility of his activity of self-dispositioning.

This in no way weakens the thesis that history is a human activity in which the historian, or if one prefers, man as historian, enters into active communication with past human beings and their thoughts and concerns, and, in so doing, moulds himself up into what he is. In fact, it enables us to maintain this thesis without the tortuous difficulties which Collingwood tried to solve by his theory of incapsulation,[321] that is, the problem as to how the same activity of thinking could be in a past historical figure and in the present historian's mind.

[318] *Ibid.*
[319] *Ibid.*, 193b 5-15.
[320] *Ibid.*, 193a 1-10.
[321] See his *Autobiography*, pp. 112-114. See also Chapter 6, above, p. 227 et seq.

There is no problem if the two (that is, the past and present) thinkers have some identity *other than their activities of thinking* and thought communication, that is, if the thinker and the thought he thinks are not entirely the same.

If the absolute in being is self-conscious, self-creative activity, then man is simply his history, and history as the story of what human consciousness has made of itself is the absolute in knowledge. But if man is something more than self-conscious, self-creative activity, then there is room for other disciplines than history to make their own autonomous contribution to answering the question, "What is man?" Man as a natural being, for instance, can be investigated as something found at the terminal of a long period of biological evolution. His evolution to his place in the natural world is part of the explanation of what he is. Thus, H.-I. Marrou distinguishes "between . . . two pasts of mankind," one pertaining "to biological evolution" and "the other to history."[322] The prehistorian not only "examines the remains of human skeletons, analyses their physical characteristics. . . . concentrates . . . on the size of the cranial brainpan or the erect posture," which pertains to palaeontology, he also "studies objects that bear the marks or traces of man's voluntary action . . ." which "pertains to archaeology, which is a branch of history . . . in the full meaning of the term."[323]

Self-consciousness, then, can be seen as something man has in addition to many other things which he has but of which perhaps he may never become conscious. It is by his activities of thinking and caring that he exercises self-conscious, self-creative activity, but he does not actively endow himself with the very ability to do these actions; rather, he finds himself endowed with them. It is not by my free, self-determining activity that I am able to act freely. If freedom were not something already in what I find that I am I could never take myself in hand in the exercise of free, self-determining activity. Man is free inasmuch as he is not fully determined to being one kind of thing, but is open, in what he is, to alternative possibilities. But he is not free about this, namely, that he is so open; as Sartre says man is condemned to freedom. But to call this openness a condemnation seems to suggest expression of displeasure and the will to punish, in a Platonic fashion, on the part of some personal being, and thus savors of myth-making. Is it not more in harmony with the facts

---

[322] H.-I. Marrou, *The Meaning of History*, Montreal, Palm Publishers, 1966, p. 36.
[323] *Ibid.*, pp. 36-37. See also *ibid.*, pp. 43-45.

of our experience of ourselves and of other human beings to say that man is 'by nature' free; that inasmuch as he belongs to the world of things which (as Aristotle says) exist 'by nature,' as having within them predetermined principles of acting, he is 'naturally,' that is, unfreely, determinately (although not by condemnation, as Sartre says) endowed with the ability to assess his situation, make decisions and take his situation, his ''thrown-ness,'' in hand. All this can be simply summarized by saying that Collingwood's assertion that ''man is what he does,'' understood, as it is by Collingwood, in an absolutized care-meaning context, has to be modified so as to allow also that ''what a man does supposes what he is,'' which assertion recognizes man's membership in the world of things which exist 'by nature.' Inasmuch as man is a historical being he is what he makes of himself by his freedom. Inasmuch as he is a natural being he acts determinately, as his nature so determines him to act. And one of the ways his nature determines him to act is to act freely.

> Ainsi la spontanéité de l'esprit n'est pas absolue. La science est bien une construction de l'esprit, mais elle n'est pas une *pure* construction de l'esprit. C'est une construction partiellement au moins dirigée du dehors. Les choses sont compréhensibles, ce qui était pour Einstein la plus incompréhensible des choses. Le progrès dans la connaissance de l'univers et de ses lois vérifie à sa manière la vieille thèse métaphysique de l'intelligibilité de l'être.[324]

If, consequently, we follow Cassirer and take perception as the phenomenological point of departure in the analysis of the source of the two orders of meaning, are we in fact faced with an either/or situation: *either* objective, physical, natural, deterministic explanation, in which the subject of concern is obliterated by being absorbed into the natural, *or* subjective, personalist, spiritual, free, self-determining explanation, in which the objective is merely a concern of the subject? It is submitted that the evidence of perception justifies the validity of both types of explanation, and that the inability of one or the other type of explanation to deal adequately with all areas of experience in supplying a rational account of experience is a warning against the validity of such a reduction. Science bears on the natural world as its object, and bears on it according to its determinate characteristics without, however, appropriating to itself the natural world as its own exclusive object. For philosophy also bears on the natural world, as

---

[324] J. de Finance, *Essai sur l'agir humain*, Rome, Presses de l'Université Grégorienne, 1962, p. 165.

also on the cultural world. Science bears on the natural world according to the determinate observable and measurable relations found therein, which it expresses as 'the laws of nature.' Philosophy, on the other hand, inquires as to the mode of being of that which exists 'by nature.'[325] Science, therefore, should not attempt to impose its method of objectivising and proceeding on history, nor should history attempt to foist its way of objectivising and proceeding onto science. Each discipline may borrow the methods of the other, as Robert Stover showed,[326] but each should honestly recognize that the borrowing is in fact a borrowing, and that what is borrowed belongs to the other discipline by right. The Michelson-Morley experiment in physics can be regarded from the point of view of answering the question which manifested the personal interests and concerns of the physicists involved. Explained in that way, that experiment has its place in the history of physics and in the unfolding of man's thinking about the world. Or the answer which that same experiment supplied can be regarded also as a proposition which says something about the natural world. Thus taken, that experiment belongs to the domain of physics, not merely to the history of physics.

From this we can draw the conclusion which, in essence, is the necessary modification which must be borne in mind in agreeing with Collingwood that the method of history is one of questioning.[327] When propositions which are answers to questions are taken precisely as

---

[325] The point made here is that the natural world and the scientific world are not synonymous, as if thinking scientifically about an object is what entitles it to be called 'natural.' This latter position seems to be that of Collingwood. Thus, in *The New Leviathan*, p. 7, par. 1.83, he says: "Man as body is *whatever the sciences of body say that he is*" (emphasis in original), whereas (1.84) "Man as mind is *whatever he is conscious of being.*" Again, in *Speculum Mentis*, p. 167, he says, "Mathematics, mechanism and materialism are the three marks of all science, a triad of which none can be separated from the others, since in fact *they all follow from the original act by which the scientific consciousness comes into being,* namely, the assertion of the abstract concept. They are all, it may be said, *products of the classificatory frame of mind . . .*" (emphasis added). The notion of the naturality of a thing as constituted by an inbuilt source of determinate activity is foreign to the modern mentality dominated by the Cartesian identification of the natural world with externality, understood spatially in terms solely of extension. Kant's subsequent understanding of the natural world as constituted by the *a priori* form of space, followed by the Hegelian reduction of the 'given' characteristic of experience to an activity of the mind which gives its object to itself, eventually leads, as in Collingwood, to the identification of the natural with the 'spatial,' that is, that which is given according to the mode of externality, and which is thus given by and in the mind thinking scientifically. Such a reduction, widespread in modern thought, evacuates from a natural world its very naturality, namely, that it is an autonomous possessor of its own source of acting in a definite way.

[326] See this chapter, above, pp. 362-363.

[327] See this chapter, above, p. 416.

answers to questions, they are taken precisely as satisfying some personal need or concern in a care-meaning context. But when these answers are taken not precisely as answers, but precisely as propositions, they are taken precisely as statements which say something about the natural world, and their meaning is therefore that of the natural order type. Collingwood strives to suppress the propositional character which bears natural order meaning to inflate and absolutize the answering character, which is the bearer of careful-concern meaning. This statement summarizes in a nutshell the core critique against the whole Collingwoodian enterprise.

One final point may be made which helps to drive home the above conclusion. Ernst Cassirer noted[328] that natural science, by its manner of conceiving, extends itself to the whole of cosmic space, whereas, due to their way of universalizing, the cultural sciences, including history, are confined to a small region of that domain, namely the human region.[329] But this implies that man is a distinguishable part of that cosmic whole, capable of being recognized by defining characteristics, which presupposes genus-species logic. Thus Dr. Leakey, looking for evidences of prehistoric man in East Africa, had to be able to recognise what were properly human artifacts from what was the result of natural forces.[330] Man's conscious reappropriation of the cosmic totality within himself, and ability to bend it to his own satisfaction, is conditioned by and presupposes his location as one among others forming part of the natural cosmic whole. Man is thus *naturally* a part of the whole before the whole becomes *consciously* part of man through man's investigative activity. The delineation of the area of self-creative, dialectical activity manifesting itself as a scale of forms in an overlap of classes thus *supposes* the availability of genus-species logic capable of recognizing man as *specifically distinct* from all other natural entities in the cosmos. Thus, even though the dialectical logic of the active self-creating concrete universal may locate genus-species thinking as a self-transcended part within itself, nevertheless the very process itself is attached to the being of man, which requires distinguishing him *as a species from other species* in the cosmos with whom he would be *contained under some common genus*. Thus there is mutual superiority and mutual subordination of one way of proceed-

---

[328] See this chapter, above, p. 411.

[329] See his *The Logic of the Humanities*, p. 144.

[330] See L. Leakey, *Adam's Ancestors*, New York, Harper and Row, 1960, Ch. 3, pp. 29-53.

ing with respect to the other, depending upon the standpoint from which one operates, namely, the scientific point of view[331] or the historical point of view.

[331] Or, more widely than in Stover's perspective, the natural point of view, that is, the point of view which visualises a thing in its determinate mode of being, whether objectivised scientifically in terms of laws of nature, which consist in universal and necessary correlations of observable and measurable phenomena, or visualized philosophically, in terms of the mode of being involved in that which acts in a determinate way. The opposition Stover recognizes is that between the predetermined and the self-determining. Stover restricts his inquiry to the pre-determined as scientifically investigated.

Chapter Ten
Conclusion

A. Summary of Foregoing

The foregoing has presented Collingwood's call for the development of a hitherto neglected logic of questioning, which he alleges to be the logical procedure proper to history (Chapter One). Chapter Two investigated Collingwood's identification of knowledge with the activity of asking and answering questions, in which he takes the distinctive position that the meaning of any proposition can only be known in relation to the question of which the proposition is intended to be the answer. This, in consequence, involved him in also taking the distinctive position that all statements are essentially historical in meaning, since every question (to which propositions are relative as regards their meaning) is asked by a definite person in a definite context of time, place and circumstances.

Chapter Three investigated the suppositional character essential to the activity of questioning according to Collingwood. Questioning as supposing is important for Collingwood since it is this characteristic that marks knowledge as an activity, as distinct from the mere copresence of the mind with an already given object simply to be contemplated. In the supposal of questioning, the mind takes an active stand with respect to its objects and is able to consider the non-existent, the ''supposed.'' Since Collingwood identifies this particular feature of questioning with artistic activity, it was necessary to examine Collingwood's position on the nature of art and its relevance to the questioning process, which consists in this: that in questioning as supposal the mind exercises its creative initiative in knowledge as distinct from mere contemplative co-presence with a datum, as Positivists maintain (this latter consideration being deferred to Appendix One).

The key logical entity in Collingwood's proposed question and answer logic was shown, in Chapter Four, to be that of the concrete universal, which is the principle of concrete unity in diversity in history and "the daily bread of every historian," in contradistinction to the abstract universal, which is the principle of intelligibility in mathematical deduction and scientific induction. Whereas the abstract universal is the principle of genus-species, propositional logic most apt for scientific thought, the concrete universal is necessary for concrete, historical thinking, which, according to Collingwood, alone could provide mankind with the intellectual instrument for the harmonious management of human affairs. The application of Renaissance science to the solution of human problems had resulted in the sundering of the various life forms, namely, art, religion, science, history and philosophy, and Collingwood sought, through history and its properly human logic, to bring about a *rapprochement* of the now departmentalized human affairs which had culminated in the disaster of the first World War. Collingwood's concrete question and answer logic was thus shown to be his answer to the problem of human ills as he saw them to flow from the practice of treating human affairs with the methods of science, which are properly applicable only to the natural external world.

Chapter Five exposed Collingwood's theory that relevant questions arise from historically determined presuppositions. According to him, Metaphysics is really the historical study of the absolute presuppositions which determine the relevant questions of this or that particular historical period. Whereas relative presuppositions are answers to previous questions, absolute presuppositions, which ultimately condition the meaning of all questions and their answers, are statements that are made not in answer to any question, but are catalytic tools that the mind forms out of itself in order to question experience and thereby convert it into the experience of a rational being. They are simply presupposed for the purpose of interrogation.

Chapter Six then dealt with Collingwood's theory of mind as the ultimate basis of all that is, and his position that mind is not a thing or substance, but self-activating and self-creating activity. The totality of that which is, or reality, is thus understandable as reduced to the dialectical self-creativity of mind becoming progressively conscious of itself through the raising and solving of problems.

The remaining three previous chapters (Seven, Eight and Nine) attempted to evaluate critically the various main aspects of Collingwood's thought on the logic of questioning. Chapters Seven and Eight concluded that although Collingwood was right in recognizing that propositions have a distinctive meaning in the concrete situation in which they function as answers to questions, he was wrong in denying that they have a meaning in themselves distinct from their function as answers to questions.

Chapter Eight took Collingwood to task for neglecting to recognize an already well-developed logic of questioning in Aristotle. In terms of the Aristotelian logic of questioning, Collingwood's theory of knowledge as question and answer was criticized as being the result of disorderly interrogation.

Chapter Nine investigated Collingwood's claim that history proceeds by a distinctive logic of questioning, and concluded by agreeing that such a logic proper to history does exist. This chapter also showed precisely how historical meaningfulness differs from scientific meaningfulness, namely, that whereas science looks at the world as intelligible in terms of impersonal deterministic laws, the historian looks at events in their relation to human, free causality, in which there is self-determination of alternative possibilities. When a proposition is taken precisely as an answer to a definite question, it is taken precisely in relation to human freedom and its projects and concerns, which the question articulates.

B. General Conclusion Regarding Collingwood's Logic of Question and Answer

The conclusion of Chapter Nine, above, that history does in fact proceed, as Collingwood says, according to a logical method of question and answer, is also the answer to one of the basic questions which has dominated this whole investigation, namely, whether Collingwood's allegation of the existence of such an undeveloped logic is to be admitted.

Having agreed on this basic point with Collingwood, subject to the important reservations made in Chapter Nine, in which Collingwood's undue subordination of induction and deduction to historical methodology was rejected, the second basic concern which has dominated this whole inquiry becomes relevant, namely, what can be said regarding the nature, structure and modes of procedure of this interrogative logic?

All that can be said in this regard, as far as the preceding investigation warrants, is that the rationale or basis which gives the concrete inquiry of history (and of science precisely in its historical dimension, that is, precisely in its function of discovery and research as distinct from its codification into established principles and laws) is the concrete universal, which has been shown (in Chapter Four above) to have the character of an overlap of classes in a scale of forms in dialectical development, and (in Chapter Nine) to be constituted by a linguistic unity in diversity, a unity in diversity of human consciousness in linguistic dialogue.

But if this linguistic unity in diversity is the logical principle, the one in the many, at the basis of history and of research, what can be said regarding its principles?

Such an investigation would involve considerable research in its own right, and could scarcely be given adequate treatment in a mere concluding chapter. But an attempt is made, in Appendix Two, below, to give some idea of the direction in which such a research could be pursued. The principal authority followed is Michael Polanyi, whose main concern is with the logic of scientific discovery. Since the research and discovery aspect of science is properly historical, what Polanyi says is easily adaptable in giving an at least *prima facie* outline of the logic of interrogation.

C. Final Remarks: Collingwood's Contribution to the History of Philosophy

The history of philosophy has provided us many times with visual aids disclosing the forms which philosophy takes when it takes its cue and inspiration from other disciplines. Pythagoras, Descartes and Spinoza show us philosophy under the spell of mathematics and its ideals and methods. The ancient atomists, Kant and A. J. Ayer show us philosophy dominated by one or other of the ideals of physical science. Bergson illustrates philosophy as elaborated from the view of evolutionary biology, and apparently, to some extent at least, so does Aristotle, who, according to E. Gilson, "exaggerated the scope of one science [biology] and the value of its method, to the detriment of others";[1] when "committing the opposite mistake to that of Des-

[1] E. Gilson, "Concerning Christian Philosophy," in *Philosophy and History: The Ernst Cassirer Festschrift*, edited by Raymond Klibansky and H. J. Paton, New York, Harper Torchbooks, 1963, p. 71.

cartes, Aristotle sets up biological method as a physical method."[2] Gilson says that thus

> to biologize the inorganic as he and the mediaeval philosophers did was to condemn oneself to ignorance about those sciences of the inorganic world whose present popularity comes chiefly from the inexhaustible fertility which they display in things practical.[3]

With Collingwood, following in the wake of Hegel and Croce, philosophy is presented to us as dominated by the thought patterns and modes of procedure proper to history, in which the explanation of the world of experience is in terms of an active striving of the conscious self in the overcoming of obstacles. If "in the cult of antinomies in modern philosophy . . . Kant comes up against them" and "Hegel lives by them,"[4] I think one may say that Collingwood, in his theory of questioning, completely radicalizes them. In presenting us with this living, historically based philosophy, Collingwood has rendered a service to philosophy itself. We can look at this living example, appraise its strengths, criticise its weaknesses and disengage its genuine contribution to the perennial philosophy.

Collingwood's most outstanding contribution to philosophy is perhaps his highlighting of philosophy in its multifarious forms as answers to pressing questions, an aspect of philosophy consistently overlooked, yet of the highest importance for the full understanding of any particular philosophy and indeed for that whole which is philosophy itself. Thus, Aristotle's hylomorphic theory is scarcely intelligible if it is not viewed against the background of problems grappled with by Parmenides, Heraclitus, the ancient Atomists, Pythagoras and Plato. His theory not only speaks to us about the changing world of experience, but also answers a definite question that was a matter of urgent personal concern in Aristotle's mind, which question arose as a result of his inheritance from his forerunners.

The philosophy of the future will have to reckon with Collingwood's contribution, and will perhaps show a greater concern for the dependence which philosophy or science, or any discipline for that matter, has upon its own history.

---

[2] *Ibid.*, pp. 71, 72.
[3] *Ibid.*, p. 73.
[4] *Ibid.*, p. 68.

APPENDIX ONE

The Nature of Supposal as an Activity of Imagination

At the conclusion of Chapter Three, above, the treatment of supposal as imaginative activity, or artistic activity, was deferred to an appendix. This treatment is now to be taken up here.

Collingwood investigates the nature of the imagination in his 1938 publication *The Principles of Art*, which, as the title indicates, aims to answer the question, ''What is Art?'' The imagination is therefore dealt with in that work inasmuch as its activity is an integral element in the answer to that question. The question, however, which is of relevance here is, ''What is the imagination inasmuch as imaginative activity is supposal or questioning?'' Thus, a somewhat different viewpoint will be taken in this textual analysis of that work than the one adopted by the work itself.

Three main points will be the focus of attention of the present inquiry. First, the imagination will be considered as the activity by which the objective order is most fundamentally constituted. By this activity the fleeting present here-and-now actuality of feeling, in which, previous to reflection, subject and object are indistinguishably blended, is primarily converted into a recognizable object for a conscious subject, and therefore the activity in which the subject-object dichotomy required for consciousness and thought is first established. That is, the activity of imagining is the activity in which the conscious subject actively disengages itself from that which it opposes to itself as its object. The very possibility of meaningfully recognizing the present fleeting moment of actual experience as something having a past, and as pointing towards an as yet unrealized but realizable future, depends upon this fundamental objectivizing function of imagination as a presupposition thereof.[1] The work of thought in all its forms and at all its levels is to convert merely felt experience into the intelligible experience of a rational being. Foundational to this, and a presupposition of any explanation and mediation, is the objectivization of the actuality of present experience, the transformation of it from merely felt actuality to objectivity. This is basically the work of the imagination.[2] It provides the primary instance of the recognition of an 'other' and the presupposition of all subsequent recognition of an 'other,' and, therefore, of all subsequent relating activity of thought, and of all subsequent disengagement from and transcendence over immediacy, and of the very possibility of establishing the alternatives which are a prerequired condition for the activity of questioning. Imagining, as an objectivizing disengagement from actuality but yet as ordered back to the actuality of experience, is the activity of supposing, an essential constituent of the activity of questioning, according to Collingwood.

Second, imagination as expression correlates the theoretical aspect of knowledge, the apprehension of unity in diversity, with the dynamism of appetite, which is the forward striving of the mind

---

[1] "The real is the present, conceived not as a mathematical point between the present and the past, but as the union of present and past in a duration or permanence that is at the same time change: the possible parting with its unnecessariness and the necessary parting with its impossibility in an actuality which is at once possible [because open to an undetermined future] *and* necessary [because brought to actuality by a determined past], not (like the abstract mathematical present) neither. Within this present there are, as really as you like, two elements (necessity and possibility), each of which taken singly or in isolation characterizes a being which is not real but ideal — the past and future respectively. Thus the past *as past* and the future *as future* do not exist at all, but are purely ideal; the past as living in the present and the future as germinating in the present are wholly real and indeed are just the present itself. It is because of the presence of these two elements *in* the present . . . that the present is a concrete and changing reality and not an empty mathematical point." R. G. Collingwood, "Some Perplexities about Time," *Proceedings of the Aristotelian Society*, XXVI (1925-1926), p. 149. For a further consideration of the necessity of the past and the contingency of the future see Dominique Parodi, "Nécessité et contingence en histoire," *Revue de Métaphysique et de Morale*, LIV (1949), pp. 273-279.

[2] What Bosanquet says in his *The Principle of Individuality and Value*, pp. 57-58, well indicates this objectivizing function of the imagination: ". . . a work of art . . . is an object in which we can realize what the Greeks meant by *Theoria*. In its essence, as a thing of beauty, and neglecting its aspect as a physical object or movement, it is self-contained and a true whole, possessing its significance in itself, and not driving our thought beyond it to a detached meaning and explanation. Every point in it carries the burden, or lives with the life of the whole."

which constitutes the open possibilities of the future and is not really distinct from the activity of questioning dynamically considered, that is, as the cutting edge of the mind in action.

Third, imaginative expression, as language, is concerned precisely with the process of the objectivizing of feeling, the very constituting of feeling into an object of imagination, that is, an idea. Language is the expressed embodiment of objective, therefore actually achieved, consciousness, which thus acquires the determinate characteristic of a past. Furthermore, as linguistically expressed, what is expressed is available not only to the speaker, but publicly. "In speech and art the individuals not only share what they already possess; it is only by virtue of this sharing process in speech and art that individuals have attained what they possess."[3]

The structure of the content of this appendix follows generally the order in which it appears in Collingwood's *The Principles of Art*. Section A exposes Collingwood's distinction between thinking, feeling and imagining. Section B deals with the problem of distinguishing imagination from sensation and Section C with Collingwood's identification of the imagination with consciousness in its wide sense and with the distinction of imagination from thought in the strict sense of intellectual thought. Section D outlines Collingwood's summary of his general theory of the imagination, and section E presents Collingwood's theory of art as expression, specifically with emphasis on linguistic expression as the basic form of art.

A. Imagination, Thinking and Feeling Distinguished

Collingwood first appeals to the commonly recognized distinction between thinking and feeling.[4] Whereas in our experience of feeling there is the experience of a perpetual flux, in that of thinking there is the experience of a genuine recurrence of something; something is recognized to stand firm amidst the flux, somewhat in the same way that in the flow of a river at least the soil and the rocks in which the river is channelled are recognized to stand firm, at least in relation to the flow of the water. In order that something be an object in the most minimal sense of the term it is necessary that it be taken out of the condition of radical flux. If feeling is to be an object, the

---

[3] Ernst Cassirer, *The Logic of the Humanities*, New Haven, Yale University Press, 1966, p. 113.

[4] See R. G. Collingwood, *The Principles of Art*, London, Oxford University Press, 1938 (first Galaxy paperback edition, 1958), p. 159.

flux which characterizes it in its own right has to be stabilized in some way.[5]

Collingwood distinguishes two recognizably distinct kinds of feeling. First there is the kind we refer to when we say we feel hot or cold, hard or soft. Then there is the kind we refer to when we say we feel pleasure, pain, anger or fear, etc. He accepts established custom in calling the first type sensations and the second type emotions. In our experience, every sensation has its corresponding distinctive emotional charge associated with it, or, more correctly, with its object, its sensum, or that which is sensed as distinct from the act of sensing it.[6]

From our experience of it "feeling appears to arise in us independently of all thinking, in a part of our nature which exists and functions below the level of thought and . . . unaffected by it."[7] This sensuous-emotional nature which we have as feeling-creatures, and independent of our thinking nature as rational creatures, constituting a level of experience below that of thought, presents itself as something basic, a foundation upon which the rational part is superstructured. This level of awareness, at which we merely feel, Collingwood calls the psychical level, which he says is the proper domain of the science of psychology.

Thought presents itself in our experience under two forms. Its primary form is exclusively occupied with its substratum of feeling, which gives to it (thought) its sole and universal subject-matter. When we think 'I am tired,' or 'It is a hot day,' or 'There is a patch of blue,' we are thinking about our feelings.

> . . . our experience of the world . . . the 'world of nature' . . . is an experience partly sensuous (strictly, sensuous-emotional) and partly intellectual: sense being concerned with the colours we see, the sounds we hear, and so forth; and thought, with the relations between these things.[8]

But in its secondary form, thought thinks not about feelings and the relations between these, but rather about our thoughts themselves.

---

[5] I understand this to mean that pure feeling as such would be too fleeting to be taken hold of and dealt with in any way; it could not be said to possess a recognizable identity, and would thus defy objectivization.

[6] See R. G. Collingwood, *The Principles of Art*, pp. 160-163. See also Collingwood's *The New Leviathan*, pp. 18-26.

[7] R. G. Collingwood, *The Principles of Art*, p. 163.

[8] *Ibid.*, p. 166. See also *The New Leviathan*, p. 16, pars. 3.54 and 3.6, for 'feeling' as one of the traditional meanings of the term 'body.'

> The propositions asserted by thought in this secondary form may be indifferently described as affirming relations between one act of thinking and another, or between one thing we think and another . . .[9]

Collingwood says that the distinction of first order thought from second order thought is the same as the traditional distinction of understanding from reason, or of science from philosophy.

The problem of imagination, or the question to which the imagination is the answer, arises in connection with thought in its primary function, namely as thought is concerned with the relations between sensa. This question arises in regard to the very possibility of making comparisons and establishing relationships between sensa, as when we say 'this is blue,' 'that is red,' and 'this blue is bluer than that blue.' To say that it is hotter now than a moment, or a minute, or an hour ago, implies that we have some means of comparing the presently felt sensum with the sensum felt a moment, or a while, or an hour ago. But the sensum of a moment, or a minute, or an hour ago is no longer present, having since been carried away by the flux of sensation. The flux of sense seems to destroy any sensum before it has lasted long enough to permit its relations with other sensa to be studied. So-called sense data are not mere data of sensation, but sensa *as given and retained*, established or fixed like a datum line in a survey, and therefore *already taken out of the pure flux of sensation*. Sense data are already thought-interpreted (mediated) and stabilized.[10]

Again, we cannot call our relationship with our sensa an 'acquaintance,' for in order to be acquainted with someone or something there has to be contact made with it on a number of occasions. It would thus have to be a recurring feature of one's experience. But sensa neither persist nor recur. Redness may recur and enable us to get acquainted with it, but this red patch does not recur so that one cannot get acquainted with it.

> [Talk about and appeal to sense data] implies that I can know what certain sensa, not present to me, would be like if they were present to me, and can say 'These are, or are not, the sensa I expected,' comparing sensa which I now have with some idea of them which I framed in advance of having them; and it should be explained how this is possible.[11]

---

[9] R. G. Collingwood, *The Principles of Art*, p. 167.

[10] See *ibid.*, p. 169. That is, 'sensa,' as data, are *established*, as well as *defined*, as objects. Note also the character of 'past' which becomes relevant in such a comparison.

[11] *Ibid.*, p. 170.

Talking about our sensa in this way is to raise the term 'sensum' and all its cognates to mean something quite different from the momentary and evanescent colours, sounds, scents and the rest which we actually feel in sensation but which we mistakenly substitute for them. If there are such things, they differ from sensa in not being wholly fluid and evanescent, since any one of them admits of retention in the mind as an object of attention after the sensation of it has occurred, or in anticipation of the sensation before it occurs. But, on the other hand, these quasi-sensa are sufficiently like sensa to be taken for them and to receive the same name.

Collingwood maintains that there are such things. Hume called them ideas as distinct from impressions, and Collingwood holds that "there is a special activity of the mind correlative to them . . . what we generally call the imagination, as distinct from sensation on the one hand, and intellect on the other."[12]

This activity, according to Collingwood, is the point at which the activity of thought makes contact with the merely psychic life of feeling. It is Kant's blind but indispensable faculty which bridges sensation and understanding, and that without which, according to Aristotle, thought would be impossible.

Collingwood draws attention to the fact that everyday language has two ways of talking about sense and sensations, one of which maximizes the similarity between (say) seen and imagined colours, the other which stresses the difference. In the first way of talking, 'really seeing' and 'imagining' are both named sensations, and the objects of both are indifferently referred to as sensa, or sense data. In the second way of talking, the words 'sensum' and 'sensation' are restricted to cases referred to as 'really seeing,' whereas the terms 'imagining' and 'what is imagined' are used in opposition to 'really seeing' and 'what is really seen.' The second way of talking does not have any generic term to cover 'sensing' and 'sensum' on the one hand and 'imagining' and 'what is imagined' on the other hand, so that there is no indication, in this way of talking, as to how the two are generically related. Collingwood proposes to use the term 'sensation' as a generic term to cover both the act of sensing and the act of imagining, and the term 'to sense' when a verb is needed, and the object sensed will be referred to as 'a sensum,' whose species are colors, sounds, etc. As specific names for the two cases, he proposes to reserve the terms 'real sensation' and 'imagination,' the species of

[12] *Ibid.*, p. 171.

real sensation being really seeing, really hearing, etc., and their objects; the species 'real sensa' are real colors, real sounds, etc., and imaginary sensa are imagined colors, imagined sounds, etc.[13]

B. Problem of the Distinction of Imagination from Sensation in the History of Modern Philosophy

Collingwood examines the opinions of philosophers throughout the history of modern philosophy on the question of the distinction of imagination from sensation.[14]

The assumption by the mediaevals, that sensation gives real acquaintance with the real world, was undermined by sixteenth-century sceptics, and the problem of guarding against illusions arising from mistaking an imaginary sensation for a real one became a major issue with Descartes, as, for example, our inability to distinguish sleeping and dreaming from the state of being awake. Descartes did not deny that there was a distinction between a real sensation and an imaginary one, between really sitting in front of the fire and only dreaming it, but he concluded that there was no available test whereby one could decide and make the distinction. Hobbes subsequently denied the reality of the distinction, since immediacy was a feature of all sense experience. Spinoza agrees with Hobbes against Descartes that in principle all sensation is imagination. For Leibnitz, sensa deserve the name 'idea,' since for him, unlike Spinoza, they are modes of thought. But they are ideas of a peculiar kind, namely confused ideas, which, if they could be brought to distinctness, would thereby lose their sensory character.[15]

Locke does not distinguish real from imaginary sensa, although his reason for doing so differs from that of Hobbes and Descartes. For Hobbes and Descartes all sensa are imaginary, whereas for Locke they are all real ("Our simple ideas are all real"). But Locke was the first to attempt to distinguish 'real ideas' from 'fantastical.' The only ideas Locke allows as fantastical are certain complex ideas formed at will by the arbitrary combination of simple ideas.[16]

Berkeley and Hume disowned Locke's position that sensa are real. Accepting Newton's world populated by bodies, Locke consequently accepted that real ideas are "such as have a conformity with the real Being and Existence of things, or with their Archetypes." Locke

[13] See *ibid.*, pp. 172-174.
[14] See *ibid.*, pp. 174-190.
[15] See *ibid.*, pp. 174-176.
[16] See *ibid.*, p. 176.

describes a 'fantastical idea' as one which the mind 'makes to itself.' Complex ideas are sometimes fantastical, as when they are voluntary combinations of simple ones in which the mind ''uses some kind of liberty in forming them.'' Simple ideas can never be fantastical, because they can never be 'Fictions at Pleasure.'[17]

Locke inadvertently provided Berkeley with a method which he (Berkeley) thought would enable him to distinguish real from imaginary sensa by introspection, namely, by distinguishing ideas voluntarily produced in us from those 'forced upon' us. Thus Berkeley distinguished 'ideas of sense' from 'ideas of imagination' on the score that 'ideas of sense' are stronger and more lively than those of imagination. But this turns out to be ambiguous. If the distinction refers to *the objects* (the sensa) it would simply mean that a real sensation, (say) sound, is louder than an imaginary one, so that to call a sound real or imaginary would refer merely to a difference in audibility. But if the distinction refers to *the acts* of sensing and imagining, it would mean that a real sound has a way of forcing itself upon us that an imaginary sound does not, so that a real sound is heard whether we will it or no, whereas an imaginary one can be summoned up at will. In the latter case, the difference is not between what we hear, the sounds, but between the experiences of hearing them, that is, a difference not recognizable by the ear but by the introspection of reflexive consciousness in which we are aware of those experiences.[18]

However, this introspectionist theory, drawn out of Locke by Berkeley, is still unable satisfactorily to distinguish real from imaginary sensations. In the hallucinations of mental disease the patient is obsessed by imaginary sights, sounds, etc., which are altogether beyond his control. Indeed, in a healthy organism, a man who has been horrified by certain sights and sounds cannot banish them from his mind for quite some time, for he continues to imagine the crash, the blood, the cries, etc., in spite of all his efforts to stop doing so. On Berkeley's principle, this should be a sure indication that he is really seeing them. But all it proves is that we have but limited control over our imaginative activity.[19]

Berkeley then attempts a second theory of distinction. Ideas of sense have a certain steadiness, order and coherence according to discoverable laws, the so-called Laws of Nature which we learn by expe-

---

[17] See *ibid.*, pp. 176-177.
[18] See *ibid.*, p. 177.
[19] See *ibid.*, pp. 178-179.

rience. For Berkeley, the Laws of Nature are not laws concerning the relations between bodies or bodily movements or bodily forces, but are laws concerning the relations between sensa, and the Laws of Nature are obeyed by 'ideas of sense' but not by 'ideas of imagination,' which are wild.[20]

Collingwood says that this theory does not hold on closer scrutiny. Either I see what looks like a black animal crouching in the half-light or I do not. No animal is found when I turn on the light and search the room, so I conclude that I was not really seeing but only imagining. But does the imaginary animal really disobey the laws of nature, as Berkeley contends it should? It may disobey some, but it obeys others. It comes under definite conditions, in the dark, when I am tired, bringing a slight but perceptible fear to one who, as a little boy, was frightened by the dark. So, although it does not belong to a family discernable in physical terms, yet it clearly does belong to one discernable in psychological terms.[21]

Can a distinction be made between real sensa and imaginary sensa, asks Collingwood, on the score that real sensa obey natural laws whereas imaginary sensa obey psychological laws appropriate to the mental world? He rejects this possibility, first, since the two orders are not absolutely distinct. Real sensa obey psychological laws as well as do imaginary sensa, and it is still a debated point whether the psychological is not reducible to the physical. Second, the laws of nature are what we learn by studying our real sensa, and the laws of psychology are what we learn by studying our imaginary sensa. But this distinction presupposes that we are already able, as a principle of that distinction, to distinguish real sensa from imaginary sensa. So,

> if we need rules in order to distinguish real sensa from imaginary ones, it cannot be sensation, the undistinguished mixture of sensation and imagination, that teaches us those rules.[22]

Hume, therefore, says Collingwood, dropped the relation theory and tried to redevelop the introspection theory. Since in his *Treatise on Human Nature* he set himself the task of showing how all our knowledge is derived from what Berkeley called 'ideas of sense,' and what he himself called 'impressions,' he says that it is necessary to be able to distinguish ideas of sense from ideas of imagination, which he calls simply 'ideas.' How, therefore, are we to place this distinction on a

---

[20] See *ibid.*, pp. 179-180.
[21] See *ibid.*, p. 181.
[22] *Ibid.*, p. 182.

firm basis? The distinction, he maintained, must depend on a difference between the two types of experience perceptible by direct inspection. ''The difference betwixt these,'' he says at the beginning of his Treatise, ''consists in the degrees of force and liveliness with which they strike upon the mind, and make their way into our thought or consciousness.'' This is the same as Berkeley's criterion ''more strong, lively, distinct.'' The distinction between real sensation and imagination is resolved into the distinction between our inability of set purpose to control, excite, suppress or modify our sensory experiences. But, when Hume admits that ideas 'in sleep, in fever, in madness, or in any very violent emotions of the soul' conform to what he calls impressions and the definition he gives of impressions, instead of concluding that they are impressions, or that his definition of impression is faulty, he excuses himself by pleading that these are exceptional cases. Collingwood says that Hume does not see that this is really an appeal to the alternative criterion which he has already rejected, namely that of the relation in which our various experiences stand to each other, since 'exceptionalness' is something we can attribute to something only when we try to think of it as an instance of a rule, which, here, is a question of a rule determining the relations which our sensa must bear to one another if they are to be regarded as real sensa. Thus, Hume's attempt to derive all knowledge from sensation has broken down on the very first page of his Treatise.[23]

But, Collingwood maintains, progress in the science of man, as in any other science, comes through taking exceptions seriously. Kant noted that if there is any distinction between real and imaginary sensa, it cannot lie in a difference of 'force and liveliness,' that is, in the involuntary or voluntary character of the acts by which we perceive them, but must lie elsewhere.

Kant distinguished thought into a primary function, called understanding, which is concerned with the relations between sensa, and thought in its secondary function, which is concerned with the relations between thoughts in the primary function. Thought in its secondary function, for Kant, is reasoning.

According to Berkeley, the laws of nature are all empirical laws learned from experience by noting the relations between sensa. Kant explicitly attacked this position by showing that these first-order laws (relations between sensa) imply second-order laws, which Kant called 'principles of the understanding.' Now, a sensum may be wild (unruly)

---

[23] See *ibid.*, p. 185.

relatively to first-order laws of nature, so far as they have been discovered at any given moment in the history of scientific discovery. The laws, as thus far known, may not yet permit classification of such sensa in any family whose law of behaviour is known. But this, according to Kant, cannot be so in regard to second-order laws, which govern the understanding itself. It is a principle of the understanding that every event has a cause, and no event that comes under our notice can escape that principle. To call a sensation wild is simply to say that we have not yet discovered what its particular cause is.[24]

According to Collingwood, Kant's discovery of second-order laws, the principles of the understanding, involves the discovery that there are no wild sensa, and at the same time tells us why we can speak of wild sensa existing. What we are really saying is that certain sensa considered in the light of second-order laws *must admit of interpretation* but *have not yet been interpreted,* and *may remain uninterpreted until certain hitherto unknown first-order laws are discovered.* Furthermore, according to Kant, 'reality' (as when we speak of 'real' sensa) is a category of the understanding, so that a sensation said to be real or not real is a sensation interpreted by the interpretative work of the understanding.[25]

Thus, Kant, instead of trying to conceive real sensa and imaginary sensa as two coordinate species of the same genus, conceived the difference between them as a difference of degree.

> For [Kant] a real sensum can only mean one which has undergone interpretation by the understanding, which alone has the power to confer the title real; an imaginary sensum will then mean one which has not yet undergone that process.[26]

So, the common sense distinction between real and imaginary sensa, although flatly denied by the Cartesians, can be justified according to Kant, but it cannot be justified as a distinction between two classes of sensa.[27]

Now, Collingwood has already drawn attention to the ambiguity in common sense dialogue when it refers to real and imaginary sensa.[28] Sometimes it refers to both under the common genus of sensa, but other times it distinguishes sensation from imagination without the mediation of a common generic term. This second way of talking is

---

[24] See *ibid.*, pp. 186-187.
[25] See *ibid.*, p. 186.
[26] *Ibid.*, p. 187.
[27] See *ibid.*, p. 188.
[28] See this appendix, pp. 444-445.

most obvious in referring to what we call illusions or illusory sensa, which common sense abruptly distinguishes from real sensa. An illusory sensum is (not merely an imagined sensum, but) an imagined sensum *taken for a real one*. There is nothing special *in the sensa themselves* which would make them illusory; to say that they are illusory is simply to say that *a mistake has been made about them*. Pride may lead us to place the blame *on them* as if our error were due *to them* and not to our faulty thinking about them.

But so-called real sensa are likewise susceptible to such mistakes about them. A child, or a primitive, looking into a mirror for the first time, may look for the source of his sensations by placing his hand behind the mirror, whereas experience will teach him that he has to place his hand in front of the mirror to interpret them correctly.[29]

Collingwood says, therefore, that it is wrong to define illusory sensa as imaginary sensa which we mistake for real ones (according to that second way in which common sense talks). He maintains that illusory sensa can be defined without any reference to the distinction between imaginary and real. ''Any sensum [so-called 'real' or so-called 'imaginary'] is illusory in so far as we make an error about it.''[30]

Collingwood clarifies by explaining that this error does not consist in mistaking one sensum for a different sensum. Such an error is impossible because all that there can ever be in a sensum is present to the act of sensation. The mistakes we make about our sensa are mistakes about their relations with other sensa, possible or anticipated. The child or primitive before the mirror is not mistaken in thinking that he sees a pattern of colours, nor that what he sees is like what he sees when he looks at someone else's face two feet away. His mistake lies in thinking that, because of these facts, he can touch the face he sees by feeling behind the glass. Further experience will teach him that in order to touch it he must feel in front of the glass.

> An illusory sensum, then, is simply a sensum as to which we make mistakes about the relation in which it stands to other sensa. The conception of illusion disappears, resolved into the conception of error.[31]

---

[29] See R. G. Collingwood, *The Principles of Art*, pp. 188-189.

[30] *Ibid.*, p. 189. This echoes Kant's ''. . . truth or illusion is not in the object, insofar as it is intuited, but in the judgment about it, insofar as it is thought.'' *Critique of Pure Reason* (B.350); see Norman Kemp Smith's translation, London, Macmillan, 1933, p. 297.

[31] R. G. Collingwood, *The Principles of Art*, p. 190.

Collingwood then applies his conclusion to such problems as the distant man looking smaller than the nearby one, and the railway lines which 'look' convergent but 'really' are parallel, etc. He rejects the commonly accepted distinction between realities (real men, real train tracks, etc.) and appearances (of men, of train tracks, etc.), according to which the men are really the same size but appear different and the train tracks are really parallel though appearing to converge. He criticises the attitude which attributes to sensation itself an inducement or temptation to us to make a mistake about it. He insists, contrarily, that "just as no sensation can force us to make a mistake about it, so none can persuade or tempt us to do the same."[32] When we say that a distant man looks smaller, or that railway tracks converge, what we mean is that we are warning ourselves or others against the error of thinking that because the pattern of colours we now see resembles the patterns we have seen on occasions of a certain kind then the further sensa which we may expect to behave in certain ways will continue to show the same kind of resemblance.

> Thus, the phrase 'illusions of sense' or 'illusory sensa' describes cases in which actual errors are made as to the relations between sensa, so 'appearances of sense' describes cases in which care is taken that errors of this kind shall not be made.[33]

Before coming to his conclusion of this long discussion, Collingwood asks us to consider three cases in which we speak of imagining something. A match box is before me; I say that I really see the three sides facing me, but that I imagine the other three, each with its appropriate color. I also say that I imagine the inside of the box and the matches in it. I also imagine the feel of the box and the smell of its phosphorous edge. These imagined things are really there, and as Collingwood, following Kant, says, it is only inasmuch as I imagine these things to be really there (though not sensed to be there) that I am aware of the box as a solid body. If a person could only see the box without imagining, he would not see a world of solid bodies, but only (as Berkeley put it) 'various colours variously disposed.' Thus, says Kant, imagination is an 'indispensable function' for our knowledge of the world around us.[34]

The second kind of imaginings Collingwood asks us to consider is illustrated by looking out of a window and seeing a rainbow. When

---

[32] See *ibid.*, p. 190.
[33] *Ibid.*, p. 191.
[34] See *ibid.*, p. 192.

I do this, he says, I do not think of an arched and painted structure, over which men might climb and swallows build their nests, standing upon two plots of ground at its two ends. I think that I am looking at rain (although I see no drops) which is lit by sunlight and whose whiteness is broken into colors. Collingwood says that in saying this he is rejecting one interpretation and embracing another. Collingwood further says that the rainbow is 'really' there not in one sense only but in two. As a sensum, or arrangement of sensa, it is really there in the sense that I see it. And, he adds, *in that sense* of being real, the imaginary beast in the twilight is really there as also are the snakes in delirium tremens. In the second sense of being 'really' there, what is really there is the rain and the sunshine, that is, the things in terms of which I interpret my sensa.

A person suffering from a bilious attack may see a zigzag pattern of lines before his eyes. Are these really there? Collingwood answers that they are really there in the first sense of being really there. That is, they are sensa actually seen. In the second sense of being really there, no answer can be given to the question until they have been interpreted in the way the rainbow was interpreted in terms of rain and sunlight. If in seeing the rainbow we see drops of rain and in seeing the zigzag lines we see the bilious attack, we are seeing what is 'really' there, just as when a man goes red in the face we are seeing his anger which is really there. But 'really there' in this second sense is a correct interpretation according as our sensa (in the first sense of being 'real') are placed in a correct relationship among themselves and with other sensa.[35]

The third case of imagining Collingwood proposes for consideration is that of a boy who dreams that a fire destroys his home while he looks on helplessly. This, says Collingwood, is a clear case of imagination, probably complicated by illusion. When the boy awakens the illusion is dispelled, whereas the imagination (if he remembers his dream) remains. The fire is 'really' there in the first sense; he 'really sees' the sensa, the colors and shapes, etc. But is it 'really there' in the second sense? The dream has to be interpreted in order to answer this question. If it is interpreted as meaning that his house will short-

---

[35] See *ibid.*, pp. 192-193. On p. 186 of *The Principles of Art*, Collingwood, following Kant, understands reality to be a category of the understanding. But here, Collingwood distinguishes two senses of 'reality' as applicable to the question, "Are these [sensa] 'really' there?" The interpretative sense of the term is quite consistent with the Kantian meaning of 'reality,' but in what sense is 'reality' to be taken when it is used to mean that the sensa are actually seen?

ly be burned down, then we must say that the fire is unreal. Modern psychologists will connect the dream, in their interpretation, with the awakening of adolescent passions, etc. If their interpretation is right, then the fire is as real as the rainbow and the zigzag lines; it is the way that the boy sees the crisis which has come upon him.[36]

The conclusion, therefore, of this long analysis of sensation and imagination, according to Collingwood, is as follows:

> Sensa cannot be divided, by any test whatever, into real and imaginary; sensations cannot be divided into real sensations and imaginations. That experience which we call sensation is of one kind only, and is not amenable to the distinctions between real and unreal, true and false, veridical and illusory. That which is true or false is thought; and our sensa are called real and illusory in so far as we think truly or falsely about them. To think about them is to interpret them, which means stating the relations in which they stand to other sensa, actual or possible. A real sensum means a sensum correctly interpreted; an illusory sensum, one falsely interpreted. And an imaginary sensum means one which has not been interpreted at all; either because we have tried to interpret it and have failed, or because we have not tried.[37]

Collingwood adds that these latter are not three kinds of sensa, nor are they sensa which, when correctly interpreted, are related to fellow sensa in three ways. They are sensa ''in respect of which the interpretative work of thought has been done well, or done ill, or left undone.''[38] He adds that the common sense distinction between real and imaginary sensa is not false, but that it is not a distinction among sensa. It is a distinction among the various ways in which sensa may be related to the interpretative work of thought.[39]

What this amounts to, as far as it is relevant to the theory of questioning in Collingwood, is that sensa are not distinguished *as sensa*, but are distinguished according to the different interpretative attitudes taken in regard to them. An imaginary sensum, for Collingwood, is one not yet interpreted, one not yet subjected to affirmation and nega-

---

[36] R. G. Collingwood, *The Principles of Art*, pp. 193-194.

[37] *Ibid.*, p. 194.

[38] *Ibid.*

[39] See *ibid.* But, what about that second sense of real referred to above (p. 452 n. 35)? A second difficulty also suggests itself as a result of this conclusion. If imaginary sensa are uninterpreted sensa, as contrasted with real sensa understood to be correctly interpreted sensa and illusory sensa understood to be incorrectly interpreted sensa, this seems to regard imaginary sensa as something more basic than real sensa. But this appears to be quite inconsistent with the basic character of actually experienced feeling (that is, sensation accompanied by its emotive charge), for imaginary sensa would now appear to be something more primordial.

tion in thought. But nevertheless, a sensum, as imaginary, is one already set before the mind as an object about which one can ask (question) what interpretative attitudes are possible and justifiable in regard to it. Thus, the world of objectivities, which, according to Meinong, constitutes a distinctive world on its own, meriting study by a distinct branch of philosophy ignored until now, is Collingwood's world of imagination and supposal.[40] "Attending to a feeling," says Collingwood, "means holding it before the mind; rescuing it from the flux of mere sensation, and conserving it for so long as may be necessary in order that we should take note of it."[41]

These imaginary, uninterpreted sensa, if considered as looking forward towards affirmational or negational interpretation which they call for, are Bradley's floating ideas. They float inasmuch as they demand of the mind that they be attached to reality somehow or another, that is, that they be interpreted through judgment.[42] So, although not yet affirmed or denied, they are nevertheless objectivized, and thus constituted to be in a proximate condition to enter into the interpretative business of affirmation and negation.

In dialectic thought, action replaces the static concept of substance as the basic reality. Action results in polarized opposites, the most basic of which is that between subject and object. Aesthetic activity

---

[40] See "The Movement Towards Objectivity," Ch. 8 of J. Passmore's *A Hundred Years of Philosophy*, Middlesex, England, Penguin, 1968, pp. 174-200. See especially pp. 180-185 for a contextualization of Meinong's problem regarding the status of the objective order as an order distinct from that of reality. Also, see J. Dawes Hicks, "The Philosophical Researches of Meinong," *Mind*, XXXI (1922), pp. 1-30.

[41] R. G. Collingwood, *The Principles of Art*, p. 209.

[42] See Bradley's "On Floating Ideas and the Imaginary," *Mind*, XV (1906), pp. 445-472, especially p. 446 and pp. 449-450; and see pp. 451-453 for Bradley's application of his theory of floating ideas and the imaginary to questions and suppositions, which, no doubt, influenced Collingwood's theory of imagination, supposal and questioning. See also G. F. Stout, "Truth and Falsity," *Mind*, XLI (1932), for a discussion of objectivity as abstracting from truth and falsity. Stout says (p. 299): "By a proposition I mean what Meinong calls 'an objective.' It is best represented in language by such phrases as 'that a is b' or 'a being b'. It is a factor common to judgments, supposals and questions. I may affirm or deny that a is b. I may inquire whether a is b; I may merely suppose a to be b. The object in such processes, considered as something proposed or capable of being proposed to a thinking mind, is a proposition. As I maintain, propositions are just possibilities considered as proposed to a thinking mind [compare Collingwood's notion of possibility as explained in *An Essay on Metaphysics*, pp. 274-276]. A proposition is the intentional term corresponding to the 'formal' term possibility." If I understand Collingwood rightly, I think he would say that an imaginary is the term of an intentional act constitutive of the objectivity of a sensum, whereby the sensum is in such a condition as *to be able to* enter into the constitution of propositions; and, rather than being capable of being proposed, is, more fundamentally, *made* an object (of imagination), which is the very *foundation of the possibility* of propositional capacity.

is *par excellence* the work of objectivization, the activity whereby the subject opposes himself to himself as subject to object.[43]

> [The] immediate origin [of the idealist conception of the absolute] was the relation between subject and object in knowledge. The opposition between subject and object, in turn, had its roots in the Cartesian dualism of mind and matter, and the group of ideas of which it formed a part. This dualism reappears in idealism as the difference between nature and spirit; and is formulated as a contrast between a dead, material substance on the one hand, and a free, creative subject on the other. In fact, the idealists often identify spirit with freedom, which they interpret as a spontaneous act of creation. This notion is best illustrated by art, because in art the self is most clearly creative. The artist objectifies his ideas in a medium. He puts himself into his works so completely that they could not have been made by anyone else. They are different from their author, yet they are himself in another form, for in them he expresses, or better, actualizes himself. Their full meaning is lost if they are taken in abstraction from him; in fact, to suggest this possibility is to imply their imperfection, by an external relationship between artist and work. Ideally he, as artist, can be adequately seen in them. The difference is that the work is finished, a created thing, while the artist is creating; it is the difference between being and becoming.[44]

The essential dialectical relationship between artistic activity and the subject-object dichotomy is further explained by Otis Lee:

> Dialectic considers the process of objectification, negation. It is variously described: creation, objectification, actualization, self-expression. But these are all ways of stating the same idea. It is the activity by which spirit objectifies itself, not an operation performed on a pre-existent substance. To continue our analogy [with art], we must suppose that the artist creates his own medium. Also, the act is double, for it both results in an object which is other than spirit, and relates this to spirit. The artist paints a picture, which is outside himself: he may sell it, or give it to a friend, and forget about it. This is the "first negation." Yet at the same time, the work is himself; it does not really have an independent being, although it may be elsewhere in space and time. This is the second negation, the "negation of the negation." The two phases are not separate, nor do they succeed one another in time. To become other is to become more completely oneself. Thus objectification is a process of differentiation and at the same time of integration; it is both exclu-

---

[43] See Otis Lee, "Dialectic and Negation," *The Review of Metaphysics*, I (1947), pp. 3-7. See also Ernst Cassirer, *The Logic of the Humanities*, p. 115.

[44] Otis Lee, "Dialectic and Negation," *The Review of Metaphysics*, I (1947), pp. 6-7.

sion and inclusion, antithesis and synthesis; the first can not occur without bringing the second along with it.[45]

Thus, the artistic dimension in conscious activity is the activity whereby objectivity is seen to arise from the creative, self-assertive activity of the conscious subject.[46] Now questioning, as Collingwood has repeatedly stated, is primarily a dialogue of the soul with itself, of the soul *as subject* with itself *as object*. The first and fundamental condition of the very possibility of a questioning situation is here posited by the aesthetic activity of the conscious subject. Collingwood says that we can speak to others only because we first have the experience of being ourselves listeners to ourselves as speakers: ". . . the speaker is his own first hearer."[47]

C. Relation Between Imagination and Consciousness

Collingwood next investigates the contrast of imagination with sensation as something active contrasted with something passive, something we do as contrasted with something we undergo, something under our control compared to something we cannot help, a making compared to a receiving.[48] Common sense makes this distinction, but is unable to give an account of it.[49]

This distinction cannot be one between activity and passivity as such, since sensation itself is an activity. Even if we are stimulated to do it by outside forces, it still remains something to do. Response to a stimulus is passive in the sense that it cannot arise without a stimulus, but it is active inasmuch as it is a response. It is not a distinction between activities we choose to do and those we cannot help doing; it is easier to stop seeing this paper (by shutting one's eyes) than it is to stop imagining the frightful accident seen yesterday.[50]

But in spite of this, in some sense or another, imagination is more free than sensation. Sensation is not entirely unfree, arising as it does from the spontaneity of a living and sentient organism. The freedom of the imagination goes a step further than this, although it is not the conscious freedom of carrying out an intention, which is freedom of choice.

---

[45] *Ibid.*, p. 7.

[46] See Ernst Cassirer, *The Logic of the Humanities*, pp. 114-115.

[47] R. G. Collingwood, *The Principles of Art*, p. 249. See *ibid.*, pp. 247-252, for discussion of the distinction between speaker and listener. But see also Chapter 9, above, p. 400, for comment on this.

[48] R. G. Collingwood, *The Principles of Art*, p. 195.

[49] See *ibid.*, p. 196.

[50] See *ibid.*, p. 197.

> Regarded as an activity or manifestation of freedom, then, imagination seems to occupy a place intermediate between the less free activity of mere feeling and the more free activity of what is generally called thought.[51]

The task, then, says Collingwood, is to define this intermediate place, between the minimal freedom of spontaneous response characteristic of sensation and the full conscious freedom of choice.

Two things which must be distinguished are confused, according to Collingwood, by modern philosophers when they talk about sensation and sensa. When they are talking about the sensa that we should or would perceive given certain circumstances, or those perceived in the past or expected to be perceived in the future, they are talking about something vastly different from the sensa we actually see or feel, and the sensations we are actually having.[52] Actually experienced sensation must be regarded as a flux of activity in which a greater or lesser number of distinct sensory acts are going on together at one time, and each is no sooner achieved than it gives place to another. What is thus sensed can only be present in the performance of the corresponding act.[53] Its *esse* is *sentiri*. But the sensa and sensations referred to above as spoken of by modern philosophers are not present, and not presently available to be spoken of in relation to one another and in relation to past or future sensations. Unless there is a distinction between these two kinds of things, the statements made about relations between sensa, quite apart from whether they are true or not, could not be made; they would be meaningless.

> There must, in other words, be a form of experience other than sensation, but closely related to it; so closely as to be easily mistaken for it, but different in that colours, sounds, and so on which in this experience we 'perceive' are retained in some way or other before the mind, anticipated, recalled, although these same colours and sounds, in their capacity as sensa, have ceased to be seen and heard.
>
> This other form of experience is what we ordinarily call imagination.[54]

Collingwood says that it was in order to distinguish this other form of experience, imagination, from sensation, that Hume distinguished ideas from impressions. What modern philosophers call relations between impressions are, according to Collingwood, really relations

---

[51] *Ibid.*, pp. 197-198.
[52] See *ibid.*, p. 202.
[53] See *ibid.*, p. 198.
[54] *Ibid.*, p. 202.

between ideas.[55] It is not, as Locke says, sensation which furnishes the empty room of the mind, but imagination.[56] It is Hume's ideas which inhabit Locke's empty room, which becomes furnished with what 'the busy and boundless Fancy of man provides.' And it is, according to Collingwood, imagination to which Empiricists really appeal when they appeal to 'experience.'[57]

It is then the work of thought to detect relations between sensa, to find similarity between this patch of colour and others, so that it can be called 'red.' But the detection of such resemblances, or of any other relationships, prerequires that each sensum must be recognized as distinct in itself. "This act of appreciating something, just as it stands, before I begin to classify it, is what we call attending to it."[58] This act of attention must be carefully distinguished from an act of sensation.

> Looking is different from seeing and listening from hearing. Seeing and hearing are species of sensation; looking and listening are the corresponding species of attention.[59]

Collingwood says that current tradition speaks of 'a red patch' as a sensum, but, he continues, what we see inasmuch as we merely see is always a visual field, more or less parti-coloured, never a red patch. A red patch is a well-defined region cut out of this field, so that to describe it as a patch implies that the field is divided into an object of attention and a background or penumbra from which attention is withdrawn. This division of the general field of feeling into foreground and background by conscious attention is the division into the conscious and the (more or less) unconscious.[60] Such a distinction does not exist at the merely psychical level, where the mind exists only in the shape of sentience.[61]

When the light of consciousness falls on sentience, what was sentient becomes imagination. Consciousness can tell us clearly only of

---

[55] See *ibid.*, p. 203.

[56] See *ibid.*, p. 200.

[57] See *ibid.*, p. 203. No doubt Bernard Bosanquet influenced this conclusion: "The object which thought in the true sense has worked upon is not a relic of decaying sense, but is a living world, analogous to a perception of the beautiful, in which every thought-determination adds fresh point and deeper bearing to every element of the whole." See Bosanquet's *The Principle of Individuality and Value*, p. 58.

[58] R. G. Collingwood, *The Principles of Art*, p. 203. See also *The New Leviathan*, p. 35, pars. 5.63, 5.65. See also Alan Donagan, *The Later Philosophy of R. G. Collingwood*, pp. 39-40.

[59] R. G. Collingwood, *The Principles of Art*, pp. 203-204.

[60] See *ibid.*, pp. 204-205.

[61] See *ibid.*, p. 205.

the things to which it attends and obscurely of those it ignores (which remain more or less unconscious). The so-called unconscious of modern psychology is not an area distinct from and independent of consciousness, but it is that which consciousness ignores in sentience by not attending to it.

Attention has a double object, whereas sentience has one only. What we attend to is not only what we sense but the act of sensing it. Sight sees only colour, but conscious attention attends both to the colour seen and to the act of seeing it. The word 'consciousness' signifies the togetherness of sensation and sensum in the object. A man conscious of his anger is aware that the anger he has is his.[62]

> Thus, the difference between seeing and looking, or hearing and listening, is that a person who is said to be looking is described as aware of his own seeing as well as of the thing he sees.[63]

When consciousness enters into experience a new principle is established. Mere presence to sense does not claim attention; it may solicit it but does not secure it. Consciousness is master in its own house and dominates feeling. Feeling thus dominated, as compelled to accept the place accorded to it by consciousness, is no longer impression but idea (in Hume's sense). Consciousness is absolutely autonomous. Its decision alone determines whether a given sensum or emotion will be attended to or not. A conscious being is not free to decide what sensations he will *have*, but he is free to decide what feeling he will *place in the focus of his consciousness*. He is not free as to whether he will decide or not; he must decide. But he is free to decide to attend to this or that feeling. And that choice, says Collingwood, is consciousness itself.[64]

This freedom of consciousness, Collingwood explains, is not a freedom of choice between alternatives; the decision is not made by reviewing the various sensations and opting for this or that one. In fact, says Collingwood, this freedom of consciousness (as distinct from freedom of choice) is a prerequired condition of the possibility of alternatives for there to be choice about. Freedom of choice is enjoyed only at the intellectual level. The freedom of mere consciousness is thus an

---

[62] But in *The New Leviathan*, Collingwood says that "in anger you have no consciousness of being angry; that comes only with reflection upon anger; what you are aware of is simply a contrast between yourself and something . . . other than yourself" (pp. 70-71, par. 10.43). See Alan Donagan, *The Later Philosophy of R. G. Collingwood*, p. 48, on Collingwood's revised conception of first order consciousness.

[63] R. G. Collingwood, *The Principles of Art*, p. 206.

[64] See *ibid.*, p. 207.

elementary kind of freedom, yet it is very real. At the level of sentience, of psychical experience, the self is dominated by its own feelings. This dominance, says Collingwood, is what Berkeley and Hume refer to as 'force' or 'liveliness.' A child screams when it feels pain; a child become conscious of its feelings screams because it wants attention. The consciousness of the self as something other than the feeling of the moment, something to which the feeling belongs, is the assertion of the self as able in principle to dominate the feeling. The self-consciousness and the self-assertion, the theoretical and the practical, form together a single indivisible experience.[65]

An effect of this is that the feelings are tempered in their strength, and thereby they more and more lose their power of determining our actions, although their quality or intensity is not thereby altered. They become domesticated and less like storms or earthquakes. They are structurally fitted into the fabric of our life instead of going their own way. They become fitted into a structure.

> [I]n asserting ourselves as against our feelings we have asserted in principle a structure of some kind, though as yet an indeterminate one. In becoming aware of myself I do not yet know at all what I am; but I do know that I am something to which this feeling belongs, not something belonging to it.[66]

Another result of this domestication of feelings is that we become able to perpetuate them, including sensa, at will.

> Attending to a feeling means holding it before the mind; rescuing it from the flux of mere sensation, and conserving it for so long as may be necessary in order that we should take note of it. This, again, means perpetuating the act by which we feel it.[67]

In the flux of sensation one pattern of the total sensory field is replaced by another. Attention focuses itself on one element of that field. By attending, say, to the scarlet and neglecting everything else, a kind of compensation for the fading of the sensation in pure flux is effected.

---

[65] See *ibid.*, p. 209. Owing to the distinction which Collingwood makes later in *The New Leviathan*, as referred to by Alan Donagan (see above, p. 459 n. 62), we have to say that consciousness has only implicit awareness of self, an awareness in which the self is experienced but not yet objectified by second order consciousness. That is, by conscious awareness, the self asserts itself in the presence of its feelings, but reflection on this assertive activity is needed in order that the self be distinguished *objectively* from the non-self.

[66] R. G. Collingwood, *The Principles of Art*, p. 209.

[67] *Ibid.*

> By . . . adjusting our attention . . . we obtain a new kind of experience by moving as it were with the flux, so that the self and the object are (so to speak) at rest relatively to each other for an appreciable time. . . . We have [thereby] liberated ourselves for a moment from the flux of sensation and kept something before us long enough to get a fair sight of it.[68]

By doing this Collingwood says we have converted a sensation into an idea. We have become conscious of ourselves as its masters, and broken its mastery over us. "We have told it to stay still, and it has stayed, though only for a moment."[69]

So long as any trace of a past feeling remains, attention is able to single it out and reconstitute the original feeling into an idea. What we call remembering an emotion is never anything but this focusing of our attention on the traces it has left in our present feeling. This, says Collingwood, is the meaning which we can attach to Hume's formula, that all ideas are derived from impressions.

Collingwood draws attention to the fact that the analysis so far yields two concepts of the imagination, both arrived at by an analysis of Hume's distinction between impression and idea.

First, the distinction between impressions and ideas was understood to be equivalent to the distinction between sensa interpreted by thought and sensa not so interpreted. Next, the analysis revealed that this same (Humean) distinction between impressions and ideas is equivalent to the distinction between sheer feeling and feeling modified by consciousness dominating and perpetuating it.

But there seems to be a double discrepancy between these two explanations stemming from the same source (the Humean distinction). First, a discrepancy between impression as real sensum (that is, sensum interpreted by thought) and impression as sheer feeling;[70] second, a discrepancy between idea as feeling not interpreted by thought and idea as feeling perpetuated and dominated by consciousness. Collingwood proposes to show that these discrepancies arise not from his own faulty analysis but from Hume's confusing of two quite different things under the one same pair of terms.[71]

Considering the second discrepancy first, Collingwood says that the work of determining the relations between things must depend

---

[68] *Ibid.*, p. 210.

[69] *Ibid.* A truly objective world is thereby constituted, a world set over against the conscious subject, which can be looked at 'from a distance,' so to speak.

[70] See *ibid.*, p. 212.

[71] *Ibid.*

on something prior, namely, that those things whose relations are to be determined must be able to be held before the mind in such a way as to make it possible to compare one with another, in order to see how they resemble one another, and so forth. Each must be known as to *what it is in itself* before being known as related.[72] Our knowledge that this is a patch of red goes far beyond knowing a patch of red as it is in itself; it is a consideration of it in relation to an established system of colours with established names. Our knowledge of what a red patch is in itself, if it could be verbalized, would be expressed something like 'This is what I see,' or 'This is how I feel.' This is the kind of thing that Collingwood maintains we must be able to say before we can begin the work of interpreting, that is, discussing relations. It is not through bare sensation, however, that we are able to do this, but through the consciousness of sensation, by its work of attention, selecting and perpetuating some element in the field of sensation and some corresponding element in the sensory act.

Collingwood reconciles the two conflicting accounts (of impressions and ideas) by recognizing that a feeling of which we have become conscious, by attending to it, is one not yet interpreted but nevertheless ready for interpretation; one which we have only begun to interpret. Conversely, an uninterpreted feeling, if it means a feeling which is ready for interpretation, can only be a feeling of which we have become conscious through attention.

As regards the first discrepancy, between imagination as real sensum (interpreted by thought) and impression as sheer feeling, the situation is different. Collingwood points out that three stages in the life of feeling have so far been distinguished: (i) feeling as bare feeling below the level of consciousness, (ii) feeling of which we have become conscious by the process of attending to it, and (iii) feeling not only brought to consciousness by attention but also placed in relation with others. The second stage of this enumeration is what, according to Collingwood, Hume means by idea. But Hume failed to distinguish impression in sense (i) from impression in sense (iii). An impression, for Hume, is distinguished from an idea only by its force or liveliness. But this force may be of two kinds: either the brute force of pure sensation not yet dominated by thought, or the solid strength of a

[72] This knowing of *what it is in itself*, previously to setting up any relational knowledge, is what I understand by saying that it is first constituted in its objectivity. That is, interpretation presupposes objectivization. But if this is to be called knowledge, then we have admitted the existence of something entitled to be called knowledge which is not the answer to a question.

sensum firmly placed in its context by the interpretative work of thought. Hume's failure to recognize this difference provided a pitfall for subsequent philosophers, especially in regard to the term 'sense datum.' The word 'sense datum' or 'sensum' is applied promiscuously not only to something given by sensation (in which case it would at once be taken away again: sense [i]), and not only to something perpetuated by consciousness or imagination (in which case the only region from which it could be called up would be that of past sensation: sense [ii]), but also to something constructed inferentially by the work of intellect (sense [iii]).

Impressions are converted into ideas, that is, crude sensations into imaginations, by the activity of consciousness. Consciousness and imagination are synonyms if taken as names for a certain kind or level of experience. But within that same experience imagination is the name for what has undergone conversion, and consciousness is the name for what effects that conversion (of sensa into ideas). Imagination is thus the new form which feeling takes when it is transformed by the (attending) activity of consciousness. Imagination is a distinct level of experience, intermediate between sensation and intellect, and the point at which the life of thought makes contact with the purely psychical experience, the world of pure feeling. So, it is not sensa as such that provide the intellect with data; it is sensa transformed into ideas of imagination by the attending work of consciousness.[73]

D. General Theory of the Imagination by Way of a Summary of the Foregoing

The discussion began with the common sense two-term distinction between feeling and thought as two distinct features of experience.[74] We now seem to have a three-term distinction, with consciousness assuming an intermediate level of experience connecting the two. But, Collingwood insists, consciousness is not something other than thought; it is a level of thought which is not yet intellect. The two-term distinction includes consciousness inasmuch as thought in that expression is taken in the wide sense, as including consciousness, and not in its strict sense, which is intellection. The work of intellect, that is, of thought par excellence, is to construct relations. But this takes two forms. Intellect in its primary function apprehends relations, not between feelings as such, as experienced psychically, but between

---

[73] R. G. Collingwood, *The Principles of Art,* p. 215.
[74] See this appendix, p. 444 above.

feelings modified by consciousness and so converted into ideas. Intellect in its secondary function apprehends relations between the acts of primary intellection, or between what we think in such acts.[75]

Consciousness, says Collingwood, is that activity without which we would have no terms between which intellect in its primary form could detect or construct relations. Thus, consciousness is thought in its absolutely fundamental or original form.[76]

> All thought presupposes feeling; and all the propositions which express the results of our thought belong to one of two types: they are either statements about feelings, in which case they are called empirical, or statements about the procedure of thought itself, in which case they are called 'a priori'. 'Thought', here, means intellect; 'feeling' means not feeling proper, but imagination.[77]

By the act of attention, or conscious awareness, which is activity distinct from mere feeling, the self asserts itself as owner of its feelings, dominating them, so that they are no longer experiences forcing themselves upon us unawares, but are experiences in which we experience our own activity. Their brute power over us has been replaced by our power of conscious control over them. From being impressions of sense they are transformed into ideas of imagination. They are still feelings of the same kind as before, but have ceased to be mere sensations and have become what we call imaginations.

The important point of this discussion is that imagination has affinity with two opposite features of experience. It shares in the characteristics of thought and also in the characteristics of feeling.[78] From one point of view imagination does not differ from sensation. What we sense and what we imagine are the same kinds of things (colors, scents, etc.). This explains why philosophers have experienced difficulties in distinguishing them, and why common sense, according to one way of speaking, calls both (sensations and imaginings) by the same name, sensation, and their respective objects, sensa. But from another point of view there is a vast difference between imagining and sensing, which is indicated by the fact that common sense has a second way of talking about them, in which imagination is sharply distinguished from sensation, with no generic term to cover both. What we imagine is tamed or domesticated, and what tames it is consciousness, which is a kind of thought, the kind of thought which stands

---

[75] R. G. Collingwood, *The Principles of Art*, p. 216.
[76] See *ibid*.
[77] *Ibid*., p. 221.
[78] See *ibid*., p. 223.

closest to sensation or to mere feeling. Every further development of thought presupposes this kind of thought and is based upon it as upon a foundation; it deals not with feeling in its raw form, but with feeling as transformed into imagination.[79]

Consciousness itself does not consider the likenesses and differences between feeling, or classify them or group them into other arrangements, such as a time series, etc., but only prepares the ground for such systematizing work of thought. In itself, consciousness does no more than attend to the here-and-now of feeling. In attending to a present feeling, it perpetuates it at the cost of turning it into something new, no longer merely crude feeling or impression, but domesticated feeling, or imagination, or idea.[80]

Consciousness does not compare one idea with another. While enjoying one idea, another may be summoned up. The new idea is not held alongside the first, as two distinct experiences between which I can detect relations. Rather the two ideas fuse into one, the new idea presenting itself as a colouring or modification of the old.

> Thus, imagination resembles feeling in this, that its object is never a plurality of terms with relations between them, but a single indivisible unity: a sheer here and now.[81]

Conceptions of past, future, the possible, the hypothetical are meaningless for imagination as for feeling. These conceptions belong to a further development of thought.[82]

"To form an idea of feeling is already to feel it in imagination," so that "imagination is blind, that is, it cannot anticipate its own results by preconceiving them as purpose in advance of executing them."[83] Its freedom is not freedom of choice to carry out a preconceived plan or to choose between alternative possible plans, which freedom belongs to a later stage of thought. Its freedom is freedom from submission to the sheer flux of feeling, whereby these can be, so to speak, 'distanced' from the self and looked at, or objectivized.

---

[79] See *ibid.*, pp. 222-223.

[80] See *ibid.*, p. 223.

[81] *Ibid.*

[82] See *ibid.*, p. 224. Nevertheless, these conceptions *suppose* imagination. Imagination, as ordered to the past, the future, the possible, and, especially, the hypothetical, is *supposal*, as looking beyond itself towards the assertion of reality, questioning activity mediating. So, although these terms are meaningless *for imagination*, imagination is not meaningless *for them*, for they *suppose* imagination as a necessary precondition.

[83] *Ibid.*

Imagination is, therefore, for Collingwood, consciousness of feeling, or, conversely, feeling become conscious. From one point of view, it is thought in affinity or continuity with the psychic, with feeling. From another point of view, it is feeling as thought. It mediates between the two levels of experience, namely thought and feeling, by partaking of both (in Kant's way of saying it, it is homogeneous with the categories on the one hand and with sensation on the other). So, the identification by philosophers sometimes of imagination with thought (Leibnitz) and sometimes with feeling (Spinoza) has a foundation in experience itself. Sensations imagined and sensations 'sensed' or felt do not differ *as sensations* (as the distinction of sensations into real and imaginary sensations implies), but they differ as to whether they have been consciously attended to or not. Brute, merely felt, feelings, such as they are at the psychic level of experience, impose themselves forcibly upon us, and we experience a certain helplessness with respect to them. But feelings imagined, or feelings attended to by being consciously aware of them, are, so to speak, put at a distance with a definite control over them on our part. But they are the same feelings in both cases.

Sensations at the brute level of feeling are uninterpreted, and indeed are not in a condition of being able to be interpreted by thought. Sensations as imagined are indeed not interpreted sensations (that is, ordered with other past, present and future sensations into a systematic unity), but are now proximately ready for the interpretative work of thought operating at a higher level of consciousness, namely that of the understanding. What are usually referred to as 'sense data' are really sensations, not merely attended to and stabilized, but also interpreted by the understanding, by being put in relational order with other sensations from which they are distinguished. They are, as such, not something absolutely 'given' to the mind, as if the mind were a mere passive recipient, but they are rather something the mind gives to itself in its systematizing function, and it gives them to itself from the storehouse of the imagination, not from brute sensa.

Imagination is therefore both 'consciousness' and 'idea' (in the Humean sense). As 'idea' it refers to feelings, but feelings now transformed, by attention, into objects of consciousness. As consciousness it is thought, not intellectual thought, but thought preliminary to intellection and ready for intellection.

What is important in this discussion, as far as Collingwood's theory of questioning is concerned, is that, just as brute sensation as such lacks objectivity, lacks that characteristic of being set over against a subject as its object, but only becomes objectivized as converted into an idea of imagination, so, correspondingly, awareness would not be a conscious awareness if that of which it was the awareness were not 'set before it' as something confronting it with which it has to deal. Thus, just as imagination renders sensa available as objects, so also it constitutes imaginative awareness to be the awareness of a conscious subject set before and over against the object of that awareness. That is, imagination first gives rise to the fundamental distinction and relationship of subject and object. It distinguishes them in the very same activity in which it unites them. In other words, the bipolarity characteristic of thought is first constituted by imagination; that is, imagination sets up both the object *as object*, and the subject *as subject*.

E. Imagination and Expression

To relate further the theory of imagination to that of questioning, it is necessary to understand that consciousness becomes, or makes itself aware of, or objectivizes, its feelings by expressing them to itself. Language comes into existence with imagination as a feature of the conscious level of experience.[84] Original or native language is imaginative or expressive. Calling it imaginative describes what it is; calling it expressive describes what it does. Language is an imaginative activity whose function is to express emotion.[85]

Below the conscious level of linguistic expression there is psychic expression, the involuntary cry of pain, the scream of fear, which rather betrays feeling than expresses it.[86] Every kind and shade of emotion which occurs at the purely psychical level of experience has its counterpart in some change of muscular or circulatory or glandular system which, in a sense, 'expresses' that emotion.[87] To express an emotion is one thing, but to observe and interpret it is another, which belongs to acts of higher levels of consciousness reflecting on acts of lower levels. Psychical expression is the only expression of which the psychical (basic feeling-level of) emotions are capable, although levels of consciousness other than the psychic may admit also of psychic

---

[84] See R. G. Collingwood, *The Principles of Art*, p. 225.

[85] See *ibid*. This identification of art, as expression, with language derives from Croce. A. Gennaro, in his "Croce and Collingwood," *The Personalist*, XLVI (1965), p. 197, writes: "Having established that art is spirituality, Croce deduces that art is language. 'If Aesthetics were a manifestation different from Linguistics, it could not have as its object the expression which is an aesthetic fact.' "

[86] R. G. Collingwood, *The Principles of Art*, pp. 228-229.

[87] See *ibid*., p. 230.

expression. Emotions such as hatred, love, anger and shame arise only through consciousness of self, and they thus presuppose self-awareness.[88] Unlike purely psychical emotions, these emotions accompanying acts of self-awareness, of consciousness, admit of expression in language, in a phrase or controlled gesture or suchlike.[89] But in addition to this linguistic mode of expression (characterized by conscious control), they have their own special psychical expressions, the blush of shame accompanied by muscular reaction, the blush of anger with muscular tension and rigidity. A psychical emotion is an emotional charge on a sensum, but an emotion of consciousness is an emotional charge, not on a sensum, but on a certain mode of consciousness.

Emotions can be expressed in two different ways because of the relation between the level of expression and the next above it. The lower level achieves a new principle of organization when, through the reflection of consciousness, it is transformed to a higher level.[90] Emotions of consciousness can be expressed either formally as modes of consciousness or materially as constellations of psychic elements.[91]

Psychical expression is uncontrollable, characterized by brute givenness. A change occurs at the level of consciousness. Consciousness transforms the emotion from the condition of brute givenness to that of being dominated by thought, so that we now do not just simply feel but feel in a new way, a way we call imagining. This change likewise affects the bodily act of expression, which is raised above the crude psychical level to the imaginative.[92]

> Bodily actions expressing certain emotions, insofar as they come under our control and are conceived by us, in our awareness of controlling them, as our way of expressing them, are language.[93]

In its wide sense language is thus simply the bodily expression of emotion as this is dominated by thought in its primitive form as consciousness. A child soon learns to distinguish the cry expressive of uncontrolled emotion from the self-conscious cry deliberately uttered to call attention to its needs. The second cry is language though not yet speech. It bears a new relation to the child's experience as a whole. It is the cry of a child aware of itself and asserting itself.[94] Thus, imag-

[88] *Ibid.*, p. 231.
[89] *Ibid.*, p. 232.
[90] *Ibid.*, p. 233.
[91] See *ibid.*, p. 234.
[92] See *ibid.*, p. 235.
[93] *Ibid.*
[94] See *ibid.*, p. 236.

inative experience creates for itself, by an infinite work of refraction, reflection, condensation and dispersal, an infinity of emotions demanding expression in an infinite subtlety of articulations in the language created in expressing them. An emotion cannot be felt at any level of consciousness without being expressed; there are no unexpressed emotions.[95] So-called unexpressed emotions are emotions felt and expressed at their appropriate level of which the one who feels them is trying to become conscious of them, that is, trying to convert them into the material of a higher level of consciousness which, when achieved, will be an emotion at this higher level, with an appropriate expression of it.[96]

Every kind of language (speech, gesture, etc.) is an offshoot from, and transformation at a higher level of, an original language of total bodily gesture. The language of total bodily gesture is thus the motor side of our total imaginative experience.[97] Language in its original nature expresses, not thought in the narrower sense of intellectual thought, but only emotions; not crude impressions of brute sensations, but sensations transmuted into ideas by the activity of consciousness. Language at the intellectual level undergoes modification to express intellectual emotions.[98]

The general distinction between imagination and intellect is that, whereas imagination presents to itself an object which it experiences as one and indivisible, intellect goes beyond that single object and presents to itself a world of many such objects with determinate kinds of relations between them.[99] However complex an imagined object may be, it is imagined as a single whole in which *relations between the parts* are *present as a quality of the whole*. Thus, a May thrush singing a May song is imagined alongside the previous imagination of a January thrush singing a January song. Insofar as the experience remains *at the imaginative level*, distinct from that of the intellect, the two songs are not *imagined separately* as two things with a relation between them. Rather, the January song coalesces with the May song, toning it with a quality of mature mellowness. In an *imagined whole* the parts do not present themselves *distinctly as parts* but merely as *qualitative modifications of the whole*. But if the experience is thought about, by strictly intellectual thinking, it is analysed into its parts, becoming a manifold,

---

[95] See *ibid.*, p. 238.
[96] See *ibid.*, p. 239.
[97] See *ibid.*, p. 246.
[98] See *ibid.*, p. 252.
[99] See *ibid.*

a network of things with relations between them. Different notes are distinguished in a definite way; each note subject to being thought about itself (one note might have been higher or lower, etc.). The difference between the two songs can be described; one has a sweeter tone, or is longer, or contains more notes than the other. This is analytic thought.

Every different kind of activity has its corresponding kind of emotion, which has its corresponding kind of expression. The emotional charges on sense experience, as broadly distinguished from thought, are expressed psychically by automatic reactions, being felt at the purely psychic level. The emotional charges on thought-experiences are expressed by the controlled activity of language. The emotions of consciousness, taking consciousness as a level of thought lower than intellect, are expressed by language in its primitive and original form. The emotions proper to intellectual experience have their appropriate expression in language in its intellectualized form. The excitement which drove Archimedes from his bath was not a mere generalized excitement, but specifically the excitement of a man who has just solved an intellectual problem.[100] In the expression of intellectualized emotion by intellectualized language, the thought is also expressed in the same linguistic expression. Symbolism is intellectualized language: language because it is expression of emotion, intellectualized because it is adapted to the expression of intellectual emotions. Original, imaginative language is expressive but lacks meaning, not distinguishing the speaker's meaning from what he says. Intellectualized language both expresses and expresses meaning. As language it is emotively expressive; as symbolic, it refers beyond that emotion to the thought whose emotional charge it is. Thus we commonly distinguish what we say from what we mean. What we say is what we immediately express: the eager, or reluctant, or triumphant, or regretful. The emotions, together with the sounds and gestures that express them, are inseparable parts of a single experience. What we mean is the intellectual activity upon which these emotions are the emotive charge, which is, so to speak, pointed to by the words expressing the emotions as a sort of finger-post indicating from which direction they came and in which direction the other person must go if he is to understand what we say, that is, if he is to reconstruct for himself and in himself the intellectual experience which has led us to say what we did.[101]

---

[100] See *ibid.*, pp. 266-267.

[101] *Ibid.*, p. 269. It is interesting to note, in regard to researches on learning dis-

Theoretically, the artist is a person who comes to know himself, his emotions, which is also to come to know his world, that is, the sights and sounds, etc., which go together to make up his total imaginative experience. Knowing himself and his world is one knowledge only, since the sights and sounds are to him steeped in the emotion by which he contemplates them; they are the language in which that emotion utters itself to his consciousness. His world is one with his language; what it says to him it says about him; his imaginative vision of it is his self-knowledge.

But this knowledge of himself is together a making of himself, a conversion of himself from mere psyche, from mere feeling, into consciousness. Coming to know his emotions is, at the same time, coming to dominate them, to assert himself as their master. This is not yet the life of morality but is an indispensable step towards that life. He has learnt to acquire by his own efforts a new set of mental endowments which must be accomplished first if later he is to acquire by his own effort those mental endowments whose possession will bring him nearer his moral ideal.

The world he has together come to know and to make is a linguis-

---

abilities in children conducted at the Montreal General Hospital, as presented by a C.B.C. telecast, that children who, without any apparent physical or psychological handicaps, but simply seemed unable to learn, were basically unable to distinguish one part of their body from another (e.g., neck from knee), and were thus unable to piece together the drawing of a human figure cut up into jigsaw puzzle pieces. They remained unable to distinguish such parts *until they were able to name each bodily part distinctly*. Once the distinctions were made by being *able to express the appropriate names* ('neck,' 'knee,' etc.) the learning process proceeded more or less normally. Ernst Cassirer, *The Logic of the Humanities*, Yale University Press, 1960, pp. 59-60, says that ". . . the language function . . . is not a mere result but a continuous self-renewing process: and the contour of man's 'world' defines itself more clearly and determinately in direct proportion to the development of this process. Thus a name is not simply tacked on to its completed and ready-to-be-used objective intuition as an *extraneous symbolizing* of knowledge; instead, in it there is expressed a determinate way, a manner, and tendency for *coming to know*. Everything we know concerning the development of language in children gives factual substantiation to this basic view. For it is manifestly not the case that in this development a fixed stage of already acquired objective intuition is strung on to a subsequent stage, in which this new acquisition is now also named, defined, and grasped by words. Quite the contrary, language awareness—the awakening symbol consciousness—impresses *its* stamp upon observation and perception in ever-increasing measure as it grows in strength and extends and clarifies itself. Both observation and perception become 'objective' just to the degree that this linguistic energy succeeds in clarifying, differentiating, and organizing the mute, undifferentiated chaos of particular circumstances. Linguistic symbolism opens up an original phase of spiritual and intellectual life. A life of 'meanings' supplants the life of mere impulses. . . ." Compare this with Collingwood's discussion of the child throwing its bonnet off its head with the exclamation 'Hattiaw'; see his *The Principles of Art*, pp. 227-228 and pp. 239-241.

tic world in which everything is an expression of emotion. To the extent to which this world is expressive, or meaningful, it was he who made it so, out of what was presented to him in the still more elementary stage of the purely psychical experience of colours, sounds, etc.

Looking at the aesthetical experience from a higher level of reflection, distinction is made between the theoretical and the practical. It is seen to embody both kinds of activity. It is both *a knowing* of one's self and one's world and *a making* of one's self and one's world. The self consists only of emotions expressed in that language which is its world, and the world is the language expressive of those emotions. The self which was psychic was remade in the shape of consciousness, and the world, previously crude sensa, was remade in the shape of language, or sensa converted into imagery and charged with emotional significance. Thus, the step forward in the development of experience that leads from the psychic level to the level of consciousness, which step is the specific achievement of art, is a step forward both in theory and in practice, though it is one step only, not two.[102]

The essence of art is that it is activity whereby we become conscious of our own emotions. As such it has nothing that is due to intellect. The proper sphere of art is, therefore, that of making conscious the previously unconscious emotions of the psychic level. On the other hand, if art is taken not merely as art but as a certain kind of art, it may express emotions of another order, such as intellectual emotions, or emotions that only an intellectual being can feel. These emotions are emotional charges, not merely upon psychical experience of bare feeling, nor upon conscious experience, but upon thought at the level of intellectual experience, or thought in the strict sense of the word.[103]

A change in the type of emotion felt takes place as a new level of conscious awareness is reached. The feeling of an intellectual being is different from that of the child or primitive; his emotions at the psychic level are modified and expressed in a different way. Higher levels of consciousness react on the basic structure, thereby modifying it. The poet converts human experience into poetry by fusing thought itself into emotion, not by first expurgating the intellectual elements and then perceiving the emotional, but by thinking in a certain way, then expressing how it feels to think in that way. Thus, Dante fused the Thomistic philosophy into a poem which expressed what it feels like to be a Thomist.[104]

---

[102] See R. G. Collingwood, *The Principles of Art*, pp. 291-292.
[103] See *ibid.*, pp. 293-294.
[104] See *ibid.*, p. 295.

To relate this theory of linguistic expression to the theory of questioning, we have to recall Collingwood's position regarding the relation of art to other aspects of human experience. Collingwood regards the experience of human life as having two aspects, the aesthetic and the logical.[105] Art indulged in as an all-absorbing form of experience mistakes the aesthetic aspect for the whole, the aspect which Collingwood had previously referred to as the cutting edge of the mind. The logical aspect is the weight of fact as something asserted which gives power to that edge to cut.[106] Collingwood's account of art in *Speculum Mentis* is an attempt to give an account of the mind's cutting edge, which he likewise refers to as the element of pure spontaneity, or inquiry, that is, questioning, which is the true beginning of knowledge. The life of art regarded as a distinctive type of life-experience is the assertion of a philosophical error which abstracts the aesthetic from the logical function and erects each into separate experiences. The so-called life of art is the persistent pursuit of this error, which increasingly becomes a futile quest at every turn.

In fact, the so-called aesthetic experience is a name for all experience insofar as this element of spontaneity or questioning or supposal enters into it.[107] Art correctly appreciated is therefore an element of all experience. It is the element of spontaneity or questioning or supposing. That is, it is identical with the very dynamism of the mind in its forward thrust. Divorced from the matrix of fact in which it is embedded, artistic activity is intuition removed from its context in experience, intuition absolutized and divorced from its logical context. But when artistic activity is seen in its concrete context in lived experience, it is seen to be the questioning or supposal aspect of conscious life.

Representing, as it does, the lowest level of human experience, art is the most rudimentary form of thought. The concept of art seems to function in Collingwood's thought something like the concept of Eros in Plato's thought. It is the name for that intrusive and dynamic power which spirits the mind forward to further adventures.[108]

This activity, absolutized as art, seems to be identical with the power of negativity, which is the dynamism of the dialectical movement of the mind. The mind is what it does; it is its activity. But this

---

[105] See *Speculum Mentis*, p. 101. This dichotomy is of Grocean derivation; see William M. Johnston, *The Formative Years of R. G. Collingwood*, p. 73.

[106] See *Speculum Mentis*, p. 101.

[107] See *ibid.*, p. 103.

[108] See L. O. Mink, *Mind, History and Dialectic*, p. 237.

activity is a dynamic striving from a prior state or condition to a subsequent state or condition. The questioning or imaginative or supposal activity which is absolutized in aesthetic activity seems in its context in lived experience to be the condition of negating the prior state of affairs (expressed by past fact, indeed constituted as past fact inasmuch as it is recognized to be prior) and the simultaneous thrusting towards a posterior state of affairs (a future which, when expressed, thereby becoming present, will function as answers fulfilling the mind's striving). We can see here the radical constitution of the historical character of human experience. Achievements of the mind's striving, answers to its questions, form the content of the past and provide the matter (the weight behind the cutting edge) for further striving in the form of questioning. A question unanswered, but with a restlessness demanding an answer, points to something as yet unrealized, but demanding to be realized; this is what we recognize as the future. Thus, the question and answer activity, dynamically and exercisedly considered, is basically constitutive of human experience as distinguishable (by thought when it reflects over the process) into a past (which is achieved) and a future (which is as yet unrealized, but straining to be realized), both features being co-actualized in the present.[109] But that is precisely what we mean by the historical character of human experience. The objectivised, the achieved, the asserted is (more or less) the past.[110] The striving to achieve, the questioning, is the present activity of the mind (and mind is what it does) reaching beyond this present actuality to what is as yet not actualized in order to actualize it; and this is the future. So the future already exists, not actually achieved, but in the dynamic tendency of the mind in the process of realizing the as yet unrealized. The intimate relationship between questioning and history becomes clear. The mind thus, so to speak, 'feeds' on its own (past) achievements (the ascertained facts which are the weight behind the cutting edge of questioning)

---

[109] See Collingwood's "Some Perplexities about Time," *Proceedings of the Aristotelian Society*, XXVI (1925-1926), p. 49.

[110] "All externality is imaginary; for externality—a mutual outsidedness in the abstract sense of the denial of a mutual insidedness—is, as such, abstraction, and abstraction is always intuition or imagination. It is only to the imagination that the mind is ever outside its object; but to the imagination it is always outside its object, even when that object is itself as in fact it always is. Therefore, since the externality of the object is only imaginary, the act by which we create the object is never capricious: we only imagine it to be capricious: in point of fact it is necessary and an integral part of the life of reason." R. G. Collingwood, *Speculum Mentis*, p. 293. Imagination and imaginative activity would therefore seem to be the foundation of that externality which is characteristic of the objects of scientific (as distinct from historical or philosophical) knowledge.

in bringing itself forth in new forms of activity (which, we must remember, are identical with the mind itself). Its past is a quasi-substance which it transforms into new forms of itself through questioning and reflection. Objectivizing is the making available of what is achieved (past facts) as a condition of the mind's questioning activity (that it *have* something to act about in a questioning way) so that it can creatively bring forth new forms of itself as successive reflections upon its (past) achievements.[111]

Towards this future affirmational state to which it is striving, the question—that is, the mind in its activity of questioning—is related both affirmatively and negatively: it affirms it, inasmuch as the very asking of the question presupposes and demands an (as yet unrealized) answer as its complement; it negates it, inasmuch as the question, being an unanswered question, is the absence (negation) of the answer. Likewise, in relation to the past, to the body of achieved factual information, it affirms it inasmuch as questioning presupposes available information; it negates it inasmuch as the mind, through the question, strives to go beyond it, and in doing so alters it; past facts look different as they are seen from greater perspective. Hence, the question is in dialectic interplay with prior fact (presupposed by the question and to some extent embodied in the question) and subsequent (future, as yet not ascertained) fact, to be made known through the process of questioning. Thus it appears how the question is a function of the dialectic of negativity, which is the lifeblood of the mind.

The questioning activity of the mind, as *Speculum Mentis* illustrates, is the dialectical dynamism whereby the mind ascends from lower to higher levels of consciousness. Collingwood denies that any level of consciousness necessitates a higher one. But, as Louis O. Mink well points out,[112] every mental activity is accompanied by its own characteristic emotion, and this emotion may *remain unexpressed at that level.* What drives the mind to a higher level is not the thirst for knowledge or novelty, but the need to express the emotion attendant on thinking at the lower. Only a higher level of conscious activity can express an emotion attendant on a lower order of activity. And here is the dynamism of the dialectical transformation from implicit (lower level) to explicit (higher level). It is in reflection that emotion is made explic-

---

[111] ". . . inasmuch as its whole life is a process of self-determination, the past in any such process is the evil which is rejected, good when it was brought into being but now outworn and therefore evil if it had been retained." *Speculum Mentis,* p. 296.

[112] See his *Mind, History, and Dialectic,* p. 237.

it, not as an object of reflection but by being expressed in the discursive act of reflection.

> The relation between emotion and thought therefore occupies in Collingwood's thought the conceptual locus which in other systems is occupied by a distinction between "subjective" and "objective." Emotion is, in all its occurrences, the felt ("subjective") quality of experience, thought the grasp of ("objective") structure in experience. Emotion never becomes an object of thought, but it is not opposed to thinking, as if every human act were a tactical victory over the other in a Manichean war for dominance.[113]

Unexpressed emotion inhibits and holds consciousness back from its forward thrust, from its work of questioning and criticism. Collingwood regards this failure to express an emotion as a corruption of consciousness. Louis O. Mink says [114] that the expression of emotion is not just an ornamental decoration on the life of reason, nor a mere catharsis which clears the way for thought, *but it is the energy for all rational inquiry*. The mind's search for truth must not only satisfy the logical criteria for knowledge, but must also express the emotions which well up through the levels of consciousness. For this end, the work of art is indispensable. The life of art may be a stage to be eventually abandoned, but it is indispensable for education in expressiveness.[115] "Thought never outgrows the need of language, never learns to live without that immediacy or intuitiveness which marks the aesthetic consciousness. To lose that would be to lose its own cutting edge."[116]

---

[113] *Ibid.*, p. 237. Felt disconformity with itself is the emotional counterpart of error, and the spur to the mind to reconstitute itself by reflection over its present actual state, and bring to (objective) clarity what is there in implicit confusion. ". . . any error [the mind] makes concerning its own nature (and every error is that) creates two conflicting results: a new state of itself, and a new notion of itself. But because its own state is a state of consciousness, this new state, even though only implicitly, contradicts the newly formed notion. Thus there are two conflicting notions of itself in the mind, and this conflict is the mark of error and the signal that a return to the road of truth is required. Thus the equilibrium of thought is a stable equilibrium like that of a gyroscope; but it is only the energy of the gyroscope that keeps it upright." *Speculum Mentis*, p. 297.

[114] See Louis O. Mink, *Mind, History, and Dialectic*, p. 238.

[115] See *ibid*.

[116] R. G. Collingwood, *Speculum Mentis*, p. 158.

Appendix Two

Further Elaboration of the Logic of Question and Answer

The conclusion of Chapter Nine, that there is a logic of questioning proper to history, leads one inevitably to ask about the principles, structure and mode of procedure of this logic.

The basic principle of this logic, the concrete universal, was shown, following Collingwood, in Chapter Four above, to have the structure of an overlap of classes in a scale of forms. Looking at this logical basis precisely as it is the principle of logical structure in history, it was shown, in Chapter Nine, with the help of Robert Stover, John Wild, and Ernst Cassirer, that this basic unity in diversity (universal) is actively created and constitutively constructed by a unity in diversity of consciousnesses unified in the activity of dialectical interrogation.

Given, then, that this is the basis of the intelligibility of history, what can we say regarding the structural outlines of this logic such as might form the subject matter of a logical textbook?

Leslie Armour feels that the elaboration of such a dialectical logic is possible.

> I can only conclude, then, that, though it is by no means easy to say what a dialectical logic is, there is, certainly, a kind of logic which one can distinguish in this way and the possibilities for its development seem to be considerable. In fact, there are important grounds for thinking that we cannot do without such a logic.

> Under the circumstances, there seems to be no reason to suppose that the demand for this kind of logic is illegitimate.[1]

But the elaboration of such a logic does seem to be frustrated in principle right from the start, since the identity of form and content is a basic tenet of dialectical thought. This seems to imply that any concrete situation embodies its own *sui generis* logic, so that nothing can be said *a priori* to the concrete situation. This seems to be confirmed by what Collingwood says about the need for insight into the concrete situation in order to know how to act[2] and also when he says that nothing is recognizable as evidence until it is used as such.[3]

But the situation is not quite as hopeless as this may make it appear to be. For, notwithstanding that Leslie Armour affirms the possibility of elaborating such a logic, he warns that such a logic is limited and kept in check by the existence of other and equally valid logical systems:

> . . . arguments in favour of a dialectical logic need not be arguments for the abolition of other sorts of logic. The relations between them are complex, but I have suggested that one may well need a "logic of limiting cases" even within the confines of a dialectical logic and that this will have important properties in common with traditional logics. The relationship in this case will simply be that of part and whole—and the worst sin of the traditional logician on this view, will have been that he made the part seem like the whole.[4]

In agreement with this observation it is submitted (conclusion to Chapter Nine above) that genus-species, abstract, propositional logic is valid and exercises its own *sui generis* autonomy. Therefore, from the point of view of abstract genus-species logic, we can look at dynamic self-developing human consciousness from the outside, not merely from inside some concrete situation of involvement.[5] We can then articulate what belongs to it *as such*.

If looked at in this way, what is said about the dialectical logic of self-constituting consciousness is going to concern more the moral qualities of the conscious subject and the insight-abilities accruing to such a conscious subject. In other words, this chapter of a logical textbook would bear upon that part of logic in which logic overlaps with

---

[1] Leslie Armour, *The Concept of Truth*, Assen, Van Gorcum, 1969, p. 115.
[2] See his *An Autobiography*, pp. 101-106.
[3] See his *The Idea of History*, p. 280. See also Chapter 1, above, pp. 32-33.
[4] Leslie Armour, *op. cit.*, p. 115.
[5] It may be remarked that Charles Reich's *The Greening of America* is a dialectical analysis of the American fact from within the American consciousness.

ethics, which is quite consistent with what Collingwood said about the various parts of philosophy.[6]

If, following Collingwood's lead, we focus our train of thought on the logic of a consciousness engaged in research of a scientific, or historical, or crime-solving nature, we can delineate, at least in a broad perspective, what some of these moral qualities are and how they are logically determining, especially if we allow a little help to come from Michael Polanyi, since Collingwood himself leaves his investigative logic little developed. The principal feature of logical import in the following analysis will be the concept of "guessing," and more specifically, of "right guessing." There are *no prescribed rules* for discovering the unknown or finding the solution to a newly arisen problem. Where prescribed rules and methods are silent, the researcher or questioner is "on his own," and his personal judgment, in which he "puts himself and his reputation on the line," is ultimately determining. But his "guesses," though prescriptively without pre-given foundation, are not arbitrary, or mere "shots in the dark." His guesses or "hunches" are "right" if he correctly follows a *directive principle* which *he himself embodies within his conscious self*. And this is nothing other than the *infixed directedness of consciousness to reality*, or to the whole of that which is. This infixed directedness is present not as something seen but as a conscious "weight," presenting itself as a feeling of concern and urgency to act. The researcher has to "feel his way" through the pathway to the solution of his problem, and the pathway becomes gradually evident only as each step is actually taken. In other words, he constructs the pathway in the very activity of "picking" his way through it. Fidelity to this tension of consciousness in search of the truth, the fully real in which his true self is also realized, is the only lodestone the researcher or problem solver has to follow in cutting new pathways where none are already provided.

A. Commitment to a Consensus: Acceptance of Authority: Aim of Each to Be His Own Authority

The first quality of a good researcher is *commitment to a tradition*. In order to learn 'the tricks of his trade' a novice researcher does not simply consult research manuals. He attaches himself to a reputable community of scientists (or historians, if his field is history), which community is defined by scientific (or historical) aims and objects and professes a responsible commitment to truth.

[6] See his *Essay on Philosophical Method*, pp. 189-190. Also see Chapter 4, above, pp. 140 et seq.

As the bearer of a tradition, this scientific (or historical) body, which may be regarded as a common consciousness shared by many individual consciousnesses (whose unity in diversity is basically linguistic, as was explained in Chapter Nine above), is defined by an accepted set of convictions about the world (natural or historical, as the case may be) as to how it looks and how it should look, or as to how it was in the past.

> Every interpretation of nature, whether scientific, non-scientific or anti-scientific, is based on some intuitive conception of the general nature of things.[7]

This corresponds to Collingwood's presuppositions, including absolute presuppositions, which determine the integrity and identity of the corporate consciousness, and whose acceptance by the members identifies them as scientists, or historians, as the case may be. Each member thus begins his career by an act of fidelity to this tradition and to its embodied values and aims, which that group as such is committed to safeguard. "The premisses of science on which all scientific teaching and research rests are the beliefs held by scientists on the general nature of things."[8] The medical profession thus has the Hippocratic Oath as something in the nature of a codified creed, much in the same way as Collingwood regarded the Catholic Faith of the early Christian Fathers as a codification of the basic commitments of Western civilization and culture.[9]

Similarly, if one aspires to become proficient in the business of crime detection, one will apprentice oneself to competent masters who embody that art, judgment and skill. Collingwood refers to two traditions (exemplified in a Kierkegaardian fashion) as typified by the fact-finding Sherlock Holmes on the one hand, and by the more rationalistic Hercule Poirot on the other hand, with Collingwood's decided preference in favor of the latter.[10]

Since discovery is more of an art, and "since an art cannot be precisely defined, it can be transmitted only by examples of the practice which embodies it."[11] This means that the novice researcher must "recognize as authoritative the art which he wishes to learn and those

---

[7] Michael Polanyi, *Science, Faith and Society*, Chicago, University of Chicago Press, 1946, p. 10.

[8] *Ibid.*, p. 11.

[9] See Collingwood's *An Essay on Metaphysics*, pp. 225-227; see also, above, Chapter 5, pp. 191-192.

[10] See Collingwood's *The Idea of History*, pp. 281-282.

[11] Michael Polanyi, *op. cit.*, p. 15.

from whom he would learn it.''[12] He must incorporate a tradition by first subjecting himself to acknowledged masters as authorities in order that he may eventually establish himself as an authority.

> To learn an art by the example of its practice is to accept an artistic tradition and to become a representative of it. Novices to the scientific profession are trained to share the ground on which their masters stand and to claim this ground for establishing their independence of it. The imitation of their masters teaches them to insist on their own originality which may oppose part of the current teachings of science.[13]

This traditionalism and respect for authority are not to be understood in the pejorative sense which one associates with conservatism and authoritarianism as quite inconsistent with critical inquiry and progress. For ''it is inherent in the nature of scientific authority that in transmitting itself to a new generation it should invite opposition to itself and assimilate this opposition in a reinterpretation of the scientific tradition.''[14] A scientific community thus delicately balances enforcement of discipline together with inducement to dissent, *which balance is realized in dialectical intercourse between the society and its members*. This balance is expressed in the way research resources are allocated, and in the acceptance of publishable material by the scholarly organs of publication. A certain tension will be inevitable in such a dialectical balance.

Scientific opinion embodies the sovereign authority of science. ''It is only by his peers that any claimant to knowledge is judged.''[15] Furthermore, ''every succeeding generation is sovereign in reinterpreting the tradition of science.''[16]

However, the sovereignty of everyone's position is more fundamental than the differences in rank between scientists.[17] Collingwood says much the same thing when he says that

> . . . the only way of knowing whether a given type of argument is cogent or not is to learn how to argue that way, and find out. Meanwhile the second best thing is to take the word of people who have done so for themselves.[18]

[12] *Ibid.*
[13] *Ibid.*
[14] *Ibid.*, pp. 15-16.
[15] R. G. Collingwood, *The Idea of History*, p. 252.
[16] M. Polanyi, *op. cit.*, p. 16.
[17] *Ibid.*
[18] R. G. Collingwood, *The Idea of History*, p. 263.

But if each aspirant is finally to become his own authority, he has to recognize the superiority of his peers when he is in the formative stage. Docility, in the ancient, traditional sense of the virtue whereby a student is a good learner through a readiness to submit himself to his masters, is thus an indispensable quality in any apprentice researcher being put through his paces in acquiring his own status as an authority.

> If any reader wishes to rise here on a point of order and protest that a philosophical question, which ought therefore to be settled by reasoning, is being illegitimately disposed of by reference to the authority of historians, and quote against me the good old story about the man who said 'I'm not arguing, I'm telling you', I can only admit that the cap fits. I am not arguing; I am telling him.[19]

The novice in scientific research will thus find himself, through commitment to the tradition, the possessor of a body of information in the form of dogma, of commonly accepted commitments about the *status quo* of his field, and convictions as to how the world ought to behave and respond. Columbus, as an explorer, was committed to a world view specifiable meaningfully in the directional terms, north, south, east and west, and likewise he was committed to the view of the global structure of the navigable world. Given these presuppositions he was right in conjecturing that the East could be reached from the West.[20] His mistaking of America for India was not due to these legitimate principles, but to his own mistaken belief that the world was much smaller than it actually is, so that he was quite unsuspecting as regards the likelihood of a large, hitherto unknown continent intervening.

> The power to expand hitherto accepted beliefs far beyond the scope of hitherto explored implications is itself a pre-eminent force of change in science. It is this kind of force which sent Columbus in search of the Indies across the Atlantic. His genius lay in taking it literally and as a guide to practical action that the earth was round, which his contemporaries held vaguely and as a mere matter for speculation.[21]

Thus ''the influence of these premisses [beliefs held regarding the nature of the world] on the pursuit of discovery is great and indispensable. They indicate to scientists the kind of questions which seem

[19] *Ibid*.

[20] See Michael Polanyi, *Personal Knowledge*, Chicago, University of Chicago Press, 1962, p. 310.

[21] *Ibid*., p. 277.

reasonable and interesting to explore."[22] Howsoever greatly our convictions may be assailed, some holding firm is essential, otherwise there is a complete fragmentation of consciousness, with a sceptical abandonment of all research and no longer a conscious domination of the situation, no longer a "self-reliance" in which the conscious self stands firm.

> The intellectual daring which impels our acts of commitment retains its dynamic character within the state of commitment, in relying on its own resourcefulness to deal with the unspecifiable implications of the knowledge acquired by the act of commitment. In this self-reliance lies our ultimate power for keeping our heads in the face of a changing world. It makes us feel at home in a universe presenting us with a succession of unprecedented situations and even makes us enjoy life best precisely on these occasions which force us to respond to novelty by reinterpreting our accepted knowledge.[23]

Such a commitment to a tradition does not eliminate doubt; otherwise research would be frustrated in principle from the start. But, no matter to what extent we encounter the unexpected, or the situation changes, we must stand firm on our beliefs, which represent nothing other than our commitment to the truth.

> A fiduciary philosophy does not eliminate doubt, but (like Christianity) says that we should hold on to what we truly believe, even when realizing the absurdly remote chances of this enterprise, trusting the unfathomable intimations that call upon us to do so.[24]

Thus, just as Jacques Maritain points out that a man sacrificing his life to save a friend or to defend his country, if he does this consciously and deliberately, must know in the depths of himself that all is not lost, even though this action seems to negate that basic conviction,[25] so also the committed researcher must hold fast to some basic conviction right to the end, in spite of all obstacles and apparent failures, if he is ever to achieve success. The Newtonian physicist, confronted with the challenging conclusions of the Michelson-Morley experiment, will not sceptically abandon science as a self-refuting enterprise, but will see rather a fuller and richer understanding of his spatio-temporal vision of the world. The world conceived spatio-temporally

---

[22] M. Polanyi, *Science, Faith and Society*, p. 11.
[23] M. Polanyi, *Personal Knowledge*, p. 317.
[24] *Ibid.*, p. 318.
[25] See J. Maritain, *Neuf leçons sur les notions premières de la philosophie morale*, Paris, Téqui, 1951, p. 96.

stands firm, whereas the conception of an immobile framework of an absolute space at rest has to be relinquished. Again, as far as crime detection is concerned, the T.V. series *Columbo* has, as perhaps its unique characteristic, the image of a man who stands firm to certain unshakeable basic insights whose dynamic logic he follows to the end, much in the same way as Columbus' unshakeable commitment to the rotundity of the earth impelled him three times to try to seek a sea route to the Indies.

> The science of today serves as a heuristic guide for its own further development. It conveys a conception about the nature of things which suggests to the enquiring mind an inexhaustible range of surmises. The experience of Columbus who so fatefully misjudged his own discovery is inherent to some extent in all discovery. The implications of new knowledge can never be known at its birth. For it speaks of something real and to attach reality to something is to express the belief that its presence will yet show up in an indefinite number of unpredictable ways.[26]

B. Relativism of World Views: Reality of the Whole of That Which Is as Arbiter Between Different World Views

But this grounding of research in the basic convictions of a community seems to involve a relativism of world views. Given a certain constellation of (absolute and relative) presuppositions, or basic convictions, "it would seem that our daily experience compels us with the force of logical necessity to accept certain natural laws as true," and that "generalizations such as 'all men must die' or 'the sun sheds daylight' seem to follow from experience without any intervention of an intuitive faculty on the part of ourselves as observers."[27] But Polanyi objects that this is due to our inclination to regard our own particular convictions (or absolute presuppositions) as inescapable. Such general convictions, however, are often denied by primitive peoples. Death may seem to them not to be a matter of natural necessity but of evil magic, or the darkness of night may be seen as due to the sun returning across the night sky without shedding any light. Thus, the smashing of a man's skull would be regarded by us as highly significant in regard to his death, but would be regarded by the magical culture as incidental and irrelevant, whereas, to them, the passage overhead of a rare bird would be ominous and seized upon as the explanatory cause. Such primitive people are of normal intelligence,

---

[26] M. Polanyi, *Personal Knowledge*, p. 311.
[27] M. Polanyi, *Science, Faith and Society*, p. 25.

yet they hold their views as consistent with daily experience and "uphold them firmly in the face of any attempt on the part of Europeans to refute them by reference to such experience."[28] Because "the terms of interpretation which we derive from our intuition of the fundamental nature of external reality cannot be readily proved inadequate by pointing at any particular new element of experience," "it would seem that we are in danger of losing sight of any difference between the rival claims of the magical and the naturalistic interpretations of events."[29]

The objection stated above is that we stand enmeshed in a radical cultural relativism, inasmuch as each culture seems to be locked in its own world view determinants as regards the meaning of events, and therefore as regards what attitude or action, together with its inherent logic, is to be taken in the presence of such events. Polanyi cites the instance of St. Augustine, who had a profound interest in science before his conversion to Christianity. But, as he approached conversion, he came to regard all scientific knowledge as barren and its pursuit as spiritually misleading. Eleven hundred years later, at the time of the Renaissance, the spell Augustine exercised over Western culture waned as the secular, critical, extrovert, rationalist spirit spread into many fields and finally revived the scientific study of nature. Today, however, a reverse process seems to be felt, as among many scientists there is felt "the balance of mental needs tilting back once again."[30] Science today seems to be not so self-confident as regards the extent to which its generalizations make sense as far as the world as a whole is concerned. How do we stand, then, when, with wavering and multiple basic commitments, we are confronted with wavering and multiple interpretations of daily experience? Polanyi answers as follows:

> . . . objective experience cannot compel a decision either between the magical and the naturalist interpretation of daily life or between the scientific and the theological interpretation of nature; it may favour one or the other, but *the decision can be found only by a process of arbitration in which alternative forms of mental satisfaction will be weighed in the balance.*[31]

But what is to be the final arbiter in determining the alternative forms of mental satisfaction? It is that of *reality* or *wholeness*. Each

---

[28] *Ibid.*
[29] *Ibid.*
[30] *Ibid.*, p. 27.
[31] *Ibid.*, p. 28. Emphasis added.

explanation is more or less satisfactory insofar as it gives us an *acceptable coherent view* of experience. Consciousness is a tendency to an integral and fully determined totality of experience, according to which integrity and determination the elements of experience are regarded as real. Unfortunately, though,

> . . . there exist . . . no explicit rules by which a scientific proposition can be obtained from observational data, and we must therefore accept also that no explicit rules can exist to decide whether to uphold or abandon any scientific proposition in face of any particular new observation.[32]

If no observation can supply explicit rules for the abandonment or upholding of any scientific position, what, then, is the role of observation? Polanyi states that the precise role of observation in scientific discovery "is to supply [not *rules*, but] *clues* for the apprehension of reality"; and he adds that this is "the process underlying scientific discovery."[33]

This apprehension of reality thus gained in its turn provides us with a *clue* to future observations, which is the process underlying verification. This corresponds to Collingwood's position that experience produces strains in our basic convictions (a constellation of absolute presuppositions) as they prove more and more inept to integrate our experience into a coherent totality. With readjustment following restructuring in the constellation of absolute presuppositions, we *then* approach experience, whose meaningfulness (i.e., implication in wholeness) is intelligible in the light of our readjusted constellation of absolute presuppositions. This also corresponds to Collingwood's mutual relation of implication that exists between the starting point and conclusion of a philosophical inquiry, which he conceives as a constant return to the beginning, combining together the movements from principle to conclusion (as in deduction) and from fact to principle (as in induction).[34]

Polanyi says that "in both processes there is involved an intuition of the relation between observation and reality: a faculty which can range over all grades of sagacity, from the highest level present in the inspired guesses of scientific genius, down to a minimum required for ordinary perception."[35] Polanyi concludes with the incisive

---

[32] *Ibid.*, p. 29.
[33] *Ibid*. Emphasis added.
[34] See R. G. Collingwood, *An Essay on Philosophical Method*, pp. 160-175; see also Chapter 4, above, pp. 133 et seq.
[35] M. Polanyi, *Science, Faith and Society*, p. 29.

statement that "*verification*, even though usually more subject to rules than *discovery*, rests ultimately on mental powers which go beyond the application of any definite rules."[36]

Three things stand out here as highly relevant to our investigation of the logic of questioning, research and discovery. First, the highest level of all grades of sagacity is accredited to be the "inspired guesses of scientific genius," and second, there are basic "mental powers which go beyond the application of any definite rules." Furthermore, third, "these mental powers" and "the inspired guesses of scientific genius" are concerned precisely with "an intuition of the relation between observation" (facts, events, fragmentary experiences, etc.) "and *reality*" (that is, universal integrity, mutual implicatedness, wholeness, the totality of all that is grasped somehow as a universal, that is, as a unity in diversity, not as an abstract, generic type of universal but as a universal whose differences are relevant to it).

Thus, says Polanyi, "in the course of any single experimental inquiry the mutual stimulus between intuition and observation goes on all the time and takes on the most varied forms."[37] He illustrates as follows:

> Most of the time is spent in fruitless efforts, sustained by a fascination which will take beating after beating for months on end, and produce ever new outbursts of hope, each as fresh as the last so bitterly crushed the week or month before. Vague shapes of the surmised truth suddenly take on the sharp outlines of certainty, only to dissolve again in the light of second thoughts or of further experimental observations. Yet from time to time certain visions of the truth, having made their appearance, continue to gain strength both by further reflection and additional evidence. These are the claims which may be accepted as final by the investigator and for which he may assume public responsibility by communicating them in print. This is how scientific propositions normally come into existence.[38]

The common consensus of the professional community will be the deciding factor as to whether the individual researcher's conclusion is incorporated into the fund of scientific knowledge as an established position that future researchers have to take into account; that is, whether science as a whole will now operate from a world view which is adjusted so as to take account of the individual researcher's findings. Thus, the common acceptance of the Einsteinian reinterpreta-

---

[36] *Ibid*. Emphasis added.
[37] *Ibid*., p. 30.
[38] *Ibid*.

tion of the spatio-temporal constitution of the world as a result of the Michelson-Morley experiment gave rise to the world view of relativity physics, which present researchers adopt as their basis for future investigations.

C. Objectivity and Universality Guaranteed by Responsible Commitment

The concept of responsible commitment discussed above[39] presents us also with the concept of consciousness establishing its autonomy and *creating itself as a true self* in the decisive act of staking its identity as a unique and original self on that very self-made decision in regard to the alternatives with which it is faced in seeking the solution to a problem. *Human consciousness as such is an innate tendency to an integral and fully determined totality of experience.* It maintains its selfness and conscious integrity in backing its decision with the whole of that self as its stake.

What finally survives in research is what is commonly accepted by reputable authorities. As Collingwood puts the matter:

> . . . it is not enough that science should be autonomous or creative, it must also be cogent or objective; it must impress itself as inevitable on anyone who is able and willing to consider the grounds upon which it is based and to think for himself what the conclusions are to which they point.[40]

Columbus, in staking his all in seeking a route to the Indies, was being *true to himself*. Notwithstanding that his empirical self went down in personal shame after three unsuccessful attempts to reach the Indies, his apparent "failure" was recognized for the success it really was when subsequent navigators correctly understood the epoch-making discovery he had made. The universal intent of his responsible self was valid, notwithstanding all appearances to the contrary. The Columbus we know would never have been incorporated into history if he had relinquished his project merely to avoid the shame of failure. His fundamental commitment was too strong for that. Being true to one's self means holding one's self *in truth*, that is, in relation to the whole of that which is, in which the self has its life, movement and mode of being. The "failures" of a truly committed man are never *mere* failures, but more often are the very occasions of that leap in the scale of forms whereby the consciousness moves from a lower to

---

[39] See this appendix, pp. 479 et seq.

[40] R. G. Collingwood, *The Idea of History*, p. 265.

a higher level. Not all is lost; the true self survives in its self-developed fulfillment.

> . . . the explorer gambles for indefinite stakes. Columbus sailed out to find a Western route to the Indies; he failed and after repeating his voyage three times to prove that he had reached the Indies, he died in shame. Still, Columbus did not merely blunder into America. He was wrong in accepting on the evidence of Esdras and, presumably, of Toscanelli's map, that the westward distance of the Indies from Spain was only about twice that of the Azores, but he was right in concluding that the East could be reached from the West. He staked his life and reputation on what appear now to be insufficient grounds for an unattainable prize, but he won another prize instead, far greater than he was ever to realize. He had committed himself to a belief which we now recognize as a small distorted fragment of the truth, but which impelled him to make a move in the right direction. Such wide uncertainties of its aims are attached to every great scientific enquiry. They are implicit in the looseness of the hold which a daring anticipation of reality has upon it.[41]

This "daring anticipation of reality" is nothing other than the responsible judgment of the researcher, which, in addition to being a daring anticipation of reality, is also a positing of his true self in putting his whole self "on the line" as the stakes. As Aristotle had said:[42] ". . . the end appears to each man in the form answering to his character"; and G. C. de Menasce incisively explains as follows:

> Choice of the end is more of a *conversion* than a simple choice. There is only an analogy between the liberty of conversion or betrayal and that little, imperious liberty of the choice of means. In love of the end, it is the *whole subject* who seeks, wills, mobilizes himself and finds. In this choice of the end, it is the whole subject who thus finds his *raison d'être*, his justification, his honor and glory. When I choose the means, it is the means that presents itself to me to be sought; when I choose the end, I myself give an account of it.[43]

Commitment is the act of a conscious self in which it *together* posits its authentic self (its self-posited self) and its relation to reality, or to the whole of that which is. It is the self-taking-in-hand act whereby the self posits its selfness in itself and in its distinction from, and in relation to, the non-self. This decisional self-positing of the self is not merely a self-isolating act but is together an *act of universal import* (therefore involving logical relevance which is our precise concern)

---

[41] M. Polanyi, *Personal Knowledge*, p. 310.

[42] In his *Nicomachean Ethics*, 1114a 30-35, b 1-5; see W. D. Ross' translation in *The Basic Works of Aristotle*, ed. Richard McKeon, p. 973.

[43] G. C. de Menasce, *The Dynamics of Morality*, New York, Sheed and Ward, 1961, p. 86. Emphasis in original.

and an act of consciusness, a grasping of a self in its opposedness to a non-self *in a whole of meaning*, an opposition which is *together* a definition of the self. Consciousness (as explicit or implicit thought) reaches out for the whole of that which is—that is, reality—but it effectively takes hold only of fragments, parts, which, as such, do not totally embody meaningfulness. Commitment, therefore, involves the desire and intent that others also ought to hold to the same, and it is therefore an appeal to the approbation of others, especially to the authorities to ratify.

D. The Researcher or Problem Solver Goes Beyond Rules and Present Achievements by Committing Himself in His Decisive Judgments

Responding to Plato's objection in the *Meno* that to search for the solution to a problem is absurd, since if you do not already know what you are looking for you would neither be moved to look for it (and so would not have a problem), nor would you be able to recognize it if you found it (so, even if you had a problem its solution would be impossible), M. Polanyi says that "we must have a foreknowledge sufficient to guide our conjecture with reasonable probability in choosing a good problem and in choosing hunches that might solve the problem."[44]

But he agrees with Plato in that "we still have no clear conception of how discovery comes about."[45] Such rules as we may have at our disposal are "merely rules of art" whose "application . . . must always rely on acts not determined by rule."[46] Polanyi does not deny the existence or usefulness of rules guiding verification, but says "only that there are none which can be relied on in the last resort." If rules guide decisions, it is equally the case that decisions are needed in the choice of the relevant guiding rule. Between whatever rules may be at hand and the solution of the problem at hand there is interposed the personal, responsible decision of the researcher or problem solver. It is precisely this which must bring the researcher out of a state of continuous suspension in doubt in the presence of the alternative courses he may take. To say that there are no established rules of procedure

> is not to say that we must always remain in doubt, but only that our decision what to accept as finally established cannot be wholly

---

[44] M. Polanyi, *Science, Faith and Society*, p. 14.
[45] *Ibid.*
[46] *Ibid.*

derived from any explicit rules but must be taken in the light of our personal judgment on the evidence.[47]

Problems that are solved by the mere application of available rules are solved by mere routine procedure. Likewise it is routine procedure (i.e., without any creative effort on the part of the researcher) to explain deviations from a rule by a deeper understanding of the rule and of its application. But some deviations from accepted rules may involve serious consequences for the rule itself, such as, for example, deviation of the movement of the planet Mercury from the rules of Galilean and Newtonian physics. Likewise, in judicial decisions "interpretation is most challenging when it involves a jump from the domain of rules to the domain of action. It then establishes the link between guidance and action."[48] According to Collingwood:

> . . . the generality of law, like the generality of archaeology, is not an end in itself but a means to an explicitly-recognized concrete end. The ideal of law is not the self-identical immobility of this or that statute, but its interpretation; and this interpretation means using the law as an aid to the determination of concrete cases, just as the archaeologist uses his generalized rules for the purpose of fixing the concrete date of a building or other object.[49]

Polanyi maintains that "the rules of scientific inquiry leave their own interpretation wide open, to be decided by the scientist's judgment."[50] This decisive judgment is the researcher's major function, and it includes "the finding of a good problem, and of the surmises to pursue it, and the recognition of a discovery that solves it."[51] If he does rely on the support of a rule, he selects the rule that applies to the case, as a golfer chooses the club most appropriate to his next stroke. The researcher does not operate like a pre-programmed machine, but "from beginning to end he is himself the ultimate judge in deciding on each consecutive step of his inquiry."[52]

This brings us to the crucial activity in research, that in which the researcher vitally and deliberately inserts his personal, decisive judg-

---

[47] *Ibid.*, p. 30. In judicial decisions "rules, principles and canons are an assortment of tools for the performance of the judicial task. The judge picks the tool he finds most suitable and proceeds to use it on the body of the text interpreted." Gidon Gottlieb, *The Logic of Choice*, London, Allen and Unwin, 1968, p. 102.

[48] Gidon Gottlieb, *The Logic of Choice*, p. 98.

[49] R. G. Collingwood, *Speculum Mentis*, p. 225.

[50] M. Polanyi, *Science, Faith and Society*, pp. 14-15.

[51] *Ibid.*, p. 15.

[52] *Ibid.*

ment into his proceedings. This is the activity of "guessing" and of "guessing rightly."

Polanyi says that, although natural laws are discoverable, this is not done by the mere application of some explicitly known operation to the given evidence of measurements and observations.[53] The way a research scientist proceeds is somewhat similar to the way one proceeds in "discovering" a burglar in a neighboring room. First, there is the registering of observed facts: certain unusual sounds are heard from the direction of the room. Then, there is questioning: "Is it wind? A burglar? A rat?" First there is guessing as to what it might be; then some definite feature, what sounds like a foot-fall, is accepted as evidence that it is a burglar. There is then *the judgment in which we decide* that it is in fact a burglar. In that judgment the adjudicator stands committed, himself to be judged right or wrong upon verification.[54] Here it is not a question of *deducing* new observational data from what is already given. The data, as data, are consistent with an infinite number of possible future observations. What was discovered is the presence of a burglar. Polanyi explains the manner in which the discovery was made as follows:

> Curious noises are noticed; speculations about wind, rats, burglars, follow, and finally *one more clue being noticed and taken to be decisive,* the burglar theory is established.[55]

Polanyi says that what we see here is "a consistent effort at guessing and at guessing right." He details as follows:

> The process starts with the very moment when, certain impressions being *felt to be unusual* and suggestive, a 'problem' is presenting itself to the mind; it continues with the collection of clues with an eye to a definite line of solving the problem; and it *culminates in the guess* of a definite solution.[56]

Now "the propositions of science . . . appear to be in the nature of guesses," and ". . . their conjectural character remains inherent in them," notwithstanding that they are "founded on the assumptions of science concerning the structure of the universe and on the evidence of observations collected by the methods of science" and "subject to a process of verification in the light of further observations according to the rules of science."[57]

---

[53] See *ibid.*, p. 22.
[54] See *ibid.*, pp. 22-23.
[55] *Ibid.*, p. 23. Emphasis added.
[56] *Ibid.*, pp. 23-24. Emphasis added.
[57] *Ibid.*, pp. 31-32.

But this "guesswork" so vital to scientific discovery, and "right-guess work," the mark of scientific genius, is not unfounded, not a mere exercise of arbitrariness. The method inherent in its operation is as follows:

> In science the process of guessing starts when the novice *feels* first attracted to science and is *then* attracted towards a certain field of problems. This guesswork involves the assessment of the young person's own yet largely undisclosed abilities, and of a scientific material yet uncollected or even unobserved to which he may later successfully apply his abilities. It involves the sensing of hidden gifts in himself and of hidden facts in nature, from which two, in combination, will spring one day his ideas that are to guide him to discovery. It is characteristic of the scientific conjecture that it can *guess*, as in this case, the several consecutive elements of a coherent sequence—even though each step guessed at a time can be justified only by the success of the further yet unguessed steps with which it will eventually combine to the final solution.[58]

It is important to recognize in this that the progressive unfolding is *both* an unfolding of the self in its own awareness of itself *as a self* and as a centre of potentialities (a future self) *and* of the unfolding of its world, which is together the shared world in which other selves participate owing to its universal intent. As Collingwood says:

> The self and its world are correlative. I am the self that I am, simply because of the nature of the world: by studying a certain kind of world and living in it as my environment, I develop my own mind in a determinate way. And conversely my world is the world of my mind . . .[59]

This process of serial "guessing" implies a certain foreknowledge of the solution to the problem.

> In order to guess a series of such steps an intimation of approaching nearer towards a solution must be received at every step. There must be a sufficient foreknowledge of the whole solution to guide conjecture with reasonable probability in making the right choice at each consecutive step.[60]

This process "resembles the creation of a work of art," about which Polanyi says that

> [it] is firmly guided by a fundamental vision of the final whole, even though that whole can be definitely conceived only in terms of its yet undiscovered particulars—with the remarkable difference,

---

[58] *Ibid.*, p. 32.
[59] R. G. Collingwood, *Speculum Mentis*, p. 248.
[60] M. Polanyi, *Science, Faith and Society*, p. 32.

> however, that in natural science the final whole lies not within the powers of our shaping, but must give a true picture of a hidden pattern of the outer world.[61]

This comparison of the groping and guessing of scientific discovery with artistic creation can be well appreciated if we consider what Jacques Maritain[62] and G. C. de Menasce[63] say regarding "the practical intelligence, in wedlock with love" initiating a project and carrying it through to completion. We are asked to reflect on how a primitive man, as yet ignorant of a boat as a means of crossing a river, may come to invent one. Such a man, standing on a river bank, *desires* to cross the river. He retains *in his memory* fragments of knowledge from past experiences, such as, for example, having seen uprooted trees carried down a stream by a current, and having seen some birds perched on these flowing trees. He likewise remembers how he bound together tree branches to make a hut. These different memories, experiences and facts become fluid in his consciousness, no longer tied together in the unity they had as originally experienced. In this free-floating state they *form a dialogue with his desire* to cross the river. This desire to cross the river is the active form of his consciousness according to which these fragmented memories recoalesce in the growing consciousness of a primitive boat.

> These recalled experiences, rubbing against each another [in dialogue with his desire to cross the river], form a new synthesis, and thus, the savage creates his first boat, which is a true *boat* because it will be the man's knowledge and experience that will enable him to cross the river.[64]

The *desire* which characterises the inventor's present conscious experience is the motor principle in the reassembling of the fragmented past experiences presently retained in memory. The reunification, in which consists precisely the invention of the man-made boat, takes place when the fragments are viewed not in relation to the unity they had in their original past experience but in relation to the presently *felt*, but as yet unfulfilled, *desire* to cross the river. The truth of the true boat consists in the conformity of what is known to the desires of the agent.

---

[61] *Ibid*.

[62] See his *Creative Intuition in Art and Poetry*, New York, World Publishing Company (Meridian Books), 1954, pp. 38-39.

[63] See his *The Dynamics of Morality*, New York, Sheed and Ward, 1961, p. 149.

[64] *Ibid*. Emphasis in original.

From this comparison of creative inventiveness with research and discovery we can see how "the historian is . . . the criterion of history," in which he is "both the architect of the past, which is the subject of his judgment, and the author of the principles by which he judges."[65] The historian will be equipped with information derived from experience and past research. He will likewise have a question in his mind, a problem calling for solution. His informational data, referred to his conscious self *in the form determined by his question*, will guide him to his solution. His skill as a historian will consist in his ability, perfected by past experience and the guidance of masters, to synthesise his material content in terms of the desire of which his question is the expression. His continuing decisions, in which there is exercise of the creative imagination in constituting a whole of parts,[66] involve that he stake his all in each succeeding decision, like a judge in applying law to a concrete case, and like a scientific researcher in solving a scientific problem. The scientist engaged in the work of research and discovery is bringing about science precisely in its historical dimension, inasmuch as the knowledge involved is taken precisely as it satisfies human concerns and answers questions which articulate the particular human interest involved.[67] Concerning the researcher in science, Polanyi says the following:

> [T]he curious fact [is] that *he* is *himself* the ultimate judge of what he accepts as true. His brain labours to *satisfy its own demands* according to criteria applied by its own judgment. It is like a game of patience in which the player has discretion to apply the rules to each run *as he thinks fit*. Or, to vary the simile, the scientist appears acting here as detective, policeman, judge and jury all rolled into one.[68]

Thus, the "scientific conscience cannot be satisfied by the fulfillment of any rules, since all rules are subject to its own interpretation."[69] Quite distinct from the routine verification of facts and references which forms part of the daily work of any scientist, "the real scientific conscience is involved in judging how far other people's data can be relied upon[70] and avoiding at the same time the dangers of either too little

---

[65] L. Rubinoff, "Introduction" to his edition of F. H. Bradley's *The Presuppositions of Critical History*, Don Mills, Ontario, Dent University Press, J. M. Dent and Sons, 1968, p. 49.

[66] See R. G. Collingwood, "The Historical Imagination," in *The Idea of History*, pp. 231-249.

[67] See Chapter 9, above, pp. 422 et seq.

[68] M. Polanyi, *Science, Faith and Society*, p. 38. Emphasis added.

[69] *Ibid.*, p. 40.

[70] "Confronted with a ready-made statement about the subject he is studying, the

or too much caution."[71] In his responsible decisions not predetermined by rules, the researcher is making *himself* the reason for his assertions and putting his reputation on the line in doing so. His conscious awareness and decision is thus an ultimate in scientific progress. "The scientist takes complete responsibility for every one of these actions and particularly for the claims he puts forward."[72] The same is true of the historian, the judge making a difficult decision and the crime solver interpreting clues in the disclosure of evidence. If his judgment is later confirmed by others *he* is said to have been right. If his work is later proven wrong *he* is said to have failed,[73] since "he is bound to no explicit rules and is entitled to accept or reject any evidence at his own discretion."[74] His job "is not to observe any allegedly correct procedure but to get the right results."[75] Thus, in discovery, "the scientist has to be judge in his own case,"[76] and, in doing so, he proceeds much in the same way as the primitive inventing a boat. The process is described as follows:

> Intuitive impulses keep arising in him stimulated by some of the evidence but conflicting with other parts of it. One half of his mind keeps putting forward new claims, the other half keeps opposing them. Both these parties are blind, as either of them left to itself would lead indefinitely astray. Unfettered intuitive speculation would lead to extravagant wishful conclusions; while rigorous fulfillment of any set of critical rules would completely paralyse discovery. The conflict can be resolved only through a judicial decision by a third party standing above the contestants. The third party in the scientists' mind which transcends both his creative impulses and his critical caution, is his scientific conscience.[77]

This central role of the conscience in forming the researcher's progressive and final decisions thus highlights the basic moral factor in the foundations of scientific research.

---

scientific historian never asks himself 'Is this statement true or false?' The question he asks himself is 'What does this statement mean?' [which] . . . is equivalent . . . to the question, 'What light is thrown on the subject in which I am interested by the fact that this person made this statement meaning by it what he did mean?' " R. G. Collingwood, *The Idea of History*, p. 275.

[71] M. Polanyi, *Science, Faith and Society*, p. 40.

[72] *Ibid.*

[73] *Ibid.*

[74] *Ibid.*

[75] *Ibid.*

[76] *Ibid.*, p. 41.

[77] *Ibid.*

E. Reality as Researcher's Aim

Sometimes the solution to a problem, or answer to a question, consists in selecting for its solution one of the known elements of reality, and such was the discovery of the presence of a burglar as solution to the problem of the noises in a neighbouring room.[78] At other times, and often in scientific research, the solution is by way of the postulation of an entirely new entity. It is this reality-orientation that distinguishes the art of scientific or historical or crime research from the art of creating fiction, such as in the writing of a novel.

Now "our principal clue to the reality of an object is its possession of a coherent outline."[79] It would take years of labour to recognize an egg-shape if it were presented as a catalogue of spacial coordinates, a feat which would rate equal to the discovery of the Copernican system as an intellectual achievement.[80] But it is precisely this capacity to grasp wholes in the presence of fragments which is at the root of discovery.[81]

> . . . the capacity of scientists to guess at the presence of shapes as tokens of reality differs from the capacity of our ordinary perception only by the fact that it can integrate shapes presented to it in terms which the perception of ordinary people cannot readily handle. The scientist's intuition can integrate widely dispersed data, camouflaged by sundry irrelevant connections, and indeed seek out such data by experiments guided by a dim foreknowledge of the possibilities which lie ahead.[82]

The most fundamental commitment, or most basic absolute presupposition, to use Collingwood's term, seems to be that of reality, or wholeness, the conviction that all our experience is "of a whole," or that our variety of experiences, to be intelligibly and meaningfully appreciated, must be reduced to the unity of a whole, to a global world view of unity in diversity.

Human consciousness, as such, seeks wholeness but everywhere encounters only particularity, and so must continue its search.

Now (and this is precisely the remarkable difference referred to by Polanyi between a work of art and a work of science),[83] there is some mysterious, pre-established harmony between human conscious-

---

[78] See this appendix, above, p. 492 et seq.
[79] M. Polanyi, *Science, Faith and Society,* p. 24.
[80] *Ibid.*
[81] Refer to Chapter 4, above, p. 86 et seq.
[82] M. Polanyi, *Science, Faith and Society,* p. 24.
[83] See above, this appendix, p. 480 et seq.

ness and reality, or the whole of that which is. According to Collingwood, consciousness is itself the whole of that which is, so that for consciousness to be consciousness of unintegrated multiples of experience is for consciousness to be not fully conscious of itself. The fulfillment of consciousness is thus, according to Collingwood, the full consciousness of experience as a whole. This is clear from what he says about philosophy as the absolute knowledge of absolute mind.[84]

> We do not assert that the trees and hills and people of our world are 'unreal', or 'mere ideas in my mind.' . . . The very essence of trees and hills and people is that they should be not myself but my objects in perception: they are not subjective but objective, not states of myself but facts that I know. None the less my knowing them is organic to them: it is because they are what they are that I can know them, because I know them that they can be what to me they really are. They and I alike are members of one whole, a whole which the destruction of one part would in a sense destroy throughout, as the death of our dearest friend darkens for us the very light of the sun.[85]

It is this commitment to reality which guides the researcher in his discoveries.

> All these processes of creative guesswork have in common that they are guided by the urge to make contact with a reality, which is felt to be there already to start with, waiting to be apprehended.[86]

It is this 'sense of reality' which validates the researcher's right to make 'hunches' and 'intelligent guesses,' even though it sometimes happens that such 'hunches' and 'guesses' turn out to be abortive (how many primitives perished in their first attempts to cross a river on improvised rafts?). In research, the realignment of the fluid fragments of past experience in the search for wholeness is articulated in the form of a question or problem.

Thus, the reality-concept functions both as the basis of criticism of different conflicting traditions, absolute presuppositions or world views and as that which guarantees the universality of the individual researcher's commitment, and the principle which determines whether his ''guesses'' are ''right'' or wrong.

F. Unpredictability of Solution: Emergent Character of Solution

Discovery of the solution to a problem, or answer to a question,

---

[84] See his *Speculum Mentis*, pp. 291-317.
[85] R. G. Collingwood, *Speculum Mentis*, p. 311.
[86] M. Polanyi, *Science, Faith and Society*, p. 35.

occurs as "a process of emergence rather than a feat of operative action."[87] This is somewhat contrary to Collingwood's position, for which the activity of questioning is unduly credited as being the sole operative force in the discovery of the answer.[88] Since Collingwood recognizes as the sole agent in knowledge the self-creative activity of self-evolving consciousness, he does not allow that rest of the mind in which the questioned other (person or object) is allowed to respond in its own right.[89]

Since, as was said above, "there are no manuals prescribing the conduct of research . . . because its method cannot be definitely set out,"[90] the rules of research, like those of all the higher arts, are embodied only in practice, and so can be learned only by going through a form of apprenticeship to someone who is already proficient in its practice. Discovery is "an extremely delicate and personal art which can be but little assisted by any formulated precepts."[91]

Polanyi holds that "the most essential phase of discovery thus represents a process of spontaneous emergence,"[92] and, contrary to Collingwood, maintains that "discovery does not usually occur at the culmination of mental effort . . . but more often comes in a flash after a period of rest or distraction."[93] Collingwood insists on the need for sustained effort.[94] But Polanyi says that after "our labours are spent as it were in an unsuccessful scramble among the rocks and in the gullies on the flanks of a hill . . . when we would give up for the moment and settle down to tea we suddenly find ourselves transported to the top."[95] Thus, "the efforts of the discoverer are but preparations for the main event of discovery, which eventually takes place—if at all—by a process of spontaneous mental reorganization uncontrolled by conscious effort."[96]

---

[87] *Ibid.*, p. 33.

[88] See above, Chapter 5, p. 158. See also criticism of Collingwood's position in this regard, Chapter 7, above, pp. 308-309.

[89] See above, Chapter 7, pp. 271-277. See also Chapter 9, above, pp. 425-427.

[90] M. Polanyi, *Science, Faith and Society*, p. 33.

[91] *Ibid.*, p. 34.

[92] *Ibid.*

[93] *Ibid.*

[94] See especially his *Essay on Metaphysics*, p. 37. "Everything that we call specifically human is due to man's power of thinking hard . . . high grade thinking . . . depends on . . . increase of mental effort and skill in the direction of that effort."

[95] M. Polanyi, *Science, Faith and Society*, p. 34.

[96] *Ibid.*

G. Research and Problem Solving as the Building Up of a Concrete Universal

It was shown above [97] that "the origin of the spontaneous coherence prevailing among scientists" is found in the fact that "they are speaking with one voice because they are informed by the same tradition."[98] Because the scientists as a corporate group are agreed in accepting one tradition, and trust each other as being informed by this tradition, they express the continued existence of science.[99] Fruitful discussion between them would become impossible if they were not able somehow to rely on each other's opinion.[100]

But Polanyi warns that this coherence of scientific opinion must not be imposed after the manner of a central authority, but rather by the spontaneous recognition and acceptance by the members of those whose judgment is the more authentic and to be followed rather than questioned.[101] And it is precisely in this that we can see the character of a concrete universal realized in the case of a scientific body informed by a scientific tradition.[102] It fits perfectly what Collingwood said regarding the concrete universal when he says that "everything in it is determined by its place in the whole."[103]

Collingwood says, further, that determinism is not involved, since "every part determines the whole and therefore by implication every other part." Thus, "each part taken separately may be regarded as the crucial determinant of everything else, just as every separate link bears the whole responsibility for keeping the chain together." And so, "everything in it is as unique as the whole, and the uniqueness of every part is based upon the uniqueness of every other."[104] Polanyi speaks similarly regarding the scientific corporation:

> Every time a scientist makes a decision in which he ultimately relies on his own conscience or personal beliefs, he shapes the substance of science or the order of scientific life as one of its sovereign rules.[105]

---

[97] See this appendix, p. 480 et seq.
[98] M. Polanyi, *Science, Faith and Society*, p. 52.
[99] See *ibid.*
[100] See *ibid.*
[101] See *ibid.*, p. 53.
[102] *Ibid.*, pp. 64-65.
[103] R. G. Collingwood, *Speculum Mentis*, p. 221.
[104] *Ibid.*
[105] M. Polanyi, *Science, Faith and Society*, p. 63.

Polanyi asserts that scientists themselves recognize this in recognizing that "inasmuch as each scientist is following the ideals of science according to his own conscience, the resultant decisions of scientific opinion are rightful."[106]

This "give and take" within the scientific corporation simply articulates from another optic all that was said previously[107] about the essentially linguistic character of the concrete universal. It is in language that is established that unity in diversity of consciousness. Language is nothing other than intersubjective objectivity or, what says the same, objective intersubjectivity, or again, one consciousness participated in by many conscious subjects.

Thus, speaking of the "logic" of institutions, Peter Berger and Thomas Luckman[108] say that this logic "does not reside in the institutions and their external functionalities, but in the way these are treated in reflection about them."[109] This means only that "reflective consciousness superimposes the quality of logic on the institutional order."[110] They explain further as follows:

> Language provides the fundamental superimposition of logic on the objectivated social world. The edifice of legitimations is built upon language and uses language as its principal instrumentality. The "logic" thus attributed to the institutional order is part of the socially available stock of knowledge and taken for granted as such.[111]

H. Overlap of Classes in a Scale of Forms

The essentially ongoing temporal character of this corporate body and its convictions is understandable as an overlap of classes in a scale of forms. The successful guidance of science by scientific opinion demands strict understanding that "this opinion represents only a temporary and imperfect embodiment of the traditional standards of science."[112] The aim of a scientific researcher, however, in soliciting the confirmation of his colleagues in the acceptance of his innovations, must not consist of merely subjective motives of personal prestige and

---

[106] *Ibid.*

[107] See Chapter 9, above, p. 399 and 410-416.

[108] See their *The Social Construction of Reality*, Garden City, New York, Doubleday-Anchor, 1967, pp. 64-67.

[109] *Ibid.*, p. 64.

[110] *Ibid.*

[111] *Ibid.*

[112] M. Polanyi, *Science, Faith and Society*, p. 53.

salary status, but "only in satisfying the standards of science,"[113] that is, as a lover of truth. Thus, "viewing oneself as a lover of truth may provide the motivation for great effort in both creative and critical thinking."[114] This forward movement of communally controlled scientific convictions takes the shape of a scale of forms in an overlap of classes, such as was explained above,[115] regarding the dynamic, dialectical, concrete character of the concrete universal.

---

[113] *Ibid*.

[114] W. Edgar Moore, *Creative and Critical Thinking*, Boston, Houghton Mifflin & Co., 1967, p. 8.

[115] See Chapter 9, above, especially pp. 422-424.

BIBLIOGRAPHY

A. COLLINGWOOD'S WORKS

A.i. Books and Booklets, Excluding Strictly Historical and Archaeological Works, Published Prior to 1943

1916 *Religion and Philosophy*, London, Macmillan, 1916.

---- Essay on "The Devil" in H. B. Streeter's and others' *Concerning Prayer: Its Nature, Its Difficulties and Its Value*, London, Macmillan, 1916.

1920 *Ruskin's Philosophy: An Address delivered at The Ruskin Centenary Conference, 8th August, 1919*, London, Titus Wilson and Son, 1920. Reprinted in Alan Donagan's edition of R. G. Collingwood's *Essays in the Philosophy of Art*, Bloomington, Indiana University Press, 1964, pp. 5-41.

1924 *Speculum Mentis*, Oxford, Clarendon Press, 1924.

1925 *Outlines of a Philosophy of Art*, London, Oxford University Press, 1925. Reprinted in Alan Donagan's edition of R. G. Collingwood's *Essays in the Philosophy of Art*, Bloomington, Indiana University Press, 1964, pp. 45-54.

1927 "Aesthetic," an essay in *The Mind*, edited by R. J. S. McDowell, London, Longmans, 1927, pp. 214-244.

1928 *Faith and Reason: A Study in the Relation between Reason and Science*, An Affirmations Series Pamphlet, London, Ernst Benn, 1928. Pp. 12-24 contain a reprint of "Reason is Faith Cultivating Itself," originally in *The Hibbert Journal*, XXVI (1927-1928), pp. 3-14. Reprinted in Lionel Rubinoff's edition of *Faith and Reason: Essays in the Philosophy of Religion by R. G. Collingwood*, Chicago, Quadrangle Books, 1968, pp. 122-147.

1930 *The Philosophy of History*, An Historical Association Leaflet, No. 79, London, G. Bell and Sons, 1930. Reprinted in William Debbins' edition of R. G. Collingwood's *Essays in the Philosophy of History*, Austin, Texas, University of Texas Press, 1965, pp. 121-139.

1933 *An Essay on Philosophical Method*, Oxford, Clarendon Press, 1933.

1935 *The Historical Imagination*, inaugural lecture delivered at Oxford, 28th October, 1935, Oxford, Clarendon Press, 1935. Reprinted in *The Idea of History*, Oxford, Clarendon Press, 1946, pp. 231-249.

1938 *The Principles of Art*, Oxford, Clarendon Press, 1938.

1939 *An Autobiography*, London, Oxford University Press, 1939.

1940 *An Essay on Metaphysics*, Oxford, Clarendon Press, 1940.

---- *The First Mate's Log of a Voyage to Greece in the Schooner ''Fleur de Lys'' in 1939*, London, Oxford University Press, 1940.

1941 *The Three Laws of Politics*, The L. T. Hobhouse Memorial Trust Lecture No. 11, London, Oxford University Press, 1941.

1942 *The New Leviathan*, Oxford, Clarendon Press, 1942.

A.ii. Posthumous Publications after 1943

1945 *The Idea of Nature*, Oxford, Clarendon Press, 1945. Posthumous publication of material compiled between August 1933 and September 1934 and revised September 1939. Minor editing by T. M. Knox.

1946 *The Idea of History*, Oxford, Clarendon Press, 1946. Posthumous publication of the following: (a) material of thirty-two lectures on the history of philosophy, divided into two parts, each part originally intended to be one book, compiled during the first six months of 1936 and revised, especially the sections on Greece and Rome, in 1940; (b) the fragmentary *Principles of History*, sketched in Java in the Spring of 1939; the whole work being compiled and edited by T. M. Knox. Contains also a valuable introduction by T. M. Knox with much useful information on the evolution of Collingwood's thought. Published 1961 as an Oxford Galaxy paperback.

A.iii. Edited Collections of Previously Published Works

1964 *Essays in the Philosophy of Art*, edited by Alan Donagan, Bloomington, Indiana University Press, 1964. A collection of Collingwood's writings on art previously separately published. Contains 'Ruskin's Philosophy' and 'Outlines of a Philosophy of Art.'

1965 *Essays in the Philosophy of History*, edited by William Debbins, Austin, University of Texas Press, 1965.

1968 *Faith and Reason: Essays in the Philosophy of Religion*, edited by Lionel Rubinoff, Chicago, Quadrangle Books, 1968.

A.iv. Strictly Historical and Archaeological Books

1923 *Roman Britain*, London, Oxford University Press, 1923 (revised 1934).

1930 *The Archaeology of Roman Britain*, London, Methuen and Co., 1930.

1934 *Roman Britain*, Oxford, Clarendon Press, 1934. Revised edition of the 1930 publication.

1936 *Roman Britain and the English Settlements*, co-authored with J. N. L. Myers, Oxford, Clarendon Press, 1936. Collingwood contributed the material of pp. 1-324 and pp. 462-478.

A.v. Collingwood's Articles, Discussions, etc., Excluding Book Reviews and Strictly Historical and Archaeological Writings

1920 "What is the Problem of Evil?" *Theology*, I (1920), pp. 66-74. Reprinted in L. Rubinoff's edition of Collingwood's *Faith and Reason: Essays in the Philosophy of Religion*, pp. 148-158.

1921 "Croce's Philosophy of History," *Hibbert Journal*, XIX (1921), pp. 263-278. Reprinted in William Debbins' edition of Collingwood's *Essays in the Philosophy of History*, pp. 5-22.

1922 "Are History and Science Different Kinds of Knowledge?" *Mind*, XXXI (1922), pp. 443-451. Reprinted in William Debbins' edition of Collingwood's *Essays in the Philosophy of History*, pp. 23-33.

1923 "Can the New Idealism Dispense with Mysticism?" A symposium with Evelyn Underhill and W. R. Inge, *Proceedings of the Aristotelian Society*, Supplementary Volume III (1923), pp. 161-175. Reprinted in Lionel Rubinoff's edition of Collingwood's *Faith and Reason: Essays in the Philosophy of Religion*, pp. 270-282.

---- "Sensation and Thought," *Proceedings of the Aristotelian Society*, XXIV (1923-1924), pp. 55-76.

1925 "The Nature and Aims of a Philosophy of History," *Proceedings of the Aristotelian Society*, XXV (1924-1925), pp. 151-174. Reprinted in William Debbins' edition of Collingwood's *Essays in the Philosophy of History*, pp. 34-56.

---- "Plato's Philosophy of Art," *Mind*, XXXIV (1925), pp. 154-172. Reprinted in Alan Donagan's edition of Collingwood's *Essays in the Philosophy of Art*, pp. 157-183.

---- "Economics as a Philosophical Science," *International Journal of Ethics*, XXXVI (1925-1926), pp. 162-185.

1926 "The Place of Art in Education," *Hibbert Journal*, XXIV (1925-1926), pp. 434-448. Reprinted in Alan Donagan's edition of Collingwood's *Essays in the Philosophy of Art*, pp. 187-207.

---- "Some Perplexities about Time," *Proceedings of the Aristotelian Society*, XXVI (1925-1926), pp. 135-150.

---- "Religion, Science and Philosophy," *Truth and Freedom*, II, 7 (1926), pp. 1-3. Reprinted in *Faith and Reason: Essays in the Philosophy of Religion*, ed. Lionel Rubinoff, pp. 89-92.

1927 "Reason Is Faith Cultivating Itself," *Hibbert Journal*, XXVI (1927-1928), pp. 3-14. Reprinted in *Faith and Reason: Essays in the Philosophy of Religion*, ed. Lionel Rubinoff, pp. 108-121.

---- "Oswald Spengler and the Theory of Historical Cycles," *Antiquity*, I (1927), pp. 311-325. Reprinted in William Debbins' edition of Collingwood's *Essays in the Philosophy of History*, pp. 57-75.

---- ''The Theory of Historical Cycles: II. Cycles and Progress,'' *Antiquity*, I (1927), pp. 435-446. Reprinted as ''The Theory of Historical Cycles'' in William Debbins' edition of Collingwood's *Essays in the Philosophy of History*, pp. 76-89.

1928 ''The Limits of Historical Knowledge,'' *Journal of Philosophical Studies*, III (1928), pp. 213-222. Reprinted in William Debbins' edition of Collingwood's *Essays in the Philosophy of History*, pp. 90-103.

1929 ''Political Action,'' *Proceedings of the Aristotelian Society*, XXIX (1928-1929), pp. 155-176.

---- ''A Philosophy of Progress,'' *The Realist*, I (1929), pp. 64-77. Reprinted in William Debbins' edition of Collingwood's *Essays in the Philosophy of History*, pp. 104-120.

---- ''Form and Content in Art,'' *Journal of Philosophical Studies*, VI (1929), pp. 332-345. Reprinted in Alan Donagan's edition of Collingwood's *Essays in the Philosophy of Art*, pp. 211-232.

1934 ''The Present Need of a Philosophy,'' *Philosophy*, IX (1934), pp. 262-265.

1936 ''Human Nature and Human History,'' *Proceedings of the British Academy*, XXII (1936), pp. 97-127. Reprinted in Collingwood's posthumous *The Idea of History*, Oxford, Clarendon Press, 1946, pp. 205-231.

1938 ''On The So-called Idea of Causation,'' *Proceedings of the Aristotelian Society*, XXXVIII (1937-1938), pp. 85-112.

1940 ''Fascism and Nazism,'' *Philosophy*, XXV (1940), pp. 168-176.

A.vi. Translations Made by Collingwood

1913 CROCE, Benedetto, *Filosofia di Giambattista Vico*, Bari, 1911, translated by R. G. Collingwood as *The Philosophy of Giambattista Vico*, London, Latimer, 1913. Reprinted by Russell and Russell, New York, 1964.

1921 RUGGIERO, Guido da, *Modern Philosophy*, translated by R. G. Collingwood and A. Howard Hanney, London, Library of Philosophy, George Allen and Unwin, 1921.

1927 CROCE, Benedetto, *Contributatio alla critica di me stesso*, Bari, 1915. Translated by R. G. Collingwood under the title of *Benedetto Croce, An Autobiography*, Oxford, Clarendon Press, 1927. Reprinted by Books for Libraries Press, Freeport, New York, 1970.

---- RUGGIERO, Guido da, *History of European Liberalism*, translated by R. G. Collingwood, London, Oxford University Press, 1927. Reprinted by The Beacon Press, Boston, 1959.

1929 CROCE, Benedetto, ''Aesthetic,'' translated by R. G. Collingwood for the 14th edition of *Encyclopaedia Brittanica*, Chicago, 1929, pp. 263-271.

A.vii. Book Reviews by Collingwood

ALEXANDER, S., *Art and Instinct*, Oxford, 1927. Reviewed in *Journal of Philosophical Studies*, III (1928), pp. 547-548.

BAILEY, Cyril, ed., *Epicurus: The Extant Remains*, Oxford, 1926. Reviewed in *The Monthly Criterion*, VI (1927), pp. 369-372.

BRIGHTFIELD, Myron F., *The Issue in Literary Criticism*, Berkeley, 1932. Reviewed in *Philosophy*, XII (1937), pp. 114-116.

BURY, J. B., *Selected Essays*, ed. Harold Temperly, Cambridge, 1930. Reviewed in *The English Historical Review*, XLVI (1931), pp. 461-465. Reprinted in part in Collingwood's posthumous *The Idea of History*, pp. 147-151.

CARR, H. W., *A Theory of Monads*, London, 1922. Reviewed in *Hibbert Journal*, XXIII (1924-1925), pp. 380-382.

CHARLETON, Walter, ed. *Epicurus, His Morals*, London, 1926. Reviewed in *The Monthly Criterion*, VI (1927), pp. 369-372.

D'ARCY, M. C., *The Nature of Belief*, London, 1931. Reviewed in *The Criterion*, XI (1931-1932), pp. 334-336.

DUCASSE, C. J., *The Philosophy of Art*, London, 1931. Reviewed in *Journal of Philosophical Studies*, IX (1931), pp. 383-386.

GORE, Charles, *The Philosophy of the Good Life*, Edinburgh, 1930. Reviewed in *The Criterion*, X (1930-1931), pp. 560-562.

LAIRD, John, *The Idea of Value*, Cambridge, 1929. Reviewed in *The Criterion*, IX (1929-1930), pp. 320-327.

LODGE, R. C., *Plato's Theory of Ethics*, London, 1928. Reviewed in *The Criterion*, VIII (1928-1929), p. 159.

REID, L. A., *A Study of Aesthetics*, London, 1931. Reviewed in *Philosophy*, VII (1932), pp. 335-337.

STACE, W. T., *The Meaning of Beauty*, London, 1929. Reviewed in *Journal of Philosophical Studies*, V (1930), pp. 460-463. Rejoinder by W. T. Stace in *Journal of Philosophical Studies*, V (1930), pp. 653-654.

TAYLOR, A. E., *Plato: The Man and His Work*, London, 1926. Reviewed in *The Monthly Criterion*, VI (1927), pp. 65-68.

TEGGART, F. J., *Theory of History*, New Haven, 1925. Reviewed in *The Journal of Philosophical Studies*, I (1926), pp. 255-256.

URBAN, W. M., *The Intelligible World*, London, 1929. Reviewed in *The Criterion*, IX (1929-1930), pp. 320-337.

WAHL, Jean, *Etude sur le Parménide de Platon*. Reviewed in *The Monthly Criterion*, VI (1927), pp. 65-68.

B. WORKS ON COLLINGWOOD

B.i. Bibliographies on Collingwood

DEBBINS, William, ed., *Essays in the Philosophy of History*, by R. G. Collingwood, Austin, University of Texas Press, 1965, pp. 141-148.

DONAGAN, Alan, *The Later Philosophy of R. G. Collingwood*, Oxford, Clarendon Press, 1962, pp. 308-310. Pp. 311-312 contain an appendix of dates in which Collingwood's philosophical works after *An Essay on Philosophical Method* were composed.

JOHNSTON, William M., *The Formative Years of R. G. Collingwood*, The Hague, Martinus Nijhoff, 1967, pp. 155-165.

Proceedings of the British Academy, XXIX (1943), accompanying obituary notice. Bibliography, excluding Book Reviews and Historical Works and Articles, including Philosophical Books and Articles, Translations and Memoirs, compiled by T. M. Knox, pp. 474-475. Bibliography of writings on Ancient History and Archaeology compiled by I. A. Richmond, pp. 481-485.

RUBINOFF, Lionel, *Collingwood and the Reform of Metaphysics*, Toronto, University of Toronto Press, 1970, pp. 394-406.

RUBINOFF, Lionel, editor, *Faith and Reason: Essays in the Philosophy of Religion*, by R. G. Collingwood, Chicago, Quadrangle Books, 1968, pp. 305-311.

SHALOM, Albert, *R. G. Collingwood: philosophe et historien*, Paris, Presses Universitaries de France, 1967. Annotated bibliography in body of text, pp. 154-157; includes both philosophical and archaeological works by Collingwood.

TOMLIN, E. W. F., *R. G. Collingwood*, London, Longmans, Green and Co., 1961. Brief bibliography on pp. 39-40.

B.ii. Books (including Unpublished Dissertation) on Collingwood and his Work

CRAGG, R. C., *Collingwood's Logic of Question and Answer*, diss. University of Toronto, 1948.

DONAGAN, Alan, *The Later Philosophy of R. G. Collingwood*, Oxford, Clarendon Press, 1962.

JOHNSTON, William M., *The Formative Years of R. G. Collingwood*, The Hague, Martinus Nijhoff, 1967.

MINK, Louis O., *Mind, History, and Dialectic*, Bloomington, Indiana University Press, 1969.

RUBINOFF, Lionel, *Collingwood and the Reform of Metaphysics*, Toronto, The University of Toronto Press, 1970.

SHALOM, Albert, *R. G. Collingwood, philosophe et historien*, Paris, Presses Universitaires de France, 1967.

TOMLIN, E. W. F., *R. G. Collingwood*, London, Longmans, Green and Co., new edition 1961 (first published 1953).

B.iii. Books with Significant Sections on Collingwood

NYGREN, Anders, *Meaning and Method, Prolegomena to a Scientic Philosophy of Religion and a Scientific Theology*, translated by Philip S. Watson, London, Epworth Press, 1972. Pp. 194-199 are on Collingwood's theory of absolute presuppositions.

POLANYI, Michael, *The Study of Man*, Chicago, University of Chicago Press, 1969. Bibliographical note, pp. 100-102, on Collingwood's *The Idea of History*.

VERSFELT, Martin, *The Perennial Order*, London, Society of St. Paul, 1954. Chapter Eleven, pp. 162-175, on Collingwood's *The Idea of History*.

B.iv. Articles on Collingwood

BUCHDAHL, Gerd, "An Assessment of R. G. Collingwood's Idea of History," *The Australasian Journal of Philosophy*, XXVI (1948), pp. 94-113.

BUCHDAHL, Gerd, "Has Collingwood Been Unfortunate in His Critics?" *The Australasian Journal of Philosophy*, XXXVI (1958), pp. 95-108.

COHEN, L. Jonathan, "Has Collingwood Been Misinterpreted?" *The Philosophical Quarterly*, VII (1957), pp. 149-150.

DONAGAN, Alan, "The Croce-Collingwood Theory of Art," *Philosophy*, XXXIII (1958), pp. 162-167.

DONAGAN, Alan, "Does Knowing Make a Difference to What is Known? (A Rejoinder to Mr. Post)," *The Philosophical Quarterly*, XVI (1966), pp. 352-355.

DONAGAN, Alan, "The Verification of Historical Theses," *The Philosophical Quarterly*, VI (1956), pp. 193-208.

DRAY, William, "R. G. Collingwood and the Acquaintance Theory of Knowledge," *Revue Internationale de Philosophie*, XI, 42 (1957), pp. 420-432.

DUCASSE, C. J., "Mr. Collingwood on Philosophical Method," *The Journal of Philosophy*, XXXIII (1936), pp. 95-106.

DYKSTRA, V., "Philosophers and Presuppositions" (discussion), *Mind*, LXIX (1960), pp. 63-68.

FRUCHON, Pierre, "Signification de l'histoire de la philosophie selon l'Autobiographie de Collingwood," *Les Etudes philosophiques*, XIII (1958), pp. 143-160.

GENNARO, Angelo de, "Croce and Collingwood," *The Personalist*, XLVI (1965), pp. 193-203.

GOLDSTEIN, Leon J., "Collingwood's Theory of Historical Knowing," *History and Theory*, IX (1970), pp. 3-36.

GRANT, C. K., "Collingwood's Theory of Historical Knowledge," *Renaissance and Modern Studies*, I (1957), pp. 65-90.

HARRIS, Errol, "Collingwood's Theory of History," *The Philosophical Quarterly*, VII (1957), pp. 35-49.

HARRIS, Errol, "Collingwood on Eternal Problems," *The Philosophical Quarterly*, I (1951), pp. 228-241.

KNOX, T. M., ''Notes on Collingwood's Philosophical Work'' (accompanying obituary notice, with attached bibliography), *Proceedings of the British Academy*, XXIX (1943), pp. 469-475.

LLEWELYN, John, ''Collingwood's Doctrine of Absolute Presuppositions,'' *The Philosophical Quarterly*, XI (1961), pp. 49-60.

MACKAY, Donald S., ''On Supposing and Presupposing,'' *The Review of Metaphysics*, II, 5 (1948), pp. 1-20.

McCALLUM, R. B., ''Robin George Collingwood 1889-1943'' (obituary notice), *Proceedings of the British Academy*, XXIX (1943), pp. 462-468.

MINK, Louis O., ''Collingwood's Dialectic of History,'' *History and Theory*, VII, 1 (1968), pp. 3-37.

MISCHEL, Theodore, ''Collingwood on Art as 'Imaginative Expression,' '' *The Australasian Journal of Philosophy*, XXXIX (1961), pp. 241-250.

POST, J. F., ''Does Knowing Make a Difference to What Is Known?'' *The Philosophical Quarterly*, XV (1965), pp. 220-228.

RICHMOND, I. A., ''An Appreciation of R. G. Collingwood as an Archaeologist'' (accompanying obituary notice), *Proceedings of the British Academy*, XXIX (1943), pp. 476-480 (with archaeological bibliography appended, pp. 481-485).

RITCHIE, A. D., ''The Logic of Question and Answer,'' *Mind*, LII (1943), pp. 24-38.

ROTENSTREICH, Nathan, ''From Facts to Thoughts: Collingwood's Views on the Nature of History,'' *Philosophy*, XXXV (1960), pp. 122-137.

ROTENSTREICH, Nathan, ''Historicism and Philosophy,'' *Revue internationale de philosophie*, XI, 42 (1957), pp. 401-419.

RUBINOFF, Lionel, ''Collingwood's Theory of the Relation between Philosophy and History: A New Interpretation,'' *The Journal of the History of Philosophy*, VI, 4 (1968), pp. 363-380.

RYNIN, David, ''Donagan on Collingwood: Absolute Presuppositions, Truth and Metaphysics,'' *The Review of Metaphysics*, XVIII (1964-1965), pp. 301-333.

SHALOM, Albert, ''R. G. Collingwood et la métaphysique,'' *Les Etudes philosophiques*, X (1955), pp. 693-711.

SHOEMAKER, Robert G., ''Inference and Intuition in Collingwood's Philosophy of History,'' *The Monist*, LIII (1969), pp. 100-115.

STRAUSS, Leo, ''On Collingwood's Philosophy of History,'' *The Review of Metaphysics*, V, 4 (1952), pp. 559-586.

WALSH, W. H., ''R. G. Collingwood's Philosophy of History'' (discussion), *Philosophy*, XXII (1947), pp. 153-160.

C. PUBLICATIONS ON QUESTIONING

C.i. Books on or with Significant Sections on Questioning

ARISTOTLE, *De Sophisticis Elenchis,* trans. W. A. Pickard-Cambridge, Book 1 only in Richard McKeon's edition of *The Basic Works of Aristotle,* New York, Random House, 1941, pp. 207-212.

ARISTOTLE, *Posterior Analytics,* trans. G. R. G. Mure, in Richard McKeon's edition of *The Basic Works of Aristotle,* New York, Random House, 1941, pp. 108-186.

ARISTOTLE, *Topica* and *De Sophisticis Elenchis,* translated by W. A. Pickard-Cambridge, in W. D. Ross' edition of *The Works of Aristotle,* Oxford, Clarendon Press, 1928, Vol. 1, 100a1-164b30; 164a1 [sic]-184b10.

ARISTOTLE, *Topics,* trans. W. A. Pickard-Cambridge, Book 1 only in Richard McKeon's edition of *The Basic Works of Aristotle,* New York, Random House, 1941, pp. 187-206.

BARDEN, Garrett, and Philip McShane, *Towards Self-Meaning,* Dublin, Gill and Macmillan, 1969. Chapters Three and Four on 'The What-Question' and 'The Is-Question.'

BELNAP, Nuel D. Jr., "Questions: Their Presuppositions and How They Can Fail to Arise," in *The Logical Way of Doing Things,* ed. Karel Lambert, New Haven, Yale University Press, 1969, pp. 23-37.

GADAMER, Hans Georg, *Wahrheit und Methode,* Tübingen, J. C. B. Mohr, 1965.

HARRAH, David O., *Communication: A Logical Model,* Cambridge, Massachusetts, M.I.T. Press, 1963.

HARRAH, David O., "Erotetic Logistics," in *The Logical Way of Doing Things,* ed. Karel Lambert, New Haven, Yale University Press, 1969.

JOHANN, Robert, *Building Up the Human,* New York, Herder and Herder, 1968. Pp. 35-53 on Inquiry.

LAMBERT, Karel, ed., *The Logical Way of Doing Things,* New Haven, Yale University Press, 1969.

LEONARD, Henry S., *Principles of Reasoning,* New York, Dover, 1967. Units Three and Four, pp. 22-43, on "The Recognition of a Problem."

PAYNE, Stanley L., *The Art of Asking Questions,* Princeton, New Jersey, Princeton University Press, 1951.

RÉGIS, L.-M., *Epistemology,* New York, Macmillan, 1959. Pp. 126-140 on the theory of interrogative methodology.

WAISMANN, F., *The Principles of Linguistic Philosophy,* ed. R. Harré, New York, St. Martin's Press, 1965. Chapter Twenty, pp. 387-417, "Towards a Logic of Questions."

WELLMAN, Francis L., *The Art of Cross-Examination,* New York, Collier Books, fourth edition, revised and enlarged, 1962.

C.ii. Articles on Questioning

BELNAP, Nuel D., "Questions, Answers and Presuppositions," *The Journal of Philosophy*, LXIII, 20 (1966), pp. 609-611.

BRADLEY, F. H., "On Floating Ideas and the Imaginary," *Mind*, XV (1906), pp. 445-472. See pp. 446, 450, and 451-452 on questions.

BROMBERGER, Sylvain, "Questions," *The Journal of Philosophy*, LXIII, 20 (1966), pp. 597-606.

COHEN, F. S., "What is a Question?" *The Monist*, XXXIX (1929), pp. 350-364.

HAMBLIN, C. L., "Discussion: Questions Aren't Statements," *Philosophy of Science*, XXX (1963), pp. 62-63.

HAMBLIN, C. L., "Questions," *The Australasian Journal of Philosophy*, XXXVI (1958), pp. 159-168.

HAMBLIN, C. L., Entry "Questions," in Vol. VII of *The Encyclopedia of Philosophy*, ed. Paul Edwards, New York, The Macmillan Company and The Free Press, 1967, pp. 49-53.

HARRAH, David, "A Logic of Questions and Answers," *Philosophy of Science*, XXVIII (1961), pp. 40-46.

HARRAH, David, "A Model for Applying Information and Utility Functions," *Philosophy of Science*, XXX (1963), pp. 267-272.

HARRAH, David, "Question Generators," *The Journal of Philosophy*, LXIII, 20 (1966), pp. 606-608.

HIZ, Henry, "Questions and Answers," *The Journal of Philosophy*, LIX (1962), pp. 253-265.

KUBINSKI, Tadeusz, "An Essay in Logic of Questions," *Atti del Congresso Internazionale di Filosofia*, Venezia, 12-18 September, 1958, Firenze, Sansoni, 1960, Vol. XV, pp. 315-322.

LEONARD, Henry S., "Interrogatives, Imperatives, Truth, Falsity and Lies," *Philosophy of Science*, XXVI (1959), pp. 172-186.

LEONARD, Henry S., "A Reply to Professor Wheatley," *Philosophy of Science*, XXVII (1961), pp. 55-64.

LLEWELYN, John E., "What Is a Question?" *Australasian Journal of Philosophy*, XLII (1964), pp. 69-85.

MAYO, Bernard, "Deliberative Questions: A Criticism," *Analysis*, XVI (1955), pp. 58-63.

PRESLEY, C. F., "A Note on Questions," *Australasian Journal of Philosophy*, XXXVII (1959), pp. 64-66.

PRIOR, Mary L., and Arthur N. Prior, "Erotetic Logic," *The Philosophical Review*, LXIV (1955), pp. 43-59.

RITCHIE, A. D., "The Logic of Question and Answer," *Mind*, LII (1943), pp. 24-38.

SPERANTIA, Eugenio, ''La métaphysique implicite dans les postulats de toute pensée possible,'' *Travaux de IXe Congrès International de philosophie*, Congrès Descartes (Actualités Scientifiques et Industrielles, Numéro 537), Vol. VIII, *Analyse réflexive et transcendence*, 1er partie, éditeurs Hermann et Cie, Paris, 1937, pp. 80-85.

SPERANTIA, Eugenio, ''Remarques sur les propositions interrogatives: Projet d'une logique du problème,'' *Actes du Congrès International de philosophie scientifique*, Paris, Sorbonne, 1935 (Actualités scientifiques et industrielles, Numéro 394), Vol. VII, *Logique*, éditeurs Hermann et Cie, Paris, 1936, pp. 18-28.

WHEATLEY, J. M. O., ''Deliberative Questions,'' *Analysis*, XV, 3 (1955), pp. 49-60.

WHEATLEY, J. M. O., ''Note on Professor Leonard's Analysis of Interrogatives,'' *Philosophy of Science*, XXVIII (1961), pp. 52-54.

C.iii. Unpublished Manuscript on Questioning

BELNAP, Nuel D., ''An Analysis of Questions: Preliminary Report,'' California, System Development Corporation, 1963.

D. BOOKS CITED NOT OTHERWISE CLASSIFIED, AND BOOKS CONSULTED

ALLARD, J.-L., *Le Mathématisme de Descartes*, Ottawa, Editions de l'Université d' Ottawa, 1963.

AQUINAS, Thomas, *Exposition of Aristotle's Metaphysics*, trans. Pierre Conway, Columbus, Ohio, College of St. Mary of the Springs, 1963.

AQUINAS, Thomas, *Exposition of Aristotle's Posterior Analytics*, trans. Pierre Conway, Québec, La Librairie philosophique, 1956.

ARMOUR, Leslie, *The Concept of Truth*, Assen, Van Gorcum, 1969.

ARMOUR, Leslie, *The Rational and the Real*, The Hague, Martinus Nijhoff, 1962.

ARON, Raymond, *Introduction to the Philosophy of History*, trans. G. J. Irwin, Boston, Beacon Press, 1961 (original French edition published 1938, revised 1948).

BACON, Francis, *The Works of Francis Bacon*, ed. J. Spedding, R. L. Ellis and D. D. Heath, London, Longman and Co., 1857-1874.

BARZUN, Jacques, *The Delights of Detection*, New York, Criterion Books, 1961.

BERGER, Peter, and Thomas Luckman, *The Social Construction of Reality*, Garden City, New York, Doubleday (Anchor paperback), 1966.

BLOCH, Marc, *The Historian's Craft*, New York, Alfred Knopf, 1963.

BLONDEL, Maurice, *L'Action: Essai d'une critique de la vie et d'une science de la pratique*, Paris, Presses Universitaires de France, 1950 (first published 1893).

BOSANQUET, Bernard, *The Essentials of Logic*, London, Macmillan and Co. Ltd., 1897.

BOSANQUET, Bernard, *Implication and Linear Inference*, London, Macmillan and Co. Ltd., 1920.

BOSANQUET, Bernard, *Knowledge and Reality*, London, Swan and Sonnenschein, 1892.

BOSANQUET, Bernard, *The Principle of Individuality and Value*, London, Macmillan and Co. Ltd. (Kraus Scholarly Reprint, New York, 1968.)

BRACELAND, F. J., and M. Stock, *Modern Psychiatry*, New York, Doubleday, 1963.

BRADLEY, F. H., *Appearance and Reality*, Oxford, Clarendon Press, 9th corrected impression 1930.

BRADLEY, F. H., *Collected Essays*, 2 vols., Oxford, Clarendon Press, 1935.

BRADLEY, F. H., *Essays on Truth and Reality*, Oxford, Clarendon Press, 1914.

BRADLEY, F. H., *The Presuppositions of Critical History*, introduced and edited by Lionel Rubinoff, Don Mills, Ontario, Dent University Press, J. M. Dent and Sons, 1968.

BRADLEY, F. H., *The Principles of Logic*, 2 vols., London, Oxford University Press, 1928.

BREASTED, James H., *A History of Egypt*, New York, Scribner's, 1959 (first published 1905).

BRUNNER, A., *La Connaissance humain*, Paris, Aubier, 1943.

BUTTERFIELD, Herbert, *Man On His Past*, London, Cambridge University Press, 1969.

CAHN, Steven M., *Fate, Logic and Time*, New Haven, Yale University Press, 1967.

CAPONIGRI, A. Robert, *Time and Idea: The Theory of History in Giambattista Vico*, Notre Dame, University of Notre Dame Press, 1968 (first published 1953).

CARR, Wildon, *What is History*? Harmondsworth, Penguin Books, 1964.

CASSIRER, Ernst, *The Logic of the Humanities*, New Haven, Yale University Press, 1960.

CERAM, C. W., *Gods, Graves and Scholars*, New York, Alfred Knopf, revised edition 1968.

COHEN, Morris R., *The Meaning of Human History*, LaSalle, The Open Court Publishing Company, 2nd edition 1961 (1st edition 1947).

CROCE, Benedetto, *History as the Story of Liberty*, translated by Sylvia Sprigge, Chicago, Henry Regnery Co. (Gateway paperback), 1970 (first published 1941).

CROCE, Benedetto, *History: Its Theory and Practice*, translated by Douglas Ainslie, New York, Russell and Russell, 1960.

CROCE, Benedetto, *Logica*, Vol. II of *Filosofia come scienza dello spirito*, Bari, Gius, Laterza and Figli, 1964.

CROCE, Benedetto, *What Is Living and What Is Dead of the Philosophy of Hegel*, translated by Douglas Ainslie, New York, Russell and Russell, 1915.

DANIEL, Glyn, *Man Discovers His Past*, New York, Thomas Y. Crowell Company, 1968.

DANIELOU, Jean, *Christ and Us* (translation by Walter Roberts of *Approches du Christ*), London, A. R. Mowbray, 1961.

D'ARCY, M. C., *The Sense of History, Secular and Sacred*, London, Faber and Faber, 1959.

DESCARTES, René, *The Philosophical Works of Descartes*, in 2 vols., ed. Elizabeth S. Haldane and G. R. T. Ross, New York, Dover, 1955.

DILTHEY, Wilhelm, *Pattern and Meaning in History: Thoughts on History and Society*, New York, Harper Torchbooks, 1962.

DIMITROFF, G., and James G. Baker, *Telescopes and Accessories*, Philadelphia, The Blakiston Company, 1945.

DRAY, William, *Laws and Explanation in History*, Oxford, Clarendon Press, 1957.

EDDINGTON, Arthur Stanley, *The Nature of the Physical World*, London, J. M. Dent and Sons, 1935.

FACKENHEIM, Emil L., *Metaphysics and Historicity*, Milwaukee, Marquette University Press, 1961.

FAIN, Haskell, *Between Philosophy and History*, New Jersey, Princeton University Press, 1970.

FINANCE, Joseph de, *Essai sur l'agir humain*, Rome, Presses de l'Université Grégorienne, 1962.

FISCHER, David Hackett, *Historians' Fallacies: Toward a Logic of Historical Thought*, New York, Harper Torchbooks, 1970.

FORBES, Duncan, *The Liberal Anglican Idea of History*, London, Cambridge University Press, 1952.

GARDINER, Patrick, *Theories of History*, Glencoe, The Free Press, 1959.

GEIGER, L.-B., *Philosophie et spiritualité*, 2 Vols. Paris, Editions du Cerf, 1963.

GENNARO, Angelo de, *The Philosophy of Benedetto Croce*, New York, The Citadel Press, 1961.

GILSON, Etienne, *Being and Some Philosophers*, Toronto, Pontifical Institute of Mediaeval Studies, 1949.

GILSON, Etienne, *The Christian Philosophy of Thomas Aquinas*, London, Victor Gollancz Ltd., 1961.

GOTTLIEB, Gidon, *The Logic of Choice*, London, George Allen and Unwin, 1968.

GOTTSCHALK, Louis, *Understanding History*, 2nd ed., New York, Alfred Knopf, 1969.

GRAY, Wood, and others, *Historian's Handbook*, 2nd ed., Boston, Houghton Mifflin Co., 1964.

GRENE, Marjorie, *A Portrait of Aristotle*, London, Faber and Faber Ltd., 1963.

GROTE, George, *Aristotle*, edited by Alexander Bain and G. Croom Robertson, 2nd ed., London, John Murray, 1880.

HABERMAS, Jürgen, *Knowledge and Human Interests*, Boston, Beacon Press, 1971.

HARVEY, Van A., *The Historian and the Believer*, New York, The Macmillan Company, 1966.

HEIDEGGER, Martin, *Being and Time*, translated by John Macquarrie and Edward Robinson, New York, Harper and Row, 1962.

HILDEBRAND, Dietrich von, *Man and Woman*, Chicago, Franciscan Herald Press, 1965.

HOENEN, Peter, *Cosmologia*, Vol. II, Rome, Gregorian University Press, 1936.

HOOK, Sidney, ed., *Philosophy and History*, New York, New York University Press, 1963.

JOACHIM, H. H., *The Nature of Truth*, New York, Greenwood Press, 1969. (Originally published by The Clarendon Press, Oxford, 1906.)

KANT, Immanuel, *The Critique of Judgment*, translated by J. H. Bernard, 2nd rev. ed., London, Macmillan, 1931.

KANT, Immanuel, *Critique of Pure Reason*, translated by Norman Kemp Smith, London, Macmillan, 1933.

KANT, Immanuel, *On History*, translated, edited and introduced by Lewis White Beck, New York, Bobbs-Merrill Co. Inc., 1963.

KEEN, Sam, *Apology for Wonder*, New York, Harper and Row, 1969.

KELLY, George Armstrong, *Idealism, Politics and History: Sources of Hegelian Thought*, London, Cambridge University Press, 1969.

KLIBANSKY, R., and H. J. Paton, editors, *Philosophy and History, The Ernst Cassirer Festschrift*, New York, Harper Torchbooks, Harper and Row, 1963.

LEAKEY, L. S. B., *Adam's Ancestors*, New York, Harper and Row, 1960.

LEVASSEUR, J. M., *Le lieu Théologique "Histoire,"* Trois Rivières, Québec, Editions du Bien Publique, 1960.

LOFTHOUSE, W. F., *F. H. Bradley*, London, Epworth Press, 1949.

LOWITH, Karl, *Meaning in History*, Chicago, University of Chicago Press (Phoenix Books), 1949.

LUCEY, W. L., *History: Methods and Interpretation*, Chicago, Loyola University Press, 1958.

LUPASCO, S., *Du devenir logique et de l'affectivité*, 2 vols., Paris, Vrin, 1935.

MARITAIN, Jacques, *Creative Intuition in Art and Poetry*, New York, World Publishing Company (Meridian Books), 1954.

MARITAIN, Jacques, *The Degrees of Knowledge*, London, G. Bles, 1959.

MARITAIN, Jacques, *Formal Logic*, New York, Sheed and Ward, 1946.

MARITAIN, Jacques, *Moral Philosophy*, London, G. Bles, 1964.

MARITAIN, Jacques, *Neuf leçons sur les notions premières de la philosophie morale*, Paris, Téqui, 1951.

MARITAIN, Jacques, *The Range of Reason*, London, G. Bles, 1953.

MARROU, Henri-Irenée, *De la connaissance historique*, 5e edn. rev. et aug., Paris, Editions du Seuil, 1966.

MARROU, Henri-Irenée, *The Meaning of History*, Montreal, Palm Publishers, 1966. (Translation by Robert J. Olson of Marrou's *De la connaissance historique*, 4th edition, 1959.)

MAZLISH, Bruce, *The Riddle of History*, New York, Harper and Row, 1966.

McNEILL, John, *The Blondelian Synthesis*, Leiden, E. Brill, 1966.

MENASCE, G. C. de, *The Dynamics of Morality*, New York, Sheed and Ward, 1961.

MERTZ, Barbara, *Temples, Tombs and Hieroglyphs*, New York, Howard McCann, 1964.

MEYERHOFF, Hans, ed., *The Philosophy of History in Our Time*, New York, Doubleday (Anchor), 1959.

MILLER, Hugh, *An Introduction to Modern Philosophy*, New York, Macmillan, 1947.

MILNE, A. J. M., *The Social Philosophy of English Idealism*, London, George Allen and Unwin, 1962.

MOORE, W. Edgar, *Creative and Critical Thinking*, Boston, Houghton Mifflin and Co., 1967.

MURE, G. R. G., *A Study of Hegel's Logic*, London, Oxford University Press, 1950.

NEVINS, Alan, *The Gateway to History*, New York, Doubleday (Anchor), 1962.

ORGAN, Troy W., *The Art of Critical Thinking*, Boston, Houghton Mifflin and Co., 1965.

PASSMORE, John, *A Hundred Years of Philosophy*, London, Penguin Books, 1968.

PLATO, *The Dialogues of Plato*, in 2 vols., trans. Benjamin Jowett, New York, Random House, 1937.

POLANYI, Michael, *Personal Knowledge*, Chicago: University of Chicago Press, 1962.

POLANYI, Michael, *Science, Faith and Society*, Chicago, University of Chicago Press (Phoenix Books), 1946.

POLANYI, Michael, *The Tacit Dimension*, New York, Doubleday (Anchor), 1967.

POWELL, Betty, *Knowledge of Actions*, London, George Allen and Unwin, 1967.

RAEYMAEKER, Louis de, *The Philosophy of Being*, St. Louis, B. Herder, 1954.

RAHNER, Karl, *Spirit in the World*, Montreal, Palm Publishers, 1968.

RÉGIS, L.-M., *L'opinion selon Aristote*, Paris, Vrin, 1935.

RENIER, G. J., *History: Its Purpose and Method*, New York, Harper and Row (Torchbooks), 1965. (Original publisher: London, George Allen and Unwin, 1965.)

RONAN, Colin, *Astronomers Royal*, New York, Doubleday, 1969.

SCHÖKEL, Luis Alonso, *Understanding Biblical Research*, Montreal, Palm Publishers, 1963.

SIMON, Yves, *The Great Dialogue of Nature and Space*, edited by G. J. Dalcourt, Albany, Magi Books, 1970.

SOMERVILLE, J. E., *Total Commitment: Blondel's L'Action*, Washington, Corpus Books, 1968.

SPIEGELBERG, Herbert, *The Phenomenological Movement: A Historical Introduction*, in 2 vols., 2nd ed., The Hague, Martinus Nijhoff, 1960.

STERN, Alfred, *La philosophie de l'histoire et le problème des valeurs*, Paris, Centre de Documentation Universitaire (Sorbonne), n.d.

STERN, Fritz, ed., *The Varieties of History from Voltaire to the Present*, New York, World Publishing Company (Meridian Books), 1956.

STOVER, Robert, *The Nature of Historical Thinking*, Chapel Hill, The University of North Carolina Press, 1967.

STRASSER, Stephan, *Phenomenology and the Human Sciences*, Pittsburgh, Duquesne University Press, 1963.

STURZO, Luigi, *Inner Laws of Society*, trans. Barbara Barclay Carter, New York, P. J. Kenedy & Sons, 1944.

TAYLOR, A. E., *Plato, The Man and His Work*, London, Methuen and Co., 1952.

TREVELYAN, G. M., *Clio: A Muse*, London, Longmans, Green and Co., 1930.

VAIHINGER, Hans, *The Philosophy of 'As-If,'* translated by C. K. Ogden, London, Routledge & Kegan Paul Ltd., 1924.

VEATCH, Henry, *Two Logics*, Evanston, Illinois, Northwestern University Press, 1969.

VICO, Giambattista, *The New Science of Giambattista Vico*, abridged trans. of the 1744 third ed. by Thomas Goddard Bergin and Max Harold Fisch, Ithaca, Cornell University Press, 1970.

WALSH, W. H., *Philosophy of History*, New York, Harper and Row, 1960.

WEISS, Paul, *History, Written and Lived*, Carbondale, Southern Illinois University Press, 1962.

WHEELER, Mortimer, *Archaeology From the Earth*, Harmondsworth, England, Pelican Books, 1956.

WILD, John, *The Challenge of Existentialism*, Bloomington, Indiana University Press, 1955.

WINKS, Robin W., ed., *The Historian as Detective*, New York, Harper and Row, 1968.

WOODBURY, David O., *The Glass Giant of Palomar*, London, Heinemann, 1940.

E. SCHOLARLY ARTICLES CITED OR CONSULTED

ACTON, H. B., "The Theory of Concrete Universals," *Mind*, XLV (1936), pp. 417-431; XLVI (1937), pp. 1-13.

ALLARD, J.-L., "The Role of Experience in the Philosophy of Descartes," unpublished paper, 1968.

BEDELL, Gary L., "Bradley's Monistic Idealism," *The Thomist*, XXXIV (1970), pp. 568-583.

BOSANQUET, Bernard, "Contradiction and Reality," *Mind*, XV (1906), pp. 1-12.

BOSANQUET, Bernard, "The Relation of Coherence to Immediacy and Specific Purpose," *The Philosophical Review* XXVI (1917), pp. 259-273.

BRADLEY, Francis Herbert, "Coherence and Contradiction," *Mind*, XVIII (1909), pp. 489-508.

BRADLEY, Francis Herbert, "On Floating Ideas and the Imaginary," *Mind*, LX (1906), pp. 445-472.

BRADLEY, Francis Herbert, "On Truth and Coherence," *Mind*, XVIII (1909), pp. 329-342.

CAPONIGRI, A. Robert, "Philosophy and History," *The Review of Metaphysics*, III, 10 (1949), pp. 119-136.

COLLINS, James, "The Role of Monistic Idealism in Croce's Aesthetic," *The New Scholasticism*, XVII (1943), pp. 32-58.

DANTO, A. C., "Mere Chronicle and History Proper," *The Journal of Philosophy*, L, 6 (1953), pp. 173-182.

DEELY, John N., "Animal Intelligence and Concept-Formation," *The Thomist*, XXXV (1971), pp. 43-93.

DIBBLE, Vernon K., "Four Types of Inference from Documents to Events," *History and Theory*, III (1964), pp. 203-219.

DIETL, Paul J., "Deduction and Historical Explanation," *History and Theory*, VII, 2 (1968), pp. 167-188.

ESLICK, Leonard J., "The Platonic Dialectic of Non-Being," *The New Scholasticism*, XXIX (1955), pp. 33-49.

FEUER, Lewis, "What is the Philosophy of History?" *The Journal of Philosophy*, XLIX, 10 (1952), pp. 329-340.

FOSTER, Michael B., "The Concrete Universal: Cook Wilson and Bosanquet," *Mind*, XL (1931), pp. 1-12.

FRIESS, Horace L., "Historical Interpretation and Critical Analysis," *The Journal of Philosophy*, XLIX, 10 (1952), pp. 340-350.

GALLIE, W. B., "The Historical Understanding," *History and Theory*, III (1964), pp. 149-202.

GARDEIL, A., "La 'Certitude Probable,' " *Revue des Sciences Philosophiques et Théologiques*, L (1911), pp. 237-266; 441-485.

GERHARD, W. A., "Plato's Theory of Dialectic," *The New Scholasticism*, XXI (1947), pp. 192-211.

GEWIRTZ, A., "Experience and the Non-Mathematical in the Cartesian Method," *Journal of the History of Ideas*, II, 2 (1941), pp. 183-210.

GRAM, Moltke S., "The Reality of Relations," *The New Scholasticism*, XLIV (1970), pp. 49-68.

GRIFFISS, James E., "The Kantian Background to Hegel's Logic," *The New Scholasticism*, XLIII (1969), pp. 509-529.

HANSON, N., "An Anatomy of Discovery," *The Journal of Philosophy*, LXIV, 11 (1967), pp. 321-352.

HANSON, N., "The Idea of a Logic of Discovery, *Dialogue*, IV (1965-1966), pp. 48-61.

HICKS, J. Dawes, "The Philosophical Researches of Meinong," *Mind*, XXXI (1922), pp. 1-30.

HODGSON, Shadworth, "The Idea of Totality," *Proceedings of the Aristotelian Society*, VIII (1907-1908), pp. 95-114.

HOERNLÉ, R. F. Alfred, "Professor Stout's Theory of Possibilities, Truth and Error," *Mind*, XL (1931), pp. 273-284.

HOGAN, James, "The Dialectic of Aristotle," *Philosophical Studies*, V (1955), pp. 3-21.

HUGHES, Percy, "Concrete Conceptual Synthesis," *The Journal of Philosophy*, IV (1907), pp. 623-630.

HUGHES, Percy, "Is There a Distinct Logic of Historical Construction?" *The Journal of Philosophy*, II (1905), pp. 54-55.

JOACHIM, H. H., "Absolute and 'Relative' Truth," *Mind*, XIV (1905), pp. 1-14.

JONES, E. E. Constance, "Logic and Identity in Difference," *Proceedings of the Aristotelian Society*, VII (1906-1907), pp. 81-92.

KOSOK, Michael, "The Formalization of Hegel's Dialectical Logic," *The International Quarterly*, VI (1966), pp. 596-631.

KRONER, Richard, "Philosophy of Life and Philosophy of History," *The Journal of Philosophy*, XXXIII, 8 (1936), pp. 204-212.

KRONER, S., "The Year 1800 in the Development of German Idealism," *The Review of Metaphysics*, I (1948), pp. 1-31.

KUNTZ, Paul G., "The Dialectic of Historicism and Anti-Historicism," *The Monist*, LIII (1969), pp. 656-669.

LAURIERS, M.-L. Guérard des, "A propos de la 'Connaissance Historique' [de H.-I. Marrou]," *Revue des sciences philosophiques et théologiques*, XXXIX (1955), pp. 569-602.

LEAHY, Daniel J., "A Pragmatic Theory of Past, Present and Future," *The Review of Metaphysics*, VI (1953), pp. 369-380.

LEE, Otis, "Dialectic and Negation," *The Review of Metaphysics*, I (1947), pp. 3-23.

LEE, Donald S., "Hypothetic Inference in Systematic Philosophy," *International Philosophical Quarterly*, IX (1969), pp. 363-390.

LOBKOWICZ, N., "Abstraction and Dialectics," *The Review of Metaphysics*, XXI (1967-1968), pp. 468-490.

MacLACHLAN, D. L. C., "Presuppositions in Bradley's Philosophy," *Dialogue*, II (1963-1964), pp. 155-169.

MANDELBAUM, Maurice, "Some Neglected Problems Regarding History," *The Journal of Philosophy*, XLIX, 10 (1952), pp. 317-329.

MARROU, Henri-Irenée, "De la logique de l'histoire à une ethique de l'histoire," *Revue de métaphysique et de morale*, LIV (1949), pp. 248-272.

MESNARD, Pierre, "L'intelligibilité en histoire," *Proceedings of the XIth International Congress of Philosophy*, Bruxelles, XIV (1953), pp. 205-215.

NIEL, H., "Le sens de l'histoire," *Recherches de science religieuse*, XLVI (1958), pp. 60-77. See also English translation in *Philosophy Today*, II (1954), pp. 230-239.

PARKER, Francis H., "Traditional Reason and Modern Reason," *Philosophy Today*, VII (1963), pp. 235-244.

PARODI, D., "Nécessité et contingence en histoire," *Revue de métaphysique et de morale*, XLIV (1949), pp. 273-279.

PETRIN, Jean, "Les modes de dire 'per se' et la démonstration," *Revue de l'Université d'Ottawa*, XXI (1951), pp. 173-192.

PETRIN, Jean, "Univocité et analogie dans les lois de la logique," *Angelicum*, XXVI-XXVII (1949-1950), pp. 233-249.

PITT, Jack, "Generalizations in Historical Explanation," *The Journal of Philosophy*, LVI, 13 (1959), pp. 578-586.

RESCHER, Charles, and Carey B. Joynt, "Evidence in History and in the Law," *The Journal of Philosophy*, LVI, 13 (1959), pp. 561-578.

RIEZLER, Kurt, "The Historian and Truth," *The Journal of Philosophy*, XLV, 14 (1948), pp. 378-388.

ROBINSON, Richard, "Cook Wilson's View of Judgment," *Mind*, XXXVII (1928), pp. 304-317.

ROBINSON, Richard, "Cook Wilson's View of the Origin of Judgment," *Mind*, XXXVII (1928), pp. 454-470.

RUSSELL, L. J., "The Basis of Bosanquet's Logic," *Mind*, XXVII (1918), pp. 432-463.

SABINE, George H., "Professor Bosanquet's Logic and the Concrete Universal," *The Philosophical Review*, XXI (1912), pp. 546-565.

SIMON, Yves, "On Order in Analogical Sets," *The New Scholasticism*, XXXIV (1960), pp. 1-42.

SMITH, John E., "The Relation of Thought to Being," *The New Scholasticism*, XXXVIII (1964), pp. 22-43.

SMITH, N. K., "The Fruitfulness of the Abstract," *Proceedings of the Aristotelian Society*, XXVIII (1927-1928), pp. 203-220.

SMITH, N. K., "The Nature of Universals," *Mind*, XXXVI (1927), pp. 137-157; 265-280; 393-422.

STERRETT, J. Macbride, "Methods of Studying the History of Philosophy," *The Journal of Philosophy*, II (1905), p. 55.

STOUT, G. F., "Bradley on Truth and Falsity," *Mind*, XXIV (1925), pp. 39-54.

STOUT, G. F., "Truth and Falsity," *Mind*, XLI (1932), pp. 297-310.

STRANGE, E. H., "Mr. Bradley's Doctrine of Knowledge," *Mind*, XX (1911), pp. 457-488.

URBAN, Wilbur M., "Metaphysics and History," *The Review of Metaphysics*, III, 11 (1950), pp. 263-299.

WADIA, A. R., "Mr. Joachim's Coherence Notion of Truth," *Mind*, XXVIII (1949), pp. 427-435.

WARD, James, "Bradley's Doctrine of Experience," *Mind*, XXIV (1925), pp. 13-38.

WILD, John, "Being, Meaning, and the World," *The Review of Metaphysics*, XVIII (1965), pp. 411-429.

INDEX OF PROPER NAMES